I0890957

Oceanographic Processes of Coral Reefs

In the last two decades since publication of the first edition, substantial advancements have been made in the science, and the need for transdisciplinary approaches to coral reef protection are greater than ever before. This new edition, now in full color with accompanying animations, goes beyond identifying foundational information and current problems to pinpoint science-based solutions for managers, stakeholders, and policy makers.

Coral reefs are connected by currents that carry plankton and the larvae of many reef-based organisms. Furthermore, they supply food to reefs. Currents also bring pollutants from the land and, together with the atmosphere, affect the surrounding ocean. The chapters in this book provide a much-needed review of the biophysics of reefs with an emphasis on the Great Barrier Reef as an ecosystem. The focus is on interactions among currents, waves, sediment, and the dynamics of coastal and reef-based ecosystems. The topographic complexity of reefs redirects mainstream currents and creates tidal eddies, mushroom jets, boundary layers, and stagnation zones, and this turbulence is enhanced by the oceanographic chaos in the adjoining Coral Sea. This is the environment in which particles and organisms of a range of sizes live, from tiny plankton to megafauna. This generates faunal connectivity at scales of meters to thousands of km within the Great Barrier Reef and with the adjoining ocean.

Pollution and other impacts from the land and the Coral Sea is increasing, and remedial measures are described both on land and on coral cays. The impact of climate change is quantified in case studies about mangroves and corals. Modelling this biophysical complexity is increasing in sophistication, and the authors suggest how the field can advance further.

Oceanographic Processes of Coral Reefs

Physical and Biological Links in the Great Barrier Reef

Second Edition

Edited by
Eric Wolanski
James Cook University, Australia
Michael J. Kingsford
James Cook University, Australia

CRC Press
Taylor & Francis Group
Boca Raton London New York

CRC Press is an imprint of the
Taylor & Francis Group, an **informa** business

Designed Cover:
Front page pictures from NASA/USGS, Norman Duke, Michael Kingsford, and Mark Hamann; Back page pictures from NASA/USGS, Peter Harrison and Great Barrier Reef Marine Park Authority.

Second edition published 2024
by CRC Press
2385 Executive Center Drive, Suite 320, Boca Raton, FL 33431

and by CRC Press
4 Park Square, Milton Park, Abingdon, Oxon, OX14 4RN

CRC Press is an imprint of Taylor & Francis Group, LLC

ISBN: 978-1-032-34069-2 (HB)
ISBN: 978-1-032-34074-6 (PB)
ISBN: 978-1-003-32042-5 (EB)

DOI: 10.1201/9781003320425

Typeset in Times
by Apex CoVantage, LLC

Access the eResources: https://www.routledge.com/9781032340692

Animations available online at https://www.routledge.com/9781032340692

Dedication

We dedicate this book to researchers on the Great Barrier Reef. May their contributions help to sustain one of the great natural wonders of the world for all including our children and grandchildren, Philippe, Oliver, Grace, and Harry Wolanski, Jessica and Sam Kingsford, and Kaitlin and Sophie Arbouin.

Contents

SECTION 2 Land-Sea Connectivity

SECTION 3 Biophysical Oceanography

SECTION 4 Consequences: Impact on GBR Water and Remediation

Foreword

The Great Barrier Reef is the world's largest and most iconic coral reef ecosystem. Visible from space, the Reef is larger than two-thirds of the world's countries and famous for its spectacular beauty and amazing biodiversity. The Reef's World Heritage listing is in recognition of its outstanding universal value. In addition to its immense natural values, the Great Barrier Reef is an economic powerhouse, contributing billions of dollars to the Australian economy each year and providing tens of thousands of jobs. It has deep cultural and spiritual significance for the First Nation's People, Traditional Owners whose connections to the Reef go back some 50 thousand years (Sutton and Walshe 2021). For many Australians, the Reef is part of their national identity. It is also a global icon that attracts many international visitors each year. Healthy reef ecosystems are essential and intrinsically linked to the wellbeing of people, underpinning economies, and jobs dependent on the Reef.

Climate change impacts to reef ecosystems have grown over the last decade. Marine heatwaves in the summers of 2016, 2017, 2020, and 2022 resulted in mass coral bleaching. Although climate change is the most significant threat to the Reef, it is also being impacted by crown-of-thorns starfish outbreaks, coastal development, land-based run-off, and threats from direct human use of the Reef, such as overfishing of species and illegal fishing activities. Conserving and restoring natural spaces, both on land and in the water including the Reef, is essential for limiting carbon emissions, adapting to an already changing climate, and ensuring that there is system resilience to threats to and pressures. The Great Barrier Reef Marine Park Authority (Reef Authority) is Australia's reef management agency, leading a global effort to preserve and protect the Great Barrier Reef by strengthening reef resilience, reducing threats, and protecting and restoring reef habitats to help ensure that the Reef copes better with cumulative threats and a changing climate.

The management of the Reef by the Reef Authority is globally recognized as a model for effective marine park management. The cornerstone is a multiple-use zoning management plan that includes a network of no-take marine reserves, known as green zones, that cover approximately one-third of the marine park. Other management tools include legislation, plans of management, Traditional Owner agreements, compliance, policies, permits, stewardship and best practice, education and community awareness, and research and monitoring and reporting. All of the Reef's management approaches and tools are evidence-based, and scientific information is a critical part of decision making and reporting. The ever-increasing knowledge base about the Great Barrier Reef is supported by a wide range of world-class science and knowledge providers including research institutions, government agencies, universities, Traditional Owners, industry, and the Reef community.

The Reef is one of the most well researched and monitored ecosystems in the world. This is due to sustained investment in science and long-term partnerships with dedicated and committed research providers enhanced by the increasing use of Traditional Owner knowledge, industry partnerships, and a growing community contribution through citizen science programs. This research has a long history of multidisciplinary approaches that not only focus on enhanced scientific knowledge but also address the needs of the Reef and associated communities. This tradition started with the 1928 Great Barrier Reef expedition, a 13-month scientific program that included geology, oceanography, meteorology, biology, and fishery assessments.

The new edition of *Oceanographic Processes of Coral Reefs: Physical and Biological Links in the Great Barrier Reef* continues and builds on this tradition by providing new scientific insights into the Reef's physical oceanography, biogeomorphology, land-sea connectivity, ecohydrology, contaminants and pesticides, and the biophysics of many Reef animals and plants. In addition, it addresses new areas of reef research including land remediation, coral restoration, and engineering measures to combat climate change. Congratulation to the editors, Professors Eric Wolanski and Michael J. Kingsford, and all the contributors for producing a book that will significantly enhance the Reef's scientific knowledge base and provide a very useful reference for the managers and users of the Reef including the implementation of Australia's Reef 2050 Plan, the nation's overarching framework for protecting and managing the Great Barrier Reef from 2015 to 2050.

Ian Poiner
Chair, Great Barrier Reef Marine Park Authority

REFERENCE

Sutton P, Walshe K (2021) Farmers or Hunter-Gathers? Appendix 1. Melbourne University Press.

Access the eResources: https://www.routledge.com/9781032340692
Animations available online at https://www.routledge.com/9781032340692

About the Editors

Adjunct Professor Eric Wolanski is an estuarine oceanographer and ecohydrologist at James Cook University. His research interests range from the oceanography of coral reefs, mangroves, and muddy estuaries to the interaction between physical and biological processes that determine ecosystem health in tropical waters. He has published 430 book chapters, scientific papers, and reports. He has a Google Scholar h-index of 88 and 25,500 citations. Eric is Fellow of the Australian Academy of Technological Sciences and Engineering, the Institution of Engineers Australia (ret.), and l'Académie Royale des Sciences d'Outre-Mer (Belgium). He was awarded an Australian Centenary medal, 2 Doctorate Honoris Causa (by the Catholic University of Louvain and the University of Hull), a Queensland Information Technology and Telecommunications Award for Excellence, and a Lifetime Achievement Award by the Estuarine & Coastal Sciences Association. Eric is Editor-in-Chief of *Wetland Ecology and Management* and of the Elsevier book series *Ecohydrology from catchment to coast*, Honorary Editor of *Estuarine, Coastal and Shelf Science*, and a member of the editorial board of 4 other journals.

Michael J. Kingsford is Distinguished Professor in the Marine Biology and Aquaculture group of the College of Science and Engineering at James Cook University (JCU), Australia. Over a 14-year period, he has held positions as Head of the School of Marine Biology and Tropical Biology and Dean of the College of Marine and Environmental Sciences. Furthermore, he was President of the Australian Coral Reef Society, Director of One Tree Island Research Station, and member of the Great Barrier Reef Research Foundation and of the Museum of Tropical Queensland advisory committees. His awards include the K. Radway Allen Award awarded for outstanding contribution in fish or fisheries science by the Australian Society for Fish Biology (2017) and the AMSA Jubilee Award (2021) for excellence in marine research and outstanding contribution to marine research in Australia. He has published extensively on the ecology of reef fishes, biological oceanography, climate change, and jellyfishes. In total, he has 220 publications including 4 major books, 42 book chapters, 165 refereed publications, and 9 refereed proceedings (h-index Google Scholar = 58). He was Chief Investigator with the ARC Centre of Excellence for Innovative Coral Reef Studies. Major focuses of his research have been on reef fish ecology and demography, connectivity of reef fish populations, the ecology and behavior of larval fishes, the utility of Marine Protected Areas, and environmental records in corals and fishes. He has 40 years of research experience in studying fishes, jellyfishes, and oceanography in the temperate and tropical regions of Australia and other parts of the world.

About the Editors

Contributors

Stephen Attard
AgriTech Solutions
Ayr, Australia

Zoe Bainbridge
James Cook University
Australia

Rebecca Bartley
CSIRO Land and Water
Australia

Ian P. Bell
Queensland Parks and Wildlife Service
Queensland Government, Australia

Stacy Bierwagen
Australian Institute of Marine Science
Australia

Michael Bode
Queensland University of Technology
Australia

Michael Bradley
North Queensland Dry Tropics
Australia

Maria Byrne
University of Sydney
Australia

Ciemon F. Caballes
University of Guam
USA

Adam D. Canning
James Cook University
Australia

Alexandra Carter
James Cook University
Australia

Paula Cartwright
James Cook University
Australia

Taha Chaiechi
James Cook University
Australia

Josie F. Chandler
James Cook University
Australia

Carla C.M. Chen
James Cook University
Australia

Andrew Chin
James Cook University
Australia

Severine M. Choukroun
James Cook University
Australia

Christophe Cleguer
James Cook University
Australia

Owen Coffee
Department of Environment and Science
Queensland Government, Australia

Robert G. Coles
James Cook University
Australia

John G. Connell
James Cook University
Australia

Kay Critchell
Deakin University
Australia

Aaron Davis
James Cook University
Australia

John Dawson
Department of Environment and Science
Queensland Government, Australia

Peter C. Doll
James Cook University
Australia

Jennifer M. Donelson
James Cook University
Australia

Norman C. Duke
James Cook University
Australia

Mike J. Emslie
Australian Institute of Marine Science
Australia

Shawna A. Foo
University of Sydney
Australia

Gabriele G. Gerlach
Helmholtz Institute for Functional Marine Biodiversity
Germany

Alana Grech
James Cook University
Australia

Mark Hamann
James Cook University
Australia

Sarah M. Hamylton
University of Wollongong
Australia

Daniel P. Harrison
Southern Cross University
Australia

Peter L. Harrison
Southern Cross University
Australia

Sebastian J. Hennige
University of Edinburgh
United Kingdom

Mia O. Hoogenboom
James Cook University
Australia

Nicholas A. Kamenos
Umeå University
Sweden

Michael J. Kingsford
James Cook University
Australia

Stuart Kininmonth
University of Queensland
Australia

Sara Kophamel
James Cook University
Australia

Sarah Kwong
James Cook University
Australia

Jonathan Lambrechts
Université Catholique de Louvain
Belgium

Bethan J. Lang
James Cook University
Australia

Stephen Lewis
James Cook University
Australia

Colin J. Limpus
Queensland Department of Environment and Science
Australia

Nicolas Lubitz
James Cook University
Australia

Jock R. Mackenzie
Earthwatch Australia
Australia

George Marmorino
Naval Research Laboratory
USA

Helene Marsh
James Cook University
Australia

Sam A. Matthews
Great Barrier Reef Marine Park Authority
Australia

Carlo Mattone
North Queensland Dry Tropics
Australia

Gillian McCloskey
Department of Environment and Science
Queensland Government, Australia

Roger McLean
Australian Defence Force Academy
University of New South Wales, Australia

Katie Motson
James Cook University
Australia

Jaelen Myers
James Cook University
Australia

Mikaela Nordborg
Australian Institute of Marine Science
Australia

Liudmila Osipova
International Council on Clean Transportation
Germany

Richard Pearson
James Cook University
Australia

Ian Poiner
Great Barrier Reef Marine Park Authority
Australia

Morgan S. Pratchett
James Cook University
Australia

Kate M. Quigley
Minderoo Foundation and James Cook University
Australia

Jean-Baptiste Raina
University of Technology Sydney
Australia

Katharine Robertson
Department of Science and Environment
Queensland Government, Australia

Mike Ronan
Department of Science and Environment
Queensland Government, Australia

Jodie L. Rummer
James Cook University
Australia

Jodie A. Schlaefer
CSIRO Environment
Australia

Justin R. Seymour
University of Technology Sydney
Australia

Marcus Sheaves
James Cook University
Australia

Rong Situ
James Cook University
Australia

Lisa Spiecker
Carl von Ossietzky University Oldenburg
Germany

Owen Stewart
Queensland University of Technology
Australia

Michelle A. Templeman
James Cook University
Australia

Colette R. Thomas
Great Barrier Reef Marine Park Authority
Australia

Samantha J. Tol
James Cook University
Australia

Ryan Turner
University of Queensland
Australia

Vinay Udyawer
Australian Institute of Marine Science
Australia

Sven Uthicke
Australian Institute of Marine Science
Australia

Ajay A. Venkataraman
James Cook University
Australia

Ana Vila-Concejo
University of Sydney
Australia

Nathan J. Waltham
James Cook University
Australia

Jane Waterhouse
James Cook University
Australia

Gregory E. Webb
University of Queensland
Australia

Christopher D. Williams
James Cook University
Australia

Sara Winter
James Cook University
Australia

Eric Wolanski
James Cook University
Australia

Kennedy Wolfe
University of Queensland
Australia

Paul H. York
James Cook University
Australia

Section 1

The Key Role of Oceanography and How It Influences Life in the GBR

1 The Emergence of Biophysical Sciences for the Great Barrier Reef

Michael J. Kingsford and Eric Wolanski

1.1. INTRODUCTION

The ocean is a three-dimensional environment that has a great influence on the movement of particulate matter, sediments, plankton, and nekton. The complexity of currents increases greatly in mosaics of coral reefs, and here, we focus on the interactions among physical oceanography, biology, and abiotic particles that include pollutants.

The scale of the Great Barrier Reef (GBR) is huge: the Great Barrier Marine Park (GBRMP) is 345,500 km^2 (Figure 1). It drains 35 watersheds with a total surface area of 423,122 km^2. The GBR, including its northern extension in Torres Strait, stretches nearly 2,000 km and is visible from space (Figure 2). The GBR is buffered by mainland Australia in the west and a shelf slope that quickly descends beyond a 2,000-m depth in the east in the Coral Sea. Its width varies greatly; in the far southern region at the Swain Reefs, it is 300 km wide, and in the north near the Flinders Group, it is only 27 km wide. Its bathymetry is extremely complex; it forms a dense but porous wall at the outer continental shelf, except in the central region where the system is much more open to the influence of the Coral Sea (Figure 3). Scattered throughout the continental shelf, numerous reefs are present, in some places as a dense reef matrix and in other places as reefs located relatively far away from one another. Remarkably, the entire shelf area occupied by the reefs was above sea level approximately 12,000 years ago. Furthermore, 5,000 years ago, the sea level was near the present one (Hopley and Smithers 2019). Accordingly, today's currents sweep through ancient river valleys and recent reefs that are largely composed of the skeletons of reef-building corals and other depositors of calcium carbonates. The reefs and shoals near the surface make up 9% of the area of the GBRMP and therefore have a great influence on local current patterns.

1.2. THE GROWTH OF BIOPHYSICAL SCIENCE OF THE GBR

The GBR coral reefs provide habitats for a host of organisms associated with reefs. The diversity of life supported by habitats in the GBR is remarkable. Species diversity includes approximately 1,500 fishes (Hutchings and Kingsford 2019), 450 species of reef-building corals (Wallace 2019), over 6,000 species of molluscs (Willan 2019), 630 species of echinoderms (Byrne 2019), 30 species of marine mammals, six species of turtles, and 32 species of sea snakes (Marsh et al. 2019). The historical focus of research in many marine environments of the GBR has been on the biota, including species richness, biology, and ecology (e.g., Hutchings et al. 2019), and the oceanography has lagged behind. Over time, it became recognised that the currents drive patterns of connectivity and especially the early life history stages of organisms. Furthermore, currents disperse freshwater nutrients, sediments, and pollutants. The GBR coral reefs compose a solid but porous body that interacts with the prevailing currents forced by the tides, the Coral Sea, and the wind to generate complex, topographically generated currents. These currents transport waterborne 'babies' (e.g., seeds, shoots, rhizome fragments, propagules, eggs, larvae, and hatchlings) and enable a connectivity of fauna and flora between reefs and regions. Some babies are passive, and they may be entirely dependent on the water currents, while others have behaviours that enable them to control their trajectories and their fate. Quantifying these processes is necessary to understand how the GBR works as an ecosystem. This is greatly complicated by the ubiquitous, intense, topographically generated sub-mesoscale turbulence in the GBR (Figure 4).

Organisms associated with the reef interact with currents in a range of ways. Many taxa spend time in the plankton as larva, that is, they are moved by the water currents. This transport influences connections among reefs and the likelihood of staying close to home and recruiting back home. Species associated with reefs often rely on currents for the supply of plankton, or the growth of plants through nutrient supply, if they are to survive. Other taxa may move as large nekton to spawning locations or hatching sites or, to follow the supply of food.

Despite this broad acknowledgment that oceanography is important for many such processes in the GBR, biophysical studies (i.e., those that combine physical and biological oceanography) have lagged until recently. The realisation that biophysics is important for the whole GBR ecosystem probably started with the first edition of this book (Wolanski 2001). There have been considerable advances in GBR biophysical knowledge since then, and this edition brings it all together in a contemporary picture.

The grand challenges, or 'wicked problems', facing reefs generally relate to how people interact with the physical environment and the biota. These challenges include dealing with global warming and associated coral bleaching,

DOI: 10.1201/9781003320425-2

FIGURE 1 A map of the GBRMP, the major watersheds, and the location of many sites discussed in this book.

Source: Courtesy of Geospatial Services, Great Barrier Reef Marine Park Authority © Commonwealth of Australia (GBRMPA) 2022.

FIGURE 2 A MERIS Envisat image of the GBR from space. The image clearly shows the nature of reefs across the shelf from inshore islands with fringing reefs that are exposed to nearshore turbidity from wind-driven resuspension of mud and riverine input to mid-shelf, and the outer-shelf reefs are in relatively clear waters. The GBR extends almost to Papua New Guinea where the Fly River inhibits coral growth. Two frontal cloud bands are visible that extend seaward from the Central GBR.

Source: www.esa.int/Applications/Observing_the_Earth/Earth_from_Space_Great_Barrier_Reef

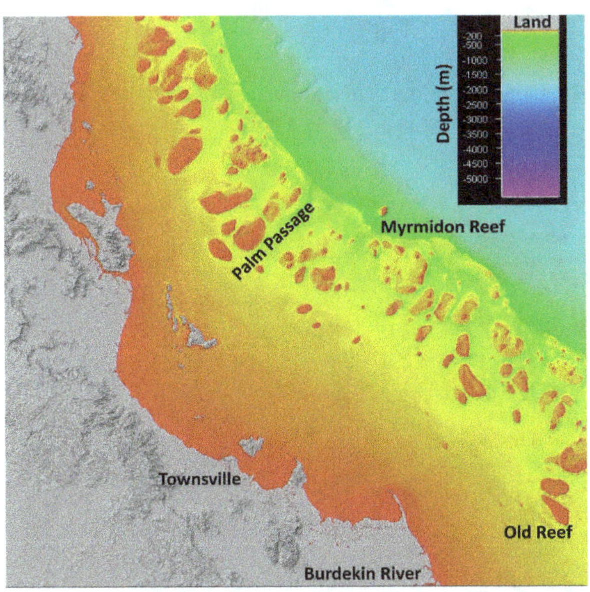

FIGURE 3 Subtidal bathymetry of the central GBR. The Palm Island Passage is the widest passage in the GBR. Ancient river beds (> 12,000 years BP) can be seen at the edge of the green highlighting. The straight-line distance from Myrmidon Reef to Townsville is approximately 119 km.

Source: www.deepreef.org/2010/07/06/3dgbr-bathy/

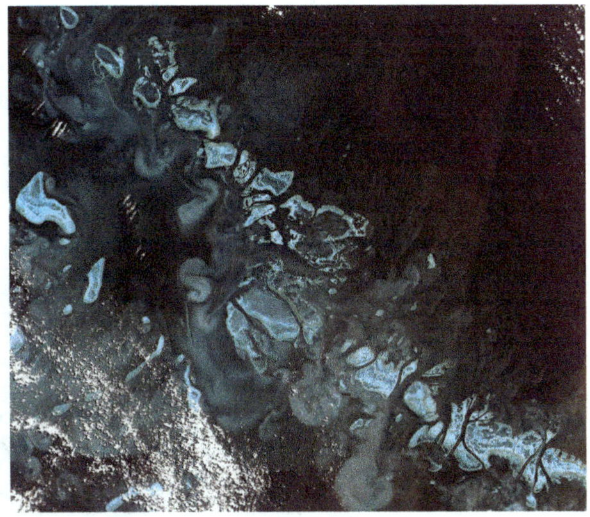

FIGURE 4 A Sentinel-2 satellite image showing the chaotic-like topographically induced turbulence in the reef matrix offshore of the Whitsunday Group, Southern GBR, opposite Proserpine.

ocean acidification, cyclones, crown-of-thorns-starfish (COTS), fisheries, coastal runoff along with associated sediment load, herbicides and pesticides, coastal impacts such as dredging, and pollutants that include plastics and coal dust that may disperse with currents (Wolanski et al. 2019). The GBR also receives pollutants such as plastics from the Coral Sea and from 500 ships that operate within the GBR that use marine fuels with sulphur levels above 0.5%. These ships remove sulphur from the exhaust gas by using seawater, and they discharge the used washwater overboard.

An understanding of oceanography is an important component of the toolkit toward solutions where point sources can be identified and broader impacts determined though an understanding of oceanographic transport.

The challenges of managers concerned with conservation and sustainable practices must cater to stakeholder needs, such as fishing and preserving habitats and biodiversity. The Great Barrier Reef Marine Park Authority (GBRMPA), for example, constantly seeks to modify marine zoning to maximise its effectiveness. An understanding of oceanography can help greatly for the identification of 'source reefs' that may transport planktonic progeny, including less appealing taxa such as COTS, to reefs downstream. Furthermore, the study of currents allows the identification of areas affected by different processes, such as runoff for coastal and some mid-shelf reefs, warming seas, acidified waters, COTS infestation, and processes that may be more sheltered because of topographically generated upwelling. Even in coastal areas, the chaotic-like features of the currents and freshwater plumes can result in reefs in close proximity and even sites within reefs being impacted differently at different times (Figure 5).

However, the GBR's ecological watershed is not restricted to the continental shelf and the adjoining river systems on

FIGURE 5 Composite Sentinel-2 satellite views of the Burdekin River plume in flood; the seaward extent of the plume changed quickly between (a) February 13, 2019 and (b) February 16, 2019. In this case, sediment is acting as a tracer.

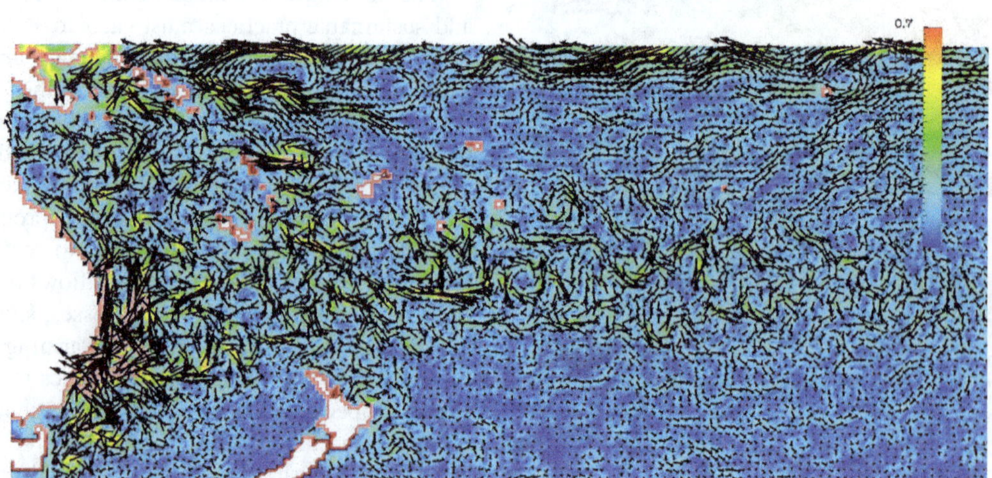

FIGURE 6 A snapshot of the satellite altimetry-derived, 5-day averaged, near-surface currents in the Coral Sea and the South Pacific Ocean, which highlights the intense mesoscale turbulence. Australia is the land mass on the left. The colour bar shows the speed in m s^{-1}.

land. It is also a node within the vast oceanic migration areas of important species such as sea turtles (Hamann et al., 2024), humpback and GBR dwarf minke whales (Andrews-Goff et al. 2018; Ramirez-Flores et al. 2019), the ornate spiny lobster (Dao et al. 2015), and black marlin billfish (Domeier and Speare 2012). These migration routes, from and to the GBR, include vast areas of the tropical Western Pacific Ocean, the coastal waters of Vietnam and the Philippines, Melanesia, Micronesia, and Polynesia, and the south Pacific Ocean south of Australia and around Antarctica. Accordingly, connectivity to and from the GBR ecosystem is considerably large. The currents in expanses of the Pacific are patchy and unsteady with intense mesoscale turbulence (Figure 6). Given that these animals migrate successfully in such a complex flow field is a feat of navigation that is addressed in this book.

1.3. GBR BIOPHYSICS IN THIS BOOK

The intent of this book is to focus on the physical and biological links of the GBR, but the messages in this work have a much broader relevance to other tropical reefs of the world and geomorphologically complex environments. We have invited experts in their respective fields to contribute to the following major sections of the book.

(i) The key role of oceanography and how it influences the GBR. Here, we cover broad- and small-scale elements of GBR oceanography, geomorphological and biophysical links, fluid dynamics, the GBR in a changing world, and how currents can influence population dynamics.
(ii) Land-sea connectivity. The land and sea are intertwined not only through riverine run off but also through the sediment/nutrient fluxes that go with it. Herbicides and pesticides are also transported from the land, and pollutants that include microplastics and coal dust can be transported through the mosaics of reefs. This section concludes with a review of the ecological consequences of currents for fishes and other fauna.
(iii) GBR biophysical oceanography. This section addresses interactions between physical oceanography and the behaviour of organisms ranging from seagrasses and mangroves to planktonic jellyfish and fish larvae; the dispersal and connectivity of COTS larvae are also reviewed. Furthermore, we describe the interactions between currents and charismatic nektonic taxa such as dugongs; in some cases, these animals migrate over thousands of kilometres.
(iv) The consequences of interactions with humans: impacts on GBR waters and remediation. Given the challenges faced by the GBR, the book addresses the impact of climate change on GBR-associated organisms and the steps taken to address threats through the restoring of corals, wetlands, and irrigation infrastructure, engineering the beach morphology for nesting turtles, and engineering proposals to combat climate change.

The book concludes with a synthesis chapter on research opportunities for biophysical science. For the science, advances in technology are increasingly allowing us to collect data to address important biophysical questions. Such advances include low-cost current meters that can be deployed in large numbers and at different spatial scales (Schlaefer et al. 2021), tags that get smaller every year, remote sensing, the development of specialist sensors from space, and drones that can provide high-resolution images of currents, sediment, and nutrients, and more sophisticated physical and biophysical models. These models will become operational and more realistic when the models fully assimilate real-time data and deal realistically with the highly turbulent flows in the reef matrix. Biologically, the increasing sophistication of genetics will allow us to test for successful connectivity. Improved models will allow, and have allowed, for smart irrigation and wetlands restoration to diminish the diversity of human impacts on the GBR. Further applications will be, among the many described in the various chapters in this book, informed genetic engineering for repopulating fish and corals, new methods to quantify pollution plumes originating from the land and from the Coral Sea, the engineering of sand profiles on coral cays to help sea turtle nesting, innovative engineering methods to combat climate change, and optimising the design of Marine Protected Areas by utilising biophysics. With existing and potential changes to atmospheric and oceanic weather and currents, the use of biophysical science to inform the management of the GBR has a promising future.

ACKNOWLEDGEMENTS

We thank all the authors and reviewers for contributing to the richness of this book. Given the diversity of the field, multiple experts were required to review the knowledge, describe gaps in our understanding and opportunities for improvement, and provide glimpses of innovation for the future.

REFERENCES

Andrews-Goff V, Bestley S, Gales NJ, Laverick SM, Paton D, Polanowski AM, Schmitt NT, Double MC (2018) Humpback whale migrations to Antarctic summer foraging grounds through the southwest Pacific Ocean. Scientific Reports 8:12333.

Byrne M (2019) Echinoderms. In: Hutchings P, Kingsford M, Hoegh-Guldberg O (eds) The Great Barrier Reef: Biology, Environment and Management. CSIRO Press, Clayton, pp. 365–376.

Dao HT, Smith-Keune C, Wolanski E, Jones CM, Jerry DR (2015) Oceanographic currents and local ecological knowledge indicate, and genetics does not refute, a contemporary pattern of larval dispersal for the Ornate Spiny Lobster, *Panulirus ornatus* in the South-East Asian archipelago. PLoS One 10(5):e0124568.

Domeier M, Speare P (2012) Dispersal of adult Black Marlin (*Istiompax indica*) from a Great Barrier Reef spawning aggregation. PLoS One 7(2):e31629

Hamann M, Limpus CJ, Kophamel S (2024) Dispersal and connectivity of marine turtles in the Great Barrier Reef and links to the south Pacific Ocean. In: Wolanski E, Kingsford MJ (eds.), Oceanographic Processes of Coral Reefs. Physical and Biological Links in the Great Barrier Reef. CRC Press, Boca Raton, FL.

Hopley D, Smithers S (2019) Geomorphology of coral reefs with special reference to the Great Barrier Reef. In: Hutchings P, Kingsford M, Hoegh-Guldberg O (eds) The Great Barrier Reef: Biology, Environment and Management. CSIRO Press, Clayton, pp. 9–24.

Hutchings P, Kingsford MJ, Hoegh-Guldberg O (2019) Great Barrier Reef: Biology, Environment and Management. CSIRO Publishing and CSIRO Press, Clayton.

Marsh H, Heatwole H, Lukoschek V (2019) Marine mammals and reptiles. In: Hutchings P, Kingsford M, Hoegh-Guldberg O (eds) The Great Barrier Reef: Biology, Environment and Management, vol. 3169. CSIRO Publishing, Clayton South, pp. 4707–4718

Ramirez-Flores O, Birtles A, Pazmino D, Zenger KR, Van-Herwerden L (2019) Dwarf minke whales from the South Pacific share a matrilineal lineage distinct from Balaenoptera acutorostrata and B. a. scammony. Australian Mammalogy. doi: 10.1071/AM18015

Schlaefer JA, Wolanski E, Lambrecht J, Kingsford MJ (2021) Behavioural and oceanographic isolation of an island-based jellyfish (*Copula sivickisi*, Class Cubozoa) population. Scientific Reports 11:10280. https://doi.org/10.1038/s41598-021-89755-7

Wallace CC (2019) Hexacorals 2: Reef-building or hard corals (Scleractinia). In: Hutchings P, Kingsford M, Hoegh-Guldberg O (eds) The Great Barrier Reef: Biology, Environment and Management. CSIRO Press, Clayton, pp. 267–282

Willan R (2019) Molluscs. In: Hutchings P, Kingsford M, Hoegh-Guldberg O (eds) The Great Barrier Reef: Biology, Environment and management. CSIRO Press, Clayton, pp. 341–356

Wolanski E (2001) Oceanographic Processes of Coral Reefs: Physical and Biological Links in the Great Barrier Reef. CRC Press, Boca Raton, FL, 356p.

Wolanski E, Day JW, Elliott M, Ramachandran R (eds) (2019) Coasts and Estuaries: The Future. Elsevier, London, 701p.

2 The Physical Oceanography of the Great Barrier Reef: *A Review*

Eric Wolanski, Michael Kingsford, Jonathan Lambrechts, and George Marmorino

2.1. INTRODUCTION

The Great Barrier Reef (GBR; Figure 1a-d) is immense. It extends over 1,840 km from Lady Elliott Island in the south to Cape York in the north. Those are the administrative limits because the GBR as an ecosystem extends another 170 km through Torres Strait to Papua New Guinea (Figure 1b). Figure 1b emphasizes the shallowness of the GBR compared to the Coral Sea. Figure 1e-f emphasizes the complex and rugged bathymetry of the GBR. Its width varies with location and is seldom larger than 200 km, with vast areas with a width less than 100 km. As a result of the reef density on the outer shelf, the GBR can be divided in four regions (Figure 1b). The Far Northern GBR (FNGBR) extends southward from the Gulf of Papua and Torres Strait, where it is under the influence of the Gulf of Carpentaria (Figure 1c). It extends to South Warden Reef where the shelf is partially blocked, creating a bottleneck for the water circulation (Figure 1d). All along the FNGBR, reefs are scattered throughout, with an inshore and an offshore longshore channel south of Cape Grenville (Figure 1b). North of Cape Grenville, the offshore channel is largely blocked by reefs scattered throughout (Figure 1e). The shelf edge is blocked for approximately 90% of the distance by elongated (ribbon) reefs located at the shelf edge and separated by narrow reef passages. The water depth is seldom greater than 30 m. The Northern GBR (NGBR) extends to the south of South Warden Reef with numerous reefs scattered on the shelf and where the fringing reefs at the shelf edge block about 90% of the distance (Pickard et al. 1977). Further south is the Central GBR (CGBR) where the reefs block only approximately 10% of the distance. The water depth seldom exceeds 40 m except near the shelf break. This area has the least reef density. Finally, to the south, starting at Old Reef and ending in the Swain Reefs, is the main Southern GBR (SGBR) where the outer reefs block about 90% of the distance, and these reefs are several tens of kilometres landward from the shelf edge (Figure 1f). The mid-shelf of the SGBR is typically 50 m deep, and both the mid-shelf and the inner shelf are reef-free. The SGBR ends at its offshore side by the Swain complex of small patch reefs and on its landside by the deep Capricorn Channel that reaches a depth of approximately 145 m at its entrance. The SGBR has a linear 'extension' in the far south, namely, the line of reefs forming the Capricorn Bunker Group that includes Lady Elliott Island.

Here, we emphasise a fundamental difference between the shelf waters of the GBR and other parts of the world. Commonly, shelf waters have a gentle slope, with few topographic anomalies until the shelf break. In contrast, the GBR has approximately 2,500 reefs scattered on the shelf, and as a result, the flows are very complex. The water circulation in the GBR is forced by the tides, the circulation in the Coral Sea, wind, oceanic swell waves, and river inflow. This results in a pattern across the shelf whereby inshore waters are often affected by riverine runoff, while outer reefs are often subjected to upwelling driven by wind, tides, and topography.

There are several small rivers all along the GBR coast. In the CGBR, there is one large river, the Burdekin River, whose discharge is seasonal and highly variable inter-annually. Its floods are short-lived (lasting typically a few days to 2–3 weeks) with a peak flood discharge of approximately 40,000 $m^3 s^{-1}$ in the recorded history starting in 1922. In the SGBR, the largest river is the Fitzroy River that has a larger catchment area than that of the Burdekin but a smaller peak discharge (15,000 $m^3 s^{-1}$), and similar to the Burdekin, these floods are short-lived. In the NGBR, there is no large river. In the FNGBR, the largest river is the perennial Fly River in Papua New Guinea (PNG) at the edge of Torres Strait, with a typical discharge of 6,000 $m^3 s^{-1}$ that varies little seasonally but that occasionally decreases by about 50% for a few months during an El Niño period. The moderate-sized Normanby River flows in Princess Charlotte Bay. These three rivers have a high suspended sediment concentration with typical values of 1,000 mg L^{-1} during floods.

2.2. THE EXTERNAL FORCING OF THE WATER CIRCULATION IN THE GBR

The depth of the well-mixed layer in the Coral Sea fringing the GBR is typically 50–70 m in summer and 90–110 m in winter, that is, it exceeds or is equivalent to the water depth on the shelf (Figure 2). The depth of the well-mixed upper layer in winter varies inter-annually by up to 50 m, possibly because of the ENSO (El Niño-La Niña) phenomenon (Sih

DOI: 10.1201/9781003320425-3

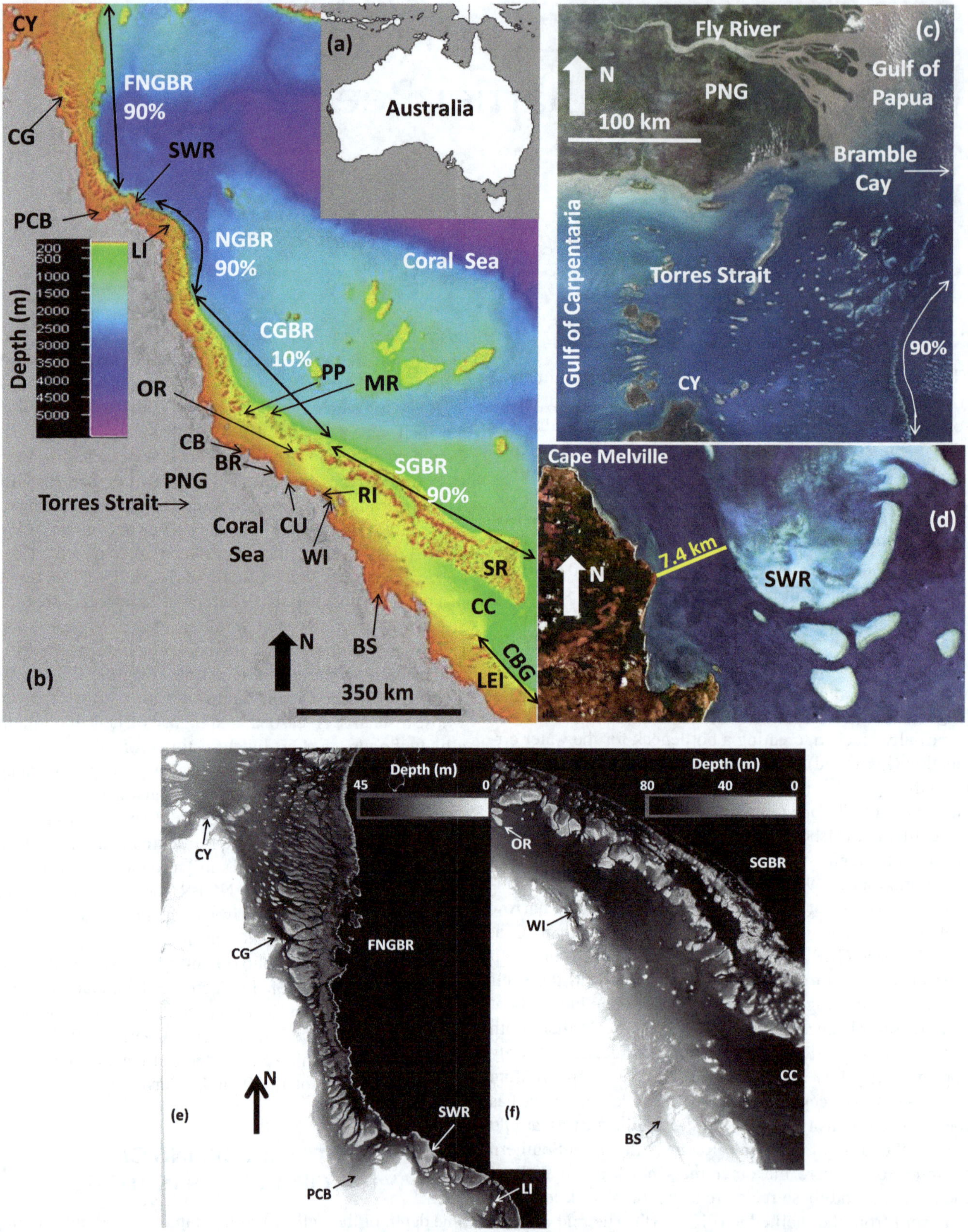

FIGURE 1 (a), (b) Location maps. From north to south: CY: Cape York. CG: Cape Grenville. PCB: Princess Charlotte Bay. SWR: South Warden Reef. LI: Lizard Island. PP: Palm Passage. MR: Myrmidon Reef. CB: Cleveland Bay. OR: Old Reef. BR: Burdekin River mouth. CU: Cape Upstart. WI: Whitsunday Islands. BS: Broad Sound. CC: Capricorn Channel. SR: Swain Reefs. CBG: Capricorn Bunker Group. LEI: Lady Elliott Island. %: the approximate level of blockage by reefs following Pickard et al. (1977). Satellite images of (c) the Torres Strait and (d) the bottleneck near South Warden Reef (SWR). The bathymetry in (e) the FNGBR and (f) the SGBR.

Source: For the bathymetry, www.deepreef.org/2010/07/06/3dgbr-bathy/

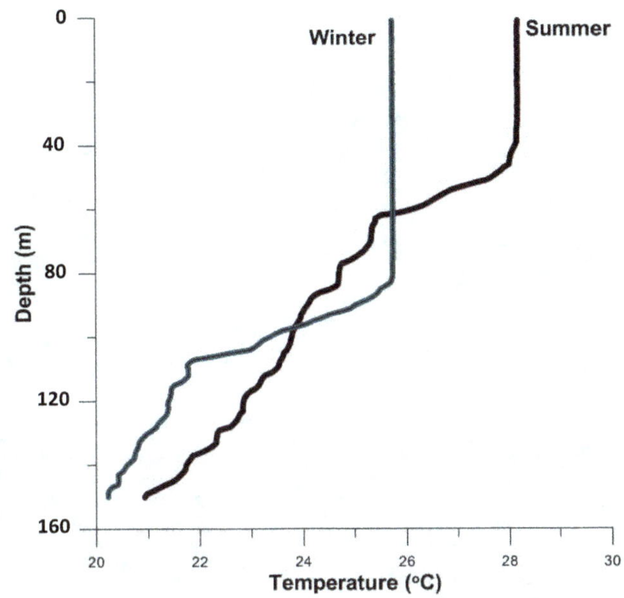

FIGURE 2 Typical temperature profiles in the Coral Sea fringing the CGBR in the summer and winter of 1983. Note the strong thermocline between 50 and 80 m.

Source: AIMS

et al. 2017). Thus, except for upwelling processes (described later), the GBR receives water only from the well-mixed surface layer of the Coral Sea. The main processes that generate this inflow/outflow of water that flushes the GBR shelf and thus prevents stagnation, are the tides, the mean sea level fluctuations in Torres Strait, the mesoscale turbulent jets and eddies in the Coral Sea, the breaking of oceanic swell waves at the outer reef crest that generates an inflow of oceanic water over the outer reef flat, and the wind. As described below, the influence of river flows is mainly inshore, local, and transient.

2.2.1. The Tides and the Mean Sea Level

The tides at sea are the sum of sinusoidal (tidal harmonic) components that vary with location. The dominant tidal harmonics are the semi-diurnal and diurnal components. The dominant tidal component at the GBR shelf break is the semi-diurnal M_2 tide, with a period of 12.42 h. Its tidal amplitude and phase are nearly constant along the shelf break (Figure 3a). The maximum tidal range, which includes all the tidal constituents, varies little north-south along the shelf break, but it varies significantly along the coast (Figure 3b). The maximum tidal range is larger at the coast than at the shelf break at the same latitude. There are also wind-driven fluctuations of the mean sea level by typically 0.2 m at the coast because of the passage of south-easterly strong wind events that generate continental shelf waves in the CGBR (Wolanski and Bennett 1983). There is no apparent mean sea level difference between the seaward and landward sides of the outer fringing reefs. There are, however, strong seasonal mean sea level fluctuations of up to 1 m east-west in the FNGBR through the Torres Strait (Animations 1a and 1b). The level is higher on the Gulf of Carpentaria side of the Strait in the monsoon season (e.g., in January). The inverse situation prevails during the south-easterly trade wind season (e.g., in August).

2.2.2. The Circulation in the Coral Sea

There is a westward (landward) current, the South Equatorial Current (SEC), in the Coral Sea fronting the NGBR to the SGBR and a northward current further north (Andrews and Clegg 1989). The SEC enters the Coral Sea from the east as a series of jets between the Solomon Islands, Vanuatu, and New Caledonia (Choukroun, 2010). The 5-day average circulation as measured by satellite altimetry is mesoscale turbulent (Animations 2a and 2b for 2019 and 2020, respectively). Mesoscale turbulent jets occur, and these fluctuate widely in location, direction, and intensity throughout the year. They are accompanied by mesoscale eddies. The points where the jets reach the GBR shelf break vary by 700 km north-south as the jets meander in the Coral Sea; thus, the location of the bifurcation points also varies continuously in time. The jets occasionally have filaments. There is no apparent repeatable pattern of the jets' meandering between seasons and between years. In the Coral Sea fringing the Capricorn Channel in the SGBR, the transient Capricorn Eddy is apparent in Animation 2.

The annual-mean circulation shows that on average, there are three main jets: one jet impacting the NGBR and the other two jets impacting the SGBR (Figure 4). The northern jet bifurcates seaward from the GBR to form two branches. One branch forms the southward

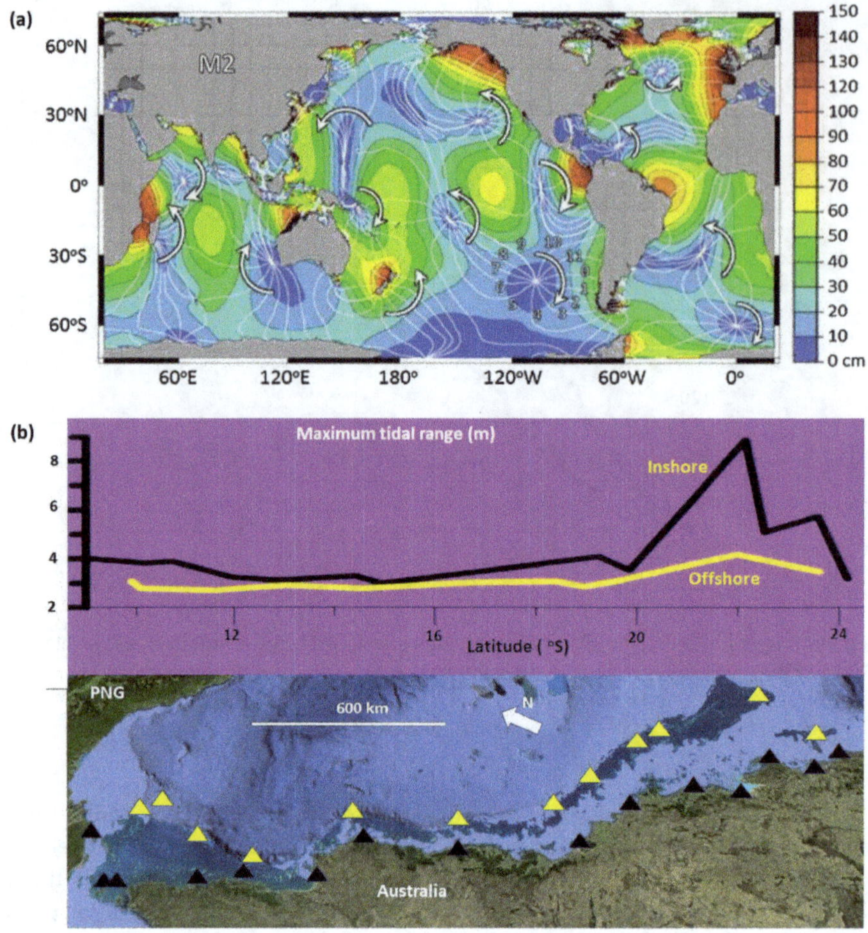

FIGURE 3 (a) The M_2 tidal constituent. The amplitude is indicated by colour, and the white lines are cotidal lines spaced at phase intervals of 30° (slightly > 1 hr). The amphidromic points are where the lines come together. The arrows indicate the direction of rotation of the tidal wave.

Source: NASA

(b) The maximum tidal range offshore and inshore along the length of the GBR.

Source: Bureau of Meteorology

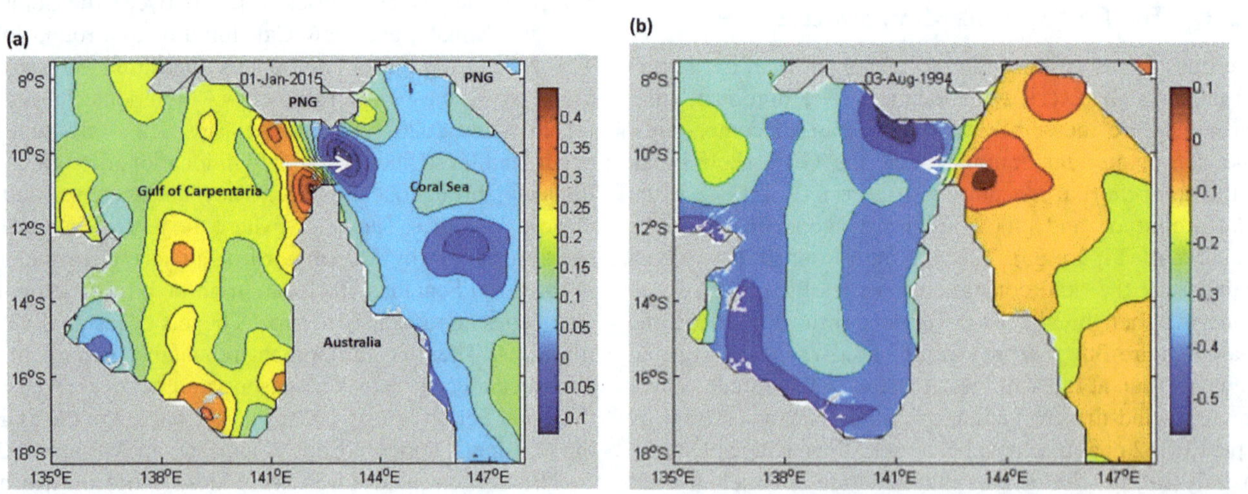

ANIMATION 1 The daily evolution of the distribution of the altimetry-derived mean sea level (in m) in the Gulf of Carpentaria, the Torres Strait, and the Coral Sea in (a) January 2015 and (b) August 1994. The arrows indicate the prevailing mean current direction in Torres Strait as determined by the mean sea level slope.

Available online at https://www.routledge.com/9781032340692

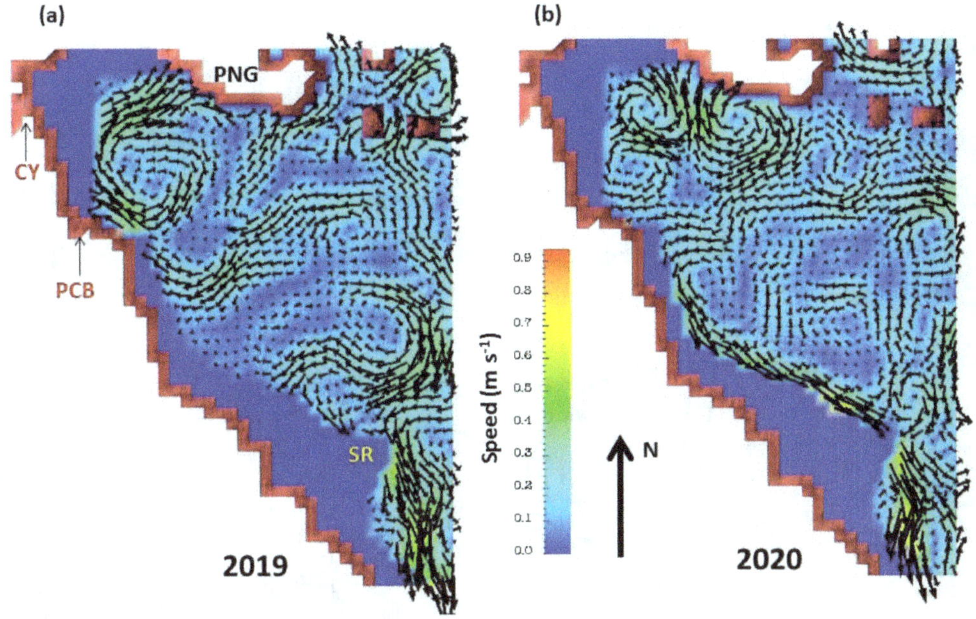

ANIMATION 2 An animation of the 5-day average near-surface water circulation in the Coral Sea fringing the GBR during (a) 2019 and (b) 2010, as measured by satellite altimetry. The colour bar shows the speed in m s^{-1}.

Available online at https://www.routledge.com/9781032340692

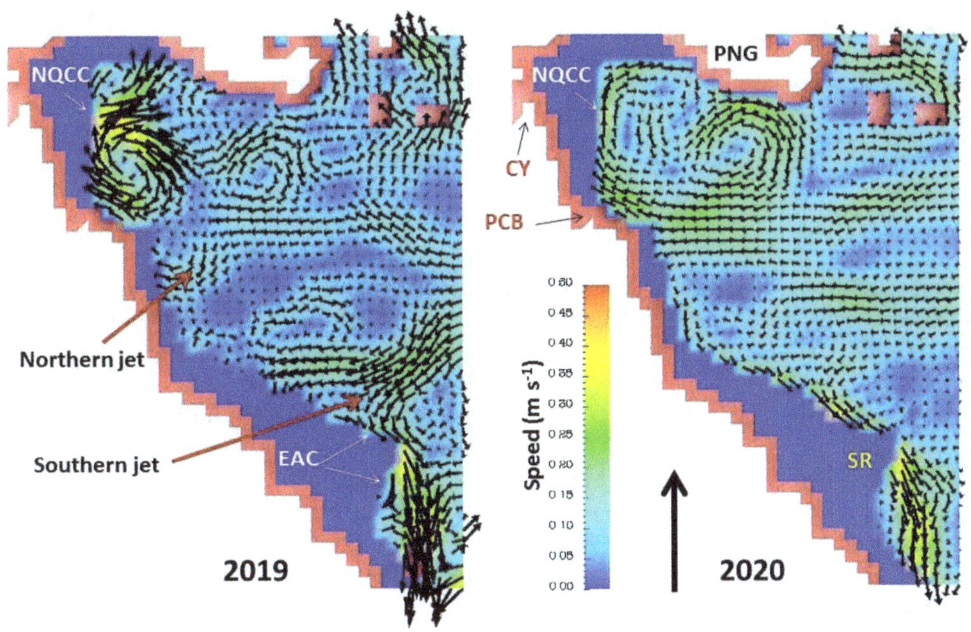

FIGURE 4 The annual mean near-surface circulation in the Coral Sea fringing the GBR in 2019 and 2020, as measured by satellite altimetry. The colour bar shows the speed in m s^{-1}.

flowing East Australian Current (EAC) along the outer shelf and shelf slope of the SGBR; the EAC is then reinforced by the jets impacting the SGBR. The other branch is the eddy-rich, generally northward flowing North Queensland Coastal Current offshore from the FNGBR. The location of the bifurcation point varies with the mesoscale turbulence. The whole flow field is highly unsteady and mesoscale turbulent.

2.2.3. The Wind

The wind histogram (Figure 5b) shows a dominance of the southeasterly wind, which includes periods of light wind and calm weather, and monsoonal wind (Figure 5b). The southeasterly wind is highly coherent within each zone (FNGBR, CGBR, and SGBR) and shows variability between zones as the GBR is long enough for its northern

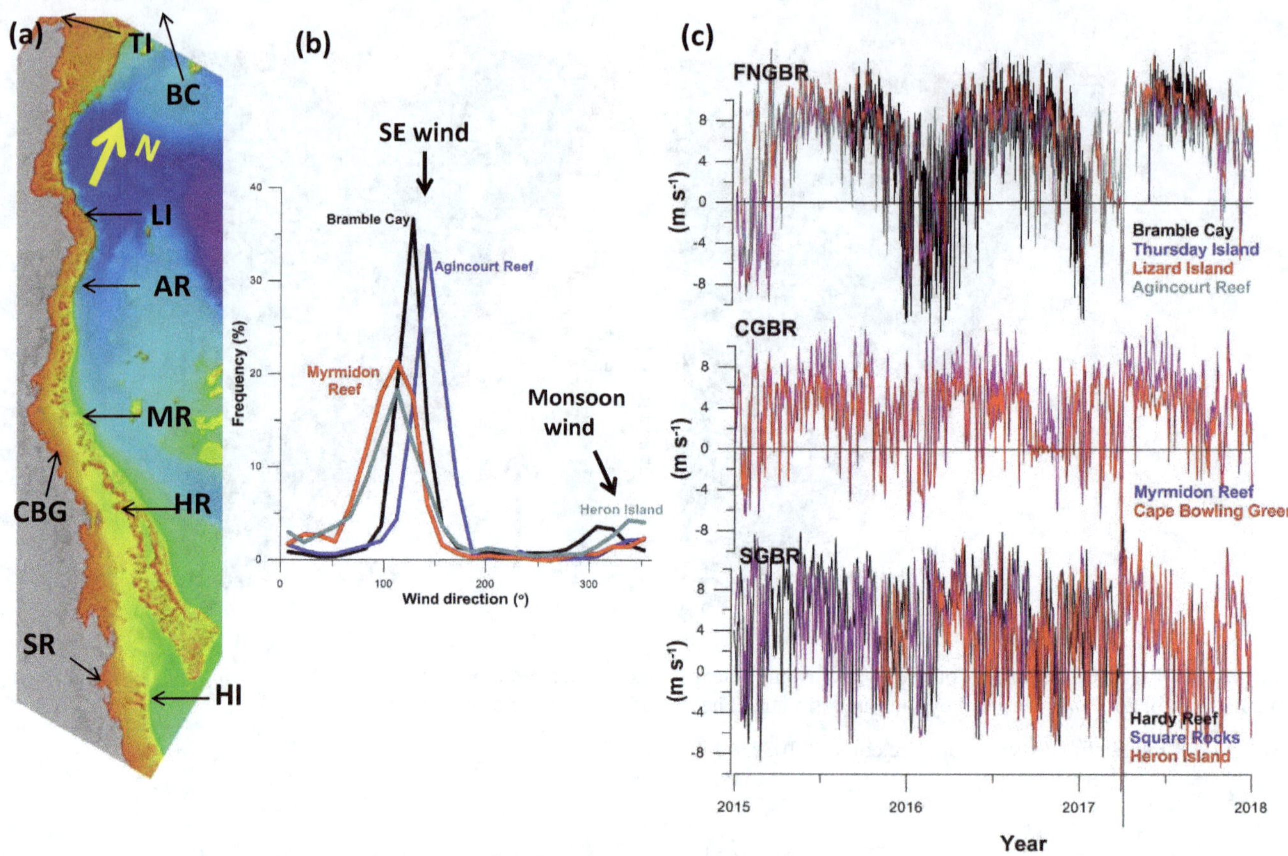

FIGURE 5 (a) A location map of the weather stations. (b) A histogram of the wind direction. (c) Time-series plot of the southeasterly wind component in the FNGBR, CGBR, and SGBR. Positive values indicate a southeasterly wind component, negative values a northwesterly wind component.

Source: AIMS

and southern boundaries to experience the influence of the different weather systems in tropical and temperate Australia (Figure 5c).

2.2.4. Waves in the Coral Sea

Wave data in the northern Coral Sea are few; until recently, the only wave data were from a wave gauge moored near Myrmidon Reef at the shelf break of the CGBR (Wolanski 1986). More recently, wave data from radar and altimetry have become available (Jaffres et al. 2010; Gallop et al. 2014). These data show a dominance of short wind-driven waves (a period of typically 4 seconds) and of swell waves (period of 8–10 seconds). Upon breaking over the outer reef crest, these waves generate an inflow of oceanic water over the reef flat.

2.3. The Rugged Bathymetry Generates a Bioengineered Water Circulation

A major theme of this chapter is that the rugged bathymetry of the GBR not only modifies the water currents but also completely transforms the water circulation at the scale of individual reefs to the point that, through energy dissipation at these small scales, both the small- and large-scale water circulation are completely reshaped. The fluid mechanics of the GBR turbulence are combined in both 2D and 3D aspects. In 3D turbulence, large eddies decay into small eddies, which decay into smaller eddies that dissipate by viscosity. In 2D turbulence, there is an inverse cascade where small eddies merge into larger eddies (see, for example, the animation at https://vimeo.com/213705279). Large eddies could thus form in the GBR as it is shallow. However, the shallowness of the GBR shelf implies a high friction that quickly dampens such eddies.

Thus, in the GBR, topographically generated eddies dissipate the tidal energy sufficiently fast such that they create a novel, bioengineered water circulation completely different from the one that would prevail in the absence of reefs. There are currently 12 known such processes of reef bioengineered water circulation in the GBR shelf (Figure 6). To explain this reef-generated physical oceanography, this chapter starts with the flow past one headland, and it progresses to increasingly topographically complex reef systems.

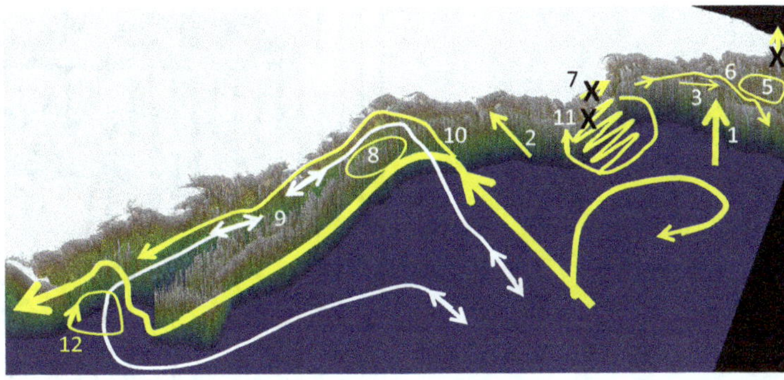

FIGURE 6 A vertically distorted 3D view of the GBR showing some of the key processes by which the reefs create a bioengineered water circulation. (1) Bernoulli tidal upwelling in reef passages. (2) Inflow of oceanic water from waves breaking on the reef crest. (3) Channelisation of the tidal flow on the shelf. (4) Tidal friction, reefs, and islands blocking the mean flow through Torres Strait. (5) Tidal friction in a high-density reef matrix deflects intruding wind-driven water (6) back out to sea. (7) Reefs create a bottleneck preventing throughflow from the NGBR to the FNGBR. (8) Patchiness in a reef matrix due to the sticky water effect. (9) Convergence of tidal waves from the north and south in the southern GBR. (10) Reefs reduce the oceanic inflow on the GBR. (11) Reefs prevent oceanic swell waves and mesoscale turbulence to spread onto the shelf. (12) To follow the bathymetry, the EAC changes direction around the Swain Reefs, and this generates transient eddies (the Capricorn Eddy) offshore from the Capricorn Channel. **X** indicates that the flow is blocked.

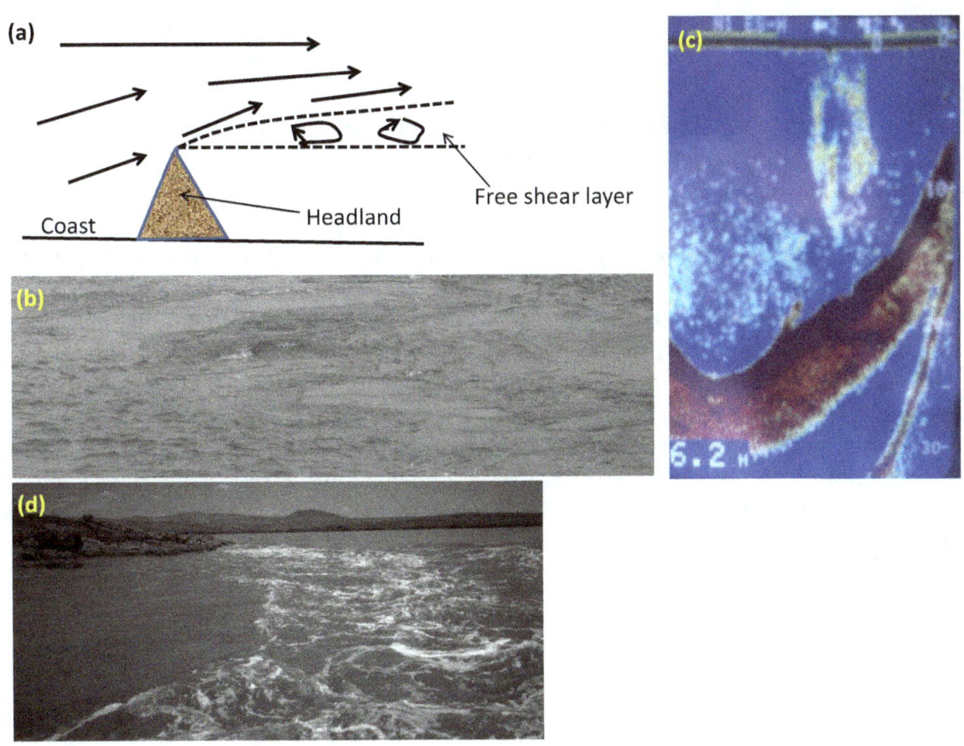

FIGURE 7 (a) Sketch of the dynamics of the free shear layer behind a headland in shallow coastal waters for moderate tidal currents. (b) A photograph showing the smooth surface of eddies in the free shear layer for moderate tidal currents. The eddies are 1–3 m wide at the surface. (c) The backscatter from an echo-sounding transect across such an eddy shows that the eddy depth is typically 5–10 times larger than its width. (d) A photograph of a highly turbulent free shear layer behind a headland.

2.3.1. Flows Past a Headland in Shallow Coastal Waters

Provided the tidal current speed is larger than approximately 0.1 m s^{-1}, which is common, the flow past a sharp headland in shallow waters separates at the tip of the headland to form a free shear layer (Figure 7a-c). The free shear layer separates the nearly stagnant waters in the lee of the headland from the free-flowing waters offshore. Within the free shear layer, there are eddies as evidenced by their smooth surface, which is due to the upwelled eddy water having just reached the surface and thus not being disturbed yet by the wind (Wolanski 1993). Echo-sounding reveals that such eddies readily reach a depth of 6 m for a tidal

ANIMATION 3 A video of the upwelling generated by a phalarope spinning.

Available online at https://www.routledge.com/9781032340692

Source: William Hamner kindly provided this video

ANIMATION 4 A computer animation of the predicted (and verified against field data; Wolanski et al. 1996) 3D flows around Rattray Island over a tidal cycle. The horizontal flow is shown by arrows, and the vertical flows are shown by colour shading; the colour bar shows the scale for the vertical velocity. Surface aggregation of floating matter occurs at the edges of the eddy where material is brought in by the currents and cannot sink with the downwelling water as it is buoyant.

Available online at https://www.routledge.com/9781032340692

current of about 0.2 m s^{-1}, and the upwelling velocity may reach 0.05 m s^{-1}. The upwelling is due to the rotation of the water around the vertical axis of the eddy. The centrifugal acceleration generates a radial sea surface slope, and, in turn, this generates a radial hydrostatic pressure gradient, which generates a horizontal converging flow of water near the bottom; this water is then upwelled.

This process brings bottom material to the surface, which, in turn, attracts seabirds. Some reef fish preferentially recruit on such headland fronts with consistent oceanographic and habitat features (Booth et al. 2000). Similar dynamics are exploited on a much smaller scale by the spinning behavior of the phalarope bird (Obst et al. 1996), as the resulting vortex upwells shrimps to the surface for the bird to feed on (Animation 3).

2.3.2. Flows Past an Island in Shallow Waters

The GBR is shallow in the fluid dynamics framework such that the depth is much smaller than the horizontal size of reefs and islands. An island or reef in such shallow waters typically has two headlands. Each headland generates a free shear layer, and these merge in the lee of the island to form an island wake (Figure 8a). The flood tide eddies behind an island do not simply transfer from one side to the island at ebb tide as the tidal current reverses direction. They cannot do this because they have a different sign of rotation. Instead, as the tide reverses, these eddies are advected away on the opposite side of the island where they dissipate their energy through bottom friction. Island wakes thus function as energy dissipators, that is, all the tidal energy facing an island or reefs is dissipated within one tidal cycle. They also function as mixers by spreading their waters over wide areas. The island wake is unsteady, and it has an upwelling velocity—this makes the eddy visible as bottom mud is upwelled to the surface. The free shear layers separating the waters inside and outside the island wake (7a and 8a) have an elevated level of turbulence because the horizontal and vertical variations of all the components of velocity contribute to the production of turbulence (Wolanski et al. 1996). In contrast, the turbulence in the rest of the flow is much smaller as it is generated by the vertical shear of the horizontal velocity (Fisher et al. 1979). As a result, the eddy viscosity in the free shear layer is large and prevents a vertical shear of the horizontal currents. This prevents an exchange of water between the waters on either side of the free shear layer, resulting in a downwelling of water along the sides of the island wake (Animation 4). This downwelling is made apparent by the aggregation of floating matter such as coral spawn (the pinkish streak in Figure 8b). The surface aggregation of floating matter occurs at the edges of the eddy where material is brought in by the currents, but it cannot sink with the downwelling water because it floats.

The shape and size of the island/reef wakes vary with the island/reef shape and depth and with the water currents. As in classical laboratory fluid mechanics, the island wakes in shallow waters attain many forms, including attached flows (i.e., no flow separation), bubble-shaped island eddies such as in Figure 8a, meandering wakes, Karman vortices, and fully turbulent wakes. Examples of such features are readily visible in Figure 9.

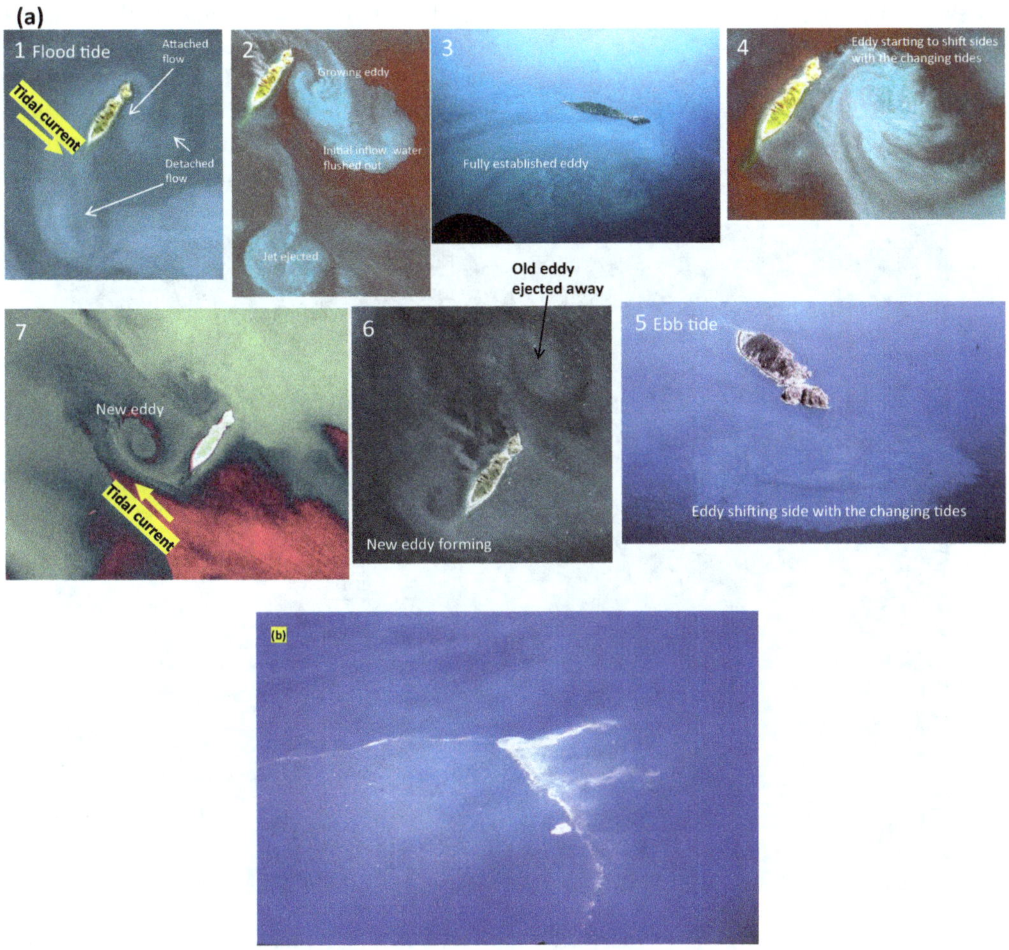

FIGURE 8 (a) Sentinel-2 and aerial images of the island wake at Rattray Island in shallow coastal waters of the SGBR over (1–3) the developing flood tide, (4–5) the changing tides from flood to ebb, and (6–7) the developing ebb tide. The island is 1.24 km long from tip to tip. (b) An oblique aerial image of floating coral spawn aggregating along the side of an island wake because of a downwelling at the outer edges of an island wake behind a reef where coral spawning occurred.

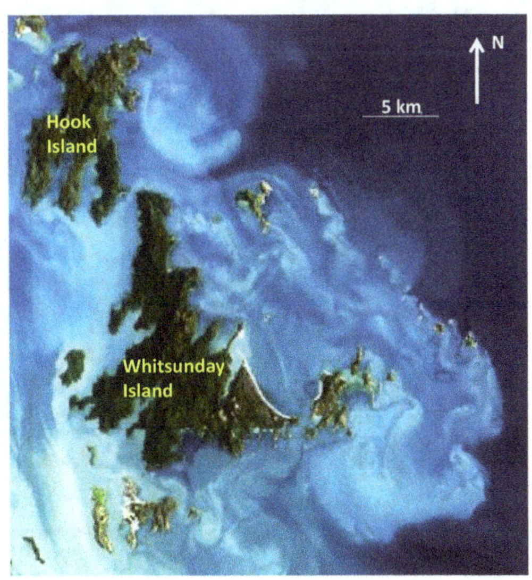

FIGURE 9 A Landsat 8 image of the Whitsunday Islands on 20 February 2019 during calm weather and spring flood tides, showing numerous island wakes and free shear layers of various shapes and sizes and the resulting widespread sub-mesoscale and macroscale turbulence.

2.3.3. Flows Past a Linear String of Islands and Reefs

Satellite images show sub-mesoscale turbulence around a linear string of islands in the SGBR (Figure 10). Two such images at 2 m horizontal resolution were obtained 54 s apart. From the displacement of the surface features, it was possible to estimate the instantaneous horizontal velocity field (Delandmeter et al. 2017). This was used to successfully verify for this instant in time a 2D model of the water circulation. The predicted flow field (Animation 5) shows both production and dissipation of sub-mesoscale and

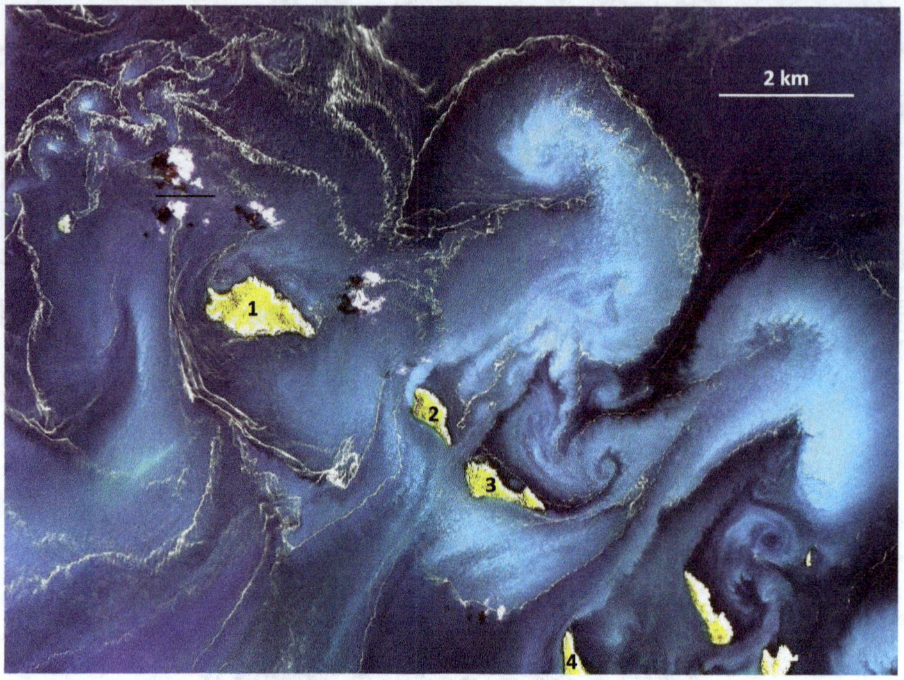

FIGURE 10 A WorldView-3 satellite image on February 27, 2015 at 10:10 AM local time at falling tide showing sub-mesoscale turbulence around a string of islands in the SGBR. (1) Minster Island. (2) Beverlac Island. (3) Hull Island. (4) Noel Island.

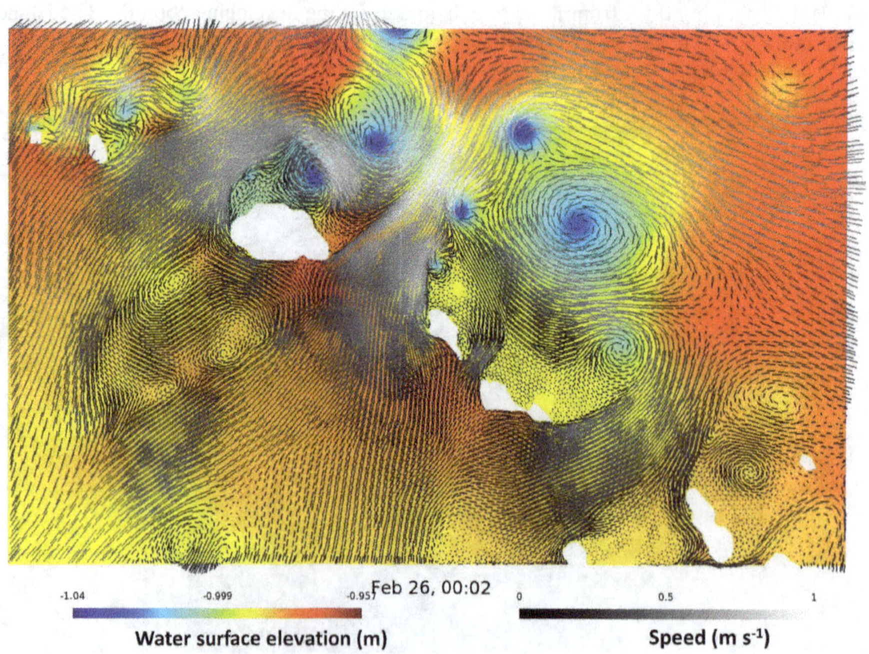

ANIMATION 5 A computer animation of the SLIM2D prediction of the horizontal flow field over 2 days around the string of islands in the SGBR shown in Figure 10. The sea level is shown by colour, and the velocity is shown by arrows.

Available online at https://www.routledge.com/9781032340692

Source: Philippe Delandmeter kindly provided the video clip.

macroscale turbulence at tidal periods, that is, the islands and reefs behave as dissipators of energy and mixers of water.

2.3.4. Flows through a Reef Matrix: The Tidal Friction and Sticky Water Effects

The tidal flow through a high-density reef matrix (i.e., numerous reefs close together) is highly dynamic with swift tidal currents forming island wakes, which makes the tidal flow field appear chaotic (Figure 11).

The mean current that faces a dense reef matrix with swift tidal currents is deflected sideways to flow around the reef matrix and not through it; this is the sticky water effect (Spagnol et al. 2001). This effect is important as it deflects the mean flow between the CGBR and the SGBR sideways around Old, Stanley, and Darley Reefs that form the boundary between these two regions (Animation 6a); thus, it limits the exchange of water between the SGBR and the CGBR to the small gap between Cape Upstart and Old Reef. The sticky water effect also prevails in the Whitsunday Islands archipelago in the SGBR where the tidal currents are strong (Animation 6b). During calm weather, the sticky water deflects the net southward coastal current to flow around the Whitsunday Islands archipelago and not through it (Animation 6c). Ecologically this effect may be important as it helps explain the genetic discontinuity of dugongs north and south of the archipelago (Marsh and Cleguer 2024). This effect is also important in the SGBR reef matrix as it leads to patchiness with zones of high connectivity between reefs and other zones of high self-recruitment (Andutta et al. 2012). This effect also inhibits the mean water circulation in the FNGBR north of Cape Grenville and south of Cape York, and this redirects seawards and landwards oceanic water imported on the shelf further south where the reef density is smaller (Wolanski and Lambrechts 2020).

The physical explanation for the sticky water effect is the tidal friction, which is due to the non-linear interaction between tidal and mean currents through the bottom friction term, τ, in the momentum equations (Wolanski and Thomson 1984). The instantaneous value of τ is

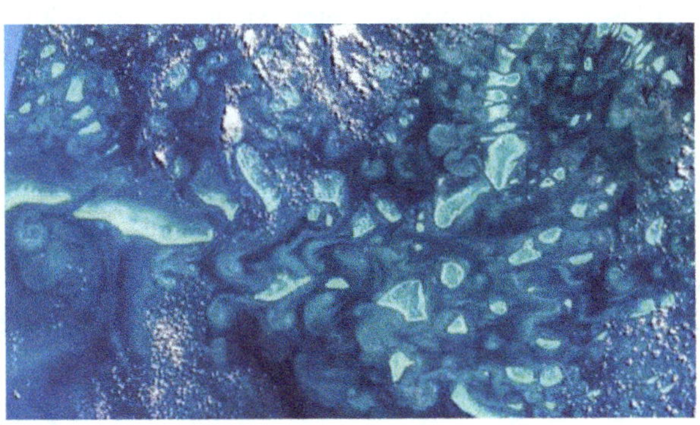

FIGURE 11 A satellite image showing the turbulent tidal flows in a reef matrix in the SGBR.

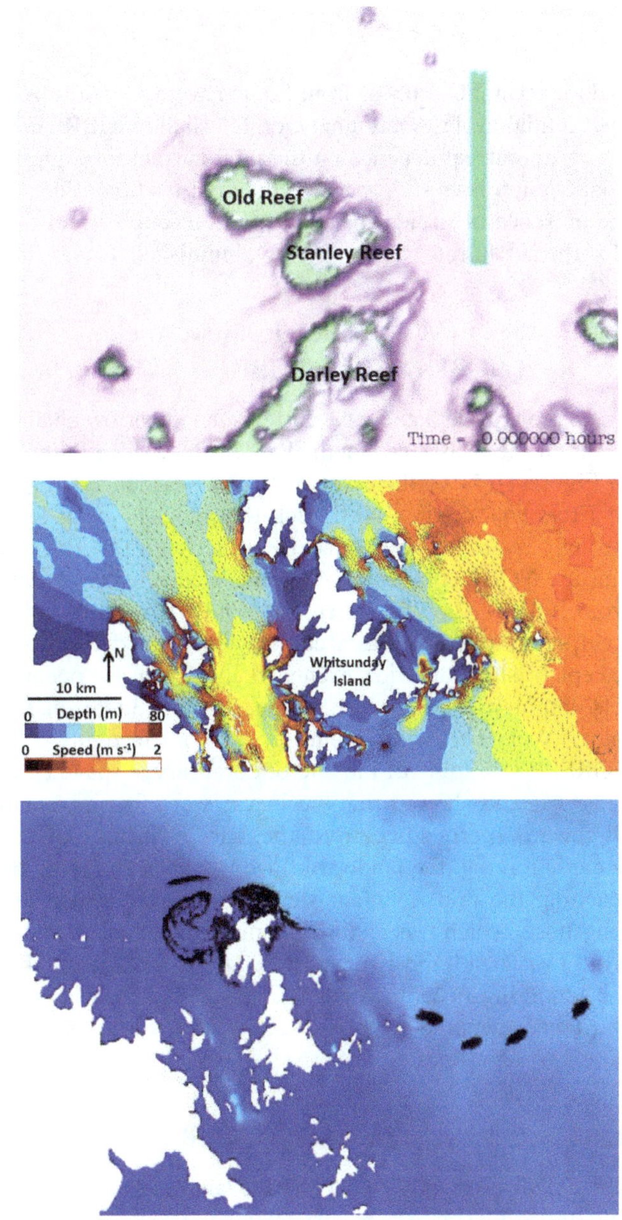

ANIMATION 6 (a) The modeled trajectory of tagged water in calm weather showing the deflection of the mean around Old, Stanley, and Darley Reefs. (b) The predicted tidal currents over one tidal cycle in the Whitsunday Islands archipelago. The colour bar shows the speed in m s^{-1} with a saturation for currents greater than 1 m s^{-1}. (c) The modeled trajectory at a 12-hr interval over 9 days during calm weather of tagged water starting upstream of the Whitsunday Islands, showing the deflection of the net southward currents around the archipelago.

Available online at https://www.routledge.com/9781032340692

$$\tau = C_d\left(\langle \mathbf{u} \rangle + \mathbf{u}'\right)^2 \tag{1}$$

where C_d is a friction coefficient, <> denotes the average over a tidal cycle, <**u**> is thus the mean current, and **u'** indicates the tidal current. Therefore, averaging τ in Eq. (1) over one tidal cycle, the friction term affecting the mean current is

$$<\tau>=C_d\left(<\mathbf{u}>^2+<\mathbf{u}'^2>\right) \quad (2)$$

and the term 2 C_d **<u> u'** from Eq. (1) averages out to zero over a tidal cycle. Accordingly, on the shallow GBR shelf, the mean current depends on the tidal current through the tidal friction term **<u'²>**; this is the tidal friction effect. In the presence of strong tidal currents **u'** through a reef matrix, the net current **<u>** is strongly diminished.

2.3.5. South Warden Reef Isolating the FNGBR from the NGBR

There are no field data on the currents in the narrow channel between the coast and South Warden Reef (location map in Figure 1). Nevertheless, two lines of evidence suggest that this channel is a bottleneck that blocks—or at least strongly restricts—the connectivity between the FNGBR and the NGBR. Firstly, six months of current meter data from 1981–1982 on the inner shelf south of South Warden Reef, suggest a small coastal current with a speed < 0.1 m s^{-1}, except in short-duration events likely driven by oceanic jets impinging on the GBR (see Figure 23d discussed later; Wolanski and Thomson, 1984). Secondly, the CORSPEX95 study of the fate of coral larvae spawned at Lizard Island showed that the larvae were first advected shoreward (perpendicular to the wind) because of the oceanic forcing (the landward flowing SEC), and then, on reaching the mid- to inner shelf they moved southward longshore, which was against the prevailing southeasterly wind (Wolanski and Kingsford 2024). This suggests that the incoming oceanic water was unable to move northward with the wind, probably because of the bottleneck at South Warden Reef.

2.3.6. Oceanic Inflow in the CGBR

The CGBR is the only area where the landward oceanic current (the SEC) can readily intrude onto the continental shelf as the reef density there is ≈ 10% (Figure 1). The passages are very wide, for example, the Palm Passage is 13 km wide. Two quantify this inflow, a current meter was deployed for several years at the shelf break near Myrmidon Reef, and two current meters were deployed for one year in the gap between Cape Upstart and Old Reef: one mooring was near Cape Upstart, and the other mooring was near Old Reef. The data at the Myrmidon Reef mooring site (Figure 12) show that in calm weather and in the monsoon season, the water either came in from the Coral Sea and flowed towards Old Reef-Cape Upstart (i.e., an oceanic inflow in the CGBR) or it left the CGBR to flow seaward northward during strong southeasterly wind (i.e., an outflow of CGBR water to the Coral Sea). The net currents in the gap between Cape Upstart and Old Reef were southward longshore during calm weather and monsoonal winds, and they were northward longshore during strong southeasterly winds (Wolanski and Bennett 1983; Wolanski and Pickard 1985). The return coefficient

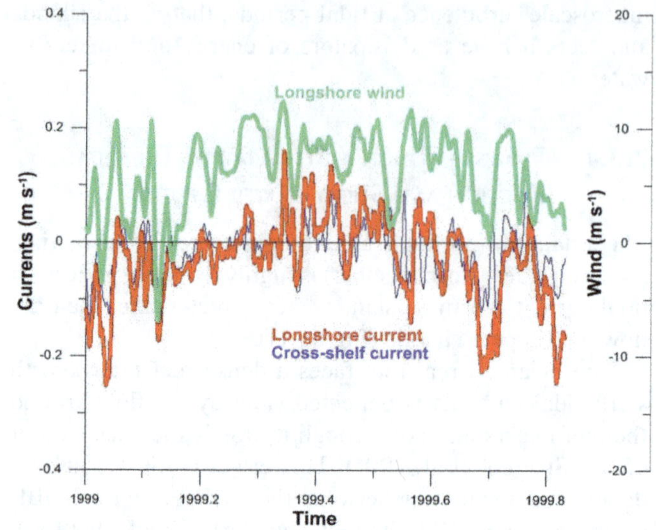

FIGURE 12 A time-series plot of the observed, longshore wind component (> 0 if northward longshore) at Myrmidon Reef in 1999 and of the observed, net longshore current (> 0 if northward) and cross-shelf current (> 0 if seaward) at the shelf break near Myrmidon Reef in the CGBR.

Source: AIMS

of this forward-backward flow between the CGBR and the SGBR was as high as 50% (Andutta et al. 2013).

Thus, during southeasterly winds, the CGBR is not a zone of oceanic inflow, and it does not provide water to the SGBR through the gap between Cape Upstart and Old Reef. To generate the observed net southward current in the central and southern regions of the SCGBR (Middleton 1987), the intrusion of oceanic water into the SGBR bypasses the CGBR and occurs further south through outer reef passages in the SGBR. There are no field data to verify this prediction. The age of oceanic waters intruding in the CGBR was estimated to be between 3 and 5 months on leaving the SGBR, depending on the location (Andutta et al. 2013).

2.3.7. Topographic Driven Upwelling

The internal tide on the continental slope has two modes (Baines 1982). Mode 1 internal tides raise and lower at tidal frequency the thermocline separating the well-mixed upper layer from the deeper water (Figure 2). Mode 2 internal tides are waves below the thermocline within the stratified waters below the thermocline without a movement of the thermocline. Only mode 1 internal waves can upwell water onto the GBR shelf and then only if the slope of the internal tide rays is larger than that of the continental shelf slope. This upwelling brings up nutrients from below the thermocline and supports the GBR ecosystem (Andrews and Gentien 1982). This is the case in the CGBR where upwelling events are common and are due to internal waves over the continental slope that spill over the continental shelf (Figure 13a; Wolanski 1986; Benthuysen et al. 2016). Indeed, 100-m internal waves have been observed there

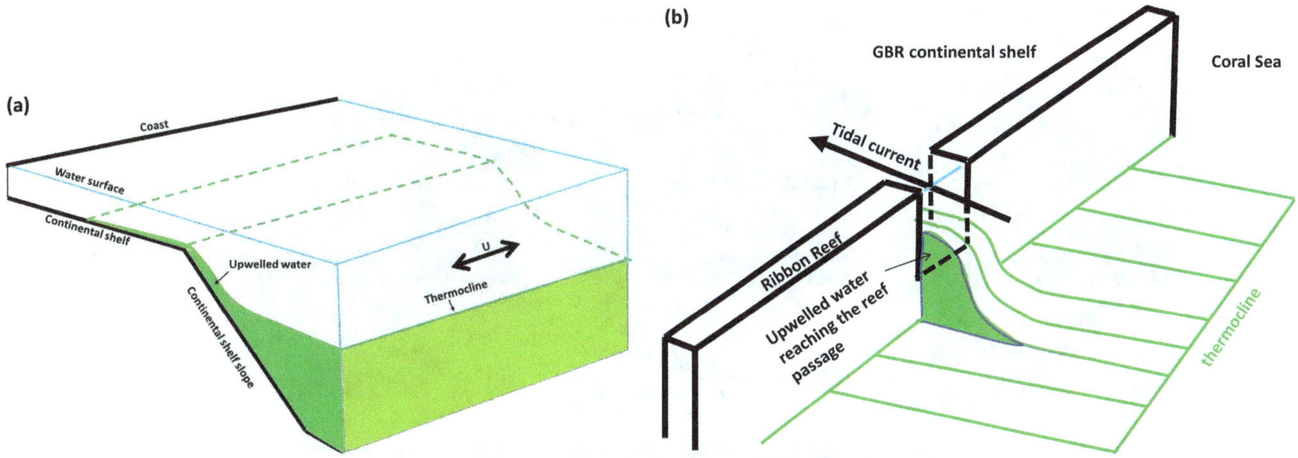

FIGURE 13 Sketch of (a) the upwelling by long internal waves in the CGBR due to the waxing and waning of the prevailing longshore current U on the continental slope and (b) the upwelling in front of a reef passage in the NGBR and the FNGBR.

because of the continental shelf slope being nearly identical to the slope of the internal tide ray in the CGBR (Wolanski and Pickard 1983). This enables the waxing and waning of the oceanic current along the shelf slope to generate occasional upwelling as bottom-tagging, long waves on the outer shelf and the shelf slope of the CGBR, a classic example of the Ekman transport mechanism, which is particularly true in summer when the thermocline is close to the shelf break (Figure 2; Andrews and Furnas 1986; Benthuysen et al. 2016). The upwelled water has been documented to occasionally travel from the shelf break to the mid-shelf of the CGBR.

In the FNGBR, internal tides on the continental shelf slope cannot spill unto the continental shelf because the continental shelf slope is much larger than the slope of the internal tide ray. Upwelling at tidal period still occurs; however, it is restricted to the vicinity of reef passages, and it is due to the strong tidal currents through reef passages (Figure 13b). There, a Bernoulli effect locally lifts the thermocline (Thompson and Golding 1981) by as much as 60 to 80 m during spring tides, which is enough for sub-thermocline water to reach the reef passage (Wolanski et al. 1988). Each reef passage along the FNGBR outer shelf thus acts like a suction pump providing nutrient-rich sub-thermocline water to the FNGBR (Thomson and Wolanski 1984). This would not occur without the reefs; thus, the FNGBR is a bioengineered ecosystem par excellence.

There are no field studies of upwelling in the SGBR, except for those of the upwelling due to the intermittent Capricorn eddy offshore from the Capricorn Channel (Weeks et al. 2013 and 2015; McDuie et al. 2018). The Capricorn eddy results from the instability of the prevailing EAC as it turns 90° around the Swain Reef complex to follow the bathymetry of the continental slope. As a result, the mean current in the ocean immediately offshore from the Capricorn channel is highly unsteady, fluctuating among periods with no eddies, periods with a transient anti-cyclonic eddy, and other periods with a transient cyclonic eddy (Figure 14 a-c). In addition, there are also internal waves in this same area, whose orientation suggests that they are propagating into the Capricorn Channel, with the prevailing tidal currents (Figure 14d).

2.3.8. Tidal Jets

On reaching the passage, the upwelled water is swiftly moved landward on the continental shelf by a tidal jet that can take many forms depending on the bathymetry and the tidal range. Tidal jets are important as they provide upwelled sub-thermocline nutrients to meadows of the calcareous green alga *Halimeda*, which are located a few kilometre landward of reef passages (Wolanski et al. 1988). The jet is attached to the reef at the start of the flood tide (Figure 15a), and it can remain attached throughout the flood tide if the tidal currents are too small to generate flow separation (Figure 15b). It can be detached and forms a thin, turbulent jet (Figure 15c) if the width of the reef passage is approximately equal to the depth so that bottom friction effects do not control the jet dynamics. A mushroom jet (Figure 15d) forms when the width of the reef passage is larger than the depth so that bottom friction effects control the dynamics. Mushroom jets are predicted to have strong 3D motions at their leading edge (Figure 15e), which deliver the upwelled nutrients to Halimeda on the seafloor and to plankton in the water column, but there are no field data to verify this prediction. Some tidal jets meet a reef when propagating shoreward, and this makes them split into two or three smaller mushroom jets (Figure 15f), thus increasing the turbulence. Two nearby parallel tidal jets can also merge (Figure 15g), a finding that extends to scales of a kilometre, this result well known in engineering fluid mechanics at scales of centimetre to metre (Lee and Hassan 2018).

In reef passages of the SGBR, tidal currents can reach 4 m s^{-1}, which is over twice that measured (but then only in a few passages) in the FNGBR (Marmorino 2022). A distinguishing characteristic of fast flowing waters is the occurrence of intermittent, upward bursts of turbulence

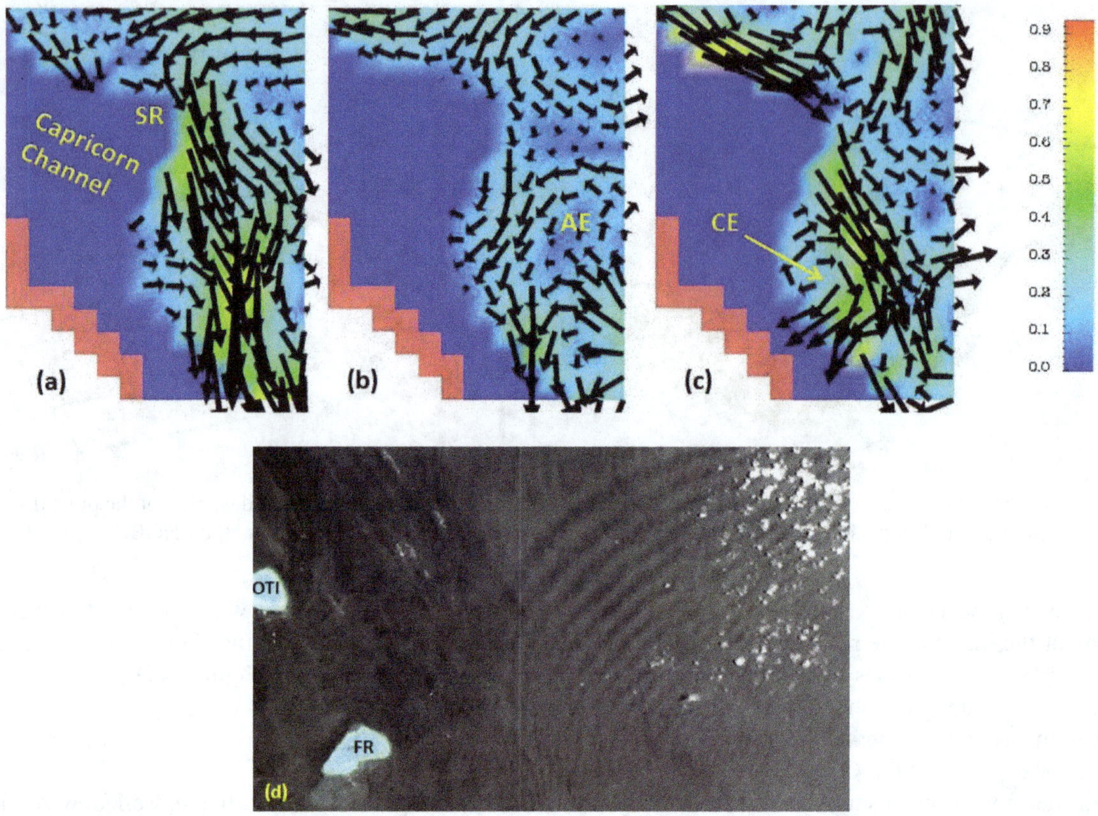

FIGURE 14 Snapshots of the altimetry-derived 5-day average near-surface currents in 2019 offshore from the Capricorn Channel showing (a) no Capricorn eddy, (b) an anticyclonic eddy (AE), and (c) a cyclonic eddy (CE). SR = Swain Reefs. (d) A Landsat-8 image of internal waves just offshore from the Capricorn Channel. The wavelength is 1.74 km. OTI = One Tree Island; FR = Fitzroy Reef.

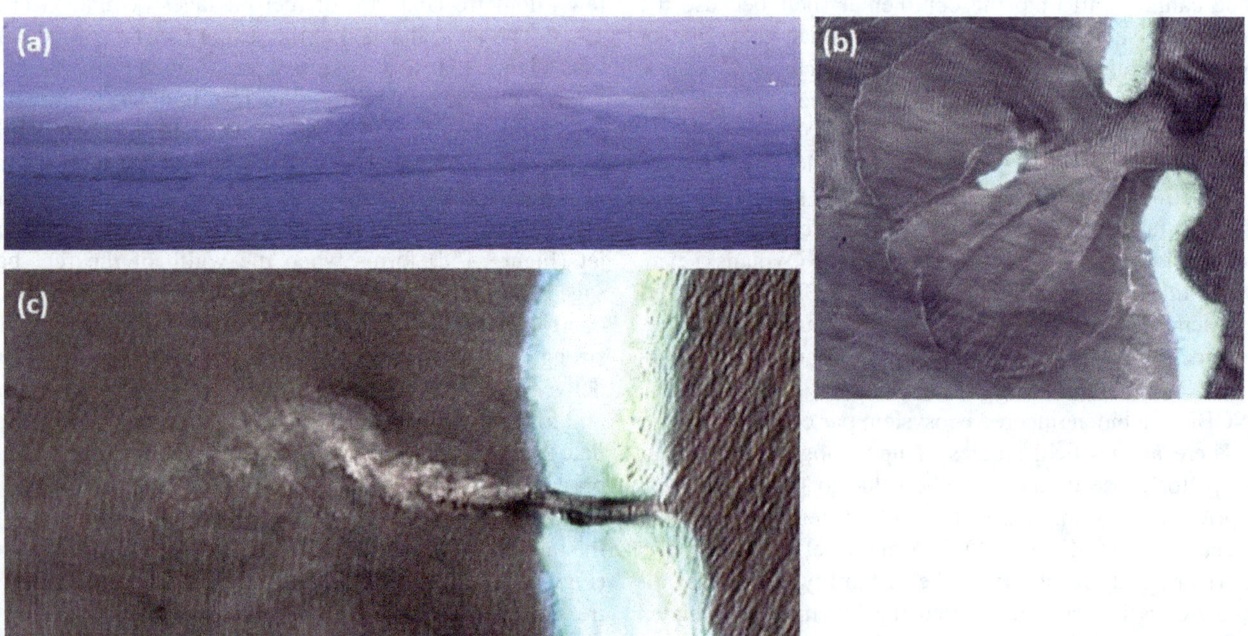

FIGURE 15 (a) An aerial photograph of an attached tidal jet at the start of the rising tide. (b) A Sentinel-2 image of an attached tidal jet. (c) A Sentinel-2 photograph of a thin turbulent tidal jet. (d) An aerial photograph of a wide, detached mushroom tidal jet. (e) The predictions by Deleersnijder et al. (1992)'s island wake model, which is adapted to the reef passage bathymetry and its open boundary conditions, of the 3-D motions of a developing mushroom tidal jet 100-m wide in the reef passage and spreading in vertically well-mixed water over a 30-m deep flat shelf. (f) A Sentinel-2 image of a tidal jet that meets a reef and splits into three mushroom jets, leading to increased turbulence. (g) A Sentinel-2 image of two merging tidal jets.

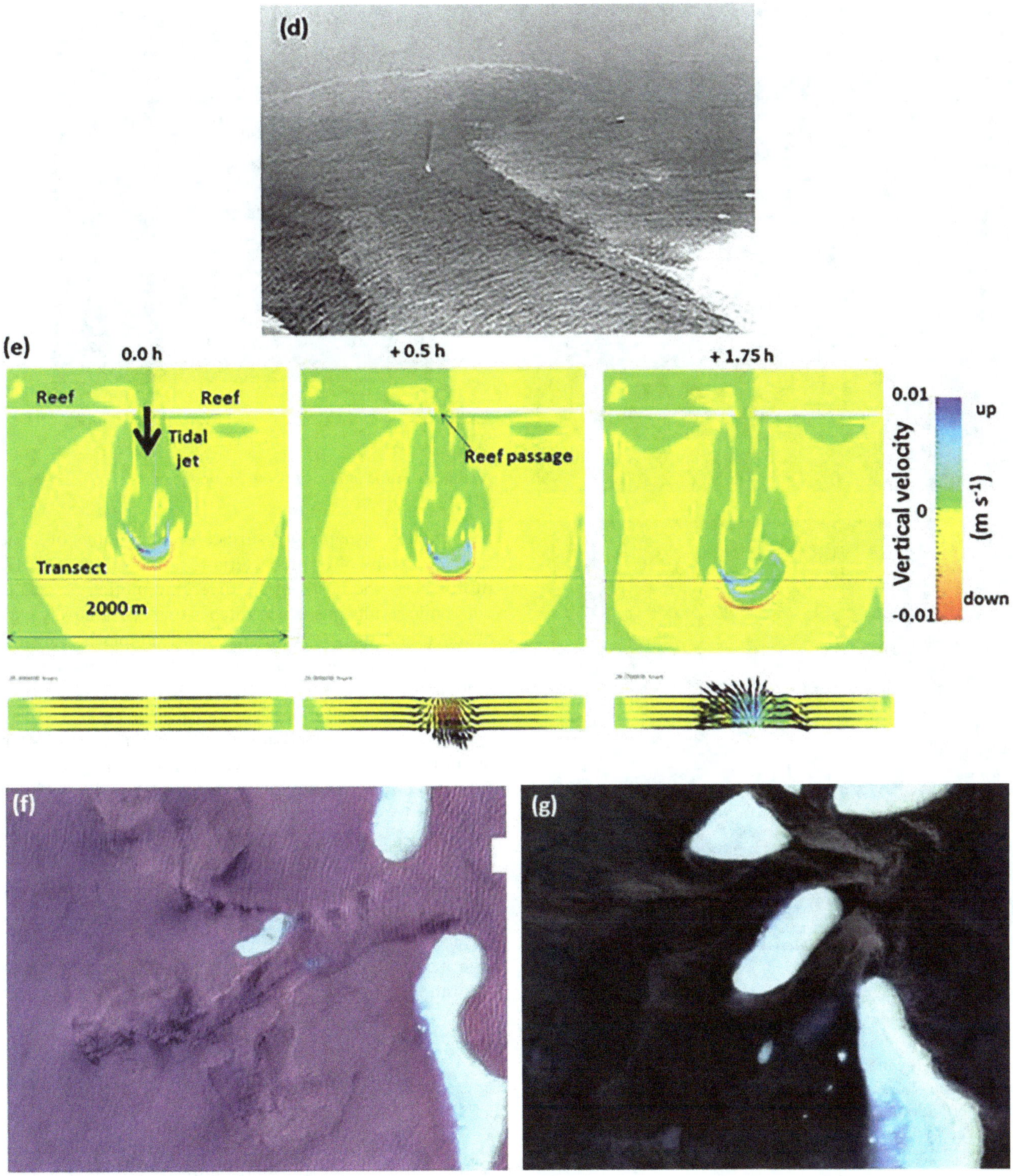

FIGURE 15 *(Continued)*

that originate from the shear layer above the seabed. In weakly stratified water, bursts impinge on the sea surface to produce boils—approximately circular patches of diverging motion where the water is brought to the surface from below and spreads on the sea surface. The spreading motion results in a surrounding band of convergence where there is a steepening and often breaking of small-scale surface waves (Longuet-Higgins 1996). As a result, the outline of a boil becomes visible as a ring (or partial ring) of enhanced sea surface roughness and increased foam and bubbles from breaking waves. An example is shown in Figure 16a. The bright features in the figure are surface roughness signatures, that is, enhanced scattering of sunlight from bubbles and foam from breaking waves; feature 'A' is a partial

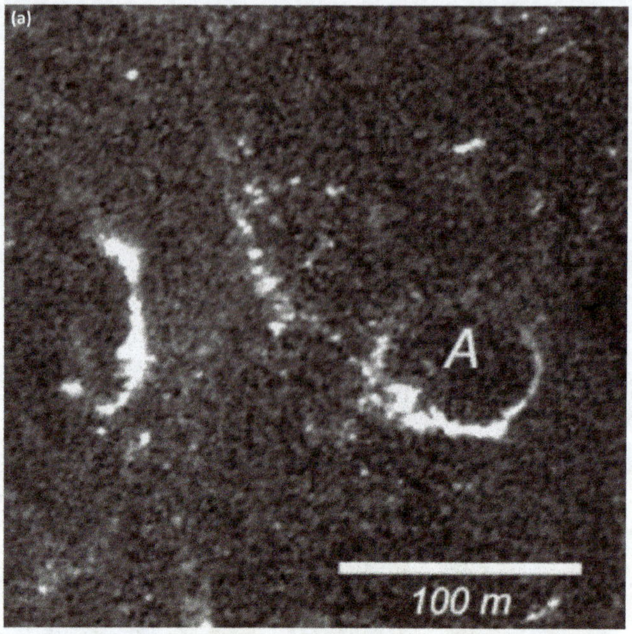

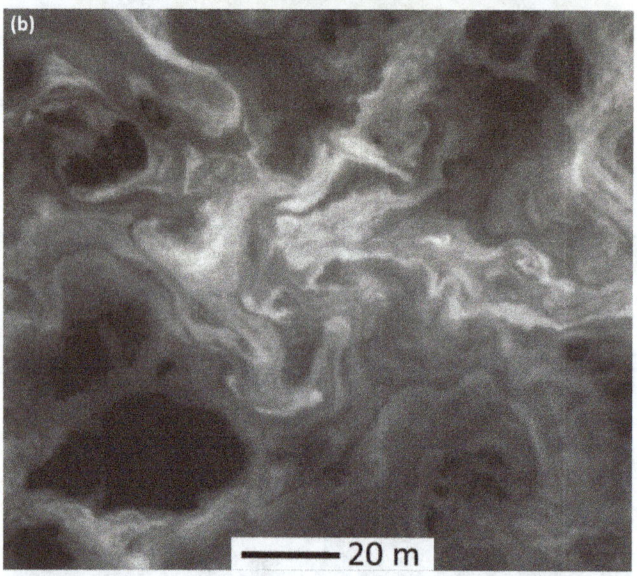

FIGURE 16 (a) WorldView-2 satellite image of a reef passage in the SGBR showing a particular boil signature (feature 'A'). Local water depth is 36 m. Data are from the WorldView-2 satellite (Maxar Technologies, Westminster, CO, USA; adapted from Marmorino 2022). (b) A WorldView-3 satellite image of cyanobacteria bloom having dark patches (no algae) believed to be produced by boils. Location is in the SGBR, north of Townsend Island (22.25°S, 150.5°E), in water approximately 30 m deep.

Source: Adapted from Marmorino et al. 2017

outline of a particular boil. Over time (Animation 7), the boil can be seen to grow and then dissipate over a period of approximately 100 seconds. From the rate of expansion of the boil signature, a radial surface velocity can be estimated as being about 0.25 m s^{-1}, or approximately 8% of the background surface flow speed in the reef passage.

In less energetic GBR shelf waters, under relatively calm wind conditions, the occurrence of boils can be revealed as

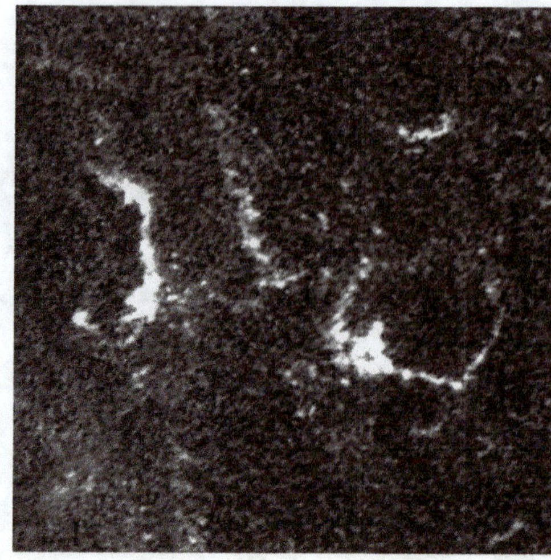

ANIMATION 7 A time-lapse video of the evolution of feature 'A' in Figure 16a.

Available online at https://www.routledge.com/9781032340692

a continuing reshaping of surface accumulations of algae, such as blooms of cyanobacteria (Marmorino et al. 2017). In this case, the horizontally diverging motion associated with boils results in surface patches that are relatively clear of algae and that grow in size. An example is shown in Figure 16b; the boil-induced divergence values are of the order of 10^{-3} s^{-1}.

Shown in Animation 7 are nine sequential views of approximately the same area of ocean surface as feature 'A' moves down the reef passage. The mean time interval between frames is 12.9 s, and each frame has a size of 256 m by 256 m (adapted from Marmorino 2022).

Flow separation can occur at each side of the entrance of a reef passage. If the passage is funnel-shaped, then this can generate a sharp convergence front (Figure 17a) that aggregates surface material in a slick line that can attract surface feeders (Wolanski and Hamner 1988). If the passage has parallel sides or if the passage is short, then the separation layers may not meet and do not form a convergence front in the passage (Figure 17b).

2.3.9. Vertical Stratification on the GBR Continental Shelf

In the FNGBR away from river plumes, no significant vertical stratification in temperature or salinity has been measured at more than a few kilometre landward from the outer fringing reefs, that is, at the limit of the intrusion on the shelf by tidal jets with their upwelled sub-thermocline Coral Sea water (EW, unpubl. data). In the CGBR away from river plumes, occasional stratification events have been noticed on the outer shelf of the GBR, and they are due to the spread of sub-thermocline Coral Sea water upwelled by low-frequency mode-1 internal waves on the continental slope (Benthuysen et al. 2016). In the SGBR, a vertical stratification occurs on the mid-shelf, although its frequency

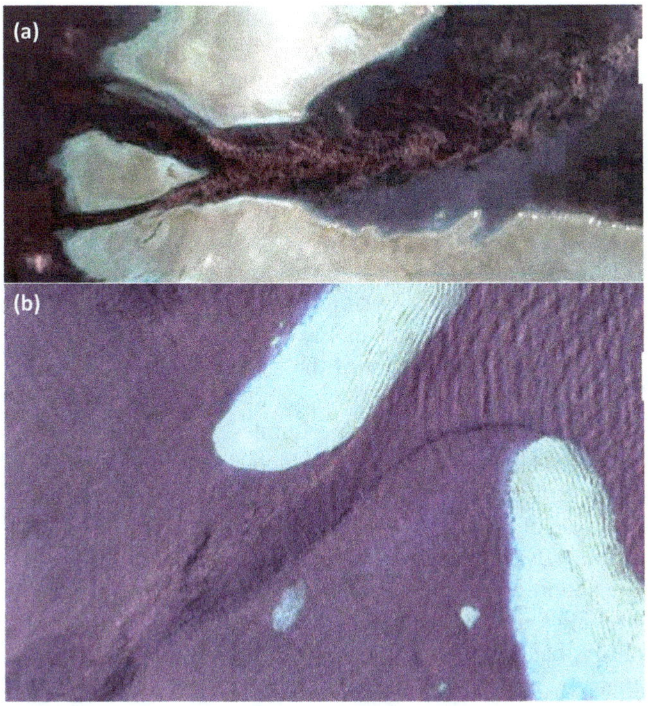

FIGURE 17 Sentinel-2 images of (a) the convergence zone in a funnel-shaped reef passage and (b) two free shear layers that do not merge in a short reef passage.

and duration are unknown. That this stratification exists is evidenced in Sentinel-2 and LANDSAT-8/9 satellite images that show expressions of internal waves. A compelling example, which has at least five distinct packets of waves, is shown in Figure 18. A prominent packet, having seven wave crests, lies just below the center of the image. A southward propagation direction can be inferred from the rank ordering of the wavelengths. The waves leading the packet have a wavelength of approximately 275 m, while that of the trailing waves is about 125 m. Wave crests are markedly curved, indicating either a local generation source or refraction by spatially varying currents. Local generation may result from direct interaction of tidal currents with bathymetry, or internal waves may be forced by intrusion of reef-mixed fluid into a stratified ambient. The present example suggests that upwelled Coral Sea water can penetrate quite far inshore within the SGBR. This finding calls for future research.

2.3.10. Wave Breaking

The fringing reefs on either side of a reef passage not only increase the speed of the tidal jet in the reef passage but also let some oceanic water (and its plankton, planktivores, and detritus) onto the continental shelf by the breaking of waves onto the reef crest. The breaking of oceanic waves at

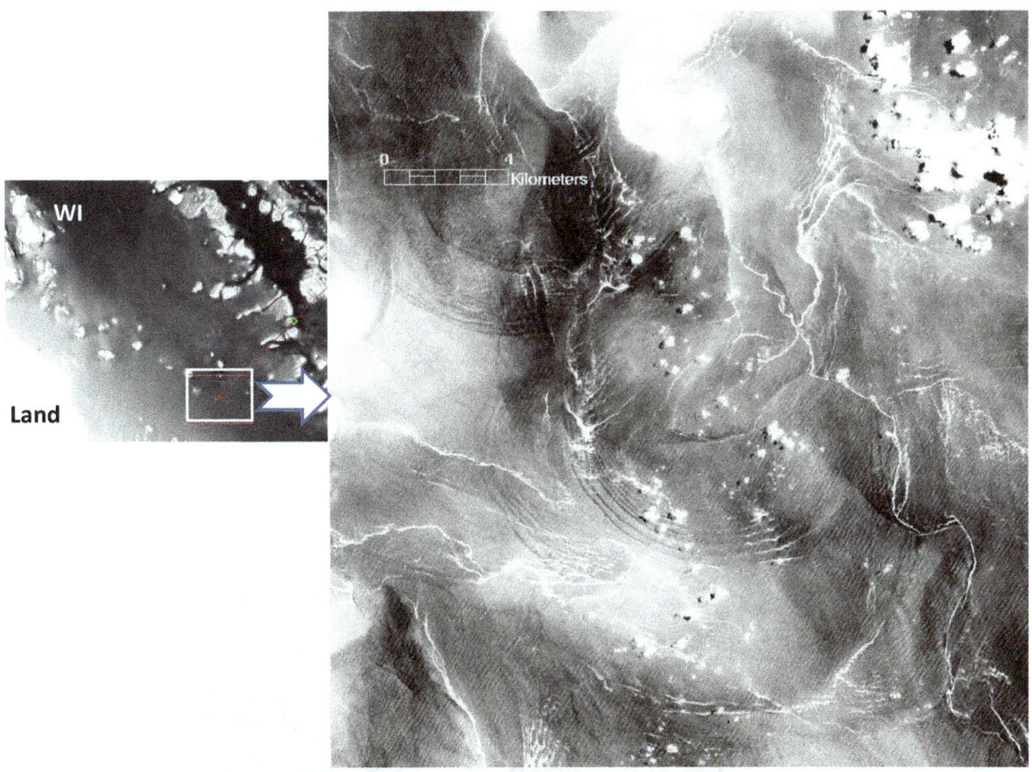

FIGURE 18 (Left) A location map. (Right) A LANDSAT-9 image showing likely surface expressions (the circular lines) of internal waves in the SGBR approximately 100 km inshore of the shelf break. The image was acquired on March 1, 2022 at 09:58:56 local time (AST). Shown are data from visible band 3 (wave lengths of 530 to 590 nm), contrast-stretched to emphasize the internal waves. A circular, white object in the northern (upper) part of the image is Sandpiper Reef (21.025°S, 150.082°E); clouds (and their shadows) are scattered across the image, appearing prominently in the upper right. Tidal phase was high-water slack (end of flood tide); winds were 4.7 m/s from the NNW. WI = Whitsunday Islands.

the reef crest transfers their momentum to the water at the reef crest, and this generates a net current over the reef flat at the back of the reef crest. Secondary waves may also be generated over the reef flat. The dynamics of wave breaking are controlled by the wave height and period and the slope of the reef on its seaward side (Massel 2013). This is strongly modulated by the presence or absence of a spur and grooves structure on the outer reef slope (Duce et al. 2022). An empirical rule from field data is that wind-driven waves drive little flow over the outer reef flat. Instead, most of the wave-driven flow over the reef flat comes from the breaking of swell waves; on breaking, a swell wave of 1 m (2 m, 3 m) exerts on the water at the reef crest a stress equivalent to that of a landward wind of 10 m s^{-1} (20 m s^{-1}, 30 m s^{-1}; Wolanski and Lambrechts 2020 and references therein). Using this rule, it is estimated that approximately half of the oceanic water intruding in the FNGBR during strong southeasterly winds comes from the breaking of oceanic swell waves over the crest of the fringing reefs, and the other half comes from an inflow through reef passages. A major influence of the fringing reefs on the oceanography is to thus prevent the oceanic swell waves from propagating onto the FNGBR shelf. Another influence is that a large part of the wave energy is converted to a landward current over the outer reef flat (Gallop et al. 2014).

The wave dynamics are responsible for the creation and maintenance of stable coral cays (Figure 19; Gourlay 1988). Other aspects of the ecological importance of waves on reefs are discussed in this book by Hamylton et al. (2024) and Robertson et al. (2024).

2.3.11. Boundary Mixing

The reef surface is very rough, and reefs thus exert a high friction on the water currents; for instance, the Manning coefficient, a parameter of the bottom friction exerted on the water by the rugosity of the surface, is typically 0.25–0.35 for a living coral surface and 0.03 for a sandy surface. Furthermore, the high roughness of the coral surface generates boundary mixing, which generates temperature microstructures in the surrounding water in the calm environment of a reef lagoon (Figure 20). Scatterers of the echo-sounder waves (presumably plankton and some sediment) are found aggregated in the micro-thermoclines, but they are not found within 1–2 m from the slopes of the coral; this is because boundary mixing creates water of intermediate buoyancy, and this water spreads horizontally away from the boundary by buoyancy-driven currents (Phillips et al. 1986). The biological impact of such microstructures on plankton dynamics in lakes is major (Prairie et al. 2012),

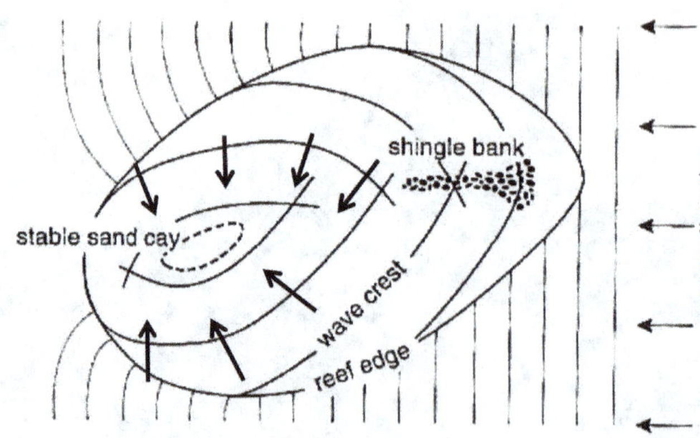

FIGURE 19 Wave refraction on a platform reef. The arrows show the wave direction. The wave convergence forms the stable sand cay.

Source: Modified from Gourlay 1988

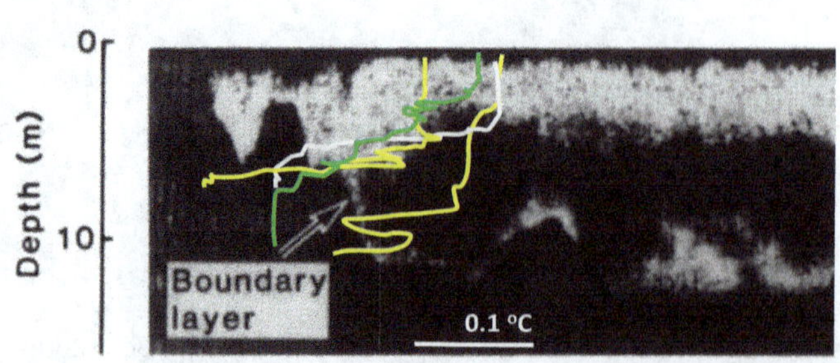

FIGURE 20 Evidence for intense boundary mixing along a coral reef in the calm water of an enclosed lagoon in Davies Reef in the CGBR. The temperature profiles near a sloping coral outcrop showed the presence of microstructures that varied hugely within a few metres and a few minutes from one another, and the echo sounding scatterers (presumably plankton) in the lagoon were absent from the boundary layer generated by boundary mixing (redrawn from Wolanski 1987). The horizontal distance is approximately 50 m.

but this effect has been ignored in studies of turbulence and plankton dynamics and supply to consumers in coral reef environments (Davis et al. 2021).

2.3.12. Steering of the Tidal Wave by the Reef

On the FNGBR continental shelf, the tidal currents are channelized by the reefs (Figure 21). As a result, the propagation of the tidal wave is also steered by the reef-generated bathymetry (Animation 8a). Two different tides meet in Torres Strait: one tide comes from the Gulf of Carpentaria, and the other tide is in the FNGBR. The east-west tidal propagation through the Torres Strait is hindered by the numerous reefs and shoals oriented north-south of Torres Strait, leading to strong currents in passages and much weaker currents in the surrounding areas (Animation 8b; Wolanski et al. 1988, 2013). In the SGBR, the blocking by the reef matrix on the outer shelf is so strong that the tide moves in from the south and from the north, which forms two tidal waves that meet near Broad Sound (Animation 8c; Middleton et al. 1984). This leads at rising tide to southward (northward) currents in the northern (southern) region of the SGBR and strong cross-shelf currents through reef passages. The currents reverse direction 180° when the tide changes from flood to ebb.

2.3.13. Forcing by the Rivers

There are several rivers all along the GBR coast, and although their discharge varies seasonally, all have short-lived (a few days to a few weeks at most) floods following heavy rainfall in the river catchment (Lewis et al. 2024). The rivers discharging into the SGBR and SGBR are subjected to stochastic rain that leads to considerable inter-annual variability of the yearly volumetric discharge, for example, 0.25–54.03 million ML year^{-1} for the Burdekin

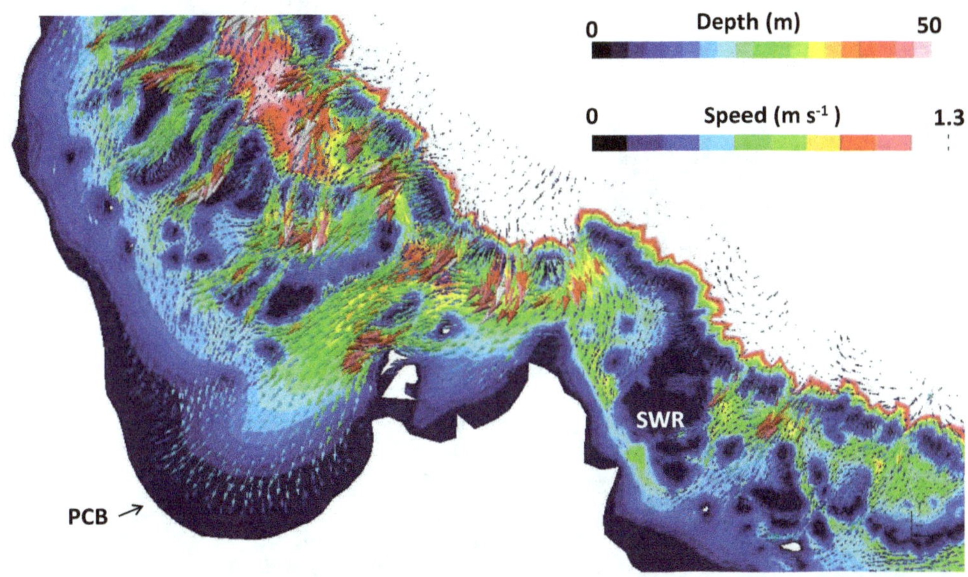

FIGURE 21 A snapshot of the SLIM2D model prediction of the bathymetry (colour) and the tidal currents (arrows) at spring flood tide in the FNGBR around Princess Charlotte Bay (PCB).

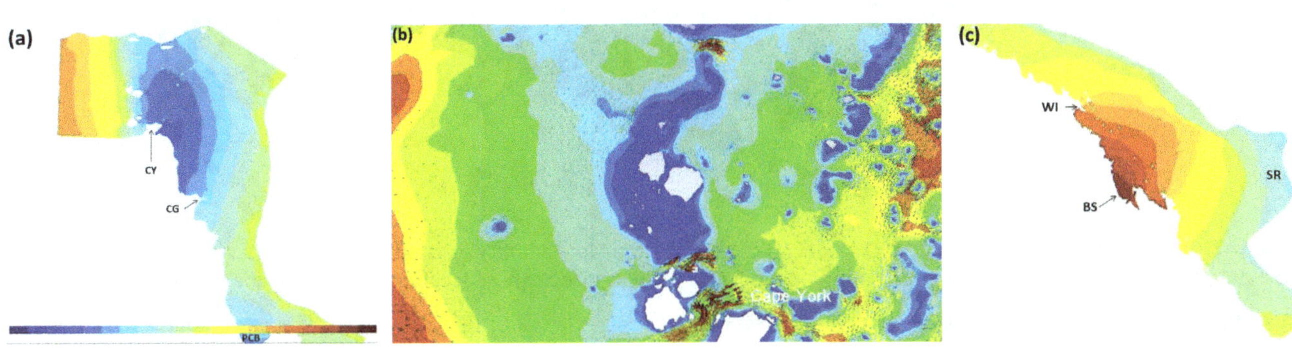

ANIMATION 8 A video of the SLIM2D model prediction of (a) the spring tidal wave propagation in the FNGBR, with a scale of 0–5 m, (b) the water circulation during one spring tidal cycle in the southern Torres Strait during calm weather and spring tide, and (c) the spring tidal wave propagation in the SGBR, with a scale of 0–8 m.

Available online at https://www.routledge.com/9781032340692

River. The discharge of the far north rivers is more predictable by season and years. This terrestrial runoff includes fine sediment, nutrients, pesticides, and herbicides, and it is harmful to the seagrass, corals, and coral reefs of the GBR (Fabricius 2005; Schlaefer et al. 2021; Carter et al. 2021).

The Burdekin River has the largest river discharge during floods, and these floods generate turbid river plumes in the CGBR. These have been extensively studied using salinity data, aerial and satellite images, and models (Animation 9a; Delandmeter et al. 2015). During particularly large floods, the Burdekin River plume tags the coast for more than 200 km (Figure 22), and all long this distance, the plume is reinforced by the outflow from smaller rivers. The plume does not strongly impact mid-shelf reefs until it reaches the latitude of Cairns, at which point river plume water spreads seaward in the Coral Sea off Cairns. This may be one reason why primary outbreaks of coral-eating crown-of-thorns starfish have occurred in reefs off Cairns (Wolanski and De'ath 2005; Pratchett et al. 2024). After the flood has peaked, the plume breaks up in large patches, and

(a)

(b)

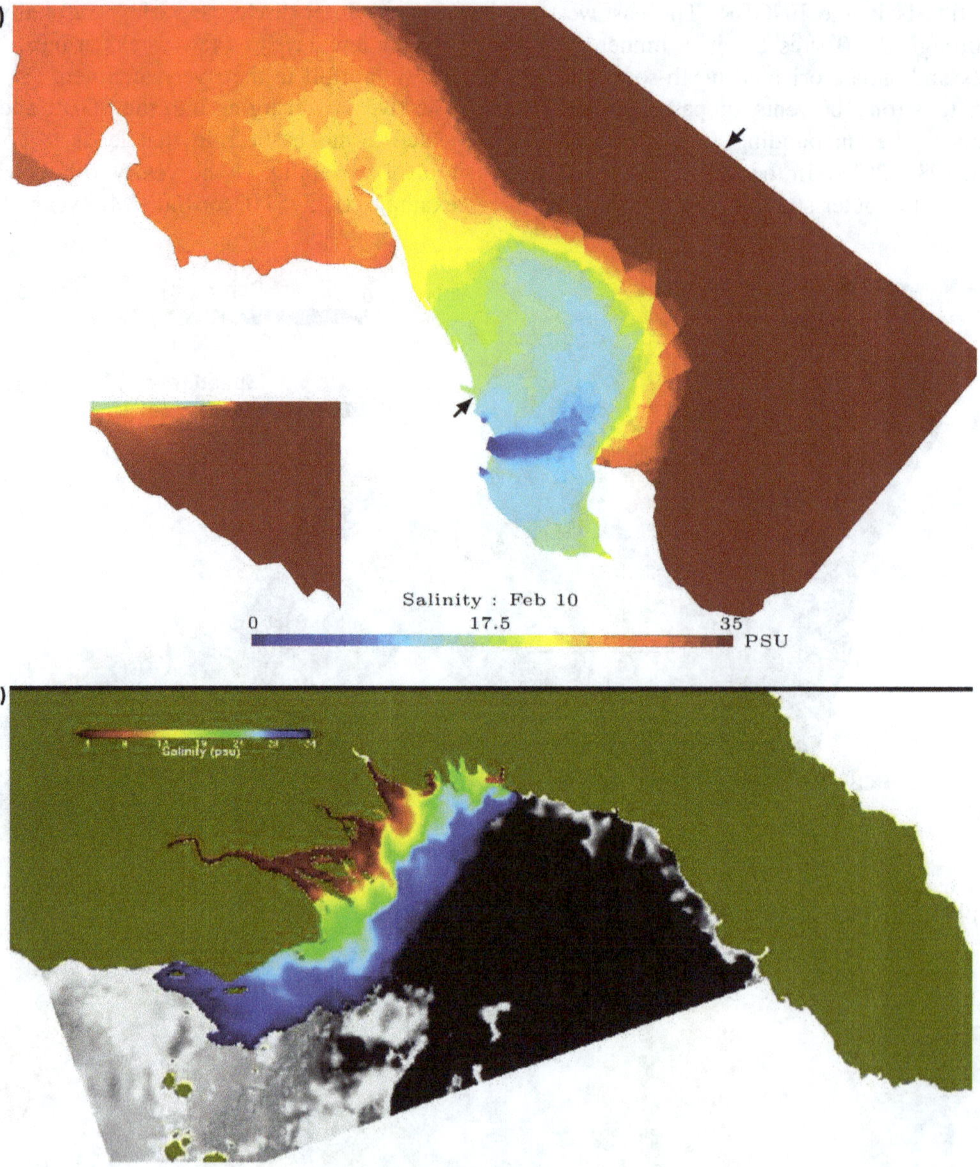

ANIMATION 9 (a) A video of the SLIM3D model prediction of the surface salinity distribution in the Burdekin River plume during the 1991 flood. The inset shows the vertical salinity distribution in the transect whose location is shown by the two arrows.

Available online at https://www.routledge.com/9781032340692

Source: Philippe Delandmeter kindly provided this video clip

(b) A video of the MOHID3D model prediction of the surface salinity in the Fly River plume during August 1994.

Source: Flavio Martins kindly provided this video clip

(c) and (d) Video clips of harpacticoid copepod—*Euterpina*—feeding on a muddy marine snow floc in the Fly River plume.

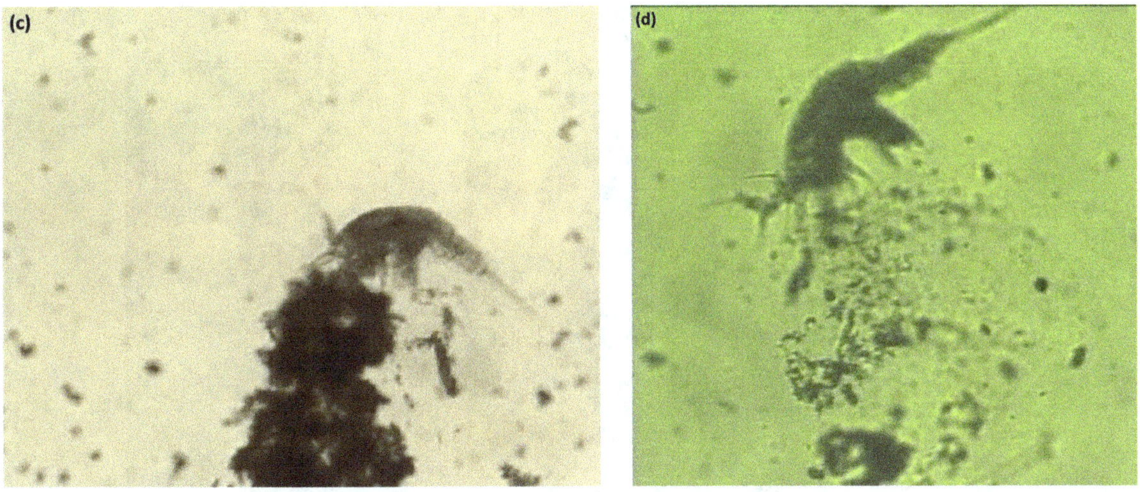

ANIMATION 9 *(Continued)*

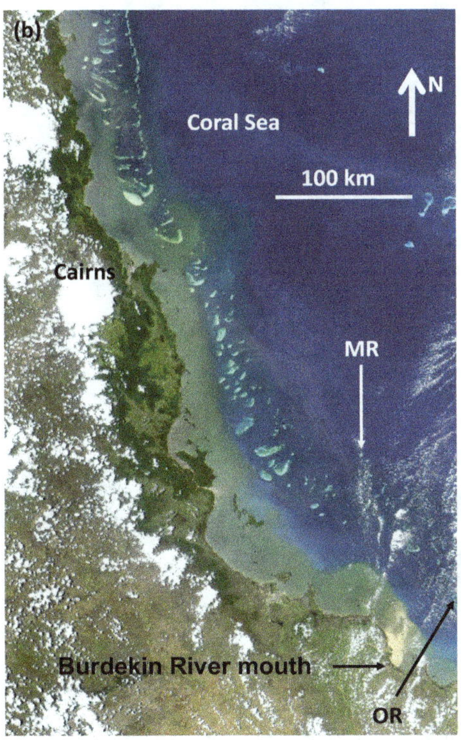

FIGURE 22 A MODIS satellite image of the turbid, coastal-hugging plume of the Burdekin River and several smaller rivers in February 2007.

it disperses widely in one week (Wolanski and van Senden 1982). The Burdekin River plume floats over the more saline water underneath; the water is strongly vertically stratified in salinity. The suspended fine sediment makes the plume highly turbid and readily visible. Because of flocculation, most of that sediment settles down quickly in the near field where it remains in the shelter of the bays for thousands of years (Delandmeter et al. 2015).

The large river impacting the FNGBR is the perennial Fly River in Papua New Guinea. Its river plume intrudes in Torres Strait permanently during the southeast trade wind season, and the plume waxes and wanes as the mean sea level slope through Torres Strait fluctuates over time (Animations 1a and b; Li et al. 2017).

Some fine sediment remains in suspension in these two river plumes as biologically rich muddy marine snow flocs that support an intense biological activity (Animations 9c and d; Ayukai and Wolanski 1997). These muddy marine snow flocs can travel long distances because the organic matter makes the flocs nearly neutrally buoyant (Bainbridge et al. 2012). Understanding and quantifying these river plumes, fine sediment, and nutrient processes are essential for science to inform managers of land use (Lewis et al. 2024).

2.4. THE MEAN CIRCULATION

The mean water circulation in the GBR, inferred from altimetry and field data, is sketched in Figure 23a-c. It is ecologically important as it determines the connectivity between reefs within the pelagic larval duration of the species (Bode et al. 2024; Kamenos and Hennige 2024; Kingsford et al. 2024; Pratchett et al. 2024). It is determined by the wind, reef density, oceanic jets in the Coral Sea, and non-linear interaction of the tidal currents with the prevailing currents. The oceanic jets in the Coral Sea are mesoscale turbulent. The outer fringing reefs form a porous wall that transforms the mesoscale oceanic turbulence into sub-mesoscale turbulence; similar grids are used to reduce the macroscale turbulence in fluid dynamics flume studies. The reefs and islands on the continental shelf transform this sub-mesoscale turbulence into macroscale turbulence. This macroscale turbulence is dissipated by friction over the shallow waters of the GBR. Tidal friction and the sticky water effect further smooth out the mean flows through the reef matrix. The prevailing southeasterly winds also strongly affect the whole northern region of the SGBR, its reef matrix, and the southern coastal region of the SGBR.

The exchange of water between the FNGBR and the NGBR through the passage at South Warden Reef alternates between northward and southward, and it is

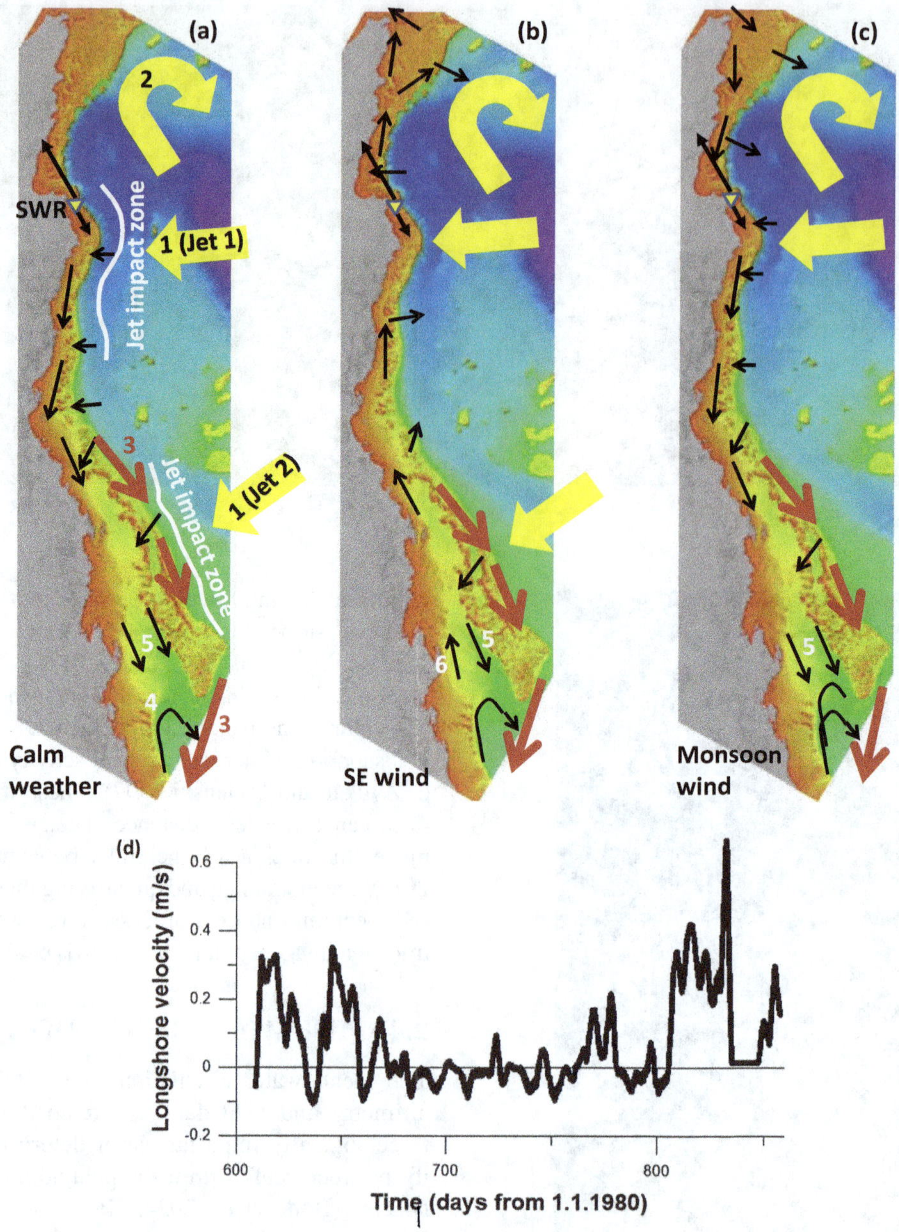

FIGURE 23 A sketch of the mean circulation on the GBR shelf under (a) calm weather, (b) strong southeasterly wind, and (c) monsoon wind. The white line 'Jet impact zone' shows where the jets of the SEC in the Coral Sea impact the GBR shelf break during their north-south meandering (Animation 2). The yellow arrows in the Coral Sea denote (1) the two dominant jets that form the SEC and (2) the highly dynamic eddies of the North Queensland Coastal Current. (3) The EAC. (4) The transient Capricorn eddy. (5) The Coral Sea Lagoonal Current (CSLC). (6) The wind-driven Coastal Boundary Layer (CBL).

Source: Modified from Wolanski et al. (2013).

(d) A time-series plot of the observed, longshore mean current (> 0 if northward) in coastal water near Linnet Reef located slightly south of the passage at South Warden Reef (SWR).

small because the mean currents are small (< 0.1 m s^{-1}; Figure 23d). However, there are occasional events of very strong (peaking at 0.6 m s^{-1}) northward mean currents. Thus, the connectivity is mainly from the NGBR to the FNGBR. These events are likely due to Coral Sea oceanic jets impacting the GBR in this area (Wolanski and Kingsford 2024).

2.5. CLIMATE CHANGE

2.5.1. The Mean Sea Level

This is measured by satellite altimetry for the world ocean and by using water level recorders at coastal ports. The data (Figure 24a) suggest a rise in the global sea level of approximately 9 cm from 1992 to 2021. The relative mean

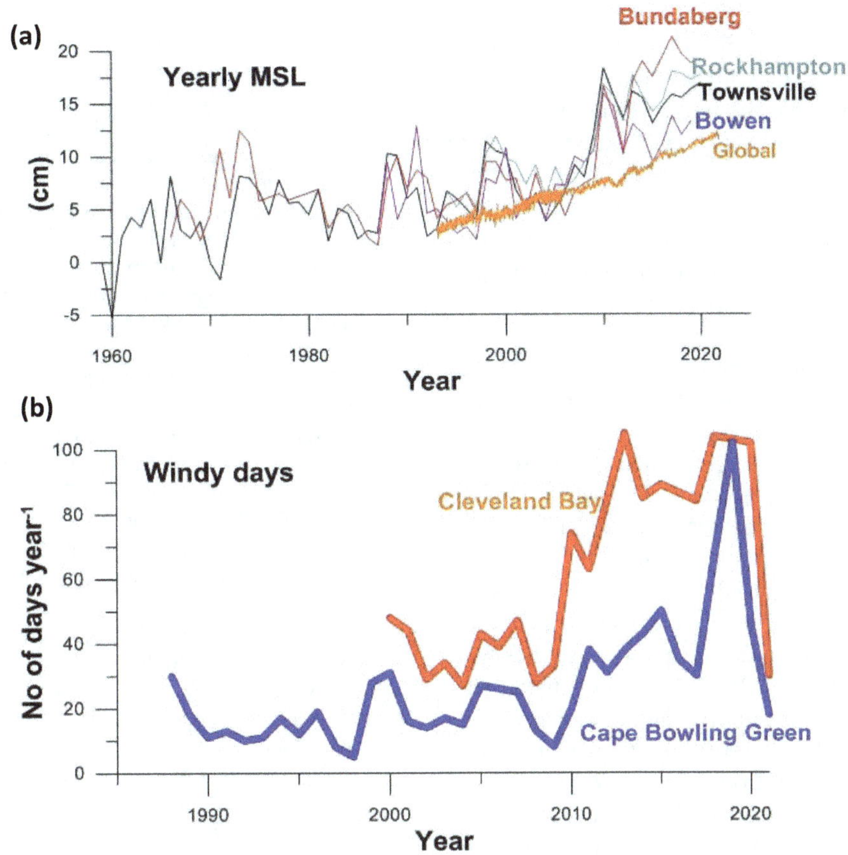

FIGURE 24 (a) Time-series plot of the global mean sea level measured by altimetry and the relative mean sea level measured at coastal ports in the CGBR and the SGBR. (b) Time-series plot of the number of days per year of moderate southeasterly wind (wind speed > 7 m s^{-1}) at Cape Bowling Bay and Cleveland Bay.

Source: AIMS

sea level rise at the coast is more variable and somewhat larger (14 cm from 1992 to 2021) than in the ocean, and this discrepancy may be due to tectonic movements.

The rise in mean sea level will increase the flow over the reef flats from waves breaking at the reef crest, which changes the dynamics of sand cays. It may also slightly change some of the tidal components at the coast (i.e., the M_2 tidal constituent may decrease, and the S_2 tidal constituent may increase), but its main impact may be to alter coastal flooding risks with a spatially varying impact (Mawson et al. 2022).

2.5.2. The Wind

The GBR wind data span a period of approximately 40 years. This is too short to reliably infer climate change effects on the wind in view of the El Niño-La Niña episodes that typically each last 9–12 months, although some La Niña events can last a few years with no regular return period. At some sites but not at others, the number of windy days (southeasterly wind speed > 7 m s^{-1}) increased from 2010 to 2020, but by 2021, the situation was back to 'normal' (Figure 24b). It is not known what drove this decadelong event.

The evidence on changes to the frequency and strength of tropical cyclones is contradictory. GBRMPA (2022) stated that there is no evidence of a change of cyclones' frequency, strength, or their location. However, Duke et al. (2024) reported that there has been a notable increase in the number of *severe* tropical cyclones, with seven recorded during the 2010–2019 decade compared to zero in the two previous decades.

2.5.3. The Temperature

Sea surface temperatures in the Australian region have warmed by approximately 1 °C since 1910, with the GBR warming by 0.8 °C in the same period (GBRMPA 2022). This warming is smaller than the warming (e.g., 2.28 °C) at the southernmost penetration of the EAC in Tasmania (Ridgway 2007). The year 2019 was Australia's hottest on record, with average air temperatures exceeding 1.5 °C above the long-term average. As global mean temperature reaches 1.5 °C above pre-industrial times, the temperatures experienced in 2019 may become typical rather than exceptional. February 2020 was the hottest month on record for the Reef, with temperatures 1.2 °C above the long-term

February average (1961–1990). This may have increased the frequency of heat-induced coral bleaching events in the GBR. Heating is due to the combined effect of local heating of water locally trapped and the import of hot water from hot spots in the ocean by water currents. The FNGBR 2015 bleaching event was facilitated by hot water brought in from the Gulf of Carpentaria via Torres Strait in January 2015 (Wolanski et al. 2017). The 2020 SGBR bleaching event was facilitated by hot water brought in from the tropics by a filament of the SEC in February (Wolanski et al. 2021). A review of historical and recent bleaching events is provided by Kamenos and Hennige (2024) who showed that some bleaching events occurred in pre-European times.

The oceanography of oceanic heat waves impacting the GBR has been greatly under-studied. The whole science of predicting changes to the physical oceanography of the GBR from climate change remains educated guesswork (Steinberg 2007). Process-based biophysical field studies are clearly needed to solve this challenge. Such studies are neglected at present in favour of model studies with insufficient data, and there is clearly a need to change this approach. To emphasise this need, Saint-Amand (2022) compared the outputs of a SLIM 2D finite-element model with five resolutions ranging from 250 m to 4 km near the reefs. The predicted tidal signal is similar for all resolutions, but the predicted currents are very different. As a result, the predictions of connectivity and larval dispersal patterns from the coarse-scale models quickly diverge significantly, that is, they become unreliable, from those predicted by the fine-scale model. Hence, the present whole-GBR models need to increase their spatial resolution to resolve the dominant fine-scale oceanography features of the GBR and to include the wave breaking on the outer reef crest and the oceanic jets in the Coral Sea.

REFERENCES

Andrews JC, Clegg S (1989) Coral Sea circulation and transport deduced from modal information models. Deep Sea Research Part A. Oceanographic Research Papers 36:957–974.

Andrews JC, Furnas M (1986) Subsurface intrusions of Coral Sea water into the central Great Barrier Reef. I. Structure and shelf-scale dynamics. Continental Shelf Research 6:491–514.

Andrews JC, Gentien P (1982) Upwelling as a source of nutrients for the Great Barrier Reef ecosystems: A solution to Darwin's question. Marine Ecology Progress Series 8:257–269.

Andutta F, Kingsford M, Wolanski E (2012) 'Sticky water' enables the retention of larvae in a reef mosaic. Estuarine, Coastal and Shelf Science 101:54–63.

Andutta F, Ridd PV, Wolanski E (2013) The age and the flushing time of the Great Barrier Reef waters. Continental Shelf Research 53:11–19.

Ayukai T, Wolanski E (1997) Importance of biologically mediated removal of fine sediments from the Fly River plume, Papua New Guinea. Estuarine, Coastal and Shelf Science 44:629–639.

Bainbridge ZT, Wolanski E, Alvarez-Romero JG, Lewis SE, Brodie JE (2012). Fine sediment and nutrient dynamics related to particle size and floc formation in a Burdekin River flood plume, Australia. Marine Pollution Bulletin 65:236–248.

Benthuysen JA, Tonin H, Brinkman R, Herzfeld M, Steinberg C (2016) Intrusive upwelling in the Central Great Barrier Reef. Journal of Geophysical Research 121:8395–8416.

Bode M, Stewart O, Choukroun S (2024) Incorporating biophysical larval dispersal simulations into marine spatial management decisions on coral reefs. In: Wolanski E, Kingsford MJ (eds.), Oceanographic Processes of Coral Reefs. Physical and Biological Links in the Great Barrier Reef. CRC Press, Boca Raton, FL.

Booth DJ, Kingsford MJ, Doherty PJ, Berretta GA (2000) Recruitment of damselfishes in One Tree Island lagoon: Persistent interannual spatial patterns. Marine Ecology Progress Series 202:212–230.

Carter AB, Collier C, Lawrence E., Rasheed MA, Robson BJ, Coles R (2021) A spatial analysis of seagrass habitat and community diversity in the Great Barrier Reef World Heritage Area. Scientific Reports 11:22344.

Choukroun S (2010). *The surface circulation of the Coral Sea and Great Barrier Reef.* PhD thesis, James Cook University, 222pp.

Davis KA, Pawlak G, Monismith SG (2021) Turbulence and coral reefs. Annual Review of Marine Science 13:343–373.

Delandmeter P, Lambrechts J, Marmorino GO, Legat V, Wolanski E, Remacle JP, Chen W, Deleersnijder E (2017) Submesoscale tidal eddies in the wake of coral islands and reefs: Satellite data and numerical modelling. Ocean Dynamics 67:897–913.

Delandmeter P, Lewis S, Lambrechts J, Deleersnijder E, Legat V, Wolanski E (2015) The transport and fate of riverine fine sediment exported to a semi-open system. Estuarine, Coastal and Shelf Science 167:336–346.

Deleersnijder E, Norro A, Wolanski E (1992) A three-dimensional model of the water circulation around an island in shallow water. Continental Shelf Research 12:891–906.

Duce S, Vila-Concejo A, Mccarroll RJ, Yiu B, Perris LA, Webster JM (2022) Field measurements show rough fore reefs with spurs and grooves can dissipate more wave energy than the reef crest. Geomorphology 413(3):108365.

Duke NC, Canning A, Mackenzie J (2024) More intense severe tropical cyclones in recent decades cause greater impacts on mangroves bordering Australia's Great Barrier Reef. In: Wolanski E, Kingsford MJ (eds.), Oceanographic Processes of Coral Reefs. Physical and Biological Links in the Great Barrier Reef. CRC Press, Boca Raton, FL.

Fabricius K (2005) Effects of terrestrial runoff on the ecology of corals and coral reefs: Review and synthesis. Marine Pollution Bulletin 50:125–146.

Fischer H, List E, Koh R, Imberger J, Books N (1979) Mixing in Inland and Coastal Waters. Academic Press Inc, New York, 483pp.

Gallop SL, Young IR, Ranasinghe R, Durrant TH, Haigh ID (2014) The large-scale influence of the Great Barrier Reef matrix on wave attenuation. Coral Reefs 33:1167–1178.

GBRMPA (Great Barrier Reef Marine Park Authority) (2022) www.gbrmpa.gov.au/our-work/threats-to-the-reef/climate-change

Gourlay MR (1988) Coral cays: Products of wave action and geological processes in a biogenic environment. In Proceedings 6th International Coral Reef Symposium, Australia 2:491–496.

Hamylton SM, McLean R, Bell IP, Thomas CR (2024) Biological and geological links on coral reef islands. In: Wolanski E, Kingsford MJ (eds.), Oceanographic Processes of Coral Reefs. Physical and Biological Links in the Great Barrier Reef. CRC Press, Boca Raton, FL.

Jaffrés J, Heron ML, Middleditch A, Steinberg CR, Durrant TH (2010) *Waves in the southern Great Barrier Reef.* In: Proceedings of Oceans 2010 IEEE, Sydney.

Kamenos NA, Hennige SJ (2024) A historic perspective to thermal and heatwave induced bleaching on the Great Barrier Reef. In: Wolanski E, Kingsford MJ (eds.), Oceanographic Processes of Coral Reefs. Physical and Biological Links in the Great Barrier Reef. CRC Press, Boca Raton, FL.

Kingsford MJ, Spiecker L, Gerlach G (2024) The interactions between larval behaviour and oceanography. In: Wolanski E, Kingsford MJ (eds.), Oceanographic Processes of Coral Reefs. Physical and Biological Links in the Great Barrier Reef. CRC Press, Boca Raton, FL.

Lee S, Hassan YA (2018) Experimental study of flow structures near the merging point of two parallel plane jets using PIV and POD. International Journal of Heat and Mass Transfer 116:871–888.

Lewis S, Bainbridge Z, Davis A, McCloskey G, Bartley R, Turner R (2024) Sediment and nutrient flux from land. In: Wolanski E, Kingsford MJ (eds.), Oceanographic Processes of Coral Reefs. Physical and Biological Links in the Great Barrier Reef. CRC Press, Boca Raton, FL.

Li Y, Martins F, Wolanski E (2017) Sensitivity analysis of the physical dynamics of the Fly River plume in Torres Strait, Estuarine. Coastal and Shelf Science 194:84–91.

Longuet-Higgins MS (1996) Surface manifestations of turbulent flow. Journal of Fluid Mechanics 308:15–29.

Marmorino GO (2022) Investigation of turbulent tidal flow in a coral reef channel using Multi-LookWorldView-2 satellite imagery. Remote Sensor 14:783.

Marmorino GO, Smith GB, Miller WD (2017) Turbulence characteristics inferred from time-lagged satellite imagery of surface algae in a shallow tidal sea. Continental Shelf Research 148:178–184.

Marsh H, Cleguer C (2024) Interactions between dugong biology and the biophysical determinants of their environment: A review. In: Wolanski E, Kingsford MJ (eds.), Oceanographic Processes of Coral Reefs. Physical and Biological Links in the Great Barrier Reef. CRC Press, Boca Raton, FL.

Massel SR (2013) Ocean Surface Waves: Their Physics and Prediction. 2nd edition. World Scientific, Singapore, 692p.

Mawson EE, Lee KC, Hill J (2022) Sea level rise and the Great Barrier Reef: the future implications on reef tidal dynamics. Journal of Geophysical Research Oceans 127:e2021JC017823.

McDuie F, Weeks SJ, Congdon BC (2018) Oceanographic drivers of near-colony seabird foraging site use in tropical marine systems. Marine Ecology-Progress Series 589:209–225.

Middleton JH (1987) Steady coastal circulation due to oceanic alongshore pressure gradients. Journal of Physical Oceanography 13:1371–1382.

Middleton JH, Buchwald VT, Huthnance JM (1984). The anomalous tides near Broad Sound. Continental Shelf Research 3:359–381.

Obst, BS, Hamner, WM, Hamner PP, Wolanski E, Rubega M, Littlehales B (1996) Kinematics of phalarope spinning. Nature 384:121.

Phillips OM, Shyu J, Salmun H (1986) An experiment on boundary mixing: Mean circulation and transport rates. Journal of Fluid Mechanics 173:473–499.

Pickard GL, Donguy JR, Henin C, Rougerie F (1977). A review of the physical oceanography of the Great Barrier Reef and western Coral Sea. Australian Institute of Marine Science, Monograph Series 2.

Prairie JC, Sutherland KR, Nickols KJ, Kaltenberg AM (2012) Biophysical interactions in the plankton: A cross-scale review. Limnology and Oceanography: Fluids and Environments 2:121–145.

Pratchett MS, Choukroun SM, Chandler J, et al. (2024) Biophysical processes involved in the initiation and spread of population irruptions of crown-of-thorns starfish on the Great Barrier Reef. In: Wolanski E, Kingsford MJ (eds.), Oceanographic Processes of Coral Reefs. Physical and Biological Links in the Great Barrier Reef. CRC Press, Boca Raton, FL.

Ridgway KR (2007) Long-term trend and decadal variability of the southward penetration of the East Australian Current. Geophysical Research Letters 34:22921–22936.

Robertson K, Dawson K, Coffee O (2024) Raine Island Recovery Project—Intervening at one of the most significant sites on the Great Barrier Reef. In: Wolanski E, Kingsford MJ (eds.), Oceanographic Processes of Coral Reefs. Physical and Biological Links in the Great Barrier Reef. CRC Press, Boca Raton, FL.

Saint-Amand A (2022). Bridging the gap between small and large spatial scales in marine dispersal modelling: Applications to coral ecosystems in the Great Barrier Reef. PhD thesis, Catholic University of Louvain. https://dial.uclouvain.be/pr/boreal/object/boreal:266081

Schlaefer JA, Tebbett SB, Bellwood DR (2021) *The study of sediments on coral reefs: A hydrodynamic perspective.* Marine Pollution Bulletin 169:112580.

Sih TL, Cappo M, Kingsford M (2017) Deep-reef fish assemblages of the Great Barrier Reef shelf-break (Australia). Scientific Reports 7:10886.

Spagnol S, Wolanski E, Deleersnjider E (2001) Steering by coral reef assemblages. In: Wolanski E (ed) Oceanographic Processes of Coral Reefs: Physical and Biological Links in the Great Barrier Reef. CRC Press, Boca Raton, FL.

Steinberg C (2007) Impacts of climate change on the physical oceanography of the Great Barrier Reef. Chapter 3. In: Johnson JE, Marshall PA (eds) Climate Change and the Great Barrier Reef. Great Barrier Reef Marine Park Authority and Australian Greenhouse Office, Clayton.

Thompson R, Golding TJ (1981) Tidally induced upwelling by the Great Barrier Reef. Journal of Geophysical Research 86:6517–6521.

Thomson RE, Wolanski E (1984) Tidal period upwelling within Raine Island Entrance, Great Barrier Reef. Journal of Marine Research 42:787–808.

Weeks SJ, Magno-Canto MM, Jaine FRA, Brodie J, Richardson AJ (2015) Unique sequence of events triggers manta ray feeding frenzy in the Southern Great Barrier Reef, Australia. Remote Sens. 7:3138–3152.

Weeks SJ, Steinberg C, Congdon BC (2013) Oceanography and seabird foraging: Within-season impacts of increasing sea-surface temperature on the Great Barrier Reef. Marine Ecology-Progress Series 490:247–254

Wolanski E (1986a) A simple analytical model of low-frequency wind-driven upwelling on a continental slope. Journal of Physical Oceanography 16:1634–1702.

Wolanski E (1986b) Observations of wind-driven surface gravity waves offshore from the Great Barrier Reef. Coral Reefs 4:213–219.

Wolanski E (1987) Some evidence for boundary mixing near coral reefs. Limnology and Oceanography 32:735–739.

Wolanski E (1993) Facts and numerical artefacts in the dispersal of crown-of-thorns starfish larvae in the Great Barrier Reef. Australian Journal of Marine and Freshwater Research 44:427–436.

Wolanski E, Andutta F, Deleersnijder E, Li Y, Thomas CJ (2017) The Gulf of Carpentaria heated Torres Strait and the Northern Great Barrier Reef during the 2016 mass coral bleaching. Estuarine, Coastal and Shelf Science 194:172–181.

Wolanski E, Asaeda T, Tanaka A, Deleersnijder E (1996) Three-dimensional island wakes in the field, laboratory, and numerical models. Continental Shelf Research 16:1437–1452.

Wolanski E, Bennett AF (1983) Continental shelf waves and their influence on the circulation around the Great Barrier Reef. Australian Journal of Marine and Freshwater Research 34:23–47.

Wolanski E, De'ath G. (2005) Predicting the present and future human impact on the Great Barrier Reef. Estuarine, Coastal and Shelf Science 64:504–508.

Wolanski E, Hamner W (1988) Topographically controlled fronts and their biological influence. Science 241:177–181.

Wolanski E, Kingsford MJ (2024). Great barrier reef biophysics: A synthesis of challenges and opportunities. In: Wolanski E, Kingsford MJ (eds.), Oceanographic Processes of Coral Reefs. Physical and Biological Links in the Great Barrier Reef. CRC Press, Boca Raton, FL.

Wolanski E, Lambrechts J (2020) The net water circulation in the Far Northern Great Barrier Reef. Estuarine, Coastal and Shelf Science 235:106569.

Wolanski E, Lambrechts J, Thomas C, Deleersnijder E (2013) The net water circulation through Torres Strait. Continental Shelf Research 64:66–74.

Wolanski E, Pickard GL (1985) Long-term observations of currents on the central Great Barrier Reef continental shelf. Coral Reefs 4:47–57.

Wolanski E, Richmond RH, Golbuu Y (2021) Oceanographic chaos and its role in larval self-recruitment and connectivity among fish populations in Micronesia. Estuarine, Coastal and Shelf Science 259:107461.

Wolanski E, Ridd P, Inoue M (1988) Currents through Torres Strait. Journal of Physical Oceanography 18:1535–1545.

Wolanski E, Thomson RE (1984) Wind-driven currents on the northern Great Barrier Reef continental shelf in summer. Estuarine, Coastal and Shelf Science 18:271–289.

Wolanski E, van Senden D (1982) Mixing of Burdekin River flood waters in the Great Barrier Reef. Australian Journal of Marine and Freshwater Research 34:49–63.

3 Biological and Geological Links on Coral Reef Islands

Sarah M. Hamylton, Roger McLean, Ian P. Bell, and Colette R. Thomas

3.1. INTRODUCTION

Coral reefs are three-dimensional platforms comprised of thin veneers of living coral and reef-associated organisms overlying previously deposited calcium carbonate that has been laid down over thousands of years (Kench 2009). The network of nearly 3,000 reef platforms that comprises the Great Barrier Reef (GBR) can be seen from space and represents the largest biological construction on Earth, running for 2,400 km in length (from 9°30'S to 24 °30'S) and stretching over 260 km offshore in places (Hopley et al. 2007). Coral reefs and the low-lying islands that develop on top of their surfaces are composed predominantly of calcium carbonate produced by calcifying animals and plants that convert calcium (Ca^{2+}) and carbonate (CO_3^{2-}) ions from seawater into calcium carbonate ($CaCO_3$) via a biological process that, over time, builds the rigid form of a coral reef platform.

Modern reefs have developed over the past few millennia. In many cases they have re-established over a reefal substrate inherited from a previous sea-level high stand, such as the last interglacial some 120,000 years ago. Over geological timescales (i.e., centuries or millennia) active calcification by a range of reef dwelling organisms leads to the build-up of substantial carbonate reef structures or platforms. The inner framework of such platforms builds up as carbonate sand and rubble become bound into interstitial spaces between skeletal structures of hard scleractinian corals and loose materials become overgrown by coralline algae into a large, solid structure (Hubbard 1997). Reefs undergo growth pulses with periods of rising sea levels that are interspersed with erosive periods of low sea level that are not conducive to coral growth (Hubbard 1985). The last post glacial submersion of the GBR shelf occurred some 5,000–12,000BP, subsequent to which the reefs that we see today developed (Henderson and Johnson 2016).

As islands are often the only offshore exposed land areas, they are important for wildlife, habitat and livelihoods. Sedimentary landforms associated with coral reefs including low islands, beaches, shingle spits or ramparts, conglomerate platforms, sand aprons and beach rock are frequently associated with the coasts of the high granitic and low limestone islands within the GBR. These accumulations of unconsolidated sediments deposited by waves and currents are initially sourced from an adjacent coral reef slope and become consolidated and lithified and therefore stabilised in position over time. Many reef sedimentary landforms provide a foundational surface for vegetation communities including seagrasses, mangroves, trees and shrubs (Heatwole 2011).

3.1.1. Physical and Biological Linkages between Coral Reefs and Islands

Within the GBR, there are two distinct types of islands, namely, the continental or high islands and the low or reef islands, which are closely linked to the adjacent mainland and the coral reef environment, respectively (Figure 1). Within the Great Barrier Reef Marine Park, almost a thousand islands are scattered among the matrix of reef platforms along the shelf, and many of these (approximately 617) are high islands with attached reefs, while over 300 of these are low-lying reef islands that rarely rise more than a few metres above sea level (Hopley et al. 2007; Hopley 2008). There is considerable variety within the islands in terms of their size, shape, surface topography, rock type, elevation, and distance from the mainland within each of these two groups. The low reef islands are distributed unevenly across the shelf and latitudinally as they are strongly influenced by spatial patterns in reef size and relative sea-level history (Figure 2). Two features that are common to virtually all GBR islands are coral reefs, such as fringing reefs, carbonate sand and shingle beaches, berms, spits and forelands, and littoral vegetation, such as *Scaevola*, *Tournefortia*, *Pemphis* and *Pisonia* thickets, stands of *Casuarina*, *Pandanus* and palms, and individual and forests of mangroves.

Islands are discrete biological and physical entities where the sharp discontinuity between land and sea is reflected in the landforms, flora, and fauna of the intertidal or littoral zone. The land-water interface that clearly delimits a subaerial island from the submarine environment is not just a single line separating the terrestrial and oceanic domains. Rather, it is a transitional zone of indeterminate width between land and sea that combines characteristics of terrestrial and marine environments to demonstrate close linkages and interactions within this insular interface.

Some littoral features owe their characteristics to the geology and topography of the land, for example, the bold bluffs, cliffs, and valleys of the granitic, sedimentary, and limestone islands on the GBR, although other features on these same high islands such as fringing reefs, shore platforms, and peripheral sedimentary accumulations are

DOI: 10.1201/9781003320425-4

FIGURE 1 Examples of high and low islands within the GBR. (A) Heron Island, (B) Hamilton Island and Whitsunday Group, and (C) Lady Musgrave Island, a low-lying island on top of a coral reef platform in the southern GBR formed from carbonate sediments derived from the surrounding coral reef (photo credit: Prithvi Bhattacharya).

more connected with the ocean. A considerable number of dynamic physical and biological interactions closely link the island and marine environments in exchanges of energy and materials. In nearshore areas, fresh rainwater, river silt, and debris passing into the ocean or onto the reef from high island slopes affect marine and reef communities, while salts from sea spray blowing onto islands affect the terrestrial plant and animal communities, soils, and substrates. Indeed, some organic material on islands ultimately has a marine origin, for example, the phosphate deposits and guano deposited by fish-eating seabirds, which have been mined on many of the islands in the GBR from Raine in the north to Lady Elliot in the south.

The insular shorelines of reef islands are laterally continuous, enclosing what is often the only subaerial landmass on a reef platform surface. Island shores face all points of the compass and therefore have the potential to be impacted by wind, waves, currents, and storms from any direction, unlike continental coastlines elsewhere in Australia. Rarely, if ever, is the distribution of processes uniform around an island. Strong variations occur in the intensity of processes, notably and consistently between the rough windward and calm leeward sides of the GBR islands in the belt of southeasterly trade winds (September to March). The result is easily recognizable, especially on the low islands with shingle deposits and boulder ramparts common on the eastern,

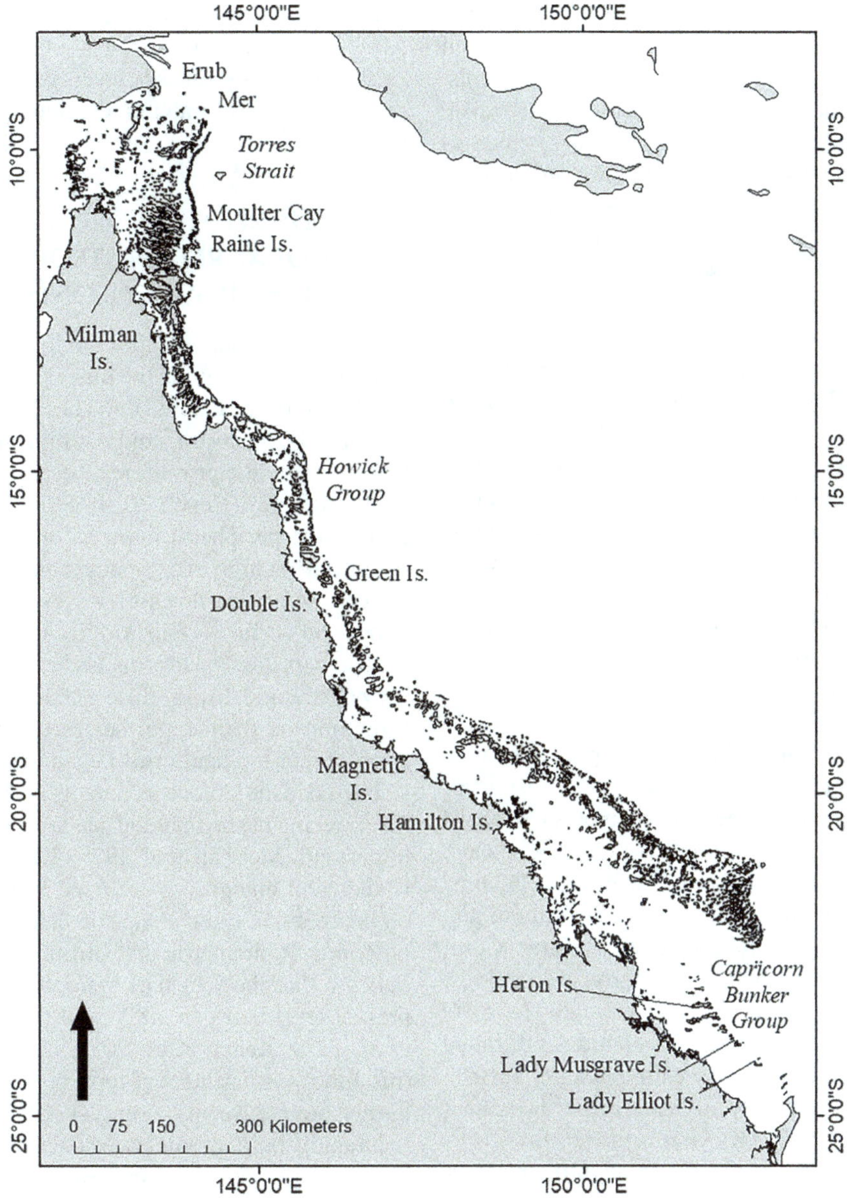

FIGURE 2 Islands within the GBR that are referred to in this chapter (image credit: Sarah M. Hamylton).

windward side of reef platforms and low sand cays on the western leeward side. This windward/leeward difference is also important for the plants and animals that inhabit sedimentary accumulations and intervening intertidal reef flat. Here, high gravel deposits mainly derived from corals afford shelter, allowing the initial colonisation then growth and expansion of mangroves with *Avicennia* occupying the exposed rocky shores and *Rhizophor*a inhabiting the sheltered interior flats (Stoddart 1980).

3.1.2. The Carbonate Factory that Underpins Coral Reef and Island Development

The structural integrity of coral reefs over time fundamentally relies on the balance between carbonate addition to and removal from the reef, which is termed the carbonate budget. A reef carbonate budget quantifies the net gain or loss of calcium carbonate to a reef system once all the additive and removal processes have been accounted for (Stearn et al. 1977; Scoffin et al. 1980; Hubbard et al. 1990). Several calcifying organisms are responsible for much of the carbonate production in coral reef environments including hard corals, coralline algae, calcareous green algae, such as *Halimeda*, and a variety of crustaceans, molluscs, and foraminifera (Perry 2007). In most reefs, hard corals that build a solid skeleton are the dominant producers of calcium carbonate, although the relative contribution of these different components of the carbonate budget varies geographically, depending on local conditions. For example, foraminifera are the source of carbonate on Raine Island (Dawson et al. 2014).

Biological, chemical, and physical processes break down and transport carbonate into lagoons and moats and in some cases, remove carbonate from reef environments. Bioerosion occurs when organisms such as parrotfish, invertebrates, and endolithic macroborers (e.g., invertebrates such as worms and sponges) break down and feed on calcifiers, digesting the organic material and excreting carbonate sediments as waste products (Hutchings 1986). This converts carbonate skeletons to sediments and transports them over reef platforms to build up sedimentary landforms over time (Perry et al. 2015). Currents and waves also mechanically break down and transport carbonate material around and off reefs. This happens on an ongoing basis at relatively low levels through tidal cycles and seasonal fluctuations in wind and waves. It also occurs sporadically with high energy events such as storms and cyclones when large quantities of sediment are transported to produce shingle spits and ramparts in a short timespan. Chemical processes can remove carbonate from a reef system, notably through acidification of ocean waters, which both reduces the rates of calcification and dissolves carbonate, thereby reducing the density of carbonate skeletal material for corals (Kleypas and Yates 2009).

Environmental, physical, and ecological processes play a major role in the production, breakdown, and transport of calcium carbonate throughout reef environments (Hubbard 1997, see also chapter 6 in this volume). At the global scale, the main environmental controls on tropical $CaCO_3$ production are sea surface temperature, light penetration, and the calcium carbonate saturation state of seawater. Coral reefs occur in shallow tropical and subtropical settings, where seawater temperatures fall in the range 17–34 °C on shallow surfaces that fall within the photic depth range to which sunlight can penetrate beneath the water surface (< 90m), although the majority of reefs thrive in waters shallower than 20 m in depth (Kleypas et al. 1999).

The GBR has developed over time through the production, breakdown, and transport of carbonate resulting in the accumulation and eventual lithification of reef platforms (Hopley 2008). Much of the carbonate produced by marine flora and fauna does, not do not remain *in situ*; rather, it is broken down into sediments that accumulate in sedimentary 'sinks', such as shingle spits, sand aprons, and islands that form across the reef surface, and in the surrounding waters of the GBR lagoon. Surficial sedimentary landforms can become colonised by vegetation over time. Once deposited sediments have raised the ground surface to an elevation within the suitable tidal range, mangroves may become established and grow. The root systems of mangroves stabilise the substrate, while the trees add organic matter to transform the reef surface over time, bringing associated communities of birds and invertebrates. Likewise, grasses, trees, and shrubs can form on sandy cays, where they stabilise unconsolidated sediments, leading to the formation of vegetated reef islands and sedimentary accumulations. Carbonate breakdown and transport around a reef system, particularly those that have occurred over the last 5,000 years since the current configuration of reefs became established at a stable sea level, governs the landforms that can be seen above water on the upper surfaces of GBR reefs today.

3.2. THE RELATIONSHIP BETWEEN CORAL REEF PLATFORM SIZE AND MANGROVE COLONISATION

The Howick group includes ten low wooded islands within an area of 1,500 km^2 on the inner shelf of the northern GBR, 100 km north of Cooktown. These low wooded islands are composed of leeward sand cays on the western side of the reef flat with mangrove forests occupying the upper surface of the reef flat (Kench 2010). Shingle ridges have formed around the peripheral margins of most of these reef platforms during high energy storm and cyclone events, which, in turn, influence flows of waves and currents across the upper reef surfaces (Fairbridge and Teichert 1948).

The coral reef platforms underlying the Howick Islands vary in size and display different degrees of mangrove forest formation on their upper surfaces (Figure 3). The present array of reef-top landforms began to develop after sea levels stabilised in the Holocene, notably in the last 7,000 years with the lowering of the regional sea level from its mid-Holocene high stand (McLean et al. 1978). Rapid development of both islands and mangroves occurred in ponded, sheltered sites on the reef flat, often alongside fields of distinctive microatoll corals. Radiometric dates from microatolls at these sites indicate that the reef flats were at, or slightly above, their present level between 4,800 and 5,800 years ago (McLean et al. 1978; Kench et al. 2012; Perry et al. 2017). During this time, the island reef-tops in the northern GBR were highly productive, providing skeletal carbonate sediments that rapidly built up sand cays and shingle ramparts and provided the surface for mangrove colonisation (McLean and Stoddart 1978; McLean et al. 1978).

Cartographic records extending across 45 years on these islands have revealed rapid recent expansion of *Rhizophora stylosa* mangrove forests (Hamylton et al. 2023). Mangrove area and rate of spread across upper reefs are thought to be functions of reef-top morphology, primarily the outcome of upward reef growth from older foundations in the Holocene, with periodic interruptions to colonisation and spread from catastrophic cyclone damage (Stoddart 1980; see also Chapter 12 in this volume). It has been argued that reef platform size influences the timing of platform filling such that larger reef platforms represent sites where mangrove forests and other landforms should be less well-developed and where expanding forests are undergoing slower rates of change because of the extended times taken for lagoon infill of these larger platforms prior to the development of sedimentary landforms and mangrove colonisation and expansion (Perry et al. 2017). Here, we test the reef platform size hypothesis by exploring both the extent of reef flat mangrove

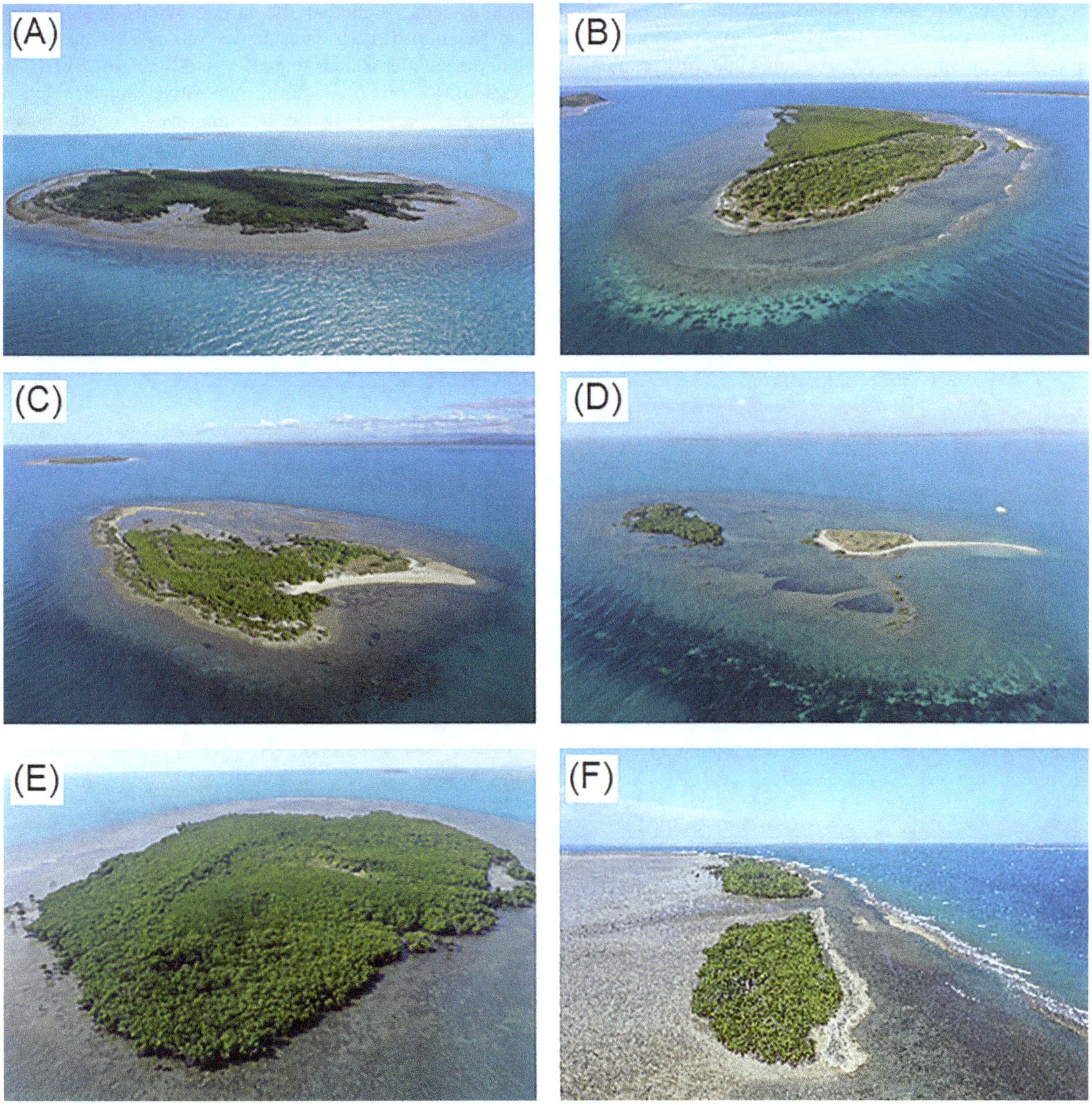

FIGURE 3 Low, wooded island of the Howick group exhibiting different proportions of reef flat mangrove cover: (A) Newton, (B) Houghton, (C) Leggatt, (D) Sinclair-Morris, (E) Hampton, and (F) Ingram-Beanley (image credit: Sarah M. Hamylton).

colonisation and expansion in relation to reef platform size for eight reef sites in the Howick Islands that were mapped in 1973 and re-surveyed in 2021: Bewick, Watson, Newton, Houghton, Coquet, Leggatt, Sinclair-Morris, and Hampton.

Images were acquired using an aerial drone in 2021, along with in situ ground referencing photographs to identify *Rhizophora* trees from the imagery. Aerial surveys were planned using the Map Pilot for DJI Application, and images were processed into a single aerial orthomosaic for each island (spatial resolution 30 cm) using the photogrammetric software Agisoft Photoscan (AgiSoft 2014). The contemporary distribution of *Rhizophora* woodland was visually assessed and digitised at a scale of 1:400 from the high-resolution aerial photomosaics. The areas of resulting vector polygons were calculated to evaluate mangrove areal changes.

The 1973 maps made by compass-traverse field survey were scanned, georeferenced, and digitised at the same scale of 1:400 (Stoddart 1978; Hamylton 2017). Changes in mangrove cover were ascertained from a comparison of vector shapefiles depicting the *Rhizophora* distribution in 1973 and 2021. Finally, the outline of the peripheral margin of the reef flats was also digitised from the high-resolution aerial imagery to enable the relationship among the extent of reef flat mangrove colonisation, mangrove expansion rate, and reef platform size to be explored.

Figure 4 illustrates the proportion of reef flat covered by mangrove in 1973 and 2021 for the eight islands for which cartographic records are available, while Figure 5 shows scatter plots of both mangrove cover and changes to mangrove cover as a function of reef platform size for these same islands.

A positive relationship between reef platform size and mangrove cover emerged from observations across the eight islands, such that the larger islands (e.g., Bewick and Newton) had a greater degree of cover by mangroves than the smaller islands (e.g., Leggatt, Sinclair-Morris, and Watson) ($R^2 = 0.86$). A similarly positive, weaker relationship was apparent between the mangrove cover change and reef platform size ($R^2 = 0.59$). Both relationships run contrary to the reef platform size hypothesis, which suggests that smaller reefs should fill with sediment more rapidly and therefore have reef flats with well-developed sedimentary landforms and larger mangrove forests on them. Many of the smaller islands appeared to have substantial areas of unoccupied reef flat, while the larger reef platforms (e.g.,

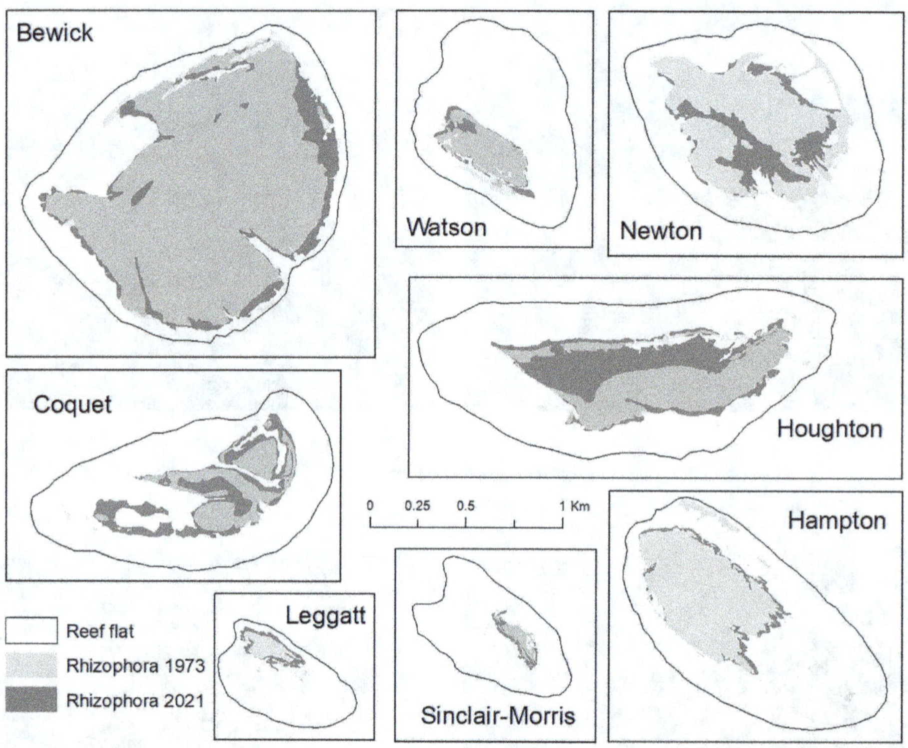

FIGURE 4 Formation and expansion of *Rhizophora* mangrove forests on the reef flats of the Howick Islands, as mapped in 1973 and 2021. All reef flats are shown at a consistent scale for comparison (image credit: Sarah M. Hamylton).

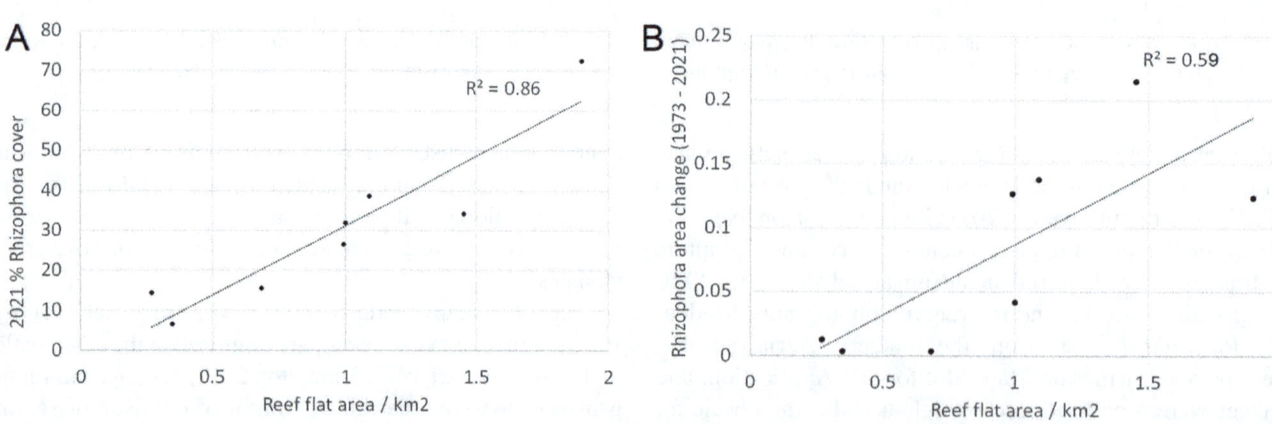

FIGURE 5 (A) Plot of reef flat area vs proportion of reef flat covered by *Rhizophora* mangrove, and (B) plot of reef flat area vs changes to mangrove area (1973–2021).

Bewick, Houghton, and Newton) were the three highest ranking sites in terms of proportion of reef flat covered by mangrove.

These observations relate only to the eight sites for which cartographic records and observational field data are available. Two other sites falling within the Howick group (Pipon and Ingram-Beanley) have large reef platforms with relatively limited mangrove coverage on their reef surfaces and would likely adhere more closely to the reef platform size hypothesis.

Notwithstanding variability in timing of landform formation on reef flats, mangrove expansion on the Howick Islands results from an interplay between sea level, reef flat sediments, and substrates. The remarkable lateral expansions of *Rhizophora* mangrove observed across all eight sites has likely been made possible through increases in the vertical elevation of the reef flat driven by deposition of sand and shingle in the lower energy and protected leeward areas of the mangrove woodland. As the reef flat gradually accretes upward into the lower intertidal, it reaches a vertical range suitable for *Rhizophora* seedlings to become established under conditions of tidal emergence and submergence that are conducive to seedling settlement and survival of juvenile saplings (Smith 1987). Once established, the mangrove roots dissipate energy, thereby facilitating further localised lateral expansion. The vertical shift into a favourable tidal range is accompanied by limited subsurface current flows over the reef flat as a result of protection by windward shingle ramparts (Baldwin et al. 1996). This collective set of incremental growth processes can rapidly replace bare reef flats with the complex array of landforms commonly seen on low wooded islands (Stoddart et al. 1978; Hamylton et al. 2019).

In the Howick group, shingle spits and ramparts formed by storms and rough seas on the windward side of reef-tops have been variously dated between 2,000 and 3,550 years old (McLean and Stoddart 1978), providing a timeframe from which the sea level height relative to the reef flat allowed relatively early growth of *Rhizophora* in the sheltered lee of such deposits. Shingle spits, ramparts, and boulder tracts have continued to develop around the windward reef-top margins, particularly because of high energy storms and cyclones, which provide ongoing opportunities for colonisation by mangroves.

Both the accumulation of sedimentary landforms and colonisation by mangroves are triggered when there is critical water depth over the reef platform that enables enhanced sediment production, which governs a process regime that can mobilize and transfer sediments around the reef surface (Kench et al. 2012). This depth window is governed by the interplay between sea-level change and reef carbonate dynamics and has driven an important transition in hydrodynamics across the upper reef flats of the Howick Islands on the northern GBR, which has provided muddy substrates for subsequent colonisation by mangroves. In turn, this has led to substantial mangrove forest expansions observed over recent, decadal stages of reef development. Our findings across the Howick Islands suggest that reef platform size is a significant control on mangrove forest development, although not in the manner previously suggested.

3.3. HOW DOES ISLAND GEOMORPHOLOGY INFLUENCE TURTLE NESTING IN THE NORTHERN GBR?

Hundreds of GBR islands provide nesting habitat for marine turtle species such as loggerheads (*Caretta caretta*), hawksbills (*Eretmochelys imbricata*), flatbacks (*Natator depressus*), and greens (*Chelonia mydas*). Here, we use examples from two of the most important marine turtle rookeries in the northern GBR to demonstrate how reef geomorphic processes can shape the accessibility, suitability, and availability of nesting habitat for *C. mydas*: Raine Island, a small, outer-reef coral cay that supports the largest extant *C. mydas* (green turtle) breeding cohort in the world; and Milman Islet, a densely wooded inner-reef sand cay surrounded by a large reef flat that is among the most significant hawksbill and green turtle rookeries in the northern GBR.

Turtles are ancient air-breathing reptiles that have thrived in aquatic environments for hundreds of millions of years. Today's marine turtle rookeries are the product of geomorphic processes that may have only been operating for several thousand years. In the northern GBR, the sea level falls and associated geomorphological processes required to support development of outer-shelf reefs and islands, including Raine Island, began around 5,000 years ago, and for inner-shelf islands such as Milman Islet, these processes began approximately 7,000–8,000 years ago (Smithers et al. 2018).

Modern marine turtles spend most of their lives in the sea and occasionally emerge onto land to nest or bask. Adult females exhibit strong homing and are believed to nest almost exclusively within ~150 km of their natal beach (Bowen and Karl 2007). Most Australian marine turtles nest offshore on islands. An adult female *C. mydas* (green turtle) may swim hundreds or thousands of kilometres from her foraging area to nest. She must find a way across the reef crest, reef flat, and intertidal beach to access the island and then find a site to dig a body pit and egg chamber above the high tide line. In their search for a suitable site, *C. mydas* frequently abandon nesting attempts and may wander a beach for the entire night, digging as many as twelve trial nests (Limpus et al. 1983). If a suitable nesting site is found, then she will complete oviposition, backfill the egg chamber, cover and camouflage the nest, then descend to the beach and return to the sea (Figure 6).

3.3.1. Accessibility of Turtle Nesting Habitat

Marine turtles are graceful and elegant in the water but cumbersome and ungainly on land. Nesting turtles must negotiate a series of geomorphological features that can either help or hinder their journey to nesting habitat. The shape, size, and configuration of the surrounding reef (e.g.,

FIGURE 6 High density nesting of *Chelonia mydas* (green turtle) at Raine Island on the northern GBR. Photo credit: Chris Sammut.

large or elevated reef flats), the location of the island within the reef (e.g., proximity of the island to the reef edge), and the presence and scale of beach landforms (e.g., phosphate cliffs and dunes) and shoreline erosion all determine nesting habitat accessibility (Figure 7).

At dusk, gravid females congregate at the deep edge of the reef and wait for the tide to be sufficiently high and ambient light to be sufficiently low to begin their night's nesting journey. Reef flats that are elevated relative to sea level can prevent turtles negotiating the reef crest and accessing the reef flat at all. Reefs with margins and channels that sit close to the shoreline can reduce the time required for turtles to cross the reef flat, decreasing predation exposure and other energetic costs. These journeys are very significant for new hatchlings, which can take over one-half hour to swim a one-kilometre-wide reef flat before reaching the safety of deeper water. Predation pressure is highest on low tides, and only 18% of hatchlings are expected to survive such a crossing (Gyuris 1994). These factors may change over time as sediment erosion and accretion processes drive islands to 'migrate' around the reef flat (Garcin et al. 2022).

From the moment they emerge to the point they start digging, many nesting turtles can be obstructed by beach landforms. After coming ashore, turtles navigate the beach to locate areas above the high tide mark. Reef islands are highly dynamic landscapes and erosion frequently undermines beaches to produce cliff-like landforms and steep dune scarps, undercut vegetated areas, and expose blocks of beach rock, which can be impossible for turtles to navigate over or around. Similarly, strong storms and tidal surges often transport logs and other large debris onto nesting beaches. This can interrupt, redirect, or prevent females from accessing nesting habitat to cause turtles to traverse long sections of beach or give up entirely on nesting that night. In extreme circumstances, storms and tidal surges can trap and kill adults and hatchlings attempting to return to the sea (Garcin et al. 2022). Turtle flippers are unsuited to navigating unstable or steep terrain. Their poor eyesight prevents them from avoiding tight spaces, hanging branches, and drop-offs. If they find themselves caught or they topple off a ledge, their inability to crawl backwards prevents them from backing out and they can become trapped.

3.3.2. Suitability of Turtle Nesting Habitat

Local geomorphic factors will influence when and how turtles access their nesting habitat, including whether they attempt to dig, complete oviposition (egg laying), or produce viable hatchlings. Nesting *C. mydas* are sensitive to beach topography and prefer to nest in steeply sloped concave dunes with moderately rugged beach and dune faces, preceded by a gradually sloped foredune and surrounded by fringing vegetation (Hays et al. 2001; Kelly et al. 2017). At some rookeries the slope and width of the intertidal beach

FIGURE 7 Geomorphic zones and their influence on turtle nesting beaches at Milman Islet, northern GBR: (A) aerial view of Milman Islet; (B) beach rock cliffs; (C) nests eroded by waves; (D) fringing low vegetation provides shade, but large branches and roots can obstruct digging; and (E) tall open woodland on gentle dunes are good habitat. Image credits: Shane Ross, Ian Bell and Colette Thomas.

before the dunes can also influence where females choose to nest.

The location of a nest relative to island and beach features also influences its susceptibility to the elements and determines whether or not it will produce viable hatchlings. On the low-lying islands common to the GBR, storm surge can erode beaches and flood or expose nests. Some islands have unique geomorphological characteristics that influence nesting. For example, Raine Island is encircled by a relatively steep beach face that then gently slopes landward to a depression before rising again to a line of phosphate rock cliffs (Limpus et al. 2003). A rising water table, heavy rainfall, and wave and tide wash-over can flood the swale and inundate nests, potentially killing embryos (Limpus et al. 2003). Nesting beaches on islands that do not have these landforms can still experience catastrophic washout. For example, at Milman Islet, strong tides periodically erode and flood nesting beaches, drowning the nests (Staines et al. 2020).

Marine turtles select where to excavate to control nest incubation temperature and gas exchange, which determine egg survival rates, embryo sex ratios, and hatchling fitness. Porewater availability and sediment grain size and variability collectively determine the compaction and cohesiveness of nesting substrates, which are important controls on nest structure and integrity. Nests constructed in dry, poorly sorted, or coarse sands are often less cohesive and more likely to collapse during excavation, whereas sands that are too fine-grained or wet are often too compacted for turtles to dig (Stewart et al. 2019). At some rookeries, a lack of fine roots can reduce nest stability and moisture, which affects embryo survival and fitness (Chen et al. 2007). Obstacles including large roots, large low branches, and buried stumps, logs, and beach rock can increase sand compaction and obstruct nest excavation and hatchling emergence (Garcin et al. 2022).

The major *C. mydas* nesting beaches of the world are typically composed of moderately sorted sands with highly spherical grains and low levels of organic carbon; however, rookeries are quite varied. Optimal conditions at any given rookery may change depending on precipitation, mineral composition, and sand sediment particle shape. In the northern GBR, major *C. mydas* rookeries including Raine Island typically have well-sorted, coarse-grained sands; Milman Islet's medium-grained sands are a rare exception (Fuentes et al. 2010a). The sediment composition of a rookery varies depending on its location relative to various geomorphological features (e.g., cliff, berm, dune, and beach toe). Rookery sediment composition reflects the environmental settings that govern the productivity of local reef-building organisms (Fuentes et al. 2010b; Hamylton and Puotinen 2015). Consequently, although Raine Island sands are dominated by foraminifera, Milman Islet sands are dominated by marine mollusc shells and the calcified green macroalgae *Halimeda*.

Temperature is a critical determinant of turtle embryo sex, survival, and fitness, and turtle nest habitat suitability. All marine turtles have temperature-dependent sex determination, whereby offspring gender is determined by the temperature of the sand around the nest during egg incubation. If temperatures rise above a 'pivotal' point (~29 °C) during the critical mid-trimester of incubation, then the resulting hatchlings will be female. In contrast, if temperatures remain below this temperature, then the hatchlings will be male (Ackerman 2017). Very high nest temperatures can result in egg mortality; for *C. mydas*, this occurs when sand temperatures consistently exceed approximately 33 °C. Sediment characteristics can affect nest temperature. Temperatures tend to be higher at open nesting areas with dark sand, which are less shaded, or that are further from the sea (Hays et al. 2001; Fuentes et al. 2010a).

3.3.3. Availability of Turtle Nesting Habitat

Turtles appear to be unaware of, or not influenced by, the presence of an existing nest when selecting a place to start digging (Chen et al. 2007). During the high-density breeding seasons observed at some *C. mydas* rookeries, reduced availability of nesting space can increase the rate at which eggs are exhumed by turtles digging new nests on or near existing nests (Bustard and Tognetti 1969). Indeed, this often occurs at Raine Island, where the beach can be completely dug over almost nightly (Limpus et al. 2003).

The interplay of geomorphological, ecological, and climatic factors has determined the conditions for successful marine turtle nesting for millennia. Islands of the northern GBR have been successful rookeries for 5,000–6,000 years. In the last two decades, feminisation ratios have climbed to 99.4%, and large-scale failures in hatchling production have been observed at major rookeries including Raine Island (Limpus et al. 2003). Increased sand temperature is strongly correlated with rises in sea surface temperature (Fuentes et al. 2009). Eggs at major *C. mydas* rookeries in the northern GBR, such as Raine Island and Moulter Cay, have been consistently incubated *above* the pivotal temperature since the early 1990s. The corresponding number of juvenile turtles recruiting into foraging populations, where they are first able to be found and measured after returning from their post-hatchling pelagic stage, has declined (Jensen et al. 2016; Jensen et al. 2018). Further temperature increases at major rookeries in the northern GBR are likely to overstep the thermal mortality threshold, which causes further losses to hatchling production. In addition, modelled projections indicate that beach inundation from sea level rise and storm wave run-up could remove 30% and > 50% of suitable turtle nesting habitat at Raine Island and Milman Islet respectively (Fuentes et al. 2011). The density-dependent population regulation processes previously described will certainly exacerbate these effects at the major northern GBR rookeries (Bustard and Tognetti 1969).

3.4. HUMAN USES OF ISLANDS IN THE GBR

The low-lying reef islands of the GBR provide important goods and services to humans, who have interacted with reef islands in ways that have modified their geomorphic behaviour through engineering and by influencing the reef carbonate dynamics on which reef landforms rely.

3.4.1. Tourism on the Great Barrier Reef

Islands have brought many national and international visitors to the Queensland coastline. The GBR tourism industry has grown steadily since the early 1900s, generating an income estimated at AUD$56 billion in 2017 (Economics 2017). Tourist activities have expanded over the last century in the GBR supported by localised development at sites close to the coastal cities of Cairns and Townsville, including the jetty construction at Double Island, Green Island, and Magnetic Island (Daley 2014). In the 1930s, resorts were established to accommodate increased numbers of visitors from outside Queensland at Green Island (1920–36), Lady Musgrave (1930–39), and Heron Island (1970).

In the early 1900s, tourists travelled by boat and landed on the islands, where they would picnic and walk fringing and near-shore reefs at low tide, often staying overnight or for a day visit. Shells, corals, and other exotic creatures were collected for souvenirs, resulting in localised mortality of marine life. Pollution from inadequately treated sewage at resort sites resulted in a proliferation of algal beds on reef flats where corals had previously thrived. The construction of infrastructure on otherwise dynamic surfaces has interacted with existing geomorphic processes. For example, hard engineering of resort infrastructure and the introduction of beach groynes has disrupted seasonal alongshore flows of beach sand around island peripheries. Furthermore, excavating boat channels through underlying reef framework using dynamite, with the associated construction of a jetty, dredging the boat harbour, moving the dredge spoil, and erecting rock and timber retaining seawalls have reduced the capacity for sediments to be transported around the reef flat by tides and currents.

3.4.2. Coastal Livelihoods in the Central Torres Strait

Between northern mainland Australia and Papua New Guinea, the Torres Strait is home to 270 islands, including low-lying sand cays, muddy inshore islands, and high volcanic and granitic islands (Figure 8a). These islands formed through geological processes that have occurred since the flooding of the Sahul Shelf that connected northern Australia to New Guinea some 20,000 years ago, and many of these islands have been inhabited for thousands of years.

In central Torres Strait, coral reefs and islands are shaped by a combination of regional geological processes, Holocene sea-level history and exposure to contemporary environmental wind and wave regimes. In geological terms, reef islands have formed relatively recently on top of platforms that have grown over topographic remnants of previous interglacial reefs (Woodroffe et al. 2000). Thus, the antecedent topography of Pleistocene reefs preserved in the geological foundations of the underlying sea shelf acts as a fundamental control on the distribution of islands across Torres Strait. Sandy cays commonly form on the leeward, northwestern side of many elongated reef platforms, which themselves are elongated, having been shaped by strong currents driven by contrasting tidal systems to the East and West in the Coral Sea and Gulf of Carpentaria, respectively (Jones 1995).

Annual seasonal patterns in wind and rainfall govern the habitability of many Torres Strait islands, with a marked summer wet season in which the northwesterly monsoon drives heavy rainfall (December to March), followed by a longer period of low rainfall often characterised by periods of drought while southeast trade winds prevail (April to November).

The availability of freshwater determines the survival of flora and fauna on low-lying islands. Groundwater aquifers on small islands are formed by a convex-shaped freshwater lens that floats above denser saltwater. Such aquifers are supplied through precipitation that infiltrates the unconsolidated limestone sediments of the island surface and percolates down until it reaches either an impermeable surface or denser saltwater. These freshwater lenses are a critical resource of water for irrigation and drinking on small islands. Their character depends on geological factors including the permeability and porosity of reef sediments and the underlying reef structure.

3.4.3. Warraber Island: A Central Hub for Maritime Communities

Central Torres Strait is populated by the Kulkalgal Islanders, who have likely occupied the region for over 2,000 years and developed specialist strategies to occupy these islands. Kulkalgal territory covers 26 islands in central Torres Strait, and 19 of these are low-lying reef islands (Figure 8b). This area includes Warraber Island, one of the most studied islands in central Torres Strait.

Drilling on the reef platform at Warraber has indicated an older Pleistocene limestone foundation at a depth of approximately 6 m, over which reefs were established around 6,700 years BP. They subsequently accreted vertically upwards to reach a sea level approximately 0.8–1.0 m above present conditions around 5,300 years BP, as indicated by microatoll dates from the reef flat (Woodroffe et al. 2000). Radiocarbon dating of sedimentary ridges suggests that the sandy island has expanded incrementally in a southeasterly direction from a northernmost initiation point approximately 4,000 to 3,000 years ago, becoming stable some 2,000 years ago (Woodroffe et al. 2007).

The capacity to move between islands is key to survival for Islanders in central Torres Strait. The Kulkalgal adopted semi-nomadic settlement patterns across central Torres Strait with communities dispersing in the dry season and congregating in the wet season to enable continual access to freshwater for sustenance throughout the year (McNiven 2015). Warraber is thought to have functioned for several

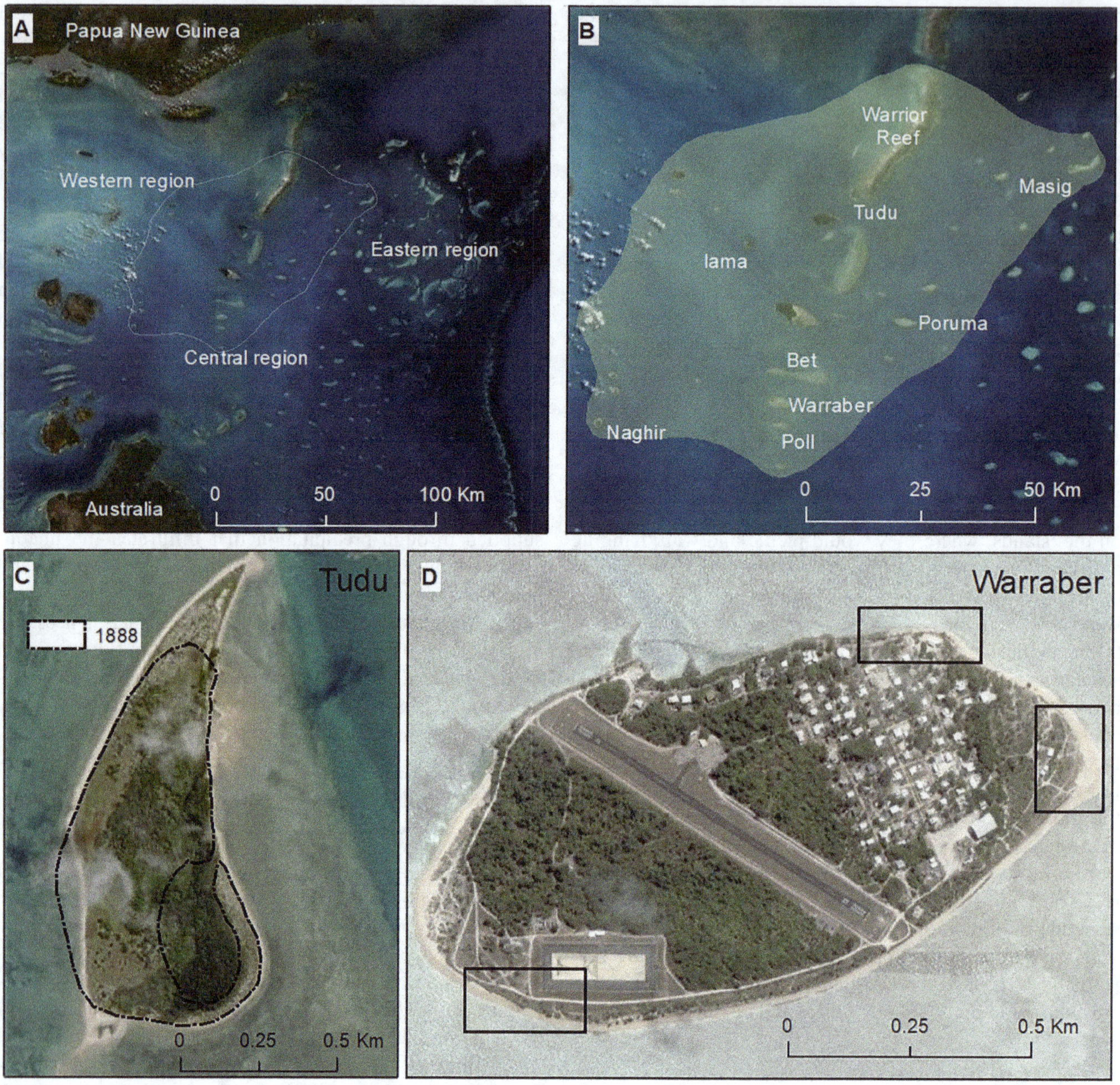

FIGURE 8 (A) The islands of Torres Strait on the Sahul Shelf between Australia and Papua New Guinea. (B) The islands of the Central Region, including low-lying reef islands populated by Kulkalgal society. (C) Shoreline changes, including lagoon infill over the past 150 years at Tudu Island. (D) Eroding shorelines leaving the water storage facility, cemetery, and church vulnerable to inundation at Warraber (image credit: Sarah M. Hamylton).

hundred years as a larger residential island because of its freshwater well that flows consistently throughout the year. Residents of neighbouring Naghir and Poruma Islands likely gathered in the dry season at Warraber when their freshwater wells had been exhausted (MacGillivray 1852). Such movement patterns of Islanders allowed food plants such as yam, taro, and sweet potato to be planted at the end of the dry season and harvested at times of communal gathering (McNiven 2015). Regional mobility on outrigger canoes also opened nearby unoccupied 'foraging islands' for day visits, where new fishing grounds were exploited, and freshwater, firewood, seabird, and turtle eggs were gathered (Fuary 1991; Crouch 2008).

Visits further afield to neighbouring New Guineans to the north and Meriam Islanders to the east enabled regional alliances to be developed for trade in cultivated plants during the dry season, when the terrestrial plant foods and drinking water resources on the reef islands had been impoverished (McNiven 2015). The pearl shell and beche-de-mer of Warrior Reef were the focal points of a regional trading network that exposed the Kulkalgal and others in central Torres Strait to broader international networks, including the adjacent mainland coastlines and extending along the Kimberley coastline and out to the Coral Sea (Lawrence 1998).

Harvesting activities throughout northern coastal Australia extend back to the 1700s, representing some of Australia's oldest international trading exchanges, when fleets of 'trepangers' sailed across northern Australia on seasonal winds from Makassar in Indonesia to collect sea cucumbers. They stopped on the low-lying reef islands for several months at a time to smoke their catches before transporting them back to Indonesia, where they would sell them to Chinese traders (Schwerdtner Máñez and Ferse 2010).

Sustainable use of the diverse array of marine resources surrounding the coral reefs of the central islands has been supported by specialist strategies for fishing, hunting dugong, and collecting shellfish at Warraber. Tidal fish traps, known as *sai*, occur across much of Torres Strait, including the reef flats adjacent to Warraber's beaches (Rowland and Ulm 2011). These traps are constructed from rock walls that stand up to a metre high and run continuously out from on the shore to enclose a portion of the reef. On a high tide, *sai* walls become submerged to allow fish and turtles to move inside them, which subsequently become trapped as the tide recedes and the tops of the walls are exposed. These traps were designed hundreds, perhaps thousands, of years ago based on a knowledge of reef flat tidal geomorphology and have been used for shell fishing for many years, particularly by women (Lui-Chivizhe 2022).

Communal ceremonial practices refined over centuries of living alongside and being nourished by reefs convey important information about ownership of community hunting grounds, which seasons to harvest or hunt, and movements of marine species. The presence of a Turtle *zogo*, or ceremonial site, at Warraber indicates the attribution of totemic importance to this marine species. Similar status has been afforded to sharks, saltwater crocodiles, turtles, and dugong by Torres Strait Islanders. Reciprocal responsibilities that require kinsmen to respect and abstain from killing or eating animal totems may mediate localised levels of exploitation (Haddon et al. 1904; McNiven 2011). The broader role played by totemism in sustaining wildlife of otherwise inhospitable regions is a recognised aspect of Australian Indigenous societies that encourages restraint and management for long-term productivity of local resources (Rose 2013).

Geological factors and biological factors combine in central Torres Strait to create dynamic conditions for coastal communities. These communities have been described as living on the margins of sustainability, where islanders can experience shoreline movements, periods of drought, and poor terrestrial resources (McNiven 2015). Sketches and notes from Alfred Haddon's 1888 visit to the low-lying island of Tudu suggest that until relatively recently, this island had an open lagoon some 250 m wide, which has now completely filled internally (Philp and Herle 2020; Figure 8c).

Climate change threatens the livelihoods and cultural identity of Torres Strait Islanders, with the inundation from rising seas and erosion from storms presenting an existential threat to those living on low reef islands. According to the Australian Bureau of Statistics, Warraber had a population of 274 people spread across 57 households on the eastern side of the island in 2021. Aerial photography indicates that the northeastern and southwestern corners of the island have eroded as sediments have been transported along the shoreline and deposited further to the east and to the west, respectively, since the 1960s (Figure 8d). Seasonal cut and fill of beach sediments give the island an appearance of rotating over the longer term (Duce et al. 2010). Shoreline erosion in the low-lying northeastern corner of the island has exposed the cemetery and church sites to inundation, while erosional shorelines close to the water facility are of concern to the community.

3.4.4. An Uncertain Future for Reef Islands

Each of the case studies linked geological and biological aspects of reefs within the GBR. We are beginning to understand how these dynamic links behave with environmental changes and the implications that follow for reef islands, although much remains to be seen.

The landforms that we see today such as the mangrove forests growing on upper reef flat surfaces in the Howick Islands are shaped by a suite of geomorphological processes occurring on the reef platforms that underlie them. Climate change impacts on the GBR will continue to significantly reduce the amount of suitable nesting habitat available for marine turtles whilst simultaneously increasing feminisation rates. Climate-driven changes to the type and amount of reef sediment produced and supplied to the islands will likely change the geomorphology of many islands in the northern GBR. However the islands most affected and the impact of such changes on the resilience of the islands and nesting beaches are unclear. There is a critical need to address these questions before GBR populations of marine turtles and other marine species collapse.

Coral reef tourism, maritime trade, and the livelihoods of Torres Strait Islanders face uncertainty about the condition and future of reefs in the GBR region. The GBR attracts many visitors. In 2015, Tourism Australia reported that 42% of international tourists to Australia ranked the GBR as Australia's most appealing tourist attraction, and it attracted approximately 2.1 million international and 1.7 million domestic visitors. It remains to be seen whether the GBR will remain a globally competitive tourist destination or whether visitors will look to alternative sites in other countries as the GBR gradually becomes degraded over time because of global heating and other anthropogenic impacts.

Severe tidal flooding of the large low-lying muddy Boigu and Saibai Islands in northern Torres Strait resulted in the voluntary relocation of community members in the 1940s (Smithers 2022). Occupants of other low-lying islands, particularly those across central Torres Strait, may need to relocate over time, either to the mainland or to the higher volcanic and granitic islands, such as Mer and Erub in eastern Torres Strait. Strategies developed by the Kulkalgal community for living sustainably across the inhospitable low-lying islands of the central region will take on a renewed importance for modern reef island communities facing the predicted impacts of global warming.

The evolution of reef islands over time will depend on a complex interplay among sea-level rise, reef flat sediment deposition, hydrodynamics, and a continued supply of calcium carbonate from producers. Alongside these geomorphic considerations, direct human interventions such as dredging for marinas and navigation channels will have localised implications. The effects of global climate change in the ocean, notably coral bleaching and ocean acidification, will influence the carbonate budget in ways that will change island formation and maintenance over time. Reef islands are isolated outposts of dynamic, habitable land subject to seasonal erosion and accretion on which communities are faced with precarious terrestrial resources including a paucity of vegetation and freshwater juxtaposed with high marine resource productivity. These circumstances have given rise to distinct strategies for living in rapidly changing environments that have emerged as important themes for human communities inhabiting reef islands.

REFERENCES

Ackerman RA (2017) The nest environment and the embryonic development of sea turtles. In: The Biology of Sea Turtles. CRC Press, pp. 83–106

AgiSoft L (2014) Agisoft Photoscan. Professional Edition, St Petersburg, Russia

Baldwin AH, McKee KL, Mendelssohn IA (1996) The influence of vegetation, salinity, and inundation on seed banks of oligohaline coastal marshes. American Journal of Botany 83:470–479

Bowen BW, Karl SJMe (2007) Population genetics and phylogeography of sea turtles. Molecular Ecology 16:4886–4907

Bustard HR, Tognetti KPJS (1969) Green sea turtles: A discrete simulation of density-dependent population regulation. Science 163:939–941

Chen H-C, Cheng I-J, Hong EJ (2007) The influence of the beach environment on the digging success and nest site distribution of the green turtle, Chelonia mydas, on Wan-an Island, Penghu Archipelago, Taiwan. Journal of Coastal Research 23:1277–1286

Crouch J, Hola J (2008) Reading between the lands: Toward an amphibious archaeological settlement model for maritime migrations. Handbook of Landscape Archaeology 131–140

Daley B (2014) The Great Barrier Reef: An Environmental History. Routledge

Dawson JL, Smithers SG, Hua QJG (2014) The importance of large benthic foraminifera to reef island sediment budget and dynamics at Raine Island, northern Great Barrier Reef. Geomorphology 222:68–81

Duce SJ, Parnell KE, Smithers SG, McNamara KE (2010) A synthesis of climate change and coastal science to support adaptation in the communities of Torres Strait. Reef and Rainforest Research Centre Report.

Deloitte Access Economics (2017) At what price? The economic, social and icon value of the Great Barrier Reef, Report Commissioned by the Great Barrier Reef Foundation.

Fairbridge RW, Teichert C (1948) The low isles of the Great Barrier Reef: A new analysis. The Geographical Journal 111:67–88

Fuary MM (1991) Fishing and its social significance on Yam Island. CSIRO Report, Melbourne.

Fuentes M, Hamann M, Limpus C (2010a) Past, current and future thermal profiles of green turtle nesting grounds: Implications from climate change. Journal of Experimental Marine Biology and Ecology 383:56–64

Fuentes M, Limpus C, Hamann M (2011) Vulnerability of sea turtle nesting grounds to climate change. Global Change Biology 17:140–153

Fuentes M, Maynard J, Guinea M, Bell I, Werdell P, Hamann M (2009) Proxy indicators of sand temperature help project impacts of global warming on sea turtles in northern Australia. Endangered Species Research 9:33–40

Fuentes MMB, Dawson J, Smithers S, Hamann M, Limpus C (2010b) Sedimentological characteristics of key sea turtle rookeries: Potential implications under projected climate change. Marine and Freshwater Research 61:464–473

Garcin M, Vende-Leclerc M, Read T, Oremus M, Bourgogne HJ (2022) Assessment method of the sea turtle-nesting habitat of small reef islands. Journal of Coastal Conservation 26:1–13

Gyuris E (1994) The rate of predation by fishes on hatchlings of the green turtle (*Chelonia mydas*). Coral Reefs 13(3):137–144.

Haddon AC, Rivers WHR, Haddon AJ (1904) Totemism. Reports of the Cambridge anthropological expedition to torres straits: Sociology. Magic and Religion of the Western Islanders 153–193

Hamylton, S, Kelleway, J, Rogers, K, McLean, R, Nagel-Tynan, Z, Repina, O (2023). Mangrove expansion on the low wooded islands of the Great Barrier Reef. Proceedings of the Royal Society B, 290(2010), 20231183

Hamylton S, McLean R, Lowe M, Adnan F (2019) Ninety years of change on a low wooded island, Great Barrier Reef. Royal Society Open Science 6:181314

Hamylton SM (2017) Mapping coral reef environments: A review of historical methods, recent advances and future opportunities. Progress in Physical Geography 41:803–833

Hamylton SM, Puotinen M (2015) A meta-analysis of reef island response to environmental change on the Great Barrier Reef. Earth Surface Processes and Landforms 8:1006–1016.

Hays G, Ashworth J, Barnsley M, Broderick A, Emery D, Godley B, Henwood A, Jones EJ (2001) The importance of sand albedo for the thermal conditions on sea turtle nesting beaches. Oikos 93:87–94

Heatwole H (2011) Coral cays, vegetational succession. In Encyclopedia of Modern Coral Reefs. Springer, pp. 256–261

Henderson R, Johnson D (2016) The Geology of Australia. Cambridge University Press

Hopley D (2008) Geomorphology of coral reefs with special reference to the Great Barrier Reef. In: The Great Barrier Reef: Biology, Environment and Management, pp. 5–16. CSIRO Publishing

Hopley D, Smithers SG, Parnell K (2007) The Geomorphology of the Great Barrier Reef: Development, Diversity and Change. Cambridge University Press

Hubbard D (1985) What do we mean by reef growth. Proceedings of the Fifth International Coral Reef Congress 6:433–438

Hubbard D (1997) Reefs as Dynamic Systems. Life and Death of Coral Reefs. Chapman & Hall, pp. 43–67

Hubbard DK, Miller AI, Scaturo D (1990) Production and cycling of calcium carbonate in a shelf-edge reef system (St. Croix, US Virgin Islands); applications to the nature of reef systems in the fossil record. Journal of Sedimentary Research 60:335–360

Hutchings P (1986) Biological destruction of coral reefs. Coral Reefs 4:239–252

Jensen MP, Allen CD, Eguchi T, Bell IP, LaCasella EL, Hilton WA, Hof CA, Dutton PH (2018) Environmental warming and feminization of one of the largest sea turtle populations in the world. Current Biology 28:154–159. e154

Jensen MP, Bell I, Limpus CJ, Hamann M, Ambar S, Whap T, David C, FitzSimmons NNJ (2016) Spatial and temporal genetic variation among size classes of green turtles (Chelonia mydas) provides information on oceanic dispersal and population dynamics. Marine Ecology Progress Series 543:241–256

Jones MJ (1995) The Torres Reefs, North Queensland, Australia—strong tidal flows a modern control on their growth. Coral Reefs14:63–69

Kelly I, Leon JX, Gilby BL, Olds AD, Schlacher TAJ (2017) Marine turtles are not fussy nesters: A novel test of small-scale nest site selection using structure from motion beach terrain information. PeerJ 5:e2770

Kench P, Perry, C, Spencer, T (2009) Coral reefs. In: Slaymaker O (ed.), Geomorphology and Global Environmental Change. Cambridge University Press, pp. 180–213

Kench P, Smithers S, McLean RJ (2012) Rapid reef island formation and stability over an emerging reef flat: Bewick Cay, northern Great Barrier Reef, Australia. Geology 40:347–350

Kench PJ (2010) Low Wooded Island. In: Encyclopedia of Modern Coral Reefs. Springer, p. 191

Kleypas JA, McManus JW, Meñez LA (1999) Environmental limits to coral reef development: Where do we draw the line? American Zoologist 39:146–159

Kleypas JA, Yates KK (2009) Coral reefs and ocean acidification. Oceanography 22:108–117.

Lawrence DJ (1998) Customary exchange in the Torres Strait. Australian Aboriginal Studies 2:13–25

Limpus, CJ, Miller, J, Baker, V, McLachlan, E (1983) The Hawksbill Turtle, *Eremochelys imbricata* (L.), in North-Eastern Australia: The Campbell Island Rookery. Wildlife Research 10(1):185–197.

Limpus CJ, Miller JD, Parmenter CJ, Limpus DJ (2003) The green turtle, Chelonia mydas, population of Raine Island and the northern Great Barrier Reef. Memoirs—Queensland Museum 149:349–440

Lui-Chivizhe L (2022) Coral reefs of Zenadth Kes. In: Hamylton SM, Hutchings, P., Hoegh-Guldberg, O. (ed) Coral Reefs of Australia: Perspectives from Beyond the Water's Edge. CSIRO Publishing.

MacGillivray J (1852) Narrative of the Voyage of HMS Rattlesnake. T&W Boone.

McLean R, Stoddart D (1978) Reef island sediments of the northern Great Barrier Reef. Philosophical Transactions of the Royal Society of London Series A, Mathematical and Physical Sciences 291:101–117

McLean R, Stoddart D, Hopley D, Polach H (1978) Sea level change in the Holocene on the northern Great Barrier Reef. Philosophical Transactions of the Royal Society of London A: Mathematical, Physical and Engineering Sciences 291:167–186

McNiven IJ (2011) Torres Strait Islanders: The 9000-year History of a Maritime People. The Torres Strait Islands, Queensland Art Gallery/Gallery of Modern Art, pp. 210–219

McNiven IJ (2015) Precarious islands: Kulkalgal reef island settlement and high mobility across 700 km of seascape, central Torres Strait and northern Great Barrier Reef. Quaternary International 385:39–55

Perry CJ (2007) Tropical coastal environments: Coral reefs and mangroves. Environmental Sedimentology 302–350.

Perry CT, Kench PS, O'Leary M, Morgan K, Januchowski-Hartley FJ (2015) Linking reef ecology to island building: Parrotfish identified as major producers of island-building sediment in the Maldives. Geology 43:503–506

Perry CT, Kench PS, Smithers SG, Riegl B, Gulliver P, Daniells JJ (2017) Terrigenous sediment-dominated reef platform infilling: An unexpected precursor to reef island formation and a test of the reef platform size–island age model in the Pacific. Coral Reefs 36:1013–1021

Philp J, Herle A (2020) Recording Kastom: Alfred Haddon's Journals From the Torres Strait and New Guinea, 1888 and 1898. Sydney University Press, Sydney.

Rose DB (2013) Common Property Regimes in Aboriginal Australia: Totemism Revisited. ANU Press, pp. 127–143

Rowland MJ, Ulm SJ (2011) Indigenous fish traps and weirs of Queensland. Queensland Archaeological Research 14:1–58

Schwerdtner Máñez K, Ferse SC (2010) The history of Makassan trepang fishing and trade. PLoS One 5:e11346

Scoffin TP, Stearn C, Boucher D, Frydl P, Hawkins C, Hunter I, MacGeachy J (1980) Calcium-carbonate budget of a fringing-reef on the west-coast of Barbados. 2. Erosion, sediments and internal structure. Bulletin of Marine Science 30:475–508

Smith TJ (1987) Effects of light and intertidal position on seedling survival and growth in tropical tidal forests. Journal of Experimental Marine Biology and Ecology 110:133–146

Smithers S, Browne, N (2022) Australia's reef islands. In: Hamylton S, Hutchings, P., Hoegh-Guldberg, O. (ed) Coral Reefs of Australia: Perspectives from Beyond the Water's Edge. CSIRO Publishing.

Smithers S, Hopley D, Mclean RJ (2018) Holocene sea levels on the Great Barrier Reef: Reflections on four decades of debate and the legacy of the 1973 expedition to the northern Great Barrier Reef. Atoll Research Bulletin 619:79–104

Staines MN, Booth DT, Madden Hof CA, Hays GC (2020) Impact of heavy rainfall events and shading on the temperature of sea turtle nests. Marine Biology 167:1–11

Stearn C, Scoffin T, Martindale W (1977) Calcium carbonate budget of a fringing reef on the west coast of barbados part I—Zonation and productivity. Bulletin of Marine Science 27:479–510

Stewart TA, Booth DT, Rusli MUJ (2019) Influence of sand grain size and nest microenvironment on incubation success, hatchling morphology and locomotion performance of green turtles (Chelonia mydas) at the Chagar Hutang Turtle Sanctuary, Redang Island, Malaysia. Australian Journal of Zoology 66:356–368

Stoddart D (1978) Mapping reef and islands. In Coral Reefs Research Methods Monographs on Oceanographic Methodology. UNESCO, pp. 17–22

Stoddart D, McLean R, Scoffin T, Gibbs P (1978) Forty-five years of change on low wooded islands, Great Barrier Reef. Philosophical Transactions of the Royal Society of London Series B, Biological Sciences 63–80

Stoddart DJ (1980) Mangroves as successional stages, inner reefs of the Northern Great Barrier Reef. Journal of Biogeography 269–284

Woodroffe CD, Kennedy D, Hopley D, Rasmussen C, Smithers SJMG (2000) Holocene reef growth in Torres Strait. Marine Geology 170:331–346

Woodroffe CD, Samosorn B, Hua Q, Hart DEJ (2007) Incremental accretion of a sandy reef island over the past 3000 years indicated by component-specific radiocarbon dating. Geophysical Research Letters 34

4 Currents Modulate the Genetic Character of Marine Populations in the Great Barrier Reef

Stuart Kininmonth

4.1. INTRODUCTION

The overwhelming impression of any visitor to the Great Barrier Reef (GBR) is the vastness and intricate nature of this collection of coral reefs. Built primarily by tiny scleractinian animals, these coral reefs occupy 7% of the associated GBR world heritage area and can be clearly observed from space (Lyons et al. 2020). However, the remarkable relationship between the coral dispersal processes and the water circulation patterns results in a complexity of species and structures along the 1,840-km span (Wolanski et al. 2024). In many respects, the creation of the GBR over the last 10,000 years can be conceptualized similar to the age-old philosophical question of 'which came first, the chicken or the egg?' The oceanic currents as they push against the Queensland coast shape the dispersal patterns of the coral species, which, in turn, gradually grow into the water column to influence the energy and direction of the very same currents. Few organisms have shaped their environment so dramatically. This chapter explores how the water circulation patterns have interacted with the growth and dispersal of marine species resulting in complex genetic patterns.

4.2. THE MARINE WORLD AS WE KNOW IT

Early biologists upon observing the GBR were impressed with the vastness of reef structures (Taylor 1931). Subsequent surveys along the entire GBR revealed a diversity that extended from the warm tropical waters of the Torres Strait to the cooler waters of the Swains complex (Fernandes et al. 2005). Furthermore, this diversity was confounded by the distance from the shore and the depth profile. Considerable research in the 1970s explored the principal drivers of this diversity based on the spatial distribution of species (Harriott and Banks 2002; Veron et al. 2015) and the topography of coral reefs (Hopley 2011). The scientific focus was on the capacity of corals to flourish in the microenvironments found across the GBR (Arias-González et al. 2006). With the more recent observation of the massive spawning events (Wallace 1985; Willis et al. 1985; Oliver and Willis 1987; Harrison et al. 2010), the focus shifted to understanding how the life cycles of corals and other marine species were impacted by the associated water currents (Wolanski 2017). Coral species do not simply disperse in a uniform manner, but instead, the propagules from sexual encounters are transported to destinations because of dynamic water currents (Willis and Oliver 1990; James et al. 2002; Bode et al. 2005; Jones et al. 2009). The modelling of these currents required advanced methods in particle transfer that could encompass tidal flows, oceanic currents, predominant wind forcing, and three-dimensional interactions against coastal and reef structures (James et al. 2002; Legrand et al. 2006). Animation 1 shows the high theoretical complexity of the particle movements and the transition to settlement success. Even more thought-provoking is the incorporation of the switch from a passive to an active propagule (Kingsford et el. 2024) within the computer simulations (Wolanski and Kingsford 2014). The challenge in understanding the impact that this water movement complexity has on seascape genetics cannot be understated. Empirical evidence of water flows combined with satellite images and particle release experiments provided the validation and calibration required for these interaction models. Interestingly, the rezoning of the GBR in 2007 used a bioregionalisation approach (Fernandes et al. 2005) that explicitly did not incorporate currents primarily because the conservation modelling (Watts et al. 2009; Friesen et al. 2019) had not caught up to the hydrodynamic modelling advances. At the same time as hydrodynamic models developed in sophistication, an explosion of genetic analytical capacity occurred. Combined with an intensive sampling program at a variety of scales (Underwood et al. 2007; Noreen et al. 2009; van Oppen et al. 2011), this genetic analysis led to a revolution in the understanding of the species taxonomy (Veron 2013; Hobbs et al. 2022; Ramírez-Portilla et al. 2022) and spatial distributions of marine species (Kininmonth et al. 2009, 2012; Harrison et al. 2012; Riginos et al. 2019). The water circulation patterns are described in detail in Wolanski et al. (2024), but the observation of this process can be dramatic when a hidden coral reef measuring many square kilometers emerges from the ocean as the daily tide drops. Within this daily dynamic context over thousands of years, coral species, amongst many others, have evolved and distributed along the continental shelf (Smith-Keune and Oppen 2006). The scientific understanding of the genetic complexity of the GBR is in its infancy, but the important role of hydrodynamics in shaping the genetic seascape is clear.

DOI: 10.1201/9781003320425-5

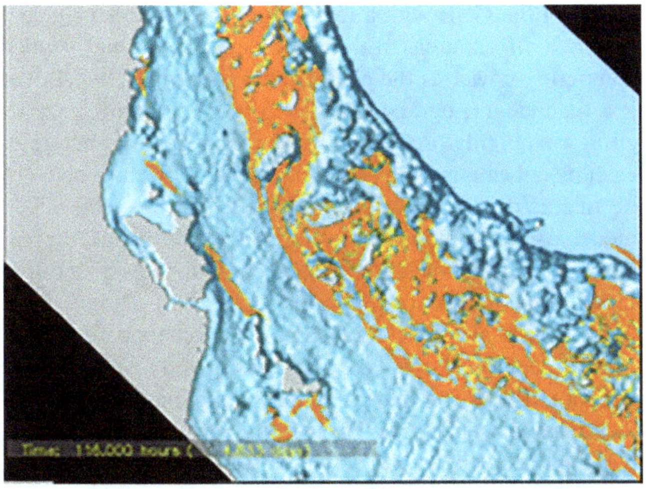

ANIMATION 1 Simulation of propagule dispersal in the GBR (based on James et al. 2002). The orange particles are released from each reef edge and are dispersed. As time goes on, many of the larvae die, but enough survive to settle on distant reefs.

Source: Permission granted by Lou Mason, animation creator

Available online at https://www.routledge.com/9781032340692

4.3. THE PASSAGE OF GENES IN THE MARINE WORLD

The lifecycle of many marine animals, including fish (Edgar and Stuart-Smith 2014; Bode et al. 2019; Lefcheck et al. 2021), corals (Treml et al. 2007; Veron et al. 2015), sea stars (MacNeil et al. 2016; Pratchett et al. 2017), and mollusks (Vanderklift et al. 2020) embraces a pelagic component at the mercy of the prevailing currents. The most well-known is the synchronized spawning events of the scleractinian coral whereby the egg and sperm from millions of colonies are released into the water column (Wallace 1985; Harrison et al. 2010). Brooding coral retain the egg and seek sperm from the water column, while broadcaster corals rely completely on the intermingling at the ocean surface. This haphazard manner of sexual reproduction where one colony of a specific species tries to exchange genetic information with another colony of the same species relies completely on the enormous volumes of gametes produced (Harrison et al. 2021) with help from the chemical recognition of conspecific gametes (Babcock 1995). These gametes rapidly disperse into the wider food chain, while the few successful fertilized propagules drift with small-scale currents within the reef complex and large-scale currents moving down the GBR. After a period referred to as the pelagic larval duration (PLD; Treml et al. 2007), the coral larvae then actively seek a suitable substrate to settle on. Despite the high failure rate, this successful colonization can disperse corals outside the GBR to distant places such as Lord Howe Island by using the East Australian Current (S. Choukroun, pers. comm.). Critically, the uncertainty of the strength and direction of these currents at any given moment makes the dispersal events appear chaotic, but when examined over many years, a strong dispersal pattern is revealed (Kininmonth et al. 2009). Some coral reefs due to their placement in the GBR act as source reefs for propagules, while others, particularly those on the fringe of coral reef complexes, are sinks (Bode et al. 2005). Over long periods, this hydrodynamically driven dispersal process directly influences the genetic character of the reef's inhabitants (van Oppen et al. 2008; Kininmonth et al. 2010b). Other species such as fish also include for dispersal an active swimming component at the adult stage (Harrison et al. 2012; Kingsford et al. 2024). However, despite the high risk of failure, the dominant currents with their associated warm waters can still disperse mature individuals to distant locations (Willis 2011; Stuart-Smith et al. 2017).

4.4. THE CHARACTERIZATION OF MARINE POPULATIONS

Early taxonomic work on coral species based on morphological traits assumed that species in one location would be similar if not identical to ones found across the GBR. Species distribution maps presented a global picture of diversity based on species counts (Veron et al. 2009). A nominal 150 coral ecoregions capture over 830 coral species (Green and Mous 2008; Veron et al. 2015). The GBR, for example, was subdivided into four regions (Veron et al. 2015): GBR Far North and Torres Strait, GBR North Central, GBR Southeast, including Pompey and Swain reefs, and the GBR South. These regions are based on species diversity, that is, the coral species that an observer might commonly see within this region. Although the GBR is certainly special, the diversity of species is higher within the equatorial regions known as the coral triangle (Veron et al. 2009). The assumption of genetic uniformity that defines the described species that make up these coral regions is now realized to be misleading (Ramírez-Portilla et al. 2022). Consequently, the region where the original type specimen (the very first sample of the species) was first collected is now an important component to describing the definition of the species. Subsequent collections of this species in other regions are then compared to the original collected specimen. For example, van Oppen et al. (2008) conducted genetic analysis of a limited number of samples of a restricted set of species and emphasized how spatial interaction of the focal species shaped the genetic structure across a dynamic water circulation pattern (van Oppen et al. 2008). At the reef scale, individual corals (*Acropora millepora*) growing on a single reef, such as Davies Reef (GBR: 18–096), are often more related to adjacent reefs than those within the reef, which indicates the dominant role of water circulation (Smith-Keune et al. 2006) and differential post-settlement mortality (Penin et al. 2010). At a larger scale, the GBR shows a genetic differentiation of northern versus southern populations of reef fish (Doherty et al. 1995) and calcareous sponges, such as *Leucetta chagosensis* (Wörheide et al. 2002). The clams *Tridacna maxima* and *Tridacna derasa* indicate genetic differentiation for the southern populations (Benzie and Williams 1997). Five reef-building corals show genetic diversity from north, central, and southern populations (Ayre and Hughes 2004). These population splits are likely to be partially because of low sea levels from the last glacial period that created the Queensland and Marion Plateaus (Davies 2011). Despite over 10,000 years of dispersal events and migration and colonization, the lack of a pan mixed population in the GBR emphasizes that hydrodynamics at a variety of scales, combined with environmental factors such as temperature (Mora et al. 2003) and post-settlement processes (Penin et al. 2010), continue to act as a barrier to gene flow (Smith-Keune and Oppen 2006). Barriers, such as the persistent frontal convergence zone south of the Whitsunday Islands (Cape Clinton; -22.5S, 150.7E) (Burrage et al. 1996; Weeks et al. 2010), continue to influence modern marine populations. Early work on the holobionts of marine organisms was limited to taxonomic differences, and for coral, this was principally focused on the symbiotic algal strains where noticeable differences were observed (Planes et al. 2019; Pollock et al. 2019; Stévenne et al. 2021). Future work in this area is urgently needed to understand the impact of climate change and the capacity to migrate to alternative habitats (Selmoni et al. 2020).

4.5. MODELLING THE COMPLEX PROPAGULE MOVEMENT

Since the days of sailing vessels and perhaps even the canoes of Australian First Nation peoples, society has noted and understood the dominant currents of the GBR. However, it was only in recent years with supercomputing that models could be developed that specifically quantify the water flow (James et al. 2002; Legrand et al. 2006). One key driver of these models was the prediction of 'particle' transfer from reef to reef for both pollution events and biological activity. Early models used a simple diffusion approach, but this was highly limited in such a complex environment as the GBR. More sophisticated approaches based on Lagrangian statistics calculated the trajectory of a particle to essentially step across discrete cells based on the forces in the model (Dight et al. 1990; James et al. 2002). To make this type of model scientifically useful, many particles were included in the release with simplified individual biological behaviour of the larvae (Treml et al. 2007). The early models by James et al. (2002) included 320 reefs located between latitudes 14 and 19° S. The tides, wind, and oceanic influences on water currents were included. To provide a long-term picture of the oceanographic influences, the particles were dispersed for each spawning season between 1967 and 1998. The results opened a novel window into the character of the GBR (Bode et al. 2005, 2008; Kininmonth et al. 2009, 2011, 2012) despite some significant simplifications. These simplifications were necessary to run the model in a sensible timeframe but importantly included being based on a two-dimensional model of the GBR with a regular grid network. This meant that the uniform areas of the GBR were consuming as much computing power as the complicated reef edges, which was less than ideal (Legrand et al. 2006). Validation of the modelling was highly problematic given the stochastic nature of each model run, but the perception of the GBR as a network of interacting reefs was established. More sophisticated models were developed especially aided by advanced computing power that included variable mesh models with three-dimensional interactions (Legrand et al. 2006; Thomas et al. 2014). Fine-scale models that captured the chaotic turbulence of coral reef interfaces were also constructed (Wolanski and Spagnol 2000). The marine propagules remained simple in that behaviors, such as active swimming to settle on a reef or moving reactively up and down the water column (Wolanski et al. 1997), were given uniform values. Subsequent improvements incorporated larval behavior and especially the concept of 'sticky water' that retained larvae in the coral reefs (Kingsford et al. 2024). Validation through genetic studies of populations, satellite imagery analysis, and particle release increased the confidence in the models (Manel et al. 2019).

4.6. THE TOPOLOGY OF DISPERSAL NETWORKS

The final output from the hydrodynamic models was essentially a large table (known as connectivity matrices) that showed the particles departing and arriving from each coral reef in the model scope (Figure 1). The same reef is shown

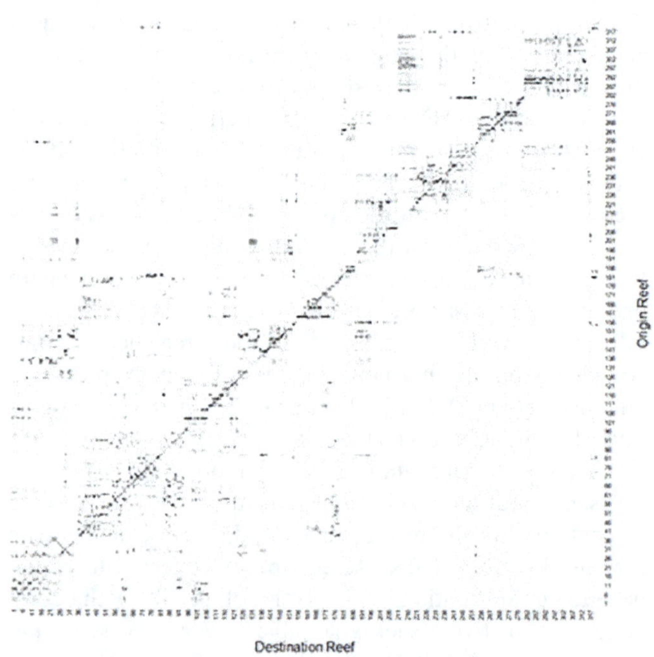

FIGURE 1 Connectivity matrix. The black dots represent a group of larvae that have travelled from the origin to the destination reef. The white areas highlight that larvae did not travel to the destination reef from the origin reef.

Source: Generated from James (2002)

on both the vertical and horizontal axes so that the diagonal represents the self-fertilization numbers. To compensate for the size variation of GBR coral reefs, the length of the coral reef polygon edge was used as a crude measure of propagule production capacity and settlement success for the entire reef. Although clearly these values are only potential, the immediate impression of the model output is that some reefs are sources of gene flow, while others are sinks (Bode et al. 2005).

This realization is critical in terms of conservation planning (Bode et al. 2024) where the impact from the decimation of a coral reef from tropical cyclones, crown-of-thorns starfish plagues (Pratchett et al. 2024), warm water bleaching, etc. is in many respects a function of where the reef is situated in the system (Fernandes et al. 2005). Transferring all of this connectivity data to a network where the nodes represent reefs and the lines indicate propagule transfer highlights both the complexity and the hidden structure of the GBR (Figure 2). Research has shown that the thousands of reefs are assembled into a small-world network (Kininmonth et al. 2009; Thomas et al. 2014) where some reefs form clusters that are connected to other clusters.

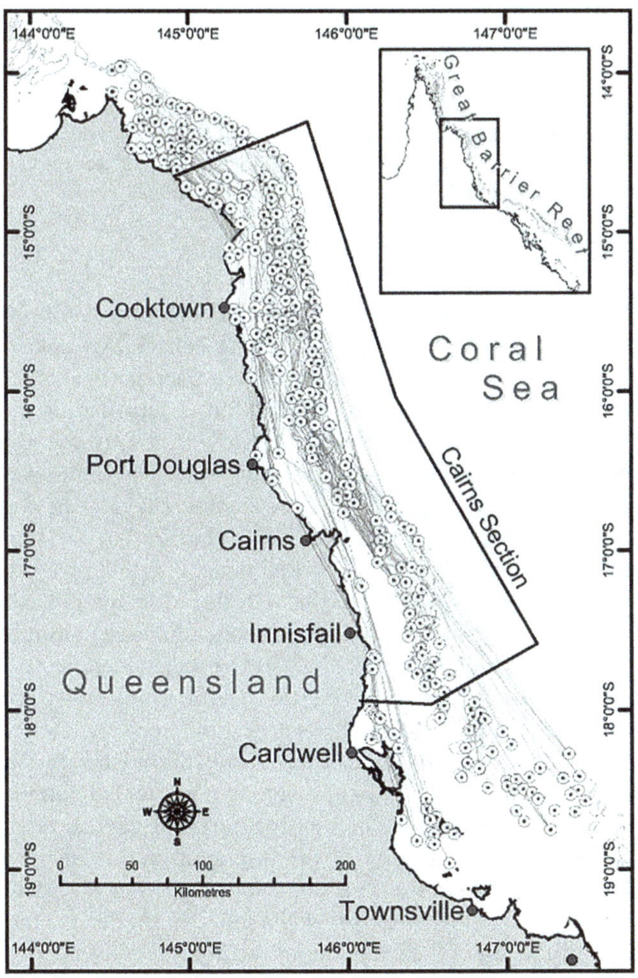

FIGURE 2 Map of the GBR showing the spatially located network (Kininmonth et al. 2009). The direction is not shown due to the clarity required at this scale, but there is a general north and south flow from the central region.

Source: Reproduced with permission from Springer Nature, license number 5479160987172

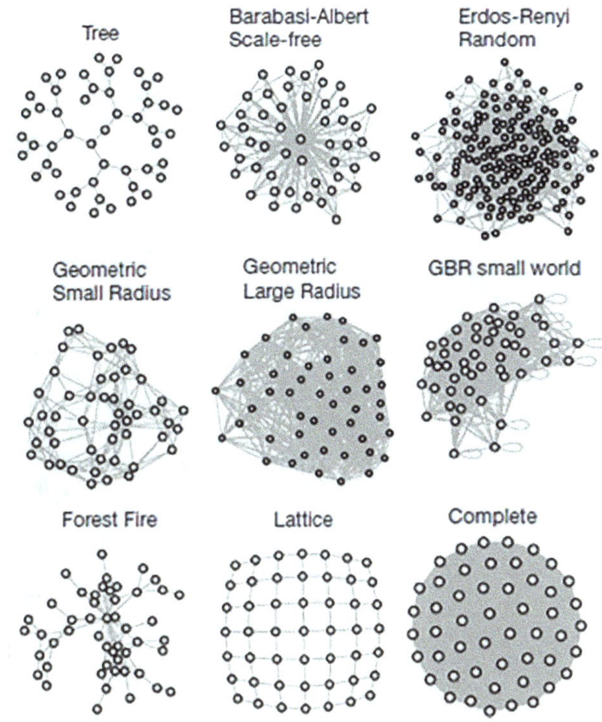

FIGURE 3 Network topologies available to genetic dispersal.

The style of networks can be described by their topology (Dunne et al. 2002; Kininmonth et al. 2010a). Some networks are very regular where the nodes (representing a reef, for example) are only connected to their nearest neighbor (lattice: Figure 3). Other networks are called random networks (Figure 3) since any node can connect to any other node. The small-world network (Figure 3) has clusters of nodes connected by long links and is commonly found in ecological systems. This facilitates material transfer such as genes to propagate rapidly through the network but without the concern of what might happen if a single key reef is damaged. The small-world topology of gene flow throughout the GBR has resulted in a series of metapopulations that regularly interact.

Marine organisms commonly face disruption at a local scale, perhaps through cyclones (Duke et al. 2024), and hence being connected to other smaller populations is very useful for recolonization. Extrapolating this connection pattern across 'geological' time ensures that the reefs that expand are dominated by the coral colonies that benefit from the water circulation patterns (Harriott and Banks 2002). The growth of the marine organisms, especially corals, then tends to trap the water through turbulence thus positively reinforcing the cycle of growth. Isolated coral reefs are very vulnerable to extinction events. The genetic diversity (Figure 4) reflects this pattern of gene flow as a function of water circulation over many years (Underwood et al. 2007; van Oppen et al. 2008; Kininmonth et al. 2010b).

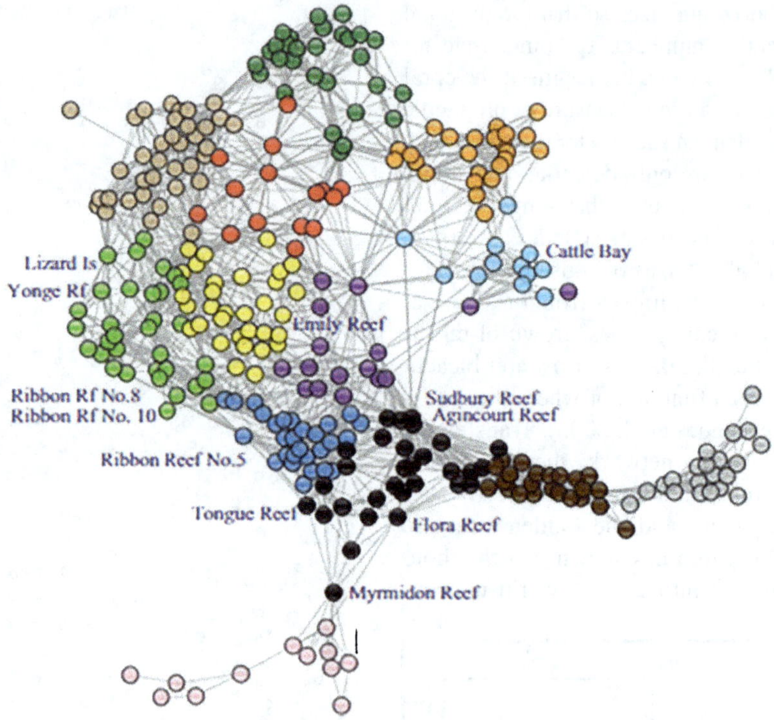

FIGURE 4 Plot showing the small-world community structure of the coral *Seriatopora hystrix* in the central GBR based on hydrodynamic model data and genetic surveys (Kininmonth et al. 2010b). Note that the position of the reefs in this plot only relates to their community connections, not the geospatial location. The named coral reefs were sampled for their genetic composition.

Source: Reproduced with permission from Elsevier, license number 5479151182191

4.7. EXTREME EVENTS

The emphasis on hydrodynamic modelling, following the increase in computational power, has been to increase resolution and confidence in predicting the character of regular water circulation patterns. Although these patterns are based on tidal recordings, wind forcing, and oceanic currents, there is a substantial gap in modelling extreme events. These events, which are caused by cyclones, create highly chaotic and extremely powerful forces in water motion. Despite the difficulty in understanding the impact as they cross a complex coral reef structure, these events could result in the transport of genetic information to distant locations (Radford et al. 2014; Vercelloni et al. 2020). The impact on genetics within a metapopulation is evident as reefs experience a cycle of destruction and recovery with propagules transported from neighboring reefs to the damaged reef (Gerlach et al. 2021). The local replenishment of propagules still remains an important factor in the population character (Game et al. 2008) despite the process of settlement for larvae arriving from distant reefs being hazardous (Keith et al. 2015).

4.8. UNDERSTANDING THE MISMATCH BETWEEN HYDRODYNAMICS AND POPULATION GENETICS

A naive perspective on hydrodynamic modelling is that the outputs will closely resemble the dispersal of a particle launched at one location and arriving at another to settle. This is simply not possible due to the complex nature of interacting scales and chaotic water behavior (Kankara et al. 2007; Jahnke and Jonsson 2022). Each particle, such as a propagule, is subject to minute forces surrounding the organism combined with large-scale forces pushing the entire water column across the reef. The resulting movements of 'surviving' actual particles over a typical 14-day period will be highly variable (North et al. 2008). The particles are also subject to a random death rate, but post-settlement mortality is not normally included in the models. The best outcome is to produce a model that can highlight the character of the water flow so that our understanding of the GBR is enhanced (Figure 5). The opposite is true for genetic studies. Here, the genetics are precisely known, and the link between populations is only diminished by the sampling intensity within and among coral reefs (Galindo et al. 2006). However, the missing parts for genetic models are the exact timing of the migration events and the likelihood of sexual interactions (van Oppen et al. 2011). The genetic models then provide a longer-term perspective and validation of the character of the mixing but very little information about an individual event since the coral reef community composition is the result of multiple recruitment and mortality events (Manel et al. 2019). Fine-scale sampling of a population can reveal parentage connections that assist with estimating the timing of genetic transfers (Harrison et al. 2012), and this has generated significant genetic advances in understanding how fish populations can 'spill' over from a protected zone to a fished zone (Bode

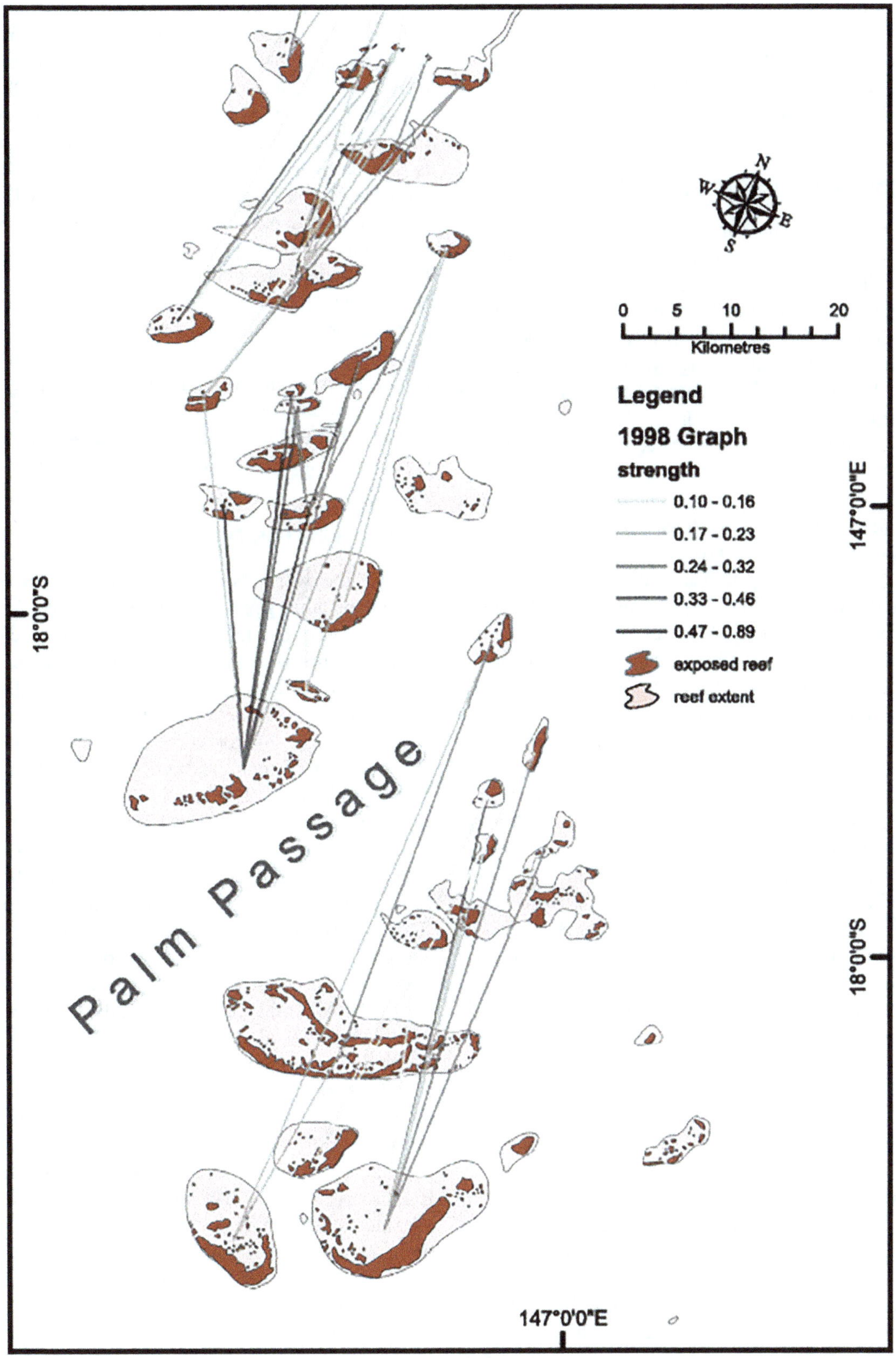

FIGURE 5 Local dispersal influences supporting genetic population patterns (Kininmonth et al. 2009) that identify the isolation of successful transfer events. The strength shown in the line thickness relates to the recruitment success from a source reef.

Source: Reprinted with permission from Springer Nature, no. 5479040419854 et al. 2019). The genetic models then provide a sense of the underlying population structure of the GBR, while the hydrodynamic models create insights into the dynamics of the interactions. Combining all datasets including environmental factors can significantly increase the predictive power of population models (Galindo et al. 2010).

4.9. CONSERVATION APPLICATIONS

With coral reefs under substantial stress, the need to develop conservation priorities and determine where action is required is critical (De'ath et al. 2009). Previous work in conservation planning was strongly focused on protecting the highest species diversity with the expectation that other areas would benefit from the health of highly diverse places (Veron et al. 2009; Klein et al. 2010). Unfortunately, threats to the marine environment are process-driven; hence, an approach that recognizes the functional role of each species is required (Stuart-Smith et al. 2013). The rezoning of the GBR in 2004 was a major attempt to capture this suite of processes with a series of planning rules designed to account for the scientific understanding of the day in biological diversity formation, major water circulation patterns, and disturbance histories (Fernandes et al. 2005). The planning methods did not explicitly try to model these processes but instead adopted an approach of placing protected zones across the bioregions and across the GBR both north and south and from the coast to the shelf. This practical solution provided an excellent example of conservation planning in action (Watts et al. 2009). Advances in conservation planning within a dynamic environment have identified the need to also protect the processes of recruitment (Treml et al. 2015; Magris et al. 2016; Olds et al. 2016; Kininmonth et al. 2019). With the realization of climate change impacts, additional actions are required to incorporate them in the modelling to better inform managers who protect the GBR (Great Barrier Reef Marine Park Authority 2009; De'ath et al. 2012; Condie et al. 2021; Kamenos 2024). One threat to coral abundance is the crown-of-thorns starfish (COTS: *Acanthaster planci*), which utilize the currents along the GBR to disperse (Pratchett et al. 2024). In plague numbers, COTS can decimate a reef through their consumption of live coral (De'ath et al. 2012; MacNeil et al. 2016; Pratchett et al. 2017, 2024). Protecting such an immense structure as the GBR is difficult and highly costly, but using the knowledge of COTS dispersal and population genetics (Babcock et al. 2016; Pratchett et al. 2017) based on hydrodynamics and long-term records has enabled a selection of key reefs to be prioritized for COTS destruction programs (Condie et al. 2021). The evidence so far suggests that this focused program can have an impact at a regional scale (Pratchett et al. 2024). Scientific discussions are also underway to develop coral reef rescue actions such as translocation of thermally resilient corals (van Oppen and Gates 2006; Selmoni et al. 2020; Suggett and van Oppen 2022; Quigley and Donelson 2024) and concentrated larvae (Boström-Einarsson et al. 2020; Harrison et al. 2021; Quigley and Donelson 2024). The exploration of a coral's genetics to determine the capacity of populations to survive through heat waves highlights the genetic diversity across the GBR (Smith-Keune et al. 2006; Jin et al. 2020). Some coral populations can thrive in waters that would rapidly lead to bleaching for southern species. Understanding just how this is possible is a dominant scientific endeavor. Undoubtedly, the future impact of recurrent heat waves will be to shape the genetic structure of the GBR. The prediction that resilient corals might migrate south as temperatures increase is fundamental to unlocking the future of the GBR (Almany et al. 2009; Riginos et al. 2019). Other efforts are focusing on coral restoration through the concentration of the spawning process and subsequent active transport to damaged regions for settlement (Harrison et al. 2021). Interfering with natural recruitment and settlement processes is fraught with danger, and we will need to significantly increase our understanding before these applications are commonly utilized.

4.10. LESSONS FROM AFAR

Around the world, scientific efforts are being made to understand the genetic assemblages that influence the marine environment. The early work on the coral triangle highlights the critical role that geological diversity plays in shaping the water circulation patterns and the settlement opportunities (Veron et al. 2011, 2015). Various models of radiating migration for coral species are confounded by the deep time of geological processes (Bagnato et al. 2004, 2005; Webb 2024). The Pacific Ocean shows reduced coral and fish diversity as the coral reefs extend both north and south and including across to the East (Hensley and Allen 1996; Stuart-Smith et al. 2013; Veron et al. 2015). The same fine-scale patterns are observed in the Indian Ocean (DiBattista et al. 2017). As noted earlier, corals and fish adapt and genetically differentiate as the populations expand into new environments. Climate change stress will challenge the survival of all coral reefs including those in pristine oceanic waters (Evans et al. 2019; Dutra et al. 2021). For example, evaluating the strength of the major oceanic currents as they fluctuate from climate-induced weather patterns will revise the viability of many coral reefs especially for those at the margins of their niche (Banks et al. 2007; Selkoe et al. 2010; Watson et al. 2012; Schill et al. 2015; Álvarez-Romero et al. 2017). The human-centric approach, which considers coral reefs a socio-ecological system, underlines the dependence that humans place on coral reefs and yet coral reefs are directly threatened by human activity (Van Putten et al. 2016). Due to over-fishing, land runoff, and climate change, many reefs around the world are denuded and on a trajectory towards decimation (Spalding et al. 2007; Burke et al. 2011). Despite being notably the best managed coral reef system in the world, the GBR is also suffering from these stresses (De'ath et al. 2012).

4.11. FUTURE DIRECTIONS IN SEASCAPE GENETICS

Seascape genetics is defined as the study of how spatially variable ocean features, including benthic composition and environmental fluctuations, influence the genetic patterns of marine organisms (Riginos and Liggins 2013). Both the hydrodynamics and genetic models of the marine world are rapidly improving in both precision and reliability as more supporting data are collected and advanced techniques utilize supercomputing capacity (Selkoe et al. 2016). Satellite information of surface water dynamics has confirmed the movement of floating particles (Kleypas et al. 2016), such as *Trichodesmium* (Subramaniam et al. 2001), while the use of high frequency radar arrays to precisely measure wave formations (Wyatt et al. 2011; Wyatt 2014) is a key aspect to circulation models. More accurate benthic mapping that uses multibeam sonar can create the underlying geological structure including sunken reefs (Arias-González et al. 2006; Bridge et al. 2011; Harris et al. 2012) on which water circulation models are constructed (Van Wynsberge et al. 2017). A particular promising area of study for geneticists is the micro turbulence of the coral reefs and how this impacts particle transfer (Davis et al. 2021). At the other end of the scale is the long-term variability of currents such as the Eastern Australian Current (Wolanski et al. 2024). Small changes in flow could create dramatic effects on coral genetics especially if combined with large sea level rises. The primary advances for the seascape genetics have been the widespread collection of samples and the establishment of linkages between thermal resilience and genetic composition (Selmoni et al. 2020). Experiments with gene manipulation are also being considered, but is likely to remain theoretical for some decades (Suggett and van Oppen 2022). Untangling the very fabric of what constitutes an organism, a recent expedition (Planes et al. 2019) involved a research vessel (RV TARA) with sufficient laboratory capacity to analyse coral holobiont composition (Stévenne et al. 2021) of collected specimens across the oceans. A coral holobiont comprises the coral organism and the dinoflagellate algae and associated viruses, bacteria, and other microeukaryotes. Understanding the biodiversity at this level reveals many linkages that commonly rely on water circulation interactions. Modelling this set of linkages among genes, individuals, populations, and water circulation is still limited in sophistication (Galindo et al. 2006).

REFERENCES

Almany GR, Connolly SR, Heath DD, Hogan JD, Jones GP, McCook LJ, Mills M, Pressey RL, Williamson DH (2009) Connectivity, biodiversity conservation and the design of marine reserve networks for coral reefs. Coral Reefs 28:339–351

Álvarez-Romero JG, Munguía-Vega A, Beger M, del Mar Mancha-Cisneros M, Suárez-Castillo AN, Gurney GG, Pressey RL, Gerber LR, Morzaria-Luna HN, Reyes-Bonilla H, Adams VM, Kolb M, Graham EM, Vanderwal J, Castillo-López A, Hinojosa-Arango G, Petatán-Ramírez D, Moreno-Baez M, Godínez-Reyes CR, Torre J (2017) Designing connected marine reserves in the face of global warming. Global Change Biology 24:e671– e691. https://doi.org/10.1111/gcb.13989

Arias-González J, Done T, Page C, Cheal A, Kininmonth S, Garza-Pérez J (2006) Towards a reefscape ecology: Relating biomass and trophic structure of fish assemblages to habitat at Davies Reef, Australia. Mar Ecol Prog Ser 320:29–41

Ayre DJ, Hughes TP (2004) Climate change, genotypic diversity and gene flow in reef-building corals. Ecol Lett 7:273–278

Babcock RC (1995) Synchronous multispecific spawning on coral reefs: Potential for hybridization and roles of gamete recognition. Reprod Fertil Dev 7:943–950

Babcock RC, Dambacher JM, Morello EB, Plagányi ÉE, Hayes KR, Sweatman HPA, Pratchett MS (2016) Assessing different causes of Crown-of-Thorns starfish outbreaks and appropriate responses for management on the Great Barrier Reef. PLoS One 11:1–20

Bagnato S, Linsley BK, Howe SS, Wellington GM (2004) Evaluating the use of the massive coral Diploastrea heliopora for paleoclimate reconstruction. Paleoceanography 19:PA1032, doi:10.1029/2003PA000935

Bagnato S, Linsley BK, Howe SS, Wellington GM (2005) Coral oxygen isotope records of interdecadal climate variations in the South Pacific Convergence Zone region. Geochemistry, Geophysics, Geosyststem 6:Q06001, doi:10.1029/2004GC000879

Banks SC, Piggott MP, Williamson JE, Bové U, Holbrook NJ, Beheregaray LB (2007) Oceanic variability and coastal topography shape genetic structure in a long-dispersing sea urchin. Ecology 88:3055–3064

Benzie JAH, Williams ST (1997) Genetic structure of giant clam (Tridacna maxima) populations in the west pacific is not consistent with dispersal by present-day ocean currents. Evolution (N Y) 51:768–783

Bode M, Bode L, Armsworth PR (2005) Larval dispersal reveals regional sources and sinks in the Great Barrier Reef. The Marine Ecology Progress Series I:17–25

Bode M, Burrage K, Possingham HP (2008) Using complex network metrics to predict the persistence of metapopulations with asymmetric connectivity patterns. Ecological Modelling 214:201–209

Bode M, Leis JM, Mason LB, Williamson DH, Harrison HB, Choukroun S, Jones GP (2019) Successful validation of a larval dispersal model using genetic parentage data. PLoS Biology 17:1–13

Bode M, Stewart O, Choukroun SM (2024) Incorporating biophysical larval dispersal simulations into coral reef conservation decision-making. In: Wolanski E, Kingsford MJ (eds.), Oceanographic Processes of Coral Reefs. Physical and Biological Links in the Great Barrier Reef. CRC Press, Boca Raton, FL.

Boström-Einarsson L, Babcock RC, Bayraktarov E, Ceccarelli D, Cook N, Ferse SCA, Hancock B, Harrison P, Hein M, Shaver E, Smith A, Suggett D, Stewart-Sinclair PJ, Vardi T, McLeod IM (2020) Coral restoration—A systematic review of current methods, successes, failures and future directions. PLoS One 15:1–24

Bridge TCL, Done TJ, Beaman RJ, Friedman A, Williams SB, Pizarro O, Webster JM (2011) Topography, substratum and benthic macrofaunal relationships on a tropical mesophotic shelf margin, central Great Barrier Reef, Australia. Coral Reefs 30:143–153

Burke L, Reytar K, Spalding MD, Perry A (2011) Reefs at Risk Revisited. World Resources Institute.

Burrage DM, Steinberg CR, Skirving WJ, Kleypas JA (1996) Mesoscale circulation features of the Great Barrier Reef region inferred from NOAA satellite imagery. Remote Sensing of Environment 56:21–41

Condie SA, Anthony KRN, Babcock RC, Baird ME, Beeden R, Fletcher CS, Gorton R, Harrison D, Hobday AJ, Plagányi ÉE, Westcott DA (2021) Large-scale interventions may delay decline of the Great Barrier Reef. Royal Society Open Science 8.

Davies PJ (2011) Great barrier reef: Origin, evolution, and modern development BT. In: Hopley D. (eds) Encyclopedia of Modern Coral Reefs: Structure, Form and Process. Springer, Dordrecht, pp. 504–534

Davis KA, Pawlak G, Monismith SG (2021) Turbulence and coral reefs. Annual Review of Marine Science 13:343–373

De'ath G, Fabricius KE, Sweatman H, Puotinen M (2012) The 27-year decline of coral cover on the Great Barrier Reef and its causes. Proceedings of the National Academy of Sciences of the United States of America 1–5

De'ath G, Lough JM, Fabricius KE (2009) Declining coral calcification on the Great Barrier Reef. Science 323:116–119

DiBattista JD, Travers MJ, Moore GI, Evans RD, Newman SJ, Feng M, Moyle SD, Gorton RJ, Saunders T, Berry O (2017) Seascape genomics reveals fine-scale patterns of dispersal for a reef fish along the ecologically divergent coast of Northwestern Australia. Mol Ecol 26:6206–6223

Dight IJ, James MK, Bode L (1990) Modelling the larval dispersal of *Acanthaster planci*. Coral Reefs 9:125–134

Doherty PJ, Planes S, Mather P (1995) Gene flow and larval duration in seven species of fish from the Great Barrier Reef. Ecology 76:2373–2391

Duke NC, Adam C, Jock M (2024) Damaging impacts of severe tropical cyclones on mangroves bordering Australia's Great Barrier Reef. In: Wolanski E, Kingsford MJ (eds.), Oceanographic Processes of Coral Reefs. Physical and Biological Links in the Great Barrier Reef. CRC Press, Boca Raton, FL.

Dunne JA, Williams RJ, Martinez ND (2002) Network structure and biodiversity loss in food webs: Robustness increases with connectance. Ecology Letters 5:558–567

Dutra LXC, Haywood MDE, Singh S, Ferreira M, Johnson JE, Veitayaki J, Kininmonth S, Morris CW, Piovano S (2021) Synergies between local and climate-driven impacts on coral reefs in the Tropical Pacific: A review of issues and adaptation opportunities. Marine Pollution Bulletin 164:111922

Edgar GJ, Stuart-Smith RD (2014) Systematic global assessment of reef fish communities by the Reef Life Survey program. Science Data 1:1–8

Evans RD, Ryan NM, Travers MJ, Feng M, Hitchen Y, Kennington WJ (2019) A seascape genetic analysis of a stress-tolerant coral species along the Western Australian coast. Coral Reefs 38:63–78

Fernandes L, Day J, Lewis A, Slegers S, Kerrigan B, Breen D, Cameron D, Jago B, Hall J, Lowe D, Innes J, Tanzer J, Chadwick V, Thompson L, Gorman K, Simmons M, Barnett B, Sampson K, De'ath G, Mapstone B, Marsh H, Possingham H, Ball I, Ward T, Dobbs K, Aumend J, Slater D, Stapleton K (2005) Establishing representative no-take areas in the great barrier reef: Large-scale implementation of theory on marine protected areas. Conservation Biology 19:1733–1744

Friesen SK, Martone R, Rubidge E, Baggio JA, Ban NC (2019) An approach to incorporating inferred connectivity of adult movement into marine protected area design with limited data. Ecological Applications 29.

Galindo HM, Olson DB, Palumbi SR (2006) Seascape genetics: A coupled oceanographic-genetic model predicts population structure of Caribbean corals. Current Biology 16:1622–1626

Galindo HM, Pfeiffer-Herbert AS, McManus MA, Chao Y, Chai F, Palumbi SR (2010) Seascape genetics along a steep cline: Using genetic patterns to test predictions of marine larval dispersal. Molecular Ecology 19:3692–3707

Game ET, McDonald-Madden E, Puotinen ML, Possingham HP (2008) Should we protect the strong or the weak? Risk, resilience, and the selection of marine protected areas. Conservation Biology 22:1619–1629

Gerlach G, Kraemer P, Weist P, Eickelmann L, Kingsford MJ (2021) Impact of cyclones on hard coral and metapopulation structure, connectivity and genetic diversity of coral reef fish. Coral Reefs 40:999–1011

Great Barrier Reef Marine Park Authority (2009) Great Barrier Reef Outlook Report 2009. Great Barrier Reef Marine Park Authority, Townsville.

Green AL, Mous PJ (2008) Delineating the Coral Triangle, its ecoregions and functional seascapes. Nature Conservation 44

Harriott V, Banks SA (2002) Latitudinal variation in coral communities in eastern Australia: A qualitative biophysical model of factors regulating coral reefs. Coral Reefs 21:83–94

Harris PT, Bridge TCL, Beaman RJ, Webster JM, Nichol SL, Brooke BP (2012) Submerged banks in the Great Barrier Reef, Australia, greatly increase available coral reef habitat. ICES Journal of Marine Science70:284–293

Harrison HB, Williamson DH, Evans RD, Almany GR, Thorrold SR, Russ GR, Feldheim KA, Van Herwerden L, Planes S, Srinivasan M, Berumen ML, Jones GP (2012) Larval export from marine reserves and the recruitment benefit for fish and fisheries. Current Biology 22:1023–1028

Harrison PL, Babcock RC, Bull GD, Oliver JK, Carden C, Willis BL (2010) Mass spawning in tropical reef corals. Advance Science 223:1186–1189

Harrison PL, dela Cruz DW, Cameron KA, Cabaitan PC (2021) Increased coral larval supply enhances recruitment for coral and fish habitat restoration. Frontiers in Marine Science 8:1–22

Hensley, D. A. (1996). Review of G. R. Allen, 1991, Damselfishes of the World. Mergus Press, Melle, Germany. Copeia 1996:229–231.

Hobbs JPA, Richards ZT, Popovic I, Lei C, Staeudle TM, Montanari SR, Di Battista JD (2022) Hybridisation and the evolution of coral reef biodiversity. Coral Reefs 41:535–549

Hopley D (2011) Encyclopedia of Modern Coral Reefs: Structure, Form and Process. edited by David Hopley. Springer, Dordrecht and London.

Jahnke M, Jonsson PR (2022) Biophysical models of dispersal contribute to seascape genetic analyses. Philosophical Transactions of the Royal Society B 377.

James MK, Armsworth PR, Mason LB, Bode L (2002) The structure of reef fish metapopulations: Modelling larval dispersal and retention patterns. Proceedings of the Royal Society B: Biological Sciences 269:2079–2286

Jin YK, Kininmonth S, Lundgren PB (2020) Predicting the spatial distribution of allele frequencies for a gene associated with tolerance to eutrophication and high temperature in the reef-building coral, Acropora millepora, on the Great Barrier Reef. Coral Reefs 39:147–158

Jones GP, Almany GR, Russ GR, Sale PF, Steneck RS, Van Oppen MJH, Willis BL (2009) Larval retention and connectivity among populations of corals and reef fishes: History, advances and challenges. Coral Reefs 28:307–325

Kamenos N (2024) Coral bleaching biophysics. In: Wolanski E, Kingsford MJ (eds.), Oceanographic Processes of Coral Reefs. Physical and Biological Links in the Great Barrier Reef. CRC Press, Boca Raton, FL.

Kankara RS, Subramanian BR, Sampath V (2007) Validation of a deterministic hydrodynamic model in the ICZM framework for Gulf of Kachchh, India. Journal of Coastal Research 23:1324–1331

Keith SA, Woolsey ES, Madin JS, Byrne M, Baird AH (2015) Differential establishment potential of species predicts a shift in coral assemblage structure across a biogeographic barrier. Ecography (Cop) 38:1225–1234

Kingsford MJ, Spiecker L, Gerlach G (2024) The interactions between larval behaviour and oceanography. In: Wolanski E, Kingsford MJ (eds.), Oceanographic Processes of Coral Reefs. Physical and Biological Links in the Great Barrier Reef. CRC Press, Boca Raton, FL.

Kininmonth SJ, Beger M, Bode M, Peterson E, Adams VM, Dorfman D, Brumbaugh DR, Possingham HP (2011) Dispersal connectivity and reserve selection for marine conservation. Ecological Modelling 222:1272–1282

Kininmonth SJ, De'ath G, Possingham HP (2009) Graph theoretic topology of the Great but small Barrier Reef world. Theoretical Ecology 3:75–88

Kininmonth SJ, Drechsler M, Johst K, Possingham H (2010a) Metapopulation mean life time within complex networks. Marine Ecology Progress Series 417:139–149

Kininmonth SJ, van Oppen MJH, Castine S, Peplow L, Lutz A (2012) The small genetic world of Seriatopora hystrix. Network Biology 2:1–15

Kininmonth SJ, van Oppen MJH, Possingham HP, Oppen MJH Van, Possingham HP (2010b) Determining the community structure of the coral Seriatopora hystrix from hydrodynamic and genetic networks. Ecological Modelling 221:2870–2880

Kininmonth SJ, Weeks R, Abesamis RA, Bernardo LPC, Beger M, Treml EA, Williamson D, Pressey RL (2019) Strategies in scheduling marine protected area establishment in a network system. Ecological Applications 29:1–9

Klein CJ, Ban NC, Halpern BS, Beger M, Game ET, Grantham HS, Green A, Klein TJ, Kininmonth S, Treml E, Wilson K, Possingham HP (2010) Prioritizing land and sea conservation investments to protect coral reefs. PLoS One 5:e12431

Kleypas JA, Thompson DM, Castruccio FS, Curchitser EN, Pinsky M, Watson JR (2016) Larval connectivity across temperature gradients and its potential effect on heat tolerance in coral populations. Global Change Biology 22:3539–3549

Lefcheck JS, Edgar GJ, Stuart-Smith RD, Bates AE, Waldock C, Brandl SJ, Kininmonth S, Ling SD, Duffy JE, Rasher DB, Agrawal AF (2021) Species richness and identity both determine the biomass of global reef fish communities. Nature Communications 12:1–9

Legrand S, Deleersnijder E, Hanert E, Legat V, Wolanski E (2006) High-resolution, unstructured meshes for hydrodynamic models of the Great Barrier Reef, Australia. Estuar Coast Shelf Sci 68:36–46

Lyons M, Roelfsema CV, Kennedy EM, Kovacs E, Borrego-Acevedo R, Markey K, Roe M, Yuwono D, Harris D, Phinn S, Asner GP, Li J, E. Knapp D, Fabina N, Larsen K, Traganos D, Murray N (2020) Mapping the world's coral reefs using a global multiscale earth observation framework. Remote Sensing in Ecology and Conservation 6:557–568

MacNeil MA, Mellin C, Pratchett MS, Hoey J, Anthony KRN, Cheal AJ, Miller I, Sweatman H, Cowan ZL, Taylor S, Moon S, Fonnesbeck CJ (2016) Joint estimation of crown of thorns (*Acanthaster planci*) densities on the Great Barrier Reef. PeerJ 4:e2310

Magris RA, Treml EA, Pressey RL, Weeks R (2016) Integrating multiple species connectivity and habitat quality into conservation planning for coral reefs. Ecography (Cop) 39:649–664

Manel S, Loiseau N, Andrello M, Fietz K, Goñi R, Forcada A, Lenfant P, Kininmonth S, Marcos C, Marques V, Mallol S, Pérez-Ruzafa A, Breusing C, Puebla O, Mouillot D (2019) Long-distance benefits of marine reserves: Myth or reality? Trends in Ecology & Evolution 34:342–354

Mora C, Chittaro PM, Sale PF, Kritzer JP, Ludsin SA, Africa S (2003) Patterns and processes in reef fish diversity. Nature 421:933–936

Noreen AME, Harrison PL, Van Oppen MJH (2009) Genetic diversity and connectivity in a brooding reef coral at the limit of its distribution. Proceedings of the Royal Society B: Biological Sciences 276:3927–3935

North EW, Schlag Z, Hood RR, Li M, Zhong L, Gross T, Kennedy VS (2008) Vertical swimming behavior influences the dispersal of simulated oyster larvae in a coupled particle-tracking and hydrodynamic model of Chesapeake Bay. The Marine Ecology Progress Series 359:99–115

Olds AD, Connolly RM, Pitt KA, Pittman SJ, Maxwell PS, Huijbers CM, Moore BR, Albert S, Rissik D, Babcock RC, Schlacher TA (2016) Quantifying the conservation value of seascape connectivity: A global synthesis. Global Ecology and Biogeography 25:3–15

Oliver JK, Willis BL (1987) Coral-spawn slicks in the Great Barrier Reef: Preliminary observations. Marine Biology 94:521–529

Penin L, Michonneau F, Baird AH, Connolly SR, Pratchett MS, Kayal M, Adjeroud M (2010) Early post-settlement mortality and the structure of coral assemblages. The Marine Ecology Progress Series 408:55–64

Planes S, Allemand D, Agostini S, Banaigs B, Boissin E, Boss E, Bourdin G, Bowler C, Douville E, Flores JM, Forcioli D, Furla P, Galand PE, Ghiglione JF, Gilson E, Lombard F, Moulin C, Pesant S, Poulain J, Reynaud S, Romac S, Sullivan MB, Sunagawa S, Thomas OP, Troublé R, De Vargas C, Thurber RV, Voolstra CR, Wincker P, Zoccola D, Armstrong E, Audrain S, Aury JM, Banaig B, Barbe V, Belser C, Beraud E, Bonnival E, Bourgois E, Carradec Q, Cassar N, Cohen NR, Conan P, Cronin DR, Da Silva O, Djerbi N, Dolan JR, Dominguez Herta G, Du J, Filée J, Friedrich R, Gorsky G, Guinther M, Haëntjens N, Henry N, Hertau M, Hochart C, Hume BCC, Iwankow G, John SG, Karp-Boss L, Kelly RL, Kitano Y, Klinges G, Koren I, Labadie K, Lancelot J, Lang-Yona N, Lê-Hoang J, Lemee R, Lin Y, Marie D, McMind R, Miguel-Gordo M, Trainic M, Monmarche D, Mucherie Y, Noel B, Ottaviani A, Paoli L, Pedrotti ML, Pogoreutz C, Pujo-Pay M, Reverdin G, Röthig T, Rottinger E, Rouan A, Ruscheweyh HJ, Salazar G, Vardi A, Vega-Thunder R, Zahed A, Zamoum T (2019) The Tara Pacific expedition—A pan-ecosystemic approach of the "-omics" complexity of coral reef holobionts across the Pacific Ocean. PLoS Biology 17.

Pollock FJ, Lamb JB, Van De Water JAJM, Smith HA, Schaffelke B, Willis BL, Bourne DG (2019) Reduced diversity and stability of coral-associated bacterial communities and suppressed immune function precedes disease onset in corals. Royal Society Open Science 6.

Pratchett MS, Caballes CF, Wilmes JC, Matthews S, Mellin C, Sweatman HPA, Nadler LE, Brodie J, Thompson CA, Hoey J, Bos AR, Byrne M, Messmer V, Fortunato SAV, Chen CCM, Buck ACE, Babcock RC, Uthicke S (2017) Thirty years of research on crown-of-thorns starfish (1986–2016): Scientific advances and emerging opportunities. Diversity 9.

Pratchett MS, Choukroun S, Chandler S, et al. (2024) Biophysical processes involved in the initiation and spread of population irruptions of crown-of-thorns starfish on the great barrier reef. In: Wolanski E, Kingsford MJ (eds.), Oceanographic Processes of Coral Reefs. Physical and Biological Links in the Great Barrier Reef. CRC Press, Boca Raton, FL.

Quigley K, Donelson J (2024) Selective breeding and promotion of naturally heat tolerant coral reef species. In: Wolanski E, Kingsford MJ (eds.), Oceanographic Processes of Coral Reefs. Physical and Biological Links in the Great Barrier Reef. CRC Press, Boca Raton, FL.

Radford B, Babcock R, Van Niel K, Done T (2014) Are cyclones agents for connectivity between reefs? Journal of Biogeography 41:1367–1378

Ramírez-Portilla C, Baird AH, Cowman PF, Quattrini AM, Harii S, Sinniger F, Flot JF (2022) Solving the coral species delimitation conundrum. Systematic Biology 71:461–475

Riginos C, Hock K, Matias AM, Mumby PJ, van Oppen MJH, Lukoschek V (2019) Asymmetric dispersal is a critical element of concordance between biophysical dispersal models and spatial genetic structure in Great Barrier Reef corals. Diversity and Distributions 25:1684–1696

Riginos C, Liggins L (2013) Seascape genetics: Populations, individuals, and genes marooned and adrift. Geography Compass 7:197–216

Schill SR, Raber GT, Roberts JJ, Treml EA, Brenner J, Halpin PN (2015) No reef is an island: Integrating coral reef connectivity data into the design of regional-scale marine protected area networks. PLoS One 10:1–24

Selkoe KA, D'Aloia CC, Crandall ED, Iacchei M, Liggins L, Puritz JB, Von Der Heyden S, Toonen RJ (2016) A decade of seascape genetics: Contributions to basic and applied marine connectivity. Marine Ecology Progress Series 554:1–19

Selkoe KA, Watson JR, White C, Horin T Ben, Iacchei M, Mitarai S, Siegel DA, Gaines SD, Toonen RJ (2010) Taking the chaos out of genetic patchiness: Seascape genetics reveals ecological and oceanographic drivers of genetic patterns in three temperate reef species. Molecular Ecology 19:3708–3726

Selmoni O, Lecellier G, Ainley L, Collin A, Doucet R, Dubousquet V, Feremaito H, Ito Waia E, Kininmonth S, Magalon H, Malimali S, Maugateau A, Meibom A, Mosese S, René-Trouillefou M, Satoh N, van Oppen MJH, Xozamé A, Yékawene M, Joost S, Berteaux-Lecellier V (2020) Using modern conservation tools for innovative management of coral reefs: The MANACO Consortium. Frontiers in Marine Science 7:1–5

Smith-Keune C, Oppen M (2006) Genetic structure of a reef-building coral from thermally distinct environments on the Great Barrier Reef. Coral Reefs 25:493–502

Smith-Keune C, Van Oppen M, Oppen M (2006) Genetic structure of a reef-building coral from thermally distinct environments on the Great Barrier Reef. Coral Reefs 25:493–502

Spalding MD, Fox HE, Allen GR, Davidson N, Ferdaña ZA, Finlayson MAX, Halpern BS, Jorge MA, Lombana AL, Lourie SA, Martin KD, Manus MC, Molnar J, Recchia CA, Robertson J (2007) Marine ecoregions of the world: A bioregionalization of coastal and shelf areas. BioScience 57(7):573–583

Stévenne C, Micha M, Plumier JC, Roberty S (2021) Corals and sponges under the light of the holobiont concept: How microbiomes underpin our understanding of marine ecosystems. Frontiers in Marine Science 8.

Stuart-Smith RD, Edgar GJ, Barrett NS, Bates AE, Baker SC, Bax NJ, Becerro MA, Berkhout J, Blanchard JL, Brock DJ, Clark GF, Cooper AT, Davis TR, Day PB, Emmett Duffy J, Holmes TH, Howe SA, Jordan A, Kininmonth S, Knott NA, Jonathan LS, Ling SD, Parr A, Strain E, Hugh S, Russell T (2017) Assessing national biodiversity trends for rocky and coral reefs through the integration of citizen science and scientific monitoring programs. Bioscience 67:134–146

Stuart-Smith RR, Bates AAE, Lefcheck JJ, Duffy JE, Baker SC, Thomson RJ, Stuart-Smith JF, Hill NA, Kininmonth SJ, Airoldi L, Becerro MA, Campbell SJ, Dawson TP, Navarrete SA, Soler GA, Strain EMA, Willis TJ, Edgar GJ (2013) Integrating abundance and functional traits reveals new global hotspots of fish diversity. Nature 501:539–542

Subramaniam A, Brown CW, Hood RR, Carpenter EJ, Capone DG (2001) Detecting Trichodesmium blooms in SeaWiFS imagery. Deep Sea Research Part II: Topical Studies in Oceanography 49:107–121

Suggett DJ, van Oppen MJH (2022) Horizon scan of rapidly advancing coral restoration approaches for 21st century reef management. Emerging Topics in Life Sciences 1–12

Taylor G (1931) A Year on the Great Barrier Reef. C. M. Yonge. The Journal of Geology 39:489–490

Thomas CJ, Lambrechts J, Wolanski E, Traag VA, Blondel VD, Deleersnijder E, Hanert E (2014) Numerical modelling and graph theory tools to study ecological connectivity in the Great Barrier Reef. Ecological Modelling 272:160–174

Treml EA, Ford JR, Black KP, Swearer SE (2015) Identifying the key biophysical drivers, connectivity outcomes, and metapopulation consequences of larval dispersal in the sea. Movement Ecology 3

Treml EA, Halpin PN, Urban DL, Pratson LF (2007) Modeling population connectivity by ocean currents, a graph-theoretic approach for marine conservation. Landscape Ecology 23:19–36

Underwood JN, Smith LD, Van Oppen MJH, Gilmour JP (2007) Multiple scales of genetic connectivity in a brooding coral on isolated reefs following catastrophic bleaching. Molecular Ecology16:771–784

van Oppen MJH, Gates RD (2006) Conservation genetics and the resilience of reef-building corals. Molecular Ecology 15:3863–3883

van Oppen MJH, Lutz A, De'ath G, Peplow L, Kininmonth S (2008) Genetic traces of recent long-distance dispersal in a predominantly self-recruiting coral. PLoS One 3:e3401

van Oppen MJH, Peplow LM, Kininmonth S, Berkelmans R (2011) Historical and contemporary factors shape the population genetic structure of the broadcast spawning coral, Acropora millepora, on the Great Barrier Reef. Molecular Ecology 20:4899–4914

Van Putten IE, Dichmont CM, Dutra LXC, Thébaud O, Deng RA, Jebreen E, Owens R, Pascual R, Read M, Thompson C (2016) Objectives for management of socio-ecological systems in the Great Barrier Reef region, Australia. Regional Environmental Change16:1417–1431

Van Wynsberge S, Andréfouët S, Gaertner-Mazouni N, Tiavouane J, Grulois D, Lefèvre J, Pinsky ML, Fauvelot C (2017) Considering reefscape configuration and composition in biophysical models advance seascape genetics. PLoS One 12:1–23

Vanderklift MA, Doropoulos C, Gorman D, Leal I, Minne AJP, Statton J, Steven ADL, Wernberg T (2020) Using propagules to restore coastal marine ecosystems. Frontiers in Marine Science 7:1–15

Vercelloni J, Liquet B, Kennedy E V., González-Rivero M, Caley MJ, Peterson EE, Puotinen M, Hoegh-Guldberg O, Mengersen K (2020) Forecasting intensifying disturbance effects on coral reefs. Global Change Biology 26:2785–2797

Veron J (2013) Overview of the taxonomy of zooxanthellate Scleractinia. Zoological Journal of the Linnean Society169:485–508

Veron J, Stafford-Smith M, DeVantier L, Turak E (2015) Overview of distribution patterns of zooxanthellate Scleractinia. Frontiers in Marine Science 2:1–19

Veron JE, DeVantier LL, Turak E, Green AL, Kininmonth S, Stafford-Smith M, Peterson N (2011) The coral triangle. In: Z. Dubinsky & N. Stambler (eds.), Coral Reefs: An Ecosystem in Transition (pp. 47–55). Springer

Veron JEN, Devantier LM, Turak E, Green AL, Kininmonth S, Stafford-Smith M, Peterson N (2009) Delineating the Coral Triangle. Galaxea, Journal of the Coral Reef Studies 11:91–100

Wallace CC (1985) Reproduction, recruitment and fragmentation in nine sympatric species of the coral genus Acropora. Marine Biology 88:217–233

Watson JR, Kendall BE, Siegel DA, Mitarai S (2012) Changing seascapes, stochastic connectivity, and marine metapopulation dynamics. The American Naturalist 180:99–112

Watts ME, Ball IR, Stewart RS, Klein CJ, Wilson K, Steinback C, Lourival R, Kircher L, Possingham HP (2009) Marxan with Zones: Software for optimal conservation based land- and sea-use zoning. Environmental Modelling & Software 24:1513–1521

Webb GE (2024) GBR climate change including longer-term perspectives from coral cores. In: Wolanski E, Kingsford MJ (eds.), Oceanographic Processes of Coral Reefs. Physical and Biological Links in the Great Barrier Reef. CRC Press, Boca Raton, FL.

Weeks SJ, Bakun A, Steinberg CR, Brinkman R, Hoegh-Guldberg O (2010) The Capricorn Eddy: A prominent driver of the ecology and future of the southern Great Barrier Reef. Coral Reefs 29:975–985

Willis BL, Babcock RC, Harrison PL, Oliver JK (1985) Patterns in the mass spawning of corals on the Great Barrier Reef from 1981 to 1984. Proceedings of the Fifth International Coral Reef Congress 4:343–348

Willis BL, Oliver JK (1990) Direct tracking of coral larvae: Implications for dispersal studies of planktonic larvae in topographically complex environments. Ophelia 32:145–162

Willis J (2011) Modelling swimming aquatic animals in hydrodynamic models. Ecol Modell 222:3869–3887

Wolanski E (2017) Bounded and unbounded boundaries—Untangling mechanisms for estuarine-marine ecological connectivity: Scales of m to 10,000 km—A review. Estuarine, Coastal and Shelf Science 198:378–392.

Wolanski E, Doherty P, Carleton J (1997) Directional swimming of fish larvae determines connectivity of fish populations on the great barrier reef. Naturwissenschaften 84:262–268

Wolanski, E, Kingsford MJ (2014) Oceanographic and behavioural assumptions in models of the fate of coral and coral reef fish larvae. Journal of the Royal Society Interface 11:20140209.

Wolanski E, Kingsford MJ, Lambrechts J, Marmorino G (2024). The physical oceanography of the great barrier reef: A review. In: Wolanski E, Kingsford MJ (eds.), Oceanographic Processes of Coral Reefs. Physical and Biological Links in the Great Barrier Reef. CRC Press, Boca Raton, FL.

Wolanski E, Spagnol S (2000) Sticky waters in the Great Barrier Reef. Estuarine, Coastal and Shelf Science 50:27–32

Wörheide G, Hooper JNA, Degnan BM (2002) Phylogeography of western Pacific Leucetta "chagosensis" (Porifera: Calcarea) from ribosomal DNA sequences: Implications for population history and conservation of the Great Barrier Reef World Heritage Area (Australia). Molecular Ecology 11:1753–1768

Wyatt LR (2014) Wave Measurements From the Australian Coastal Ocean Radar Network. Oceans 2014 – Taipei, Taiwan, pp. 1–5. doi: 10.1109/OCEANS-TAIPEI.2014.6964286.

Wyatt LR, Green JJ, Middleditch A (2011) HF radar data quality requirements for wave measurement. Coastal Engineering 58:327–336

5 Advances in Understanding Climate Change on the Great Barrier Reef Using Coral-Based Proxies

Gregory E. Webb

5.1. INTRODUCTION

Coral reefs have been resilient over many timeframes despite the paradox that reef-building scleractinian corals thrive best at several 'edges' of existence. Reef corals are innately fragile but grow best in high-energy, shallow water, extending vertically to low tide levels only centimetres from exposure and death. Their strength is tested daily by prevailing wave climate and commonly by extreme storm events, disturbance regimes to which reef communities are well-adapted (Connell 1978). Reef corals thrive best at temperatures just below those that cause decreased extension rates (Lough and Cooper 2011), bleaching, and death (Hoegh-Guldberg 2005). Although late Pleistocene reef communities have tracked both falling and rising sea levels to reconstitute and re-establish themselves through each glacial-interglacial cycle (Pandolfi 1996; Webster et al. 2018), reefs now face anthropogenic climate threats (Hoegh-Guldberg et al. 2017). Increased thermal bleaching, ocean acidification (OA) (Eyre et al. 2018), and increasing cyclone intensity (Knutsen et al. 2020; Duke et al. 2024) have altered reef communities, both marine and terrestrial (Hughes et al. 2018; Fine et al. 2019; Hamylton et al. 2024), in what are thought to be unprecedented ways. To make meaningful predictions of the effects of projected climate change on reef communities, we must better understand both palaeoclimate itself (Carré et al. 2021; Emile-Geay et al. 2021) and its specific effects on particular reefs and coral reef communities.

Although the Great Barrier Reef (GBR) was resilient through previous glacial-interglacial cycles (Davies et al. 1989; Webster et al. 2018), at timescales of only weeks, corals may bleach and die with ephemeral elevated sea surface temperature (SST). Major gaps exist in our ability to understand and predict the effects of changing climate on coral reefs between these temporal extremes because of the short timeframes for quantitative instrumental observations and uncertainties relating to the current unusually high rates of change (Pörtner et al. 2014). In inshore GBR reefs, anthropogenic land-use change confounds natural secular trends since European settlement of the adjoining catchments (Brodie et al. 2012; D'Olivo and McCulloch 2022). Hence, there is great need to better document climate variability and its effects on reef environments over decadal, multi-century, millennial, and especially pre-anthropogenic timeframes (Ramos et al. 2020). As our understanding of the variable tolerance of corals to environmental stressors improves, we can utilise historical records to better target conservation efforts (Barkley and Cohen 2016). Historical (Clark et al. 2017; Fine et al. 2019) and even palaeontological (Pandolfi and Kiessling 2014; Brenner et al. 2020; Leonard et al. 2020b; Kiessling et al. 2023) scales are increasingly relevant.

This review provides a brief update on climatically sensitive data recoverable from reef corals. The topic has exploded since the review of Lough (2010), including major advances in some fields. Therefore, this contribution presents a guide to recent reviews and selected specific advances (Table 1) with a bias towards coral-based palaeoclimate records for the GBR.

5.2. CORAL-BASED PALAEOCLIMATE DATA ARCHIVES

Coral skeletons record their local environment in many ways, from differences in extension and calcification rates to direct geochemical proxies for ambient SST, sea surface salinity (SSS), and pH (Table 1). Relative precipitation can be recorded by $\delta^{18}O$ and geochemical proxies for terrigenous runoff, including trace elements and luminescence bands (Lough et al. 2015; Saha et al. 2016; Lewis et al. 2024). Coral-based geochemical proxies also exist for upwelling (Saha et al. 2016; Erler et al. 2020b; Rahaman et al. 2022) and with adequate context, wind direction (Chapmann et al. 2022). Secular changes in such proxies inform broader climate dynamics, including the Australian-Indonesian Summer Monsoon (AISM), El Niño-Southern Oscillation (ENSO), Pacific Decadal Oscillation (PDO), Interdecadal Pacific Oscillation, Western Pacific Gradient, and Indian Ocean Dipole (e.g., Deng et al. 2014; Rodriquez-Ramirez et al. 2014; Zinke et al. 2014; Reed et al. 2019; Abram et al. 2020; Mohtar et al. 2021). Even secular trends in marine radiocarbon reservoir corrections vary with broader climate-oceanographic changes (Hua et al. 2015, 2020).

5.2.1. Advances in Coral Archive Studies

Thompson (2022) reviewed many aspects of environmental archives in corals (Table 1). Although different coral genera have been analysed for palaeoclimate archives (e.g., Felis et al. 2014; Sadler et al. 2014, Saha et al. 2018b,

DOI: 10.1201/9781003320425-6

TABLE 1

Selected recent reviews and advances by topic and author. Large Xs represent a larger source than small xs.

Reference	D'Olivo and Mcculloch 2022	Thompson 2022	Wu et al. 2021	Brenner et al. 2020	Felis 2020	Hernández et al. 2020	Hua et al. 2020	Cuny-Guirriec et al. 2019	D'Olivo et al. 2018	Decarlo et al. 2018	Decarlo and Cohen 2017	Hoegh-Guldberg et al. 2017	De Carlo et al. 2016	Erler et al. 2016	Saha et al. 2016	Deng et al. 2014	Sadler et al. 2014	Lough and Cooper 2011
TOPIC																		
PALAEOCLIMATE CONTEXT		X		X	X	X	X					X						
SYNTHESIS IN MODELLING		X				X												
SAMPLE VETTING		x													X		X	
GEOCHRONOLOGY		X					X											
RADIOCARBON RESERVOIR							X											
CORAL EXTENSION											X							X
CORAL CALCIFICATION		X								X			X		x		X	x
CORAL MICROSTRUCTURE															X		x	
BIOSMOOTHING EFFECT		x															X	
CRYSTALLOGRAPHY		X													X			
STRESS BANDS											X							X
CORAL BLEACHING		x									X	X						x
PROXY UPTAKE IN CORALS		X													X			
PROXY TRANSFER EQUATIONS		x	X						X				X				X	
CALIBRATION VALIDATION		x											x					
PALAEOTHERMOMETRY (SST)			X						X				X			X	X	
O ISOTOPES																X		
SST SR/AA PROXY			X					x	x				X			x		
SST MG/CA PROXY			X					x	x							x		
SST U/CA PROXY			X						x				X					
SST B/MG PROXY			X															
SST LI/MG PROXY			X					X	x									
SST SR-U PROXY													X					
SST MULTI-PROXIES			X						X				X					
WATER QUALITY	X													X	X			
LUMINESCENCE BANDS																		
BA/CA	X														x			
MN/CA															x			
V/CA															x			

REES													x			
N ISOTOPES		X										X				
ANTHROPOGENIC POLLUTION	X												X			
OCEAN ACIDIFICATION (OA)										X						x
PH B ISOTOPE PROXY		X							X		X					
NON-*PORITES* CORAL ARCHIVES				X				x							X	
GBR HOLOCENE CLIMATE				X	X	X	X									
GBR DEGLACIAL CLIMATE				X	X											
GBR CLIMATE PROXIES	X		X	x			X					X		X		

Cuny-Guirriec et al. 2019, Razak et al. 2020), *Porites* is the dominant genus used in the Indo-Pacific because of its abundant, large colonies, slow growth rates, long survival times, and well-developed annual growth bands, which allow sclerochronology (Figure 1). Annual growth bands reflect interactions among coral tissue thickness, extension rate, calcification rate, and dissepiment emplacement (Figure 2). Many advances in understanding coral

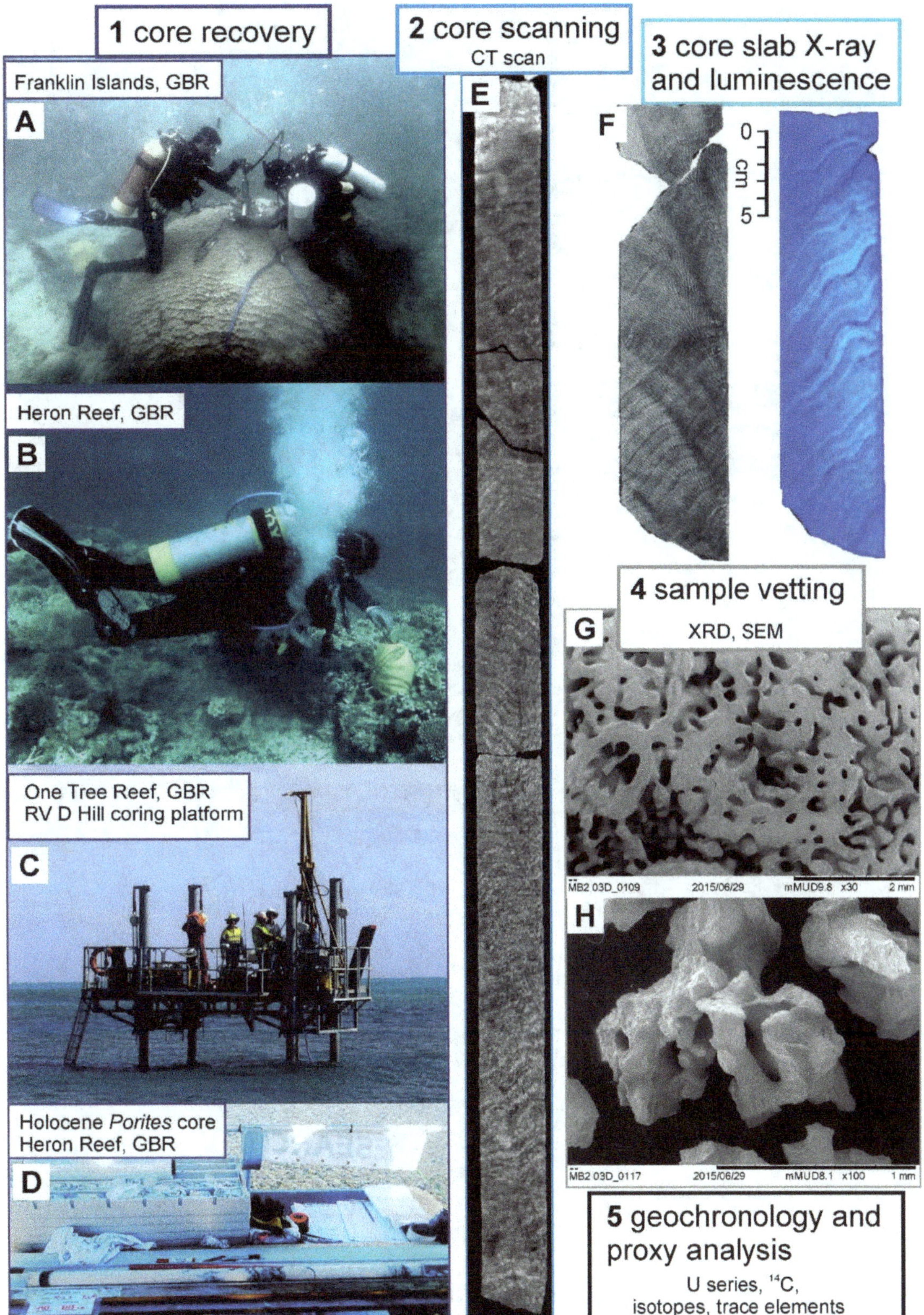

FIGURE 1 Sequence of necessary processes for the recovery of palaeoclimate data from cores.

Source: Photos A and F from Nicole Leonard; other photos from author

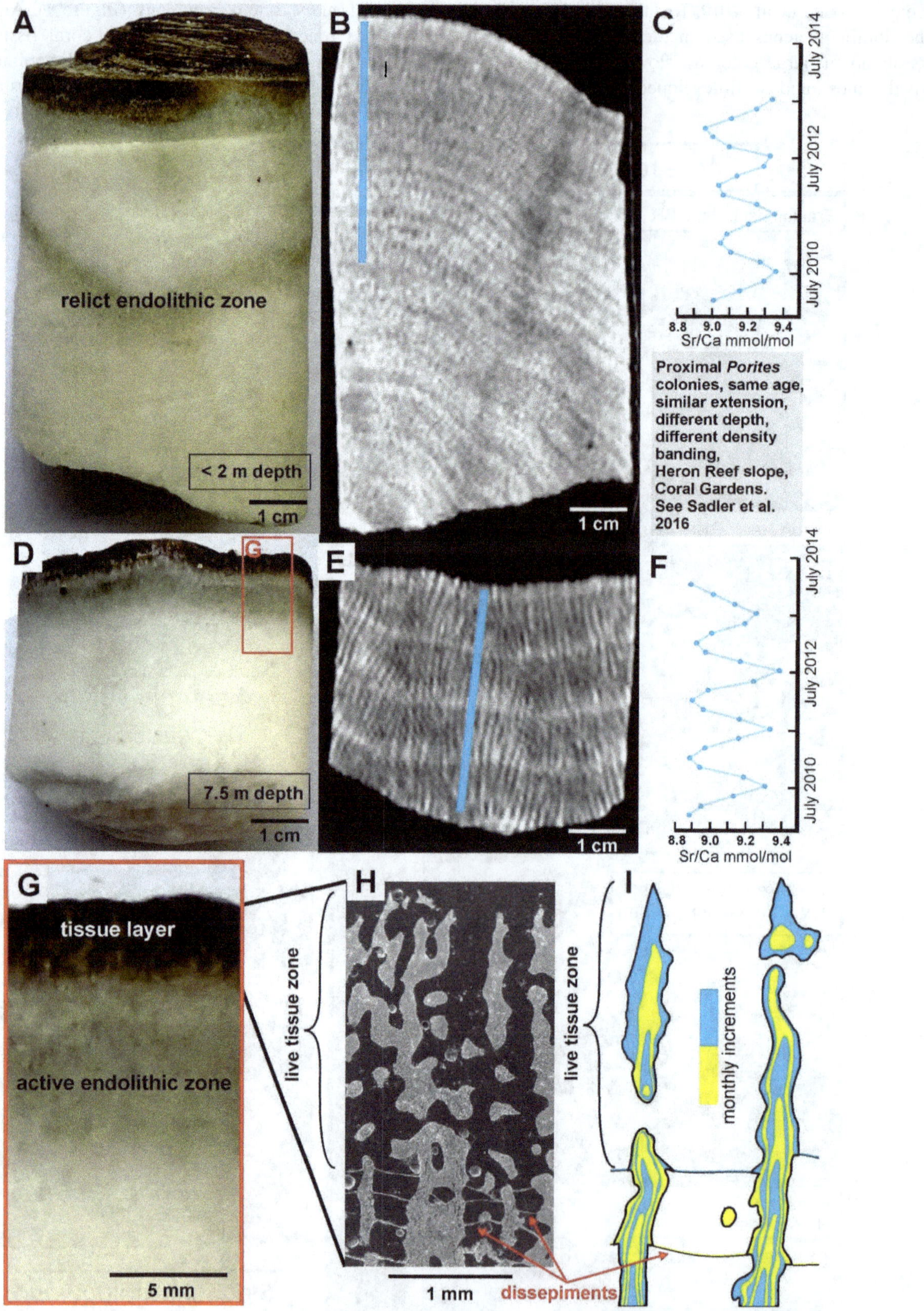

FIGURE 2 The nature of coral growth. A, B, D, E, G, H, *Porites* cores from Heron Reef. B, E, CT scans of adjacent corals. C, F, curves showing Sr/Ca ratios and matching seasons along blue trajectories in each coral. Despite the corals being sampled near one another, the growth banding apparent in CT scans is very different, possibly because of their different depths. H, scanning electron micrograph cross section of *Porites* surface. I, schematic diagram showing monthly growth increments assuming each dissepiment is emplaced on the lunar month (DeCarlo and Cohen 2017).

Source: All photos from the author

density bands have occurred since the review of Lough and Cooper (2011), but significant gaps remain, and the exact nature of the mechanisms that control banding remains unclear (DeCarlo and Cohen 2017). Improvements in computed tomography (CT) densitometry emphasise the need for careful calibration as instrument settings make greater differences than even coral slab thickness (DeCarlo 2017). Coral orientation and internal growth trajectory are critical for internal data consistency (e.g., Reed et al. 2021) because topography of the growing coral surface controls the diameter, morphology, and skeletal density of adjacent corallites.

Density banding in *Porites* is clearly annual with lower density reflecting greater extension rates based on correlation with lunar monthly dissepiment emplacement in some corals (DeCarlo and Cohen 2017). However, regional (e.g., latitude) or site-specific factors (e.g., depth, on-shore versus offshore, wave action, freshwater influx, and upwelled nutrients) contribute significantly (Lough and Cooper 2011; Spreter et al. 2022) (Figure 2), even potentially dominating the timing of bands where seasonality is weaker near the tropics (Reed et al. 2019). Tidal range may also be a factor but is rarely discussed (Schoepf et al. 2020). Even where skeletal density correlates with SST, calcification and linear extension may not (Razak et al. 2020). Some inconsistency may relate to varying tissue thickness associated with biosmoothing (Nothdurft and Webb 2007; Gagan et al. 2012; Sadler et al. 2014). The calice depth (= tissue thickness = TTL) controls how long the calicoblastic layer is in contact with a given portion of skeleton (Figure 2). New skeleton is layered on old, which increases skeletal density and causes time-averaging of the skeleton until emplacement of a dissepiment isolates it from soft tissues. Thinner TTL should provide more accurate seasonal proxies (Nothdurft and Webb 2007), but TTL varies systematically as a function of position on a colony, colony size, season, nutrient levels, and position/depth on a reef (Lough and Cooper 2011). Even proximal contemporary *Porites* from the same setting may record different density patterns (DeCarlo and Cohen 2017) (e.g., Figure 2A-F). These authors (2017, p. 759) concluded that 'Ultimately, skeletal banding patterns reflect the sensitivity of the coral holobiont to environmental variability, but these sensitivities—and the environmental forcing involved—may vary among reef locations, species, phenotypes, and potentially even over time'. OA provides an additional challenge for comparing corals over longer timespans (Mollica et al. 2018).

To place environmental proxies into a temporal context, coral chronologies are critical. Sclerochronology is used on cores collected from living colonies, but radiometric dating is required for longer and older coral archives. Prior to dating or geochemical proxy analysis, coral samples must be vetted to assure that they are unaltered. Sample vetting (Figure 3) has become increasingly sophisticated with new non-destructive spectroscopic approaches (x-ray fluorescence—Sadler et al. 2014; hyperspectral—Murphy et al. 2017; mRaman—Cuny-Guirriec et al. 2019; Stolarski et al. 2021) and greater awareness of marine diagenetic alteration (e.g., Nothdurft and Webb 2009b; Sayani et al. 2011). Spectroscopic techniques allow pre-vetting of entire cores (Figure 3). Vetting through scanning electron microscopy (SEM) outperforms the petrographic and x-ray diffraction approaches aimed primarily at recognising meteoric diagenesis in older samples. Early marine aragonite cement may contain significantly higher Sr/Ca ratios than contemporary coral aragonite (Sayani et al. 2011; Sadler et al. 2014), thus mooting the utility of x-ray diffraction for vetting samples for palaeothermometry. As cements, microbialites, and bioerosion products present different problems (Figure 4), vetting using SEM is now widely advocated (Hibbert et al. 2016; Webb et al. 2016). Improved radiometric dating techniques have also advanced coral U-series geochronology (e.g., improved decay constants and $^{234}U/^{238}U_{seawater}$ measurements and high precision, high throughput multicollector (MC)-ICP-MS and laser ablation) (Zhao et al. 2009; Clark et al. 2012; Shen et al. 2012; Cheng et al. 2013; Chen et al. 2016; Chutcharavan et al. 2018). Radiocarbon dating now has more detailed, temporally constrained reservoir corrections (Grothe et al. 2016), which also inform palaeoclimate and palaeoceanography (Hua et al. 2020).

Interpretation of direct geochemical proxies in coral skeleton depends in part on understanding the sources of individual elements (Figure 5), how they are incorporated into coral aragonite, and what physical and biological factors control this incorporation (e.g., Thompson 2022 her Figure 4). Such knowledge is particularly important for direct quantification using proxy data (e.g., relating elemental ratios to SST) by elucidating the following: degrees of Rayleigh fractionation; any critical physico-chemical thresholds; and potential sources of noise in datasets. Elemental data must be understood in light of skeletal microstructure in the coral skeleton (Farfan et al. 2022) and its broader three-dimensional growth architecture (Nothdurft and Webb 2007). The relationship of elemental concentrations and isotopic ratios in coral skeleton relative to those in seawater is affected by the site of precipitation and degree to which particular elements are enriched in, or excluded from, calcifying fluids by passive or active transport processes. Some elements are more affected by 'vital affects' and may be enhanced in intraskeletal organic matter rather than skeletal aragonite (e.g., Li, Mg and U) (Brahmi et al. 2012; DeCarlo et al. 2016; Frankowiak et al. 2016; Raddatz et al. 2014; Wu et al. 2021). Hence, centres of rapid accretion contain higher concentrations of some metals than proximal thickening deposits (Figure 5) (e.g., Brahmi et al. 2012). Extraneous organic matter from endoliths may also host additional or exotic metal concentrations (Cuny-Guirriec et al. 2019). Internal variations in metal ratios can be accommodated by sampling adequate amounts of skeleton, but this reduces temporal resolution of the proxy (D'Olivo et al. 2018). As with densitometry, the exact trajectory along which a proxy is sampled is critical (Delong et al. 2013; Sayani et al. 2019; Grothe et al. 2020).

Improved sampling and analytical techniques allow greater control of spatial (temporal) resolution. Laser ablation (LA) can sample smaller volumes of skeleton than micro-milling but with the trade-off that lower

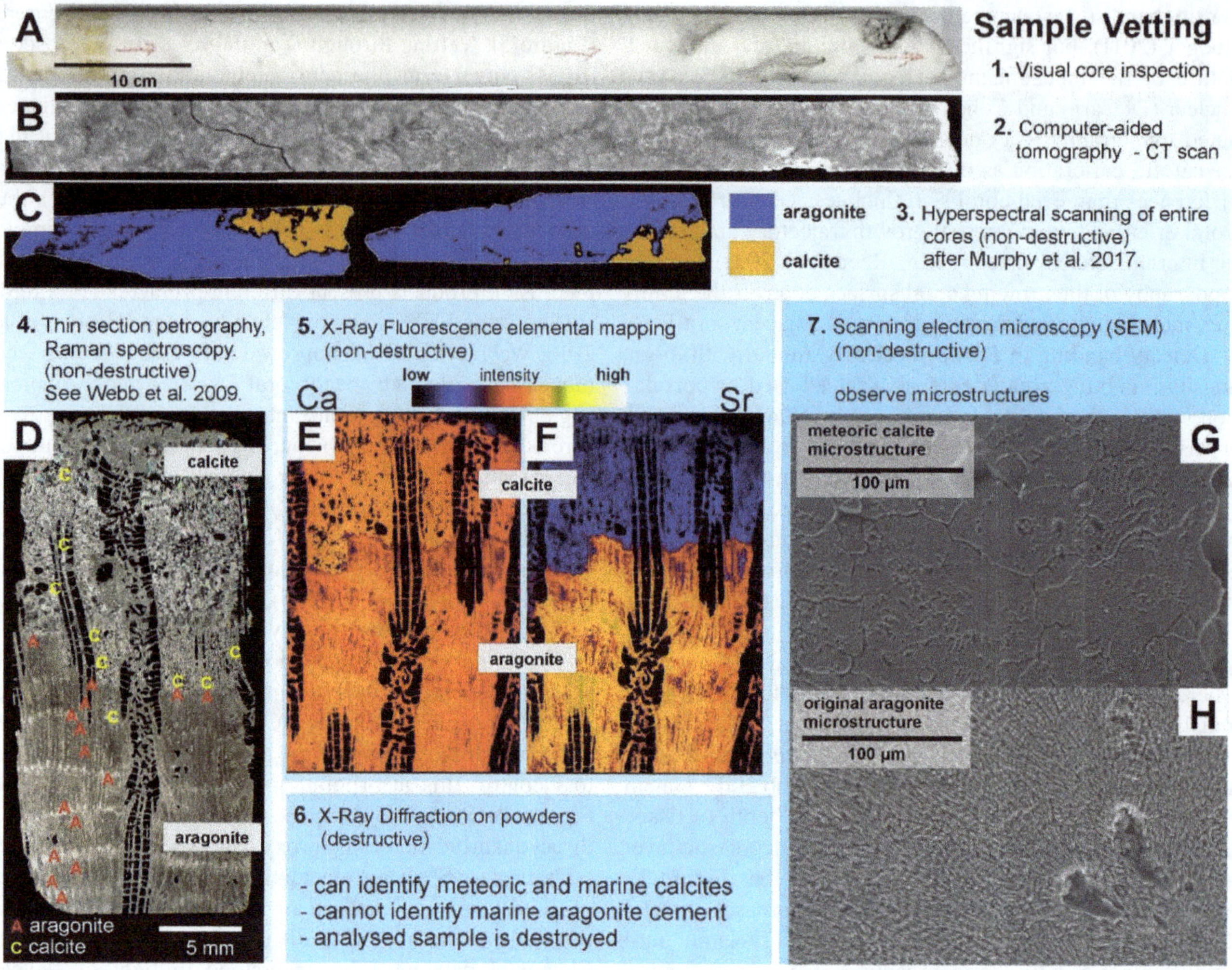

FIGURE 3 Sequence of processes for diagenetic vetting before subsampling for dating and geochemical analysis. Figures E and F are elemental maps generated using the Maia detector on the XFM beamline at the Australian Synchrotron (project AS133_XFM).

Source: Image C from Richard Murphy; Photo D from Luke Nothdurft; other images from the author

concentration trace elements are measured with less precision. Regardless, sub-annual resolution is easily obtainable for many coral proxies (e.g., McCulloch et al. 2003; Sadekov et al. 2019). LA mapping of larger areas of coral skeleton (Chalk et al. 2021) also accounts for biosmoothing and aids selection of appropriate trajectories for secular trend data. Secondary ion mass spectrometry (SIMS) is providing new insights into very high-resolution coral proxy data (Jurikova et al. 2019; Sayani et al. 2022). SIMS Sr/Ca data provide robust sub-seasonal SST reconstructions (Sayani et al. 2022), but greater noise inherent to high sample resolution still requires smoothing to establish secular trends (e.g., Brahmi et al. 2012; Domart-Coulon et al. 2014). Regardless, SIMS may allow analysis of diagenetically affected coral aragonite by sampling only from unaltered areas of pristine aragonite (Sayani et al. 2022). A similar approach may work for LA-ICP-MS. As individual metal-oxygen bonds in coral aragonite are investigated by Raman spectroscopy, additional environmental proxies may become available (Farfan et al. 2022).

5.2.2. Palaeothermometry, Palaeosalinity, and pH

Thompson (2022) reviewed the use of stable isotopes and metal/Ca ratios in corals to reconstruct secular trends in SST, SSS, and pH. Despite great interest, broad availability of mass spectrometry, and an explosion of published works (e.g., Walter et al. 2022), interpretation of coral-based proxies remains non-trivial. The primary palaeothermometres in corals are Sr/Ca and $\delta^{18}O$ (e.g., Hetzinger et al. 2016; Sayani et al. 2019), and the latter is also sensitive to $\delta^{18}O_{seawater}$ and, thus, to changes in SSS. Both proxies have sensitivities to drivers other than temperature. Replication among local colonies allows a more consistent climate signal to be isolated (Sayani et al. 2019), but the regional context is critical for monitoring broader climate trends (Hetzinger et al. 2016; Mohtar et al. 2021). In some regions, coral $\delta^{18}O$ primarily reflects temperature, but in others, SSS is a significant driver. Thus, temporal changes in climatic systems that vary both precipitation-evaporation and temperature (e.g., ENSO) require replication, cross referencing

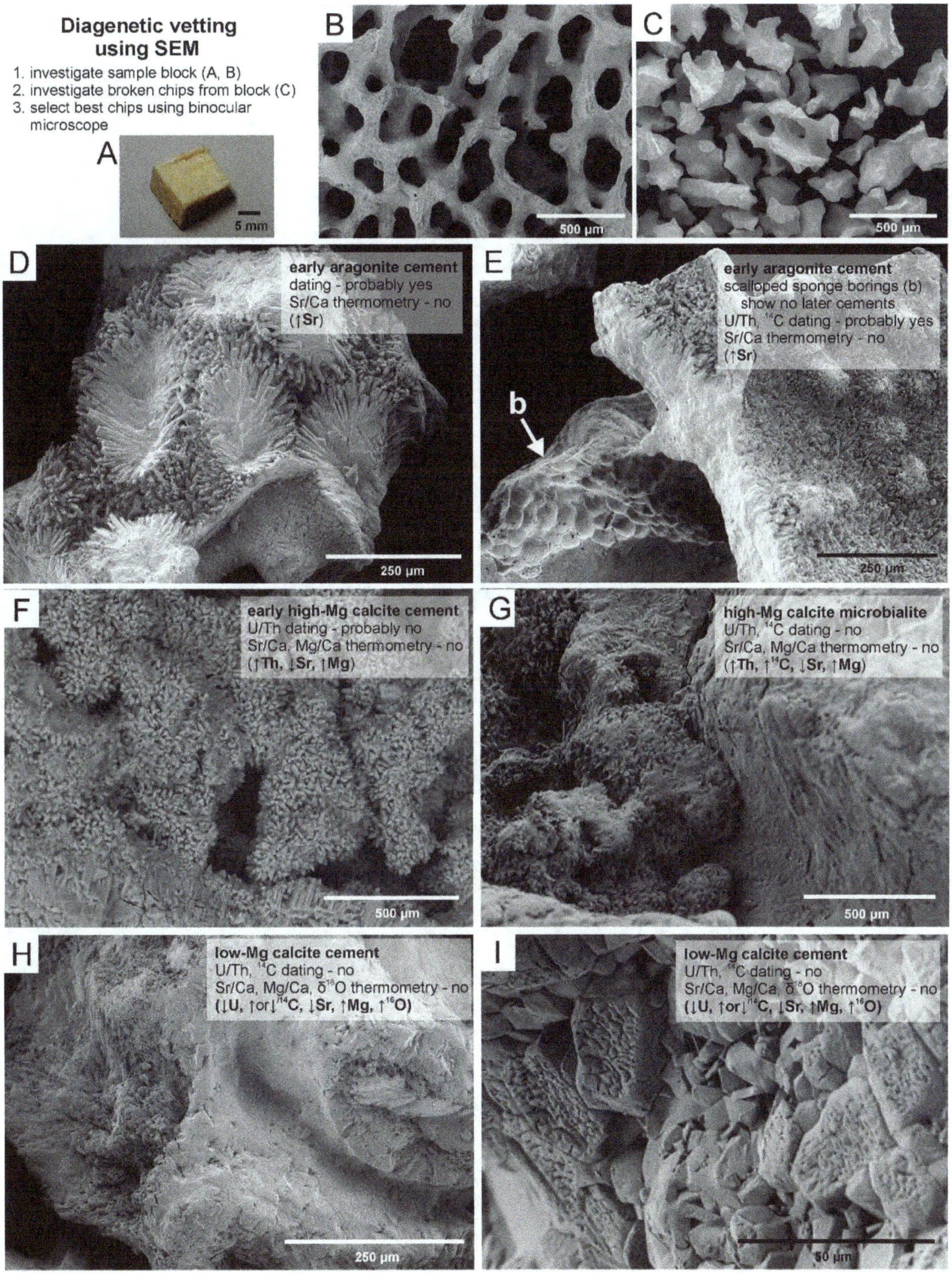

FIGURE 4 Procedure for vetting samples for diagenesis by using SEM from small blocks to chips showing common diagenetic precipitates and the problems that they cause for analyses (all samples from Heron Reef, GBR).

Source: All images from the author

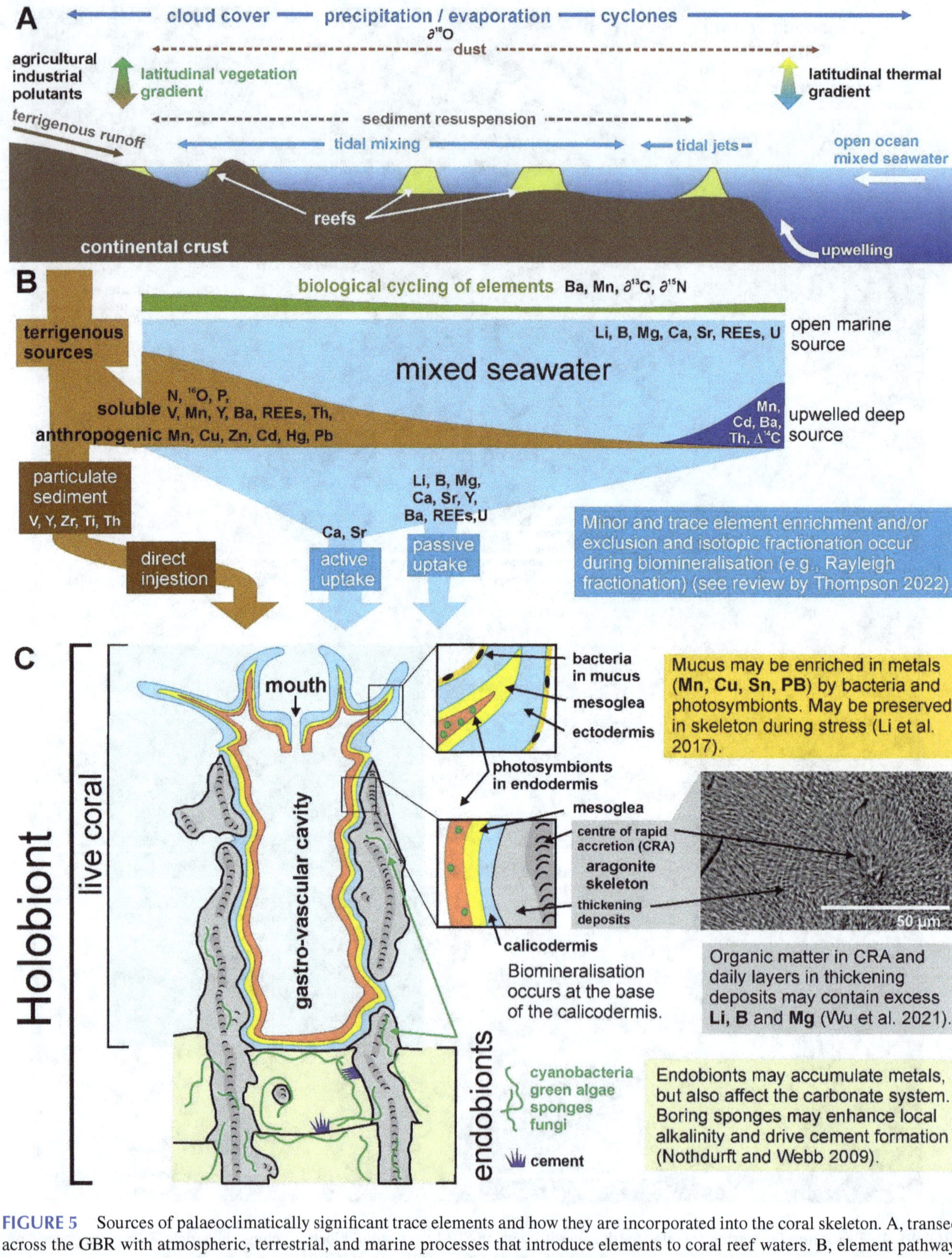

FIGURE 5 Sources of palaeoclimatically significant trace elements and how they are incorporated into the coral skeleton. A, transect across the GBR with atmospheric, terrestrial, and marine processes that introduce elements to coral reef waters. B, element pathways from source to coral skeleton. C, affinities for different elements within the coral holobiont.

between techniques, and careful site-specific calibration against instrumental records (Sayani et al. 2019; Mohtar et al. 2021).

Multiproxies are important for developing the best secular records of SST as each proxy has different positive and negative attributes (Thompson 2022). DeCarlo et al. (2016) combined Sr/Ca and U/Ca data to reflect internal biomineralisation processes and created a Sr/U thermometer that outperformed Sr/Ca alone. Wu et al. (2021) found that although Li/Mg and B/Mg ratios could cancel out the effects of Rayleigh fractionation and outperform Sr/Ca and U/Ca proxies, the affinity of Li, B and Mg for organic matter introduced significant anomalies in long-term reconstructions. Sr/Ca ratios had lower temperature sensitivity, and the U/Ca proxy was also found to be subject to additional controls (Wu et al. 2021). To circumvent problems associated with individual proxies, D'Olivo et al. (2018) suggested a 'universal multi-trace element calibration scheme' for *Porites* utilising Mg/Ca, Sr/Ca, U/Ca, Li/Mg, B/Ca, and Li/Ca data. They concluded that Sr/Ca and Li/Mg were the best proxies and that a weighted least squares method applied to multiple proxies could provide an SST estimate free from many individual colony variations and vital affects. However, different species require different calibrations.

OA is monitored in corals primarily by using B isotopes (see review of DeCarlo et al. 2018). Although B uptake and isotope fractionation are intimately tied to vital effects in the calcifying fluid, comparison of seasonal and annual B data in a 75-year *Porites* record from the GBR showed that the long-term secular trend in decreasing $pH_{seawater}$ was captured by the coral (D'Olivo et al. 2019). However, inshore records are affected by variable terrestrial runoff (D'Olivo et al. 2015), and shelf waters have reduced pH relative to the adjacent Coral Sea (Wei et al. 2009). New techniques aimed at better understanding OA in coral archives include observation of microcrystallographic effects of precipitation at lower pH, which involve the incorporation of increased organic matrix and water and better constrained crystal orientation (Coronado et al. 2019). Chalk et al. (2021) used LA-ICP-MS to map the spatial distribution of both element concentrations and boron isotopes in *Porites* skeleton to better discriminate the internal heterogeneity of geochemistry and thus calcification processes within the skeleton.

5.2.3. Terrestrial Runoff as a Precipitation Proxy

Terrestrial runoff reflective of precipitation can be monitored in corals by using luminescent bands and various trace elements and isotopes (Lough 2011; Yu et al. 2022; Lewis et al. 2024) (Figure 5). Luminescence bands in corals have proven to be a robust proxy for terrigenous runoff in the inner GBR (Lough et al. 2015) and have demonstrated utility for monitoring and potentially improving terrestrial management to support coral reef health (e.g., Kaushal et al. 2021). Nitrogen isotopes in coral skeletal-bound organic matter provide a developing tool for tracking terrestrial flux, nutrient cycling, and upwelling in the GBR (Erler et al. 2016, 2020a, 2020b).

Element/Ca ratios (e.g., Ba/Ca) in corals were used early on to infer trends in terrigenous runoff in the GBR pre- and post-European colonisation (McCulloch et al. 2003) with additional proxies developed since then. Importantly, interpreting secular trends in precipitation from terrestrial runoff proxies is not straightforward. Different elements, or even the same element, may travel in different ways, behave differently within local water bodies, or have differing sources (Saha et al. 2016). Most metals used to trace terrestrial sediment flux enter the marine environment through exchange reactions from fluvial sediment in the estuaries and nearshore plumes (Walther and Nims 2015). However, these processes are inadequately understood for many elements (Lewis et al. 2018; Saha et al. 2021). Vanadium (V), for example, is sourced primarily from terrestrial sediment but may be quantitatively decoupled from sediment flux through chemical reactions in the catchment (Saha et al. 2019a). Intense heating of V-bearing lateritic soils during anthropogenic clearing and burning releases V from insoluble Fe-oxyhydroxides as highly mobile vanadyl ions that then provide a better secular proxy for clearing and burning than for precipitation, runoff, or sediment flux (Saha et al. 2019a). The behaviour of Ba, Mn, and Y in the fluvial and estuarine waters of the Burdekin and Fitzroy systems and the Ba/Ca, Mn/Ca and Y/Ca ratios in associated inshore coral archives differ because of very different desorption behaviours in the estuaries (Lewis et al. 2018; Saha et al. 2021). Some coral-based Ba/Ca archives did not correlate well with Burdekin discharge, despite luminescent bands being consistent (Lewis et al. 2018); Mn/Ca and Y/Ca archives did not reflect proximal discharge at all (Lewis et al. 2018; Saha et al. 2021). Multi-century GBR coral records of Ba/Ca are useful for tracking terrestrial plumes and correlate relatively well with luminescence data (D'Olivo and McCulloch 2022). However, Ba/Ca ratios reflect total sediment flux within the plume rather than the volume of freshwater runoff and, therefore, precipitation. Thus, coral archives of sediment flux (Ba/Ca) may be decoupled from precipitation volume, the humic substances that cause luminescence, and independent SSS proxies ($\delta^{18}O$). Biological cycling of Ba in inshore waters may also obscure temporal runoff patterns (Saha et al. 2018b), whereas farther offshore, Ba/Ca archives may be dominantly sourced from upwelling (Walther et al. 2013; Spreter et al. 2022) (Figure 5). Unravelling Ba sources may be possible with Ba isotope studies (Hsieh et al. 2022), but more work is required. Understanding the sources and marine cycling behaviour of a given proxy element and the spatial distribution of the proxy archive in relation to local geography is critical (Lewis et al. 2018). Strontium isotopes also vary systematically with ENSO because of differences in groundwater mixing with seawater (Rahaman et al. 2022), but interpretation depends on the local geographic and hydrologic context.

Coral-based rare earth element (REE) distributions are useful proxies for terrigenous flux and are superior to individual element/Ca ratios where direct comparisons are made (Saha et al. 2019b, 2021). REEs are useful proxies because of their coherent chemical behaviour with key differences in surface complexation across their mass range. These characteristics allow multiple parameters (e.g., light REE (LREE) depletion; Y/Ho ratios) to develop in tandem in the estuary (Saha et al. 2016). These parameters are decoupled from total REE concentration and thus reflect terrigenous runoff-seawater mixing regardless of the volume of sediment flux, which controls single element concentrations. Importantly, REE proxies can be related directly to the terrigenous source and may allow 'fingerprinting' of runoff from particular catchments (Saha et al. 2021). However, as with other terrigenous metal/Ca runoff proxies, they can be affected by sediment resuspension during cyclones and other large wave events, so they may be decoupled from precipitation volumes (Saha et al. 2019b).

Regardless of the coral-based proxy used, a variety of complications exist when inferring palaeoclimate from secular trends in proxies. Evidence of runoff is spatially variable depending on flood plume distribution, which reflects the volume of runoff, geography, depth, currents, and tides (Devlin et al. 2012; Walther et al. 2013; Wolanski et al. 2024). Even proximal records may vary because of local hydrodynamics (Leonard et al. 2019), but despite this, individual flood plumes can be traced remotely (Petus et al. 2019), although the time of monitoring may cause bias (Petus et al. 2022). Regardless, broader water quality gradients have been inferred from coral-based proxies (Lewis et al. 2018; Leonard et al. 2019; Saha et al. 2021). More generally, latitudinal differences in coastal vegetation, especially between the wet and dry tropics, modulate the relationship between terrestrial sediment flux and discharge (Neil et al. 2002) (Figure 5). Where rainfall is more persistent, sediment is better flushed from local catchments, and more sediment can be trapped by estuarine vegetation. Thus, the same degree of rainfall and discharge may cause different levels of sediment flux with latitude. Temporal mismatches between precipitation and sediment flux also occur. Subtle geochemical signals may be masked because of routinely high inshore concentrations of some elements (Saha et al. 2018b), and water and/or sediment retention in a given catchment may vary significantly over time (Lough et al. 2015; D'Olivo and McCulloch 2022). Sediment flux can be disproportionately enhanced by early rains following a drought interval just as sediment flux may decrease relative to precipitation and runoff through successive wet years (Saha et al. 2018a). Thus, coral-based sediment proxies cannot provide a strictly quantitative record of precipitation, but secular trends in terrigenous flux may provide useful data at the seasonal to multiannual, multidecadal, and centennial levels. As a complicating factor, the decoupling of sediment flux from precipitation has increased with anthropogenic catchment modification (D'Olivo and McCulloch 2022). Land clearing and European agricultural practices have systematically increased terrestrial sediment flux in many major catchments of the GBR (McCulloch et al. 2003; D'Olivo and McCulloch 2022), which means that pre- and post-European colonisation data are not directly comparable, and the date of transformation varies geographically. Coastal development and particularly port dredging activities also obscure the temporal record of precipitation in inshore settings of the GBR (Saha et al. 2019b). Despite these problems, terrestrial runoff proxies are still useful for palaeoclimate studies with the proper context, and over much longer timeframes, changes in offshore water quality based on REE proxies in reefal microbialites have informed changes in the AISM and ENSO over > 6,000 years of the Holocene for the southern GBR (Salas-Saavedra et al. 2022).

5.3. SUBLETHAL STRESS INDICATORS IN CORALS

Climate-controlled stressors on corals include thermal bleaching, cyclones, and reduced water quality (e.g., turbidity, low salinity, and increased nutrients) associated with continental runoff and sediment resuspension from storms (Figure 6). Although reefs are well-adapted to storms and cyclones (Connel 1978), storm intensity has varied over longer (Nott and Hayne 2001) and shorter (Haig et al. 2014; Wolff et al. 2016; Duke et al. 2024) timescales. Importantly, cyclones affect reefs in a multitude of ways. Aside from the direct damage to corals and increased rubble generation (Wolfe et al. 2021) and reduced water quality from sediment resuspension, increased cyclone activity is putting additional stress on inshore mangrove environments (Duke et al. 2024). Reduced mangrove systems increase the potential for greater terrestrial sediment flux in the future, particularly in the northern GBR where these environments sequester significant sediment (Neil et al. 2002). Some mid-Holocene hiatuses in reef growth may reflect increased cyclone activity (Ryan et al. 2016). However, reef turn-off events may also reflect local factors (e.g., terrigenous runoff) (Perry and Smithers 2011) and minor sea level fluctuations (Dechnik et al. 2017; Leonard et al. 2020b), which may also indicate changing climate. Improvements in precision and the cost of U-series dates for corals have provided an approach to unravel this record and a means of relating mass coral die-offs to specific causes. Coral death assemblage analysis uses the precise dating of bulk dead coral and debris from reef flats or forereef slopes (Clark et al. 2014a; Leonard et al. 2020a) to statistically link mass death events to individual storms (Yu et al. 2012a; Liu et al. 2014), bleaching events (Clark et al. 2014b, 2017), or anthropogenic factors (Chen et al. 2021).

Although mass coral die-offs can be associated with specific causes in the past, recognition of sub-lethal stress events is more challenging but potentially more important for identifying specific thresholds and stress responses in corals (DeCarlo and Cohen 2017). The sub-annual geochemical context can be recovered from death assemblage

FIGURE 6 Stressors that cause diebacks in colonies. A, B, exposure at low tide levels causes microatolls. C, sedimentation can cause false microatolls in deeper water when sediment is deposited on top of a colony and cannot be removed. D, sediment may partially kill massive colonies from the base. E, poor water quality can cause die-backs. F, a variety of corallivores, such as this crown-of-thorns starfish can kill sections of colonies. G, thermal stress may cause bleaching and death or just anomalous high-density bands, but interpretations may remain speculative without corroborating geochemical evidence. All photos are from Heron Reef, except for E, which is from Moreton Bay, Queensland.

Source: All photos from the author

corals (Clark et al. 2014b), but the lethal event may have been too brief to be captured by skeletal growth, as precipitation may decrease abruptly during stress (DeCarlo and Cohen 2017). Soft tissue predation, burial in sand, or brief episodes of anoxia or thermal stress may not be recorded by skeletal growth, so the cause of death may remain unknown.

Many sublethal stress indicators have been identified. Individual bite scars from fish are recognised in x-radiographs (Glynn and Manzello 2015), and broader partial colony mortality may leave localised die-back surfaces followed by regrowth from adjacent polyps. Determining the cause of diebacks in massive *Porites* cores (Hendy et al. 2003b; Cantin and Lough 2014) is difficult, especially without access to lateral areas that survived the stress. Where entire colonies can be observed, the source of the stress may be more obvious as in the case of microatolls, false microatolls with sediment-smothered tops, and edge diebacks because of sedimentation events (Figure 6). Bands of increased microbial bioerosion preserved beneath live tissues in massive corals (Figure 2A) may also indicate stress intervals with lower coral extension rates (Le Campion-Alsumard et al. 1995; Fordyce et al. 2021), but they may be difficult to recognize and quantify in core because of the small dimensions of the borings (diameters commonly < 10 μm). Relating a bioerosion 'band' to the age when it formed is also difficult, as their depth may vary within the skeleton. Additionally, cyanobacterial bioerosion intensity varies on reefs with nutrients and light penetration through depth and turbidity gradients (Perry and Macdonald 2002). Although the endolithic biome is extensive and pervasive in many shallow corals and although endolithic metabolites may affect the overlying live corals, even affecting resilience to bleaching (Ricci et al. 2019; Cárdenas et al. 2022), quantification is rare (Perry and Hepburn 2008; Nothdurft and Webb 2009b).

Sublethal stress indicators include changes in calcification-extension rates, skeletal density, and dissepiment spacing (De Carlo and Cohen 2017). Where associated specifically with other proxies, such as luminescence, trace elements, N isotopes, and palaeothermometry, identified stress bands may be linked to specific types of stress (Figure 6). Proxies for water quality and palaeothermometry were discussed earlier. The following section focuses specifically on proxies for thermal bleaching.

5.3.1. Coral Bleaching

The bleaching history of the GBR is of great interest (Kamenos and Hennige 2024), but the topic is controversial because of the difficulty of attributing pre-instrumental stress events specifically to marine heat waves (e.g., Kamenos and Hennige 2018, 2020; Hoegh-Guldberg et al. 2019; DeCarlo 2020), for reasons discussed later, and the sparsity and temporal and geographic unevenness of the proxy records and correlative heatwave reconstructions (Kamenos and Hennige 2024). Radiometric dating has allowed coral mortality events to be attributed to possible thermal stress back in time (e.g., Holocene, Wang et al. 2021; 1860s, 1890s, Yu et al. 2012b; 1930s, Clark et al. 2017; 1950s, Yu et al. 2006), but the earliest scientifically observed coral bleaching on the GBR was in 1929 by the Yonge Expedition to Low Isles (Dunne et al. 2022). In this case, summer bleaching in February resulted from high SST combined with reef flat exposure (i.e., low tide) and high solar insolation. Bleaching during winter (June) in the same year reflects only high solar irradiance at low tide (Dunne et al. 2022). Although bleaching has been well documented in much of the GBR over the last several decades (Hoegh-Guldberg et al. 2017), it was first documented in the Torres Strait as late as 2010 (Bainbridge and Berkelmans 2014).

As with the formation of annual density bands, individual corals behave very differently during the same heating event. Bleaching not only reflects increased regional SST but also local factors such as differing symbionts, normal daily temperature ranges, tidal exposure, cloud cover, nutrient levels, turbidity, and physical shading (Silverstein et al. 2015; Barkley and Cohen 2016; McGowan and Theobald 2017; Safaie et al. 2018; Baumann et al. 2019). The absence of expected bleaching in some GBR corals during the 2004 heating event may reflect lower nutrient loads caused by decreased upwelling associated with a slow Eastern Australian Current (EAC) at the time (DeCarlo and Harrison 2019). Corals in other settings (e.g., lagoons, shallow inshore) may be better acclimated to increased SSTs and high nutrient loads and are thus less susceptible to thermal bleaching (Cantin and Lough 2014; Barkley and Cohen 2016; Mollica et al. 2019). Corals also become more resilient to heating events through successive heat waves (DeCarlo et al. 2019; Hughes et al. 2021). Thus, individual coral responses to heating events are highly variable and relate to regional, local, and individual factors, some of which are temporally variable in themselves (e.g., acclimatization Hughes et al. 2021). Hence, bleaching is not uniquely related to elevated SST even for well-documented modern events, and both spatial and temporal variations complicate the interpretation of past mass bleaching. To obtain longer-term records of thermal stress in corals, we must understand individual responses to multiple stressors for each coral and obtain adequate geographic coverage to elucidate regional drivers (e.g., marine heat waves). Even with reliable records of past thermal stress, the relation of this record to palaeoclimate is nontrivial, as it is not clear in the GBR how much of a given anomalous SST is the result of El Niño, local patterns of cloud cover (e.g., weather), or proximal terrestrial heat waves (McGowan and Theobald 2017; Karnauskas 2020; Zhao et al. 2021). Regardless, independent proxies to relate specific density bands in corals to thermal stress are needed.

Apart from anomalous density bands themselves and zones of intensified microbioerosion (e.g., Fordyce et al. 2021), geochemical palaeothermometres may associate specific stress bands with a given season or broader temperature anomaly (i.e., circumstantial evidence) (Figure 6). DeCarlo and Cohen (2017) documented a stress band during a high temperature interval with reduced carbonate ion abundance based on U/Ca data, which is consistent with bleaching, but a skeleton was not produced during the bleaching event itself because of low calcification. Regardless, the occurrence of a stress band during summer or an interval of unseasonably high SST based on the coral's local secular trend provides support for an interpretation of thermal stress if not bleaching. Wang et al. (2021) associated mortality surfaces and stress bands in Holocene *Porites* with anomalously high SSTs (based on Sr/Ca) and decreased $\delta^{13}C$ suggestive of reduced photosynthesis. Such concurrence of independent proxies strengthens the support for interpretations. Li et al. (2017) compared proximal bleached and unbleached *Porites* and found major increases in Mn, Cu, Sn, and Pb within a thin band of skeletal aragonite that they attributed to polyp retraction and incorporation into the skeleton of mucous containing metals accumulated by both coral and excluded symbionts. However, transition metals in corals commonly have anthropogenic sources, providing significant noise in recent times (Saha et al. 2016; Bolouki Kourandeh et al. 2021). Nevertheless, the ease of analysing such trace metals using ICP-MS provides an exciting avenue for additional study.

Changes in $\delta^{15}N$ in coral skeleton also might provide a proxy for bleaching with additional work because of disruption of symbiont metabolic stability (Thompson 2022). Decreased photosynthesis during bleaching was hypothesised to lower intracoralline pH so that reduced $\delta^{11}B$ might serve as a proxy (Wei et al. 2009; Dishon et al. 2015), but the results of similar studies have varied, and Ross et al. (2022) concluded that pH within coral calcifying fluid depends on complex, site-specific interactions of many factors including light and temperature. Although heat stress may not affect the calcifying fluid pH, it may disrupt typical metal/Ca proxies for many months following the stress (Cheung et al. 2021; Schoepf et al. 2021). If recognisable in stable secular trends, then such disruptions themselves might serve as proxies for thermal events, but they also increase noise. Attribution uniquely to thermal bleaching is still difficult.

5.4. APPLICATION OF CORAL-BASED PALAEOCLIMATE PROXIES IN THE GBR

The majority of secular coral environmental proxy records have fewer than 50 years of data younger than the 1970s and have focused on calibration and validation using instrumental records in order to document anthropogenic warming, OA, and inshore water quality (see reviews in Table 1). Fewer records investigate the secular trends over centennial and longer timescales that are appropriate to understand natural climate trends and internal forcing relationships for key climate systems (Emile-Geay et al. 2021). In the GBR, most data represent inshore to mid-shelf reefs (Table 2, Figure 7) where 'noise' is associated with continental runoff and anthropogenic factors. Multi-century to millennial studies use long or multiple overlapping *Porites* records aimed at documenting the Medieval Climate Anomaly and warming since the Little Ice Age (LIA) (Zinke et al. 2014; Chen et al. 2018), terrigenous runoff (Lough et al. 2015), and in the GBR, European colonisation (D'Olivo and McCulloch 2022). Rare studies have focused on isolated time intervals within the Holocene and deglacial interval, but even Last Interglacial palaeothermometry is becoming available (Felis 2020).

The most abundant, longest and best integrated GBR coral proxy archives represent inshore to mid-shelf reefs of the central GBR (Table 2, Figure 7) by utilising the AIMS Coral Core Archive as supported by a > 300-year master luminescence chronology (Hendy et al. 2003a). Coral luminescence bands suggest significant interannual and decadal variation of precipitation in the central GBR since ~1650 (Lough 2007, 2011; Lough et al. 2015) (Figure 8). An interval of relatively dry climate occurred between the 1760s and 1850s when peak flows increased significantly (Figure 8D). However, this same time corresponds to settlement of the Burdekin catchment by Europeans (McCulloch et al. 2003), so some increased runoff may relate to changed land use (Lewis et al. 2007; Lough et al. 2015; D'Olivo and McCulloch 2022). The Fitzroy River had a similar but lower amplitude record of discharge (Lough 2007). The record relates in part to ENSO and PDO dynamics with a recent shift towards greater amplitude ENSO variability since the 19th century, and although no general trend towards wetter or drier climate is indicated, the abundance of extreme wet and dry events is increasing (Lough et al. 2015). Other regions are covered by more scattered data (Table 2). The least well-studied area is at the far north of the GBR in the Torres Strait, where Lough (2016) reported *Porites* luminescence from Bramble Cay and Darnley Island, including a multi-century (1781–1993) record from Bramble Cay. *Porites* calcification rates from both localities did not appear to respond to increased SSTs but Bramble Cay cores showed long-term annual luminescence pulses related to the Fly River outflow to the northwest throughout the record. Farther south, Ba/Ca ratios provide the longest terrigenous runoff archives (D'Olivo and McCulloch 2022) with multi-century records for the northern, central, and southern regions of the GBR all emphasising the effects of European settlement and land-use change (D'Olivo and McCulloch 2022). However, drier versus wetter intervals are evident. Inclusion of Y and Mn with Ba allowed documentation of specific effects of sheep and cattle grazing in a 174-year archive (Lewis et al. 2007). However, a 168-year record of central GBR *Porites* $\delta^{15}N$ showed stability through European settlement, against expectation, suggesting very efficient N dynamics within the coastal ecosystem that may have 'buffered' the anthropogenic signature (Erler et al. 2016). Additional longer records are required to better distinguish climate signals in runoff.

Despite the large number of global studies on coral-based palaeothermometry, multi-century records are surprisingly rare, and no GBR record was used in the Western Pacific multi-century reconstruction of Tierney et al.

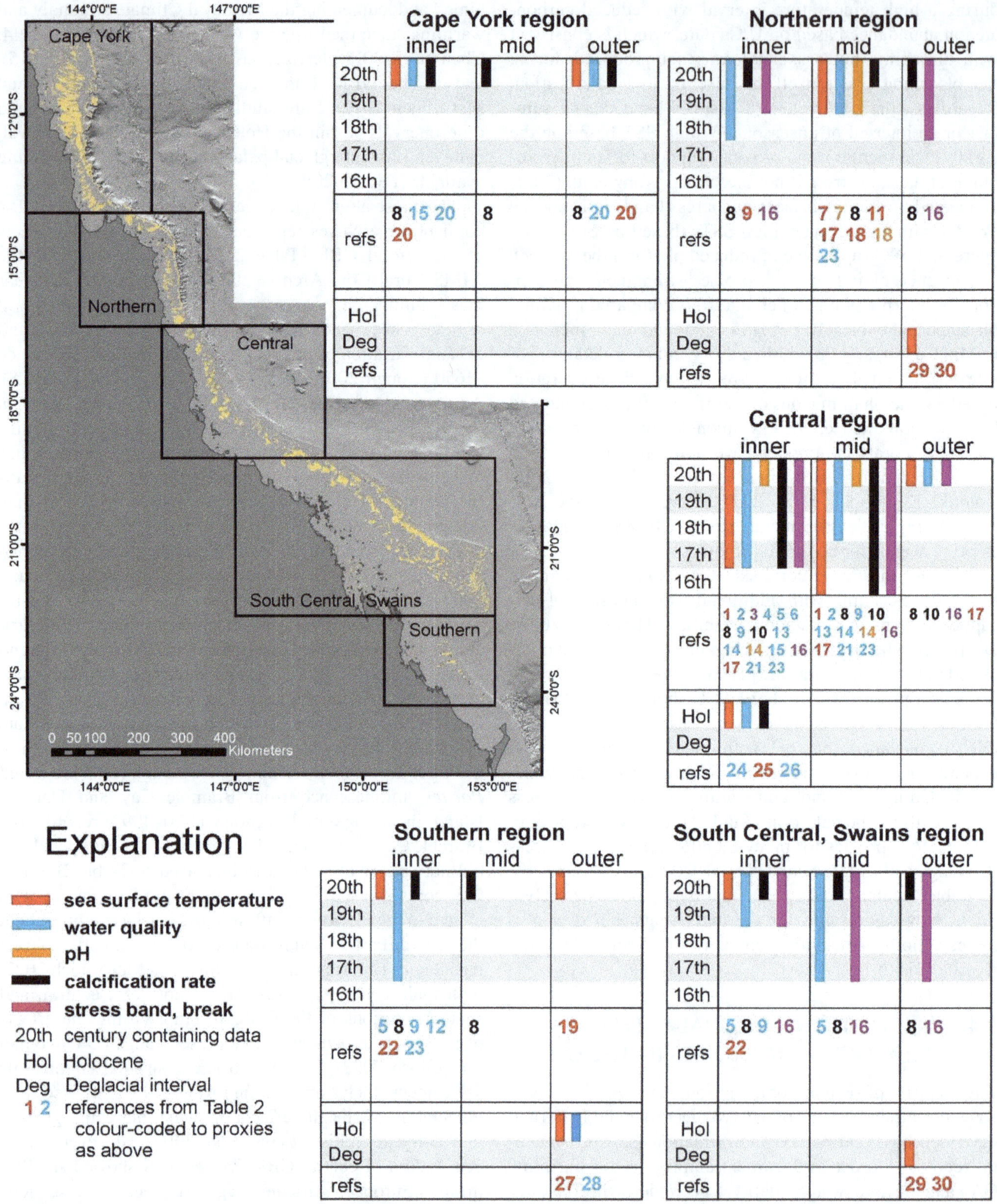

FIGURE 7 Map showing five regions of the GBR with palaeoclimate data listed for inner, mid-shelf, and outer shelf reefs. Vertical coloured bars show the temporal distribution of different climate proxy data for the 16th to 20th centuries. However, data may not extend completely through a given century. Data for the Holocene and deglacial interval are shown at the bottom. Reference numbers are from Table 2 and are also colour-coded by proxy type.

Source: Map kindly provided by Atefeh Sansoleimani

TABLE 5.2
Selected references with secular trends for climate proxies in the GBR.

Number	Reference	source	region*	Inshore	mid-shelf	offshore	start	end	range	# cores	Sr/Ca	Mg/Ca	U/Ca	Li/Mg	B/Mg	B/Ca	Ba/Ca	Y/Ca	Mn/Ca	REEs	del11B	del^{15}N	del^{18}O	Luminescence	density bands	calcification	growth breaks
1	Hendy et al. 2002	*Porites*	C	X	X		1565	1985	421	8	X		X										X				
2	Hendy et al. 2003a	*Porites*	C	X	X		1615	1985	371	8														X			
3	Hendy et al. 2003b	*Porites*	C	X	X		1612	1985	374	8																	X
4	McCulloch et al. 2003	*Porites*	C	X			1750	1998	249	2							X							X			
5	Lough 2007	*Porites*	C	X	X		1631	1996	366	25														X			
6	Lewis et al. 2007	*Porites*	C	X			1813	1986	174	1							X	X	X								
7	Wei et al. 2009	*Porites*	N		X		1800	2004	205	1	X	X					X				X		X				
8	De'ath et al. 2009	*Porites*	CY, N, C, SCS, S	X	X	X	1572	2005	434	328																X	
9	Lough 2011	*Porites*	N, C, SCS, S	X	X		1639	1995	357	20														X			
10	D'Olivo et al. 2013	*Porites*	C	X	X	X	1930	2008	79	41																X	
11	Deng et al. 2014	*Porites*	N		X		1937	2009	73	2	X												X				
12	Rodriguez-Ramirez et al. 2014	*Porites*	S	X			1921	2011	91	6														X			
13	Lough et al. 2015	*Porites*	C	X	X		1648	2011	364	6														X			
14	D'Olivo et al. 2015	*Porites*	C	X	X		1940	2009	70	5											X						
15	Erler et al. 2016	*Porites*	C	X			1820	1987	168	1												X	X				
15	Erler et al. 2016	*Porites*	CY	X	X		1973	1990	18	3												X	X				
16	Kamenos and Hennige 2018	*Porites*	N, C, SCS, S	X	X	X	1620	2001	382	44															X		
17	D'Olivo et al. 2018	*Porites*	N, C	X	X	X	1990	2013	24	12	X	X	X	X													
18	D'Olivo et al. 2019	*Porites*	N		X		1939	2013	75	1						X					X						
19	Ellis et al. 2019	*Porites*	S			X	1990	2012	23	1	X																
20	Reed et al. 2019	*Porites*	CY	X	X	X	1972	2008	37	5	X												X	X	X		
21	Erler et al. 2020b	*Porites*	C	X	X		1997	2004	8	18												X					
22	Wu et al. 2021	*Porites*	SCS, S	X			1898	2016	119	7	X		X	X	X												
23	D'Olivo and McCulloch 2022	*Porites*	N, C, S	X	X		1759	2008	250	10							X										
	Holocene records																										
24	Lough et al. 2014	*Porites*	C	X			~6.2 ky	~5.9 ky	na	5														X		X	
25	Roche et al. 2014	*Porites*	C	X			~4.7 ky	~4.66 ky	na	1	X												X				
26	Leonard et al. 2016	*Porites*	C	X			~6.0 ky	~3.8 ky	na	8														X			
27	Sadler et al. 2016	*Porites, Acropora*	S			X	~7.0 ky	~5.2 ky	na	3	X																
28	Salas-Saavedra et al. 2022	microbialite	S			X	~8.2 ky	~1.8 ky	na	na								X	X	X							
	Deglacial records																										
29	Felis et al. 2014	*Isopora*	N, SCS			X	~25 ky	~11 ky	na	25	X												X				
30	Brenner et al. 2020	*Isopora*	N, SCS			X	~25 ky	~11 ky	na	25	X												X				

Source: *CY—Cape York region; N—Northern region; C—Central region; SCS—South-Central, Swains region; S—Southern region

(2015). Of the rare GBR records, a 420-year central GBR record of coral Sr/Ca and U/Ca (Hendy et al. 2002) does not correlate well with the Tierney et al. (2015) reconstruction, showing reduced temperatures from 1565 to 1700 followed by elevated temperatures from 1700 until the 1870s when they fell again before rising to modern levels (Figure 8B). Importantly, although the temperature anomalies in Figures 8A and B are plotted against different averages (as shown), the trending in the two curves is almost antithetical between ~1650 and 1900. Hendy et al. (2002) suggested that warmer temperatures during part of the LIA could reflect increased latitudinal temperature gradients at that time. However, comparable data from a greater latitudinal spread along the GBR are currently not available to test this hypothesis. Interestingly, the trend of inferred bleaching frequency (Kamenos and Hennige 2024, their Figure 1) from a similar range of GBR corals as those of Hendy et al. (2002) better matches the Tierney et al. (2015) curve. An abrupt freshening of water (lower $\delta^{18}O$) from the mid-19th century (Figure 8C) corresponds in part to

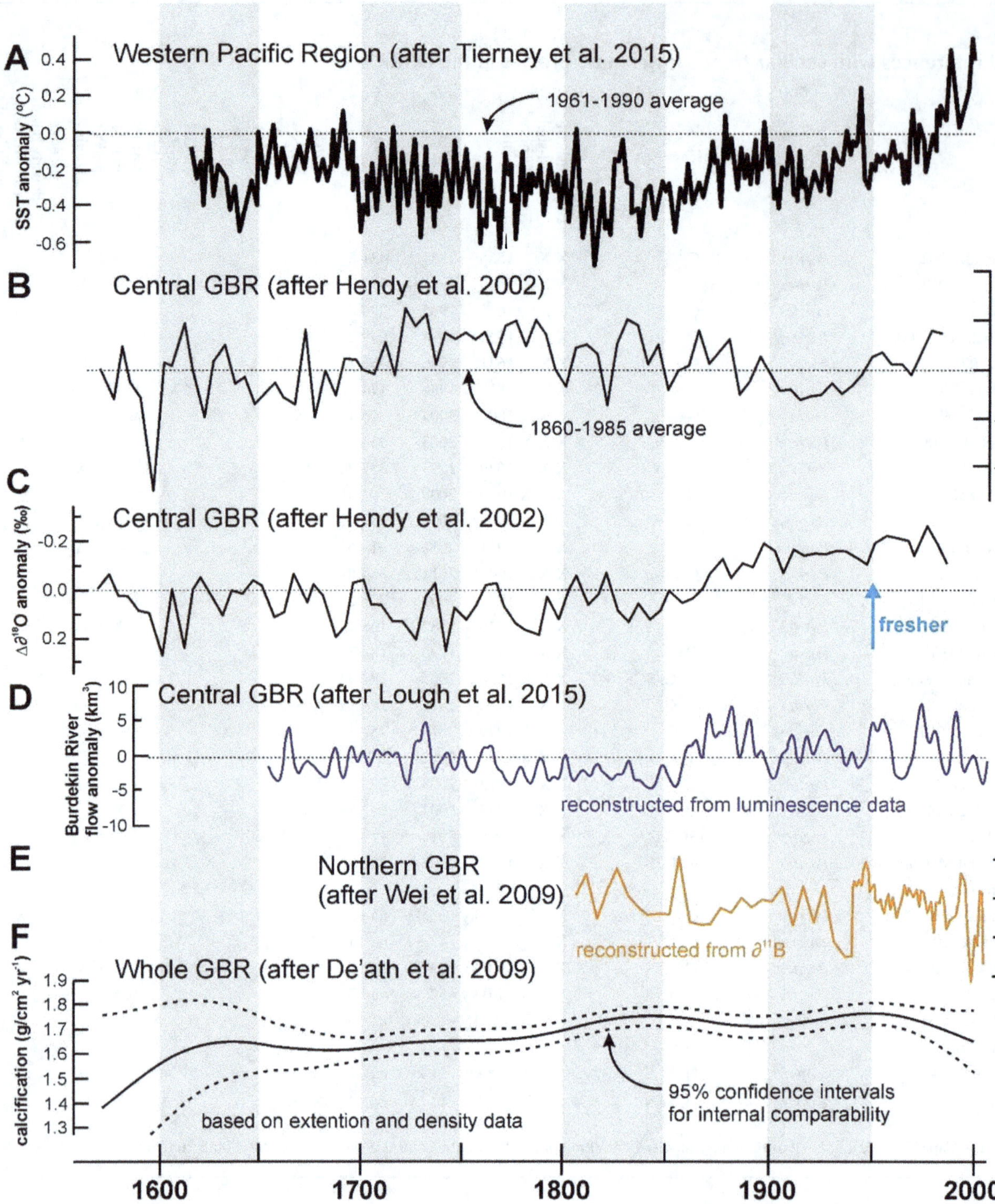

FIGURE 8 Coral based palaeoclimate archives for the GBR. A, SST anomaly for the Western Pacific region. B, SST anomaly based on Sr/Ca palaeothermometry for the central GBR region. C, salinity anomaly measured as $\delta^{18}O$ adjusted for temperature for the central GBR region. D, reconstructed Burdekin River flow based on coral luminescence banding from the central GBR. E, pH reconstructed from $\delta^{11}B$ data in the northern GBR region. F, calcification rate trends for the GBR.

late 19th-century cooling in the GBR (Hendy et al. 2002) but corresponds better to increasing temperatures in the regional average (Tierney et al. 2015) (Figure 8A). This increase in freshwater input was also documented by Lough et al. (2015) based on luminescence records of Burdekin River flux (Figure 8D). However, the GBR fresh-water flux also correlates with proximal anthropogenic land-use change, so this may not be indicative only of climate. More recently, post-1900 temperature records of Wu et al. (2021) are mostly, but are not all, consistent with the 20th-century increase in SST (Tierney et al. 2015), but GBR proxy records of SST varied geographically (Wu et al. 2021). Therefore, studies of broader geographic effects on the distribution of SST over time are needed in the GBR, although none of the palaeothermometres appear to be affected by anthropogenic differences in coastal runoff.

Kamenos and Hennige (2018) interpreted anomalous density bands (reduced linear extension) as thermal bleaching stress bands in 44 GBR cores spanning from 1620–2001 and suggested that bleaching increased (1620–1753), decreased (1754–1820), and then increased (1821–2001) in the 380-year record. An initial negative response to this interpretation (Hoegh-Guldberg et al. 2019; DeCarlo 2020) illustrates the many difficulties with interpreting the causes of individual stress bands in corals and in having adequate samples for robust statistical analysis (e.g., Hendy et al. 2003b). Kamenos and Hennige (2024) pointed out that where validation is easiest using the Tierney et al. (2015) curve, there is still a good case for thermal bleaching in the GBR beginning in the 19th century. Although acknowledging the increase in thermal bleaching observed in recent decades, they pointed out that there is little evidence to suggest that thermal bleaching was not responsible for observed stress bands in earlier times when direct bleaching observations were lacking. They highlight the need to investigate if coral acclimatization may have already caused shifting baselines in coral bleaching thresholds. Regardless, adequate geographic and temporal coverage to confidently identify mass bleaching events in the past may not be feasible at this time. Earlier centuries require significantly more cores to be analysed, especially with an independent palaeotemperature proxy context for individual stress bands (e.g., DeCarlo and Cohen 2017) (Figure 6G). De'ath et al. (2009) summarised long-term coral calcification rates for the GBR (Figure 8F) and noted an increase from the early 1600s followed by a decline since 1990. However, the record may be biased towards inshore reefs. Outer and mid-shelf reefs show less of a decline since 1990 (D'Olivo et al. 2013). A multi-century record of pH using $\delta^{11}B$ suggests decadal and interdecadal cycles until the 1940s when it began to decline more consistently (Figure 8E) (Wei et al. 2009; see also D'Olivo et al. 2019). Correlation with decreasing $\delta^{13}C$ suggests that increasing anthropogenic CO_2 is the driver (Wei et al. 2009). However, the pattern at Flinders Reef in the open Coral Sea is more stable over time, suggesting that nearshore waters may be more subject to reduced pH (Wei et al. 2009) and to terrigenous runoff and biological cycling effects in the plankton (D'Olivo et al. 2015). Combined, the palaeoclimate records for the GBR over the last several centuries are difficult to unify into a single climate history, most likely because of the large size and geomorphological complexity of the GBR combined with the uneven distribution and general sparsity of records.

Older, Holocene coral-based palaeoclimate archives are required to understand the drivers and dynamics of many key climate systems, such as ENSO (Emile-Geay et al. 2021), and to make better predictions of climate change going forward (Lawman et al. 2022). Such archives also provide direct evidence of the effects of specific changes on the corals and reef communities themselves (Webb and Webster 2022). Corals have provided significant information about snapshots of Holocene climate in many reefs around the world (e.g., McGregor and Gagan 2004; McGregor et al. 2013; Toth and Aronson 2019; Wang et al. 2021), but data from the GBR are few. Lough et al. (2014) suggested a weaker summer monsoon and possibly reduced ENSO variability during the mid-Holocene (~6,000 years before the present—yBP) of the central GBR based on two > 100-year records of coral luminescence. Leonard et al. (2016) also suggested weak ENSO variability between ~5,200 and 4,300 yBP based on *Porites* in the same region. They suggested that SSTs at that time were as much as 2.5 °C cooler than modern SSTs and those in the previous millennium (i.e., ~6,000 yBP, Lough et al. 2014), both of which accord with palaeo-SST proxy data from the southern GBR at Heron Reef (Sadler et al. 2016). A 32-year *Porites* dataset that indicates warmer temperatures at that time in the central GBR (Roche et al. 2014) may have been affected by reef flat ponding (Sadler et al. 2016). Southern GBR SSTs were also cooler than the present at ~7,000 yBP, thus constraining the mid-Holocene climate optimum in the GBR to ~6,800 to 6,000 yBP (Sadler et al. 2016). Although existing coral proxies are very informative for this interval, more core analyses are needed to improve the temporal and geographic coverage.

Even older data are required to improve modelling of glacial-interglacial scale climate and particularly the specific effects of warming on reefs, increasing atmospheric pCO_2, and sea level rise, which are major features of deglaciation. As coral records are analysed from Expedition 325 of the Integrated Ocean Drilling Program (Webster et al. 2018), details of GBR palaeoclimate from the last glacial maximum through the interglacial are becoming known. Data (Sr/Ca, $\delta^{18}O$) obtained from glacial and deglacial *Isopora* colonies at two Expedition 325 GBR sites separated by 3° of latitude suggest a steeper latitudinal SST gradient during the deglacial than at present (Felis et al. 2014). These authors attributed this to a possible weakening of the South Pacific gyre and expansion of cooler southern waters northward into the GBR. Brenner et al. (2020) used new calibration curves on the same *Isopora* data to suggest that the Younger Dryas (~12.9–11.7 yBP) was not cooler in the GBR, as in the northern hemisphere, but had similar SSTs to those of today. The results serve as a reminder of both the physical scale of the GBR and the complexities in the ocean-climate system over glacial cycles and the need to obtain broad geographic datasets for modelling.

5.5. CONCLUSIONS AND THE WAY FORWARD

Tremendous progress has been made in our scientific understanding of coral biomineralisation, elemental and isotopic cycles in terrigenous-marine and atmosphere-marine processes (e.g., Figure 6), and in our ability to analyse corals and recover useful environmental proxies. However, due in part to the great size and geographic differentiation of the GBR along latitude and from onshore to offshore, our record of climate over the last many centuries is limited to rather sparse datasets that may better reflect small, localised regions of the GBR; in some cases, these provide more questions than answers. For example, why are existing palaeo-temperature records for the last centuries so out of step with the southwest Pacific regional average of Tierney et al. (2015) (Figures 8A, B)? If changes in latitudinal heat gradients are responsible (Hendy et al. 2002), which were also thought to have affected SSTs in the GBR during the deglacial (Felis et al. 2014), then large-scale oceanographic and climatic drivers may enhance highly regionalised differences in the temporal trends and values within the greater GBR. A key take-home message of this chapter should be that the GBR is too large and complex to be treated as a single entity where palaeoclimate studies are concerned. The following points are offered as a way forward to guide future research.

1. Coral-based palaeoenvironmental archives need to be extended latitudinally along the GBR, both inshore and offshore, in the adjoining Coral Sea, and back in time through the Holocene. This will require the targeting of new cores, both from individual coralla and from deeper reef coring, and additional types of geochemical analysis on many existing cores, which have so far been analysed only for calcification rates and luminescence.
2. At the other end of the spatial scale, continued improvement of our understanding of the biochemistry of element uptake and isotope fractionation in corals is required, especially as it relates to the local context and vital effects where they govern the distribution of elements within the skeleton and/or organic matrix. It is critical to try to match understanding of coral biomineralisation with physical scales appropriate for our increasing resolution of measurement (e.g., LA-ICP-MS, SIMS). Accordingly, we also must ensure that amount of time averaging within the sample is appropriate for the problem being addressed. Sub-weekly to annual scales of analytical measurement (i.e., averaging) are useful for addressing different questions.
3. Improved integration of datasets (extension, density, calcification, luminescence, trace elements, and isotopes) for individual contemporary corals in close proximity and in slightly different settings will aid the unravelling of the effects of the local context (i.e., noise and bias) on secular records. At the same time, new experimental approaches will help isolate variables and allow better modelling of environmental parameters from proxies in complex settings.
4. Finally, and of a timely nature, the many advances in understanding coral proxies now make it possible to increase the resolution of geochemical analyses associated with individual stress bands in corals in order to interpret their individual causes. This should allow us to better evaluate the history of, and resilience to, different specific types of non-lethal and lethal environmental stressors. Improved understanding of sublethal stress effects in corals will greatly aid our ability to predict the potential outcomes of projected climate change scenarios.

REFERENCES

Abram NJ, Hargreaves JA, Wright NM, et al. (2020) Palaeoclimate perspectives on the Indian Ocean Dipole. Quaternary Science Reviews 237:106302.

Bainbridge S, Berkelmans R (2014) The use of climatologies and Bayesian models to link observations to outcomes; an example from the Torres Strait. Environmental Science: Processes & Impacts 16:1041–1049

Barkley HC, Cohen AL (2016) Skeletal records of community-level bleaching in *Porites* corals from Palau. Coral Reefs 35:1407–1417.

Baumann JH, Ries JB, Rippe JP, et al. (2019) Nearshore coral growth declining on the Mesoamerican Barrier Reef System. Global Change Biology 25:3932–3945.

Bolouki Kourandeh M, Nabavi SMB, Shokri MR, et al. (2021) Trace metal content in annually banded scleractinian coral '*Porites lobata*' across the northern Persian Gulf. Environmental Science and Pollution Research 28:61008–61020.

Brahmi C, Kopp C, Domart-Coulon I, et al. (2012) Skeletal growth dynamics linked to trace-element composition in the scleractinian coral *Pocillopora damicornis*. Geochimica et Cosmochimica Acta 99:146–158.

Brenner LD, Linsley BK, Webster JM, et al. (2020) Coral record of Younger Dryas Chronozone warmth on the Great Barrier Reef. Paleoceanography and Paleoclimatology 35:e2020 PA003962.

Brodie J, Kroon F, Schaffelke B, et al. (2012) Terrestrial pollutant runoff to the Great Barrier Reef: An update of issues, priorities and management responses. Marine Pollution Bulletin 65:81–100.

Cantin NE, Lough JM (2014) Surviving coral bleaching events: *Porites* growth anomalies on the Great Barrier Reef. PLoS One 9(2): e88720.

Cárdenas A, Raina J-B, Pogoreutz C, et al. (2022) Greater functional diversity and redundancy of coral endolithic microbiomes align with lower coral bleaching susceptibility. ISME Journal 16:2406–2420.

Carré M, Braconnot P, Elliot M, et al. (2021) High-resolution marine data and transient simulations support orbital forcing of ENSO amplitude since the mid-Holocene. Quaternary Science Reviews 268:107125.

Chalk TB, Standish CD, D'Angelo C, et al. (2021) Mapping coral calcification strategies from in situ boron isotope and trace element measurements of the tropical coral *Siderastrea siderea*. Scientific Reports 11:472.

Chapman AU, Thompson DM, Hlohowskyj SR, et al. (2022) A mechanistic investigation of the coral Mn/Ca-based trade-wind proxy at Kiritimati. Geochimica et Cosmochimica Acta 328:58–75.

Chen T, Chen T, Robinson LF, et al. (2016) Ocean mixing and ice-sheet control of seawater $^{234}U/^{238}U$ during the last deglaciation. Science 1015:626–629.

Chen T, Cobb KM, Roff G, et al. (2018) Coral-derived western Pacific tropical sea surface temperatures during the last millennium. Geophysical Research Letters 45: 3542–3549.

Chen T, Li S, Zhao JX, Feng YX (2021) Uranium-thorium dating of coral mortality and community shift in a highly disturbed inshore reef (Weizhou Island, northern South China Sea). Science of the Total Environment 752:141866.

Cheng H, Lawrence Edwards R, Shen CC, et al. (2013) Improvements in ^{230}Th dating, ^{230}Th and ^{234}U half-life values, and U-Th isotopic measurements by multi-collector inductively coupled plasma mass spectrometry. Earth and Planetary Science Letters 371–372:82–91.

Cheung AH, Cole JE, Thompson DM, et al. (2021) Fidelity of the Coral Sr/Ca paleothermometer following heat stress in the Northern Galápagos. Paleoceanography and Paleoclimatology 36:e2021PA004323.

Chutcharavan PM, Dutton A, Ellwood MJ (2018) Seawater $^{234}U/^{238}U$ recorded by modern and fossil corals. Geochimica et Cosmochimica Acta 224:1–17.

Clark TR, Roff G, Zhao J, et al. (2017) U-Th dating reveals regional-scale decline of branching *Acropora* corals on the Great Barrier Reef over the past century. Proceedings of the National Academy of Sciences of the United States of America 114:10350–10355.

Clark TR, Roff G, Zhao JX, et al. (2014a) Testing the precision and accuracy of the U-Th chronometer for dating coral mortality events in the last 100 years. Quaternary Geochronology 23:35–45.

Clark TR, Zhao J, Roff G, et al. (2014b) Discerning the timing and cause of historical mortality events in modern *Porites* from the Great Barrier Reef. Geochimica et Cosmochimica Acta 138:57–80.

Clark TR, Zhao JX, Feng YX, et al. (2012) Spatial variability of initial $^{230}Th/^{232}Th$ in modern *Porites* from the inshore region of the Great Barrier Reef. Geochimica et Cosmochimica Acta 78:99–118.

Connell JH (1978) Diversity in tropical rain forests and coral reefs. Science 199:1302–1310.

Coronado I, Fine M, Bosellini FR, Stolarski J (2019) Impact of ocean acidification on crystallographic vital effect of the coral skeleton. Nature Communication 10:2896.

Cuny-Guirriec K, Douville E, Reynaud S, et al. (2019) Coral Li/Mg thermometry: Caveats and constraints. Chemical Geology 523:162–178.

Davies PJ, Symonds PA, Feary DA, Pigram CJ (1989) The evolution of the carbonate platforms of northeast Australia. In Crevello PD, Wilson JL, Sarg JF. Read JF (eds.), Controls on Carbonate Platforms and Basin Development, SEPM Special Publication 44, SEPM Society for Sedimentary Geology, Tulsa, OK, 233–258.

De'ath G, Lough JM, Fabricius KE (2009) Declining coral calcification on the Great Barrier Reef. Science 323:116–119.

DeCarlo TM (2017) Deriving coral skeletal density from computed tomography (CT): Effects of scan and reconstruction settings. Matters Select 3:e201706000005.

DeCarlo TM (2020) Commentary: Reconstructing four centuries of temperature-induced coral bleaching on the Great Barrier Reef. Frontiers in Marine Science 7:30.

DeCarlo TM, Cohen AL (2017) Dissepiments, density bands and signatures of thermal stress in *Porites* skeletons. Coral Reefs 36:749–761.

DeCarlo TM, Gaetani GA, Cohen AL, et al. (2016) Coral Sr-U thermometry. Paleoceanography 31:626–638.

DeCarlo TM, Harrison HB (2019) An enigmatic decoupling between heat stress and coral bleaching on the Great Barrier Reef. PeerJ 7:e7473.

DeCarlo TM, Harrison HB, Gajdzik L, et al. (2019) Acclimatization of massive reef-building corals to consecutive heatwaves. Proceedings of the Royal Society B: Biological Sciences 286:20190235.

DeCarlo TM, Holcomb M, McCulloch MT (2018) Reviews and syntheses: Revisiting the boron systematics of aragonite and their application to coral calcification. Biogeoscience 15:2819–2834.

Dechnik B, Webster JM, Webb GE, et al. (2017) Successive phases of Holocene reef flat development: Evidence from the mid- to outer Great Barrier Reef. Palaeogeography, Palaeoclimatology, Palaeoecology 466:221–230.

DeLong KL, Quinn TM, Taylor FW, et al. (2013) Improving coral-base paleoclimate reconstructions by replicating 350 years of coral Sr/Ca variations. Palaeogeography, Palaeoclimatology, Palaeoecology 373:6–24.

Deng W, Wei G, McCulloch M, et al. (2014) Evaluation of annual resolution coral geochemical records as climate proxies in the Great Barrier Reef of Australia. Coral Reefs 33:965–977.

Devlin MJ, McKinna LW, Álvarez-Romero JG, et al. (2012) Mapping the pollutants in surface riverine flood plume waters in the Great Barrier Reef, Australia. Marine Pollution Bulletin 65:224–235.

Dishon G, Fisch J, Horn I, et al. (2015) A novel paleo-bleaching proxy using boron isotopes and high-resolution laser ablation to reconstruct coral bleaching events. Biogeoscience 12:5677–5687.

D'Olivo JP, Ellwood G, DeCarlo TM, McCulloch MT (2019) Deconvolving the long-term impacts of ocean acidification and warming on coral biomineralisation. Earth and Planetary Science Letters 526:115785.

D'Olivo JP, McCulloch MT (2022) Impact of European settlement and land use changes on Great Barrier Reef river catchments reconstructed from long-term coral Ba/Ca records. Science of the Total Environment 830:154461.

D'Olivo JP, McCulloch MT, Eggins SM, Trotter J (2015) Coral records of reef-water pH across the central Great Barrier Reef, Australia: Assessing the influence of river runoff on inshore reefs. Biogeosci 12:1223–1236.

D'Olivo JP, McCulloch MT, Judd K (2013) Long-term records of coral calcification across the central Great Barrier Reef: Assessing the impacts of river runoff and climate change. Coral Reefs 32:999–1012.

D'Olivo JP, Sinclair DJ, Rankenburg K, McCulloch MT (2018) A universal multi-trace element calibration for reconstructing sea surface temperatures from long-lived *Porites* corals: Removing 'vital-effects'. Geochimica et Cosmochimica Acta 239:109–135.

Domart-Coulon I, Stolarski J, Brahmi C, et al. (2014) Simultaneous extension of both basic microstructural components in scleractinian coral skeleton during night and daytime, visualized by in situ ^{86}Sr pulse labeling. Journal of Structural Biology 185:79–88.

Duke NC, Canning A, Mackenzie A (2024) More intense severe tropical cyclones in recent decades cause greater impacts on mangroves bordering Australia's Great Barrier Reef. In Wolanski E, Kingsford MJ (eds.), Oceanographic Processes of Coral Reefs: Physical and Biological Links in the Great Barrier Reef. CRC Press, Boca Raton, FL.

Dunne RP, Brown BE, Spencer T, Woodworth PL (2022) Coral bleaching at Low Isles during the 1928–1929 Great Barrier Reef Expedition. Coral Reefs. doi: 10.1007/s00338-022-02327-7.

Ellis B, Grant K, Mallela J, Abram N (2019) Is XRF core scanning a viable method for coral palaeoclimate temperature reconstructions? Quaternary International 514:97–107.

Emile-Geay J, Cobb KM, Cole JE, et al. (2021) Past ENSO Variability: Reconstructions, models, and implications (In El Niño Southern Oscillation in a Changing Climate, eds. McPhaden MJ, Santoso A, Cai W). Geophysical Monograph 253:87–118.

Erler DV, Farid HT, Glaze TD, et al. (2020a) Coral skeletons reveal the history of nitrogen cycling in the coastal Great Barrier Reef. Nature Communications 11(1):1500.

Erler DV, Rangel MS, Tagliafico A, et al. (2020b) Can coral skeletal-bound nitrogen isotopes be used as a proxy for past bleaching? Biogeochem 151:31–41.

Erler DV, Wang XT, Sigman DM, et al. (2016) Nitrogen isotopic composition of organic matter from a 168 year-old coral skeleton: Implications for coastal nutrient cycling in the Great Barrier Reef Lagoon. Earth and Planetary Science Letters 434:161–170.

Eyre BD, Cyronak T, Drupp P, et al. (2018) Coral reefs will transition to net dissolving before end of century. Science 359:908–911.

Farfan GA, Apprill A, Cohen A, et al. (2022) Crystallographic and chemical signatures in coral skeletal aragonite. Coral Reefs 41:19–34.

Felis T (2020) Extending the instrumental record of ocean-atmosphere variability into the last interglacial using tropical corals. Journal of Oceanography 33:68–79.

Felis T, McGregor HV, Linsley BK, et al. (2014) Intensification of the meridional temperature gradient in the Great Barrier reef following the last glacial maximum. Nature Communications 5(1):4102.

Fine M, Hoegh-Guldberg O, Meroz-Fine E, Dove S (2019) Ecological changes over 90 years at Low Isles on the Great Barrier Reef. Nature Communications 10:4409.

Fordyce AJ, Ainsworth TD, Page CE, et al. (2021) Microalgal blooms in the skeletons of bleached corals during the 2020 bleaching event on Heron Island, Australia. Marine and Freshwater Research 72:1689–1694.

Frankowiak K, Kret S, Mazur M, et al. (2016) Fine-scale skeletal banding can distinguish symbiotic from asymbiotic species among modern and fossil scleractinian corals. PLoS One 11(1):e0147066.

Gagan MK, Dunbar GB, Suzuki A (2012) The effect of skeletal mass accumulation in *Porites* on coral Sr/Ca and $\delta^{18}O$ paleothermometry. Paleoceanography 27.

Glynn PW, Manzello DP (2015) Bioerosion and coral reef growth: A dynamic balance. In Birkeland C (ed.), Coral Reefs in the Anthropocene, Springer, Dordrecht, 67–97.

Grothe PR, Cobb KM, Bush SL, et al. (2016) A comparison of U/Th and rapid-screen ^{14}C dates from Line Island fossil corals. Geochemistry, Geophysics, Geosystems 17(3):833–845.

Grothe PR, Cobb KM, Liguori G, et al. (2020) Enhanced El Niño–Southern Oscillation variability in recent decades. Geophysical Research Letters 47:e2019GL083906.

Haig J, Nott J, Reichart G-J (2014) Australian tropical cyclone activity lower than at any time over the past 550–1,500 years. Nature 505:667–671.

Hamylton SM, McLean R, Bell IP, Thomas CR (2024) Biological and geological links on coral reef islands. In Wolanski E, Kingsford MJ (eds.), Oceanographic Processes of Coral Reefs: Physical and Biological Links in the Great Barrier Reef. CRC Press, Boca Raton, FL.

Hendy EJ, Gagan MK, Alibert CA, et al. (2002) Abrupt decrease in tropical Pacific sea surface salinity at end of Little Ice Age. Science 295:1511–1514.

Hendy EJ, Gagan MK, Lough JM (2003a) Chronological control of coral records using luminescent lines and evidence for nonstationary ENSO teleconnections in northeast Australia. Holocene 13:187–199.

Hendy EJ, Lough JM, Gagan MK (2003b) Historical mortality in massive *Porites* from the central Great Barrier Reef, Australia: Evidence for past environmental stress? Coral Reefs 22:207–215.

Hernández A, Martin-Puertas C, Moffa-Sánchez P, et al. (2020) Modes of climate variability: Synthesis and review of proxy-based reconstructions through the Holocene. Earth-Science Reviews 209:103286.

Hetzinger S, Pfeiffer M, Dullo W-C, et al. (2016) A change in coral extension rates and stable isotopes after El Niño-induced coral bleaching and regional stress events. Science Reports 6:32879.

Hibbert FD, Rohling E J, Dutton A, et al. (2016) Coral indicators of past sea-level change: A global repository of U-series dated benchmarks. Quaternary Science Reviews 145:1–56.

Hoegh-Guldberg O (2005) Low coral cover in a high-CO_2 world. Journal of Geophysical Research 110:C09S06.

Hoegh-Guldberg O, Poloczanska ES, Skirving W, Dove S (2017) Coral reef ecosystems under climate change and ocean acidification. Frontiers in Marine Science 4:158.

Hoegh-Guldberg O, Skirving WJ, Lough JM, et al. (2019). Commentary: Reconstructing four centuries of temperature-induced coral bleaching on the Great Barrier Reef. Frontiers in Marine Science 6:86.

Hsieh YT, Paver R, Tanzil JTI, et al. (2022) Multi-colony calibration of barium isotopes between shallow-water coral skeletons and in-situ seawater: Implications for paleo proxies. Earth and Planetary Science Letters 580:117369.

Hua Q, Ulm S, Yu K, et al. (2020) Temporal variability in the Holocene marine radiocarbon reservoir effect for the Tropical and South Pacific. Quaternary Science Reviews 249:106613.

Hua Q, Webb GE, Zhao J-x, et al. (2015) Large variations in the Holocene marine radiocarbon reservoir effect reflect ocean circulation and climatic changes. Earth Planet Science Letters 422:33–44.

Hughes TP, Kerry JT, Baird AH, et al. (2018) Global warming transforms coral reef assemblages. Nature 556:492–496.

Hughes TP, Kerry JT, Connolly SR, et al. (2021) Emergent properties in the responses of tropical corals to recurrent climate extremes. Current Biology 31:5393–5399.

Jurikova H, Libetrau V, Raddatz J, et al. (2019) Boron isotope composition of the cold-water coral *Lophelia pertusa* along the Norwegian margin: Zooming into a potential pH-proxy by combining bulk and high-resolution approaches. Chemical Geology 513:143–152.

Kamenos NA, Hennige SJ (2018) Reconstructing four centuries of temperature-induced coral bleaching on the Great Barrier Reef. Frontiers in Marine Science 5:283.

Kamenos NA, Hennige SJ (2020) Commentary: Commentary on Reconstructing Four Centuries of Temperature-Induced Bleaching on the Great Barrier Reef by Hoegh-Guldberg et al. 2019 and DeCarlo 2020. Frontiers in Marine Science 7:570620.

Kamenos NA, Hennige SJ (2024) A historical perspective on thermal- and heatwave-induced bleaching on the Great Barrier Reef. In Wolanski E, Kingsford MJ (eds.), Oceanographic Processes of Coral Reefs: Physical and Biological Links in the Great Barrier Reef. CRC Press, Boca Raton, FL.

Karnauskas KB (2020) Physical diagnosis of the 2016 Great Barrier Reef bleaching event. Geophysical Research Letters 47:e2019GL086177.

Kaushal N, Sanwlani N, Tanzil JTI, et al. (2021) Coral skeletal luminescence records changes in terrestrial chromophoric dissolved organic matter in tropical coastal waters. Geophysical Research Letters 48:e2020GL092130.

Kiessling W, Smith JA, Raja NB (2023) Improving the relevance of paleontology to climate change policy. Proceedings of the National Academy of Sciences of the United States of America 120:e2201926119.

Knutson T, Camargo SJ, Chan JC, et al. (2020) Tropical cyclones and climate change assessment: Part II: Projected response to anthropogenic warming. Bulletin of the American Meteorological Society 101:E303–E322.

Lawman AE, Di Nezio PN, Partin JW, et al. (2022) Unraveling forced responses of extreme El Niño variability over the Holocene. Science Advances 8:eabm4313.

Le Campion-Alsumard T, Golubic S, Hutchings P (1995) Microbial endoliths in skeletons of live and dead corals: *Porites lobata* (Moorea, French Polynesia). Marine Ecology Progress Series 117:149–157.

Leonard ND, Lepore ML, Zhao JX, et al. (2020a) A U-Th dating approach to understanding past coral reef dynamics and geomorphological constraints on future reef growth potential; Mazie Bay, Southern Great Barrier Reef. Paleoceanography and Paleoclimatology 35:e2019PA003768.

Leonard ND, Lepore ML, Zhao JX, et al. (2020b) Re-evaluating mid-Holocene reef "turn-off" on the inshore Southern Great Barrier Reef. Quaternary Science Reviews 244:106518.

Leonard ND, Welsh KJ, Lough JM, et al. (2016) Evidence of reduced mid-Holocene ENSO variance on the Great Barrier Reef, Australia. Paleoceanography 31:1248–1260.

Leonard ND, Welsh KJ, Nguyen AD, et al. (2019) High resolution geochemical analysis of massive *Porites* spp. corals from the Wet Tropics, Great Barrier Reef: Rare earth elements, yttrium and barium as indicators of terrigenous input. Marine Pollution Bulletin 149:110634.

Lewis SE, Lough JM, Cantin NE, et al. (2018) A critical evaluation of coral Ba/Ca, Mn/Ca and Y/Ca ratios as indicators of terrestrial input: New data from the Great Barrier Reef, Australia. Geochimica et Cosmochimica Acta 237:131–154.

Lewis SE, McCloskey G, Bainbridge Z, et al. (2024) Sediment and nutrient flux from land. In Wolanski E, Kingsford MJ (eds.), Oceanographic Processes of Coral Reefs: Physical and Biological Links in the Great Barrier Reef. CRC Press, Boca Raton, FL.

Lewis SE, Shields GA, Kamber BS, Lough JM (2007) A multitrace element coral record of land-use changes in the Burdekin River catchment, NE Australia. Palaeogeography, Palaeoclimatology, Palaeoecology 246:471–487.

Li S, Yu K, Zhao J, et al. (2017) Trace metal anomalies in bleached *Porites* coral at Meiji Reef, tropical South China Sea. Chinese Journal of Oceanology and Limnology 35:115–121.

Liu ET, Zhao JX, Clark TR, et al. (2014) High-precision U–Th dating of storm-transported coral blocks on Frankland Islands, northern Great Barrier Reef, Australia. Palaeogeography, Palaeoclimatology, Palaeoecology 414:68–78.

Lough JM (2007) Tropical river flow and rainfall reconstructions from coral luminescence: Great Barrier Reef, Australia. Paleoceanography 22:PA2218.

Lough JM (2010) Climate records from corals. Wiley Interdisciplinary Reviews Climate Change 1:318–331.

Lough JM (2011) Great Barrier Reef coral luminescence reveals rainfall variability over northeastern Australia since the 17th century. Paleoceanography 26:PA2201.

Lough JM (2016) *Coral Core Records of the North-east Torres Strait*. Report to the National Environmental Science Programme. Reef and Rainforest Research Centre Limited, Cairns, 24pp.

Lough JM, Cooper TF (2011) New insights from coral growth band studies in an era of rapid environmental change. Earth-Science Reviews 108:170–184.

Lough JM, Lewis SE, Cantin NE (2015) Freshwater impacts in the central Great Barrier Reef: 1648–2011. Coral Reefs 34:739–751.

Lough JM, Llewellyn LE, Lewis SE, et al. (2014) Evidence for suppressed mid-Holocene northeastern Australian monsoon variability from coral luminescence. Paleoceanography 29:581–594.

McCulloch M, Fallon S, Wyndham T, et al. (2003) Coral record of increased sediment flux to the inner Great Barrier Reef since European settlement. Nature 421:727–730.

McGowan H, Theobald A (2017) ENSO weather and coral bleaching on the Great Barrier Reef, Australia. Geophysical Research Letters 44:10601–10607.

McGregor HV, Fischer MJ, Gagan MK, et al. (2013). A weak El Niño/Southern Oscillation with delayed seasonal growth around 4,300 years ago. Nature Geoscience 6:949–953.

McGregor HV, Gagan MK (2004) Western Pacific coral $\delta^{18}O$ records of anomalous Holocene variability in the El Niño--Southern Oscillation. Geophysical Research Letters 31:L11204.

Mohtar AM, Hughen KA, Goodkin NF, et al. (2021) Coral-based proxy calibrations constrain ENSO-driven sea surface temperature and salinity gradients in the Western Pacific Warm Pool. Palaeogeography, Palaeoclimatology, Palaeoecology 561:110037.

Mollica NR, Cohen AL, Alpert AE, et al. (2019) Skeletal records of bleaching reveal different thermal thresholds of Pacific coral reef assemblages. Coral Reefs 38:743–757.

Mollica NR, Guo W, Cohen AL, et al. (2018) Ocean acidification affects coral growth by reducing skeletal density. Proceedings of the National Academy of Sciences of the United States of America 115:1754–1759.

Murphy R, Webster JM, Nothdurft L, et al. (2017) High-resolution hyperspectral imaging of diagenesis and clays in fossil coral reef material: A nondestructive tool for improving environmental and climate reconstructions. Geochemistry, Geophysics, Geosystems 18:3209–3230.

Neil DT, Orpin AR, Ridd PV, Yu B (2002) Sediment yield and impacts from river catchments to the Great Barrier Reef lagoon: A review. Marine and Freshwater Research 53:733–752.

Nothdurft LD, Webb GE (2007) Microstructure of common reef-building coral genera *Acropora*, *Pocillopora*, *Goniastrea* and *Porites*: Constraints on spatial resolution in geochemical sampling. Facies 53:1–26.

Nothdurft LD, Webb GE (2009) Earliest diagenesis in scleractinian coral skeletons: Implications for palaeoclimate-sensitive geochemical archives. Facies 55:161–201.

Nott J, Hayne M (2001) High frequency of 'super-cyclones' along the Great Barrier Reef over the past 5000 years. Nature 413:508–512.

Pandolfi JM (1996) Limited membership in Pleistocene reef coral assemblages from the Huon Peninsula, Papua New Guinea: Constancy during global change. Paleobiology 22:152–176.

Pandolfi JM, Kiessling W (2014) Gaining insights from past reefs to inform understanding of coral reef response to global climate change. Current Opinion in Environmental Sustainability 7:52–58.

Perry CT, Hepburn LJ (2008) Syn-depositional alteration of coral reef framework through bioerosion, encrustation, and cementation: Taphonomic signatures of reef accretion and reef depositional events. Earth-Science Reviews 86:106–144.

Perry CT, Macdonald IA (2002) Impacts of light penetration on the bathymetry of reef microboring communities: Implications for the development of microendolithic trace assemblages. Palaeogeography, Palaeoclimatology, Palaeoecology 186:101–113.

Perry CT, Smithers SG (2011) Cycles of coral reef 'turn-on', rapid growth and 'turn-off' over the past 8500 years: A context for understanding modern ecological states and trajectories. Global Change Biology 17:76–86.

Petus C, Waterhouse J, Lewis S., et al. (2019) A flood of information: Using Sentinel-3 water colour products to assure continuity in the monitoring of water quality trends in the Great Barrier Reef (Australia). Journal of Environmental Management 248:109255.

Petus C, Waterhouse J, Tracey D, Wolanski E, Brodie J (2022) Using optical water-type classification in data-poor water quality assessment: A case study in the Torres Strait. Remote Sensing 14:22212.

Pörtner HO, Karl D, Boyd PW, et al. (2014). Ocean systems. pp. 411–484 in Field CB, Barros VR, Dokken DJ, et al. (eds.), Climate Change 2014: Impacts, Adaptation, and Vulnerability Part A: Global and Sectoral Aspects Contribution of Working Group II to the Fifth Assessment Report of the Intergovernmental Panel of Climate Change. Cambridge University Press, Cambridge, NY.

Raddatz J, Ruggeberg A, Flogel S, et al. (2014) The influence of seawater pH on U/Ca ratios in the scleractinian cold-water coral *Lophelia pertusa*. Biogeoscience 11:1863–1871.

Rahaman W, Tarique M, Fousiya AA, et al. (2022) Tracing impact of El Niño Southern Oscillation on coastal hydrology using coral $^{87}Sr/^{86}Sr$ record from Lakshadweep, South-Eastern Arabian Sea. Science Total of the Environment. doi: 10.1016/j.scitotenv.2022.157035

Ramos RD, Goodkin NF, Fan T-Y (2020) Coral records at the Northern Edge of the Western Pacific warm pool reveal multiple drivers of sea surface temperature, salinity, and rainfall variability since the end of the little ice age. Paleoceanography and Paleoclimatology 35:e2019PA003826.

Razak TB, Roff G, Lough JM, Mumby PJ (2020) Growth responses of branching versus massive corals to ocean warming on the Great Barrier Reef, Australia. Science of the Total Environment 705:135908.

Reed EV, Cole JE, Lough JM, Cantin NE (2019) Linking climate variability and growth in coral skeletal records from the Great Barrier Reef. Coral Reefs 38:29–43.

Reed EV, Thompson DM, Cole JE, et al. (2021) Impacts of coral growth on geochemistry: Lessons from the Galápagos Islands. Paleoceanography and Paleoclimatology 36:e2020PA004051.

Ricci F, Marcelino VR, Blackall LL, et al. (2019) Beneath the surface: Community assembly and functions of the coral skeleton microbiome. Microbiome 7:1–10.

Roche RC, Perry CT, Smithers SG, et al. (2014) Mid-Holocene sea surface conditions and riverine influence on the inshore Great Barrier Reef. Holocene 24:885–897.

Rodriguez-Ramirez A, Grove CA, Zinke J, et al. (2014) Coral luminescence identifies the pacific decadal oscillation as a primary driver of river runoff variability impacting the southern Great Barrier Reef. PLoS One 9:e84305.

Ross CL, Warnes A, Comeau S, et al. (2022) Coral calcification mechanisms in a warming ocean and the interactive effects of temperature and light. Communications Earth & Environment 3:72.

Ryan EJ, Smithers SG, Lewis SE, et al. (2016) The influence of sea level and cyclones on Holocene reef flat development: Middle Island, central Great Barrier Reef. Coral Reefs 35:805–818.

Sadekov A, Lloyd NS, Misra S, et al. (2019) Accurate and precise microscale measurements of boron isotope ratios in calcium carbonates using laser ablation multicollector-ICPMS. Journal of Analytical Atomic Spectrometry 34:550–560.

Sadler J, Webb GE, Leonard ND, et al. (2016) Reef core insights into mid-Holocene water temperatures of the southern Great Barrier Reef. Paleoceanography 31:1395–1408.

Sadler J, Webb GE, Nothdurft LD, Dechnik B (2014) Geochemistry-based coral palaeoclimate studies and the potential of 'non-traditional' (non-massive *Porites*) corals: Recent developments and future progression. Earth-Science Reviews 139:291–316.

Safaie A, Silbiger NJ, McClanahan TR, et al. (2018) High frequency temperature variability reduces the risk of coral bleaching. Nature Communication 9:1671.

Saha N, Rodriguez-Ramirez A, Nguyen AD, et al. (2018a) Seasonal to decadal scale influence of environmental drivers on Ba/Ca and Y/Ca in coral aragonite from the southern Great Barrier Reef. Science Total of the Environment 639:1099–1109.

Saha N, Webb GE, Christy AG, Zhao J-X (2019a) Vanadium in the massive coral *Porites*: A potential proxy for historical wood clearing and burning. Earth and Planetary Science Letters 527:115793.

Saha N, Webb GE, Zhao J-X (2016) Coral skeletal geochemistry as a monitor of inshore water quality. Science Total of the Environment 566–567:652–684.

Saha N, Webb GE, Zhao J-X, et al. (2018b) Influence of marine biochemical cycles on seasonal variation of Ba/Ca in the near-shore coral *Cyphastrea*, Rat Island, southern Great Barrier Reef. Chemical Geology 499:71–83.

Saha N, Webb GE, Zhao J-X, et al. (2019b) Coral-based high-resolution rare earth element proxy for terrestrial sediment discharge affecting coastal seawater quality, Great Barrier Reef. Geochimica et Cosmochimica Acta 254:173–191.

Saha N, Webb GE, Zhao J-X, et al. (2021) Spatiotemporal variation of rare earth elements from river to reef continuum aids monitoring of terrigenous sources in the Great Barrier Reef. Geochimica et Cosmochimica Acta 299:85–112.

Salas-Saavedra M, Webb GE, Sanborn K, et al. (2022) Holocene microbialite geochemistry records > 6000 years of secular influence of terrigenous flux on water quality for the southern Great Barrier Reef. Chemical Geology 604:120871.

Sayani HR, Cobb KM, Cohen AL, et al. (2011) Effects of diagenesis on paleoclimate reconstructions from modern and young fossil corals. Geochimica et Cosmochimica Acta 75:6361–6373.

Sayani HR, Cobb KM, DeLong K, et al. (2019) Intercolony $\delta^{18}O$ and Sr/Ca variability among *Porites* spp. corals at Palmyra Atoll: Toward more robust coral-based estimates of climate. Geochemistry, Geophysics, Geosystems 20: 5270–5284.

Sayani HR, Cobb KM, Monteleone B, Bridges H. (2022) Accuracy and reproducibility of coral Sr/Ca SIMS timeseries in modern and fossil corals. Geochemistry, Geophysics, Geosystems 23:e2021GC010068.

Schoepf V, D'Olivo JP, Rigal C., et al. (2021) Heat stress differentially impacts key calcification mechanisms in reef-building corals. Coral Reefs 40:459–471.

Schoepf V, Jung MU, McCulloch MT, et al. (2020) Thermally variable, macrotidal reef habitats promote rapid recovery from mass coral bleaching. Frontiers in Marine Science 7:245.

Shen C-C, Wu C-C, Cheng H, et al. (2012) High-precision and high-resolution carbonate ^{230}Th dating by MC-ICP-MS with SEM protocols. Geochimica et Cosmochimica Acta 99:71–86.

Silverstein RN, Cunning R, Baker AC (2015) Change in algal symbiont communities after bleaching, not prior heat exposure, increases heat tolerance of reef corals. Global Change Biology 21:236–249.

Spreter PM, Reuter M, Mertz-Kraus R, et al. (2022) Calcification response of reef corals to seasonal upwelling in the northern Arabian Sea (Masirah Island, Oman). Biogeoscience 19:3559–3573.

Stolarski J, Coronado I, Murphy JG, et al. (2021) A modern scleractinian coral with a two-component calcite–aragonite skeleton. Proceedings of the National Academy of Sciences of the United States of America 118:e2013316117.

Thompson DM (2022) Environmental records from coral skeletons: A decade of novel insights and innovation. Wiley Interdisciplinary Reviews Climate Change 13:e745.

Tierney JE, Abram NJ, Anchukaitis KJ, et al. (2015) Tropical sea surface temperatures for the past four centuries reconstructed from coral archives. Paleoceanography 30:226–252.

Toth TL, Aronson RB (2019) The 4.2ka event, ENSO, and coral reef development. Climate of the Past 15:105–119.

Walter RM, Sayani HR, Felis T, et al. (2022) The CoralHydro2k Database: A global, actively curated compilation of coral δ^{18}O and Sr/Ca proxy records of tropical ocean hydrology and temperature for the Common Era. Earth System Science Data. doi: 10.5194/essd-2022-172.

Walther BD, Kingsford MJ, McCulloch MT (2013) Environmental records from Great Barrier Reef corals: Inshore versus offshore drivers. PLoS One 8:e77091.

Walther BD, Nims MK (2015) Spatiotemporal variation of trace elements and stable isotopes in subtropical estuaries: I. Freshwater endmembers and mixing curves. Estuaries Coasts 38:754–768.

Wang H, Yu K, Tao S, et al. (2021) New evidence for the periodic bleaching and recovery of *Porites* corals during the mid-late Holocene in the northern South China Sea. Global Planet Change 197:103387.

Webb GE, Nothdurft LD, Kamber BS, et al. (2009) Rare earth element geochemistry of scleractinian coral skeleton during meteoric diagenesis: A sequence through neomorphism of aragonite to calcite. Sedimentology 56:1433–1463.

Webb GE, Nothdurft LD, Zhao J-X, et al. (2016) Significance of shallow core transects for reef models and sea-level curves, Heron Reef, Great Barrier Reef. Sedimentology 63:1396–1424.

Webb GE, Webster J (2022) Scientific drilling on the Great Barrier Reef: Unlocking the history of the reef. In Hamylton SM, Hutchings P, Hoegh-Guldberg O (eds.), Coral Reefs of Australia: Perspectives from beyond the Water's Edge, CSIRO Publishing, Clayton, Vic, 124–130.

Webster JM, Braga JC, Humblet M, et al. (2018) Response of the Great Barrier Reef to sea-level and environmental changes over the past 30,000 years. Nature Geoscience 11:426–432.

Wei G, McCulloch MT, Mortimer G, et al. (2009) Evidence for ocean acidification in the Great Barrier Reef of Australia. Geochimica et Cosmochimica Acta 73:2332–2346.

Wolanski E, Kingsford M, Lambrechts J, Marmorino G (2024) The physical oceanography of the Great Barrier Reef: A review. In Wolanski E and Kingsford MJ (eds.), Oceanographic Processes of Coral Reefs: Physical and Biological Links in the Great Barrier Reef. CRC Press, Boca Raton, FL.

Wolfe K, Kenyon TM, Mumby PJ (2021) The biology and ecology of coral rubble and implications for the future of coral reefs. Coral Reefs 40:1769–1806.

Wolff NH, Wong A, Vitolo R, et al. (2016) Temporal clustering of tropical cyclones on the Great Barrier Reef and its ecological importance. Coral Reefs 35:613–623.

Wu Y, Fallon SJ, Cantin NF, Lough JM (2021) Assessing multiproxy approaches (Sr/Ca, U/Ca, Li/Mg, and B/Mg) to reconstruct sea surface temperature from coral skeletons throughout the Great Barrier Reef. Science of the Total Environment 786:147393.

Yu K-F, Zhao J-X, Roff G, et al. (2012a) High-precision U-series ages of transported coral blocks on Heron Reef (southern Great Barrier Reef) and storm activity during the past century. Palaeogeography, Palaeoclimatology, Palaeoecology 337–338:23–36.

Yu K-F, Zhao J-X, Shi Q, et al. (2006) U-series dating of dead *Porites* corals in the South China sea: Evidence for episodic coral mortality over the past two centuries. Quaternary Geochronology 1:129–141.

Yu K-F, Zhao J-X, Shi Q, Price GJ (2012b) Recent massive coral mortality events in the South China Sea: Was global warming and ENSO variability responsible? Chemical Geology 320–321:54–65.

Yu Y, Hathorne E, Siebert C, et al. (2022) Monthly resolved coral barium isotopes record increased riverine inputs during the South Asian summer monsoon. Geochimica et Cosmochimica Acta 329:152–167.

Zhao JX, Yu KF, Feng YX (2009) High-precision U-238-U-234-Th-230 disequilibrium dating of the recent past: A review. Quaternary Geochronology 4:423–433.

Zhao W, Huang Y, Siems S, Manton M (2021) The role of clouds in coral bleaching events over the Great Barrier Reef. Geophysical Research Letters 48:e2021GL093936.

Zinke J, Hoell A, Lough JM, et al. (2014) Coral record of southeast Indian Ocean marine heatwaves with intensified Western Pacific temperature gradient. Nature Communication 6:8562.

6 From the Microscale to the Reef

The Role of Microorganisms in the Chemical Ecology and Gaseous Emissions of the Great Barrier Reef

Jean-Baptiste Raina and Justin R. Seymour

6.1. INTRODUCTION

With a length greater than 2,300 km, the Great Barrier Reef (GBR) is the largest coral reef system on Earth (Wolanski et al. 2024). Comprised of 2,900 individual reefs, it is the largest biological structure on the planet, affecting physical, biological, and chemical processes far beyond its boundaries. The colossal size of the GBR means that it can influence atmospheric processes, and surveys over Australia identified it as a significant emitter of aerosol particles (Bigg and Turvey 1978; Modini et al. 2009), which are fine solid particles or liquid droplets that remain suspended in the air. These aerosols can be introduced into the atmosphere by bursting bubbles at the air-sea interface (Lewis et al. 2004) or through volatile organic compounds emitted by benthic invertebrates and microorganisms (Mahilang et al. 2021). Marine unicellular eukaryotes, bacteria, and archaea are especially important regulators of these gaseous emissions as they can act as both sources and sinks for these compounds (Peñuelas et al. 2014; Kramshøj et al. 2018; Tang et al. 2019). As some of these gases can impact the climatic processes of the GBR region, understanding their emission requires delving into the biology of the microorganisms involved in their release and the physics constraining their activity.

Each litre of seawater contains approximately 1 billion prokaryotic cells (bacteria and archaea) (Kallmeyer et al. 2012), and taken together, these microorganisms account for approximately 70% of the total marine biomass in the world's oceans (Bar-On Yinon et al. 2018). Because of their sheer abundance and metabolic versatility, these microorganisms play tremendously important roles in the health and ecology of marine systems by directly impacting primary productivity (Azam and Malfatti 2007), biogeochemical cycling (Azam 1998), and the production of climate-active molecules (Oremland Ronald 1975; Curson et al. 2011). However, contrary to our intuition, planktonic microorganisms and solutes are not homogeneously mixed in the ocean. Indeed, the seascape experienced by marine microorganisms is, in fact, quite heterogeneous and interspersed by chemical hotspots derived from planktonic cells, pelagic and benthic organisms, along with sinking and suspended organic particles (Stocker 2012; Clerc et al. 2022). This microscale heterogeneity, shaped by diffusion and flow (Guasto et al. 2012), influences the behaviour, physiology, and trophic interactions of these microorganisms and ultimately impacts their growth and the chemical transformations that they mediate (Azam and Malfatti 2007; Stocker et al. 2008; Smriga et al. 2016). Yet, the recognition of this microscale heterogeneity in the marine environment and its importance for microbial life is quite recent. As a result, the ecology of marine bacterial and archaeal communities on the GBR has rarely been studied at scales that properly consider the microenvironments they experience.

In coral reefs, the roles and impacts of marine prokaryotes have primarily been studied in the context of their association with benthic organisms (Bourne et al. 2016; Thomas et al. 2016). However, studies focusing on microbially mediated chemical transformations in the water column (Haas et al. 2016; Gao et al. 2020), at the water-sediment interface (Rusch and Gaidos 2013; Dong et al. 2022), or on the volatiles emitted into the atmosphere (Lawson et al. 2020) have started to unravel the roles that microbes play in the chemical ecology of coral reefs. In this chapter, we describe the biophysical processes that structure microenvironments at the scale most relevant for marine microbes and their influence on large-scale biogeochemical processes that affect the GBR. We highlight the impact of coral reef microbes on the cycling of carbon, nitrogen, and sulphur and their emission of volatile organic compounds, which can affect atmospheric processes beyond GBR boundaries.

6.2. FROM THE MACRO TO THE MICROSCALE

Microbial processes in the pelagic ocean have traditionally been investigated over large spatial scales, under the common (but incorrect) assumption that water motion and turbulence homogenise the distribution of microorganisms and solutes. Consequently, microbial oceanography is most often investigated within the context of large oceanographic features (e.g., currents, eddies, and gyres). Although these large-scale processes certainly determine the dynamics of marine microbial communities, the life of individual planktonic cells is most strongly influenced by

DOI: 10.1201/9781003320425-7

processes occurring at the microscale. At this very small scale, the environment experienced by these cells is controlled by physical rules that are extremely different from those that we experience and can therefore sometimes seem counterintuitive.

To understand why microorganisms are not homogenised by turbulence, it is important to consider how energy is transferred from large scales to smaller scales. Turbulent flow is composed of eddies of different sizes; the largest ones (i.e., several kilometres in size) have the largest amount of energy, while smaller eddies have very little energy. As large eddies break up, the energy that they contain is transferred to smaller eddies (Kolmogorov 1941), and this process continues until the smallest turbulent eddy size is reached (~10–30 mm in the upper mixed layer in the ocean) (Guasto et al. 2012) (Figure 1). Below this scale, viscous forces dominate over inertial forces, and the remaining energy is dissipated as heat (Figure 1). Importantly, the vast majority of planktonic microorganisms are smaller than the smallest turbulent eddies, which means that they typically do not experience chaotic changes in fluid velocity (i.e., turbulences).

Living in a world dominated by viscous forces means that water feels like honey for a bacterial cell. Most motile bacteria swim using helical flagella, and the prevalence of viscous forces is central to the way that these cells swim and

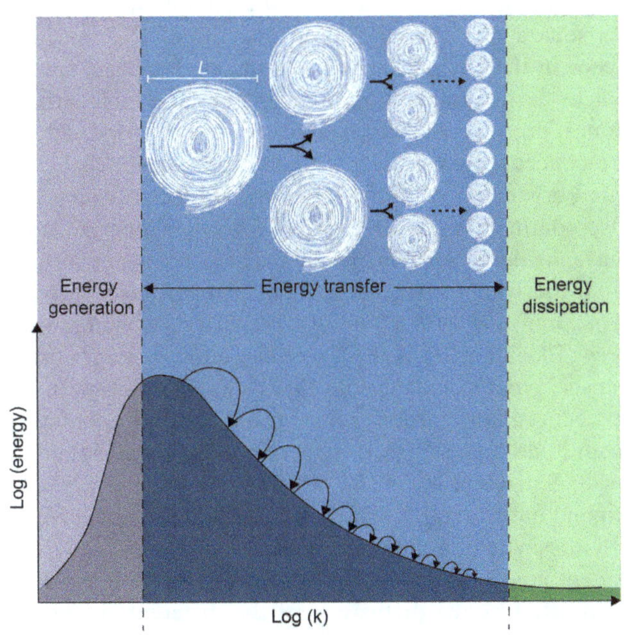

FIGURE 1 **Large eddies transfer their kinetic energy to smaller eddies through an energy cascade.** Three phases compose this energy cascade: (i) the injection range where eddies form (through instability processes); (ii) the inertial range where transfers of energy from large eddies (tens of km) to smaller ones (tens of mm) occur; and (iii) the dissipation range where viscous forces take over and convert energy into heat through friction. Parameter k represents the wave number ($k = 1/L$, where L is the size of the eddy).

to the evolution of their flagella (Lauga 2016). If a swimming cell stops rotating its flagellum, its movement will stop instantly (Purcell 1977). This is perhaps not intuitive, because at the human scale, when a boat stops its propeller, it keeps moving because of inertia. Importantly, the lack of inertia at the microscale means that non-motile microorganisms are transported by fluid flow as if they were water parcels. Therefore, unless they are motile or buoyant, marine microorganisms behave as true plankton or "drifters". As a consequence, fluid flow alone (including turbulence) cannot affect the distribution of non-motile microorganisms because water is incompressible, and microbes travel with the water.

6.3. THE MARINE ENVIRONMENT FROM THE PERSPECTIVE OF A MICROORGANISM

Physical processes occurring at the micrometre scale structure the local environments experienced by microorganisms and dictate how they can interact with one another. As a result, the chemical seascape experienced by microorganisms is not homogenous but is instead a mosaic of transient hotspots scattered in nutrient-poor waters. Here, we focus on the physical drivers of microscale heterogeneity in the ocean, namely, diffusion and fluid flow, to better emphasize their impact on the chemical transformations mediated by marine microorganisms. We then focus on the cell distance and encounter rates between microbial cells in the ocean, as these distances directly affect the interaction rates of these microorganisms.

6.3.1. Molecular Diffusion and Fluid Flow Shape Life at the Microscale

At the scale of marine microorganisms, the transport of dissolved molecules (including nutrients) is dominated by diffusion. This is in stark contrast with processes occurring over large spatial scales, where fluid flow is mostly responsible for transport. Although the effect of turbulence is intuitively associated with the mixing of water masses that ultimately generates homogeneous conditions, the real action of turbulence is to stir. This stirring increases the contact area between two water masses, but it is molecular diffusion that ultimately homogenises the two fluids. Indeed, at small scales, turbulence stirs, stretches, and folds a patch of nutrients (for example, from a fish excretion) into thin sheets and filaments (Figure 2a-b) (Taylor and Stocker 2012). These thin filaments become progressively thinner down to the Batchelor scale (the smallest scale at which turbulence can affect the distribution of dissolved molecules). In the marine environment, the Batchelor scale typically ranges from 30 to 300 μm (Guasto et al. 2012), which means that below this scale, molecular diffusion completes the homogenisation of these Batchelor-scale filaments.

The effect of turbulence not only decreases with the length scale but also is quenched by surfaces. The diffusion

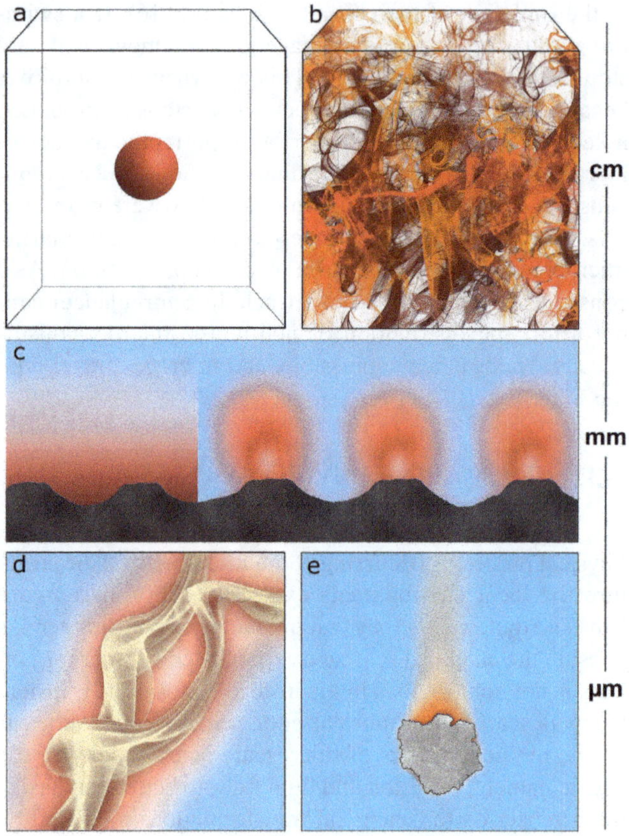

FIGURE 2 **Selected sources of patchiness and heterogeneity affecting marine microorganisms.** (a) The effect of turbulence on a small patch of nutrient creates (b) an intricate web of enriched filaments down to the Batchelor's scale (inspired from a simulation derived from (Stocker 2012)). (c) Diffusion of solute from the DBL of a coral with (right side) and without (left side) ciliary flow. (d) Coral mucus strings and filaments can vary in sizes from cm to mm and sometimes smaller. They contain high amounts of carbohydrates that diffuse out and create concentration gradients. (e) The sinking of particles creates comet-shaped gradients.

Source: Artwork produced by Paige Strudwick

boundary layer (DBL) is the region near a surface that is not affected by turbulence and where molecular diffusion dominates transport (Thar and Kühl 2003). DBLs are present around all surfaces in the ocean, ranging from small phytoplankton and marine snow particles to benthic invertebrates, sediments, and even large animals (Figure 2c). The thickness of the DBL around a surface is strongly affected by the strength of the surrounding flow. For example, the DBL around corals can be as thin as 200 µm under high flow velocity but larger than 1 cm in stagnant waters (Kühl et al. 1995).

In addition to affecting the width of the DBL, when the fluid flow is stronger, the solute transport to and from a surface is greater. This process enhances the photosynthetic rates of benthic organisms, such as corals or seagrasses, by increasing the oxygen flux from the organisms to the water (Mass et al. 2010). In addition to increasing photosynthesis, flow also positively affects multiple processes in corals, such as nutrient uptake (Baird and Atkinson 1997), respiration (Patterson et al. 1991), growth (Nakamura and Yamasaki 2005), and calcification (Schutter et al. 2011).

In some cases, the physical processes near the surfaces of benthic organisms are complicated further by the behaviour of the organisms. For instance, corals have the unique ability to actively stir their DBL using epidermal cilia, which means that they do not rely solely on external flow to enhance the transport of solutes from their boundary layer (Shapiro et al. 2014) (Figure 2c). This stirring ability is possibly an adaptative mechanism to overcome the detrimental health effects caused by weak ambient flow (Shapiro et al. 2014) or to mitigate oxidative stress (Pacherres et al. 2022). The vortices generated by ciliary flow on the coral surface impact the gradients arising from the DBL, and the resulting flow can reach speeds that far exceed the swimming capabilities of marine microorganisms (Shapiro et al. 2014); therefore, they likely affect how planktonic microorganisms interact with coral surfaces. However, for benthic organisms other than corals, the diffusive transport arising from DBLs results in steady gradients of solutes and thus contributes to the spatial heterogeneity at the microscale.

Other steady gradients can occur around mucus strings that are still attached to coral colonies (Figure 2d), or to positively buoyant mucus aggregates at the surface (Huettel et al. 2006). Mucus is released continuously by reef-building and soft corals, with *Acropora* exuding nearly 5 litres per square meter per day (Wild et al. 2004). Despite differences in the chemical compositions among coral species, mucus is invariably viscous and composed of carbohydrates, lipids, and proteins (Huettel et al. 2006), which create concentration gradients that can be exploited by reef microorganisms (Azam et al. 1983). Pulses of solutes are also common in coral reef waters and can generate strong but ephemeral nutrient patches. These patches can come from the lysis of microorganisms through viral infection (Riemann and Middelboe 2002), partial consumption of prey (Blackburn et al. 1997), or excretion by larger organisms (Lehman and Scavia 1982) and often contain dissolved organic carbon concentrations several orders of magnitude higher than the surrounding seawater (Fenchel 1988; Smriga et al. 2016). As time passes, these patches expand outwards through molecular diffusion and eventually dissipate as solute concentrations progressively reach background levels. The rate of expansion of the patches is dictated by the diffusivity of the different compounds released. Indeed, compounds with a high molecular weight diffuse more slowly and generate more persistent gradients than those with a low molecular weight. This difference can be substantial because the diffusivity of different cell constituents released during a lysis event can vary by two to four orders of magnitude.

Other unsteady gradients can arise when a nutrient source moves relative to the water. Common examples include the sinking of faecal pellets, mucus flocs, and particles (Silver et al. 1978; Kiørboe and Jackson 2001; Huettel et al. 2006). Unlike spherical gradients that arise from static diffusive sources (such as phytoplankton cells), the sinking of particles deforms their solute emissions into a plume that looks like a comet (Figure 2e). This means that the plume generated is highly asymmetrical, with steep gradients in front of the particles and a long tail of weaker gradients in their wake (Kiørboe and Jackson 2001; Stocker et al. 2008). As it extends through the water column, this sinking plume is subjected to turbulence and diffusion and ultimately becomes diluted in the background.

6.3.2. Cell Density and Encounter Rates at the Microscale

The mean abundance of bacterial cells is relatively constant in oceanic waters, ranging between 10^5 and 10^6 cells per millilitre (including on the GBR (Patten et al. 2006)). If we distribute 10^6 bacteria uniformly in the space taken by 1 ml (i.e., 1 cm^3), then the distance between a bacterium and its nearest neighbour will be approximately 100 µm (or ~100 bacterial body lengths). This spacing is not drastically affected by small changes in cell concentrations, as it is proportional to the cubic root of the cell density (for example, a 10-fold difference in cell concentration causes only a 2-fold change in cell separation). Similar considerations apply to the distance between bacteria and phytoplankton (which constitute one of the sources of nutrient for heterotrophic bacteria). For example, for 10^6 bacteria per millilitre and a typical phytoplankton density of 10^3 per millilitre, each bacterium would be at approximately 1 mm to the nearest phytoplankton cell when both populations are randomly distributed. As an analogy, this distance is equivalent to 1,000 human body lengths or 1.7 kilometres. Overall, these considerations reveal that despite the high concentration of microorganisms present in seawater, the ocean at the microscale is a dilute suspension of microorganisms, with bacteria separated by 10–100 body lengths from their nearest neighbour and by 100–1,000 body lengths from nutrient sources such as phytoplankton cells. However, despite these distances, microorganisms will encounter one another more often than is perhaps expected during their lifetime.

Non-motile bacterial cells are subjected to Brownian motion, the small fluctuations in position driven by random collisions with water molecules. The Brownian motion of a bacterium can be quantified by its diffusivity, which depends on the bacterial radius and the temperature and dynamic viscosity of seawater. For example, for a seawater temperature of 26 °C (the average temperature on the GBR), the diffusivity of a bacterium of radius 0.4 µm is approximately 5.5×10^{-13} m^2/s. This means that because of the random path they follow by Brownian motion, non-motile bacteria will cover approximately 35 µm in 10 minutes or 450 µm in one day and will on average encounter 0.01 algal cells over a day (Clerc et al. 2022), which means that random Brownian motion alone is not an effective strategy to interact with point sources of nutrients, such as phytoplankton cells, in the ocean.

Many marine bacteria are motile and can swim between 10 to 100 body lengths per second (Mitchell et al. 1995a; Mitchell et al. 1995b). Motile marine bacteria often alternate straight "runs" with random reorientations, which results in a random walk that vastly increases not only the volume that they can explore but also their chances of encountering resources. A motile bacterium can explore a domain of ~2 mm in 10 min or ~2 cm in a day (Kiørboe 2018), which is nearly 50 times larger than the distance reached by non-motile cells through Brownian diffusion. The volume explored by motile bacteria is correspondingly greater by a factor of 50^3, which translates in greater encounter rates: a motile bacterium will encounter approximately 25 algal cells per day (Clerc et al. 2022). In addition, most motile cells are also chemotactic (Raina et al. 2019), which means that they can alter their swimming patterns based on chemical gradients. This behaviour can further enhance the encounter rate of microorganisms, altering their spatial distribution and, thus, the frequency and intensity of their interactions. Chemotaxis also significantly enhances the ability of microorganisms to exploit the many sources of microscale heterogeneity previously described to thus promote the survival of these cells.

Considering the importance of microscale physical processes is paramount to understand the activity and interaction of microorganisms in the marine environment. Here, we presented only a non-technical overview of these processes. We refer interested readers to more in-depth and mathematical accounts of the effect of flow and other physical processes on the life of marine microorganisms (Guasto et al. 2012; Stocker and Seymour 2012; Rusconi and Stocker 2015; Clerc et al. 2022).

6.4. CHEMICAL TRANSFORMATION MEDIATED BY MICROBES ON THE GREAT BARRIER REEF

The biophysical processes described above dictate the biochemical activity of microbial cells. Here, we review the chemical transformation mediated by microorganisms on the GBR by focusing on processes occurring in the water column, in the sediments, and within benthic hosts.

6.4.1. Water Column

Despite the very high levels of gross productivity that coral reef ecosystems are renowned for (Crossland et al. 1991), the pelagic waters overlaying coral reefs are generally characterised by nutrient-poor conditions and low rates of planktonic primary productivity. Indeed, phytoplankton production above most coral reefs contributes only a very small fraction to overall coral reef productivity. For example, phytoplankton production accounted for only 37% of total primary production in the central GBR (Furnas and Mitchell 1987), while it was 60 times lower than benthic production within a Red Sea coral reef (Cardini et al. 2016). Consistent with the oligotrophic conditions of the water column, phytoplankton communities overlaying coral reefs are also typically dominated by small picophytoplankton (Furnas and Mitchell 1986), which results in a reduced potential for vertical carbon flux in the water column (De Martini et al. 2018). Furthermore, phytoplankton-derived organic carbon makes only a very small contribution to the total dissolved organic carbon (DOC) pool and secondary production within coral reef environments (Alldredge et al. 2013; Cardini et al. 2016). Thus, it is not surprising that the growth of heterotrophic bacteria in coral reef waters is often poorly correlated with planktonic primary production (Rochelle-Newall et al. 2008). Nevertheless, the metabolic activity and carbon assimilation rates of bacterioplankton overlaying coral reefs are often very high relative to surrounding waters (Moriarty et al. 1985). This is a consequence of bacterioplankton utilization of organic carbon released from the benthic environment (Silveira et al. 2017).

Heterotrophic bacterioplankton that inhabit the water column overlaying reefs experience significant fluxes of DOC and particulate organic carbon (POC) from the reef environment, with bacterial assimilation of this organic matter playing an important role in coral reef carbon cycling (Silveira et al. 2017; Nelson et al. 2023). Planktonic microbes reliably obtain a diverse pool of DOC exuded from benthic reef organisms (Nelson et al. 2023) and POC-rich mucus shed by corals (Wild et al. 2004) and detritus from other reef organisms (Alldredge et al. 2013). This results not only in a rich, and heterogenous, seascape of organic carbon that sustains high levels of bacterial metabolism within the water column (Moriarty et al. 1985) but also in microscale (Seymour et al. 2005) and metre-scale (Gast et al. 1998) gradients in bacterioplankton abundance and activity in the waters overlaying reefs. Indeed, there is evidence that the metabolism of bacterioplankton assemblages overlaying coral reefs is finely tuned to exploit a combination of both reef-derived and oceanic DOC (Nelson et al. 2013) and is likely very sensitive to changes in the nature of organic matter delivered from the reef. For example, seasonal shifts observed in the relative proportion of Alphaproteobacteria and Bacteroidota observed in GBR waters were reflective of an increased genomic potential to degrade macroalgal-derived polysaccharides (Glasl et al. 2020), and both bacterioplankton community composition and growth rates have been shown to vary according to coral or macroalgal sources of organic matter (Nelson et al. 2013; Silveira et al. 2017).

Coral mucus sloughed into reef waters as ropes and flocs is rich in carbohydrates (Coffroth 1990) and proteins (Krupp 1985), and it becomes heavily colonized by heterotrophic bacteria (Rublee et al. 1980) that can metabolize it (Sharon and Rosenberg 2008) and in some cases, hone in on it by using chemotaxis (Garren et al. 2014; Tout et al. 2015). Notably, the subsequent degradation of flocs of coral mucus and the directed colonization and passive trapping of microbial biomass within mucus suspended in the water column may also regulate coral reef carbon cycling by influencing POC deposition on reefs (Figure 3). Physical action (e.g., waves, currents) breaks mucus down into microscopic marine snow particles (Huettel et al. 2006), which both fuels bacterial metabolism in the water column (Simon et al. 1990) and influences carbon flux dynamics (Omand et al. 2020). Mucus fractions that resist physical breakdown trap microorganisms and other material, leading to orders of magnitude increases in dry weight, microalgae, bacteria, nitrogen, and carbon content (Huettel et al. 2006) and much higher levels of pelagic-benthic C flux than would otherwise be expected in a picophytoplankton-dominated ecosystem (Naumann et al. 2009). This subsequently catalyses benthic remineralization of deposited mucus flocs (Huettel et al. 2006), which, in turn, enhances benthic-pelagic nutrient exchange, including the flux of organic and inorganic species of nitrogen into the water column and further fuels biological activity and nitrogen cycling processes.

Benthic metabolism is largely responsible for catalysing several key steps of the marine nitrogen cycle, including ammonification and nitrification, within the waters surrounding coral reefs (O'Neil and Capone 2008). However, pelagic nitrogen fixation, whereby some species of diazotrophic bacteria and archaea reduce atmospheric N_2 gas to ammonium by using the nitrogenase enzyme (Capone and Carpenter 1982), independently delivers a sometimes quantitatively important amount of biologically available nitrogen to coral reef ecosystems (Garcia et al. 2007; Benavides et al. 2017). Evidence from both the GBR (Bell et al. 1999) and the Red Sea (Tilstra et al. 2018) indicate that pelagic diazotrophy can support a significant fraction of primary production in the water column overlaying coral reefs. Although nitrogen fixation rates within the water column are typically lower than within other reef microenvironments (Benavides et al. 2017), sometimes very high rates of pelagic nitrogen fixation (e.g., 68 nmol L^{-1} d^{-1}) have been observed in reef waters (Messer et al. 2017). The identity of diazotrophs in the water column is also distinct from those in other (e.g., benthic and host-associated) reef microenvironments. In the GBR and coral reefs in the south Pacific Ocean, the filamentous cyanobacterium *Trichodesmium* is often the dominant and most active planktonic diazotroph (Bell et al. 1999; Garcia et al. 2007; Hewson et al. 2007; Messer et al. 2017). However, other planktonic diazotrophic groups including heterotrophic proteobacteria (Hewson et al. 2007; Messer et al. 2017) and unicellular cyanobacteria (Biegala and Raimbault 2008) can also be abundant and contribute to water column nitrogen fixation in coral reefs (Figure 3). Together, these diverse communities of planktonic diazotrophs (Hewson et al. 2007; Messer et al. 2017) are believed to deliver an important share of biologically available nitrogen to coral reef nitrogen budgets and thereby contribute to the overall productivity of coral reef ecosystems (O'Neil and Capone 2008; Benavides et al. 2017).

Similarly to the carbon and nitrogen cycles, the microbial processes that drive sulphur cycling within coral reefs have been most extensively studied within the coral holobiont (Gardner et al. 2022) and benthic environment

FIGURE 3 Diverse populations of marine microorganisms mediate key chemical transformations on the GBR. These include microbial communities present in bloom of the cyanobacterium *Trichodesmium* (top); in the water column; sediments; or those associated with benthic hosts. Carbon cycling: red; nitrogen cycling: blue; sulphur cycling: purple. Org N: organic nitrogen compounds; Org C: organic carbon compounds; POC: particulate organic carbon; DOC: dissolved organic carbon; Met: methionine; DMSP: dimethylsulfoniopropionate; DMS: dimethyl sulphide.

Source: Artwork produced by Paige Strudwick

(Broadbent and Jones 2004) rather than the water column (see sections 3.3 and 3.4 below). However, largely as a consequence of the substantial production of dimethylsulfoniopropionate (DMSP) and its breakdown product dimethyl sulfide (DMS) by benthic organisms (Broadbent et al. 2002; Raina et al. 2013), very high concentrations of these ecologically and climatically important organosulphur compounds are commonly recorded in the waters overlaying and surrounding coral reefs (Broadbent and Jones 2006). In addition to elevated dissolved concentrations of these chemicals, coral reef waters also experience localised hotspots of DMSP and its derivatives in the form of shed coral mucus. Concentrations of DMSP and DMS measured in coral mucus and mucus ropes on the GBR are the highest recorded in any marine environment (Broadbent and Jones 2004; Gardner et al. 2022). These extremely high concentrations are notable given that DMSP is a very important metabolite within microbial foodwebs, whereby it delivers a significant fraction of the carbon and sulphur demands of bacterioplankton populations (Kiene et al. 2000), can be used as an antioxidant by non-DMSP-producing phytoplankton (Theseira et al. 2020), and governs ecological interactions among planktonic microbes (Strom et al. 2003; Seymour et al. 2010). Furthermore, there is evidence that marine microbes are particularly attuned to exploiting microscale chemical hotspots of DMSP (Gao et al. 2020), akin to suspended coral mucus particles and flocs. Coral reef bacterioplankton exhibit high levels of chemotaxis towards the DMSP in coral mucus (Garren et al. 2014). Moreover, a recent study in GBR waters demonstrated that DMSP is assimilated by both bacterioplankton and phytoplankton and that dissolved DMSP concentrations govern the abundance and community composition of both groups in the water column (Fernandez et al. 2021). Given the very high concentrations of DMSP produced within coral reefs, it is perhaps unsurprising that this chemical and its products so profoundly shape microbiological processes within the surrounding pelagic environment.

Notably, the important microbiological and biogeochemical processes that occur within the water column overlaying coral reefs are not stable, and environmental perturbations, including anthropogenically driven declines in water quality and climate change, are causing changes to this part of the reef ecosystem, which may ultimately have profound implications for coral reef health. Many coastal coral reef environments are experiencing eutrophication as a consequence of increased nutrient run-off from terrestrial environments (Lesser 2021), which will likely reshape the balance of pelagic production and biochemistry in the water surrounding reefs (D'Angelo and Wiedenmann 2014). Elevated nutrient levels catalyse increased phytoplankton production and biomass in the water column, which has the potential to drive a number of potentially detrimental ecosystem cascades, including light limitation of Symbiodiniaceae through shading, increased prey for the larvae of the predatory crown-of-thorns starfish, release of phytoplankton toxins, and deoxygenation of the water column and benthic environment when dead phytoplankton biomass is metabolised by heterotrophic bacteria (D'Angelo and Wiedenmann 2014). Furthermore, shifts in the concentrations and balances of nutrients (DOC, DIN, DON, and DIP) will also alter biogeochemical cycling rates and directions within reef waters, with unpredictable impacts on the health of the reef below (O'Neil and Capone 2008). Finally, the combined impacts of nutrient pollution, over-fishing, and declines in coral cover due to increased seawater temperatures can drive significant increases in benthic macroalgal communities (Zaneveld et al. 2016), which leads to an increased flux of often highly labile DOC into the water column (Manikandan et al. 2021). This fuels elevated pelagic bacterial metabolism (Haas et al. 2011) and potentially leads to deoxygenation events (Silveira et al. 2019) and increases in copiotrophic bacterial taxa, which can often include opportunistic pathogens (Nelson et al. 2013; Glasl et al. 2020). Clearly, although the nature of the pelagic environment overlaying healthy coral reefs is primarily mediated by benthic processes, changing environmental conditions can potentially shift this balance and processes occurring within the water column to negatively impact reef health.

6.4.2. Reef Sediments

The sediments surrounding coral reefs comprise fine and coarse sands, dead coral rubble, other reef detritus, and sometimes dense layers of microalgal and cyanobacterial mats (Heil et al. 2004; Werner et al. 2006; Brocke et al. 2015). This environment experiences significant levels of organic matter deposition from the reef (Silveira et al. 2017) and as a consequence, supports very high rates of microbiological activity and biogeochemical cycling, thereby playing an important role in benthic-pelagic chemical and biological coupling and supporting the overall productivity of coral reef ecosystems (Garren and Azam 2012).

Large quantities of POC, in the form of sloughed coral mucus (Naumann et al. 2009), detritus from other benthic reef plants and animals (Hansen et al. 1992), and fish faeces (Meyer and Schultz 1985), sink to the surface of coral reef sediments, which fuels hotspots of heterotrophic activity (Clavier and Garrigue 1999). A further substantial contribution of organic material comes from the significant levels of benthic primary production occurring on the upper surface layer of the sediments. Benthic diatoms, green microalgae, and cyanobacteria are responsible for very high photosynthesis rates in the surface of reef sediments, which may in some cases be comparable to, or even exceed, those occurring on the reef itself (Werner et al. 2008).

Because of these two reliable sources of organic matter (i.e., deposition of POC and benthic photosynthesis), both the rates of heterotrophy and bacterial biomass are extremely high within coral reef sediments (Figure 3). Bacterial abundances reaching up to 10^{10} cells ml^{-1} have been observed in coral reef sediments, which are among the highest concentrations recorded in any marine sediments, while rates of microbial production and metabolism are often comparable to those observed in the sediments of highly eutrophic environments (Sorokin 1973). These high levels of bacterial production on coral reef surface sediments lead not only to rapid and highly efficient (up to 99%) remineralization of organic material (Alongi et al. 2008) but also to the depletion of oxygen (Werner et al. 2006). While the coarse sandy nature of coral reef sediments allows for higher levels of diffusion and oxygen penetration into the sediments than in other marine ecosystems, oxygen levels still decay with depth into the sediment, leading to microscale oxygen gradients (Johnstone et al. 1990; Werner et al. 2006). Similarly to the benthos in other marine ecosystems (Jørgensen et al. 2019), sulphate reduction often becomes an important microbial metabolism below the top 1–2 cm of reef sediment (Pigott and Land 1986; Nedwell and Blackburn 1987), whereby organic matter is degraded by heterotrophic bacteria that use sulphate as a terminal electron acceptor under anoxic conditions (Jørgensen and Kasten 2006). However, in contrast to many other benthic marine habitats, aerobic respiration rather than sulphate reduction often remains responsible for the bulk (55–98%) of organic matter remineralization in coral reef sediments (Alongi et al. 2008), presumably due to greater oxygen penetration into the highly permeable, coarse sands often found on reefs (Werner et al. 2006). In line with this pattern, there is often little evidence for notable levels of methanogenesis within coral reef sediments (Nedwell and Blackburn 1987; Alongi et al. 2008).

Benthic microbial remineralization of organic matter deposited from the reef and related microscale oxygen gradients also govern benthic nitrogen cycling processes within coral reef sediments. High nitrogen concentrations, strong microscale gradients of ammonium, nitrate and nitrite, and highly dynamic turnover of the sediment nitrogen pool are indicative of the substantial nitrogen recycling potential within this habitat (Williams et al. 1985; Johnstone et al.

1989; Capone et al. 1992). Indeed, significant levels of ammonification, ammonia uptake, nitrification, and denitrification have all been observed in coral reef sediments, with rates of these processes varying according to sediment characteristics (e.g., coarse sandy vs muddy) and depth relative to microscale gradients of O_2 and different inorganic nitrogen species (Capone et al. 1992). Very high levels of nitrogen fixation have also been recorded in coral reef sediments, with evidence that sediment diazotrophy contributes a significant fraction of bioavailable nitrogen to the reef environment (Capone et al. 1992; Charpy-Roubaud et al. 2001; Cardini et al. 2014).

Unsurprisingly, given the substantial microscale heterogeneity in chemical concentrations (e.g., O_2, NO_2, NO_3, and NH_4^+) and related microbial metabolisms (Johnstone et al. 1989; Capone et al. 1992) within coral reef sediments, the microbial assemblages within this habitat are generally highly diverse (Rusch et al. 2009; Dong et al. 2022). The bacterial and archaeal communities within sediments also display pronounced microscale vertical transitions in composition related to the sediment redox state and chemical gradients (Hewson and Fuhrman 2006; Rusch et al. 2009), and they are enriched in taxa and metabolic potential consistent with key nitrogen and sulphur cycling processes (Rusch and Gaidos 2013; Dong et al. 2022). Cumulatively, the metabolic activity of these microbial communities plays fundamentally important roles in the regulation of benthic reef biogeochemistry and ultimately the delivery of limiting nutrients to coral reef ecosystems.

6.4.3. Host Association

Many benthic invertebrates, such as reef-building corals or sponges, harbour specific, diverse, and abundant consortia or microorganisms, which have the capacity to carry out important functions supporting the health of their hosts (Rosenberg et al. 2007; Webster and Taylor 2012; van Oppen and Blackall 2019; Pogoreutz et al. 2020). Indeed, these microorganisms can extend the metabolic repertoire, immunity, and environmental adaptation capacities of their hosts (Rosenberg et al. 2007; Rädecker et al. 2015; Ziegler et al. 2017). These complex symbiotic systems are now called "holobionts", and are considered discrete ecological units formed by benthic hosts and their microbial partners (Rohwer et al. 2002).

Specific bacteria and archaea in coral and sponge holobionts have the genetic capacity to fix carbon through multiple pathways (Botté et al. 2019; Robbins et al. 2019), which means that these microorganisms may actively contribute to carbon acquisition in the holobiont. Symbiotic interactions between corals and bacteria also involve the cycling of essential compounds, such as B vitamins. Indeed, both corals and Symbiodiniaceae require exogenous sources of vitamin B_{12} (Robbins et al. 2019; Matthews et al. 2020), which is required for protein synthesis and for a range of other important metabolic pathways. Concentrations of vitamin B_{12} in the coral gastrovascular cavity (GVC) can be 35 times higher than in surrounding reef waters (Agostini et al. 2009). This essential vitamin is mostly produced by bacteria and archaea living in anoxic or low oxygen environments, and the GVC, with its strong oxygen gradients, may be an ideal niche for these microorganisms (Hughes et al. 2022).

Corals are also known to harbour nitrogen-fixing bacteria (Lesser et al. 2004; Rädecker et al. 2015), which can provide "new" biologically available nitrogen to both coral and Symbiodiniaceae (Benavides et al. 2017). These nitrogen-fixing bacteria may substantially support the functioning of the holobiont under low nitrogen availability; however, altered nutrient cycling during heat stress (Rädecker et al. 2021) reduces the assimilation of diazotroph-derived nitrogen in the coral host and Symbiodiniaceae (Rädecker et al. 2022). Other key nitrogen cycling processes carried out by microbes (e.g., nitrification, denitrification, and anaerobic ammonium oxidation) in corals have received less research attention (Wegley et al. 2007; Robbins et al. 2019) but have been more thoroughly studied in sponges (Webster and Taylor 2012; Zhang et al. 2019). Indeed, waste products derived from the sponge host fuels complex interactions among aerobic, anaerobic, autotrophic, and heterotrophic microbial processes, allowing for an efficient recycling of nutrients within the holobiont (Hoffmann et al. 2009; Moitinho-Silva et al. 2017).

Corals also harbour diverse bacteria that can cycle methylated sulphur compounds. These bacteria can produce DMSP from methionine (Kuek et al. 2022), degrade DMSP into DMS (Raina et al. 2009), demethylate DMSP to produce methanethiol (Frade et al. 2016), or further catabolise DMS (Raina et al. 2009), thus reducing the emissions of this gas from corals. Increases in water temperature and other stressors cause an increase in DMSP concentrations in corals (Raina et al. 2013; Deschaseaux et al. 2014b; Gardner et al. 2017) and a concomitant increase in the abundance of bacterial genes involved in the catabolism of this compound (Frade et al. 2016; Gardner et al. 2022). Importantly, during heat stress, the cleavage pathway (producing DMS) becomes more abundant than the demethylation pathway (Gardner et al. 2022), which suggests that a larger fraction of the DMSP pool is catabolised into DMS by coral-associated bacteria.

6.4.4. Gaseous Emissions

Some molecules produced by microorganisms on the GBR have a high vapour pressure at ambient temperature, which means that they are typically released as gases after being produced. These are called biogenic volatile organic compounds (BVOCs) and have been well-studied in terrestrial ecosystems, where plants and microorganisms are recognised as the largest emitters (Peñuelas et al. 2014). BVOCs fulfil many functions, from protection against pathogens to inter-species signalling (Lawson et al. 2022), but as they often are the end products of metabolic pathways, they can also be excellent stress markers (Loreto and Schnitzler 2010).

In coral reefs, most of the research on BVOC emissions has focused on the sulphur-containing compound DMS. On a global scale, emissions of DMS represent by far the largest natural flux of sulphur between the ocean and the atmosphere (Lovelock et al. 1972), accounting for 50% of natural sulphur emissions (Barnes et al. 2006), or 27 Tg of sulphur per year (Hulswar et al. 2022). The highest concentration of DMS ever recorded in the environment came from coral mucus ropes on the GBR (i.e., 18.7 μM) (Broadbent and Jones 2004). DMS can be produced by Symbiodiniaceae (Steinke et al. 2011; Deschaseaux et al. 2014a), corals (Alcolombri et al. 2015), and coral-associated bacteria (Raina et al. 2009), which contributes to the large concentrations measured above corals and coral reefs. However, other BVOCs released by coral reefs have also recently been characterised. For example, other sulphur volatiles have been detected from corals, such as methanethiol, carbon disulphide, dimethyl trisulphide, or thiirane (Swan et al. 2016). In addition, recent untargeted chemical analyses applied to Symbiodiniaceae cultures, Symbiodiniaceae-associated bacteria, or corals have identified nearly 200 BVOCs belonging to 20 chemical classes including hydrocarbons (e.g., alkanes, alkenes, aromatics, and halogenated hydrocarbons), organosulphurs, monoterpenes, ketones, esters, or ethers (Lawson et al. 2019; Lawson et al. 2020; Lawson et al. 2021; Olander et al. 2021). Specifically, one of these studies revealed that Symbiodiniaceae exudates impact the composition and increase the chemical diversity of the BVOC emissions produced by their associated bacteria (Lawson et al. 2020).

Environmental factors affect the BVOC emissions from coral reefs. For example, DMS emissions are much larger at low and incoming tides during the day, when the corals are exposed to air (Jones and Trevena 2005; Jones et al. 2007; Hopkins et al. 2016). In addition, environmental stressors, such as heat stress, can negatively impact DMS emissions (Jones et al. 2007; Fischer and Jones 2012) and the diversity, abundance, and functional potential of other BVOCs emitted from corals in the GBR (Lawson et al. 2021).

6.5. THE IMPACT OF GASEOUS EMISSIONS FROM THE GBR

A fraction of the BVOCs emitted by benthic holobionts, unicellular algae, and other microorganisms on the GBR reaches the atmosphere, where they can form secondary aerosols. Aerosols are small particles (typically less than 1 μm) suspended in the atmosphere. Two major types of aerosols are formed in the marine environment: primary sea spray aerosols, which are caused by bubbles bursting at the air-sea interface through wave action (containing both organic matter and inorganic salts) (Wilson et al. 2015); and secondary aerosols, which are produced via atmospheric oxidation of BVOCs (Figure 3). All atmospheric aerosols scatter incoming solar radiation and can therefore have a cooling effect on climate by enhancing the total amount of solar radiation reflected from the planet (Twomey 1977; Jackson et al. 2020). However, aerosols can also increase in size (Riipinen et al. 2011), which affects climatic processes even more. Indeed only particles larger than 40 nm can act as cloud condensation nuclei (CCN), providing support for water vapour to condense and form cloud droplets (Dusek et al. 2006) (Figure 4). The abundance of aerosol particles large enough to act as CCN therefore impacts the number and size of cloud droplets, which ultimately affect precipitation and cloud reflectivity (Dusek et al. 2006) (Figure 4). Although both primary and secondary aerosols can be a source of CCN (Quinn and Bates 2011), secondary aerosols seem to play a dominant role in their formation in the marine environment (Mayer et al. 2020).

Natural aerosol emissions have been identified as the largest source of uncertainty in our current estimates of the Earth's radiative budget (Carslaw et al. 2013). Early surveys above Australia identified the GBR as a significant source of aerosol particles, with levels seven times higher on average than in typical marine air (Bigg and Turvey 1978). More recently, the formation of aerosol particles has been recorded over the GBR (Modini et al. 2009; Vaattovaara et al. 2013), and their concentration increases in spring and summer (Leck and Bigg 2008; Cropp et al. 2018; Jackson et al. 2018), which strongly suggests that their origin is biological.

One of the most scrutinised candidates underpinning the formation of these aerosol particles is DMS. The fate of DMS in the atmosphere has received significant attention because of its hypothesised role in regulating a process that was proposed to cool oceanic surface waters through the formation of CCN (Charlson et al. 1987). Atmospheric DMS has a lifetime of one to two days and is oxidised to produce methane sulfonic acid (MSA), sulphur dioxide, and ultimately, sulfuric acid, leading to the formation of new aerosol particles (Kreidenweis and Seinfeld 1988). DMS emissions may have a sizable (but variable among studies) effect on the amount of solar radiation received by the planet (Six et al. 2013; Schwinger et al. 2017; Fiddes et al. 2018). However, when focusing specifically on the importance of coral-derived DMS, recent modelling studies have indicated that the estimated emissions from the GBR are insufficient, compared to the anthropogenic aerosol emissions along the east coast of Australia, to have a significant influence on regional climatic processes (Fiddes et al. 2021; Fiddes et al. 2022; Jackson et al. 2022).

6.6. CONCLUSION: FROM THE MICROSCALE TO THE REEF

Microorganisms are diverse and extremely abundant on the GBR. The processes that they mediate have tremendous and multifaceted impacts on the chemical ecology of this ecosystem, which range from the recycling of limiting elements (i.e., nitrogen) to the synthesis of essential molecules (i.e., B vitamins) and the production of volatile compounds. However, these key processes are rarely investigated at a scale that matches the activity of the microorganisms

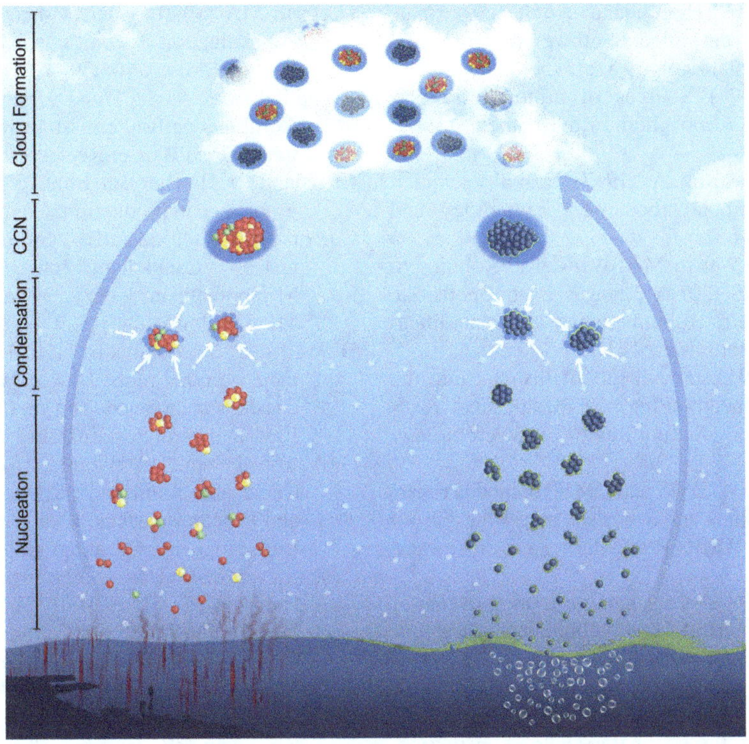

FIGURE 4 **Routes of aerosol formation and growth above coral reefs that can lead to cloud condensation nuclei (CCN).** BVOCs' emission from coral reefs (left side) and sea spray (right side) can lead to new particle formation in the atmosphere (in green: surface active organic material forming the sea surface microlayer). These newly formed particles can grow in size to act as CCN, providing a substrate onto which water vapour may condense.

Source: Artwork produced by Paige Strudwick

involved. Indeed, because of the importance of the microenvironments experienced by these cells on the rate of chemical transformation that they mediate, or the symbiotic interactions that they may be involved in, it appears that the scale at which microorganisms are typically sampled (e.g., collecting litres of seawater using Niskin bottles or homogenising the tissues of benthic organisms) is inadequate. These large-scale sampling strategies blend microorganisms and chemicals that are usually separated, breaking up the spatial structure of these communities and increasing the risk of erroneously characterising their biology and ecology. Although technical bottlenecks have long prevented the study of microorganisms at the appropriate scale, numerous techniques are now applicable to coral reefs (Engelberts et al. 2021; Hughes et al. 2022; van Oppen and Raina 2023). These new capabilities will undoubtedly contribute to determining how the consequences of these microscale processes scale up to affect the GBR.

REFERENCES

Agostini S, Suzuki Y, Casareto BE, Nakano Y, Hidaka M, Badrun N (2009) Coral symbiotic complex: Hypothesis through vitamin B12 for a new evaluation. Galaxea, Journal of Coral Reef Studies 11:1–11

Alcolombri U, Ben-Dor S, Feldmesser E, Levin Y, Tawfik DS, Vardi A (2015) Identification of the algal dimethyl sulfide–releasing enzyme: A missing link in the marine sulfur cycle. Science 348:1466–1469

Alldredge AL, Carlson CA, Carpenter RC (2013) Sources of organic carbon to coral reef flats. Oceanography 26:108–113

Alongi DM, Trott LA, Pfitzner J (2008) Biogeochemistry of inter-reef sediments on the northern and central Great Barrier Reef. Coral Reefs 27:407–420

Azam F (1998) Microbial control of oceanic carbon flux: The plot thickens. Science 280:694–696

Azam F, Fenchel T, Field JG, Gray JS, Meyer-Reil LA, Thingstad F (1983) The ecological role of water-column microbes in the Sea. Marine Ecology Progress Series 10:257–263

Azam F, Malfatti F (2007) Microbial structuring of marine ecosystems. Nature Reviews Microbiology 5:782–791

Baird ME, Atkinson MJ (1997) Measurement and prediction of mass transfer to experimental coral reef communities. Limnology and Oceanography 42:1685–1693

Barnes I, Hjorth J, Mihalopoulos N (2006) Dimethyl sulfide and dimethyl sulfoxide and their oxidation in the atmosphere. Chemical Reviews 106:940–975

Bar-On Yinon M, Phillips R, Milo R (2018) The biomass distribution on Earth. Proceedings of the National Academy of Sciences 115:6506–6511

Bell PRF, Elmetri I, Uwins P (1999) Nitrogen fixation by Trichodesmium spp. in the Central and Northern Great Barrier Reef Lagoon: Relative importance of the fixed-nitrogen load. Marine Ecology Progress Series 186:119–126

Benavides M, Bednarz VN, Ferrier-Pagès C (2017) Diazotrophs: Overlooked key players within the coral symbiosis and tropical reef ecosystems? Frontiers in Marine Science 4

Biegala IC, Raimbault P (2008) High abundance of diazotrophic picocyanobacteria (<3 μm) in a Southwest Pacific coral lagoon. Aquatic Microbial Ecology 51:45–53

Bigg EK, Turvey DE (1978) Sources of atmospheric particles over Australia. Atmospheric Environment (1967) 12:1643–1655

Blackburn N, Azam F, Hagström Å (1997) Spatially explicit simulations of a microbial food web. Limnology and Oceanography 42:613–622

Botté ES, Nielsen S, Abdul Wahab MA, Webster J, Robbins S, Thomas T, Webster NS (2019) Changes in the metabolic potential of the sponge microbiome under ocean acidification. Nature Communications 10:4134

Bourne DG, Morrow KM, Webster NS (2016) Insights into the coral microbiome: Underpinning the health and resilience of reef ecosystems. Annual Review of Microbiology 70:317–340

Broadbent AD, Jones GB (2004) DMS and DMSP in mucus ropes, coral mucus, surface films and sediment pore waters from coral reefs in the Great Barrier Reef. Marine and Freshwater Research 55:849–855

Broadbent AD, Jones GB (2006) Seasonal and diurnal cycles of dimethylsulfide, dimethylsulfoniopropionate and dimethylsulfoxide at one tree reef lagoon. Environmental Chemistry 3:260–267

Broadbent AD, Jones GB, Jones RJ (2002) DMSP in corals and benthic algae from the Great Barrier Reef. Estuarine, Coastal and Shelf Science 55:547–555

Brocke HJ, Polerecky L, de Beer D, Weber M, Claudet J, Nugues MM (2015) Organic matter degradation drives benthic cyanobacterial mat abundance on caribbean coral reefs. PLoS One 10:e0125445

Capone DG, Carpenter EJ (1982) Nitrogen fixation in the marine environment. Science 217:1140–1142

Capone DG, Dunham SE, Horrigan SG, Duguay LE (1992) Microbial nitrogen transformations in unconsolidated coral reef sediments. Marine Ecology Progress Series 80:75–88

Cardini U, Bednarz VN, Foster RA, Wild C (2014) Benthic N2 fixation in coral reefs and the potential effects of human-induced environmental change. Ecology and Evolution 4:1706–1727

Cardini U, Bednarz VN, van Hoytema N, Rovere A, Naumann MS, Al-Rshaidat MMD, Wild C (2016) Budget of primary production and dinitrogen fixation in a highly seasonal red sea coral reef. Ecosystems 19:771–785

Carslaw KS, Lee LA, Reddington CL, Pringle KJ, Rap A, Forster PM, Mann GW, Spracklen DV, Woodhouse MT, Regayre LA, Pierce JR (2013) Large contribution of natural aerosols to uncertainty in indirect forcing. Nature 503:67–71

Charlson RJ, Lovelock JE, Andreae MO, Warren SG (1987) Oceanic phytoplankton, atmospheric sulphur, cloud albedo and climate. Nature 326:655–661

Charpy-Roubaud C, Charpy L, Larkum A (2001) Atmospheric dinitrogen fixation by benthic communities of Tikehau Lagoon (Tuamotu Archipelago, French Polynesia) and its contribution to benthic primary production. Marine Biology 139:991–998

Clavier J, Garrigue C (1999) Annual sediment primary production and respiration in a large coral reef lagoon (SW New Caledonia). Marine Ecology Progress Series 191:79–89

Clerc EE, Raina J-B, Peaudecerf FJ, Seymour JR, Stocker R (2022) Survival in a sea of gradients: Bacterial and archaeal foraging in a heterogeneous Ocean. In: Stal LJ, Cretoiu MS (eds), The Marine Microbiome. Springer International Publishing, Cham, 47–102

Coffroth MA (1990) Mucous sheet formation on poritid corals: An evaluation of coral mucus as a nutrient source on reefs. Marine Biology 105:39–49

Cropp R, Gabric A, van Tran D, Jones G, Swan H, Butler H (2018) Coral reef aerosol emissions in response to irradiance stress in the Great Barrier Reef, Australia. Ambiology 47:671–681

Crossland CJ, Hatcher BG, Smith SV (1991) Role of coral reefs in global ocean production. Coral Reefs 10:55–64

Curson ARJ, Todd JD, Sullivan MJ, Johnston AWB (2011) Catabolism of dimethylsulphoniopropionate: Microorganisms, enzymes and genes. Nature Reviews Microbiology 9:849–859

D'Angelo C, Wiedenmann J (2014) Impacts of nutrient enrichment on coral reefs: New perspectives and implications for coastal management and reef survival. Current Opinion in Environmental Sustainability 7:82–93

De Martini F, Neuer S, Hamill D, Robidart J, Lomas MW (2018) Clade and strain specific contributions of Synechococcus and Prochlorococcus to carbon export in the Sargasso Sea. Limnology and Oceanography 63:S448–S457

Deschaseaux ESM, Beltran VH, Jones GB, Deseo MA, Swan HB, Harrison PL, Eyre BD (2014a) Comparative response of DMS and DMSP concentrations in Symbiodinium clades C1 and D1 under thermal stress. Journal of Experimental Marine Biology and Ecology 459:181–189

Deschaseaux ESM, Jones GB, Deseo MA, Shepherd KM, Kiene RP, Swan HB, Harrison PL, Eyre BD (2014b) Effects of environmental factors on dimethylated sulfur compounds and their potential role in the antioxidant system of the coral holobiont. Limnology and Oceanography 59:758–768

Dong X, Lan H, Huang L, Zhang H, Lin X, Weng S, Peng Y, Lin J, Wang J-h, Peng J, Yang Y (2022) Metagenomic views of microbial communities in sand sediments associated with coral reefs. Microbial Ecology 85:465–477

Dusek U, Frank GP, Hildebrandt L, Curtius J, Schneider J, Walter S, Chand D, Drewnick F, Hings S, Jung D, Borrmann S, Andreae MO (2006) Size matters more than chemistry for cloud-nucleating ability of aerosol particles. Science 312:1375–1378

Engelberts JP, Robbins SJ, Damjanovic K, Webster NS (2021) Integrating novel tools to elucidate the metabolic basis of microbial symbiosis in reef holobionts. Marine Biology 168:175

Fenchel T (1988) Marine Plankton food chains. Annual Review of Ecology and Systematics 19:19–38

Fernandez E, Ostrowski M, Siboni N, Seymour JR, Petrou K (2021) Uptake of Dimethylsulfoniopropionate (DMSP) by Natural Microbial Communities of the Great Barrier Reef (GBR), Australia. Microorganisms 9:1891

Fiddes SL, Woodhouse MT, Lane TP, Schofield R (2021) Coral-reef-derived dimethyl sulfide and the climatic impact of the loss of coral reefs. Atmospheric Chemistry and Physics 21:5883–5903

Fiddes SL, Woodhouse MT, Nicholls Z, Lane TP, Schofield R (2018) Cloud, precipitation and radiation responses to large perturbations in global dimethyl sulfide. Atmospheric Chemistry and Physics 18:10177–10198

Fiddes SL, Woodhouse MT, Utembe S, Schofield R, Alexander SP, Alroe J, Chambers SD, Chen Z, Cravigan L, Dunne E, Humphries RS, Johnson G, Keywood MD, Lane TP, Miljevic B, Omori Y, Protat A, Ristovski Z, Selleck P, Swan HB, Tanimoto H, Ward JP, Williams AG (2022) The contribution of coral-reef-derived dimethyl sulfide to aerosol burden over the Great Barrier Reef: A modelling study. Atmospheric Chemistry and Physics 22:2419–2445

Fischer E, Jones G (2012) Atmospheric dimethysulphide production from corals in the Great Barrier Reef and links to solar radiation, climate and coral bleaching. Biogeochemistry 110:31–46

Frade PR, Schwaninger V, Glasl B, Sintes E, Hill RW, Simó R, Herndl GJ (2016) Dimethylsulfoniopropionate in corals and its interrelations with bacterial assemblages in coral surface mucus. Environmental Chemistry 13:252–265

Furnas MJ, Mitchell AW (1986) Phytoplankton dynamics in the central Great Barrier Reef—I. Seasonal changes in biomass and community structure and their relation to intrusive activity. Continental Shelf Research 6:363–384

Furnas MJ, Mitchell AW (1987) Phytoplankton dynamics in the central Great Barrier Reef—II. Primary production. Continental Shelf Research 7:1049–1062

Gao C, Fernandez VI, Lee KS, Fenizia S, Pohnert G, Seymour JR, Raina J-B, Stocker R (2020) Single-cell bacterial transcription measurements reveal the importance of dimethylsulfoniopropionate (DMSP) hotspots in ocean sulfur cycling. Nature Communications 11:1942

Garcia N, Raimbault P, Sandroni V (2007) Seasonal nitrogen fixation and primary production in the Southwest Pacific: Nanoplankton diazotrophy and transfer of nitrogen to picoplankton organisms. Marine Ecology Progress Series 343:25–33

Gardner SG, Nitschke MR, O'Brien J, Motti CA, Seymour JR, Ralph PJ, Petrou K, Raina JB (2022) Increased DMSP availability during thermal stress influences DMSP-degrading bacteria in coral mucus. Frontiers in Marine Science 9.

Gardner SG, Raina J-B, Nitschke MR, Nielsen DA, Stat M, Motti CA, Ralph PJ, Petrou K (2017) A multi-trait systems approach reveals a response cascade to bleaching in corals. BMC Biology 15:117

Garren M, Azam F (2012) New directions in coral reef microbial ecology. Environmental Microbiology 14:833–844

Garren M, Son K, Raina J-B, Rusconi R, Menolascina F, Shapiro OH, Tout J, Bourne DG, Seymour JR, Stocker R (2014) A bacterial pathogen uses dimethylsulfoniopropionate as a cue to target heat-stressed corals. The ISME Journal 8:999–1007

Gast GJ, Wiegman S, Wieringa E, van Duyl FC, Bak RPM (1998) Bacteria in coral reef water types: Removal of cells, stimulation of growth and mineralization. Marine Ecology Progress Series 167:37–45

Glasl B, Robbins S, Frade PR, Marangon E, Laffy PW, Bourne DG, Webster NS (2020) Comparative genome-centric analysis reveals seasonal variation in the function of coral reef microbiomes. The ISME Journal 14:1435–1450

Guasto JS, Rusconi R, Stocker R (2012) Fluid mechanics of planktonic microorganisms. Annual Review of Fluid Mechanics 44:373–400

Haas AF, Fairoz MFM, Kelly LW, Nelson CE, Dinsdale EA, Edwards RA, Giles S, Hatay M, Hisakawa N, Knowles B, Lim YW, Maughan H, Pantos O, Roach TNF, Sanchez SE, Silveira CB, Sandin S, Smith JE, Rohwer F (2016) Global microbialization of coral reefs. Nature Microbiology 1:16042

Haas AF, Nelson CE, Wegley Kelly L, Carlson CA, Rohwer F, Leichter JJ, Wyatt A, Smith JE (2011) Effects of coral reef benthic primary producers on dissolved organic carbon and microbial activity. PLoS One 6:e27973

Hansen JA, Klumpp DW, Alongi DM, Dayton PK, Riddle MJ (1992) Detrital pathways in a coral reef lagoon. Marine Biology 113:363–372

Heil CA, Chaston K, Jones A, Bird P, Longstaff B, Costanzo S, Dennison WC (2004) Benthic microalgae in coral reef sediments of the southern Great Barrier Reef, Australia. Coral Reefs 23:336–343

Hewson I, Fuhrman JA (2006) Spatial and vertical biogeography of coral reef sediment bacterial and diazotroph communities. Marine Ecology Progress Series 306:79–86

Hewson I, Moisander PH, Morrison AE, Zehr JP (2007) Diazotrophic bacterioplankton in a coral reef lagoon: Phylogeny, diel nitrogenase expression and response to phosphate enrichment. The ISME Journal 1:78–91

Hoffmann F, Radax R, Woebken D, Holtappels M, Lavik G, Rapp HT, Schläppy M-L, Schleper C, Kuypers MMM (2009) Complex nitrogen cycling in the sponge Geodia barretti. Environmental Microbiology 11:2228–2243

Hopkins FE, Bell TG, Yang M, Suggett DJ, Steinke M (2016) Air exposure of coral is a significant source of dimethylsulfide (DMS) to the atmosphere. Scientific Reports 6:36031

Huettel M, Wild C, Gonelli S (2006) Mucus trap in coral reefs: Formation and temporal evolution of particle aggregates caused by coral mucus. Marine Ecology Progress Series 307:69–84

Hughes DJ, Raina J-B, Nielsen DA, Suggett DJ, Kühl M (2022) Disentangling compartment functions in sessile marine invertebrates. Trends in Ecology & Evolution 37:740–748

Hulswar S, Simó R, Gali Tapias M, Bell TG, Lana A, Inamdar S, Halloran PR, Manville G, Mahajan AS (2022) Third revision of the global surface seawater dimethyl sulfide climatology (DMS-Rev3). Earth System Science Data 14:2963–2987

Jackson RL, Gabric AJ, Cropp R (2018) Effects of ocean warming and coral bleaching on aerosol emissions in the Great Barrier Reef, Australia. Scientific Reports 8:14048

Jackson RL, Gabric AJ, Cropp R (2020) Coral reefs as a source of climate-active aerosols. PeerJ 8:e10023

Jackson RL, Woodhouse MT, Gabric AJ, Cropp RA, Swan HB, Deschaseaux ES, Trounce H (2022) Modelling the influence of coral reef-derived dimethylsulfide on the atmosphere of the Great Barrier Reef, Australia. Frontiers in Marine Science:1476

Johnstone RW, Koop K, Larkum AWD (1989) Fluxes of inorganic nitrogen between sediments and water in a coral reef lagoon. Proceedings of the Linnean Society of New South Wales 110:219–227

Johnstone RW, Koop K, Larkum AWD (1990) Physical aspects of coral reef lagoon sediments in relation to detritus processing and primary production. Marine Ecology Progress Series 66:273–283

Jones GB, Curran M, Broadbent A, King S, Fischer E, Jones R (2007) Factors affecting the cycling of dimethylsulfide and dimethylsulfoniopropionate in coral reef waters of the Great Barrier Reef. Environmental Chemistry 4:310–322

Jones GB, Trevena AJ (2005) The influence of coral reefs on atmospheric dimethylsulphide over the Great Barrier Reef, Coral Sea, Gulf of Papua and Solomon and Bismarck Seas. Marine and Freshwater Research 56:85–93

Jørgensen BB, Findlay AJ, Pellerin A (2019) The biogeochemical sulfur cycle of marine sediments. Frontiers in Microbiology 10

Jørgensen BB, Kasten S (2006) Sulfur Cycling and Methane Oxidation. In: Schulz HD, Zabel M (eds) Marine Geochemistry. Springer Berlin Heidelberg, Berlin, Heidelberg, 271–309

Kallmeyer J, Pockalny R, Adhikari Rishi R, Smith David C, D'Hondt S (2012) Global distribution of microbial abundance and biomass in subseafloor sediment. Proceedings of the National Academy of Sciences 109:16213–16216

Kiene RP, Linn LJ, Bruton JA (2000) New and important roles for DMSP in marine microbial communities. Journal of Sea Research 43:209–224

Kiørboe T (2018) A Mechanistic Approach to Plankton Ecology: A Mechanistic Approach to Plankton Ecology. Princeton University Press, Princeton.

Kiørboe T, Jackson GA (2001) Marine snow, organic solute plumes, and optimal chemosensory behavior of bacteria. Limnology and Oceanography 46:1309–1318

Kolmogorov AN (1941) The local structure of turbulence in incompressible viscous fluid for very large Reynolds numbers. Comptes rendus de l'Académie des sciences de l'URSS 30:301–305

Kramshøj M, Albers CN, Holst T, Holzinger R, Elberling B, Rinnan R (2018) Biogenic volatile release from permafrost thaw is determined by the soil microbial sink. Nature Communications 9:3412

Kreidenweis SM, Seinfeld JH (1988) Nucleation of sulfuric acid-water and methanesulfonic acid-water solution particles: Implications for the atmospheric chemistry of organosulfur species. Atmospheric Environment (1967) 22:283–296

Krupp DA (1985) An immunochemical study of the mucus from the solitary coral Fungia scutaria (Scleractinia, Fungiidae). Bulletin of Marine Science 36:163–176

Kuek FWI, Motti CA, Zhang J, Cooke IR, Todd JD, Miller DJ, Bourne DG, Raina J-B (2022) DMSP Production by Coral-Associated Bacteria. Frontiers in Marine Science 9

Kühl M, Cohen Y, Dalsgaard T, Jørgensen BB, Revsbech NP (1995) Microenvironment and photosynthesis of zooxanthellae in scleractinian corals studied with microsensors for O2, pH and light. Marine Ecology-Progress Series 117:159–172

Lauga E (2016) Bacterial hydrodynamics. Annual Review of Fluid Mechanics 48:105–130

Lawson CA, Camp E, Davy SK, Ferrier-Pagès C, Matthews J, Suggett DJ (2022) Informing coral reef conservation through metabolomic approaches. In: van Oppen MJH, Aranda Lastra M (eds) Coral Reef Conservation and Restoration in the Omics Age. Springer International Publishing, Cham, 179–202

Lawson CA, Possell M, Seymour JR, Raina J-B, Suggett DJ (2019) Coral endosymbionts (Symbiodiniaceae) emit species-specific volatilomes that shift when exposed to thermal stress. Scientific Reports 9:17395

Lawson CA, Raina J-B, Deschaseaux E, Hrebien V, Possell M, Seymour JR, Suggett DJ (2021) Heat stress decreases the diversity, abundance and functional potential of coral gas emissions. Global Change Biology 27:879–891

Lawson CA, Seymour JR, Possell M, Suggett DJ, Raina J-B (2020) The volatilomes of Symbiodiniaceae-associated bacteria are influenced by chemicals derived from their algal partner. Frontiers in Marine Science 7

Leck C, Bigg KE (2008) Comparison of sources and nature of the tropical aerosol with the summer high Arctic aerosol. Tellus B: Chemical and Physical Meteorology 60:118–126

Lehman JT, Scavia D (1982) Microscale nutrient patches produced by zooplankton. Proceedings of the National Academy of Sciences 79:5001–5005

Lesser MP (2021) Eutrophication on coral reefs: What is the evidence for phase shifts, nutrient limitation and coral bleaching. BioScience 71:1216–1233

Lesser MP, Mazel CH, Gorbunov MY, Falkowski PG (2004) Discovery of symbiotic nitrogen-fixing cyanobacteria in corals. Science 305:997–1000

Lewis ER, Lewis R, Schwartz SE (2004) Sea Salt Aerosol Production: Mechanisms, Methods, Measurements, and Models. American Geophysical Union, New York.

Loreto F, Schnitzler J-P (2010) Abiotic stresses and induced BVOCs. Trends in Plant Science 15:154–166

Lovelock JE, Maggs RJ, Rasmussen RA (1972) Atmospheric dimethyl sulphide and the natural sulphur cycle. Nature 237:452–453

Mahilang M, Deb MK, Pervez S (2021) Biogenic secondary organic aerosols: A review on formation mechanism, analytical challenges and environmental impacts. Chemosphere 262:127771

Manikandan B, Thomas AM, Shetye SS, Balamurugan S, Mohandass C, Nandakumar K (2021) Macroalgal release of dissolved organic carbon in coral reef and its interaction with the bacteria associated with the coral Porites lutea. Environmental Science and Pollution Research 28:66998–67010

Mass T, Genin A, Shavit U, Grinstein M, Tchernov D (2010) Flow enhances photosynthesis in marine benthic autotrophs by increasing the efflux of oxygen from the organism to the water. Proceedings of the National Academy of Sciences 107:2527–2531

Matthews JL, Raina J-B, Kahlke T, Seymour JR, van Oppen MJH, Suggett DJ (2020) Symbiodiniaceae-bacteria interactions: Rethinking metabolite exchange in reef-building corals as multi-partner metabolic networks. Environmental Microbiology 22:1675–1687

Mayer KJ, Wang X, Santander MV, Mitts BA, Sauer JS, Sultana CM, Cappa CD, Prather KA (2020) Secondary marine aerosol plays a dominant role over primary sea spray aerosol in cloud formation. ACS Central Science 6:2259–2266

Messer LF, Brown MV, Furnas MJ, Carney RL, McKinnon AD, Seymour JR (2017) Diversity and activity of diazotrophs in great barrier reef surface waters. Frontiers in Microbiology 8

Meyer JL, Schultz ET (1985) Migrating haemulid fishes as a source of nutrients and organic matter on coral reefs. Limnology and Oceanography 30:146–156

Mitchell JG, Pearson L, Bonazinga A, Dillon S, Khouri H, Paxinos R (1995b) Long lag times and high velocities in the motility of natural assemblages of marine bacteria. Applied and Environmental Microbiology 61:877–882

Mitchell JG, Pearson L, Dillon S, Kantalis K (1995a) Natural assemblages of marine bacteria exhibiting high-speed motility and large accelerations. Applied and Environmental Microbiology 61:4436–4440

Modini RL, Ristovski ZD, Johnson GR, He C, Surawski N, Morawska L, Suni T, Kulmala M (2009) New particle formation and growth at a remote, sub-tropical coastal location. Atmospheric Chemistry and Physics 9:7607–7621

Moitinho-Silva L, Díez-Vives C, Batani G, Esteves AIS, Jahn MT, Thomas T (2017) Integrated metabolism in sponge–microbe symbiosis revealed by genome-centered metatranscriptomics. The ISME Journal 11:1651–1666

Moriarty DJW, Pollard PC, Hunt WG (1985) Temporal and spatial variation in bacterial production in the water column over a coral reef. Marine Biology 85:285–292

Nakamura T, Yamasaki H (2005) Requirement of water-flow for sustainable growth of Pocilloporid corals during high temperature periods. Marine Pollution Bulletin 50:1115–1120

Naumann MS, Richter C, el-Zibdah M, Wild C (2009) Coral mucus as an efficient trap for picoplanktonic cyanobacteria: Implications for pelagic–benthic coupling in the reef ecosystem. Marine Ecology Progress Series 385:65–76

Nedwell DB, Blackburn TH (1987) Anaerobic metabolism in lagoon sediments from Davies Reef, Great Barrier Reef. Estuarine, Coastal and Shelf Science 25:347–353

Nelson CE, Goldberg SJ, Wegley Kelly L, Haas AF, Smith JE, Rohwer F, Carlson CA (2013) Coral and macroalgal exudates vary in neutral sugar composition and differentially enrich reef bacterioplankton lineages. The ISME Journal 7:962–979

Nelson CE, Wegley Kelly L, Haas AF (2023) Microbial Interactions with dissolved organic matter are central to coral reef ecosystem function and resilience. Annual Review of Marine Science 15:431–460

Olander A, Lawson CA, Possell M, Raina J-B, Ueland M, Suggett DJ (2021) Comparative volatilomics of coral endosymbionts from one- and comprehensive two-dimensional gas chromatography approaches. Marine Biology 168:76

Omand MM, Govindarajan R, He J, Mahadevan A (2020) Sinking flux of particulate organic matter in the oceans: Sensitivity to particle characteristics. Scientific Reports 10:5582

O'Neil JM, Capone DG (2008) Nitrogen cycling in coral reef environments. In: Capone D, Bronk D, Mullhollan MC (eds) Nitrogen in the Marine Environment. Elsevier, London, 949–989

Oremland Ronald S (1975) Methane production in shallow-water, tropical marine sediments. Applied Microbiology 30:602–608

Pacherres CO, Ahmerkamp S, Koren K, Richter C, Holtappels M (2022) Ciliary flows in corals ventilate target areas of high photosynthetic oxygen production. Current Biology 32:4150–4158

Patten NL, Seymour JR, Mitchell JG (2006) Flow cytometric analysis of virus-like particles and heterotrophic bacteria within coral-associated reef water. Journal of the Marine Biological Association of the United Kingdom 86:563–566

Patterson MR, Sebens KP, Olson RR (1991) In situ measurements of flow effects on primary production and dark respiration in reef corals. Limnology and Oceanography 36:936–948

Peñuelas J, Asensio D, Tholl D, Wenke K, Rosenkranz M, Piechulla B, Schnitzler JP (2014) Biogenic volatile emissions from the soil. Plant, Cell & Environment 37:1866–1891

Pigott JD, Land LS (1986) Interstitial water chemistry of Jamaican reef sediment: Sulfate reduction and submarine cementation. Marine Chemistry 19:355–378

Pogoreutz C, Voolstra CR, Rädecker N, Weis V, Cardenas A, Raina J-B (2020) The Coral Holobiont Highlights the Dependence of Cnidarian Animal Hosts on Their Associated Microbes Cellular Dialogues in the Holobiont. CRC Press, London, 91–118

Purcell EM (1977) Life at low Reynolds number. American Journal of Physics 45:3–11

Quinn PK, Bates TS (2011) The case against climate regulation via oceanic phytoplankton sulphur emissions. Nature 480:51–56

Rädecker N, Pogoreutz C, Gegner HM, Cárdenas A, Perna G, Geißler L, Roth F, Bougoure J, Guagliardo P, Struck U, Wild C, Pernice M, Raina J-B, Meibom A, Voolstra CR (2022) Heat stress reduces the contribution of diazotrophs to coral holobiont nitrogen cycling. The ISME Journal 16:1110–1118

Rädecker N, Pogoreutz C, Gegner HM, Cárdenas A, Roth F, Bougoure J, Guagliardo P, Wild C, Pernice M, Raina J-B, Meibom A, Voolstra CR (2021) Heat stress destabilizes symbiotic nutrient cycling in corals. Proceedings of the National Academy of Sciences 118:e2022653118

Rädecker N, Pogoreutz C, Voolstra CR, Wiedenmann J, Wild C (2015) Nitrogen cycling in corals: The key to understanding holobiont functioning? Trends in Microbiology 23:490–497

Raina J-B, Fernandez V, Lambert B, Stocker R, Seymour JR (2019) The role of microbial motility and chemotaxis in symbiosis. Nature Reviews Microbiology 17:284–294

Raina J-B, Tapiolas DM, Forêt S, Lutz A, Abrego D, Ceh J, Seneca FO, Clode PL, Bourne DG, Willis BL, Motti CA (2013) DMSP biosynthesis by an animal and its role in coral thermal stress response. Nature 502:677–680

Raina J-B, Tapiolas DM, Willis Bette L, Bourne David G (2009) Coral-associated bacteria and their role in the biogeochemical cycling of sulfur. Applied and Environmental Microbiology 75:3492–3501

Riemann L, Middelboe M (2002) Viral lysis of marine bacterioplankton: Implications for organic matter cycling and bacterial clonal composition. Ophelia 56:57–68

Riipinen I, Pierce JR, Yli-Juuti T, Nieminen T, Häkkinen S, Ehn M, Junninen H, Lehtipalo K, Petäjä T, Slowik J, Chang R, Shantz NC, Abbatt J, Leaitch WR, Kerminen VM, Worsnop DR, Pandis SN, Donahue NM, Kulmala M (2011) Organic condensation: A vital link connecting aerosol formation to cloud condensation nuclei (CCN) concentrations. Atmospheric Chemistry and Physics 11:3865–3878

Robbins SJ, Singleton CM, Chan CX, Messer LF, Geers AU, Ying H, Baker A, Bell SC, Morrow KM, Ragan MA, Miller DJ, Forêt S, Ball E, Beeden R, Berumen M, Aranda M, Ravasi T, Bongaerts P, Hoegh-Guldberg O, Cooke I, Leggat B, Sprungala S, Fitzgerald A, Shang C, Lundgren P, Fyffe T, Rubino F, van Oppen M, Weynberg K, Robbins SJ, Singleton CM, Xin Chan C, Messer LF, Geers AU, Ying H, Baker A, Bell SC, Morrow KM, Ragan MA, Miller DJ, Foret S, Voolstra CR, Tyson GW, Bourne DG, Voolstra CR, Tyson GW, Bourne DG, ReFuGe C (2019) A genomic view of the reef-building coral Porites lutea and its microbial symbionts. Nature Microbiology 4:2090–2100

Rochelle-Newall EJ, Torréton JP, Mari X, Pringault O (2008) Phytoplankton-bacterioplankton coupling in a subtropical South Pacific coral reef lagoon. Aquatic Microbial Ecology 50:221–229

Rohwer F, Seguritan V, Azam F, Knowlton N (2002) Diversity and distribution of coral-associated bacteria. Marine Ecology Progress Series 243:1–10

Rosenberg E, Koren O, Reshef L, Efrony R, Zilber-Rosenberg I (2007) The role of microorganisms in coral health, disease and evolution. Nature Reviews Microbiology 5:355–362

Rublee PA, Lasker HR, Gottfried M, Roman MR (1980) Production and bacterial colonization of mucus from the soft coral briarium asbestinum. Bulletin of Marine Science 30:888–893

Rusch A, Gaidos E (2013) Nitrogen-cycling bacteria and archaea in the carbonate sediment of a coral reef. Geobiology 11:472–484

Rusch A, Hannides AK, Gaidos E (2009) Diverse communities of active bacteria and archaea along oxygen gradients in coral reef sediments. Coral Reefs 28:15–26

Rusconi R, Stocker R (2015) Microbes in flow. Current Opinion in Microbiology 25:1–8

Schutter M, Kranenbarg S, Wijffels RH, Verreth J, Osinga R (2011) Modification of light utilization for skeletal growth by water flow in the scleractinian coral Galaxea fascicularis. Marine Biology 158:769–777

Schwinger J, Tjiputra J, Goris N, Six KD, Kirkevåg A, Seland Ø, Heinze C, Ilyina T (2017) Amplification of global warming through pH dependence of DMS production simulated with a fully coupled Earth system model. Biogeosciences 14:3633–3648

Seymour JR, Patten N, Bourne DG, Mitchell JG (2005) Spatial dynamics of virus-like particles and heterotrophic bacteria within a shallow coral reef system. Marine Ecology Progress Series 288:1–8

Seymour JR, Simó R, Ahmed T, Stocker R (2010) Chemoattraction to dimethylsulfoniopropionate throughout the marine microbial food web. Science 329:342–345

Shapiro O, Fernandez V, Garren M, Guasto JS, Debaillon-Vesque F, Kramarsky-Winter E, Vardi A, Stocker R (2014) Vortical ciliary flows actively enhance mass transport in reef corals. Proceedings of the National Academy of Sciences 111:13391–13396

Sharon G, Rosenberg E (2008) Bacterial growth on coral mucus. Current Microbiology 56:481–488

Silveira CB, Cavalcanti GS, Walter JM, Silva-Lima AW, Dinsdale EA, Bourne DG, Thompson CC, Thompson FL (2017) Microbial processes driving coral reef organic carbon flow. FEMS Microbiology Reviews 41:575–595

Silveira CB, Luque A, Roach TNF, Villela H, Barno A, Green K, Reyes B, Rubio-Portillo E, Le T, Mead S, Hatay M, Vermeij MJA, Takeshita Y, Haas A, Bailey B, Rohwer F (2019) Biophysical and physiological processes causing oxygen loss from coral reefs. eLife 8:e49114

Silver MW, Shanks AL, Trent JD (1978) Marine snow: Microplankton habitat and source of small-scale patchiness in pelagic populations. Science 201:371–373

Simon M, Alldredge AL, Azam F (1990) Bacterial carbon dynamics on marine snow. Marine Ecology Progress Series 65:205–211

Six KD, Kloster S, Ilyina T, Archer SD, Zhang K, Maier-Reimer E (2013) Global warming amplified by reduced sulphur fluxes as a result of ocean acidification. Nature Climate Change 3:975–978

Smriga S, Fernandez Vicente I, Mitchell James G, Stocker R (2016) Chemotaxis toward phytoplankton drives organic matter partitioning among marine bacteria. Proceedings of the National Academy of Sciences 113:1576–1581

Sorokin YI (1973) Microbiological Aspects of the Productivity of Coral Reefs. Academic Press, New York

Steinke M, Brading P, Kerrison P, Warner ME, Suggett DJ (2011) Concentrations of dimethylsulfoniopropionate and dimethyl sulfide are strain-specific in symbiotic dinoflagellates (Symbiodinium sp., Dinophyceae). Journal of Phycology 47:775–783

Stocker R (2012) Marine microbes see a sea of gradients. Science 338:628–633

Stocker R, Seymour JR (2012) Ecology and physics of bacterial chemotaxis in the ocean. Microbiology and Molecular Biology Reviews 76:792–812

Stocker R, Seymour JR, Samadani A, Hunt Dana E, Polz Martin F (2008) Rapid chemotactic response enables marine bacteria to exploit ephemeral microscale nutrient patches. Proceedings of the National Academy of Sciences 105:4209–4214

Strom S, Wolfe G, Slajer A, Lambert S, Clough J (2003) Chemical defense in the microplankton II: Inhibition of protist feeding by β-dimethylsulfoniopropionate (DMSP). Limnology and Oceanography 48:230–237

Swan HB, Crough RW, Vaattovaara P, Jones GB, Deschaseaux ESM, Eyre BD, Miljevic B, Ristovski ZD (2016) Dimethyl sulfide and other biogenic volatile organic compound emissions from branching coral and reef seawater: Potential sources of secondary aerosol over the Great Barrier Reef. Journal of Atmospheric Chemistry 73:303–328

Tang J, Schurgers G, Rinnan R (2019) Process Understanding of soil BVOC fluxes in natural ecosystems: A review. Reviews of Geophysics 57:966–986

Taylor JR, Stocker R (2012) Trade-offs of chemotactic foraging in turbulent water. Science 338:675–679

Thar R, Kühl M (2003) Bacteria are not too small for spatial sensing of chemical gradients: An experimental evidence. Proceedings of the National Academy of Sciences 100:5748–5753

Theseira AM, Nielsen DA, Petrou K (2020) Uptake of dimethylsulphoniopropionate (DMSP) reduces free reactive oxygen species (ROS) during late exponential growth in the diatom Thalassiosira weissflogii grown under three salinities. Marine Biology 167:127

Thomas T, Moitinho-Silva L, Lurgi M, Björk JR, Easson C, Astudillo-García C, Olson JB, Erwin PM, López-Legentil S, Luter H, Chaves-Fonnegra A, Costa R, Schupp PJ, Steindler L, Erpenbeck D, Gilbert J, Knight R, Ackermann G, Victor Lopez J, Taylor MW, Thacker RW, Montoya JM, Hentschel U, Webster NS (2016) Diversity, structure and convergent evolution of the global sponge microbiome. Nature Communications 7:11870

Tilstra A, van Hoytema N, Cardini U, Bednarz VN, Rix L, Naumann MS, Al-Horani FA, Wild C (2018) Effects of water column mixing and stratification on planktonic primary production and dinitrogen fixation on a Northern Red Sea Coral Reef. Frontiers in Microbiology 9

Tout J, Jeffries TC, Petrou K, Tyson GW, Webster NS, Garren M, Stocker R, Ralph PJ, Seymour JR (2015) Chemotaxis by natural populations of coral reef bacteria. The ISME Journal 9:1764–1777

Twomey S (1977) The influence of pollution on the shortwave albedo of clouds. Journal of the Atmospheric Sciences 34:1149–1152

Vaattovaara P, Swan HB, Jones GB, Deschaseaux E, Miljevic B, Laaksonen A, Ristovski ZD (2013) The contribution of sulfate and oxidized organics in climatically important ultrafine particles at a coral reef environment. International Journal of Environmental, Chemical, Ecological, Geological and Geophysical Engineering 7:720–724

van Oppen MJH, Blackall LL (2019) Coral microbiome dynamics, functions and design in a changing world. Nature Reviews Microbiology 17:557–567

van Oppen MJH, Raina J-B (2023) Coral holobiont research needs spatial analyses at the microbial scale. Environmental Microbiology 25(1):179–183.

Webster NS, Taylor MW (2012) Marine sponges and their microbial symbionts: Love and other relationships. Environmental Microbiology 14:335–346

Wegley L, Edwards R, Rodriguez-Brito B, Liu H, Rohwer F (2007) Metagenomic analysis of the microbial community associated with the coral Porites astreoides. Environmental Microbiology 9:2707–2719

Werner U, Bird P, Wild C, Ferdelman T, Polerecky L, Eickert G, Jonstone R, Hoegh-Guldberg O, de Beer D (2006) Spatial patterns of aerobic and anaerobic mineralization rates and oxygen penetration dynamics in coral reef sediments. Marine Ecology Progress Series 309:93–105

Werner U, Blazejak A, Bird P, Eickert G, Schoon R, Abed RMM, Bissett A, de Beer D (2008) Microbial photosynthesis in coral reef sediments (Heron Reef, Australia). Estuarine, Coastal and Shelf Science 76:876–888

Wild C, Huettel M, Klueter A, Kremb SG, Rasheed MYM, Jørgensen BB (2004) Coral mucus functions as an energy carrier and particle trap in the reef ecosystem. Nature 428:66–70

Williams SL, Yarish SM, Gill IP (1985) Ammonium distributions, production, and efflux from backreef sediments, St. Croix, US Virgin Islands. Marine Ecology Progress Series 24:57–64

Wilson TW, Ladino LA, Alpert PA, Breckels MN, Brooks IM, Browse J, Burrows SM, Carslaw KS, Huffman JA, Judd C, Kilthau WP, Mason RH, McFiggans G, Miller LA, Nájera JJ, Polishchuk E, Rae S, Schiller CL, Si M, Temprado JV, Whale TF, Wong JPS, Wurl O, Yạkobi-Hancock JD, Abbatt JPD, Aller JY, Bertram AK, Knopf DA, Murray BJ (2015) A marine biogenic source of atmospheric ice-nucleating particles. Nature 525:234–238

Wolanski E, Kingsford M, Lambrechts J, Marmorino G (2024) The physical oceanography of the great barrier reef: A review. In: Wolanski E, Kingsford MJ (eds.), Oceanographic Processes of Coral Reefs. Physical and Biological Links in the Great Barrier Reef. CRC Press, Boca Raton, FL.

Zaneveld JR, Burkepile DE, Shantz AA, Pritchard CE, McMinds R, Payet JP, Welsh R, Correa AMS, Lemoine NP, Rosales S, Fuchs C, Maynard JA, Thurber RV (2016) Overfishing and nutrient pollution interact with temperature to disrupt coral reefs down to microbial scales. Nature Communications 7:11833

Zhang F, Jonas L, Lin H, Hill RT (2019) Microbially mediated nutrient cycles in marine sponges. FEMS Microbiology Ecology 95

Ziegler M, Seneca FO, Yum LK, Palumbi SR, Voolstra CR (2017) Bacterial community dynamics are linked to patterns of coral heat tolerance. Nature Communications 8:14213

Section 2

Land-Sea Connectivity

7 Great Barrier Reef Ecohydrology

Jane Waterhouse, Richard Pearson, Stephen Lewis, Aaron Davis, and Nathan Waltham

7.1. INTRODUCTION

The Great Barrier Reef (GBR) is an iconic and highly valued marine protected area in Queensland, Australia that is recognised internationally for its natural and cultural values. It is one of the largest coral reef systems in the world (348,000 km^2) and contains extensive areas of soft-bottom, seagrass, and mangrove communities. It hosts a range of megafauna including whales, dolphins, dugongs, turtles, sharks, and large fish and supports both commercial and recreational fisheries (GBRMPA 2019). The adjacent catchment area (424,000 km^2; McCloskey et al. 2021) includes an extensive network of freshwater ecosystems. The total system incorporates the landscape and seascape from the catchment to the reef and beyond, including the terrestrial catchment, rivers, streams, coastal floodplain wetlands, estuaries, islands, reefs and inter-reefal areas, the continental slope, and deep oceanic waters (Creighton et al. 2021) (Figure 1). This interconnected system is critical to the continued condition and resilience of the GBR and is described by Brodie and Pearson (2016) as the 'GBR province'.

The GBR has been managed as a marine park since 1975 and was listed as a World Heritage Area in 1981 (Day and Dobbs 2013). Joint management arrangements have existed between the Australian and Queensland State Governments since the establishment of the protected area and have focused on managed use and ecosystem protection.

Ecohydrology integrates different branches of environmental science, including ecology and hydrology, to determine the effects of hydrological processes on the distribution, structure, and function of ecosystems (Nuttle 2002). It merges this understanding with management actions to develop practical solutions to increase the resilience of the aquatic environment to human impacts (Wolanski 2007).

The GBR catchment area comprises 35 major basins (Figure 2). The basins vary in size, climate (500–4,500 mm/yr rainfall; McCloskey et al. 2021), land use, flow, drainage, and socioeconomic characteristics (Waterhouse et al. 2017). The basins are grouped into six natural resource management (NRM) regions (Figure 1). The GBR catchment area has been modified since European settlement (Lewis et al. 2021), with the current predominant land uses of cattle grazing (75%), nature conservation (13%), cropping (4%; including sugar cane, 1.3%), and other uses, including urbanisation, industries, and mining (8%) (McCloskey et al. 2021; Figure 2).

These land-use developments have led to major changes to the connected ecosystems between the land and the sea and, at the largest scale, in the quantity and quality of water discharges into the GBR (Furnas 2003). Monitored and modelled estimates have identified several-fold increases in pollutant delivery from the GBR catchment area to the GBR since pre-development (Bartley et al. 2017). The broad implications of these changes to contaminant delivery between the catchment and the reef are discussed in this chapter.

Physical connections between the GBR catchment area and marine waters are largely the result of water flow derived from precipitation and delivered through run-off, groundwater, streams, rivers, estuaries, and wetlands to coastal areas, where marine currents facilitate connections. Wonky holes, where submarine groundwater discharge from paleo river channels formed during lower sea levels seeps into lagoon waters, have also been shown to be of local importance in some areas (Cappo and Kelley 2000; Stieglitz 2005). Many other connections, including ecological links, occur at multiple scales from the landscape to individual species' movements (Figure 3), which is also discussed in Sheaves et al. (2024). Rivers, forested floodplains, wetlands, saltmarshes, and mangrove ecosystems provide not only habitat for many species that move to a greater or lesser extent through the system but also nutrient and sediment distribution and processing. These connections contribute to the trophic linkages that are essential to the ecosystems of the GBR World Heritage Area (Great Barrier Reef Marine Park Authority 2012). Organisms move actively and passively through the water and the air, but their movement can be impaired by natural and man-made barriers such as dams, weirs, and bunds. Cultural connections are also strong, and the health of land and water together is central to the culture of First Nations people in Australia and the broader community.

This chapter describes the key features of the land-sea connections in the GBR and the interdependencies among the catchment, coastal, and marine ecosystems, which are underpinned by a whole-of-system conceptual framework. It also summarises the importance of land-sea connections in terms of ecosystem functions, how these have been modified over time, and the implications of these changes to the GBR. Finally, the management approaches required to address multi-ecosystem issues are outlined, emphasising a whole-of-system approach and how this can be achieved for the GBR.

DOI: 10.1201/9781003320425-9

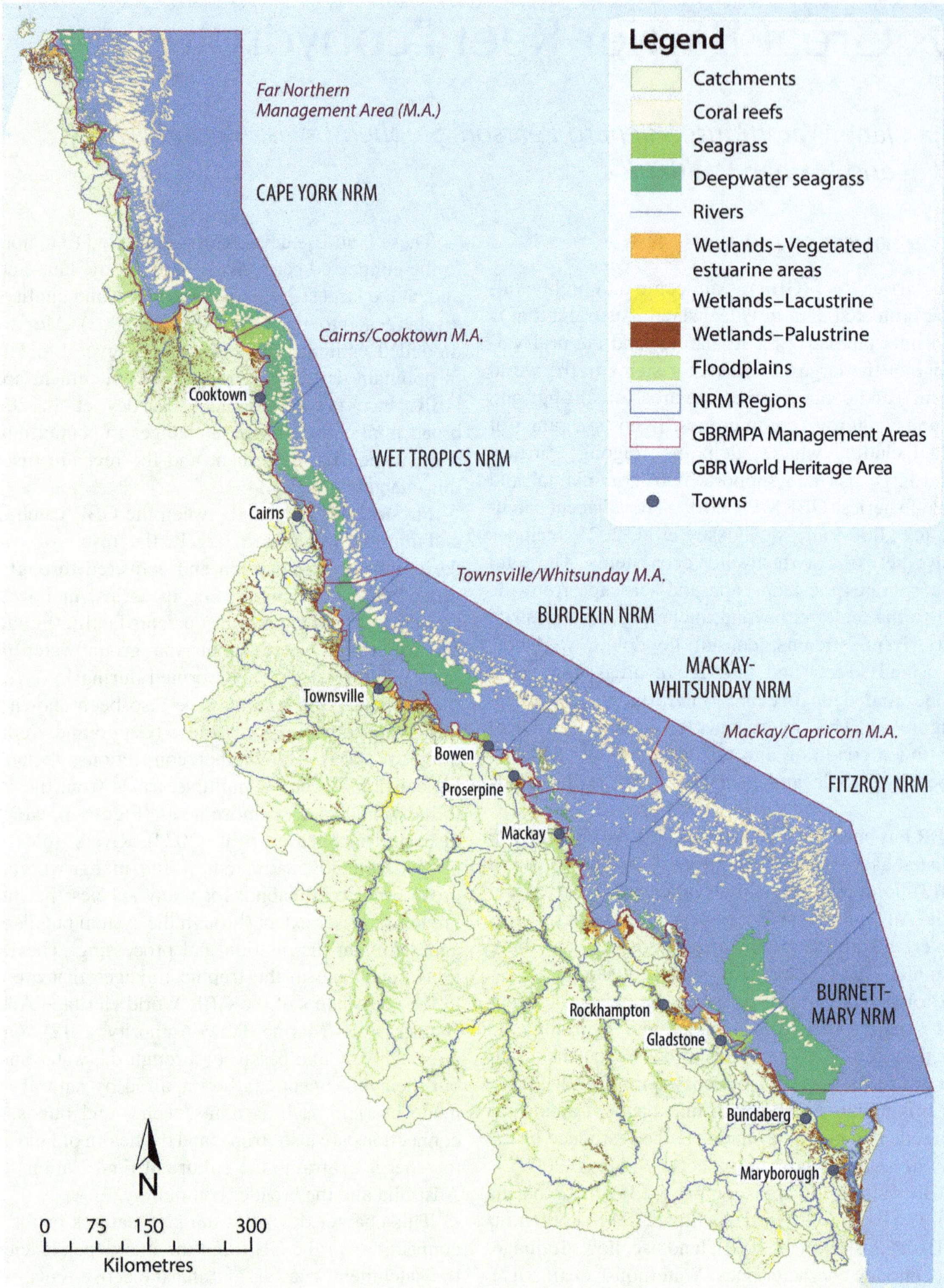

FIGURE 1 The GBR and its catchments, the six natural resource management (NRM) regions, and key ecosystems.

Source: Reproduced from Schaffelke et al. (2017)

7.2. GBR HYDROLOGY: RIVER DISCHARGE AND MAJOR FLOW PATHWAYS

The primary factor driving land-sea connections in the GBR is water; therefore, the hydrological connectivity among freshwater, estuarine, and coastal ecosystems is critical for the continued existence of many species (Wolanski 2007; Devlin and Schaffelke 2009; Waltham et al. 2019; Canning et al. 2022). For example, the mangrove jack (*Lutjanus argentimaculatus*) has a life cycle

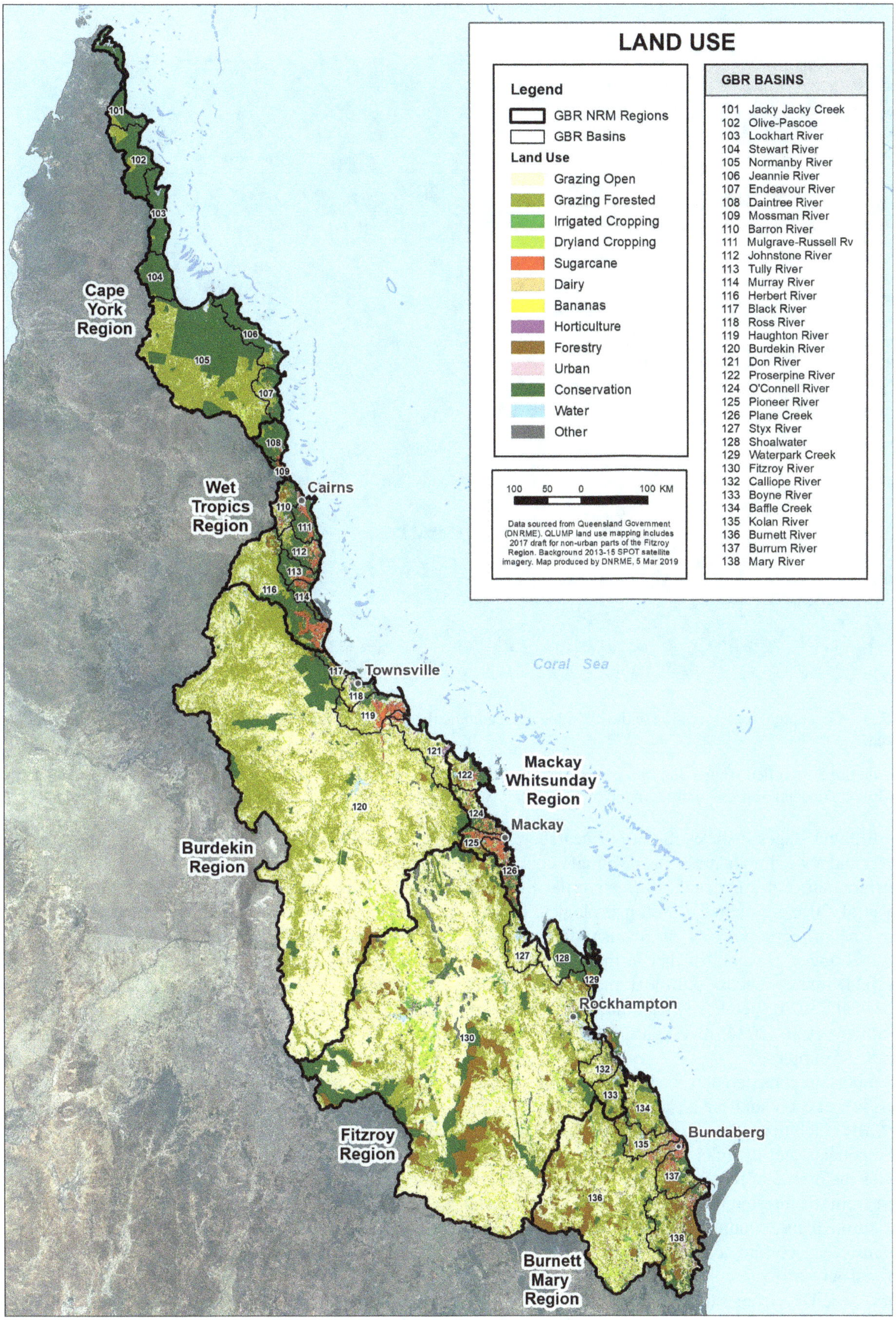

FIGURE 2 Land use in the GBR catchment area, showing basins listed from north to south by their Australian Water Resources Council (AWRC) basin number.

Source: McCloskey et al. (2021)

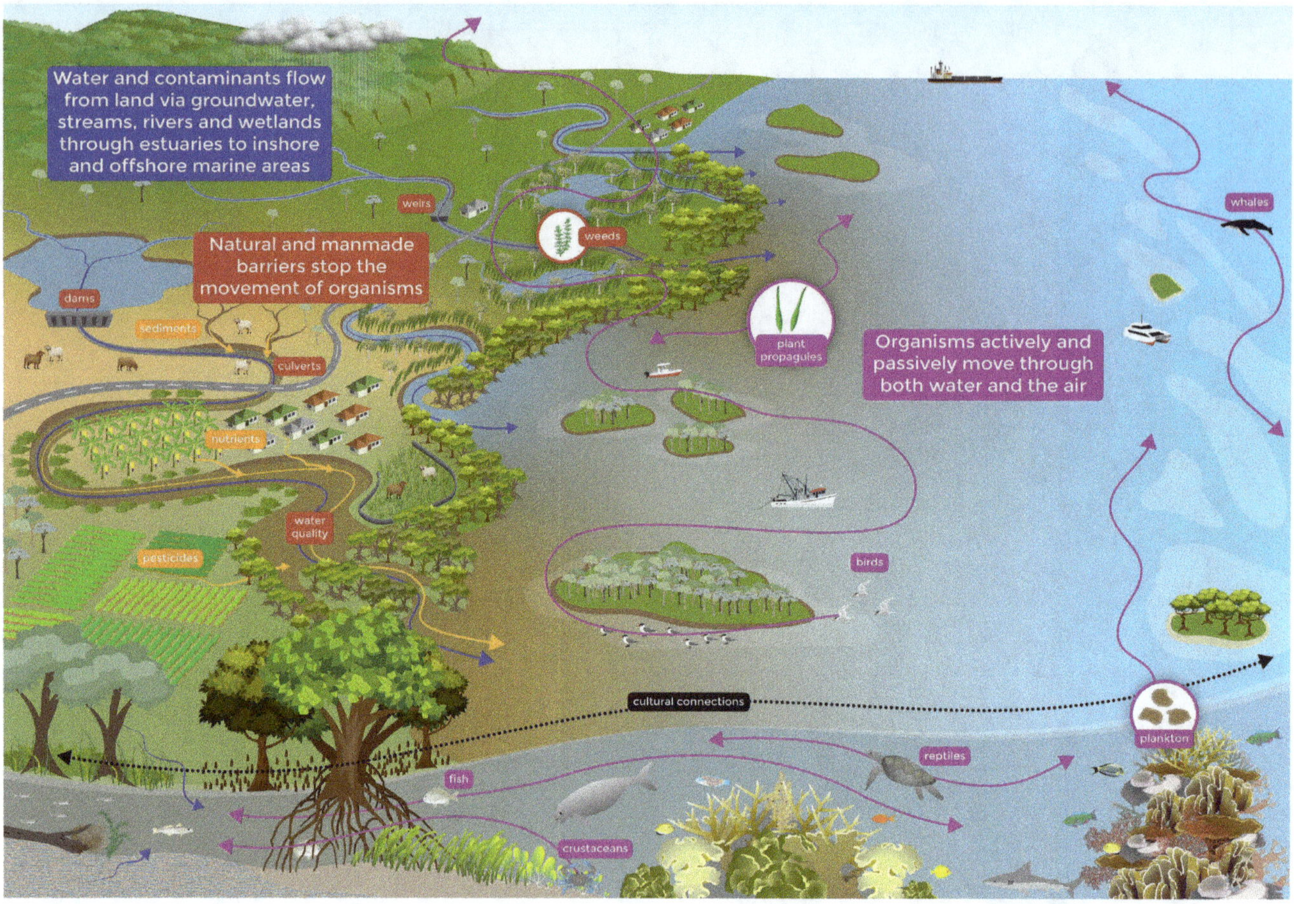

FIGURE 3 Conceptual representation of the hydrological, ecological, and social connections between the GBR and its adjacent catchment area.

Source: Figure 3 modified from a diagram prepared for the 2022 Scientific Consensus Statement for Land-based Runoff to the Great Barrier Reef. Conceptual presentation guided by the authors and prepared by Kate Hodge, Hodge Environmental

that at different stages includes the use of nearshore reefs, estuaries, and lower floodplain areas (Waltham et al. 2019). The barramundi (*Lates calcarifer*) has a similar life cycle (Crook et al. 2016). As noted above, hydrological connectivity is the primary delivery mechanism for pollutants to the GBR and is explored further in this chapter. Lewis et al. (2024) further discuss sediment and nutrient inputs, Sheaves et al. (2024) specifically examine fish populations, and Waltham et al. (2024) investigate wetland restoration options and examples.

An important consideration for these hydrological connections is that each land-sea or 'catchment to reef' profile in the GBR catchment area is unique in terms of ecosystem components and hydrology. The example in Figure 4 illustrates the profile of the Tully River, located in the Wet Tropics: from the forested headwater streams of the coastal ranges, through the floodplain with its remnant wetlands and extensive agriculture, to the shallow waters of the GBR lagoon, and outward to the continental shelf break. The profile of rivers of the Dry Tropics (e.g., the Burdekin and Fitzroy) is greatly extended to the landward side, with little forest and extensive woodland grazing country (Pearson et al. 2021). These characteristics significantly influence system processes, functions, and connectivity and are explained further in the following.

The most recent estimates of river discharge from the 35 major basins in the GBR catchment area are presented in Puignou Lopez et al. (in review). This highlights the variability of mean annual flow between basins and their inter-annual variability (Figure 5). The mean annual discharge exhibits a general north-south gradient with the wettest basins in the Cape York and Wet Tropics NRM regions (south to the Herbert River). These northern rivers, and the Wet Tropics rivers in particular, tend to have more consistent flows between seasons and contribute a much greater discharge per unit area than the rivers of the Dry Tropics (Puignou-Lopez et al. in review). The Wet Tropics rivers occupy approximately 5% of the GBR catchment area (McCloskey et al. 2021) but contribute around 32% of the mean total annual flow that reaches the GBR. In contrast, the larger Dry Tropics Burdekin and Fitzroy basins tend to deliver larger flows every 3 to 10 years (Furnas 2003; Lough et al. 2015). Although these basins comprise 68% of the total catchment area, together, they contribute only 21% of total freshwater runoff. Rainfall in these basins is influenced by the El Niño Southern Oscillation (ENSO) with the largest

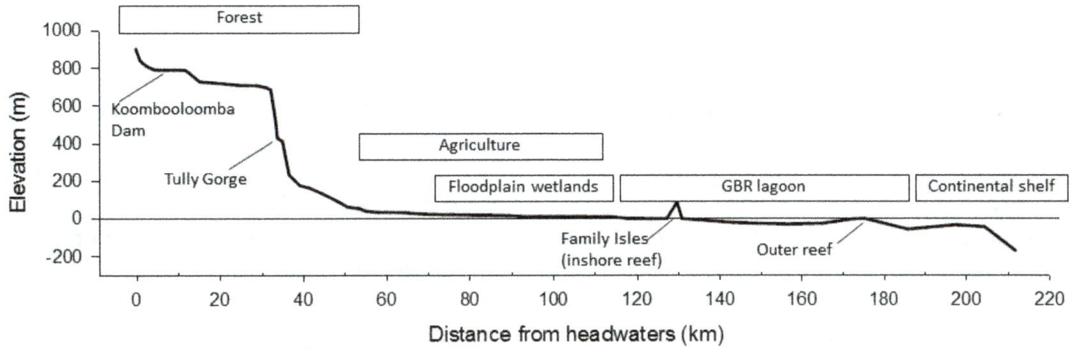

FIGURE 4 Connectivity of the GBR catchment to the reef.

Source: Brodie and Pearson (2016)

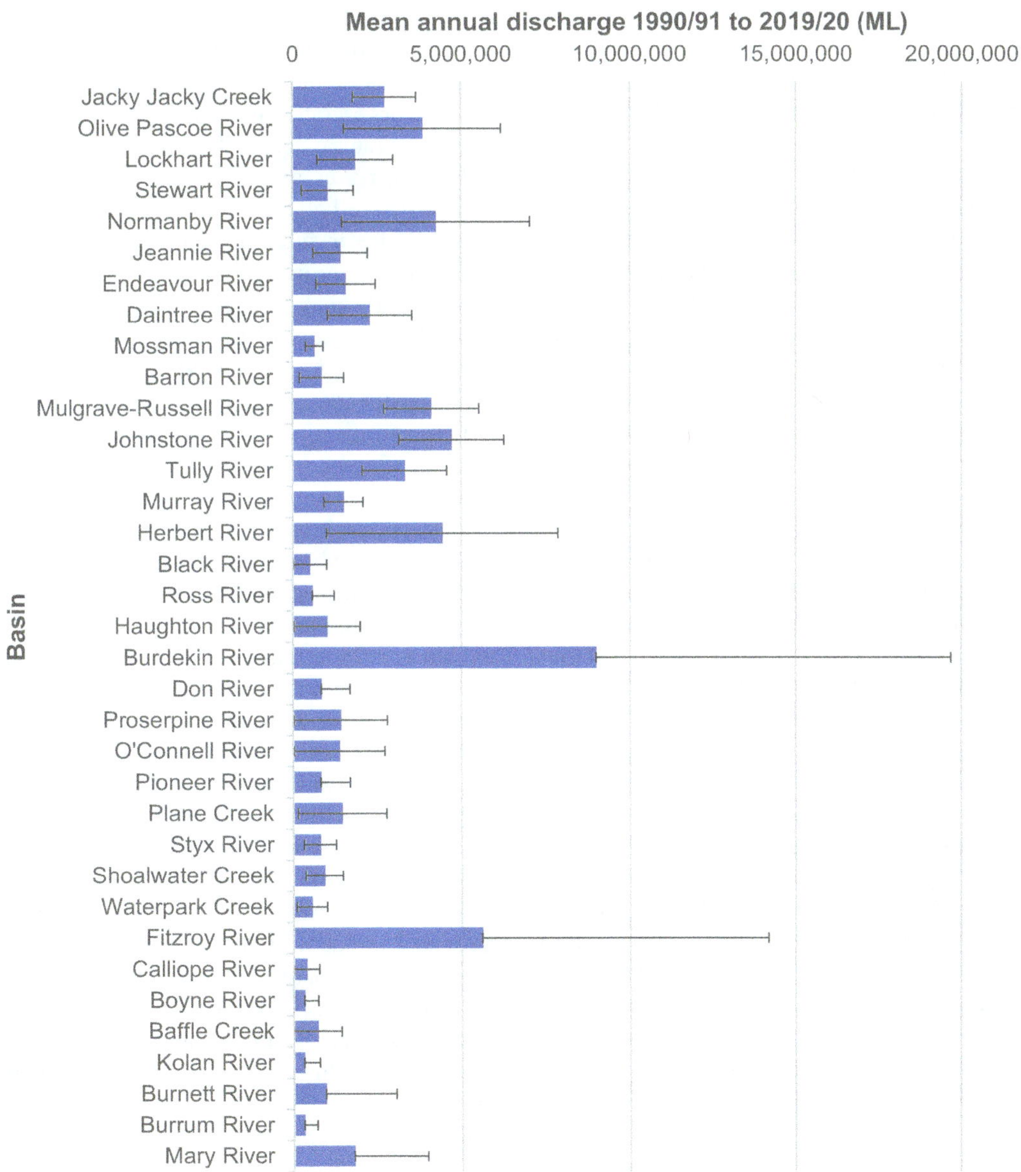

FIGURE 5 Mean annual discharge with 1 standard deviation for the 35 basins in the GBR catchment area.

Source: Puignou Lopez et al. (in review)

flows generated by major events such as rain depressions, the northern Australian monsoon, and tropical cyclones. Long proxy records of river discharge from luminescent lines in coral cores show the changing influence of ENSO in the GBR region over the Holocene period (e.g., Lough et al. 2014; Roche et al. 2014; Leonard et al. 2016). The most recent 50-to-100-year period of the 363-year (1648 to 2011) discharge reconstruction for the Burdekin River shows a near doubling of the extreme 90th percentile flows; such flow events also become much more frequent over this same period, which is potentially exacerbated in some basins by land clearing and catchment hardening (Lough et al. 2015). A 592-year (1413 to 2005) tree-ring proxy record of discharge from the Daly River (Northern Territory) displays similar trends, with streamflow increasing since the 1800s. The most recent 40-year period is unprecedented in terms of extreme flow events (Higgins et al. 2022), and this trend is likely widespread across northern Australia.

Hydrological connectivity is not restricted to the consideration of surface waters. Groundwater is vital to the health of freshwater systems, but accessing it for human supply applies pressure on the resource and the systems that depend on it (Nevill et al. 2010). Although much GBR water quality research has focussed on surface-water processes, and groundwater contaminant transport remains generally poorly understood (Brodie et al. 2008), a small number of studies have examined these interactions (including Hunter 2012, Connolly et al. 2015; Shishaye et al. 2020). There is considerable diversity of aquifer size, complexity, and groundwater-surface water connectivity across the Wet Tropics, Lower Burdekin, and Mackay-Whitsunday areas (Hunter 2012), with unconfined alluvial aquifers widely represented and confined or semi-confined aquifers present in some areas. There is a high degree of connectivity between groundwater and surface waters in all three areas. This has led to a diversity of management strategies across the GBR catchment area with the construction of artificial drainage networks in some areas to lower water tables and prevent waterlogging of sugarcane (Hunter 2012). This is particularly evident in the Wet Tropics region (Waltham et al. 2017), where groundwater discharge to waterways far exceeds groundwater discharge directly to the coast; in contrast, groundwater discharge to the coast via distributaries in the Lower Burdekin and Mackay-Whitsunday areas represents approximately 40% of total groundwater discharge from each aquifer (Hunter 2012). In the Lower Burdekin area, a history of groundwater use for irrigation and overdraft of shallow coastal aquifers have led to rising water tables and saltwater intrusion (Narayan et al. 2007; Lenahan and Bristow 2010; McMahon et al. 2012; Perna et al. 2012b; Shishaye et al. 2020). The challenges associated with water management in the Lower Burdekin are discussed further in Attard et al. (2024).

Wonky holes, referred to earlier, were initially studied in Halifax Bay and have since been identified in offshore areas from most rivers of the central and northern GBR (Stieglitz 2005). The major paleo-channel for the Burdekin region, near the mouth of the Haughton River, has been identified from seismic surveys (Fielding et al. 2003) and marks possible submarine groundwater discharge sites from the Burdekin region. Although the amount of freshwater discharged into the marine environment from wonky holes is largely unquantified, there is some evidence that groundwater discharge to the inshore GBR may have increased since European settlement. This increase is related to the clearing of the landscape for agriculture, which may have resulted in higher land-sea hydraulic gradients (Gagan et al. 2002).

7.3. GBR ECOLOGICAL CONNECTIVITY WITHIN THE CATCHMENT AND TO THE REEF

The physical hydrological connectivity of surface and ground water within the GBR catchment area and across the marine environment is reflected in many ecological processes and at a variety of scales, ranging from localised nutrient supply and chemical interactions to animal migrations across biomes. It includes active or passive breeding movements among habitats, dispersal, trophic interconnections, and interactions within and between species (Sheaves 2009; Harris et al. 2017). Connectivity provides access to alternative habitats within and between marine, estuarine, and freshwater environments that may be obligatory or opportunistic in the life cycles of many animal species. Connectivity may be linear along streamlines, driven by coastal currents, or distributed, such as on floodplains and in the sea. Connectivity is also important in waterway- or marine-dependent terrestrial species that may travel overland or aerially along waterways or between habitat patches such as islands, including species of birds (Buelow et al. 2018) and fruit bats (Reef et al. 2014).

The focus here is on ecological connectivity at the macro-scale, which is defined as the biota's continual or intermittent access within and between water bodies and their associated terrestrial habitats, ranging from headwater streams to estuaries, freshwater and saline wetlands, and marine waters, including the GBR lagoon and beyond. Connectivity for different species may be active or passive and facultative or obligatory. Active animal movements are typically two-way, namely, with or against the direction of water flow. Examples of aquatic animals, especially crustaceans and fish, which require system connectivity to sustain populations and complete their life cycles, are the main focus of this discussion. The extent of connectivity for these animals is influenced by directional currents and barriers to dispersal, both natural (e.g., waterfalls) and anthropogenic (e.g., dams). These notions of connectivity were embodied in the Australian Rainforest and Reef Cooperative Research Centres' "Catchment to Reef" research program (Pearson and Stork 2008), and subsequent similar programs, and are captured in the conceptual model shown in Figure 3.

Natural connectivity in the GBR catchment area varies substantially with the degree of seasonality and the total amount of rainfall, which affects many processes, with particular contrast between the Wet and Dry Tropics (Davis et al. 2017), as noted previously. The contrast is reflected in the terrestrial environment, which ranges from rainforest to open savanna woodland and grassland. During large floods, the capacity of lowland rivers is exceeded, and the floods extend on to the floodplain. This flood-pulse is crucial to the ecology of floodplain wetlands (Junk et al. 1989) in the global tropics. Connectivity on floodplains is often regarded as two-way, with wetlands replenished by seasonal flooding that then gradually drain back to the river system during the dry season (e.g., Junk et al. 1989; Karim et al. 2012, 2014; Molinari et al. 2022). However, many rivers in the GBR catchment area do not have a two-way connection as floodwaters drain to the sea via wetland linkages and local floodplain distributaries (Brizga et al. 2006; Rayner et al. 2008).

In the Dry Tropics, there may be sustained dry season base flow in larger rivers and those fed from basaltic aquifers, or there may be progressive disconnection (Blanchette and Pearson 2013). This pattern is reflected in some standing freshwater systems (Pearson et al. 2022a). The flows and connections between wetlands and whether they are permanent or intermittent (depending on run-off and groundwater dynamics) are major determinants of the species composition of a freshwater habitat. Throughout the GBR catchment area, these relationships affect the nature of estuaries, floodplain wetlands, the species composition, and extent of mangroves and saltpans in a similar way.

Many species use both saline and freshwater systems during their life cycles (Table 1). For example, freshwater prawns (*Macrobrachium* species) and jungle perch (*Kuhlia rupestris*) breed in estuaries (Pusey et al. 2004; Short 2004); barramundi (*Lates calcarifer*) breed at sea but undertake much of their development in rivers and wetlands (Halliday et al. 2010); and many species, such as mangrove jack (*Lutjanus argentimaculatus*), opportunistically access rivers (Pearson et al. 2010; GBRMPA 2012; Waltham et al. 2019). Cling gobies (e.g., *Stiphodon* species) live in short streams, but their larvae require saline waters (Thuesen et al. 2011), while the eel (*Anguilla reinhardtii*) migrates from headwater streams and rivers towards New Caledonia to breed, with juveniles swimming and using GBR currents to return to rivers (Pusey et al. 2004). Even some small freshwater shrimp and snails have estuarine larvae (Page et al. 2007; Davis and Ponder 2019).

Connectivity is also important within floodplain wetlands. In the Wet Tropics' Tully-Murray basins, floodplain lagoons support the early life-history stages of 15 of the 21 native species present, including 4 species that require access to saline habitats for larval development (Godfrey et al. 2017). Connectivity to the rivers, distance from the coast, and flood dynamics influence the fish assemblage. Similarly, on the Fitzroy floodplain in the Dry Tropics, connections among individual wetlands and adjacent habitats allow fish to access them as nurseries (Sheaves and Johnston 2008).

When considering the factors that can, or have, influenced ecological connectivity in the GBR basins, it is also important to maintain a long-term view of the ecological role of connectivity—for example, it is important in biogeographic and evolutionary processes. This may be illustrated by the Terapontidae fishes (grunters), several species of which occur in the GBR catchment area. Ancestrally, they appear to have been euryhaline, with a single transition to freshwater environments producing all Australasian freshwater species with subsequent dietary radiation and lineage diversification (Davis et al. 2012). Natural lack of connectivity can also be important in evolutionary processes and for maintaining particular biotic assemblages (Pearson et al. 2010). For example, isolation on mountain tops has led to speciation of crayfish, but even within basins, genetic differentiation within species occurs and may lead to speciation in the long term (Hurwood and Hughes 1998; Ponniah and Hughes 2004).

7.3.1. Passive Connectivity and Dispersal

Passive connectivity includes natural interchange of materials and species between basins and waterways and within waterways, such as organic input to fresh waters and estuaries, and insect emergence into terrestrial environments, providing subsidies to food webs (Cheshire et al. 2005; Leigh et al. 2013; Molinari et al. 2022). Passive dispersal includes the drifting of plankton, plant propagules, and animal larvae within currents in basins, estuaries, and coastal zones, which provides for colonisation of suitable habitat. For example, many invertebrate species drift in streams and rivers (Connolly and Pearson 2018), and larvae of the eastern Australian shrimp *Australatya* spp. drift as much as 160 km from headwater streams to estuaries, which they require for further development (Smith 1987). Others in the GBR catchment area include snails (Neritidae; Davis and Ponder 2019) and cling gobies (*Stiphodon* species; Thuesen et al. 2011), whose larvae develop in estuaries or at sea before migrating back upstream. Zooplankton move up and down estuaries with tides and stream currents (Godfrey et al. 2020). Seeds of some plants are also propagated along streams (Hyslop and Trowsdale 2012; Traveset et al. 2014), while seeds of some riparian species may be dispersed by the wind. Seeds of wetland plants are widely transported by birds (Soons et al. 2016), and birds can disperse other materials, such as nutrients from mangroves to islands (Buelow et al. 2018). Fruit bats may also transport nutrients from forests to mangroves (Reef et al. 2014). Aquatic passive dispersal may be arrested by large dams, but aerial dispersal is independent of such barriers.

7.3.2. Active Connectivity, Dispersal, and Migration

Many species actively move within freshwater environments and to estuarine or marine environments (Table 1). Fishes that actively move between fresh and saline waters during their life cycle include catadromous species that move downstream to breed in estuaries or at sea (e.g., jungle perch, *Kuhlia rupestris*, and the eel, *Anguilla reinhardtii*, respectively; Pusey et al. 2004). In the Burdekin system, at least eight species are critically dependent on access to estuarine and marine areas for spawning (Brizga et al. 2006). Other

TABLE 7.1
Native fish and crustacean species of the GBR catchment area that move between marine/estuarine environments and fresh waters.

Class/Family	Species	Common name
Chondrichthyes		
Carcharhinidae	*Carcharhinus leucas*	River whaler
Osteichthyes		
Ambassidae	*Ambassis agrammus*	Glassfish[1]
Anguillidae	*Anguilla obscura*	Pacific short-finned eel
	Anguilla marmorata	Giant mottled/marbled eel
	Anguilla reinhardtii	Long-finned eel
Centropomidae	*Lates calcarifer*	Barramundi
Chanidae	*Chanos chinos*	Milkfish
Cichlidae	*Oreochromis mossambicus*	Tilapia
	Tilapia mariae	Tilapia
Clupeidae	*Herklotsichthys castenaui*	River herring
Eleotridae	*Bunaka gyrinoides*	Greenback gauvina
	Eleotris acanthopoma	Spine-cheek gudgeon
	Eleotris fusca	Brown gudgeon
	Eleotris melanosoma	Ebony gudgeon
	Hypseleotris compressa	Empire gudgeon
Elopidae	*Elops hawaiensis*	Giant herring
Engraulidae	*Thryssa hamiltoni*	Anchovy[1]
Gerreidae	*Gerres filamentosus*	Silver biddy
Gobiidae	*Awaous acritosus*	Roman-nosed goby[2]
	Glossogobius aureus	Golden goby[2]
	Glossogobius bicirrhosus	Bearded goby[2]
	Glossogobius giurus	Flathead goby[2]
	Glossogobius sp.	Mountain goby[2]
	Mugilogobius notospilus	Pacific mangrove goby[2]
	Redigobius bikolanus	Speckled goby[2]
	Schismatogobius sp.	Scaleless goby[2]
	Sicyopterus lagocephalus	Rabbithead goby[2]
	Sicyopterus sp.	Cling goby
	Sicyopus sp.	Cling goby
	Smilosicyopus sp.	Cling goby
	Stenogobius cf. watsoni	(goby)
	Stiphodon spp.	Cling goby
Hemiramphidae	*Arramphis sclerolepis*	Snub-nose gar
Kuhliidae	*Kuhlia marginata*	Spotted flagtail
	Kuhlia rupestris	Jungle perch
Lutjanidae	*Lutjanus argentimaculatus*	Mangrove jack
Megalopidae	*Megalops cyprinoides*	Tarpon
Mugilidae	Ellochelon *vaigiensis*	Diamond scale mullet[2]
	Mugil cephalus	Sea mullet
	Crenimugil *buchanani*	Blue tailed mullet[2]
	Planiliza subviridis	Greenback mullet
Muraenidae	*Gymnothorax polyuranodon*	Freshwater moray
Pristidae	*Pristis* sp.	Freshwater sawfish
Scatophagidae	*Scatophagus argus*	Spotted butterfish
	Selenotoca multifasciata	Striped butterfish
Sciaenidae	*Nibea soldado*	Silver jewfish
Scorpaenidae	*Notesthes robusta*	Bullrout[2]
Sparidae	*Acanthopagrus berda*	Pikey bream[1]
Syngnathidae	*Doryichthys* sp.	River pipefish[2]

TABLE 1 *(Continued)*

Native fish and crustacean species of the GBR catchment area that move between marine/estuarine environments and fresh waters.

Class/Family	Species	Common name
	Hippichthys heptagonus	Reticulated freshwater pipefish[2]
	Microphis brachyrus	Short-tailed river pipefish[2]
Decapoda		
Atyidae	*Australatya* spp.	Freshwater shrimp: two species
	Caridina spp.	Freshwater shrimp: up to 8 species
	Paratya australiensis	Freshwater shrimp
Palaemonidae	*Macrobrachium* spp.	Freshwater prawn: 10 species

Source: Information sources: Palaemonidae, Short (2004); Atyidae, Davie et al. (2002), Page et al. (2007) and Choy et al. (2019); fish, Brizga et al. (2006), Pearson et al. (2010), Thuesen et al. (2011) and Marcus Sheaves (pers. com.).

Notes: [1] Vagrant, [2] Uncertain.

species are anadromous, spending much of their juvenile lives in fresh waters or estuaries, including barramundi, *Lates calcarifer* (Russell and Garrett 1985), mangrove jack, *Lutjanus argentimaculatus* (Waltham et al. 2019), and marine prawns, *Penaeus* species (Coles and Greenwood 1983). Many species use fresh waters and estuaries facultatively but may be considered important components of the assemblages (Pearson et al. 2010; GBRMPA 2012). In the Tully-Murray basin, connectivity of wetlands is variable in duration according to the distance from the coast or rivers, the drainage network (including artificial drains and wetlands), and flood magnitude barriers (Karim et al. 2012, 2014) but is important for a number of fish species (Godfrey et al. 2017).

7.4. THREATS: BARRIERS TO CONNECTIVITY AND LANDSCAPE MODIFICATION

7.4.1. Natural Barriers within the GBR Catchment Area

Natural barriers to connectivity in the GBR catchment area include waterfalls, rapids, and seasonal drying of habitats (Pearson et al. 2021). River levees may also create one-way barriers that block the return of floodwaters to the main river, which occurs in many GBR systems such as the Burdekin River in the Dry Tropics (Pearson et al. 2022a) and the Mulgrave River in the Wet Tropics (Rayner et al. 2008). Many waterfalls in the GBR catchment area prevent movement of most fish and prawn species, although the shrimp (*Australatya* spp.) and the eel (*Anguilla reinhardtii*) can climb or circumvent waterfalls (Smith 1987; Brizga et al. 2006). In many of the rivers of the Dry Tropics, seasonal drying leads to disconnection of permanent waterholes, which tend to develop their own character in terms of water quality and animal and plant assemblages (Blanchette and Pearson 2013; Marshall et al. 2016; Davis et al. 2017; Preite and Pearson 2017, 2021).

Nevertheless, genetic studies suggest that, even in the absence of obvious barriers, movements may be restricted in various species of shrimps, insects, and rainbowfish in headwater networks in the Wet Tropics (Hughes et al. 1996; Bunn and Hughes 1997; Thuesen et al. 2008). Additionally, although waterfalls typically act as barriers, they may be beneficial to some species—for example, the Bloomfield cod, *Guyu wujalwujalensis*, is restricted to upstream of a waterfall, which may protect it from predators (Pusey et al. 2004), and freshwater prawns (*Macrobrachium* species) in Wet Tropics streams may be seen during the day grazing over rocks upstream of barriers to large fish but not downstream (unpublished data). Similarly, fish assemblages upstream of the Burdekin Falls did not include large predators until the artificial introduction of the piscivorous sleepy cod (*Oxyeleotris lineolatus*), which resulted in negative impacts on a small gudgeon (*Mogurnda adspersa*) (Pusey et al. 2006).

7.4.2. Anthropogenic Barriers within the Catchment Area

A large proportion of the world's coastal systems and their land-sea interconnection are exposed to high human pressure (Williams et al. 2022). Increasingly, biological reserves throughout the world are threatened by alterations in hydrologic connectivity that cause extirpation of migratory species, food-web effects, and genetic isolation (Pringle 2001). Declining fish communities characterise global and Australian freshwater environments, particularly

FIGURE 6 Connectivity issues: (a) dams (Burdekin Falls), (b) weirs (Ross River), (c) drop boards, (d) fish way around drop boards, (e) and (f) blockage of waterways by invasive plants, (c)-(f) located on the Burdekin floodplain.

Source: Image source: Richard Pearson

through lost river connectivity due to water resource development (Harris et al. 2017), which isolates populations and reduces gene flow (Machado et al. 2022). The ecological impacts of dams and flow management are generally well understood and include severance of instream and riparian connectivity, reduction of seasonality, and major habitat changes upstream and downstream (e.g., Kingsford 2000; Geist 2021).

Figure 6 presents a number of examples. Dams and other water storage infrastructure on major rivers are obvious potential barriers to catchment to reef connectivity (Waterhouse et al. 2016), and they modify the hydrological environment of downstream wetland and marine ecosystems and vegetation communities (Robertson et al. 1996).

Dams and large weirs have been built on many rivers, both large and small, in the GBR catchment area; Table 2 shows the major dams in the GBR catchment area. Although these structures block the easy passage of most aquatic fauna, both within rivers and between rivers, estuaries, and the sea, in some cases, dams may be built on sites that were previously impassable by fish. For example, the

TABLE 7.2
Major dams in the GBR catchment area—location by basin, year of completion, and capacity.

River basin	Reservoir	Year completed	Capacity (ML)	End of catchment mean gauge flow (ML)	Year record	C:Q ratio	Gauging station used
Barron	Tinaroo Falls Dam	1958	407,000	840,000	1916–1958	0.54	Barron River at Kuranda
	Copperlode Falls Dam	1976	45,560				
Tully	Koombooloomba Dam	1961	212,020	3,100,000	1973–2012	0.07	Tully River at Euramo
Ross	Ross River Dam	1974	417,000	280,000	1916–1974	1.49	Ross River at Gleeson's Weir
Burdekin	Burdekin Falls Dam	1987	1,860,000	8,300,000	1921–1969	0.24	Burdekin River at Clare
	Eungella Dam	1969	131,000				
Proserpine	Peter Faust Dam	1990	500,000	92,000	1957–1990	5.43	Proserpine River at Peter Faust Tailwater
Sandy Creek	Kinchant Dam	1977	62,800	170,000	1967–2012	0.37	Sandy Creek at Homebush
Pioneer	Teemburra Dam	1997	148,000	830,000	1917–1997	0.18	Pioneer River at Pleystowe
Fitzroy	Fairbairn Dam	1972	1,440,000	5,600,000	1915–1970	0.27	Fitzroy River at Yaamba
	Fitzroy River Barrage	1970	65,920				
Boyne	Awoonga High Dam	1984	250,000	380,000	1939–1983	0.66	Boyne River at Annondale
Kolan	Fred Haigh Dam	1975	586,000	420,000	1967–1975	1.40	Kolan River at Kinagin
Burnett	Boondooma Dam	1983	212,000	1,300,000	1911–1998	0.39	Burnett River at Walla
	Paradise Dam	2005	300,000				

Source: Reproduced from Waterhouse et al. (2016)

site of the Burdekin Falls Dam always prevented passage of fish upstream, except for *A. rheinhardtii* (Brizga et al. 2006). Some dams are built upstream of the falls that have always prevented upstream movement of fish (e.g., Tinaroo Dam, upstream of the Barron Falls on the Barron River). Depending on where such dams are located, however, they may preclude normal, within-river movement for a major portion of the river thereby blocking necessary habitat in species' life cycles. Weirs may similarly block passage of fish for much of the time but allow passage during floods that overtop weirs. However, not all species can move under high flow conditions (Brizga et al. 2006), although side channels and riparian zones may facilitate passage because of reduced current velocities.

The presence of water storages can also have a substantial influence on material delivery. For example, Lewis et al. (2013) showed that sediment trapping in the Burdekin Falls Dam is typically between 60–70% during moderate to very large flow events with a more consistent trapping efficiency around ~60% (± 10%). Importantly, this sediment trapping efficiency differs depending on particle size, with a large proportion (> 50%) of fine particles (< 20 μm) readily passing through the dam, in contrast to sediment particles > 30 μm, which are almost totally trapped (Lewis et al. 2013). These fine particles are of greatest concern because of deleterious effects on marine ecosystems (Bainbridge et al. 2012, 2018, 2021). Hence, sediment trapping by the Burdekin Falls Dam has only a moderate effect on reducing the risk from total suspended solids in the GBR ecosystems.

On floodplains in the GBR, natural connectivity has been disrupted by construction of drains, levee banks, bunds, tidal barrages, culverts, drop-boards, and low flows and by excessive plant growth and poor water quality (Brizga et al. 2006; Carter et al. 2007; Veitch et al. 2008; Perna et al. 2012a; Sheaves et al. 2014; Waterhouse et al. 2016; Waltham et al. 2019). Bunds consist of earth walls that are constructed to retain water in low-lying coastal areas to increase agricultural production. However, they have also restricted movement of fish, although efforts are now underway to remove or modify these structures to increase habitat values (Abbott et al. 2020; Waltham et al. 2019; Waltham et al. 2024). Even simple culverts can prevent upstream movement of fish and are very abundant—for example, over 5,000 possible barriers have been identified in the Wet Tropics region (Lawson et al. 2010; Wet Tropics Management Authority 2013; Kroon and Phillips 2016; Wet Tropics Waterways 2022), and in the Fitzroy region, Moore and Marsden (2008) identified approximately half of a total of 10,632 potential barriers as being actual barriers to fish migration. The absence of effective fishways to circumvent many barriers in many GBR basins is noteworthy. Flow regulation that restricts flooding may also prevent dispersal of organisms between rivers and wetlands and within rivers (Bunn and Arthington 2002).

7.4.3. Landscape Modification

GBR basins and coastal ecosystems have been highly modified through changes in land use at a landscape scale, resulting in significant modification of natural ecosystem processes. However, the extent or consequences of these changes have not been fully quantified. Waterhouse et al. (2016) discussed the loss of coastal ecosystems, including wetlands, and the presence of water storages across the GBR catchment area. This information has been reviewed and updated where possible here. Other datasets that are consistently available across the GBR basins are limited, although the type and scale of information available in this field is continuously improving through NRM bodies, research institutions, and government organisations. A particularly relevant example is the Queensland Government's WetlandInfo website.

Many of the riparian and wetland areas in the GBR catchment area have been significantly reduced in function and output. In their natural state, these systems have the capacity to capture and process basin outflows into essential inputs to marine productivity from diatoms and phytoplankton to higher trophic levels that include crustaceans and fish (Sheaves et al. 2014). A comparison of 2017 and pre-clear wetland extent indicates that the area of wetlands remaining in GBR basins has declined substantially in some areas (Table 3; DES 2015). The loss of wetlands has been most significant in the Wet Tropics region (30.5%) and Burnett Mary region (28.5%). The greatest losses are in palustrine wetlands across all regions (except for Cape York), particularly in the Wet Topics and Mackay Whitsunday regions (an approximate 49% and 43% loss, respectively). Riverine wetlands are also showing greater losses, ranging between approximately 10% and 36% (excluding Cape York). The condition and values of the remaining wetlands across the GBR basins are not well documented. This is an important gap in the knowledge required to inform a more comprehensive whole-of-system approach to GBR management.

The implications of these changes have not been quantified and are very difficult to assess. However, there is sufficient conceptual understanding to suggest that the modification of ecosystems and the loss of ecosystem function have modified the biological, biogeochemical, and physical functions that would have supported the health of the GBR and the connected coastal ecosystems (Great Barrier Reef Marine Park Authority 2012; Creighton et al. 2021). The work of the Great Barrier Reef Marine Park Authority (GBRMPA) in assessing the status and function of coastal ecosystems in the GBR catchment area is a noteworthy step towards establishing methods to quantify hydrological connectivity and associated functions. A key outcome of this program was the Coastal Ecosystem Assessment Framework and the 'Blue Maps' products (GBRMPA 2017a), which identify areas in the GBR catchment area that hydrologically connect coastal ecosystem functions to the GBR. Several regional NRM bodies have used this information to help prioritise areas where ecosystem repair may be advantageous for improving water quality and other values such as biodiversity, but there has not been widespread recognition of this framework across the GBR. Further evaluation of this approach in conjunction with field assessments could lead to a much greater understanding of the connection between the catchment and reef ecosystems.

Many of these changes have been associated with vegetation clearing or the instalment of drainage networks in the lower parts of the catchment area, which has broader implications for the quality and quantity of surface runoff from the basins to the GBR. In the upstream reaches, it is well-established from plot-scale trials that the clearing and compaction of the soil results in less infiltration and greater (and faster) surface runoff (e.g., McIvor et al. 1995; Roth

TABLE 7.3

Percentage of wetlands remaining in each NRM region based on the 2017 wetland extent/pre-clear wetland extent, presented by wetland type and total wetland extent. The areas do not include marine or estuarine waters but do include estuarine wetland vegetation (e.g., mangroves and tidal flats) and exclude artificial and highly modified wetlands. As a relative measure, red cells indicate areas with less than 60% remaining, orange indicate 61–80%, yellow indicate 81–95%, and green indicate greater than 95%.

NRM region	Total area 2017 (km²)	2017 area/pre-clear wetland extent (% remaining)					Total % loss
		Estuarine	Lacustrine (e.g., lakes)	Palustrine (e.g., swamps)	Riverine	Total	
Cape York	136,521	100	99.2	99.3	99.8	**99.6**	**0.4**
Wet Tropics	22,230	93.4	99.3	51.4	70.7	**69.5**	**30.5**
Burdekin	140,874	89.6	99.8	91	90.9	**90.8**	**9.2**
Mackay Whitsunday	9,335	97.1	n/a	56.4	88.2	**89.3**	**10.7**
Fitzroy	156,762	85.7	87.8	80.8	81.8	**83.2**	**16.8**
Burnett Mary	55,768	98.1	100	68.5	63.8	**71.5**	**28.5**

Source: DES (2015)

2004). At a larger paddock scale (10 to 20 ha), data from the Brigalow Catchment Study show that discharge has doubled to tripled (with higher peak runoff) when a virgin Brigalow catchment is compared to surrounding catchments that were cleared for grazing and cropping (Thornton et al. 2007; Thornton and Yu 2016; Thornton and Elledge 2021). The clearing of Brigalow has also led to increased losses of sediment and phosphorus from the landscape (Bartley et al. 2017; Elledge and Thornton 2017). However, at a larger sub-catchment to catchment scale (> 1,000,000 ha), the increase in runoff because of the Brigalow clearing is more difficult to quantify due to climate variability, rainfall patterns, and the catchment response (e.g., Siriwardena et al. 2006; Peña-Arancibia et al. 2012). An approach based on multiple lines of evidence has recently untangled these other influences to reveal an increase at a sub-catchment scale since the clearing of the Brigalow between 1960 and 1980 (Cheng and Yu 2019). Bartley et al. (2022) demonstrated that regenerative grazing practices (i.e., allowing the landscape to 'rest' from cattle) will result in improvements to vegetation (plant biomass) and soil and land condition but may take up to 20 years before significant improvements can be measured.

In coastal areas, drainage and modification of coastal plains can lead to acidification, metal contamination, deoxygenation, and iron precipitation in reef-receiving waters (Powell and Martens 2005). This is clearly documented in a case study at East Trinity Inlet near Cairns where disturbance of coastal areas resulted in discharge of acid into Trinity Inlet, resulting in periodic fish kills (Powell and Martens 2005). However, despite good progress in the broad identification of potential areas of acid sulphate soils (coastal areas below 20 m Australian Height Datum), there is no clear knowledge of the extent of the impacts of acid sulphate soils in the GBR catchment area and coastal zones.

7.5. CONSEQUENCES OF MODIFICATIONS TO THE GBR CATCHMENT AREA

The loss of ecological connectivity, both within the GBR catchment area and from the catchment to the reef, has highlighted a number of consequences for particular species and catchment ecosystems. At the landscape scale, an important influence of modifications is on the increases in the delivery of land-derived pollutants to the GBR (Brodie et al. 2013; Waterhouse et al. 2017), especially during floods, while direct impacts of pollutant exposure are especially important for freshwater ecosystems (e.g., Davis et al. 2017).

In the GBR, dissolved nutrients promote the growth of benthic macroalgae, which may smother suitable settlement sites for many coral species, thereby affecting coral establishment and growth (Fabricius 2011; McCook et al. 2001). Dissolved nutrients may promote proliferation of planktonic algae, which facilitate the growth and survival of crown-of-thorns starfish larvae; this is one of several drivers of population outbreaks of the starfish, which can severely damage coral reefs (Brodie et al. 2005; Fabricius et al. 2010). Fine sediments (less than 20 μm) reduce the light available to seagrass ecosystems (e.g., Collier et al. 2012, 2016; Chartrand et al. 2018) and inshore coral reefs (Wolanski et al. 2003; Fabricius et al. 2014, 2016), and pesticides pose a toxicity risk to freshwater ecosystems and some inshore and coastal habitats (Lewis et al. 2009, 2012; Templeman and Williams 2024).

Assessments of risk of land-based pollutant inputs to GBR coral reef and seagrass areas have consistently identified areas of higher exposure for each pollutant (Brodie et al. 2013; Waterhouse et al. 2017). The total area of potential influence has been defined as approximately 91,000 km^2, or one-quarter of the total area of the GBR Marine Park (Waterhouse et al. 2017). The extent of exposure varies among basins and pollutants, and heavily depends on the volume of river discharge, distance from the coast, weather conditions, and currents (typically, there is northward movement of river discharge along the GBR coast) (Waterhouse et al. 2017). Waterhouse et al. (2017) concluded that exposure to dissolved inorganic nitrogen (DIN) is significant for all inshore areas, with the highest end-of-catchment DIN loads, including the basins of the Wet Tropics and Burdekin regions and the mid-shelf areas between Lizard Island and Townsville. Exposure to fine sediment is significant in areas of shallow seagrass and coral reefs on inshore areas adjacent to basins with high fine sediment loads, particularly the Burdekin (south of Townsville), Fitzroy (Rockhampton), and Mary (Maryborough) basins. Large areas of the inshore coral reefs and seagrass meadows are in the highest exposure categories (Figure 7). Notably, the areas of highest exposure have high recreational and commercial use and include highly valued tourism areas including Low Isles, the Frankland Island group, Dunk Island, Hinchinbrook Island, Magnetic Island, and many of the Whitsunday Island group. Therefore, it is clearly of great importance to understand the factors that influence the delivery of land-based pollutants to these areas of the GBR.

Pesticides are a potential major problem for organisms in the GBR (Waterhouse et al. 2017). The highest pesticide concentrations are typically found adjacent to and downstream of their source (see Templeman and Williams 2024), including areas of intensive cropping, which are largely sugarcane cultivation and horticulture (Brodie and Landos 2019; Water Quality and Investigations 2020). Freshwater wetlands, rivers, and estuaries are exposed to the highest concentrations, followed by coastal ecosystems, seagrass, and coral reefs. In the assessment of pesticide risk, Waterhouse et al. (2017) concluded that only a few basins present a very high to moderate risk to end-of-catchment ecosystems from Photosystem II herbicides, with diuron presenting the highest risk. These basins are generally characterised as smaller coastal catchments with high proportions of sugarcane land use (i.e., the basins within the Mackay-Whitsunday, Lower Burdekin, and Wet Tropics regions). The capacity to assess the spatial exposure of pesticides in the marine environment has recently been developed in the eReefs modelling suite (Skerrat et al. 2023), which will assist in updating the risk of pesticide exposure in the GBR.

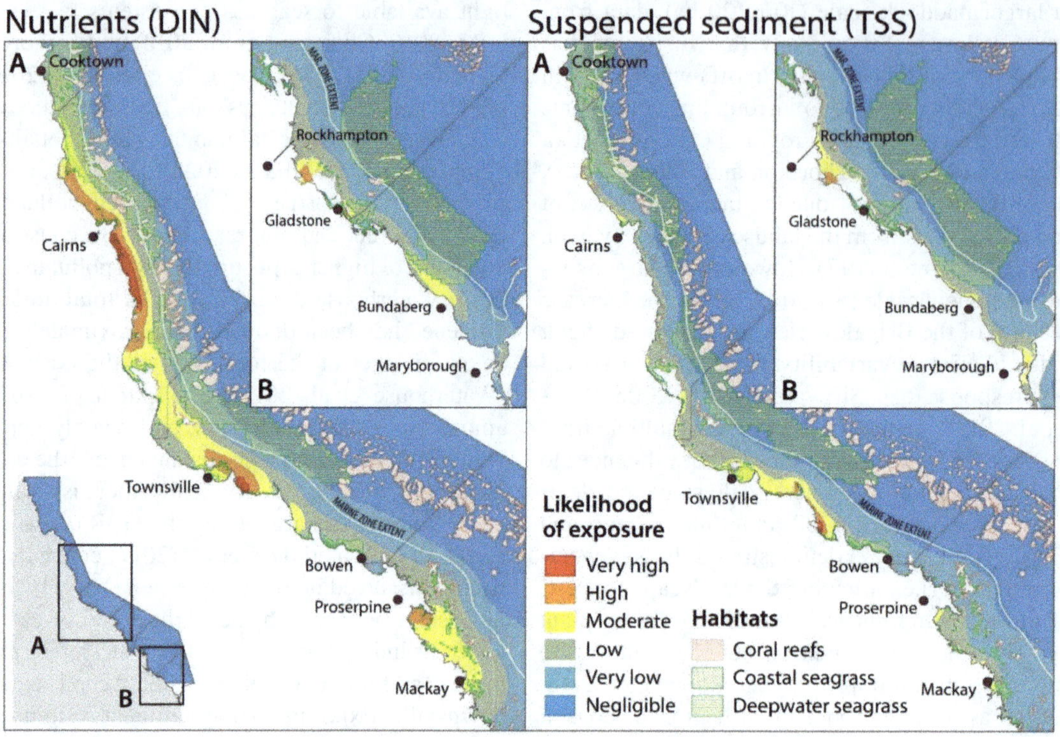

FIGURE 7 Likelihood of exposure of DIN and total suspended sediment in the GBR. Cape York is not shown here as the assessment identified a low or very low likelihood of exposure north of Cooktown.

Source: Reproduced from Waterhouse et al. (2017); refer to the full publication for the description of the methods and results at the entire GBR scale and the following notes for data sources.

The timing of inputs is important and differs between freshwater and marine ecosystems. The greatest risk in fresh waters is often from acute contamination during early wet-season 'pre-flush' run-off, transporting contaminants that have built up during the dry season, or continuous input of contaminants over long periods of base flow (Pearson and Stork 2008; Brodie et al. 2009; Davis et al. 2017), whereas the estuarine and marine areas are at greatest risk during the floods of the wet season (November to May). This is also when seawater temperature is highest, which potentially results in greater combined impacts, such as increased bleaching susceptibility with elevated nutrients (Wooldridge 2009; Vega Thurber et al. 2013; Wiedenmann et al. 2013; D'Angelo and Wiedenmann 2014; Wooldridge et al. 2017). This is also a period of other adverse stressors, such as larval crown-of-thorns starfish survivorship (Pratchett et al. 2023). Outside of the wet season, land-based influences on the broader GBR are comparatively smaller, and nutrient requirements for productivity are dominated by recycling in the GBR lagoon, although nutrient upwelling can be important in some outer shelf areas (e.g., Swains, Palm Passage, and far northern GBR) (Furnas et al. 2011). However, chronic inputs from rivers and floodplains can be important, especially in irrigated areas (see Attard et al. 2024). Turbidity from the presence of fine sediment is heavily influenced by wind- and wave-driven resuspension and supplemented by river inputs (Lewis et al. 2020). These differences are important to recognise in the design and implementation of management options across the landscape.

7.6. FUTURE CONNECTIVITY: DEVELOPMENT AND MANAGEMENT

Further construction of water infrastructure has long been discussed in the GBR catchment area, with numerous proposals for dams and associated infrastructure to service new agricultural areas, mines, and urban development. The most recent of these has been raising the Burdekin Falls Dam and two major dams higher in the Burdekin catchment, on the upper Burdekin River itself and on the Broken River (Pearson et al. 2022b). All proposals present a high risk of substantial degradation of habitats, connectivity, and ecological processes in large proportions of the river lengths, including migrations by freshwater fish and coastal fishery production (Brizga et al. 2006). Whatever developments may take place, it would be expected from current regulations that impact on the aquatic system associated with land-based development would be incorporated in environmental impact statements (e.g., Urannah Project, Queensland Government 2021). It might also be expected that new infrastructure would be designed to make proper use of environmental flows (i.e., supplements or reductions to mirror the natural hydrograph) and fish passage devices to maintain connectivity, which mostly appears not to be the case currently.

Future management strategies for the GBR need to protect, maintain, and restore coastal ecosystems and their functions to maintain and restore land-sea connectivity, if the values of the GBR and its resilience are to be retained in the long term (Waterhouse et al. 2016). Maintaining normal biodiversity and ecosystem processes means maintaining natural connectivity or its absence (i.e., natural barriers, which may be important in the evolution of isolated populations of species). This means maintaining natural flow regimes and linkages in rivers, wetlands, and estuaries, mitigating the effects of anthropogenic barriers by incorporating environmental flows and fish passage mechanisms, reducing weed infestations and water quality barriers, and possibly supplementing flows through distributary channels (Pearson et al. 2010; GBRMPA 2013; Waltham et al. 2019). The possible effects of climate change are also a major consideration as they may affect flows and within-river disconnections, potentially reducing biotic diversity (James et al. 2017). Rising sea levels will cause incursion of saline water in the lowlands with negative impact on wetlands and their biota (Pearson et al. 2010). Management strategies for more specific consequences of landscape modification are discussed in other chapters—for example, sediment management in Lewis et al. (2024) and pesticides in Templeman and Williams (2024).

7.7. IMPLICATIONS FOR THE GBR AND FUTURE APPROACHES

Because of the close links between the GBR and its catchment area, these landscapes need collective management as the 'Greater Barrier Reef' region (Brodie and Pearson 2016). Although this chapter has emphasised that connections are most prevalent between the catchment and inshore coastal waters, these connections extend through both physical and ecological processes to reefs and beyond. Although there is no overarching authority to undertake this task, the Queensland and Australian governments and regional NRM bodies have commenced programs to improve water quality reaching the GBR lagoon, which are underpinned by the primary joint governmental water quality policy and planning document—the Reef 2050 Water Quality Improvement Plan (and preceding policies from 2003; Australian and Queensland Government 2018). The actions within this Plan and associated initiatives (e.g., Queensland Government Reef Water Quality Program, the Australian Government Reef Trust Program, and the Reef Trust Partnership coordinated by the Great Barrier Reef Foundation) can be generally beneficial to ecological processes, although water quality itself is only one of the issues affecting connectivity and reef health. Management actions in catchments and on floodplains, as outlined previously, need major attention to alleviate anthropogenic impact on habitats and broad catchment to reef connections.

There has been some positive progress in this area of policy and management for the GBR in recent years. Of particular relevance, the Queensland Wetlands Program has recently published the Whole-of-System, Values-Based Framework (DES 2022). The Framework *'draws explicit connections between the biophysical environment, the beneficiaries of the services provided by that ecosystem and their values'* and *'is designed to ensure that outcomes are achieved for the entire socio-ecological system'*. It provides the rationale to underpin management approaches that are relevant to the GBR catchment to reef landscape and guides users in assessing key elements of the system to direct optimal management outcomes, including the assessment of drivers, system parts, and processes; intrinsic values and ecosystem service; stakeholder values and beneficiaries; potential threats/pressures and opportunities; the definition of management objectives; the identification of a mix of management interventions; detailed design and analysis; and implementation and adaptive management. This Framework is directly applicable to the challenges raised in this chapter and provides a strong foundation for investors in strategic design and implementation of catchment to reef management for the GBR. It is, however, not formally funded and will require adoption by governments to provide confidence that it is being applied at a broad scale and across major initiatives.

The Regional Report Card program is also relevant here, and it is supported by solid partnerships among key stakeholders, the government, and community in many of the NRM regions in the GBR catchments. These programs incorporate specific indicators of river flows, fish movement, and estuarine health (e.g., Wet Tropics Waterways 2022) and are collecting valuable information to inform future investment options and management decisions.

The Coastal Ecosystem Assessment Framework developed by the GBRMPA has real potential in prioritising areas of coastal ecosystems to maximise water quality and biodiversity outcomes for the GBR. However, the current recognition of this framework is limited. Further reference to this work in new policies such as the *Reef Blueprint for Resilience* (GBRMPA 2017b) would ensure that a whole-of-system approach to management of the GBR and its catchments is formally valued in future management strategies. Despite the lack of widespread adoption of this work, the Blue Maps products were adopted in a number of Regional Water Quality Improvement Plans between 2014 and 2016 (e.g., Wet Tropics, Burdekin, and Fitzroy). A key example is the comprehensive assessment of options for restoring catchment waters and ecological function in the Wet Tropics basins undertaken as part of the Wet Tropics Water Quality Improvement Plan (Terrain NRM 2015). This was supported by a specifically designed Wet Tropics System Repair Spatial Planning Framework that aggregates values, threats, and the management response at the regional scale, with further details on the impacts of one or more threats within a basin. The assessment derives management objectives for specific areas (based on the outcomes of the spatial analysis) framed around 'intact and natural catchments' including actions to *protect* and *restore* assets, services, and processes and 'disturbed and modified catchments' including actions to *maintain* and *adapt* assets, services, and processes.

The underpinning concepts of these frameworks align with the key criteria for managing complex catchment to reef landscapes proposed by Creighton et al. (2021). The suggested criteria for success are (i) the lead role of the community, (ii) the need for a systems approach, (iii) the need for targeted, cost-effective, and sustainable long-term investment, and (iv) building knowledge by maintaining consensus and commitment in the face of politically motivated interference. Criteria (i) and (ii) around community involvement and the need for a systems approach are recognised in the examples above, whereas criteria (iii) and (iv) warrant further reference in relation to this chapter.

Regarding criterion (iii), Creighton et al. (2021) suggest three key principles including the need for refined investment prioritisation analyses at smaller scales than previously conducted (e.g., within catchments), allocation of high investment priority for protective management and natural area acquisition rather than restorative actions (if possible), and the need for long-term investment options. Long-term investment is vital for projects that explicitly adopt a whole-of-system approach that aims to protect, maintain, restore, or adapt different parts of an entire system. This approach can take many years to design and implement. Short-term funding in the GBR investment portfolio has tended to lead to a piecemeal approach to investment, which means that critical parts of programs are often left out (Eberhard et al. 2017).

Finally, criterion (iv) refers to building knowledge, building and maintaining consensus, and political commitment. These aspects are critically important for the future management of the GBR and rely on an ongoing commitment to filling priority knowledge gaps and the development of policy and management strategies that are underpinned by strong evidence. The physical and ecological connections between the GBR catchment area and the reef are strong, and although there is a substantial foundational knowledge base to support this, there are ongoing challenges in overcoming the jurisdictional issues of the joint State and Federal government management arrangements from the catchment to the reef. There is a need for renewed recognition from both governments of the importance of these connections in maintaining the Outstanding Universal Values of the GBR World Heritage Area.

Accordingly, it is evident that the design and delivery of more sustainable social, ecological, and economic outcomes for the GBR and its catchment area will accrue from a number of highly significant actions. These include further development of the science underpinning ecological connectivity between the GBR catchment area and the reef; better understanding of the likely implications of climate change to these relationships; greater integration of indigenous and community values and knowledge into the catchment to reef management frameworks; expanded efforts in monitoring and evaluation to allow for adaptive management; and greater education and awareness of the key issues among the broader community, including politicians. With strategic implementation of these actions and recognition of the need for a whole-of-system approach to the management of the GBR and its catchments, there is a higher likelihood that the important connections between the catchment and the reef and the benefits that they provide can be protected, maintained, and in many cases, restored.

ACKNOWLEDGEMENTS

The authors would like to thank Dr. Marcus Sheaves for his valuable input on the finalisation of Table 1 and the associated text. We are also grateful for the input provided by Dr. Mari-Carmen Pineda in the development of the conceptual diagram in Figure 3 and to Kate Hodge for preparing the Figure.

REFERENCES

Abbott BN, Wallace J, Nicholas DM, Karim F, Waltham NJ (2020) Bund removal to re-establish tidal flow, remove aquatic weeds and restore coastal wetland services—North Queensland, Australia. PLoS One, 15:e0217531.

Attard S, Chaiechi T, Connell JG (2024) Pathways to improved water quality in the GBR lagoon—exploring opportunities for broadscale application of low-risk practices in the Lower Burdekin irrigated agriculture areas. In: Wolanski E, Kingsford MJ (eds.), Oceanographic Processes of Coral Reefs. Physical and Biological Links in the Great Barrier Reef. CRC Press, Boca Raton, FL.

Australian and Queensland government (2018) Reef 2050 Water Quality Improvement Plan 2017–2025. State of Queensland, Queensland.

Bainbridge ZT, Lewis SE, Bartley R, Fabricius KE, Collier CJ, Waterhouse J, Garzon-Garcia A, Robson BJ, Burton JM, Wenger A, Brodie JE (2018) Fine sediment and particulate organic matter: A review and case study on ridge-to-reef transport, transformations, fates, and impacts on marine ecosystems. Marine Pollution Bulletin 135:1205–1220.

Bainbridge ZT, Lewis SE, Stevens T, Petus C, Lazarus E, Gorman J, Smithers S (2021) Measuring sediment grain size across the catchment to reef continuum: Improved methods and environmental insights. Marine Pollution Bulletin 168:112339.

Bainbridge ZT, Wolanski E, Álvarez-Romero JG, Lewis SE, Brodie JE (2012) Fine sediment and nutrient dynamics related to particle size and floc formation in a Burdekin River flood plume, Australia. Marine Pollution Bulletin 65:236–248.

Bartley R, Abbott BN, Ghahramani A, Ali A, Kerr R, Roth CH, Kinsey-Henderson A (2022) Do regenerative grazing management practices improve vegetation and soil health in grazed rangelands? Preliminary insights from a space-for-time study in the Great Barrier Reef catchments, Australia. The Rangeland Journal 44(4):221–246. doi:10.1071/RJ22047

Bartley R, Waters D, Turner R, Kroon FJ, Wilkinson SN, Garzon-Garcia A, Kuhnert PM, Lewis SE, Smith RA, Bainbridge ZT, Olley J, Brooks AP, Burton JM, Brodie JE, Waterhouse J (2017) Scientific Consensus Statement 2017: A Synthesis of the Science of Land-based Water Quality Impacts on the Great Barrier Reef, Chapter 2: Sources of Sediment, Nutrients, Pesticides and Other Pollutants to the Great Barrier Reef. State of Queensland, Brisbane.

Blanchette ML, Pearson RG (2013) Dynamics of habitats and macroinvertebrate assemblages in rivers of the Australian dry tropics. Freshwater Biology 58:742–757.

Brizga SO, Lait R, Butler B, Cappo M, Connolly N, Kapitzke IR, Pearson RG, Post D, Pusey B, Smithers S, Werren G (2006) Burdekin Basin Draft Water Resource Plan. Environmental Assessment Report Phase 1—Current Environmental Condition. Queensland Department of Natural Resources, Mines and Water, Brisbane.

Brodie J, Binney J, Fabricius K, Gordon I, Hoegh-Guldberg O, Hunter H, O'Reagain P, Pearson R, Quirk M, Thorburn P, Waterhouse J, Webster I, Wilkinson S (2008) Synthesis of Evidence to Support the Scientific Consensus Statement on Water Quality in the Great Barrier Reef. The State of Queensland (Department of the Premier and Cabinet), Brisbane. https://www.reefplan.qld.gov.au/__data/assets/pdf_file/0020/46118/reef-consensus.pdf.

Brodie J, Fabricius KE, De'ath G, Okaji K (2005) Are increased nutrient inputs responsible for more outbreaks of crown-of-thorns starfish? An appraisal of the evidence. Marine Pollution Bulletin 51:266–278.

Brodie J, Landos M (2019) Pesticides in Queensland and Great Barrier Reef waterways-potential impacts on aquatic ecosystems and the failure of national management. Estuarine, Coastal and Shelf Science 230:106447.

Brodie J, Pearson RG (2016) Ecosystem health of the Great Barrier Reef: Time for effective management action based on evidence. Estuarine, Coastal and Shelf Science 183:438–451.

Brodie J, Pearson RG, Lewis S, Bainbridge Z, Waterhouse J, Prange J (2009) Water Quality Research: Baseline Synthesis and Year 1 Summary. Report to the Marine and Tropical Sciences Research Facility. Reef and Rainforest Research Centre Limited, Cairns, 124p.

Brodie J, Waterhouse J, Maynard J, Bennett J, Furnas M, Devlin M, Lewis S, Collier C, Schaffelke B, Fabricius K, Petus C, da Silva E, Zeh D, Randall L, Brando V, McKenzie L, O'Brien D, Smith R, Warne MStJ, Brinkman R, Tonin H, Bainbridge Z, Bartley R, Negri A, Turner RDR, Davis A, Bentley C, Mueller J, Álvarez-Romero JG, Henry N, Waters D, Yorkston H, Tracey D (2013a) Assessment of the Relative Risk of Water Quality to Ecosystems of the Great Barrier Reef. A Report to the Department of the Environment and Heritage Protection, Queensland Government, Brisbane. TropWATER Report 13/28, Townsville.

Buelow CA, Baker R, Reside AE, Sheaves M (2018) Nutrient subsidy indicators predict the presence of an avian mobile-link species. Ecological Indicators 89:507–515.

Bunn SE, Arthington AH (2002) Basic principles and ecological consequences of altered flow regimes for aquatic biodiversity. Environ Manage 30:492–507.

Bunn SE, Hughes JM (1997) Dispersal and recruitment in streams: Evidence from genetic studies. Journal of the North American Benthological Society 16(2):338–346.

Canning AD, Smart JC, Dyke J, Curwen G, Hasan S, Waltham NJ (2022) Constructed wetlands suitability for sugarcane profitability, freshwater biodiversity and ecosystem services. Environmental Management 1–17.

Cappo M, Kelley R (2000) Connectivity in the great barrier reef world heritage area—An overview of pathways and processes. Oceanographic Processes of Coral Reefs 20:181–208.

Carter J, Tait J, Kapitzke R, Corfield J (2007) Burdekin Dry Tropics NRM Region Fish Passage Study. Alluvium Consulting. Available at: www.nqdrytropics.com.au/publications/. Accessed July 25, 2022.

Chartrand KM, Szabó M, Sinutok S, Rasheed MA, Ralph PJ (2018) Living at the margins—The response of deep-water seagrasses to light and temperature renders them susceptible to acute impacts. Marine Environmental Research 136:126–138. doi: 10.1016/j.marenvres.2018.02.006

Cheng Z, Yu B (2019) Effect of land clearing and climate variability on streamflow for two large basins in Central Queensland, Australia. Journal of Hydrology 578:124041.

Cheshire K, Boyero L, Pearson RG (2005) Food webs in tropical Australian streams: Shredders are not scarce. Freshwater Biology 50:748–769.

Choy S, Page TJ, Mos B (2019) Taxonomic revision of the Australian species of Australatya Chace 1983 (Crustacea, Decapoda, Atyidae), and the description of a new species. Zootaxa 4711(2). doi: 10.11646/zootaxa.4711.2.8

Coles RG, Greenwood JG (1983) Seasonal movement and size distribution of three commercially important Australian prawn species (Crustacea: Penaeidae) within an estuarine system. Marine and Freshwater Research 34(5):727–743.

Collier CJ, Adams MP, Langlois L, Waycott M, O'Brien KR, Maxwell PS, McKenzie L (2016) Thresholds for morphological response to light reduction for four tropical seagrass species. Ecological Indicators 67:358–366.

Collier CJ, Waycott M, McKenzie LJ (2012) Light thresholds derived from seagrass loss in the coastal zone of the northern Great Barrier Reef, Australia. Ecological Indicators 23(0):211–219. doi: 10.1016/j.ecolind.2012.04.005

Connolly NM, Pearson RG (2018) Colonisation, emigration and equilibrium of stream invertebrates in patchy habitats. Freshwater Biology 63:1446–1456.

Connolly NM, Pearson RG, Loong D, Maughan M, Brodie J (2015) Water quality variation along streams with similar agricultural development but contrasting riparian vegetation. Agriculture, Ecosystems & Environment 213:11–20.

Creighton C, Waterhouse J, Day JC, Brodie J (2021) Criteria for effective regional scale catchment to reef management: A case study of Australia's Great Barrier Reef. Marine Pollution Bulletin 173:112882.

Crook DA, Buckle DJ, Allsop Q, Baldwin W, Saunders TM, Kyne PM, Woodhead JD, Maas R, Roberts B, Douglas MM (2016) Use of otolith chemistry and acoustic telemetry to elucidate migratory contingents in barramundi Lates calcarifer. Marine and Freshwater Research, 68(8):1554–1566.

D'Angelo C, Wiedenmann J (2014) Impacts of nutrient enrichment on coral reefs: New perspectives and implications for coastal management and reef survival. Current Opinion in Environmental Sustainability 7:82–93.

Davie PJF (2002) Atyidae. In: Davie PJF. (ed.), Crustacea: Malacostraca: Phyllocarida, Hoplocarida, Eucarida (Part 1) Vol. 19.3A. CSIRO, Melbourne, 222–229.

Davis AM, Pearson RG, Brodie JE, Butler B (2017) Review and conceptual models of agricultural impacts and water quality in waterways of the Great Barrier Reef catchment area. Marine and Freshwater Research 68:1–19.

Davis AM, Ponder W (2019) Biogeographic conundrum: Why so few stream nerite species (Gastropoda: Neritidae) in Australia?. Freshwater Biology 64(11):2084–2088.

Davis AM, Unmack P, Pusey BJ, Pearson RG, Johnson J (2012) Marine-freshwater transitions are associated with the evolution of dietary diversification in terapontid grunters (Teleostei: Terapontidae). Journal of Evolutionary Biology 25: 1163–1179.

Day JC, Dobbs K (2013) Effective governance of a large and complex cross-jurisdictional marine protected area: Australia's Great Barrier Reef. Marine Policy 41:14–24.

Department of Environment and Science Queensland (2015) Regional natural resource management (NRM) body 2017/pre-clear wetland extent percentages. WetlandInfo. Available at: https://wetlandinfo.des.qld.gov.au/wetlands/ecology/statistics/wetland-pre-clear-percent/nrm.html. Accessed January 21, 2023.

Department of Environment and Science Queensland (2022) Whole-of-System, Values-Based Framework—Process Diagram (version 1.0). WetlandInfo Website. Available at: https://wetlandinfo.des.qld.gov.au/wetlands/management/whole-system-values-framework/

Devlin M, Schaffelke B (2009) Spatial extent of riverine flood plumes and exposure of marine ecosystems in the Tully coastal region, Great Barrier Reef. Marine and Freshwater Research 60:1109–1122.

Eberhard R, Thorburn P, Rolfe J, Taylor B, Ronan M, Weber T, Flint N, Kroon F, Brodie J, Waterhouse J, Silburn M, Bartley R, Davis A, Wilkinson S, Lewis S, Star M, Poggio M, Windle J, Marshall N, Hill R, Maclean K, Lyons P, Robinson C, Adame F, Selles A, Griffiths M, Gunn J, McCosker K (2017) Scientific Consensus Statement 2017: A Synthesis of the Science of Land-based Water Quality Impacts on the Great Barrier Reef, Chapter 4: Management Options and Their Effectiveness. State of Queensland, Brisbane. https://www.reefplan.qld.gov.au/__data/assets/pdf_file/0033/45996/2017-scientific-consensus-statement-summary-chap04.pdf.

Elledge A, Thornton C (2017) Effect of changing land use from virgin brigalow (Acacia harpophylla) woodland to a crop or pasture system on sediment, nitrogen and phosphorus in runoff over 25 years in subtropical Australia. Agriculture, Ecosystems & Environment 239:119–131.

Fabricius KE (2011) Factors determining the resilience of coral reefs to eutrophication: A review and conceptual model. In: Dubinsky Z, Stambler N (eds.), Coral Reefs: An Ecosystem in Transition, Springer Press, New York, 493–506.

Fabricius KE, Logan M, Weeks S, Brodie J (2014) The effects of river run-off on water clarity across the central Great Barrier Reef. Marine Pollution Bulletin 84(1–2):191–200. doi: 10.1016/j.marpolbul.2014.05.012

Fabricius KE, Logan M, Weeks SJ, Lewis SE, Brodie J (2016) Changes in water clarity in response to river discharges on the Great Barrier Reef continental shelf: 2002–2013. Estuarine, Coastal and Shelf Science 173:A1–A15. doi: 10.1016/j.ecss.2016.03.001

Fabricius KE, Okaji K, De'ath G (2010) Three lines of evidence to link outbreaks of the crown-of-thorns seastar *Acanthaster planci* to the release of larval food limitation. Coral Reefs 29:593–605.

Fielding CR, Trueman JD, Dickens GR, Page M (2003) Anatomy of the buried Burdekin River channel across the Great Barrier Reef shelf: How does a major river operate on a tropical mixed siliciclastic/carbonate margin during sea level lowstand? Sedimentary Geology 157(3–4):291–301.

Furnas M (2003) Catchments and Corals: Terrestrial Runoff to the Great Barrier Reef. Australian Institute of Marine Science, Townsville, 334pp.

Furnas M, Alongi D, McKinnon D, Trott L, Skuza M (2011) Regional-scale nitrogen and phosphorus budgets for the northern (14°S) and central (17°S) Great Barrier Reef shelf ecosystem. Continental Shelf Research 31:1967–1990.

Gagan MK, Ayliffe LK, Opdyke BN, Hopley D, Scott-Gagan H, Cowley J (2002) Coral oxygen isotope evidence for recent groundwater fluxes to the Australian Great Barrier Reef. Geophysical Research Letters 29(20):41–43.

Geist J (2021) Editorial: Green or red: Challenges for fish and freshwater biodiversity conservation related to hydropower. Aquatic Conservation: Marine and Freshwater Ecosystems. doi: 10.1002/aqc.3597.

Godfrey PC, Arthington AH, Pearson RG, Karim FJW, Wallace J (2017) Fish larvae and recruitment patterns in floodplain lagoons of the Australian Wet Tropics. Marine and Freshwater Research 68:964–979.

Godfrey PC, Pearson RG, Pusey BJ, Arthington AH, (2020), Drivers of zooplankton dynamics in a small tropical lowland river. Marine and Freshwater Research 71:1–13.

Great Barrier Reef Marine Park Authority (2012) Informing the Outlook for Great Barrier Reef Coastal Ecosystems. Great Barrier Reef Marine Park Authority, Townsville.

Great Barrier Reef Marine Park Authority (2013) Coastal Ecosystems Management—Case Study: Water Management, Great Barrier Reef Marine Park Authority, Townsville.

Great Barrier Reef Marine Park Authority (2017a) A Method for Identifying and Prioritising Coastal Ecosystem Functional Connections to the Great Barrier Reef World Heritage Area and Great Barrier Reef Marine Park, Great Barrier Reef Marine Park Authority, Townsville.

Great Barrier Reef Marine Park Authority (2017b) Great Barrier Reef Blueprint for Resilience, Great Barrier Reef Marine Park Authority, Townsville.

Great Barrier Reef Marine Park Authority (2019) Great Barrier Reef Outlook Report 2019. Great Barrier Reef Marine Park Authority, Townsville.

Halliday IA, Robins JB, Mayer DG, Staunton-Smith J, Sellin MJ (2010) Freshwater flows affect the year-class strength of 'Barramundi lates calcarifer' in the Fitzroy River estuary, Central Queensland. Proceedings of the Royal Society of Queensland 116:1–0.

Harris J, Kingsford RT, Peirson W, Baumgartner L (2017) Mitigating the effects of barriers to freshwater fish migrations: The Australian experience, Marine and Freshwater Research 68(4):614–628.

Higgins PA, Palmer JG, Rao MP, Andersen MS, Turney CS, Johnson F (2022) Unprecedented High Northern Australian Streamflow Linked to an Intensification of the Indo-Australian Monsoon. Water Resources Research, 58(3):e2021WR030881.

Hughes JM, Bunn SE, Hurwood DA, Choy S, Pearson RG (1996) Genetic differentiation among different populations of Caridina zebra (Decapoda: Atyidae) in tropical rainforest streams, northern Australia. Freshwater Biology 36:289–296.

Hunter HM (2012) Nutrients and Herbicides in Groundwater Flows to the Great Barrier Reef Lagoon Processes, Fluxes and Links to On-farm Management. Queensland Government, Townsville.

Hurwood DA, Hughes JM (1998) Phylogeography of the freshwater fish, Mogurnda adspersa, in streams of northeastern Queensland, Australia: Evidence for altered drainage patterns. Molecular Ecology 7:1507–1517.

Hyslop J, Trowsdale S (2012) A review of hydrochory (seed dispersal by water) with implications for riparian rehabilitation. Journal of Hydrology (New Zealand) 137–152.

James CS, Reside AE, VanDerWal J, Pearson RG, Burrows D, Capon SS, Harwood TD, Hodgson L and Waltham N (2017) Sink or swim? Potential for high faunal turnover in Australian rivers under climate change. Journal of Biogeography. doi: 10.1111/jbi.12926.

Junk WJ, Bayley PB, Sparks RE (1989) The flood pulse concept in river-floodplain systems. Canadian Special Publication of Fisheries and Aquatic Sciences 106(1):110–127.

Karim F, Kinsey-Henderson A, Wallace J, Arthington AH, Pearson RG (2012) Modelling wetland connectivity during overbank flooding in a tropical floodplain in north Queensland, Australia. Hydrological Processes 26:2710–2723.

Karim F, Kinsey-Henderson A, Wallace J, Godfrey P, Arthington AH, Pearson RG (2014) Modelling hydrological connectivity of tropical floodplain wetlands via the stream network. Hydrological Processes 28:5696–5710.

Kingsford RT (2000) Ecological impacts of dams, water diversions and river management on floodplain wetlands in Australia. Austral Ecology 25:109–127.

Kroon FJ, Phillips S (2016) Identification of human-made physical barriers to fish passage in the Wet Tropics region, Australia. Marine and Freshwater Research 67:677–681. doi: 10.1071/MF14397

Lawson T, Kroon F, Russell J, Thuesen P (2010) Audit and Prioritisation of Physical Barriers to Fish Passage in the Wet Tropics Region. Milestone report, MTSRF Project 262. Marine and Tropical Science Research Facility/Reef and Rainforest Research Centre, Cairns.

Leigh C, Reis TM, Sheldon F (2013) High potential subsidy of dry-season aquatic fauna to consumers in riparian zones of wet–dry tropical rivers. Inland Waters 3:411–420.

Lenahan MJ, Bristow KL (2010) Understanding sub-surface solute distributions and salinization mechanisms in a tropical coastal floodplain groundwater system. Journal of Hydrology 390:131–142.

Leonard ND, Welsh KJ, Lough JM, Feng YX, Pandolfi JM, Clark TR, Zhao JX (2016) Evidence of reduced mid-Holocene ENSO variance on the Great Barrier Reef, Australia. Paleoceanography 31(9):1248–1260.

Lewis S, Bainbridge ZT, Stevens T, Garzon-Garcia A, Chen C, Bahadori M, Burton JM, Rezaei Rashti M, James C, Smithers S, Olley J (2020) What's really damaging the Reef?: Tracing the origin and fate of the environmentally detrimental sediment and associated bioavailable nutrients. Report to the National Environmental Science Program. Reef and Rainforest Research Centre Limited, Cairns.

Lewis S, McCloskey G, Bainbridge ZT, Davis A, Bartley R, Turner R (2024). Sediment and nutrient flux from land. In: Wolanski E, Kingsford MJ (eds.), Oceanographic Processes of Coral Reefs. Physical and Biological Links in the Great Barrier Reef. CRC Press, Boca Raton, FL.

Lewis SE, Bainbridge ZT, Kuhnert PM, Sherman BS, Henderson BS, Dougall C, Cooper M, Brodie JE, Kuhnert ZT, Sherman PM, Dougall B, Cooper C, M. Brodie JE (2013) Calculating sediment trapping efficiencies for reservoirs in tropical settings: A case study from the Burdekin Falls Dam, NE Australia. Water Resources Research 49:1017–1029.

Lewis SE, Bartley R, Wilkinson SN, Bainbridge ZT, Henderson AE, James CS, Irvine SA, Brodie JE (2021) Land use change in the river basins of the Great Barrier Reef, 1860 to 2019: A foundation for understanding environmental history across the catchment to reef continuum. Marine Pollution Bulletin 166:112193.

Lewis SE, Brodie JE, Bainbridge ZT, Rohde KW, Davis AM, Masters BL, Maughan M, Devlin MJ, Mueller JF, Schaffelke B (2009) Herbicides: A new threat to the Great Barrier Reef. Environmental Pollution 157:2470–2484.

Lewis SE, Schaffelke B, Shaw M, Bainbridge ZT, Rohde KW, Kennedy KE, Davis AM, Masters BL, Devlin MJ, Mueller JF, Brodie JE (2012) Assessing the risks of PS-II herbicide exposure to the Great Barrier Reef. Marine Pollution Bulletin 65:280–291.

Lough JM, Lewis SE, Cantin NE (2015) Freshwater impacts in the central Great Barrier Reef: 1648–2011. Coral Reefs 34:739–751.

Lough JM, Llewellyn LE, Lewis SE, Turney CSM, Palmer JG, Cook CG, Hogg AG (2014) Evidence for suppressed mid-Holocene northeastern Australian monsoon variability from coral luminescence. Paleoceanography 29. doi: 10.1002/2014PA002630.

Machado CB, Braga-Silva A, Freitas PD, Galetti Jr PM (2022) Damming shapes genetic patterns and may affect the persistence of freshwater fish populations. Freshwater Biology 67(4):603–618.

Marshall JC, Menke N, Crook DA, Lobegeiger JS, Balcombe SR, Huey JA, Fawcett JH, Bond NR, Starkey AH, Sternberg D, Linke S (2016) Go with the flow: The movement behaviour of fish from isolated waterhole refugia during connecting flow events in an intermittent dryland river. Freshwater Biology 61(8):1242–1258.

McCloskey GL, Baheerathan R, Dougall C, Ellis R, Bennett FR, Waters D, Darr S, Fentie B, Hateley LR, Askildsen M (2021) Modelled estimates of fine sediment and particulate nutrients delivered from the Great Barrier Reef catchments. Marine Pollution Bulletin 165:112–163.

McCook LJ, Jompa J, Diaz-Pulido G (2001) Competition between corals and algae on coral reefs: A review of evidence and mechanisms. Coral Reefs 19(4):400–417.

McIvor JG, Williams J, Gardener CJ (1995) Pasture management influences runoff and soil movement in the semi-arid tropics. Australian Journal of Experimental Agriculture 35:55–65.

McMahon GA, Reading L, Foy Z, Wang J, Bajracharya K, Corbett N, Gallagher M, Lenahan MJ, Gurieff L (2012) Development of a Hydrological Modelling Toolkit to Support Sustainable Development in the Lower Burdekin Groundwater System: Conceptualisation of the Lower Burdekin Aquifer. Department of Science, Information Technology, Innovation and the Arts, Queensland Government, Brisbane.

Molinari B, Stewart-Koster B, Malthus TJ, Bunn SE (2022) Impact of water resource development on connectivity and primary productivity across a tropical river floodplain. Journal of Applied Ecology 59(4):1013–1025.

Moore M, Marsden TJ (2008) Fitzroy Basin Fish Barrier Prioritisation Project. Queensland Department of Primary Industries, Fisheries, Queensland, 51pp.

Narayan KA, Schleeberger C, Bristow KL (2007) Modelling seawater intrusion in the Burdekin Delta Irrigation Area, North Queensland, Australia. Agricultural Water Management 89:217–228.

Nevill JC, Hancock PJ, Murray BR, Ponder WF, Humphreys WF, Phillips ML, Groom PK (2010) Groundwater-dependent ecosystems and the dangers of groundwater overdraft: A review and an Australian perspective. Pacific Conservation Biology 16(3):187–208.

Nuttle WK (2002) Is ecohydrology one idea or many? Hydrological Sciences Journal 47(5):805–807.

Page T, von Rintelen K, Hughes JM (2007) An island in the stream: Australia's place in the cosmopolitan world of Indo-West Pacific freshwater shrimp (Decapoda: Atyidae: Caridina). Molecular Phylogenetics and Evolution 43:645–659.

Pearson RG, Connolly NM, Davis AM, Brodie JE (2021) Aquatic systems of the Great Barrier Reef catchment: Impacts and management of anthropogenic disturbance on freshwater and estuarine biodiversity, ecology and connectivity. Marine Pollution Bulletin 166:112–194. doi: 10.1016/j.marpolbul.2021.112194

Pearson RG, Davis AM, Birtles RA (2022a) The Burdekin River: A Review of its Ecology, Conservation and Management. TropWater report 22/06, James Cook University, Townsville. Available at: www.tropwater.com/publications/technical-reports/

Pearson RG, Davis AM, Birtles RA (2022b) Enhancing whole-of-river conservation. Marine and Freshwater Research 73:729–741. doi: 10.1071/MF21287.

Pearson RG, Godfrey PC, Arthington AH (2010) Freshwater-marine Connectivity in the Great Barrier Reef Catchment: A Review of Current Knowledge and a Case Study of the Tully-Murray Floodplain. Marine and Tropical Sciences Research Facility. Reef and Rainforest Research Centre Limited, Cairns, 61pp.

Pearson RG, Stork NE (2008) Catchment to reef: Water quality and ecosystem health in tropical streams. In: Stork N., Turton S. (eds) Living in a Dynamic Tropical Forest Landscape. Blackwells Publishing, Carlton, 557–576.

Peña-Arancibia JL, van Dijk AI, Guerschman JP, Mulligan M, Bruijnzeel LA, McVicar TR (2012) Detecting changes in streamflow after partial woodland clearing in two large catchments in the seasonal tropics. Journal of Hydrology 416:60–71.

Perna CN, Cappo M, Pusey BJ, Burrows DW, Pearson RG (2012a) Removal of aquatic weeds greatly enhances fish community richness and diversity: An example from the Burdekin River floodplain, tropical Australia. River Research and Applications 28:1093–1104.

Perna CN, O'Connor R, Cook B (2012b) Hydroecology of the Lower Burdekin River Alluvial Aquifer and Associated Groundwater Dependent Ecosystems. Department of Environment and Resource Management, Brisbane.

Ponniah M, Hughes JM (2004) The evolution of Queensland spiny mountain crayfish of the genus *Euastacus*. I. Testing vicariance and dispersal with interspecific mitochondrial DNA. Evolution 58:1073–1085.

Powell B, Martens M (2005) A review of acid sulfate soil impacts, actions and policies that impact on water quality in Great Barrier Reef catchments, including a case study on remediation at East Trinity. Marine Pollution Bulletin 51:149–164.

Pratchett MS, Chandler J, Choukroun SM, Doll PC, Land BJ, Kwong S, Emslie M, Caballes CF, Uthicke S, Matthews S (2024) Biophysical processes involved in the initiation and spread of population irruptions of crown-of-thorns starfish on the great barrier reef. In: Wolanski E, Kingsford MJ (eds.), Oceanographic Processes of Coral Reefs. Physical and Biological Links in the Great Barrier Reef. CRC Press, Boca Raton, FL.

Preite CK, Pearson RG (2017) Water quality variability in dryland riverine waterholes: A challenge CSIRO publishing for ecosystem assessment. Annales de Limnologie—International Journal of Limnology 53:221–232.

Preite CK, Pearson RG (2021) Phytoplankton in dryland riverine waterholes: Environmental drivers, variability, and ecosystem-monitoring potential using different levels of taxonomic resolution and dataset reduction. Marine and Freshwater Research 72:244–255. doi: 10.1071/MF19343.

Pringle CM (2001) Hydrologic connectivity and the management of biological reserves: A global perspective. Ecological Applications 11(4):981–998.

Puignou Lopez O, Lewis, SE, James CS, Davis AM, Mackay SJ (in review) Hydrology of the Great Barrier Reef catchment area along a latitudinal gradient: Implications for estimating discharge. Journal of Hydrology.

Pusey B, Burrows D, Arthington A, Kennard M (2006) Translocation and spread of piscivorous fishes in the Burdekin River, north-eastern Australia. Biological Invasions 8(4):965–977.

Pusey B, Kennard MJ, Arthington AH (2004) Freshwater Fishes of North-eastern Australia. CSIRO Publishing, London, 2004.

Queensland Government (2021) Urannah Project. Available at: www.statedevelopment.qld.gov.au/coordinator-general/assessments-and-approvals/coordinated-projects/current-projects/. Accessed August 10, 2022.

Rayner TS, Pusey BJ, Pearson RG (2008) Seasonal flooding, instream habitat structure and fish assemblages in the Mulgrave River, north-east Queensland: Towards a new conceptual framework for understanding fish-habitat dynamics in small tropical rivers. Marine and Freshwater Research 59:97–116.

Reef R, Feller IC, Lovelock CE (2014) Mammalian herbivores in Australia transport nutrients from terrestrial to marine ecosystems via mangroves. Journal of Tropical Ecology 30(3):179–188.

Robertson AI, Lee Long WJ, Coles RG, Pearson RG, Hegerl EJ (1996) Impacts of land and water use practices on riparian areas, wetlands, estuaries and coastal seagrasses. In: Hunter, HM, Eyles, AG, and Rayment, GE (eds) Downstream Effects of Land Use. Department of Natural Resources, Queensland, 85–92.

Roche RC, Perry CT, Smithers SG, Leng MJ, Grove CA, Sloane HJ, Unsworth CE (2014) Mid-Holocene sea surface conditions and riverine influence on the inshore Great Barrier Reef. The Holocene 24(8):885–897.

Roth C (2004) A framework relating soil surface condition to infiltration and sediment and nutrient mobilisation in grazed rangelands of north-eastern Queensland. Earth Surface Processes and Landforms 29:1093–1104.

Russell DJ, Garrett RN (1985) Early life history of barramundi, Lates calcarifer Bloch. in north-eastern Queensland. Australian Journal of Marine and Freshwater Research 36:191–201.

Schaffelke B, Collier C, Kroon F, Lough J, McKenzie L, Ronan M, Uthicke S, Brodie J (2017) Scientific Consensus Statement 2017. Scientific Consensus Statement 2017: A Synthesis of the Science of Land-based Water Quality Impacts on the Great Barrier Reef, Chapter 1: The Condition of Coastal and Marine Ecosystems of the Great Barrier Reef and Their Responses to Water Quality and Disturbances. State of Queensland, Brisbane.

Sheaves M (2009) Consequences of ecological connectivity: The coastal ecosystem mosaic. Marine Ecology Progress Series 391:107–115.

Sheaves M, Bradley M, Mattone C (2024) The influence of the spatio-temporal dynamics of fish populations on the outcomes of land-sea connectivity. In: Wolanski E, Kingsford MJ (eds.), Oceanographic Processes of Coral Reefs. Physical and Biological Links in the Great Barrier Reef. CRC Press, Boca Raton, FL.

Sheaves M, Brookes J, Coles R, Freckelton M, Groves P, Johnston R, Winberg P, Freckleton M (2014) Repair and revitalisation of Australia's tropical estuaries and coastal wetlands: Opportunities and constraints for the reinstatement of lost function and productivity. Marine Policy 47:23–38.

Sheaves M, Johnston R (2008) Influence of marine and freshwater connectivity on the dynamics of subtropical estuarine wetland fish metapopulations. Marine Ecology Progress Series 357:225–243.

Shishaye HA, Tait DR, Befus KM, Maher DT, Reading MJ (2020) New insights into the hydrogeology and groundwater flow in the Great Barrier Reef catchment, Australia, revealed through 3D modelling. Journal of Hydrology: Regional Studies 30:100708.

Short JW (2004) A revision of Australian river prawns, Macrobrachium (Crustacea: Decapoda: Palaemonidae) Hydrobiologia 525:1–100.

Siriwardena L, Finlayson BL, McMahon TA (2006) The impact of land use change on catchment hydrology in large catchments: The Comet River, Central Queensland, Australia. Journal of Hydrology 326(1–4):199–214.

Skerratt JH, Baird ME, Mongin M, Ellis R, Smith RA, Shaw M, Steven AD (2023) Dispersal of the Pesticide Diuron in the Great Barrier Reef. Science of the Total Environment, Available at: SSRN 4216259

Smith REW (1987) The ecology of Australatya striolata (McCulloch and McNeill) (Decapoda: Atyidae). PhD thesis, James Cook University, Townsville.

Soons MB, Brochet AL, Kleyheeg E, Green AJ (2016) Seed dispersal by dabbling ducks: An overlooked dispersal pathway for a broad spectrum of plant species. Journal of Ecology 104(2):443–455.

Stieglitz T (2005) Submarine groundwater discharge into the nearshore zone of the Great Barrier Reef, Australia. Marine Pollution Bulletin 51:51–59.

Templeman MA, Williams CD (2024) Jellyfish: A Window into Pesticide Distribution and Risks on the great barrier reef. In: Wolanski E, Kingsford MJ (eds.), Oceanographic Processes of Coral Reefs. Physical and Biological Links in the Great Barrier Reef. CRC Press, Boca Raton, FL.

Terrain NRM (2015) Wet Tropics Water Quality Improvement Plan 2015–2020. Terrain NRM, Innisfail. Available at: www.terrain.org.au/Projects/Water-Quality-Improvement-Plan.

Thornton CM, Cowie BA, Freebairn DM, Playford CL (2007) The Brigalow Catchment Study: II*. Clearing brigalow (Acacia harpophylla) for cropping or pasture increases runoff. Soil Research 45(7):496–511.

Thornton CM, Elledge AE (2021) Heavy grazing of buffel grass pasture in the Brigalow Belt bioregion of Queensland, Australia, more than tripled runoff and exports of total suspended solids compared to conservative grazing. Marine Pollution Bulletin 171:112704.

Thornton CM, Yu B (2016) The Brigalow Catchment Study: IV.* Clearing brigalow (Acacia harpophylla) for cropping or grazing increases peak runoff rate. Soil Research, 54(6):749–759.

Thuesen PA, Ebner BC, Larson H, Keith P, Silcock RM, Prince J, Russell DJ (2011) Amphidromy links a newly documented fish community of continental Australian streams, to oceanic islands of the West-Pacific. PLoS One 6:e26685. doi: 10.1371/journal.pone.0026685

Thuesen PA, Pusey BJ, Peck DJ, Pearson RG, Congdon B (2008) Genetic differentiation over small spatial scales in the absence of physical barriers in an Australian rainforest stream fish. Journal of Fish Biology 72:1174–1187.

Traveset A, Heleno R, Nogales, M (2014) The ecology of seed dispersal. In: Gallagher RS (ed.), Seeds: The Ecology of Regeneration in Plant Communities. 3rd edition. CAB International, Wallingford, 62–93.

Vega Thurber RL, Burkepile DE, Fuchs C, Shantz AA, McMinds R, Zaneveld JR (2013) Chronic nutrient enrichment increases prevalence and severity of coral disease and bleaching. Global Change Biology 20(2):544–554.

Veitch V, Tait J, Burrows D (2008) Fish Passage Connectivity Issues, Lower Sheep Station Creek -Dry and Wet Season Investigation of Water Quality and Fish Assemblages. Report number 08/02, Australian Centre for Tropical Freshwater Research, James Cook University, Townsville.

Waltham NJ, Burrows DW, Wegscheidl C, Buelow C, Ronan M, Connolly NM, Groves P, Marie-Audas D, Creighton C, Sheaves M (2019) Lost floodplain wetland environments and efforts to restore connectivity, habitat, and water quality settings on the Great Barrier Reef. Frontiers in Marine Science 6:1–14.

Waltham NJ, Cartwright P, Sheaves, M, Ronan M (2024) Coastal wetlands restoration: Case studies from great barrier reef catchment. In: Wolanski E, Kingsford MJ (eds.), Oceanographic Processes of Coral Reefs. Physical and Biological Links in the Great Barrier Reef. CRC Press, Boca Raton, FL.

Waltham NJ, Wegscheidl C, Smart JCR, Volders A, Hasan S, Waterhouse J (2017) Scoping land conversion options for high DIN risk, low-lying sugarcane, to alternative use for water quality improvement in Wet Tropics catchments. Report to the National Environmental Science Program. Reef and Rainforest Research Centre Limited, Cairns (141pp).

Water Quality & Investigations (2020) Catchment Loads Monitoring Program Pesticide Reporting Portal, Department of Environment and Science, Brisbane.

Waterhouse J, Brodie JE, Lewis S, Audas DM (2016) Land-sea connectivity, ecohydrology and holistic management of the Great Barrier Reef and its catchments: Time for a change. Ecohydrology and Hydrobiology 16:45–57.

Waterhouse J, Brodie JE, Tracey D, Smith R, Vandergragt M, Collier C, Petus C, Baird M, Kroon F, Mann R, Sutcliffe T, Waters D, Adame F (2017) Scientific Consensus Statement 2017: A Synthesis of the Science of Land-based Water Quality Impacts on the Great Barrier Reef, Chapter 3: The Risk from Anthropogenic Pollutants to Great Barrier Reef Coastal and Marine Ecosystems. State of Queensland, Brisbane.

Wet Tropics Management Authority (2013) Annual Report and State of Wet Tropics Report 2012–13. Wet Tropics Management Authority, Cairns, Queensland.

Wet Tropics Waterways (2022) Wet Tropics Report Card 2022 (reporting on data 2020–21). Waterway Environments: Results. Wet Tropics Waterways and Terrain NRM, Innisfail.

Wiedenmann J, D'Angelo C, Smith EG, Hunt AN, Legiret F-E, Postle AD, Achterberg EP (2013) Nutrient enrichment can increase the susceptibility of reef corals to bleaching. Nature Climate Change 3(5):160–164.

Williams BA, Watson JE, Beyer HL, Klein CJ, Montgomery J, Runting RK, Roberson LA, Halpern BS, Grantham HS, Kuempel CD, Frazier M (2022) Global rarity of intact coastal regions. Conservation Biology e13874. doi: 10.1111/cobi.13874

Wolanski E (2007) Chapter 7—Ecohydrology Solutions. In: Wolanski E (eds.), Estuarine Ecohydrology Elsevier, 125–137. doi: 10.1016/B978-044453066-0.50008-8.

Wolanski E, Richmond R, McCook L, Sweatman H (2003) Mud, marine snow and coral reefs: The survival of coral reefs requires integrated watershed-based management activities and marine conservation. American Scientist 91(1):44–51.

Wooldridge SA (2009) Water quality and coral bleaching thresholds: Formalising the linkage for the inshore reefs of the Great Barrier Reef, Australia. Marine Pollution Bulletin 58(5):745–751. doi: 10.1016/j.marpolbul.2008.12.013.

Wooldridge SA, Heron SF, Brodie JE, Done TJ, Masiri I, Hinrichs S (2017) Excess seawater nutrients, enlarged algal symbiont densities and bleaching sensitive reef locations: 2. A regional-scale predictive model for the Great Barrier Reef, Australia. Marine Pollution Bulletin 114(1):343–354.

8 Sediment and Nutrient Flux from Land

Stephen Lewis, Gillian McCloskey, Zoe Bainbridge, Aaron Davis, Rebecca Bartley, and Ryan Turner

8.1. INTRODUCTION

The interactions between terrestrial runoff from the adjacent landscape and tropical marine ecosystems, such as coral reefs and seagrass meadows, have been subject to intense research activity over several decades. Indeed, the internal structures of inshore or nearshore coral reefs of the Great Barrier Reef (GBR) exposed to terrestrial runoff reveal a long, millennial-scale history of coral growth coupled with the deposition of fine-grained terrigenous sediments (Smithers et al. 2006; Browne et al. 2012; Perry et al. 2012; Ryan et al. 2016, 2018). It is also well-recognised that nutrients such as carbon, nitrogen, and phosphorus help support the productivity of marine ecosystems and particularly inshore fisheries (Devlin et al. 2001; Furnas et al. 2011). However, land runoff of freshwater, suspended sediments, and nutrients can disrupt the dynamic balance of receiving marine ecosystems and cause both acute and chronic effects such as burial or smothering, reduced water clarity, increased growth of macroalgae, coral bleaching, and increased coral bioerosion and disease (Fabricius 2005; Erftemeijer et al. 2012; Risk 2014; Thompson et al. 2014; Lough et al. 2015; Bainbridge et al. 2018). The attribution of enhanced land-based runoff to changes in the health and condition of marine ecosystems first requires a fundamental understanding of (1) the level of increase of terrestrial pollutants exported from the adjacent catchment following catchment development for livestock, cropping, mining, and urban land use in the GBR and (2) the transformations of the terrestrial pollutants across the catchment to marine continuum.

Since the Furnas and Mitchell (2000) review of "Runoff of terrestrial sediment and nutrients into the Great Barrier Reef World Heritage Area" in the first edition, there have been considerable developments in the quantification, source attribution, and management of sediment and nutrient loads from the Great Barrier Reef Catchment Area (GBRCA) and their transformations through the catchment and in the GBR lagoon. These improvements have been fostered through coordinated monitoring and modelling efforts and the development of proxy-based records. The following review summarises this new knowledge gathered over the past two decades and is covered by five key objectives:

1. The history of research on understanding sediment and nutrient fluxes to the GBR;
2. The sources and fluxes of sediment to the GBR;
3. The sources and fluxes of nitrogen and phosphorus to the GBR;
4. The transformations of sediment and nutrients across the catchment to marine continuum; and
5. Management options to reduce sediment and nutrient loads to the GBR.

The term sediment is generically used in the environmental management space and depends on what aspects or types of sediment are measured/monitored, and/or modelled, which can result in varying outcomes and impacts. We define suspended fine sediment as terrigenous mineral particles that are less than 20 μm in diameter; this is the fraction that is preferentially transported in the GBR lagoon (Bainbridge et al. 2021). The nutrients covered in this review include the more biologically active dissolved inorganic and particulate forms of nitrogen (i.e., dissolved inorganic nitrogen (DIN) and particulate nitrogen (PN)) and phosphorus (i.e., particulate phosphorus (PP)). Although the dissolved organic forms of nitrogen and phosphorus (i.e., dissolved organic nitrogen (DON) and dissolved organic phosphorus (DOP)) can be bioavailable (e.g., Lonborg et al. 2018), the best available evidence is that these constituents contain more recalcitrant forms and that their loads have likely not changed greatly since the arrival of Europeans. Therefore, DON and DOP are not considered further in this chapter; dissolved inorganic phosphorus (DIP) is bioavailable, but a greater understanding of its sources, transport, and fate is required before it can be considered in a review.

8.2. HISTORY OF RESEARCH ON UNDERSTANDING SEDIMENT AND NUTRIENT FLUXES TO THE GBR

The GBRCA (423,000 km^2) of northeastern Australia stretches along nearly 2,000 km of coastline, contains 35 separate river basins (Figure 1), and is part of the North East Coast topographic drainage division of Australia. The basins contain a vast mix of different geologies and soil types (e.g., McCulloch et al. 2003a; Douglas et al. 2006; Bainbridge et al. 2014) and various climatic zones across seven distinct bioregions (Furnas 2003; Peel et al. 2007; Petheram et al. 2008). These basins have been subjected to varying levels of intensive land modification such as livestock grazing (predominately cattle), cropping, mining, and urban development since the arrival of Europeans around 1850 (Lewis et al. 2021). It is well-recognised that such land modifications have led to increased exports of sediment and nutrients to the GBR, although there is some variability in the magnitude of increase of these estimates. For

 DOI: 10.1201/9781003320425-10

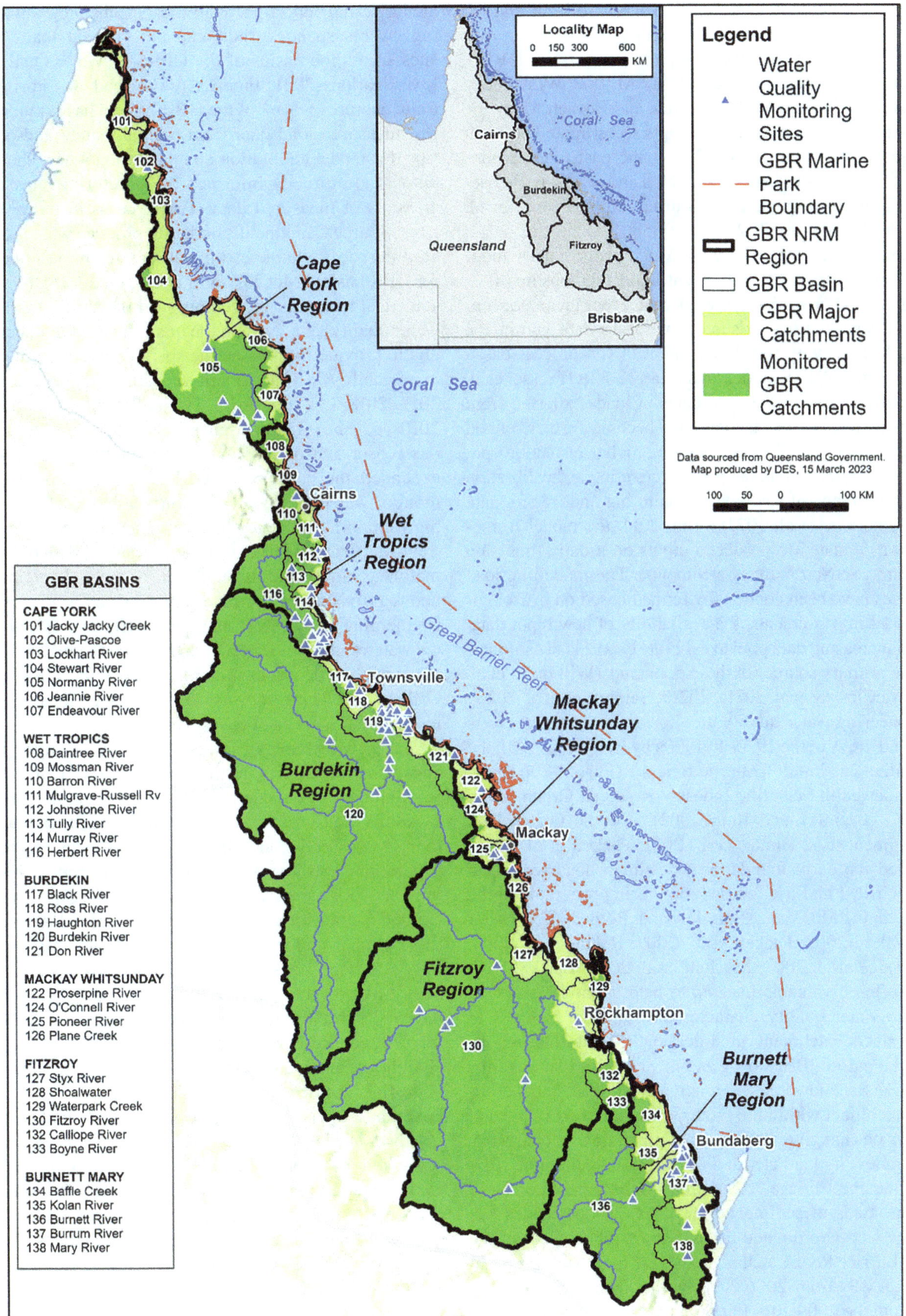

FIGURE 1 GBR catchment area showing the 35 river basins and the monitoring sites of the GBR Catchment Loads Monitoring Program (GBRCLMP).

management purposes, the GBRCA has also been divided into six natural resource management (NRM) regions including Cape York, Wet Tropics (or Terrain NRM), Burdekin (or NQ Dry Tropics NRM), Mackay Whitsunday (or Reef Catchments NRM), Fitzroy, and Burnett Mary.

There have been several attempts to estimate sediment and nutrient loads to the GBR over the past four decades (Belperio 1983; Moss et al. 1992; Neil and Yu 1996; Prosser et al. 2001; Neil et al. 2002; Furnas 2003; Brodie et al. 2003; McKergow et al. 2005a, 2005b; Kroon et al. 2012; McCloskey et al. 2021a, 2021b). Much of the earlier modelled estimates used flow-concentration relationships (i.e., rating curves) or site/land use-specific event mean concentrations derived from existing monitoring data within the GBRCA. Following the development of the process-based SedNet and ANNEX models in the early 2000s (Prosser et al. 2001; Brodie et al. 2003), estimates of loads exported from the basins to the GBR lagoon have become further refined with improved model input data (i.e., higher resolution spatial data) and refined model parameterization and calibration by using a range of catchment monitoring and tracing data (see Bainbridge et al. 2023). From 2014, the model framework was updated to produce outputs on a daily time step (Dynamic SetNet/Source Catchments). The modelling load estimates have been continually refined based on hydrological model improvements, the availability of new input data, and an increasing data pool to calibrate export load estimates and the primary sources of these pollutants (Wilkinson et al. 2014; McCloskey et al. 2021a, 2021b; Bainbridge et al. 2023).

Pioneering water quality monitoring of the major rivers of the GBRCA in the 1970s to 1990s by Belperio (1979), the Australian Institute of Marine Science (Furnas 2003), the Commonwealth Scientific Industry Research Organisation (CSIRO; Bramley and Roth 2002), and the Queensland Government (e.g., Hunter et al. 2001; Packett et al. 2009) provided the early foundation to estimate loads exported to the GBR. Land-use focused monitoring programs in the 1990s and 2000s (e.g., Bramley and Roth 2002; Hunter and Walton 2008; Packett et al. 2009; Rohde et al. 2008; Bainbridge et al. 2009; Mitchell et al. 2009; Bartley et al. 2012) offered valuable insights to help quantify land use-specific water quality characteristics and the changes to loads since catchment modification. Monitoring results from the paired 'Brigalow Catchment Study' that contains a pristine Brigalow forest, a grazing catchment, and a cropping catchment within the Fitzroy Basin, present important insights on changing hydrology and water quality characteristics following land clearing (Elledge and Thornton 2017; Thornton and Elledge 2022). More recently, monitoring of riverine loads of sediments, nutrients, and pesticides has become well-coordinated through the establishment of the Great Barrier Reef Catchment Loads Monitoring Program (GBRCLMP) from 2006/2007, with annual reports available from the 2009/2010 financial year (e.g., Joo et al. 2012; Turner et al. 2012; Water Quality and Investigations 2021). The program has a rigorous quality assurance and quality control approach around sample collection, processing, analysis, and reporting, with the current program including river suspended sediment and nutrient loads for 56 sites across the basins of the GBRCA (Water Quality and Investigations 2021; Figure 1). The load monitoring program captures > 90% of the GBRCA and has provided critical measured load data from predominately end-of-basin sites for model calibration and validation where the model outputs are directly informed by monitoring sites with > 10 years of measured data. This process has allowed the latest Source Catchments models to produce what is considered the most reliable estimation of current baseline loads for GBR basins (McCloskey et al. 2021a, 2021b). Moreover, statistical techniques have improved the ability to interpret water quality data through the quantification of uncertainty and to provide insights on the spatial and temporal drivers of variability in water quality constituents (Kuhnert et al. 2012, 2018; Liu et al. 2018, 2021a, 2021b). Since the mid-2010s, advances in fit-for-purpose, field deployable continuous sensor and digital imagery technologies have greatly expanded the capacity of GBR monitoring programs to monitor water quality and streamflow-based parameters at high temporal frequency (i.e., 15 minute intervals) in nearly 'real-time' (Davis et al. 2021). In particular, with the rapidly maturing nitrate and turbidity sensor technology, instantaneous provision of accurate, high-resolution observations is emerging as an increasingly attractive and effective tool for water quality monitoring and associated water resource management decisions across multiple spatial scales in GBR basins. These sensors are most suited for finer scales in the upper reaches of basins (Figure 2).

The new technology offers greater potential for farmer engagement regarding fertiliser use with a greater temporal frequency of data collection and near real time delivery of results. Sensor-based technologies have the potential for more cost-effective identification of sediment and nutrient generation 'hotspots' when sensing equipment is installed in a network array, at a range of spatial scales, within a defined catchment area (e.g., Bartley et al. 2014). Ignoring these critical source areas can have profound ramifications for cost effectiveness of water quality improvement practices (Sharpley and Smith 1994). These considerations have received substantial longer-term recognition in minimizing sediment losses in grazing catchments of the GBRCA (Bainbridge et al. 2014) both at the sub-catchment scale and at the finer scale hydro-geomorphological features (i.e., gullies) that are recognised as suspended sediment 'hotspots' (Wilkinson et al. 2018). Identified through existing technologies (remote sensing, historic aerial photography, LiDAR, etc.), remediating these small spatial scale features (at least at the entire catchment scale) is a more recent focus of water quality management in GBR grazing catchments (Doriean et al. 2021; Koci et al. 2021).

Nutrient (DIN) generation 'hotspots' are certainly implicit in some aspects of predictive water quality modelling of farming practice changes in the GBRCA for agricultural commodities such as sugarcane (McCloskey et al. 2017). However, these are based primarily on coarse-scale

FIGURE 2 Alternative optical lens cleaning options utilised throughout Project 25. (a) Compressed air blast. (b) Automated, externally mounted lens wiper being trialled. (c) Real-time monitoring station during wet season flood events.

Source: Davis et al. (2020)

data relating to catchment farm management practices rather than the actual identification of high nutrient contributing zones within the landscape. The on-ground identification and subsequent targeting of intervention effort at nutrient loss 'hotspots' are, accordingly, still in a state of relative infancy in the more intensively developed agricultural catchment areas of the GBR (Webster et al. 2012). However, in recent years, multiple programs have rolled out that combine sensor-based nutrient monitoring with automated Doppler, or camera-based streamflow quantification techniques, to identify high DIN exporting hotspots in GBR agricultural catchments (Davis et al. 2020).

High frequency, sensor-based monitoring programs will also increase in importance for future GBR catchment water quality sampling as the need to identify catchment hotspots and monitor the success of fine spatial scale remediation activities increases. As basin size decreases, it is well-established that high-flow events typically become of shorter duration (i.e., 'flashier'), and changes in discharge and pollutant concentrations also occur over shorter periods (Davis et al. 2012; Horowitz 2013; Kaushal et al. 2014; Novic et al. 2018). Accordingly, sampling frequency has to increase to generate acceptable flux estimates and/or concentration ranges to avoid missing the 'hot moments' (McClain et al. 2003), which are episodic events that carry a disproportionate amount of catchment flux (Koci et al. 2020). This certainly includes the 'first flush' events documented in many tropical, agricultural sub-catchments when highly elevated concentrations (and loads) are transported through the catchment (Davis et al. 2020). Here, streamflow variations and associated sediment and nitrate-nitrogen (N) concentration fluxes documented with turbidity and nitrate sensor data at finer sub-catchment scales are often pronounced, with order-of-magnitude changes occurring in the space of hours (Bartley et al. 2006; Wilkinson et al. 2013; Figure 3). The sample collection frequency required to accurately capture this scale of variability in a traditional automated, discrete water sample collection approach is impractical and cost prohibitive, particularly with a high number of sampling sites across multiple sub-catchments. With water quality sensor data progressively becoming less expensive per data point as the sampling frequency of discrete samples increases, there comes a point where despite initial instrumentation costs, sensor technologies become a more cost-effective option. Therefore, it seems inevitable that many GBR water quality monitoring programs in the future will shift towards more sensor-based technologies.

Despite numerous investigations of N dynamics from GBR basins, detailed studies on the temporal aspects of N fluxes and stream discharge in agricultural areas of the GBRCA are currently scarce. Quantification of nutrient fluxes in sewage systems and urban watersheds has identified initial catchment 'first flushes' as driving a significant proportion of nutrient export from watersheds in many scenarios (Lee et al. 2002; Li et al. 2007). The first flush (Figure 4) phenomenon has been most frequently

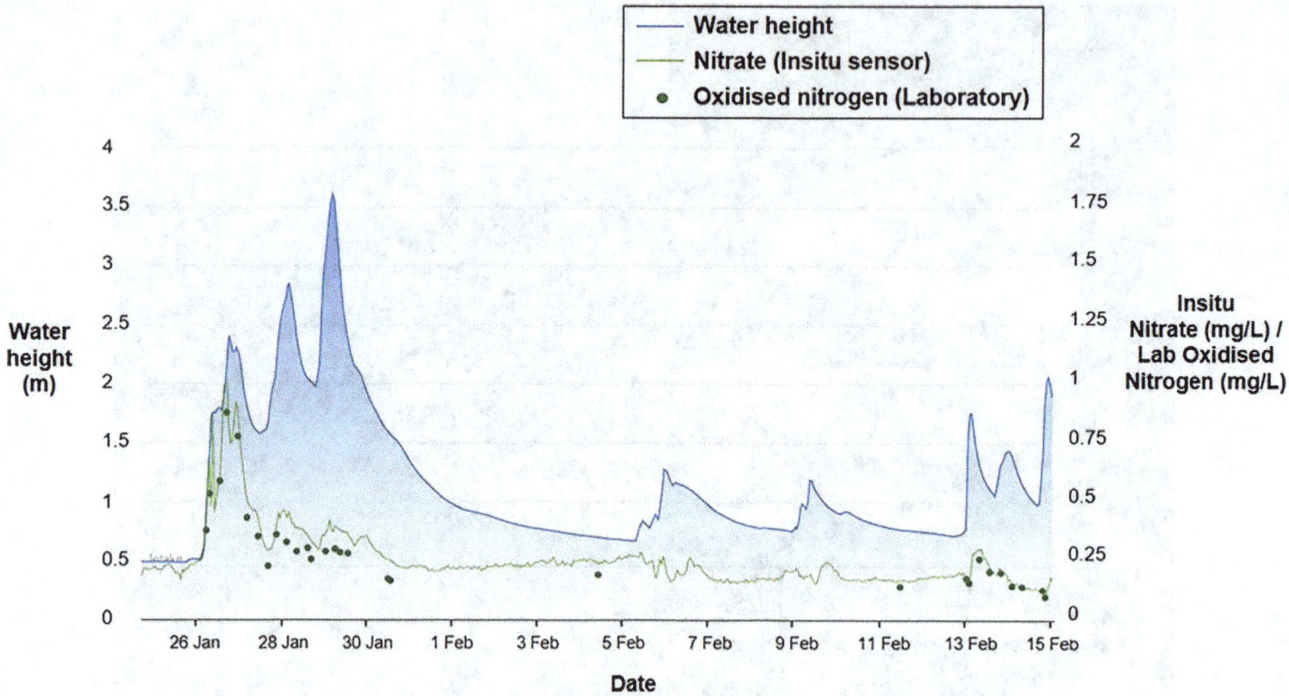

FIGURE 3 Example of high-resolution sensor data available from the GBRCLMP in a GBR catchment.

Source: GBRCLMP, QDES

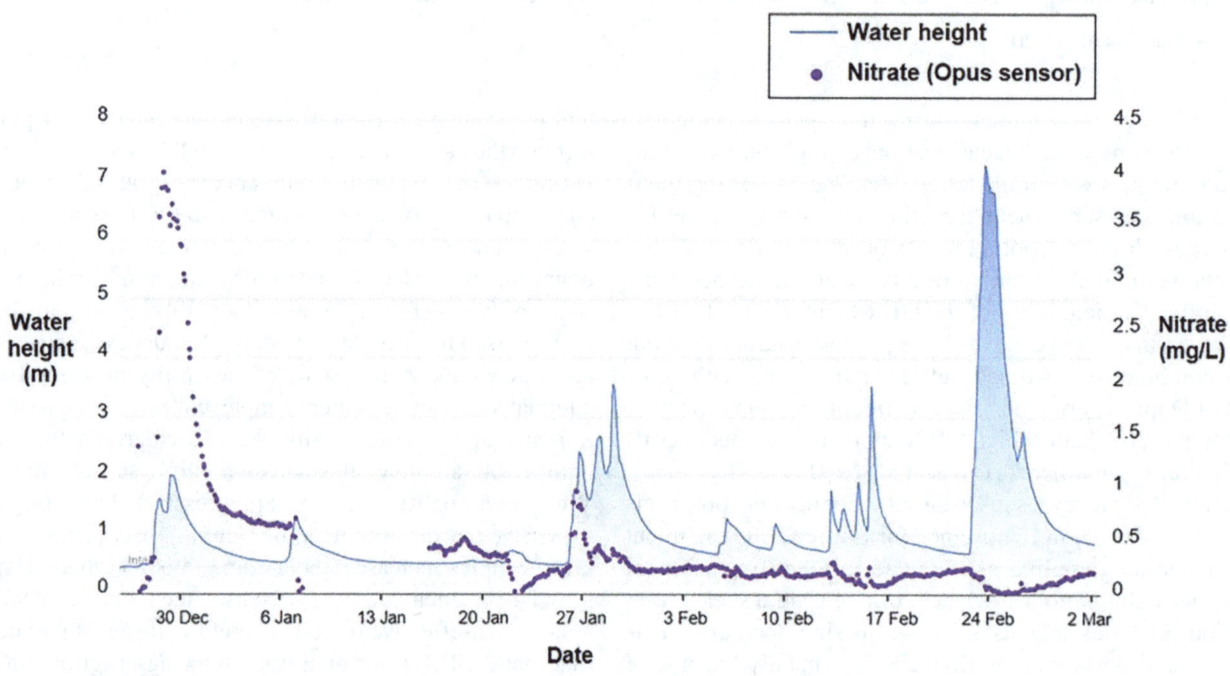

FIGURE 4 High-resolution nitrate in situ sensor data showing highly elevated concentrations during the first flush in a GBR catchment.

Source: GBRCLMP, QDES

demonstrated in small urban catchments characterised by large catchment proportions of impervious area, from which surface flow dominates the runoff process. A growing body of research from agricultural catchments is also suggesting elements of first flush behaviours in water quality dynamics from farming land uses (Obermann et al. 2007; Yang et al. 2015). Recent water quality monitoring of nitrate loss dynamics in the GBRCA at both paddock and sub-catchment scales indicates that some of the highest nitrate-N concentrations and a significant proportion of annual nitrate load losses from sugarcane-dominated catchments can often occur in the first 3–4 significant rainfall-runoff events of the year (Masters et al. 2017; Davis et al. 2021), which is often prior to the largest wet season flood events. The implications of these concentrated initial load loss processes in terms of potential intervention points for the management of diffuse pollution and downstream ecosystem impacts remain to be studied.

Another critical aspect of the calculation of GBR river loads is the estimation of the pre-development or pre-European settlement loads. Best estimates of pre-development load coupled with the current load provide the 'anthropogenic load', which is a key measure of the change in delivery to the GBR lagoon and is used for target setting and basin prioritization exercises (e.g., Brodie et al. 2017). Modelling pre-development suspended sediment and associated particulate nutrient loads has typically been performed by applying dry-weather/event mean concentrations from 'reference or pristine' sites coupled with discharge (i.e., to produce a rating curve) and considering various bio-geomorphological features of the landscape to calculate a Pre-European load (e.g., Neil and Yu 1996; Furnas 2003; Brodie et al. 2003; McKergow et al. 2005a; McCloskey et al. 2021a). Similarly for dissolved nutrient loads, data from reference or pristine sites are used coupled with discharge to estimate a 'natural' pre-development load (e.g., Furnas 2003; Brodie et al. 2003; McKergow et al. 2005b; McCloskey et al. 2021b).

Other insights on relative changes in terrestrial runoff since the arrival of Europeans are provided in various proxy records. For example, luminescent lines in coral cores provide a valuable proxy of annual riverine discharge from the adjacent catchment. Long (> 300 year) records have been produced for some of the larger GBR rivers, which show an increase in annual average discharge since 1850 and an increase in the frequency and magnitude (~2 fold) of 90th-percentile discharge events (Lough 2007; Lough et al. 2015). Notably, this change in discharge is not incorporated into the modelling of GBR pollutant loads for the pre-development period (McCloskey et al. 2021a). The geochemistry of coral core records, such as the Ba/Ca ratio, provides additional insights on changes to terrestrial runoff, where a record from Havannah Island on the inshore GBR indicates a 5- to 10-fold increase following the arrival of Europeans (McCulloch et al. 2003b). Although this record was originally interpreted as a proxy of increased sediment flux from the Burdekin River (McCulloch et al. 2003b), there is currently dispute whether the record reflects changes in sediment flux, freshwater discharge, or a combination of the two (Lewis et al. 2018; D'Olivo and McCulloch 2022). The rare earth element composition within the coral skeleton shows considerable promise as a proxy of sediment exposure in the GBR lagoon (Saha et al. 2019, 2021), while Mn and Y also provide some indication of changes in top soil erosion and terrestrial runoff, respectively (Lewis et al. 2007; Saha et al. 2021). Measurements of the N isotope composition in coral cores have also shown promise for reconstructing long-term changes in nutrient cycling at coral reef sites (Erler et al. 2020; Marion et al. 2021), while phosphorus measurements in coral cores may provide an indication of changes in loads over time (Mallela et al. 2013). Coral core records also offer important insights on changing climates across the GBR (see Webb 2024). Other useful proxies to examine changes in sediment fluxes include the examination of accumulation rates in sediment core records (Douglas et al. 2010; Lewis et al. 2014), Be-10-derived land denudation rates measured in the catchment (Nichols et al. 2014; Croke et al. 2015; Bartley et al. 2015; Mariotti et al. 2021), and changes in fluvial (stream) bench deposition (Hughes et al. 2009, 2010; Pietsch et al. 2015; Leonard and Nott 2016; Bartley et al. 2018).

8.3. SOURCES AND FLUXES OF SEDIMENT TO THE GBR

The latest best available modelled estimates of mean river basin discharge and mean sediment loads for the 35 basins of the GBRCA (McCloskey et al. 2021a; Puignou Lopez et al. in review) are presented in Table 1. These records encompass slightly different time periods but represent long-term (over 30 years) averages. In general, the basins with the largest areas (i.e., Burdekin and Fitzroy) and/or within the wetter climatic regions (i.e., Wet Tropics and Cape York NRM regions) have the highest mean annual discharge (Puignou Lopez et al. in review). The top five basins with the highest mean annual discharge are Burdekin, Fitzroy, Johnstone, Herbert, and Normanby Basins (Table 1). For further details on the hydrology of the GBR including changes in the connection between the catchment and marine zones, the reader is referred to Waterhouse et al. (2024).

The top five basins in terms of mean annual sediment export include the Burdekin, Fitzroy, Mary, Herbert, and Burnett Basins (Table 1). In general, the basin area and the basins that are within the drier climate areas largely explain these contributions, although other factors such as land use/management and the geomorphology of the streams (i.e., soil types and steepness) also play a role (e.g., see Bainbridge et al. 2014).

Over the past two decades, fine-scale monitoring, sediment tracing, and catchment mapping exercises have helped to better identify the key sediment sources (i.e., hillslope, gully, and streambank sources) within some of the main contributing basins. In particular, the use of fallout radionuclides such as Cs-137 has shown that sub-soil erosion (i.e., either gully or streambank sources) largely dominates the sediment loads exported from the catchment (Tims et al. 2010; Olley et al. 2013; Hancock et al. 2014; Wilkinson et al. 2013, 2015). Such atmospheric fallout radionuclides 'coat' the surface of the top ~15 cm of the soil profile, and accordingly, suspended sediment that contains little to no radionuclides must be derived from the subsoil. The latest Source Catchments modelling has been informed from these data and improved gully density mapping (Darr et al. 2017), which show that gully erosion dominates in the Burdekin Basin, contributing 54% of the sediment load (McCloskey et al. 2021a), while streambank and hillslope sources contribute the remaining 23% and 22%, respectively (Figure 5). In comparison, streambank erosion is the dominant sediment source for the other main contributing basins (Fitzroy, Mary, and Herbert), while the contribution

TABLE 1

Latest modelled estimates of mean annual discharge and fine sediment loads for the river basins of the GBRCA.

Basin	Basin area (km²)	Mean discharge (ML) 1990/91 to 2019/20[1]	Mean fine sediment load (kt/y) 1986/87 to 2013/14[2]
Jacky Jacky Creek	2,963	2,740,000	20
Olive Pascoe River	4,180	3,860,000	45
Lockhart River	2,883	1,870,000	61
Stewart River	2,743	1,050,000	22
Normanby River	24,399	4,260,000	156
Jeannie River	3,638	1,430,000	31
Endeavour River	2,182	1,590,000	81
Cape York NRM	**42,988**	**16,800,000**	**415**
Daintree River	2,107	2,300,000	44
Mossman River	473	648,000	10
Barron River	2,188	865,000	148
Mulgrave-Russell River	1,983	4,120,000	92
Johnstone River	2,325	4,720,000	187
Tully River	1,683	3,330,000	115
Murray River	1,107	1,510,000	40
Herbert River	9,844	4,440,000	518
Wet Tropics NRM	**21,710**	**21,900,000**	**1,153**
Black River	1,057	514,000	49
Ross River	1,707	583,000	35
Haughton River	4,051	1,020,000	145
Burdekin River	130,120	9,000,000	3,335
Don River	3,736	844,000	213
Burdekin NRM	**140,671**	**12,000,000**	**3,777**
Proserpine River	2,494	1,420,000	72
O'Connell River	2,387	1,380,000	212
Pioneer River	1,572	825,000	149
Plane Creek	2,539	1,460,000	88
Mackay-Whitsunday NRM	**8,992**	**5,090,000**	**522**
Styx River	3,013	801,000	81
Shoalwater Creek	3,601	927,000	55
Waterpark Creek	1,836	556,000	33
Fitzroy River	142,552	5,600,000	1,369
Calliope River	2,241	388,000	42
Boyne River	2,496	324,000	14
Fitzroy NRM	**155,740**	**8,590,000**	**1,594**
Baffle Creek	4,085	718,000	65
Kolan River	2,901	316,000	34
Burnett River	33,207	955,000	328
Burrum River	3,362	318,000	20
Mary River	9,466	1,800,000	582
Burnett-Mary NRM	**53,021**	**4,110,000**	**1,030**
GBR Total	**423,122**	**68,500,000**	**8,490**

Source: [1]Puignou Lopez et al. (in review); [2]McCloskey et al. (2021a)

of gully and streambank sources is fairly evenly spread for the Burnett Basin (Figure 5; McCloskey et al. 2021a).

The latest available Source Catchments modelling suggests that sediment loads have increased by more than 3-fold for most basins within the Burdekin, Mackay Whitsunday, Fitzroy, and Burnett Mary NRM regions (exceptions include the Black, Proserpine, Shoalwater, and Waterpark Basins) (McCloskey et al. 2021a). The basins of the Cape York NRM region have modelled load increases of 1.3-fold or less except for the Normanby Basin from which loads have been

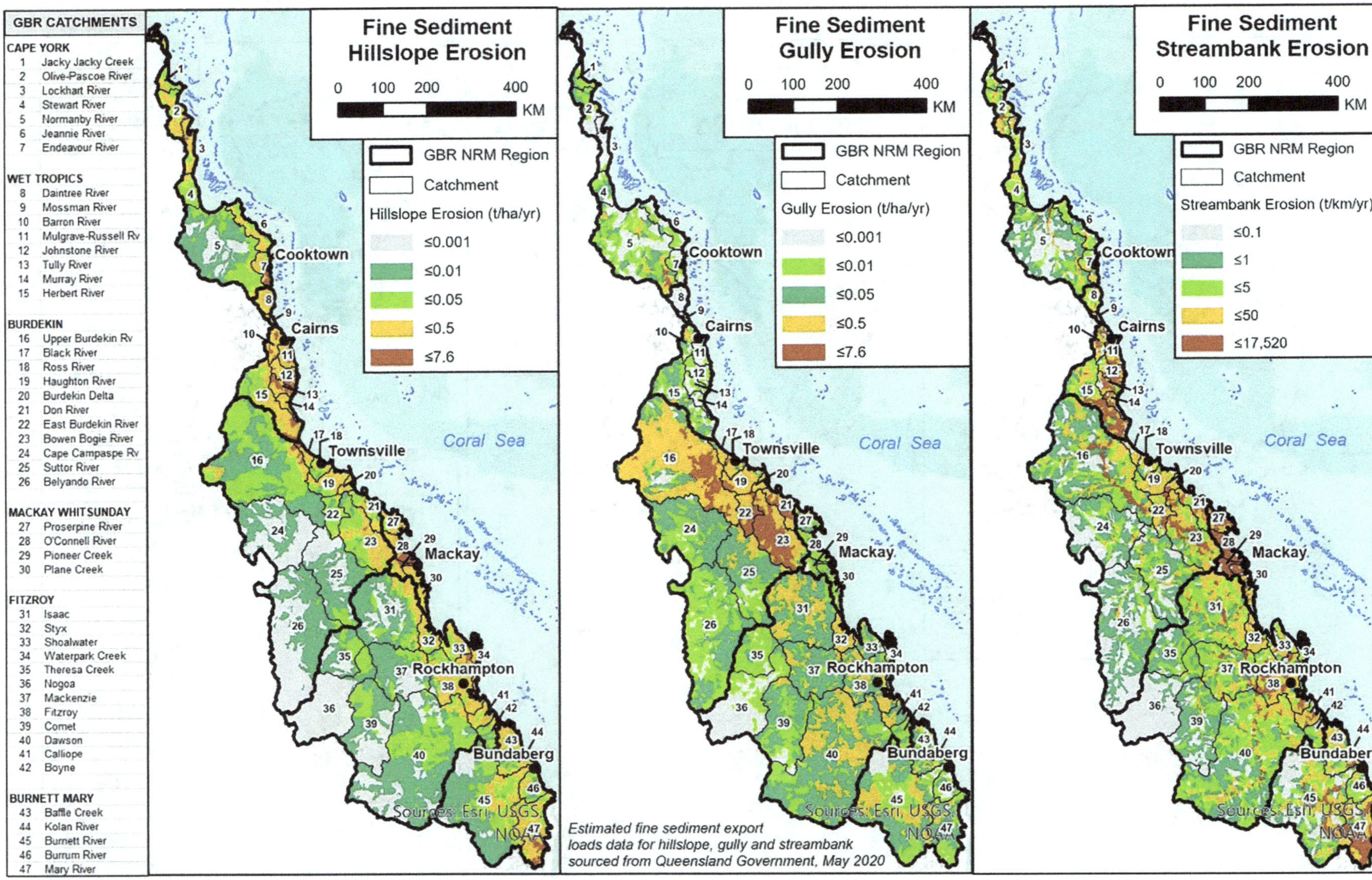

FIGURE 5 Modelled contribution of different erosion processes within the GBRCA.

Source: Modified from McCloskey et al. (2021a)

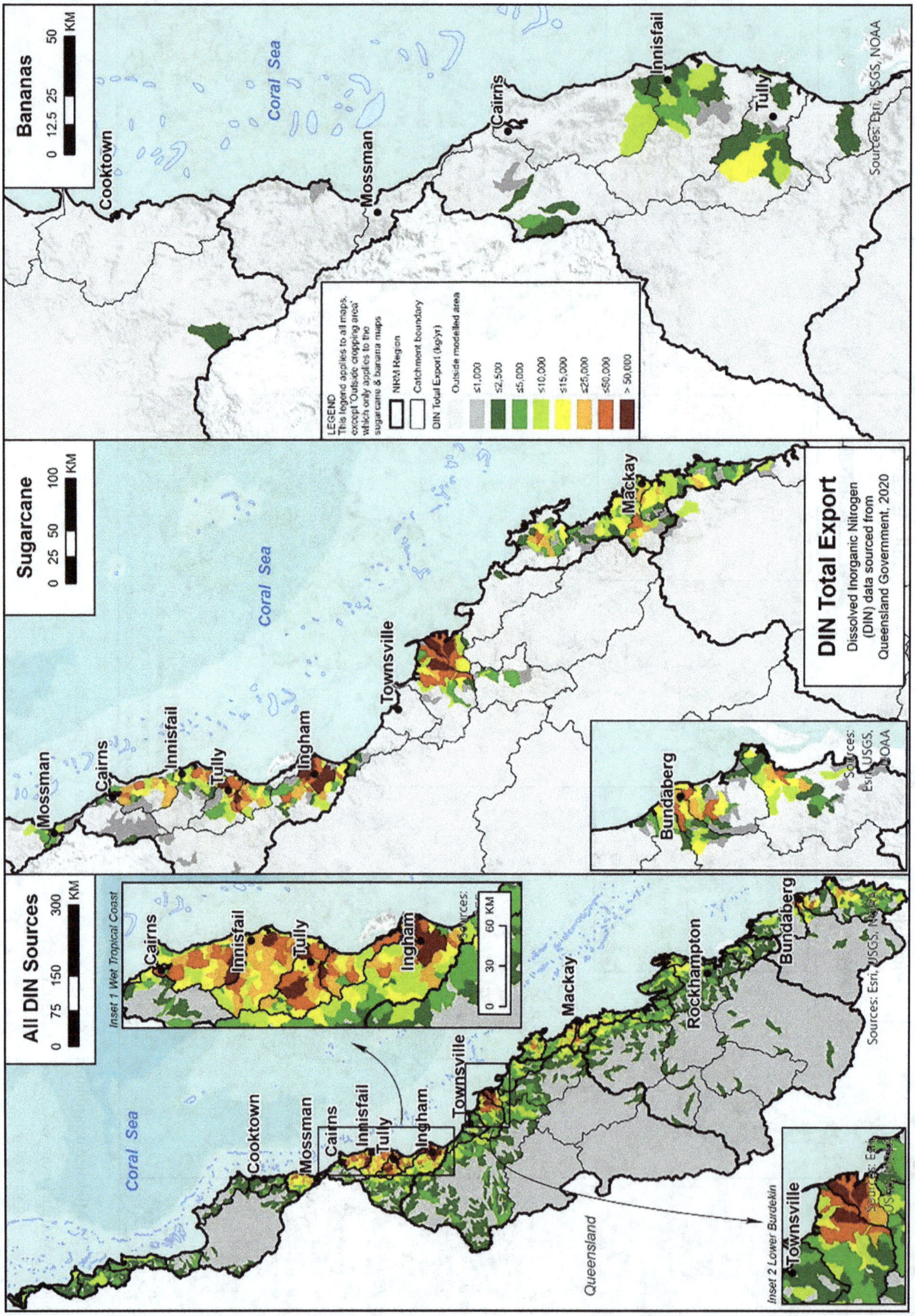

FIGURE 6 Modelled sources of DIN from the GBRCA highlighting key hotspots for DIN from cropping lands.

Source: Modified from McCloskey et al. (2021b)

projected to have increased by 3- to 5-fold (McCloskey et al. 2021a; G. McCloskey unpublished data). The magnitude of increase in sediment loads across the Wet Tropics basins ranges from 1.2-fold (Daintree Basin) to 2.3-fold (Herbert Basin) (McCloskey et al. 2021a). For the highest sediment contributing basins, the modelled load increases since the arrival of Europeans exceed 5-fold for the Burdekin, Fitzroy, and Mary basins, while increases are 2.3-fold for the Herbert and 3.8-fold for the Burnett (McCloskey et al. 2021a). In general, these increase factors are supported by the available proxy data derived mostly from the Burdekin and Fitzroy Basins (Douglas et al. 2010; Hughes et al. 2010; Thompson et al. 2011; Lewis et al. 2014; Bartley et al. 2015, 2018; Mariotti et al. 2021), although monitoring data from the Normanby Basin suggest that a large proportion of the load may be deposited on the floodplain and is not exported from the catchment (Brooks et al. 2013; Pietsch et al. 2015; Howley et al. 2021).

8.4. SOURCES AND FLUXES OF NITROGEN AND PHOSPHORUS TO THE GBR

The latest best available modelled estimates of mean GBR river basin loads of PN, PP, and DIN are presented in Table 2 (McCloskey et al. 2021a, 2021b). The top five contributing basins in terms of mean annual PN (Burdekin, Fitzroy, Herbert, Mary, and Johnstone Basins) and PP (Burdekin, Fitzroy, Herbert, Mary, and Burnett Basins) loads generally follow the sediment load rankings, with some exceptions that are likely related to variability in soil composition and erosional sources (i.e., surface versus sub-surface). The latest Source Catchments modelling suggests that hillslope erosion is the greatest contributor of both PN and PP loads for most of the GBR basins with some exceptions (McCloskey et al. 2021a). The most notable exceptions include the dominance of gully erosion for the PN and PP contribution to the Burdekin and Don Basins and for PP in the Normanby Basin (McCloskey et al. 2021a).

Our knowledge of PN and PP contributions has been improved from catchment monitoring and increased soil profile data from the river basins. For example, soil profile sampling from the Burdekin catchment reveals that certain soil types contribute much greater proportions of PN and PP than others, and the relative contributions from the surface and sub-surface can also be variable depending on the soil type (Burton et al. 2015; Garzon-Garcia et al. 2018). In addition, recent tracing research that utilises a combination of stable isotope tracers ($\delta^{15}N$ and $\delta^{13}C$) and trace element geochemistry has provided new insights on PN contributions in catchments (Bahadori et al. 2019, 2020). In this regard, tracing data from the Johnstone Basin show that although rainforest lands were the largest contributor of PN, the proportional contribution from bananas was much greater than the area of bananas (proportion enrichment 4.5) within the basin (Bahadori et al. 2020); this finding agrees with historical land-use monitoring data from the Johnstone Basin (Hunter and Walton 2008).

The top five modelled estimates of mean annual DIN loads (Herbert, Burdekin, Johnstone, Fitzroy, and Mulgrave-Russell Basins) generally reflect the basin size (and in particular, the discharge volume) and the area of intensive fertilised cropping (Table 2). The latest modelling suggests that nearly half of the total DIN load exported from the GBRCA is from the Wet Tropics NRM region, and half of this total DIN load is derived from the sugarcane cropping area (Figure 6; McCloskey et al. 2021b). In general, the basins with intensive cropping have seen DIN loads increase by factors between 1.5 and 3.0 (McCloskey et al. 2021b).

8.5. TRANSFORMATIONS OF SEDIMENT AND NUTRIENTS ACROSS THE CATCHMENT TO MARINE CONTINUUM

Targeted research over the past two decades has revealed new insights on the transformations of sediment and nutrients as they travel through the catchment network and are exported to the GBR lagoon. For example, monitoring to determine the sediment trapping efficiency of the Burdekin Falls Dam was used to produce revised equations for application in the Source Catchments modelling that better predict the trapping efficiencies for dams in tropical settings (Lewis et al. 2013; Waters and Lewis 2017; McCloskey et al. 2021a). Sediment tracing that uses geochemistry or clay mineralogy has provided evidence for the preferential transport of sediment through the catchment and out into the GBR lagoon (e.g., Smith et al. 2008; Bainbridge et al. 2016; Furuichi et al. 2016). Geochemical tracing in the Fitzroy catchment suggests that basaltic soils are preferentially transported within the Fitzroy River and out into the coastal zone (Smith et al. 2008), while clay mineralogy tracing indicates that the expandable clay fraction is preferentially transported in Burdekin River flood plumes (Bainbridge et al. 2016). Particle size measurements of suspended sediment in Burdekin River flood plumes also reveal that the < 20 µm terrigenous fraction travels farthest in the marine environment and forms the nucleus of large sediment floc aggregates composed of terrigenous particles, bacterial and fungal mucus, and phytoplankton remains and excrement; hence, this terrigenous < 20 µm fraction should be the primary focus for catchment management efforts (Bainbridge et al. 2012, 2018, 2021). The exposure of terrestrial materials within the GBR lagoon is dependent on the spatial movement of the riverine flood plumes, which tend to move northwards along the coast and are therefore largely confined to the inner GBR (Devlin and Brodie 2005). On occasion, during large river discharge events coupled with slack winds, flood plumes can move further offshore and impinge on the middle and outer areas of the GBR lagoon. Aerial photographs and satellite images capture such exposure events and provide valuable insights on the transformation of suspended sediments and nutrients in the GBR lagoon (Figure 7).

TABLE 2

Latest modelled estimates of particulate N and phosphorus and DIN loads for the river basins of the GBRCA.

Basin	Mean PN load (t/y) 1986/87 to 2013/14[1]	Mean PP load (t/y) 1986/87 to 2013/14[1]	Mean DIN load (t/y) 1986/87 to 2013/14[2]
Jacky Jacky Creek	326	45	67
Olive Pascoe River	540	80	98
Lockhart River	422	102	49
Stewart River	142	32	31
Normanby River	403	139	107
Jeannie River	289	43	35
Endeavour River	572	164	40
Cape York NRM	**2,695**	**604**	**425**
Daintree River	160	45	378
Mossman River	47	12	92
Barron River	507	135	100
Mulgrave-Russell River	462	130	787
Johnstone River	1,041	357	857
Tully River	495	126	706
Murray River	171	31	458
Herbert River	1,416	607	1,747
Wet Tropics NRM	**4,299**	**1,443**	**5,124**
Black River	181	63	55
Ross River	92	31	70
Haughton River	313	163	516
Burdekin River	3,928	2,161	1,189
Don River	370	133	111
Burdekin NRM	**4,884**	**2,551**	**1,941**
Proserpine River	330	108	307
O'Connell River	964	318	256
Pioneer River	553	173	262
Plane Creek	483	160	403
Mackay-Whitsunday NRM	**2,329**	**760**	**1,228**
Styx River	391	213	95
Shoalwater Creek	320	158	105
Waterpark Creek	375	158	58
Fitzroy River	3,228	1,890	799
Calliope River	271	140	44
Boyne River	66	35	35
Fitzroy NRM	**4,650**	**2,595**	**1,136**
Baffle Creek	155	100	64
Kolan River	65	29	109
Burnett River	959	397	245
Burrum River	36	19	206
Mary River	1,405	547	490
Burnett-Mary NRM	**2,620**	**1,093**	**1,113**
GBR Total	**21,477**	**9,045**	**10,967**

Source: [1]McCloskey et al. (2021a); [2]McCloskey et al. (2021b)

Recent research has identified that PN is highly reactive within the stream network and within river flood plumes and can provide an important source of bioavailable nutrients to the GBR through both desorption of ammonium ions attached to the sediment and microbial mineralisation (Burton et al. 2015; Garzon-Garcia et al. 2018, 2021). Field studies show that the fine (< 10 μm) fraction of surface soils is preferentially enriched in bioavailable N and phosphorus, while laboratory studies on the soil samples indicate that large proportions of PN are readily mineralised to DIN up

FIGURE 7 Sentinel 2 satellite image of the plume from the Daintree River during the 2021/22 season (image processed by Caroline Petus, TropWATER) (left panel) and aerial photograph of the Burdekin River plume impinging over Old Reef on the GBR mid-shelf during the 2018/19 season (right panel image: Matt Curnock).

to 7 days after incubation (Burton et al. 2015). Therefore, surface soils (and surface erosion processes) are likely to make a disproportionate contribution to PN and PP loads relative to sub-surface soils. Notably, however, in instances where sub-surface erosion is dominant (i.e., > 80%), this erosion process will still be the dominant contributor to PN and PP loads (Burton et al. 2015). Research on soil samples shows that colloidal N can contribute up to 45% of the total soil N pool and may explain previous monitoring data from unfertilised grazing lands that found higher than expected concentrations of DIN (Judy et al. 2018). Sediment tracing studies suggest that terrestrially derived P in seafloor sediments in the GBR lagoon off the Wet Tropics region is predominantly sourced to basaltic soils (McCulloch et al. 2003a). The concentration of terrestrially derived P in the marine sediments is much lower than in the adjacent catchment, which indicates that a considerable fraction of PP can become bioavailable in the GBR lagoon, although it is likely contributed through longer-term desorption processes under anoxic conditions (McCulloch et al. 2003a).

Dissolved inorganic nutrients follow a distinctive conservative mixing pattern until the 15 to 25 PSU salinity zone where they then rapidly deplete, which generally coincides with increasing chlorophyll a concentrations (Devlin et al. 2001; Devlin and Brodie 2005; Bainbridge et al. 2012). These results indicate that dissolved inorganic nutrients are completely bioavailable and are rapidly consumed by primary productivity in the GBR lagoon. Recent research has suggested that runoff of undegraded fertiliser, urea, can represent not only a significant but formerly unrecognised proportion of the total DON pool leaving intensive agricultural land uses but also an important form of nitrogenous export from fertilised agricultural land uses (Davis et al. 2016). Situations where substantial rainfall or irrigation-driven surface water runoff occurs within 1–2 weeks of fertiliser application particularly provide the scope for major, and in some cases dominant, losses of undegraded urea from paddocks (Rohde et al. 2013; MacDonald et al. 2020). Fertiliser-derived urea can be a significant, bioavailable, and anthropogenic form of dissolved 'organic' N export and warrants further attention in many field and catchment scale research applications. Notably, this 'organic' pool could be confounded by disproportionally high DON originating from natural forested areas.

8.6. MANAGEMENT OPTIONS TO REDUCE SEDIMENT AND NUTRIENT LOADS TO THE GBR

Over the past two decades, the state and federal governments have implemented policy instruments such as increased agricultural extension efforts, financial incentive packages for farmers, the regulation of farming practices, and the setting of basin-specific water quality targets to manage sediment and nutrient runoff to the GBR. These policy instruments are perceived to have varying degrees of success but are difficult to quantify (Eberhard et al. 2021). Mechanisms to better evaluate the cost-effectiveness of such policies are currently being developed (e.g., Star et al. 2021), although the assignment of economic values to the various ecosystems of the GBR that are highly complex comes with great uncertainty (Rolfe and De Valck 2021). Over the past decade, large government on-ground investments have occurred through the Reef Trust Program (federal government) in partnership with the Great Barrier Reef Foundation, and Major Integrated Projects (MIP) (Queensland state government) in an attempt to reduce the loads of sediments, nutrients, and pesticides to the GBR lagoon.

The management of soil erosion in grazing lands has resulted not only in several major programs to rehabilitate severely gullied landscapes but also in engineering measures to stabilise streambanks. One major area of government investment is in the Bowen-Broken-Bogie 'BBB' catchments of the Burdekin Basin where a grassroots 'Landholders Driving Change' Queensland Government-funded MIP project is focussed on building extension capacity in the area for improved holistic land management and gully remediation (see https://ldc.nqdrytropics.com.au). The Reef Trust Program, via the Great Barrier Reef Foundation, has also focussed on gully rehabilitation efforts in the BBB, which includes evaluating the effectiveness (both in terms of cost and landscape repair) of different techniques (https://ldc.nqdrytropics.com.au; Bainbridge et al. 2023).

The federal and Queensland state governments have (including the partnership with the Great Barrier Reef Foundation) also invested heavily in management to reduce fertiliser runoff from cropping lands through their Reef Trust Program and Wet Tropics MIP (run by Terrain NRM, see https://terrain.org.au/what-we-do/waterway-health/major-integrated-project/), respectively. Both programs contain a strong extension component that utilises local agronomic practitioners to refine and implement improved fertiliser application to either meet or exceed industry standards. Many of the projects embedded within these programs utilise water quality monitoring as a form of direct farmer engagement including the instalment of near time nitrate loggers (Davis et al. 2021). Other options to manage runoff from coastal cropping lands include the restoration of (or conversion of land to) coastal wetlands and land retirement (Wallace and Waltham 2021; Waltham et al. 2021). Indeed, a timely, up-scaled and multi-pronged approach to reduce sediment and nutrient loads to the GBR will be required to achieve water quality targets and help provide crucial resilience to the coastal and marine ecosystems also threatened by a rapidly changing climate.

8.7. CONCLUSIONS

Overall, in the past two decades, vast improvements have been made in the methods available to refine load estimates of sediment and nutrients delivered from the river basins to the GBR lagoon and to identify hotspot areas for management prioritisation. The monitoring of river basin loads has become coordinated through the state government, while other research monitoring has become more targeted to address specific questions such as erosion processes, preferential transport of sediments, bio-availability assessment of particulate nutrients, and evaluation of management practices and to identify catchment hotspots. Modelling of river basin loads has likewise become more robust and has seen continual improvement over this same period. These improvements have, in turn, allowed governments to make better policy decisions and investments to begin to reduce the loads of sediments and nutrients delivered to the GBR lagoon. Although some success has occurred on localised scales from such improvements, the ability to achieve larger-scale reductions from broader best management practice uptake across catchment scales remains elusive and requires further considerable investment (Attard et al. 2024). Innovative and robust water quality monitoring programs in the future will need to continually adapt and evolve to respond to the increasing pressure on coastal and marine ecosystems from climate change. Such programs will likely require high temporal resolution measurements of multiple parameters across wide spatial scales to help inform management interventions.

ACKNOWLEDGEMENTS

We are most grateful to Shuci Liu (Department of Environment and Science; DES) for the production of Figure 1, Angela Pollett (DES) for the production of Figure 5 and 6, Caroline Petus for processing the Sentinel 2 image in Figure 7, and Matt Curnock for the Burdekin plume image (Figure 7). We acknowledge the DES Water Quality and Investigation's GBRCLMP for supplying the high-resolution nitrate sensor data plots and the CSIRO-JCU Partnership (Catchment Water Quality Science) for supporting this chapter.

REFERENCES

Attard S, Connell JG, Chaiechi T (2024) Chapter 25: Pathways to improved water quality in the GBR lagoon—exploring opportunities for broadscale application of low-risk practices in the Lower Burdekin irrigated agriculture areas. In Wolanski E, Kingsford MJ (eds.), Oceanographic Processes of Coral Reefs. Physical and Biological Links in the Great Barrier Reef. CRC Press.

Bahadori M, Chen C, Lewis S, Rezaei Rashti M, Cook F, Parnell A, Esfandbod M, Boyd S (2019) A novel approach of combining isotopic and geochemical signatures to differentiate the sources of sediments and particulate nutrients from different land uses. Sci Total Environ 655:129–140.

Bahadori M, Chen C, Lewis S, Rezaei Rashti M, Cook F, Parnell A, Esfandbod M, Stevens T (2020) Tracing the sources of sediment and associated particulate nitrogen from different land uses in the Johnstone River catchment, Wet Tropics, northeastern Australia. Mar Pollut Bull 157:111344.

Bainbridge ZT, Brodie JE, Faithful JW, Sydes DA, Lewis SE (2009) Identifying the land-based sources of suspended sediments, nutrients and pesticides discharged to the Great Barrier Reef from the Tully Basin, Queensland, Australia. Mar Freshwater Res 60:1081–1090.

Bainbridge ZT, Lewis S, Bartley R, Fabricius K, Collier C, Waterhouse J, Garzon-Garcia A, Robson B, Burton J, Wenger A, Brodie J (2018) Fine sediment and particulate organic matter: A review and case study on ridge-to-reef transport, fates, and impacts on marine ecosystems. Mar Pollut Bull 135:1205–1220.

Bainbridge ZT, Lewis S, Smithers S, Kuhnert PM, Henderson BL, Brodie JE (2014) Fine-suspended sediment and water budgets for a large, seasonally dry tropical catchment: Burdekin River catchment, Queensland, Australia. Water Resources Res 50:9067–9087.

Bainbridge ZT, Lewis S, Smithers S, Wilkinson S, Douglas G, Hillier S, Brodie J (2016) Clay mineral source tracing and characterisation of Burdekin River (NE Australia) and flood plume fine sediment. J Soils Sediments 16:687–706.

Bainbridge ZT, Lewis S, Stevens T, Petus C, Lazarus E, Gorman J, Smithers S (2021) Measuring sediment grain size across the catchment to reef continuum: Improved methods and environmental insights. Mar Pollut Bull 168:112339.

Bainbridge ZT, Olley J, Wilkinson S, Bartley R, Lewis S, Dougall C, Khan S, Kuhnert P, Burton J (2023) Refining fine sediment source identification through integration of spatial modelling, concentration monitoring and source tracing: A case study in the Great Barrier Reef catchments. Sci Total Environ 892:164731.

Bainbridge ZT, Wolanski E, Álvarez-Romero JG, Lewis SE, Brodie JE (2012) Fine sediment and nutrient dynamics related to particle size and floc formation in a Burdekin River flood plume, Australia. Mar Pollut Bull 65:236–248.

Bartley R, Corfield JP, Hawdon AA, Kinsey-Henderson AE, Abbott BN, Wilkinson SN, Keen RJ (2014) Can changes to pasture management reduce runoff and sediment loss to the Great Barrier Reef? The results of a 10-year study in the Burdekin catchment, Australia. Rangeland J 36:67–84.

Bartley R, Croke J, Bainbridge ZT, Austin JM, Kuhnert PM (2015) Combining contemporary and long-term erosion rates to target erosion hot-spots in the Great Barrier Reef, Australia. Anthropocene 10:1–12.

Bartley R, Roth CH, Ludwig J, McJannet D, Liedloff A, Corfield J, Hawdon A, Abbott B (2006) Runoff and erosion from Australia's tropical semi-arid rangelands: Influence of ground cover for differing space and time scales. Hydrological Processes 20:3317–3333.

Bartley R, Speirs WJ, Ellis TW, Waters DK (2012) A review of sediment and nutrient concentration data from Australia for use in catchment water quality models. Mar Pollut Bull 65:101–116.

Bartley R, Thompson C, Croke J, Pietsch T, Baker B, Hughes K, Kinsey-Henderson A (2018) Insights into the history and timing of post-European land use disturbance on sedimentation rates in catchments draining to the Great Barrier Reef. Mar Pollut Bull 131:530–546.

Belperio AP (1979) The combined use of wash load and bed material load rating curves for the calculation of total load: An example from the Burdekin River, Australia. Catena 6:317–329.

Belperio AP (1983) Late Quaternary terrigenous sedimentation in the Great Barrier Reef lagoon. In: Baker JT, Carter RM, Sammarco PW, Stark KP (eds) Proceedings: Inaugural Great Barrier Reef Conference, Townsville, August 28 to September 2, 1983. JCU Press, 71–76.

Bramley RGV, Roth CH (2002) Land-use effects on water quality in an intensively managed catchment in the Australian humid tropics. Mar Freshwater Res 53:931–940.

Brodie J, Baird M, Waterhouse J, Mongin M, Skerratt J, Robillot C, Smith R, Mann R, Warne M (2017) Development of basin-specific ecologically relevant water quality targets for the Great Barrier Reef. TropWATER Report No. 17/38, James Cook University, Published by the State of Queensland, Brisbane, Australia, 68pp.

Brodie J, McKergow LA, Prosser IP, Furnas M, Hughes AO, Hunter H (2003) Sources of sediment and nutrient exports to the Great Barrier Reef World Heritage Area. ACTFR Report No. 03/11, Australian Centre for Tropical Freshwater Research, James Cook University, Townsville, 208p.

Brooks A, Spencer J, Olley J, Pietsch T, Borombovits D, Curwen G, Shellberg J, Howley C, Gleeson A, Simon A, Bankhead N, Klimetz D (2013) An empirically-based sediment budget for the Normanby Basin: Sediment sources, sinks, and drivers on the Cape York savannah. Griffith University, 506pp.

Browne NK, Smithers SG, Perry CT (2012) Coral reefs of the turbid inner-shelf of the Great Barrier Reef, Australia: An environmental and geomorphic perspective on their occurrence, composition and growth. Earth Sci Rev 115:1–20.

Burton J, Moody P, DeHayr R, Chen C, Lewis S, Olley J (2015). Sources of bioavailable particulate nutrients: Phase 1 (RP128G). Department of Science, Information Technology and Innovation.

Croke J, Bartley R, Chappell J, Austin JM, Fifield K, Tims SG, Thompson CJ, Furuichi T (2015) 10Be-derived denudation rates from the Burdekin catchment: The largest contributor of sediment to the Great Barrier Reef. Geomorphology 241:122–134.

Darr SD, Pringle MJ (2017) Improving gully density maps for modelling water quality within Great Barrier Reef Catchments. In: Syme G, Hatton MacDonald D, Fulton B, Piantadosi J (eds.) MODSIM2017, 22nd International Congress on Modelling and Simulation. Modelling and Simulation Society of Australia and New Zealand, Hobart, 1920–1926.

Davis AM, Lewis SE, Bainbridge ZT, Glendenning L, Turner R, Brodie JE (2012) Dynamics of herbicide transport and partitioning under event flow conditions in the lower Burdekin region, Australia. Mar Pollut Bull 65:182–193

Davis AM, Taylor B, Fielke S (2020) 'Project 25': Engaging with farmers and demonstrating water quality outcomes to create confidence in on-farm decision making. In: Report to the National Environmental Science Program. Reef and Rainforest Research Centre Limited, Cairns, 83pp.

Davis AM, Tink M, Rohde K, Brodie JE (2016) Urea contributions to dissolved 'organic' nitrogen losses from intensive, fertilised agriculture. Agric Ecosyst Environ 223:190–196. doi: 10.1016/j.agee.2016.03.006.

Davis AM, Webster AJ, Fitch P, Fielke S, Taylor BM, Morris S, et al. (2021) The changing face of science communication, technology, extension and improved decision-making at the farm water quality interface. Mar Pollut Bull 169:112534.

Devlin M, Waterhouse J, Taylor J, Brodie J (2001) Flood plumes in the Great Barrier Reef: Spatial and temporal patterns in composition and distribution. GBRMPA Publication No. 68. Great Barrier Reef Marine Park Authority, Townsville, Australia.

Devlin MJ, Brodie J (2005) Terrestrial discharge into the Great Barrier Reef Lagoon: Nutrient behaviour in coastal waters. Mar Pollut Bull 51:9–22.

D'Olivo JP, McCulloch M (2022) Impact of European settlement and land use changes on Great Barrier Reef river catchments reconstructed from long-term coral Ba/Ca records. Sci Total Environ 830:154461.

Doriean NJC, Bennett WW, Spencer JR, Garzon-Garcia A, Burton JM, Teasdale PR, Welsh DT, Brooks AP (2021) Intensive landscape-scale remediation improves water quality of an alluvial gully located in a Great Barrier Reef catchment. Hydrol Earth Syst Sci 25:867–883.

Douglas GB, Ford PW, Palmer M, Noble RM, Packett R (2006) Fitzroy River Basin, Queensland, Australia. I. Identification of sediment sources in impoundments and flood events. Environ Chem 3:364–376.

Douglas GB, Kuhnen M, Radke LC, Hancock G, Brooke B, Palmer MR, Pietsch T, Ford PW, Trefry MG, Packett R (2010) Delineation of sediment sources to a coastal wetland in the Great Barrier Reef catchment: Influence of climate variability and land clearing since European arrival. Environ Chem 7:190–206.

Eberhard R, Coggan A, Jarvis D, Hamman E, Taylor B, Baresi U, Vella K, Dean AJ, Deane F, Helmstedt K, Mayfield H (2021) Understanding the effectiveness of policy instruments to encourage adoption of farming practices to improve water quality for the Great Barrier Reef. Mar Pollut Bull 172:112793.

Elledge A, Thornton C (2017) Effect of changing land use from virgin brigalow (Acacia harpophylla) woodland to a crop or pasture system on sediment, nitrogen and phosphorus in runoff over 25 years in subtropical Australia. Ag Ecosyst Environ 239:119–131.

Erftemeijer PL, Riegl A, Hoeksema BW, Todd PA (2012) Environmental impacts of dredging and other sediment disturbances on corals: A review. Mar Pollut Bull 64:1737–1765.

Erler DV, Farid HT, Glaze TD, Carlson-Perret NL, Lough JM (2020) Coral skeletons reveal the history of nitrogen cycling in the coastal Great Barrier Reef. Nature Comm. 11:1–8.

Fabricius KE (2005) Effects of terrestrial runoff on the ecology of corals and coral reefs: Review and synthesis. Mar Pollut Bull 50:125–146.

Furnas M (2003) Catchments and corals: Terrestrial runoff to the Great Barrier Reef. Australian Institute of Marine Science, Townsville, 334pp.

Furnas M, Mitchell A (2000) Runoff of terrestrial sediment and nutrients into the Great Barrier Reef World Heritage Area. In Wolanski E (ed) Oceanographic processes of coral reefs (pp. 57–72). CRC Press.

Furnas M, Alongi D, McKinnon D, Trott L, Skuza M (2011) Regional-scale nitrogen and phosphorus budgets for the northern (14°S) and central (17°S) Great Barrier Reef shelf ecosystem. Cont Shelf Res 31:1967–1990.

Furuichi T, Olley J, Wilkinson S, Lewis S, Bainbridge Z, Burton J (2016) Paired geochemical tracing and load monitoring analysis for identifying sediment sources in a large catchment draining into the Great Barrier Reef Lagoon. Geomorphology 266:41–52.

Garzon-Garcia A, Burton, J, Ellis R, Askildsen M, Finn L, Moody P, DeHayr R (2018) Sediment particle size and contribution of eroded soils to dissolved inorganic nitrogen export in great barrier reef catchments. Department of Environment and Science, Queensland Government.

Garzon-Garcia A, Burton JM, Lewis S, Bainbridge Z, De Hayr R, Moody P, Brodie J (2021) The bioavailability of nitrogen associated with sediment in riverine plumes of the Great Barrier Reef. Mar Pollut Bull 173:112910.

Hancock GJ, Wilkinson SN, Hawdon AA, Keen RJ (2014) Use of fallout tracers 7Be, 210Pb and 137Cs to distinguish the form of sub-surface soil erosion delivering sediment to rivers in large catchments. Hydrol Process 28:3855–3874.

Horowitz AJ (2013) A review of selected inorganic surface water quality-monitoring practices: Are we really measuring what we think, and if so, are we doing it right? Environ Sci Technol 47:2471–2486. doi: 10.1021/es304058q.

Howley C, Shellberg J, Olley J, Brooks A, Spencer J, Burford M (2021) Sediment and nutrient sources and sinks in a wet-dry tropical catchment draining to the Great Barrier Reef. Mar Pollut Bull 165:112080.

Hughes AO, Croke JC, Pietsch TJ, Olley JM (2010) Changes in the rates of floodplain and in channel bench accretion in response to catchment disturbance, central Queensland, Australia. Geomorphology 114:338–347.

Hughes AO, Olley JM, Croke JC, McKergow LA (2009) Sediment source changes over the last 250 years in a dry-tropical catchment, central Queensland, Australia. Geomorphology 104:262–275.

Hunter HM, Sologinkin CJ, Choy SC, Hooper AR, Allen WS, Raymond MAA, Peeters J (2001) Water management in the Johnstone Basin, Final Report, Natural Heritage Trust Project No. 952194. Department of Natural Resources and Mines, Brisbane, Queensland, p. 117.

Hunter HM, Walton RS (2008) Land-use effects on fluxes of suspended sediment, nitrogen and phosphorus from a river catchment of the Great Barrier Reef, Australia. J Hydrol 356:131–146.

Joo M, Raymond MAA, McNeil VH, Huggins R, Turner RDR, Choy S (2012) Estimates of sediment and nutrient loads in 10 major catchments draining to the Great Barrier Reef during 2006–2009. Mar Pollut Bull 65:150–166.

Judy JD, Kirby JK, Farrell M, McLaughlin MJ, Wilkinson SN, Bartley R, Bertsch PM (2018) Colloidal nitrogen is an important and highly-mobile form of nitrogen discharging into the Great Barrier Reef lagoon. Sci Rep 8:1–11.

Kaushal SS, Mayer PM, Vidon PG, Smith RM, Pennino MJ, Newcomer TA et al. (2014) Land use and climate variability amplify carbon, nutrient, and contaminant pulses: A review with management implications. J Am Water Res Assoc 50:585–614.

Koci J, Sidle RC, Kinsey-Henderson AE, Bartley R, Wilkinson SN, Hawdon AA, Jarihani B, Roth CH, Hogarth L (2020) Effect of reduced grazing pressure on sediment and nutrient yields in savanna rangeland streams draining to the Great Barrier Reef. J Hydrology 582:124520. doi: 124510.121016/j.jhydrol.122019.124520.

Koci J, Wilkinson SN, Hawdon AA, Kinsey-Henderson AE, Bartley R, Goodwin NR (2021) Rehabilitation effects on gully sediment yields and vegetation in a savanna rangeland. Earth Surf Process Landform 46:1007–1025. doi: 1010.1002/esp.5076.

Kroon FJ, Kuhnert PM, Henderson BL, Wilkinson SN, Kinsey-Henderson A, Brodie JE, Turner RDR (2012) River loads of suspended solids, nitrogen, phosphorus and herbicides delivered to the Great Barrier Reef lagoon. Mar Poll Bull 65:167–181.

Kuhnert PM, Henderson BL, Lewis SE, Bainbridge ZT, Wilkinson SN, Brodie JE (2012) Quantifying total suspended sediment export from the Burdekin River catchment using the loads regression estimator tool. Water Res Farahani 48:W04533.

Kuhnert PM, Pagendam DE, Bartley R, Gladish DW, Lewis SE, Bainbridge ZT (2018) Making management decisions in the face of uncertainty: A case study using the Burdekin catchment in the Great Barrier Reef. Mar Freshwater Res 69:1187–1200.

Lee JH, Bang KW, Ketchum LH, Choe JS, Yu MJ (2002) First flush analysis of urban storm runoff. Sci Total Environ 293:163–175

Leonard S, Nott J (2016) A Holocene fluvial archive of the Mulgrave River, Notheastern Australia: Influence of tropical cyclones and sediment delivery to the Great Barrier Reef. Palaeogeog Palaeoclimatol Palaeoecol 441:845–853.

Lewis SE, Bainbridge ZT, Kuhnert PM, Sherman BS, Henderson B, Dougall C, Cooper M, Brodie JE (2013) Calculating sediment trapping efficiencies for reservoirs in tropical settings: A case study from the Burdekin Falls Dam, NE Australia. Water Resources Res 49:1017–1029.

Lewis SE, Bartley R, Wilkinson SN, Bainbridge ZT, Henderson AE, James CS, Irvine SA, Brodie JE (2021) Land use change in the river basins of the Great Barrier Reef, 1860 to 2019: A foundation for understanding environmental history across the catchment to reef continuum. Mar Pollut Bull 166:112193.

Lewis SE, Lough JM, Cantin NE, Matson EG, Kinsley L, Bainbridge ZT, Brodie JE (2018) A critical evaluation of coral Ba/Ca, Mn/Ca and Y/Ca ratios as indicators of terrestrial input: New data from the Great Barrier Reef, Australia. Geochim Cosmochim Acta 237:131–154.

Lewis SE, Olley J, Furuichi T, Sharma A, Burton J (2014) Complex sediment deposition history on a wide continental shelf: Implications for the calculation of accumulation rates on the Great Barrier Reef. Earth Planet Sci Lett 393:146–158.

Lewis SE, Shields GA, Kamber BS, Lough JM (2007) A multi-trace element coral record of land-use changes in the Burdekin River catchment, NE Australia. Palaeogeog Palaeoclimatol Palaeoecol 246:471–487.

Li LQ, Yin CQ, He QC, Kong LL (2007) First flush of storm runoff pollution from an urban catchment in China. J Environ Sci 19:295e299.

Liu S, Ryu D, Webb JA, Lintern A, Guo D, Waters D, Western AW (2021a) A Bayesian approach to understanding the key factors influencing temporal variability in stream water quality–a case study in the Great Barrier Reef catchments. Hydrol Earth Syst Sci 25:2663–2683.

Liu S, Ryu D, Webb JA, Lintern A, Guo D, Waters D, Western AW (2021b) A multi-model approach to assessing the impacts of catchment characteristics on spatial water quality in the Great Barrier Reef catchments. Environ Pollut 288:117337.

Liu S, Ryu D, Webb JA, Lintern A, Waters D, Guo D, Western AW (2018) Characterisation of spatial variability in water quality in the Great Barrier Reef catchments using multivariate statistical analysis. Mar Pollut Bull 137:137–151.

Lønborg C, Álvarez–Salgado XA, Duggan S, Carreira C (2018) Organic matter bioavailability in tropical coastal waters: The Great Barrier Reef. Limnol Oceanograph 63:1015–1035.

Lough JM (2007) Tropical river flow and rainfall reconstructions from coral luminescence: Great Barrier Reef, Australia. Paleoceanography 22:PA2218.

Lough JM, Lewis SE, Cantin NE (2015) Freshwater impacts in the central Great Barrier Reef: 1648–2011. Coral Reefs 34:739–751.

MacDonald BCT, Nachimuthu G, Chang YF, Nadelko AJ, Tuomi S, Watkins M (2020) Nitrogen composition in furrow irrigated run-off water. Ag Water Manag 242:106399.

Mallela J, Lewis SE, Croke B (2013) Coral skeletons provide historical evidence of phosphorus runoff on the Great Barrier Reef. PLoS One 8:e75663.

Marion GS, Jupiter SD, Radice VZ, Albert S, Hoegh-Guldberg O (2021) Linking isotopic signatures of nitrogen in nearshore coral skeletons with sources in catchment runoff. Mar Pollut Bull 173:113054.

Mariotti A, Croke J, Bartley R, Kelley SE, Ward J, Fülöp RH, Rood AH, Rood DH, Codilean AT, Wilcken K, Fifield K (2021) Pre-development denudation rates for the Great Barrier Reef catchments derived using 10Be. Mar Pollut Bull 172:112731.

Masters B, Tahir N, Fries J, Mortimore C, Enderlin N, Armour J (2017) Paddock-scale water quality monitoring of nitrogen fertiliser management practices in sugarcane cropping 2014–2016 technical report, Wet Tropics region. Department of Natural Resources and Mines, Brisbane, Queensland 73 pp.

McClain ME, Boyer EW, Dent CL, Gergel SE, Grimm NB, Groffman PM, et al. (2003) Biogeochemical hot spots and hot moments at the interface of terrestrial and aquatic ecosystems. Ecosystems 6:301–312.

McCloskey GL, Baheerathan R, Dougall C, Ellis R, Bennett FR, Waters D, Darr S, Fentie B, Hateley LR, Askildsen M (2021a) Modelled estimates of fine sediment and particulate nutrients delivered from the Great Barrier Reef catchments. Mar Pollut Bull 165:112163.

McCloskey GL, Baheerathan R, Dougall C, Ellis R, Bennett FR, Waters D, Darr S, Fentie B, Hateley LR, Askildsen M (2021b) Modelled estimates of dissolved inorganic nitrogen exported to the Great Barrier Reef lagoon. Mar Pollut Bull 171:112655.

McCloskey GL, Waters D, Baheerathan R, Darr S, Dougall C, Ellis R, Fentie B, Hateley LR (2017) Modelling reductions of pollutant loads due to improved management practices in the Great Barrier Reef catchments: Updated methodology and results. Technical Report for Reef Report Card 2015, Queensland Department of Natural Resources and Mines

McCulloch M, Pailles C, Moody P, Martin CE (2003a) Tracing the source of sediment and phosphorus into the Great Barrier Reef lagoon. Earth Planet Sci Lett 210:249–258.

McCulloch MT, Fallon S, Wyndham T, Hendy E, Lough J, Barnes D (2003b) Coral record of increased sediment flux to the inner Great Barrier Reef since European settlement. Nature 421:727–730.

McKergow LA, Prosser IP, Hughes AO, Brodie J (2005a) Sources of sediment to the Great Barrier Reef world heritage area. Mar Pollut Bull 51:200–211.

McKergow LA, Prosser IP, Hughes AO, Brodie J (2005b) Regional scale nutrient modelling: Exports to the Great Barrier Reef world heritage area. Mar Pollut Bull 51:186–199.

Mitchell A, Reghenzani J, Faithful J, Furnas M, Brodie J (2009) Relationships between land use and nutrient concentrations in streams draining a 'wet-tropics' catchment in northern Australia. Mar Freshwater Res 60:1097–1108.

Moss AJ, Rayment GE, Reilly N, Best EK (1992) A preliminary assessment of sediment and nutrient exports form Queensland coastal catchments. Department of Primary Industries, Queensland.

Neil DT, Orpin AR, Ridd PV, Yu B (2002) Sediment yield and impacts from river catchments to the Great Barrier Reef lagoon. Mar Freshwater Res 53:733–752.

Neil DT, Yu B (1996) Fluvial sediment yield to the Great Barrier Reef Lagoon: Spatial patterns and the effect of land use. In: Hunter HM, Eyles AG, Rayment GE (eds) Downstream effects of land use. Department of Natural Resources, Queensland, Australia, 281–286.

Nichols KK, Bierman PR, Rood DH (2014) 10Be constrains the sediment sources and sediment yields to the Great Barrier Reef from the tropical Barron River catchment, Queensland, Australia. Geomorphology 224:102–110.

Novic AJ, Ort C, O'Brien DS, Lewis SE, Davis AM, Mueller JF (2018) Understanding the uncertainty of estimating herbicide and nutrient mass loads in a flood event with guidance on estimator selection. Water Res 132:99–110.

Obermann M, Froebrich J, Perrin J-L, Tournoud M-G (2007) Impact of significant floods on the annual load in an agricultural catchment in the Mediterranean. J Hydrol 334:99–108.

Olley J, Brooks A, Spencer J, Pietsch T, Borombovits D (2013) Subsoil erosion dominates the supply of fine sediment to rivers draining into Princess Charlotte Bay, Australia. J Environ Radioact 124:121–129.

Packett R, Dougall C, Rhode K, Noble R (2009) Agricultural lands are hot-spots for annual runoff polluting the southern Great Barrier Reef lagoon. Mar Pollut Bull 58:976–986.

Peel MC, Finlayson BL, McMahon TA (2007) Updated world map of the Koppen-Geiger climate classification. Hydrol Earth Syst Sci 11:1633–1644.

Perry CT, Smithers SG, Gulliver P, Browne NK (2012) Evidence of very rapid reef accretion and reef growth under high turbidity and terrigenous sedimentation. Geology 40:719–722.

Petheram C, McMahon TA, Peel MC (2008) Flow characteristics of rivers in northern Australia: Implications for development. J. Hydrol 357:93–111.

Pietsch TJ, Brooks AP, Spencer J, Olley JM, Borombovits D (2015) Age, distribution, and significance within a sediment budget, of in-channel depositional surfaces in the Normanby River, Queensland, Australia. Geomorphology 239:17–40.

Prosser IP, Rustomji P, Young WJ, Moran CJ, Hughes A (2001) Constructing river basin sediment budgets for the National Land and Water Resources Audit. CSIRO Land and Water Technical Report 15/01, CSIRO, Canberra.

Puignou Lopez O, Lewis S, James C, Davis A, Mackay S (in review) Hydrology of the Great Barrier Reef catchment area along a latitudinal gradient: Upscaling discharge to reflect catchment inputs. J Hydrology.

Risk MJ (2014) Assessing the effects of sediments and nutrients on coral reefs. Curr Opin Environ Sustain 7:108–117.

Rohde K, Masters B, Fries N, Noble R, Carroll C (2008) Fresh and Marine Water Quality in the Mackay Whitsunday Region 2004/05 to 2006/07. Queensland Department of Natural Resources and Water for the Mackay Whitsunday Natural Resource Management Group, Australia.

Rohde K, McDuffie K, Agnew J (2013) Paddock to Sub-catchment Scale Water Quality Monitoring of Sugarcane Management Practices. Final Report 2009/10 to 2011/12 Wet Seasons, Mackay Whitsunday Region. Department of Natural Resources and Mines, Queensland Government for Reef Catchments (Mackay Whitsunday Isaac) Limited, Australia.

Rolfe J, De Valck J (2021) Values for protecting the Great Barrier Reef: A review and synthesis of studies over the past 35 years. Mar Pollut Bull 169:112531.

Ryan EJ, Smithers SG, Lewis SE, Clark TR, Zhao J-x (2016) Chronostratigraphy of Bramston Reef reveals a long-term record of fringing reef growth under muddy conditions in the central Great Barrier Reef. Palaeogeog Palaeoclimatol Palaeoecol 441:734–747.

Ryan EJ, Smithers SG, Lewis SE, Clark TR, Zhao J-x (2018) The variable influences of sea level, sedimentation and exposure on Holocene reef development over a cross-shelf transect, central Great Barrier Reef. Diversity 10:110.

Saha N, Webb GE, Zhao J-x, Lewis SE, Nguyen A, Feng Y (2021) Spatiotemporal variation of rare earth elements from river to reef continuum aids monitoring of terrigenous sources in the Great Barrier Reef. Geochim Cosmochim Acta 29:85–112.

Saha N, Webb GE, Zhao J-x, Nguyen A, Lewis SE, Lough J (2019) Coral-based high-resolution rare earth element proxy for terrestrial sediment discharge affecting coastal seawater quality, Great Barrier Reef. Geochim Cosmochim Acta 254:173–191.

Sharpley AN, Smith SJ (1994) Wheat tillage and water quality in the Southern Plains. Soil Tillage Res 30:33–48.

Smith J, Douglas GB, Radke LC, Palmer M, Brooke BP (2008) Fitzroy River Basin, Queensland, Australia. III. Identification of sediment sources in the coastal zone. Environ Chem 5:231–242.

Smithers SG, Hopley D, Parnell KE (2006) Fringing and nearshore coral reefs of the Great Barrier Reef: Episodic Holocene development and future prospects. J Coastal Res 22:175–187.

Star M, Rolfe J, Farr M, Poggio M (2021) Transferring and extrapolating estimates of cost-effectiveness for water quality outcomes: Challenges and lessons from the Great Barrier Reef. Mar Pollut Bull 171:112870.

Thompson A, Schroeder T, Brando VE, Schaffelke B (2014) Coral community responses to declining water quality: Whitsunday Islands, Great Barrier Reef, Australia. Coral Reefs 33:923–938.

Thompson CJ, Croke JC, Purvis-Smith D (2011) Floodplain sediment disconnectivity at a tributary junction and valley constriction site in the Fitzroy River basin, Queensland, Australia. Geomorphology 125:293–304.

Thornton CM, Elledge AE (2022) Leichhardt, land clearing and livestock: The legacy of European agriculture in the Brigalow Belt bioregion of central Queensland, Australia. Ani Product Sci. doi: 10.1071/AN21468

Tims S, Everett S, Fifield L, Hancock G, Bartley R (2010) Plutonium as a tracer of soil and sediment movement in the Herbert River, Australia. Nucl Instrum Method Phys Res Sec B. Beam Int Mater Atoms 268:1150–1154.

Turner R, Huggins R, Wallace R, Smith R, Vardy S, Warne M (2012) Sediment Nutrient and Pesticide Loads Great Barrier Reef Catchment Loads Monitoring 2009–2010. Brisbane. www.reefplan.qld.gov.au/__data/assets/pdf_file/0027/45981/2009-2010-gbr-catchment-loads-report.pdf

Wallace J, Waltham NJ (2021) On the potential for improving water quality entering the Great Barrier Reef lagoon using constructed wetlands. Mar Pollut Bull 170:112627.

Waltham NJ, Wegscheidl C, Volders A, Smart JC, Hasan S, Lédée E, Waterhouse J (2021) Land use conversion to improve water quality in high DIN risk, low-lying sugarcane areas of the Great Barrier Reef catchments. Mar Pollut Bull 167:112373.

Waterhouse J, Pearson R, Lewis S, Davis A, Waltham N (2024) Chapter 7: Great Barrier Reef ecohydrology. In: Wolanski E, Kingsford MJ (eds.), Oceanographic Processes of Coral Reefs. Physical and Biological Links in the Great Barrier Reef. CRC Press.

Waters DK, Lewis SE (2017) Calculating sediment trapping efficiency for reservoirs in series. In: Syme G, Hatton MacDonald D, Fulton B, Piantadosi J (eds) MODSIM2017, 22nd International Congress on Modelling and Simulation. Modelling and Simulation Society of Australia and New Zealand, pp. 1990–1996. www.mssanz.org.au/modsim2017/L22/waters.pdf

Water Quality & Investigations (2021) Great barrier reef catchment loads monitoring program: Loads and yields for sediment and nutrients, and Pesticide metric results (2019–2020) for rivers that discharge to the great barrier reef. Department of Environment and Science, Brisbane.

Webb G (2024) Advances in understanding climate change on the great barrier reef using coral-based proxies. In: Wolanski E, Kingsford MJ (eds.), Oceanographic Processes of Coral Reefs. Physical and Biological Links in the Great Barrier Reef. CRC Press.

Webster AJ, Bartley R, Armour JD, Brodie JE, Thorburn PJ (2012) Reducing dissolved inorganic nitrogen in surface runoff water from sugarcane production systems. Mar Pollut Bull 65:128–135.

Wilkinson SN, Dougall C, Kinsey-Henderson AE, Searle RD, Ellis RJ, Bartley R (2014) Development of a time-stepping sediment budget model for assessing land use impacts in large river basins. Sci Total Environ 468:1210–1224.

Wilkinson SN, Hancock GJ, Bartley R, Hawdon AA, Keen R (2013) Using sediment tracing to assess processes and spatial patterns of erosion in grazed rangelands, Burdekin River basin, Queensland, Australia. Agric Ecosyst Environ 180:90–102.

Wilkinson SN, Kinsey-Henderson AE, Hawdon AA, Hairsine PB, Bartley R, Baker B (2018) Grazing impacts on gully dynamics indicate approaches for gully erosion control in northeast Australia. Earth Surf Proc Landforms 43:1711–1725.

Wilkinson SN, Olley JM, Furuichi T, Burton J, Kinsey-Henderson AE (2015) Sediment source tracing with stratified sampling and weightings based on spatial gradients in soil erosion. J Soils Sediments 15:2038–2051.

Yang X, Li T, Hua K, Zhang Y (2015) Investigation of first flushes in a small rural-agricultural catchment. Polish J Environ Studies 24:381–389.

9 Dispersal and Environmental Impacts of Pan-Oceanic Contaminants

Mia O. Hoogenboom, Liudmila Osipova, Mikaela Nordborg, Jodie A. Schlaefer, and Kay Critchell

9.1. INTRODUCTION

Understanding the structure, function, and dynamics of coral reef ecosystems requires knowledge of the physical processes that link reefs to the open ocean. For instance, ocean currents carry a flux of plankton together with particulate and dissolved organic matter that are important to the nutritional requirements of reef-associated plants and animals (e.g., Hamner et al. 1988; Morais and Bellwood 2019). At the same time, however, ocean currents can also carry pollutants that have negative impacts on marine life. Notable examples include plastic debris, which has become ubiquitous in our global oceans (Cozar et al. 2014) and can lead to mortality of wildlife and habitat degradation around the world (Jang et al. 2014; Krelling et al. 2017), including coral reefs (Reichert et al. 2019). In addition, sediments and a range of different chemicals (e.g., agricultural fertilizers and pesticides) are also transported from land to sea through terrestrial runoff processes (Bainbridge et al. 2018; Lewis et al. 2021; Lewis et al. 2024), disperse within nearshore environments according to prevailing winds and currents (Wolanski et al. 2008), and can be harmful to marine life (e.g., Fabricius 2005). Finally, a wide range of other contaminants, including fuel oils, ghost-nets, antifouling chemicals, grey and black water, bilge water, lubricants, and exhaust gas cleaning system (EGCS) washwater enter the marine environment from shipping and resource extraction activities (Moldanová et al. 2022).

Environmental risk assessments provide an objective framework to understand the risks that pollutants pose to the environment and prioritise management and spill response actions (Bottrill et al. 2008). Such risk assessments combine the likelihood of an identified hazard (e.g., the likelihood of a pollutant being spilled) and the consequences to the environment should this hazard occur (e.g., how populations would be affected by the pollutant). In the case of marine pollution, the severity of the environmental impact depends on the concentrations of pollutants present in the environment, longevity of pollutants remaining in a toxic form, local environmental conditions during the exposure, timing and duration of the exposure, and sensitivity of the local fauna and flora (e.g., Nordborg et al. 2020). Moreover, the consequences of exposure can depend on the size and age of organisms due to varying sensitivities to exposure because of differing habitat use, diet, or biological development (e.g., Critchell and Hoogenboom 2018; Nordborg et al. 2021; Duncan et al. 2021). To perform risk assessments, reliable toxicity thresholds are required for relevant taxa and environmental conditions, but these are not well-characterised for many ecosystems and pollutants (French-McCay et al. 2018; Parkerton et al. 2018). Even when precise toxicity thresholds are known, the likelihood of pollutant-environment interactions in coastal marine systems is highly variable because of the spatial and temporal variability in the composition and concentrations of contaminants (e.g., Wilcox et al. 2013) and in the physical environmental conditions (e.g., temperature, light, wave height, tidal state, and currents). Quantifying how contaminant concentrations change as the contaminants are dispersed in dynamic marine environments is fundamental to assessing the risk that these contaminants pose to marine life. Hydrodynamic models can be used in this capacity and are therefore important tools in environmental risk assessment (e.g., French-McCay et al. 2021).

Hydrodynamic models coupled with sub-models representing the properties of contaminants have been applied to advance the understanding of contaminant dispersal (e.g., James 2002). Complex physical properties (including particle size and shape, hydrophobicity, and specific gravity), chemical degradation processes, and environmental factors such as light and temperature, affect how contaminants move through the marine environment and must be incorporated into these contaminant modules. For example, parameterisation must change over time to represent how contaminants change via chemical degradation (e.g., aromatic hydrocarbons that change form when exposed to air and water), bio-fouling (e.g., microbial film that grows on the surface of floating debris), and physical degradation (e.g., mechanical breakdown because of wave action). Coupled hydrodynamic-contaminant models have improved over recent decades to include more accurate representations of contaminant dynamics and high-resolution representations of ocean processes (e.g., Anselain et al. 2023; Skerratt et al. 2023; Turrell 2020). However, numerous knowledge gaps remain as to how many contaminants interact with seawater and how environmental factors such as light, temperature, and pH alter these interactions (e.g., Nordborg et al. 2020). In addition to representing dynamic processes, it is also important to understand and correctly parameterise where pollutants enter the environment (i.e., their source).

The sources of contaminants are key determinants of the risks that they pose to marine life. Some contaminants

DOI: 10.1201/9781003320425-11

enter the marine environment at specific, fixed points of origin and often result in a gradient of concentration/exposure with distance from the point source. Common examples include sediment and nutrient contamination carried by flood-waters exiting major rivers (Bainbridge et al. 2018) or oil spills that spread out from the site of a shipping incident or oil and gas extraction facility (Beyer et al. 2016). However, other contaminants can enter the marine environment in more complex ways or with a mobile point of origin. Some examples include (1) marine plastic debris that enters the marine environment via rivers, coasts, and ships and the breakdown of large items that fragment, float, or sink depending on biofouling and (2) fuel oil residues and EGCS washwater released from ships while underway. Plastic and shipping pollution is an issue in all regions of the globe, including the Great Barrier Reef (GBR). The GBR is connected not only to the Coral Sea via the shelf break on its eastern boundary but also to the Arafura Sea via the Torres Strait, which joins the Northern GBR. Additionally, there are numerous, and heavily used, shipping channels throughout the GBR. This means that there are complex pathways for marine plastics and shipping by-products to enter the GBR system. Here, we present new perspectives on how to assess the risks that contaminants with pan-oceanic sources and complex behaviours in water pose to marine life. We focus on plastic debris and pollution from shipping, including fuel oils and EGCS washwater, and identify key information required for further development of the environmental risk assessment framework for application to pan-oceanic contaminants. We discuss global issues but highlight their complexities with GBR-specific examples.

9.2. PAN-OCEANIC DISPERSAL AND ACCUMULATION OF PLASTIC POLLUTION

Marine debris that washes up on beaches has implications for human health (e.g., Campbell et al. 2019), for ecosystem functioning, and for local economies because of, for example, lost revenue through decreased tourism (Jang et al. 2014). The accumulation of plastics on local beaches is influenced by the particle supply rate, as determined by global, local, and seasonal hydrodynamic processes (Ryan et al. 2014; Hu et al. 2016). The supply rate is intrinsically variable because tides and winds can change the location of convergence zones at a scale of hundreds of meters (Hu et al. 2016). However, accumulation also depends on particle residence time and export, with resuspension being a key determinant of residence time (Hinata et al. 2017). In developing dispersal models to understand the environmental risks associated with plastic pollution, there are multiple considerations for capturing these processes. When focusing on accumulation-resuspension dynamics, aside from estimates for large buoyant objects such as fishing floats (Johnson and Eiler 1999; Hinata et al. 2017), the resuspension rate and volume are not well-characterised (Turrell 2020). However, plastic dispersal predictions are sensitive

FIGURE 1 Plastic pollution accumulates on a rocky beach on the Central GBR.

Source: Photo credit: Annie Bauer-Civiello, Orpheus Island

to resuspension dynamics (Critchell and Lambrechts 2016; Patterson et al. 2023); therefore, this process should be incorporated in dispersal models. Additionally, we recommend that a wave field be incorporated into models that predict resuspension of plastics from the coast (Isobe et al. 2014). Furthermore, resuspension estimates need to account for particle size, because the residence times of microplastics can be 4x lower than that of macroplastics (Hinata et al. 2017).

In addition to accumulation-resuspension dynamics, the oceanic dispersal of plastic pollution is affected by plastic supply from global sources, degradation of macroplastics into microplastics, wind and Stokes drifts, particle settling rates, diffusivity, and in complex coastal environments, effects of wind shadows caused by land masses (e.g., Critchell and Lambrechts 2016; Chubarenko et al. 2018). Recent work on the GBR shows that model predictions of plastic accumulation vary up to 5-fold in response to relatively small changes in the parameter estimates for these processes (Critchell and Lambrechts 2016). For example, understanding marine plastic dispersal requires knowledge of how rapidly large plastic items degrade into microplastics. However, very few studies have quantified this degradation for different plastics under open ocean or beached conditions to allow for parameterisation in a model (but see Zhang et al. 2021). Further studies are required to understand how physical properties of beaches (e.g., sand grain size, temperature, and UV intensity) and mechanical breakdown driven by wave action affect plastic fragmentation (e.g., Weinstein et al. 2016; Welden and Cowie 2017). The parameterisation of wind drift processes also needs further research because it is often neglected in models of plastic distributions (Reisser et al. 2013; Isobe et al. 2014; Maes and Blanke 2015). The wind drift that plastic particles experience in nature is dependent on the cross-sectional area of the object exposed to the wind, which, in turn, is influenced by object size, shape, and buoyancy such that each floating object would experience different wind drift effects. Applying a single average wind drift value to all simulated particles in a model will capture broad patterns in plastic dispersal (as per Dunkin 2015) but could underestimate the spatial extent to which particles can spread. Obtaining more precise parameter estimates for each of these processes is critical to produce more accurate dispersal estimates and better assessment of environmental risk from plastic pollutants (van Sebille et al. 2020).

The GBR World Heritage Site is a complex hydrodynamic environment that poses challenges to modelling plastic dispersal at scales that are relevant to local management. In addition to the global challenges identified above, lack of knowledge of the input of plastics from external sources (e.g., large cities upstream, shipping lanes, and/or via the Coral Sea) is a key limitation to understanding plastic dispersal within the GBR. The GBR's northern boundary is also open to the Torres Strait, which represents another potential source of plastics. However, this is probably not a major transport pathway, given that the water currents through the Strait generally run east to west from the GBR to the Arafura Sea (Wolanski et al. 2013). Further south, initial modelling (Critchell 2018) suggests that offshore sources are dominant in the Whitsunday region of the central GBR for both micro- and macro-plastics, although additional research is needed to distinguish near shore from oceanic (Coral Sea) sources. Alongside more field sampling to test model predictions, especially for microplastics, experiments and field data (e.g., the Australian Marine Debris Initiative database complied by Tangaroa Blue) should be used to quantify plastic hydraulic behaviours and improve parameter estimates of hydrodynamic models. A similar parameterisation appoach has been implemented to model the dispersal of plastics from a cargo spill (Ebbesmeyer et al. 2007), dispersal of plastic bottle caps (Kako et al. 2011), and dispersal of crab pots (Ebbesmeyer et al. 2011).

9.3. PAN-OCEANIC POLLUTION BY EGCS WASHWATER AND FUEL OILS

Much of the world's trade goods are transported via oceanic commercial shipping vessels, and the risks posed to marine environments from shipping-associated pollutants are complex. Although international shipping is regulated by a designated UN agency, the International Maritime Organization (IMO), the movement of over 390,000 ships operating globally remains a pan-oceanic source of marine pollution (Olmer et al. 2017). The operations of these ships result in the production of 12 recognised types of water pollution, including solid waste, atmospheric pollution, noise pollution, and nine types of wastewater (Jalkanen et al. 2020; Moldanová et al. 2022). Even though most of these pollutants are regulated by the IMO, there are limitations on the capacity of local management authorities to enforce compliance with IMO regulations. For example, independent compliance tests conducted on ballast water in Singapore, Canada, and Australia revealed that only 48% of the samples met IMO requirements, despite high compliance reported in official inspections (99% overall; MEPC 78/4/1).

EGCS, also known as scrubbers, are systems that remove sulphur from ship engine exhaust gas by using seawater, with the resulting washwater typically discharged overboard. The washwater is more acidic than seawater and contains nitrates, particulate matter, polycyclic aromatic hydrocarbons (PAHs), and heavy metals that can be toxic to marine organisms (Comer et al. 2020; Teuchies et al. 2020). However, there are major knowledge gaps about dispersal of this pollutant from ships and its effects on marine organisms. The use of EGCS has increased dramatically since January 2020, when the IMO prohibited the use of marine fuels with sulphur levels above 0.5%, except for ships using EGCS. Because of this change in global shipping regulations, there are now > 4,300 ships worldwide that use EGCS, which is up from less than 250 in 2015 (Osipova et al. 2021). Approximately 500 of these ships regularly operate within the GBR Particularly Sensitive Sea Area (PSSA; Osipova 2021). Globally, at least 10 gigatonnes (Gt) of EGCS washwater are estimated to be released annually (see Figure 2), including in PSSAs such as the GBR (~32 MT p.a.), Hawaii, Papua New Guinea, Japan, and parts of the Caribbean (Osipova et al. 2021). In Australian waters, the release of washwater from EGCS is permitted under the Commonwealth Protection of the Sea Act 1983, although some Port Authorities have introduced local regulations that prevent the release of washwater within port limits. These discharges are contingent on continual washwater monitoring and reporting by each ship and compliance with threshold levels of nitrates, PAHs, turbidity, and pH (MEPC 77–16/Add.1). Current regulations do not require direct measurement of heavy metal contamination in EGCS washwater, although analysis of a limited number of samples from washwater in Europe shows concentrations of copper and zinc in washwater of up to 260 and 450 $\mu g\ L^{-1}$, respectively (Turner et al. 2017). Consequently, in areas with high shipping traffic where there are repeated instances of washwater release, there is potential for the accumulation of contaminants from multiple ships, or over time, and the breaching of water quality guideline values.

In the last few years, multiple studies have demonstrated the toxic effects of EGCS washwater on marine organisms. Severe toxic effects have been revealed for the microplankton community of the Baltic Sea (Ytreberg et al. 2019); for zooplankton, including copepods (Thor et al. 2021); and for algae, shrimps, and fish (Ji et al. 2022). The most recent study implemented by a coalition of European research institutions reported a strong toxic effect of EGCS washwater on phytoplankton and planktonic invertebrates, even at very low concentrations (Magnusson and Granberg 2022). For example, fertilisation of sea urchin eggs was significantly impeded at 0.0001% washwater concentration (1 mL washwater per m^3 of seawater), whereas for other organisms, negative effects were observed at concentrations between 0.001%–0.1%. To date, however, the potential

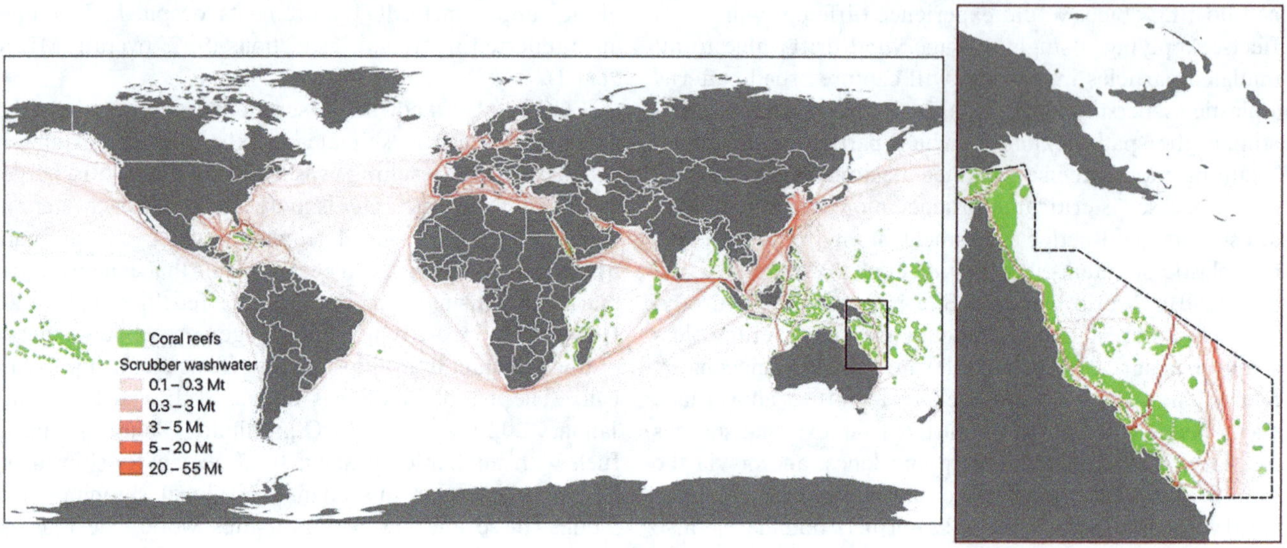

FIGURE 2 Estimates of global washwater discharge based on global shipping data (from Osipova et al. 2021).

effects of EGCS washwater on corals and other coral reef organisms have not been investigated. Scientific research is critically required to understand the chemical composition of washwater released in close proximity to coral reefs and to establish no-effect concentrations of washwater on key coral reef organisms to guide further development of regulations for washwater release in PSSAs.

Research into the toxicity of fuel oils on reef-building corals provides some insight into the potential effects of EGCS washwater on reefs. The ecological outcomes for oil spills from ship groundings on or near coral reefs have varied, but investigations of and reporting on environmental impacts have typically been limited. However, widespread mortality (Jackson et al. 1989), changes to community structure (e.g., Bak 1987), and inhibited recovery (e.g., Green et al. 1997) have been observed. Longer-term spills and chronic pollution (Rinkevich and Loya 1977; Bak 1987) with heavier oils appear to cause the most severe impacts, but the lack of information on specific exposure concentrations and environmental conditions prevents direct comparison among studies. Recent experimental studies reveal high sensitivity of some early life stages of coral to oil. For example, 69–100% of the toxicity thresholds recently derived for *Acropora millepora* corals were lower than the measured concentrations of dissolved aromatic hydrocarbons reported from the field following larger or repeated spills of other oil types (< 385 μg PAH L^{-1}; Diercks et al. 2010; Spies et al. 2017). Recent examples of spills also highlight the likelihood of extended exposures (> 48 h), particularly following ship groundings in reef environments (Daley 2019; Rajendran et al. 2021).

Most existing data for oil toxicity towards marine and aquatic organisms have been derived under temperate conditions (French-McCay 2002; McGrath et al. 2018), and if applied in risk assessments, the potential consequences of spills in tropical marine environments may be underestimated. Tropical marine organisms often live close to their upper thermal limit, which makes them particularly vulnerable to environmental pressures such as marine heatwaves and high ultraviolet radiation (UVR; Nguyen et al. 2011; Rummer et al. 2014), and these pressures are concurrently likely to influence oil toxicity during spill events and affect coral reefs. Most of the available information indicates that the impacts of oil spills are magnified when organisms are exposed to environmental co-factors, with UVR highly likely to coincide with spills in tropical marine ecosystems and expected to cause the largest increases in toxicity (Nordborg et al. 2021). Because of the correlation between UVR intensity and the magnitude of phototoxicity for oils (Barron 2017), toxicity increases may be substantially larger under natural field conditions than under controlled laboratory conditions. Application of toxicity thresholds generated from temperate and freshwater species to risk assessments for tropical reefs assumes that ecologically important taxa such as corals have similar sensitivities. However, there have been previous suggestions that the application of sensitivity or toxicity data from other climates or environments may be inappropriate (Leung et al. 2001; Kwok et al. 2007). As the early life history stages of *A. millepora* have been identified as very sensitive, further oil toxicity studies on other coral reef taxa are urgently needed. Moreover, studies of the effects of EGCS washwater on corals are also important for risk management of these pollutants within the GBR.

The transport and accumulation of EGCS washwater, fuel oils, and other pollutants with origins on vessels can be modelled to help understand their cumulative impacts and risks. Modelling for oil spills is arguably the most advanced pollution modelling field as the methods are well-established (e.g., Keramea et al. 2021; French-McCay et al. 2021), and the management implications are clear. These models include sophisticated parametrisations of physical and chemical transport and degradation processes, including oil dispersal, evaporation, and emulsification, that are monitored during the management of and responses to oil spills (Ross et al. 2016). Models are used in a predictive capacity after a ship grounding or other spill incident to anticipate the most appropriate clean-up locations or to identify infrastructure or habitats at risk from the spill (e.g., Janeiro et al. 2014). Satellite data can be assimilated into model outputs to improve these forecasts (Fingas and Brown 2018). Oil spill models can also be used during the planning stage of projects to understand the risks of new or proposed infrastructure or transport routes. Furthermore, where model outputs inform decision makers, communicating the uncertainty associated with the outputs could help to inform their responses. However, the standards around providing uncertainty estimates are not well-defined (Keramea et al. 2021). The modelling of oil spills is an advanced field, but some physio-chemical processes, such as dissolution, vertical mixing, photo-oxidation, and biodegradation, are generally not well-represented in the research, and further investigation of the dynamics and parameterisation of these process could improve spill modelling outcomes (e.g., Socolofsky et al. 2019; Ward et al. 2018).

The principles from oil modelling can be applied to other pollutants of emerging concern, such as EGCS washwater, but will require adaptation to account for their varying physical and chemical properties (see Parkerton et al. 2018 for an example of risk modelling for produced water). For example, models could be run to estimate dispersion from vessel paths for consecutive vessels to understand cumulative concentrations in the water column. However, as these pollutants comprise complex mixtures of different chemicals that may not be well-characterised, further information is needed on their specific degradation, evaporation, and sinking processes so that dispersal models can be appropriately parameterised. Such models would then require ground-truthing through the use of field collected data prior to application in risk assessments.

9.4. CONCLUSIONS

The case studies presented above bracket the spectrum of issues that present obstacles for environmental risk assessments of pan-oceanic contaminants. For marine plastics, a major challenge to understanding environmental risk is linked to difficulties in modelling the dispersal of particles that have different sizes, shapes, and buoyancy, that accumulate on and resuspend from beaches, and that undergo complex mechanical breakdown over long timescales. For EGCS, highly detailed ship movement information is available, together with robust methods for estimating the volume of washwater released, by using information on engineering and travel speed. However, there is limited publicly available data describing the composition of EGCS washwater in tropical regions in general, and, specifically, for the GBR, and a deficiency of knowledge of how washwater exposure impacts tropical marine organisms. Finally, for fuel oils, toxicity thresholds have primarily been quantified in temperate waters and for temperate species, and it remains unclear how toxicity could change under the higher temperatures and higher light levels present in tropical marine environments. For all three pollutant categories, data quantifying rates of breakdown of materials and compounds in the natural environment are also needed to enable the development and accurate parameterisation of particle dispersal models that can incorporate these complex behaviours.

REFERENCES

Anselain T, Heggy E, Dobbelaere T, Hanert E (2023) Qatar Peninsula's vulnerability to oil spills and its implications for the global gas supply. Nature Sustainability. doi: 10.1038/s41893-022-01037-w.

Bainbridge Z, Lewis S, Bartley R, Fabricius K, Collier C, Waterhouse J, Garzon-Garcia A, Robson B, Burton J, Wenger A, (2018). Fine sediment and particulate organic matter: A review and case study on ridge-to-reef transport, transformations, fates, and impacts on marine ecosystems. Marine Pollution Bulletin 135:1205–1220. doi: 10.1016/j.marpolbul.2018.08.002.

Bak RPM (1987). Effects of chronic oil pollution on a Caribbean coral reef. Marine Pollution Bulletin 18:534–539. doi: 10.1016/0025-326X(87)90537-6.

Barron MG (2017). Photoenhanced toxicity of petroleum to aquatic invertebrates and fish. Archives of Environmental Contamination and Toxicology 73:40–46.

Beyer J, Trannum HC, Bakke T, Hodson PV, Collier TK (2016) Environmental effects of the Deepwater Horizon oil spill: A review. Marine Pollution Bulletin 110:28–51. doi: 10.1016/j.marpolbul.2016.06.027.

Bottrill MC, Joseph LN, Carwardine J, Bode M, Cook C, Game ET, Grantham H, Kark S, Linke S, McDonald-Madden E, Pressey RL, Walker S, Wilson KA, Possingham HP (2008) Is conservation triage just smart decision making? Trends in Ecology & Evolution 23(12):12. doi: 10.1016/j.tree.2008.07.007.

Campbell ML, Peters L, McMains C, de Campos MCR, Sargisson RJ, Blackwell B, Hewitt CL (2019) Are our beaches safe? Quantifying the human health impact of anthropogenic beach litter on people in New Zealand. Science of The Total Environment 651:2400–2409. doi: 10.1016/j.scitotenv.2018.10.137.

Chubarenko I, Esiukova E, Bagaev A, Isachenko I, Demchenko N, Zobkov M, Efimova I, Bagaeva M, Khatmullina L (2018) Behavior of microplastics in coastal zones. In E. Y. Zeng (Ed.), Microplastic Contamination in Aquatic Environments: An Emerging Matter of Environmental Urgency (pp. 175–223). Elsevier. doi: 10.1016/B978-0-12-813747-5.00006-0.

Comer B, Georgeff E, Osipova L (2020) Air emissions and water pollution discharges from ships with scrubbers. International Council on Clean Transportation. https://theicct.org/publication/air-emissions-and-water-pollution-discharges-from-ships-with-scrubbers/.

Cózar A, Echevarría F, González-Gordillo JI, Irigoien X, Úbeda B, Hernández-León S, Palma ÁT, Navarro S, García-de-Lomas J, Ruiz A, Fernández-de-Puelles ML, Duarte CM (2014) Plastic debris in the open ocean. Proceedings of the National Academy of Sciences 111(28):28. doi: 10.1073/pnas.1314705111.

Critchell K (2018) Using hydrodynamic models to understand the impacts and risks of plastic pollution. PhD Thesis, James Cook University. doi: 10.25903/5bd25335639a2.

Critchell K, Hoogenboom MO (2018) Effects of microplastic exposure on the body condition and behaviour of planktivorous reef fish (*Acanthochromis polyacanthus*). PLoS One 13(3). doi: 10.1371/journal.pone.0193308.

Critchell K, Lambrechts J (2016) Modelling accumulation of marine plastics in the coastal zone; what are the dominant physical processes? Estuarine, Coastal and Shelf Science 171:111–122. doi: 10.1016/j.ecss.2016.01.036.

Daley J (2019) Month-long oil spill in the Solomon Islands threatens world's largest coral reef atoll. Smart News, Smithsonian Magazine. www.smithsonianmag.com/smart-news/month-long-oil-spill-threatens-world-heritage-site-solomon-islands-180971674/. Accessed March 12, 2019.

Diercks AR, Highsmith RC, Asper VL, Joung D, Zhou Z, Guo L, et al. (2010) Characterization of subsurface polycyclic aromatic hydrocarbons at the Deepwater Horizon site. Geophysical Research Letters 37:L20602 1–6. doi: 10.1029/2010GL045046.

Duncan EM, Broderick AC, Critchell K, Galloway TS, Hamann M, Limpus CJ, Lindeque PK, Santillo D, Tucker AD, Whiting S, Young EJ, Godley BJ (2021) Plastic pollution and small juvenile marine turtles: A potential evolutionary trap. Frontiers in Marine Science 8. www.frontiersin.org/article/10.3389/fmars.2021.699521.

Dunkin M (2015) A Comparison of Analytical, Experimental, and Numerical Formulations of the Wind-drift Coefficient for Buoyant Marine Litter. Bachelor of Engineering (Mechanical Engineering), James Cook University.

Ebbesmeyer CC, Ingraham WJ, Jones JA, Donohue MJ (2011) Marine debris from the Oregon Dungeness crab fishery recovered in the Northwestern Hawaiian Islands: Identification and oceanic drift paths. Marine Pollution Bulletin 65(1). doi: 10.1016/j.marpolbul.2011.09.037.

Ebbesmeyer CC, Ingraham WJ, Royer TC, Grosch CE (2007) Tub toys orbit the Pacific subarctic gyre. Eos, Transactions American Geophysical Union 88(1). doi: 10.1029/2007EO010001.

Fabricius KE (2005) Effects of terrestrial runoff on the ecology of corals and coral reefs: Review and synthesis. Marine Pollution Bulletin 50:125–146. doi: 10.1016/j.marpolbul.2004.11.028.

Fingas M, Brown CE (2018) A review of oil spill remote sensing. Sensors 18:91. doi: 10.3390/s18010091.

French-McCay DP (2002) Development and application of an oil toxicity and exposure model, OilToxEx. Environmental Toxicology and Chemistry 21:2080–2094. doi: 10.1002/etc.5620211011.

French-McCay DP, Crowley D, Rowe JJ, Bock M, Robinson H, Wenning R, et al. (2018) Comparative Risk Assessment of spill response options for a deepwater oil well blowout: Part 1. Oil spill modeling. Marine Pollution Bulletin 133:1001–1015. doi: 10.1016/j.marpolbul.2018.05.042.

French-McCay DP, Spaulding ML, Crowley D, Mendelsohn D, Fontenault J, Horn M (2021) Validation of oil trajectory and fate modelling of the Deepwater Horizon oil spill. Frontiers in Marine Science 8:618463. doi: 10.3389/fmars.2021.618463

Green A, Burgett J, Molina M, Palawski D, Gabrielson P (1997) The impact of a ship grounding and associated fuel spill at Rose Atoll National Wildlife Refuge, American Samoa. U.S. Fish and Wildlife Service, Honolulu, Hawaii. https://catalog.data.gov/dataset/the-impact-of-a-ship-grounding-and-associated-fuel-spill-at-rose-atoll-national-wildlife-r. Accessed June 9, 2018.

Hamner WM, Jones MS, Carleton JH, Hauri IR, Williams D (1988) Zooplankton, Planktivorous Fish, and Water Currents on a Windward Reef Face: Great Barrier Reef, Australia. Bulletin of Marine Science 42:459–479.

Hinata H, Mori K, Ohno K, Miyao Y, Kataoka T (2017) An estimation of the average residence times and onshore-offshore diffusivities of beached microplastics based on the population decay of tagged meso- and macrolitter. Marine Pollution Bulletin 122(1):1. doi: 10.1016/j.marpolbul.2017.05.012.

Hu Z, Pan D, He X, Bai Y (2016) Diurnal variability of turbidity fronts observed by geostationary satellite ocean color remote sensing. Remote Sensing 8(2):2. doi: 10.3390/rs8020147.

Isobe A, Kubo K, Tamura Y, Kako S, Nakashima E, Fujii N (2014) Selective transport of microplastics and mesoplastics by drifting in coastal waters. Marine Pollution Bulletin 89(1–2):1–2. doi: 10.1016/j.marpolbul.2014.09.041.

Jackson JB, Cubit JD, Keller BD, Batista V, Burns K (1989) Ecological effects of a major oil spill on Panamanian coastal marine communities. Science 243:37. doi: 10.1126/science.243.4887.37.

Jalkanen J-P, Johansson L, Wilewska-Bien M, Granhag L, Ytreberg E, et al. (2020) Modeling of discharges from Baltic Sea shipping. Ocean Science Discussion 17:699–728. doi: 10.5194/os-2020-99.

James ID (2002) Modelling pollution dispersion, the ecosystem and water quality in coastal waters: A review. Environmental Modelling & Software 17(4):363–385. doi: 10.1016/S1364-8152(01)00080-9.

Janeiro J, Zacharioudaki A, Sarhadi E, Neves A, Martins F (2014) Enhancing the management response to oil spills in the Tuscany Archipelago through operational modelling. Marine Pollution Bulletin 85(2):574–589. doi: 10.1016/j.marpolbul.2014.03.021.

Jang YC, Hong S, Lee J, Lee MJ, Shim WJ (2014) Estimation of lost tourism revenue in Geoje Island from the 2011 marine debris pollution event in South Korea. Marine Pollution Bulletin 81(1):1. doi: 10.1016/j.marpolbul.2014.02.021.

Ji Z, Yang Y, Zhu Y, Ling Y, Ren D, Huo Z, Zhang N (2022) Toxic Effects on Dunaliella salina, Mysidopsis bahia, and Mugilogobius chulae from Ship Exhaust Gas Closed-loop Scrubber Wash Water [Preprint]. In Review. doi: 10.21203/rs.3.rs-1324578/v1.

Johnson SW, Eiler JH (1999) Fate of radio-tagged trawl web on an Alaskan beach. Marine Pollution Bulletin 38(2):2. doi: 10.1016/S0025-326X(98)00109-X.

Kako S, Isobe A, Magome S, Hinata H, Seino S, Kojima A (2011) Establishment of numerical beach-litter hindcast/forecast models: An application to Goto Islands, Japan. Marine Pollution Bulletin 62(2). doi: 10.1016/j.marpolbul.2010.10.011

Keramea P, Spanoudaki K, Zodiatis G, Gikas, G, Sylaios G (2021) Oil spill modeling: A critical review on current trends, perspectives, and challenges. Journal Marine Science and Engineering 9:181. doi: 10.3390/jmse9020181.

Krelling AP, Williams AT, Turra A (2017) Differences in perception and reaction of tourist groups to beach marine debris that can influence a loss of tourism revenue in coastal areas. Marine Policy 85(Supplement C), Article Supplement C. doi: 10.1016/j.marpol.2017.08.021

Kwok KWH, Leung KMY, Lui GSG, Chu VKH, Lam PKS, Morritt D, et al. (2007) Comparison of tropical and temperate freshwater animal species' acute sensitivities to chemicals: Implications for deriving safe extrapolation factors. Integrated Environmental Assessment and Management 3:49–67. doi: 10.1002/ieam.5630030105

Leung KMY, Morritt D, Wheeler JR, Whitehouse P, Sorokin N, Toy R, et al. (2001) Can saltwater toxicity be predicted from freshwater data? Marine Pollution Bulletin 42:1007–1013. doi: 10.1016/S0025-326X(01)00135-7.

Lewis SE, Bartley R, Wilkinson SN, Bainbridge ZT, Henderson AE, James CS, Irvine SA, Brodie JE (2021) Land use change in the river basins of the Great Barrier Reef, 1860 to 2019: A foundation for understanding environmental history across the catchment to reef continuum. Marine Pollution Bulletin 166:112193. doi: 10.1016/j.marpolbul.2021.112193.

Lewis SE, McCloskey G, Bainbridge Z, Davis A, Bartley R, Turner R (2024) Sediment and nutrient flux from land. In Wolanski E, Kingsford MJ (eds.), Oceanographic Processes of Coral Reefs. Physical and Biological Links in the Great Barrier Reef. CRC Press, Boca Raton, FL.

Maes C, Blanke B (2015) Tracking the origins of plastic debris across the Coral Sea: A case study from the Ouvéa Island, New Caledonia. Marine Pollution Bulletin 97(1–2):160–168. doi: 10.1016/j.marpolbul.2015.06.022.

Magnusson K, Granberg M (2022) Report on scrubber water whole effluent toxicity testing, at different geographical regions (Deliverable D2.3; Evaluation, Control and Mitigation of the Environmental Impact of Shipping Emissions (H2020/EMERGE Project)).

McGrath JA, Fanelli CJ, Di Toro DM, Parkerton TF, Redman AD, Paumen ML, et al. (2018) Re-evaluation of target lipid model–derived HC5 predictions for hydrocarbons. Environmental Toxicology and Chemistry 37:1579–1593. doi: 10.1002/etc.4100.

Moldanová J, Hassellöv I-M, Matthias V, Fridell E, Jalkanen J-P, Ytreberg E, Quante M, Tröltzsch J, et al. (2022) Framework for the environmental impact assessment of operational shipping. Ambio 51(3):754–769. doi: 10.1007/s13280-021-01597-9.

Morais RA, Bellwood DR (2019) Pelagic subsidies underpin fish productivity on a degraded coral reef. Current Biology 29(9):1521–1527.e6. doi: 10.1016/j.cub.2019.03.044

Nguyen KDT, Morley SA, Lai CH, Clark MS, Tan KS, Bates AE, Peck LS (2011) Upper temperature limits of tropical marine ectotherms: Global warming implications. PLoS One 6(12):e29340. doi: 10.1371/journal.pone.0029340

Nordborg FM, Brinkman DL, Ricardo GF, Agusti S, Negri AP (2021) Comparative sensitivity of the early life stages of a coral to heavy fuel oil and UV radiation. Science of the Total Environment 781:146676. doi: 10.1016/j.scitotenv.2021.146676.

Nordborg FM, Jones RJ, Oelgemoller M, Negri AP (2020) The effects of ultraviolet radiation and climate on oil toxicity to coral reef organisms—A review. Science of the Total Environment 720:137486. doi: 10.1016/j.scitotenv.2020.137486.

Olmer N, Comer B, Roy B, Mao X, Rutherford D (2017) Greenhouse gas emissions from global shipping, 2013–2015. International Council on Clean Transportation. https://theicct.org/publication/greenhouse-gas-emissions-from-global-shipping-2013–2015/

Osipova L (2021), May 12. Coral reefs face a new threat from international shipping. *ICCT Blog Post*. https://theicct.org/coral-reefs-face-a-new-threat-from-international-shipping/.

Osipova L, Georgeff E, Comer B (2021) Global scrubber washwater discharges under IMO's 2020 fuel sulfur limit. International Council on Clean Transportation. https://theicct.org/publication/global-scrubber-washwater-discharges-under-imos-2020-fuel-sulfur-limit/.

Parkerton TF, Bok M, Ireland AW, Prosser CM (2018) An evaluation of cumulative risks from offshore produced water discharges in the Bass Strait. Marine Pollution Bulletin 126:610–621. doi: 10.1016/j.marpolbul.2017.10.003

Patterson RG, Wolanski E, Groom R, Critchell K, Playford L, Grubert M, Kennett R, et al. (2023) Improving certainty in marine ecosystems: A biophysical modelling approach in the remote, data-limited Gulf of Carpentaria. Estuarine, Coastal and Shelf Science 108254. doi: 10.1016/j.ecss.2023.108254.

Rajendran S, Vethamony P, Sadooni FN, Al-Kuwari HAS, Al-Khayat JA, Seegobin VO, et al. (2021) Detection of Wakashio oil spill off Mauritius using Sentinel-1 and 2 data: Capability of sensors, image transformation methods and mapping. Environmental Pollution 274:116618. doi: 10.1016/j.envpol.2021.116618.

Reichert J, Arnold AL, Hoogenboom MO, Schubert P, Wilke T (2019) Impacts of microplastics on growth and health of hermatypic corals are species-specific. Environmental Pollution 254:113074. doi: 10.1016/j.envpol.2019.113074.

Reisser J, Shaw J, Wilcox C, Hardesty BD, Proietti M, Thums M, Pattiaratchi C (2013) Marine plastic pollution in waters around Australia: Characteristics, concentrations, and pathways. PLoS One 8(11):11. doi: 10.1371/journal.pone.0080466.

Rinkevich B, Loya Y (1977) Harmful effects of chronic oil pollution on a Red Sea scleractinian coral population. Proceedings of the Third International Coral Reef Symposium Miami, Florida, pp. 585–591.

Ross A, Revill A, Qi X, Stalvies C, Griffin D, Hook S (2016) Response phase monitoring. In Oil Spill Monitoring Handbook. CSIRO Publishing, pp. 107–140.

Rummer JL, Couturier CS, Stecyk JAW, Gardiner NM, Kinch JP, Nilsson GE, Munday PL (2014) Life on the edge: Thermal optima for aerobic scope of equatorial reef fishes are close to current day temperatures. Global Change Biology 20:1055–1066. doi: 10.1111/gcb.12455.

Ryan JP, Harvey JBJ, Zhang Y, Woodson CB (2014) Distributions of invertebrate larvae and phytoplankton in a coastal upwelling system retention zone and peripheral front. Journal of Experimental Marine Biology and Ecology 459:51–60. doi: 10.1016/j.jembe.2014.05.017.

Skerratt JH, Baird ME, Mongin M, Ellis R, Smith RA, Shaw M, Steven ADL (2023). Dispersal of the pesticide diuron in the great barrier reef. Science of the Total Environment 879:163041. doi: 10.1016/j.scitotenv.2023.163041.

Socolofsky SA, Gros J, North E, Boufadel MC, Parkerton TF, Adams EE (2019) The treatment of biodegradation in models of sub-surface oil spills: A review and sensitivity study. Marine Pollution Bulletin 143:204–219. doi: 10.1016/j.marpolbul.2019.04.018.

Spies RB, Mukhtasor M, Burns KA (2017). The Montara oil spill: A 2009 well blowout in the Timor Sea. Archives of Environmental Contamination and Toxicology 73:55–62. doi: 10.1007/s00244-016-0356-7.

Teuchies J, Cox TJS, Van Itterbeeck K, Meysman FJR, Blust R (2020) The impact of scrubber discharge on the water quality in estuaries and ports. Environmental Sciences Europe 32(1):103. doi: 10.1186/s12302-020-00380-z.

Thor P, Granberg ME, Winnes H, Magnusson K (2021) Severe toxic effects on pelagic copepods from maritime exhaust gas scrubber effluents. Environmental Science & Technology. doi: 10.1021/acs.est.0c07805.

Turner DR, Hassellöv I-M, Ytreberg E, Rutgersson A (2017) Shipping and the environment: Smokestack emissions, scrubbers and unregulated oceanic consequences. Elementa: Science of the Anthropocene 5(0):0. doi: 10.1525/elementa.167.

Turrell WR (2020) How litter moves along a macro tidal mid-latitude coast exposed to a coastal current. Marine Pollution Bulletin 160:111600. doi: 10.1016/j.marpolbul.2020.111600.

van Sebille E, Aliani S, Law KL, Maximenko N, Alsina JM, Bagaev A, et al. (2020) The physical oceanography of the transport of floating marine debris. Environmental Research Letters 15(2):2. doi: 10.1088/1748-9326/ab6d7d.

Ward CP, Sharpless CM, Valentine DL, French-McCay DP; Aeppli C, White HK, et al. (2018) Partial photochemical oxidation was a dominant fate of Deepwater Horizon surface oil. Environmental Science and Technology 52:1797–1805. doi: 10.1021/acs.est.7b05948.

Weinstein JE, Crocker BK, Gray AD (2016) From macroplastic to microplastic: Degradation of high-density polyethylene, polypropylene, and polystyrene in a salt marsh habitat. Environmental Toxicology and Chemistry 35:1632–1640. doi: 10.1002/etc.3432.

Welden NA, Cowie PR (2017) Degradation of common polymer ropes in a sublittoral marine environment. Marine Pollution Bulletin 118(1):1. doi: 10.1016/j.marpolbul.2017.02.072.

Wilcox C, Hardesty BD, Sharples R, Griffin DA, Lawson TJ, Gunn R (2013) Ghostnet impacts on globally threatened turtles, a spatial risk analysis for Northern Australia. Conservation Letters 1:1–8. doi: 10.1111/conl.12001

Wolanski E, Fabricius KE, Cooper TF, Humphrey C (2008) Wet season fine sediment dynamics on the inner shelf of the Great Barrier Reef. Estuarine. Estuarine, Coastal and Shelf Science 77:755–762. doi: 10.1016/j.ecss.2007.10.014.

Wolanski E, Lambrechts J, Thomas C, Deleersnijder E (2013) The net water circulation through Torres Strait. Continental Shelf Research 64:66–74. doi: 10.1016/j.csr.2013.05.013.

Ytreberg E, Hassellöv IM, Nylund AT, Hedblom M, Al-Handal AY, Wulff A (2019) Effects of scrubber washwater discharge on microplankton in the Baltic Sea. Marine Pollution Bulletin 145:316–324. doi: 10.1016/j.marpolbul.2019.05.023.

Zhang K, Hamidian AH, Tubić A, Zhang Y, Fang JKH, Wu C, Lam PKS (2021) Understanding plastic degradation and microplastic formation in the environment: A review. Environmental Pollution 274:116554. doi: 10.1016/j.envpol.2021.116554.

10 Jellyfish

A Window into Pesticide Distribution and Risks on the Great Barrier Reef

Michelle A. Templeman and Christopher D. Williams

10.1. GENERAL INTRODUCTION

'Pesticide' is a generic term for any substance that is used to control animals, plants, or other organisms regarded as pests (NIEHS 2022). This definition includes herbicides, insecticides, fungicides, algaecides, rodenticides, disinfectants, anti-fouling treatments, and timber preservatives. The earliest records of pesticide use date back 3,000–4,500 years with sulphur compounds used by the Sumerians to control insects and compounds of arsenic and mercury used by the Chinese to control body lice (Unsworth 2010). Chemical coatings to limit the build-up of marine organisms on the hulls of ships were recorded more than 2,500 years ago (WHOI 1952). Pesticide innovations in the late 1700s and early 1800s were driven by naval and merchant shipping concerns and included biocidal anti-fouling paints containing copper, arsenic, and mercury compounds and copper sheathing of ship hulls (WHOI 1952). Widespread use of inorganic metal salts and acids as controls on agricultural weeds began in the late 1800s (Barrett et al. 2021; Chauvel et al. 2022). Modern, synthetic pesticides first came into significant use during World War II as a means of combatting malaria, typhus, and other insect-borne human diseases (NPIC 1999).

Pesticides currently play a critical role in maintaining food production and human health and well-being in many parts of the world. Globally, it is estimated that almost 80% of fruit production, over 50% of vegetable production, and more than 30% of cereal production currently relies on pesticides for crop protection (Tudi et al. 2021). In Australia, the combination of improved farming practices and pesticide use has led to a 250% increase in productivity since the 1960s (Productivity Commission 2005). Effective management of malaria, dengue, and other tropical diseases also remains reliant on chemical vector control measures including residual spraying of insecticides and insecticide impregnated nets (WHO 2015; PAHO 2019). The global impact of termite damage to building infrastructure amounts to approximately US $40 billion per year, and application of termiticides to building pad soils and chemical treatment of structural timbers are still the dominant means of preventing termite attack (Oi 2022). The use of anti-fouling systems on the hulls of ships is considered of critical importance to efficient commerce and to impede the spread of harmful aquatic organisms and pathogens (IMO 2001).

However, it has also been observed that the only thing that separates a remedy from a poison is the dose (California DPR 2017). Pesticides, by definition, are environmentally active compounds, and many can persist in soil, water, and tissue for a long time. Residual effects caused by gradual uptake and accumulation in non-target plants and animals can take many years to be recognised. As a result, the history of pesticide use has a parallel history of lagged understanding, regulation, and ultimately, withdrawal or bans on different formulations. Commercial imperatives mean that each pesticide formulation withdrawn from the market is quickly replaced by new-generation products. These are invariably bound by more stringent regulation and toxicological testing criteria, but history suggests that the full environmental consequences may still not be apparent for some time. This reflects the very low effect-concentrations for many pesticide compounds, variable break-down trajectories and mixture implications in natural waters, and a lag in the ability to test for many compounds in environmental samples.

Dichlorodiphenyltrichloroethane (DDT) provides the clearest example of this evolution. It initially presented as a highly effective control on insect pests in crop and livestock production, buildings, and gardens, and use grew rapidly after World War II (NPIC 1999). Other organochlorine compounds soon followed DDT into agricultural and domestic usage as insecticides and herbicides, although increasing evidence of health and environmental issues resulted in greater regulation imposed through the late 1950s and 1960s (NPIC 1999). All registered uses for DDT in the US were eventually cancelled in 1972, with further restrictions and bans on organochlorine compounds implemented across the globe in the 1980s and 1990s, which culminated in the Stockholm Convention on persistent organic pollutants (POPs) in 2001 and the Rotterdam Convention on the trade of hazardous chemicals in 2004 (UNEP 2023). These developments caused a dramatic decline in organochlorine pesticide use, although the residual effects of past practices continue to be observed. The estimated half-life of DDT in the aquatic environment is around 150 years (NPIC 1999), and despite the 20–50 years elapsed time since bans were implemented on most of its applications, it is still found in many freshwater and marine environments across the globe (e.g., Cavanagh et al. 1999; Zheng et al. 2020).

DOI: 10.1201/9781003320425-12

10.2. HISTORY OF PESTICIDE USE IN NORTHERN AUSTRALIA

Early pest control in Australian agriculture was mostly physical and utilised ploughs, hoes, fire, grazing, and crop rotation to minimize weed growth and insect damage. Early termite protection for buildings in Northern Australia was also physical, with barriers provided by raised stumps and 'ant' caps. The first agricultural pesticides introduced in the early 20th century comprised inorganic salts of arsenic, copper, and sulphur (Barrett et al. 2021; Thompson and Chauhan 2022). Many of these products required high rates of application and were toxic to humans and farm animals (Barrett et al. 2021; Unsworth 2010). Synthetic organochlorine pesticides displaced many inorganic salts after World War II, and compounds such as DDT, dieldrin, chlordane, and heptachlor were widely used between the 1950s and 1980s. DDT was registered for use as an herbicide, insecticide, and anti-fouling agent across many different sectors of the economy (Table 1). Other organochlorine compounds were mainly applied as insecticides, with a predominant focus on termite control. A key functional aspect for these organochlorine pesticides was their long persistence in the environment; however, this was also a major factor in the environmental risks that they presented. Most organochlorine products were deregistered and phased out of use by the late 1980s, although dieldrin continued to be utilised as a termiticide until 1994 (ANZG 2018), and endosulphans remained in use until 2010 (APVMA 2022). Organochlorine residuals have been found in many soils in agricultural areas, around cattle dips, and urban environments where historical usage was greatest, with downstream signatures often associated with sediment mobilisation and transport patterns (Cavanagh et al. 1999; ATSE 2002).

Subsequent generations of pesticides have been more strongly defined by mode-of-action, allowing a more selective targeting of pest species (Unsworth 2010). There are currently more than 8,000 products registered in Australia for use as herbicides, fungicides, insecticides, parasiticides, miticides, molluscicides, vertebrate poisons, and wood preservatives (APVMA 2022). These products incorporate some 600–700 separate chemical compounds as the active agent. On-going research and development are concentrated on finding new members of existing chemical families with better economic and/or environmental profiles and new chemical formulations with alternate modes of action or selective capacities (Unsworth 2010). Integrated management strategies that employ a range of techniques to dissuade pests and reduce the use of pesticides and other interventions have also been encouraged.

Notwithstanding the technological developments in pesticide formulation and use, the parallel 'races' among pesticide effectiveness, economic viability, and environmental sustainability continue to play out in Northern Australia. Growing weed resistance to selectively targeted formulations means more and varied herbicide usage is being required over time (Thompson and Chauhan 2022). Concerns for the health of aquatic environments have led to many diuron uses being curtailed in Australia in 2012 and no-spray windows imposed in various drainage systems for parts of the year (Table 1; APVMA 2012). Methoxyethyl mercury chloride, a potent fungicide applied to sugar cane crops, was deregistered in Australia in 2020 due to human and environmental health risks, although the use of this fungicide continued long after the bans on mercury pesticides had been imposed in most other countries (APVMA 2020; Schneider 2021).

Tributyltin (TBT) paints were introduced in the 1960s as a less harmful alternative to anti-fouling systems based on DDT, arsenic, and mercury but have subsequently been described as the most toxic substance that mankind has deliberately introduced into the marine environment (IMO 2002). The International Convention on the Control of Harmful Anti-fouling Systems on Ships has introduced prohibitions on the use of organotin coatings for ocean-going vessels without additional barriers to prevent leaching into the water column (IMO 2001). However, industry concerns regarding acceptable alternatives and the unknown environmental risks associated with other biocides and their metabolites have impeded a wider phase-out (IMO 2002). Elevated concentrations of TBT and other antifouling pesticides are an increasing concern in areas of the GBR, especially those most closely associated with shipping lanes and activities (Kroon et al. 2020).

10.3. PESTICIDE INPUTS TO THE GBR

Terrestrial exports of pesticides from GBR catchments derive from agricultural activities, urban runoff, and wider landscape management practices such as weed control on roads, railways, and power line corridors (Kroon et al. 2016; Warne et al. 2020a). Major agricultural activities in the GBR catchment area include grazing, grain, sugarcane, horticulture, tree fruit, bananas, and cotton (Thorburn et al. 2013). Grazing is the predominant land use (97% of the agricultural footprint) (Thorburn et al. 2013), although the most intensive pesticide usage is generally associated with grain and sugar cane cropping (Davis et al. 2008; Thorburn et al. 2013).

The extent of pesticide runoff from any given drainage catchment will be a complex derivative of land-use patterns and crop stage; rainfall and temperature conditions; topography; riparian extent; the formulation, soil sorption, and breakdown profiles of applied pesticides; and wider pest management strategies (e.g., Thorburn et al. 2013; Devlin et al. 2015; Rippy et al. 2017). Farm and catchment scale assessments of pesticide partitioning between dissolved and particulate phases found that the majority of pesticide load was transported in the dissolved fraction, but there was relatively wide variability in load transport, even for the same pesticide (Devlin et al. 2015). Variability in pesticide runoff appears more closely related to the residual pesticide concentration at the time of a rainfall event than the particular properties of the pesticide (Thorburn et al. 2013).

TABLE 1

Common uses for selected pesticides in the Australian context (APVMA 2022; ANZG 2018). Usage coded as 'A' = algicide, 'F' = fungicide, 'H' = herbicide, 'I' = insecticide, 'M' = marine antifouling biocide, and 'R' = rodenticide. Black shading identifies current registered use (APVMA 2022). Grey shading identifies historical use. Included pesticides comprise the 16 pesticides with the greatest contribution to the toxicity of pesticide mixtures in rivers discharging to the GBR (Spilsbury et al. 2020), widely used organochlorines and metals, and selected other compounds addressed in this discussion. 'MCPA' = 2-methyl-4-chlorophenoxyacetic acid. 'MEMC' = methoxyethyl mercury chloride. 'MSM' = metsulfuron-methyl.

Pesticide	Usage	Tree Fruit	Bananas	Cotton	Sugar Cane	Horticulture	Grains / Pulses	Grass / Seed Crops	Forestry	Grazing Pasture	Cattle Tick	Channels / Drains	Roads / Rights of Way	Commercial / Industrial	Marine Anti-Fouling	Termites	Timber Preservatives	Garden / Household
2,4-D	H	■	■	■	■	■	■	■	■	■		■	■	■				■
Ametryn	H				■	■						■	■	■				
Arsenic	HIMR	□									□		□		□	□	■	□
Atrazine	H				■		■	■	■				■					
Bifenthrin	I	■	■	■	■	■	■							■		■		■
Bromacil	H	■				■							■	■				
Chlordane	I															□		
Chlorpyrifos	I	■	■	■	■	■	■	■		■	■			■				
Copper	FAM	■	■			■									■		■	■
DDT	HIM	□	□	□	□	□					□				□	□		□
Diazinon	I	■	■	■	■		■	■		■	■	■		■				■
Dieldrin	I	□	□	□	□		□				□					□		□
Diuron	HM	□	■	■	■	■	■	□				■	□	□	■			■
Endosulfan	I			□		□	□	□		□								
Flumethrin	I										■							■
Heptachlor	I					□	□			□								
Hexazinone	H				■				■	■			■	■				
Imazapic	H				■		■											
Imidacloprid	I	■		■		■		■										■
Isoxaflutole	H				■	■	■											
MCPA	H						■	■		■								■
MEMC	F				□													
Metolachlor	H			■	■	■	■			■								
Metribuzin	H					■	■											
MSM	H					■	■		■	■			■	■				
Tebuthiron	H								■	■								
Tributyltin	FM														■		■	■
Triclopyr	FH						■		■	■			■	■				■

Upstream catchment size and associated topography can influence patterns of downstream pesticide delivery. High rainfall, coastal catchments in the Wet Tropics have year-round base flows, more frequent runoff events, and typically short flow transit times compared to dryland grazing catchments in the Burdekin and Fitzroy River basins. Solute and sediment exports from catchments are also influenced by climate patterns and associated effects on vegetation cover, runoff intensity, and streamflow thresholds (Brodie et al. 2012; Lewis et al. 2023). The extent of pesticide migration via groundwater pathways (Devlin et al. 2015) can also vary with the long-term climate. Overall, across GBR catchments, pesticide export loads tend to be higher in wetter-than-average years, while pesticide export concentrations tend to be higher across average to drier years (Devlin et al. 2015).

The last peak of rainfall excess in much of Northern Australia was in mid-2012, with subsequent rainfall patterns presenting as a drying phase of the climate cycle (Figure 1). In GBR catchments, these underlying climate drivers map with changes in cropping area, cattle numbers, and human population through the 20th century (Lewis et al. 2021, 2023; Figure 1). This creates synergies between pesticide use and land-use intensity that are linked to climate patterns.

Major urban conurbations in the GBR catchment also have coastal locations with short runoff times. This response is exacerbated by the large extent of impervious surfaces in urban environments (Chen et al. 2019). A wide variety of dissolved pesticides have been found in urban runoff from major Australian cities; the most common are herbicides associated with weed control in home gardens, urban waterways, road verges, and industrial and commercial settings including diuron, MCPA, 2,4-D, simazine, and triclopyr (Vincente-Becket et al. 2006; Rippy et al. 2017). The pesticide mixtures observed by Rippy et al. (2017) were stable over time but strongly catchment-specific. Urban contributions are relatively low by global standards and exceed environmental benchmarks only in outlier samples. Other international studies have found elevated concentrations of imidacloprid in urban wastewater discharges (Sadaria et al. 2016) and pyrethroid insecticides associated with urban stormwater particulates (Gorgoglione et al. 2018). Assessment of pesticide residues in sewage discharges to the GBR in the Wet Tropics region has found relatively low concentrations and loads of pesticides compared to agricultural inputs although point-source, press-effects on downstream environments were not ruled out (Devlin et al. 2015; Kroon et al. 2015).

Inputs of antifouling pesticides to the GBR system predominantly reflect leaching of biocides from antifouling paints and groundings of large vessels (Kroon et al. 2020). Historical accumulations of antifouling pesticides have also been found in most of the major port facilities on the Queensland coast (Kroon et al. 2020). Environmental loads are a function of vessel size and duration of passage, and they increase with greater marine traffic movements and mooring times. Commercial shipping traffic through the major ports on the GBR broadly doubled from 2000 to 2012 with a further 2–3-fold increase projected between 2012 and 2032 (Kroon et al. 2020; BTRE 2002; BITRE 2013). The potential risks of contamination from grounding of ships will also increase with ship size and traffic.

The Development of a Pesticide Risk Baseline for the Reef 2050 Water Quality Improvement Plan has assessed toxicity mixtures for 50 common pesticides in terrestrial

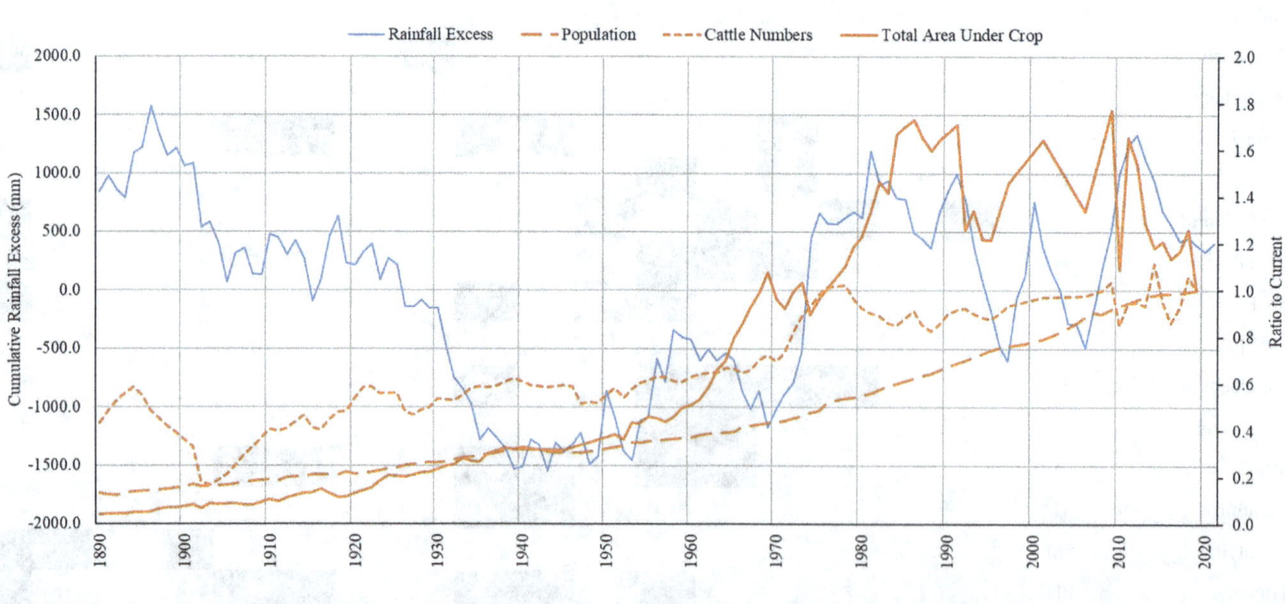

FIGURE 1 Cumulative variation from average rainfall at Charters Towers in the Burdekin Basin between 1886 and 2021 (BOM Stations 34002 and 34084). Data calculated daily from monthly average rainfall. Initial bound adjusted to balance total water excess and deficit over the period of record. Population, cattle numbers, and total area under crops in the GBR catchment area are from Lewis et al. (2021). Total area under crops predominantly comprises sugarcane and grains.

runoff from the GBR catchment area (Warne et al. 2020a; Spilsbury et al. 2020). The majority of pesticide risk in major rivers discharging to the GBR (92%) is attributed to the herbicides diuron, atrazine, metolachlor, and hexazinone and the insecticide imidacloprid. The clearest land-use association is with sugarcane, and seasonal occurrences are more frequent in the wet season than in the dry season (Warne et al. 2020b). However, other pesticides are also major contributors to the risk quotient of individual samples including 2,4-D ametryn, atrazine, isoxaflutole, MCPA, and MSA (Spilsbury et al. 2020). This reinforces the diversity of pesticide exports to the GBR. Herbicide formulations that inhibit photosynthesis by targeting the Photosystem II complex (PSII herbicides) are considered to be the strongest drivers of potential toxicity (Warne et al. 2020a). These include ametryn, atrazine, diuron, metribuzin, and tebuthiuron (Table 1). However, the relative contributions to pesticide mixture toxicity from different groups of herbicides and insecticides are highly variable across different drainage systems in the GBR catchment (Warne et al. 2020a, 2020b). Other herbicides and insecticides jointly contributed approximately 50% of the estimated total pesticide toxicity, and multiple modes of pesticide action were identified in most samples.

Pesticides are found in almost all surface waters draining the GBR catchment (Brodie and Landos 2019; Warne et al. 2020b). Most samples contain mixtures of 2–20 quantifiable pesticide residues, although only a limited range of formulations have been assessed to date, and most fungicides and chemical breakdown products have not yet been addressed (Warne et al. 2020b). Other point sources and potential input pathways (coastal urban discharges, marine pesticide sources, and reservoirs) also largely lie outside of the current assessment framework. The relative load and toxicity contributions currently reported for pesticides reflect the available monitoring data and the history of land and pesticide use (Brodie and Landos 2019). The risks posed to the GBR by different pesticide groups will likely change as more pesticide formulations are included in risk assessments (Warne et al. 2020a, 2020b) and land and pesticide use patterns continue to evolve (Brodie and Landos 2019).

10.4. MEASUREMENT OF PESTICIDES ON THE GBR

It is considered that much of the inshore GBR is exposed to low levels of pesticides, which are contributing to a chronic decline in coral and seagrass cover (Devlin et al. 2015; King et al. 2013). The highest water-borne exposures are associated with wet season flood plumes, with concentrations and detection rates tending to decrease with distance from source (Devlin et al. 2015; Kennedy et al. 2012; Lewis et al. 2009). These pulse exposures occur concurrently with other stressors including high suspended sediment and nutrient loads (Devlin et al. 2015). Terrestrial pesticide exports and coastal measures are recorded year-round, however, and chronic press effects of pesticide release are also likely (Shaw et al. 2010; Devlin et al. 2015; Brodie and Landos 2019). Assessments of common herbicide solutes in the marine environment have found that degradation processes are accelerated in the presence of a diurnal light cycle and suspended sediments, but expected half-lives in the GBR lagoon are typically more than 100 days (Mercurio et al. 2016). The effects of light and sediments on herbicide breakdown are attributed to their influence on the microbial community and the ability to utilise the herbicides as a carbon source (Mercurio et al. 2016).

The distribution of particulate pesticide loads on the GBR is dependent on local sediment transport, remobilisation, and water column mixing factors downstream of the source reservoirs associated with ports, shipping lanes, and historical grounding sites (Kroon et al. 2020). The sediment fans beyond river mouths can also be a reservoir of historical terrestrial exports. Seabed particulates are likely to continue releasing solutes into the surrounding environment (Negri and Marshall 2009).

Past studies on the pesticide risks to the GBR have primarily focused on five priority PSII herbicides, namely, ametryn, atrazine, diuron, hexazinone, and tebuthiuron (Negri et al. 2020), and these are the most commonly recorded pesticides in the GBR lagoon (Lewis et al. 2009; Devlin et al. 2015). However, a wide variety of other pesticide formulations have been recorded in the waters and sediments of the GBR at both inshore and offshore locations (Haynes and Johnson 2000; Lewis et al. 2009; Gallen et al. 2019; Thai et al. 2020). Relatively limited information is available on historical organochlorine residues on the GBR (Haynes and Johnson 2000). They are known to persist in local biota but are considered to pose a low risk to marine ecosystems (Douglas et al. 2005; Lewis et al. 2009; Cagnazzi et al. 2013).

Sampling techniques used to develop the pesticide dataset in GBR waters include both 'point-in-time' grab sampling and, more recently, the use of passive samplers that provide a proxy temporal exposure measure. Grab sampling offers the opportunity for a wide suite of laboratory analyses, but logistics often constrain spatial and temporal coverage. Large sample volumes may also be necessary to detect compounds at very low concentrations. Passive samplers are an attractive alternative for monitoring low concentration contaminants by providing in situ enrichment, integrated exposure concentrations, and selective sampling of targeted analytes (Raub et al. 2015; Becker et al. 2021). However, passive sampling approaches are not suitable for assessing peak, short-term variations in contaminant concentrations and present other potential limitations (Caban et al. 2022). These include the need for intensive calibration assessments to understand the uptake of each additional analyte, sensitivity of the uptake rate to changes in environmental conditions, and a lack of reference calibration methods.

In this space, the use of biological monitors or 'biomonitors' has been investigated to provide a sensitive and time-integrated measure of contaminant exposure and the potential risk posed to local ecologies. A range of organisms are now used to monitor contaminant exposure under a series of spatial and temporal regimes, with certain criteria now defined to assess the potential utility of a given species as a biomonitor of specific contaminants (Rainbow and Phillips 1993). With increasing knowledge, toxicokinetic and toxicodynamic models are assisting in the identification of key monitoring traits to understand ecological risk from contaminant exposure (Spurgeon et al. 2020).

10.5. JELLYFISH LIFE HISTORY

Jellyfish are ubiquitous inhabitants of all oceans and seas globally. A number of species, including the venomous cubozoans, Box Jellyfish, Irukandji, and a number of Scyphozoan jellyfish have been found all along the GBR (Schlaefer and Kingsford 2024). They have generally been considered part of the local ecology, but there has been little interest in them ecologically, except when negative interactions with humans occur, such as jellyfish stings and the currently well-recognised 'stinger season' in the northern Australian summer season (Schlaefer and Kingsford 2024).

Despite this lack of historical research interest, jellyfish have a number of physiological and behavioural features that make them potentially very effective contaminant monitors. Structurally, they are very simple animals; however, they possess complex life histories with both sexual and asexual reproductive strategies. The juvenile polyp form is small, benthic, and sessile and is typically found anchored to hard substrates, such as rocks, buoys, pontoons, etc. Polyps reproduce asexually in multiple ways through budding (producing new sessile polyps), strobilation (producing free-swimming ephyra), or encystment (Figure 2). Encystment occurs when polyps are exposed to poor environmental conditions such as hypoxia or starvation. They develop a hard-shelled cyst (podocyst) that can remain dormant for days to years until environmental conditions become more favourable to survival and growth

FIGURE 2 Life stages of the *Cassiopea* jellyfish. A) Cassiopea polyp with planuloid bud; B) Polyp undergoing strobilation to ephyra (photographer: Pinetreella CC BY 4.0.) C) Newly released ephyra; D) Adult medusae (© WoRMS Photogallery, Photographer: André Morandini, CC BY 4.0).

(e.g., Thein et al. 2012; Kawahara et al. 2013). In contrast, the adult medusae reproduces sexually by producing eggs and sperm that when fertilised, produce free-swimming planula larvae that migrate to settle on a suitable hard substrate. The planula then develops into a polyp, which begins the reproductive cycle again. The combination of podocyst formation and asexual budding not only allows the polyps to remain viable for long time periods when environmental conditions are poor but also allows for rapid responses to optimum conditions, a feature that allows jellyfish to take rapid advantage of 'boom-bust cycles'.

In addition to high reproductive rates in polyps, growth rates in jellyfish medusa can be very high with instantaneous growth rates reported to be as high as 25–30% day^{-1} in some species (e.g., Pitt and Kingsford 2003; Hirst et al. 2003; Riisgard and Larsen 2022). The combination of the high strobilation rates and growth rates allows jellyfish to exploit favourable environmental conditions to a level that most other marine organisms are unable to achieve (e.g., Stoner et al. 2011). Of the two phases, the polyps play the key role in determining the numbers of individuals that can drive development of a jellyfish bloom (Lucas et al. 2012; Thein et al. 2012).

Development of jellyfish blooms can be challenging, particularly when there are negative impacts on economies including tourism, fishing, and power generation industries. Expanded media coverage has triggered concerns regarding the increasing frequency in bloom formation and associated economic and ecological impacts on coastal ecosystems (Hay 2006; Ghermandi et al. 2015). These concerns coupled with the perception that jellyfish are trophic dead-ends have also raised questions about anthropogenic drivers of blooms and the slide to gelatinous-dominated marine systems (Hay 2006; Lynam et al. 2006).

These perceptions regarding jellyfish as trophic dead-ends and drivers of ecosystem switches have been questioned to a large degree (Young et al. 2015; Hays et al. 2018; Pitt et al. 2018). New research is demonstrating that jellyfish not only fill a range of ecological niches but also are key components in marine food webs and biogeochemical cycles. Although lower in calorific content than some other prey species, jellyfish biomass and abundance can provide adequate nutrition to a range of predators. Recent studies have shown that numerous marine species, from bristle-worms and nudibranchs to apex predators including turtles, swordfish, and tuna, will regularly target jellyfish as prey (Stoner and Layman 2015; Young et al. 2015; Guzman et al. 2020). Furthermore, they are considered important structures in the pelagic environment for invertebrates and fishes to shelter, feed, and are sometimes commensal (Kingsford 1993; Browne and Kingsford 2005).

Many other perceptions regarding jellyfish physiology and behaviour are also being revisited due to knowledge advances. This includes the view that medusae are largely passive drifters with limited ability to control or maintain local populations. Field studies have demonstrated that a range of jellyfish species have the capacity to both orient their movement and positions within currents and maintain localized populations, particularly within shallow coastal environments (e.g., Fossette et al. 2015; Schlaefer et al. 2018; Kingsford et al. 2021; Schlaefer and Kingsford 2024). For the benthic *Cassiopea* in particular, their sedentary habit ensures limited spatial movement unless environmental conditions become unsuitable.

One of the key physiological features of jellyfish is their capacity to act as biological pumps. This results in high prey clearance and water turnover rates. Studies on 4-cm *Aurelia aurita* reported clearance rates ranging from 0.6–3.8 L hr^{1} $individual^{-1}$ (Riisgard and Madsen 2011), with similar results reported for 6-cm *A. coerulea* (Wang et al. 2020). This is even higher in *Cassiopea* where estimates for 6-cm medusa range from 9–24 L hr^{1} (Santhanakrishnan et al. 2012).

Accordingly, although jellyfish have historically been considered pests with little in the way of redeeming features, the combination of their simple structure, physiological resilience, and complex life history provides a number of criteria that can be usefully harnessed for monitoring environmental contaminants at both small and large spatial and temporal scales. Both the adult medusae and polyps of a range of species have been assessed for their capacity to provide environmental markers of disturbance, but in many cases, due to size, clearance rates, and ease of identification, the medusae provide greater monitoring opportunities in this space.

10.6. CASSIOPEA

One genus with an increased research focus recently as an environmental monitor has been the Upside-Down Jellyfish (*Cassiopea* spp.). A member of the Cassiopeidae family, the genus comprises twelve species along with five synonyms, although this continues to be under revision as new populations are discovered (Collins et al. 2023). Of the known *Cassiopea* spp. globally, five species have been identified in tropical Australian waters with populations currently extending as far south as the mid-coast of New South Wales, where they are considered to be potentially invasive (Keable and Ahyong 2016). *Cassiopea* spp. are found all along the Queensland coast including in inshore and mid-shelf reef lagoons, mangroves, urban canals, sand flats, and seagrass meadows from Lizard Island to the Gold Coast (Templeman and Kingsford 2010, 2012).

Unlike most other jellyfish, *Cassiopea* demonstrates atypical behaviour by resting upside down on soft substrates rather than having a truly pelagic existence (Figure 2). As such, they are generally found in shallow, sheltered locations in coastal areas. A study by Epstein et al. (2016) established that elemental concentrations in *Cassiopea maremetens* tissues not only reflected a degree of coastal development in signatures but also showed site discrimination at the scale of hundreds of metres between sites.

Cassiopea are present in many urban marine environments including urban marine lakes and canal systems.

These systems often tend to be spatially stratified with poor mixing and high variability in physico-chemical parameters across different parts of the lakes/canals (Dunn et al. 2007, among others). This can lead to zones of hypoxia, eutrophication, salinity stratification, aggregations of debris, and other anthropogenic wastes in parts of the waterbodies. Under these conditions, water quality is also considered to be variable at small spatial scales. A study on the variation in elemental signatures in *Cassiopea* across multiple locations within three urban marine lakes and canals found variation in bioconcentration both within and between the lakes and canals (Templeman and Kingsford 2010). Variation was greater between locations but was also measurable at the scale of hundreds of metres. This further emphasises the capacity for this genus to be effective biomonitors in systems as large and diverse as the GBR and across small spatial scales found in urban marine systems.

The reef patchiness and complexity of marine structures within the GBR mean that currents, tidal flows, and water mixing create heterogeneity in water quality and assemblages. Ecological hot spots can be created by eddies, etc. and lead to patchiness in community ecology and biotic assemblages across both coastal and lagoonal parts of the GBR (Kingsford et al. 2021; Schlaefer et al. 2022; Benthuysen et al. 2022; Wolanski et al. 2024). These same factors can also concentrate sediments, nutrients, and contaminants in localised areas (Wolanski et al. 2024). This also contributes to variations in water residence times within and across reefs and can have a major influence on contaminant exposure regimes as toxicity is a function not only of concentration but also of exposure time.

Rather than being uniformly distributed across the GBR, *Cassiopea* tend to aggregate in specific locations. This combined with their sedentary nature allow them to act in a manner similar to passive samplers and thus provide a time-integrated measure of contaminant exposure. Templeman and Kingsford (2012) determined that *Cassiopea* could be regularly found within specific locations on the GBR across a four-year monitoring program and that cohorts of the jellyfish had elemental signatures that were specific to their location. The same study also demonstrated significant cross-shelf elemental signatures for two other zooxanthellate Scyphozoan jellyfish species, specifically, *Mastigias* sp. and *Netrostoma* sp. (Templeman and Kingsford 2012).

The capacity to bioconcentrate elements has been further quantified with laboratory studies showing that *Cassiopea* concentrate aqueous copper and zinc within their tissues in a defined manner. Bioconcentration rates for copper and zinc were 99 and 104 times exposure and biological half-lives of 1.7 and 9.1 days, respectively (Templeman and Kingsford 2015). This was further supported by Epstein et al. (2016) who found that increased nutrient status increased the bioconcentration rates for both copper and zinc under laboratory conditions. In the same study, field deployment of animals showed site specific accumulation of copper and zinc in a six-day exposure period across a coastal development gradient (Epstein et al. 2016). The capacity to establish concentration gradients across such short temporal and spatial scales provides additional capacity for effectively monitoring contaminants in habitats subjected to patchy flows and exposures.

Due to their sedentary, benthic nature, *Cassiopea* have been shown to be very efficient biogeochemical cyclers of nutrients through a combination of vertical mixing between the sediment and the water column and in situ exudate production (Jantzen et al. 2010; Niggl et al. 2010; Durieux et al. 2021). The study by Niggl et al. (2010) estimated the organic carbon contribution from *Cassiopea* exudates was 2–15 times greater than estimated for hermatypic corals. In oligotrophic waters associated with coral reef systems, this behaviour may be a significant driver of primary productivity for the wider reefal system (Jantzen et al. 2010; Niggl et al. 2010).

Similarly, in low energy environments such as those found in mangrove systems, poor mixing between the benthos and overlying water can lead to the stratification of nutrients and other elements. *Cassiopea* with their biological pump mechanism actively remobilise nutrients from porewater into the water column under these conditions and are considered to be an important driver of pelagic productivity (Durieux et al. 2021). This allows the jellyfish to modify conditions to suit their needs and create a higher nutrient environment for their own benefit and can also drive increased productivity into the surrounding water column (Jantzen et al. 2010; Niggl et al. 2010; Santhanakrishnan et al. 2012; Durieux et al. 2021). A side effect of this activity, however, can be the remobilisation of sequestered sediment-bound or porewater contaminants. Such remobilisation effects have the potential to re-introduce contaminants into the pelagic system where they can influence pelagic ecology.

Like many of their Anthozoan cousins, *Cassiopea* have an endosymbiotic relationship with a range of *Symbiodinium* spp., a dinoflagellate algae. Due to this symbiotic relationship, *Cassiopea* have recently been put forward as a model organism for understanding symbiotic mechanisms and are also considered to be a good proxy for hermatypic corals (Ohdera et al. 2018). Unlike many other species, however, the host-symbiont relationship in *Cassiopea* is considered unique in that it requires symbiont colonisation for the polyp to strobilate to the adult medusa (Ohdera et al. 2018).

The criticality of the symbiosis in *Cassiopea* is reflected in the dependence of the host jellyfish on nutrient contributions from symbiont to host. The majority of the carbon required for growth and reproduction is supplied by the endosymbiont, with bleached animals reducing their body mass even with adequate heterotrophic nutrition (Ohdera et al. 2018). This demonstrates the essentiality of the host-symbiont relationship in *Cassiopea*. This also provides the capability for monitoring contaminant effects by using both endosymbiont autotrophic and host heterotrophic responses simultaneously.

Pesticides have been identified as one of the key threats to the GBR ecology (Lewis et al. 2009; Negri et al. 2020). With a range of pesticide contributions from multiple sources, challenges lie in effectively monitoring both their presence and effects on GBR ecology (Negri et al. 2020). One of the benefits of using *Cassiopea* as a pesticide monitor is due to their physiology and life history. The symbiotic relationship with zooxanthellae and sedentary nature allow them to act as proxies for hermatypic corals. *Cassiopea* are also tolerant of fluctuating salinities as low as 12 ppt salinity. They are also adaptable to higher water temperatures that at times exceed 35 °C in parts of the Red Sea (Klein et al. 2019). This flexibility allows them to be used as environmental monitors across inshore marine systems that are routinely subject to greater salinity and temperature fluctuations.

Inorganic arsenic and copper compounds have long been used as antifoulants on ship hulls and as termiticide treatments for timber, despite increasing concerns regarding their toxicity to aquatic systems (Weis et al. 1992; Kroon et al. 2020). The capacity for *Cassiopea* to bioconcentrate a range of elements including copper, arsenic, and chromium makes them ideal for monitoring inorganic antifoulants such as copper complexes and urban termiticides such as copper-chrome-arsenic treatments in urban marine systems. Although no specific field studies have investigated the relationship between antifoulant and termiticide leaching on *Cassiopea*, the demonstrated accumulation capacity of the contributing metals indicates that they could be effective and relevant monitors (Templeman and Kingsford 2010, 2012).

The increase in synthetic organic pesticides in agriculture has led to increased detection of these pesticides in GBR catchments and coastal systems (Lewis et al. 2009). In particular, PSII herbicides have been an effective herbicide group across many agricultural catchments on the GBR (Table 1). Consequently, PSII herbicides are currently the most commonly detected group of herbicides in the GBR (Shaw et al. 2010). PSII herbicides have demonstrated impacts at low concentrations on corals and microalgae under laboratory conditions (Jones 2005; Negri et al. 2005; Magnusson et al. 2008; among others).

Cassiopea has also shown to be sensitive to the PSII herbicides diuron and hexazinone. Growth of the jellyfish was significantly affected by both herbicides at concentrations as low as 0.1 ug L^{-1} after seven days, indicating that *Cassiopea* is very sensitive to PSII exposure (Rowen et al. 2017, Figure 3). McKenzie et al. (2020) demonstrated similar responses to hexazinone exposure, but the intensity of response differed. It was suggested that the use of analytical-grade herbicides in this study were the driver of the variation in results compared to the agricultural-grade herbicides with their associated adjuvants in the Rowen et al. (2017) study.

Simulation of wet season conditions has also shown that herbicide impacts can be magnified in jellyfish. Combined exposure of the PSII herbicide atrazine under a range of salinities to simulate light to heavy rainfall events caused mortality in the highest atrazine/lowest salinity concentrations (Klein et al. 2016). Decreased salinity in conjunction with herbicide exposure also caused synergistic decreases in photosynthetic yields. The outcomes of this study indicate that the combination of herbicide exposure with decreased salinity from rainfall events can increase impacts on jellyfish (Klein et al. 2016). As the majority of herbicide dispersion into the GBR is associated with wet season rainfall events, this has implications for coastal systems. Collectively, both the consistency of response by *Cassiopea* and their endosymbiont to herbicide exposure, as well as the influences of confounding effects (chemical grade, salinity), show that there is a very real herbicide inhibition to both host jellyfish and endosymbiont under a range of environmental conditions (Figure 3).

More widely, studies on the effect of insecticides on the moon jellyfish (*Aurelia aurita*) ephyra found that both serine (carbamate) and chlorpyrifos (organophosphorus) insecticides were toxic to survival and behavior (Costa

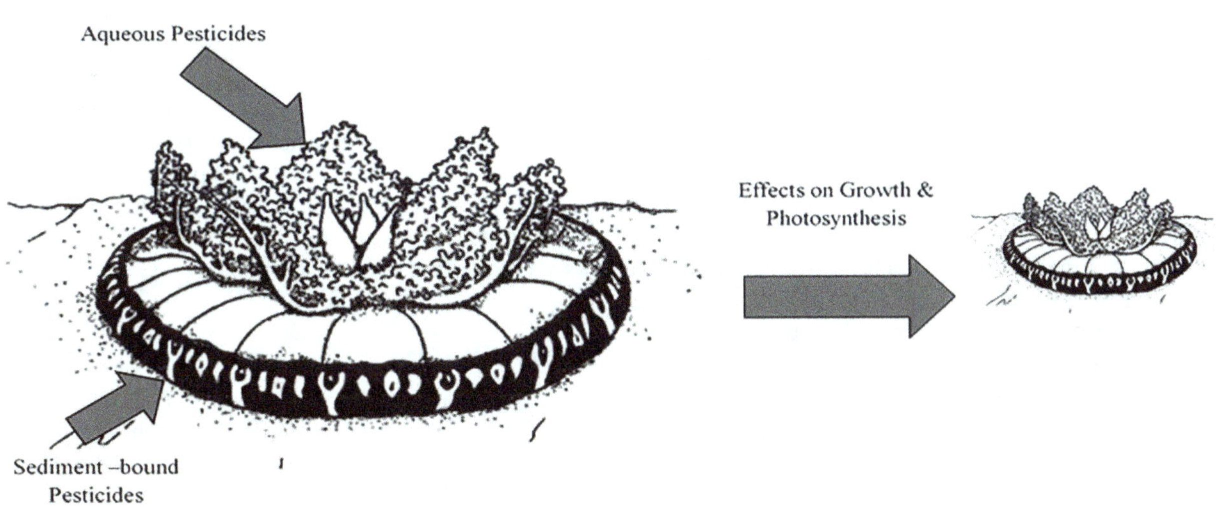

FIGURE 3 Sources and pesticide processes influencing *Cassiopea* sp. health. Pesticide exposure causing reductions in bell size and photosynthetic efficiency.

Source: Image produced by J Stuart

et al. 2015). Of the two compounds, chlorpyrifos was the more toxic, inhibiting pulse frequency at 0.1 mg L^{-1} after 48 hours of exposure. Comparison of inhibition from the two insecticides with other marine invertebrates showed that *A. aurita* was more sensitive to both insecticides than other reported studies (Costa et al. 2015).

Interestingly, some studies have shown that although the ephyra and medusa tend to be very sensitive to contaminant exposure, polyps may have a greater level of resilience. Olguin-Jacobson et al. (2020) reported no inhibition in polyp survival over a nine-week period in *A. aurita* exposed to the insecticide chlorpyrifos and the herbicide atrazine both individually and as a mixture. This contrasts with the results in Costa et al. (2015), although exposure concentrations differed between the studies. Interestingly, although *Aurelia coerulea* polyps also displayed resilience to both atrazine and chlorpyrifos exposure, ephyra from chlorpyrifos-exposed polyps exhibited a greater proportion of deformities and a reduction in some key metabolites (Olguin-Jacobson et al. 2021). Again, this suggests that even when direct effects are not identified, chronic pesticide exposure can have longer-term physiological effects on different life stages.

Variability in sensitivity to contaminant exposure is common across age in most species, with Almeda et al. (2013) showing both inter-age and inter-species specificity in sensitivity to polycyclic aromatic hydrocarbons. Generally, when the life history is more complex, the likelihood of variability in sensitivity to contaminant exposure across different life stages is greater (Spurgeon et al. 2020). In particular, shifts in the toxicodynamics of target site interactions, diversity between different life stages, and simple physiological factors such as size can all influence sensitivity to contaminant exposure (Spurgeon et al. 2020; Olguin-Jacobson et al. 2021).

Variability in response to contaminant exposure among jellyfish species is also expected given the differences in distribution, physiology, and ecology (Arai 1997; Almeda et al. 2013). Symbiotic jellyfish such as *Cassiopea* with their photosynthetic zooxanthellae could be expected to be more sensitive to herbicide exposure than the non-symbiotic *Aurelia* jellyfish. There is potential for inconsistency between pesticides in this response, however, with Olguin-Jacobson and Pitt (2021) who showed similar responses between zooxanthellate and azooxanthellate *Cassiopea xamachana* to atrazine exposure. This is despite numerous studies that link the majority of the carbon assimilation in *Cassiopea* to photosynthates produced by the endosymbiont zooxanthellae (Ohdera et al. 2018). Although this contradiction highlights the complexity of the symbiotic relationship, the breadth of studies indicates a close relationship between host and symbiont.

10.7. CHALLENGES AND FUTURE DIRECTIONS

At present, despite a growing body of research into pesticide risk, there are still major knowledge gaps in understanding the challenges that pesticides pose. For the GBR, periodicity of exposure and seasonality of exposure for some compounds make monitoring of concentrations challenging. This combined with the hydrodynamic complexity of the GBR, which causes localised hot spots in fronts and eddies, create difficulties in identifying priority locations for contaminant monitoring, particularly at small spatial scales.

Spatial and temporal complexity in monitoring is further challenged by the continuing evolution of pesticide formulations being released to the market. Realistically, these issues make it impossible to effectively manage chemical risk in real-time in a system as large and diverse as the GBR. Historically, the risk-based approach has been driven by an 'as needed' basis, with the greatest focus directed to the chemicals perceived to be of greatest risk to the community—both human and ecological (Spurgeon et al. 2020). Ideally, understanding the key physiological traits that contribute to species sensitivity can shortcut some of these processes (Spurgeon et al. 2020).

Cassiopea offer a level of robustness for pesticide monitoring within the GBR, in particular due to their rapidity of response to pesticide and other contaminant exposures. Chronic reactions to exposure, both in terms of endosymbiont (e.g., photosynthetic yield) and host responses (e.g., pulse behaviours and growth) can be identified in hours to days, which is also reflective of in situ exposure regimes (Templeman et al. 2021). Questions still remain on whether they can accumulate pesticides in their tissues, although examples from other studies (e.g., Almeda et al. 2103; Kingsford and Templeman 2015; Epstein et al. 2016) suggest that they may offer potential in this space.

At present, there is limited information on *Cassiopea* responses to multiple pesticides, both inorganic and organic, but the opportunity remains to assess their effectiveness. The capacity to tolerate both variable thermal and salinity regimes in estuaries also provides an opportunity to measure pulse-driven pesticides delivered from catchments. This prospect may be able to guide a biological 'early warning' system of increasing threat to the wider GBR.

Ultimately, the capacity to accurately determine risk and provide a clear relationship between contaminant cause and response in a model organism in real-time under real-world conditions is the goal for ecosystem protection for the GBR and more widely. Whether *Cassiopea* holds potential as a model in this area is still to be determined, but early indications suggest that it holds promise.

BIBLIOGRAPHY

Almeda R, Wambaugh Z, Chai C, Wang Z, Liu Z, Buskey EJ (2013). Effects of crude oil exposure on bioaccumulation of polycyclic aromatic hydrocarbons and survival of adult and larval stages of gelatinous zooplankton. PLoS One, 8:e74476.

ANZG (2018). Australian and New Zealand guidelines for fresh and marine water quality. Australian and New Zealand Governments and Australian state and territory governments, Canberra ACT, Australia. www.waterquality.gov.au/anz-guidelines

APVMA (2012). Diuron, Final Review Report. The reconsideration of the registrations of selected products containing diuron and their associated labels. Aust Pest Vet Med Author, 2012.

APVMA (2020). Regulatory operations-chemical reviews-label reviews-mercury fungicides. https://apvma.gov.au/node/75941

APVMA (2022). Australian pesticides and veterinary medicines authority, public chemical registration information system search. https://portal.apvma.gov.au/pubcris?p_auth=8RLgByYh&p_p_id=pubcrisportlet_WAR_pubcrisportlet&p_p_lifecycle=1&p_p_state=normal&p_p_mode=view&p_p_col_id=column-1&p_p_col_pos=2&p_p_col_count=4&_pubcrisportlet_WAR_pubcrisportlet_javax.portlet.action=search. Accessed December, 2022.

Arai M (1997). A functional biology of Scyphozoa. Chapman & Hall, London.

ATSE (2002). Pesticide use in Australia. Australian Academy of Technological Sciences and Engineering. www.atse.org.au/wp-content/uploads/2019/01/pesticide-use-australia.pdf

Barrett L, Macfadyen S, Williams, Parry H (2021). A short history of agricultural chemical usage and development. ECOS, 278. https://ecos.csiro.au/history-of-agricultural-chemicals/

Becker B, Kochleus C, Spira D, Mohlenkamp C, Bachtin J, Meinecke S, Vermeirssen (2021). Passive sampler phases for pesticides: Evaluation of AttractSPE™ SDB-RPS and HLB versus Empore™ SDB-RPS. Environ Sci Poll Res, 28:11697–11707.

Benthuysen JA, Emslie MJ, Currey-Randall LM, Cheal AJ, Heupel MR (2022). Oceanographic influences on reef fish assemblages along the Great Barrier Reef. Prog Ocean, 208:102901.

BITRE (2013). Statistical report, Australian sea freight 2011–12. Bureau of Infrastructure, Transport and Regional Economics. Department of Infrastructure and Transport, Canberra, Australia. www.bitre.gov.au/publications/2013/asf_2011_12

Brodie J, Landos M (2019). Pesticides in Queensland and Great Barrier Reef waterways—Potential impacts on aquatic ecosystems and the failure of national management. Est Coast Shelf Sci, 230:106447.

Brodie JE, Kroon FJ, Schaffelke B, Wolanski EC, Lewis SE, Devlin MJ, Bohnet IC, Bainbridge ZT, Waterhouse J, Davis AM (2012). Terrestrial pollutant runoff to the Great Barrier Reef: An update of issues, priorities and management responses. Mar Poll Bull, 65:81–100.

Browne JG, Kingsford MJ (2005). A commensal relationship between the scyphozoan medusae *Catostylus mosaicus* and the copepod *Paramacrochiron maximum*. Mar Biol, 146:1157–1168.

BTRE (2002). Australian Sea Freight, 1999–2000. Bureau of Transport and Regional Economics. Information Paper 47. www.bitre.gov.au/publications/2002/ip_047

Caban M, Lis H, Stepnowski P (2022). Limitations of integrative passive samplers as a tool for the quantification of pharmaceuticals in the environment—A critical review with the latest innovations. Crit Rev Anal Chem, 52:1386–1407.

Cagnazzi D, Fossi MC, Parra GJ, Harrison PL, Maltese S, Coppola D, Soccodato A, Bent M, Marsili L (2013). Anthropogenic contaminants in Indo-Pacific humpback and Australian snubfin dolphins from the central and southern Great Barrier Reef. Environ Poll, 182:490–494.

California DPR (2017). A brief history of pesticide regulation. California Department of Pesticide Regulation, February 2017. www.cdpr.ca.gov/docs/pressrls/dprguide/historical_timeline.pdf

Cavenagh JE, Burns KA, Brunskill GJ, Coventry RJ (1999). Organochlorine pesticide residues in soils and sediments of the Herbert and Burdekin River regions, North Queensland—Implications for contamination of the Great Barrier Reef. Mar Poll Bull, 39:367–375.

Chauvel B, Gauvrit C, Guillemin J-P (2022). From sea salt to glyphosate salt: A history of herbicide use in France. Adv Weed Sci, 40(Spec1):e020220015.

Chen C, Guo W, Ngo HH (2019). Pesticides in stormwater runoff–A mini review. Front Environ Sci Eng, 13:72.

Collins AG, Jarms G, Morandini AC (2023). World list of Scyphozoa. *Cassiopea* Péron & Lesueur, 1810. Accessed through: World Register of Marine Species. https://www.marinespecies.org/aphia.php?p=taxdetails&id=135253. Accessed January 5, 2023.

Costa E, Gambardella C, Piazza V, Greco G, Lavorano S, Beltrandi M, Bongiovanni E, Gnone G, Faimali M, Garaventa F (2015). Effect of neurotoxic compounds on ephyrae of *Aurelia aurita* jellyfish. Hydrobiologia, 759:75–84.

Davis A, Lewis S, Bainbridge Z, Brodie J, Shannon E (2008). Pesticide residues in waterways of the lower Burdekin region: Challenges in ecotoxicological interpretation of monitoring data. Aust J Ecotox, 14:89–108.

Devlin M, Lewis S, Davis A, Smith R, Negri A, Thompson M, Poggio M (2015). Advancing our understanding of the source, management, transport and impacts of pesticides on the Great Barrier Reef. A report for the Queensland Department of Environment and Heritage Protection. Tropical Water & Aquatic Ecosystem Research (TropWATER) Publication, James Cook University, Cairns, 134pp.

Douglas G, Ford P, Mossa A, Noble B, Packett B, Palmere M, Revilla A, Robson B, Tillman P, Webster I. (2005). Carbon and nutrient cycling in a Subtropical Estuary (The Fitzroy), Central Queensland. Final Report of Coastal CRC Project FH1 (2000–2003).

Dunn RJK, Ali A, Lemckert CJ, Teasdale PR, Welsh DT (2007). Short-term variability of physico-chemical parameters and the estimated transport of filterable nutrients and chlorophyll-a in the urbanised Coombabah Lake and Coombabah Creek system, southern Moreton Bay, Australia. J Coast Res, SI50:1062–1068.

Durieux DM, Du Clos KT, Lewis DB, Gemmell BJ (2021). Benthic jellyfish dominate water mixing in mangrove ecosystems. PNAS, 118:e2025715118.

Epstein HE, Templeman MA, Kingsford MJ (2016). Fine-scale detection of pollutants by a benthic marine jellyfish. Mar Poll Bull, 107:340–346.

Fossette S, Gleiss AC, Chalemeau J, Bastian T, Armstrong CD, Vandenabeele S, Karpytchev M, Hays GC (2015). Current-oriented swimming by jellyfish and its role in bloom maintenance. Curr Biol, 25:342–347.

Gallen C, Thai P, Paxman C, Prasad P, Elisei G, Reeks T, Eaglesham G, Yeh R, Tracey D, Grant S, Mueller J (2019). Marine monitoring program: Annual report for inshore pesticide monitoring 2017–18. Report for the Great Barrier Reef Marine Park Authority, Great Barrier Reef Marine Park Authority, Townsville, 118pp.

Ghermandi A, Galil B, Gowdy J, Paulo ALD, Nunes PALD (2015). Jellyfish outbreak impacts on recreation in the Mediterranean Sea: Welfare estimates from a socioeconomic pilot survey in Israel. Ecosyst Serv, 11:140–147.

Gorgoglione A, Bombardelli FA, Pitton BJ, Oki LR, Haver DL, Young TM (2018). Role of sediments in insecticide runoff from urban surfaces: Analysis and modeling. Int J Environ Res Public Health, 15:1464.

Guzman HM, Kaiseer S, van Hinsberg VJ (2020). Accumulation of trace elements in leatherback turtle (*Dermochelys coriacea*) eggs from the south-western Caribbean indicates potential health risks to consumers. Chemosphere, 243:125424

Hay S (2006). Marine ecology: Gelatinous bells may ring change in marine ecosystems. Curr Biol, 16:R679–R682.

Haynes D, Johnson JE (2000). Organochlorine, heavy metal and polyaromatic hydrocarbon pollutant concentrations in the Great Barrier Reef (Australia) environment: A review. Mar Pollut Bull, 41:267–278.

Hays GC, Doyle TK, Houghton JDR (2018) A paradigm shift in the trophic importance of jellyfish? Trends Ecol Evol, 33:874–884.

Hirst AG, Roff JC, Lampitt RS (2003). A synthesis of growth rates in marine epipelagic invertebrate zooplankton. Adv Mar Biol, 44:1–142.

IMO (2001). International convention on the control of harmful anti-fouling systems on ships. International Maritime Organization. https://www.imo.org/en/About/Conventions/Pages/International-Convention-on-the-Control-of-Harmful-Anti-fouling-Systems-on-Ships-(AFS).aspx

IMO (2002). Background on anti-fouling systems. International Maritime Organization. https://wwwcdn.imo.org/localresources/en/OurWork/Environment/Documents/FOULING2003.pdf

Jantzen C, Wild C, Rasheed M, El-Zibdah M, Richter C (2010). Enhanced pore-water nutrient fluxes by the upside-down jellyfish *Cassiopea* sp. in a Red Sea coral reef. Mar Ecol Prog Ser, 411:117–125.

Jones R (2005). The ecotoxicological effects of Photosystem II herbicides on corals. Mar Pollut Bull, 51:495–506.

Kawahara M, Ohtsu K, Uye S (2013). Bloom or non-bloom in the giant jellyfish *Nemopilema nomurai* (Scyphozoa: Rhizostomeae): Roles of dormant podocysts. J Plankt Res, 35:213–217.

Keable SJ, Ahyong ST (2016). First records of the invasive "Upside-down jellyfish", *Cassiopea* (Cnidaria: Scyphozoa: Rhizostomae: Cassiopeidae), from coastal lakes of New South Wales, Australia. Rec Aust Mus, 68:23–30.

Kennedy K, Devlin M, Bentley C, Lee-Chue K, Paxman C, Carter S, Lewis SE, Brodie J, Guy E, Vardy A, Martin KC, Jones A, Packett R, Mueller JF (2012). The influence of a season of extreme wet weather events on exposure of the World Heritage Area Great Barrier Reef to pesticides. Mar Pollut Bull, 64:1495–1507.

King J, Alexander F, Brodie J (2013). Regulation of pesticides in Australia: The Great Barrier Reef as a case study for evaluating effectiveness. Agric Ecosys Environ, 180:54–67.

Kingsford M (1993). Biotic and abiotic structure in the pelagic environment: Importance to small fishes. Bull Mar Sci, 53:393–415.

Kingsford MJ, Schlaefer JA, Morrissey SJ (2021). Population structures and levels of connectivity for Scyphozoan and Cubozoan jellyfish. Diversity, 13:174.

Klein SG, Pitt KA, Carroll AR (2016). Reduced salinity increases susceptibility of zooxanthellate jellyfish to herbicide toxicity during a simulated rainfall event. Environ Poll, 209:79–86.

Klein SG, Pitt KA, Lucas CH, Hung S-H, Schmidt-Roach S, Aranda M, Duarte CM (2019). Night-time temperature reprieves enhance the thermal tolerance of a symbiotic Cnidarian. Front Mar Sci, 6:453.

Kroon F, Berry K, Brinkman D, Davis A, King O, Kookna R, Lewis S, Leusch F, Makarynskyy O, Melvin S, Müller J, Neale P, Negri A, O'Brien D, Puotinen M, Smith R, Tsang J, van de Merwe J, Warne M, Williams M (2016). Identification, impacts, and prioritization of emerging contaminants present in the GBR and Torres Strait marine environments. Report to the National Environmental Science Programme. Reef and Rainforest Research Centre Limited, Cairns, 138pp.

Kroon FJ, Berry KLE, Brinkman DL, Kookna R, Leusch FDL, Melvin SD, Neale PA, Negri AP, Puotinen M, Tsang JJ, van de Merwe JP, Williams M (2020). Sources, presence and potential effects of contaminants of emerging concern in the marine environments of the Great Barrier Reef and Torres Strait, Australia. Sci Tot Environ, 719:13514.

Kroon FJ, Hook SE, Jones D, Metcalfe S, Henderson B, Smith R, Warne MStJ, Turner RD, McKeown A, Westcott DA (2015). Altered transcription levels of endocrine associated genes in two fisheries species collected from the Great Barrier Reef catchment and lagoon. Mar Environ Res, 104:51–61.

Lewis S, Bainbridge Z, Davis A, et al. (2023). Sediment and nutrient flux from land. *In*: Oceanographic processes of coral reefs: Physical and biological links in the Great Barrier Reef. CRC Press, London.

Lewis SE, Bartley R, Wilkinson SN, Bainbridge ZT, Henderson AE, James CS, Irvine SA, Brodie JE (2021). Land use change in the river basins of the Great Barrier Reef, 1860 to 2019: A foundation for understanding environmental history across the catchment to reef continuum. Mar Poll Bull, 166:112193.

Lewis SE, Brodie JE, Bainbridge ZT, Rohde KW, Davis AM, Masters BL, Maughan M, Devlin MJ, Mueller JF, Schaffelke B (2009). Herbicides: A new threat to the Great Barrier Reef. Environ Poll, 157:2470–2484.

Lucas CH, Graham WM, Widmer C (2012). Chapter 3—Jellyfish life histories: Role of polyps in forming and maintaining Scyphomedusa populations. Adv Mar Biol, 63:133–196.

Lynam CP, Gibbons MJ, Axelsen BE, Sparks CA, Coetzee J, Heywood BG, Brierley AS (2006). Jellyfish overtake fish in a heavily fished ecosystem. Curr Biol, 16:R492–R493.

Magnusson M, Heimann K, Negri AP (2008). Comparative effects of herbicides on photosynthesis and growth of tropical estuarine microalgae. Mar Pollut Bull, 56:1545–1552.

McKenzie MR, Templeman MA, Kingsford MJ (2020). Detecting effects of herbicide runoff: The use of *Cassiopea maremetens* as a biomonitor to hexazinone. Aquat Toxic, 221:105442.

Mercurio P, Mueller JF, Eaglesham G, O'Brien J, Flores F, Negri AP (2016). Degradation of herbicides in the tropical marine environment: Influence of light and sediment. PLoS One, 11:e0165890.

Negri A, Marshall P (2009). TBT contamination of remote marine environments: Ship groundings and ice-breakers as sources of organotins in the Great Barrier Reef and Antarctica. J Environ Manage, 90:S31–S40.

Negri A, Vollhardt C, Humphrey C, Heyward A, Jones R, Eaglesham G, Fabricius K (2005). Effects of the herbicide diuron on the early life history stages of coral. Mar Pollut Bull, 51:370–383.

Negri AP, Templeman S, Flores F, van Dam J, Thomas M, McKenzie M, Stapp L, Kaserzon S, Mann RM, Smith R, Warne M StJ, Mueller J (2020). Ecotoxicology of pesticides on the Great Barrier Reef for guideline development and risk assessments. Final report to the National Environmental Science Program. Reef and Rainforest Research Centre Limited, Cairns.

NIEHS (2022). Environmental health topics—Environmental agents—Pesticides. United States National Institute of Environmental Health Sciences website, www.niehs.nih.gov/health/topics/agents/pesticides/index.cfm. Accessed December, 2022.

Niggl W, Naumann MS, Struck U, Manasrah R, Wild C (2010). Organic matter release by the benthic upside-down jellyfish *Cassiopea* sp. fuels pelagic food webs in coral reefs. J Exp Mar Biol Ecol, 384:99–106.

NPIC (1999). DDT general fact sheet. United States National Pesticide Information Center. Oregon State University. http://npic.orst.edu/factsheets/ddtgen.pdf

Ohdera A, Abramson M, Ames C, Baker D, Suescún-Bloívar L, Collins A, Freeman C, Gamero-Mora E, Goulet T, Hofmann D, Jaimes-Becerra A, Long P, Marques A, Miller L, Mydlarz L, Morandini A, Newkirk C, Putri S, Samson J, Stampar S, Steinworth B, Templeman M, Thom´e P, Vlok M, Woodley C, Wong J, Martindale M, Fitt W, Medina M (2018). Upside-down but headed in the right direction: Review of the highly versatile *Cassiopea xamachana* system. Front Ecol Evol, 6:35.

Oi F (2022). A review of the evolution of termite control: A continuum of alternatives to termiticides in the United States with emphasis on efficacy testing requirements for product registration. Insects, 13:50.

Olguin-Jacobson C, Pitt KA (2021). Symbiotic microalgae do not increase susceptibility of zooxanthellate medusae (*Cassiopea xamachana*) to herbicides. Aquat Toxic, 236:105866.

Olguin-Jacobson C, Pitt KA, Carroll AR, Melvin SD (2020). Polyps of the jellyfish *Aurelia aurita* are unaffected by chronic exposure to a combination of pesticides. Environ Tox Chem, 39:1685–1692.

Olguin-Jacobson C, Pitt KA, Carroll AR, Melvin SD (2021). Chronic pesticide exposure elicits a subtle carry-over effect on the metabolome of *Aurelia coerulea* ephyrae. Environ Poll, 275:116641.

PAHO (2019). Technical document for the implementation of interventions based on generic operational scenarios for *Aedes aegypti* control. Pan American Health Organization. https://idl-bnc-idrc.dspacedirect.org/bitstream/handle/10625/60176/a7443bba-a3b7-450d-ae4e-032843973aec.pdf

Pitt KA, Kingsford MJ (2003). Temporal and spatial variation in recruitment and growth of medusae of the jellyfish, *Catostylus mosaicus* (Scyphozoa: Rhizostomeae). Mar Fresh Res, 54:117–125.

Pitt KA, Lucas CH, Condon RH, Duarte CM, Stewart-Koster B (2018). Claims that anthropogenic stressors facilitate jellyfish blooms have been amplified beyond the available evidence: A systematic review. Front Mar Sci, 5:451.

Productivity Commission (2005). Trends in Australian agriculture, Research Paper, Canberra.

Rainbow PS, Phillips DJH (1993). Cosmopolitan biomonitors of trace metals. Mar Poll Bull, 26:593–601.

Raub KB, Vlahos P, Whitney M (2015). Comparison of marine sampling methods for organic contaminants: Passive samplers, water extractions, and live oyster deployment. Mar Environ Res, 109:148–158.

Riisgard HU, Larsen PS (2022). Bioenergetic model and specific growth rates of jellyfish *Aurelia* spp. Mar Ecol Prog Ser, 688:49–56.

Riisgard HU, Madsen CV (2011). Clearance rates of ephyrae and small medusae of the common jellyfish *Aurelia aurita* offered different types of prey. J Sea Res, 65:51–57.

Rippy MA, Deletic A, Black J, Aryal R, Lampard J-L, Tang JY-M, McCarthy D, Kolotelo P, Sidhu J, Gernjak W (2017). Pesticide occurrence and spatio-temporal variability in urban run-off across Australia. Water Res, 115:245–255.

Rowen DJ, Templeman MA, Kingsford MJ (2017). Herbicide effects on the growth and photosynthetic efficiency of *Cassiopea maremetens*. Chemosphere, 182:143–148.

Sadaria AM, Sutton R, Moran KD, Teerlink J, Brown JV, Halden RU (2016). Passage of fiproles and imidacloprid from urban pest control uses through wastewater treatment plants in Northern California, USA. Environ Tox Chem, 36:1473–1482.

Santhanakrishnan A, Dollinger M, Hamlet CL, Colin SP, Miller LA (2012). Flow structure and transport characteristics of feeding and exchange currents generated by Upside-Down Cassiopea Jellyfish. J Exp Biol, 215:2369–2381.

Schlaefer J, Wolanski E, Kingsford MJ (2018). Swimming behaviour can maintain localised jellyfish (*Chironex fleckeri*: Cubozoa) populations. Mar Ecol Prog Ser, 591:287–302.

Schlaefer JA, Kingsford MJ (2024). Bio-physical interactions of jellyfish on the great barrier reef. In Wolanski E, Kingsford MJ (eds.), Oceanographic Processes of Coral Reefs. Physical and Biological Links in the Great Barrier Reef. CRC Press, Boca Raton, FL.

Schlaefer JA, Tebbett SB, Bowden CL, Collins WP, Duce S, Hemingson CR, Huertas V, Mihalitsis M, Morais J, MoraisRA, Siqueira AC, Streit RP, Swan S, Valenzuela J, Bellwood DR (2022). A snapshot of sediment dynamics on an inshore coral reef. Mar Environ Res, 181:105763.

Schneider L (2021). When toxic chemicals refuse to die—An examination of the prolonged mercury pesticide use in Australia. Sci Anth, 9:1.

Shaw M, Furnas MJ, Fabricius K, Haynes D, Carter S, Eaglesham G, Mueller JF (2010). Monitoring pesticides in the Great Barrier Reef. Mar Pollut Bull, 60:113–122.

Spilsbury FD, Warne MStJ, Backhaus T (2020). Risk assessment of pesticide mixtures in Australian rivers discharging to the Great Barrier Reef. Environ Sci Technol, 54:14361–14371.

Spurgeon D, Lahive E, Robinson A, Short S and Kille P (2020). Species sensitivity to toxic substances: Evolution, ecology and applications. Front Environ Sci, 8:588380.

Stoner EW, Layman CA (2015). Bristle worms attack: Benthic jellyfish are not trophic dead ends. Front Ecol Environ, 13:226–227.

Stoner EW, Layman CA, Yeager LA, Hassett HM (2011). Effects of anthropogenic disturbance on the abundance and size of epibenthic jellyfish *Cassiopea* spp. Mar Poll Bull, 62:1109–1114.

Templeman MA, Kingsford MJ (2010). Trace element accumulation in *Cassiopea* sp. (Scyphozoa) from urban marine environments in Australia. Mar Environ Res, 69:63–72.

Templeman MA, Kingsford MJ (2012). Variation in soft tissue chemistry among scyphozoan and cubozoan jellyfishes from the Great Barrier Reef, Australia. Hydrobiologia, 690:279–290.

Templeman MA, Kingsford MJ (2015). Predicting aqueous copper and zinc accumulation in the upside-down jellyfish *Cassiopea maremetens* through the use of biokinetic models. Environ Monit Assess, 187:416–424.

Templeman MA, McKenzie MM, Kingsford MJ (2021). The utility of jellyfish as marine biomonitors. Mar Poll Bull, 173:113056.

Thai P, Paxman C, Prasad P, Elisei G, Reeks T, Eaglesham G, Yeh R, Tracey D, Grant S, Mueller J (2020). Marine Monitoring Program: Annual report for inshore pesticide monitoring 2018–19. Report to the Great Barrier Reef Marine Park Authority, Great Barrier Reef Marine Park Authority, Townsville, 69pp.

Thein H, Ikdea H, Uye S (2012). The potential role of podocysts in perpetuation of the common jellyfish *Aurelia aurita* s.l. (Cnidaria: Scyphozoa) in anthropogenically perturbed coastal waters. Hydrobiology, 690:157–167.

Thompson M, Chauhan BS (2022). History and perspective of herbicide use in Australia and New Zealand. Adv Weed Sci, 40:e20210075.

Thorburn PJ, Wilkinson SN, Silburn DM (2013). Water quality in agricultural lands draining to the Great Barrier Reef: A review of causes, management and priorities. Agric Ecosys Environ, 180:4–20.

Tudi M, Daniel Ruan DH, Wang L, Lyu J, Sadler R, Connell D, Chu C, Phung DT (2021). Agriculture development, pesticide application and its impact on the Environment. Int J Environ Res Public Health, 18:1112.

UNEP (2023). Stockholm Convention at a glance. www.pops.int/Portals/0/download.aspx?d=UNEP-POPS-PAWA-OVERV-AtaGlance.En.pdf

Unsworth J (2010). History of pesticide use. https://agrochemicals.iupac.org/index.php?option=com_sobi2&sobi2Task=sobi2Details&catid=3&sobi2Id=31

Vicente-Beckett V, Noble R, Packett R, Verwey P, Ruddle L, Munksgaard N, Morrison H (2006). Pesticide, polycyclic aromatic hydrocarbon and metal contamination in the Fitzroy Estuary, Queensland. Technical Report 37, Cooperative Research Centre for Coastal Zone, Estuary and Waterway Management, Indooroopilly, Qld.

Wang P, Zhang F, Sun S, Wang W, Wan A, Li C (2020). Experimental clearance rates of *Aurelia coerulea* ephyrae and medusae, and the predation impact on zooplankton in Jiaozhou Bay. J Ocean Limn, 38:1256–1269.

Warne MS, Neelamraju C, Strauss J, Smith RA, Turner RDR, Mann RM (2020a). Development of a method for estimating the toxicity of pesticide mixtures and a Pesticide Risk Baseline for the Reef 2050 Water Quality Improvement Plan. Department of Environment and Science, Queensland Government, Brisbane.

Warne MS, Smith RA, Turner RDR (2020b). Analysis of pesticide mixtures discharged to the Great Barrier Reef, Australia. Environ Poll, 265:114088.

Weis P, Weis JP, Greenberg A, Nosker TJ (1992). Toxicity of construction materials in the marine environment: A comparison of chromated-copper-arsenate-treated wood and recycled plastic. Arch Environ Contam Toxicol, 22:99–106.

WHO (1952). The history of the prevention of fouling. Chapter 11, Marine fouling and its prevention. Contribution No. 580 Woods Hole Oceanographic Institute. https://darchive.mblwhoilibrary.org/bitstream/handle/1912/191/chapter%2011.pdf

WHO (2015). World Health Organization global technical strategy for malaria 2016–2030. www.who.int/docs/default-source/documents/global-technical-strategy-for-malaria-2016-2030.pdf

Wolanski E, Kingsford MJ, Lambrechts J, Marmorino G (2024). The physical oceanography of the great barrier reef: A review. In Wolanski E, Kingsford MJ (eds.), Oceanographic Processes of Coral Reefs. Physical and Biological Links in the Great Barrier Reef. CRC Press, Boca Raton, FL.

Young JW, Hunt BPV, Cook TR, Llopiz JK, Hazen EL, Pethybridge HR, Ceccarelli D, Lorrain A, Olsen RJ, Allain V, Menkes C, Patterson T, Nicol S, Lehodey P, Kloser RJ, Arrizabalaga H, Choy CA (2015). The trophodynamics of marine top predators: Current knowledge, recent advances and challenges. Deep-Sea Res II, 113:170–187.

Zheng H, Gao Y, Xia Y, Yang H, Cai M (2020). Seasonal variation of legacy organochlorine pesticides (OCPs) from East Asia to the Arctic Ocean. Geophys Res Lett, 47:e2020GL089775.

11 The Influence of the Spatiotemporal Dynamics of Fish Populations on the Outcomes of Land-Sea Connectivity

Marcus Sheaves, Michael Bradley, Nicolas Lubitz, Carlo Mattone, Jaelen Myers, Ajay A. Venkataraman, Nathan J. Waltham, and Sara Winter

11.1. INTRODUCTION

The Great Barrier Reef (GBR) region comprises a diversity of ecosystems including the reefs themselves, inter-reefal habitats, and a mosaic of coastal systems. Here, land and sea are strongly interconnected at multiple spatial and temporal scales by diverse patterns of fish movement. This functional interlinking enables a range of vital ecological processes (Sheaves and Johnston 2008; Sheaves 2009). Among the most important contributors to reef function are the mosaics of coastal ecosystems (Figure 1) that are diverse in nature and function; they support vital links between marine environments and nutritionally rich terrestrial ecosystems (Waterhouse et al. 2023). This complex interlinking of land and sea via coastal ecosystems provides the GBR region with its unique ecological character. The interplay of processes ranges from small-scale interactions within the daily life of organisms (Sheaves 2005; Nagelkerken et al. 2015), which are often focused on the scale of an individual habitat, to the large-scale spatiotemporal dynamics (Jones et al. 2010) of fish undergoing major ontogenetic migrations. Together, these interact to influence the outcomes of land-sea connectivity (Sheaves 2009). Understanding the role of fish in land-sea connectivity is central to understanding their importance in the region, both because of the diversity of their linkages (Sheaves et al. 2007a; Nagelkerken et al. 2015) and because of the economic, social, and cultural importance of fish to people of the GBR region (Hutchings et al. 2019).

The coastal zone of the GBR region consists of a diversity of ecosystems and habitats that interact to function as a Coastal Ecosystem Mosaic (CEM) (Sheaves 2009). The component habitats and ecosystems of the CEM are many and varied (Figure 2) and include mangroves, seagrass, tidal flats, estuarine/fresh transitional wetlands, saltmarsh, oyster reefs, coastal littoral forests, and freshwater streams, among many others. The roles played by these individual components are vital; the connections between the components that support the diverse physical and biological interactions that occur in the CEM are also vital (Sheaves and Johnston 2008; Davis et al. 2012; Nagelkerken et al. 2015). Although many factors combine to enable this connectivity, freshwater flows and tidal pulses dominate these systems.

Freshwater flows, in the form of stream, overland, and ground water flows, mediate largely unidirectional connectivity from terrestrial to marine environments (Figure 2a). These flows are influential at connecting 'upstream' and terrestrial components to more 'downstream' components at local and large scales, but they also influence the nature of connectivity within marine areas because net seasonal outwelling is driven by freshwater flows (Ridd et al. 1997). Particularly in the dry tropical areas that comprise most of the GBR coast, these flows are very seasonal, to the extent that there are major shifts in the influence of terrestrial- versus marine-derived productivity on coastal marine food webs (Abrantes and Sheaves 2010). Tides are the second major facilitator of connectivity in the CEM (Figure 2b), with the regular pulsing of the tide particularly important for lateral connectivity (Wolanski et al. 2024), enabling marine organisms and water chemistries to enter intertidal areas on a predictable schedule. Both freshwater and tidal forces are critical in determining the physical conditions (e.g., dissolved oxygen, temperature, and salinity) throughout the various components of the CEM that produce conditions that either promote or impede connectivity at particular points in space and time.

The interaction of rainfall/streamflow and tides results in a complex fish assemblage that varies in context-specific ways across coastal ecosystems (Sheaves et al. 2007a; Bradley et al. 2017) and influences the way that fish use available coastal habitats (Bradley et al. 2020). Moreover, given the importance of rainfall and tides as major drivers of connectivity and the substantial differences in rainfall and tidal range along the coast of the GBR region (Figure 3), the details of connectivity vary substantially among locations.

Fish have always provided important ecosystem services to the people of the GBR region. For millennia, fish provided key sources of protein for indigenous people and feature prominently in their culture (Smith 1989; Cadet-James et al. 2017). Although these roles continue to be important, the increase in human population densities and the new social-economic landscape following European colonisation have added a range of new values of fish to people living in the GBR region. Fish now support a valuable commercial fishing industry servicing domestic and

DOI: 10.1201/9781003320425-13

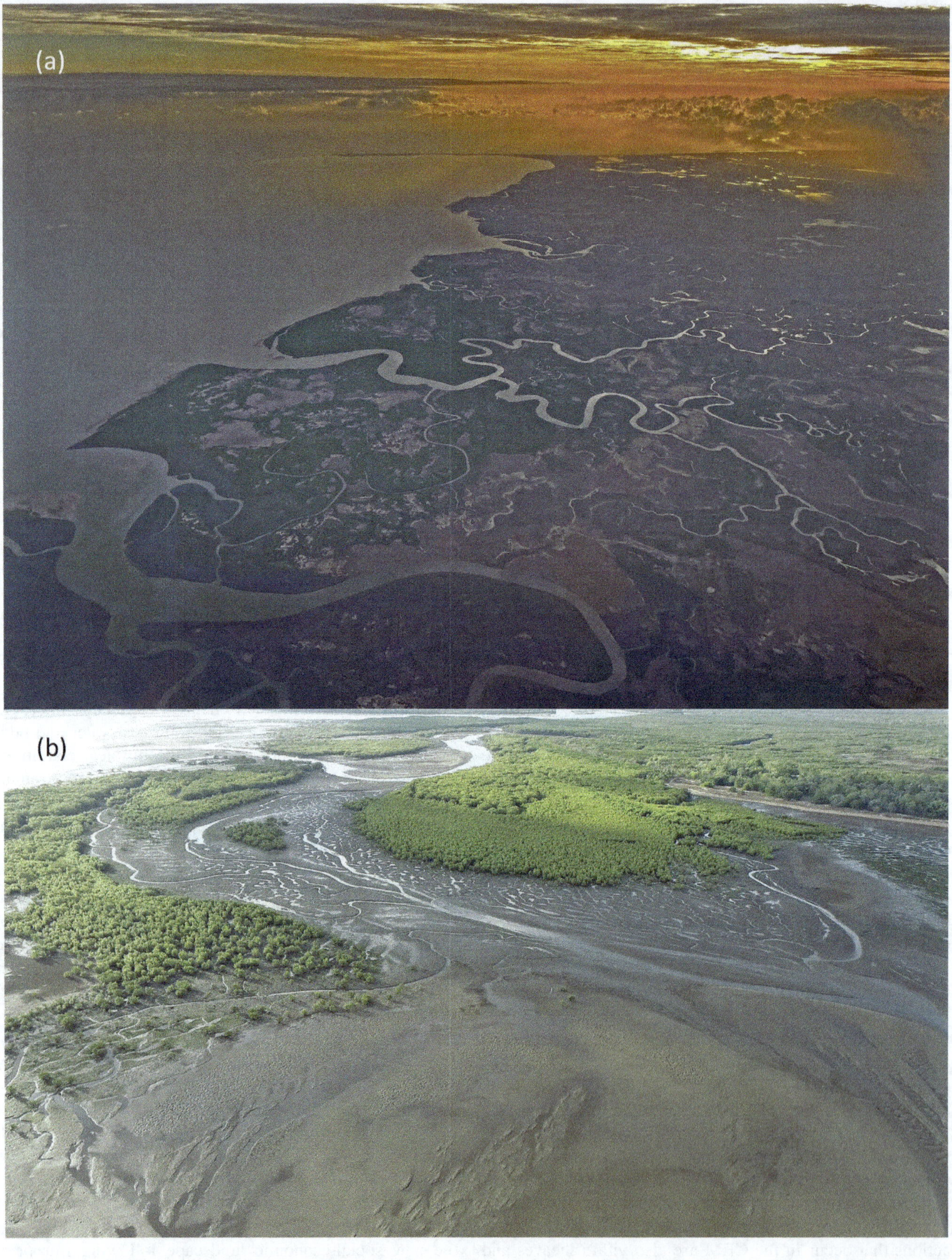

FIGURE 1 The GBR region's complex coastal ecosystems. (a) The mosaic of waterways, mangrove forests, saltmarshes, and saltpans comprising the coast of Queensland's Bowling Green Bay. (b) A close-up of the intertidal flats of Bowling Green Bay (lower left of part a) showing how the complexity of coastal ecosystems is manifest across different spatial scales.

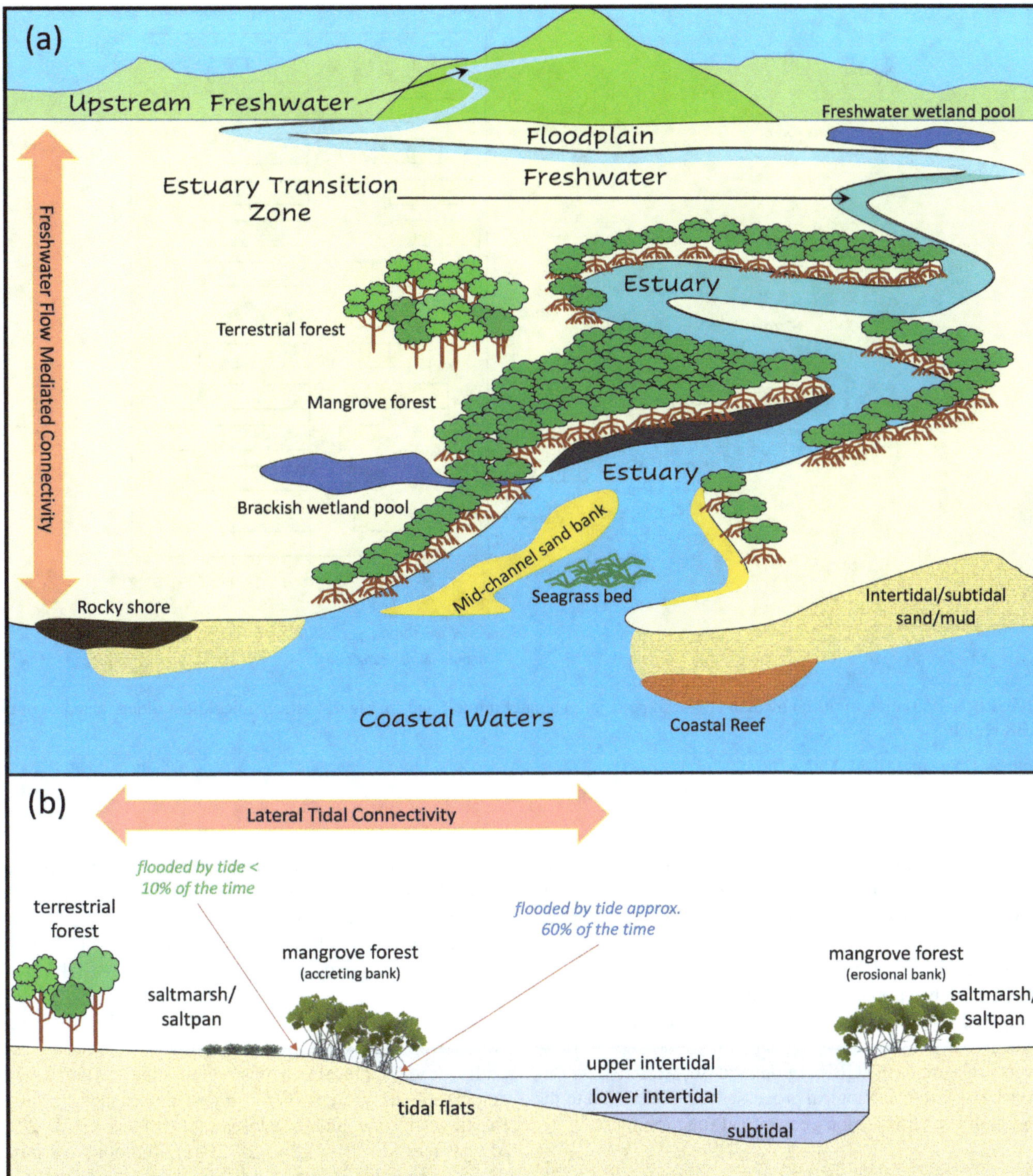

FIGURE 2 Two axes of connectivity across the diversity of ecosystems and habitats that comprise the GBR region's coastal ecosystem mosaic. (a) Unidirectional connectivity of terrestrial to coastal ecosystems created by seasonal freshwater flows. (b) Bidirectional connectivity across intertidal and tidal areas mediated by tidal patterns. Most components of the diagram are self-explanatory except for the 'Estuary Transition Zone', which comprises shallow upstream areas at the limit of tidal incursion where physical conditions vary between freshwater and marine flows dominated under the influence of northern Australia's strongly seasonal rainfall.

international export markets, a substantial international aquarium fish trade, and ever-expanding recreational fisheries (Chin et al. 2019). The narrative of the future value of fish in the GBR region becomes even more complex as the potential impact of anthropogenic activities (Kroon and Phillips 2015) and climate change (Sheaves et al. 2007c) on coastal fish populations becomes evident, as the relationship of fish with the region's expanding urban-industrial

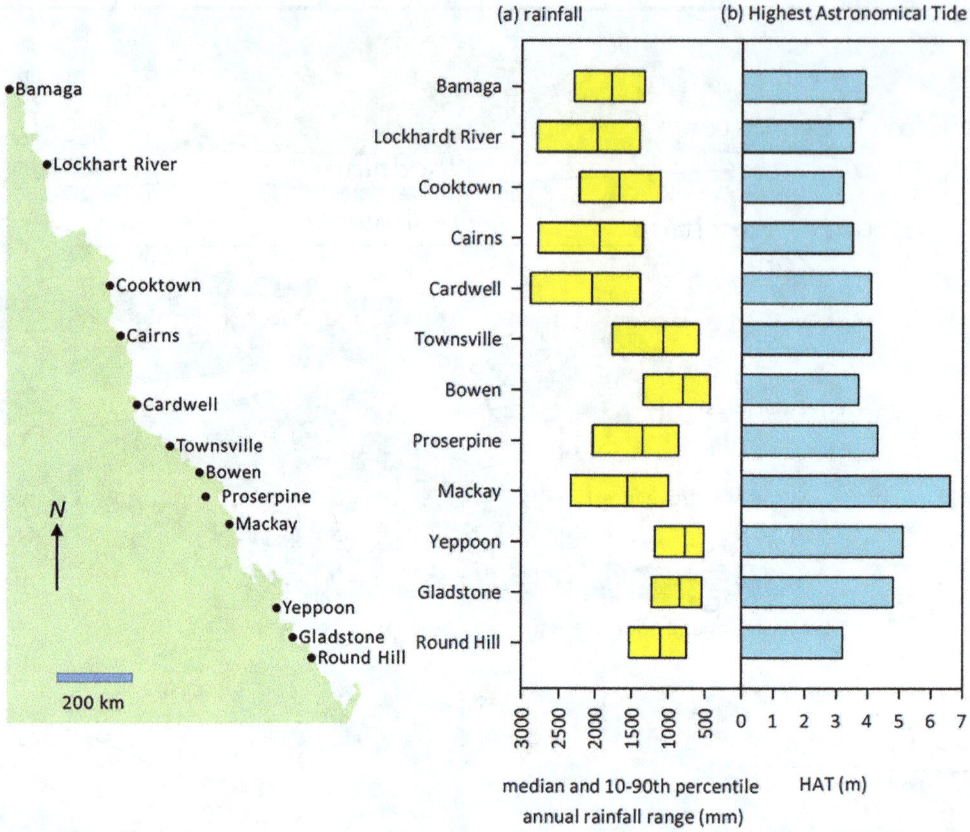

FIGURE 3 Variation in the (a) median (10–90th percentile) annual rainfall and (b) highest astronomical tides along the coast of the GBR region.

Source: Data from BOM 2022

seascapes becomes clearer (Bradley et al. 2023), and as the co-benefits of coastal restoration and blue carbon projects to support fisheries' stocks are considered (Waltham et al. 2019, 2024).

In this chapter, we (i) explore and document the nature of events in fish populations that influence land-sea connectivity, (ii) provide a comprehensive, integrated overview of the current state of knowledge about the outcomes of these events relative to the GBR region, and (iii) explore how certain events affect ecological processes that are crucial to the health of the GBR region's ecological functioning.

11.2. SOURCES OF ENERGY AND NUTRIENTS

Fish in coastal and inshore areas derive their energy and nutrients from both autochthonous and allochthonous sources (Abrantes and Sheaves 2010; Hitchcock et al. 2016) (Figure 4). Autochthonous sources involve production from both phytoplankton and from benthic producers (e.g., microphytobenthos (MPB), seagrass, and algae). Allochthonous inputs into coastal and inshore water are derived from terrestrial or freshwater environments (Abrantes and Sheaves 2008, 2010; Bell et al. 2014; Al-Najjar et al. 2020). To a large extent, allochthonous material is delivered via stream flow or seasonal flooding (Waterhouse et al. 2024), but it may also result from the tidal flooding of coastal wetlands (Abrantes and Sheaves 2008), from dust blown by offshore winds (Al-Najjar et al. 2020), or from groundwater flow (Valiela et al. 1978).

The balance of autochthonous and allochthonous inputs varies over space and time. Allochthonous sources are generally more important in upstream estuarine areas, with autochthonous sources increasing in dominance in downstream estuarine and coastal areas (Franca et al. 2011; Mwijage et al. 2018; Li et al. 2022). Temporal variation is common and complex; allochthonous inputs are particularly high in dry tropical coastal systems during flood pulse events in the wet season (Abrantes and Sheaves 2010), and autochthonous inputs vary depending on water clarity, salinity levels, and the availability of nutrients, which is itself affected by variations in stream flow (Hitchcock et al. 2016). The effect of hydrologic connectivity on energy and nutrients (Waterhouse et al. 2024) makes it a key factor in regulating the delivery of nutrients to coastal systems and in determining the level and composition of the energy and nutrients available for upload and offshore translocation (Abrantes and Sheaves 2010).

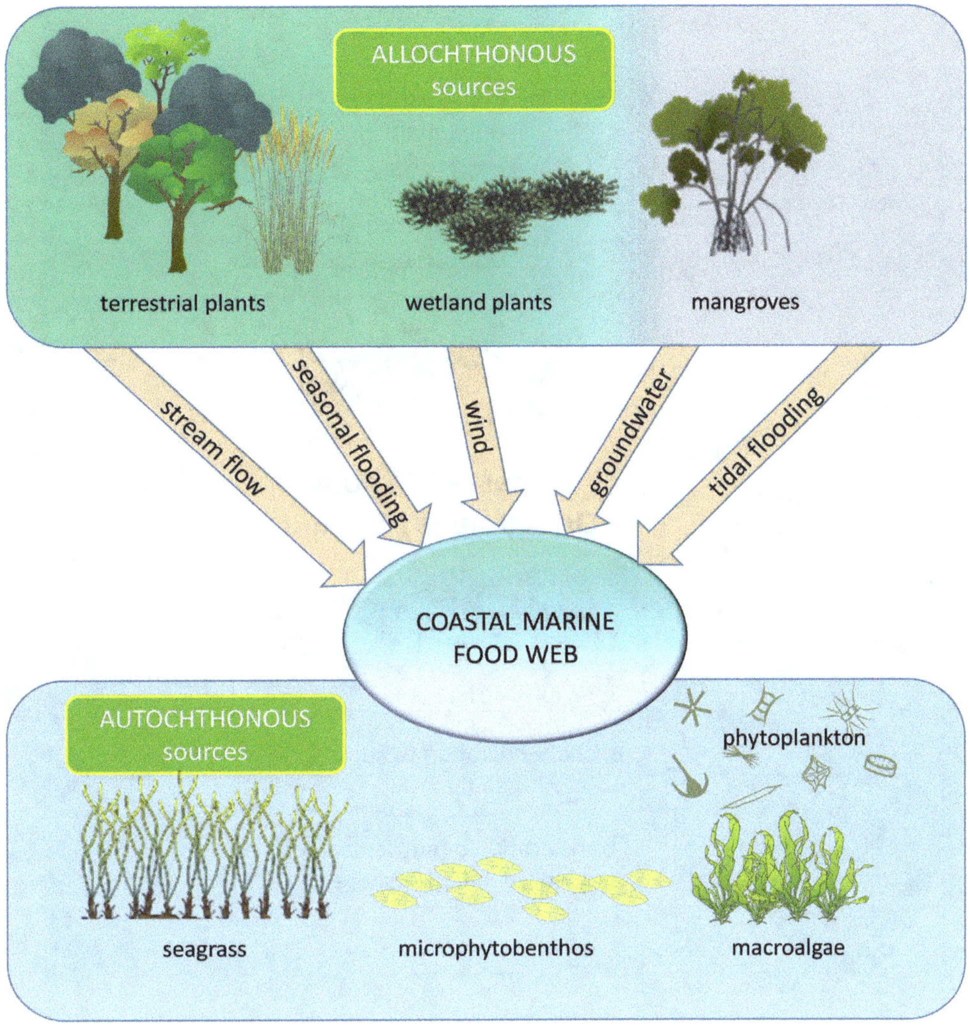

FIGURE 4 Allochthonous and autochthonous inputs to coastal marine food webs along GBR coasts.

11.3. EVENTS IN FISH POPULATIONS THAT INFLUENCE LAND-SEA CONNECTIVITY

Fish are mobile consumers that are extensively engaged in cross-boundary connectivity and nutrient/energy translocation within and among ecosystems (Nagelkerken 2009; Pittman 2017). Fish-enabled links between coastal and offshore waters are pervasive, with up to 80% of commercial species and their key prey around the world dependent on coastal systems (Blaber et al. 1995; Lellis-Dibble et al. 2008; Litvin et al. 2018). These linkages can be thought of as migrations at a variety of conceptual scales: ontogenetic migrations, spawning migrations, and intermittent migrations.

One of the strongest and most ubiquitous manifestations of fish-mediated land-sea connectivity takes the form of life-history or ontogenetic migration (Figure 5). These can be defined as gradual, predictable spatial displacements as individuals move to satisfy continually changing food and refuge requirements. Fish undergo massive changes in size as they grow through their life stages, from larvae a few millimetres long, through a series of juvenile stages, to adults that may be metres in length. Different stages can have particular habitat and environmental requirements (Lecchini and Galzin 2005b; Switzer et al. 2015) that drive movements ranging from local habitat shifts to large-scale cross-ecosystem migrations (Eggleston 1995; Lecchini and Galzin 2005a; Heupel and Simpfendorfer 2008; Heupel et al. 2015; Kingsford et al. 2024). These ontogenetic shifts mean that populations are often distributed across different habitats based on their size and age/life stage. Although some of this habitat differentiation can occur over local scales (e.g., from shallow to deeper waters across a coastal continuum), in many instances, the adults of a species inhabit deep coral reefs, while juveniles are confined to coastal and freshwater nursery grounds that are often tens of kilometres away from the adult habitats (Simpfendorfer and Milward 1993). These nursery grounds are vital to fish populations, providing food, protection from predators, and appropriate abiotic conditions. Large scale movements offer obvious connections that underpin land-sea connectivity, but even smaller scale movements play important roles in

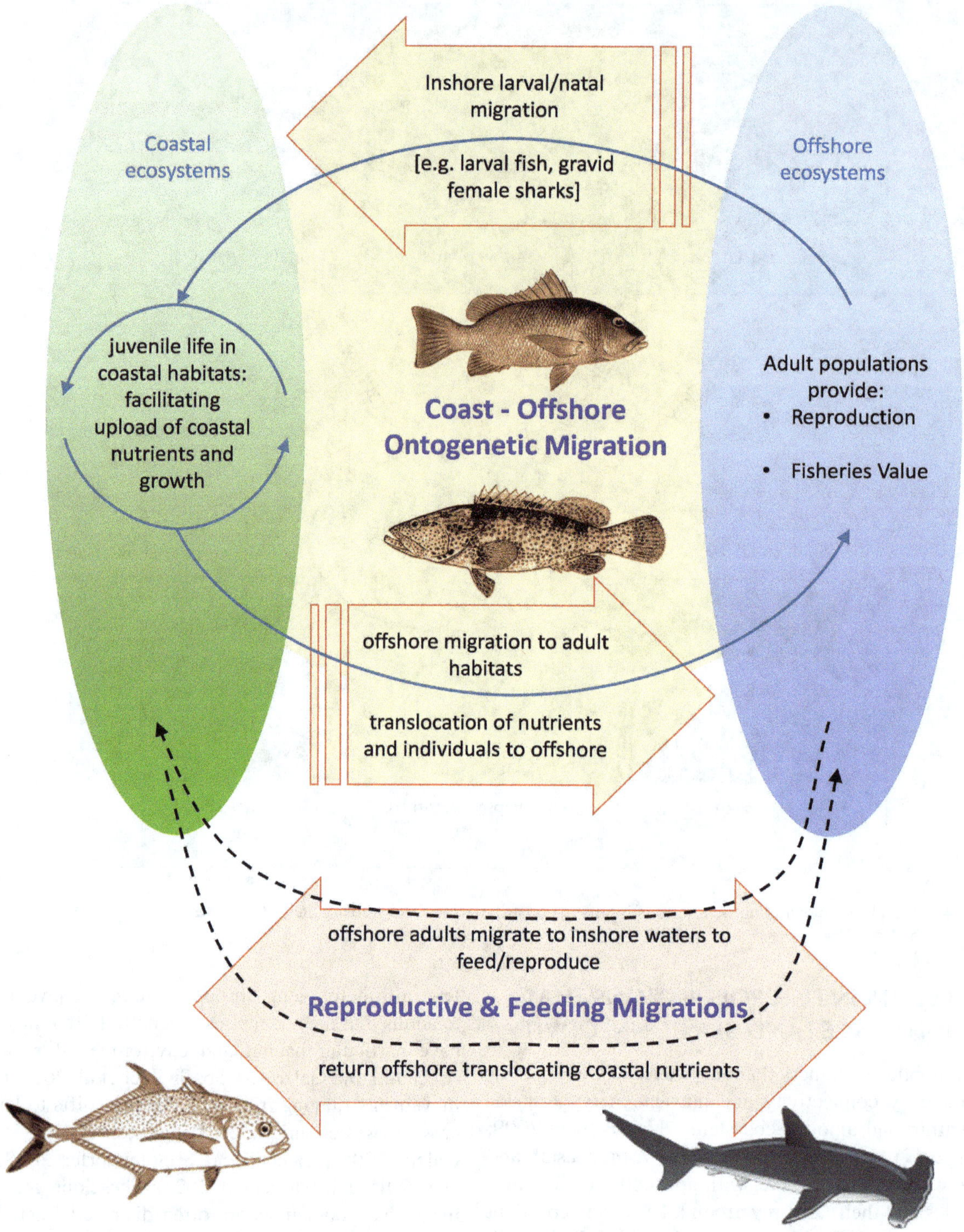

FIGURE 5 Fish-mediated land-sea connections.

determining exactly how the values of land-sea connectivity are derived (Sheaves et al. 2015).

The use of nursery grounds involves complex patterns of migration. For instance, tropical snappers of the genus *Lutjanus* commonly feature substantial spatial separations, with adults and spawning grounds on reefs and juveniles using a range of different coastal ecosystems at a range of scales (Baker et al. 2019). In the GBR region, Mangrove Jack, *Lutjanus argentimaculatus*, uses upper tidal estuaries as juvenile habitats to feed mostly on crabs and other small crustaceans (Sheaves and Molony 2000) before migrating to the reef as adults (Sheaves 1995). This strategy is also utilised by related species such as *Lutjanus fulvus* (Nakamura et al. 2008). Although the offshore-to-reef migration pattern of post-settlement fish is common when life history is viewed as a whole, the pattern may be reversed sometimes when viewed at shorter temporal scales. For instance, at some locations, during early juvenile stages, *Lutjanus johnii* move from coastal food webs to estuarine mangrove food webs to access their preferred prey (Tanaka et al. 2011). Indeed, at local scales, nursery value is rarely a function of a single habitat but rather conferred via a mosaic of interacting habitats (Nagelkerken et al. 2015; Litvin et al. 2018).

Elasmobranchs also commonly display ontogenetic migrations that facilitate land-sea connectivity. Along the coast of Queensland, large numbers of juvenile stingrays, such as the Australian whipray, Himantura australis, the cowtail ray *Pastinachus ater*, the mangrove whipray *Urogymnus granulatus*, and the giant shovelnose *Glaucostegus typus*, make extensive use of shallow coastal nurseries (Crook et al. 2022; Martins et al. 2022), a pattern observed around the world in which stingrays provide important links between shallow coastal waters and offshore ecosystems (Aguiar et al. 2009; Ajemian and Powers 2016). In a similar way, juvenile bull sharks make extensive use of freshwater and estuarine nurseries (Heupel et al. 2010) before taking up a roving coastal lifestyle as adults (Heupel and Simpfendorfer 2011; Heupel et al. 2015), while blacktip reef sharks move from coastal habitats to coral reefs as they grow (Chin et al. 2013). In addition, key prey species, such as prawns, rely on coastal ecosystems for nutrition (Sheaves et al. 2012; Abrantes et al. 2019) and undergo regular offshore ontogenetic migrations (Balston 2008).

Although critically intertwined with life-history connectivity, spawning migrations represent specific, short-term events that can connect coastal and offshore ecosystems. Spawning migrations are predictable and range from short movements to particular spawning locations that maximise the opportunity for offshore dispersal of larvae (Sheaves et al. 1999; Kingsford et al. 2024) to large-scale movements to access the specific conditions required for juvenile development. Spawning aggregations can influence offshore reef ecosystems in a variety of ways, including producing eggs and larvae that provide vital prey resources for planktivores and concentrating adult spawners (Fulton et al. 1999) that provide a prey resource for large predators (Espinoza et al. 2016).

Intermittent migrations include the activities of roving predators that move between coastal and offshore areas to take advantage of assemblages of preferred prey. This includes predators that target predictable offshore spawning aggregations and seasonal migrations (Graham and Castellanos 2012). Although we do not fully understand the drivers of these intermittent migrations, they probably include both purposive migrations to arrive at advantageous locations at specific times (Lubitz et al. in review) and more random movements to find assemblages of prey whose spatiotemporal occurrence is more haphazard.

Predictable connectivity events have obvious underlying drivers, for instance, the need to access stage/size-appropriate resources or the need to move to a site of mass spawning. For instance, the release of crab zoeae in saltmarshes during periods of inundation and cue feeding migrations of planktivorous glass fish (Ambassidae) to the flooded saltmarsh (McPhee et al. 2015) link marine ecosystems to terrestrial productivity. The details and drivers of connection events vary and are modified by the duration and frequency of tidal connections. Despite these differences, connection events serve to fulfil a species' particular needs. The ability of different species to access these events is constrained at some level by the range of physical conditions under which the species can operate. For example, the occurrence of juvenile snappers and groupers in tropical estuaries that feature extreme seasonal physical variability (Sheaves 1996, 1998) indicates formidable abilities to live under extreme physical conditions that would exclude less-tolerant species. Similarly, the distributions of closely related species within estuarine nurseries are modified by species-specific differences in physiological tolerances to such things as salinity (Sheaves 1996) and dissolved oxygen (Dubuc et al. 2017; Mattone et al. 2022). Consequently, even if some areas seem to be suitable nurseries for juveniles, physical constraints can limit occupancy, which means that a potentially valuable area for terrestrial nutrient uptake can actually act as a nutrient sink.

Although the sub-adults of many species move from coastal nurseries to the GBR, thus taking the coastal and terrestrial nutrients with them in the form of biomass export, individuals and nutrients also move in the reverse direction. For instance, fish larvae (Sheaves et al. 2015; Kingsford et al. 2024) and sharks undertake reproductive migrations (Espinoza et al. 2016) from the GBR to coastal systems, creating an influx of marine offshore individuals and nutrients into coastal systems.

11.4. ECOLOGICAL OUTCOMES OF LAND-SEA CONNECTIVITY FOR THE GBR

Land-sea connectivity results in the complex interchange of individuals and biomass between terrestrial/coastal ecosystems and offshore waters that mediates the translocation of productivity within and among ecosystems (Nelson et al. 2013; Litvin et al. 2018). Ecological outcomes of land-sea connectivity are substantially interlinked, with outcomes from the provision of critical nurseries as part of life-history connectivity, spawning migrations, and the intermittent and seasonal movements of roving species inextricably bound to trophic connectivity. Although this connectivity is not unidirectional, it is the movement of individuals, biomass, and nutrients from coastal/terrestrial ecosystems to offshore areas, which is often most pronounced and has the most obvious and far-reaching ecological consequences. This outwelling is dependent on the complex interlinking of physical and biological processes within and among the component habitats of coastal/terrestrial ecosystems. This leads to emergent properties that magnify the individual values of mangroves, seagrass, tidal flats, saltmarsh, oyster reefs, littoral forests, freshwater streams, and other habitats.

The movement of fish from coastal marine areas to reefal systems generates a diversity of outcomes for individuals, populations, and ecosystems. These movements are facilitated by events such as the supply of new individuals to adult populations from coastal nursery grounds (Sheaves 1995, 2009) and the ecological linking of coastal and reef waters by the seasonal migration of large predators such as sharks (Espinoza et al. 2016) that prey upon aggregations of spawning reef fish. This interlinking provides for the varied life-history needs of individuals, allowing different species to utilise resources that provide for particular stage-specific requirements for refuge, food, and physical conditions. At the same time, the interlinking supports the resilience and dispersal of populations and enables species with diverse physical and biological needs, trophic identities, and functional roles to co-exist across space and time. However, the specific outcomes of interlinking are regulated by cross-shelf position, with the details of reef fish communities influenced by their proximity to the coast (Hoey and Bellwood 2008; Hoey et al. 2013). Thus, land-sea connectivity mediated by fish and the proximity to a large continental land mass result in the GBR possessing biological, ecological, and functional profiles much richer, more diverse, and more robust than would be possible if dependent on reefal ecosystems alone.

11.4.1. Trophic Outcomes

Trophic connections are a pervasive and obvious outcome of fish-mediated land-sea connectivity in GBR waters. Nitrogen and phosphorous availability drive phytoplankton and algal growth on the GBR. However, reef communities exhibit high productivity despite low nutrient concentration in the surrounding waters (Bell et al. 2014). This productivity is sustained by high rates of nutrient recycling and resupply from external sources. Coastal ecosystems also make substantial trophic contributions to off-shore waters (Eyre and McKee 2002; Herrmann et al. 2015). These take the form of both nutrients derived from benthic and planktonic productivity in intertidal and shallow subtidal waters (Leguerrier et al. 2003; Abrantes and Sheaves 2009; Lee et al. 2019; Kwon et al. 2020; Baker et al. 2021) and the incorporation of terrestrial and freshwater productivity into marine systems (Caddy and Bakun 1994; Abrantes and Sheaves 2008). The biological transfer process is vital to the extent that biologically mediated organic carbon fluxes can dominate physical transport fluxes (Kemp et al. 1997). These biological transfers are facilitated by the feeding activities of fish. Much of this feeding occurs in biologically rich intertidal areas that are utilised by a variety of fish by using a range of different foraging strategies cued to the rise and fall of the tide (Figure 6).

FIGURE 6 Common tidally referenced spatiotemporal distribution patterns of fish that use intertidal areas, based on 5 representative species, with arrowed lines indicating temporal movements with the tide: (i) the mangrove crab feeder *Lutjanus argentimaculatus* that makes tidal migrations between low tide refuge structures once the mangrove forest is tidally inundated (solid black line), (ii) the intertidal benthos feeder *Sillago ciliata* that migrates with the tide, remaining in water shallow enough to limit risk from most piscivorous nekton throughout the tide (dashed black line), (iii) the structure-associated *Epinephelus coioides* that remains associated with complex structures along steep accreting banks moving vertically up bank with the tide (yellow lines), (iv) the cursorial predator *Caranx papuensis* that roves in search of prey aggregations and targets predictable 'hot spots' (e.g., the mangrove edge as the falling tide forces fish to move from cover) (blue dashed line), and (v) the lie-in-wait predator *Platycephalus fuscus* that moves into the intertidal, buries in an advantageous position, and attacks prey with a sudden lunge when they approach in range (black dashed line).

11.4.2. The Specific Role of Fish in Nutrient Translocation

The movement of fish across spatial gradients is a key facilitator of the transfer of nutrients and energy throughout seascapes (Heupel et al. 2014), promoting the upload and integration of coastal/terrestrial productivity into marine food webs (Davis et al. 2014; Sheaves et al. 2015). This translocation is obvious during ontogenetic migrations, where juveniles in coastal nurseries consume location-specific prey at each life-history stage (Nakamura et al. 2008) before transporting biomass, energy, and nutrients accumulated in coastal environments to offshore waters.

Some offshore translocation pathways are more subtle as illustrated by the example of juvenile stingrays, *Himantura australis* and *Pastinachus ater*, in coastal GBR waters, which are involved in offshore nutrient translocation in a variety of ways. These juvenile stingrays feed extensively on coastal intertidal sand flats, where they excavate the sediment to extract invertebrate prey (Crook et al. 2022). In addition to the direct nutritional support for the juvenile stingrays, this excavation feeding resuspends large areas of the sediment by freeing buried organic carbon that can then be exported by tides and currents or provide nourishment for organisms that reintegrate it into food-webs (Flowers et al. 2021). At the same time, the growing stingrays represent a substantial biomass of nutrients and energy extracted from coastal food webs that can be consumed by roving predators, such as Great Hammerheads sharks, *Sphyrna mokarran*. These predators undertake regular foraging migrations between coral reef flats and shallow inshore habitats to feed on juvenile rays (Lubitz et al. in review), thus facilitating additional transfers of energy and nutrients (Cliff 1995; Dale et al. 2011) in coastal and reef waters and resuspending buried organic material.

In a parallel way, the roles of large migratory marine fauna in coastal-offshore nutrient and energy transfer may be pervasive and complex (Bauer and Hoye 2014). For example, bull sharks, *Carcharhinus leucas*, are coastal top predators that make extensive migrations along Australia's East Coast, inhabiting reefs in the GBR region ((Heupel et al. 2015; Espinoza et al. 2016). Migratory females move to coastal nurseries to pup each reproductive cycle (Espinoza et al. 2016), with juveniles utilising estuarine and freshwater nursery grounds (Heupel and Simpfendorfer 2011). Being an oviparous species, bull shark embryos receive nutrients directly from the mother via a placental-like connection (Wourms and Demski 1993). The deposition of these nutrients is evident in the isotopic values from the tissue of small juvenile sharks that resemble those of the mother (Olin et al. 2011). Consequently, nutrients acquired in offshore systems are directly transferred to coastal systems. However, the nutrient translocation goes both ways as juvenile bull sharks gradually move to occupy areas further offshore over ontogeny, providing a pathway for offshore trophic transfer. To add to the complexity, the use of coral reefs by bull sharks has been linked to seasonally abundant food availability in the form of fish spawning aggregations (Espinoza et al. 2016). Thus, adult bull sharks continue to be involved in complex trophic translocations by bringing nutrients into reefal environments as they migrate in and removing nutrients as they feed on spawning fish.

11.4.3. Trophic Importance of Fish beyond Energy Translocation

The trophic importance of fish goes far beyond nutrient and energy translocation. Fish perform vital roles in food web interactions (Meziane and Tsuchiya 2000; Leguerrier et al. 2003), and by migrating between ecosystems at a variety of scales (Krumme 2009; Nelson et al. 2013), they connect food webs. In this way, fish not only bring with them nutrients and embodied productivity but also perform important functions in the ecosystem, such as herbivory and predation (Hoey and Bellwood 2008; Olds et al. 2012). This results in ecosystems that are functionally linked across space (sometimes called 'meta-ecosystems'). It is thought that this functional connectivity is a key aspect of ecosystem resilience (Cumming 2011) and of particular importance in coastal seascapes (Olds et al. 2018). This means that land-sea connectivity at all levels is a key enabler of ecological outcomes (Sheaves 2009).

11.5. CONSEQUENCES OF CHANGES TO LAND-SEA CONNECTIVITY FOR ECOLOGICAL FUNCTIONING

The proximity of coastal areas to human activities, their low-lying geography, and the direct influence of tide and climate on them (Figures 2 and 3) make coastal ecosystems particularly vulnerable to the joint influences of climate and anthropogenic change (Lotze et al. 2006; Seneviratne et al. 2012; Sheaves et al. 2016). Because of the key roles that coastal ecosystems play in the lives of fish, the increasing pace of anthropogenic disturbance and climate change (Syvitski et al. 2005) is likely to present serious challenges to the integrity of fish-mediated land-sea connectivity in the GBR region. The challenges are interactive, with anthropogenic factors likely to exacerbate the effects of climate change (Lotze et al. 2006; Seneviratne et al. 2012) but also providing opportunities to mitigate its impacts. However, a dearth of comprehensive research means that we currently lack adequate understanding to either precisely predict the results of change or respond to these challenges with certainty. These challenges to land-sea connectivity in fish population, together with stressors that act directly on the GBR (e.g., bleaching, crown-of-thorns starfish outbreaks, and temperature rise), can substantially impair the GBR ecosystem health and its overall resilience in the long term (GBRMPA 2019).

11.5.1. Major Challenges

11.5.1.1. Climate Change

Climate change is likely to affect coastal habitats directly and rapidly. Sea level rise will profoundly change the current distribution and extent of many key coastal habitats, with habitats such as mangrove and saltmarsh dependent on the duration and depth of tidal flooding (Bearup and Blasius 2017). Without human-constructed barriers, intertidal habitats, such as mangroves, would be able to gradually migrate landwards in many areas and likely maintain ecological functioning over the long term (Borchert et al. 2018). However, barriers are common in the region (Sheaves et al. 2007b; Sheaves et al. 2014; Kroon and Phillips 2015), and in many areas, steep topography will limit landward habitat migration. Moreover, impacts to supralittoral habitats, such as the dieback of coastal Melaleuca forests (Saintilan et al. 2019), pose poorly assessed threats to nutrient flows into coastal waters.

Elevated temperatures due to climate change will inevitably increase stress on fish and their prey, particularly in shallow subtidal and intertidal areas (Hobbs and McDonald 2010) either as a direct effect of heat stress or because of the effect of temperature on dissolved oxygen solubility (Iriarte et al. 2010). Changes in the amount, periodicity, predictability, and interannual variability of rainfall will profoundly influence the physical nature of coastal marine environments, with the potential of shifting the balance between hyposaline- and hypersaline-adapted fish assemblages (Sheaves et al. 2007b). Variations in rainfall will also influence the inflow of allochthonous nutrients and anthropogenic pollutants.

Together, the cumulative effects of sea level rise, increasing temperatures, and more variable rainfall are likely to change the nature of coastal habitats (Hughes 2003; Sheaves et al. 2007c; Saintilan et al. 2019), induce range shifts (Sorte et al. 2010; Bond et al. 2011), change the viability of nursery grounds (Hughes et al. 2015), food-web efficiency, and fish production (Lefebure et al. 2013), and influence the spawning of temperature-sensitive species (Sheaves 2006). These effects may be moderated for many of the coastal species involved in coastal-reef connectivity because of their tolerance to extreme environmental conditions (Molony and Parry 2006; Dubuc et al. 2019; Mattone et al. 2022). However, tolerance to different parameters is species-specific (e.g., salinity, Molony and Parry 2006; temperature, Waltham and Sheaves 2017), with variations in response to the mix of environmental conditions faced determining the details of the distribution of co-occurring species (Mattone et al. 2022).

11.5.1.2. Anthropogenic Change

Queensland's population is clustered along its coast, exposing coastal waters to a variety of anthropogenic impacts ranging from dredging and landscape modification (Burt 2014; Martín-Antón et al. 2016), pollution (Shriadah 2000; Fong et al. 2020), and water quality impacts from human industry (Ibrahim and Eltahir 2019) to coastal eutrophication (Nicastro and Bishop 2013) and overfishing. Sixty percent of human-engineered structures in the GBR region are sited close to the coast, while breakwaters, training walls, and jetties occupy approximately 10% of the coastline's linear extent (Waltham and Sheaves 2015). In addition, Queensland's Wet Tropics World Heritage Area has some 130,000 hectares of its coastal catchments converted to sugar cane (Dale et al. 2017), although the rate of wetland conversion to agricultural use has slowed in recent decades (Canning and Waltham 2021). In addition, there are several large urban areas situated close to the coast. Together, these human activities, in close proximity to the coast, exert substantial pressures on coastal ecosystems (Nicholls et al. 2007; Espinoza et al. 2016). Coastal aquaculture is also increasing rapidly, which brings with it loss of habitat and the potential for effluents to reduce water quality (Wolanski et al. 2000; Pearson et al. 2021).

Anthropogenic challenges to the coastal ecosystems of the GBR are diverse and include such things as the following.

Overfishing: Overfishing also poses risks for the functioning of land-sea connectivity. Fishing-related declines in key fish stocks in the GBR region are increasing, with notable declines in the stocks of important fisheries species such as Pink Snapper and Chrysophrys auratus (Wortmann et al. 2018) in the southern GBR and Spanish Mackerel, *Scomberomorus commerson* (Tanimoto et al. 2021). These are large, long-living species, and declines in their populations pose risks for the nature and intensity of predator-prey relationships, trophic cascades (Ripple et al. 2016), and ultimately, the proper functioning of fish-mediated land-sea connectivity. Impacts can be particularly severe if fishing is focused on spawning aggregations (Fulton et al. 1999) or if fisheries activities impact on early life-history stages (Sumpton and Jackson 2005).

Direct habitat destruction, modification: Intense human activity results in substantial direct habitat destruction and modification (Walker and Kendrick 1998), which leads to changes in the availability of key habitats and to food web instability. Coastal habitat destruction or disturbance often results in substantial flow-on effects, such as the widespread development of acid sulphate soils that, when disturbed, can lead to coastal acidification resulting in fish kills and dissolved oxygen depression (Cook et al. 2000).

Road and infrastructure construction—connectivity barriers: Coastal urbanisation, industrialisation, and agriculture require substantial infrastructure development, ranging from road networks to dams, bunds, and weirs (Hyland 2002; Boys et al. 2012; Kroon and Phillips 2015), all with the potential to modify the connectivity needed for the successful utilisation of coastal waters by fish (Boys et al. 2012; Sheaves et al. 2014; Kroon and Phillips 2015). Substantial barriers can have far-reaching impacts on coastal fish assemblages, particularly in the dry tropical climate that characterises much of the GBR region. Dams and weirs can reduce or eliminate stream flow during dry years, accentuating differences in flow regimes between wet and dry climatic cycles that lead to sustained changes to fish assemblages and ecological processes (Sheaves et al. 2007b). Moreover, a substantial part of the vulnerability of coastal ecosystems to climate change is attributable to human-emplaced barriers that prevent self-adaptation to sea-level rise (Sheaves et al. 2016).

Siltation due to run-off: Agriculture and erosion from mine sites also significantly contribute to stream sediment loads, with an approximately six-fold increase in the sediment load of some major rivers since regional development began (Bartley et al. 2003). In unregulated rivers, this can lead to substantial changes in the distribution and movement of sediments by altering both the relative mix of habitat types and the progression and development of habitats over time. In contrast, rivers, dams, and weirs trap sediments (Pringle 1991), often alleviate additional sediment loads, but bring a different set of changes as sediments are washed from systems without being replaced, which fundamentally alters the nature and mix of habitat types.

Pollutants (agricultural and domestic): Freshwater flows also inject extraneous nutrients (Mitchell et al. 1997) and pollutants (O'Brien et al. 2013) into GBR catchments. In the heavily agriculturalized catchments, such flows can contain substantial levels of pesticides and herbicides (Haynes et al. 2000; Johnson and Ebert 2000; O'Brien et al. 2013), but pollutants from many other sources, such as coal dust, heavy/trace metals, seeps from acid soils, marine debris, microplastics, pharmaceuticals, petroleum hydrocarbons (Kroon et al. 2020), and aquaculture (Pearson et al. 2021), are also present. Increasingly, human activities such as agriculture, development, aquaculture, and waste-water treatment are all adding to allochthonous nutrient flows. Substantial additional nutrient inputs can lead to coastal eutrophication (Fabricius 2005) that can adversely impact coastal food webs, impacting fish habitats and damaging land-sea connectivity. These pollutants can impact pelagic food webs that support coastal fish (Lefebure et al. 2013). For instance, they can greatly reduce seagrass extent and biomass, which directly impacts fish that rely directly on seagrass (Nelson et al. 2012) and greatly reduces fish-mediated energy and nutrient exports (Heck et al. 2008). There have been extensive negative impacts on the GBR region's seagrass beds in recent years, with severe tropical storms and associated floods exacerbating existing human-induced stressors (Coles et al. 2015).

Together, these many potential sources of anthropogenic impacts represent a broad spectrum of risks for coastal marine ecosystems and the fish that they support. These risks range from habitat loss and degradation, food web impairment, pollution, and the loss of key connectivities to fundamental changes in the complex physical and biological factors that support the coastal phases of the life cycles of many fish species that support fish and the coastal-offshore connectivity (Sheaves et al. 2021a). These risks interact with, and are likely to be intensified by, the diver effects of climate change on the region's coastal ecosystems (Sheaves et al. 2007c).

11.6. ENSURING THE FUTURE OF FISH-MEDIATED LAND-SEA CONNECTIVITY

11.6.1. Problems with the Current Biological Knowledge Base

Despite the importance of fish-mediated land-sea connectivity to the flow of terrestrial/coastal nutrients and energy and to the resupply of juveniles to fish populations of offshore waters, many details of the way fish use coastal ecosystems and the factors that constrain this use are poorly understood. For instance, many species targeted by fisheries recruit to juvenile nursery areas at around 10 mm in length and leave a few years later as a sub-adult of 250–500 mm. This rapid, orders-of-magnitude growth means that juveniles need access to a constantly changing suite of prey specific to their stage/size-specific requirements (Baker and Sheaves 2005) and the availability of refuges at scales appropriate to their changing physical size; all the time, they need to remain in areas that they are physiologically adapted to occupy. These changes require subtle, or sometimes extensive, changes of spatial location and habitat. We know in a general sense that many species require shallow water habitats during their early life history, but there are very few cases where the specifics of stage-specific utilisation have been detailed (Litvin et al. 2018).

Queensland's population and associated urban, agricultural, and industrial activities are concentrated along its coastline. This concentration of anthropogenic pressures, together with the incipient risks of climate change, mean that species that engage in ontogenetic migration beginning in shallow coastal waters have at-risk life cycles (Figure 7), which makes the development of detailed stage-specific understandings critical for these species. This knowledge is vital to those managing the GBR region's coastal ecosystems, to the success of restoration and revitalisation projects, to developing a precise understanding of the scale and consequences of coastal habitat and environmental change, and ultimately, to the resilience of coastal-offshore connectivity. Ensuring that fish life histories, particularly

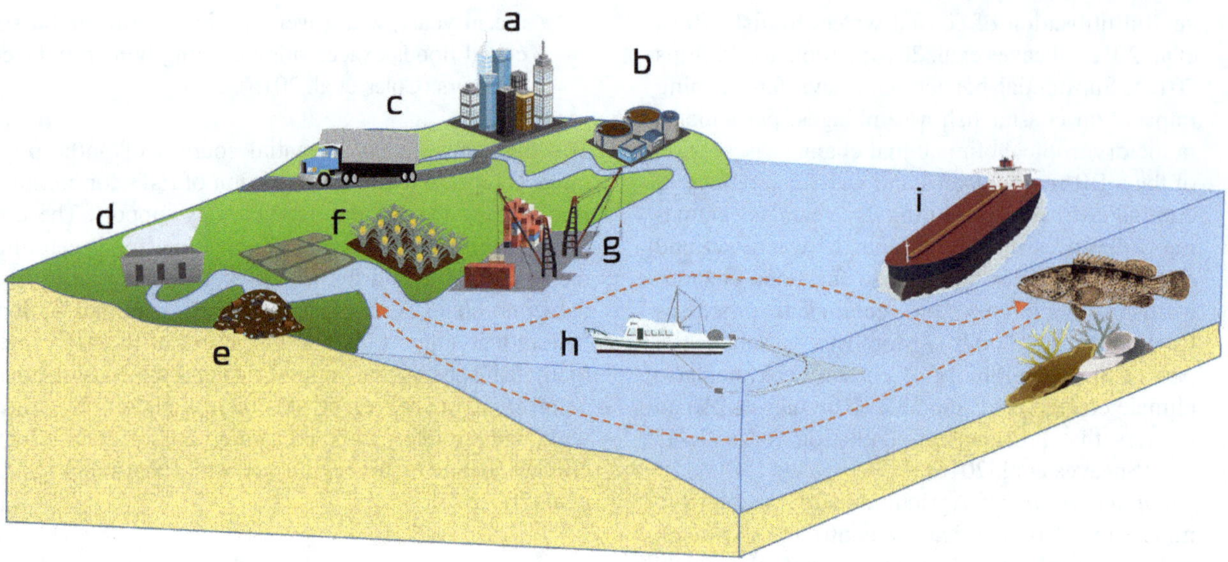

FIGURE 7 At-risk life cycles. Species such as the estuary cod, *Epinephelus coioides*, inhabit offshore waters as adults but rely on shallow coastal waters and estuaries as nursery grounds. This life-cycle reliance on coastal nurseries places vulnerable juvenile stages in habitats that are exposed to numerous risks from many anthropogenic activities. These include (a) urban development, (b) urban wastes, (c) road and rail transport and the ecosystem impacts of road and rail crossings, (d) the loss of upstream connectivity and altered hydrology due to dam construction, (e) rubbish disposal, (f) aquaculture and agriculture, (g) port construction and operation, (h) habitat destruction and catch during commercial and recreational fishing activities, and (i) shipping.

the requirements of different life-history stages, are understood, recognised in management planning, and included in management strategies, is a fundamental need. Full understanding represents a major knowledge gap, so it is imperative that research resources are focused on developing this fundamental knowledge.

11.6.2. Approaches to Support Functionally Robust and Dynamic Fish-Mediated Land-Sea Connectivity

Supporting functionally robust and dynamic fish-mediated land-sea connectivity goes beyond the development of precise and detailed scientific understanding. In the broadest sense, it is vital that the importance of land-sea links to the health of the GBR are appreciated by managers and the general public alike and that the maintenance of robust land-sea linkages becomes a central tenant of GBR management. Fish have an important role to play in this; they are both crucial components and vectors of land-sea connectivity and a highly recognisable group of organisms that can stand as impactful icons for publicising its importance. Accordingly, it is important that the role that fish play in land-sea connectivity is fully understood, clearly communicated to all sectors, and included in decision making. Although little information on connectivity was available to support rezoning in 2003, the approach used was at least partially successful in maintaining connectivity (Bode and Day 2020), but a more complete understanding is vital to validating current zoning decisions and fine-tuning them in the future.

Along with this knowledge and appreciation of fish-mediated land-sea connectivity, it is important to develop a detailed and spatially explicit understanding (sensu Nagelkerken et al. 2015) of the likely impacts of increased urbanisation and industrialisation of coastal seascapes on this connectivity. This will provide the basis for understanding how urban-industrial seascapes can be managed to optimise environmental outcomes. Knowledge of land-sea connectivity is also crucial in determining the most appropriate responses to the predicted impacts of climate change.

Ecosystem restoration is seen as a key to reversing environmental change (Creighton et al. 2015). However, success is often tenuous (Sheaves et al. 2014), so precise, well-considered and accountable decision making that optimises the benefits of restoration projects is required (Sheaves et al. 2021b). An important consideration is that co-benefits of restoration projects are maximised. For example, mangrove restoration aimed at blue carbon benefits can provide far-reaching benefits for coastal fish populations if care is taken to ensure that restoration results in quality juvenile fish habitats. Ensuring the integrity and health of coastal nursery habitats is one of the most prominent of these goals of restoration, especially when the role of these habitats in supporting vital land-sea connectivity is considered.

REFERENCES

Abrantes K, Sheaves M (2008) Incorporation of terrestrial wetland material into aquatic food webs in a tropical estuarine wetland. Estuarine Coastal and Shelf Science 80:401–412

Abrantes K, Sheaves M (2009) Food web structure in a near-pristine mangrove area of the Australian Wet Tropics. Estuarine Coastal and Shelf Science 82:597–607

Abrantes K, Sheaves M (2010) Importance of freshwater flow in terrestrial-aquatic energetic connectivity in intermittently connected estuaries of tropical Australia. Marine Biology 157:2071–2086

Abrantes KG, Sheaves M, Fries J (2019) Estimating the value of tropical coastal wetland habitats to fisheries: Caveats and assumptions. PLoS One 14:e0215350

Aguiar AA, Valentin JL, Rosa RS (2009) Habitat use by Dasyatis americana in a south-western Atlantic oceanic island. Journal of the Marine Biological Association of the United Kingdom 89:1147–1152

Ajemian MJ, Powers SP (2016) Seasonality and ontogenetic habitat partitioning of cownose rays in the Northern Gulf of Mexico. Estuaries and Coasts 39:1234–1248

Al-Najjar MA, Munday C, Fink A, Abdel-Moati MA, Hamza W, Korte L, Stuut J-B, Al-Ansari IS, Al-Maslamani I, de Beer D (2020) Nutritive effect of dust on microbial biodiversity and productivity of the Arabian Gulf. Aquatic Ecosystem Health and Management 23:122–135

Baker R, Abrantes K, Feller IC (2021) Stable isotopes suggest limited role of wetland macrophyte production supporting aquatic food webs across a mangrove-salt marsh ecotone. Estuaries and Coasts 44:1619–1627

Baker R, Barnett A, Bradley M, Abrantes K, Sheaves M (2019) Contrasting seascape use by a coastal fish assemblage: A multi-methods approach. Estuaries and Coasts 42:292–307

Baker R, Sheaves M (2005) Redefining the piscivore assemblage of shallow estuarine nursery habitats. Marine Ecology-Progress Series 291:197–213

Balston J (2008) Climate impacts on barramundi and banana prawn fisheries of Queensland tropical east coast. Environmental flows for sub-tropical estuaries: Understanding the freshwater needs of estuaries for sustainable fisheries production and assessing the impacts of water regulation. Final Report FRDC Project (2001/022), Fisheries Research and Development Corporation, 63–77

Bartley R, Henderson A, Prosser I, Hughes A, McKergow L, Lu H, Brodie J, Bainbridge Z, Roth C (2003) Patterns of erosion and sediment and nutrient transport in the Herbert River catchment, Queensland. Consultancy Report, CSIRO Land and Water, Clayto.

Bauer S, Hoye BJ (2014) Migratory animals couple biodiversity and ecosystem functioning worldwide. Science 344:1242552

Bearup D, Blasius B (2017) Ecotone formation induced by the effects of tidal flooding: A conceptual model of the mud flat-coastal wetland ecosystem. Ecological Complexity 32:217–227

Bell PRF, Elmetri I, Lapointe BE (2014) Evidence of large-scale chronic eutrophication in the Great Barrier Reef: Quantification of Chlorophyll a thresholds for sustaining Coral Reef Communities. Ambio 43:361–376

Blaber S, Brewer D, Salini J (1995) Fish communities and the nursery role of the shallow inshore waters of a tropical bay in the Gulf of Carpentaria, Australia. Estuarine, Coastal and Shelf Science 40:177–193

Bode M, Day C (2020) The Great Barrier Reef -Systematically protecting connectivity without connectivity data IUCN guidance for conserving connectivity through ecological corridors and networks. IUCN, Gland

BOM (2022) Australian Government Bureau of Meteorology. http://bom.gov.au

Bond N, Thomson J, Reich P, Stein J (2011) Using species distribution models to infer potential climate change-induced range shifts of freshwater fish in south-eastern Australia. Marine and Freshwater Research 62:1043–1061

Borchert SM, Osland MJ, Enwright NM, Griffith KT (2018) Coastal wetland adaptation to sea level rise: Quantifying potential for landward migration and coastal squeeze. Journal of Applied Ecology 55:2876–2887

Boys CA, Kroon FJ, Glasby TM, Wilkinson K (2012) Improved fish and crustacean passage in tidal creeks following floodgate remediation. Journal of Applied Ecology 49:223–233

Bradley M, Baker R, Sheaves M (2017) Hidden components in tropical seascapes: Deep-estuary habitats support unique fish assemblages. Estuaries and Coasts 40:1195–1206

Bradley M, Nagelkerken I, Baker R, Sheaves M (2020) Context-dependence: A conceptual approach for understanding the habitat relationships of coastal marine fauna. BioScience 70:986–1004.

Bradley M, Sheaves M, Waltham NJ (2023) Urban-industrial seascapes can be abundant and dynamic fish habitat. Frontiers in Marine Science 9:1034039

Burt JA (2014) The environmental costs of coastal urbanization in the Arabian Gulf. City 18:760–770

Caddy J, Bakun A (1994) A tentative classification of coastal marine ecosystems based on dominant processes of nutrient supply. Ocean & Coastal Management 23:201–211

Cadet-James Y, James RA, McGinty S, McGregor R (2017) Yaru Gugubadhungu Gugu Badhun: People of the Valley of Lagoons. Aboriginal Studies Press, Canberra, ACT, pp. 1–13

Canning AD, Waltham NJ (2021) Ecological impact assessment of climate change and habitat loss on wetland vertebrate assemblages of the Great Barrier Reef catchment and the influence of survey bias. Ecology and Evolution 11:5244–5254

Chin A, Cameron D, Saunders R (2019) Fisheries of the Great Barrier Reef. The Great Barrier reef: Biology. Environment and Management:117–130

Chin A, Heupel M, Simpfendorfer C, Tobin A (2013) Ontogenetic movements of juvenile blacktip reef sharks: Evidence of dispersal and connectivity between coastal habitats and coral reefs. Aquatic Conservation: Marine and Freshwater Ecosystems 23:468–474

Cliff G (1995) Sharks caught in the protective gill nets off KwaZulu-Natal, South Africa. 8. The Great hammerhead shark Sphyrna mokarran(Rüppell). South African Journal of Marine Science 15:105–114

Coles RG, Rasheed MA, McKenzie LJ, Grech A, York PH, Sheaves M, McKenna S, Bryant C (2015) The Great Barrier Reef World Heritage Area seagrasses: Managing this iconic Australian ecosystem resource for the future. Estuarine, Coastal and Shelf Science 153:A1–A12

Cook F, Hicks W, Gardner E, Carlin GD, Froggatt D (2000) Export of acidity in drainage water from acid sulphate soils. Marine Pollution Bulletin 41:319–326

Creighton C, Boon P, Brookes J, Sheaves M (2015) Repairing Australia's estuaries for improved fisheries production – What benefits, at what cost? Marine and Freshwater Research 66(6):493–507

Crook KA, Sheaves M, Barnett A (2022) Species-specific foraging behaviors define the functional roles of sympatric stingrays. Limnology and Oceanography 67:219–230

Cumming GS (2011) Spatial resilience: Integrating landscape ecology, resilience, and sustainability. Landscape Ecology 26:899–909

Dale A, Ryan S, Broderick K (2017) Natural resource management as a form of multi-level governance: The impact of reform in Queensland and Tasmania. Multi-level Governance 327

Dale JJ, Wallsgrove NJ, Popp BN, Holland KN (2011) Nursery habitat use and foraging ecology of the brown stingray Dasyatis lata determined from stomach contents, bulk and amino acid stable isotopes. Marine Ecology Progress Series 433:221–236

Davis B, Johnston R, Baker R, Sheaves M (2012) Fish utilisation of wetland nurseries with complex hydrological connectivity. PLoS One 7:e49107

Davis JP, Pitt KA, Fry B, Olds AD, Connolly RM (2014) Seascape-scale trophic links for fish on inshore coral reefs. Coral Reefs 33:897–907

Dubuc A, Baker R, Marchand C, Waltham NJ, Sheaves M (2019) Hypoxia in mangroves: Occurrence and impact on valuable tropical fish habitat. Biogeosciences 16:3959–3976

Dubuc A, Waltham N, Malerba M, Sheaves M (2017) Extreme dissolved oxygen variability in urbanised tropical wetlands: The need for detailed monitoring to protect nursery ground values. Estuarine, Coastal and Shelf Science 198:163–171

Eggleston DB (1995) Recruitment in Nassau grouper Epinephelus striatus: Post-settlement abundance, microhabitat features, and ontogenetic habitat shifts. Marine Ecology Progress Series 124:9–22

Espinoza M, Heupel MR, Tobin AJ, Simpfendorfer CA (2016) Evidence of partial migration in a large coastal predator: Opportunistic foraging and reproduction as key drivers? PLoS One 11:e0147608

Eyre BD, McKee LJ (2002) Carbon, nitrogen, and phosphorus budgets for a shallow subtropical coastal embayment (Moreton Bay, Australia). Limnology and Oceanography 47:1043–1055

Fabricius KE (2005) Effects of terrestrial runoff on the ecology of corals and coral reefs: Review and synthesis. Marine Pollution Bulletin 50:125–146

Flowers KI, Heithaus MR, Papastamatiou YP (2021) Buried in the sand: Uncovering the ecological roles and importance of rays. Fish and Fisheries 22:105–127

Fong CR, Gaynus CJ, Carpenter RC (2020) Extreme rainfall events pulse substantial nutrients and sediments from terrestrial to nearshore coastal communities: A case study from French Polynesia. Scientific Reports 10:1–8

Franca S, Vasconcelos RP, Tanner S, Maguas C, Costa MJ, Cabral HN (2011) Assessing food web dynamics and relative importance of organic matter sources for fish species in two Portuguese estuaries: A stable isotope approach. Marine Environmental Research 72:204–215

Fulton E, Kault D, Mapstone B, Sheaves M (1999) Spawning season influences on commercial catch rates: Computer simulations and Plectropomus leopardus, a case in point. The Canadian Journal of Fisheries and Aquatic Sciences 56:1096–1108

GBRMPA (2019) Great Barrier Reef Outlook Report 2019. Great Barrier Reef Marine Park Authority, Townsville

Graham R, Castellanos D (2012) Apex predators target mutton snapper spawning aggregation. Coral Reefs 31:1017–1017

Haynes D, Müller J, Carter S (2000) Pesticide and herbicide residues in sediments and seagrasses from the Great Barrier Reef World Heritage Area and Queensland coast. Marine Pollution Bulletin 41:279–287

Heck KL, Carruthers TJ, Duarte CM, Hughes AR, Kendrick G, Orth RJ, Williams SW (2008) Trophic transfers from seagrass meadows subsidize diverse marine and terrestrial consumers. Ecosystems 11:1198–1210

Herrmann M, Najjar RG, Kemp WM, Alexander RB, Boyer EW, Cai WJ, Griffith PC, Kroeger KD, McCallister SL, Smith RA (2015) Net ecosystem production and organic carbon balance of US East Coast estuaries: A synthesis approach. Global Biogeochemical Cycles 29:96–111

Heupel M, Knip D, Simpfendorfer C, Dulvy N (2014) Sizing up the ecological role of sharks as predators. Marine Ecology Progress Series 495:291–298

Heupel MR, Simpfendorfer CA (2008) Movement and distribution of young bull sharks Carcharhinus leucas in a variable estuarine environment. Aquatic Biology 1:277–289

Heupel MR, Simpfendorfer CA (2011) Estuarine nursery areas provide a low-mortality environment for young bull sharks Carcharhinus leucas. Marine Ecology Progress Series 433:237–244

Heupel MR, Simpfendorfer CA, Espinoza M, Smoothey AF, Tobin A, Peddemors V (2015) Conservation challenges of sharks with continental scale migrations. Frontiers in Marine Science 2:12

Heupel MR, Yeiser BG, Collins AB, Ortega L, Simpfendorfer CA (2010) Long-term presence and movement patterns of juvenile bull sharks, Carcharhinus leucas, in an estuarine river system. Marine and Freshwater Research 61:1–10

Hitchcock JN, Mitrovic SM, Hadwen WL, Growns IO, Rohlfs A-M (2016) Zooplankton responses to freshwater inflows and organic-matter pulses in a wave-dominated estuary. Marine and Freshwater Research 67:1374–1386

Hobbs JP, McDonald C (2010) Increased seawater temperature and decreased dissolved oxygen triggers fish kill at the Cocos (Keeling) Islands, Indian Ocean. Journal of Fish Biology 77:1219–1229

Hoey A, Bellwood D (2008) Cross-shelf variation in the role of parrotfishes on the Great Barrier Reef. Coral Reefs 27:37–47

Hoey A, Brandl S, Bellwood D (2013) Diet and cross-shelf distribution of rabbitfishes (f. Siganidae) on the northern Great Barrier Reef: Implications for ecosystem function. Coral Reefs 32:973–984

Hughes BB, Levey MD, Fountain MC, Carlisle AB, Chavez FP, Gleason MG (2015) Climate mediates hypoxic stress on fish diversity and nursery function at the land–sea interface. Proceedings of the National Academy of Sciences 112:8025–8030

Hughes L (2003) Climate change and Australia: Trends, projections and impacts. Austral Ecology 28:423–443

Hutchings P, Kingsford M, Hoegh-Guldberg O (2019) The Great Barrier Reef: Biology, environment and management. CSIRO Publishing, Melbourne, 378pp

Hyland S (2002) An investigation of the impacts of ponded pastures on barramundi and other finfish populations in tropical coastal wetlands. Department of Primary Industries, Queensland, 146pp

Ibrahim HD, Eltahir EA (2019) Impact of brine discharge from seawater desalination plants on Persian/Arabian gulf salinity. Journal of Environmental Engineering 145:04019084

Iriarte A, Aravena G, Villate F, Uriarte I, Ibáñez B, Llope M, Stenseth NC (2010) Dissolved oxygen in contrasting estuaries of the Bay of Biscay: Effects of temperature, river discharge and chlorophyll A. Marine Ecology Progress Series 418:57–71

Johnson A, Ebert S (2000) Quantifying inputs of pesticides to the Great Barrier Reef Marine Park–A case study in the Herbert River catchment of north-east Queensland. Marine Pollution Bulletin 41:302–309

Jones DL, Walter JF, Brooks EN, Serafy JE (2010) Connectivity through ontogeny: Fish population linkages among mangrove and coral reef habitats. Marine Ecology Progress Series 401:245–258

Kemp W, Smith E, Marvin-DiPasquale M, Boynton W (1997) Organic carbon balance and net ecosystem metabolism in Chesapeake Bay. Marine Ecology Progress Series 150:229–248

Kingsford M, Spiecker L, Gerlach G (2024) The interactions between larval behaviour and oceanography. In Wolanski E, Kingsford MJ (eds.), Oceanographic Processes of Coral Reefs. Physical and Biological Links in the Great Barrier Reef. CRC Press, Boca Raton, FL.

Kroon FJ, Berry KL, Brinkman DL, Kookana R, Leusch FD, Melvin SD, Neale PA, Negri AP, Puotinen M, Tsang JJ (2020) Sources, presence and potential effects of contaminants of emerging concern in the marine environments of the Great Barrier Reef and Torres Strait, Australia. Science of the Total Environment 719:135140

Kroon FJ, Phillips S (2015) Identification of human-made physical barriers to fish passage in the Wet Tropics region, Australia. Marine and Freshwater Research 67:677–681

Krumme U (2009) Diel and tidal movements by fish and decapods linking tropical coastal ecosystems Ecological connectivity among tropical coastal ecosystems. Springer, New York, pp. 271–324

Kwon B-O, Kim H, Noh J, Lee SY, Nam J, Khim JS (2020) Spatiotemporal variability in microphytobenthic primary production across bare intertidal flat, saltmarsh, and mangrove forest of Asia and Australia. Marine Pollution Bulletin 151:110707

Lecchini D, Galzin R (2005a) Spatial repartition and ontogenetic shifts in habitat use by coral reef fishes (Moorea, French Polynesia). Marine Biology 147:47–58

Lecchini D, Galzin R (2005b) Spatial repartition and ontogenetic shifts in habitat use by coral reef fishes (Moorea, French Polynesia). Marine Biology 147:47–58

Lee SL, Chong VC, Then AY-H (2019) Fish trophodynamics in tropical mudflats: A dietary and isotopic perspective. Estuaries and Coasts 42:868–889

Lefebure R, Degerman R, Andersson A, Larsson S, Eriksson LO, Båmstedt U, Byström P (2013) Impacts of elevated terrestrial nutrient loads and temperature on pelagic food-web efficiency and fish production. Global Change Biology 19:1358–1372

Leguerrier D, Niquil N, Boileau N, Rzeznik J, Sauriau P-G, Le Moine O, Bacher C (2003) Numerical analysis of the food web of an intertidal mudflat ecosystem on the Atlantic coast of France. Marine Ecology Progress Series 246:17–37

Lellis-Dibble KA, McGlynn K, Bigford TE (2008) Estuarine fish and shellfish species in US commercial and recreational fisheries: Economic value as an incentive to protect and restore estuarine habitat. NOAA Technical Memorandum NMFS-F/SPO-90. https://spo.nmfs.noaa.gov/sites/default/files/TM90.pdf

Li XX, Yang W, Sun T, Yang ZF (2022) Trophic diversity and food web structure of vegetated habitats along a coastal topographic gradient. Frontiers in Marine Science 9

Litvin SY, Weinstein MP, Sheaves M, Nagelkerken I (2018) What makes nearshore habitats nurseries for nekton? An emerging view of the nursery role hypothesis. Estuaries and Coasts 41:1539–1550

Lotze HK, Lenihan HS, Bourque BJ, Bradbury RH, Cooke RG, Kay MC, Kidwell SM, Kirby MX, Peterson CH, Jackson JB (2006) Depletion, degradation, and recovery potential of estuaries and coastal seas. Science 312:1806–1809

Lubitz N, Daly R, Filmalter J, Sheaves M, Cowley P, Naesje T, Barnett A (in review) Context drives movement patterns in a mobile marine predator. Movement Ecology 11(1):28

Martín-Antón M, Negro V, del Campo JM, López-Gutiérrez JS, Esteban MD (2016) Review of coastal land reclamation situation in the world. Journal of Coastal Research:667–671

Martins A, Matley J, Heupel M, Fisk A, Chin A, Simpfendorfer C (2022) Trophic ecology of sympatric juvenile stingrays within a nursery area. Marine and Freshwater Research 73:678–688

Mattone C, Bradley M, Barnett A, Konovalov DA, Sheaves M (2022) Environmental conditions constrain nursery habitat value in Australian sub-tropical estuaries. Marine Environmental Research 175:105568

McPhee JJ, Freewater P, Gladstone W, Platell ME, Schreider MJ (2015) Glassfish switch feeding from thalassinid larvae to crab zoeae after tidal inundation of saltmarsh. Marine and Freshwater Research 66:1037–1044

Meziane T, Tsuchiya M (2000) Fatty acids as tracers of organic matter in the sediment and food web of a mangrove/intertidal flat ecosystem, Okinawa, Japan. Marine Ecology Progress Series 200:49–57

Mitchell AW, Bramley R, Johnson A (1997) Export of nutrients and suspended sediment during a cyclone-mediated flood event in the Herbert River catchment, Australia. Marine and Freshwater Research 48:79–88

Molony B, Parry G (2006) Predicting and managing the effects of hypersalinity on the fish community in solar salt fields in north-western Australia. Journal of Applied Ichthyology 22:109–118

Mwijage AP, Shilla DA, Machiwa JF, Zhang J, Wang FQ, Ying W (2018) Important organic matter sources and trophic pathways for the nutrition of Hilsa kelee (Cuvier, 1829) and Valamugil buchanani (Bleeker, 1853) in Pangani macrotidal estuary, Tanzania. Chemistry and Ecology 34:941–963

Nagelkerken I (2009) Ecological connectivity among tropical coastal ecosystems. Springer, New York, 615pp

Nagelkerken I, Sheaves M, Baker R, Connolly RM (2015) The seascape nursery: A novel spatial approach to identify and manage nurseries for coastal marine fauna. Fish and Fisheries 16:362–371

Nakamura Y, Horinouchi M, Shibuno T, Tanaka Y, Miyajima T, Koike I, Kurokura H, Sano M (2008) Evidence of ontogenetic migration from mangroves to coral reefs by black-tail snapper Lutjanus fulvus: Stable isotope approach. Marine Ecology Progress Series 355:257–266

Nelson J, Wilson R, Coleman F, Koenig C, DeVries D, Gardner C, Chanton J (2012) Flux by fin: Fish-mediated carbon and nutrient flux in the northeastern Gulf of Mexico. Marine Biology 159:365–372

Nelson JA, Stallings CD, Landing WM, Chanton J (2013) Biomass transfer subsidizes nitrogen to offshore food webs. Ecosystems 16:1130–1138

Nicastro A, Bishop MJ (2013) Weak and habitat-dependent effects of nutrient pollution on macrofaunal communities of southeast Australian estuaries. PLoS One 8:e65706

Nicholls RJ, Wong PP, Burkett VR, Codignotto JO, Hay JE, McLean RF, Ragoonaden S, Woodroffe CD (2007) Coastal systems and low-lying areas. In Parry ML, Canziani OF, Palutikof JP, van der Linden PJ, Hanson CE (eds.) Climate change 2007: Impacts, adaptation and vulnerability (Contribution of working group II to the fourth assessment report of the intergovernmental panel on climate change). Cambridge University Press, Cambridge, pp. 315–356

O'Brien D, Davis A, Nash M, Di Bella L, Turner R, Reghenzani J, Brodie J (2013) Water quality within the Herbert River Catchment associated with specific land use. In Proceedings of the 35th Conference of the Australian Society of Sugar Cane Technologists, pp. 1–14

Olds AD, Nagelkerken I, Huijbers CM, Gilby BL, Pittman SJ, Schlacher TA (2018) Connectivity in coastal seascapes. Seascape Ecology:261–292

Olds AD, Pitt KA, Maxwell PS, Connolly RM (2012) Synergistic effects of reserves and connectivity on ecological resilience. Journal of Applied Ecology 49:1195–1203

Olin JA, Hussey NE, Fritts M, Heupel MR, Simpfendorfer CA, Poulakis GR, Fisk AT (2011) Maternal meddling in neonatal sharks: Implications for interpreting stable isotopes in young animals. Rapid Communications in Mass Spectrometry 25:1008–1016

Pearson RG, Connolly NM, Davis AM, Brodie JE (2021) Fresh waters and estuaries of the Great Barrier Reef catchment: Effects and management of anthropogenic disturbance on biodiversity, ecology and connectivity. Marine Pollution Bulletin 166:112194

Pittman SJ (2017) Seascape ecology. John Wiley & Sons, Hoboken

Pringle AW (1991) Fluvial sediment supply to the north-east Queensland coast, Australia. Australian Geographical Studies 29:114–138

Ridd P, Sam R, Hollins S, Brunskill G (1997) Water, salt and nutrient fluxes of tropical tidal salt flats. Mangroves and Salt Marshes 1:229–238

Ripple WJ, Estes JA, Schmitz OJ, Constant V, Kaylor MJ, Lenz A, Motley JL, Self KE, Taylor DS, Wolf C (2016) What is a trophic cascade? Trends in Ecology & Evolution 31:842–849

Saintilan N, Rogers K, Kelleway J, Ens E, Sloane D (2019) Climate change impacts on the coastal wetlands of Australia. Wetlands 39:1145–1154

Seneviratne S, Nicholls N, Easterling D, Goodess C, Kanae S, Kossin J, Luo Y, Marengo J, McInnes K, Rahimi M (2012) Changes in climate extremes and their impacts on the natural physical environment. In Chapter 3 in Field BC, Barros V, Stocker TF, Dahe Q (eds.), Managing the risks of extreme events and disasters to advance climate change adaptation. Special Report of the intergovernmental panel on climate change, Cambridge University Press, Cambridge, pp. 109–230

Sheaves M (1995) Large lutjanid and serranid fishes in tropical estuaries: Are they adults or juveniles? Marine Ecology-Progress Series 129:31–40

Sheaves M (1996) Do spatial differences in the abundance of two serranid fishes in estuaries of tropical Australia reflect long term salinity patterns? Marine Ecology-Progress Series 137:39–49

Sheaves M (1998) Spatial patterns in estuarine fish faunas in tropical Queensland: A reflection of interaction between long-term physical and biological processes? Marine and Freshwater Research 49:31–40

Sheaves M (2005) Nature and consequences of biological connectivity in mangrove systems. Marine Ecology-Progress Series 302:293–305

Sheaves M (2006) Is the timing of spawning in sparid fishes a response to sea temperature regimes? Coral Reefs 25:655–669

Sheaves M (2009) Consequences of ecological connectivity: The coastal ecosystem mosaic. Marine Ecology-Progress Series 391:107–115

Sheaves M, Baker R, Nagelkerken I, Connolly RM (2015) True value of estuarine and coastal nurseries for fish: Incorporating complexity and dynamics. Estuaries and Coasts 38:401–414

Sheaves M, Brodie J, Brooke B, Dale P, Lovelock C, Waycott M, Gehrke P, Johnston R, Baker R (2007c) Chapter 19 Vulnerability of coastal and estuarine habitats in the GBR to climate change. In Johnson J, Marshal P (eds.), Climate change and the Great Barrier Reef: A vulnerability assessment. GBRMPA, Townsville, pp. 593–620

Sheaves M, Brookes J, Coles R, Freckelton M, Groves P, Johnston R, Winberg P (2014) Repair and revitalisation of Australia's tropical estuaries and coastal wetlands: Opportunities and constraints for the reinstatement of lost function and productivity. Marine Policy 47:23–36

Sheaves M, Johnston R (2008) Influence of marine and freshwater connectivity on the dynamics of subtropical estuarine wetland fish metapopulations. Marine Ecology-Progress Series 357:225–243

Sheaves M, Johnston R, Abrantes K (2007a) Fish fauna of dry tropical and subtropical estuarine floodplain wetlands. Marine and Freshwater Research 58:931–943

Sheaves M, Johnston R, Connolly R, Baker R (2012) Importance of estuarine mangroves to juvenile banana prawns. Estuarine, Coastal and Shelf Science 114:208–219

Sheaves M, Johnston R, Molony B, Shepard G (2007b) The effect of impoundments on the structure and function of fish fauna in a highly regulated dry tropics estuary. Estuaries and Coasts 30:507–517

Sheaves M, Mattone C, Connolly RM, Hernandez S, Nagelkerken I, Murray N, Ronan M, Waltham NJ, Bradley M (2021a) Ecological constraint mapping: Understanding outcome-limiting bottlenecks for improved environmental decision-making in marine and coastal environments. Frontiers in Marine Science 8:717448

Sheaves M, Molony B (2000) Short-circuit in the mangrove food chain. Marine Ecology Progress Series 199:97–109

Sheaves M, Molony B, Tobin A (1999) Spawning migrations and local movements of a tropical sparid fish. Marine Biology 133:123–128

Sheaves M, Sporne I, Dichmont CM, Bustamante R, Dale P, Deng R, Dutra LX, Van Putten I, Savina-Rollan M, Swinbourne A (2016) Principles for operationalizing climate change adaptation strategies to support the resilience of estuarine and coastal ecosystems: An Australian perspective. Marine Policy 68:229–240

Sheaves M, Waltham N, Benham C, Bradley M, Mattone C, Diedrich A, Sheaves J, Sheaves A, Hernandez S, Dale P (2021b) Restoration of marine ecosystems: Understanding possible futures for optimal outcomes. Science of the Total Environment 796:148845

Shriadah MA (2000) Contamination by petroleum hydrocarbons in two semi-enclosed areas along the Arabian Gulf Coast of the United Arab Emirates. OEbalia (Tarauto) 26:49–62

Simpfendorfer CA, Milward NE (1993) Utilisation of a tropical bay as a nursery area by sharks of the families Carcharhinidae and Sphyrnidae. Environmental Biology of Fishes 37:337–345

Smith A (1989) Usage of marine resources by Aboriginal communities on the east coast of Cape York Peninsula. Great Barrier Reef Marine Park Authority, Townsville, p. 137. http://hdl.handle.net/11017/209

Sorte CJ, Williams SL, Carlton JT (2010) Marine range shifts and species introductions: Comparative spread rates and community impacts. Global Ecology and Biogeography 19:303–316

Sumpton W, Jackson S (2005) The effects of incidental trawl capture of juvenile snapper (Pagrus auratus) on yield of a subtropical line fishery in Australia: An assessment examining habitat preference and early life history characteristics. Fisheries Research 71:335–347

Switzer TS, Chesney EJ, Baltz DM (2015) Habitat use by juvenile red snapper in the northern Gulf of Mexico: Ontogeny, seasonality, and the effects of hypoxia. Transactions of the American Fisheries Society 144:300–314

Syvitski JP, Vorosmarty CJ, Kettner AJ, Green P (2005) Impact of humans on the flux of terrestrial sediment to the global coastal ocean. Science 308:376–380

Tanaka K, Hanamura Y, Chong VC, Watanabe S, Man A, Mohd Kassim F, Kodama M, Ichikawa T (2011) Stable isotope analysis reveals ontogenetic migration and the importance of a large mangrove estuary as a feeding ground for juvenile John's snapper Lutjanus johnii. Fisheries Science 77:809–816

Tanimoto M, Fox A, O'Neill MF, Langstreth J (2021) Stock assessment of Australian east coast Spanish mackerel (Scomberomorus commerson). Technical Report, State of Queensland, Brisbane. https://era.daf.qld.gov.au/id/eprint/8226/

Valiela I, Teal JM, Volkmann S, Shafer D, Carpenter EJ (1978) Nutrient and particulate fluxes in a salt marsh ecosystem: Tidal exchanges and inputs by precipitation and groundwater 1. Limnology and Oceanography 23:798–812

Walker DI, Kendrick GA (1998) Threats to macroalgal diversity: Marine habitat destruction and fragmentation, pollution and introduced species. Botanica Marina 41:105–112.

Waltham N, Cartwright P, Motson K, Sheaves M, Ronan M (2024) Coastal wetland restoration: Case studies from Great Barrier Reef catchments. In Wolanski E, Kingsford MJ (eds.), Oceanographic Processes of Coral Reefs. Physical and Biological Links in the Great Barrier Reef. CRC Press, Boca Raton, FL.

Waltham NJ, Burrows D, Wegscheidl C, Buelow C, Ronan M, Connolly N, Groves P, Marie-Audas D, Creighton C, Sheaves M (2019) Lost floodplain wetland environments and efforts to restore connectivity, habitat, and water quality settings on the Great Barrier Reef. Frontiers in Marine Science 6:71

Waltham NJ, Sheaves M (2015) Expanding coastal urban and industrial seascape in the Great Barrier Reef World Heritage Area: Critical need for coordinated planning and policy. Marine Policy 57:78–84

Waltham NJ, Sheaves M (2017) Acute thermal tolerance of tropical estuarine fish occupying a man-made tidal lake, and increased exposure risk with climate change. Estuarine, Coastal and Shelf Science 196:173–181

Waterhouse J, Pearson R, Lewis S, Davis A, Waltham N (2024) Great Barrier Reef ecohydrology. In Wolanski E, Kingsford MJ (eds.), Oceanographic Processes of Coral Reefs. Physical and Biological Links in the Great Barrier Reef. CRC Press, Boca Raton, FL.

Wolanski E, Kingsford M, Lambrechts J, Marmorino G (2024) The physical oceanography of the Great Barrier Reef: A review. In Wolanski E, Kingsford MJ (eds.), Oceanographic Processes of Coral Reefs. Physical and Biological Links in the Great Barrier Reef. CRC Press, Boca Raton, FL.

Wolanski E, Spagnol S, Thomas S, Moore K, Alongi D, Trott L, Davidson A (2000) Modelling and visualizing the fate of shrimp pond effluent in a mangrove-fringed tidal creek. Estuarine, Coastal and Shelf Science 50:85–97

Wortmann J, O'Neill MF, Sumpton WD, Campbell MJ, Stewart J (2018) Stock assessment of Australian east coast snapper, Chrysophrys auratus. Predictions of stock status and reference points for 2016. Technical Report, State of Queensland. https://era.daf.qld.gov.au/id/eprint/6341/

Wourms JP, Demski LS (1993) The reproduction and development of sharks, skates, rays and ratfishes: Introduction, history, overview, and future prospects. Environmental Biology of Fishes 38:7–21

Section 3

Biophysical Oceanography

12 Estimates of Wind Drift Coefficient to Inform Biophysical Models of Seagrass Dispersal in the Great Barrier Reef

Samantha J. Tol, Alexandra Carter, Paul H. York, Alana Grech, Rong Situ, and Robert G. Coles

12.1. INTRODUCTION

The rate of exchange, or connectivity, among populations affects population and metapopulation dynamics and genetics, responses to species invasions and disease transmission, species expansion, and population growth (Cowen and Sponaugle 2009). In the case of marine aquatic plants such as seagrasses, dispersal of propagules (seeds, fruit, or viable plant fragments) is critical to the distribution, structure, and resilience of populations (Kendrick et al. 2012). The convective forces of ocean waves and currents are efficient vectors for dispersal of seagrass and facilitate the replenishment and recovery of populations after disturbance at multiple spatial and temporal scales (McMahon et al. 2014; Grech et al. 2016; Kendrick et al. 2017). Connectivity underpins the persistence of seagrass populations, so knowledge on mechanisms and pathways for seagrass dispersal are integral to the design of management and conservation actions.

Seagrasses have multiple modes of dispersal (Kendrick et al. 2012). Seagrass meadows can expand locally by clonal growth through rhizome extension but only at small scales (hundreds of metres) (Kendrick et al. 2012; McMahon et al. 2014). Some seagrass species produce propagules that detach from the parent plant and can grow in the water column and remain positively buoyant for long periods of time (months and potentially years), which allows for long-distance dispersal (Thomson et al. 2015; Smith et al. 2018). Many species can produce fruits that can float on the water's surface for short timescales (days, weeks) or have fruiting bodies that are attached to floating propagules, enabling the movement of new genetic material outside the source meadow (Källström et al. 2008; van Dijk et al. 2009). Seagrass seeds can also disperse tens of meters to tens of kilometres depending on whether they are positively, neutrally, or negatively buoyant (Harwell and Orth 2002; McMahon et al. 2014). Strong waves or currents can cause rhizomes to break to create a propagule, which depending on the species, can persist in the ocean for many months and facilitate long-distance dispersal (Harwell and Orth 2002; Smith et al. 2018). These many dispersal modes make empirical measures of seagrass dispersal challenging to obtain, especially given the logistics and costs of tracking propagules in a dynamic three-dimensional fluid environment (Saunders et al. 2015).

An alternative approach to empirical measures is using biophysical models to predict the dispersal of seagrass propagules. This approach combines the simulation of movement by using numerical models to describe the motion of waters and biological parameters to define the reproductive attributes of seagrass (e.g., Jahnke et al. 2017; Sinclair et al. 2018; Jahnke et al. 2020; Evans et al. 2021; Jackson et al. 2021). Modelling the potential dispersal distance and paths of seagrass propagules can provide vital information for management and conservation decision making at relevant spatial scales. It can also assist in mitigation efforts against coastal development and climate change (Lai et al. 2018) and inform the design of spatial management and environmental impact assessment (Swearer et al. 2019). The accuracy of models is dependent on the accuracy of the parameters used, as the persistence of plant species in marine environments is influenced by how propagule characteristics interact with disperser attributes, such as water movement (McConkey et al. 2012). Unfortunately, there is little empirical information on propagule density, viability, predation, drift potential, and species-specific sensitivity to wind effects for tropical seagrasses, which ultimately limits the predictive power of biophysical models (Wolanski 2017). This is particularly important for parameters such as wind drift coefficients where species morphology can influence the model values and where temperate species traits (such as fragment size) (e.g., Harwell and Orth 2002) are not appropriate for tropical species (Lai et al. 2018).

Empirical studies to fill the knowledge gaps in model parameters have been trailing the use of models, especially for seagrasses (Lai et al. 2018). Data on the movement of seagrass vegetative fragments are lacking for many parameters used in biophysical models, such as float survival time and wind drift effect (Lai et al. 2018; Lai et al. 2020). A recent methodological study by Schlaefer et al. (2022) found that the long dispersal time and reproductive period of seagrasses and their complex life history necessitate the use of sensitivity analysis to determine the number of particles (i.e., 'virtual' propagules) required to robustly simulate dispersal in biophysical models. The optimal number

DOI: 10.1201/9781003320425-15

of particles was influenced by three factors: seagrass habitat type and location, month of release, and wind drift of the 'virtual' particles. Wind drift coefficients (*Cw*) are influenced by a propagule's mass, shape, and buoyancy and by the angle that it faces in relation to the wind direction. Schlaefer et al. (2022) used two values *Cw* (1% and 2%) to represent the variety of structurally diverse forms of seagrass propagules. To simulate the dispersal of propagules below the water surface, *Cw* was set to 0%, so dispersal was solely determined by the water currents. Schlaefer et al. (2022) concluded that capturing the variability of dispersal trajectories for seagrasses requires releasing enough particles to robustly represent dispersal across the full range of wind drift coefficients. The importance of propagules to the connectivity of seagrasses and the year-round availability of these propagules (Kendrick et al. 2012; McMahon et al. 2014) mean that it is important to improve the parameterisation of biophysical models to provide a more accurate simulation of the wind drift of seagrass vegetative fragments.

In marine conservation areas such as Australia's Great Barrier Reef World Heritage Area (GBRWHA), understanding how the components of seagrass ecosystems interact at broad scales has become important for understanding cycles of recovery and loss. The GBRWHA includes one of the world's largest seagrass habitats with around 20% of the world's seagrass species—as many as 15 different species (Short et al. 2011; Coles et al. 2015). Seagrass meadows extend from the intertidal zone down to nearly 70 m deep. Modelling indicates that there is likely to be at least 35,000 km^2 of seagrass habitat (Coles et al. 2009; Coles et al. 2015) and as many as 36 different seagrass communities (Carter et al. 2021a). Our knowledge of seagrass ecology and distribution comes from extensive research starting in the early 1980s, which has been collated in a publicly available database (Carter et al. 2021b).

For tropical northeastern Australia, water flows have been extensively modelled (Wolanski et al. 2024), and information on water movements is readily available. Wind speed and direction are also publicly available, but modelling of seagrass movement (Grech et al. 2016; Jackson et al. 2021) has had to rely on published wind drag coefficients developed for species not found in the tropics (Erftemeijer et al. 2008). Biophysical modelling indicates seagrass communities in the GBRWHA are connected through propagule exchange, whose extent influences resilience to environmental disturbance and influences long-term trends in meadow parameters (Grech et al. 2016; Grech et al. 2018; Carter et al. 2022). To maximise the value of these modelling approaches, it is important to refine model parameters to closely reflect the behaviour and survival of propagules in the water for the morphology of each species and for the environmental conditions likely to be experienced.

The aim of the study presented in this chapter is to determine the wind drift coefficient of vegetative fragments that contribute to long-distance dispersal in the tropical seagrass species of the GBRWHA to address the problem of inappropriate input parameters in biophysical models. We chose to study *Halophila ovalis* (a round leaf plant) and *Zostera muelleri* (a strap leafed plant), as they are morphologically different and two of the most common species found in eastern Queensland tropical waters (Carter et al. 2021a; Figure 3). Quantifying the wind drift coefficient will improve the accuracy of seagrass dispersal biophysical models and improve understanding of the role of vegetative fragments in seagrass connectivity.

12.2. METHODS

1. Study site

Seagrass fragments were harvested from an intertidal seagrass meadow at Cape Pallarenda, Townsville (-19.186, 146.772) in August 2015 (Figure 1). This seagrass meadow consists of a mixed species assemblage dominated by *Halophila ovalis*, *Halodule uninervis*, and *Zostera muelleri* (Wells and Rasheed 2017).

2. Flume Tank

The flume tank consisted of three compartments: a wind tunnel containing a fan and a bundle tube flow straightener, a water channel on top of a steel frame, and the wind flume above the water (Figure 2). The flow straightener comprised 144 individual PVC (20-mm diameter) piping joined into a square (12 × 12) and placed 1 m away from the fan. The fan was an Aerovent AM030–201 plate-mounted axial fan. The water channel consisted of clear acrylic 500 cm long, 30 cm wide, and 50 cm high with an acrylic cover to contain airflow and a hatch after the flow straightener to allow the seagrass fragments to be added. The flume tank was filled with tap water, with aquarium salt (Red Sea Salt) added to create 34 ppt salt water.

3. Evaluating fragment travel speed

A total of 135 *H. ovalis* fragments and 174 *Z. muelleri* fragments were placed individually into the flume tank under varying wind speeds. Harvested fragments were cut-up to ensure a mixture of two-, four-, and six-leaf shoots (Table 1). All fragments were stored in sea water collected from the study site for a maximum of 10 h before completion of the experiment.

Individual fragments were timed for how long it took (s = seconds) them to move 1 m under varying wind speeds: low (< 15 knots, equivalent to < 7.7 m/s); medium (15 to 20 knots, equivalent to between 7.8–10.1 m/s); and high (> 20 knots, equivalent to > 10.2 m/s). Each fragment was placed into the flume tank the same distance from the wind source (approximately 15 cm away from the flow straightener) and allowed to drift until they reached the first 1-m point in the water channel. The time taken for the seagrass fragment to drift 1 m (between the 1-m and 2-m point on the water channel) was measured with a stopwatch. Wind speed was measured (± 0.01 m/s) with a manometer (Dwyer 475–000) once the individual fragments were placed into the flume tank. To reduce the influence of water currents, the water

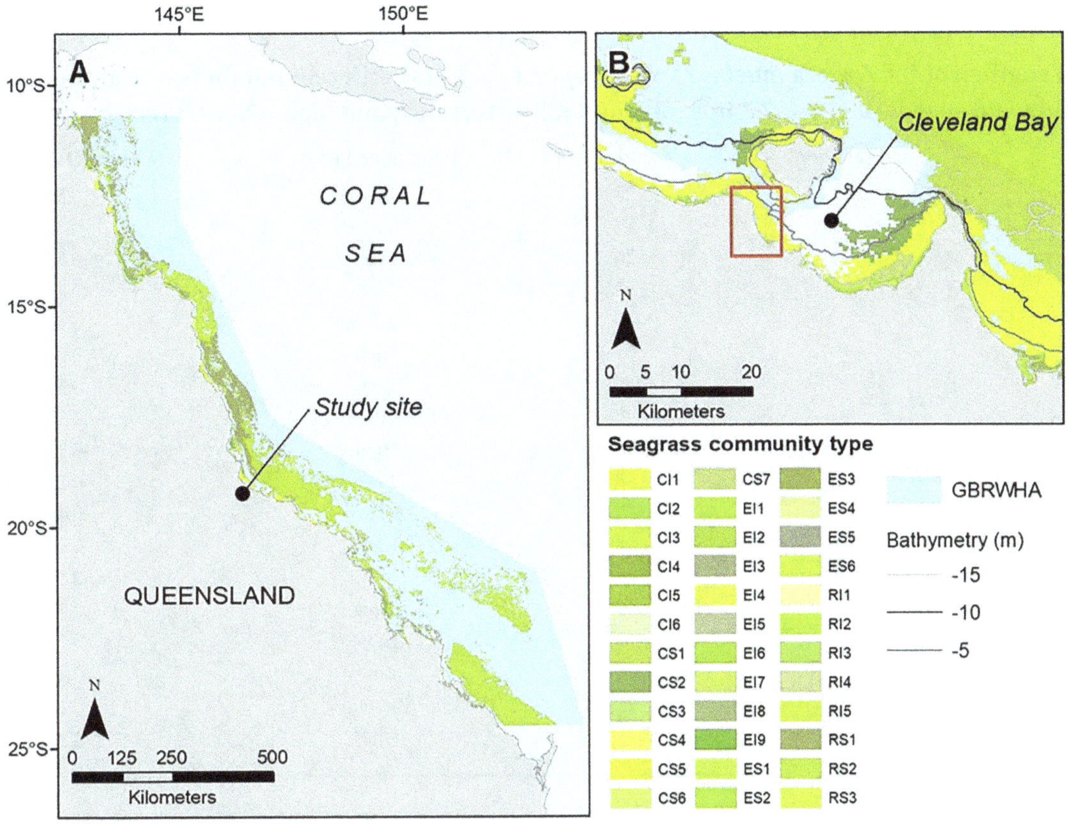

FIGURE 1 (A) Thirty-six seagrass communities predicted for the GBRWHA and adjacent estuaries: estuary intertidal (EI1–EI9), estuary subtidal (ES1–ES6), coastal intertidal (CI1–CI6), coastal subtidal (CS1–CS7), reef intertidal (RI1–RI5), and reef subtidal (RS1–RS3) communities (Carter et al. 2021b), (B) Red box indicates the location of seagrass meadow at the study site, Cape Pallarenda, Townsville Queensland, Australia. This intertidal seagrass meadow contains *Halophila ovalis*, *Halodule uninervis*, and *Zostera muelleri* species.

FIGURE 2 Flume tank used in the experiment, located in Townsville, Queensland Australia.

within the flume tank was allowed to become stationary before a new fragment was added.

4. Fragment measurements

The number of leaf shoots was recorded for all fragments. Leaves were then removed, placed flat on a piece of paper, and photographed to determine the surface area. Rhizome length (± 0.01 mm) and fragment surface area (± 0.01 mm^2) were recorded using ImageJ software (v. 1.53e; Rasband, 1997–2018). Individual fragments were then dried in an oven at 60 °C until they reached a constant dry weight and fragment biomass (± 0.001 g) could be measured.

TABLE 1

Wind drift coefficient for *Zostera muelleri* and *Halophila ovalis* for different numbers of leaf shoots (two, four, and six) and wind speeds (low = < 7.7 m/s, medium = 7.8–10.1 m/s, and high = > 10.2 m/s).

Species	Leaf Shoot	n	Wind Speed (m/s)	Wind Drift Coefficient (*Cw*)
Zostera muelleri	2	28	Low	0.52% (± 0.02%)
		12	Medium	0.32% (± 0.03%)
		20	High	0.31% (± 0.02%)
	4	15	Low	0.41% (± 0.04%)
		21	Medium	0.35% (± 0.02%)
		22	High	0.31% (± 0.02%)
	6	19	Low	0.36% (± 0.03%)
		17	Medium	0.32% (± 0.02%)
		20	High	0.24% (± 0.02%)
Halophila ovalis	2	28	Low	0.59% (± 0.04%)
		7	Medium	0.44% (± 0.09%)
		28	High	0.24% (± 0.02%)
	4	22	Low	0.52% (± 0.03%)
		12	Medium	0.24% (± 0.03%)
		11	High	0.24% (± 0.02%)
	6	8	Low	0.45% (± 0.02%)
		11	Medium	0.36% (± 0.04%)
		11	High	0.21% (± 0.04%)

5. Statistical analysis

All data prior to analysis were examined for outliers and normality, and all predictor variables were assessed for collinearity using variance inflation factors (Zuur et al. 2010). One variable had a high level of collinearity and was subsequently dropped from the main model—the fragment surface area was highly collinear with biomass. We ran the model after substituting the collinear variable and returned similar results.

Statistical analyses for the effect of wind speed on the travel time of seagrass fragments were performed using a generalized linear model (GLM) with a Gamma distribution to account for our data not containing any values below zero (Manning et al. 2005). A set of models was developed as follows to determine the best prediction for the wind drift coefficient (*Cw*) of fragments.

Equation 1:

$$Cw \cdot = \cdot \text{fn} \cdot (\text{WS} \cdot + \cdot \text{Sp} \cdot + \cdot \text{LS} \cdot + \cdot \text{RL} \cdot + \cdot \text{B} \cdot + \cdot \text{WS:Sp} \cdot + \cdot \text{WS:B} \cdot + \cdot \text{WS:LS} \cdot + \cdot \text{WS:RL})$$

where WS = wind speed (m/s)
Sp = species
LS = number of leaf shoots
RL = rhizome length (mm)
B = biomass (g/DW)

Subsets of each model were ranked using the dredge function in the 'MuMIn' package (Barton 2019). The inferred best-fit model was the simplest model within two points of the lowest AIC corrected for small sample sizes (AICc) (Burnham and Anderson 2002); the simplest model was *Cw* = WS + B + LS + Sp + WS:Sp. Residuals were visually inspected for patterns by plotting the fitted versus the response variables. A Tukey post hoc analysis for pairwise comparisons was conducted using the 'emmeans' package (Lenth 2020). All statistical analyses were conducted in the statistical software environment R (R version 4.0.3; R Core Team 2020).

6. Evaluating wind drift coefficient

The wind drift (*Cw*) coefficient was calculated using the following equation:

Equation 2: $u = \frac{d}{t}$

Equation 3: $Cw = \frac{u}{V} \times 100\%$

Where: u = drift velocity (m/s)

V = air velocity ($\frac{m}{s}$) (a.k.a WS)

d = distance travelled by seagrass fragment (1 m)

t = time (s) taken to travel distance d

12.3. RESULTS

1. Fragment speeds

Fragment speed decreased with increasing fragment biomass irrespective of seagrass species (Gamma GLM: $df_{1:307}$,

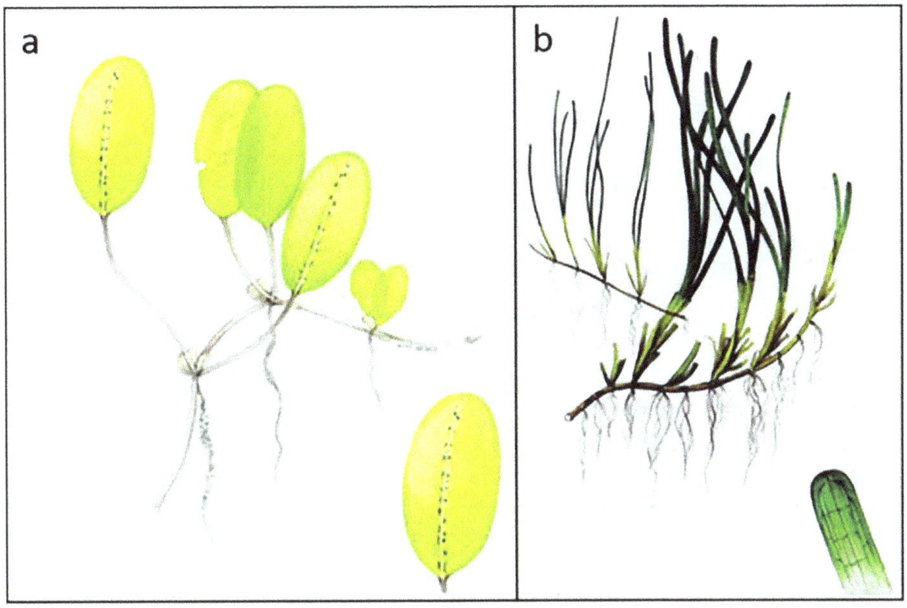

FIGURE 3 Illustration of two tropical seagrass species **a)** *Halophila ovalis*; and **b)** *Zostera muelleri*.

Source: Illustrations by Ruth Berry—Art and Science.

FIGURE 4 **a)** A two-leaf shoot *Halophila ovalis* fragment; and **b)** A six-leaf shoot *Zostera muelleri* fragment float neutrally buoyant when travelling across the flume tank.

$F = 29.60$, $p < 0.0001$; Figure 5a). Two-leaf shoot fragments moved significantly faster than six-leaf shoot fragments (Tukey post hoc: $Z = 3.180$, $p = 0.004$; Figure 5b). The relationship between increasing wind speeds and fragment speed was negative for *H.* ovalis but positive for *Z. muelleri* (Gamma GLM: $df_{1:302}$, $F = 24.31$, $p < 0.0001$; Figure 5c).

All seagrass fragments maintained neutral buoyancy across all wind speeds (all fragments floated just below the water's surface, with no part breaking the surface; Figure 4). *Z. muelleri* fragments would maintain a stiff and straight shape when placed in the flume tank. *H. ovalis* fragments would regularly collapse in on themselves, especially during faster wind speeds. This collapsing would cause the fragments to randomly 'tumble'.

2. Wind drift coefficient

Wind drift coefficients were calculated based on significant factors in the statistical model (Table 1).

The wind drift coefficient value varied between 0.34% and 0.52% for *Z. muelleri* and between 0.21% and 0.59% for *H. ovalis*. The wind drift coefficient for *H. ovalis* fragments decreased with increasing leaf shoots and wind speed. The wind drift coefficient for *Z. muelleri* also decreased with increasing leaf shoots but not for medium wind speeds. For *Z. muelleri*, only low wind speeds caused a decrease in the wind drift coefficient, with no difference between medium and high wind speeds.

12.4. DISCUSSION

We used an experimental approach to measure the values of a biological parameter critical in biophysical models of seagrass dispersal—the wind drift coefficient. Our study revealed that the wind drift coefficient for *H. ovalis* and *Z. muelleri* were 4–10 times lower than the values of the parameter used in a seagrass biophysical model of the GBRWHA (i.e., 2%; Grech et al. 2016). Other tropical

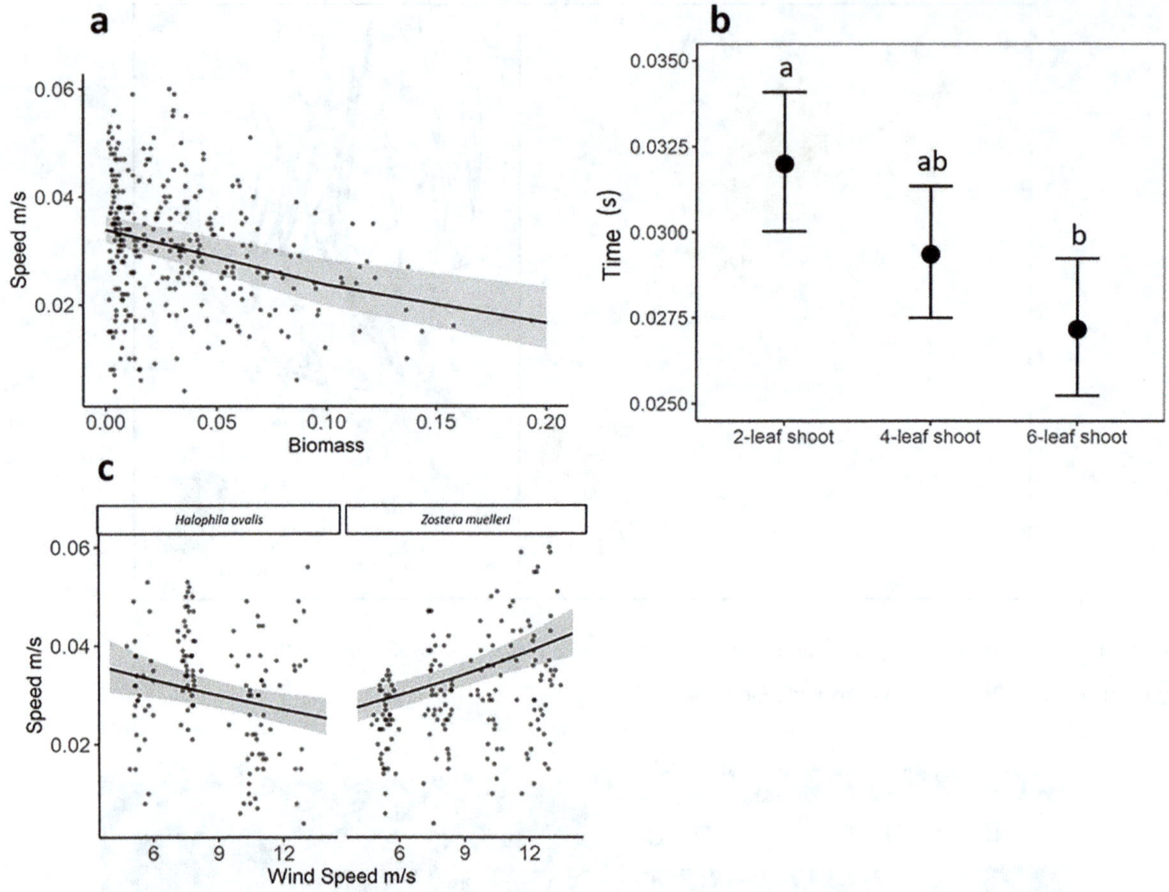

FIGURE 5 **a)** Relationship between fragment biomass and speed (metres/second) for both seagrass species; **b)** Time taken for two-, four-, and six-shoot fragments to move 1 m for both seagrass species (letters symbolise which number of leaf shoots are significantly different); and **c)** Relationship between wind speed and fragment speed for *Halophila ovalis* and *Zostera muelleri* fragments.

seagrass biophysical models (i.e., 1.2%; Ruiz-Montoya et al. 2015; Sinclair et al. 2018) also use a high wind drift parameter. We found that fragments of both species maintained only neutral buoyancy—the seagrass fragment did not protrude out of the water. This result contrasts with previous seagrass biophysical models that simulate fragments as positively buoyant particles.

1. Seagrass dispersal ecology

The wind drift coefficient was significantly impacted by the structure of the two tropical seagrass species and the complexity of the fragment. An increased number of leaf shoots at higher wind speeds caused the motion of fragments to change and resulted in a smaller wind drift value. This is likely due to the morphology of the seagrass species and the shape of the fragments. *H. ovalis* fragments have malleable rhizomes that do not hold their original shape when floating free in the water (more prominent with increasing rhizome length), which causes the fragment to be caught in turbulent water movements. *Z. muelleri* fragments have robust rhizomes that hold their shape, remain close to the water's surface, and are not as affected by turbulence. As fragments of both species are neutrally buoyant, the fragments are impacted only by wave motions when wind speeds increase, creating a larger turbulent effect. These results highlight the need for biophysical models to use specific wind drift parameters based on species type and where practical, on actual wind speeds.

Our results correspond with the findings of Lai et al. (2018) who calculated the wind drift coefficient for tropical seagrass species *Cymodocea rotundata*, *H. ovalis*, *Halodule uninervis*, and *Thalassia hemprichii*. Lai et al. (2018) found wind drift coefficients within the same range as our two species and that morphology and wave movement impacted fragment travel speed. However, our results differ from Weatherall et al. (2016) and Harwell and Orth (2002) who observed positively buoyant fragments. Weatherall et al. (2016) used an indoor seawater mesocosm to measure settling velocity of *H. ovalis* and *Z. muelleri* seagrass fragments. Their study tested the influence of temperature stress on settling velocities but did not examine the effects associated with physical disturbance, such as in a flume tank experiment used in this study. Harwell and Orth (2002) measured the dispersal potential of *Zostera marina* fragments with fruiting bodies attached (spathes) and found them to maintain positive buoyancy for up to two weeks. This lack of positive buoyancy favours fragment dispersal at smaller spatial scales than currently presumed and greater self-recruitment and bounding of seagrass ecosystems via this dispersal mechanism (Wolanski 2017).

The wind drift coefficient discrepancy between what we found and that which has been implemented in previous tropical seagrass biophysical models in the GBRWHA and elsewhere will impact maximum dispersal estimates and have a flow-on effect in connectivity analyses. As most biophysical models select a percentage of fragments to be settled per day (Swearer et al. 2019), a reduction in the wind drift parameter can have an exponential difference in maximum dispersal distance. Therefore, overestimating the wind drift parameter can greatly impact dispersal and connectivity model outputs, potentially undermining conservation and management advice. Grech et al. (2016) used a wind drift coefficient of 2% and estimated a maximum dispersal potential of 900 km when released in the central GBRWHA. This high wind drift parameter could result in the overestimation of long-distance dispersal and the scale of connectivity where a wind effect was included. Fortunately, half of the particles released by Grech et al. (2016) were modelled with 0% wind drift to simulate the dispersal of fragments floating below the surface (negatively buoyant). The connectivity analysis of Grech et al. (2018) used both floating and suspended particle dispersal data and concluded that seagrass meadows in the central GBRWHA were primarily self-recruiting. This conclusion corresponds with the outcomes of this chapter, that is, that tropical seagrass fragments have low to moderate fragment dispersal potential and are most likely dispersed at smaller spatial scales.

The low wind drift and neutral buoyancy of seagrass fragments found in this study suggest that fragment dispersal potential and connectivity of tropical seagrass populations are lower than previously predicted by biophysical models. However, the long-distance dispersal of seagrass is not facilitated by abiotic processes alone. Seeds are also transmitted in the faecal content of seagrass grazers, such as green sea turtles and dugongs (Tol et al. 2017). Movement in water by grazing animals facilitates connectivity over hundreds of kilometres (Kendrick et al. 2012; McMahon et al. 2014). Tol et al. (2017) demonstrated that the availability of seagrass propagules in the GBRWHA may assist the natural recovery of seagrass meadows because green sea turtles and dugong can move viable seagrass seeds to new locations. The directional transport of seeds by green sea turtles and dugongs is independent of wind- and water-driven currents, enabling connectivity to occur at different directions and scales than abiotic processes.

2. *Limitations to seagrass dispersal*

This chapter found that as wind speed increases, the wind drift coefficient decreases, leading to a decrease in the dispersal of seagrass fragments. Increased wind speeds are also known to increase sediment suspension in intertidal and subtidal coastal habitats (Wolanski et al. 2024) and can directly impact seagrass health by lowering light availability (Chartrand et al. 2012). Increased sediment loads could also reduce the survival time of seagrass fragments that are negatively buoyant, as we found in this chapter. Within the GBR, sediment load has increased between 1.2- and 5.3-fold since the arrival of Europeans (Lewis et al. 2024). Information on sediment load impacts to seagrass fragments would greatly enhance dispersal model predictions by refining the parameters associated with their survival time and mortality.

12.5. RESEARCH GAPS

This chapter demonstrates that when using models to determine dispersal potential and connectivity, the species of the seagrass fragment and the wind speed need to be considered for correct wind drift parameters. However, our experimental approach was limited to single fragments and did not consider floating rafts of multiple seagrass fragments. A lack of knowledge on the density and description of rafts of floating viable tropical seagrass fragments makes it difficult to ascertain whether they are common or an important form of fragment movement. It is possible that rafts have a higher wind drift parameter compared to individual fragments and are therefore likely to disperse further than what we would predict. Work is also needed to determine fragment float survival time and the wind drift coefficient for other seagrass species not included in this or Lai et al. (2020) study. Furthermore, increased run-off during storms can greatly affect the salinity of catchment zones (Lewis et al. 2024). Our study did not investigate the impact that lower salinity levels could have on the buoyancy of seagrass fragments. Filling in these gaps will assure the accuracy of parameters for dispersal and connectivity models. Until these research gaps are resolved, we believe that sensitivity analyses such as those conducted in Schlaefer et al. (2022) are necessary to robustly predict dispersal by ensuring that biophysical models are parameterised using a suite of values and that the effects of changes to parameters are transparently and systematically assessed.

12.6. CONCLUSION

This chapter contributes to a growing body of literature that improves the implementation of biophysical models of seagrass dispersal by refining biological parameters (e.g., Weatherall et al. 2016; Lai et al. 2018; Lai et al. 2020) and testing the sensitivity of model inputs (e.g., Schlaefer et al. 2022). Our flume tank experiment gives insight as to how seagrass fragments move in the water under a variety of wind and water conditions. Other influences will play a part in the natural world of seagrass ecosystems. Fragments in nature may be impacted by herbivory, damage from wave action, and loss of buoyancy from deteriorating tissue structures. They may also clump into rafts and present an enhanced profile to the prevailing winds, and rafts may provide conditions that enhance fragment survival times.

Although these stochastic processes are difficult to model and predict, our chapter gives guidance on the most likely outcomes for the species included and supports the prediction of the dispersal of wind-influenced viable fragments in biophysical models. Tol et al. (2017) demonstrated that vast numbers of viable seagrass fragments are moving in the nearshore water column at any time. A better understanding of the biophysical constraints to fragment movement improves our understanding of seagrass meadow connectivity and resilience and the management approaches needed to ensure that meadows are adequately protected.

REFERENCES

Barton K (2019) MuMIn: Multi-model inference. R Package Version 0.12.2/r18

Burnham K, Anderson D (2002) Model selection and multimodel inference: a practical information-theoretic approach, 2nd edition. Springer-Verlag

Carter A, Collier C, Coles R, Lawrence E, Rasheed M (2022) Community-specific "desired" states for seagrasses through cycles of loss and recovery. Journal of Environmental Management 314:115059

Carter A, Collier C, Lawrence E, Rasheed M, Robson B, Coles R (2021a) A spatial analysis of seagrass habitat and community diversity in the Great Barrier Reef World Heritage Area. Scientific Reports 11:1–19

Carter A, McKenna S, Rasheed M, Collier C, McKenzie L, Pitcher R, Coles R (2021b) Synthesizing 35 years of seagrass spatial data from the Great Barrier Reef World Heritage Area, Queensland, Australia. Limnology and Oceanography Letters 6:216–226

Chartrand K, Rasheed M, Petrou K, Ralph P (2012) Establishing tropical seagrass light requirements in a dynamic port environment. In Proceedings of the 12th International Coral Reef Symposium. Seagrasses and seagrass ecosystems, Cairns, Australia

Coles R, McKenzie L, De'ath G, Roelofs A, Lee Long W (2009) Spatial distribution of deepwater seagrass in the inter-reef lagoon of the Great Barrier Reef World Heritage Area. Marine Ecology Progress Series 392:57–68

Coles R, Rasheed M, McKenzie L, Grech A, York P, Sheaves M, McKenna S, Bryant C (2015) The Great Barrier Reef World Heritage Area seagrasses: Managing this iconic Australian ecosystem resource for the future. Estuarine, Coastal and Shelf Science 153:A1–A12

Cowen R, Sponaugle S (2009) Larval dispersal and marine population connectivity. Annual Review of Marine Science 1:443–466

Erftemeijer P, van Beek J, Ochieng C, Jager Z, Los H (2008) Eelgrass seed dispersal via floating generative shoots in the Dutch Wadden Sea: A model approach. Marine Ecology Progress Series 358:115–124

Evans R, McMahon K, Van Dijk K-J, Dawkins K, Jacobi M, Vikrant A (2021) Identification of dispersal barriers for a colonising seagrass using seascape genetic analysis. Science of the Total Environment 763:143052

Grech A, Hanert E, McKenzie L, Rasheed M, Thomas C, Tol S, Wang M, Waycott M, Wolter J, Coles R (2018) Predicting the cumulative effect of multiple disturbances on seagrass connectivity. Global Change Biology:1–12

Grech A, Wolter J, Coles R, McKenzie L, Rasheed M, Thomas C, Waycott M, Hanert E (2016) Spatial patterns of seagrass dispersal and settlement. Diversity and Distributions 22:1150–1162

Harwell M, Orth R (2002) Long-distance dispersal potential in a marine macrophyte. Ecology 83:3319–3330

Jackson E, Smith T, York P, Nielsen J, Irving A, Sherman C (2021) An assessment of the seascape genetic structure and hydrodynamic connectivity for subtropical seagrass restoration. Restoration Ecology 29:e13269

Jahnke M, Casagrandi R, Melià P, Schiavina M, Schultz S, Zane L, Procaccini G (2017) Potential and realized connectivity of the seagrass Posidonia oceanica and their implication for conservation. Diversity and Distributions 23:1423–1434

Jahnke M, Moksnes P, Olsen J, Serra oceanic N, Nilsson Jacobi M, Kuusemäe K, Corell H, Jonsson P (2020) Integrating genetics, biophysical, and demographic insights identifies critical sites for seagrass conservation. Ecological Applications 30:e02121

Källström B, Nyqvist A, Åberg P, Bodin M, André C (2008) Seed rafting as a dispersal strategy for eelgrass (*Zostera marina*). Aquatic Botany 88:148–153

Kendrick G, Orth R, Statton J, Hovey R, Ruiz Montoya L, Lowe R, Krauss S, Sinclair E (2017) Demographic and genetic connectivity: The role and consequences of reproduction, dispersal and recruitment in seagrasses. Biological Reviews 92:921–938

Kendrick G, Waycott M, Carruthers T, Cambridge M, Hovey R, Krauss S, Lavery P, Les D, Lowe R, I Vidal O (2012) The central role of dispersal in the maintenance and persistence of seagrass populations. BioScience 62:56–65

Lai S, Cheok Z, Bouma T, Yaakub S, Todd P (2020) The relative importance of light and hydrodynamics in the decay and transport of vegetative seagrass fragments. Estuarine, Coastal and Shelf Science 232:106430

Lai S, Yaakub S, Poh TS, Bouma T, Todd P (2018) Unlikely Nomads: Settlement, establishment, and dislodgement processes of vegetative seagrass fragments. Frontiers in Plant Science 9:1–11

Lenth R (2020) Emmeans: Estimated marginal means, aka least-squares means. R Package Version 1.4.8. https://CRAN.R-project.org/package=emmeans

Lewis S, Bainbridge Z, Davis A (2024) Sediment and nutrient flux from land. In Wolanski E, Kingsford MJ (eds.), Oceanographic Processes of Coral Reefs. Physical and Biological Links in the Great Barrier Reef. CRC Press

Manning W, Basu A, Mullahy J (2005) Generalized modeling approaches to risk adjustment of skewed outcomes data. Journal of Health Economics 24:465–488

McConkey K, Prasad S, Corlett R, Campos-Arceiz A, Brodie J, Rogers H, Santamaria L (2012) Seed dispersal in changing landscapes. Biological Conservation 146:1–13

McMahon K, van Dijk K, Ruiz-Montoya L, Kendrick G, Krauss S, Waycott M, Verduin J, Lowe R, Statton J, Brown E, Duarte C (2014) The movement ecology of seagrasses. Proceedings of the Royal Society B 281:1–9

R Core Team (2020) R: A language and environment for statistical computing. Vienna, Austria. https://www.R-project.org/

Ruiz-Montoya L, Lowe R, Kendrick G (2015) Contemporary connectivity is sustained by wind-and current-driven seed dispersal among seagrass meadows. Movement Ecology 3:1–9

Saunders M, Brown C, Foley M, Febria C, Albright R, Mehling M, Kavanaugh M, Burfeind D (2015) Human impacts on connectivity in marine and freshwater ecosystems assessed using graph theory: A review. Marine and Freshwater Research 67:277–290

Schlaefer J, Carter A, Choukroun S, Coles R, Critchell K, Lambrechts J, Rasheed M, Tol S, Grech A (2022) Marine plant dispersal and connectivity measures differ in their sensitivity to biophysical model parameters. Environmental Modelling & Software 149:105313

Short F, Polidoro B, Livingstone S, Carpenter K, Bandeira S, Bujang J, Calumpong H, Carruthers T, Coles R, Dennison W (2011) Extinction risk assessment of the world's seagrass species. Biological Conservation 144:1961–1971

Sinclair E, Ruiz-Montoya L, Krauss S, Anthony J, Hovey R, Lowe R, Kendrick G (2018) Seeds in motion: Genetic assignment and hydrodynamic models demonstrate concordant patterns of seagrass dispersal. Molecular ecology 27:5019–5034

Smith T, York P, Broitman B, Thiel M, Hays G, van Sebille E, Putman N, Macreadie P, Sherman C (2018) Rare long-distance dispersal of a marine angiosperm across the Pacific Ocean. Global Ecology and Biogeography 27:487–496

Swearer S, Treml E, Shima J (2019) A review of biophysical models of marine larval dispersal. CRC Press

Thomson A, York P, Smith T, Sherman C, Booth D, Keough M, Ross J, Macreadie P (2015) Seagrass viviparous propagules as a potential long-distance dispersal mechanism. Estuaries and Coasts 38:927–940

Tol S, Jarvis J, York P, Grech A, Congdon B, Coles R (2017) Long distance biotic dispersal of tropical seagrass seeds by marine mega-herbivores. Scientific Reports 7:1–8

van Dijk J, van Tussenbroek B, Jiménez-Durán K, Márquez-Guzmán G, Ouborg J (2009) High levels of gene flow and low population genetic structure related to high dispersal potential of a tropical marine angiosperm. Marine Ecology Progress Series 390:67–77

Weatherall E, Jackson E, Hendry R, Campbell M (2016) Quantifying the dispersal potential of seagrass vegetative fragments: A comparison of multiple subtropical species. Estuarine, Coastal and Shelf Science 169:207–215

Wells J, Rasheed M (2017) Port of Townsville annual seagrass monitoring and baseline survey: September—October 2016. In James Cook University Publication, Centre for Tropical Water & Aquatic Ecosystem Research (TropWATER), Cairns, p. 54

Wolanski E (2017) Bounded and unbounded boundaries–untangling mechanisms for estuarine-marine ecological connectivity: Scales of m to 10,000 km–a review. Estuarine, Coastal and Shelf Science 198:378–392

Wolanksi E, Kingsford M, Lambrechts J, Marmorino G (2024) The physical oceanography of the Great Barrier Reef: A review. In Wolanski E, Kingsford MJ (eds.), Oceanographic Processes of Coral Reefs. Physical and Biological Links in the Great Barrier Reef. CRC Press

Zuur A, Ieno E, Elphick C (2010) A protocol for data exploration to avoid common statistical problems. Methods in Ecology and Evolution 1:3–14

13 Interactions between Dugong Biology and the Biophysical Determinants of Their Environment

A Review

Helene Marsh and Christophe Cleguer

13.1. INTRODUCTION

The Great Barrier Reef (GBR) region supports globally significant populations of the dugong (*Dugong dugon*), a medium-sized coastal marine mammal and a seagrass community specialist (Marsh et al. 2011). The dugong's huge range spans ~135 degrees of longitude in the Indo-West Pacific and > 50 degrees of latitude, ~25 degrees both north and south of the equator. The dugong has high biodiversity value because it occupies a unique niche as the only extant member of the family Dugongidae, one of only five members of the order Sirenia (sea cows), and the only herbivorous mammal that spends its entire life in the sea (Marsh et al. 2011).

The International Union for Conservation of Nature lists the dugong as 'Vulnerable' because anecdotal evidence suggests that their numbers have decreased throughout most of their range (Marsh and Sobtzick 2015). Australian waters now support most of the world's dugongs, and the dugong is the marine mammal that is most abundant throughout most of the year in the inshore waters of northern Australia, including the GBR ecosystem. Dugongs are the most abundant mammal in the GBR, except during winter when approximately 40,000 humpback whales undergo their winter migration to and from the Antarctic (Noad et al. 2019).

For the purpose of this chapter, the GBR region includes Torres Strait, which supports the world's largest dugong population (102,600 ± se 20,150; 2013 data; Hagihara et al. 2018). Within the Great Barrier Reef World Heritage Area (GBRWHA, see Figure 1), the remote area north of Cooktown (NGBR) supported an estimated 7,000 ± se 600 dugongs in 2018–2019 (Marsh et al. 2020); the coastal waters of the remainder of the region (SGBR) supported 3,400 ± se 650 in 2016–18 (Sobtzick et al. 2017). These populations are linked to other globally important dugong populations in Southeast Queensland: Hervey Bay (2,100±se 380 in 2016; Sobtzick et al. 2017) and Moreton Bay (600±80 in 2016; Sobtzick et al. 2017); see Figure 1 and Clark et al. (2021) for details of these relative population estimates, which are from standardised aerial surveys).

This time series of surveys suggested that the outlook of the dugong population is stable or increasing in Torres Strait, NGBR, Hervey Bay, and Moreton Bay but declining in the SGBR (Clark et al. 2021). Outside the NGBR, this assessment is based on out-of-date data (> 5 years since last survey), which weakens confidence in the assessment. The effects on dugongs from the 2022 floods in Southeast Queensland (York et al. 2022) and the declines in subtidal seagrass throughout the Western and Central Clusters and intertidal seagrasses in the Western Cluster in Torres Strait (Carter et al. 2021a) are unquantified.

Dugongs move between the GBR and adjacent regions. Genetic pedigree evidence (Cope et al. 2015) shows that individual dugongs move between Moreton Bay and Hervey Bay in Southeast Queensland. Some satellite-tracked dugongs have also moved between Moreton Bay and Hervey Bay (Zeh et al. 2016), and others have moved between Hervey Bay and the GBR as far north as Clairview, north of Shoalwater Bay (Sheppard et al. 2006). Anecdotal information suggests that dugongs also move between other areas as shown in Figure 1, including between Torres Strait and both the NGBR and the Northeastern Gulf of Carpentaria.

Australia has international obligations to conserve dugongs in its waters. The significance of the GBRWHA for dugongs was among the reasons for its World Heritage listing (GBRMPA 1981). The Convention of Migratory Species (CMS; Marsh et al. 2011) lists the dugongs in its Appendix 11. Australia is a signatory to these Conventions and the CMS Dugong Memorandum of Understanding. Consequently, the dugong is listed as a Matter of National Environmental Significance under the Commonwealth Environment Protection and Biodiversity Conservation Act 1999 (EPBC Act).

In 2020, globally Important Marine Mammal Areas (IMMAs; Tetley et al. 2022), with dugongs as a qualifying species, were identified using an internationally accepted, expert process in Central and Western Torres Strait, NGBR, Hinchinbrook to Round Hill in the SGBR, Hervey Bay, Great Sandy Strait, and Moreton Bay (see www.marinemammalhabitat.org/immas/immas-searchable-database/).

Given the importance of dugongs in the GBR and ongoing concerns about their status, especially in the SGBR, it is important to understand the interactions between dugong

DOI: 10.1201/9781003320425-16

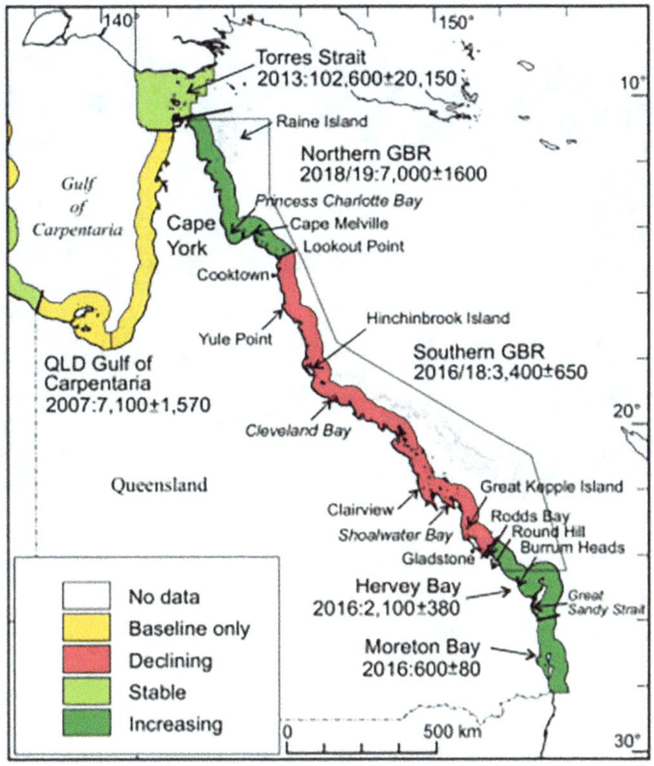

FIGURE 1 Map showing the estimated size of the dugong population of the GBR and adjacent areas based on the most recent aerial surveys at the time of writing, plus the place names mentioned in the text. The GBRWHA is outlined in grey.

Source: Adella Edwards developed the figure, reproduced with permission

biology and the biophysical determinants of their environment. Thus, the objectives of this chapter are to

a. synthesise information on the dugong's morphology, physiology, and traditions relevant to its ecology;
b. explore how these biophysical attributes interact with the biophysical attributes on the dugong's environment;
c. consider how climate change and its co-stressors might affect these interactions, especially in the GBR including Torres Strait; and
d. provide suggestions for future research.

13.2. INFORMATION ON THE DUGONG'S MORPHOLOGY, PHYSIOLOGY, AND TRADITIONS RELEVANT TO ITS ECOLOGY

13.2.1. Challenges

Three major challenges influence each dugong's interactions with its biophysical environment:

- as a homeotherm, it must maintain its core body temperature around the mammalian norm of 35–38 °C, despite the conductivity of water being 25 times greater than air at 25 °C and sea water temperatures almost always being lower than their core body temperature throughout their range;
- as an obligate air breather, benthic feeder, and seagrass community specialist (Keith-Diagne et al. 2022), it must be able to transition efficiently between the seafloor and the sea surface hundreds of times per day in shallow coastal and island waters;
- as an herbivore, it must spend a large proportion of time feeding (Marsh et al. 2011; Keith-Diagne 2022), as adults consume approximately 7% of their body weight per day (or up to ~35 kg wet weight of seagrass for a 500-kg dugong); thus, it must be able to navigate around fragmented intertidal and subtidal seagrass meadows adversely impacted by extreme weather events (Marsh et al. 2022).

13.2.2. Morphological and Sensory Adaptations

The dugong has evolved numerous morphological and sensory adaptations to enable it to meet these challenges (Marshall et al. 2022). We summarise these adaptations, some of which are inferred from research on another sirenian, the Florida manatee, to address the first objective of this chapter.

FIGURE 2 Dugong using its vibrissae-covered, hydrostatic muzzle or oral disc to find and excavate its food in a sparse seagrass community. Note (1) the other visible sensory structures—the small left eye and the left earhole posterior to the eye; (2) the excavated sediment being swept away by the current; and (3) the remoras on the dugong's back.

Source: Ahmed Shawky, photograph, reproduced with permission

The morphological and physiological adaptations that assist dugongs to maintain their core body temperature include (1) their relatively large body size (maximum body length up to 3.3 m; maximum weight up to 570 kg, Marsh et al. 2011); (2) a blubber layer in the hypodermis; (3) dense heavy skin; (4) a series of counter-current heat exchangers in their circulatory system to balance both heat loss and heat gain; and (5) a metabolic rate that is likely higher than predicted from their phylogeny (Lanyon et al. 2006 and *pers comm*., Marshall et al. 2022). The effectiveness of these adaptations means that behavioural thermoregulation mostly applies at the high latitude limits to their range in winter, including Moreton and Hervey Bays south of the GBR rather than in the GBRWHA or Torres Strait.

Dugongs propel themselves through the water with movements using caudal oscillations of their tail flukes, like dolphins and whales (Marsh et al. 2011). The arrangement of their skeletal, respiratory, integumentary, and digestive systems enables them to maintain their horizontal trim in the water (hydrostatsis). They have precise buoyancy control, which reduces the energetic costs of transiting from the benthic substrate, where their food is located, to the surface where air is inhaled (Marshall et al. 2022).

All mammals use at least the following five senses to varying degrees: touch (active and passive), hearing, vision, smell (olfaction), and taste (gustation) (Yong 2022). Although some marine mammals use other senses (e.g., odontocete cetaceans echolocate), there is no evidence that sirenians use more than the basic five. Nonetheless, along with manatees (Bauer and Reep 2022; Marshall et al. 2022), the dugong's perceptions of their environment are very different from those of other marine mammals. Research with Florida manatees indicates that sirenians are 'tactile auditory specialists with complementary visual and chemosensory abilities' and 'good spatial cognition' (Bauer and Reep 2022).

The sense most important to the dugong is somatosensation (touch, pressure, pain, temperature, position, movement, and vibration). All hairs on its body are sensory (Bryden et al. 1978; Kamiya and Yamasaki 1981). In the Florida manatee, the primary somatosensory complex, the touch centre of the brain, occupies approximately 24% of the total cortical area or ~70% of the area occupied by the primary sensory areas (Sarko and Reep 2007); this proportion is similar to that of terrestrial mammalian somatosensory specialists (Marshall et al. 2022). The dugong is likely to be similar. The Florida manatee has around 2,000 vibrissae on its face and head where its density is ~30 times greater than that of the ~3,300 hairs on the rest of its body (Reep et al. 2001, 2002). The sensory and motor functions of these facial vibrissae enable dugongs to locate and acquire food. (Figure 2).

The dugong's vibrissae-covered muzzle or oral disc is capable of numerous varied and detailed movements due to the hypertrophy of its muscles into a muscular hydrostat or shortened elephantine trunk-like structure (Marshall et al. 2022). Its 70° downwardly deflected snout makes it an obligate bottom feeder (Domning 1982).

Like those of the Florida manatee, the post-cranial hairs of the dugong likely operate like the lateral lines of fishes, serving to detect underwater features such as water movements and approaching animals (Reep et al. 2001, 2002; Gaspard et al. 2017). Thus, these hairs allow the detection of tides and currents that are important determinants of dugong movements as discussed subsequently.

Indigenous hunters of dugongs consider that dugongs have acute hearing (see Ponnampalam et al. 2022), and present-day dugong hunters in Torres Strait allow their boats to drift with the wind under a make-shift sail to avoid detection and to increase their chances of getting within harpooning distance (Kwan et al. 2006). The shallow acoustic environment in which most dugongs live is complex, and there is considerable uncertainty about their responses to most sounds including vessel noise.

Modern research on hearing in sirenians has mainly focussed on Florida manatees because of the availability of fresh carcasses, captive animals, and advanced techniques and equipment (Marshall et al. 2022). These studies show that (1) the neuroanatomical structures related to hearing are well-developed (2) at frequencies up to 20 kHz. Florida manatees have comparable or better hearing sensitivity than pinnipeds (seals and sea lions) and odontocete (toothed) cetaceans. Between 20 and 46 kHz, manatees exhibit greater auditory sensitivity than pinnipeds but less than toothed cetaceans (Gerstein et al. 1999). Manatee peak hearing sensitivity is at 8–20 kHz (Gaspard et al. 2012; Gerstein et al. 1999), which is higher than their fundamental vocalisation frequency of 2.5–5 KHz (O'Shea et al. 2022). The reasons for this difference are unknown.

The peak hearing sensitivity of the dugong is likely to be similar to that of the Florida manatee. Researchers have described their single call notes as 'bird-like' chirps, trills, and barks with frequencies in the 3–18 kHz range (see Anderson and Barclay 1995 and Marshall et al. 2022; and O'Shea et al. 2022, who provide a complete list of studies on dugong vocalisations). In shallow waters, such as most dugong habitats in the GBRWHA, dugong calls attenuate over distances of 50–150 m depending on salinity, bathymetry, sediment type, and ambient noise (Parsons et al. 2013).

Dugong vocalisations are social communication signals rather than navigational aids (O'Shea et al. 2022). Captive male dugongs in Toba Aquarium, Japan appear to produce more and longer calls when they show active movement, including certain sexual activities. Hishimoto et al. (2005) hypothesized that an increased rate of call production, especially calls of long durations, could be related to mating behaviour. Tanaka et al. (2017, 2023) documented 'vocal hotspots': areas of ~1 km^2 where dugongs showed elevated levels of vocal behaviour in the waters of Talibon Island, Thailand. The function of these hotspots is unknown; no cow-calf pairs have been observed there. Vocalisation locations were relatively consistent in both the rainy and dry seasons (Tanaka et al. 2023. Vocal hotspots have not been investigated in the GBR.

Dugongs use vision, but its usefulness is presumably limited at night and in the typically turbid environments of most seagrass meadows in the coastal GBR. Sirenian eyes and the central nervous system structures related to vision are relatively small (Marshall et al. 2022). The anatomy of their eyes indicates that dugongs are far-sighted underwater (Petit and Rochon-Duvigneaud 1929). Dugongs approach boats and inspect them visually. Based on anecdotal observations in the relatively clear waters of Shark Bay in Western Australia, Anderson (1979) rated the dugongs' underwater visual acuity as equivalent to that of a human diver wearing goggles and concluded that dugongs can see in the air when their heads are raised above water (Anderson 1981). Dugongs probably use vision at intermediate or longer distances for their initial orientation to an environment (at least during the day) and use direct tactile contact for further exploration (Marshall et al. 2022).

As with the other sensory modalities, almost all of our knowledge of gustation (taste) comes from captive Florida manatees. Yamasaki et al. (1980) described taste buds on the tongues of dugongs and West Indian and West African manatees to indicate that they can taste their food. Barboza and Larkin (2020) found that manatees have a relatively large number of taste buds along the sides of the root of their tongue, an arrangement that would enable them to taste food as they masticate and expel it (see Yong 2022) before it is swallowed. If dugongs are similar, then this arrangement may partially explain why there is so little sediment in most dugong's stomachs on necropsy (Heinsohn and Birch 1972), except when they are suffering food shortage (Spain and Heinsohn 1973; Preen and Marsh 1995). In addition, excavating dugongs tend to feed into the current, which washes the sediment away from their feeding apparatus (Figures 2 and 3).

Both male and female Florida manatees possess anal scent glands, which may be used to communicate their reproductive status to other manatees via their taste buds (Bills et al. 2013). Whether the same structures are present in dugongs is unknown. Like other marine mammals that keep their nostrils closed most of the time (Bauer and Reep 2022), dugongs are likely to have little use for olfactory clues. Sirenians have no vomeronasal organ, and the olfactory system of Florida manatees is reduced (Marshall et al. 2022). Thus, dugongs are likely to rely more on taste than on smell for chemoreception.

FIGURE 3 Aerial photograph of a large herd of dugongs feeding in Moreton Bay. Many animals are feeding into the current (from the top of the image), which is washing the disturbed sediment away from them as they move forward to feed.

Source: Susan Sobtzick, photograph, reproduced with permission

We conclude that a dugong's perception of its environment or 'umwelt' (Yong 2022) is very different from a human's perception of this environment. This difference is very important in interpreting how dugongs respond to their biophysical environment.

13.2.3. Socially Transmitted Knowledge

In addition to morphological and sensory adaptions, Florida manatees and likely dugongs rely on socially transmitted knowledge from their mothers during the years of calf dependency (Deutsch et al. 2022a; O'Shea et al. 2022). The individual differences in observed dugong movement behaviour in both two and three dimensions (Sheppard et al. 2006; Deutsch et al. 2022a, 2022b; Zeh 2022) suggest a similar learned component to their behaviour. Some satellite-tracked dugongs undertake movements of several hundred kilometres, often bypassing other seagrass meadows en route, suggesting that spatial memory informs long-range movement. For example, Sheppard et al. (2006) recorded two dugongs moving north from Hervey Bay to Great Keppel Island and Clairview in the SGBR, which passed by Rodds Bay, an area of high seagrass density 185 km north of Burrum Heads in Hervey Bay. Such knowledge of the spatiotemporal distribution of their food resources should enable dugongs to effectively exploit their heterogeneous environments. Sheppard et al. (2006) suggested that dugongs maintain a spatial memory of patches of seagrass, which they visit periodically, possibly to monitor the quantity and quality of accessible food resources. Manatees also appear to have excellent spatial cognition (Bauer and Reep 2022; Deutsch et al. 2022a, 2022b). Dugongs also appear to use a combination of spatial memory and knowledge of habitat morphology to mitigate risks from predation. For example, in places where the coral reef lagoons of New Caledonia are narrow, dugongs have learned to use the fore reef shelf to travel from one lagoonal bay to another, presumably as a strategy to avoid being ambushed by sharks in the confined spaces and dead ends created by the reticulated reefs inside the lagoon (Cleguer et al. 2020).

13.3. INTERACTIONS BETWEEN DUGONG BIOLOGY AND THE BIOPHYSICAL ATTRIBUTES OF THEIR ENVIRONMENT

13.3.1. Seagrasses

Dugongs are seagrass community specialists (Marsh et al. 2011). Since the 1980s, large-scale aerial surveys have provided temporal and spatial information on dugong distribution and abundance in most of their Australian range including the GBRWHA and Torres Strait (see Cleguer and Marsh (2023) for a complete list of surveys). The aerial surveys suggest that the most important seagrass communities for dugongs in the GBRWHA are likely to be intertidal and shallow subtidal, estuarine, and coastal communities, in addition to the reef communities in the NGBR, especially the large offshore reefs in Princess Charlotte Bay. The areas where dugongs are most abundant in the region are between ~15° S and ~14° S in the NGBR (Marsh et al. 2020) and the Hinchinbrook region and Cleveland Bay and Shoalwater Bay in the SGBR (Sobztick et al. 2017). The biophysical reasons why these areas are such important dugong habitats are poorly understood.

Carter et al. (2021b, 2021c) modelled the extent and community composition of potential seagrass communities in the GBRWHA based on a dataset sourced from multiple studies that included environmental data. They estimated (1) a total of 88,331 km^2 of potential seagrass habitat in intertidal and subtidal areas, namely, 1,111 km^2 in estuaries, 16,276 km^2 in coastal areas, and 70,934 km^2 in reef areas and (2) 36 seagrass community types defined by species assemblages within these habitat types. Their models indicate that the structure, location, and distribution of these communities are associated with complex environmental interactions including water depth, tidal exposure, latitude, current speed, benthic light, proportion of mud in the sediment, water type, water temperature, salinity, and wind speed.

The dugong aerial surveys have covered the coastal distribution of seagrass in the GBRWHA since the 1980s. However, the vast offshore subtidal communities in the SGBR (Carter et al. 2021b,2021c) have never been surveyed for dugongs. Deeper water seagrasses can recover more quickly from disturbance than seagrasses in shallow water sites (Preen et al. 1995; Rasheed et al. 2014). Thus, these offshore seagrass meadows may be important for dugongs when severe storms and/or floods damage their coastal feeding grounds. The strong spatial discontinuities identified in the modelling of offshore seagrass meadows between Clairview (north of Shoalwater Bay) and Gladstone might be inhibiting north-south movements of dugongs offshore and contributing to the deep genetic break at the Whitsundays (McGowan et al. 2023). The furthest offshore that dugongs have been confirmed in the GBRWHA is around Raine Island, on the outer edge of the northern GBR approximately 130 km off the Cape York coast, where a dugong was filmed with a drone in the summer of 2014–2015 at the western end of the reef flat, which supports seagrass (Raine Island Recovery Project *pers. comm.* 2023). The role that offshore seagrass has on dugong distribution, abundance, and movements in the GBRWHA merits further investigation.

In Torres Strait, surveys over 20 years have mapped ~13,500 km^2 of seagrass (see eAtlas:https://eatlas.org.au/data/uuid/d9a67694-d803-46ca-8f4b-63fc12d51bd1, accessed February 7, 2023). These data underestimate the area of intertidal and subtidal seagrass, especially in Central and Eastern Torres Strait because surveys are limited. Dedicated aerial surveys indicate the importance of Central and Western Torres Strait for dugongs, a region estimated to support more than 100,000 animals (Hagihara et al. 2018). Unlike the GBRWHA, most dugong sightings in Torres Strait have been in the 5 > 20 m depth zone

(66% of animals sighted in 2006, 60% in 2011, and 78% in 2013; Hagihara et al. 2018). Despite the estimated more than 2,000 km^2 of seagrass in deeper waters in Torres Strait, dugongs fitted with timed-depth reorders have not been detected in water > 20 m deep (which may be a bias resulting from the difficulty of catching dugongs in deeper water), and few dugongs have been sighted in this depth category during aerial surveys (1% in 2006, 4% in 2011, and 3% in 2013),

Dugongs probably eat all 12 species of seagrass that occur in the GBR (Waycott et al. 2007). Much has been claimed about some seagrass genera being 'preferred' by dugongs, but rigorous studies in which seagrass consumption is compared with availability are limited (but see Sheppard et al. 2010 and Tol et al. 2016 for work relevant to the GBR). No food-choice experiments have been conducted. Nonetheless, dugongs have limited capacity to masticate large, fibrous seagrasses because they cannot effectively break seagrass fragments into small digestible pieces (Lanyon and Sanson 2006a, 2006b). The dugong's diet can include infaunal invertebrates and algae in addition to seagrasses, especially when seagrasses are scarce after extreme weather events (Spain and Heinsohn 1973) or at the high latitude limits of their range in winter, where they have been recorded eating thin-shelled, burrowing mussels (*Botula vagina*) and possibly sea pens (*Virgularia*), ascidians (especially *Sycozoa pulchra*; Anderson 1998), and unidentified chaetopterid worms (Preen 1995a). Dugongs have not been recorded targeting invertebrates in the GBR, although they eat them incidentally (Marsh et al. 1982). Dugongs obtain all the freshwater that they need from their food, and there is no evidence that they drink freshwater (Smoll et al. 2020).

Dugongs use two modes of feeding: excavating and cropping (Wirsing et al. 2007a). They crop seagrass leaves when the roots and rhizomes are not accessible. In the GBR, dugongs likely crop *Thalassia hempricchii* and *Enhalus acoroides* or seagrass growing on compacted sediments (André et al. 2005; Domning and Beatty 2007).

Excavating dugongs leave characteristic feeding trails in the sediment (Figure 4), signs that have been used to study excavating patterns in intertidal meadows near Gladstone in the GBR (Rasheed et al. 2017).

Dugong feeding also affects seagrass communities in ways that depend on their feeding mode (excavating or cropping) and intensity of feeding, the season, species composition, location within the meadow, and status of the other seagrass meadows in the region (see Marsh et al. 2011 for a detailed account). Most studies in eastern Australia have been on excavating dugongs. In Moreton Bay, intensive excavating increases the rates of microbial nutrient cycling in seagrass sediments, which results in increased productivity (Perry and Dennison (1996, 1999). Aragones and Marsh (2000) experimentally showed in the GBR that intensive excavating results in seagrass meadows with a lower detrital load, presumably because most of the seagrass is eaten rather than left to die and decay.

Both low and high intensity excavating change the species composition of multispecies experimental plots, favouring the growth of early pioneer genera of seagrass such as *Halophila* and *Halodule* (Preen 1995b; Aragones and Marsh 2000). Simulated intensive excavating affects

FIGURE 4 Feeding trails produced by excavating dugongs in an intertidal seagrass bed at Yule Point in the GBR. Note the ungrazed seagrass between the feeding trails, which enables vegetative recovery.

Source: Len McKenzie, photograph, reproduced with permission the nutritional value of *Halophila* and *Halodule* by increasing the nitrogen value of the whole plant and reducing starch and fibre concentrations (Aragones and Marsh 2000).

The impacts of dugongs excavating on a seagrass meadow can be significant. Preen (1995b) estimated that a herd of over 460 dugongs consumed > 151,000 tonnes wet weight of seagrass from 41 hectares in Moreton Bay in < 17 days, removing 73–96% of the above-ground and 31–71% of the below-ground biomass. McMahon (2005) estimated that herds of 50–150 excavating dugongs in Moreton Bay removed an average of 23% of the seagrass biomass each month in a *Halophila ovalis* meadow. Removal peaked at 65% in August at the end of winter. When large areas of seagrass are lost because of seagrass dieback from unknown causes, sediment smothering, or extreme weather (or combinations of these factors), dugongs can 'overgraze' the remainder. This phenomenon has been described in Torres Strait on several occasions since the 1970s (e.g., Johannes and MacFarlane 1991; Carter et al. 2021a).

Tol et al. (2017) produced evidence from the GBRWHA suggesting that dugongs are likely to assist seagrass dispersal by eating seeds and passing the viable seeds of at least three seagrass species (*Zostera muelleri, Halodule uninervis* and *Halophila decipiens*) in their faeces. The spatial scale over which such dispersal might occur is large given the dugong's slow gut passage rate (Lanyon and Marsh 1995) and the hundreds of kilometres over which some dugongs travel in a few days (Sheppard et al. 2006; Deutsch et al. 2022a).

13.3.2. Water Depth and Bottom Type

Dugongs dive hundreds of times per day and spend a considerable portion of their daily time budget (18%; Hodgson 2004) rising to the surface and submerging to resume their previous activity. When surfacing between dives, dugongs spend a very short time breathing with only their paired nostrils above the surface (Marsh et al. 2011). Dive duration varies among individuals (e.g., Zeh 2022) and among studies (Keith-Diagne et al. 2022). At least some of this variation is likely a methodological artefact. It can be surprisingly difficult to identify a dive in a shallow diving animal living in coastal waters in variable environmental conditions, especially sea state (Hagihara et al. 2011). Nonetheless, all evidence indicates that maximum dive times (12.3 min; Chilvers et al. 2004) are short compared with other marine mammals and vary with water depth and bottom type (Keith-Diagne et al. 2022), especially the presence/absence of seagrass communities.

Most dugong dives are relatively shallow (< 5 m Chilvers et al. 2004; Hagihara 2015). Dugongs spend a high proportion of their time (1/3–1/2) within 1.5 m of the surface, although this proportion varies among locations with different bathymetries (Hagihara et al. 2014, 2018). Dugongs spend relatively more time away from the surface when they are over seagrass beds in deeper water (5 m to < 20 m), presumably to maximize their feeding time.

Hagihara's (2015) models indicate that dugongs transit quickly between the surface and the seafloor in locations supporting seagrass communities. Dugongs undertaking seafloor dives in locations apparently without seagrass transit more slowly between the surface and the bottom but may spend considerable time on the bottom, presumably resting. Dugongs in Moreton Bay rested at the bottom, mid-water, and on the surface (Hodgson 2004). Surface resting is common at the reef lagoon edge in New Caledonia (Hagihara et al. 2018) and in Shark Bay (Anderson 1982).

Dugongs can dive to depths of more than 30 m (Hagihara 2015; Sheppard et al. 2006). Lee Long et al. (1996) recorded dugong feeding trails to depths of 36 m during seagrass surveys in the GBR. There are no records of dugongs diving beyond 40 m, despite records of seagrass communities to 76 m in the GBR (Carter et al. 2021b).

Little is known about the dugong's habitat preference for mating and birthing in the GBR. Anderson (1997, 2002) recorded behaviour that he interpreted as lek mating at one site in eastern Shark Bay during spring-summer over two years, but this behaviour has never been reported elsewhere. Mating herds observed in the GBR have been consistent with those recorded in other locations including Shark Bay (see Marsh et al. 2011 and O'Shea et al. 2022 for overviews). Shallow inshore waters are likely important for birthing. Marsh et al. (1984) summarise anecdotal accounts including in the GBR that suggest that dugongs give birth in shallow coastal areas and in estuaries, perhaps to avoid shark predation (see Marsh et al. 2011). Traditional hunters confirm these observations (Johannes and MacFarlane 1991).

13.3.3. Waves

Dugongs adjust their local distribution and movement patterns to avoid rough sea conditions (Jarman 1966; Kingdon 1971; Nietschmann 1984; Johannes and MacFarlane 1991), sheltering in the lee of islands and reefs where they may form large transient groups. The pattern of these local movements interacts with seasonal wind and tidal conditions in monsoonal environments, such as the northern GBR and Torres Strait, to maximise shelter. Intense storms can result in large waves that strand dugongs in supra-tidal areas (Marsh 1989). For example, after cyclone Kathy crossed the coast of the Northern Territory near Borroloola in 1983, dugongs were stranded over an area of 250 km^2 on supra-tidal mud flats up to 9 km from the coast and behind stands of mangroves up to 3 m high (Marsh 1989).

13.3.4. Tides

The availability of the light required to support benthic marine plants restricts dugong's seagrass habitats to relatively shallow coastal and island waters. Consequently, most regional dugong populations are exposed to strong tidal influences, whose magnitude depends on the tidal regime. In the GBR, tides are diurnal, and the tidal range varies from 2.1 m to 6.7 m (Steinberg 2007). Low tides expose many seagrass meadows, which forces dugongs to commute between intertidal and sub-tidal areas to feed. Dugongs use local tidal flows and tidal channels to obtain high tide access to areas that are inaccessible at low tides (Anderson and Birtles 1978; Lanyon 2003; Sheppard et al. 2009). When dugongs undertake thermoregulatory movements in and out of Moreton Bay in winter, they save energy by taking advantage of both slack water and tidal flow. Eighty-nine percent of trips in and out of Moreton Bay were in the same direction as the tide (Zeh et al. 2018). Dugongs presumably use their sensory hairs to sense the direction of tidal flow and time their movements to maximise the opportunity for intertidal foraging while minimising the risk of being stranded.

13.3.5. Currents

McGowan et al. (2023) documented a deep genetic break in the dugong populations along the east coast of Australia at the Whitsundays. The reasons for this break are puzzling given the lack of historical or contemporary physical barriers. Wolanski et al. (2024) describe the 'sticky waters' in the Whitsunday region, the only place in the inshore waters of eastern Queensland where this effect has been documented. Although these strong current and tidal mixes would not prevent dugong movements through the region, they might be disorientating, given that hydrodynamic reception plays such a key role in how dugongs perceive their environment.

After extreme weather resulted in the loss of ~1,000 km^2 of seagrass in Hervey Bay in early 1992, at least 15 dugong carcasses were recovered in New South Wales including four south of Sydney, presumably because starving animals moved out of Hervey Bay and were caught in the southward East Australian current (Preen and Marsh 1995).

Dugongs have been recorded making extralimital movements across deep ocean trenches. For example, a subadult female dugong was stranded on Fiji, more than 600 km from the nearest known dugong population and beyond the eastern edge of the species' range (Hill-Lewenilovo et al. 2018). These records have led to speculation about the role of ocean currents in dugong dispersal and range expansion over evolutionary time.

13.3.6. Coastline Features

With the exception of thermoregulatory local scale movements in and out of Moreton Bay via South Passage (Preen 1992; Sheppard et al. 2006; Zeh et al. 2018), it has been difficult to identify travel corridors for dugongs, even though their large-scale movements (15 km > 560 km) *sensu* Sheppard et al. (2006) tend to be slow, steady and directed. Satellite tracking locations are biased against traveling dugongs because the signal is attenuated when the towed transmitter is underwater, and their travel 'paths' often have long gaps between fixes (Sheppard et al. 2006; Zeh et al. 2016; Cleguer et al. 2020).

Despite these difficulties, evidence suggests that some dugongs use coastline features to navigate, presumably with their somatosenses and hearing along pathways possibly learned from their mothers. For example, Zeh et al. (2016) tracked three dugongs moving between Moreton Bay and Hervey Bay that transited close (< 5 km) to a surf coast where there is a history of dugongs being caught in shark nets that are set parallel and close to the shore for bather protection (Marsh et al. 2005), despite the absence of seagrass along the surf coast. Along the central-west coast of New Caledonia, where the narrow lagoon contains dense reticulated reefs that impede movement, Cleguer et al. (2020) tracked four dugongs that occasionally moved outside of the lagoon to the fore reef shelf to travel from one bay to another.

Coastline features cannot aid navigation when dugongs move away from the coast. One male tracked by Sheppard et al. (2006) made a 290-km return trip from Burrum Heads in Hervey Bay to Great Keppel Island in the GBRWHA. He made several bouts of consecutive 5-min deep dives of up to 27 m during the 15 days that it took to make the initial journey and dived to 33.5 m on the 12-day return trip. The regular U-shaped profile of these dives suggests that they were to the seafloor. Sheppard et al. (2006) speculated that these dives might aid navigation, a conclusion supported by Hagihara's (2015) detailed studies of the diving behaviour of three telemetered dugongs making thermoregulatory movements in and between Moreton Bay and the adjacent oceanic waters. Water currents are always smaller close to the sea bottom due to bottom friction (*pers comm.* Eric Wolanski), and it would be possible for a dugong to sense this difference using its sensory hairs, even if it did not touch the bottom. Nonetheless, dugongs cannot always track the bottom to navigate. Occasional animals (likely vagrants) move across ocean trenches that are much deeper than their diving capability (Hobbs et al. 2007; Hill-Lewenilovo et al. 2018).

13.3.7. Light

Visual clues are presumably of limited importance to dugongs diving at night or in turbid waters. The proportion of time that they spent close to the surface varied consistently with the time of day in three locations (Moreton Bay, Torres Strait, and New Caledonia), as studied by Hagihara et al. (2018), and was least in daylight hours (0800–1600). Zeh et al. (2018) found that the diel cycle influenced thermoregulatory movements. Dugongs were more likely to travel out of Moreton Bay between midnight and noon and to return to the bay between noon and midnight. Based on data collected from 16 tracked dugongs, Derville et al. (2022) found that dugongs in the lagoons of New Caledonia used shallower waters near the mainland or near islets within the lagoon more at dusk and night than during the day and at dawn, and this effect was particularly obvious at high tide. Dugongs that were tracked outside the lagoon swam back inside the lagoon more frequently at dusk or night (n = 6), but this pattern was not statistically significant (p = 0.2).

Records from timed depth recorders demonstrate that dugongs dive throughout the diel cycle (Chilvers et al. 2004). Deutsch et al. (2022b) concluded that diel patterns in dugong activity vary among locations depending on habitat type, water depth, sea state, availability of intertidal and subtidal seagrass, and other factors including vessel traffic and predation risk. The relative importance of these factors is not known. Nothing is known about dugong sleep; sleep during resting has been monitored in a captive Amazonian manatee using electrophysiology (Marsh et al. 2011).

13.3.8. Water Temperature

The difference between the dugong's core temperature and that of seawater does not appear to be a problem for dugongs in the GBR. In contrast, anecdotal sightings of dugongs in New South Wales, which is well south of their accepted range on the east coast of Australia in summer (Allen et al. 2004), suggest that these animals may have undertaken seasonal long-distance movements from Queensland. Studies of telemetered dugongs indicate that they undertake seasonal migrations and/or local scale behavioural thermoregulation at the high-latitude limits to their range in Shark Bay on the west coast of Australia (Anderson 1982, 1986); Marsh et al. 1994; Preen et al. 1997; Holley 2006; Holley et al. 2006) and in Moreton Bay (Preen 1992; Zeh et al. 2018) and Hervey Bay on the east coast (Sheppard et al. 2006). In New Caledonia (20–22°S), dugongs rest in warmer water outside the reef lagoon at Cap Goulvain in the winter (known locally as the 'cool season'; Cleguer 2015). Although it is possible that dugongs undertake thermoregulatory movements in the SGBR, periodic seagrass loss driven by extreme weather events is likely be a more important driver of their movements.

13.3.9. Predation Risk

As long-lived slow breeding mammals (Marsh et al. 2011), dugongs are expected to undertake behaviours that maximise survival by actively reducing their risk from predation. Tiger sharks (Heithaus 2001; Simpfendorfer et al. 2001; Marsh et al. 2011), killer whales (Anderson and Prince 1985), and saltwater crocodiles (Marsh et al. 2011) have been documented eating dugongs, but how much of this consumption results from active predation rather than scavenging is not known. Some adult dugongs bear healed scars indicating that they have survived attack(s) by sharks (see Marsh et al. 2011), and Anderson and Prince (1985) documented killer whales attacking a herd of dugongs. We do not know how dugongs detect predators, but they apparently modify their behaviour to reduce predation risk.

There is anecdotal evidence that dugongs give birth in shallow water (Marsh et al. 2011). At other times, they apparently perceive deeper waters as areas of lower predation risk than shallow water and are often seen at the edges of seagrass meadows where they can easily escape to deeper water. Most of the claims that dugongs modify their behaviour to reduce predation risk come from Shark Bay in Western Australia, where Wirsing and his co-workers (Wirsing et al. 2007a, 2007b, 2007c; Wirsing and Heithaus 2012) interpreted seasonal habitat use as a tactic to avoid sharks. However, confounding correlates such as water temperature and seagrass growth may also play a role in dugong habitat selection and foraging behaviour. In tropical coral reef ecosystems in New Caledonia, dugongs venture outside of lagoons to transit from one inshore foraging habitat to another as explained in Section 13.3.6 to minimise the risk of ambush by sharks in shallow reticulated reefs (Garrigue et al. 2008; Cleguer et al. 2020). Nonetheless, dugongs are not always safe from predators in deeper water. The killer whale attack on a herd of dugongs reported by Anderson and Prince (1985) was in waters 6–7 m deep.

13.4. HOW MIGHT THESE INTERACTIONS BE AFFECTED BY CLIMATE CHANGE IN THE GBR?

In the GBR, the most important effects of climate change on dugongs is likely to be from changes to their seagrass habitats as detailed in Marsh et al. (2022). These changes will exacerbate the ongoing loss of seagrass caused by anthropogenic pressures in the coastal zone (Waycott et al. 2009). Seagrass biomass and distribution are both expected to reduce. In early 2022, floods in Hervey Bay reduced the biomass of intertidal seagrasses, and dugongs concentrated their feeding in the middle of the bay (York et al. 2022; Cleguer *per obs*). Any reduction in subtidal seagrasses will be particularly important to dugongs in Torres Strait where animals use deeper water than on the east coast (Hagihara et al. 2018). This change will presumably result in animals spending more time in intertidal waters where they are more accessible to anthropogenic sources of mortality and habitat loss (Marsh et al. 2022). Threats from continued agricultural activities and coastal development in the SGBR are also likely to reduce the resilience of seagrass meadows to climate change. Even though seagrass habitats may expand with sea level rise in some locations in the NGBR where coastal armouring is likely to remain minimal, an overall loss of estuarine and coastal seagrass is expected to further reduce the carrying capacity of the GBR for dugongs. Projected shifts in habitat use by dugongs as a result of climate change should be considered in future marine spatial planning.

Climate change is expected to alter the species composition of many seagrass meadows by favouring tropical species. Fortunately, GBR seagrasses largely comprise tropical species, Thus, the impact of marine heatwaves on seagrass communities is likely to be less than in Shark Bay in Western Australia. In 2011, a marine heatwave resulted in the decline of over 1,000 km^2 of meadows in Shark Bay, following extensive defoliation and rhizome death of the temperate *Amphibolis antarctica*, the dominant seagrass.

This loss enabled small, fast-growing tropical species eaten by dugongs such as *Halodule uninervis* to expand in the Bay (Kendrick et al. 2019).

The most relevant unknown is how climate-induced changes in seagrass community composition will affect dugong food quality, biomass, and preferred feeding locations. The basis for the dugong's food choice is poorly understood, and determinants of food quality are unknown (Aragones et al. 2006, 2012). We do not know whether the dugong's preferred feeding locations reflect environmental parameters, food quality or quantity, socially transmitted knowledge of resources, or some combination of these factors or how changes in food quality and/or biomass change will affect their time budgets.

Climate change is predicted to increase the intensity of tropical cyclones (see Duke et al. 2024) and the frequency of marine heatwaves, both of which can cause extensive loss of seagrass (see Marsh et al. 2022). Seagrass losses from tropical storms have resulted in increased dugong mortality (Heinsohn and Spain 1974; Preen and Marsh 1995; Meager and Limpus 2014). Spain and Heinsohn (1973) reported that after cyclone Althea damaged seagrass beds near Townsville in 1971, the surviving dugongs changed their diet by consuming more algae and invertebrates and unusually large amounts of sand and sediment. There are oral reports of changes in the habitat use of Torres Strait dugongs unable to find sufficient amounts of food after a seagrass dieback of unknown cause in the 1970s (Nietschmann and Nietschmann 1981; Nietschmann 1984; Johannes and MacFarlane 1991). Hunters described an unusually high proportion of 'wati dangal', lethargic animals with limited and poor-tasting fat. The 'wati dangal' had eaten larger amounts of brown and green algae than the 'malu dangal' (deeper water dugongs), which were fatter and good to eat (Nietschmann 1984).

Deeper water seagrasses tend to recover more quickly than intertidal and shallow, subtidal seagrasses after losses caused by cyclones (Preen et al. 1995; Rasheed et al. 2014) because their seed banks are less likely to have been damaged by sheer stress. Thus, any reduction in the depth range of seagrasses is likely to delay the recovery of dugong populations from climate-induced extreme weather events.

Weather events such as prolonged rain associated with the La Niña phenomenon, floods, cyclones, and marine heat waves that have caused seagrass loss (Carter et al. 2022) have also reduced fecundity in dugongs. For example, after the 1992 seagrass loss in Hervey Bay, the proportion of dugongs classified as calves fell from 22% in 1988 to 2.2% in 1993 (Preen and Marsh 1995). Sobtzick et al. (2012) did not see any calves during an aerial survey of the SGBR region in late 2011 after extreme weather events in the previous summer, including major floods and three cyclones. The calving rate recovered to normal by 2016 (Sobztick et al. 2017). The timing of this reduction in percentage of calves suggests a reduction in calf survivorship. After the seagrass dieback in Torres Strait in the 1970s, dugongs were unable to accumulate the fat stores required to ovulate and become pregnant (Marsh and Kwan 2008), suggesting that recruitment failure after seagrass loss may also lead to changes in fecundity lasting several years.

Some dugongs temporarily migrate after large-scale loss of the seagrasses communities on which they depend for food. For example, eight months after the 1992 Hervey Bay seagrass loss, the estimated population of dugongs in southern Hervey Bay was only 71 ± se 40 compared with 1753 ± se 388 in 1988 (Preen and Marsh 1995). The population had recovered to an estimated 2077 ± se 543 by 2005 (see Sobtzick et al. 2017), a recovery that can only be explained by temporary migration. Similarly, less than one year after the extreme weather on the east coast of Australia in 2010/2011, the dugong population of the SGBR survey areas was estimated to be only 478 ± se 253 compared with 1451 ± se 288 in 2005 and 2822 ± se 600 in 2016. In contrast, Bayliss et al. (2018) found no evidence of dugongs moving from Shark Bay in Western Australia after a marine heatwave in 2010/2011 killed the temperate seagrass *Amphibolis antarctica* (Kendrick et al. 2019), even though the percentage of dugongs classified as calves declined. This marine heatwave did not kill the low density tropical species that are a major predictor of dugong distribution and abundance in Shark Bay (Bayliss et al. 2019).

We do not know where individual dugongs go when their food community is destroyed because it is not ethically acceptable to fit stressed animals with tracking devices. Researchers (e.g., Gales et al. 2004) have assumed that animals move to regions where coastal seagrass has not been lost, which is a problem when losses are widespread, such as in eastern Queensland in 2011. In some areas, it is possible that dugongs move offshore. The dugong's options are likely to diminish as the depth range of seagrass contracts with climate change along with the likely ongoing loss of inshore seagrass from other anthropogenic stressors.

There is increasing evidence that sirenian calves learn the locations of key resources and the timing and direction of movements during their years of calf dependency. An important aspect of such social learning is the speed at which new behaviours can spread through populations (Brakes et al. 2019). Dugongs seem able to hedge their bets with only some animals undertaking temporary emigration following extensive seagrass loss (Preen and Marsh 1995; Sobtzick et al. 2012). Sheppard et al. (2006) and Deutsch et al. (2022a) suggested that the large-scale movements of individual dugongs may be a means of assessing the status of food sources at a scale larger than that required for their immediate needs. The locations of essential resources and cues triggering the timing of movements are predicted to rapidly alter as the climate changes, and this practice of undertaking 'exploratory' forays appears advantageous.

13.5. FUTURE DIRECTIONS

Multidisciplinary research will be required to progress our understanding of the interactions between dugongs and their biophysical environment. We outline some ideas about directions for such research as follows.

13.5.1. Modelling

Extensive long-term spatial datasets are available for both seagrasses and dugongs in the GBR. Modelling the relationships between the distribution and relative abundance of dugongs in the coastal waters and seagrass presence, community type, species, and their physical determinants should provide important insights. This information should also be used to (1) interpret temporal changes in dugong population size and distribution, 2) model the growth of different species of seagrass, and identify possible alternative protected areas if seagrass meadows in current protected areas are damaged by extreme weather events that force dugongs to move.

13.5.2. Anatomical Studies

Specimens should be salvaged from the carcasses of stranded dead dugongs that are otherwise in good condition for anatomical studies of their brain anatomy and sense organs, especially their sensory hairs, taste buds, and anal scent glands, to compare with the results of the extensive studies using Florida manatees (see Marshall et al. 2022 for references).

13.5.3. Captive Animals

Experimental studies on captive animals have provided most of the detailed information on the sensory capacities of Florida manatees (Marshall et al. 2022). Replicating this work on dugongs will be difficult because there are so few animals in captivity world-wide (Marsh 2022). Nonetheless, such experiments could provide important information on dugong touch, hearing, vision, taste, and nutritional needs, including food choice and orientation.

13.5.4. Field Studies: Seagrasses

Field experiments using enclosures and/or simulated feeding have provided important information on the effects of dugongs and other megaherbivores on seagrasses (e.g., Preen 1995b; Aragones and Marsh 2000; Scott et al. 2020, 2021). Replicating these experiments in dugong hotspots in the GBR would provide important new information on the impacts of dugongs on seagrasses. Many of the experiments to date have been in meadows where excavating is the dominant form of dugong feeding, and priority should be given to areas where cropping is also used. It is encouraging to see current work in the reefs of Torres Strait by Abbi Scott and her co-workers.

Rasheed et al. (2017) mapped dugong feeding trails in intertidal areas in the Gladstone region using photogrammetry. Extending this work across intertidal habitats in the GBR would provide insights on the spatial pattern of dugong excavating behavior within intertidal meadows.

13.5.5. Field Studies: Application of New Technologies to Dugong Studies

The dugong is a relatively cryptic animal that mostly occurs in turbid waters in the GBR and is therefore difficult to study. Developments in custom-designed, animal-borne, multi-sensor tags (which record light sensitivity, pressure, temperature, and speed and can be coupled with accelerometers, magnetometers, video cameras, and fast-acquisition Iridium/GPS-satellite tracking equipment) are revolutionizing field studies of marine mammals and have the potential to greatly improve our understanding of fine-scale, three-dimensional diving behaviour, diel patterns in activity budgets, buoyancy, habitat use, movement corridors, animal interactions, and the biophysical drivers influencing these aspects of the dugong's behavioral ecology.

Limitations in the use of tracking apparatus on wild dugongs will need to be considered. For example, current tracking techniques require the capture of animals in the wild, which can be challenging and stressful for the animals. Multi-sensor tags remain attached to the animals only for hours to days and must be retrieved to access data. To date, few GPS-satellite tags have remained attached for more than 3–6 months (see Deutsch et al. 2022a; Supplementary Information). Thus, long-term studies of individual dugongs will be very difficult. Nonetheless, such studies, particularly of mothers and their attendant calves, would provide new insights into socially transmitted knowledge.

Environmental DNA (eDNA) has emerged as a promising tool for both species detection and biodiversity monitoring (Thomsen and Willerslev 2015). When breathing and diving, dugongs leave a 'footprint' in surface waters including genetic materials from their skin, mucus, and/or faeces. Once the method is developed for dugongs, eDNA could be used to detect dugong presence is areas of the GBR that have not been surveyed from the air, including the offshore seagrass beds and reefs in the SGBR. Areas where their presence is detected using eDNA could then be prioritised for aerial surveys to estimate abundance.

Off-the-shelf small drones, which are relatively cheap and easy to use, are increasingly deployed by citizen scientists and communities across the GBR, for example, collaborations among Traditional Owners, Traditional Use of Marine Resources Agreement (TUMRA) representatives, land and sea ranger programs, and scientists conducting local scale seagrass and dugong surveys. This approach has potential to inform the understanding of the relationship between the dugong's fine-scale habitat use and their biophysical environment. Images collected from small drones can also be used to measure body size and estimate the body condition of marine mammals. These methods have been developed for manatees (e.g., Ramos et al. 2022) and could be used to assess the effects of extreme weather on the health of dugongs in the GBR.

The use of acoustic receivers to record dugong vocalizations and investigate the functions of the calls has largely been limited to Thailand and Okinawa, Japan. Combined with additional investigations on the hearing abilities of dugongs outlined above, this approach could provide new insights to inform dugong management within the GBRWHA and beyond. In association with animal-borne acoustic transmitters, acoustic receivers also have the potential to understand the dugongs' use of three-dimensional space and their interactions with biophysical and anthropogenic environments, especially near ports (see Zeh 2022 for details).

The research outlined here has the potential to provide further insights into the interactions between dugongs and their biophysical environment and to inform management responses to anthropogenic threats to both dugongs and their seagrass habitats, particularly the threat of climate change.

ACKNOWLEDGMENTS

We thank Rob Coles, Alexandra Carter, Michael Kingsford, Rachel Miller, and Eric Wolanski for their helpful comments on an earlier version of this manuscript. We also thank Len McKenzie, Ahmed Shawky, and Susan Sobtzick for permission to reproduce their photographs and Adella Edwards for developing Figure 1.

REFERENCES

Allen S, Marsh H, Hodgson A (2004) Occurrence and conservation of the dugong (Sirenia: Dugongidae) in New South Wales. Proc Linn Soc NSW 125:211–216

Anderson PK (1979) Dugong behaviour: On being a marine mammalian grazer. The Biologist 61:113–144

Anderson PK (1981) Dugong behaviour: Observations, extrapolations, and speculations. In Marsh H (ed) The Dugong. Proceedings of a seminar/workshop held at James Cook University, Townsville, 8–13 May 1979, pp. 91–111

Anderson PK (1982) Studies of dugongs at Shark Bay, Western Australia. II. Surface and subsurface observations. Aust Wildl Res 9:85–99. doi: 10.1071/WR9820085

Anderson PK (1986) Dugongs of Shark Bay, Australia—Seasonal migration, water temperature, and forage. Natl Geogr Res 2:473–490

Anderson PK (1997) Shark Bay dugongs in summer. I: Lek mating. Behaviour 134:433–462

Anderson PK (1998) Shark Bay dugongs (Dugong dugon) in summer II: Foragers in a *Halodule*-dominated community. Mammalia 62:409–425. doi: 10.1515/mamm.1998.62.3.409

Anderson PK (2002) Habitat, niche, and evolution of sirenian mating systems. J Mamm Evol 9:55–98. doi: 10.1023/A:1021383827946

Anderson PK, Barclay RMR (1995) Acoustic signals of solitary dugongs: Physical characteristics and behavioral correlates. J Mammal 76:1226–1237. doi: 10.2307/1382616

Anderson PK, Birtles A (1978) Behavior and ecology of the dugong, *Dugong dugon* (Sirenia): Observations in Shoalwater and Cleveland Bays, Queensland. Aust Wildlife Res 5:1–23. doi: 10.1071/WR9780001

Anderson PK, Prince RI (1985) Predation on dugongs: Attacks by killer whales. J Mammal 66:554–556. doi: 10.2307/1380933

André J, Gyuris E, Lawler IR (2005) Comparison of the diets of sympatric dugongs and green turtles on the Orman Reefs, Torres Strait, Australia. Wildl Res 32:53–62. doi: 10.1071/WR04015

Aragones L, Marsh H (2000) Impact of dugong grazing and turtle cropping on tropical seagrass communities. Pacific Conserv Biol 5:277–288. doi: 10.1071/PC000277

Aragones LV, Lawler IR, Foley WJ et al. (2006) Dugong grazing and turtle cropping: Grazing optimization in tropical seagrass systems. Oecologia 149:635–647. doi: 10.1007/s00442-006-0477-1

Aragones LV, Lawler IR, Marsh H et al. (2012) The role of sirenians in aquatic ecosystems. In Hines E, Reynolds J, Mignucci-Giannoni A et al. (eds) International strategies for manatee and dugong conservation. University Press of Florida, Gainesville, pp. 4–11

Barboza MLB, Larkin IV (2020) Functional morphology of the taste buds of Florida manatee, *Trichechus manatus latirostris*. Mar Mamm Sci 36:939–952. doi: 10.1111/mms.12692

Bauer GB, Reep RL (2022) Manatee cognition in the wild: An exploration of the manatee mind and behavior through neuroanatomy, psychophysics, and field observations. Anim Cogn 25:1161–1182. doi: 10.1007/s10071-022-01686-6.

Bayliss P, Raudino HC, Hutton M (2018) Dugong (Dugong dugon) population and habitat survey of Shark Bay Marine Park, Ningaloo Reef Marine Park and Exmouth Gulf. Report 1 to DBCA Marine Science Program and National Environmental Science Program, DBCA, Perth, 51pp.

Bayliss P, Raudino HC, Hutton M (2019) Modelling the spatial relationship between dugong (Dugong dugon) and their seagrass habitat in Shark Bay Marine Park, before and after the marine heatwave of 2010/11. Report 2 to DBCA Marine Science Program and National Environmental Science Program, DBCA, Perth, 55pp.

Bills ML, Samuelson DA, Larkin ILV (2013) Anal glands of the Florida manatee, *Trichechus manatus latirostris*: A potential source of chemosensory signal expression. Mar Mamm Sci 29:280–292. doi: 10.1111/j.1748-7692.2012.00564.x

Brakes P, Dall SRX, Aplin LM et al. (2019) Animal cultures matter for conservation. Science 363:1032–1034. doi: 10.1126/science.aaw3557

Bryden MM, Marsh H, MacDonald BW (1978) The skin and hair of the dugong, *Dugong dugon*. J Anat 126:637–638

Carter A, David M, Whap T, Hoffman L et al. (2021a) Torres Strait Seagrass 2021 Report card. TropWater Report. James Cook University, Cairns

Carter AB, Collier C, Coles R et al. (2022) Community-specific "desired" states for seagrasses through cycles of loss and recovery. J Environ Manage 314:115059. doi: 10.1016/j.jenvman.2022.115059.

Carter AB, Collier C, Lawrence E et al. (2021b) A spatial analysis of seagrass habitat and community diversity in the Great Barrier Reef World Heritage Area. Sci Rep 11:22344 doi: 10.1038/s41598-021-01471-4

Carter AB, McKenna SA, Rasheed MA et al. (2021c) Synthesizing 35 years of seagrass spatial data from the Great Barrier Reef World Heritage Area, Queensland, Australia. Limn Oceanog 6:216–226. doi: 10.1002/lol2.10193

Chilvers BL, Delean S, Gales NJ et al. (2004) Diving behavior of dugongs, *Dugong dugon*. J Exp Mar Bio Ecol 304:203–224. doi: 10.1016/j.jembe.2003.12.010

Clark G, Fischer M, Hunter C (2021) Coasts. In Australia state of the environment 2021. Australian Government Minister for the Environment, Commonwealth of Australia. doi: 10.26194/AANZ-RF46

Cleguer C (2015) Informing dugong conservation at several spatial and temporal scales in New Caledonia. PhD thesis, Université Pierre et Marie Curie, Paris, and James Cook University, Townsville

Cleguer C, Garrigue C, Marsh H (2020) Dugong (*Dugong dugon*) movements and habitat use in a coral reef lagoonal ecosystem. Endang Species Res 43:167–181. doi: 10.3354/esr01061

Cleguer C, Marsh H (2023) An inventory of dugong aerial surveys in Australia. Report to the National Environmental Science Program. Centre for Tropical Water and Aquatic Ecosystem Research (TropWATER), Report 23/15, James Cook University, Townsville, 49pp

Cope RC, Pollett PK, Lanyon JM et al. (2015) Indirect detection of genetic dispersal (movement and breeding events) through pedigree analysis of dugong populations in southern Queensland, Australia. Biol Conserv 181:91–101. doi: 10.1016/j.biocon.2014.11.011

Derville S, Cleguer C, Garrigue C (2022) Ecoregional and temporal dynamics of dugong habitat use in a complex coral reef lagoon ecosystem. Sci Rep 12:1–15. doi: 10.1038/s41598-021-04412-3

Deutsch CJ, Castelblanco-Martínez D, Groom R et al. (2022a) Movement behavior of manatees and dugongs: I. Environmental challenges drive diversity in migratory patterns and other large-scale movements. In Marsh H (ed) Ethology and behavioral ecology of sirenia. Springer International Publishing, Cham, pp. 155–231

Deutsch CJ, Castelblanco-Martínez D, Groom R et al. (2022b) Movement behavior of manatees and dugongs: II. Small-scale movements reflect adaptations to dynamic aquatic environments. In Marsh H (ed) Ethology and behavioral ecology of sirenia. Springer International Publishing, Cham, pp. 233–298

Domning DP (1982) Evolution of manatees: A speculative history. J Paleontol 56:599–619

Domning DP, Beatty BL (2007) Use of tusks in feeding by dugongid sirenians: Observations and tests of hypotheses. Anat Rec 290:523–538. doi: 10.1002/ar.20540

Duke NC, Canning A, Mackenzie J (2024) More intense severe tropical cyclones in recent decades cause greater impacts on mangroves bordering Australia's Great Barrier Reef. In Wolanski E, Kingsford MJ (eds) Oceanographic Processes of Coral Reefs Physical and Biological Links in the Great Barrier Reef. CRC Press, Boca Raton, FL.

Gales N, McCauley RD, Lanyon J et al. (2004) Change in abundance of dugongs in Shark Bay, Ningaloo and Exmouth Gulf, Western Australia: Evidence for large-scale migration. Wildl Res 31:283–290. doi: 10.1071/WR02073

Garrigue C, Patenaude N, Marsh H (2008) Distribution and abundance of the dugong in New Caledonia, Southwest Pacific. Mar Mamm Sci 24:81–90. doi: 10.1111/j.1748-7692.2007.00173.x

Gaspard JC, Bauer GB, Mann DA, et al. (2017) Detection of hydrodynamic stimuli by the postcranial body of Florida manatees (*Trichechus manatus latirostris*). J Comp Physiol A 203:111–120. doi: 10.1007/s00359-016-1142-8

Gaspard JC, Bauer GB, Reep RL et al. (2012) Audiogram and auditory critical ratios of two Florida manatees (*Trichechus manatus latirostris*). J Exp Biol 215:1442–1447. doi: 10.1242/jeb.089201

GBRMPA (1981) Nomination of the Great Barrier Reef by the Commonwealth of Australia for inclusion on the World Heritage List. UNESCO, Paris, 37pp

Gerstein ER, Gerstein L, Forsythe SE et al. (1999) The underwater audiogram of the West Indian manatee (*Trichechus manatus*). J Acoust Soc Am 105:3575–3583. doi: 10.1121/1.424681

Hagihara R (2015) Linking wildlife tracking data with environmental features to improve understanding of dugong diving ecology and population size estimates. PhD thesis, James Cook University, Australia.

Hagihara R, Jones RE, Grech A et al. (2014) Improving population estimates by quantifying diving and surfacing patterns: A dugong example. Mar Mamm Sci 30:348–366. doi: 10.1111/mms.12041

Hagihara R, Jones RE, Sheppard JK et al. (2011) Minimizing errors in the analysis of dive recordings from shallow-diving animals. J Exp Mar Biol Ecol 399:173–181. doi: 10.1016/j.jembe.2011.01.001

Hagihara R, Jones RE, Sobtzick S et al. (2018) Compensating for geographic variation in detection probability with water depth improves abundance estimates of coastal marine megafauna. PLoS One 13:e0191476. doi: 10.1371/journal.pone.0191476

Heinsohn GE, Birch WR (1972) Foods and feeding habits of the dugong, *Dugong dugon* (Erxleben), in northern Queensland Australia. Mammalia 36:414–422. doi: 10.1515/mamm.1972.36.3.414

Heinsohn GE, Spain AV (1974) Effects of a tropical cyclone on littoral and sub-littoral biotic communities and on a population of dugongs (*Dugong dugon* [Müller]). Biol Conserv 6:143–152. doi: 10.1016/0006-3207(74)90026-3

Heithaus MR (2001) The biology of tiger sharks, *Galeocerdo cuvier*, in Shark Bay, Western Australia: Sex ratio, size distribution, diet, and seasonal changes catch rates. Environ Biol Fish 61:25–36. doi: 10.1023/A:1011021210685

Hill-Lewenilovo R, Vuiyasawa R, Piovano S (2018) First record of a dugong (*Dugong dugon* Müller, 1776) in Fiji. Pac Conserv Biol 25:324–325. doi: 10.1071/PC18059

Hishimoto Y, Ichikawa K, Akamatsu T et al. (2005) The acoustical characteristics of dugong calls and the behavioral correlation observed in Toba aquarium. In Proceedings of the 2nd International Symposium on SEASTAR2000 and Asian Biologging Science (The 6th SEASTAR2000 Workshop). pp. 25–28, SEASTAR, Bangkok

Hobbs J-PA, Frisch AJ, Hender J et al. (2007) Long-distance oceanic movement of a solitary dugong (*Dugong dugon*) to the Cocos (Keeling) Islands. Aquat Mamm 33:175–178. doi: 10.1578/AM.33.2.2007.175

Hodgson A (2004) Dugong behavior and responses to human influences. Dissertation, James Cook University, Townsville

Holley DK (2006) Movement patterns and habitat usage of Shark Bay dugongs. MS Thesis, Edith Cowan University, Joondalup

Holley DK, Lawler IR, Gales NJ (2006) Summer survey of dugong distribution and abundance in Shark Bay reveals additional key habitat area. Wildl Res 33:243–250. doi: 10.1071/WR05031

Jarman PJ (1966) The status of the dugong (*Dugong dugon* Muller); Kenya, 1961. East Afr Wildl J 4:82–88. doi: 10.1111/j.1365–2028.1966.tb00885.x

Johannes RE, MacFarlane W (1991) Traditional fishing in Torres Strait Islands. CSIRO Division of Fisheries, Hobart

Kamiya T, Yamasaki F (1981) A morphological note on the sinus hair of the dugong. In Marsh H (ed) The dugong: Proceedings of a seminar/workshop held at James Cook University 8–13 May 1979, Townsville. James Cook University, Townsville

Keith-Diagne LW, Barlas ME, Reid JP et al. (2022) Diving and foraging behaviors. In Marsh H (ed) Ethology and behavioral ecology of sirenians. Springer, Berlin, pp. 67–100.

Kendrick GA, Nowicki RJ, Olsen YS et al. (2019) A systematic review of how multiple stressors from an extreme event drove ecosystem-wide loss of resilience in an iconic seagrass community. Front Mar Sci 6:455. doi: 10.3389/fmars.2019.00455

Kingdon J (1971) Dugong. In East African mammals: An Atlas of evolution in Africa, vol. 1. Academic Press, London, pp. 391–432

Kwan D, Marsh H, Delean S (2006) Factors affecting the customary hunting of a threatened marine mammal by a remote Indigenous community. Environ Conserv 33:164–171. doi: 10.1017/S0376892906002992

Lanyon JM (2003) Distribution and abundance of dugongs in Moreton Bay, Queensland, Australia. Wildl Res 30:397–409. doi: 10.1071/WR98082

Lanyon JM, Marsh H (1995) Digesta passage times in the dugong. Australian J Zool 43:119–127. doi: 10.1071/ZO9950119

Lanyon JM, Newgrain K, Alli TSS (2006) Estimation of water turnover rate in captive dugongs (*Dugong dugon*). Aquat Mamm 32:103–108. doi: 10.1578/AM.32.1.2006.103

Lanyon JM, Sanson GD (2006a) Degenerate dentition of the dugong (*Dugong dugon*), or why a grazer does not need teeth: Morphology, occlusion and wear of mouthparts. J Zool 268:133–152. doi: 10.1111/j.1469-7998.2005.00004.x

Lanyon JM, Sanson GD (2006b) Mechanical disruption of seagrass in the digestive tract of the dugong. J Zool 270:277–289. doi: 10.1111/j.1469-7998.2006.00135.x

Lee Long WJ, McKenzie LJ, Coles RG (1996) Deepwater seagrasses in northeastern Australia—How deep, how meaningful? In Kuo J, Phillips RC, Walker DI et al. (eds) Seagrass Biology: Proceedings of an international workshop, University of Western Australia, Rottnest Island, Perth, Western Australia, 25–29 January 1996

Marsh H (1989) Mass stranding of dugongs by a tropical cyclone in northern Australia. Mar Mamm Sci 5:78–84. doi: 10.1111/j.1748-7692.1989.tb00215.x

Marsh H (2022) Options for handling a stranded orphaned dugong calf. CMS Technical Series Publication No.44, 10pp. UNEP, Kenya.

Marsh H, Albouy C, Arraut E et al. (2022) How might climate change affect the ethology and behavioral ecology of dugongs and manatees? In Marsh H (ed) Ethology and behavioral ecology of sirenians. Springer, Berlin, pp. 351–406

Marsh H, Channells PW, Heinsohn GE et al. (1982) Analysis of stomach contents of dugongs from Queensland. Aust Wildl Res 9:55–67. doi: 10.1071/WR9820055

Marsh H, Collins K, Grech A et al. (2020) An assessment of the distribution and abundance of dugongs ad in-water, large marine turtles along the Queensland coast from Cape York to Hinchinbrook Island. A report to the Great Barrier Reef. Marine Park Authority, May 2020, GBRMPA, Townsville

Marsh H, De'ath G, Gribble N et al. (2005) Historical marine population estimates: Triggers or targets for conservation? The dugong case study. Ecol Appl 15:481–492 www.jstor.org/stable/4543368

Marsh H, Heinsohn GE, Marsh LM (1984) Breeding cycle, life history and population dynamics of the dugong, *Dugong dugon* (Sirenia: Dugongidae). Austr J Zool 32:767–788. doi: 10.1071/ZO9840767

Marsh H, Kwan D (2008) Temporal variability in the life history and reproductive biology of female dugongs in Torres Strait: The likely role of sea grass dieback. Cont Shelf Res 28:2152–2159. doi: 10.1016/j.csr.2008.03.023

Marsh H, O'Shea TJ, Reynolds JE III (2011) The ecology and conservation of Sirenia: Dugongs and manatees. Cambridge University Press, Cambridge, p. 521

Marsh H, Prince RIT, Saalfeld WK et al. (1994) The distribution and abundance of the dugong in Shark Bay, Western Australia. Wildl Res 21:149–161. doi: 10.1071/WR9940149

Marsh H, Sobtzick S (2015) *Dugong dugon*. The IUCN Red List of Threatened Species 2015:e.T6909A43792211. doi: 10.2305/IUCN.UK.2015-4.RLTS.T6909A43792211.en

Marshall CD, Sarko DK, Reep RL (2022) Morphological and sensory innovations for an aquatic lifestyle. In Marsh H (ed) Ethology and behavioral ecology of sirenians. Springer, Berlin, pp. 19–65

McGowan AM, Lanyon JM, Clark N et al. (2023) Cryptic marine barriers to gene flow in a vulnerable coastal species, the dugong (*Dugong dugon*). Mar Mamm Sci 2023:1–22. Doi: 10.1111/mms.13021

McMahon K (2005) Recovery of subtropical seagrasses from natural disturbance. PhD Thesis, University of Queensland, Brisbane, 224pp.

Meager JJ, Limpus C (2014) Mortality of inshore marine mammals in eastern Australia is predicted by freshwater discharge and air temperature. PLoS One 9:e94849. doi: 10.1371/journal.pone.0094849

Nietschmann B (1984) Hunting and ecology of dugongs and green turtles in Torres Strait. Natl Geogr Soc Res Rep 17:625–651.

Nietschmann B, Nietschmann J (1981) Good dugong, bad dugong; good turtle, bad turtle. Nat Human 90:54–62.

Noad MJ, Kniest E, Dunlop RA (2019) Boom to bust? Implications for the continued rapid growth of the eastern Australian humpback whale population despite recovery. Popul Ecol 61:198– 209

O'Shea TJ, Beck CA, Hodgson AJ et al. (2022) Social and reproductive behaviors. In Marsh H (ed) Ethology and behavioral ecology of sirenians. Springer, Berlin, pp. 101–154

Parsons MJG, Holley D, McCauley RD (2013) Source levels of dugong (*Dugong dugon*) vocalizations recorded in Shark Bay. J Acoust Soc Amer 134:2582–2588. doi: 10.1121/1.4816583

Perry CJ, Dennison WC (1996) Effects of dugong grazing on microbial processes in seagrass sediments. In: Kuo J, Walker D, Kirkman H (eds) Seagrass Biology: Scientific discussion from an international workshop, Rottnest Island, Western Australia, 25–29 January 1996. University of Western Australia, Perth

Perry CJ, Dennison WC (1999) Microbial nutrient cycling in seagrass sediments. AGSO J Austra Geo Geoph 17:227–231.

Petit G, Rochon-Duvigneaud A (1929) L'oeil et la vision de L'Halicore dugong ERXL. Bull Soc Zool de Fr 54:129–138

Ponnampalam. LS, Keith-Diagne LW, Marmontel M, et al. (2022) Historical and current interactions with humans. In Marsh H (ed) Ethology and behavioral ecology of sirenians. Springer, Berlin, pp. 299–349

Preen AR (1992) Interactions between dugongs and seagrasses in a subtropical environment. PhD thesis, James Cook University of North Queensland, Townsville

Preen AR (1995a) Diet of dugongs: Are they omnivores? J Mammal 76:163–171. doi: 10.2307/1382325

Preen AR (1995b) Impacts of dugong foraging on seagrass habitats: Observational and experimental evidence for cultivation grazing. Mar Ecol Prog Ser 124:201–213

Preen AR, Lee Long WG, Coles RG (1995) Flood and cyclone related loss, and partial recovery, of more than 1000 km2 of seagrass in Hervey Bay, Queensland, Australia. Aquat Bot 52:3–17. doi: 10.1016/0304-3770(95)00491-H

Preen AR, Marsh H (1995) Response of dugong to large-scale loss of seagrass from Hervey Bay, Queensland, Australia. Wildl Res 22:507–519. doi: 10.1071/WR9950507

Preen AR, Marsh H, Lawler IR et al. (1997) Distribution and abundance of dugongs, turtles, dolphins and other megafauna in Shark Bay, Ningaloo Reef and Exmouth Gulf, Western Australia. Wildl Res 24:185–208. doi: 10.1071/WR95078

Ramos EA, Landeo-Yauri S, Castelblanco-Martinez N et al. (2022) Drone-based photogrammetry assessments of body size and body condition of Antillean manatees. Mamm Biol 102:765–779. doi: 10.1007/s42991-022-00228-4

Rasheed M, McKenna S, Carter A et al. (2014) Contrasting recovery of shallow and deep water seagrass communities following climate associated losses in tropical north Queensland, Australia. Mar Pollut Bull 83:491–499. doi: 10.1016/j.marpolbul.2014.02.013

Rasheed MA, O'Grady D, Scott E et al. (2017) Dugong feeding ecology and habitat use on intertidal banks of Port Curtis and Rodds Bay—Final Report. Report produced for the ecosystem research and monitoring program advisory panel as part of gladstone ports corporation's ecosystem research and monitoring program. Centre for Tropical Water and Aquatic Ecosystem Research (TropWATER) Publication 16/14, James Cook University, Cairns, 68pp

Reep RL, Marshall, CD, Stoll ML (2002) Tactile hairs on the postcranial body in Florida manatees: A mammalian lateral line?. Brain Behav Evol 59:141–154. doi: 10.1159/000064161

Reep RL, Marshall CD, Stoll ML et al. (2001) Microanatomy of facial vibrissae in the Florida manatee: The basis for specialized sensory function and oripulation. Brain Behav Evol 58:1–14. doi: 10.1159/000047257

Sarko DK, Reep RL (2007) Somatosensory areas of manatee cerebral cortex: Histochemical characterization and functional implications. Brain Behav Evol 69:20–36. doi: 10.1159/000095028

Scott AL, York PH, Rasheed MA (2020) Herbivory has a major influence on structure and condition of a Great Barrier Reef subtropical seagrass meadow. Estuaries Coasts 44:506–521. doi: 10.1007/s12237-020-00868-0.

Scott AL, York PH, Rasheed MA (2021) Spatial and temporal patterns in macroherbivore grazing in a multi-species tropical seagrass meadow of the Great Barrier Reef. Diversity 13:12. doi: 10.3390/d13010012

Sheppard JK, Jones RE, Marsh H et al. (2009) Effects of tidal and diel cycles on dugong habitat use. J Wildl Manag 73:45–59. doi: 10.2193/2007-468

Sheppard JK, Marsh H, Jones RE et al. (2010) Dugong habitat use in relation to seagrass nutrients, tides, and diel cycles. Mar Mamm Sci 26:855–879. doi: 10.1111/j.1748-7692.2010.00374.x

Sheppard JK, Preen AR, Marsh H et al. (2006) Movement heterogeneity of dugongs, *Dugong dugon* (Müller), over large spatial scales. J Exp Mar Biol Ecol 334:64–83. doi: 10.1016/j.jembe.2006.01.011

Simpfendorfer CA, Goodreid AB, McAuley RB (2001) Size, sex and geographic variation in the diet of the tiger shark, *Galeocerdo cuvier*, from Western Australian waters. Environ Biol Fish 61:37–46. doi: 10.1023/A:1011021710183

Smoll L, Beard L, Lanyon J (2020) Osmoregulation and electrolyte balance in a fully marine mammal, the dugong (*Dugong dugon*). J Comp Physiol B 190:139–148. doi: 10.1007/s00360-019-01250-8

Sobtzick S, Cleguer C, Hagihara R et al. (2017) Distribution and abundance of dugong and large marine turtles in Moreton Bay, Hervey Bay and the southern Great Barrier Reef. A report to the Great Barrier Reef Marine Park Authority. Centre for Tropical Water and Aquatic Ecosystem Research (TropWATER) Publication 17/21, James Cook University, Townsville, 95pp

Sobtzick S, Hagihara R, Grech A et al. (2012) Aerial survey of the urban coast of Queensland to evaluate the response of the dugong population to the widespread effects of the extreme weather events of the summer of 2010–11. Final report to the Australian Marine Mammal Centre and the National Environmental Research Program, Hobart

Spain AV, Heinsohn GE (1973) Cyclone associated feeding changes in the dugong (Mammalia: Sirenia). Mammalia 37:678–680

Steinberg C (2007) Impacts of climate change on the physical oceanography of the Great Barrier Reef. In Johnson J, Marshall P (eds) Climate change and the Great Barrier Reef: A vulnerability assessment. Great Barrier Reef Marine Park Authority, Townsville, pp. 51–74

Tanaka K, Ichikawa K, Kittiwattanawong K et al. (2023) Spatial variation of vocalising dugongs around Talibong Island, Thailand. Bioacoustics 32:33–47. doi: 10.1080/09524622.2022.2058614

Tanaka K, Ichikawa K, Nishizawa H et al. (2017) Differences in vocalisation patterns of dugongs between fine-scale habitats around Talibong Island, Thailand. Acoust Aust 45:243–251. doi: 10.1007/s40857-017-0094-7

Tetley MJ, Braulik GT, Lanfredi C et al. (2022) The important marine mammal area network: A tool for systematic spatial planning in response to the marine mammal habitat conservation crisis. Front Mar Sci Sec. Marine Conservation and Sustainability 9. doi: 10.3389/fmars.2022.841789

Thomsen PF, Willerslev E (2015) Environmental DNA—An emerging tool in conservation for monitoring past and present biodiversity. Biol Conserv 183:4–18. doi: 10.1016/j.biocon.2014.11.019

Tol SJ, Coles RG, Congdon BC (2016) *Dugong dugon* feeding in tropical Australian seagrass meadows: Implications for conservation planning. PeerJ 4:e2194. doi: 10.7717/peerj.2194

Tol SJ, Jarvis JC, York PH et al. (2017) Long distance biotic dispersal of tropical seagrass seeds by marine mega-herbivores. Sci Rep 7:4458. doi: 10.1038/s41598-017-04421-1

Waycott M, Collier CJ, McMahon K et al. (2007) The vulnerability of seagrasses of the great barrier reef to climate change. In Johnson J, Marshall P (eds) Climate change and the Great Barrier Reef: A Vulnerability Assessment. Great Barrier Reef Marine Park Authority, Townsville, pp. 193–233. http://hdl.handle.net/11017/137

Waycott M, Duarte CM, Carruthers TJB et al. (2009) Accelerating loss of seagrasses across the globe threatens coastal ecosystems. PNAS USA:12377–12381. doi: 10.1073/pnas.0905620106

Wirsing AJ, Heithaus MR (2012) Behavioural transition probabilities in dugongs change with habitat and predator presence: Implications for sirenian conservation. Mar Fresh Res 63:1069–1076. doi: 10.1071/MF12074

Wirsing AJ, Heithaus MR, Dill LM (2007a) Can you dig it? Use of excavation, a risky foraging tactic, by dugongs is sensitive to predation danger. Anim Behav 74:1085–1091. doi: 10.1016/j.anbehav.2007.02.009

Wirsing AJ, Heithaus MR, Dill LM (2007b) Living on the edge: Dugongs prefer to forage in microhabitats that allow escape from rather than avoidance of predators. Anim Behav 74:93–101. doi: 10.1016/J.Anbehav.2006.11.016

Wirsing AJ, Heithaus MR, Dill LM (2007c) Fear factor: Do dugongs (*Dugong dugon*) trade food for safety from tiger sharks (*Galeocerdo cuvier*)? Oecologia 153:1031–1040. doi: 10.1007/S00442-007-0802-3

Wolanski E, Choukroun S, Kingsford M et al. (2024) The bioengineered water circulation in the great barrier reef: A review. In Wolanski E, Kingsford MJ (eds) Oceanographic Processes of Coral Reefs Physical and Biological Links in the Great Barrier Reef. CRC Press, Boca Raton, FL.

Yamasaki F, Komatsu S, Kamiya T (1980) A comparative morphological study on the tongues of manatee and dugong (Sirenia). Sci Rep Whales Res Inst 32:127–144

Yong E (2022) An immense world: How animal senses reveal the hidden realms around us. The Bodley Head, London, p. 449.

York PH, Bryant CV, Rasheed MA (2022) Post-flood seagrass monitoring in Hervey Bay in May 2022. James Cook University Centre for Tropical Water & Aquatic Ecosystem Research, Publication 22/31, Cairns, 24pp

Zeh DR (2022) Using acoustic and satellite tracking data to better understand the factors associated with marine mammal habitat use and movements within urban coastal environments: Dugongs in Moreton Bay, Queensland. PhD thesis, James Cook University, Australia.

Zeh DR, Heupel MR, Hamann M et al. (2016) Quick Fix GPS technology highlights risk to marine animals moving between protected areas. Endanger Species Res 30:37–44. doi: 10.3354/esr00725

Zeh DR, Heupel MR, Hamann M et al. (2018) Evidence of behavioural thermoregulation by dugongs at the high latitude limit to their range in eastern Australia. J Exp Mar Biol Ecol 508:27–34. doi: 10.1016/j.jembe.2018.08.004

14 Biophysical Interactions of Jellyfish on the Great Barrier Reef

Jodie A. Schlaefer and Michael J. Kingsford

14.1. INTRODUCTION

Jellyfish are classified as medusozoans, a subphylum within the phylum Cnidaria (Figure 1). The term jellyfish has also been used to describe the grouping of medusozoans with comb jellies (phylum Ctenophora) and thaliaceans (i.e., pyrosomes, salps, and doliolids; phylum Chordata, subphylum Tunicata, class Thaliacea, orders Pyrosomida, Salpida, and Doliolida) that outwardly resemble medusozoans. In discussing jellyfish in this chapter, we focus only on the monophyletic medusozoans. As Cnidarians, jellyfish possess stinging cells called cnidocytes that distinguish the phylum, and cnidarians are the simplest organisms to have cells organised into true tissues. Cnidarians also have a relatively simple, radially symmetric morphology with two layers of epithelial cells (an outer ectoderm and an inner endoderm) and a jelly-like, non-living substance called mesoglea that is contained within the two cell layers.

Modern Cnidaria is a diverse phylum. In addition to jellyfish/Medusozoa, it also contains the subphylum Endocnidozoa, a group of small, highly derived, obligate parasites (Foox 2021), and the class Anthozoa, which includes hard corals, soft corals, and sea anemones (Quattrini et al. 2021). Adult anthozoans live as colonial or singular benthic polys, and the calcium carbonate skeletons of hard coral polyps form the foundations of coral reefs, including the Great Barrier Reef (GBR). Medusozoans are distinguished from anthozoans as, in addition to benthic polyps, their lifecycles generally include mobile medusae that are essentially swimming polyps. Specifically, in the ancestral medusozoan lifecycle, asexually reproducing benthic polyps bud-off free-swimming medusae (Collins 2002; Marques and Collins 2004). Mature medusae then reproduce sexually to produce ciliated planulae larvae that settle and develop into polyps. Medusozoans are an ancient group in evolutionary time; there is fossil evidence that they had emerged by the Precambrian Ediacaran Period 635 to 541 million years ago (Van Iten et al. 2006; Liu et al. 2014). With their simple anatomy, complex life histories, and ancient ancestry, jellyfish are adept at surviving in a wide range of ecosystems, from freshwater to marine habitats and from polar deep seas to shallow tropical nearshore waters, including the GBR (Avian and Ramšak 2021; Blackstone and Cartwright 2021).

Jellyfish can have a marked impact on the range of socio-ecological systems that they inhabit. Ecologically, medusae are important planktonic predators (Lynam et al. 2006) that can significantly alter pelagic food webs (Brodeur et al. 2002; Möller 1984). Medusae are also important prey items for many species (Brodeur et al. 2021). Additionally, in the pelagic desert devoid of structure, medusae can provide structure for planktonic species such as fish larvae (Kingsford 1993). For human societies, jellyfish can represent both an opportunity and a nuisance. Their rapid growth rates make them ideal targets for sustainable fisheries, although this resource is not widely exploited in Australia (Kingsford et al. 2000). Conversely, their capacity to bloom in great numbers can be problematic where blooms clog fishing nets or block the saltwater intakes of power plants (Purcell et al. 2007; Uye 2014; Bosch-Belmar et al. 2021). Other nuisance species have highly venomous stings, and their presence endangers water users (Crowley-Cyr and Gershwin 2021) and can negatively impact tourism industries (Kingsford et al. 2018). Clearly, jellyfish can be both ecologically and socially significant.

Jellyfish are classified in the classes Staurozoa, Hydrozoa, Scyphozoa, and Cubozoa. Staurozoa are known as stalked jellyfish, and there are approximately 50 described species (Avian and Ramšak 2021). They live a mostly sessile existence attached to substrata via a peduncle but can detach for short periods, and float or swim by summersaulting, before reattaching (Miranda et al. 2018). Staurozoa have a largely anti-tropical distribution (Miranda et al. 2018), and no recorded occurrences in the GBR region (OBIS; https://obis.org/). Consequently, we do not discuss the biophysical interactions of Staurozoa in this chapter. Representatives of all major clades of Hydrozoa, Scyphozoa, and Cubozoa were recorded in the GBR in OBIS, reflecting the variety in jellyfish species found in the GBR (Figures 1–4). These species inhabit all ecoregions in the GBR system including estuaries, mangrove creeks, beaches, seagrass meadows, shallow reefal waters, and open waters (OBIS; Kingsford et al. 2012; Kramp 1965). The distinguishing features of hydrozoans, scyphozoans, and cubozoans influence their biophysical interactions within these habitats.

Hydrozoa are the largest and most diverse of the medusozoan classes (Collins 2009), with roughly 3,800 described species (Blackstone and Cartwright 2021; Santander et al. 2022). Hydromedusae come in a large variety of forms (Figure 2), and most fall within the size range of 0.2 to 20 cm bell diameter (BD; World Register of Marine Species; www.marinespecies.org/). The vast majority of hydrozoans are marine, but a few species are specialised for freshwater (Blackstone and Cartwright 2021). Hydrozoans

DOI: 10.1201/9781003320425-17

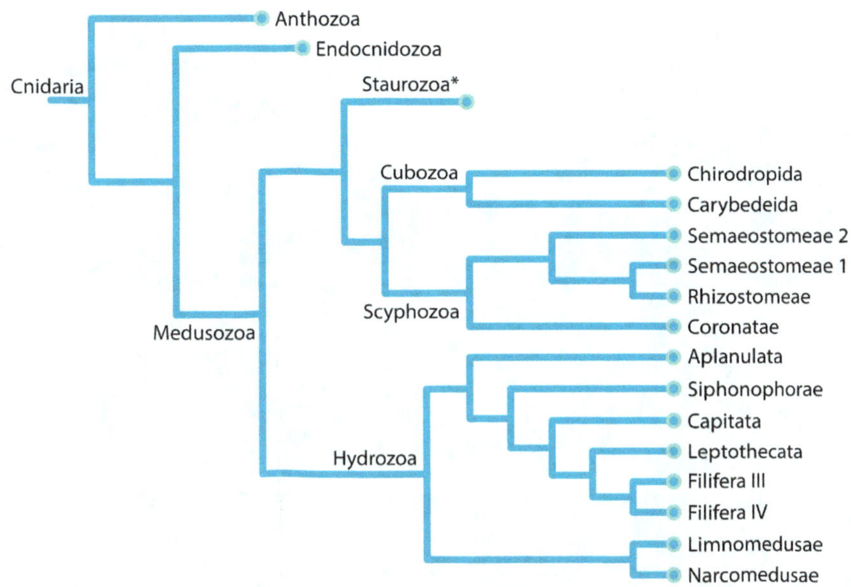

FIGURE 1 Phylogeny of phylum Cnidaria, showing all major clades within the medusozoan classes Cubozoa, Scyphozoa, and Hydrozoa. These clades all occur within the GBR, as recorded in the Ocean Biodiversity Information System (OBIS; https://obis.org/). Note, there were no recorded occurrences of Staurozoa (*), so the major clades within Staurozoa are not shown here.

Source: Adapted from Kayal et al. (2018) and Open Tree of Life's synthetic tree (OpenTree et al. 2021)

possess an array of life history strategies (Gibbons et al. 2010). Although many display a classic medusozoan bipartite lifecycle, there is extensive variation, and some lineages completely lack either a polyp or a medusa phase (e.g., Cornelius 1992). Additionally, rather than being singular, some hydrozoans are colonial organisms made up of integrated zooids that perform specialised functions (e.g., order Siphonophorae; Dunn and Wagner 2006). In fact, the longest organism in the world is thought to be a colonial hydrozoan in the order Siphonophorae that forms chains over 40 m long (Dunn 2009). Furthermore, although most hydrozoans are found below the surface, others are pleustonic, drifting on the surface where both winds and currents affect transport, dispersal, and their interactions with other zooplankton (e.g., by-the-wind-sailors *Velella velella* and Portuguese-man-of-war, *Physalia physalis*; Rayner 1986). Overall, the diversity of forms and functions in Hydrozoa translate to a diversity in the ways that hydromedusae interact with their physical environment.

Scyphozoans are known as the 'true jellyfish' (Figure 3). There are approximately 230 scyphozoan species that range in size from quite small (2 cm BD; e.g., the thimble jellyfish *Linuche unguiculata*) to the largest diameter of all jellyfish (> 2 m BD; the lion's mane jellyfish *Cyanea sp.*; Figure 3A; Avian and Ramšak 2021). Most scyphozoans have the ancestral medusozoan bipartite lifecycle; however, the polyp phase of the life cycle is highly specialised. Scyphozoan polyps develop into a strobila that successively strobilates multiple asexually produced free-swimming pre-medusae forms called ephyra. The strobilation process means that scyphozoan species can respond to favourable environmental conditions and quickly produce many ephyra when the conditions are right (Lucas 2001; Chi et al. 2019). Ephyra are tiny (a few mm) and are poor swimmers compared to medusae, but they quickly develop into juvenile and then adult medusae (Higgins et al. 2008; Zimmerman et al. 2019). Growth at the medusa phase is also rapid. For example, the Australian blubber (*Catostylus mosaicus*) can grow from 40–50 mm (< 10 g) to over 250 mm (~5 kg) in a 3–4 month period (Pitt and Kingsford 2003). Consequently, scyphozoans have boom and bust population dynamics (in both numbers and biomass) and therefore have the capacity to form large blooms of medusae (e.g., Figure 3B). Blooms of these effective planktivores can have substantial ecological ramifications (Alldredge 1984; Oguz et al. 2001).

Cubozoans are outwardly similar to scyphozoans but with more cubiform bells and greater sensory and morphological complexity (Satterlie 2011; Picciani et al. 2018). There are around 47 described species in the class, which are organised into two orders: Chirodropida and Carybdeida (Figure 4; Avian and Ramšak 2021). Chirodropids (5 to 32.5 cm bell height [BH], Figure 4A) tend to grow much larger than Carybdeids (2 to 25 cm BH, Figure 4BC; Kingsford and Mooney 2014). In contrast to scyphozoans, cubozoan polyps do not strobilate, and the metamorphosis from polyps to medusae is so complete that the resulting juvenile medusae are nearly fully formed (Werner 1983). Notably, the class Cubozoa contains the most venomous of jellyfish species, including the box jellyfish (*Chironex fleckeri*, Figure 4A) and the infamous Irukandji jellyfishes (e.g., *Carukia barnesi*, Figure 4B), which are

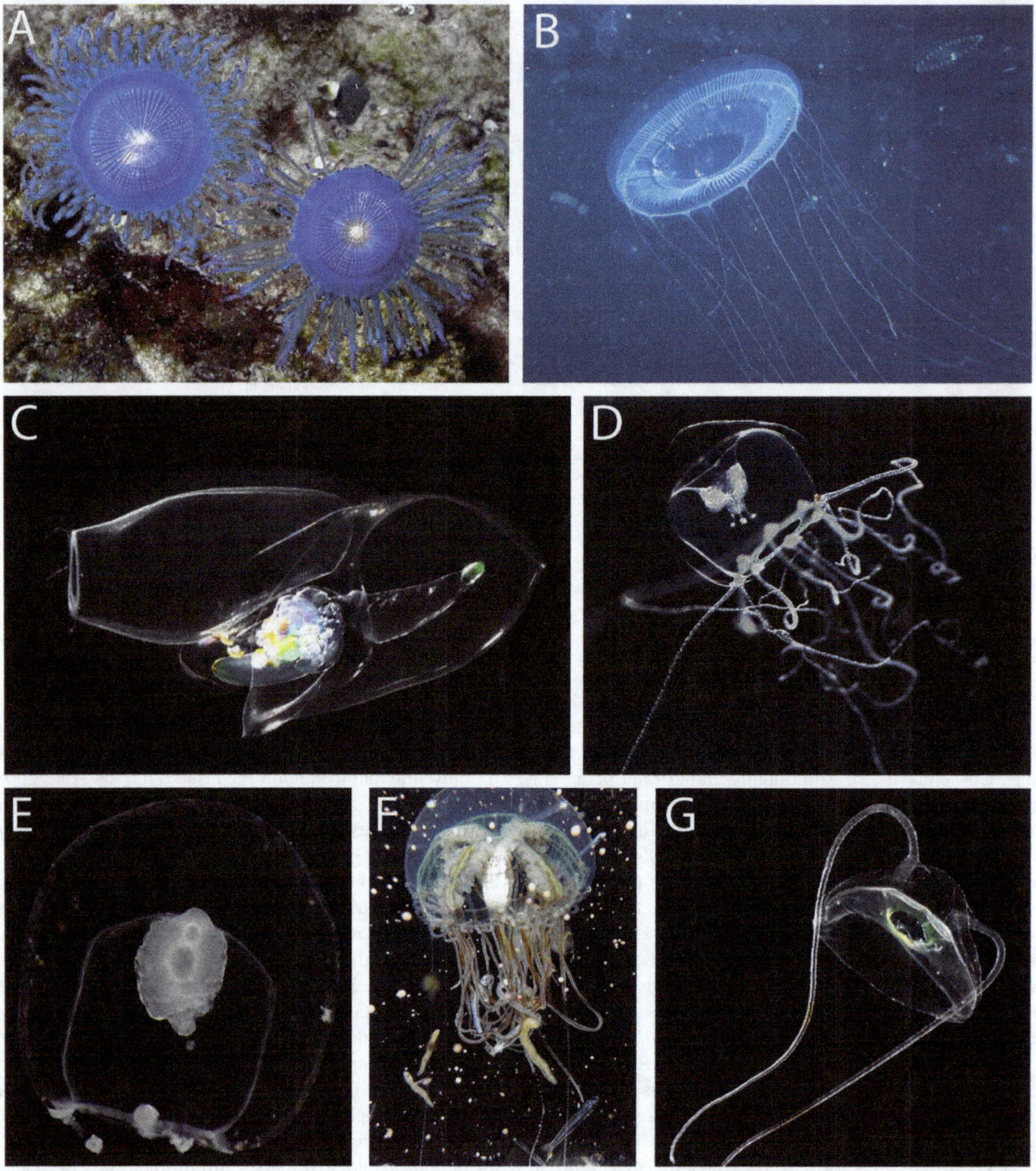

FIGURE 2 Diversity in jellyfish recorded in the GBR. Representatives from all major clades in Class Hydrozoa. Photos were either taken from within the GBR or are of species with presences recorded in the GBR region in the OBIS (https://obis.org/). A) *Porpita porpita* (suborder Capitata). Maximum diameter = 0.3 cm[1]. B) *Aequorea sp.* (order Leptothecata). Individual approximately 5 cm in diameter[2]. C) *Diphyes chamissonis* (order Siphonophorae). Individual approximately 0.6 cm long[2]. D) *Lizzia blondina* (suborder Filifera). Maximum diameter = 0.2 cm[1]. E) *Plotocnide borealis* (suborder Aplanulata). Maximum diameter = 0.3 cm[1]. F) *Olindias malayensis* (order Limnomedusae). Maximum diameter = 3.5 cm[1]. G) *Solmundella bitentaculata* (order Narcomedusae). Maximum diameter = 1.2 cm[1].

Source: Photo credits: A, C, G) ©Denis Riek, all rights reserved. B) ©Anne Hoggett, all rights reserved. D) ©Peter Schuchert, all rights reserved. E) photographer: Aino Hosia, *CC BY-SA 4.0*. F) ©Lyle Vail, all rights reserved. Size data were sourced from SeaLifeBase[1] (www.sealifebase.org, version 12/2022; Palomares and Pauly [Ed]) or from the photographer[2] if size information was not available in SeaLifeBase.

FIGURE 3 Diversity in jellyfish recorded in the GBR. Representatives from all major clades in Class Scyphozoa. Photos were either taken from within the GBR or are of species with presences recorded in the GBR region in the OBIS (https://obis.org/). A) *Cyanea sp.* (order Semaeostomeae). Maximum diameter = 229 cm. B) *Catostylus mosaicus* (order Rhizostomeae). Maximum diameter = 35 cm. C) *Atolla vanhoeffeni* (order Coronatae). Maximum diameter = 5 cm.

Source: Photo credits: A) ©Robert Streit, all rights reserved. B) ©Michael J. Kingsford, all rights reserved. C) ©Kawabata et al. (2013), photographer: James C. Hunt, CC BY 4.0. Size data were sourced from SeaLifeBase (www.sealifebase.org, version 12/2022; Palomares and Pauly [Ed]).

FIGURE 4 Diversity in jellyfish recorded in the GBR. Representatives from all major clades in Class Cubozoa. Photos were either taken from within the GBR or are of species with presences recorded in the GBR region in the OBIS (https://obis.org/). A) *Chironex fleckeri* (order Chirodropida). Maximum diameter = 30 cm. B) *Carukia Barnesi* (order Carybdeida). Maximum diameter = 2 cm. C) *Copula sivickisi* (order Carybdeida). Maximum diameter = 1.4 cm. D) A patrolled beach that was closed to swimmers due to the presence of Cubozoan jellyfish.

Source: Photo credits: A) ©Commonwealth of Australia Great Barrier Reef Marine Park Authority, photographer: M. Jones, CC BY 4.0. B, D) ©eAtlas, photographer: Lisa-ann Gershwin, CC BY 3.0 AU. C) photographer: Patrick Randall, *CC BY-SA 4.0*. Size data were sourced from SeaLifeBase (www.sealifebase.org, version 12/2022; Palomares and Pauly [Ed]).

a group of approximately 16 species with stings that elicit Irukandji syndrome (Gershwin et al. 2013). Consequently, cubozoans represent a danger to swimmers in the GBR, and their presence can trigger the closing of patrolled beaches (Figure 4D). Overall, the different features of the classes Hydrozoa, Scyphozoa, and Cubozoa underpin how they interact with their environments.

The main objective of this chapter is to provide a synthesis of the available information on jellyfish biophysical interactions, with a focus on the different types of jellyfish found in the GBR. We first discuss the sensory and nervous systems of Hydrozoa, Scyphozoa, and Cubozoa before overviewing their mobility. We then detail the biophysical interactions of jellyfish species, as evident through their behaviours and ecology; some emphasis is given to the Cubozoa. Finally, we discuss future research directions in jellyfish biophysical interactions. Notably, due to the paucity of research into jellyfish biophysical interactions conducted within the GBR, we have drawn from the broader literature; however, where possible, examples from the GBR are provided.

14.2. SENSORY AND NERVOUS SYSTEMS

Morphologically, jellyfishes are relatively simple organisms; however, they contain surprisingly complex sensory systems, and multiple senses have been reported in jellyfish species. For example, jellyfish can have mechanoreceptors and chemoreceptors (Garm et al. 2006; Skogh et al. 2006), and cues from both of these senses may be required to initiate stinging cell (cnidocyte) discharge (Kass-Simon and Scappaticci 2002; Scappaticci and Kass-Simon 2008). There is also emerging evidence that they can be sensitive to sound (Solé et al. 2016). Furthermore, jellyfish in the classes Scyphozoa and Cubozoa have sensory stalks called rhopalia that contain statoliths, hard structures made up of calcium-containing minerals (Figure 5; Boßelmann et al. 2007). In scyphozoans, the rhopalia are dispersed around the bell margin, and they are typically present in multiples of four (Figure 5A; Hyman 1940). Cubozoans have four rhopalia, with one located on each of their four sides (Figure 5B; Garm et al. 2008). The rhopalia and statoliths have been associated with a sense of gravity, which aides in behaviours such are righting (Arai 1997; Boßelmann et al. 2007). The rhopalia also house eyes, defined as structures including at least photoreceptor cells and pigment cells (Picciani et al. 2018). Although hydrozoans lack rhopalia, eyes are present in some hydrozoan clades (Capitata, Filifera, and Leptothecata; Picciani et al. 2018). Indeed, all Cnidarians appear to be photosensitive, possessing at least simple photoreceptors, but eyes of higher complexity have arisen multiple times in Medusozoa (Picciani et al. 2018). Although jellyfish are clearly capable of perceiving their environment with inputs from multiple senses, the most compelling evidence is for perception and orientation through photoreception.

The eyes of medusozoans have different structures, depending on their evolutionary origin (Picciani et al. 2018). Consequently, there is great diversity in jellyfish eyes (Picciani et al. 2018). Among hydrozoans, eyes range in complexity from eyespots made up of multiple cell types to pigment cups with compact lenses (Weber 1981; Picciani et al. 2018). In scyphozoans, pigmented photosensory cells without lenses are organised in cups and spots (Nakanishi et al. 2009; Picciani et al. 2018; Figure 5B). By contrast, the most comprehensive jellyfish visual system is found in cubozoans. In cubozoans, each rhopalium contains six eyes, giving cubomedusae 24 eyes in total (Coates and Theobald 2003; Nilsson et al. 2005; Figure 5D). Four of the six eyes per rhopalium are relatively simple eye spots, but the remaining two have lenses and form colourless images (Garm et al. 2007a). Jellyfish, particularly cubozoans, are clearly able to effectively perceive their surrounding light environment.

The signals received by jellyfish sensory structures, including their eyes, are then transmitted through their

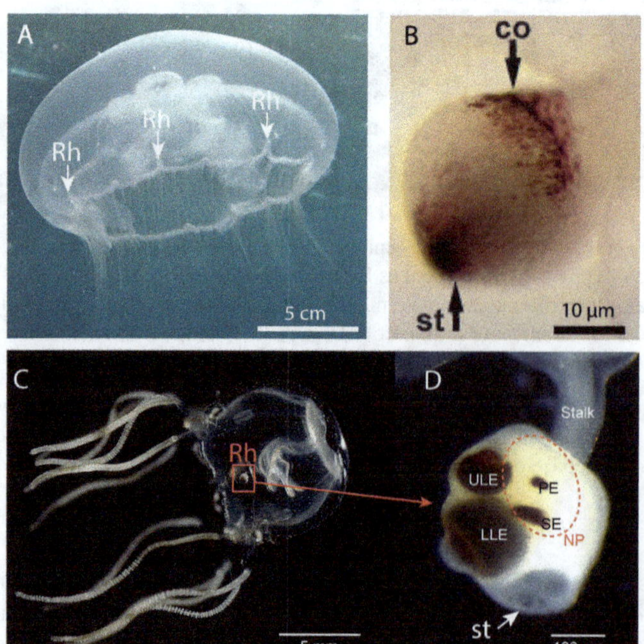

FIGURE 5 Sensory stalks (Rhopalia, Rh) in scyphozoans and cubozoans. A) Locations of a subset of Rh, around the bell margin of an *Aurelia aurita* medusa (class Scyphozoa, order Semaeostomeae). B) Zoomed view of an *A. aurita* Rh, showing the locations of the pigment cup ocellus/eye (co) and the statolith (st). C) Location of a Rh, on the side of a *Tripedalia cystophora* medusa (class Cubozoa, order Carybdeida). D) Zoomed view of a *T. cystophora* Rh, showing the locations of the upper lens eye (ULE), the lower lens eye (LLE), a pit eye (PE), a slit eye (SE), a light sensitive neuropil (NP, red broken line) and the st.

Source: Photo credits: A) ©Lyle Vail, all rights reserved. B) ©Solé et al. (2016), CC BY 4.0. C, D) ©Bielecki et al. (2014), CC BY 4.0.

well-developed nervous systems. The nervous systems of medusae generally include nerve nets that are diffusely organised (i.e., evenly distributed across their bells) and concentrated nerve nets that represent areas of centralised organisation (Satterlie 2011, 2017). In hydrozoans, concentrations of nerves form interconnected inner and outer nerve rings that transmit information around the bell margin (Mackie 2004). The inner nerve ring also serves as a swim pacemaker that modulates motor outputs (Satterlie 2002). Additionally, some species of hydrozoans have sensory hubs (e.g., eyes) that are encircled by consolidated nerve nets (Koizumi et al. 2015). In contrast, ganglia-like concentrations of nerves communicate the sensory inputs from Scyphozoa rhopalia to swim pacemakers that are also located at the rhopalia and regulate motor activity (Passano 1973). However, as the rhopalia are connected to one another by a diffuse nerve net, not a condensed one, information is not expressly communicated between them (Satterlie 2011). Cubozoans also have ganglia-like nerve concentrations and swim pacemakers at their rhopalia, but information is transmitted between their rhopalia via a condensed nerve net that forms a nerve ring around the bell margin (Garm et al. 2007c). Therefore, cubozoans have the most complex nervous systems among the medusozoans (Garm et al. 2007c; Satterlie 2011, 2017). The location of scyphozoan and cubozoan eyes and swim pacemakers on their rhopalia and the ganglia-like network of nerves that connect the structures suggest that their visual inputs are particularly important for informing their swimming behaviour (Garm and Bielecki 2008), and this suggestion has been confirmed in behavioural studies (e.g., Buskey 2003; Petie et al. 2013, 2011). Overall, in jellyfish, sensory inputs inform swimming behaviour through sophisticated and class-specific nerve pathways.

14.3. MOBILITY

To complement their sensory capabilities, jellyfish are efficient swimmers, and many species can attain relatively fast speeds. Most jellyfish swim through repeated cycles of actively contracting their bell and then relaxing to let their bell passively expand through elastic recoil (Figure 6). Notably, as the bell expands and water moves into the subumbrella cavity, vortices are generated within the cavity (Gemmell et al. 2013, 2018). If jellyfish pause for a time after the relaxation phase and before starting the next swim cycle, then these vorticities can provide an additional thrust without additional muscle movement; this mechanism is termed passive energy recapture (PER; Gemmell et al. 2013). PER is the reason why propulsion in jellyfish is so energy-efficient compared to other swimmers (Gemmell et al. 2013). Indeed, jellyfish PER was found to reduce the energy costs associated with movement by 48% (Gemmell et al. 2013). PER is widespread among diverse jellyfish taxa with different swimming modes (Gemmell et al. 2018).

For jellyfish, the swimming mode has been related to bell shape and feeding strategy. In more oblate jellyfish (e.g., most scyphozoans [Figure 3] and some hydrozoans [Figure 2ABFG]), the bell contractions and relaxations cause the bell margin to undulate in a rowing-like motion that propels the jellyfish forward (Figure 6AC; Colin and Costello 2002; Gemmell et al. 2018; Gladfelter 1972). The rowing motion can enable feeding, whereby the vortex currents that are generated during rowing move prey into the tentacles (D'Ambra et al. 2001; Katija et al. 2011). However, in an apparent trade-off between feeding and propulsion,

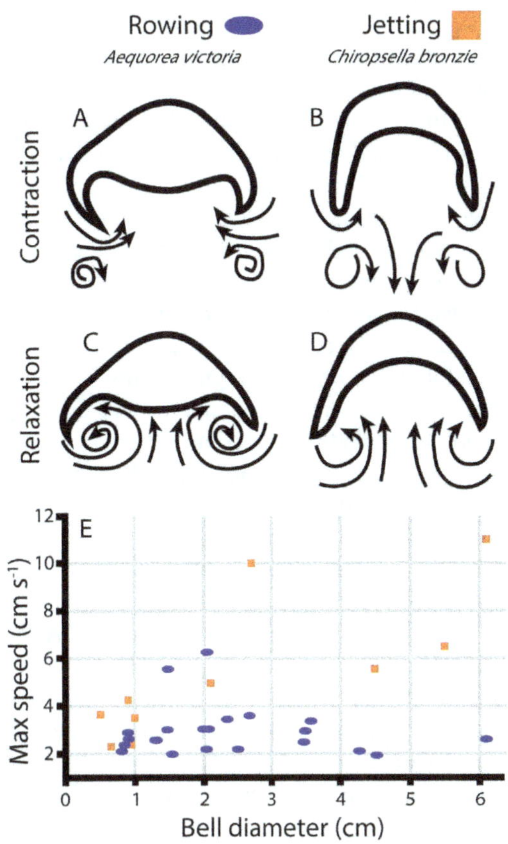

FIGURE 6 Jetting versus rowing modes of propulsion. A schematic of the bell movement (thick outlines) and resulting water movement (arrows) during the contraction and relaxation phases of a swim cycle in rowing (A, C) and jetting (B, D) propulsion. The bell outlines and water vorticity/velocity vectors have been traced from Figure 1 of Gemmell et al. (2018). E) The BD (cm) versus maximum (max) swim speed (cm s^{-1}) of medusae with rowing (blue oval) and jetting (orange square) swim modes. The data were sourced from Gemmell et al. (2018). Rowers and jetters were distinguished based on their reported bell fineness, which is the ratio between bell length to width at relaxation (Costello et al. 2008). Lower values indicate rowers with more oblate bells, while higher values indicate jetters with more prolate bells; a cut-off of 0.72 was used. The rowers include the scyphozoans *Mastigias papua*, *Catostylus mosaicus*, and *Phyllorhiza punctata*, and the hydrozoans include *Aequorea victoria*, *Clytia gregaria*, and *Stomotoca atra*. The jetters include the cubozoans *Chiropsella bronzie* and *Chironex fleckeri* and the hydrozoan *Sarsia tubulosa*. Note that the speed of the largest *C. fleckeri* was not visualised as its BD is much larger than the other jellyfish in the study (BD = 14.2 cm, maximum speed = 11.7 cm s^{-1}).

dense tentacle masses can suppress the formation of propelling vortices and therefore hamper propulsion (Miles and Battista 2019). By contrast, more prolate jellyfish (e.g., some hydrozoans [Figure 2CDE] and smaller cubozoans [Figure 4BC]) tend to swim through jet propulsion and capture prey in their proportionately long tentacles by ambushing them with targeted fast swimming bursts (Costello et al. 2008; Dabiri et al. 2010). In jetting propulsion, during the contraction phase, the water in the cavity under the bell is pushed out at force, and the resulting jet of water propels the jellyfish forward (Figure 6BD; Colin and Costello 2002; Dabiri et al. 2010; Gemmell et al. 2018). Unsurprisingly, considering the diversity in bell shapes among jellyfish, propulsion can be achieved through a combination of rowing and jetting, in a hybrid swimming mode (e.g., larger cubozoans [Figure 4A], Colin et al. 2013). Overall, jellyfish swimming is highly efficient across different swimming modes and feeding strategies.

Some members of the morphologically diverse Hydrozoa order Siphonophorae diverge from these typical swimming modes. Siphonophores are colonial organisms made up of functionally specialised zooids that are physiologically integrated (Dunn and Wagner 2006). Zooids called nectophores can propel siphonophores via jet propulsion (Dunn and Wagner 2006). In some siphonophores, nectophores are organised in chains, and colony propulsion is achieved through a division of labour in the chain, where the younger, smaller nectophores at the top use their less powerful jets to steer the colony, and the older, larger nectophores toward the bottom contribute the bulk of the forward thrust (e.g., *Nanomia bijuga*, Figure 7A; Costello et al. 2015). Furthermore, other siphonophores achieve propulsion through sail power, where a specialised gas-filled zooid called a pneumatophore protrudes above the water surface and catches the wind (e.g., blue bottles [*Physalia physalis*, Figure 7B] and by-the-wind sailors [*Velella velella*, Figure 7C]; Clark and Lane 1961; Iosilevskii and Weihs 2009). The filling of the pneumatophore with gas is actively controlled via muscle contractions, and as the muscle contraction is the only active component of propulsion (Mackie 1960), sailing is probably highly energy-efficient. Notably, *P. physalis* and *V. velella* have distinct morphotypes with pneumatophores that have either a left or right sailing orientation (Bieri 1959; Ferrer and González 2021). *V. velella* can also sail in directions ranging from downwind to nearly across-wind (Francis 1991). Clearly, the unique colonial siphonophores possess equally distinctive modes of propulsion.

Jellyfish can swim surprisingly fast, with speeds related to ontogeny, size, and swim mode. Maximum swimming speeds generally increase with age and size whereby younger, smaller medusae attain slower maximum speeds than older, larger medusae of the same species (Katija et al. 2015). However, this speed with size relationship does not hold across species with different swim modes, as evident from the data of Gemmel et al. (2018; Figure 6E). In their investigation of PER, Gemmel et al. (2018) measured the sizes, swim modes, and swim speeds of an array of diverse jellyfish species in a laboratory setting using digital video analysis, with representatives from Hydrozoa, Scyphozoa, and Cubozoa. There was considerable variability in their data at the individual level; however, all medusae tested were competent swimmers, with swim speeds ranging from 4 to 72 BDs s^{-1}. Interestingly, across the measured size range, jetting medusae tended to attain maximum speeds that were faster than rowing jellyfish of a comparable size (maximum speed ranges: 1.9 to 6.2 cm s^{-1} for rowers with BDs of 0.8 to 6.1 cm and 2.3 to 11.0 cm s^{-1} for jetters with BDs of 0.5 to 6.1 cm; Figure 6E). Relatedly, the fastest recorded burst of speed from a jetting/hybrid swimming medusae of 16.6 cm s^{-1} is among the fastest recorded for any jellyfish and was taken from a *Chironex fleckeri* medusae with an interpedalial distance of just 6 cm (Schlaefer et al. 2018). However, the maximum size of a jetting/hybrid swimming jellyfish is in the range of 35 cm BH (*C. fleckeri*, class Cubozoa; Kingsford and Mooney 2014), which is much smaller than the maximum size of a rowing scyphomedusae of approximately 2-m bell diameter (Avian and Ramšak 2021). Consequently, the fastest jellyfish swim speeds have been recorded in large scyphomedusae. For example, giant

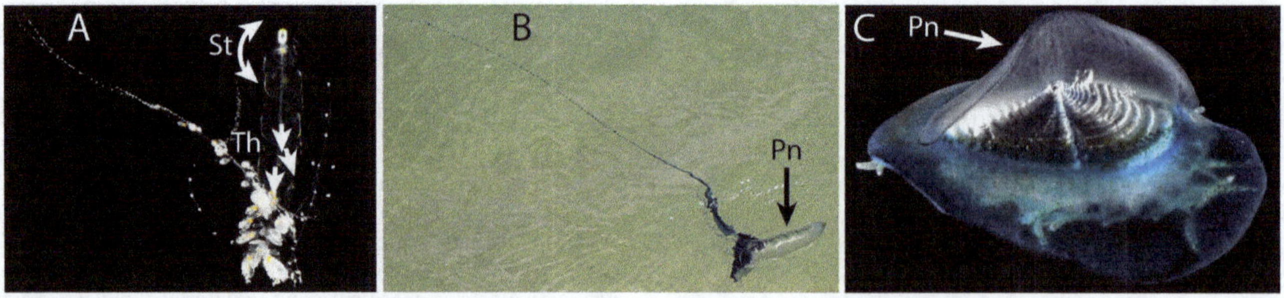

FIGURE 7 Diverse modes of propulsion in class Hydrozoa, order Siphonophorae. A) *Nanomia bijuga* as an example of a siphonophore with a chain of nectophores for propulsion, where the smaller, steering (St) nectophores are at the top of the chain, and the larger, more powerful, thrust (Th)-generating nectophores are at the bottom. B) Blue bottle (*Physalia physalis*) and C) by-the-wind sailor (*Velella velella*) as examples of siphonophores that have a pneumatophore (Pn) that catches the wind, enabling them to sail.

Source: Photo credits: A, C) ©Denis Riek, all rights reserved. B) ©Anne Hoggett, all rights reserved.

jellyfish (*Nemopilema nomurai*) with BDs from 24.5 to 68.0 cm have been measured to maintain speeds of 10.6 to 52.9 cm s^{-1} (Lee et al. 2010).

High speeds have also been recorded in hydrozoans with diverse swimming modes. Siphonophores with chains of nectophores (e.g., Figure 7A) have been recorded moving at maximum speeds of 16.2 cm s^{-1} (Sutherland et al. 2019). Furthermore, blue bottles sail at speeds of at least one order of magnitude less than the wind speeds that they sail in, which translates to sailing at speeds of tens of cm s^{-1}, with a maximum recorded speed of 40 cm s^{-1} (Totton and Mackie 1960). Clearly many jellyfish can swim or sail with, across, or down current and are far from passive drifters that are totally at the whim of currents.

14.4. BEHAVIOURS AND ECOLOGY

The multiple senses and swimming modes amongst jellyfish translate to an array of ways in which jellyfish interact with their environments. These interactions, including the formation of dense aggregations and reacting to visual stimuli, are linked to the ecological significance of jellyfish and are key determinants of population structures, as discussed in the following.

14.4.1. Aggregations

Reports of aggregation formation are widespread among medusozoans, and these aggregations are maintained through medusae behaviour in combination with oceanographic processes. In open waters commonly inhabited by scyphozoan and hydrozoan species, there can be strong currents. Although jellyfish in such environments may not be able to swim against peak tidal currents, sustained swimming against the current can facilitate the maintenance of aggregations and reduce the probability of getting stranded (Fossette et al. 2015). Similarly, bay-based jellyfish can swim horizontally away from bay mouths (Pitt and Kingsford 2000) or shift their vertical distribution to avoid strong surface ebb currents (Kopacz 1994) to help maintain population aggregations within bays. Jellyfish can even swim to maintain aggregations in bay subareas, for example, in shallow, nearshore waters close to beaches where current speeds are reduced (Schlaefer et al. 2018). Jellyfish concentrations can also be elevated within current features such as tidal eddies and slicks that are generated as waters move from open to structured habitats and interact with the topography (Chapter 3; Kingsford et al. 1991).

The maintenance of aggregations is biologically advantageous and ecologically significant. Maintaining aggregations may optimise foraging and sexual reproduction opportunities (Hamner and Dawson 2009). For example, in eddies and slicks with high jellyfish concentrations, there can be elevated concentrations of their zooplankton prey, representing a concentrated food source (Wolanski and Hamner 1988; Kingsford et al. 1991; Whitney et al. 2021). Jellyfish are important predators of zooplankton and can consume large portions of prey populations (Alldredge 1984; Oguz et al. 2001). Conversely, jellyfish aggregations themselves potentially represent important but variable pelagic subsidies to coastal and reefal ecosystems. Indeed, jellyfish are important prey to fish in temperate regions (Brodeur et al. 2021), and gelatinous tissue is a good a source of productivity-driving nitrogen (Pitt et al. 2009).

In addition to their trophic significance, jellyfish in aggregations can also significantly influence the physical dynamics of systems. For example, aggregations of golden jellyfish (*Mastigias sp.*) in Jellyfish Lake, Palau bio-engineer the deep lake system by stirring and aerating the upper layer of the lake as they migrate through it en masse (Katija and Dabiri 2009; Wolanski 2017). Similarly, aggregations of the upside-down jellyfish (*Cassiopea spp.*) rest with their bells on the bottom and tentacles oriented up in the water column in coastal habitats, including reef lagoons, mangroves, urban canals, sand flats, and seagrass meadows in the GBR and can significantly mix bottom waters as they pulse to generate feeding currents (Santhanakrishnan et al. 2012; Durieux et al. 2021; Templeman and Williams 2024). Their pulsing action can also disturb sediments, which can facilitate the biogeochemical cycling of nutrients (Jantzen et al. 2010; Durieux et al. 2021; Templeman and Williams 2024). The ecological and bio-engineering significance of jellyfish appears to be coupled with their tendency to aggregate.

14.4.2. Visually Guided Behaviours

Light is a common navigational cue in jellyfish and is associated with an array of ecologically significant behaviours. For example, responding to light has been linked to predator avoidance in jellyfish, where jellyfish can have a flight response in reaction to a sudden decrease in light intensity, representing a predator's shadow (Anderson and Mackie 1977). Reactions to changes in light intensity have also been linked to vertical migration (Arkett 1985), which could be advantageous for following prey (e.g., Canepa et al. 2014; Spadinger and Maier 1999), spawning (e.g., Canepa et al. 2014), or avoiding strong currents (e.g., Kopacz 1994). Light can also play a role in horizontal navigation, as heliotaxis (i.e., orientation to sunlight; Hamner et al. 1994; Hamner and Hauri 1981) has been reported in jellyfish species. Heliotaxis may be particularly important for species that derive energy from photosynthetic symbiotic algae that live in their mesoglea (e.g., *Mastigias sp.;* Hamner and Hauri 1981; Muscatine et al. 1986).

In structured nearshore environments, jellyfish have additional visual cues to direct their orientation. For example, the cubozoan *Carybdea rastonii* has been found to avoid dark benthic patches such as seagrass and algal beds to maintain positions on sand patches (Matsumoto 1995). *Tripedalia cystophora*, a cubozoan species that inhabits mangrove channels, has even been found to look up to and navigate by the mangrove canopy (Garm et al. 2011) and to swim through light shafts to target the prey that congregate there (Stewart 1996; Buskey 2003). Although *T. cystophora* is not found within the GBR, cubozoan species that inhabit mangroves on the GBR coast (e.g., *C. fleckeri*) could potentially perform similar behaviours. Finally, in structured habitats, obstacle avoidance is a common trait (Hamner et al. 1995; Matsumoto 1995; Garm et al. 2007b; Schlaefer et al. 2018) and is probably important for avoiding body damage. Overall, jellyfish commonly navigate using visual cues, and visually guided behaviours are entwined in predator-prey dynamics.

14.4.3. Behaviours and Population Structures

When considered together, a species' suite of behaviours can have important implications for their population structures, and good examples of this are found in cubozoan species. A population structure is the nested organisation of population units, where the broadest level is a species distribution that represents a metapopulation (Sinclair 1988; Kingsford and Battershill 1998). Metapopulations are then made up of largely self-contained mesopopulations/stocks with little genetic connectivity between them, and stocks are further divided into local populations that are connected but inhabit distinct areas (Kingsford and Battershill 1998; Kingsford et al. 2000, 2021). *Chironex fleckeri* are relatively large, strong swimming, and highly venomous cubozoans (Figure 4A). They maintain nearshore distributions in environments such as coasts and mangroves (Gordon and Seymour 2009; Kingsford et al. 2012; Schlaefer et al. 2018) and, crucially, can perform fine swimming manoeuvres to avoid the obstacles in these structurally complex environments (Hamner et al. 1995; Schlaefer et al. 2018). Additionally, currents in the nearshore shallow water environments inhabited by *C. fleckeri* are often reduced due to current shear (the slowing of current by friction; Fischer et al. 1979) and the sticky water effect (the diversion of current around structures; Andutta et al. 2012), which is likely to be very high in and among complex habitats such as mangrove stands. Given their habitat specificity and swimming ability and the reduced currents in their habitats, it is unsurprising that *C. fleckeri* stocks can be maintained across exceptionally small spatial scales of kilometres to tens of kilometres (Mooney and Kingsford 2016a, 2017; Schlaefer et al. 2018). *Copula sivickisi* (Figure 4C) are cubozoans with similarly fine-scale distributions and spatially confined stocks; however, the biophysics of how they maintain their fine-scale distributions differs from *C. fleckeri*.

C. sivickisi can inhabit bands of reef habitat in relatively high densities, and these bands can be narrow, with widths in the range of 100 m (Schlaefer et al. 2020). To investigate how *C. sivickisi* maintain this fine distribution pattern, Schlaefer et al. (2021) developed a biophysical model of *C. sivickisi* at Magnetic Island, Queensland, Australia (Figure 8ABCD, Figure 9AB). Biophysical models contain hydrodynamic modules that capture the physical environment, coupled with biological modules that can effectively represent a species biology, and for jellyfish, this can include medusae swimming behaviours. The model of Schlaefer et al. (2021) captured the complexity of the behavioural suite in *C. sivickisi* medusae and the complexity of their hydrodynamic environment, whereby current speeds were reduced in shallow, nearshore waters (Fischer et al. 1979; Andutta et al. 2012). The model included a medusae behaviour module with a nocturnal activity regime (Figure 8A), medusae maintaining positions near the bottom of the water column when they were active (Figure 8B), medusae swimming toward their habitat (Figure 8C), and medusae attaching to habitat to avoid being pushed back by the flow when the current speed exceeded a moderate threshold of 6 cm s^{-1} (Figure 8D). In a series of model runs, these researchers demonstrated that without behaviour, passive medusae were quickly lost from the habitat (Figure 9D). By contrast, medusae programmed with the full set of model behaviours were able to maintain positions on the habitat bands where the current speeds were reduced (Figure 9C; Schlaefer et al. 2021), which is in line with the quantified distribution of medusae (Schlaefer et al. 2020). Thereby, the ability of *C. sivickisi* medusae to maintain their highly restricted distribution is linked to their complex behaviours and the reduced current speeds present over their preferred habitat (Schlaefer et al. 2021).

14.5. FUTURE RESEARCH DIRECTIONS

There is a large scope for the expansion of research into jellyfish biophysical interactions. Missing information on biological processes in jellyfish at both the polyp and medusae life stages has hampered our understanding of biophysical links (Kingsford and Mooney 2014). The limited data on where to find polyps and accordingly, on the sources of medusae is notably problematic. These knowledge gaps are being addressed with both established and emergent research techniques. Environmental DNA (eDNA) analysis is a new technique particularly well-suited to jellyfish research, including for identifying sources of medusae, as the technique can overcome some of the challenges associated with sampling both polyps and medusae (e.g., hard to find, heterogenous distributions, and proximity to other dangerous organisms; Bolte et al. 2021; Morrissey et al. 2022). New biological information can be used to develop more sophisticated representations of jellyfish biology in biophysical models (e.g., Schlaefer et al. 2021, 2020). Concurrently, the physical components of biophysical models (i.e., the hydrodynamic modules) have generally increased in complexity (e.g., Swearer et al. 2019). Sophisticated biophysical models have been and will continue to be applied to investigate population structures and to predict interactions between jellyfish and humans. Examples of how fundamental biological questions are being addressed and of state-of-the-art biophysical modelling applications are discussed in the following.

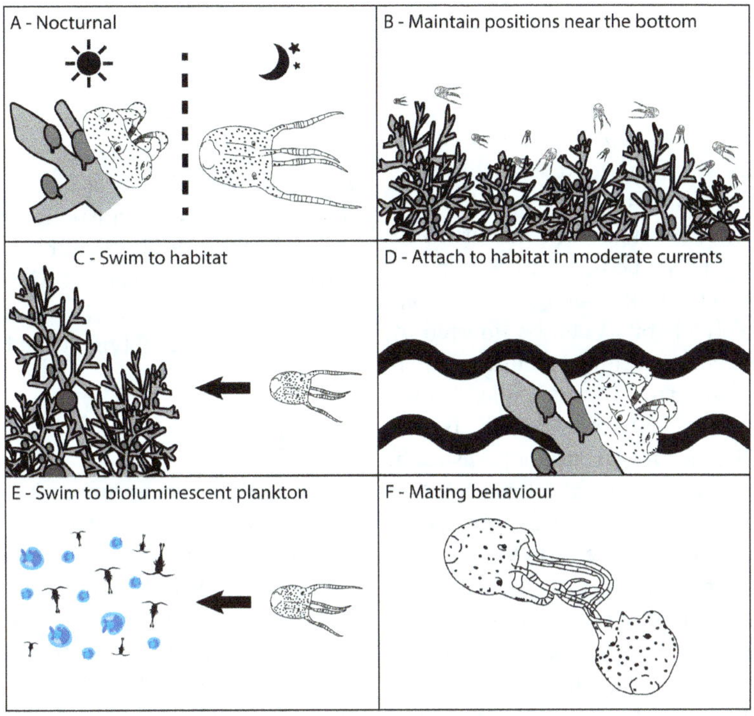

FIGURE 8 Behaviours documented in *Copula sivickisi* medusae (Class Cubozoa; Figure 4C). A) A nocturnal activity regime (Garm et al. 2012; Schlaefer et al. 2020). B) Maintaining positions nearshore and near the bottom of the water column where current speeds are reduced (Schlaefer et al. 2020). C) Swimming toward preferred habitats (Garm et al. 2012; Schlaefer et al. 2020). D) Attaching to hard structures in moderate currents (i.e., 3 to 6 cm s^{-1}) to avoid being pushed back by the flow (Schlaefer et al. 2020). E) Swimming toward bioluminescent plankton (e.g., *Noctiluca sp.*; Garm et al. 2016). F) Copulation in wedding dance mating behaviour (Lewis and Long 2005). *Sargassum sp.* (creator: Tracey Saxby) and copepod (creator: Jane Hawkey) vectors were sourced from the Integration and Application Network (https://ian.umces.edu/media-library/).

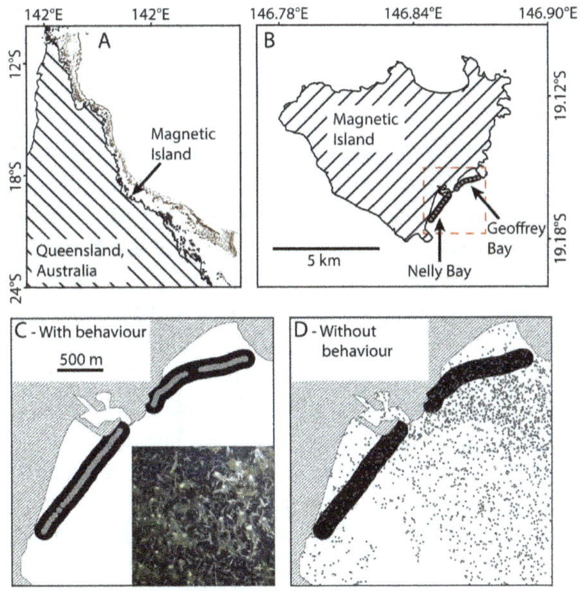

FIGURE 9 Biophysical modelling of *Copula sivickisi* (Class Cubozoa) from Schlaefer et al. (2021). A) The location of Magnetic Island, off the coast of Townsville, Queensland, Australia. Land is shown with a hatch pattern. The reefs of the GBR are shown in brown. B) Nelly Bay and Geoffrey Bay, Magnetic Island. The grey dots indicate the locations that virtual *C. sivickisi* medusae were released from in the biophysical model. The black shapes show the simulated reef habitat, which reflected the spatial extent that was measured in Schlaefer et al. (2020). The red dashed rectangle shows the extent of panels C and D. The distributions of virtual *C. sivickisi* medusae simulated C) with behaviour and D) without behaviour (passive), 7 days after their initial release. The behavioural model of Schlaefer et al. (2021) included behaviours A to D from Figure 8. The inset in panel C is a frame of underwater video footage from Nelly Bay, showing a high abundance (i.e., tens to hundreds per m^3) of *C. sivickisi* medusae over reef habitat (Schlaefer et al. 2020). Note that the dot size has been reduced in D to better show the dispersion of medusae without behaviour.

14.5.1. Polyps

There are particularly pronounced gaps in our understanding of the polyp stage of the jellyfish lifecycle. Historically, there has been great difficulty in finding cryptic polyp populations (but see Hartwick 1991). As juvenile medusae first enter the environment from polyp beds, identifying polyp habitats is key to advancing our knowledge of jellyfish biophysical interactions. Polyp habitats can be inferred by mapping the more conspicuous juvenile medusae if their distributions are highly restricted. For example, Brown (1973) repeatedly sampled the bays of Magnetic Island, off the coast of Townsville, North Queensland, Australia. The island is ~46 km in circumference, and Brown (1973) only found juvenile *Chironex fleckeri* medusae near a creek mouth in one of the bays and therefore concluded that polyps inhabited the creek. More recently, emergent methods have been utilized in the search for polyp habitats.

eDNA analysis focuses on the detection of DNA that organisms shed in the environment, and the technique has been used to help find polyp beds. As eDNA analysis involves collecting and filtering water samples, the technique is particularly well-suited to detecting the presence of polyps and medusae that are transparent, small, can inhabit murky waters, have spatially and temporally heterogenous distributions, and are sometimes found with other dangerous organisms (e.g., crocodiles). Medusae can be seasonally absent (e.g., Kingsford and Mooney 2014; Toyokawa et al. 2000), so the presence of jellyfish DNA in samples taken outside of a medusae season would suggest that sampling occurred in the vicinity of a polyp population (Bolte et al. 2021). eDNA analysis could therefore be of great utility for polyp detection as it continues to evolve in the future.

Elemental chemistry analysis has also been utilized to locate polyp habitats. The statoliths of jellyfish can record environmental signals that reflect the chemical composition of their environment, similar to fish otoliths (Secor et al. 1995; Bath et al. 2000). In scyphozoans and cubozoans, statoliths are present during metamorphosis from polyp to ephyra or juvenile medusae (Holst et al. 2016). Therefore, the elemental composition of the statolith core can give an indication of the environmental conditions experienced by the polyps, such as variation in temperature and salinity (e.g., Mooney and Kingsford 2012). However, a firm understanding of how the environment effects the uptake of elements is required before elemental composition can be used (Mooney and Kingsford 2016b; Morrissey et al. 2020). Overall, the application of emergent tools has great potential for building our understanding of polyp habitats.

Identifying the conditions that trigger polyps to bud medusae is also an active area of research. In terms of biophysical links, budding triggers determine the timing of when juveniles first enter and interact with the environment. Having information on where (i.e., polyp habitat) and when (i.e., budding triggers) medusae are likely to emerge would aid in predicting and reacting to nuisance jellyfish occurrences (Kingsford et al. 2018). The budding of medusae has been linked to physical cues such as changes in temperature, salinity, and food availability (Purcell et al. 1999; Liu et al. 2009; Lucas et al. 2012), but details are lacking for most species. Promisingly, progress has been made in determining budding triggers in controlled laboratory settings (e.g., Loveridge et al. 2021; Treible and Condon 2019). Further targeted research is needed to establish the conditions that cause polyps to bud medusae across different species.

14.5.2. Medusae

There are also gaps in our knowledge of medusae biology that hamper our understanding of biophysical links. Specifically, data on the behaviour of medusae are lacking for many jellyfish species. The behaviour of medusae in their interaction with the currents determines the fine-scale distributions of medusae within their habitats and their movements between habitats (e.g., Fossette et al. 2015; Schlaefer et al. 2018, 2021; Rahi et al. 2020). Medusae behaviour is therefore critical to determining jellyfish population structures (Schlaefer et al. 2018, 2021). Targeted laboratory studies have been instrumental in revealing the range of behaviours that medusae can perform, including studies on obstacle avoidance (e.g., Garm et al. 2007b; Hamner et al. 1995), swimming to prey (e.g., Garm et al. 2016), preferential selection of habitats (e.g., Garm et al. 2012; Schlaefer et al. 2020), avoidance of currents (e.g., Schlaefer et al. 2020), swimming faster against currents (Malul et al. 2019) and diel activity patterns (e.g., Garm et al. 2012). Field sampling and experiments can also provide insights into medusae behaviours. For example, depth preferences can be inferred from vertically stratified plankton tows (e.g., Schlaefer et al. 2020), within habitat residency preferences can be revealed by mapping distributions over small spatial scales (e.g., Pitt and Kingsford 2000; Schlaefer et al. 2020), and orientation behaviours can be discerned from translocation experiments (e.g., Garm et al. 2011). Additional laboratory and field-based studies could help to improve our understanding of the behavioural repertoires of a broader range of jellyfish species.

Elucidating medusae movements between habitats (e.g., from estuarine creeks to open waters) requires observations or sampling over long time periods. These periods range from hours of diurnal or nocturnal activity to the months that can constitute a medusa's lifespan. Electronic tags have been used to track jellyfish movements (e.g., Fossette et al. 2015; Moriarty et al. 2012); however, the technology currently has major limitations. It can be difficult to attach tags to jellyfishes' gelatinous bodies, and the tags used on jellyfish have worked only over hourly timescales in the field (Fossette et al. 2016). Tags must also be unobtrusive and lightweight relative to the tracked jellyfish to avoid impeding their swimming motion, which means that they are a viable option only for larger individuals (Fossette et al. 2016). Innovations into longer lasting electronic tags that are smaller and easier to attach would be of great utility (Gordon and Seymour 2009; Fossette et al. 2016). Drones have also been used to sample and track jellyfish, but there are similar constraints on tracking time and jellyfish size that need to be overcome (e.g., Raoult and Gaston 2018; Schaub et al. 2018; Rowley et al. 2020). In contrast, jellyfish movements can be elucidated using statolith elemental chemistry without the same size and time constraints, where elemental changes in statoliths across the axis of growth/time can reveal movements between habitats at timescales of lifespans (Mooney and Kingsford 2012). However, as mentioned previously, a comprehensive understanding of how environmental conditions impact the uptake of elements is needed before this technique can be used to infer jellyfish movements (Mooney and Kingsford 2016b; Morrissey et al. 2020). Quantifying the movements of medusae between habitats is challenging, and progress in this space may require technological advancements and a deeper knowledge of statolith elemental chemistry.

14.5.3. Biophysical Modelling

New information on polyp and medusa biology can be integrated into biophysical models of increasing sophistication with the aim to provide management-relevant information. Management-focused biophysical modelling applications in the GBR region are likely to target venomous jellyfish species that threaten the safety of water users and impact the tourism industry (e.g., Kingsford et al. 2018). Biophysical models can be used to hindcast past nuisance jellyfish interactions, including interactions with venomous species, to diagnose the drivers of the interactions (e.g., Rahi et al. 2020). Building on this knowledge, jellyfish biological modules could be paired with short-term hydrodynamic forecasts that predict within-season variability in physical conditions (e.g., wind, temperature, salinity, and currents) to provide risk outlooks for water users and managers days to weeks ahead (e.g., Brown et al. 2013; Laudy et al. 2020). There is also considerable variability in physical conditions across seasons, and deepening our understanding of how these seasonal dynamics influence jellyfish abundance patterns could be particularly useful for managers; for example, it would be invaluable to understand the variability in the onset and duration of stinging jellyfish seasons (e.g., Gordon and Seymour 2012). Seasonal hydrodynamic forecasts are operationally viable (Hobday et al. 2016), and future research could utilize such forecasts to predict nuisance jellyfish interactions months in advance. Biophysical models could also be used to look decades ahead, as hydrodynamic projections can propagate the physical signal of climate change from global to regional scales (e.g., Cheng et al. 2021; Hermann et al. 2021). In terms of jellyfish, hydrodynamic projections could be utilized to investigate habitat shifts driven by climate change impacts, as has been done for other taxa (e.g., arctic whales, Chambault et al. 2022; fish, Payne et al. 2022). Projections of habitat shifts may be most powerful when paired with field validation data, for example, eDNA monitoring for new species occurrences. Overall, hindcasts, forecasts, and projections of jellyfish distributions would be of great utility, but to get robust results, we generally need more data on species' fundamental ecology.

14.6. CONCLUSIONS

There are a wide range of biophysical interactions among the diverse assemblage of jellyfish found on the GBR. Despite their relatively simple morphology, jellyfish have multiple senses that can be used for orientation. This includes photoreception with eyes, some of which include lenses for high-resolution image acquisition. Their sensory capabilities are complemented by surprisingly complex nervous systems. The environmental cues that jellyfish receive inform their swimming behaviours, and they are capable swimmers. They can attain considerable speeds, whether swimming by rowing or jetting or by sailing. By using their abilities to sense and respond to stimuli, jellyfish can perform species-specific complex behaviours to exert some control over their positions within their environments. These behaviours include responding to currents to form ecologically significant aggregations and maintaining positions in structurally complex nearshore environments. The capacity for jellyfish to alter their distributions has implications for their population structures, and this is most evident in cubozoans. Cubozoans have the best developed senses among jellyfish and perform sophisticated behaviours to maintain stocks over small spatial scales (a few kilometres). Moving forward, we could deepen our understanding of jellyfish biophysical interactions by building our knowledge of jellyfish ecology at all life stages. Understanding jellyfish biophysical interactions is fundamental to discerning population and community ecology and reducing the risk of envenomation to humans.

REFERENCES

Alldredge AL (1984) The quantitative significance of gelatinous zooplankton as pelagic consumers. In: Fasham MJR (eds) Flows of energy and materials in marine ecosystems: Theory and practice. Springer US, pp. 407–433

Anderson PAV, Mackie GO (1977) Electrically coupled, photosensitive neurons control swimming in a Jellyfish. Science 197:186–188

Andutta FP, Kingsford MJ, Wolanski E (2012) 'Sticky water' enables the retention of larvae in a reef mosaic. Estuar Coast Shelf Sci 101:54–63

Arai MN (1997) A functional biology of Scyphozoa. Springer Science & Business Media

Arkett SA (1985) The Shadow Response of a Hydromedusan (Polyorchis penicillatus): Behavioral Mechanisms Controlling Diel and Ontogenic Vertical Migration. Biol Bull 169:297–312

Avian M, Ramšak A (2021) Phylum Cnidaria: Classes scyphozoa, cubozoa, and staurozoa. In: Schierwater B, DeSalle R (eds) Invertebrate zoology: A Tree of Life Approach. CRC Press, pp. 149–172

Bath GE, Thorrold SR, Jones CM, Campana SE, McLaren JW, Lam JWH (2000) Strontium and barium uptake in aragonitic otoliths of marine fish. Geochim Cosmochim Acta 64:1705–1714

Bielecki J, Zaharoff AK, Leung NY, Garm A, Oakley TH (2014) Ocular and extraocular expression of opsins in the Rhopalium of Tripedalia Cystophora (Cnidaria: Cubozoa). PLoS One 9:e98870

Bieri R (1959) Dimorphism and size distribution in Velella and Physalia. Nature 184:1333–1334

Blackstone NW, Cartwright P (2021) Phylum Cnidaria: Class Hydrozoa. In: Schierwater B, DeSalle R (eds) Invertebrate zoology: A tree of life approach. CRC Press, pp. 193–204

Boßelmann F, Epple M, Sötje I, Tiemann H (2007) Statoliths of calcium sulfate hemihydrate are used for gravity sensing in Rhopaliophoran Medusae (Cnidaria). Handbook of Biomineralization. WILEY-VCH Verlag GmbH & Co. KGaA, pp. 261–272

Bolte B, Goldsbury J, Huerlimann R, Jerry D, Kingsford M (2021) Validation of eDNA as a viable method of detection for dangerous cubozoan jellyfish. Environ DNA 3:769–779

Bosch-Belmar M, Milisenda G, Basso L, Doyle TK, Leone A, Piraino S (2021) Jellyfish Impacts on Marine Aquaculture and Fisheries. Rev Fish Sci Aquac 29:242–259

Brodeur R, Buckley T, Lang G, Draper D, Buchanan J, Hibpshman R (2021) Demersal fish predators of gelatinous zooplankton in the Northeast Pacific Ocean. Mar Ecol Prog Ser 658:89–104

Brodeur RD, Sugisaki H, Hunt GLJr (2002) Increases in jellyfish biomass in the Bering Sea: Implications for the ecosystem. Mar Ecol Prog Ser 233:89–103

Brown CW, Hood RR, Long W, Jacobs J, Ramers DL, Wazniak C, Wiggert JD, Wood R, Xu J (2013) Ecological forecasting in Chesapeake Bay: Using a mechanistic–empirical modeling approach. J Mar Syst 125:113–125

Brown TW (1973) Chironex fleckeri: Distribution and movements around magnetic Island, North Queensland. James Cook University, p. 26.

Buskey E (2003) Behavioral adaptations of the cubozoan medusa Tripedalia cystophora for feeding on copepod (Dioithona oculata) swarms. Mar Biol 142:225–232

Canepa A, Fuentes V, Sabatés A, Piraino S, Boero F, Gili J-M (2014) Pelagia noctiluca in the Mediterranean Sea. In: Pitt KA, Lucas CH (eds) Jellyfish blooms. Springer, pp. 237–266

Chambault P, Kovacs KM, Lydersen C, Shpak O, Teilmann J, Albertsen CM, Heide-Jørgensen MP (2022) Future seasonal changes in habitat for Arctic whales during predicted ocean warming. Sci Adv 8:eabn2422

Cheng W, Hermann AJ, Hollowed AB, Holsman KK, Kearney KA, Pilcher DJ, Stock CA, Aydin KY (2021) Eastern Bering Sea shelf environmental and lower trophic level responses to climate forcing: Results of dynamical downscaling from CMIP6. Deep Sea Res Part II Top Stud Oceanogr 193:104975

Chi X, Mueller-Navarra DC, Hylander S, Sommer U, Javidpour J (2019) Food quality matters: Interplay among food quality, food quantity and temperature affecting life history traits of *Aurelia aurita* (Cnidaria: Scyphozoa) polyps. Sci Total Environ 656:1280–1288

Clark FE, Lane CE (1961) Composition of float gases of physalia physalis. Proc Soc Exp Biol Med 107:673–674

Coates M, Theobald J (2003) Optimal visual parameters for a cubozoan jellyfish in the mangrove environment. Integ Comp Biol 43:1016–1016

Colin SP, Costello JH (2002) Morphology, swimming performance and propulsive model of six co-occurring hydromedusae. J Exp Biol 205:427–437

Colin SP, Costello JH, Katija K, Seymour J, Kiefer K (2013) Propulsion in Cubomedusae: Mechanisms and Utility. PLoS One 8:e56393

Collins AG (2002) Phylogeny of Medusozoa and the evolution of cnidarian life cycles. J Evol Biol 15:418–432

Collins AG (2009) Recent Insights into Cnidarian Phylogeny. Smithsonian Contributions to the Marine Sciences 38:141–149

Cornelius PF (1992) Cnidarian medusa suppression, hydroid rafting, and leptolids on remote islands. Sci Mar 56:245–261

Costello JH, Colin SP, Dabiri JO (2008) Medusan morphospace: Phylogenetic constraints, biomechanical solutions, and ecological consequences. Invertebr Biol 127:265–290

Costello JH, Colin SP, Gemmell BJ, Dabiri JO, Sutherland KR (2015) Multi-jet propulsion organized by clonal development in a colonial siphonophore. Nat Commun 6:8158

Crowley-Cyr L, Gershwin L (2021) Protecting the public from hazardous jellyfish: A wicked problem for regulators and operators? In: Mariottini GL, Killi N, Xiao L (eds) The cnidaria: Only a problem or also a resource? Nova Science Publishers, pp. 99–122

Dabiri JO, Colin SP, Katija K, Costello JH (2010) A wake-based correlate of swimming performance and foraging behavior in seven co-occurring jellyfish species. J Exp Biol 213:1217–1225

D'Ambra I, Costello JH, Bentivegna F (2001) Flow and prey capture by the scyphomedusa Phyllorhiza punctata von Lendenfeld, 1884. Hydrobiologia 451:223–227

Dunn C (2009) Siphonophores. Curr Biol 19:R233–R234

Dunn CW, Wagner GP (2006) The evolution of colony-level development in the Siphonophora (Cnidaria: Hydrozoa). Dev Genes Evol 216:743–754

Durieux DM, Du Clos KT, Lewis DB, Gemmell BJ (2021) Benthic jellyfish dominate water mixing in mangrove ecosystems. Proc Natl Acad Sci 118:e2025715118

Ferrer L, González M (2021) Relationship between dimorphism and drift in the Portuguese man-of-war. Cont Shelf Res 212:104269

Fischer H, List E, Koh R, Imberger J, Books N (1979) Mixing in Inland and Coastal waters. Academic Press Inc., 111 Fifth Avenue, p. 483

Foox J (2021) Phylum cnidaria: Lineage myxozoa. In: Schierwater B, DeSalle R (eds) Invertebrate zoology: A tree of life approach. CRC Press, p. 644.

Fossette S, Gleiss AC, Chalumeau J, Bastian T, Armstrong CD, Vandenabeele S, Karpytchev M, Hays GC (2015) Current-oriented swimming by jellyfish and its role in bloom maintenance. Curr Biol 25:342–347

Fossette S, Katija K, Goldbogen JA, Bograd S, Patry W, Howard MJ, Knowles T, Haddock SHD, Bedell L, Hazen EL, Robison BH, Mooney TA, Shorter KA, Bastian T, Gleiss AC (2016) How to tag a jellyfish? A methodological review and guidelines to successful jellyfish tagging. J Plankton Res 38:1347–1363

Francis L (1991) Sailing downwind: Aerodynamic performance of the Velella Sail. J Exp Biol 158:117–132

Garm A, Andersson F, Nilsson D-E (2008) Unique structure and optics of the lesser eyes of the box jellyfish Tripedalia cystophora. Vision Res 48:1061–1073

Garm A, Bielecki J (2008) Swim pacemakers in box jellyfish are modulated by the visual input. J Comp Physiol A 194:641–651

Garm A, Bielecki J, Petie R, Nilsson D-E (2012) Opposite patterns of diurnal activity in the box Jellyfish Tripedalia cystophora and Copula sivickisi. Biol Bull 222:35–45

Garm A, Bielecki J, Petie R, Nilsson D-E (2016) Hunting in Bioluminescent light: Vision in the Nocturnal Box Jellyfish Copula sivickisi. Front Physiol 7:99

Garm A, Coates MM, Gad R, Seymour J, Nilsson D-E (2007a) The lens eyes of the box jellyfish Tripedalia cystophora and Chiropsalmus sp. are slow and color-blind. J Comp Physiol A 193:547–557

Garm A, Ekström P, Boudes M, Nilsson D-E (2006) Rhopalia are integrated parts of the central nervous system in box jellyfish. Cell Tissue Res 325:333–343

Garm A, O'Connor M, Parkefelt L, Nilsson D-E (2007b) Visually guided obstacle avoidance in the box jellyfish Tripedalia cystophora and Chiropsella bronzie. J Exp Biol 210:3616–3623

Garm A, Oskarsson M, Nilsson D-E (2011) Box Jellyfish use terrestrial visual cues for navigation. Curr Biol 21:798–803

Garm A, Poussart Y, Parkefelt L, Ekström P, Nilsson D-E (2007c) The ring nerve of the box jellyfish Tripedalia cystophora. Cell Tissue Res 329:147–157

Gemmell BJ, Colin SP, Costello JH (2018) Widespread utilization of passive energy recapture in swimming medusae. J Exp Biol 221:168575

Gemmell BJ, Costello JH, Colin SP, Stewart CJ, Dabiri JO, Tafti D, Priya S (2013) Passive energy recapture in jellyfish contributes to propulsive advantage over other metazoans. Proc Natl Acad Sci 110:17904–17909

Gershwin L, Richardson AJ, Winkel KD, Fenner PJ, Lippmann J, Hore R, Avila-Soria G, Brewer D, Kloser RJ, Steven A (2013) Biology and ecology of Irukandji jellyfish (Cnidaria: Cubozoa). In Advances in marine biology. Elsevier, pp. 1–85

Gibbons MJ, Janson LA, Ismail A, Samaai T (2010) Life cycle strategy, species richness and distribution in marine Hydrozoa (Cnidaria: Medusozoa). J Biogeogr 37:441–448

Gladfelter WB (1972) Structure and function of the locomotory system of polyorchis montereyensis (Cnidaria, Hydrozoa). Helgoländer Wiss Meeresunters 23:38–79

Gordon M, Seymour J (2009) Quantifying movement of the tropical Australian cubozoan Chironex fleckeri using acoustic telemetry. Hydrobiologia 616:87–97

Gordon M, Seymour J (2012) Growth, development and temporal variation in the onset of six Chironex fleckeri medusae seasons: A contribution to understanding jellyfish ecology. PLoS One 7:e31277

Hamner W, Jones M, Hamner P (1995) Swimming, feeding, circulation and vision in the Australian box jellyfish, Chironex fleckeri (Cnidaria: Cubozoa). Mar Freshw Res 46:985–990

Hamner WM, Dawson MN (2009) A review and synthesis on the systematics and evolution of jellyfish blooms: Advantageous aggregations and adaptive assemblages. Hydrobiologia 616:161–191

Hamner WM, Hamner PP, Strand SW (1994) Sun-compass migration by *Aurelia aurita* (Scyphozoa): Population retention and reproduction in Saanich Inlet, British Columbia. Mar Biol 119:347–356

Hamner WM, Hauri IR (1981) Long-distance horizontal migrations of zooplankton (Scyphomedusae: Mastigias). Limnol Oceanogr 26:414–423

Hartwick RF (1991) Distributional ecology and behaviour of the early life stages of the box-jellyfish Chironex fleckeri. Hydrobiologia 216:181–188

Hermann AJ, Kearney K, Cheng W, Pilcher D, Aydin K, Holsman KK, Hollowed AB (2021) Coupled modes of projected regional change in the Bering Sea from a dynamically downscaling model under CMIP6 forcing. Deep Sea Res Part II Top Stud Oceanogr 194:104974

Higgins JE, Ford MD, Costello JH (2008) Transitions in morphology, nematocyst distribution, fluid motions, and prey capture during development of the Scyphomedusa Cyanea capillata. Biol Bull 214:29–41

Hobday AJ, Spillman CM, Paige Eveson J, Hartog JR (2016) Seasonal forecasting for decision support in marine fisheries and aquaculture. Fish Oceanogr 25:45–56

Holst S, Michalik P, Noske M, Krieger J, Sötje I (2016) Potential of X-ray micro-computed tomography for soft-bodied and gelatinous cnidarians with special emphasis on scyphozoan and cubozoan statoliths. J Plankton Res 38:1225–1242

Hyman LH (1940) The invertebrates: Protozoa. McGraw-Hill Book Company.

Iosilevskii G, Weihs D (2009) Hydrodynamics of sailing of the Portuguese man-of-war *Physalia physalis*. J R Soc Interface 6:613–626

Jantzen C, Wild C, Rasheed M, El-Zibdah M, Richter C (2010) Enhanced pore-water nutrient fluxes by the upside-down jellyfish Cassiopea sp. in a Red Sea coral reef. Mar Ecol Prog Ser 411:117–125

Kass-Simon G, Scappaticci Jr AA (2002) The behavioral and developmental physiology of nematocysts. Can J Zool 80:1772–1794

Katija K, Beaulieu W, Regula C, Colin S, Costello J, Dabiri J (2011) Quantification of flows generated by the hydromedusa Aequorea victoria: A Lagrangian coherent structure analysis. Mar Ecol Prog Ser 435:111–123

Katija K, Colin SP, Costello JH, Jiang H (2015) Ontogenetic propulsive transitions by Sarsia tubulosa medusae. J Exp Biol 218:2333–2343

Katija K, Dabiri JO (2009) A viscosity-enhanced mechanism for biogenic ocean mixing. Nature 460:624–626

Kawabata T, Lindsay DJ, Kitamura M, Konishi S, Nishikawa J, Nishida S, Kamio M, Nagai H (2013) Evaluation of the bioactivities of water-soluble extracts from twelve deep-sea jellyfish species. Fish Sci 79:487–494

Kayal E, Bentlage B, Sabrina Pankey M, Ohdera AH, Medina M, Plachetzki DC, Collins AG, Ryan JF (2018) Phylogenomics provides a robust topology of the major cnidarian lineages and insights on the origins of key organismal traits. BMC Evol Biol 18:68

Kingsford MJ (1993) Biotic and abiotic structure in the pelagic environment: Importance to small fishes. Bull Mar Sci 53:393–415

Kingsford MJ, Battershill CN (1998) Studying temperate marine environments: A handbook for ecologists. University of Canterbury Press, p. 335.

Kingsford MJ, Becken S, Bordehore C, Fuentes VL, Pitt KA, Yangihara AA (2018) Empowering stakeholders to manage stinging jellyfish: A perspective. Coast Manag 46:1–18

Kingsford MJ, Mooney CJ (2014) The ecology of box Jellyfishes (Cubozoa). In: Pitt KA, Lucas CH (eds) Jellyfish blooms. Springer, pp. 267–302

Kingsford MJ, Pitt KA, Gillanders BM (2000) Management of jellyfish fisheries, with special reference to the O. Rhizostomeae. Oceanogr Mar Biol 38:85–156

Kingsford MJ, Schlaefer JA, Morrissey SJ (2021) Population structures and levels of connectivity for scyphozoan and cubozoan jellyfish. Diversity 13:174

Kingsford MJ, Seymour JE, O'Callaghan MD (2012) Abundance patterns of cubozoans on and near the Great Barrier Reef. In: Purcell J, Mianzan H, Frost JR (eds) Jellyfish blooms IV: Interactions with humans and fisheries. Springer, pp. 257–268

Kingsford MJ, Wolanski E, Choat JH (1991) Influence of tidally induced fronts and Langmuir circulations on distribution and movements of presettlement fishes around a coral reef. Mar Biol 109:167–180

Koizumi O, Hamada S, Minobe S, Hamaguchi-Hamada K, Kurumata-Shigeto M, Nakamura M, Namikawa H (2015) The nerve ring in cnidarians: Its presence and structure in hydrozoan medusae. Zoology 118:79–88

Kopacz U (1994) Evidence for tidally-induced vertical migration of some gelatinous zooplankton in the Wadden Sea area near Sylt. Helgoländer Meeresunters 48:333–342

Kramp P (1965) Some medusae (mainly scyphomedusae) from Australian coastal waters. Trans R Soc S Aust 89:257–278

Laudy C, Mészáros L, Wanke S, Juan de M (2020) Mixing social media analysis and physical models to monitor invasive species. In: 2020 IEEE 23rd international conference on information fusion (FUSION). IEEE, pp. 1–8

Lee K, Bae B-S, Kim I-O, Yoon W-D (2010) Measurement of swimming speed of giant jellyfish Nemopilema nomurai using acoustics and visualization analysis. Fish Sci 76:893–899

Lewis C, Long TAF (2005) Courtship and reproduction in Carybdea sivickisi (Cnidaria: Cubozoa). Mar Biol 147:477–483

Liu AG, Matthews JJ, Menon LR, McIlroy D, Brasier MD (2014) Haootia quadriformis n. gen., n. sp., interpreted as a muscular cnidarian impression from the Late Ediacaran period approx. 560 Ma). Proc R Soc B Biol Sci 281:20141202

Liu W-C, Lo W-T, Purcell JE, Chang H-H (2009) Effects of temperature and light intensity on asexual reproduction of the scyphozoan, *Aurelia aurita* (L.) in Taiwan. In: Pitt KA, Purcell JE (eds) Jellyfish blooms: Causes, consequences, and recent advances: Proceedings of the second international jellyfish blooms symposium, held at the gold coast, Springer, pp. 247–258

Loveridge A, Lucas CH, Pitt KA (2021) Shorter, warmer winters may inhibit production of ephyrae in a population of the moon jellyfish *Aurelia aurita*. Hydrobiologia 848:739–749

Lucas, CH (2001) Reproduction and life history strategies of the common jellyfish, Aurelia aurita, in relation to its ambient environment. Hydrobiologia 451:229–246. https://doi.org/10.1023/A:1011836326717

Lucas CH, Graham WM, Widmer C (2012) Jellyfish life histories: Role of polyps in forming and maintaining scyphomedusa populations. Adv Mar Biol 63:133–196

Lynam CP, Gibbons MJ, Axelsen BE, Sparks CAJ, Coetzee J, Heywood BG, Brierley AS (2006) Jellyfish overtake fish in a heavily fished ecosystem. Curr Biol 16:R492–R493

Mackie GO (1960) Studies on Physalia physalis (L.) Part 2. Behaviour and histology. Discov Rep 30:369–407

Mackie GO (2004) Central neural circuitry in the jellyfish Aglantha. Neurosignals 13:5–19

Malul D, Lotan T, Makovsky Y, Holzman R, Shavit U (2019) The levantine jellyfish Rhopilema nomadica and Rhizostoma pulmo swim faster against the flow than with the flow. Sci Rep 9:20337

Marques AC, Collins AG (2004) Cladistic analysis of Medusozoa and cnidarian evolution. Invertebr Biol 123:23–42

Matsumoto GI (1995) Observations on the anatomy and behaviour of the cubozoan Carybdea rastonii haacke. Mar Freshw Behav Physiol 26:139–148

Miles JG, Battista NA (2019) Naut your everyday jellyfish model: Exploring how tentacles and oral arms impact locomotion. Fluids 4:169

Miranda LS, Mills CE, Hirano YM, Collins AG, Marques AC (2018) A review of the global diversity and natural history of stalked jellyfishes (Cnidaria, Staurozoa). Mar Biodivers 48:1695–1714

Möller H (1984) Reduction of a larval herring population by jellyfish predator. Science 224:621–622

Mooney C, Kingsford M (2012) Sources and movements of Chironex fleckeri medusae using statolith elemental chemistry. Hydrobiologia 690:269–277

Mooney CJ, Kingsford MJ (2016a) Discriminating populations of medusae (Chironex fleckeri, Cubozoa) using statolith microchemistry. Mar Freshw Res 68:1144–1152

Mooney CJ, Kingsford MJ (2016b) The influence of salinity on box jellyfish (Chironex fleckeri, Cubozoa) statolith elemental chemistry. Mar Biol 163:103

Mooney CJ, Kingsford MJ (2017) Statolith morphometrics as a tool to distinguish among populations of three cubozoan species. Hydrobiologia 787:111–121

Moriarty PE, Andrews KS, Harvey CJ, Kawase M (2012) Vertical and horizontal movement patterns of scyphozoan jellyfish in a fjord-like estuary. Mar Ecol Prog Ser 455:1–12

Morrissey SJ, Jerry DR, Kingsford MJ (2022) Genetic detection and a method to study the ecology of deadly Cubozoan Jellyfish. Diversity 14:

Morrissey SJ, Schlaefer JA, Kingsford MJ (2020) Experimental validation of the relationships between cubozoan statolith elemental chemistry and salinity and temperature. J Exp Mar Biol Ecol 527:151375

Muscatine L, Wilkerson FP, McCloskey LR (1986) Regulation of population density of symbiotic algae in a tropical marine jellyfish (Mastigias sp.). Mar Ecol Prog Ser 32:279–290

Nakanishi N, Hartenstein V, Jacobs DK (2009) Development of the rhopalial nervous system in Aurelia sp.1 (Cnidaria, Scyphozoa). Dev Genes Evol 219:301–317

Nilsson D-E, Gislén L, Coates MM, Skogh C, Garm A (2005) Advanced optics in a jellyfish eye. Nature 435:201–205

Oguz T, Ducklow HW, Purcell JE, Malanotte-Rizzoli P (2001) Modeling the response of top-down control exerted by gelatinous carnivores on the Black Sea pelagic food web. J Geophys Res Oceans 106:4543–4564

OpenTree et al. (2021) Open tree of life synthetic tree (Version 13.4). https://tree.opentreeoflife.org/opentree/opentree14.7@mrcaott431ott3524/Hydrozoa--Stauromedusae

Passano L (1973) Behavioral control systems in medusae; a comparison between hydro-and scyphomedusae. Publ Seto Mar Biol Lab 20:615–645

Payne MR, Danabasoglu G, Keenlyside N, Matei D, Miesner AK, Yang S, Yeager SG (2022) Skilful decadal-scale prediction of fish habitat and distribution shifts. Nat Commun 13:2660

Petie R, Garm A, Nilsson D-E (2011) Visual control of steering in the box jellyfish Tripedalia cystophora. J Exp Biol 214:2809–2815

Petie R, Garm A, Nilsson D-E (2013) Contrast and rate of light intensity decrease control directional swimming in the box jellyfish Tripedalia cystophora (Cnidaria, Cubomedusae). Hydrobiologia 703:69–77

Picciani N, Kerlin JR, Sierra N, Swafford AJM, Ramirez MD, Roberts NG, Cannon JT, Daly M, Oakley TH (2018) Prolific origination of eyes in cnidaria with co-option of non-visual opsins. Curr Biol 28:2413–2419.e4

Pitt KA, Kingsford MJ (2000) Geographic separation of stocks of the edible jellyfish Catostylus mosaicus (Rhizostomeae) in New South Wales, Australia. Mar Ecol Prog Ser 196:143–155

Pitt KA, Kingsford MJ (2003) Temporal and spatial variation in recruitment and growth of medusae of the jellyfish, Catostylus mosaicus (Scyphozoa: Rhizostomeae). Mar Freshw Res 54:117–125

Pitt KA, Welsh DT, Condon RH (2009) Influence of jellyfish blooms on carbon, nitrogen and phosphorus cycling and plankton production. Hydrobiologia 616:133–149

Purcell J, Uye S, Lo W (2007) Anthropogenic causes of jellyfish blooms and their direct consequences for humans: A review. Mar Ecol Prog Ser 350:153–174

Purcell JE, White JR, Nemazie DA, Wright DA (1999) Temperature, salinity and food effects on asexual reproduction and abundance of the scyphozoan Chrysaora quinquecirrha. Mar Ecol Prog Ser 180:187–196

Quattrini AM, DeLeo DM, Brugler MR (2021) Phylum cnidaria: Class anthozoa. In: Schierwater B, DeSalle R (eds) Invertebrate zoology: A tree of life approach. CRC Press

Rahi JE, Weeber MP, Serafy GE (2020) Modelling the effect of behavior on the distribution of the jellyfish Mauve stinger (Pelagia noctiluca) in the Balearic Sea using an individual-based model. Ecol Model 433:109230

Raoult V, Gaston TF (2018) Rapid biomass and size-frequency estimates of edible jellyfish populations using drones. Fish Res 207:160–164

Rayner JMV (1986) Pleuston: Animals which move in water and air. Endeavour 10:58–64

Rowley OC, Courtney RL, Browning SA, Seymour JE (2020) Bay watch: Using unmanned aerial vehicles (UAV's) to survey the box jellyfish Chironex fleckeri. PLoS One 15:e0241410

Santander MD, Maronna MM, Ryan JF, Andrade SCS (2022) The state of Medusozoa genomics: Current evidence and future challenges. GigaScience 11:giac036

Santhanakrishnan A, Dollinger M, Hamlet CL, Colin SP, Miller LA (2012) Flow structure and transport characteristics of feeding and exchange currents generated by upside-down Cassiopea jellyfish. J Exp Biol 215:2369–2381

Satterlie R (2017) Cnidarian neurobiology. Oxf Handb Invertebr Neurobiol 185–218

Satterlie RA (2002) Neuronal control of swimming in jellyfish: A comparative story. Can J Zool 80:1654–1669

Satterlie RA (2011) Do jellyfish have central nervous systems? J Exp Biol 214:1215–1223

Scappaticci AA, Kass-Simon G (2008) NMDA and GABAB receptors are involved in controlling nematocyst discharge in hydra. Comp Biochem Physiol A Mol Integr Physiol 150:415–422

Schaub J, Hunt B, Pakhomov E, Holmes K, Lu Y, Quayle L (2018) Using unmanned aerial vehicles (UAVs) to measure jellyfish aggregations. Mar Ecol Prog Ser 591:29–36

Schlaefer JA, Wolanski E, Kingsford MJ (2018) Swimming behaviour can maintain localised jellyfish (Chironex fleckeri: Cubozoa) populations. Mar Ecol Prog Ser 591:287–302

Schlaefer JA, Wolanski E, Lambrechts J, Kingsford MJ (2021) Behavioural and oceanographic isolation of an island-based jellyfish (Copula sivickisi, Class Cubozoa) population. Sci Rep 11:10280

Schlaefer JA, Wolanski E, Yadav S, Kingsford MJ (2020) Behavioural maintenance of highly localised jellyfish (Copula sivickisi, class Cubozoa) populations. Mar Biol 167:40

Secor DH, Henderson-Arzapalo A, Piccoli PM (1995) Can otolith microchemistry chart patterns of migration and habitat utilization in anadromous fishes? J Exp Mar Biol Ecol 192:15–33

Sinclair M (1988) Marine populations: An essay on population regulation and speciation. University of Washington Press

Skogh C, Garm A, Nilsson D-E, Ekström P (2006) Bilaterally symmetrical rhopalial nervous system of the box jellyfish Tripedalia cystophora. J Morphol 267:1391–1405

Solé M, Lenoir M, Fontuño JM, Durfort M, van der Schaar M, André M (2016) Evidence of Cnidarians sensitivity to sound after exposure to low frequency noise underwater sources. Sci Rep 6:37979

Spadinger R, Maier G (1999) Prey selection and diel feeding of the freshwater jellyfish, Craspedacusta sowerbyi. Freshw Biol 41:567–573

Stewart SE (1996) Field behavior of Tripedalia cystophora (class cubozoa). Mar Freshw Behav Physiol 27:175–188

Sutherland KR, Gemmell BJ, Colin SP, Costello JH (2019) Propulsive design principles in a multi-jet siphonophore. J Exp Biol Jeb.198242

Swearer SE, Treml EA, Shima JS (2019) A review of biophysical models of marine larval dispersal. Oceanogr Mar Biol 57:325–356

Templeman MA, Williams CD (2024) Jellyfish: A window into pesticide distribution and risks on the Great Barrier Reef. In Wolanski E, Kingsford MJ (eds.), Oceanographic Processes of Coral Reefs. Physical and Biological Links in the Great Barrier Reef. CRC Press

Totton AK, Mackie GO (1960) Studies on physalia physalis. Cambridge University Press

Toyokawa M, Furota T, Terazaki M (2000) Life history and seasonal abundance of *Aurelia aurita* medusae in Tokyo Bay, Japan. Plankton Biol Ecol 47:48–58

Treible LM, Condon RH (2019) Temperature-driven asexual reproduction and strobilation in three scyphozoan jellyfish polyps. J Exp Mar Biol Ecol 520:151204

Uye S-I (2014) The Giant Jellyfish Nemopilema nomurai in East Asian Marginal Seas. In: Pitt KA., Lucas CH (eds) Jellyfish blooms. Springer, pp. 185–205

Van Iten H, de Moraes Leme J, Simões MG, Marques AC, Collins AG (2006) Reassessment of the phylogenetic position of conulariids (?Ediacaran-Triassic) within the subphylum Medusozoa (phylum cnidaria). J Syst Palaeontol 4:109–118

Weber C (1981) Structure, histochemistry, ontogenetic development, and regeneration of the ocellus of Cladonema radiatum dujardin (cnidaria, hydrozoa, anthomedusae). J Morphol 167:313–331

Werner B (1983) Die Metamorphose des Polypen vonTripedalia cystophora (Cubozoa, Carybdeidae) in die Meduse. Helgoländer Meeresunters 36:257–276

Whitney JL, Gove JM, McManus MA, Smith KA, Lecky J, Neubauer P, Phipps JE, Contreras EA, Kobayashi DR, Asner GP (2021) Surface slicks are pelagic nurseries for diverse ocean fauna. Sci Rep 11:3197

Wolanski E (2017) Bounded and unbounded boundaries—Untangling mechanisms for estuarine-marine ecological connectivity: Scales of m to 10,000 km—A review. Estu Coastal Shelf Sci 198:378–392

Wolanski E, Hamner WM (1988) Topographically controlled fronts in the ocean and their biological influence. Science 241:177–181

Zimmerman KL, Jamshidi AD, Buckenberger A, Satterlie RA (2019) Organization of the subumbrellar musculature in the ephyra, juvenile, and adult stages of *Aurelia aurita* Medusae. Invertebr Biol 138:e12260

15 More Intense Severe Tropical Cyclones in Recent Decades Cause Greater Impacts on Mangroves Bordering Australia's Great Barrier Reef

Norman C. Duke, Adam D. Canning, and Jock R. Mackenzie

15.1. INTRODUCTION

Coastal and estuarine mangrove forests bordering the Great Barrier Reef (GBR) region play a vital role in the health and well-being of extensive coral reefs along Australia's north-eastern coast (GBRMPA 2011; Fabricius et al. 2008; De'ath et al. 2019; Fine et al. 2019; Duke and Larkum 2019). From various studies (e.g., Duke et al. 2007), we understand that mangroves provide multiple benefits to coral reefs including those associated with shared trophic interactions (Mumby et al. 2004), coastal catchment runoff (Asbridge et al. 2016), more extreme fluctuations in mean sea level (Duke et al. 2022), rapidly rising sea levels (Lovelock et al. 2015), more intense storms (Asbridge et al. 2018), and generally altered climatic conditions (Gabler et al. 2017).

Tropical cyclones are known to have severe impacts on coastal mangrove communities (Krauss and Osland 2020; Asbridge et al. 2018; Beeden et al. 2015). It is generally understood that the damage inflicted is proportionate to the intensity of the cyclone, which is chiefly defined by its high wind speeds and large storm surge waves. Although the impacts of cyclones on mangroves are often abrupt and sometimes catastrophic (see Figures 1 and 2), they are subject to the prevailing global climate and weather processes with changes in the El Niño Southern Oscillation (ENSO) atmospheric conditions (Bell et al. 2019, 2020). These ENSO forces appear to be worsening with global climate change (Holbrook et al. 2021). However, it is considered critical to appreciate that such processes are occurring with the additional damaging pressures of direct and indirect human impacts, such as large marine oil spills (Duke 2016) and generally, with increasing human development influences (Babcock et al. 2019; Duke 2020).

The types of abrupt damage inflicted by tropical cyclones on mangrove forests include canopy defoliation, limb breakage, stem breakage, below ground root exposure, above-ground root burial, and tree death (Krauss and Osland 2020). However, observations of cyclone impacts in the GBR region, while at times quite extreme (GBRMPA 2011; Fine et al. 2019), have often been reported only on an *ad hoc* basis (see also Figures 1 and 2). Furthermore, and despite the generally accepted worsening trajectory of climate change, there has been uncertainty regarding the impacts from current trends in tropical cyclone intensity and frequency for the GBR region (Callaghan et al. 2011; Chand et al. 2019, 2020; Ramsay et al. 2018; Dowdy 2019).

For these reasons, our aim has been to undertake a systematic assessment that specifically looks at severe tropical cyclones and their damaging impacts on coastal mangrove communities of the GBR area since 1970. For this purpose, we relied mostly on remote sensing satellite images and data in combination with detailed observations and records of individual tropical cyclones held by the Australian Bureau of Meteorology (BOM 2023). These observations were supported and enhanced from our field and aerial survey observations of the same region (e.g., Duke and Mackenzie 2018).

Furthermore, we focused on tropical cyclones in the GBR region with intensities of category 3 and greater (Table 1). These cyclones were generally referred to as severe tropical cyclones (STCs) having atmospheric pressures below 970 hPa and maximal sustained mean wind speeds greater than 117 km/h (Table 2). We further focused on the STCs that had made landfall and crossed the coastline, and we defined the intensity and potential damaging influences of each cyclone by its low atmospheric pressure and maximal wind speeds on landfall. In quantifying the area of damage for each STC, we measured the spatial extent of severely impacted mangroves based on specific levels of abrupt decline in canopy condition post-event. We also quantified the intensity of canopy damage by the average loss in canopy cover at the representative sites, along with estimates of longer-term recovery to pre-impact condition.

Using these resources and data, we specifically addressed a number of key questions regarding cyclone-impacted mangroves of the GBR area:

1) Have STCs become more or less intense, or frequent in recent decades?
2) Had the intensity of STCs influenced the spatial extent of damage and loss inflicted on mangrove communities?
3) Had the intensity of STCs been consistent with the levels of damage to mangrove canopies?

DOI: 10.1201/9781003320425-18

FIGURE 1 Shoreline mangroves near the Starcke River area, on the far northeastern coast of Australia, were struck by category 4 STC Ita in 2014 and again by category 3 STC Nathan in 2015. The severely impacted area of mangroves was estimated to be in excess of 600 ha when first recorded in 2017 (Duke and Mackenzie 2018). The actual extent of damage and the longer-term consequences had remained unknown and otherwise unreported.

FIGURE 2 Estuarine mangroves near Night Island, on the far northeastern coast of Australia, were struck by category 4 STC *Ingrid* in 2005 and again by category 3 STC *Monica* in 2006. Severely impacted mangroves were first recorded in 2017 (Duke and Mackenzie 2018). As noted in Figure 1 for the Starcke site, the extent of damage at this site near Night Island also remained generally unknown and otherwise unreported.

TABLE 1

Chronological listing of STCs having made landfall on the coastline bordering the GBR region since 1970. Data include range of intensity categories at landfall, cyclone name, date of landfall, barometric pressure at landfall, and location coordinates at landfall.

Cyclone Category at Landfall	Severe Tropical Cyclone (STC)	Landfall Date	Air Pressure hPa	Max Wind km/h	Landfall Latitude deg S	Landfall Longitude deg E
2 (-3)	*Ada*	18-Jan-1970	973	151	-20.3405	148.8238
4	*Althea*	24-Dec-1971	950	167	-18.9798	146.3434
3	*Aivu*	4-Apr-1989	955	157	-19.6352	147.5692
4	*Ingrid*	10-Mar-2005	953	176	-13.2871	143.5197
4	*Larry*	20-Mar-2006	940	176	-17.5270	146.0617
3	*Monica*	19-Apr-2006	965	139	-13.1721	143.5184
2 (-3)	*Ului*	21-Mar-2010	982	111	-20.2767	148.6716
5	*Yasi*	2-Feb-2011	929	204	-17.8920	146.0939
3	*Ita*	11-Apr-2014	948	176	-14.9492	145.2761
4 (-5)	*Marcia*	20-Feb-2015	933	204	-22.5655	150.4922
3	*Nathan*	20-Mar-2015	969	148	-14.6727	144.9441
3	*Debbie*	28-Mar-2017	957	148	-20.2060	148.5901
3	*Trevor*	19-Mar-2019	967	157	-12.9385	143.4842

Source: Data sources mostly from BOM (2023)

TABLE 2

Category parameters of tropical cyclones according to the Australian BoM applied in this study.

Category of Tropical Cyclone	Air Pressure (hPa)	10-Min Average Wind Speed (km/h)	Maximum Wind Gusts (km/h)
Tropical Low	> = 996	< 63	
1	986–995	63–88	< 125
2	971–985	89–117	125–164
3*	956–970	118–159	165–224
4*	930–955	160–199	225–279
5*	< 930	= > 200	=> 280

Source: Note the designation of STC categories marked by *; see also https://australiasevereweather.com/cyclones/tropical_cyclone_intensity_scale.htm

4) Had recovery occurred in mangroves damaged by STCs, how long had this taken, and had the rate differed for mangrove forests at low, mid, and high intertidal zones?
5) Were there locations with repeated impacts of STCs, and was the damage accumulative?

We believe the answers to these questions are fundamental to better understanding the impacts of STCs on mangroves and their recovery. This knowledge is needed for future evaluations and management of the threats to and consequences on these highly beneficial coastal marine ecosystems. The risks faced by such vulnerable shoreline ecosystems are reportedly increasing (cs. Gabler et al. 2017; Babcock et al. 2019; Dowdy 2019). Such a trend is in keeping with predictions from climate models of increased threats and impacts as damage levels escalate with global climate change coupled with ever-expanding, direct human pressures generally.

15.2. METHODS

15.2.1. The GBR Study Area and Data about Cyclones

STCs that made landfall on the mainland coast of the GBR region are shown in Figure 3. Tracks and damaged areas observed with each of the STCs since 1987 are further shown

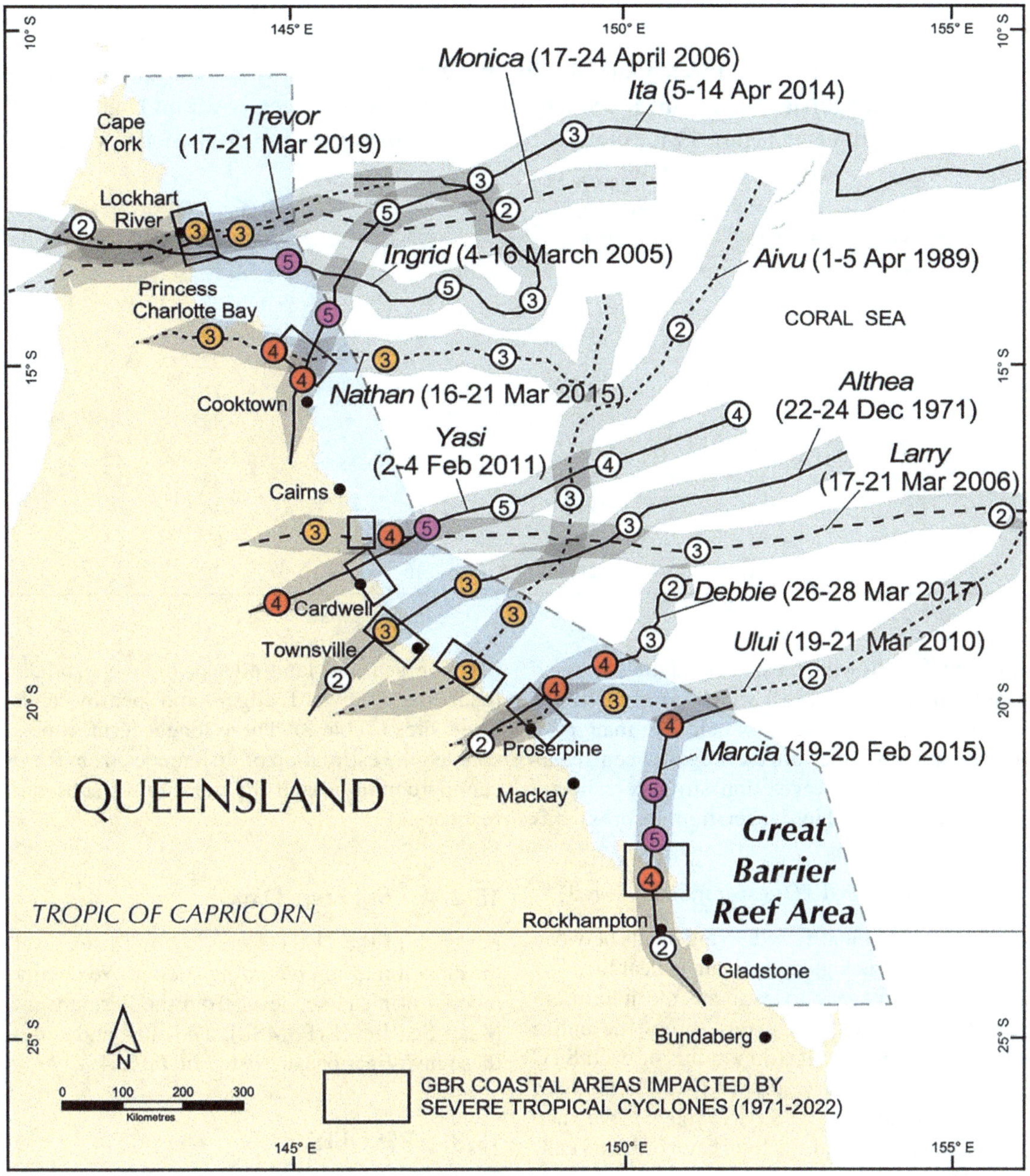

FIGURE 3 The tracks of the STCs having made landfall and crossed the GBR coast of northeastern Australia since 1970.

Source: Data sources from BOM (2023)

in the 11 GBR-specific case studies (see Case Studies 1–11). In addition to the impacts on mangroves, more detailed general knowledge about each STC was summarised in individual accounts, combined with historical records held by the Australian Bureau of Meteorology (BoM).

One intense cyclone, STC *Hamish*, category 3 to 5, during 4–14 March 2009, was not included in this assessment because it did not cross the coast. Its track and further details are available at the BoM website (www.bom.gov.au/cyclone/history/hamish09.shtml). We focused our assessment of the impact on mangroves from associated gale-force winds in this study (Table 2). We are aware of further likely impacts on mangroves located on the many islands of the GBR area (cs. Hamylton et al. 2024); however, these islands were not part of the current study.

We recognised that the linear relationship between the ten-minute average maximal wind speed (km/h) and air pressure (hPa) was significant ($r^2 = 0.8957$, $n = 12$, $P < 0.001$). Throughout this study, we mostly refer to the maximal average wind speed in our assessment.

15.2.2. Areas of Damaged Mangroves

Evaluations of mangrove areas impacted by individual cyclones were quantified using the area of substantive declines in the normalised difference vegetation index

TABLE 3

STCs (1987 onwards) and numbers of treatment sites for mangroves and nearby terrestrial forested areas. Mangrove sites were categorised at Low (near sea edge), Mid (midway between Low and High) and High (near land edge) intertidal zones (see Figure 15 for plots of treatment groupings, see Supplementary Data).

Severe Tropical Cyclone (STC)	Total Sites Surveyed	Mangrove Low Showing STC Impact	Mangrove Mid Showing STC Impact	Mangrove High Showing STC Impact	Mangrove TOTAL Showing STC Impact	Terrestrial Forest Showing STC Impact
Aivu	65	7	7	3	17	3
Ingrid	66	9	9	9	27	8
Larry	68	16	17	15	48	16
Monica	66	9	9	9	27	8
Ului	105	9	7	5	21	2
Yasi	61	15	15	13	43	15
Ita	64	16	14	6	36	4
Marcia	60	14	13	12	39	13
Nathan	64	16	14	6	36	4
Debbie	105	25	25	19	69	13
Trevor	44	11	11	10	32	11

(NDVI) immediately following the date of landfall for each STC (see Table 1). The NDVI is a widely used metric to assess the 'greenness' of vegetation, including mangrove forests. Mathematically, the difference between near-infrared reflectance (which vegetation strongly reflects) and red-light reflectance (which vegetation absorbs) was calculated as

$$\text{NDVI} = (\text{near-infrared} - \text{red}) / (\text{near-infrared} + \text{red})$$

Scores ranged between −1 and 1, with living plants between 0 and 1. When the value is higher, the plant is healthier.

We used Landsat imagery to estimate the maximum NDVI scores for 25 m × 25 m pixels across the entire extent of mangroves within a 100-km radius of each STC coastal intersection for the dry seasons (May to October inclusive) before and after each STC (Table 3). We then quantified areas of loss based on four categories of canopy damage severity: (1) no damage; (2) minimal decline (> 0–5% loss); (3) moderate decline (> 5–25% loss); and (4) severe decline (> 25% loss). These measures of loss in NDVI were consistent with the areas of impact and damage to mangrove trees observed in satellite imagery and field observations (e.g., Duke and Mackenzie 2018; Duke et al. 2021).

15.2.3. Severity of Damage to Mangroves

The severity of damage and condition of mangrove canopies were derived from NDVI measures of mangrove areas depicted in the mapping of tidal wetland areas.

Landsat image data were used to calculate NDVI monthly values between 1987 and 2022 (inclusive) at individual points at locations surrounding each STC landfall locality for 4 treatment types: low intertidal mangroves (near to the sea edge), mid intertidal mangroves (midway between sea and land edges), high intertidal mangroves (near the landward edge), and nearby terrestrial forested sites (Table 3). These longer-term, time series plots enabled the estimation of differences from the pre-cyclone canopy condition and the duration of subsequent canopy recovery.

15.2.4. Sea Level Data

For each of the 11 STC sites, we used sea level data from the closest relevant recording station. We obtained monthly records of mean sea level from the Permanent Service for Mean Sea Level (PSMSL). The sites and stations used in these investigations are listed in Table 4.

15.3. RESULTS

15.3.1. Number and Intensity of STCs

At least 12 STCs made landfall in the GBR region during the five decades since 1970 (Table 1; Figure 3; Case Studies 1–11). Over this period, there were notable increases in both the frequency and intensity of STCs. In Figure 4, the number of STCs shown had doubled, with 10 recorded during the last decade (2010–2019). Furthermore, the most intense STC *Yasi* (category 5; Case Study 7) was recorded making landfall during the recent decade.

Overall the higher wind speeds associated with the more severe STCs in recent decades were consistent with higher levels of damage to mangroves. In Figure 5, the graph shows mean maximum recorded wind speeds during 2010–2019 had exceeded 200 km/h for the first time during the last half century (also see Table 1).

In view of these findings, we review the key implications and consequences for mangrove habitat affected by such

TABLE 4

Information about sea level conditions relevant to each STC landfall location listed in Table 3. Chronological listing of 12 STCs in the GBR area since 1970, noting the respective tide gauge monitoring stations (Source: PSMSL, www.psmsl.org/). Data depicted are mean tidal range, sea level at landfall, detrended sea level at landfall, sea level anomaly (SLA) at landfall and sea level stress index (SLSI) at landfall.

STC	Sea Level Gauge Station	Tidal Range m	Landfall SL % of Longer-Term Mean	SL at Landfall De-trended for SLR	Landfall SLA Index	Landfall SLSI Index
Aivu	BOWEN II 2074	3.73	0.79	1.36	12	25
Ingrid	CAIRNS 953	3.5	0.84	0.86	74	-30
Larry	CAIRNS 953	3.6	2.39	2.73	190	33
Monica	CAIRNS 953	3.5	1.22	1.20	109	67
Ului	MACKAY 564	4.33	2.50	2.20	190	52
Yasi	TOWNSVILLE I 637	3.96	3.25	2.86	208	112
Ita	CAIRNS 953	3.4	1.20	0.97	35	17
Marcia	GLADSTONE 825	5.14	2.64	2.07	127	-16
Nathan	CAIRNS 953	3.4	0.73	0.49	7	-64
Debbie	MACKAY 564	4.33	1.87	1.26	90	30
Trevor	CAIRNS 953	3.5	1.34	0.98	94	46

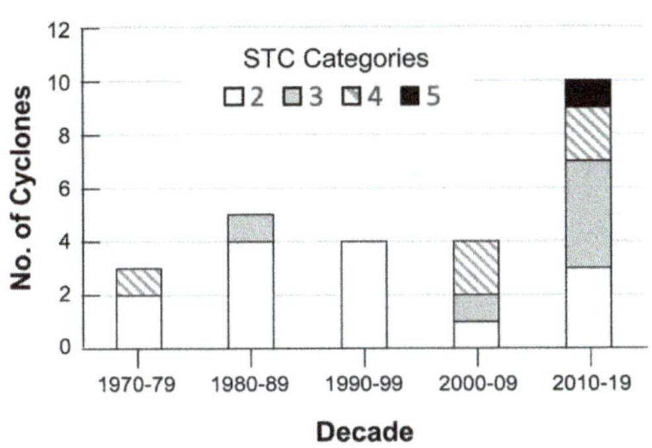

FIGURE 4 Five decades of STCs in the GBR region showing their marked increase in frequency and intensity in the most recent decade.

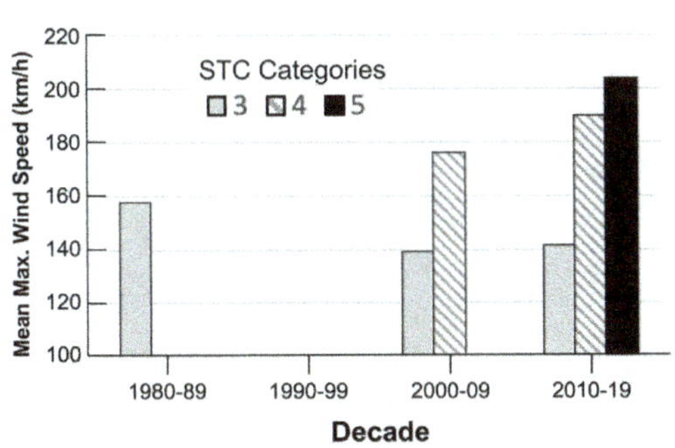

FIGURE 5 Maximal sustained wind speeds associated with STCs on landfall in the GBR region over the last four decades were notably higher in the most recent decade; notably exceeding 200 km/h. The stronger winds were consistent with greater damage to coastal mangroves in this area (see Figure 6).

extreme weather conditions. Our investigations utilised satellite vegetation indices (NDVI) applying two methods including assessments of 1) areal extent and mapping and 2) time series of localised impacts between 1987 and 2022 for analysis of impacted locations. For each location, the first objective was to quantify the relative impacts on mangrove vegetation of the respective STC(s) affecting each site.

15.3.2. Spatial Extent of Damage to Mangrove Communities

Areas of mangrove damaged by severe cyclones were assessed using change in NDVI before and after each of the 11 STCs making landfall between 1987 and 2022 (see Case Studies 1–11). In Table 5, we list the extent of impacted mangroves within 100 km from the point of landfall for each STC. We further categorised the levels of impact as minimal, moderate, and severe.

For the coastline of northeastern Australia and the GBR region, there were notable increases in the area of severe damage to mangrove areas during 2010–2019 (Figure 6). There were around 200 times more damaged mangroves in the recent decade.

15.3.3. Damage at Specific Landfall Sites of STCs in the GBR Region

Damage at specific mangrove and terrestrial sites was quantified from monthly time series plots of NDVI levels (1987–2022) in the vicinity of each STC at landfall. In Figure 7, we present a series of plots displaying the coincident impacts for each of the STCs. The abrupt impacts are apparent despite the overall longer-term fluctuations in NDVI indicative of annual and other oscillations in mean sea level reported earlier for the region (Duke et al. 2022). For the respective STC event, these plots show the timing of impact and the amount (proportion)

TABLE 5

Areas damaged (as measured by loss in NDVI levels) for each of 11 STCs having made landfall since 1987 on the coastline bordering the GBR region. Data assessments are displayed for each STC in a series of maps in Case Studies 1–11.

Severe Tropical Cyclone (STC)	Total Area of Mangrove (ha)	Minimal Damage (0–5% Loss) (ha)	Moderate Damage (> 5–25% Loss) (ha)	Severe Damage (> 25% Loss)	Minimal Damage % Total Area	Moderate Damage % Total Area	Severe Damage % Total Area
Aivu	16348	4941	4987	922	30.2	30.5	5.6
Ingrid	11826	4797	2699	477	40.6	22.8	4.0
Larry	10709	2495	3616	1880	23.3	33.8	17.6
Monica	11745	4030	3357	570	34.3	28.6	4.8
Ului	19216	7043	2260	142	36.7	11.8	0.7
Yasi	30691	3019	11843	14987	9.8	38.6	48.8
Ita	12271	3806	3191	1191	31.0	26.0	9.7
Marcia	41327	10942	15494	11834	26.5	37.5	28.6
Nathan	10986	1776	6476	2342	16.2	58.9	21.3
Debbie	18586	3875	9591	3248	20.8	51.6	17.5
Trevor	11840	1715	3773	5169	14.5	31.9	43.7

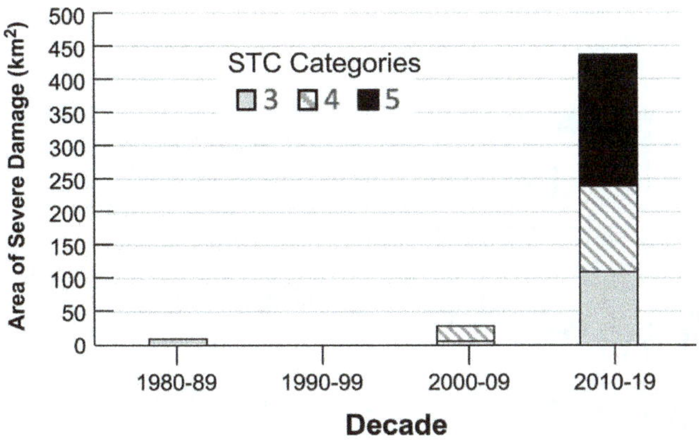

FIGURE 6 The area of severely damaged mangroves (> 25% loss in NDVI) had increased in proportion to the higher intensity and more frequent STCs over the last decade (see Figures 4 & 5).

of canopy loss inflicted. The loss in canopy condition (NDVI) was taken as a measure of the severity of impact for each event at the respective site locations.

These plots displayed the timing of impact and the amount of canopy loss inflicted. Furthermore, the plots showed the recovery from each damaging event at the respective sites.

15.3.4. Severity of Canopy Damage and Loss to Mangroves

Severity of damage was determined for each STC based on time series plots of NDVI levels from point locations near each landfall area (Table 3). We quantified the abrupt loss in NDVI following the respective STC events. As noted, there was a range of impact severity (see Figure 8). As another measure of damage severity, we calculated the average of these abrupt declines in canopy condition for each STC event.

In the example shown in Figure 8 for STC *Yasi*, canopy declines were observed at five levels of damage of approximately 10, 20, 30, 40, and 50% measures of canopy loss.

Although overall damage levels varied among STC events, they also varied within individual events. Table 6 shows estimates of mean canopy loss (damage) for mangrove and terrestrial sites for respective STC events. A notable observation was that damage to mangrove forests was predominantly higher than in nearby terrestrial forests (Figure 9). Moreover, while damage to mangroves had increased, this was less apparent in terrestrial forests nearby.

Overall, there was a significant negative relationship between canopy damage and air pressure ($r^2 = 0.6584$, $n = 11$, $P < 0.05$), which equates with cyclone intensity. This was further consistent with changes over time where greater mangrove canopy losses were observed during the recent decade (2010–2019) with the higher levels of cyclone intensity as events with decreased air pressure (Figure 9A).

We further investigated whether there were differences among three mangrove zones characterised by differences in tree species, biomass, and inundation frequency (Figure 10). For impacts on canopy condition, there were no substantive differences among low, mid, and high mangrove zones (Table 7). However, the overall difference between mangroves and terrestrial forests was significant.

15.3.5. Recovery Rates of Impacted Mangroves

We observed that different initial canopy losses had a predictable influence on recovery time. For example, recovery rates for STC *Yasi* differed for each of the mangrove zones and terrestrial forests (Figure 11).

Mean recovery rates mostly varied slightly among mangrove zones and compared to nearby terrestrial forests (Table 8) where apparent recovery rates were approximately 36% higher.

As is further evident in Table 8, there were slower recovery rates (a decline of 33% between category 3 and category 5 STCs) of impacted mangroves following progressively more intense STCs.

15.4. DISCUSSION

Our observations quantified the frequency and intensity of STCs on landfall along the GBR coast of northeastern Australia over the last half century. Both measures of cyclone impacts (frequency and intensity) were consistent with, and proportionate to, the observed levels of damage to mangrove forests. Accordingly, the more frequent and higher intensity tropical cyclones caused significantly greater harm and damage to mangrove forest canopies in recent decades. We further observed that rates of recovery became slower following more intense cyclonic events. In the following sections, we expand on these findings and briefly discuss their implications.

15.4.1. STCs Making Landfall Were More Intense and More Frequent Over Recent Decades

STCs impacting GBR coastal areas (at landfall) have increased in number in recent decades with 0–2/decade from 1970 to 1999 and up to 7/decade from 2010 to 2019. The more recent cyclones had also become more severe with two category 4 STCs first recorded in the 2000–2009 decade and three category 4 STCs in the 2010–2019 decade. Furthermore, one category 5 STC was first recorded during the 2010–2019 decade. In Figure 3, we show the number and intensity of STCs occurring in each of the five decades since 1970. This plot identifies the doubling of STCs in the most recent decade 2010–2019.

Furthermore, category 5 STC *Yasi* in 2011 (Case Study 7) was the highest intensity cyclone to have made landfall on the GBR coastline over the last half century. This STC had significantly higher wind speeds (exceeding 200 km/h) at landfall than any previous cyclone during the decades since 1970 (Figure 4). This STC was further accompanied

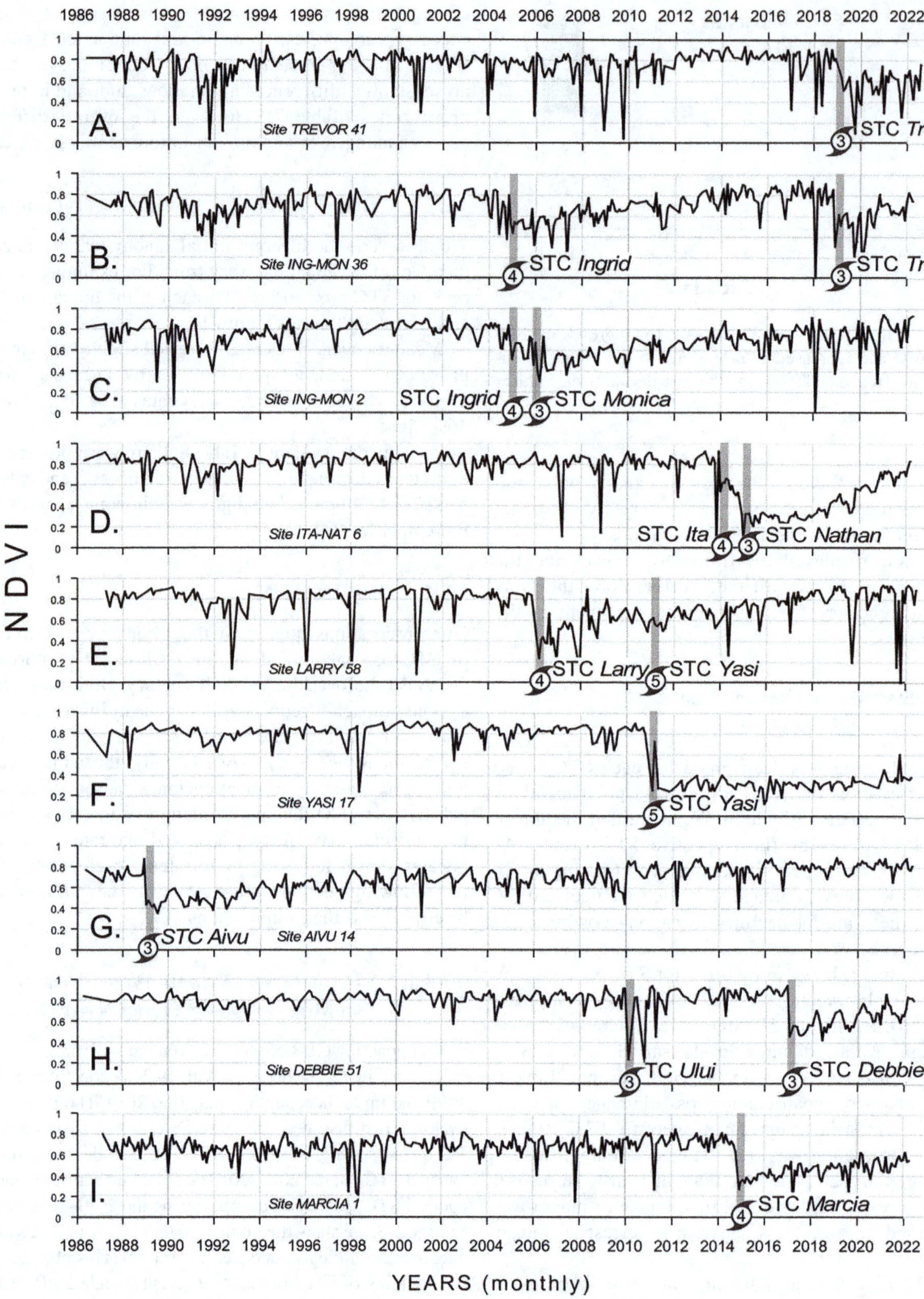

FIGURE 7 Series of nine time series plots (A to I) showing coincidental abrupt declines in NDVI (as canopy condition), indicative of each of the 11 STCs known to make landfall along the GBR coastline on specific dates. Note the occurrence of multiple STC events impacting some sites, for example, plots B to E and H. Some had accumulative impacts (C to E), while others displayed full recovery before subsequent events (B and H).

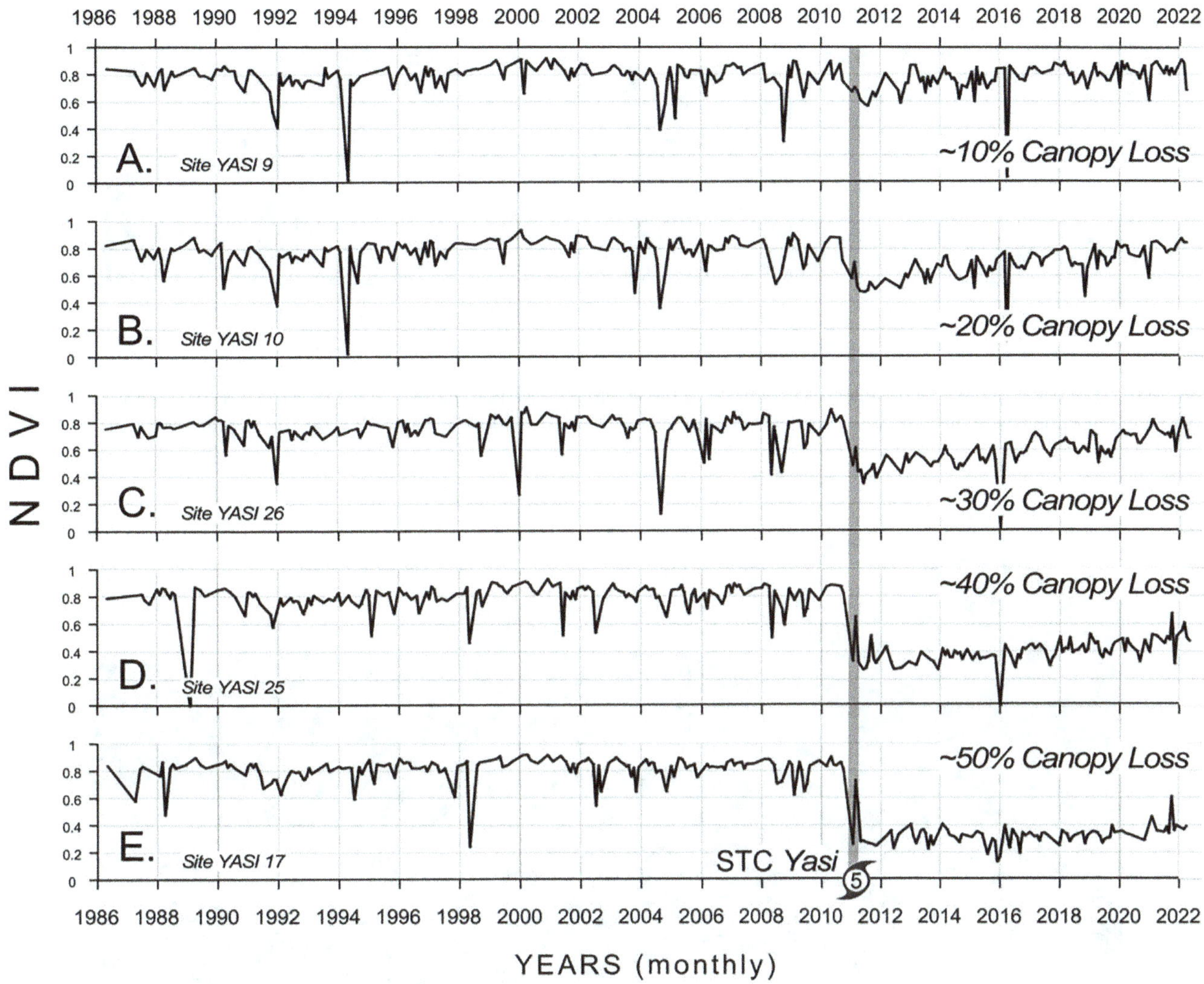

FIGURE 8 A representative set of five time series plots from STC *Yasi* showing impact severity levels ranging from approximately 10 to 50% canopy loss. The abrupt declines in NDVI levels post-event in 2011 were notably followed by proportionately longer recovery periods (if at all).

TABLE 6

Mean canopy damage of mangrove and nearby terrestrial forests for each of the 11 STCs having made landfall since 1987 on the coastline bordering the GBR Area. Refer to Table 3 for numbers of sampled sites, and 'SE x1' refers to the relevant standard error terms.

Severe Tropical Cyclone (STC)	Mean Mangrove Initial Damage %	SE x1	Mean Terrestrial Initial Damage %	SE x1
Aivu	13.2	1.7	8.3	2.0
Ingrid	23.9	2.5	14.1	1.9
Larry	26.6	1.9	25.0	1.9
Monica	10.1	0.8	9.1	1.6
Ului	10.2	1.1	10.5	4.5
Yasi	28.8	3.8	13.8	2.3
Ita	12.9	3.7	0	0
Marcia	25.6	1.9	13.3	1.6
Nathan	17.7	4.4	8.3	1.2
Debbie	18.1	3.1	10.0	0.8
Trevor	11.3	1.2	12.0	1.5

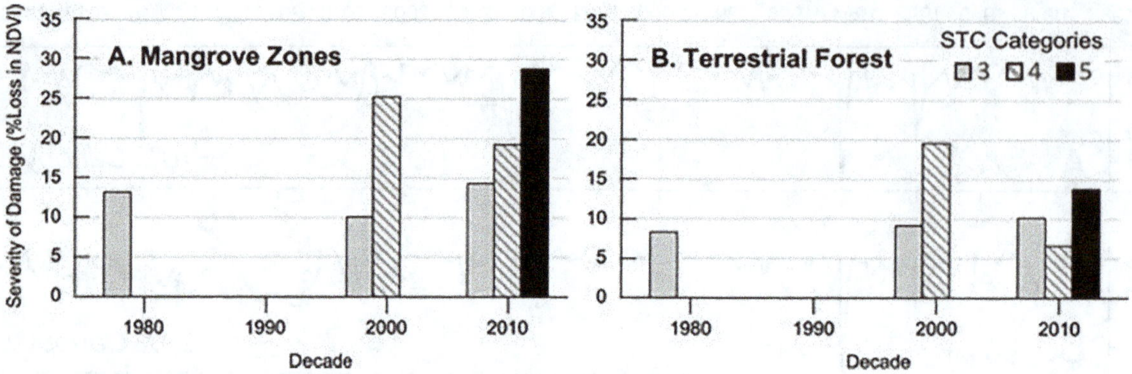

FIGURE 9 Four decades of STC severe damage affecting forest canopies of mangrove (A) and terrestrial (B) sites in the GBR region, showing greater impacts on mangrove forests with more intense STCs (especially Category 5) in the most recent decade (2010–2019).

FIGURE 10 Damage, such as tree death, breakage, and defoliation, observed in three mangrove intertidal zones of low (A), mid (B), and high (C) and in terrestrial forests (C) near Starcke River by category 4 STC *Ita* in 2014 and category 3 STC *Nathan* in 2015.

TABLE 7

Overall levels of abrupt damage by STCs for three severity categories where damage levels to mangrove and terrestrial forests was proportionate to cyclone intensity. Damage levels recorded as mean values of multiple sites (n) with standard error terms in brackets.

Severe Tropical Cyclone (STC) Category (n)	Terrestrial Forest	Mangrove High	Mangrove Mid	Mangrove Low	Mangrove Forest (TOTAL)
STC Cat. 3 (6)	9.7 (±0.6)	11.8 (±1.1)	14.9 (±2.2)	13.6 (±2.6)	13.4 (±1.5)
STC Cat. 4 (4)	13.1 (±5.1)	17.3 (±4.0)	24.7 (±2.8)	24.7 (±2.7)	22.2 (±3.2)
STC Cat. 5 (1)	13.8	21.7	29.9	34.7	28.8
ALL STCs (11)	12.2 (±1.3)	14.7 (±1.8)	19.8 (±2.3)	19.6 (±2.8)	18.0 (±2.1)

by two category 4 cyclones, STC *Ita* in 2014 (Case Study 4) and STC *Marcia* in 2015 (Case Study 11), and four category 3 cyclones, STC *Ului* in 2010 (Case Study 9), STC *Nathan* in 2015 (Case Study 5), STC *Debbie* in 2017 (Case Study 10), and STC *Trevor* in 2019 (Case Study 3). The occurrence of seven STCs in a single decade was unprecedented for the GBR area. Although there are understandable questions about cyclones prior to 1970, the trend regarding STCs in recent decades remains significant, especially for mangrove-lined shorelines currently threatened also by rapidly rising sea levels (Lovelock et al. 2015).

Notably, during the recent decade, there was also the coincident occurrence of extreme oscillations in mean sea levels, where sustained low sea levels caused the mass dieback of more than 8,000 ha of shoreline mangroves in northern Australia in 2015 (Duke et al. 2022). Additional studies are needed to test the suggestion that such coincident damaging occurrences might be related.

15.4.2. Increased Intensity of STCs Making Landfall Influenced the Area of Impacted Mangroves

The area of severely impacted mangroves correlated with STC intensity (Figure 12), but it is likely that this relationship was influenced by the small number of cyclones considered in this study. For instance, the two largest areas of mangroves on the GBR coast (see Table 5, STC *Yasi* and STC *Marcia*, Case Studies 7 and 11, each being more than twice the mean area) were struck by the two most intense STCs (wind speeds > 200 km/h) recorded since 1980. Furthermore, the percentages of severely damaged mangroves compared to the total areas of mangroves were not correlated with STC intensity (see Table 5). However, it is relevant that these two STCs both occurred during the recent decade from 2010–2019.

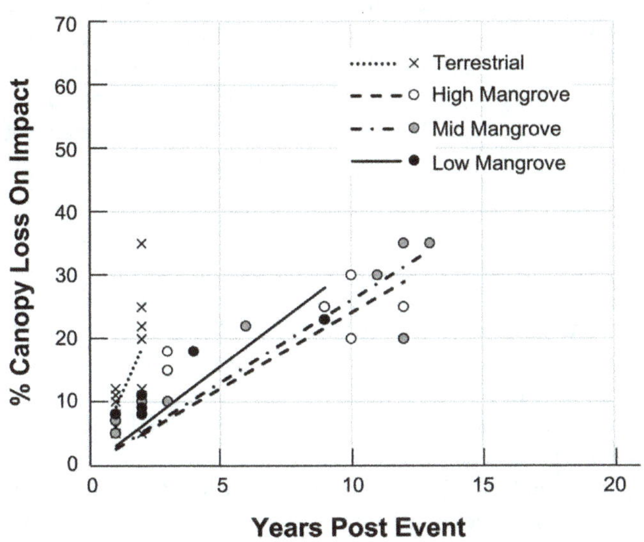

FIGURE 11 Recovery plots for STC *Yasi*, as the years taken to restore prior canopy condition (see Figure 8), showed similar rates between mangrove zones, while the rates for nearby terrestrial forests were relatively rapid.

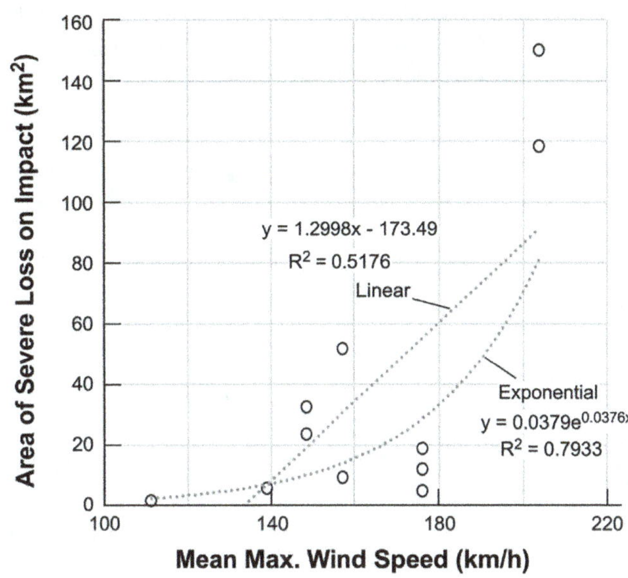

FIGURE 12 The intensity of STCs, quantified as maximal sustained wind speed, correlated with the mapped area of severe canopy damage.

TABLE 8

Mean rates of canopy recovery as percent gain in canopy condition per year (in brackets: numbers of observations or ± standard error) as determined for changes in NDVI levels before and after the respective STC events at three severity categories. See also Figure 15.

Severe Tropical Cyclone (STC)	Terrestrial Forest	Mangrove High	Mangrove Mid	Mangrove Low	Mangrove Forest (TOTAL)
STC Cat. 3	5.15 (28)	3.74 (34)	3.68 (40)	4.77 (44)	4.06 (±0.35)
STC Cat. 4	5.45 (40)	3.54 (37)	3.77 (37)	3.92 (37)	3.75 (±0.11)
STC Cat. 5	9.15 (15)	2.41 (11)	2.61 (9)	3.12 (6)	2.71 (±0.21)
ALL STCs	5.66 (±1.29)	3.21 (±0.41)	3.48 (±0.37)	4.09 (±0.48)	3.60 (±0.26)

15.4.3. The Increased Intensity of STCs Making Landfall Affected the Severity of Mangrove Canopy Loss

We quantified abrupt declines in canopy condition (as NDVI levels) at the time when the respective STCs made landfall for both mangrove and nearby terrestrial forests (Table 6). The primary finding was that these canopy losses for mangrove forests were correlated with STC intensity quantified as both air pressure ($r^2 = 0.6584$; $n = 11$; $P < 0.05$) and maximal wind speed ($r^2 = 0.6200$; $n = 11$; $P < 0.05$; see Figure 13). This observation is further depicted in Table 7 where category 3 STCs had an average canopy loss of 13.4% (± 1.5 SE), category 4 STCs had an average canopy loss of 22.2% (± 3.2 SE), and the one category 5 STC (STC *Yasi;* see Case Study 7) had an average canopy loss of 28.8%.

In Table 7, we further measured that canopy losses to mangroves (18.0% (± 2.1 SE) were notably greater than those in nearby terrestrial forests (12.2% (± 1.3). In addition, the two mangrove zones closer to the sea edge had greater levels of damage (~19.7%) compared to high intertidal zone sites (14.7% ± 1.8 SE).

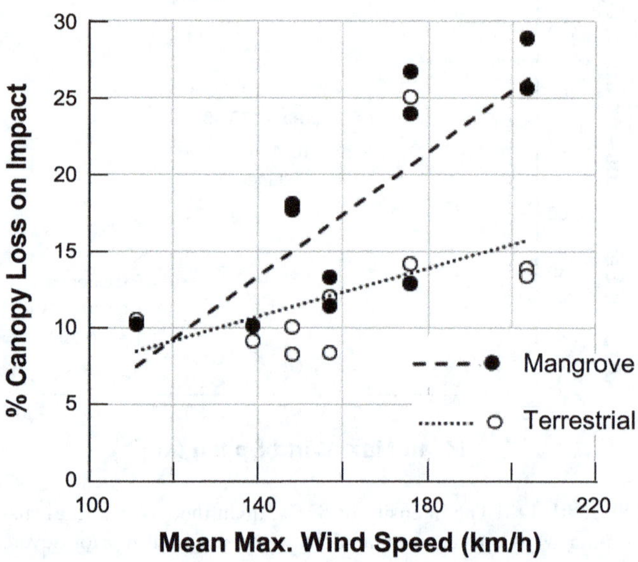

FIGURE 13 The intensity of STCs, quantified as maximal wind speed, correlated with the severity of canopy damage at impacted mangrove sites, and less so with terrestrial sites. The linear trend for impacted mangroves was significant (P<0.05).

15.4.4. Canopy Recovery Rates Were Dependent on Forest Type

Rates of recovery of mangrove canopies were largely constant and predictable. Recovery rates took ~2 years from ~10% loss, and more than 10 years for replacement of ~50% canopy loss (Figures 8 and 14). Accordingly, recovery time depended on the severity of abrupt canopy loss, as shown for STC *Yasi* (Figure 11). Overall, mean rates of apparent recovery varied with mangrove forests being ~36% slower than nearby terrestrial forests (Table 8). By contrast, rates among different mangrove zones showed little if any significant differences. However, there was a trend in the different STC categories with mangrove damage from higher intensity cyclones taking longer to recover—namely, 4.04, 3.75, and 2.71 gains in percent canopy condition (%NDVI) per year for categories 3, 4, and 5 tropical cyclones, respectively.

An earlier assessment of recovery for mangroves nearby in Australia's Gulf of Carpentaria from a different type of abrupt damaging event showed comparable recovery rates (Duke et al. 2022, Fig. 6). This observation highlighted an additional important point regarding the apparent lack of dependence of mangrove recovery on the dominant species. For instance, those in the Gulf were dominated by *Avicennia marina*, while those in the GBR area were dominated by *Rhizophora* species. Irrespective of these major differences in the dominant mangrove species, the rates were comparable between different mangrove forest biodiversity types.

By contrast, rates of recovery for terrestrial forests were substantially different. We found that terrestrial forests had rates that were faster, taking 1–2 years compared to 2–3 years from 10% loss, and 8–9 years compared to 13–14 years from 50% loss. The mean rates for each of these major forest types are listed in Table 8.

15.4.5. The Impacts of Successive STCs Making Landfall Were Accumulative

Our investigations documented multiple instances of accumulative impacts from closely occurring STCs in both space and time. In Figure 7, the series of time series plots display the process and circumstances by which accumulative impacts occurred.

Accumulative impacts occurred during periods of canopy recovery delaying full recovery more or less according to the rates shown in Table 8. Therefore, subsequent impacts added to, and seemingly reset, the recovery process already in progress. In such instances, added declines prolonged the return to maximal canopy condition. The most recognisable cases were those STCs that struck the same shoreline one year apart (see Figure 14), including STCs *Ingrid* and *Monica* (Figure 7C; Case Studies 1 and 2) and STCs *Ita* and *Nathan* (Figure 7D; Case Studies 4 and 5). In at least one area, there were STC events years later, like that for STC *Larry* (Case Study 6), which was followed by STC *Yasi* (Case Study 7) around 5 years later (Figure 7E), and STC *Trevor* (Case Study 3) that struck where STCs *Ingrid* and *Monica* (Case Studies 1 and 2) had struck around 13 years earlier (Figures 7B and C). In each case, each time series plot provided a longer term account of the damage and recovery processes associated with STCs and other damaging events at the respective sites.

Although most impacted sites had recovered, a small number did not. A further assessment is needed to quantify whether such a 'collapse' outcome might be a function of

FIGURE 14 Severely impacted low intertidal mangroves in two areas of northeastern Australia from a 2017 aerial survey (Duke and Mackenzie 2018): (A) the coast near Night Island struck by category 4 STC *Ingrid* in 2005 and category 3 STC *Monica* in 2006 (see Figure 2); and (B) near the Starcke River struck by category 4 STC *Ita* in 2014 and category 3 STC *Nathan* in 2015 (see Figure 1). Note the degraded state of standing dead *Rhizophora* trees and the apparent lack of recovery either as resprouting trees or seedlings.

more severe damaging conditions or something else. In this study, most damaged mangrove stands showed full recovery to prior canopy condition in the absence of additional abrupt impacts, as shown with STC *Aivu* (Figure 7G; Case Study 8).

These plots also showed responses in mangrove canopies to other influences. A notable component of the 'noise' in these plots was explained by Duke et al. (2022) as annual periodicity of small-scale adjustments in the canopy condition in response to annual oscillations in mean sea level. These influences were consistent with phenological observations of seasonal patterns in new leaf production and senescent leaf fall. In this way, canopy losses were driven by moisture stress that effectively limited and defined the maximal canopy state. Furthermore, more extreme maxima and minima mean sea levels were observed on occasion to result in damaging events of abrupt canopy loss (Duke et al. 2022). Such events were also evident in these plots, but their presence requires further investigation to both identify and explain such widespread, unexplained abrupt loss events in mangrove canopies, like those in 1991 (see Figures 7A-F). It seems these might be synchronous with extreme sea level events in combination with climatic conditions.

15.4.6. Cyclone Damage and Recovery of Mangrove Canopies Is an On-Going Natural Process, but Can Recovery Be Sustainable as Impacts Escalate with Global Climate Change?

In this study, we quantified key aspects of STCs in the GBR area. We investigated the impacts on mangrove forests of 11 STCs having made landfall on the GBR coastline since 1987. The impacts from these STCs were distributed across northeastern Australia from Torres Strait to the Tropic of Capricorn. Of the 1,956 km^2 of mangroves within 100 km from all landfall locations, around 428 km^2 were severely damaged (Table 5).

We documented the broad descriptors of the impacts of each STC and its impacts on coastal mangrove forests. These were based on both the spatial extent of damage and impact, combined with specific measures of severity and recovery. Each of these parameters were derived from the 35-year pool of NDVI data from Landsat satellite imagery collected since 1987. In this way, we identified the impacts of STCs having made landfall in the GBR area of northeastern Australia, and we quantified their respective rates of recovery (Figure 15). The latter was found to have been relatively constant, except where rates decreased with increased intensity of STCs.

Based on these observations, we further framed and defined the paradoxical dilemma of the sustainability of mangrove forests reliant on long-effective but fixed-rate natural recovery processes in the face of ever-increasing damaging influences of ever-more intense storms with climate change. In the situation where threats have increased unabated, the predicted outcome appears inevitable with eventual collapse of vulnerable tidal wetland ecosystems (cs. Duke 2001; Duke et al. 2021; Bergstrom et al. 2021). Such influences and dire threats have also been shown to apply to mangroves on the islands and coral communities throughout the GBR region (Hamylton et al. 2024).

In summary, we present the following key observations from this study:

1) STC events have occurred widely across the GBR area from Torres Strait to the Tropic of Capricorn (Figure 3);
2) the number and intensity of STCs increased in the recent decade (2010–2019) with 7 of the 11 STCs occurring during this decade (Figure 4) when also the mean maximum recorded wind speeds

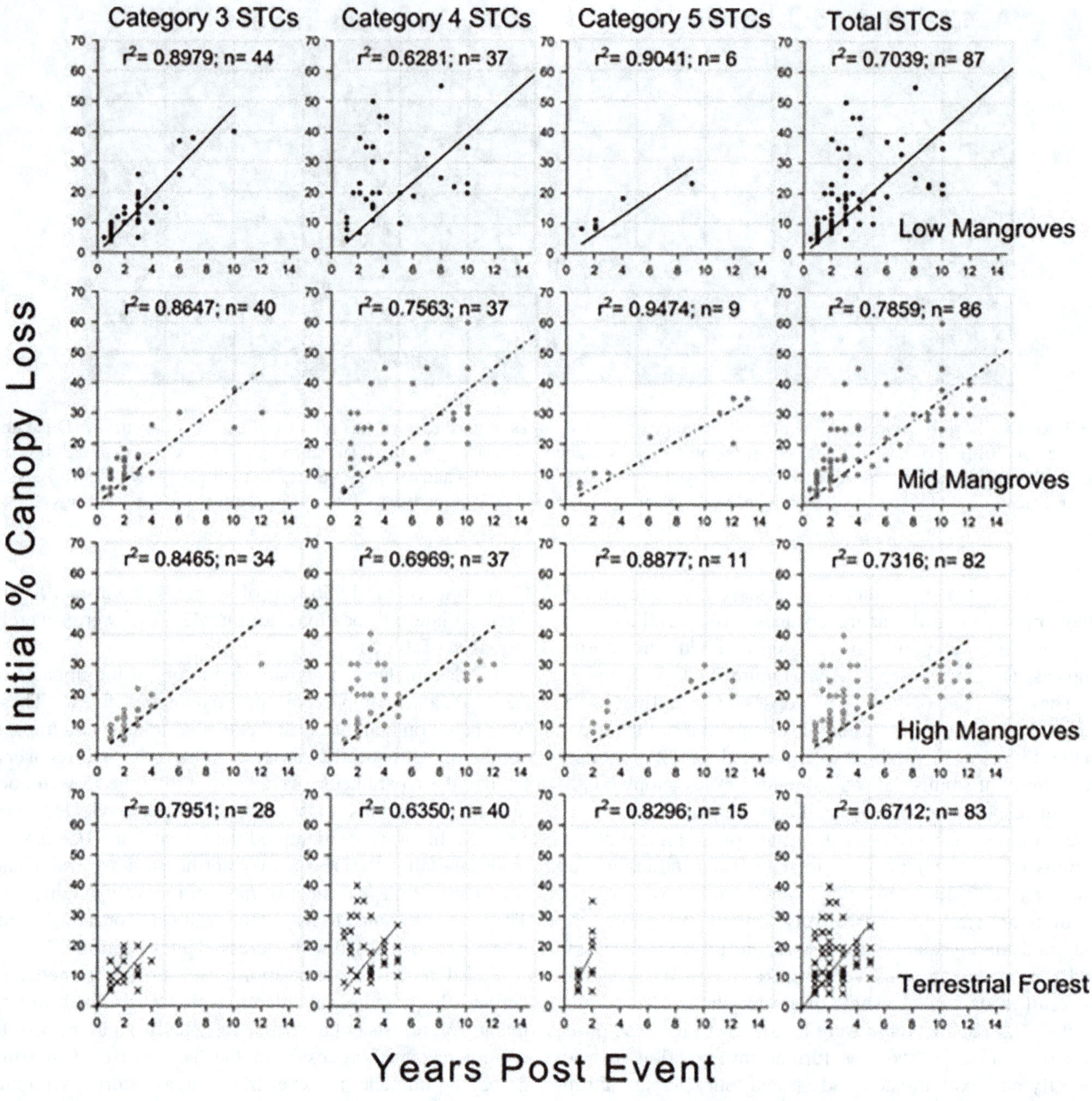

FIGURE 15 Graphs of recovery from canopy damage to mangrove zones and terrestrial forested areas following severity groupings of STC events, including categories 3, 4, and 5 and totalled categories. See Table 8 for a listing of respective recovery rates (as % canopy gain per year).

exceeded 200 km/h for the first time in the last half century (Table 1);

3) the damaging impacts on mangroves increased considerably with increased STC intensity (Table 7, Figures 5 and 6);
4) the area and severity of cyclone-damaged mangroves increased in the recent decade with the largest area of severely damaged mangroves of 38,913 ha recorded during the most recent decade up 13-fold compared to 2,926 ha maximum of previous decades (Figure 6);
5) damage to mangrove forests was greater than damage to terrestrial forests nearby, notably with % overall canopy decline for mangroves in the 11 STCs of 18 ±2.1 compared to 12.2 ±1.3 for terrestrial forests (Table 7, Figure 9);
6) overall, rates of recovery were proportionate to the amount of canopy loss, where losses of 10%, 20%, and 30% had average recovery periods of 2.2 (39 plots), 5.1 (21 plots), and 7.3 (9 plots) years, respectively (Figures 8 and 15);
7) recovery was slower from damage caused by more intense STCs, recorded by the gains in mean canopy cover (%) being 4.1, 3.8, and 2.7 per year for STC Categories 3, 4, and 5, respectively (Table 8), and these observations were consistent with the notion

of more severe storm damage further impeding natural regenerative processes;

8) damage to mangrove canopies was accumulative with recovery processes delayed proportionately whenever there was overlap; we observed a number of sites in the GBR area having repeated impacts from 2–3 STCs during the 35-year study period (Figure 7); and
9) recovery of prior canopy condition was achieved naturally in most cases, but since the rate of recovery is fixed and dependent on innate reproductive and regenerative processes, eventual ecosystem collapse appears inevitable (cs. Duke 2001), should the intensity and frequency of damaging events worsen.

ACKNOWLEDGEMENTS

We acknowledge the Traditional Owners of land and sea country where we conducted these studies. We pay our respects to their Elders—past, present, and emerging.

SUPPLEMENTARY DATA FILE

GF plot sites ALL.xlsx

EXCEL spreadsheet listing 533 sites evaluated for the 11 STCs in the GBR region by STC site code, latitude, longitude, nominal profile position, and representative zone.

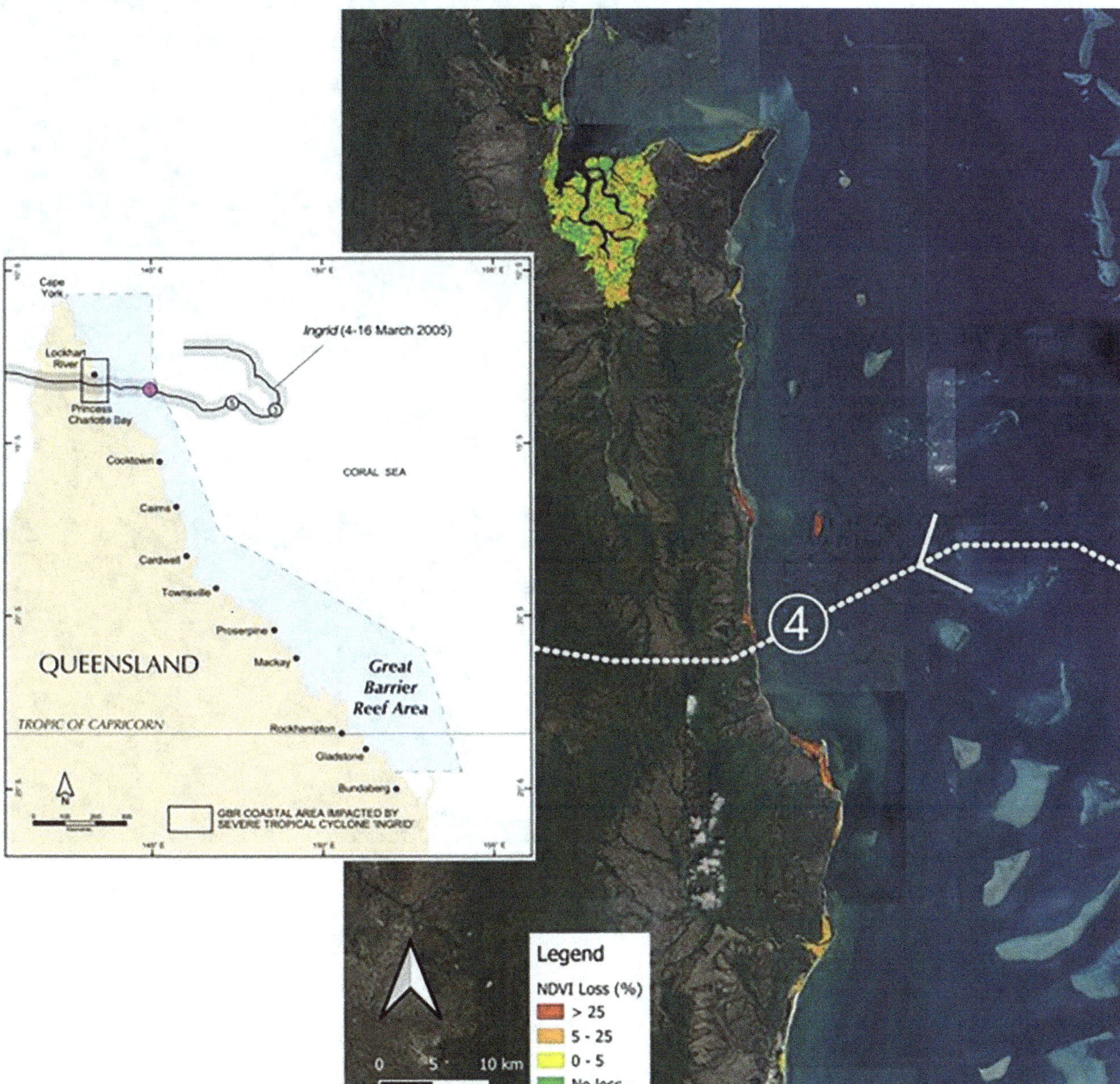

CASE STUDY 1 Impact on shoreline mangroves by STC *Ingrid* in the GBR region. Map shows the damage to mangrove areas by STC *Ingrid* at landfall on 10 March 2005. At the time, the cyclone was at category 4 intensity after tracking west across the Coral Sea. For additional detailed information about this cyclone, visit www.bom.gov.au/cyclone/history/ingrid.shtml.

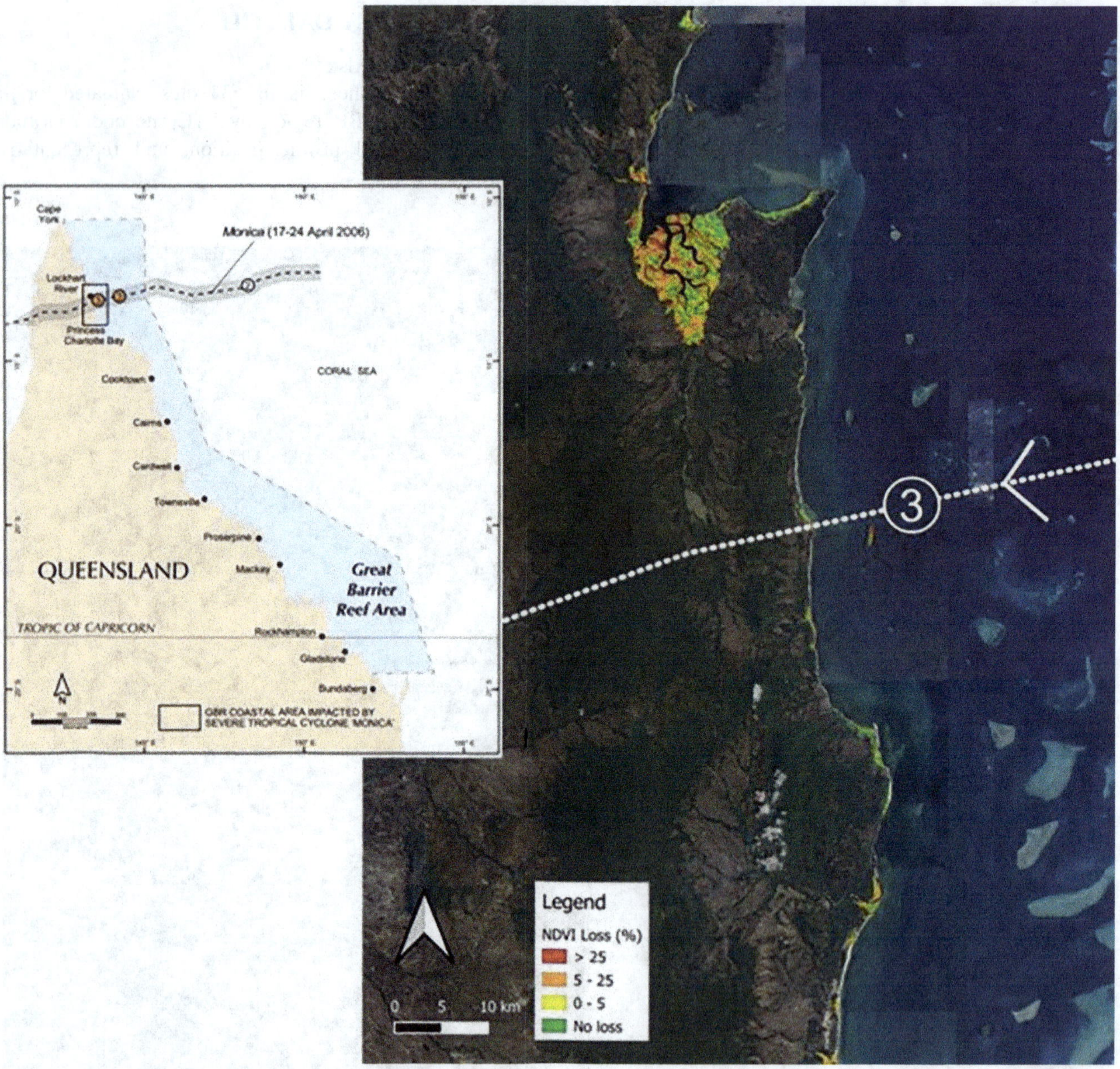

CASE STUDY 2 Impact on shoreline mangroves by STC *Monica* in the GBR region. Map shows the damage to mangrove areas by STC *Monica* at landfall on 19 April 2006. At the time, the cyclone was at category 3 intensity after tracking west across the Coral Sea. For additional detailed information about this cyclone, visit www.bom.gov.au/cyclone/history/monica.shtml.

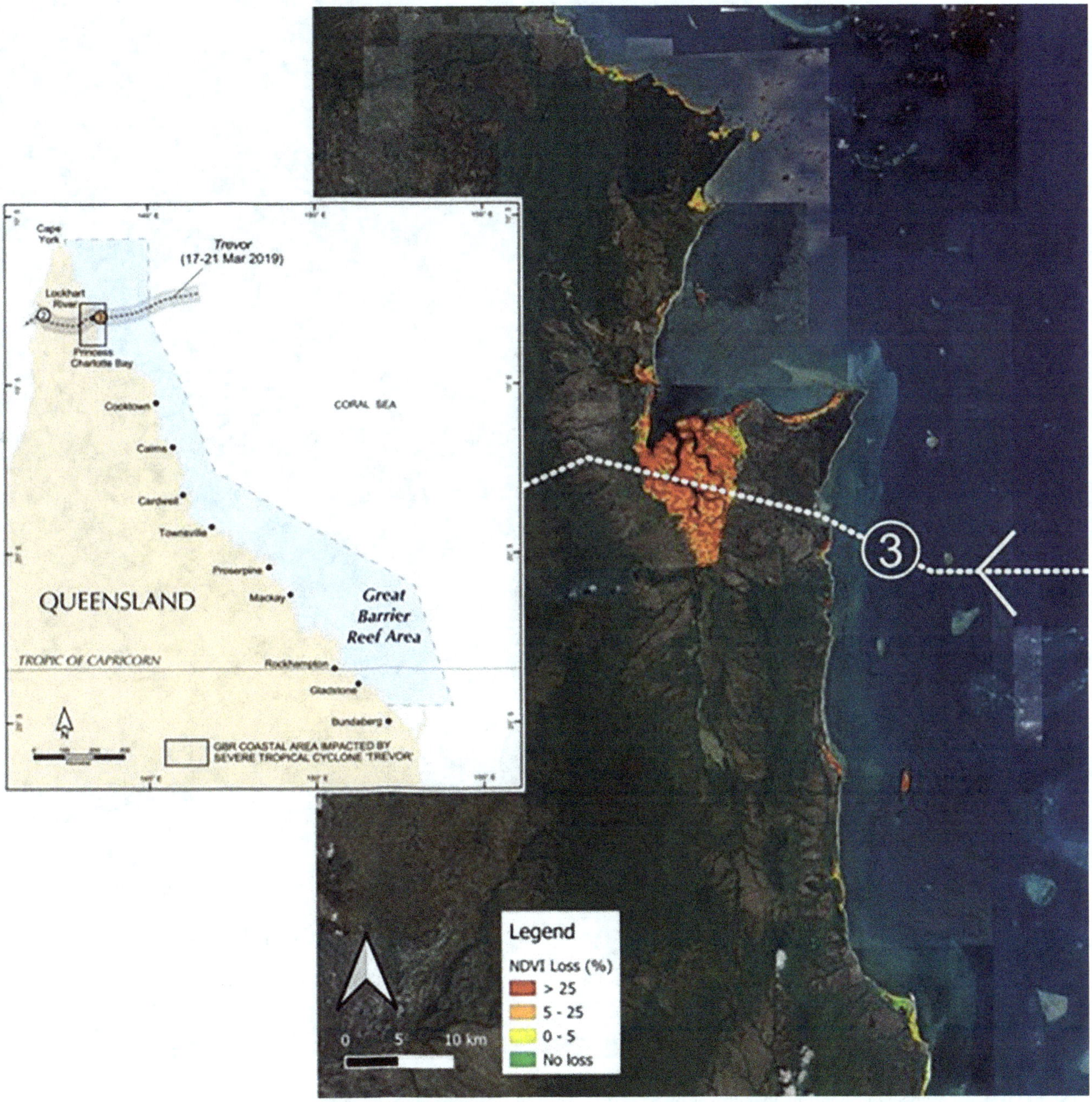

CASE STUDY 3 Impact on shoreline mangroves by STC *Trevor* in the GBR region. Map shows the damage to mangrove areas by STC *Trevor* at landfall on 19 March 2019. At the time, the cyclone was at category 3 intensity after tracking west across the Coral Sea. For additional detailed information about this cyclone, visit www.bom.gov.au/cyclone/history/Trevor.shtml.

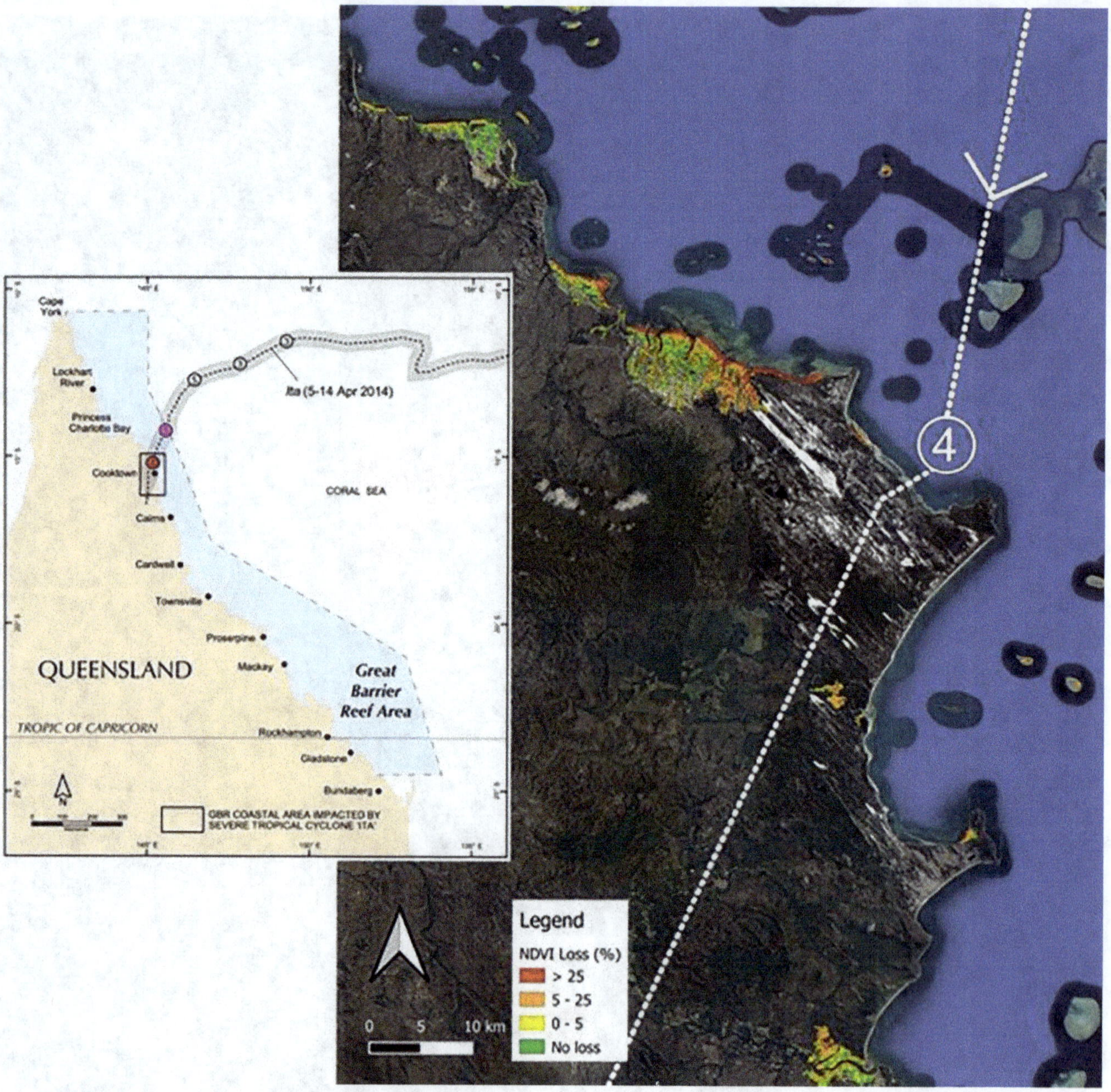

CASE STUDY 4 Impact on shoreline mangroves by STC *Ita* in the GBR region. Map shows the damage to mangrove areas by STC *Ita* at landfall on 11 April 2014. At the time, the cyclone was at category 4 intensity after tracking south across the Coral Sea. For additional detailed information about this cyclone, visit www.bom.gov.au/cyclone/history/Ita.shtml.

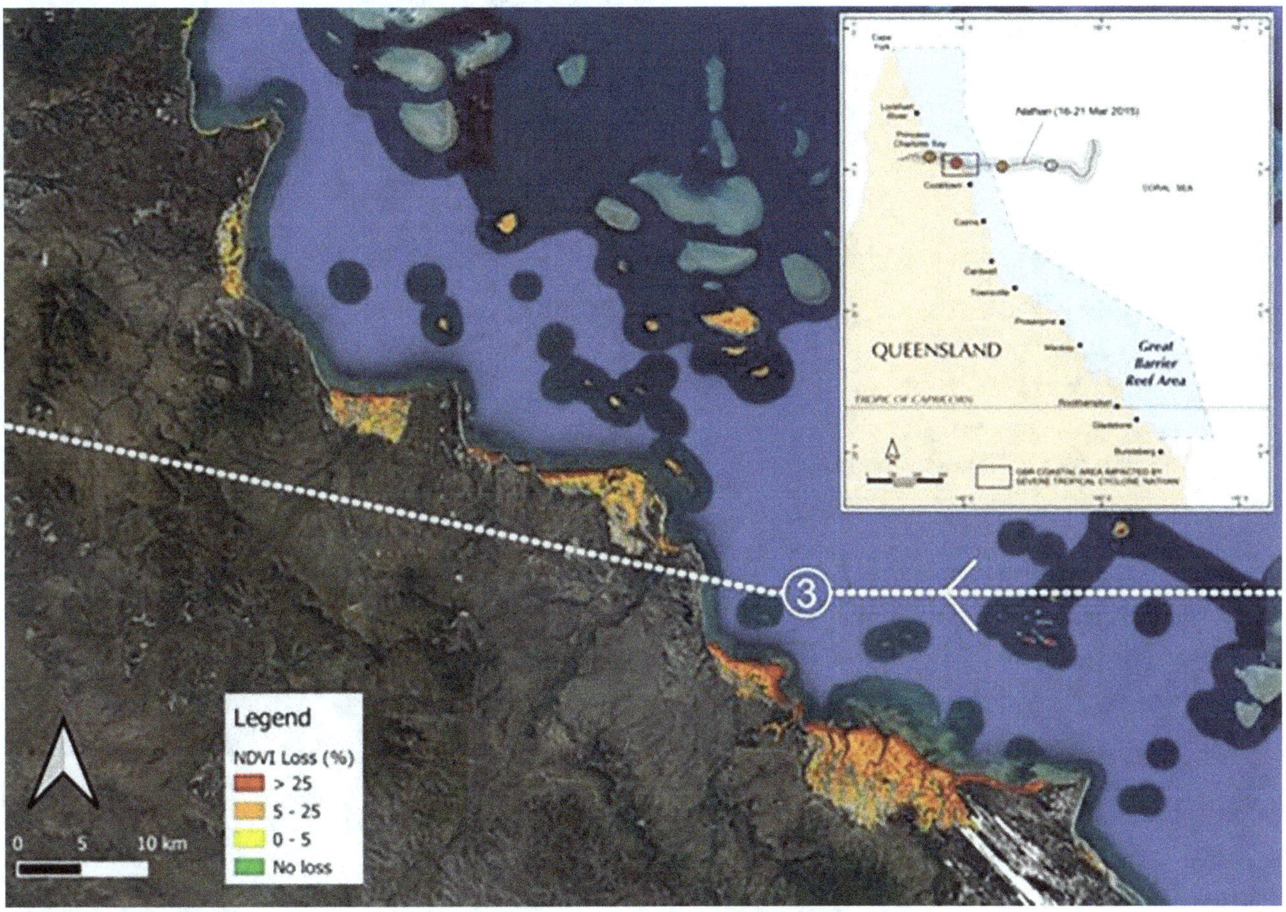

CASE STUDY 5 Impact on shoreline mangroves by STC *Nathan* in the GBR region. Map shows the damage to mangrove areas by STC *Nathan* at landfall on 20 March 2015. At the time, the cyclone was at category 3 intensity after tracking west across the Coral Sea. For additional detailed information about this cyclone, visit www.bom.gov.au/cyclone/history/Nathan-2015.shtml.

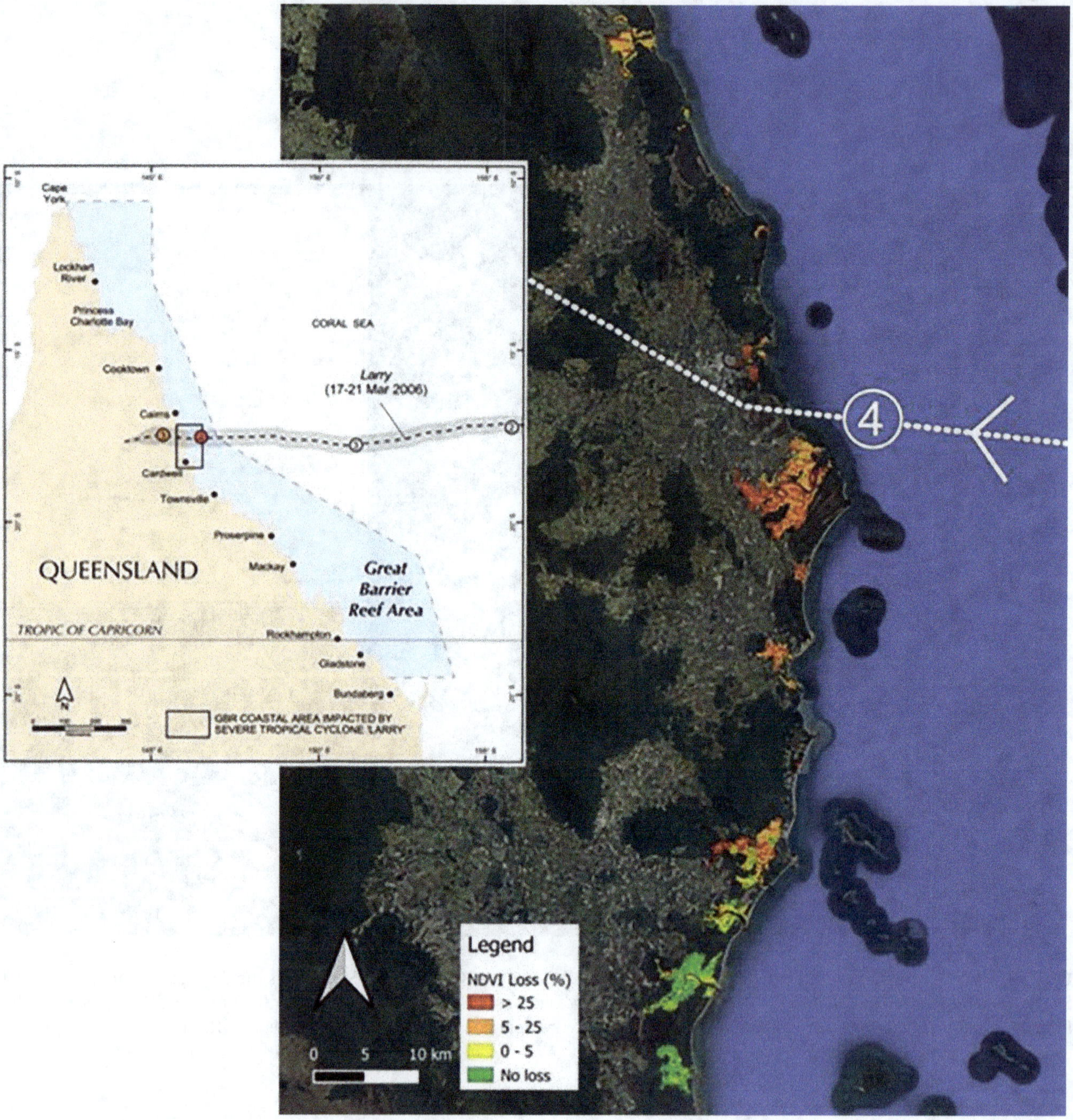

CASE STUDY 6 Impact on shoreline mangroves by STC *Larry* in the GBR region. Map shows the damage to mangrove areas by STC *Larry* at landfall on 20 March 2006. At the time, the cyclone was at category 4 intensity after tracking west across the Coral Sea. For additional detailed information about this cyclone, visit www.bom.gov.au/cyclone/history/larry.shtml.

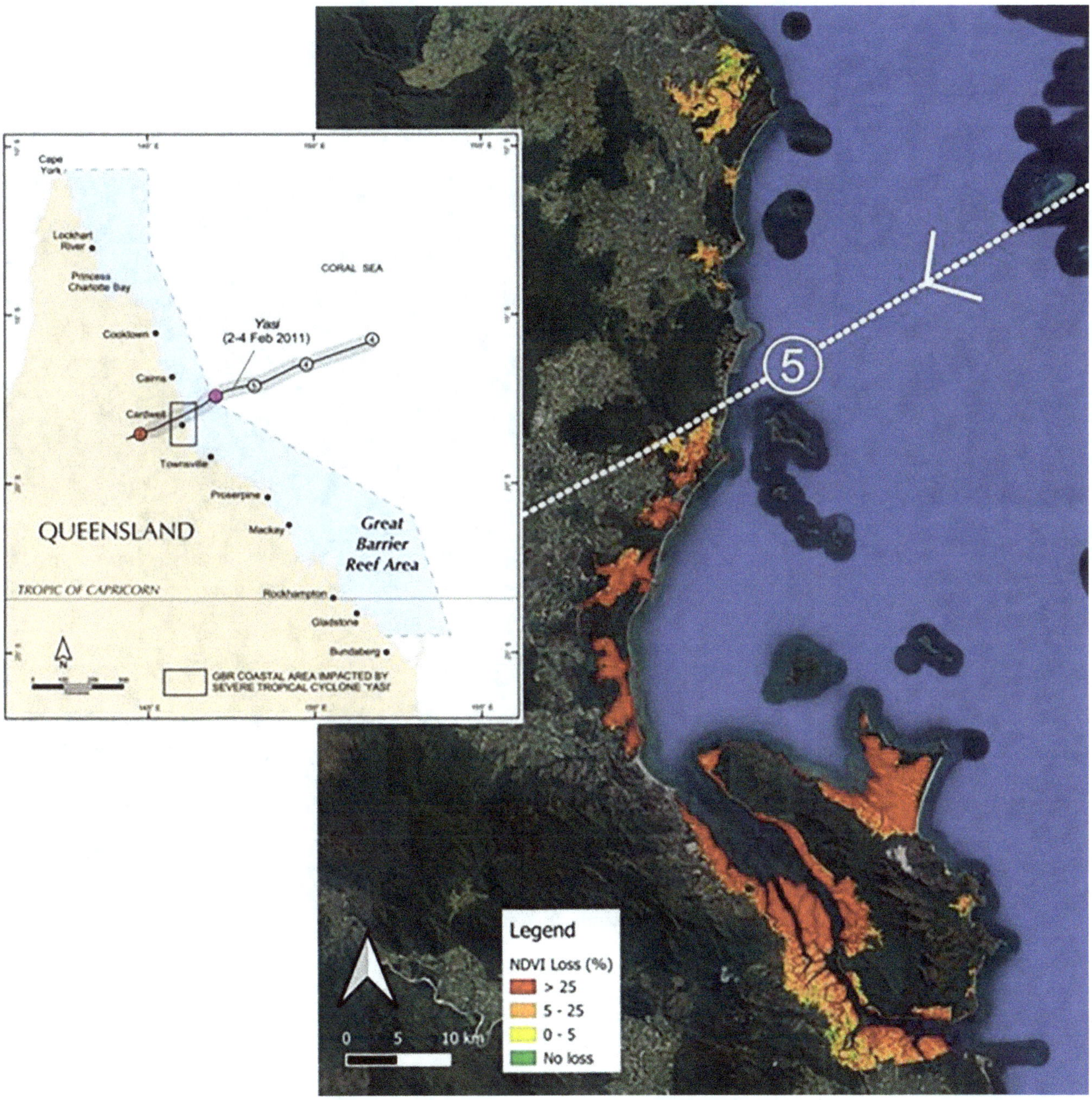

CASE STUDY 7 Impact on shoreline mangroves by STC *Yasi* in the GBR region. Map shows the damage to mangrove areas by STC *Yasi* at landfall on 2 Feb 2011. At the time, the cyclone was at category 5 intensity after tracking southwest across the Coral Sea. For additional detailed information about this cyclone, visit www.bom.gov.au/cyclone/history/yasi.shtml.

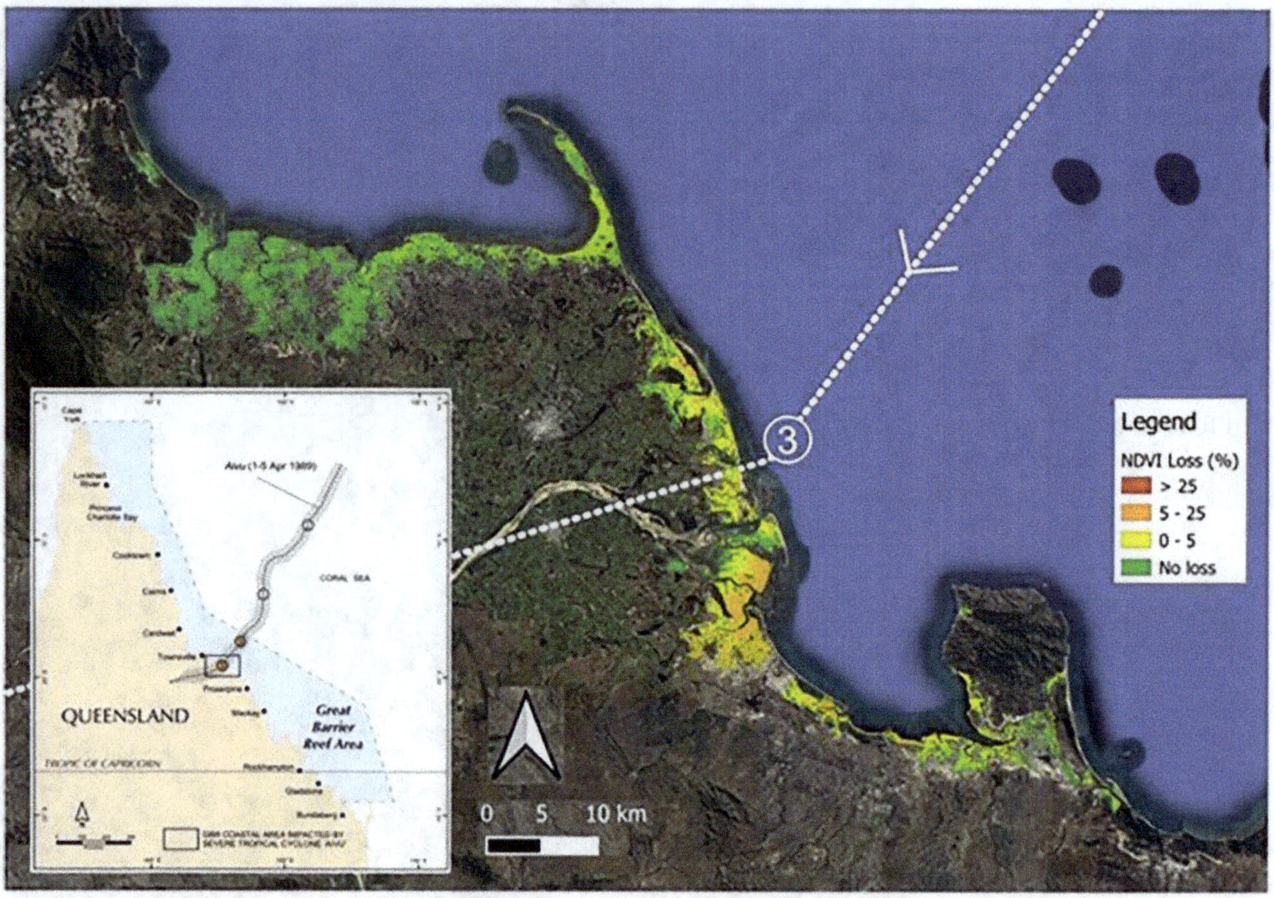

CASE STUDY 8 Impact on shoreline mangroves by STC *Aivu* in the GBR region. Map shows the damage to mangrove areas by STC *Aivu* at landfall on 4 April 1989. At the time, the cyclone was at category 3 intensity after tracking southwest across the Coral Sea. For additional detailed information about this cyclone, visit www.bom.gov.au/cyclone/history/aivu.shtml

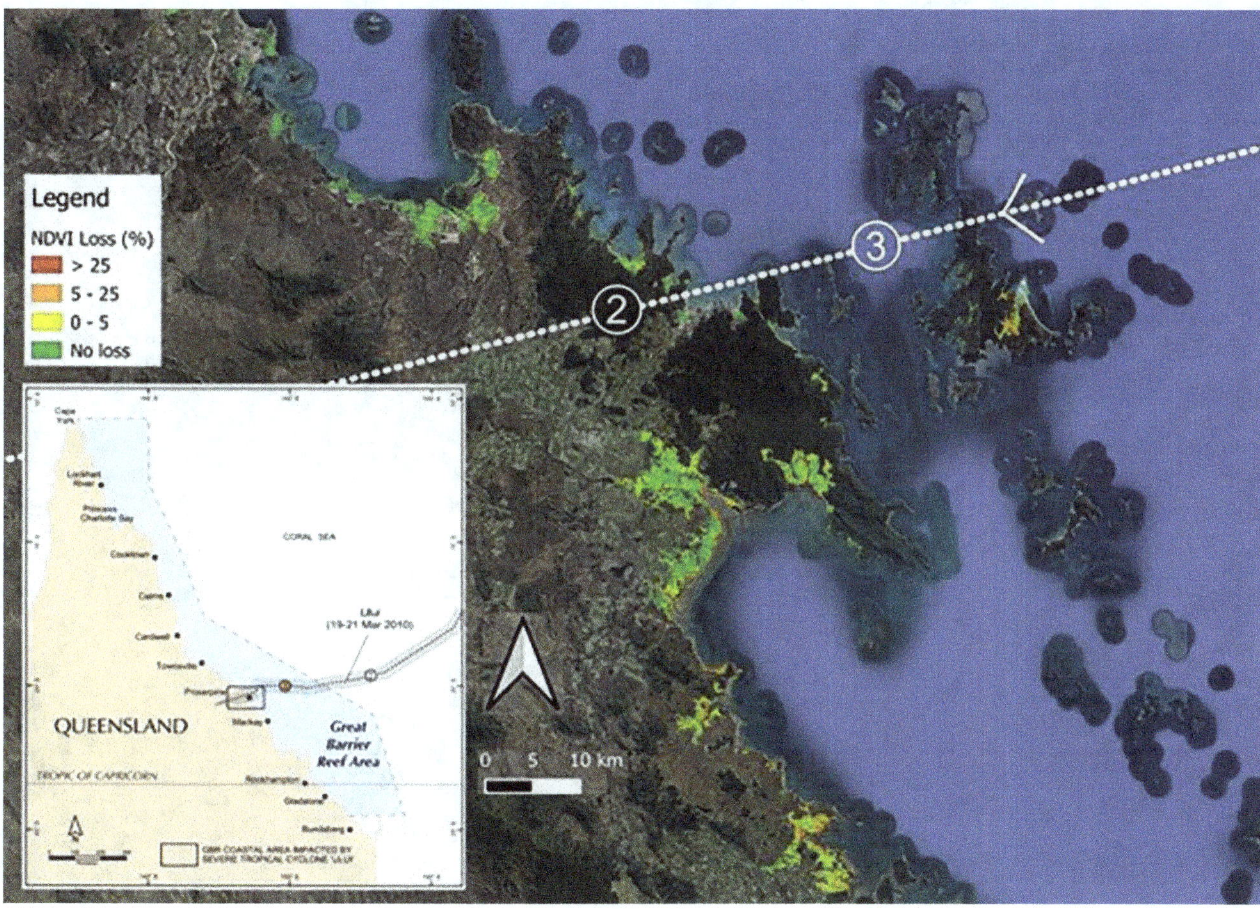

CASE STUDY 9 Impact on shoreline mangroves by STC *Ului* in the GBR region. Map shows the damage to mangrove areas by STC *Ului* at landfall on 21 March 2010. At the time, the cyclone was at category 3 to 2 intensity after tracking west across the Coral Sea. For additional detailed information about this cyclone, visit www.bom.gov.au/cyclone/history/ului.shtml.

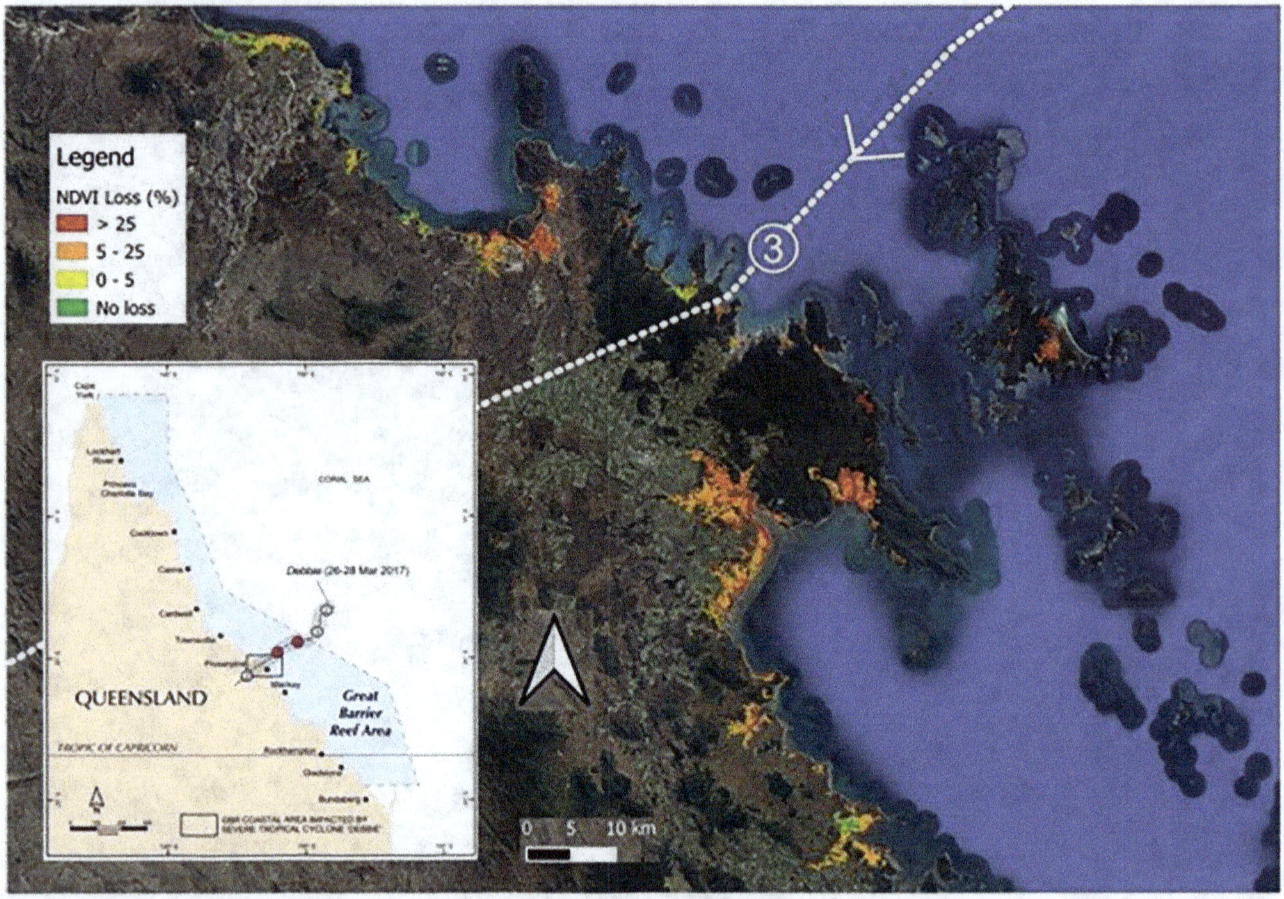

CASE STUDY 10 Impact on shoreline mangroves by STC *Debbie* in the GBR region. Map shows the damage to mangrove areas by STC *Debbie* at landfall on 28 March 2017. At the time, the cyclone was at category 3 intensity after tracking southwest across the Coral Sea. For additional detailed information about this cyclone, visit www.bom.gov.au/cyclone/history/debbie17.shtml.

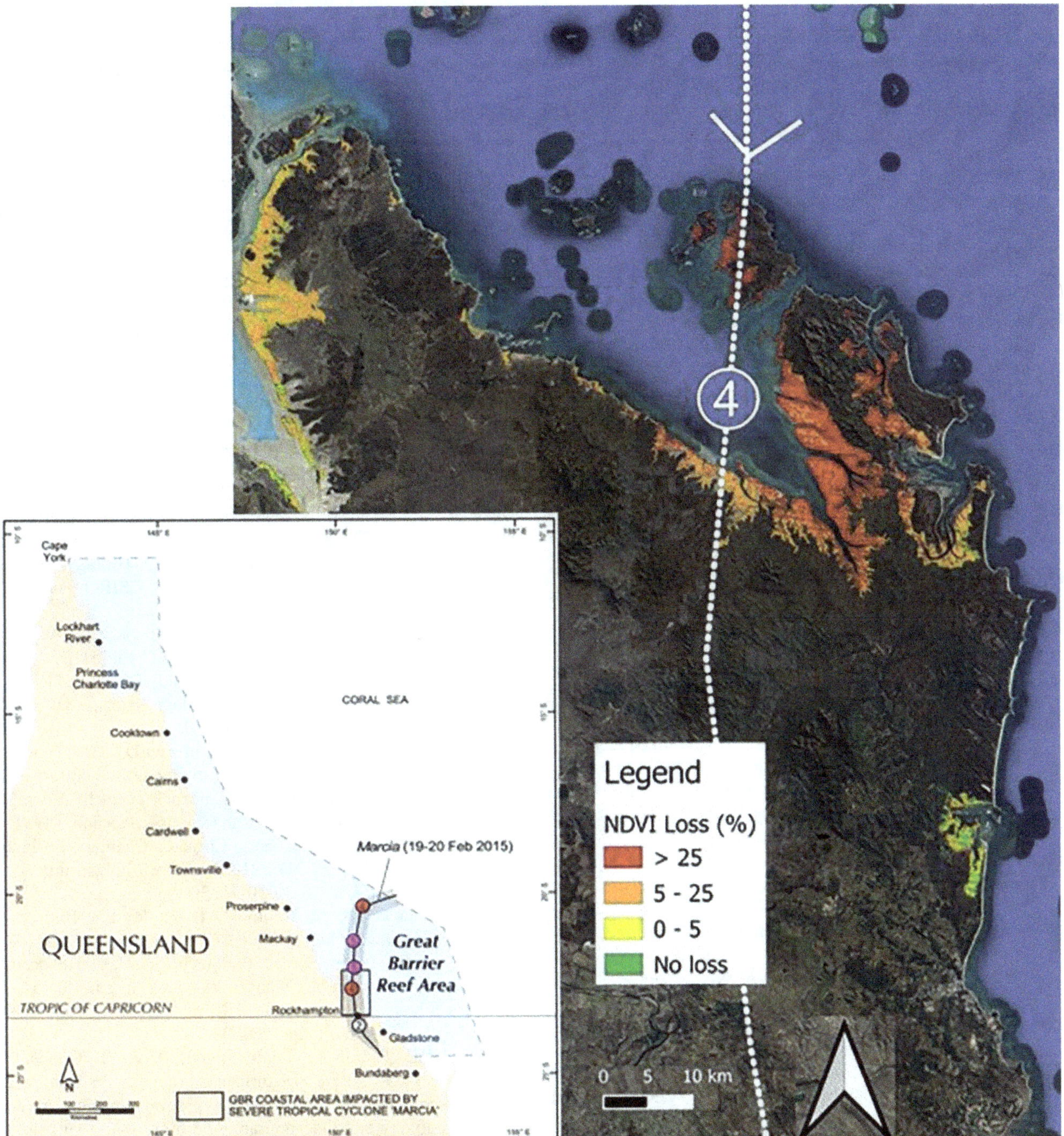

CASE STUDY 11 Impact on shoreline mangroves by STC *Marcia* in the GBR region. Map shows the damage to mangrove areas by STC *Marcia* at landfall on 20 February 2015. At the time, the cyclone was at category 4 intensity after tracking south from across the Coral Sea. For additional detailed information about this cyclone, visit www.bom.gov.au/cyclone/history/marcia.shtml.

REFERENCES

Asbridge E, Lucas R, Ticehurst C, Bunting P (2016) Mangrove response to environmental change in Australia's Gulf of Carpentaria. Ecology and Evolution 6:3523–3539. Doi: 10.1002/ece3.2140

Asbridge E, Lucas R, Rogers K, Accad A (2018) The extent of mangrove change and potential for recovery following severe Tropical Cyclone Yasi, Hinchinbrook Island, Queensland, Australia. Ecology and Evolution 8:10416–10434. Doi: 10.1002/ece3.4485

Babcock RC, Bustamante RH, Fulton EA, Fulton DJ, Haywood MDE, Hobday AJ, Kenyon R, Matear RJ, Plaganyi EE, Richardson, AJ Vanderklift MA (2019) Severe continental-scale impacts of climate change are happening now: Extreme climate events impact marine habitat forming communities along 45% of Australia's coast. Frontiers in Marine Science 6:411.

Beeden R, Maynard J, Puotinen M, Marshall P, Dryden J, Goldberg J, Willams G (2015) Impacts and Recovery from Severe Tropical Cyclone Yasi on the Great Barrier Reef. PLoS One 10(4):e0121272. Doi: 10.1371/journal.pone.0121272

Bell SS, Chand SS, Tory KJ, Dowdy AJ, Turville C, Ye H (2019). Projections of southern hemisphere tropical cyclone track density using CMIP5 models. Climate Dynamics 52(9–10). doi: 10.1007/s00382-018-4497-4

Bell SS, Chand SS, Turville C (2020). Projected changes in ENSO driven regional tropical cyclone tracks. *Climate Dynamics* 54(3–4):2533–2559. Doi: 10.1007/s00382-020-05129-1

Bergstrom DM, Wienecke BC, van der Hoff J, Hughes L, Lindemayer DL, Ainsworth TD, Baker CM, Bland L, Bowman DMJS, Brooks ST, Canadell JG, Constable A, Dafforn KA, Depledge MH, Dickson CR, Duke NC, Helmstedt KJ, Johnson CR, McGeoch MA, Melbourne-Thomas J, Morgain R, Nicholson EN, Prober SM, Raymond B, Ritchie EG, Robinson SA, Ruthrof KX, Setterfield SA, Sgro CM, Stark JS, Travers T, Trebilco R, Ward DFL, Wardle GM, Williams KJ, Zylstra PJ, Shaw, JD (2021) Ecosystem collapse from the tropics to the Antarctic: An assessment and response framework. Global Change Biology 27:1692–1703. Doi: 10.1111/gcb.15539

BOM (2023). Australian Bureau of meteorology. www.bom.gov.au/

Callaghan J, Power SB (2011) Variability and decline in the number of severe tropical cyclones making land-fall over eastern Australia since the late nineteenth century. Climate Dynamics 37(3–4):647+. Gale Academic OneFile, link.gale.com/apps/doc/A354662180/AONE?u=anon~f2967c72&sid=googleScholar&xid=0934b4e0

Chand SS, Dowdy AJ, Ramsay HA, Walsh KJE, Tory KJ, Power SB, Bell SS, Lavender SL, Ye H, Kuleshov Y (2019) Review of tropical cyclones in the Australian region: Climatology, variability, predictability, and trends. WIREs Climate Change 10. Doi: 10.1002/wcc.602

Chand SS, Dowdy AJ, Bell SS, Tory K (2020) A review of south pacific tropical cyclones: Impacts of natural climate variability and climate change (Chapter 6). In: *Climate change and impacts in the pacific*. Springer Nature, pp. 251–273. Doi: 10.1007/978-3-030-32878-8_6

De'ath G, Fabricius KE, Sweatman H, Puotinen M (2019) The 27-year decline of coral cover on the Great Barrier Reef and its causes. Proceedings of the National Academy of Sciences of the United States of America 109. Doi: 10.1073/pnas.1208909109

Dowdy A (2019) Tropical cyclones and climate change in the Great Barrier Reef region. National Environmental Science Program (NESP) Earth Systems and Climate Change (ESCC) Hub, Brochure, 6 pages. http://nespclimate.com.au/extremeweather-hazards-in-a-changing-climate-project-5-5/

Duke NC (2001) Gap creation and regenerative processes driving diversity and structure of mangrove ecosystems. Wetlands Ecology Management 9:257–269.

Duke NC (2016) Oil spill impacts on mangroves: Recommendations for operational procedures and planning based on a global review. Marine Pollution Bulletin 109(2):700–715. Doi: 10.1016/j.marpol.2016.06.082

Duke NC (2020) Mangrove harbingers of coastal degradation seen in their responses to global climate change coupled with ever-increasing human pressures. Human Ecology Journal of the Commonwealth Human Ecology Council—Mangrove Special Issue 30:19–23. www.checinternational.org/wp-content/uploads/2020/06/Journal-30-Mangroves.pdf

Duke NC, Mackenzie JR (2018) East Cape York report 2017. Final report: East Cape York shoreline environmental surveys. Report to the commonwealth of Australia. Centre for Tropical Water and Aquatic Ecosystem Research (TropWATER) Publication 17/67, James Cook University, Townsville, 144pp.

Duke NC, Larkum AWD (2019) Mangroves and seagrasses. In: Hutchings PA, Kingsford MJ, Hoegh-Guldberg O (Eds) Chapter 18, The Great Barrier Reef: Biology, environment and management, 2nd Edition. CSIRO Publishing, Collingwood, VIC, pp. 219–228.

Duke NC, Meynecke J-O, Dittmann S, Ellison AM, Anger K, Berger U, Cannicci S, Diele K, Ewel KC, Field CD, Koedam N, Lee SY, Marchand C, Nordhaus I, Dahdouh-Guebas F (2007) A world without mangroves? Science 317:41–42. Doi: 10.1126/science.317.5834.41b

Duke NC, Hutley LB, Mackenzie JR, Burrows D (2021) Processes and factors driving change in mangrove forests—An evaluation based on the mass dieback event in Australia's Gulf of Carpentaria (In: Canadell JG, Jackson RB (Eds) Ecosystem Collapse—and Climate Change. Springer). Ecological Studies 241:221–264. https://link.springer.com/chapter/10.1007/978-3-030-71330-0_9

Duke NC, Mackenzie JR, Canning AD, Hutley LB, Bourke AJ, Kovacs JM, Cormier R, Staben G, Lymburner L, Ai E (2022). ENSO-driven extreme oscillations in mean sea level destabilise critical shoreline mangroves—an emerging threat with greenhouse warming. PLoS Climate 1(8):e0000037.23pp. doi: 10.1371/journal.pclm.0000037.

Fabricius KE, De'ath G, Poutinen ML, Done T, Cooper TF, Burgess SC (2008) Disturbance gradients on inshore and offshore coral reefs caused by a severe tropical cyclone. Limnology and Oceanography 53:690–704. www.aslo.org/lo/toc/vol_53/issue_2/0690.pdf (10 MB)

Fine M, Hoegh-Guldberg O, Dove S (2019) Ecological changes over 90 years at low Isles on the Great Barrier Reef. Nature Communication 10:4409. www.nature.com/articles/s41467-019-12431-y

Gabler CA, Osland MJ, Grace JB, Stagg CL, Day RH, Hartley SB, Enwright NM, From AS, McCoy ML, McLeod JL (2017) Macroclimatic change expected to transform coastal wetland ecosystems this century. Nature Climate Change 7:142–149. Doi: 10.1038/NCLIMATE3203

GBRMPA (Great Barrier Reef Marine Park Authority) 2011. Impacts of tropical cyclone Yasi on the Great Barrier Reef: A report on the findings of a rapid ecological impact assessment, July 2011. Great Barrier Reef Marine Park Authority. https://hdl.handle.net/11017/447

Hamylton SM, McLean R, Bell IP, Thomas CR (2024) Biological and geological links on coral reef islands. In: Wolanski E, Kingsford MJ (eds) Chapter 3, Oceanographic Processes of Coral Reefs. Physical and Biological Links in the Great Barrier Reef. CRC Press, Boca Raton, FL, pp. 35–49.

Holbrook NJ, Claar DC, Hobday AJ, McInnes KL, Oliver ECJ, Gupta AS, Widlansky MJ, Zhang X (2021) ENSO-driven ocean extremes and their ecosystem impacts. In: McPhaden MJ, Santoso A, Cai W (eds) El Niño Southern Oscillation in a changing climate (Geophysical Monograph Series 253). John Wiley & Sons, Inc, Hoboken, pp. 409–428.

Krauss KW, Osland, MJ (2020) Tropical cyclones and the organization of mangrove forests: A review. Annals of Botany 125:213–234.

Lovelock CE, Cahoon DR, Friess DA, Guntenspergen GR, Krauss KW, Reef R, Rogers K, Saunders ML, Sidik F, Swales A, Saintilan N, Thuyen LX, Tiet T (2015) The vulnerability of Indo-Pacific mangrove forests to sea-level rise. Nature 526:559–563.

Mumby PJ, Edwards AJ, Arias-Gonzalez JE, Lindeman KC, Blackwell PG, Gall A, Gorczynska MI, Harborne AR, Pescod CL, Renken H, Wabnitz CCC, Llewellyn G (2004) Mangroves enhance the biomass of coral reef fish communities in the Caribbean. Nature 427:533–536.

Ramsay HA, Chand SS, Camargo SJ (2018) A statistical assessment of southern hemisphere tropical cyclone tracks in climate models. Journal of Climate 31(24). doi: 10.1175/JCLI-D-18-0377.1

16 Dispersal and Connectivity of Marine Turtles in the Great Barrier Reef and Links to the South Pacific Ocean

Mark Hamann, Colin J. Limpus, and Sara Kophamel

16.1. INTRODUCTION

Marine turtle migration has fascinated the research community for decades, from the 1950s when researchers first started to realise that turtles were migrating long distances and across international borders (e.g., Carr and Hirth 1962; Limpus et al. 1992) to recent years when researchers mapped key ocean currents, tracked turtles, and used genetic tools to identify stopovers and waypoints on the turtles' journeys (e.g., Hays et al. 2002; Luschi et al. 2003; Shimada et al. 2016 etc.). Although molecular data, satellite tracking and flipper tag recoveries give us considerable insight (e.g., Boyle et al. 2009; Perez et al. 2022), the dispersal patterns and migration routes remain part of an integral mystery about sea turtle ecology, especially while turtles are young and in their pelagic phase (Putman and Naro-Maciel 2013; Mansfield et al. 2017). Ocean currents have been increasingly mapped in recent years, and we generally understand broadscale hydrodynamic patterns, thus making it possible to investigate oceanic dispersal patterns and routes and infer the influence of behaviour on dispersal (e.g., Putman and Naro-Maciel 2013). It was broadly believed that oceanic dispersal was passive, but it is clear from recent studies that hatchling and young juvenile marine turtles use combinations of passive drifting and directional swimming behaviour linked to environmental conditions to maintain themselves in a dispersal trajectory (Putman and Mansfield 2015; Briscoe et al. 2016; Wildermann et al. 2017).

Nesting patterns of marine turtles are well-defined along the Great Barrier Reef (GBR) region. We can tell with some certainty that green turtles nest in both the northern and southern parts of the GBR (and rarely the central GBR), hawksbill turtles nest on the islands and cays north of the GBR and into Torres Strait, loggerhead turtles nest predominately in the southern part of the GBR and on the mainland coast of southeast Queensland, and flatback turtles have nesting sites spread across the mainland and inshore islands of much of the GBR coast, predominantly between Mackay and Gladstone. Although we know little about the determining factors that dictate where a particular species nests and does not nest (Hamann et al. 2010), the selection of nesting sites is likely to be influenced by bio-physical and oceanographic features (Hays et al. 1995; Santana-Garcon et al. 2010). Oceanographic features are essential for enabling hatchlings to disperse from the beaches and into nearshore or ocean currents, which are important for initiation of the pelagic phase of the post-hatchling and young turtles, their eventual recruitment to neritic coastal areas, and coupled with magnetic in-printing, their migration to and from nesting and breeding sites. In this chapter, we consider the role that oceanographic features play in each of these life history stages.

16.2. BEACH DISPERSAL

Hatchling sea turtles emerge from the sand after around eight weeks of incubation and make their way into the ocean. At this stage, they are small (4 to 5 cm in length), vulnerable, and easy prey for marine predators. Hence, their dispersal from the beach through the nearshore waters and into offshore currents relies on speed and efficiency. When they can get to the water faster, the chance of terrestrial predation is lower (e.g., Gatto and Reina 2020), and when they can move through the nearshore habitats faster, the chance of in-water predation is lower (e.g., Gyuris 1994). Once in the water, navigating the nearshore environment is challenging because the tidal-driven currents in GBR coastal waters are often more dominant than the mean surface currents. Additionally, the tidal flow rate and direction vary daily and with the lunar cycles and do not always move the water in favourable or consistent directions for hatchling dispersal. Therefore, moving through the nearshore environment relies on directional hyper-active swimming behaviour, often referred to as the swimming frenzy (e.g., Wyneken and Salmon 1992).

16.2.1. Case Study 1: Nearshore Dispersal

Flatback turtles are the only marine turtle species without a pelagic dispersal phase (Walker and Parmenter 1990; Wildermann et al. 2017), and recent research demonstrates that flatback turtle hatchlings are active rather than passive dispersers to ensure that they remain within the coastal areas (Wildermann et al. 2017). In particular, oceanographic advection-dispersal models developed to simulate dispersal from two major rookeries in the GBR indicate that flatback turtles need to swim, predominantly north-northwest, to maintain themselves within the inshore waters of the GBR and not be advected out into the oceanic currents (see Figures 6 and 7 of Wildermann et al. 2017). Their inshore

DOI: 10.1201/9781003320425-19

nesting sites, larger hatchling body size relative to the other marine turtle species, and hyper-active swimming behaviour aid in facilitating navigation within the coastal waters and prevent them from being caught in the south-flowing East Australian Current (EAC; Salmon et al. 2009; Salmon et al. 2010; Salmon et al. 2016; Wildermann et al. 2017).

16.2.2. Case Study 2: Pelagic Dispersal from the Southern GBR

In contrast to flatback turtles, green and loggerhead turtles aim to disperse from nearshore waters into the offshore ocean currents to initiate a pelagic phase within the broader South Pacific Ocean. However, evidence indicates that green and loggerhead turtles use different behaviours and biophysical cues to enable this dispersal and duration of the pelagic phase. Knowledge of swimming behaviour and movement routes are unknown because turtles of this size cannot be tracked using satellite telemetry, but data from post-hatchling stranding events and juvenile turtle bycatch indicate that these two species behave differently. Sick or injured post-hatchling green turtles generally strand along southeast Queensland and northern New South Wales (NSW) beaches. In comparison, post-hatchling loggerhead turtles have been stranded as far south as central NSW (~6 cm in length) and New Zealand (~18 cm in length), indicating entrainment in the EAC and a more southern movement trajectory (Gill 1997; Limpus 2009a, 2009b). In addition, small juvenile loggerhead turtles (~54 cm in length) caught as bycatch in Peru have been genetically linked to loggerhead turtles breeding in Queensland (Boyle et al. 2009). Therefore, loggerhead turtles likely have a more southerly and easterly pelagic phase within the South Pacific Ocean than green turtles.

The EAC is a well-described continuous western boundary current of the South Pacific Ocean. Initiating in the South Coral Sea where the South Equatorial Current (SEC) reaches the Australian continental shelf, it flows southward along Australia's east coast and into the Tasman Sea (see Figure 1). The EAC varies in strength and width along the latitudinal gradient, across seasons, and between years (Oke et al. 2019; Kerry and Roughan 2020). Among the key features of the EAC are the mesoscale eddy field south of 30°S and the two EAC extensions: the eastern Extension, which flows between eastern Australia and New Zealand (also referred to as the Tasman Front), and the EAC southern Extension, which flows south of the Tasman Sea (Kerry and Roughan 2020) (Figure 1). The influence of the EAC on marine species distribution is well-documented.

Green turtles have a general thermal preference for water above 25 °C (Spotila and Standora 1985; Seminoff et al. 2008). Because the EAC is a fast southerly flowing current, if post-hatching turtles become entrained within the EAC, then they would encounter water temperatures below their thermal threshold as they reach latitudes of 27°S to 30°S. Instead, to maintain a presence in waters with favourable temperatures, oceanographic models built with satellite-derived altimetry data demonstrate that green turtles would need to swim across the EAC before they reach approximately latitude 28°S to avoid being swept southwards into cooler waters (Figures 1 and 2) (Wolanski 2017). The behavioural components of this dispersal are not known and remain to be examined; however, oceanographic data indicate that the EAC flows near the nesting rookeries of the GBR. To cross the EAC, the hatchlings would need to leave the beach and swim eastwards, perpendicular to the EAC, throughout their swimming frenzy. The limited data on swimming behaviour of hatchlings and post-hatchlings support this possibility. In particular, laboratory and field-based experiments on green turtles in Malaysia and Florida both demonstrate that green turtle hatchlings have a one- to two-day swimming frenzy and at least six days of daytime-oriented swimming (Wyneken and Salmon 1992; Chung et al. 2009), which is likely to enable the hatchlings

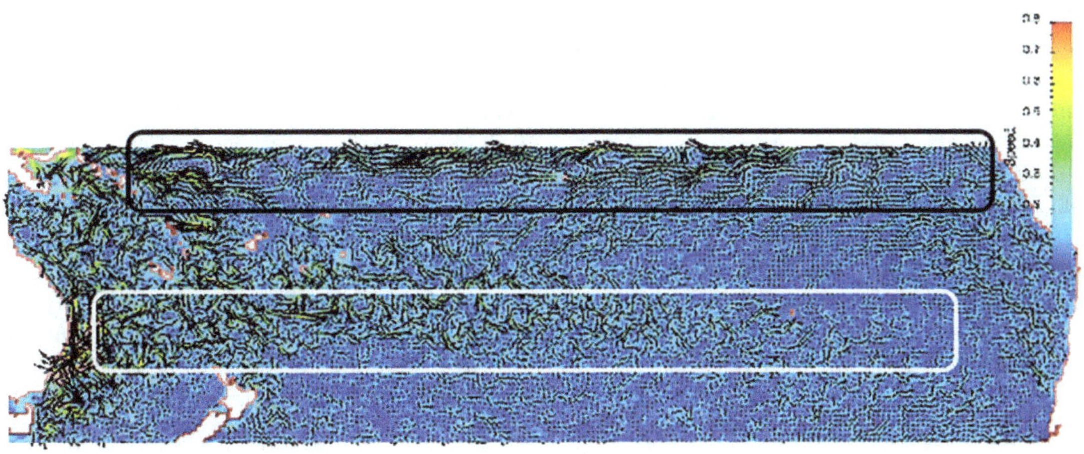

FIGURE 1 Satellite altimetry map illustrating the variable speed of the hydrodynamic condition in the South Pacific Ocean. The black box illustrates the west-flowing south equatorial current, and the white box illustrates the east-flowing current originating with the eastern Extension of the EAC.

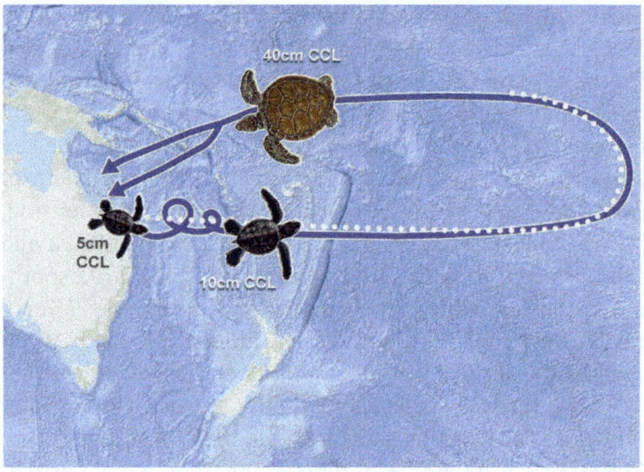

FIGURE 2 Generalised migration path of dispersal stage green turtles. The dotted white line approximates the annual average 24 °C thermocline.

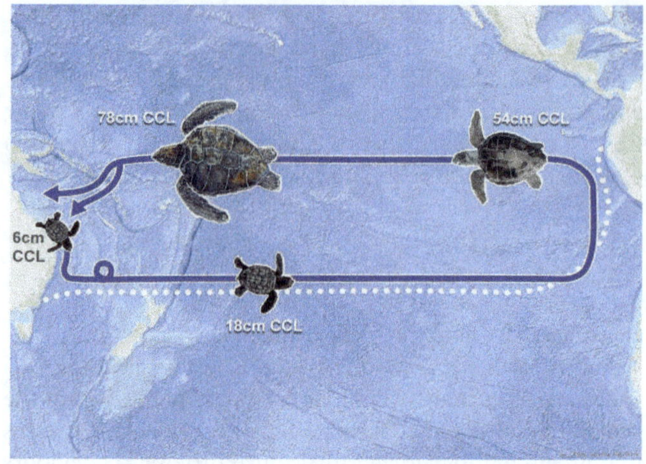

FIGURE 3 Generalised migration path of dispersal stage loggerhead turtles originating from rookeries in southeast Queensland. The dotted white line approximates the annual average 15 °C thermocline.

to reach target currents. This strategy would likely favour hatchlings emerging from rookeries closer to the EAC, such as those in the southern GBR islands. Furthermore, this behaviour, although untested, could mean that years with a narrower-than-average width or speed of the EAC convey a survival advantage to that year's cohort of turtles. Once across the EAC, oceanographic models indicate that green turtles would need to maintain an easterly/northeasterly movement pattern within a large (unnamed) easterly flowing ocean current at around latitude 28°S (Wolanski 2017). The oceanographic data indicate that the young oceanic green turtles would likely spend considerable time in eddies, which could greatly accelerate their travel if they adjust their directional swimming to use positive velocities (i.e., swim with the favourable eddy-driven currents) and swim against the unfavourable eddy-driven currents. Given that the passive drift time (i.e., excluding swimming across the EAC) would be around 2.5 years and that the age estimates of green turtles recruiting into the GBR is 5–10 years (Limpus and Chaloupka 1997), perhaps the turtles are unable to take advantage of the eddies, but this remains to be explored and is a valuable line of inquiry for future research. Then, the pelagic phase loop is completed by following the SEC back towards the Australian coastline and the GBR region (Figure 2). This behaviour-inferred oceanographic model supports the field-based data, which suggests that juvenile green turtles complete their pelagic phase in approximately 5 years and self-recruit into the waters of the central to southern GBR, with some recruitment possible to other reef areas in the northern GBR, Coral Sea, and NSW (Jensen et al. 2016).

In comparison, loggerhead turtles have a thermal threshold closer to 10 to 15 °C (Davenport 1997; Schwartz 1978) and could thus tolerate a more southerly dispersal route. Hence, it is likely that loggerhead turtles head further south in the EAC (Harrison et al. 2021) (Figures 1 and 3). Loggerhead turtles have a similar swimming frenzy to green turtles; however, their nesting sites are spread across mainland and island rookeries, differing in distance from the EAC. The farther travel distance could assist in entraining them in the EAC and enable a more passive southwards path (Harrison et al. 2021). As water temperatures cool south of 30°S they would presumably initiate active swimming behaviour to enable them to take a broader route across the southern Pacific Ocean using the EAC's eastern Extension (Tasman Front) (Tomczak and Godfrey 1994; Qiu and Chem 2004; Kerry and Roughan 2020). The turtles' orientation would possibly be guided by using a thermal gradient of around 15 °C. This hypothesis is plausible, as active dispersal was demonstrated for loggerhead turtles in the northern Pacific Ocean (Briscoe et al. 2016), and in the South Pacific Ocean, the tracking of small juveniles demonstrated that they can actively swim against currents or change the speed of dispersal in relation to environmental gradients (Kobayashi et al. 2014). Once in the southern Pacific Ocean, loggerhead turtles would presumably follow a thermal gradient as they move eastwards towards the coast of South America (e.g., like the movement of loggerheads across the northern Pacific Ocean, Briscoe et al. 2021) (Figure 3), where genetic data have linked loggerhead turtles caught by Peruvian fishers to eastern Australian nesting beaches (Boyle et al. 2009). Oceanographic particle modelling indicates that a passive transport time between the southern GBR coast and Peru would be around 8 years. The pelagic loggerhead turtles would then presumably return to the Southwest Pacific Ocean and the Australian continental shelf using the westerly flowing SEC (Figure 3) (Tomczak and Godfrey 1994; Qiu and Chem 2004) for which the passive transport time would be ~2.5 years. Although there are no coupled biological-oceanographic models to support the entire dispersal route (but see Harrison et al. 2021) or its duration, capture records and incidental sightings of transit places where turtles are found support this hypothesis.

Using back-calculated size and growth methods and information from recoveries of tagged hatchlings, loggerhead turtles likely complete their pelagic phase at approximately 78 cm in length and around 15 years old (Limpus 2009a).

16.2.3. Case Study 3: Hawksbill and Green Turtle Dispersal from the Northern GBR

In the northern GBR and Torres Strait, green turtles predominantly nest on the coral cays and islands along the outer edge of the GBR, and hawksbill turtles nest on the coral cays and islands along the inshore coast of the GBR and central islands of Torres Strait. Both species differ in their behaviour as they move from the beach into offshore currents. Green turtles have an active swimming frenzy period, and hawksbill turtles are believed to be more passive dispersers (Wyneken and Salmon 1992; Chung et al. 2009). These behaviours and locations of key nesting sites set the species up for different oceanic dispersal. Green turtle nesting sites are typically closer to the eastern edge of the GBR (Figure 4), and the dispersal of post-hatchling green turtles is likely to be strongly influenced by the clockwise circulation of the Coral Sea Coastal Current and its associated eddies (Wolanski et al. 1995). Green turtles then return to the continental shelf waters using westerly flowing eddies and currents, such as those associated with the SEC. However, no biological-oceanographic models have been developed to test these suggestions.

Given the inshore locations of hawksbill nesting sites and the spatial spread of nesting locations from the northern GBR into central Torres Strait, the dispersal patterns of post-hatchling hawksbill turtles are more difficult to predict and are likely to be spatially variable. In particular, hawksbill turtle nesting sites do not have large oceanic currents nearby (Figure 4), and the species is more likely to have a less-active frenzy swimming period than green turtles. Dispersal of hatchling hawksbill turtles is thus more likely reliant on wind-generated surface currents. However, this is difficult to predict using models because the nesting/hatching season spans from January to April, which covers the transition from dominant northerly winds to dominant southerly winds. In addition, the Torres Strait oceanographic models indicate that the tidal currents peak at 2 m/s (Figure 5) and the average currents under southeast winds peak at 0.3 m/s; thus, the difference between the currents from the tides and the averaged current is a factor of 7. Adding to the complexity of trying to elucidate patterns of hatchling dispersal and the influence of behaviour is that within the Torres Strait region, three tidal waves meet: one from the east (Coral Sea), one from the Northeast (Gulf of Papua), and one from the west (Gulf of Carpentaria) (Figure 5). Thus, these influences and uncertainty in turtle behaviour strategies mean that predictions of nearshore dispersal routes are challenging to make for hawksbill turtles, and this would be a useful line of inquiry for future research.

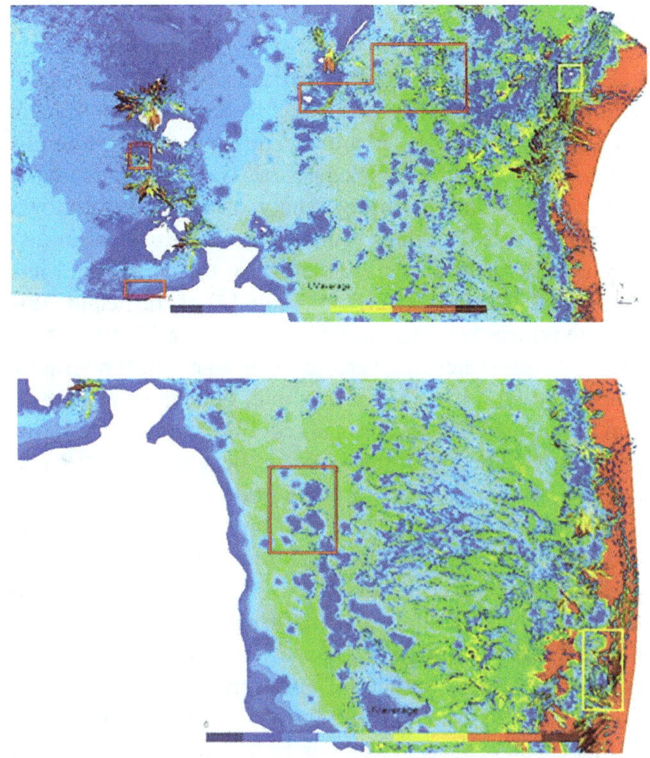

FIGURE 4 Hydrodynamic environment in Torres Strait (top) and the northern Great Barrier Reef (bottom) in relation to hawksbill turtle nesting sites. Red boxes indicate the locations of the main hawksbill turtle nesting areas. Yellow boxes indicate two of the main green turtle nesting areas (the top panel contains Mer and Dauer, and the bottom panel contains Raine Island and Moulter Cay). Scale 0 to 0.5 m/s.

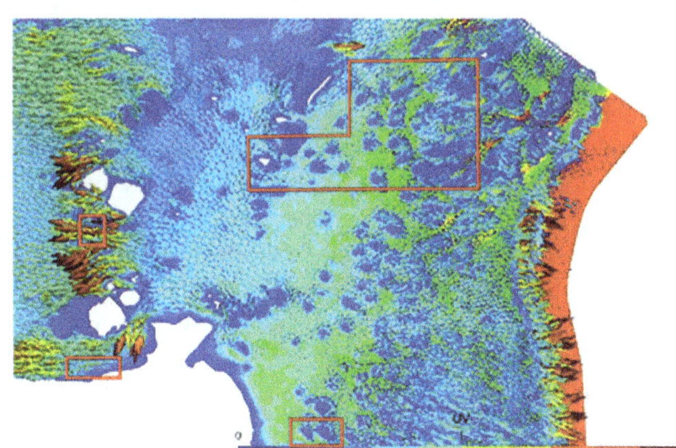

FIGURE 5 Examples of instantaneous tidal environments in Torres Strait in relation to hawksbill turtle nesting sites. Red boxes indicate the locations of the main hawksbill turtle nesting areas. Scale 0 to 2 m/s.

16.3. Partitioning of Foraging Habitats

There are two genetically distinct populations of green turtles in the GBR: a northern population that breeds on islands north of Cooktown and a southern population that breeds on the mainland and coral cay beaches south of Queensland and the GBR (Dethmers et al. 2006). Genetic studies conducted on foraging turtles in eastern Queensland have demonstrated that the pelagic turtles originating from the northern population are recruiting to GBR habitats north of Cooktown, and pelagic turtles originating from the southern GBR are recruiting to habitats south of Cooktown (Jensen et al. 2016; Jones et al. 2018). This differential distribution is likely driven by the use of different zonal jets associated with the westerly flowing SEC as the pelagic turtles are returning to the Australian continental shelf waters. The SEC bifurcates in several jets caused by the topography associated with islands, for example, Fiji, Vanuatu, Tonga, and New Caledonia (Webb et al. 2000). Although it remains to be tested, the green turtles from the southern GBR and those returning from the South Coral Sea are more likely to use the jets associated with the southern extension of the SEC and thus end up arriving in the central to the southern region of the GBR. In comparison, green turtles from the northern GBR and those returning from the northern Coral Sea are more likely to use jets associated with the northern extension of the SEC and arrive in the more northern sections of the GBR and into the Torres Strait region. Although this concept is supported by genetic-derived data (Jensen et al. 2016) and similar patterns have recently been shown for fishes (Benthuysen et al. 2022), the mechanisms, turtle behaviour, and oceanographic influences remain to be tested.

16.4. KEY RESEARCH PATHWAYS FOR OCEANOGRAPHY

Over the past 10 years, data from oceanographic modelling has demonstrated that turtle dispersal, both from inshore and oceanic environments, depends on both passive drifting and active swimming. However, because we are not yet able to track small turtles, in most cases, we can only make inferences about their behaviour in relation to the biophysical conditions that they experience. Thus, useful avenues for future work include (1) developing field and laboratory experiments to investigate the swimming activity and orientation of young pelagic-sized turtles and how behaviour could be influenced by factors such as temperature (e.g., similar to what Briscoe et al. 2021 demonstrated for the North Pacific Ocean), (2) tracking hatchling turtles to understand how or if they can adapt to their nearshore dispersal in response to variation in tidal currents, and (3) developing biological-coupled oceanographic models to understand dispersal patterns and important oceanic habitat areas (for example, by building on models developed by Harrison et al. 2021). Finally, it is clear from the literature that key parameters of ocean currents vary across temporal and spatial scales and that these parameters are likely to change under future climate scenarios. A useful avenue for future research is to examine the consequences of changing oceanographic parameters on ocean productivity and turtle dispersal.

REFERENCES

Benthuysen JA, Emslie MJ, Currey-Randall LM, Cheal AJ, Heupel MR (2022) Oceanographic influences on reef fish assemblages along the Great Barrier Reef. *Progress in Oceanography*, 102901.

Boyle MC, FitzSimmons NN, Limpus CJ, Kelez S, Velez-Zuazo X, Waycott M (2009) Evidence for transoceanic migrations by loggerhead sea turtles in the southern Pacific Ocean. *Proceedings of the Royal Society B: Biological Sciences*, *276*(1664), 1993–1999.

Briscoe DK, Parker DM, Balazs GH, Kurita M, Saito T, Okamoto H, Rice MR, Polovina JJ, Crowder LB (2016) Active dispersal in loggerhead sea turtles (*Caretta caretta*) during the 'lost years'. *Proceedings of the Royal Society B: Biological Sciences*, *283*(1832), 20160690.

Briscoe DK, Turner Tomaszewicz CN, Seminoff JA, Parker DM, Balazs GH, Polovina JJ, Kurita M, Okamoto H, Saito T, Rice MR, Crowder LB (2021) Dynamic thermal corridor may connect endangered loggerhead sea turtles across the Pacific Ocean. *Frontiers in Marine Science*, *8*, 630590.

Carr AF, Hirth HF (1962) The ecology and migrations of sea turtles. 5. Comparative features of isolated green turtle colonies. *American Museum Novitates*, *2091*.

Chung FC, Pilcher NJ, Salmon M, Wyneken J (2009) Offshore migratory activity of hawksbill turtle (*Eretmochelys imbricata*) hatchlings, I. Quantitative analysis of activity, with comparisons to green turtles (*Chelonia mydas*). *Chelonian Conservation and Biology*, *8*(1), 28–34.

Davenport J (1997) Temperature and the life-history strategies of sea turtles. *Journal of Thermal Biology*, *22*, 479–488. Doi: 10.1016/S0306-4565(97)00066-1

Dethmers KE, Broderick D, Moritz C, et al. (2006) The genetic structure of Australasian green turtles (*Chelonia mydas*): Exploring the geographical scale of genetic exchange. *Molecular Ecology*, *15*(13), 3931–3946.

Gatto CR, Reina RD (2020) Sea turtle hatchling locomotor performance: Incubation moisture effects, ontogeny and species-specific patterns. *Journal of Comparative Physiology B*, *190*(6), 779–793.

Gill BJ (1997) Records of turtles and sea snakes in New Zealand, 1837–1996. *New Zealand Journal of Marine and Freshwater Research*, *31*(4), 477–486.

Gyuris E (1994) The rate of predation by fishes on hatchlings of the green turtle (*Chelonia mydas*). *Coral Reefs*, *13*(3), 137–144.

Hamann M, Godfrey, MH, Seminoff JA, et al. (2010) Global research priorities for sea turtles: Informing management and conservation in the 21st century. *Endangered Species Research*, *11*(3), 245–269.

Harrison CS, Luo JY, Putman NF, et al. (2021) Identifying global favourable habitat for early juvenile loggerhead sea turtles. *Journal of the Royal Society Interface*, *18*, 20200799. Doi: 10.1098/rsif.2020.0799

Hays GC, Mackay A, Adams CR, Mortimer JA, Speakman JR, Boerema M (1995) Nest site selection by sea turtles. *Journal of the Marine Biological Association of the United Kingdom*, *75*(3), 667–674.

Hays GC, Broderick AC, Godley BJ, et al. (2002) Biphasal long-distance migration in green turtles. *Animal Behaviour*, *64*(6), 895–898.

Jensen MP, Bell I, Limpus CJ, et al. (2016) Spatial and temporal genetic variation among size classes of green turtles (*Chelonia mydas*) provides information on oceanic dispersal and population dynamics. *Marine Ecology Progress Series*, *543*, 241–256.

Jones K, Jensen M, Burgess G, et al. (2018) Closing the gap: Mixed stock analysis of three foraging populations of green turtles (*Chelonia mydas*) on the Great Barrier Reef. *PeerJ*, *6*, e5651.

Kerry C, Roughan M (2020) Downstream evolution of the East Australian Current system: Mean flow, seasonal, and intra-annual variability. *Journal of Geophysical Research: Oceans*, *125*(5), e2019JC015227.

Kobayashi DR, Farman R, Polovina JJ, et al. (2014) "Going with the flow" or not: Evidence of positive rheotaxis in oceanic juvenile loggerhead turtles (*Caretta caretta*) in the South Pacific Ocean using satellite tags and ocean circulation data. *PloS One*, *9*(8), e103701.

Limpus CJ (2009a) *A biological review of Australian marine turtles, 1. Loggerhead turtle, Caretta (Linneaus)*. Brisbane, Australia: Environmental Protection Agency. Brisbane.

Limpus CJ (2009b) *A biological review of Australian marine turtles, 2. Green turtle, Chelonia mydas, (Linneaus)*. Brisbane, Australia: Environmental Protection Agency. Brisbane.

Limpus C, Chaloupka M (1997) Nonparametric regression modelling of green sea turtle growth rates (southern Great Barrier Reef). *Marine Ecology Progress Series*, *149*, 23–34.

Limpus CJ, Miller JD, Paramenter CJ, et al. (1992) Migration of green (*Chelonia mydas*) and loggerhead (*Caretta caretta*) turtles to and from eastern Australian rookeries. *Wildlife Research*, *19*(3), 347–357.

Luschi P, Hays GC, Papi F (2003) A review of long-distance movements by marine turtles, and the possible role of ocean currents. *Oikos*, *103*(2), 293–302.

Mansfield KL, Mendilaharsu ML, Putman NF, et al. (2017) First satellite tracks of South Atlantic sea turtle 'lost years': Seasonal variation in trans-equatorial movement. *Proceedings of the Royal Society B: Biological Sciences*, *284*(1868), 20171730.

Oke PR, Roughan M, Cetina-Heredia P, et al. (2019) Revisiting the circulation of the East Australian Current: Its path, separation, and eddy field. *Progress in Oceanography*, *176*, 102139.

Perez MA, Limpus CJ, Hofmeister K, et al. (2022) Satellite tagging and flipper tag recoveries reveal migration patterns and foraging distribution of loggerhead sea turtles (*Caretta caretta*) from eastern Australia. *Marine Biology*, *169*(6), 1–15.

Putman NF, Mansfield KL (2015) Direct evidence of swimming demonstrates active dispersal in the sea turtle "lost years". *Current Biology*, *25*(9), 1221–1227.

Putman NF, Naro-Maciel E (2013) Finding the 'lost years' in green turtles: Insights from ocean circulation models and genetic analysis. *Proceedings of the Royal Society B: Biological Sciences*, *280*(1768), 20131468.

Qiu B, Chen S (2004) Seasonal modulations in the eddy field of the South Pacific Ocean. *Journal of Physical Oceanography*, *34*(7), 1515–1527.

Salmon M, Hamann M, Wyneken J (2010) The development of early diving behavior by juvenile flatback sea turtles (*Natator depressus*). *Chelonian Conservation and Biology*, *9*(1), 8–17.

Salmon M, Hamann, M, Wyneken J, Schauble C (2009) Early swimming activity of hatchling flatback sea turtles *Natator depressus*: A test of the 'predation risk' hypothesis. *Endangered Species Research*, *9*(1), 41–47.

Salmon M, Wyneken J, Hamann M, Whiting S (2016) Early growth and development of morphological defenses in post-hatchling flatbacks (*Natator depressus*) and green turtles (*Chelonia mydas*). *Marine and Freshwater Behaviour and Physiology*, *49*(6), 421–435.

Santana-Garcon J, Grech A, Moloney J, Hamann M (2010) Relative Exposure Index: An important factor in sea turtle nesting distribution. *Aquatic Conservation: Marine and Freshwater Ecosystems*, *20*(2), 140–149.

Schwartz FJ (1978). Behavioral and tolerance responses to cold water temperatures by three species of sea turtles (Reptilia, Cheloniidae) in North Carolina. *Florida Department of Natural Resources, 33*, 16–18.

Seminoff JA, Zárate P, Coyne M, et al. (2008) Post-nesting migrations of Galápagos green turtles *Chelonia mydas* in relation to oceanographic conditions: Integrating satellite telemetry with remotely sensed ocean data. *Endangered Species Research*, *4*(1–2), 57–72.

Shimada T, Limpus C, Jones R, et al. (2016) Sea turtles return home after intentional displacement from coastal foraging areas. *Marine Biology*, *163*(1), 1–14.

Spotila JR, Standora E (1985) Environmental constraints on the thermal energetics of sea turtles. *Copeia*, 694–702

Tomczak M, Godfrey JS (1994) Regional oceanography: An introduction. *Pergamon*, 422.

Walker TA, Parmenter CJ (1990) Absence of a pelagic phase in the life cycle of the flatback turtle, *Natator depressus* (Garman). *Journal of Biogeography*, 275–278.

Webb DJ (2000) Evidence for shallow zonal jets in the South Equatorial Current region of the southwest Pacific. *Journal of Physical Oceanography*, *30*(4), 706–720.

Wildermann N, Critchell K, Fuentes MM, et al. (2017). Does behaviour affect the dispersal of flatback post-hatchlings in the Great Barrier Reef?. *Royal Society Open Science*, *4*(5), 170164.

Wolanski E (2017) Bounded and unbounded boundaries–untangling mechanisms for estuarine-marine ecological connectivity: Scales of m to 10,000 km–a review. *Estuarine, Coastal and Shelf Science*, *198*, 378–392.

Wolanski E, Norro A, King B (1995) Water circulation in the Gulf of Papua. *Continental Shelf Research*, *15*(2–3), 185–212.

Wyneken J, Salmon M (1992). Frenzy and postfrenzy swimming activity in loggerhead, green, and leatherback hatchling sea turtles. *Copeia*, 478–484.

17 Interactions between Oceanography and the Behaviour of Larval Reef Fish

Michael J. Kingsford, Lisa Spiecker, and Gabriele G. Gerlach

17.1. INTRODUCTION

17.1.1. Dispersal Distances of Passive Larvae

Strong interactions between oceanographic processes of the GBR and the early life history stages of fish made intuitive sense, especially given that larvae were initially considered to be meroplankton that 'wander' or drift. In the 1980s, a paradigm developed where larvae could disperse over vast distances: the so-called 'teleplanic' larvae of invertebrates were thought to disperse over entire ocean basins (Scheltema 1988). At the other extreme, lecithotrophic invertebrate larvae that do not feed may disperse only centimeters to meters from their parental colony (e.g., ascidians, Davis 1987).

Fish larvae feed on plankton, and their length of larval life (termed planktonic larval duration, PLD; Victor 1991) varies from approximately nine days to three months (Brothers et al. 1983). PLD varies by group, where for damselfish of the genus *Amphiprion*, it is nine days, while the labrid *Thalassoma* has a PLD of 57 days, and the acanthurid *Naso*'s is 85 days (Brothers et al. 1983). Many larvae have a PLD of 15–30 days (Brothers et al. 1983; Victor 1986; Kingsford et al. 2014). Using a PLD of 22 days (Williams et al. 1984) predicted the dispersal patterns of fish larvae on the Great Barrier Reef (GBR). The passive larval analogs released from the mid to outer shelf reefs near Cairns drifted over 300 km in 22 days, which suggests a high level of connectivity among reefs.

Part of the perception that larvae were free drifting over great distances and with little control over their own destiny was based on two sets of evidence. Firstly, larvae from the northern hemisphere was the initial focus of research. For example, the eyesight of clupeid larvae is so poor that they can barely see the length of their own bodies until they are approximately 15 mm long, and they have a swimming motion that is slow and serpentine. Meanwhile, few gadoid or pleuronectid larvae can swim only at > 1 cm sec^{-1} at a length of 10 mm (Blaxter 1986). Accordingly, the perception of passive drift for fish larvae on the GBR seemed highly likely, especially given that uninterrupted mainstream flows on the GBR are 35–55 cm sec^{-1}, and even in reef mosaics, mean currents of 9–14 cm sec^{-1} are common (Andutta et al. 2012). Secondly, fish larvae that were collected for taxonomy (Leis and Rennis 1983; Leis and Carson-Ewart 1997) or ecology tended to be small (e.g., Leis 1986). There was generally a large gap in size between larvae caught in nets and newly settled juveniles on reefs; large presettlement juveniles were rarely caught.

Three methods detected much larger pelagic juveniles that did not align with the definition of larvae (Kingsford 1988) but were clearly presettlement individuals that had the form, and sometimes colour, of settled juveniles. Anecdotally, Doherty observed pelagic juveniles of reef fishes on the GBR that were attracted to light at night. This resulted in the design of light traps that were highly effective in catching late-stage presettlement reef fish (Doherty 1987). This was not only useful for ecological studies (e.g., Doherty et al. 1996) but also provided an abundance of specimens for laboratory studies on the mobility and orientation of these fishes. Concurrently, pelagic juvenile reef fishes were found around drifting objects (Kingsford and Choat 1985; Kingsford 1992), and large reef fish were collected in the crest nets at night (Dufour and Galzin 1993; Figure 1). There was the potential, therefore, for presettlement reef fish to be highly mobile and not fully at the mercy of currents. Here, we use the term 'larval fish' to encompass all stages of fishes from newly hatched preflexion larvae to postflexion forms (*sensu* Leis and Carson-Ewart 1997) and pelagic juveniles that are potential settlers.

Our goal is to review both the interactions between larval reef fishes and their orientation capabilities in connection to oceanography to build accurate biophysical models of dispersal and connectivity between reefs. Our specific aims are to (1) review what is known about the swimming ability of larvae and their sensory capabilities to detect reefs, (2) recognise how the nature of reef mosaics and associated 'sticky water' (Andutta et al. 2012) can influence the likelihood that larvae will 'catch a reef', and (3) determine how we should choose the correct oceanographic conditions to model connectivity through ways of 'asking' the larvae. Finally, (4) we recognise the need to incorporate more sophisticated larval behaviour into models. However, 3D oceanographic models will help with the accuracy of biophysical models as GBR waters are sometimes stratified, and Langmuir wind-driven cells and the salient topography of reefs can generate complex 3D currents even in the absence of a stratification (Wolanski and Hamner 1988; Kingsford et al. 1991)

DOI: 10.1201/9781003320425-20

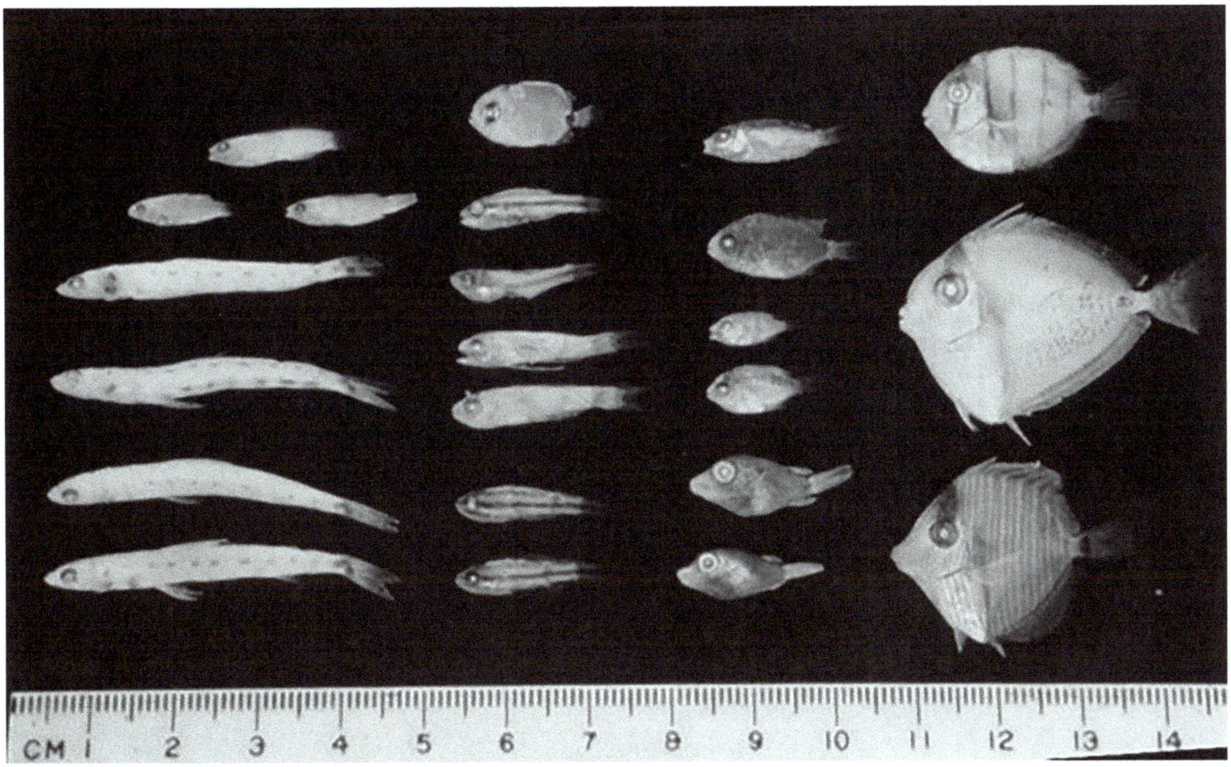

FIGURE 1 A collection of presettlement reef fish collected in a channel by MJK in Moorea, French Polynesia in 1994. From the top left and down by column. Column 1: three Labriformes, four Synodontidae. Column 2: one Chaetodontidae, six Apogonidae. Column three: four Pomacentridae, two Tetradondiformes. Column four: three Acanthuridae.

17.2. SWIMMING AND SENSORY ABILITIES OF LARVAE

17.2.1. Swimming Ability of Larvae

Multiple studies have demonstrated that presettlement reef fishes are capable swimmers and that sustained 'critical' swimming speeds of 37–46 cm sec^{-1} are common (e.g., three species of pomacentrids, Stobutzki and Bellwood 1994; Stobutzki and Bellwood 1997) and over greater distances when larvae are feeding (Fisher and Bellwood 2001). Moreover, these speeds equate with the mean currents that fish would encounter as mainstream flows in the GBR and much higher that the currents experienced in reef mosaics (Andutta et al. 2012). High swimming speeds are not just restricted to potential settlers. Leis et al. (2009) tested serranids from approximately 5 mm to 23 mm and measured swimming speeds of 2 to 30 cm sec^{-1}, and there was a clear ontogenetic increase in swimming ability where increases in swimming speed were about 1.4 cm sec^{-1} per 1 mm increase in size, while lutjanids of 12–18 mm recorded speeds of 11–16 cm sec^{-1}. *In situ* comparisons where larvae were followed have usually detected lower swimming speeds than the critical swimming speed recorded in the laboratory (Leis et al. 2009), though it is unclear how larval fish behaviour is affected by a SCUBA diver.

Larvae hatch from pelagic or demersal eggs at around 3 mm, so for a period of days, they are very small. However, even at this stage, they can swim. Fisher et al. (2000) reared two species of damselfish and a species of apogonid. Twelve hours after hatching, the three taxa had a mean critical swimming speed of 1.8 cm sec^{-1}, while approximately 20 days later and prior to settlement, swimming speeds ranged from 10 to 30.3 cm sec^{-1}.

17.2.2. Sensory Abilities and Orientation of Larvae

A series of remarkable papers that used intergenerational tags such as genetic assignment, otolith tagging, and microchemistry studies have demonstrated that a high percentage of fish return to their natal reefs (Thorrold et al. 2006; Swearer et al. 1999; Jones et al. 1999; Almany et al. 2007; Gerlach et al. 2007). This can only be explained as either larvae have orientation capabilities that guide them back to their natal reefs or larvae stay in close proximity to their natal reef. There is evidence for both.

The level of mobility and sensory ability of larvae to orientate will influence how they interact with currents (Leis 2007; Sponaugle et al. 2002). Casual inspection of the morphology of potential settlers indicates that they have

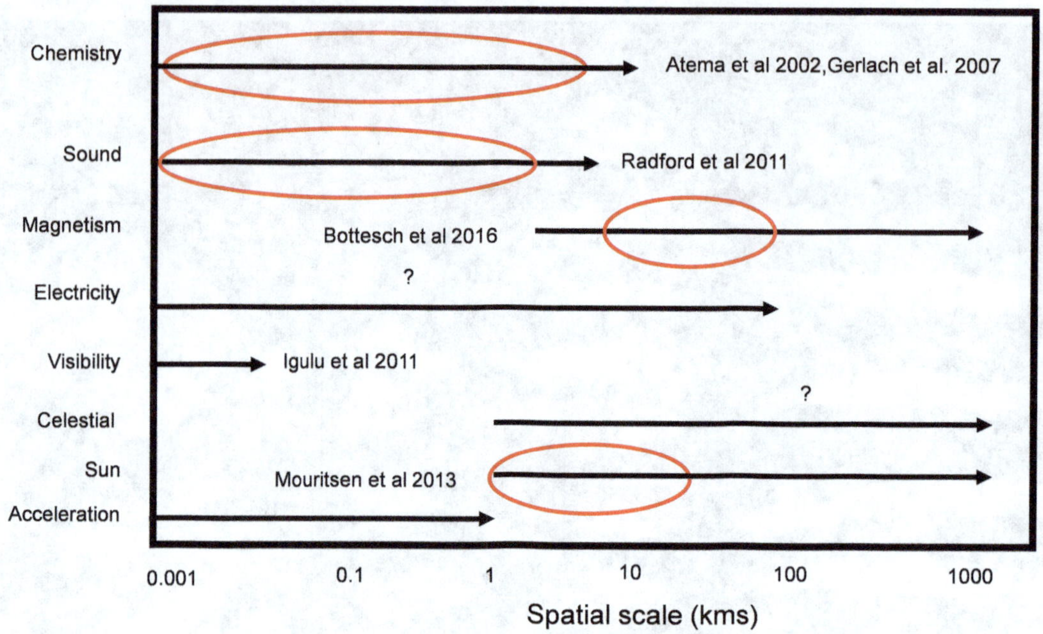

FIGURE 2 The maximum estimated range of distances from which fish larvae can detect reefs and/or orientate toward them (redrawn from Kingsford et al. 2002).

well-developed eyes, nostrils leading to olfactory lobes (Atema et al. 2002, 2015), a well-developed lateral line, and dissection of the inner ear shows three pairs of otoliths that could detect sound. Given that salmonids use a variety of sensory cues to find their natal rivers (Thorpe 1988), the potential for more evolutionarily advanced reef fishes (Choat and Bellwood 1991) to do the same seems likely. Salmonids use a hierarchy of senses to detect rivers that include magnetic, chemosensory (for river chemistry and conspecifics), and positive rheotaxis. Focused research on the sensory abilities of larval reef fish has demonstrated an astonishing set of sensory abilities that could be used for orientation to reefs and currents. Reefs are by nature noisy from the clicks, honks, and scrapings of invertebrates and fishes (Leis et al. 2002). Fishes can hear over a range of 100 to 1,000 Hz (Leis et al. 2011), and hearing may be enhanced with the swim bladder (Atema et al. 2015). Experiments on tropical (Simpson et al. 2005) and temperate (Tolimieri et al. 2000) reef fishes have demonstrated greater attraction to experimental units with sound than without sound. Settling fish can not only use the directional information contained in acoustic cues but also interpret the content of the acoustic signals for relevant habitat information, which is then used in their decision making for orientation (Radford et al. 2011).

Olfactory-guided orientation was first observed in salmon (Hasler and Wisby 1951); their peripheral olfactory system becomes physiologically tuned to population-specific (Bett et al. 2015) and even artificially added (Rehnberg et al. 1985) home stream odours so that mature fish can find their natal river to spawn after several years in the open ocean (Dittman and Quinn 1996; Scholz et al. 1976). Although many of the larval reef fishes tested for sensory responses are about 1 cm long, the use of multiple sense has been demonstrated (Figure 2). Chemotaxis by larvae to different water masses from individual reefs and the presence of conspecifics have been demonstrated. However, olfactory cues are detectable only within a few kilometres off the reef, depending on the currents. Atema et al. (2002) and in more detail Gerlach et al. (2007) and Coppock et al. (2016) showed that different species of settling coral reef fish larvae could differentiate between the olfactory cues of open ocean and reefs and also preferred the odour of their natal reef versus other nearby reefs. The importance of conspecific odour was shown in adult cardinalfish (*Cheilodipterus quinquelineatus*), which preferred chemical cues of their own versus conspecifics of a foreign reef (Doving et al. 2006). Cardinalfish larvae did not change their olfactory preference for their home reef even when they were exposed to different reef odours indicating olfactory imprinting (Miller-Sims et al. 2011). It has also been demonstrated experimentally that potential settlers can detect conspecifics in microhabitats at night. Sweatman (1985) pumped the smell of conspecifics to experimental reefs and detected much higher settlement on these reefs when compared with controls. Coral reef fish can also determine and choose the quality of settlement habitat; therefore, cues could play an important role in directing pelagic larval stage fishes to a suitable reef on which they can settle (Lecchini et al. 2013). Imprinting on olfactory cues of the natal habitat was shown in different species of anemonefish that preferred their specific anemone (Arvedlund et al. 2000). These studies show that olfactory discrimination of different habitats and related conspecifics and olfactory imprinting can contribute to homing behaviour in fish.

To distinguish between the natal reef and other reefs, hatchlings would need to imprint on the scent of their natal reef before being flushed out of the reef by the current. Miller-Sims et al. (2011) showed that the cardinal fish *Ostorhinchus doederleini* did not change their olfactory preference for One Tree Island water even if they were kept in water from a different reef, namely, Heron Reef, for several days. Imprinting can therefore contribute to homing behaviour in fish.

The neuronal basis of olfactory imprinting close to hatching was studied in detail in zebrafish, which are much easier to manipulate during rearing than coral reef fishes (Gerlach et al. 2019). These studies show that larval zebrafish can imprint during a 24-hour period on day six post fertilization shortly after hatching on olfactory and visual cues of relatives with whom they share major histocompatibility alleles (MHC class II) (Hinz et al. 2012), which influences their olfactory social preferences later in life (Gerlach and Lysiak 2006). Using neuronal activity marker PERK (phosphorylated extracellular signal regulated kinase), Biechl et al. (2016) showed that crypt cells, specific olfactory sensory neurons in the olfactory epithelium, play a role in detecting kin odour in imprinted but not in non-imprinted zebrafish larvae. Zebrafish larvae responded to very low concentrations (10^{-12} M) of MHC peptides with increased neuronal activity as shown by Ca imaging studies (Gerlach et al. 2019).

17.2.3. Longer-Distance Orientation Capabilities

Although potential abilities such as sun compass orientation and magnetic responses were based on speculation (Kingsford et al. 2002; Leis et al. 2011), responses to the sun and magnetic field have been demonstrated in coral reef fish. Mouritsen et al. (2013) showed that juvenile cardinal fish *Ostorhinchus doederleini* can use a time-compensated sun compass to orient back towards their natal reef during the day. In circular tanks with a clear view to the sky, the fish oriented towards the southeast, opposing the main northwest drift current present around their home reef, and after the fish were clock-shifted for 6 h, they chose a northwesterly direction (a 180-degree change that fit with the change in sun azimuth over six hours at the testing location and time of year). These larval cardinalfish oriented to the sun, and orientation is strongest when the sun is low on the horizon (Mouritsen et al. 2013). Field tests using a DISC have also been undertaken at Lizard Island, where Leis et al. (2014) demonstrated strong orientation in the damselfish *Chromis atripectoralis*; it was concluded that orientation was best on sunny days and at lower angles of the sun, and groups of fish showed the same orientation as individuals.

Since *O. doederleini* larvae are presumed to feed in low light (Job and Bellwood 2000), orientation must be performed in low light. As for experiments on *Amphiprion ocellaris* (Schalm et al. 2021), juvenile fourline cardinalfish were placed in circular tanks inside Helmholtz coils during the night (Bottesch et al. 2016). In the unchanged, natural geomagnetic field, the fish oriented SE, which corresponded to the sun compass data observed during the day. When magnetic north was rotated 120 degrees clockwise, the fish adjusted their orientation accordingly. Both of these experiments on cardinal fish were conducted on the Southern GBR at One Tree Island where the south easterly orientation correlates with contranatant movement against the local residual currents. Similar conclusions were drawn by another group testing individual black-axil chromis larvae (*Chromis atripectoralis*) in their presettlement phase in the absence of visual cues (O'Connor and Muheim 2017). Individual larvae responded to a 90-degree shift of the horizontal component of the magnetic field within a Helmholtz coil with a comparable shift in orientation, demonstrating that they could use a magnetic compass for orientation. These examples show that coral reef fish larvae/juveniles can orient themselves to both the sun and magnetic cues, but we still do not know whether this orientation approach is exclusively based on an inherited clock-and-compass system or on a map-based strategy that requires previous imprinting, like the geomagnetic imprinting of natal areas shown in salmon (Putman et al. 2013) or turtles (Lohmann et al. 2008).

Although compass orientation provides directional information, which enables fish to maintain a consistent bearing, only map-based orientation would offer positional information. Since landmarks are not available in the open ocean, this would require a magnetic map like the one known by experienced birds (Chernetsov et al. 2017; Pakhomov et al. 2018; Wynn et al. 2022). Juvenile *O. doederleini* were tested regarding whether they have positional information available during their pelagic dispersal (Spiecker et al. 2023). Their orientation tendencies were tested before and after physical displacement from Sweetlip lagoon to One Tree Island, GBR by 180 kilometres. At the displacement site, the fish showed a swimming direction that was indistinguishable from the control directions that they demonstrated near the capture site, which suggests that coral reef fish, similar to songbirds during their initial migration, rely on innate or learned compass orientation but have no 'true navigation', that is, they do not change orientation direction to compensate for displacement. However, whether they imprint on the magnetic cues of their natal reefs and perform homing is not known.

17.3. SENSORY FIELDS

17.3.1. Spatial Scale of Cues

The spatial scale of cues and how senses are used will vary. This, in turn, is relevant to how the movement of larvae is modelled (Pineda et al. 2007; Sponaugle et al. 2002). We suggest that a likely hierarchy of responses would be magnetism, a sun compass and polarised light (Berenshtein et al. 2014), followed by attraction to preferred reefs through chemotaxis and sound, and finally, by visual observation of the reef (Igulu et al. 2011) and settlement at night (Kingsford 2001) to avoid the 'wall of mouths', and finally use chemotaxis to find a suitable habitat and the presence of conspecifics (Figure 2).

At spatial scales below 10 kilometres, reef noise and chemical cues can play roles. Reefs are noisy (Leis et al. 2002), but as a cue, the signal 'red noise' cannot be measured beyond the 'white noise' at distances beyond 4 km (Radford et al. 2011). However, the sensory abilities of fish may limit the range to a few hundred metres (Wright et al. 2010). It has been demonstrated that fish can detect the smell of individual reefs (Gerlach et al. 2007) and can imprint to their natal reefs (Arvedlund et al. 1999). Water masses directly linked to reefs, such as ebb-tide plumes, can be viewed at distances of up to five kilometres from reefs (Booth et al. 2000; Doherty et al. 1996). We even have some evidence that larvae can swim up currents to suitable settlement sites (Booth et al. 2000). Leis et al. (2011) argued that olfaction is current- and location-dependent and that fish must swim upstream to locate the source. Although larvae would have to follow a gradient of olfactory cues on the GBR reefs, they are generally subjected to reversing tidal current and phase eddies (Figure 3—Wolanski case2.avi). Fish may detect a plume, and if they do not swim directly to the reef, then they are likely to be swept to-and-fro past the reef multiple times; this, in turn, gives them the opportunity to stay near the reef until the time is right to settle. This may also mean that the swimming of fish larvae can be more relaxed rather than just swimming the critical speed defined in laboratory tests (Leis 2010). The flushing of plumes with the tide may generate a concentration-dependent odour map (*sensu* Atema et al. 2015). At the smaller spatial scales of metres to hundreds of metres, fish would be able to see the reef, respond to mosaics of sound and olfactory cues from habitat-forming organisms such as corals, microhabitats (Figure 4), conspecific fish, and even kin (Atema et al. 2015).

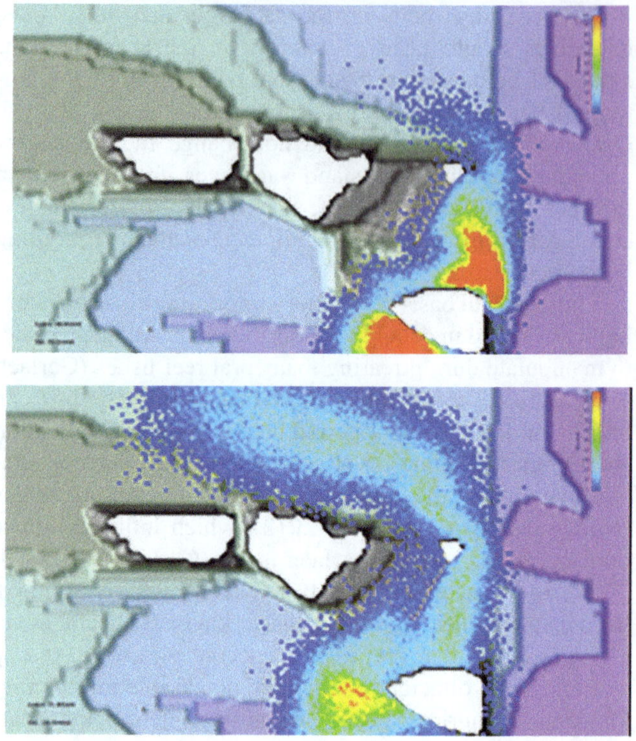

FIGURE 3 Reversing tides, the development of phase eddies, and a dispersal plume downstream of One Tree Island, GBR. Even passive particles can be advected back and forth near natal reefs and other reefs, but their longer-term fate will depend on the interaction with mainstream currents, which in this case, moves the plume of particles northward. The movement of passive particles can be viewed as a model from Animation 1. The following images are extracts from Animation 1 at 54 and 122 hours from the time of 'virtual spawning', ebb tide 54 hours from spawning, and flood tide 122 hours from spawning.

17.3.2. The Time Larvae Spend in Near Reef Waters

Given the mobility and sensory abilities of larval fish, the 'capture fields' used in biophysical models for the settlement of larvae must be considered. For example, larvae's functional attraction to reefs may often be greater than some of the distances that have been used to date (e.g., 1 to 2 km, Wolanski and Kingsford 2014). Attraction may have a number of steps as follows: (1) detection of the natal reef, or other reef with suitable habitats, swimming either as individual larvae or in groups (Kingsford and Tricklebank 1991); (2) aggregation near the reef of choice and a search for suitable habitat with conspecifics in which to settle; and (3) direct movement to settle, probably at night, and to settle in a broad habitat or microhabitat that may include conspecifics (e.g., Sweatman 1985).

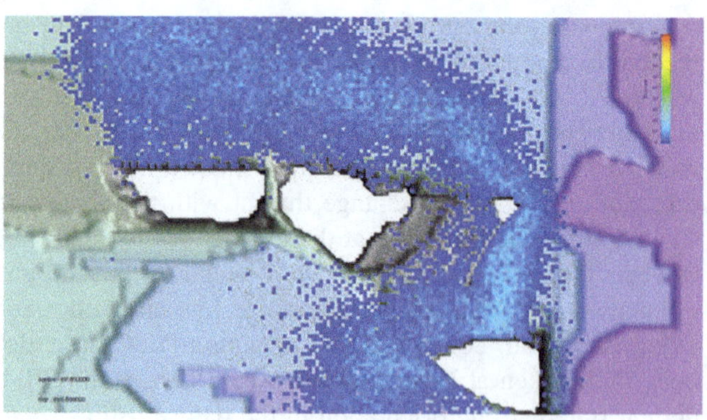

ANIMATION 1 Animation of the passive fish larva plume originating from two spawning sites at One Tree Island. The time is shown in hours next to 'dsp' on the bottom left. Model duration = 131 hours. Available online at *https://www.routledge.com/9781032340692*

There is an increasing body of evidence that fish larvae have a 'search phase' near reefs that may last hours to days. Firstly, above-average concentrations of late-stage larvae are found near reefs (Doherty et al. 1996), and phase eddies may enhance this accumulation (Burgess et al. 2007). Secondly, otoliths provide individual chronologies of growth based on the spacing of daily increments, and it has been shown in damselfishes that growth decreases within 4–6 days from settlement perhaps as fish reduce the time feeding in the quest to find a suitable settlement site. Thirdly, the chemistry of otoliths changes a few

FIGURE 4 Newly settled *Ostorhinchus doederleini* (~1 cm SL and ~20 d old, Kingsford et al. 2014) on an isolated patch reef of coral rubble that is overgrown with soft corals (*Xenia* sp.). Initially, many new settlers avoid adult and juvenile habitat on continuous reefs (Finn and Kingsford 1996).

days before settlement (Patterson et al. 2005); this may be due to physiological changes close to settlement rather than differences in water chemistry nearshore (Sih and Kingsford 2016). Fourthly, when fish larvae are placed inside a drifting *in situ* larval chamber, their behaviour changes when they are exposed to near-reef water masses (Paris et al. 2013). These findings, therefore, suggest that larvae are likely to be captured by reefs in a halo of cues up to five days prior to average PLD. Furthermore, the maintenance of position near a reef may be relatively easy where currents are reversing and phase eddies are forming (Burgess et al. 2007).

17.4. BIOPHYSICAL MODELLING

17.4.1. Biophysical Modelling and Determining Relevant Timeframes for Dispersal?

Scientists and managers require accurate biophysical models for reasons that include the following: effective marine planning (see Bode et al. 2024); ecological connectivity to determine variation in the size of reef-based populations and their resilience to changing conditions; the identification of source and sink reefs; the likelihood of successful larval dispersal among reefs from estimates of dispersal kernels (Bode et al. 2019; Buston et al. 2012; Jones 2015); how insular populations are and; the colonisation of reef following catastrophic events such as cyclones.

The use of oceanographic modelling to make predictions on the dispersal of fish larvae and connectivity among reefs and habitats has been used internationally (e.g., Cowen et al. 2000; Kobayashi 2006; Melbourne-Thomas et al. 2011; Treml et al. 2015) and on the GBR (Williams et al. 1984; Wolanksi et al. 1997; Andutta et al. 2012; Wolanski and Kingsford 2014) for decades. The reliability of oceanographic models depends on accurate timeframes that incorporate local currents. Currents can change through 'out-of-domain' physical forcing from major currents such as the East Australian Current (EAC), upwelling, storm events, eddy formation, and local winds (Wolanski et al. 2024). That is, there is a need to 'interview' successful recruits to understand the exact timeframes and physical conditions that they experience in the pelagic environment. Where physical conditions are modelled over inappropriate periods of time, inaccuracies of predicted dispersal patterns are likely to increase. The ability to age settlers to the day with daily increments in otoliths is a solution to this problem. For example, Schlaefer et al. (2018) sampled successful recruits of the snapper *Lutjanus carponatatus* and determined the exact time that fish spent in the plankton. It transpired that recruitment to One Tree Island was most likely when currents atypically alternated from north and south and from east to west, thus minimising the advection of larvae away from One Tree Island when they were at an early stage of development. Furthermore, when larval swimming was included in the model, the frequency of conditions with suitable currents explained over 29% of the variation in recruitment over 22 years. Similarly, in temperate waters, Shima and Swearer (2016) tracked the fate of individual cohorts of tripterygiid fish larvae and concluded that the temporal nature of oceanographic conditions influenced the fate of larvae. Accordingly, using generalised seasonal conditions in biophysical models may contribute greatly to inaccuracies in predicted patterns of dispersal.

17.4.2. Topography Influences Dispersal Trajectories

We have argued that the dispersal of fish larvae will be strongly influenced by swimming ability and orientation behaviour. However, the reef topography combined with level of flushing by currents will alter the outcomes of connectivity. In some cases where reef mosaics are complex, the result is decreased flushing that results in 'sticky water'. Accordingly, in these circumstances, even weak swimming larvae could find a reef or even the natal reef to settle. Andutta et al. (2012) tested the likely retention of particles in a range of geomorphological contexts in the GBR that ranged from archipelagos of small reefs with poor flushing to exposed ribbon reefs with high flushing. The self-seeding potential of even passive particles was shown to be high where the sticky water effect was also high (**Box 1**). This does not mean that all highly flushed reefs have low natal recruitment. Wolanski and Kingsford (2014) demonstrated that most drifting larvae of low mobility would indeed have a low probability of natal recruitment. In contrast, ~20% of larvae that could swim and had excellent sensory abilities could settle on their natal reefs, and approximately 8% could settle on other local reefs even when the level of flushing was high.

17.5. CURRENTS INFLUENCE DISPERSAL AND POPULATION GENETICS

There is growing evidence that fish larvae stay in groups during their pelagic phase. Firstly, there is great variation in the numbers of conspecifics collected in light traps and crest nets that are closely spaced. For example, the variation in catches of *Pomacentrus coelestis* from six traps off Fitzroy Reef in 2018 within 24 hours were [mean(range)SE], 16.5(5–31)4.4 per trap. Secondly, many larvae in the same trap were closely related as siblings or half-siblings (G. Gerlach unpubl. data). Thirdly, larvae of other species as small as 5-mm long have been observed swimming in small groups (Kingsford and Tricklebank 1991). Fourthly, small groups of gobies that have settled together are often more closely related than those among large groups (Rueger et al. 2021). Accordingly, rather than orienting to reefs as individuals, reef fish larvae may need to be modelled in groups where the integrated choices of orientation (Simpson et al. 2013) and potential group leaders might provide greater accuracy and improved levels of successful dispersal.

Population genetics can provide insight on interactions among currents, larval behaviour, and connectivity. Genetics can determine the robustness of reef-based populations and infer the likely paths of larval dispersal that could be tested with biophysical models. Gerlach et al. (2007) demonstrated very low connectivity for cardinal fish among reefs that were only separated by kilometres to tens of kilometres. Furthermore, some of these differences in 'genetic clades' were persistent over years (Gerlach et al. 2016). This corroborates with intergenerational tagging where settlement to natal reefs is often high (Jones 2015). That is, behaviour and favourable currents influence the level of recruitment to natal reefs. Major environmental perturbations, such as cyclones, have allowed the identification of source reefs of the damselfish *Pomacentrus coelestis* and a demonstration with a sufficiently large spawning stock, the larvae colonization of nearby and down current reefs can occur incrementally over years (Gerlach et al. 2021). The 'take over' by a dominant clade does not happen quickly on down current reefs, which suggests that the long tail of dispersal can be slow. The speed of such take overs also depends on the life span of the species. At the GBR, based on otolith ring counts, *O. doederleini* live approximately for a year (Kingsford et al. 2014), and *P. coelestis* have even shorter lives; their age maxima varied between 126 and 160 days (Kingsford et al. 2017). When disturbances such as cyclones alter habitats or currents, they prevent natal homing of the recruits, and the parents have no chance to produce larval batches in the next year. Accordingly, successful settlers from a different reef origin will take over empty habitat. This take over by settlers from foreign reefs can only be detected if they are of a different genotype. Overall, this has led to changes in the genetic pattern of a reef population as we observed in our long-term study on *O. doederleini* and *P. coelestis* (Gerlach et al. 2016; Gerlach et al. 2021). Such changes in the genetic population dynamics are too rapid to be driven by selection for adaptations to specific environmental conditions. They can be interpreted as 'genetic patchiness' among populations, which has also been observed in other fishes (Selwyn et al. 2016; Selkoe et al. 2006; Pujolar et al. 2006). This idea states that a large variance in reproductive success results in the surviving offspring exhibiting some of the characteristics of bottleneck populations, namely, that larvae are derived from only a small fraction of the adult population (Selkoe et al. 2010). We extend this idea to suggest that variance in the recruitment success of different genotypes that find a reef by chance or through good orientation can also lead to such 'genetic patchiness'. If such a small group of relatives, which thus consists of similar genotypes, settles together, then the genetic differences from other groups living there are due to the different degree of relatedness. This leads to patchiness of different genotypes within the reef or it even allows the spread of a new genotype over the whole area (the long tail of the dispersal kernel, Jones, 2015), especially when disturbances have created new habitats. The survival of immigrants may also be influenced by predation and intra- and interspecific interactions with resident fishes.

17.6. BIOPHYSICAL MODELLING: WHERE TO GO FROM HERE?

Accordingly, biophysical models are excellent for hypothesis generation, and some can make accurate predictions of high utility. Oceanographic models have become increasingly sophisticated, and there is often a high degree of alignment between the predictions of the models and the field measurements of current velocity and eddy formation (Delandmeter et al. 2017; Wolanski and Lambrechts 2020). A potential concern regarding the input of fish behaviour to biophysical models is that the 'rules' for fish larvae behaviour are typically simple. For example, distances of 2 to 25 kilometres from a target reef have been used to 'capture' larvae as having successfully dispersed to settlement (e.g., Kobayashi 2006; Melbourne-Thomas et al. 2011; Wolanski and Kingsford 2014). Mosaics of reefs that are typical of the GBR are often spaced at hundreds of metres to kilometres; therefore, the probability of detection of a reef must be improved, and capture distances would have to be altered in models.

Advances in the accuracy of biophysical modelling require the incorporation of more sophisticated larval parameters. Larvae have at least six senses that we know of (Figure 2), and these senses will influence the orientation of fish with distance from the targeted reef. Furthermore, currents and turbulence may be unkind to larvae that are trying to maintain a favourable orientation for settlement. How quickly larvae can reorientate to the 'correct path' and how important group interactions are where the average decision making of the group is better than that of individuals (Simpson et al. 2013) still requires further study. Estimates of detection distances are still in their infancy and are rarely based on measurement in the field (but see Radford et al. 2011 and Wright et al. 2010) for sound or observations of chemically rich plumes from reefs that can stretch kilometres on ebb tides (Booth et al. 2000). The altitude of the sun and the Earth's magnetic field are constant, and fish respond to these cues, but tests have only been performed with larvae near the end of their PLD.

Importantly, the physical conditions that are relevant to larval dispersal can be fine-tuned by examining successful settlers. Settlers contain information on date of hatching and PLD, and the spacing of increments in otoliths provide a proxy for how well larvae have fed on the plankton. Accordingly, models can be refined to the conditions experienced during the pelagic stage and validated with field measurements. The resolution of dispersal distances and the calculation of dispersal kernels are now greatly enhanced with the benefits of parentage analysis (e.g., Bode et al. 2019) and intergenerational tagging (Almany et al. 2007; Jones 2015). This type of information could also enable Bayesian priors in models that limit dispersal to measured distances.

Unstructured mesh-based oceanographic modelling allows for multiple mesh sizes that can best deal with the complexity of near-shore currents and complex eddies (Wolanski and Lambrechts 2020). Because the water column of the GBR is generally well-mixed, this has allowed for accurate 2D estimates of dispersal. However, the development of these models to incorporate a 3D understanding of the water column is required. For example, close to the mainland, adjacent to the GBR, stratification of the water column is not unusual (Figure 5). Potentially, therefore, this stratification could facilitate retention of larvae through larval 'ratcheting' as observed in, for example, prawns and flatfish. In addition, the 3D currents generated by wind-driven Langmuir cells and the salient topography of reefs, even in the absence of a stratification, can prevent the dispersion of buoyant larvae; indeed, they aggregate buoyant larvae in slick lines (Wolanski and Hamner 1988; Kingsford et al. 1991). This small-scale process needs to be incorporated in the much larger-scale 3D models of GBR oceanography.

There are many reasons why scientists and managers want accurate biophysical models that are suited to local conditions. There is great diversity not only in the configuration of reefs on the GBR (e.g., Box 1) and in other tropical locations of the world but also in the currents that affect them. The geomorphology differs greatly, from isolated reefs to complex mosaics of reefs (Hopley and Smithers 2019). Accordingly, patterns of dispersal will vary with the configuration of reefs, tidal amplitude, the nature of mainstream currents, and how they change over time. Furthermore, the mobility and sensory abilities of larvae clearly influence the nature of dispersal and how larvae near reefs should be captured in models. The most complex models would be required to determine how populations vary over time and space as this requires information on larval production, oceanography, larval behaviour, and larvae survivorship that may depend on their dispersal pathway and conditions that they experience in different water masses. Although we have not yet tagged and tracked larvae, oceanographic and behavioural models are increasingly accurate, and the ability to check their predictions has improved greatly with intergenerational tagging and high-resolution genetics.

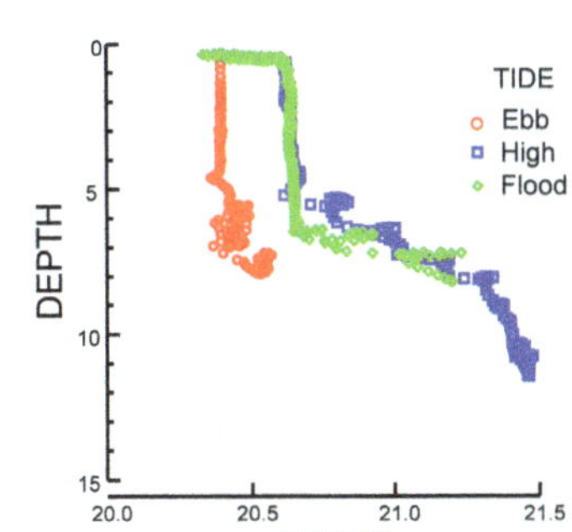

FIGURE 5 A depth profile of water density (Sigma-t) in an 8–12 m water column near fringing reefs of Magnetic Island, Inner GBR, March 5, 2022. Clear stratification of water density can be seen with a pycnocline at 5 to 8 m regardless of the state of the tide. Sigma-T is a measure of water density, which is a combination (~proportion of the relationship) of temperature (0.75) and salinity (0.25). Warm, lower salinity water is found near the surface.

BOX 1.

Sticky water effect on connectivity (Andutta et al. 2012). The movement of water becomes more restricted and the dispersal of passive particles slows as the density of reefs increases and between narrow reef channels. In this study, passive particles were released in 10 ×10 km cells, and the percentage remaining in each cell was determined over 25 days. Scenario A—dense reef mosaic; B and C—increasingly spaced reefs and shelf break was where the particles were released in unrestricted mainstream flow. Reefs were on the GBR in the Swains group 21–22°S. Green boxes are 10 × 10 km.

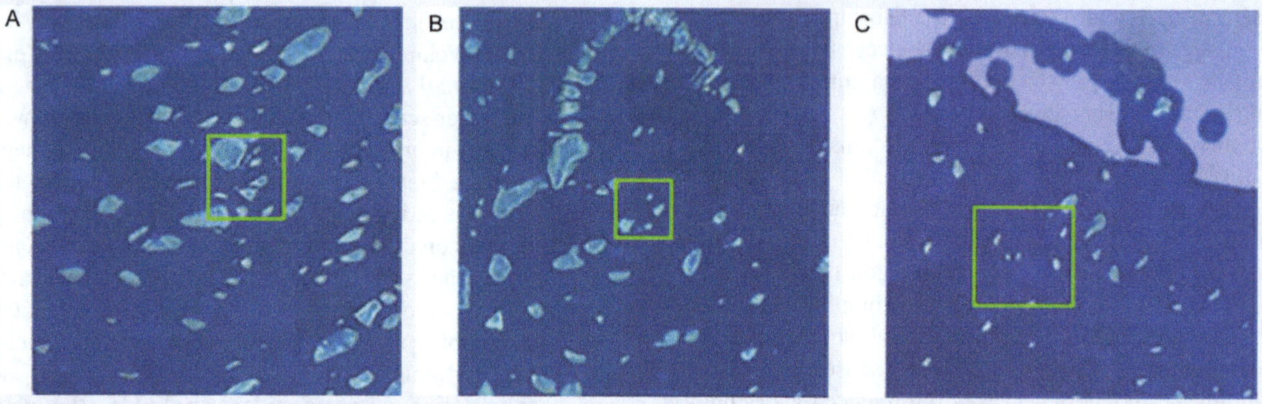

Percentage Retained

Scenario	10 days	20 days
A	75	38

Percentage Retained

Scenario	10 days	20 days
B	22	6

Percentage Retained

Scenario	10 days	20 days
C	3	< 0.5

Percentage Retained (no image provided)

Scenario	10 day	20 days
Shelf Break	< 0.1	< 0.1

REFERENCES

Almany GR, Berumen ML, Thorrold SR, Planes S, Jones GP (2007) Local replenishment of coral reef fish populations in a marine reserve. Science 316:742–744

Andutta FP, Kingsford MJ, Wolanski E (2012) 'Sticky water' enables the retention of larvae in a reef mosaic. Estuarine, Coastal and Shelf Science 101:54–63

Arvedlund M, Bundgaard I, Nielsen LE (2000) Host Imprinting in Anemonefishes (Pisces: Pomacentridae): Does it Dictate Spawning site Preferences? Environmental Biology of Fishes 58:203–213

Arvedlund M, McCormick MI, Fautin DG, Bildsoe M (1999) Host recognition and possible imprinting in the anemonefish Amphiprion melanopus (Pisces: Pomacentridae). Marine Ecology Progress Series 188:207–218

Atema J, Gerlach G, Paris CB (2015) Sensory biology and navigation behaviour of reef fish larvae. In: Moro C (ed) Ecology of fishes on coral reefs. Cambridge University Press, Cambridge, pp. 3–15

Atema J, Kingsford MJ, Gerlach G (2002) Larval reef fish could use odour for detection, retention and orientation to reefs. Marine Ecology Progress Series 241:151–160

Berenshtein I, Kiflawi M, Shashar N, Wieler U, Agiv H, Paris CB (2014) Polarized light sensitivity and orientation in coral reef fish post-larvae. PloS One 9:e88468

Bett NN, Hinch SG (2015) Attraction of migrating adult sockeye salmon to conspecifics in the absence of natal chemical cues. Behavioral Ecology 26:1180–1187

Biechl D, Tietje K, Gerlach G, Wullimann MF (2016) Crypt cells are involved in kin recognition in larval zebrafish. Science Reports 6:24590

Blaxter JHS (1986) Development of sense organs and behaviour of teleost larvae with special reference to feeding and predation avoidance. Transactions of the American Fisheries Society 115:98–114

Bode M, Leis JM, Mason LB, Williamson DH, Harrison HB, Choukroun S, Jones GP (2019) Successful validation of a larval dispersal model using genetic parentage data. PLoS Biology 17:e3000380

Bode M, Stewart O, Choukroun S (2024) Incorporating biophysical larval dispersal simulations into coral reef conservation decision-making. In Wolanski E, Kingsford MJ (eds.), Oceanographic Processes of Coral Reefs. Physical and Biological Links in the Great Barrier Reef. CRC Press, Boca Raton, FL

Booth DJ, Kingsford MJ, Doherty PJ, Berretta GA (2000) Recruitment of damselfishes in One Tree Island lagoon: Persistent interannual spatial patterns. Marine Ecology Progress Series 202:212–230

Bottesch M, Gerlach G, Halbach M, Bally A, Kingsford MJ, Mouritsen H (2016) Magnetic compass orientation helps coral reef fish larvae return to their natal reef. Current Biology December 19, 2016 R1267:R1247–R1271

Brothers EB, Williams DM, Sale PF (1983) Length of larval life in twelve families of fishes at "One Tree Lagoon", Great Barrier Reef, Australia. Marine Biology 76:319–324

Burgess SC, Kingsford MJ, Black KP (2007) Influence of tidal eddies and wind on the distribution of presettlement fishes around One Tree Island, Great Barrier Reef. Marine Ecology Progress Series 341:233–242

Buston PM, Jones GP, Planes S, Thorrold SR (2012) Probability of successful larval dispersal declines fivefold over 1 km in a coral reef fish. Proceedings of the Royal Society B: Biological Sciences 279:1883–1888

Chernetsov N, Pakhomov A, Kobylkov D, Kishkinev D, Holland RA, Mouritsen H (2017) Migratory Eurasian reed warblers can use magnetic declination to solve the longitude problem. Current Biology, 27:2647

Choat JH, Bellwood DR (1991) Reef fishes: Their history and evolution. In: Sale PF (ed) The Ecology of fishes on coral reefs. Academic Press, San Diego, pp39–66

Coppock AG, Gardiner NM, Jones GP (2016) Olfactory responses of coral-reef fishes to coral degradation and crown-of-thorns (*Acanthaster planci*). Marine and Freshwater Research 67:605–611

Cowen RK, Lwiza KMM, Sponaugle S, Paris CB, Olson DB (2000) Connectivity of marine population: Open or closed? Science 287:857–859

Davis AR (1987) Variation in recruitment of the subtidal colonial ascidian *Podoclavella cylindrica* (Quoy & Gaimard): The role of substratum choice and early survival. Journal of Experimental Marine Biology and Ecology 106:57–71

Delandmeter P, Lambrechts J, Marmorino GO, Legat V, Wolanski E, Remacle J-F, Chen W, Deleersnijder É (2017) Submesoscale tidal eddies in the wake of coral islands and reefs: Satellite data and numerical modelling. Ocean Dynamics 67:897–913

Dittman AH, Quinn TP (1996) Homing in Pacific Salmon: Mechanisms and ecological basis. The Journal of Experimental Biology 199:83–91

Doherty PJ (1987) Light-traps: Selective but useful devices for quantifying the distributions and abundance of larval fishes. Bulletin of Marine Science 41:423–431

Doherty PJ, Kingsford MJ, Booth D, Carleton J (1996) Habitat selection before settlement by *Pomacentrus coelestis*. Marine and Freshwater Research 47:391–399

Doving KB, Stabell OB, Ostlund-Nilsson S, Fisher R (2006) Site fidelity and homing in tropical coral reef cardinalfish: Are they using olfactory cues? Chemical Senses 31:265–272

Dufour V, Galzin R (1993) Colonisation patterns of reef fish larvae to the lagoon at Moorea Island, French Polynesia. Marine Ecology Progress Series 102:143–152

Finn MD, Kingsford MJ (1996) Two-phase recruitment of a coral reef fish (Pisces: Apogonidae) on the Great Barrier Reef. Marine and Freshwater Research 47:423–432

Fisher R, Bellwood DR (2001) Effects of feeding on the sustained swimming abilities of late-stage larval *Amphiprion melanopus*. Coral Reefs 20:151–154

Fisher R, Bellwood DR, Job SD (2000) Development of swimming abilities in reef fish larvae. Marine Ecology Progress Series 202:163–173

Gerlach G, Atema J, Kingsford MJ, Black KP, Miller-Sims V (2007) Smelling home can prevent dispersal of reef fish larvae. PNAS 104:858–863

Gerlach G, Atema J, Raupach MJ, Deister F, Muller A, Kingsford MJ (2016) Cryptic species of cardinalfish with evidence for old and new divergence. Coral Reefs 35:437–450

Gerlach G, Kraemer P, Weist P, Eickelmann L, Kingsford MJ (2021) Impact of cyclones on hard coral and metapopulation structure, connectivity and genetic diversity of coral reef fish. Coral Reefs 40:999–1011

Gerlach G, Lysiak N (2006) Kin recognition and inbreeding avoidance in zebrafish, Danio rerio, is based on phenotype matching. Animal Behaviour 71:1371–1377

Gerlach G, Tietje K, Biechl D, Namekawa I, Schalm G, Sulmann A (2019) Behavioural and neuronal basis of olfactory imprinting and kin recognition in larval fish. The Journal of Experimental Biology 222:jeb189746

Hasler AD, Wisby WJ (1951) Discrimination of stream odors by fishes and relation to parent stream behavior. American Natural 85:223–238

Hinz C, Gebhardt K, Hartmann AK, Sigman L, Gerlach G (2012) Influence of kinship and MHC class II genotype on visual traits in zebrafish larvae (*Danio rerio*). PloS One 7:e5118210.5118137

Hopley D, Smithers S (2019) Geomorphology of coral reefs with special reference to the Great Barrier Reef. In: Hutchings P, Kingsford M, Hoegh-Guldberg O (eds) The Great Barrier Reef: Biology, Environment and Management. CRC Press, Boca Raton, FL, pp. 9–24

Igulu MM, Nagelkerken I, Fraaije R, R, Ligtenberg H, Mgaya YD (2011) The potential role of visual cues for microhabitat selection during the early life phase of a coral reef fish (*Lutjanus fulviflamma*). Journal of Experimental Marine Biology and Ecology 401:118–125

Job S, Bellwood DR (2000) Light sensitivity in larval fishes: Implications for vertical zonation in the pelagic zone. Limnology and Oceanography 45:362–371

Jones GP (2015) Mission impossible: Unlocking the secrets of coral reef fish dispersal. In: Mora C (ed) Ecology of fishes on coral reefs. Cambridge University Press, Cambridge.

Jones GP, Milicich MJ, Emslie MJ, Lunow C (1999) Self-recruitment in a coral reef fish population. Nature 402:802–804

Kingsford MJ (1988) The early life history of fish in coastal waters of northern New Zealand: A review. New Zealand Journal of Marine and Freshwater Research 22:463–479

Kingsford MJ (1992) Drift algae and small fish in coastal waters of northeastern New Zealand. Marine Ecology Progress Series 80:41–55

Kingsford MJ (2001) Diel patterns of arrival of reef fish to a coral reef: One Tree Island, Great Barrier Reef. Marine Biology 138:853–867

Kingsford MJ, Choat JH (1985) The Fauna Associated with Drift Algae Captured with a Plankton-Mesh Purse Seine Net. Limnology and Oceanography 30:618–630

Kingsford MJ, Finn MD, O'Callaghan MD, Atema J, Gerlach G (2014) Planktonic Larval Duration, age and growth of *Ostorhinchus doederleini* (Pisces: Apogonidae) on the southern Great Barrier Reef, Australia. Marine Biology 161:245–259

Kingsford M, Leis J, Shanks A, Lindeman K, Morgan S, Pineda J (2002) Sensory environments, larval abilities and local self-recruitment. Bulletin of Marine Science 70:309–340

Kingsford MJ, O'Callaghan MD, Liggins L, Gerlach G (2017) The short-lived neon damsel *Pomacentrus coelestis*: Implications for population dynamics. Journal of Fish Biology 90:2041–2059

Kingsford MJ, Tricklebank KA (1991) Ontogeny and behavior of (*Aldrichetta forsteri*) (Teleostei: Mugilidae). Copeia:9–16

Kingsford MJ, Wolanski E, Choat JH (1991) Influence of tidally induced fronts and langmuir circulations on distribution and movements of presettlement fishes around a coral reef. Marine Biology 109:167–180

Kobayashi DR (2006) Colonization of the Hawaiian Archipelago via Johnstone Atoll: A characterization of oceanographic transport corridors for pelagic larvae using computer simulation. Coral Reefs 25:407–417

Lecchini D, Waqalevu VP, Parmentier E, Radford CA, Banaigs B (2013) Fish larvae prefer coral over algal water cues: Implications of coral reef degradation. Marine Ecology Progress Series 475:303–307

Leis J, Rennis DS (1983) The larvae of Indo-Pacific coral reef fishes. New South Wales University Press, New South Wales

Leis JM (1986) Vertical and horizontal distribution of fish larvae near coral reefs at Lizard Island, Great Barrier Reef. Marine Biology 90:505–516

Leis JM (2007) Behaviour as input for modelling dispersal of fish larvae: Behaviour, biogeography, hydrodynamics, ontogeny, physiology and phylogeny meet hydrology. Marine Ecology Progress Series 347:185–193

Leis JM (2010) Ontogeny of behaviour in larvae of marine demersal fishes. Ichthyological Research 57:325–342

Leis JM, Carson-Ewart BM (1997) In situ swimming speeds of the late pelagic larvae of some Indo-Pacific coral-reef fishes. Marine Ecology Progress Series 159:165–174

Leis JM, Carson-Ewart BM, Cato DH (2002) Sound detection *in situ* by the larvae of a coral reef damselfish (Pomacentridae). Marine Ecology Progress Series 232:259–268

Leis JM, Hay AC, Howarth GJ (2009) Ontogeny of in situ behaviours relevant to dispersal and population connectivity in larvae of coral-reef fishes. Marine Ecology Progress Series 379:163–179

Leis JM, Paris CB, Irisson JO, Yerman MN, Siebeck UE (2014) Orientation of fish larvae in situ is consistent among locations, years and methods, but varies with time of day. Marine Ecology Progress Series 505:193–208

Leis JM, Siebeck U, Dixson DL (2011) How Nemo Finds Home: The Neuroecology of Dispersal and of Population Connectivity in Larvae of Marine Fishes. Integrative and Comparative Biology 51:826–843

Lohmann KJ, Putman NF, Lohmann CMF (2008) Geomagnetic imprinting: A unifying hypothesis of long-distance natal homing in salmon and sea turtles. Proceedings of the National Academy of Sciences of the United States of America 105:19096–19101

Melbourne-Thomas J, Johnson CR, Aliño PM, Geronimo RC, Villanoy CL, Gurney GG (2011) A multi-scale biophysical model to inform regional management of coral reefs in the western Philippines and South China Sea. Environmental Modelling & Software 26:66–82

Miller-Sims VC, Atema J, Gerlach G, Kingsford MJ (2011) How stable are the reef odor preferences of settling reef fish larvae? Marine and Freshwater Behaviour and Physiology 44:133–141

Mouritsen H, Atema J, Kingsford MJ, Gerlach G. (2013). Sun compass orientation helps coral reef fish larvae return to their natal reef. PLoS One 8(6):1–9

O'Connor J, Muheim, R (2017) Pre-settlement coral-reef fish larvae respond to magnetic field changes during the day. The Journal of Experimental Biology 220:2874–2877

Pakhomov A, Anashina A, Heyers D, Kobylkov D, Mouritsen H, Chernetsov N (2018) Magnetic map navigation in a migratory songbird requires trigeminal input. Scientific Reports 8.

Paris CB, Atema J, Irisson J-O, Kingsford M, Gerlach G, Guigand CM (2013) Reef odor: A wake up call for navigation in reef fish larvae. Plos One 8:e72808–e72808

Patterson HM, Kingsford MJ, McCulloch MT (2005) Resolution of the early life history of a reef fish using otolith chemistry. Coral Reefs 24:222–229

Pineda J, Hare JA, Sponaugle S (2007) Larval transport and dispersal in the coastal ocean and consequences for population connectivity. Oceanography 20:22–39

Pujolar JM, Maes GE, Volckaert FAM (2006) Genetic patchiness among recruits in the European eel Anguilla anguilla. Marine Ecology Progress Series 307:209–217

Putman NF, Lohmann KJ, Putman EM, Quinn TP, Klimley AP, Noakes DLG (2013) Evidence for geomagnetic imprinting as a homing mechanism in Pacific salmon. Current Biology 23:312–316

Radford CA, Stanley JA, Simpson SD, Jeffs AG (2011) Juvenile coral reef fish use sound to locate habitats. Coral Reefs 30:295–305

Rehnberg BG, Curtis L, Schreck CB (1985) Homing of coho salmon (*Oncorhynchus kisutch*) exposed to morpholine. Aquaculture 44:253–255

Rueger T, Buston PM, Bogdanowicz SM, Wong MY (2021) Genetic relatedness in social groups of the emerald coral goby Paragobiodon xanthosoma creates potential for weak kin selection. Molecular Ecology 30:1311–1321

Schalm G, Bruns K, Drachenberg N, Geyer N, Foulkes NS, Bertolucci C, Gerlach G (2021) Finding Nemo's clock reveals switch from nocturnal to diurnal activity. Scientific Reports 11:6801

Schlaefer JA, Wolanski E, Lambrechts J, Kingsford MJ (2018) Wind conditions on the Great Barrier Reef influenced the recruitment of snapper (*Lutjanus carponotatus*). Frontiers in Marine Science 5:Article 193 doi: 110.3389/fmars.2018.00193

Scheltema RS (1988) Initial evidence for the transport of teleplanic larvae of benthic invertebrates across the east Pacific barrier. Biological Bulletin 174:145–156

Scholz AT, Horrall RM, Cooper JC, Hasler AD (1976) Imprinting to chemical cues: The basis for home stream selection in salmon. Science 192:1247–1249

Selkoe KA, Gaines SD, Caselle JE, Warner RR (2006) Current shifts and kin aggregation explain genetic patchiness in fish recruits. Ecology 87:3082–3094

Selkoe KA, Watson JR, White C, Ben Horin T, Iacchei M, Mitarai S, Siegel DA, Gaines SD, Toonen RJ (2010) Taking the chaos out of genetic patchiness: Seascape genetics reveals ecological and oceanographic drivers of genetic patterns in three temperate reef species. Molecular Ecology 19:3708–3726

Selwyn JD, Hogan JD, Downey-Wall AM, Gurski LM, Portnoy DS, Heath DD (2016) Kin-aggregations explain chaotic genetic patchiness, a commonly observed genetic pattern in a marine fish. PloS One 11

Shima JS, Swearer SE (2016) Evidence and population consequences of shared larval dispersal histories in a marine fish. Ecology 97:25–31

Sih TL, Kingsford MJ (2016) Near-reef elemental signals in the otoliths of settling *Pomacentrus amboinensis* (Pomacentridae) Coral Reefs 35:303–315

Simpson S, Piercy JJB, King J, Codling EA (2013) Modelling larval dispersal and behaviour of coral reef fishes. Ecological Complexity 16:68–76

Simpson SD, Meekan M, Montgomery J, McCauley R, Jeffs A (2005) Homeward Sound. Science 308:221

Spiecker L, Curdt F, Bally A, Janzen N, Kraemer P, Leberecht B, Kingsford MJ, Mouritsen H, Winklhoefer M, Gerlach G. (2023) Coral reef fish larvae show no evidence for map-based navigation after physical displacement. iScience 26(6):106950

Sponaugle S, Cowen RK, Shanks A, Morgan SG, Leis JM, Pineda J, Boehlert GW, Kingsford MJ, Lindeman K, Grimes C, Munro JL (2002) Predicting self-recruitment in marine populations: Biophysical correlates and mechanisms. Bulletin of Marine Science Special Publications 70:341–375

Stobutzki IC, Bellwood DR (1994) An analysis of sustained swimming abilities of pre- and postsettlement coral reef fishes. Journal of Experimental Marine Biology and Ecology 175:275–286

Stobutzki IC, Bellwood DR (1997) Sustained swimming abilities of the late pelagic stages of coral reef fishes. Marine Ecology Progress Series 149:35–41

Swearer SE, Caselle JE, Lea DW, Warner RR (1999) Larval retention and recruitment in an island population of a coral-reef fish. Nature 402:799–802

Sweatman HPA (1985) The influence of adults on some coral reef fishes on larval recruitment. Ecological Monographs 55:469–485

Thorpe JE (1988) Salmon migration. Science Program Oxford 72:345–370

Thorrold SR, Jones GP, Planes S, Hare JA (2006) Transgenerational marking of embryonic otoliths in marine fishes using barium stable isotopes. Canadian Journal of Fisheries and Aquatic Sciences 63:1193–1197

Tolimieri N, Jeffs A, Montgomery JC (2000) Ambient sound as a cue for navigation by pelagic larvae of reef fishes. Marine Ecology Progress Series 207:219–224

Treml EA, Ford JR, Black KP, Swearer SE (2015) Identifying the key biophysical drivers, connectivity outcomes, and metapopulation consequences of larval dispersal in the sea. Movement Ecology 3:17

Victor BC (1986) Duration of the planktonic larval stage of one hundred species of Pacific and Atlantic wrasses (family Labridae). Marine Biology 90:317–326

Victor BC (1991) Settlement strategies and biogeography of reef fishes. In: Sale PF (ed) The Ecology of Fishes on Coral Reefs. Academic Press, San Diego, pp231–260

Williams DM, Wolanski E, Andrews JC (1984) Transport mechanisms and the potential movement of planktonic larvae in the central region of the Great Barrier Reef. Coral Reefs 3:229–236

Wolanski E (2001) Oceanographic Processes of Coral Reefs: Physical and Biological Links in the Great Barrier Reef. CRC Press, Boca Raton, FL

Wolanski E, Doherty P, Carleton J (1997) Directional swimming of fish larvae determines connectivity of fish populations on the Great Barrier Reef. Naturwissenschaften 84:262–268

Wolanski E, Hamner WH (1988) Topographically controlled fronts in the ocean and their biological influence. Science 241:177–181

Wolanski E, Kingsford MJ (2014) Oceanographic and behavioural assumptions in models of the fate of coral and coral reef fish larvae. J Roy Soc Int 11:1–12

Wolanski E, Kingsford MJ, Lambrechts J, Marmorino GO (2024) The Physical Oceanography of the Great Barrier Reef: A review. In Wolanski E, Kingsford MJ (eds.), Oceanographic Processes of Coral Reefs. Physical and Biological Links in the Great Barrier Reef. CRC Press, Boca Raton, FL

Wolanski E, Lambrechts J (2020) The net water circulation in the far Northern Great Barrier Reef. Estuarine, Coastal and Shelf Science 235:106569

Wright KJ, Higgs D, Cato DH, Leis JM (2010) Auditory sensitivity in settlement-stage larvae of coral reef fishes. Coral Reefs 29:235–243

Wynn J, Padget O, Mouritsen H, Morford J, Jaggers P, Guilford T (2022) Magnetic stop signs signal a European songbird's arrival at the breeding site after migration. Science 375:446–449.

18 Incorporating Biophysical Larval Dispersal Simulations into Coral Reef Conservation Decision-Making

Michael Bode, Owen Stewart, and Severine M. Choukroun

18.1. INTRODUCTION

Conservation resources are always constrained (Wilson et al. 2006; Bode et al. 2008; McCarthy et al. 2012); decision-makers therefore need to identify which subset of locations and actions will deliver the highest return on investment (Wilson et al. 2006; Murdoch et al. 2007). Since ecosystems are highly spatiotemporally heterogeneous (Levin 1992), optimal conservation actions require an understanding of how ecosystems vary in space and time.

The heterogeneity of marine ecosystems is principally driven by changes in the physical and environmental characteristics of the ocean, which are just as variable as those of terrestrial systems (Woolley et al. 2016). However, marine ecosystems are also subject to the unique added complexity of ocean currents, which can alter thermal and chemical conditions, and disperse organisms over very large distances (Miller 2009; Jones et al. 1999; Almany et al. 2007, 2017; D'Aloia et al. 2015). Marine conservation, including coral reef conservation, therefore has to consider variation in the local environment and the variation driven by the interaction between ecological and oceanographic processes (Almany et al. 2013; Williamson et al. 2016; Bode et al. 2018).

A key connection between oceanography and coral reef ecology is the process of pelagic larval dispersal (Jones 2015; Almany et al. 2017). Most coral reef organisms have a two-stage lifecycle, with a benthic adult phase and a pelagic larval phase. Since reproduction and subsequent recruitment depend on successful larval dispersal, dispersal plays a critical role in driving population dynamics and governing the distributions of coral reef species (Leis 1991). As a consequence, larval dispersal is enormously important for coral reef conservation and can determine the success or failure of conservation actions.

18.2. BIOPHYSICAL LARVAL DISPERSAL SIMULATION MODELS

Unfortunately, the nature of pelagic larval dispersal makes it extremely difficult to empirically measure. Pelagic larvae are tiny, often centimetres or less in length, and must be tracked across distances measured in kilometres, even hundreds of kilometres (Williamson et al. 2016; Almany et al. 2017). In a large system of coral reefs like the Great Barrier Reef (GBR), a single reef could produce millions of larvae, and the resulting dispersal patterns can create millions of inter-reef linkages. Even with startling recent advances in genetic techniques of mark-recapture and assignment (Jones et al. 1999; Swearer et al. 1999; Planes et al. 2009; Bode et al. 2019), it is impossible to empirically measure larval dispersal patterns at the scale and accuracy required to support coral reef management decisions. Two methods are commonly used to extrapolate limited empirical data to predict both unobserved dispersal connections and future dispersal patterns.

The first method is to fit "dispersal kernels" to observed dispersal data. This method proposes a parametric relationship between distance and dispersal strength (Buston et al. 2012; Almany et al. 2013; D'Aloia et al. 2015; Jones 2015; Bode et al. 2017), which is then extrapolated across space and time. Dispersal kernels provide useful insights into the declining strengths of connections with distance but cannot capture the fundamentally anisotropic and advective nature of hydrodynamics. The second method uses "biophysical" simulation models, which contain both biological and physical oceanographic components, to estimate large-scale dispersal patterns on coral reefs (Dight et al. 1990a, 1990b; Cowen et al. 2000; James et al. 2002).

Biophysical models are based on hydrodynamic models—high-resolution regional hydrodynamic models that simulate oceanic flow conditions within the reef-shelf complex (Figure 1). These models are used to simulate currents throughout a period of spawning and dispersal. For example, in the southern GBR, populations of the leopard coral grouper *Plectropomus leopardus* spawn throughout the entire year but primarily in the austral spring-summer (Samoilys 1997). The larvae then spend approximately a month dispersing (Figure 2). To understand the annual larval dispersal pattern of this population for any given year, it is therefore necessary to model the east Australian ocean currents across a 13-month period (Bode et al. 2019).

Once ocean currents have been generated by the hydrodynamic model, the resulting flow field is used as an input into an agent-based biological model to simulate the transport of the pelagic larvae. A very large number of particles—often hundreds of millions—are released from the reef habitat and allowed to move with the ocean currents. Each particle represents either a single larvae or a

DOI: 10.1201/9781003320425-21

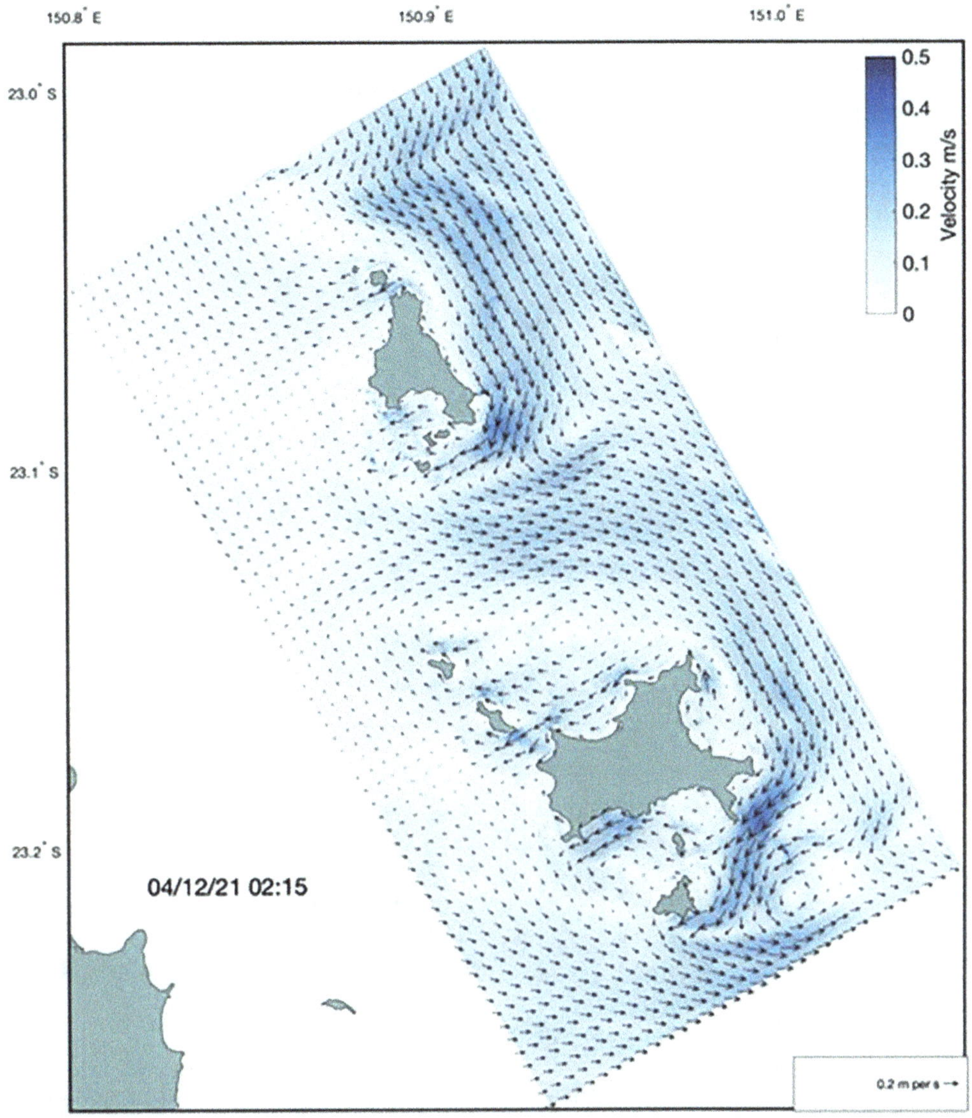

FIGURE 1 A snapshot of simulated hydrodynamic conditions around the Keppel Island group in the southern GBR, at 70-m resolution. Colours indicate the magnitude of the velocity, and arrows indicate the vector direction.

small group of larvae, and they begin their dispersal as passive particles, with specific buoyancy (e.g., coral or fish eggs). As time passes, these particles begin to change their behaviour, based on our (often limited) knowledge of larval ontogeny (Leis 2007; Bode et al. 2019). As they move across the ocean, their buoyancy changes, their swimming abilities strengthen, and newly developed sensory faculties start to guide the larvae towards suitable settlement habitat (Sponaugle et al. 2012; Staaterman et al. 2012). Eventually, the larvae develop "competency", the ability to recruit into coral reef habitat. At this point, they will try to find and reach a new coral reef.

Throughout the dispersal phase, larval mortality reduces the number of larvae that remain; eventually, all the larvae settle, senesce, or are carried by currents outside the modelling domain and are lost to the simulation. Once the fate of all the larvae has been reached, the results are summarised in a connectivity matrix **C** (Bode et al. 2006), whose elements $[\mathbf{C}]_{ij} \in [0,1]$ indicate the proportion of larvae who left reef i and settled successfully on reef j. Note that because some larvae become lost across the model boundaries, and because many larvae die during dispersal (perhaps the majority (Almany and Webster 2006)), the sum of the rows of the connectivity matrix do not add up to one, even though they represent proportions (i.e., $\sum_j [C]_{ij} < 1$). These connectivity matrices are equivalent to, and can be used to define, connectivity networks (Figure 3). Together, connectivity matrices and networks are the most common mathematical elements used to represent regional-scale larval dispersal patterns and to plan their management.

In this chapter, we examine and critique the primary ways in which larval dispersal patterns are currently incorporated in conservation planning in coral reef ecosystems.

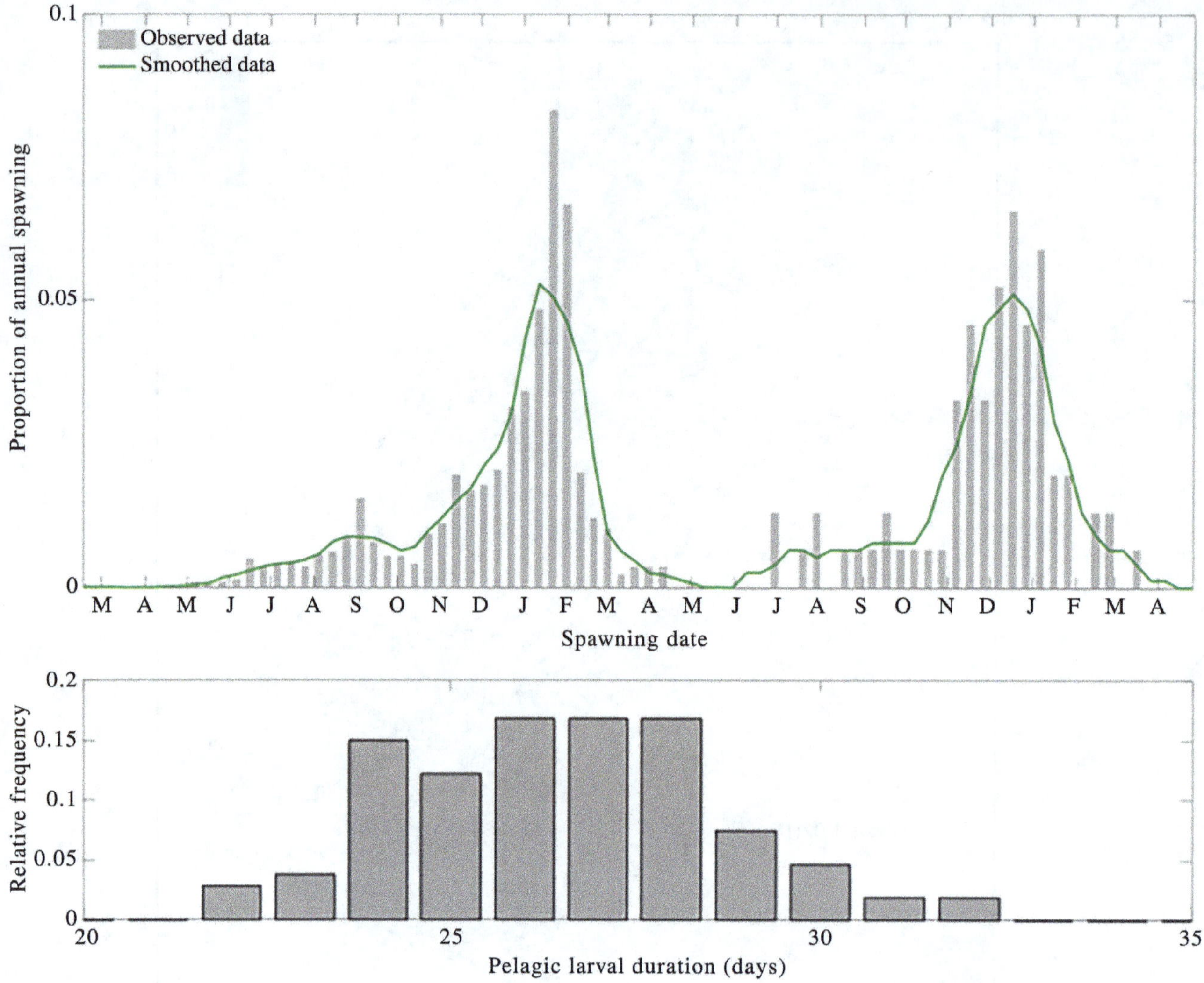

FIGURE 2 Larval spawning and dispersal data for the leopard coral grouper in the southern GBR. Upper panel shows the density of spawning across a two-year period. Spawning occurs throughout the year, except for a few months in the austral autumn. Lower Figure 2 shows pelagic larval duration of multiple animals, as estimated from otolith microchemistry.

Source: Data sourced from (Bode et al. 2019)

We consider three diverse conservation management problems where larval dispersal plays a key role and evaluate the methods used to incorporate larval dispersal into the management decision. Finally, we propose and justify a new method for considering larval connectivity in conservation decisions, based on expected utility theory. We apply this to the case study of crown-of-thorns starfish control on the GBR.

18.3. INCORPORATING LARVAL DISPERSAL INTO CONSERVATION DECISIONS

Larval dispersal patterns drive the metapopulation dynamics of most species on coral reefs, so it is unsurprising that dispersal patterns are often included in decision support tools for coral reef management. We have chosen three common conservation management decisions, each of which incorporates larval dispersal patterns in a different manner. In the examples that follow, we describe why and how larval dispersal patterns have been included in the decision.

18.3.1. Example 1: Designing Adequate Networks of No-Take Marine Protected Areas

The populations of many marine organisms are overexploited. No-take marine protected areas (MPAs) have emerged as a common and relatively effective solution to the problem, particularly in coral reef ecosystems. Unlike the (apparently) featureless open ocean, the patchy nature of coral reef habitat provides easily understood landmarks for MPA boundaries, and the relatively small size of coral reefs often means that entire reefs can be designated as protected. No-take marine reserves also have a long cultural

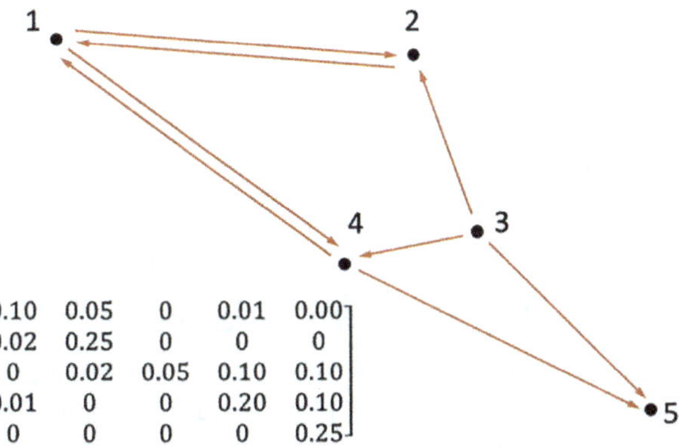

$$\mathbf{C} = \begin{bmatrix} 0.10 & 0.05 & 0 & 0.01 & 0.00 \\ 0.02 & 0.25 & 0 & 0 & 0 \\ 0 & 0.02 & 0.05 & 0.10 & 0.10 \\ 0.01 & 0 & 0 & 0.20 & 0.10 \\ 0 & 0 & 0 & 0 & 0.25 \end{bmatrix}$$

FIGURE 3 Two representations of larval dispersal patterns among a group of 5 reefs. The orange lines form a weighted directed network among the 5 nodes, indicating proportion of larvae who leave the row reef and survive to attempt settlement on the column reef. The matrix **C** shows the same network as a binary connectivity matrix. Note that self-recruitment (i.e., connection from a node to itself) has not been included in the diagram, but it represents some of the strongest dispersal connections in the reef system.

and economic history in many coastal communities, making them familiar as management tools.

However, individual coral reefs are quite demographically open, which means that they may not be able to persist without larval dispersal arriving from other reefs. MPA's must therefore be designed as networks, with larval dispersal connections that are sufficiently strong and consistent to offer reliable and mutual demographic support. Decision-makers therefore need to be able to identify particular combinations of reefs (or parts of reefs) that can deliver this mutual support.

To achieve this, researchers have proposed the application of tools from systematic conservation planning (Moilanen et al. 2009), specifically the use of "boundary length modifiers" (BLM) (Makino et al. 2013; Green et al. 2009; Magris et al. 2014, 2015; Schill et al. 2015; Weeks 2017), sometimes called "connectivity strength modifiers". Systematic conservation planning tools define an MPA vector x, whose binary elements x_i are equal to one if the reef is protected and zero if the reef is not protected. The goal of the tool is to choose a vector that minimises the total cost of an MPA network:

$$G = \sum_{i=1}^{R} k_i x_i - \alpha \sum_{j=1}^{R} \sum_{k=1}^{R} [C]_{jk} x_j x_k \quad \text{[Equation 1]}$$

In this equation, the first summation aggregates the cost of protecting each reef k_i, which is often the cost of forgone fishing opportunities. The second summation reduces this cost if connectivity occurs between protected reefs, weighted by the constant α. Although this constant suggests some equivalence between the cost of protection and the connectivity benefits, in practice, α is assigned an arbitrary value and is calibrated by visual inspection. That is, the user repeatedly runs the optimisation algorithm with differing values of α, searching for a value that provides a "compact-looking" reserve system (Ardron et al. 2010). At very high values of α, only a single large marine reserve will be created; this is normally considered "too compact". Note also that R refers to the total number of reefs present in the system, which will be used in subsequent sections.

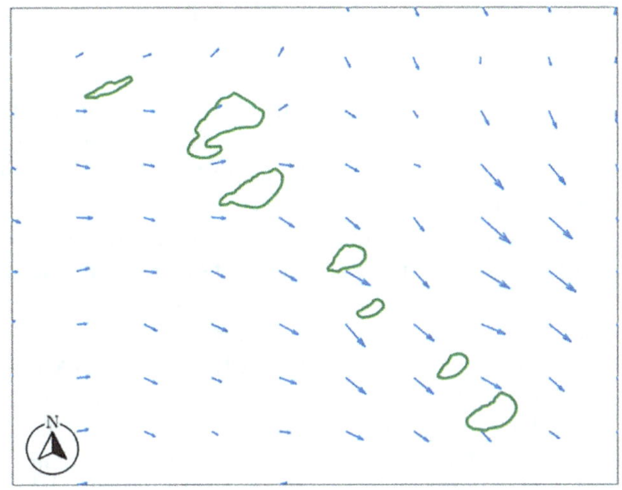

FIGURE 4 Example reef system (green polygons) with static hydrodynamic conditions superimposed (blue arrows).

18.3.2. Example 2: Spatial Planning of Coral Reef Restoration Actions

Restoration is a central pillar of conservation action (Possingham et al. 2015) and is increasingly undertaken on coral reefs. However, marine restoration is generally more expensive than terrestrial restoration, because of the remoteness of the locations, the costs of labour-intensive underwater actions, and a lack of automated or large-scale actions (Saunders et al. 2017). Coral planting, for example, costs millions of dollars per hectare (Bayraktarov et al. 2016) compared with the restoration of forests and grasslands, which cost only a few percent of these amounts.

Larval connectivity offers a justification for the high price of restoration. One hectare of corals, if placed on a reef that has been carefully chosen for its location in the larval dispersal pattern, can soon become many hectares of coral, as the offspring of the planted corals settle in large numbers on downstream reefs. All else being equal, the best location for restoration is therefore on demographically important "upstream" reefs. In the hypothetical example in Figure 4, for example, the reefs in the north-west of the chain would enable larval dispersal to all other reefs.

If the purpose of coral restoration is to promote naturally or artificially heat-tolerant coral genotypes (see Chapter 25), then larval dispersal patterns may be just as important. Once again, deployment to upstream reefs would allow heat-tolerant genotypes to disperse across other populations, amplifying the benefits and increasing the return-on-investment of this action. Conversely, during experimental deployment of artificially created heat-tolerant genotypes, regulators might aim to reduce the risk of a genotype escaping before the consequences of the *in situ* experiment can be assessed. In such situations, deployment to downstream reefs could help ensure that the new individuals do not escape outside the experimental reefs. In Figure 4, for example, the reefs in the south-east of the chain would be suitable for this purpose.

Practically, researchers propose that decision-makers identify reefs on the basis of their location in the connectivity network. Specifically, they recommend that particular network centrality metrics are used to rank coral reefs in order of their importance, particularly the PageRank metric (Wegley Kelly et al. 2020). This metric essentially calculates the eigenvector $\boldsymbol{v}$ of the connectivity matrix that is associated with the dominant eigenvalue. Each element v_i of this eigenvector is the probability that a particle, moving randomly according to the connectivity matrix, would be found on coral reef i in a given timestep. If a decision-maker was able to restore coral on a subset of coral reefs in a metapopulation, and wanted to ensure that larvae from the restored corals could colonise as many new reefs as possible, then this approach to including dispersal recommends that they prioritise reefs with high v_i values.

18.3.3. Example 3: Controlling Crown-of-Thorns Populations

The final example involves the control of an irruptive pest species with a pelagic larval dispersal phase. The crown-of-thorns starfish (COTS) *Acanthaster planci*, is an invertebrate coral-eating predator, and a major driver of coral reef degradation on the GBR (De'Ath et al. 2012). There have been four outbreaks of the species since 1962—approximately every 15 years—beginning in the northern GBR and moving slowly southwards over a period of 10–15 years (Reichelt et al. 1990; Vanhatalo et al. 2016). COTS starfish have very high fecundity (Babcock et al. 1992), and their larvae disperse from outbreak locations to downstream reefs, driving a moving wave of outbreaks. See Chapter 21 for more detail on the dynamics of COTS dispersal and outbreaks.

The most direct control action currently available to managers is the manual culling of individual starfish by human divers (Westcott et al. 2018). On individual reefs, this action has successfully driven populations below the level that causes coral cover decline. However, manual control is costly, and funding is inadequate to target every reef experiencing an outbreak (Westcott et al. 2016).

Researchers have proposed that the allocation of funding be partly determined by the structure of larval dispersal patterns. A COTS outbreak on a particular reef can cause a loss of local coral cover, and the larvae produced by those starfish can cause further downstream coral losses if they successfully recruit to a large number of downstream reefs. In a sense, the justification for priority-setting is the same as the restoration example above—those reefs that are upstream of other reefs, particularly important reefs, should receive priority attention.

Recent papers have therefore used network theory metrics to classify "key sources" among the GBR reefs, which they claim contribute disproportionately to future outbreaks (Hock et al. 2017; Bode et al. 2018). Key source reefs are defined as those that satisfy all of the following criteria. First, a high out-degree:

$$d_i = \sum_{j=1}^{R} e_{ij}, \text{ where } e_{ij} = \begin{cases} 0 & \text{if } [\boldsymbol{C}]_{ij} = 0 \\ 1 & \text{otherwise} \end{cases} \qquad \text{[Equation 2]}$$

The out-degree of a node in a network is the number of nonzero connections with other nodes (including self-connections). The second criterion is a high node strength:

$$s_i = \sum_{j=1}^{R} [\boldsymbol{C}]_{ij} \qquad \text{[Equation 3]}$$

which is simply the sum of all connections from a given node. For COTS modelling, the elements $[\boldsymbol{C}]_{ij}$ are each probabilities of successful settlement on another reef. As a consequence, s_i is the probability that a COTS larvae leaving reef i will settle successfully on any other reef in the system. The third criterion is a large number of relatively strong connections. For reef i, this is defined as:

$$f_i = \sum_{i=1}^{R} e_{ij} y_{ij} \qquad \text{[Equation 4]}$$

where e_{ij} is defined in Equation 2, and the indicator variable y_{ij} is defined as:

$$y_{ij} = \begin{cases} 0 & \text{if } [\boldsymbol{C}]_{ij} > 0.1 \sum_k [\boldsymbol{C}]_{kj} \\ 1 & \text{otherwise} \end{cases} \qquad \text{[Equation 5]}$$

A connection from reef i to reef j is therefore defined as "strong" if it represents more than 10% of the total probabilities connected to reef j.

It is worth noting that these three metrics all summarise similar quantities, and will therefore be naturally quite correlated.

18.3.4. Limitations of Existing Methods

Each of these prioritisation approaches has a series of specific limitations. For example, all of them are based on proportional connectivity—the proportion of larvae from a given reef that land on other reefs. This metric is useful and important, but it does not consider the relative size of the source reef, or the size of the population on that reef. Larger reefs, and/or reefs with larger populations, will contribute more propagules than smaller reefs—a connection worth 10% of coral larvae from a 1-hectare reef is very different from 10% of the larvae from a 1,000-hectare reef. To give another example limitation, none of the methods consider uncertainty in the larval dispersal patterns, even though we know that biophysical models are inaccurate. Of course, such omissions could be remedied with more complex analyses. Centrality metrics could be weighted by the size of node (i.e., the habitat area on the reefs), and the values could be re-calculated across multiple years (Hock et al. 2017; Bode et al. 2018) to capture variation.

More importantly, however, each of these proposed prioritisation methods shares a fundamental conceptual problem—they are all *ad hoc* constructions. The methods and metrics being used are adopted wholesale from other applications of operations research or network theory—normally from ecology but sometimes from fields as diverse as physics or electrical engineering—and applied to solve the problem of resource allocation in coral reef conservation. None of them are solutions to explicitly formulated conservation optimisation problems.

This type of *ad hoc* decision is a common flaw in conservation decision-making (Game et al. 2013). The objective of the decision-making process is stated in a very broad, verbal manner (e.g., "reduce COTS outbreaks"), but it is never quantified and never explicitly related to the decision-making process. Here, we illustrate the appropriate way to approach this problem: with an explicit and quantitative statement of the conservation objective from which the optimal solution is derived.

18.4. OPTIMAL CONTROL STRATEGY FOR CROWN-OF-THORNS STARFISH CONTROL

18.4.1. Notation

Note that lowercase and capital bolded variables (e.g., $\mathbf{p}$ and $\mathbf{C}$) correspond to vectors and matrices, respectively, with their unbolded counterparts using subscripts for indexation. Note also that $\underline{1}$ indicates a vector of ones, and $\mathbf{I}$ indicates the identity matrix. Finally, note that indexation of vector and matrix variables is applied via a subscript, for example, $x_r, \left[\mathbf{p}^{(ET)}\right]_r, [\mathbf{C}]_{ij}$, with the first index i indicating the row and the second j indicating the column.

18.4.2. Objective Function

The first essential step, in any application of conservation decision theory, is to define the management objective. For COTS control on the GBR, many different objectives could reasonably be set. We could aim to reduce the number of "outbreak" reefs in the system, which are defined by the Great Barrier Reef Marine Park Authority (GBRMPA) as reefs with COTS densities above 10,000 per hectare (Westcott et al. 2016). Alternatively, we could aim to minimise the COTS populations on the small subset of reefs that are visited by commercial tour operators—mostly off the coast of Cairns. There are likely as many objective functions as there are stakeholders and rights holders in any coral reef system, and the right choice of objective function is absolutely critical to good decision-making.

For the purposes of this analysis, we choose a relatively complicated objective: minimise the expected number of COTS, on ecologically and economically important reefs, one year in the future:

$$\min_x U = \sum_{r\in R} [\boldsymbol{p}]_r h_r v_r \qquad \text{[Equation 6]}$$

In Equation 6, h_r is the proportional live coral cover on reef r, and v_r is a binary variable representing whether substantial tourism occurs on reef r, to account for a reef's economic value. The vector elements $[\boldsymbol{p}]_r$ denote the number of COTS on reef r in the next year, which is predicted (see below) by Equation 9 using the column vector $\boldsymbol{p}^{(0)}$, which represents the COTS present on each reef in the current year. Finally, the manager seeks to minimise this utility function by choosing the best subset of reefs upon which to control COTS this year, according to the binary control vector $\boldsymbol{x}$. In this case, controlling COTS refers to culling until the ecological threshold density is reached. This threshold is defined as a density of COTS below which the annual growth of the coral population exceeds the annual consumption of coral by COTS. Note that this is a fairly "myopic" strategy and does not consider the consequences of a control decision 2 or more years in the future.

Instead of using the objective function in Equation 6 directly, we calculate instead the benefit of acting on each reef, which we define as

$$B_r = \begin{cases} E[U_0] - E[U_r] & \text{if } \left[\boldsymbol{p}^{(0)}\right]_r > \left[\boldsymbol{p}^{(ET)}\right]_r \\ 0 & \text{otherwise} \end{cases} \qquad \text{[Equation 7]}$$

where U_0 is the utility (Equation 6) under a "no action" scenario (i.e., $\boldsymbol{x} = \mathbf{0}$), and U_r is the utility when acting on reef r only. The expectation $\left(E[U_i]\right)$ is included to account for temporal variation in the connectivity dynamics. We simulate multiple connectivity matrices for different years and assume that these matrices all have an equal probability of occurring in the future. Finally, $\boldsymbol{p}^{(ET)}$ is a vector containing the COTS counts at each reef that would correspond to an ecological threshold density. As such, we set the benefit function for reefs below this threshold to 0, given that we previously defined our actions as culling COTS until reaching the ecological threshold from above. We then solve the decision problem using the following objective function:

$$\max_x F = \sum_{r\in R} B_r \qquad \text{[Equation 8]}$$

18.4.3. Constraints

Equation 8 is only one part of the optimisation decision for two critical reasons. First, the control vector $\boldsymbol{x}$ does not appear within the objective function; second, the decision is not yet constrained in any way. The optimal unconstrained solution would be to control COTS on all reefs in the system simultaneously—which is clearly not logistically possible. Each of these problems is solved by the inclusion of constraints.

The first and most obvious constraint is logistical—the COTS control program has only a limited number of boats and a limited number of divers. Of all the possible control vectors that the managers can choose in Equation 8, only a few will be affordable. We therefore set the inequality constraint:

$$\sum_{r\in R} x_r g\left(\left[p^{(0)}\right]_r, a_r\right) \le A \qquad \text{[Equation 9]}$$

where $g\left(\left[\boldsymbol{p}^{(0)}\right]_r, a_r\right)$ is the number of "person-dive-hours" that is required to control the COTS population on reef r to below a critical ecological threshold, which is defined based on an empirical formula as follows (Westcott et al. 2016):

$$g\left(\left[\boldsymbol{p}^{(0)}\right]_r, a_r\right) = 10a_r e^{5.116}\left(\frac{\left[\boldsymbol{p}^{(0)}\right]_r}{100a_r}\right)^{0.665} \qquad \text{[Equation 10]}$$

where a_r is the area of reef r in km². In Equation 9, A is the total control budget available to the managers in a given year, which is also defined in person-dive-hours.

The second constraint—a dynamical constraint—is defined by the population dynamics of the COTS population. We assume a linear model of the COTS population in the future:

$$\boldsymbol{p} = \left(\mathbf{I} + f(1-l)\mathbf{C}'\right)(1-m)\left((1-\mathbf{x})\boldsymbol{p}^{(0)} + \mathbf{x}\,\boldsymbol{p}^{(ET)}\right) \qquad \text{[Equation 11]}$$

The above model is a simple, discrete time model of COTS populations under control action. It first applies both an average mortality rate m and control action vector $\boldsymbol{x}$ to the COTS populations at each reef in the previous timestep, $\boldsymbol{p}^{(0)}$. This term ensures reefs in which action is taken (i.e., $x_r = 1$) are set to their ecological threshold in the current year, with those remaining simply left at their current levels $\boldsymbol{p}^{(0)}$. To account for dispersal, the model applies an average per-capita fecundity f and larval survival rate l, before multiplying by the larval connectivity matrix $\boldsymbol{C}$. COTS populations at each reef in the next year $\boldsymbol{p}$ are then determined by summing the results of the above two operations. (Note that $\boldsymbol{C}$ must be transposed to account for the fact that element $[\boldsymbol{C}]_{ij}$ is defined as the proportion of larvae produced at reef r_i that arrive at r_j.)

18.4.4. Model Parameterisation

To apply this model to the case study, we need to estimate the value of each of the parameters. We estimate the density (and, therefore, counts) of COTS on each reef using data gathered by the Australian Institute of Marine Science (AIMS) Long-Term Monitoring Program. For reefs in the sampled set $\mathbf{S}$, we use the most recent observed density of COTS. For reefs in the unsampled set $\boldsymbol{S}'$, we interpolate the value from nearby sampled sites using a weighted spatial average. The conservation planning area consisted of 320 reefs in the northern section of the GBR between Cairns and Townsville, between latitudes 19°S and 14°S.

For this example, we adopt the ecological threshold as 0.04 COTS per minute bottom time, as estimated by Babcock and colleagues (Babcock et al. 1992). This is equivalent to a COTS density of 285.26 COTS/km² (Westcott et al. 2016). The ecological threshold parameter used in the model, $\boldsymbol{p}^{(ET)}$, was calculated by multiplying the above density by reef area in km². We estimate the binary variable for tourism at each reef v_r using published maps of tourism use (Spalding et al. 2017). The total annual resource budget A was assumed to be $A = 2.88\times10^6$ diver hours, which is enough to control COTS populations below the ecological threshold on a moderate number (5–10) of reefs.

The demographic parameters m, l, and f used in the COTS population model are difficult to estimate. An adult COTS, with a diameter of 300 mm, produces approximately f = 8.2 million larvae annually (Pratchett et al. 2021), and we assumed that 10% of these are successfully fertilised and that a small proportion $l = 4.6\times10^{-5}$ of these survive the 24-day pelagic dispersal stage to recruit to a new reef (Almany and Webster 2006). Following Morello and colleagues (Morello et al. 2014), we assume that annual adult mortality is approximately m=0.07 y^{-1}.

18.4.5. Identifying the Optimal Control Strategy

To identify the optimal allocation of control resources, we choose a control vector $\boldsymbol{x}$, determine whether it satisfies our logistical constraint (Equation 9), calculate its effect on the following year's COTS population (Equation 11), and then calculate the impact on our utility function (Equation 8).

In our GBR case study, we consider 320 different reefs in and around the Cairns Management Section. There are therefore 2^{320} different control vectors that should be considered—far too many to assess exhaustively. As a consequence, we use integer linear programming methods to identify the optimal solution.

18.4.6. Comparison with a Network Theory Approach

We compare our expected utility theory prioritisation method to a prioritisation method based on network metrics. First, the centrality metrics described in Equations 2–4 are calculated, and reefs are then ranked for each metric, based on their average score across all connectivity matrices. These ranks are then summed and averaged across all metrics and connectivity matrices to produce a unique rank for each reef. We then choose the highest-ranked reefs, apply control until they fall below the ecological threshold, and move onto the next-highest reef until our management budget (Equation 9) is expended. This approach effectively maximises the number of "important" reefs (based on centrality metrics) where COTS are controlled below the ecological threshold, given the constrained total resources available.

Figure 5 compares the reefs targeted by the two meth ods. The left-hand panel shows the near-optimal solution identified using expected utility theory, and the right-hand panel shows the reefs targeted for COTS control when resource

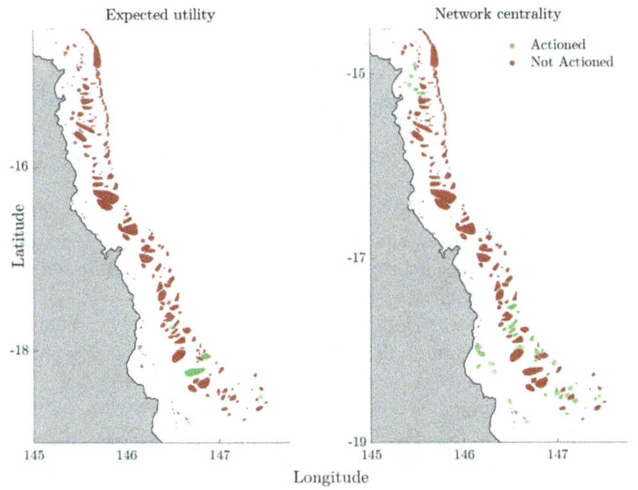

FIGURE 5 Reefs chosen by the two approaches to COTS control prioritisation. Reefs selected as high priorities are shown in green. Left panel shows the reefs selected by the expected utility approach, clustered in the southern part of the domain, where most tourism occurs. Right panel shows reefs selected by the centrality approach.

allocation is determined by network metrics. Both methods operated under the same resource constraints, but the centrality method chose a large number of smaller reefs (47 in total) compared to the utility approach (7 in total). The methods had five reefs in common—we would expect an overlap of about 4 reefs by random chance—but the decisions recommended made by each method are markedly different not only in the choices of the specific reefs themselves but also their composition and spatial location. The centrality approach chose reefs across a larger area, selecting reefs in the northern part of the domain, while the utility method targeted the southern part, where most of the tourism occurs.

18.5. DISCUSSION

Because every ecological process in a coral reef ecosystem is affected—either directly or indirectly—by the processes of pelagic larval dispersal, connectivity patterns must be included in the conservation decision-making process. The past two decades have seen impressive advances on this front, starting with the simulation and measurement of larval dispersal patterns (Dight et al. 1990a; Jones et al. 1999; James et al. 2002; Bode et al. 2006; Cowen et al. 2006; Treml et al. 2008; Harrison et al. 2012, 2020), and moving into their inclusion into conservation planning techniques (Beger et al. 2010; Treml and Halpin 2012; Schill et al. 2015; Bode et al. 2016b, 2016a; Hock et al. 2017). There is a strong consensus that conservation planning in coral reef ecosystems needs to include connectivity. However, two additional, connected facts must be admitted.

First, there are an effectively infinite number of methods for including connectivity in the conservation decision-making process. There are dozens of different ways, for example, to measure even a single network metric from landscape ecology, the "centrality" of a reef in its connectivity network (Li and Wu 2004; Baker et al. 2015). Each of the different methods for calculating network centrality will change the priority ranking of reefs. Second, as a consequence, almost all of the methods that *could* be used to include connectivity in decisions will be incorrect in that they will disagree with the true (but unknown) optimal prioritisation and therefore lead to inefficient or non-robust decisions. It is even possible that many of them will lead to worse decisions and worse conservation outcomes than ignoring connectivity entirely.

It is therefore important that the discussion on coral reef conservation moves beyond "have we included connectivity in our decisions?" and starts to ask "have we appropriately included connectivity in our decisions?" Our argument in this chapter is that, following Game and colleagues (Game et al. 2013), we should assess the answer to the second question on the basis of defensibility. That is, although we may not know if the techniques that we are using are optimal, we should at least expect there to be a logical, quantitative, and explicit connection between how we include connectivity and a clearly stated conservation objective. Our conclusion in this chapter is that, when reviewing the incorporation of connectivity into coral reef conservation decisions to date, the answer to the latter question is almost entirely negative. In almost all cases, the methods used to incorporate connectivity into planning are *ad hoc*. In almost all cases, the conservation objective is never even stated explicitly, let alone quantitatively.

In this chapter, we have outlined a method for including connectivity into the decision-making process for COTS control on the GBR. Our method has a range of important limitations. The data that we use are inaccurate—our layer for tourism usage, for example, is out-of-date (e.g., it was estimated before the global pandemic of COVID-19). Our COTS population model looks only a single year ahead, and optimal decision-making should consider longer-term consequences of immediate actions. Hydrodynamic models have come a long way in the last few decades, but they will always represent an approximation. Models cannot perfectly reflect the true hydrodynamics, particularly around the complex arrangement of reefs in the GBR lagoon (Andutta et al. 2012), and they cannot perfectly reflect the true behaviour of the dispersing organisms (Leis 2020). Even if they could do so, simulated connectivity patterns from any given year (or any given series of years) will not necessarily predict, or even encompass, next year's dispersal patterns. Finally, our characterisation of the conservation objective and utility function (Equation 6) was not developed in collaboration with the stakeholders and rights holders that it purports to represent.

These limitations are serious, but they can be rectified. Improved datasets, better treatments of uncertainty, and more accurate estimates of utility can all be incorporated into the framework that we outlined earlier. Moreover, limitations are ever-present in conservation decision-making methods—their existence does not invalidate an approach.

Although the details of our approach contain flaws, the overarching framework makes a clear connection between conservation actions and conservation objectives—avoiding the most important and common error of conservation planning tools.

Going forward, we therefore recommend that all attempts to integrate larval connectivity patterns into conservation planning tools follow this central tenet of our method. The specific approach will vary with the location and species and with the conservation goals and actions. The utility function and conservation goal (Equation 6) may change, and the connectivity patterns and population dynamics (Equation 11) certainly will. However, the logical connection between what we aim to achieve and how we believe that larval connectivity will affect this goal must still be stated explicitly. It can no longer be acceptable to simply choose a particular network metric from among the thousands available and then incorporate it into an arbitrary ranking procedure. To continue to do so will not only generate inefficient conservation decisions but also retard progress towards efficient decisions.

REFERENCES

Almany GR, Berumen ML, Thorrold SR, Planes S, Jones GP (2007) Local Replenishment of Coral Reef Fish Populations in a Marine Reserve. Science (1979) 316:742–744

Almany GR, Hamilton RJ, Bode M, Matawai M, Potuku T, Saenz-Agudelo P, Planes S, Berumen ML, Rhodes KL, Thorrold SR, Russ GR, Jones GP (2013) Dispersal of Grouper Larvae Drives Local Resource Sharing in a Coral Reef Fishery. Current Biology 23:626–630

Almany GR, Planes S, Thorrold SR, Berumen ML, Bode M, Saenz-Agudelo P, Bonin MC, Frisch AJ, Harrison HB, Messmer V, Nanninga GB, Priest MA, Srinivasan M, Sinclair-Taylor T, Williamson DH, Jones GP (2017) Larval Fish Dispersal in a Coral Reef Seascape. Nature Ecology Evolution 1:1–7

Almany GR, Webster MS (2006) The predation gauntlet: early post-settlement mortality in reef fishes. Coral Reefs 25:19–22

Andutta FP, Kingsford MJ, Wolanski E (2012) 'Sticky water' enables the retention of larvae in a reef mosaic. Estuarine, Coastal and Shelf Science 101:54–63

Ardron J, Possingham HP, Klein CJ (2000) Marxan good practices handbook. https://pacmara.org/wp-content/uploads/2010/07/Marxan-Good-Practices-Handbook-v2–2010.pdf

Babcock R, Research CM-M and F, 1992 undefined (1992) Reproductive biology, spawning and field fertilization rates of Acanthatser planci. CSIRO Publishing 43:525–559

Baker CM, Hughes BD, Landman KA (2015) Length-based connectivity metrics and their ecological interpretation. Ecological Indicators 58:192–198

Bayraktarov E, Saunders M, Abdullah S, Mills M (2016) The cost and feasibility of marine coastal restoration. Researchgate.net 26:1055–1074

Beger M, Simon L, Game E, Ball I, Treml E, Watts M, Possingham HP (2010) Incorporating asymmetric connectivity into spatial decision making for conservation. Conservation Letter 3:359–368

Bode M, Bode L, Armsworth PR (2006) Larval dispersal reveals regional sources and sinks in the Great Barrier Reef. Marine Ecology Progress Series 308:17–25

Bode M, Bode L, Choukroun S, James MK, Mason LB (2018) Resilient reefs may exist, but can larval dispersal models find them? PloS Biology 16:e2005964

Bode M, Leis JM, Mason LB, Williamson DH, Harrison HB, Choukroun S, Jones GP (2019) Successful validation of a larval dispersal model using genetic parentage data. PloS Biology

Bode M, Sanchirico JN, Armsworth PR (2016a) Returns from matching management resolution to ecological variation in a coral reef fishery. Proceedings of the Royal Society B: Biological Sciences 283:20152828

Bode M, Williamson DH, Harrison HB, Outram N, Jones GP (2017) Estimating dispersal kernels using genetic parentage data. Methods in Ecology and Evolution 9:490–501

Bode M, Williamson DH, Weeks R, Jones GP, Almany GR, Harrison HB, Hopf JK, Pressey RL (2016b) Planning Marine Reserve Networks for Both Feature Representation and Demographic Persistence Using Connectivity Patterns. PloS One 11:e0154272

Bode M, Wilson KA, Brooks TM, Turner WR, Mittermeier RA, McBride MF, Underwood EC, Possingham HP (2008) Cost-effective global conservation spending is robust to taxonomic group. Proceedings of the National Academy of Sciences of the United States of America 105:6498–6501

Buston PM, Jones GP, Planes S, Thorrold SR (2012) Probability of successful larval dispersal declines fivefold over 1 km in a coral reef fish. Proceedings of the Royal Society B: Biological Sciences 279:1883–1888

Cowen RK, Lwiza KMM, Sponaugle S, Paris CB, Olsen DB (2000) Connectivity of marine populations: Open or closed? Science 287:857–859

Cowen RK, Paris CB, Srinivasan A (2006) Scaling of connectivity in marine populations. Science 311:522–527

D'Aloia CC, Bogdanowicz SM, Francis RK, Majoris JE, Harrison RG, Buston PM (2015) Patterns, causes, and consequences of marine larval dispersal. Proceedings of the National Academy of Sciences of the United States of America 112:13940–13945

De'ath G, Fabricius KE, Sweatman H, Puotinen M (2012) The 27-year decline of coral cover on the Great Barrier Reef and its causes. Proceedings of the National Academy of Sciences of the United States of America 109 (44):17995–17999

Dight IJ, Bode L, James MK (1990a) Modelling the larval dispersal of *Acanthaster planci*—I. Large scale hydrodynamics, Cairns Section, Great Barrier Reef Marine Park. Coral Reefs 9:115–123

Dight IJ, James MK, Bode L (1990b) Modelling the larval dispersal of *Acanthaster planci*—II. Patterns of reef connectivity. Coral Reefs 9:125–134

Game ET, Kareiva P, Possingham HP (2013) Six common mistakes in conservation priority setting. Conservation Biology 27:480–485

Green A, Smith SE, Lipsett-Moore G, Groves C, Peterson N, Sheppard S, Lokani P, Hamilton R, Almany J, Aitsi J, Bualia L (2009) Designing a resilient network of marine protected areas for Kimbe Bay, Papua New Guinea. Oryx 1–11

Harrison HB, Bode M, Williamson DH, Berumen ML, Jones GP (2020) A connectivity portfolio effect stabilizes marine reserve performance. Proceedings of the National Academy of Sciences of the United States of America 117 (41):25595–25600

Harrison HB, Williamson DH, Evans RD, Almany GR, Thorrold SR, Russ GR, Feldheim KA, van Herwerden L, Planes S, Srinivasan M, Berumen ML, Jones GP (2012) Larval export from marine reserves and the recruitment benefit for fish and fisheries. Current Biology 22:1023–1028

Hock K, Wolff NH, Ortiz JC, Condie SA, Anthony KRN, Blackwell PG, Mumby PJ (2017) Connectivity and systemic resilience of the Great Barrier Reef. PloS Biol 15(11):e2003355.

James MK, Armsworth PR, Mason LB, Bode L (2002) The structure of reef fish metapopulations: Modelling larval dispersal and retention patterns. Proceedings of the Royal Society B: Biological Sciences 269:2079–2086

Jones GP (2015) Mission impossible: Unlocking the secrets of coral reef fish dispersal. In: Mora C. (eds) Ecology of Fishes on Coral Reefs. Cambridge University Press, Cambridge, pp. 16–27

Jones GP, Milicich MJ, Emslie MJ, Lunow C (1999) Self-recruitment in a coral reef fish population. Nature 402:802–804.

Leis JM (1991) The pelagic stage of reef fishes: The larval biology of coral reef fishes. In: Sale PF (eds), The Ecology of Coral Reef Fishes. Academic Press, San Diego.

Leis JM (2007) Behaviour as input for modelling dispersal of fish larvae: Behaviour, biogeography, hydrodynamics, ontogeny, physiology and phylogeny meet hydrography. The Marine Ecology Progress Series 347:185–193

Leis JM (2020) Perspectives on larval behaviour in biophysical modelling of larval dispersal in marine, demersal fishes. Oceans 2:1–25

Levin SA (1992) The problem of pattern and scale in ecology: The {Robert} {H}. {MacArthur} award lecture. Ecology 73:1943–1967

Li H, Wu J (2004) Use and misuse of landscape indices. Landscape Ecology 19:389–399

Magris R, Treml E, Pressey R, Ecography RW, et al. (2015) Integrating multiple species connectivity and habitat quality into conservation planning for coral reefs. Wiley Online Library 39:649–664

Magris RA, Pressey RL, Weeks R, Ban NC (2014) Integrating connectivity and climate change into marine conservation planning. Biological Conservation 170:207–221

Makino A, Beger M, Klein C, et al. (2013) Integrated planning for land–sea ecosystem connectivity to protect coral reefs. Biological Conservation 165:35–42

McCarthy DP, Donald PF, Scharlemann JPW, Buchanan GM, Balmford A, Green JMH, Bennun LA, Burgess ND, Fishpool LDC, Garnett ST, Leonard DL, Maloney RF, Morling P, Schaefer HM, Symes A, Wiedenfeld DA, Butchart SHM (2012) Financial Costs of Meeting Global Biodiversity Conservation Targets: Current Spending and Unmet Needs. Science (1979) 338:946–949

Miller C (2009) Biological oceanography. Wiley-Blackwell, 480pp.

Moilanen A, Wilson KA, Possingham HP (2009) Spatial conservation prioritization. Oxford University Press.

Morello EB, Plagányi ÉE, Babcock RC, Sweatman H, Hillary R, Punt AE (2014) Model to manage and reduce crown-of-thorns starfish outbreaks. The Marine Ecology Progress Series 512:167–183

Murdoch W, Polasky S, Wilson KA, Possingham HP, Kareiva P, Shaw R (2007) Maximizing return on investment in conservation. Biological Conservation 139:375–388

Planes S, Jones G, Thorrold S (2009) Larval dispersal connects fish populations in a network of marine protected areas. Proceedings of the National Academy of Sciences of the United States of America 106:5693–5700

Possingham HP, Bode M, Klein CJ (2015) Optimal Conservation Outcomes Require Both Restoration and Protection. PloS Biology 13:e1002052

Pratchett MS, Nadler LE, Burn D, Lang BJ, Messmer V, Caballes CF (2021) Reproductive investment and fecundity of Pacific crown-of-thorns starfish (Acanthaster cf. solaris) on the Great Barrier Reef. Marine Biology 168:1–8

Reichelt RE, Bradbury RH, Moran PJ (1990) Distribution of *Acanthaster planci* outbreaks on the Great Barrier Reef between 1966 and 1989. Coral Reefs 9:97–103

Samoilys MA (1997) Periodicity of spawning aggregations of coral trout Plectropomus leopardus (Pisces: Serranidae) on the northern Great Barrier Reef. The Marine Ecology Progress Series 160:149–159

Saunders MI, Bode M, Atkinson S, Klein CJ, Metaxas A, Beher J, Beger M, Mills M, Giakoumi S, Tulloch V, Possingham HP (2017) Simple rules can guide whether land- or ocean-based conservation will best benefit marine ecosystems. PloS Biology 15.

Schill S, Raber G, Roberts JJ, Treml EA, Brenner J, Halpin PN (2015) No Reef Is an Island: Integrating Coral Reef Connectivity Data into the Design of Regional-Scale Marine Protected Area Networks. PloS One 10:e0144199

Spalding M, Burke L, Wood S, Ashpole J, Policy JH-M (2017) Mapping the Global Value and Distribution of Coral Reef Tourism. Elsevier

Sponaugle S, Paris C, Walter KD, Kourafalou V, Dalessandro E (2012) Observed and modeled larval settlement of a reef fish to the Florida Keys. The Marine Ecology Progress Series 453:201–212

Staaterman E, Paris CB, Helgers J (2012) Orientation behavior in fish larvae: A missing piece to Hjort's critical period hypothesis. Journal of Theoretical Biology 304:188–196

Swearer SE, Caselle JE, Lea DE, Warner RR (1999) Larval retention and recruitment in an island population of coral reef fish. Nature 402:799–802

Treml EA, Halpin PN (2012) Marine population connectivity identifies ecological neighbours for conservation planning in the Coral Triangle. Conservation Letters 5:441–449

Treml EA, Halpin PN, Urban DL, Pratson LF (2008) Modeling population connectivity by marine currents, a graph theoretic approach for marine conservation. Landscape Ecology 23:19–36

Vanhatalo J, Hossack GR, Sweatman H (2016) Spatiotemporal modelling of crown-of-thorns starfish outbreaks on the Great Barrier Reef to inform control strategies. Wiley Online Library 54:188–197

Weeks R (2017) Incorporating seascape connectivity in conservation prioritisation. PloS One 12(7):e0182396

Wegley Kelly L, Delaney Blakeway R, Mendes T, Frys C, Saint-Amand A, Le Hénaff M, Figueiredo J, Kuba A, Walker B, Lambrechts J, Vallaeys V, Vincent D, Hanert E (2020) Fine-Scale Coral Connectivity Pathways in the Florida Reef Tract: Implications for Conservation and Restoration. Frontiers in Marine Science | www.frontiersin.org 1:312

Westcott DA, Fletcher CS, Babcock RC, Plagányi ÉE (2016) A strategy to link research and management of crown-of-thorns starfish on the Great Barrier Reef: An integrated pest management approach.

Westcott DA, Quality CF-R for NTW, 2018 undefined (2018) How effective are management responses in controlling crown of-thorns starfish and their impacts on the Great Barrier Reef. Nesptropical.edu.au

Williamson DH, Harrison HB, Almany GR, Berumen ML, Bode M, Bonin MC, Choukroun S, Doherty PJ, Frisch AJ, Saenz-Agudelo P, Jones GP (2016) Large-scale, multidirectional larval connectivity among coral reef fish populations in the Great Barrier Reef Marine Park. Molecular Ecology 25:6039–6054

Wilson KA, McBride MF, Bode M, Possingham HP (2006) Prioritizing global conservation efforts. Nature 440:337–340

Woolley S, Tittensor D, Nature PD (2016) Undefined Deep-sea diversity patterns are shaped by energy availability. Nature.com

19 A Historical Perspective on Thermal- and Heatwave-Induced Bleaching on the Great Barrier Reef

Nicholas A. Kamenos and Sebastian J. Hennige

19.1. BACKGROUND

Mass coral bleaching events during the last 30 years have caused concern over the future of coral reefs worldwide (Hughes et al. 2017). In particular, marine heatwaves—extended periods of warming—have been linked to mass coral bleaching on the Great Barrier Reef (GBR) (e.g., Hughes et al. 2017). In addition to contemporary bleaching events, there is also historical evidence of bleaching events on the GBR, yet little is known about bleaching events before the 1970s. This is primarily because prior to the 1970s, systematic observations on the GBR were not regular but rather opportunistic (for example, repeated bleaching observed in 1929 (e.g., Dunne et al. 2022; Yonge and Nicholls 1931). This incomplete bleaching record thus makes it difficult to place modern bleaching patterns in a historical context.

19.1.1. Thermally Induced Coral Bleaching

Coral bleaching is characterised by a breakdown of the coral-algae symbiosis (Glynn 1991). Bleaching may cause coral mortality, although sub-lethal bleaching can also occur (e.g., in thermally acclimatized coral colonies). This possibly acts as a mechanism enabling coral hosts to survive via symbiont shuffling (Suggett and Smith 2011), which confers acclimatization and adaptation on the coral holobiont (Kenkel and Matz 2016). Thermal acclimation by coral hosts can also occur with little symbiont replacement (Palumbi et al. 2014).

The bleaching process is caused by the physiological stress generated by single and multiple stressor agents including thermal and irradiance extremes, disease, and terrestrial runoff, which can cause local or widespread coral bleaching and result in reduced coral growth where sub-lethal bleaching occurs (Baumann et al. 2019; Fabricius 2005; Hendy et al. 2003; Kamenos and Hennige 2018; Kamenos and Hennige 2020; Wiedenmann et al. 2013). The effects of warmer sea surface temperatures (SSTs), particularly those associated with widespread heatwaves, have been directly observed to cause GBR mass bleaching events in at least 1998, 2002, 2016, 2017, and 2020 (Hughes et al. 2021). Thermal bleaching is the main driver of widespread, mass bleaching events because of its geographical extent and is the primary stressor (variable) used to forecast coral reefs at risk of bleaching over several months (Liu et al. 2006; Logan et al. 2014; NOAA 2018). Light stress is increasingly recognised as a multiple stressor in combination with temperature, which further contributes to bleaching susceptibility, mediated by the photoacclimatory response of the Symbiodinaecea to high light stress (e.g., Mieog et al. 2009; Rosenberg et al. 2007).

19.2. GREAT BARRIER REEF BLEACHING

Marine heatwaves can thus act as key agents in mass coral bleaching events. Marine ecosystems are currently experiencing marine heatwaves more frequently, and when considering the period of 1925–2016, heatwaves have become more frequent in the last 2 decades of this period with their presence being driven by increasing SST trends (Frölicher et al. 2018; Oliver et al. 2018a). Compared to a short SST spike (e.g., day-long), marine heatwaves are defined as prolonged periods of high temperature (at least 5 days long and lasting up to months) with temperatures in the 90th percentile compared with a 30-year-long baseline of data (Hobday et al. 2016). Mirroring global trends (Frölicher et al. 2018), GBR heatwaves have also become more frequent, with the northern GBR particularly affected due to the thermal gradient present across the GBR (Hughes et al. 2017). Bleaching events associated with these heatwaves have led to widespread coral death in some cases (Hughes et al. 2017); however, more recent evidence also points towards coral recovery dominated by fast-growing species (AIMS 2022a).

19.2.1. Recent Bleaching and Heatwaves (Post-1970s)

Between the 1970s and 2022, the GBR underwent five known marine heatwaves (Table 1); these occurred in the austral summers of 1998, 2002, 2016, 2017, and 2020 (Hughes et al. 2021). All heatwaves were accompanied by a degree of coral bleaching that resulted in severe bleaching in 10–38% of GBR reefs (severe bleaching includes > 60% of the corals at each reef), with bleaching in 2017 and 2020 associated with higher category heatwaves (more intense) than previous years (Table 1). Furthermore, GBR bleaching was also observed during 1982 and again in 1987 also

DOI: 10.1201/9781003320425-22

TABLE 1

Summary of observed major bleaching events on the GBR for timepoints at which extensive observations exist (observations from (AIMS 2016, 2017, 2018, 2019, 2020, 2021, 2022a, 2022b; Dunne et al. 2022; Hughes et al. 2017)). Maximum Degree Heating Weeks (DHWs) for Low Isles, GBR (16.375°S 145.575 °E) calculated using the 1985–2022 integration period. Predominant heatwave category (v1.01) for the GBR. Both calculated as per Coral Reef Watch from the CoralTemp v3.1 data product (NOAA 2018). Heatwave categories are defined as No Marine Heatwave (0), Moderate (1), Strong (2), Severe (3), Extreme (4), and Beyond extreme (5) as per (Hobday et al. 2016; NOAA 2018).

Year	Timescale	Description of event	Areas affected	Further details	Maximum daily DHW (for Low Isles)	Predominant GBR heatwave category
1929	February and June	February event was likely restricted to Acroporids, while the June event, which occurred in deeper water, likely included some massives. February event was caused by a combination of high DHWs and aerial exposure; June event was primarily caused by exposure- and insolation-induced stress.	Low Isles (this is the only area that the Expedition was observing)	Observations were made only over the 1928–1929 period during the GBR Expedition to Low Isles.	5.5 (but integrating 30 rather than 84 days of heat stress)	Data not available for full GBR
1998	Started late January and intensified by February/March	Aerial surveys of 654 reefs showed that 74% of inshore and 21% of offshore reefs had some level of bleaching. 11% of the reefs surveyed were documented to have severe levels of community bleaching.	Primarily coastal and most severe in the central and southern regions	Most reefs recovered fully, with fewer than 5% of inshore reefs suffering high coral mortality. The most severely affected reefs were in the Palm Island area, where up to 70% of corals died.	0.31	0 & 1
2002	Started in January with the worst over by April	This mass bleaching event was slightly more severe than the 1998 event. Aerial surveys of 641 reefs revealed bleaching in 54%. Nearly 41% of offshore and 72% of inshore reefs had some level of bleaching. Only 10% of reefs surveyed were documented to have severe levels of community bleaching.	Bleaching was more widespread than in 1998, and affected offshore reefs in the central region that were not impacted in 1998	Reef recovery was generally good, with fewer than 5% of the reefs suffering high mortality. The worst affected reefs were in the Bowen area, where about 70% of corals died.	1.16	1
2016	February and May	Bleaching intensity was severe (more than 60% of coral cover) at 38% of surveyed reefs. Although most reefs exhibited some degree of bleaching, this bleaching varied in intensity (from less than 10% to more than 90% of coral cover) and was patchy throughout most of the central and southern management areas. Intense prolonged heat stress capable of causing coral mortality affected more than 30% of the reef. Overall coral mortality was 22% for the entire GBR. Coral mortality was highest in the northern section, where heat stress was most intense.	Bleaching was even more extensive and much more severe, especially in the northern regions, and to a lesser extent in the central regions, where many coastal, mid-shelf, and offshore reefs were affected. Bleaching intensity decreased along a southern gradient.	The proportion of reefs with severe bleaching (> 60% of corals bleached) was over four times higher compared to 1998 or 2002.	2.62	1

(Continued)

TABLE 1 (Continued)

Summary of observed major bleaching events on the GBR for timepoints at which extensive observations exist (observations from AIMS 2016, 2017, 2018, 2019, 2020, 2021, 2022a, 2022b; Dunne et al. 2022; Hughes et al. 2017). Maximum Degree Heating Weeks (DHWs) for Low Isles, GBR (16.375°S 145.575 °E) calculated using the 1985–2022 integration period. Predominant heatwave category (v1.01) for the GBR. Both calculated as per Coral Reef Watch from the CoralTemp v3.1 data product (NOAA 2018). Heatwave categories are defined as No Marine Heatwave (0), Moderate (1), Strong (2), Severe (3), Extreme (4), and Beyond extreme (5) as per (Hobday et al. 2016; NOAA 2018).

Year	Timescale	Description of event	Areas affected	Further details	Maximum daily DHW (for Low Isles)	Predominant GBR heatwave category
2017	March	Intense heat stress capable of causing widespread coral mortality occurred during the 2017 event, mostly in the central GBR. Bleaching intensity was severe at 20% of the surveyed reefs. Some bleaching was observed in the southern region, but Long-Term Monitoring Program surveys during the next few years detected only declines in the coral cover in reef communities in the northern region (during the 2015–17 surveys) and central regions (during the 2017–19 surveys), aligning with the intense heat stress footprint.	Severe coral bleaching affected the central one-third of the GBR.	This back-to-back mass bleaching in 2016 and 2017 was unprecedented and collectively affected two-thirds of the GBR. The southern sector avoided intense heat stress and substantial bleaching in both years.	11.5	1 & 2
2020	SSTs in February were the highest recorded in 120 years of observations.	Heat stress extended into the southern GBR and caused high levels of community coral bleaching in this region compared to 2016 and 2017. 25% of reefs surveyed had severe levels of coral bleaching. Nearly 80% of the reefs throughout the GBR were exposed to heat stress levels that cause bleaching (above 5 °C for weeks). The majority of the reefs (about 60%) were exposed to moderate heat stress (4–8 °C for weeks).	Bleaching observed in northern, central, and southern regions.	Third mass bleaching event in five years to occur throughout the GBR. AIMS Long-Term Monitoring Program found that this did not cause substantial mortality, with ongoing recovery of coral cover detected in communities across the reef.	5.54	1 & 2

likely due to increased temperatures, but less observational data are available for these events (Coffroth et al. 1990; Glynn 1991; Oliver et al. 2018b). Where bleaching events were temporally close (within 3 years of each other), the probability of bleaching was reduced possibly due to coral acclimation. This was particularly the case during the 2002, 2017, and 2020 bleaching events but not intervening events in 2016 (Hughes et al. 2021). Such acclimation occurring between closely spaced heatwaves may be mechanistically explained via symbiont shuffling that occurs during sub-lethal bleaching, which is thought to promote host fitness via a preference for heat-tolerant symbiont species (Suggett and Smith 2011). Thus, field observations have been critical in developing our understanding of bleaching, as the presence of acclimation potential suggests that there may be variability in coral bleaching thresholds (the temperature at which thermally induced coral bleaching initiates) at sub-decadal timescales rather than longer-term static thresholds. This is important when considering historical bleaching records, as present-day bleaching thresholds may differ from those decades or hundreds of years ago; that is, the modern thermal thresholds at which corals bleach may be different from those in the recent and geologic past.

The sources of data used for such assessments are thus critical, particularly prior to 1998 when the onset of more detailed monitoring assessment began (including aerial surveys), with comprehensive reports being more common from 2015 (e.g., AIMS 2016, 2017, 2018, 2019, 2020, 2021, and 2022a). Moreover, records generally do not contain detailed data on bleaching event anatomy at wide spatial scales. In particular, the paucity of detailed bleaching event anatomy metrics is often due in part to the observational bleaching record being semi-quantitative, not always weighted for observer effort, or separated by taxon with little or no metric for species contribution (or contribution stability) over time (Kamenos and Hennige 2020). This, in turn, hinders a fully quantitative comparison of reef responses to particular heatwaves at longer timescales, as, for example, the coral community structure may have changed between heatwaves to confer a different widescale response. Smaller spatial scale assessments can thus be critical for a better understanding of local-level responses where such detailed assessments are possible (e.g., Hughes et al. 2021).

19.2.2. Approaches to Obtaining Pre-1970s Bleaching Records

Prior to the 1970s when more consistent records of GBR status began, few validated observational records were available. However, during an expedition to the GBR in 1928–1929, detailed observational records of two bleaching events in 1929 were made (e.g., Dunne et al. 2022; Oliver et al. 2009; Stephenson and Stephenson 1933; Yonge and Nicholls 1931; Yonge 1930). There is also inferred evidence of historical coral bleaching that is derived using corals as palaeoenvironmental and palaeobiotic proxies. This record is expanding and is fully considered in Webb (2024). That chapter also highlights the role that corals may perform in enabling an historical assessment of bleaching frequency and prevalence, particularly before the observational record, which is sparce prior to the 1970s. Although more recent observational data provide us with a short time window within which to place observations of recent rapid bleaching increases (e.g., Bellwood et al. 2004), they cannot provide a longer context of bleaching change (e.g., Baumann et al. 2019; Dishon et al. 2015; Kamenos and Hennige 2018; Kamenos and Hennige 2020). Here, it is pertinent to consider that absence of a consistent observational bleaching record prior to the 1970s could be the result of various, sometimes interacting scenarios: that bleaching did not occur prior to the 1970s, bleaching did occur but was not observed, bleaching occurred and was observed but not recorded, or bleaching did occur and was recorded, but data were not preserved as bleaching was not the aim of the study. Thus, when considering current bleaching trajectories, an unbiased metric of bleaching is required that is not driven solely by the start of complete visual observations.

This can, in part, be achieved by supplementing opportunistic observational records with proxy-derived records of coral bleaching (e.g., Baumann et al. 2019). Such proxy records are extracted from the structural and geochemical signatures of coral banding structures (e.g., Webb 2024) and allow us to place the bleaching patterns of recent heatwaves into a longer context. However, in such records, consideration is also needed with respect to bleaching signatures left by other localised stressors in addition to temperature, which include terrestrial runoff, turbidity, and nutrient supply changes (e.g., Carricart-Ganivet et al. 2012; D'Olivo and McCulloch 2017; Fabricius 2005; Goreau and Macfarlane 1990; Hendy et al. 2003; Hetzinger et al. 2016; Leder et al. 1991; Wiedenmann et al. 2013).

19.2.3. Pre-1970s Bleaching Events—The 1929 GBR (Low Isles) Observational Record

A detailed observational record exists of the 1929 bleaching events due the presence of scientists on Low Island (GBR — 16.375°S 145.575 °E) as part of the GBR Expedition from 1928–1929 (Stephenson and Stephenson 1933; Yonge 1930; Yonge and Nicholls 1931), which has recently received attention particularly by Dunne et al. (2022). During 1928–1929, two bleaching events were observed by members of the GBR Expedition: one during February 1929 followed by a second during June 1929. The February event appears to be restricted to Acroporids, while the June event, which occurred in deeper water, also included some massives (Dunne et al. 2022). The shallow-water February event was likely caused by a combination of high degree heating weeks (DHWs) and aerial exposure (Dunne et al. 2022; Yonge and Nicholls 1931), whereas the June event—a month of generally low bleaching on the reef slope—was likely primarily due to exposure and insolation-induced stress (Dunne et al. 2022). We note that the observations were restricted to Low Isles, so it is difficult to consider the wider extent of these bleaching events on the GBR. Critically, these events highlight the roles of multiple stressors including high DHWs as bleaching drivers even prior to the consistent observational record beginning in the 1970s (Dunne et al. 2022; Kamenos and Hennige 2018; Kamenos and Hennige 2020). We must remember that during the 1929 to 1998 interlude, an absence of observed bleaching events is unlikely to indicate that bleaching did not occur during this period but rather that 1929 happened to be one of the two years (1928–1929) that the GBR Expedition spent on the GBR (e.g., Dunne et al. 2022). Accordingly, this represents an opportunistic sampling time point prior to consistent observational records beginning in the 1970s.

19.2.4. Pre-1970s Bleaching Events—Global and GBR Proxy-Based Record

Globally, pre-1970 bleaching observations and reconstructions are less spatially consistent but can provide a wider context regarding historical, widespread, and possibly global bleaching. Bleaching reconstructions from the Red Sea suggest two mass bleaching events even since the last glacial maximum at ~20 kyr BP and prior to the Industrial Revolution (Dishon et al. 2015). Furthermore, in the South China Sea, coral deaths from 1865–1979 may have been caused by El Niño-associated warming (Yu et al. 2006; Yu et al. 2012).

Evidence of reef-wide stress on the GBR prior to the consistent observational record is present for specific years including 1929 (e.g., Dunne et al. 2022; Oliver et al. 2009) with regard to bleaching and 1877 with regard to stress (DeCarlo et al. 2019) (Figure 1). However, centennial-length bleaching reconstructions on the GBR also evidence bleaching prior to the onset of a consistent observational record in the 1970s. Multi-core decadal resolution reconstruction demonstrates GBR-wide bleaching trends with increasing (1620–1753), decreasing (1754–1820) and increasing (1821–2001) bleaching frequency (Kamenos and Hennige 2018; Kamenos and Hennige 2020). Bleaching prevalence (the proportion of coral cores exhibiting bleaching) decreased (1670–1774) before increasing since the late 1790s in step with positive temperature anomalies (Kamenos and Hennige 2018; Kamenos and Hennige 2020) (Figure 1).

19.3. MOVING BLEACHING BASELINES

When comparing contemporary and historical bleaching events in the context of heatwaves, we should consider the capacity of the coral holobiont for acclimatization and adaptation over time. This is because exposure to repeated heatwaves during which a coral colony bleaches but survives can lead to enhanced coral resilience (Hughes et al. 2021), likely through acclimatization and adaptation towards a

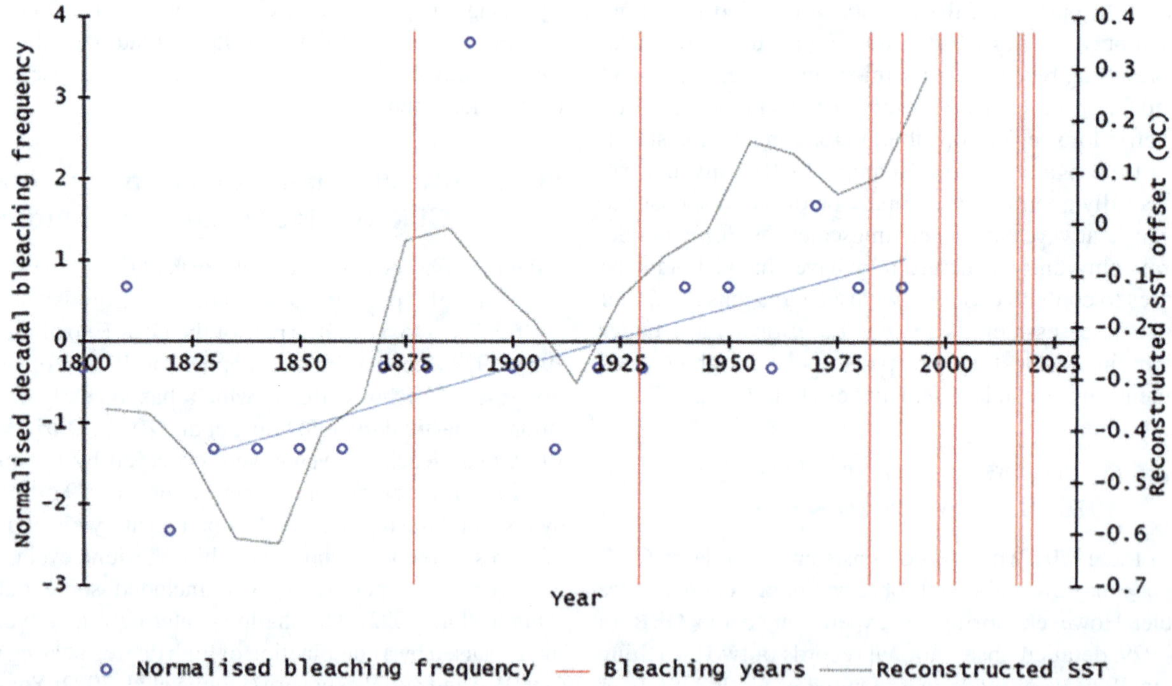

FIGURE 1 Observed and reconstructed individual bleaching/heat stress events on the GBR (red vertical lines (e.g., Coffroth et al. 1990; DeCarlo et al. 2019; Dunne et al. 2022; Glynn 1991; Oliver et al. 2009, 2018b)). Reconstructed normalised decadal bleaching frequency (blue dots, a high number indicates more bleaching is likely) and reconstructed normalised bleaching frequency trend (blue line) from Kamenos and Hennige (2018). Reconstructed SST for northern GBR from coral cores (Tierney et al. 2015) (black dashed line).

physiological ceiling beyond which resilience becomes uncertain (e.g., Kamenos and Hennige 2018; Kamenos and Hennige 2020; Kenkel and Matz 2016; Suggett and Smith 2011). We should therefore endeavour to incorporate the coral adaptive capacity when considering past bleaching events or making projections of future bleaching. For example, a heatwave during the 1800s could have caused mass coral bleaching events even though the heatwave reached a lower absolute thermal level than present-day heatwaves. Such a scenario could occur as historically, it is likely that the coral holobiont was not acclimatised to heatwave SSTs of the 1800s that were lower than those occurring in the present day (Figure 1—reconstructed SST from (Tierney et al. 2015)). This suggests that bleaching thresholds may have also changed over time as they are based on the maximum monthly mean temperature concept (bleaching thresholds are 1 °C above the summertime maxima (Glynn and D'Croz 1990)).

19.3.1. Great Barrier Reef Bleaching Thresholds over Time

Consequently, quantifying and comparing widespread detailed bleaching anatomy among heatwaves can be challenging, particularly where bleaching events are temporally dispersed or happen in temporally dispersed clusters (e.g., Hughes et al. 2021). It is, however, possible to obtain some insight into the possible roles of acclimation and adaptation by comparing bleaching thresholds (and to a lesser extent DHWs) for the Low Isles, which have an observational record from all observed bleaching events including the 1929 event. We generated bleaching thresholds using CoralTemp v3.1 with 0.5 degree pixels (NOAA 2018) and ERSST v.6 with 2.5 degree pixels (Huang et al. 2017) centred on Low Isles (16.375°S 145.575 °E). We calculated the bleaching thresholds using temporal MMM integrations of 1) 3 years (as data in Hughes et al. (2021) suggest a 3-year resilience pattern for the GBR, and for Caribbean corals as short as 56 days was observed in ca. 1978 (Coles and Jokiel 1978)) and 2) the 1985–2012 window as used by Coral Reef Watch (CRW). Maximum daily DHWs during mass bleaching years after 1985, from when data are available, were from CRW calculated using CoralTemp v3.1 data products (integrating stress over the 84 previous days) (NOAA 2018), while maximum DHW during the 1929 mass bleaching year was from Dunne et al. (2022) and calculated from instrumental records plus the CoralTemp v3.1 bleaching threshold for 1985–2012, but this integrates stress over the 30 previous days.

As expected, when calculated over the 3-year integration period, bleaching thresholds vary over time but show a general increase in more recent times (CoralTemp-derived 1987–2020 and ERSST-derived 1929–2020). The maximum increase present was 0.83 °C (CoralTemp-derived) over a 15-year period from 2002–2017. Notably, the 2017 bleaching temporally followed the 2016 bleaching event and appears to have imparted a 0.45 °C rise in thermal resilience in just one year (Figure 2). The bleaching threshold fell again between the 2017 and 2020 events by 0.15 °C as the area moved out of the 3-year window, which imparts a thermal resilience advantage. A similar pattern was generated in ERSST-derived bleaching thresholds, which although integrating notably larger spatial areas, was offset from the 5km CoralTemp product by between 0.01 and 0.56 °C. For

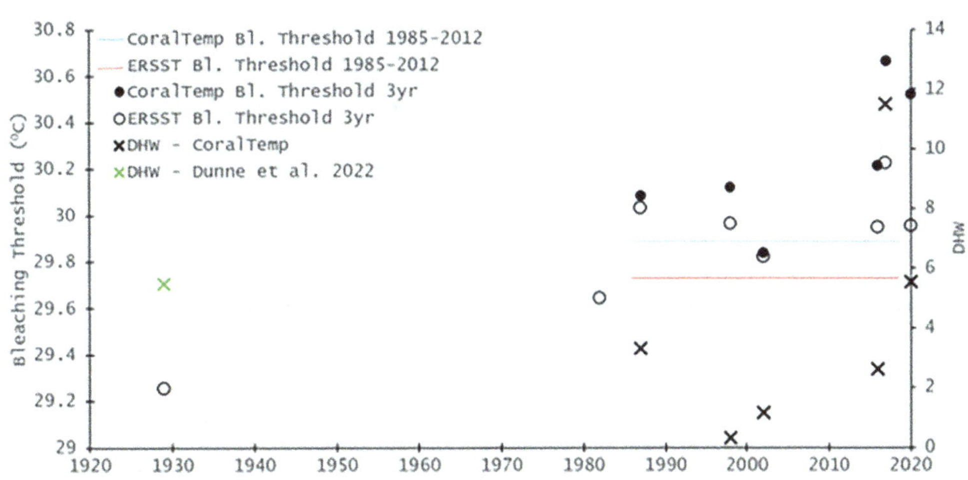

FIGURE 2 Bleaching thresholds and DHW for Low Isles, GBR. Blue line: bleaching threshold calculated using the 1985–2012 period as per CRW from the CoralTemp v3.1 data product at 5k resolution (ca. 0.5 degrees) (NOAA 2018). Red line: bleaching threshold calculated using the 1985–2012 period from the ERSST v.6 data product at 2.5 degree resolution. Bleaching threshold calculated using the MMM during the bleaching year and the two previous years (3-year window) using CoralTemp v3.1 (black dot 1985–2020) or ERSST v.6 (open dot 1929–2020). Black cross: daily DHW during heatwave mass bleaching years after 1987 from CRW calculated from CoralTemp v3.1 data products (integrates stress over the 84 previous days) (NOAA 2018). Green cross: daily DHW during the 1929 mass bleaching year (Dunne et al. 2022) calculated from instrumental records and CoralTemp v3.1 (integrates stress over the 30 previous days).

the Low Isles, the maximum daily DHW experienced in 1929 was likely similar to that experienced in 2020 with a lower maximum daily DHW during 1998, 2002, and 2016 (Figure 2, Table 1). We provide a caveat to this observation by noting that the 1929 DHW integrates only 30 days of accumulated heat stress versus the 84 days of accumulated heat stress in the 1998–2020 DHW event. Together, these observations on bleaching threshold and DHW suggest that acclimation and even adaptation can occur over short timescales given the right combination of mild thermal stress intensity and stressor frequency. Furthermore, heat stress (as indicated by DHW) has caused previous (1929) bleaching events on the GBR especially when combined with another stressor (Dunne et al. 2022).

19.4. CONCLUSIONS

Placing recently observed increases in GBR coral bleaching into a wider context is critical; acclimatisation and adaptation processes via bleaching and subsequent increases in thermal resilience (at least over 3-year timescales) mean that 1998 was highly likely not the first mass bleaching event on the GBR; this is evidenced in both the opportunistic observational record but also the inferred reconstructed bleaching record. We propose that when plotting current trajectories of coral bleaching, we need to consider increases in bleaching events over a historical timescale rather than just in recent decades, and we should thus not designate bleaching events numerically. Therefore, contrary to current practices, we propose that 1998 should not be labelled as the first mass bleaching event on the GBR. Importantly, the presence of bleaching prior to the 1970s does not detract from the depth of evidence that anthropogenic warming drives increasing trends in coral bleaching. Rather, longer time series of bleaching enable a more robust approach for evidencing that bleaching prevalence and frequency are increasing in response to global change.

REFERENCES

AIMS (2016) Annual Summary Report on Coral Reef Condition for 2015/16. Australian Institute of Marine Science, Government of Australia.

AIMS (2017) Annual Summary Report on Coral Reef Condition for 2016/17. Australian Institute of Marine Science, Government of Australia.

AIMS (2018) Annual Summary Report on Coral Reef Condition for 2017/18. Australian Institute of Marine Science, Government of Australia.

AIMS, (2019) Annual Summary Report on Coral Reef Condition for 2018/19. Australian Institute of Marine Science, Government of Australia.

AIMS (2020) Annual Summary Report on Coral Reef Condition for 2019/20. Australian Institute of Marine Science, Government of Australia.

AIMS (2021) Annual Summary Report on Coral Reef Condition for 2020/21. Australian Institute of Marine Science, Government of Australia.

AIMS, (2022a) Annual Summary Report on Coral Reef Condition for 2021/22. Australian Institute of Marine Science, Government of Australia.

AIMS (2022b) Australian Institute of Marine Science, Government of Australia. www.aims.gov.au. Accessed December, 2022.

Baumann JH, Ries JB, Rippe JP, Courtney TA, Aichelman HE, Westfield I, Castillo KD (2019) Nearshore coral growth declining on the Mesoamerican Barrier Reef System. Global Change Biology 25:3932–3945.

Bellwood DR, Hughes TP, Folke C, Nyström M (2004) Confronting the coral reef crisis. Nature 429:827.

Carricart-Ganivet JP, Cabanillas-Terán N, Cruz-Ortega I, Blanchon P (2012) Sensitivity of calcification to thermal stress varies among genera of massive reef-building corals. PloS One 7:e32859.

Coffroth MA, Lasker HR, Oliver JK, (1990) Coral mortality outside of the Eastern Pacific during 1982–1983: Relationship to El Niño, in: Glynn PW (ed.), Elsevier Oceanography Series. Elsevier, pp. 141–182.

Coles S, Jokiel PL (1978) Synergistic effects of temperature, salinity and light on the hermatypic coral Montipora verrucosa. Marine Biology 49:187–195.

DeCarlo TM, Harrison HB, Gajdzik L, Alaguarda D, Rodolfo-Metalpa R, D'Olivo J, Liu G, Patalwala D, McCulloch MT (2019) Acclimatization of massive reef-building corals to consecutive heatwaves. Proceedings of the Royal Society B: Biological Sciences 286:20190235.

Dishon G, Fisch J, Horn I, Kaczmarek K, Bijma, J, Gruber DF, Nir O, Popovich Y, Tchernov D (2015) A novel paleo-bleaching proxy using boron isotopes and high-resolution laser ablation to reconstruct coral bleaching events. Biogeosciences 12:5677–5687.

D'Olivo JP, McCulloch MT (2017) Response of coral calcification and calcifying fluid composition to thermally induced bleaching stress. Scientific Reports 7:2207.

Dunne RP, Brown BE, Spencer T, Woodworth PL (2022) Coral bleaching at Low Isles during the 1928–1929 Great Barrier Reef Expedition. Coral Reefs 41:1595–1610.

Fabricius KE (2005) Effects of terrestrial runoff on the ecology of corals and coral reefs: Review and synthesis. Marine Pollution Bulletin 50:125–146.

Frölicher TL, Fischer EM, Gruber N (2018) Marine heatwaves under global warming. Nature 560:360–364.

Glynn PW (1991) Coral reef bleaching in the 1980s and possible connections with global warming. Trends in Ecology & Evolution 6:175–179.

Glynn PW, D'Croz L (1990) Experimental evidence for high temperature stress as the cause of El Niño coincident coral mortality. Coral Reefs 8:181–191.

Goreau TJ, Macfarlane AH (1990) Reduced growth rate of *Montastrea annularis* following the 1987–1988 coral-bleaching event. Coral Reefs 8:211–215.

Hendy EJ, Lough JM, Gagan MK (2003) Historical mortality in massive *Porites* from the central Great Barrier Reef, Australia: Evidence for past environmental stress? Coral Reefs 22:207–215.

Hetzinger S, Pfeiffer M, Dullo WC, Zinke J, Garbe-Schönberg D (2016) A change in coral extension rates and stable isotopes after El Niño-induced coral bleaching and regional stress events. Scientific Reports 6:32879.

Hobday AJ, Alexander LV, Perkins SE, Smale DA, Straub SC, Oliver ECJ, Benthuysen JA, Burrows MT, Donat MG, Feng M, Holbrook NJ, Moore PJ, Scannell HA, Sen Gupta A, Wernberg T (2016) A hierarchical approach to defining marine heatwaves. Progress in Oceanography 141:227–238.

Huang BX, Thorne PW, Banzon VF, Boyer T, Chepurin G, Lawrimore J, Thomas M, Vose R, Zhang H, et al. (2017) NOAA Extended Reconstructed Sea Surface Temperature (ERSST), Version 5. NOAA National Centers for Environmental Information.

Hughes TP, Kerry JT, Álvarez-Noriega M, Álvarez-Romero JG, Anderson KD, Baird AH, Babcock RC, et al. (2017) Global warming and recurrent mass bleaching of corals. Nature 543:373–377.

Hughes TP, Kerry JT, Connolly SR, Álvarez-Romero JG, Eakin CM, Heron SF, Gonzalez MA, Moneghetti J (2021) Emergent properties in the responses of tropical corals to recurrent climate extremes. Current Biology 31:5393–5399. E5393.

Kamenos NA, Hennige SJ (2018) Reconstructing four centuries of temperature-induced coral bleaching on the Great Barrier Reef. Frontiers in Marine Science 5:283.

Kamenos NA, Hennige SJ (2020) Commentary: Commentary on Reconstructing Four Centuries of Temperature-Induced Bleaching on the Great Barrier Reef by Hoegh-Guldberg et al. 2019 and DeCarlo 2020. Frontiers in Marine Science 7.

Kenkel CD, Matz MV (2016) Gene expression plasticity as a mechanism of coral adaptation to a variable environment. Nature Ecology & Evolution 1:0014.

Leder JJ, Szmant AM, Swart PK (1991) The effect of prolonged "bleaching" on skeletal banding and stable isotopic composition in *Montastrea annularis*. Coral Reefs 10:19–27.

Liu G, Strong AE, Skirving WJ, Arzayus LF (2006) Overview of NOAA Coral Reef Watch Program's Near-Real-Time Satellite Global Coral Bleaching Monitoring Activities. Proceedings of the 10th International Coral Reef Symposium:1783–1793.

Logan CA, Dunne JP, Eakin CM, Donner SD (2014) Incorporating adaptive responses into future projections of coral bleaching. Global Change Biology 20:125–139.

Mieog JC, Olsen JL, Berkelmans R, Bleuler-Martinez SA, Willis BL, van Oppen MJH (2009) The roles and interactions of symbiont, host and environment in defining coral fitness. PloS One 4:e6364.

NOAA (2018) NOAA Coral Reef Watch. 2018, updated daily. NOAA Coral Reef Watch Version 3.1 Daily Global 5km Satellite Coral Bleaching Degree Heating Week Product, June 3, 2013-June 2, 2014. College Park, MD: NOAA Coral Reef Watch. Data Accessed using ERDDAP January, 2023.

Oliver ECJ, Donat MG, Burrows MT, Moore PJ, Smale DA, Alexander LV, Benthuysen JA, Feng M, Sen Gupta A, Hobday AJ, Holbrook NJ, Perkins-Kirkpatrick SE, Scannell HA, Straub SC, Wernberg T (2018a) Longer and more frequent marine heatwaves over the past century. Nature Communications 9:1324.

Oliver JK, Berkelmans R, Eakin CM (2009) Coral Bleaching in Space and Time, in: Lough J, van Oppen MJH (eds.), Coral Bleaching Patterns, Processes, Causes and Consequences. Springer-Verlag.

Oliver JK, Berkelmans R, Eakin CM (2018b) Coral Bleaching in Space and Time 2nd Edition, in: Lough J, van Oppen MJH (eds.), Coral Bleaching Patterns, Processes, Causes and Consequences. Springer-Verlag.

Palumbi SR, Barshis DJ, Traylor-Knowles N, Bay RA (2014) Mechanisms of reef coral resistance to future climate change. Science 344:895–898.

Rosenberg E, Koren O, Reshef L, Efrony R, Zilber-Rosenberg I (2007) The role of microorganisms in coral health, disease and evolution. Nature Reviews Microbiology 5:355–362.

Stephenson TA, Stephenson A (1933) Growth and asexual reproduction in corals. Scientific Reports Great Barrier Reef Expedition 1928–29. British Museum (Nat. Hist.) 3:167–217.

Suggett DJ, Smith DJ (2011) Interpreting the sign of coral bleaching as friend vs. foe. Global Change Biology 17:45–55.

Tierney JE, Abram NJ, Anchukaitis KJ, Evans MN, Giry C, Kilbourne KH, Saenger CP, Wu HC, Zinke J (2015) Tropical sea surface temperatures for the past four centuries reconstructed from coral archives. Paleoceanography 30:226–252.

Webb GE (2024) Advances in understanding climate change on the Great Barrier Reef using coral-based proxies. In: Wolanski E, Kingsford MJ (eds.), Oceanographic Processes of Coral Reefs. Physical and Biological Links in the Great Barrier Reef. CRC Press.

Wiedenmann J, D'Angelo C, Smith EG, Hunt AN, Legiret FE, Postle AD, Achterberg EP (2013) Nutrient enrichment can increase the susceptibility of reef corals to bleaching. Nature Climate Change 3:160–164.

Yonge CM (1930) A year on the Great Barrier Reef: The story of corals and of the greatest of their creations. Putnam.

Yonge CM, Nicholls A (1931) The structure, distribution and physiology of the zooxanthellae. Scientific Reports Great Barrier Reef Expedition 1928–29. British Museum (Nat. Hist.) 1:135–176.

Yu KF, Zhao JX, Shi Q, Chen T-G, Wang P-X, Collerson KD, Liu T-S (2006) U-series dating of dead *Porites* corals in the South China sea: Evidence for episodic coral mortality over the past two centuries. Quaternary Geochronology 1:129–141.

Yu K, Zhao JX, Shi Q, Price GJ (2012) Recent massive coral mortality events in the South China Sea: Was global warming and ENSO variability responsible? Chemical Geology 320–321:54–65.

20 Biophysical Processes Involved in the Initiation and Spread of Population Irruptions of Crown-of-Thorns Starfish on the Great Barrier Reef

Morgan S. Pratchett, Josie F. Chandler, Severine M. Choukroun, Peter C. Doll, Bethan J. Lang, Sarah Kwong, Carla C.M. Chen, Mike J. Emslie, Ciemon F. Caballes, Sven Uthicke, and Sam A. Matthews

20.1. INTRODUCTION

Population irruptions (also referred to as outbreaks; Birkeland 1982) of the western Pacific crown-of-thorns starfish (*Acanthaster* cf. *solaris*) remain one of the foremost causes of coral mortality on Australia's Great Barrier Reef (De'ath et al. 2012; Mellin et al. 2019; Pratchett et al. 2021a) and have contributed to coral loss and reef degradation at many locations throughout the Indo-West Pacific (Pratchett et al. 2014; Figure 1). During population irruptions, densities of crown-of-thorns starfish (CoTS) may exceed 10,000 starfish per km^2, and given their large size and very effective mode of feeding, often cause extensive coral loss (e.g., Chesher 1969; Kayal et al. 2012). During the most recent population irruptions of CoTS at Moorea in French Polynesia, for example, the cover of hard corals (order Scleractinia) declined from > 40% in 2005 to < 5% in 2010, with disproportionate effects on coral taxa (e.g., *Acropora* spp.) that are known to be highly preferred prey species of CoTS (Kayal et al. 2012). Although coral loss was partly compounded by other disturbances (e.g., cyclones), spatio-temporal variation in coral loss was strongly correlated with the size and abundance of CoTS (Kayal et al. 2012).

Population irruptions of CoTS have been reported from the Red Sea (Ormond et al. 1973) to Hawaii (Kenyon and Aeby 2009) but tend to be most pronounced and have the greatest ecological impacts within the western and central Pacific (e.g., Japan, Australia, Fiji, and French Polynesia). This area represents the geographic range of the western Pacific CoTS, *Acanthaster* cf. *solaris* (Haszprunar et al. 2017). This species is morphologically and genetically distinct from the three different species of CoTS (including *Acanthaster planci* (Linnaeus, 1958)) that occur in the Indian Ocean (Haszprunar et al. 2017). It is also possible that there is a distinct eastern Pacific species, nominally, *Acanthaster ellis* (Gray, 1850), but more work is required to establish the phylogenetic affinity and taxonomic status of Pacific CoTS species (Haszprunar et al. 2017). Regional differences in the occurrence and ecological impacts of population irruptions of *Acanthaster* spp. may relate to inter-specific differences in the biology or behaviour of different CoTS species (Pratchett et al. 2017a), though this is yet to be tested.

Although the ecological effects and broader ecosystems consequences of population irruptions of CoTS are well-documented (Pratchett et al. 2017a), considerable uncertainty remains about what causes or initiates population irruptions (e.g., Babcock et al. 2016a). Several alternative, not necessarily mutually exclusive, hypotheses have been proposed to account for population irruptions of CoTS (Table 1). These alternative hypotheses have been extensively discussed and debated (e.g., Birkeland and Lucas 1990; Babcock et al. 2016a), though this discourse (and much of the corresponding research) has largely focussed on biological processes (e.g., changes in reproductive output and larval survival of CoTS) or intrinsic bio-physical processes (*sensu* Treml et al. 2015), such as the availability of prey that directly affects growth and survival during successive life stages. Conversely, very limited consideration has been given to the environmental or habitat characteristics (but see Wooldridge and Brodie 2015) or extrinsic bio-physical process (*sensu* Treml et al. 2015). The majority of research into the biology and ecology of CoTS has also been conducted in the western Pacific and on *A.* cf. *solaris* (Pratchett et al. 2017a), including extensive research conducted on Australia's Great Barrier Reef (GBR). Recurrent population irruptions of CoTS recorded on the GBR appear to be somewhat unique in the global context (Pratchett et al. 2014) but may nonetheless provide a very important case study for improved understanding of the biological and physical processes that contribute to the initiation or occurrence of population irruptions of *Acanthaster* spp. more generally.

DOI: 10.1201/9781003320425-23

FIGURE 1 Elevated densities of Pacific CoTS (*Acanthaster* cf. *solaris*) observed in the southern Great Barrier Reef (Swains) in February 2021. Most hard corals (order Scleractinia) observed herein (most of which are *Acropora* sp.) are dead and were presumably eaten by the starfish.

Source: Photograph by Morgan S. Pratchett

20.2. CROWN-OF-THORNS STARFISH AND THE GREAT BARRIER REEF

20.2.1. Documented History of Population Irruptions

Population irruptions of *A.* cf. *solaris* were first documented on the GBR in the early 1960s (Barnes 1966; Pearson and Endean 1969), though there are anecdotal reports of much earlier irruptions or infestations (e.g., Vine 1973). The prevalence of CoTS spines in sediment cores on the GBR has also been presented as evidence for periodic population irruptions over thousands of years (e.g., Walbran et al. 1989); however, the utility of these data for demonstrating fluctuations in abundance of CoTS is very limited (Keesing et al. 1992; Pandolfi 1992). It is well-recognised that population irruptions of CoTS may well have occurred and gone largely unnoticed before the 1960s (Fabricus et al. 2010), but the specific timing and intensity of these events is uncertain. As such, this study (like most other analyses, e.g., Moran 1986; Pratchett et al. 2014) focusses on the well-documented occurrence of population irruptions on the GBR during the last 60 years and preferences information provided from systematic monitoring that commenced in the 1980s (Moran et al. 1988; Sweatman et al. 2011).

Four distinct episodes of population irruptions have been documented on the GBR since the early 1960s, starting in approximately 1962, 1979, 1995, and 2010 (Pratchett et al. 2014). The initiation and progression of each successive population irruption appears to have been generally similar (Kenchington 1977; Moran 1986; Moran et al. 1992; Reichelt et al. 1990; Babcock et al. 2020), starting in the northern GBR and then propagating southwards. Sequential progression of population irruptions is attributed to the successive establishment and spawning of distinct cohorts at increasing distances from the original source (or primary location) of population irruptions (Kenchington 1977; Reichelt et al. 1990; Vanhatalo et al. 2017). This infers however, that larval dispersal is relatively constrained or that subsequent (secondary) population irruptions are likely to occur within a relatively finite distance (200 km) of established population irruptions. Dispersal of CoTS may be highly constrained due to their limited planktonic larval duration (Hock et al. 2014), although Pratchett et al. (2017b) showed that CoTS larvae are competent to settle for 17–43 days post-fertilization. Larval dispersal is also influenced by directional water movement from reefs with significant reproductive potential (due to the size, distribution, and fecundity of resident CoTS; Babcock et al. 2016b; Rogers et al. 2017), which aids in the delivery of large numbers of CoTS larvae to nearby reefs that ultimately contribute to the establishment of further population irruptions (e.g., Dight et al. 1990; Hock et al. 2014).

20.2.2. The "Initiation Box"

Population irruptions of CoTS may originate independently at multiple locations across the GBR (e.g., Reichelt et al. 1990; Miller et al. 2015), although the successive occurrence of population irruptions on nearby reefs, especially

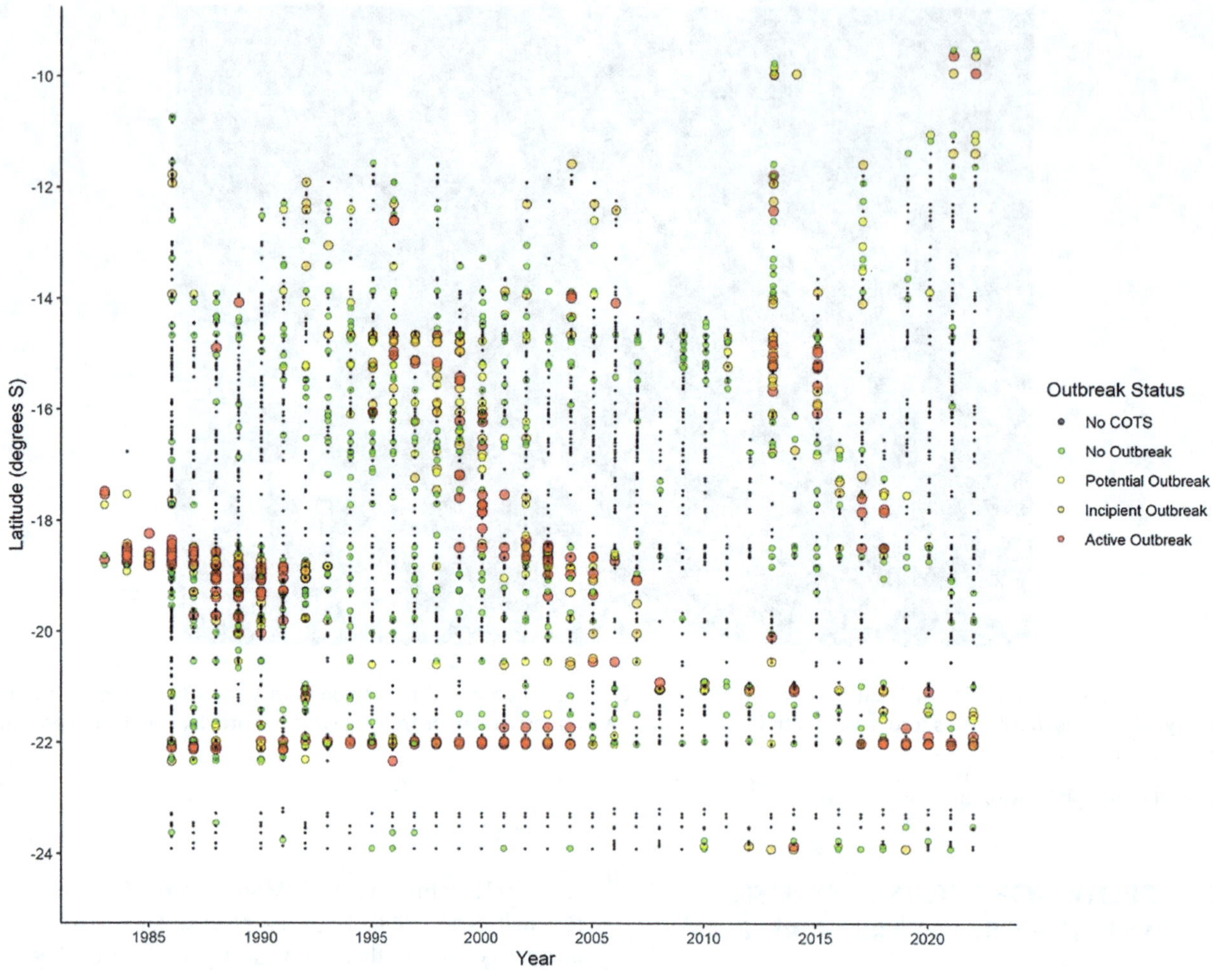

FIGURE 2 Spatiotemporal variation in the incidence and severity of CoTS outbreaks on the GBR, as documented using manta tow surveys by the Australian Institute of Marine Science Long-Term Monitoring Program (*sensu* Emslie et al. 2020). Each dot represents a survey conducted at a single individual reef in a given year.

through the central section of the GBR from 16–20°S (Figure 2), is often ascribed to the inherent spread of population irruptions in the direction of prevailing currents (e.g., Kenchington 1977; Reichelt et al. 1990; Vanhatalo et al. 2017). The critical question, therefore, is when and where do population irruptions originate? The location and timing of the initiation of population irruptions is particularly important for understanding and identifying potential causes of these recurrent disturbances on the GBR (e.g., Pratchett et al. 2014).

During the first two documented population irruptions on the GBR (in 1962 and 1979), elevated densities of CoTS were initially detected on reefs close to Cairns (e.g., Green Island); however, it is acknowledged that these infestations may have been facilitated by preceding increases in the densities of CoTS at other reefs (e.g., Pearson and Endean 1969; Kenchington 1977). Critically, there was limited broad-scale monitoring to assess the spatial patterns in the occurrence of population irruptions until after population irruptions were reported from Green Island (Vanhatalo et al. 2017). In subsequent population irruptions (in 1995 and 2010), elevated densities of CoTS were reported at reefs in the vicinity of Lizard Island at or before being reported further south at reefs near Green Island (Figure 2; Sweatman et al. 1998; Pratchett 2005, but see Wooldridge and Brodie 2015). Population irruptions of CoTS on the GBR are therefore assumed to originate on mid-shelf reefs in the area between Lizard Island (14°40'S) and Moore Reef (16o53'S), referred to as the *initiation box* (e.g., Pratchett et al. 2014). Recent and extensive research into putative causes of population irruptions of CoTS on the GBR has hence focussed on reefs within the broadly designated initiation box, but there is persistent controversy as to whether population irruptions originate mainly in the northern- (Sweatman et al. 1998; Pratchett 2005; Vanhatalo et al. 2017) or southern-most reefs (e.g., Fabricius et al. 2010; Wooldridge and Brodie 2015) within this region.

Recurrent annual monitoring across a broad cross-section of reefs on the GBR, which was established in 1986 (Emslie et al. 2020), provides important insights into spatiotemporal patterns of the occurrence of population irruptions among individual reefs and years (Figure 2; Vanhatalo et al. 2017). These monitoring data clearly indicate the southward progression of active outbreaks from reefs in the vicinity of Lizard Island (14.6°S) during 1995–1996, potentially representing the initiation of the third documented population irruption on the GBR (Figure 2; Vanhatalo et al. 2017). Notably, however, incipient outbreaks of CoTS were recorded north of 14°S in the preceding years (1992–1994). There is also evidence of independent or persistent population irruptions on reefs in the southern GBR (at approximately 22°S), though CoTS densities are consistently low further south in the Capricorn Bunkers (Figure 2; Miller et al. 2015).

20.2.3. Current Status of Population Irruptions

The current population irruption of CoTS on the GBR started in approximately 2010 (e.g., Vanhatalo et al. 2017), although there were constraints to the spatial and temporal extent of recurrent monitoring around this time that limits insights into the origination and initial spread of the fourth documented population irruption. There were not, for example, any surveys undertaken north of 14°S from 2007–2012 and no sampling undertaken within the putative initiation box (14–17°S) in 2012 (Figure 2). By 2013, population irruptions of CoTS were already widespread throughout this region (Figure 2; Harrison et al. 2017; Westcott et al. 2020) and also apparent on reefs to the north (at 12°S). From 2016–2018, population irruptions of CoTS occurred predominantly on reefs between 17 and 19°S (between Cairns and Townsville) in the central section of the GBR (Figure 2). These population irruptions almost certainly resulted from the southward propagation of population irruptions that were previously reported (2013–2015) on reefs north of Cairns (e.g., Hock et al. 2014), but these links were not clearly apparent from monitoring. Notably, the documented incidence of population irruptions north of Cairns was very limited after 2015 (Figure 2), partly due to limited monitoring in this region.

Despite the limited capacity of monitoring to clearly establish relative changes in the abundance (specifically the occurrence of outbreaks) in different regions during the latest (fourth) documented population irruption, extensive sampling of individual CoTS throughout this area (in 2013–2014) allowed for comparisons of population structure and demography (e.g., MacNeil et al. 2017). CoTS were collected across multiple reefs at four distinct regions: i) the northern sector of the initiation box (reefs around Lizard Island), ii) the central sector of the initiation box (reefs off Cooktown), iii) the southern sector of the initiation box (reefs off Cairns), and iv) to the southern edge of the outbreak in 2014 (reefs off Townsville). CoTS ranged in size from 50- to 510-mm diameter and suggest that there was a prolonged build-up in starfish numbers through multiple successive recruitment events at all reefs (e.g., Stump 1996; Pratchett 2005). Based on quantification of spine banding, CoTS collected from 2013–2014 are estimated to have recruited between 2005 and 2013. Notwithstanding potential development delays (Deaker et al. 2020), very few (< 7%) of the individuals recruited before 2007, and the majority (83.54%) are estimated to have recruited between 2007 and 2011. The average estimated year of recruitment in the northern sector (adjusted mean = 2008.53) was not significantly different from the central sector (adjusted mean = 2009.16, t-value = 1.861, $p = 0.090$) but was significantly earlier than in the southern sector (adjusted mean = 2009.48, t-value = 2.811, $p = 0.017$) and the Townsville region (adjusted mean = 2011.05, t-value = 6.622, $p < 0.001$). These data show that population irruptions were becoming established 1–2 years earlier on reefs in the northern sector of the initiation box (reefs around Lizard Island) than on reefs in the southern sector (Figure 3; Pratchett e al. 2015). However, given the almost simultaneous build-up of CoTS densities across all reefs within the broadly defined initiation box, it is not clear that population irruptions in the southern sector (at reefs off Cairns) necessarily arose from the preceding establishment of population irruptions on reefs to the north.

Extensive monitoring along the length of the GBR has been undertaken in recent years (2021–2022) and revealed limited incidence of population irruptions of CoTS (Figure 2). Most notably, established outbreaks in the central GBR have waned since 2015, potentially due to extensive culling undertaken on reefs with known outbreaks of CoTS (e.g., Westcott et al. 2020). The apparent suppression of population irruptions since 2015 also coincides with the incidence of recurrent and severe marine heatwaves that have caused extensive coral bleaching and coral loss (Hughes et al. 2017; Pratchett et al. 2021a), especially in the northern and central GBR. Although adult CoTS are sensitive to elevated temperatures (Lang et al. 2022), the links among severe marine heatwaves, extensive coral loss, and the apparent demise of population irruptions of CoTS are unknown, but do appear more than coincidence. The only population irruptions reported from 2021–2022 were on reefs in far northern GBR (north of 12°S), potentially representing the early initiation of renewed population irruptions that are expected to occur within coming years (Babcock et al. 2020), and on reefs in the southern GBR (21–24°S), which appear to have arisen prior to 2017 independently of population irruptions in the northern and central GBR.

20.3. Putative Causes of Population Irruptions

When population irruptions were documented in the late 1960s, it was generally assumed that these were new and unprecedented phenomena (e.g., Pearson and Endean 1969). As such, researchers looked to recent and marked changes in environmental or biological conditions to explain population irruptions of CoTS and inevitably focussed on increasing anthropogenic pressures (e.g., over-fishing; Endean 1969). It is possible, however, that population irruptions of CoTS are a natural phenomenon (Weber and Woodhead

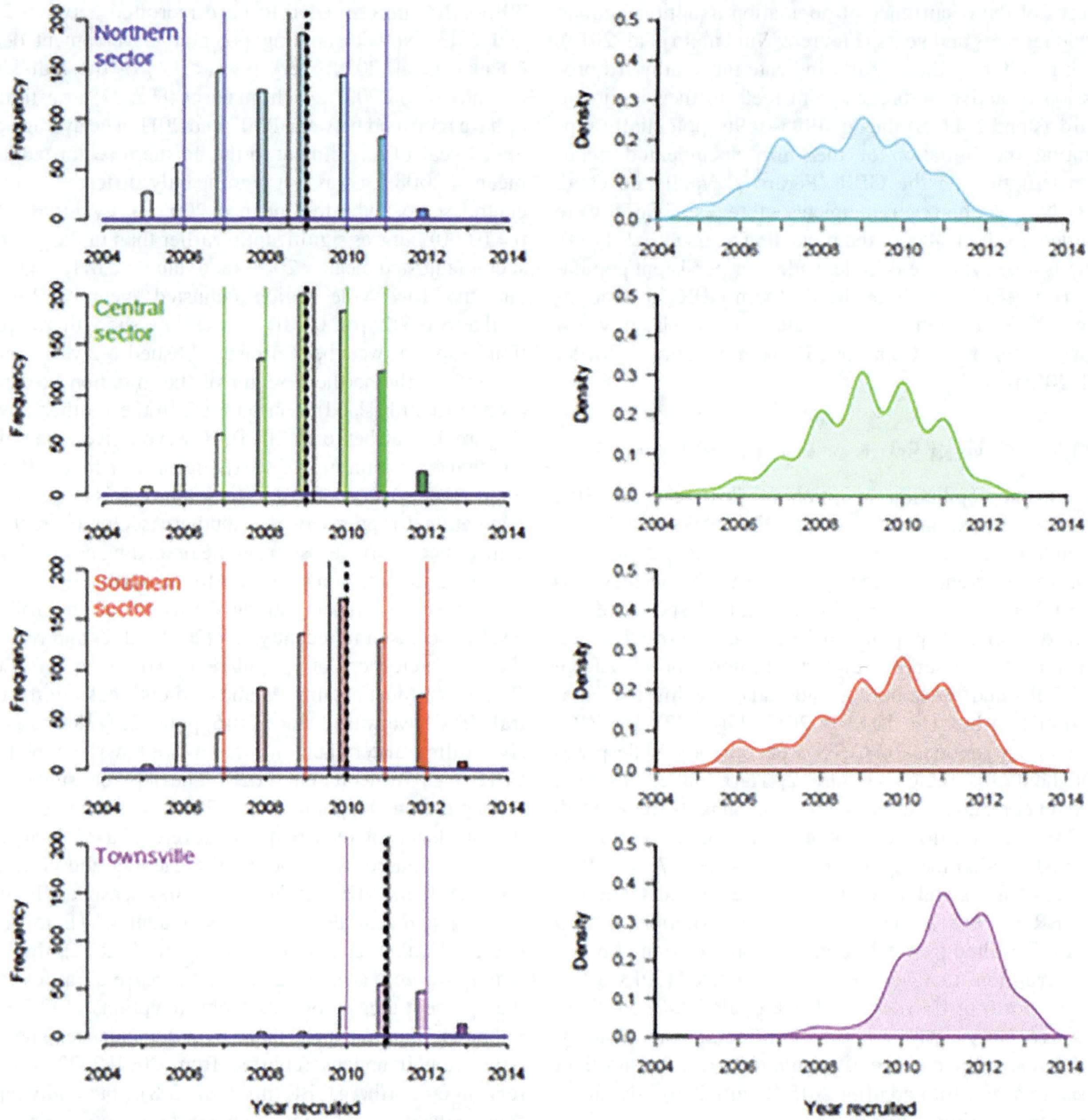

FIGURE 3 Frequency distribution (left) and density curves (right) of the estimated year of recruitment of CoTS collected (from 2013–2014) from the northern sector (Lizard Island, MacGillivray Reef, North Direction Island, and South Direction Island), central sector (Emily Reef, Endeavour Reef, Pickersgill Reef, and Spitfire Reef) and southern sector (Arlington Reef, Elford Reef, McCulloch Reef, and Hedley Reef) of the putative initiation box and the region off Townsville (Bramble Reef, Centipede Reef, Davies Reef, Rib Reef, and Trunk Reef). Solid black lines represent the mean, dashed lines indicate the median, and coloured lines identify the 10th, 25th, 75th, and 90th percentiles. For more information, see Pratchett et al. (2015).

1970; Vine 1973; Moore 1978) that simply went unnoticed prior to the 1960s (Vine 1973). Importantly, CoTS have a unique combination of life-history characteristics (e.g., very high fecundity, short generation times, developmental plasticity, generally high mortality during their early life history, and generalised patterns of prey and habitat use) that are inevitably linked to instabilities in population size (Birkeland 1989; Uthicke et al. 2009; Pratchett et al. 2014). Even if CoTS are pre-disposed to population irruptions, this does not explain specific spatiotemporal patterns in the abundance of CoTS, such as the recurrent onset of population irruptions at seemingly regular (15–17 years) intervals on the GBR (e.g., Babcock et al. 2020). A range of critical life-history processes might account for inherent population increases (Table 1) and population collapse in the aftermath of population irruptions (e.g., increased susceptibility to disease among high densities of adults; Rivera-Posada et al. 2011). However, increased consideration of extrinsic factors is needed. Oscillatory dynamics in the abundance of pest species are often moderated by changes in resource availability (Pratchett and Cumming 2019). For CoTS, resource limitation may occur not only among coral-feeding adults, whereby fecundity (if not growth and survival) is constrained by availability of specific coral prey (Caballes et al. 2016) but also during settlement, where juvenile CoTS initially feed on crustose coralline algae (Wilmes et al. 2020).

The potential role of predatory release in causing population irruptions of CoTS was initially ascribed to intensive depletion of Giant Triton (*Charonia tritonis*) in the northern GBR preceding seemingly unprecedented population irruptions of CoTS in the 1960s (e.g., Endean 1969). There is also evidence that the incidence or severity of population irruptions of CoTS is greatest in areas subject to sustained fishing pressure (Dulvy et al. 2004; Sweatman 2008; Westcott et al. 2020; Kroon et al. 2021), suggesting that normally high biomass of predominant fishery targets may contribute to suppressing population irruptions of CoTS (Kroon et al. 2021). An increasing diversity of fishes have been reported to feed on CoTS (Cowan et al. 2017; Kroon et al. 2020), which may greatly reduce individual fitness and have important non-consumptive effects (e.g., Budden et al. 2019), even if these fishes are not capable of killing CoTS outright. Many reef species also readily feed on CoTS eggs, larvae, or newly settled juveniles (e.g., Cowan et al. 2017). However, there has yet to be any explicit research that quantifies variation in predation rates or risk over relevant spatial and temporal scales, other than showing the putative benefits of no-take marine reserves in moderating the incidence and severity of CoTS outbreaks (Sweatman 2008; Kroon et al. 2021; Caballes et al. 2022). Putative predators on the early life-stages of CoTS are not generally subject to direct fisheries exploitation. However, anthropogenic degradation of reef habitats and/or trophic cascades resulting from the removal of keystone predators may have reduced the abundance of smaller-bodied predators (Dulvy et al. 2004; Pratchett et al. 2014) that might otherwise limit larval or juvenile survival (Wilmes et al. 2018).

Another pervasive hypothesis linking population irruptions of CoTS to anthropogenic activities is the terrestrial runoff hypothesis (*sensu* Fabricius et al. 2010). The central tenet of this hypothesis is that larval CoTS are generally food-limited and have very low survivorship during most years and under "normal" conditions (Lucas 1982). Elevated nutrients and thereby increased availability of plankton prey may, however, occur periodically due to terrestrial runoff (Birkeland 1982; Wooldridge et al. 2006) or specific oceanographic features (Houk et al. 2007; Miller et al. 2015). Corresponding increases in food availability and the nutritional quality of CoTS larvae are then expected to translate to greatly increased survival and developmental success of CoTS larvae (Lucas 1982; Fabricius et al. 2010). However, runoff and associated plumes also lower salinity, which have deleterious effects on larval survival of CoTS (Clements et al. 2022). Moreover, several studies have suggested that larval CoTS may be very resilient to food-limitation (Olson 1987; Wolfe et al. 2015, 2017; Caballes et al. 2017). Moderate increases in nutrient conditions and food availability may therefore increase larval survival and exacerbate the severity of outbreaks (Pratchett et al. 2014), though it is unresolved whether low-nutrient conditions would sufficiently constrain larval survival to prevent the initiation of population irruptions (Olson 1987; Wolfe et al. 2017). Periods of flooding rain and terrestrial run-off or upwellings that lead to elevated nutrients may therefore increase the probability that population irruptions become established or exacerbate the severity of population irruptions (Pratchett et al. 2014).

The initiation of population irruptions of CoTS (or *Primary Outbreaks*; Moran et al. 1992) is often characterised by progressive increases in the local abundance of CoTS (e.g., Pratchett 2005), which is attributed to the accumulation of CoTS from multiple cohorts over several successive years of elevated recruitment. It is also possible that successive cohorts of juvenile CoTS may defer transitioning from algal feeding to coral feeding if there is insufficient coral prey (Deaker et al. 2020). Although CoTS larvae have considerable potential for dispersal (Doll et al. 2022), population irruptions are most likely to originate in specific regions of the GBR where there are high levels of larval retention (Dight et al. 1990; Hock et al. 2014; Wooldridge and Brodie 2015), which facilitate high levels of self-recruitment and thereby reinforce feedback between the local abundance and reproductive potential of established populations of CoTS and successive levels of recruitment or replenishment. High levels of larval retention (combined with increased larval survival and settlement success) are fundamental in sustained increases in local densities of CoTS, but larval dispersal is necessary for the subsequent spread or propagation of population irruptions (Wooldridge and Brodie 2015) that affect the GBR more broadly. Wooldridge and Brodie (2015) suggested that relevant shifts in hydrodynamic conditions occur among reefs (including Green Island) directly offshore from Cairns, coinciding with the El Niño-Southern Oscillations (ENSO). Critically, high levels of larval retention occur during neutral ENSO phases, whereas high levels of larval dispersal occur during extreme La Niña and El Niño phases, to the north and south, respectively. The successive occurrence of conditions that enhance larval survival and promote larval retention, versus high levels of larval dispersal, is considered important in explaining why population irruptions originate and subsequently spread from reefs near Cairns (Wooldridge and Brodie 2015). Moreover, reefs in this region (e.g., Green Island) are also exposed to enhance phytoplankton concentrations, as reflected in elevated chlorophyll concentrations, because of flooding that occurs predominantly during La Niña phases (Wooldridge and Brodie

TABLE 1

Alternative (not necessarily mutually exclusive) hypotheses put forward to account for the initiation and or spread of population irruptions of *Acanthaster* spp. and the key biological and physical (including environmental) processes that underpin these hypotheses.

Hypotheses	Biological processes	Physical processes
Natural causes (Vine 1973), including marked changes in reproductive potential based on the proximity of spawning adults and spawning synchrony (Babcock et al. 1994)	Very high reproductive potential (Moore 1978; Pratchett et al. 2021b) and increased fecundity and larval quality for adult CoTS that feed on preferred coral prey (Caballes et al. 2016) and/or re-distribution of CoTS in relation to patchy availability of preferred prey (Dana et al. 1972)	High levels of larval retention, which allows for successive increases in local abundance of CoTS over multiple reproductive events (Pratchett 2005); Additionally, environmental cues or triggers for spawning (Caballes et al. 2017) and conditions that affect fertilisation success (Babcock et al. 1994)
Predator removal (Pearson and Endean 1969)	Increased size and/or abundance of adult starfish due to predatory release (Endean 1969) and/or increased survival of eggs and larvae due to a lack of planktonic predators (*sensu* Cowan et al. 2017)	Habitat complexity and the availability of physical refugia, which moderate risk of predation for adult CoTS while allowing continued feeding on coral prey (Endean and Stablum 1975)
Terrestrial runoff (Birkeland 1982) or larval starvation (Lucas 1982)	Marked increases in larval survival due to increased food availability (Lucas 1982) caused by terrestrial runoff (Birkeland 1982; Wooldridge et al. 2006) or oceanographic features (Houk et al. 2007; Miller et al. 2015); larval survival may, however, be compromised due to low salinity (Clements et al. 2022)	Drought-breaking rains (Birkeland 1982) and major flood events (Fabricius et al. 2010) or oceanic fronts and upwellings (Houk et al. 2007; Miller et al. 2015)
High levels of larval retention within specific clusters of reefs (Hock et al. 2014; Wooldridge and Brodie 2015)	Constraints on larval survival and replenishment due to food limitation (*sensu* Lucas et al. 1982) and strong stock-recruitment relationships that allow for localised population growth (Wooldridge and Brodie 2015); may also be compounded by accumulation of juveniles across multiple cohorts, which defer development until there is sufficient coral prey (e.g., Deaker et al. 2020)	ENSO and corresponding hydrodynamic conditions that influence larval retention versus dispersal (Wooldridge and Brodie 2015)

2015). Given increasing evidence that population irruptions of CoTS originate on reefs in the northern, if not far northern, section of the GBR (see Section 2), which are much less exposed to enhanced nutrient and phytoplankton concentrations, changes in larval survival may be less important than larval retention in allowing for initial increases in the local abundance of CoTS. Larval retention is likely to occur due to the "sticky water" effect (*sensu* Andutta et al. 2012), whereby there is limited flushing of water in areas with high densities of reefs. Critically, the larval retention hypothesis is one of the only hypotheses that invokes key biophysical processes (that might vary on relevant spatial and temporal scales) to explain the specific location and timing of population irruptions (e.g., Wooldridge and Brodie 2015).

20.4. BIOPHYSICAL PROCESSES AFFECTING POPULATION IRRUPTIONS OF CROWN-OF-THORNS STARFISH

The numerous hypotheses that have been put forward to explain the initiation or occurrence population irruptions of CoTS tend to focus on intrinsic biological factors, such as the aggregation and reproductive success of adults or the developmental success and survival of larvae (reviewed by Moran 1986; Birkeland and Lucas 1990; Pratchett et al. 2014; Table 1). These hypotheses are not necessarily mutually exclusive, and it is increasingly recognised that multiple factors contribute to the initiation or occurrence of population irruptions (Table 1; Reichelt et al. 1990; Babcock et al. 2020). Critically, intrinsic biological processes that allow for the periodic proliferation of adult CoTS are likely moderated by various extrinsic factors, not the least of which are ocean currents that influence the localised retention versus dispersal of CoTS larvae (Wooldridge and Brodie 2015). Moreover, environmental and habitat conditions almost certainly constrain local abundance and population growth for CoTS (e.g., availability of preferred coral prey for adult CoTS; Caballes et al. 2016), thereby highlighting the increased need for more holistic research and monitoring of intrinsic and extrinsic processes (Figure 4).

20.4.1. Ocean Currents

The circulation in the GBR is driven by the winds, tides, river inflow, and the oceanic inflow from the Coral Sea (Wolanski et al. 2024) via the South Equatorial Current (SEC). The SEC forms three zonal jets flowing westward through the Coral Sea (Couvelard et al. 2008; Gourdeau

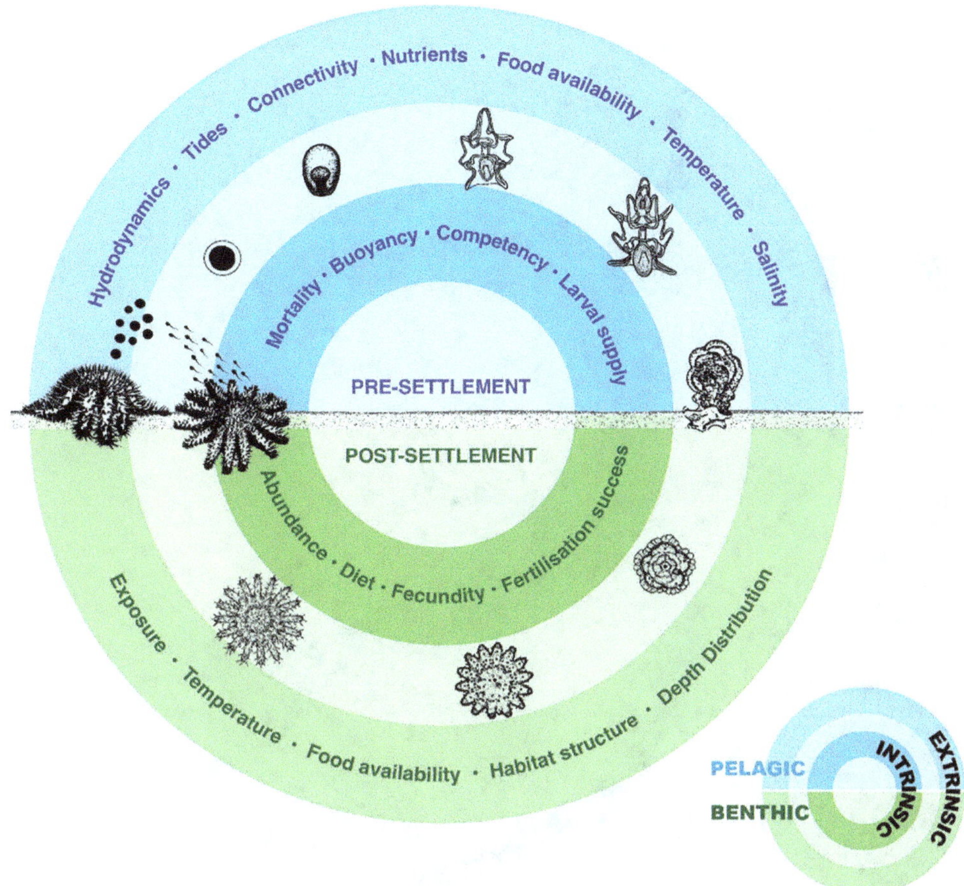

FIGURE 4 Extrinsic (outer circle) and intrinsic (inner circle) biophysical processes and factors driving variation during the pelagic pre-settlement (blue) and benthic post-settlement (green) stages in the life history of CoTS.

et al. 2008; Webb 2000), including the North Vanuatu Jet (NVJ) that intersects the GBR between 14.4°S and 14.7°S (Choukroun et al. 2010). The NVJ bifurcates to form the south-flowing East Australian Current (EAC) and the Hiri Gyre in the Gulf of Papua to the north (Ridgway et al. 2018), which are reflected in strong directional currents along the outermost edge of the GBR (Figure 5). Oceanic inflows do penetrate the perforate margin of the outer GBR and generally influence the directionality of flow within the GBR lagoon, especially among the outermost reefs. However, the rugged bathymetry of the GBR and the winds, tides, and river inflow strongly modulate oceanic flows in the vicinity of reefs and close to the mainland (Wolanski et al. 2024; Figure 5).

The prevailing longshore current within the putative initiation box (14–20°S) flows from north to south, which is consistent with southward propagation of population irruptions that originated in the vicinity of Lizard Island (Sweatman et al. 1998; Pratchett 2005; Pratchett et al. 2015; Vanhatalo et al. 2017). Similarly, population irruptions on reefs in the far northern GBR may arise from the northerly spread of larvae from reefs in the vicinity of Lizard Island. There are, however, considerable complexities and potential reversals in currents among mid-shelf reefs, especially during summer (Wolanski et al. 2024), which would modulate advection of CoTS larvae among reefs and influence fertilisation and spawning success.

20.5. MANAGING POPULATION IRRUPTIONS OF CROWN-OF-THORNS STARFISH

Population irruptions of CoTS are inexorably linked to their specific life-history characteristics, especially their exceptional reproductive capacity (Moore 1978; Uthicke et al. 2009). However, anthropogenic and environmental degradation of coral reef ecosystems may have disrupted ecological processes that otherwise regulate CoTS populations (e.g., Sweatman 2008; Fabricius et al. 2010; Kroon et al. 2021). For example, localised depletion in the abundance or activity of predators (e.g., due to overfishing) may have exacerbated population irruptions (e.g., Endean 1969; McCallum 1987), thereby accounting for documented increases in the incidence or severity of outbreaks in fished areas compared to no-take marine reserves (Sweatman 2008; Kroon et al. 2021). It is also possible that population irruptions of CoTS represent a natural and perennial threat to coral reef ecosystems (e.g., Vine 1973) that are now much more vulnerable to these disturbances given the

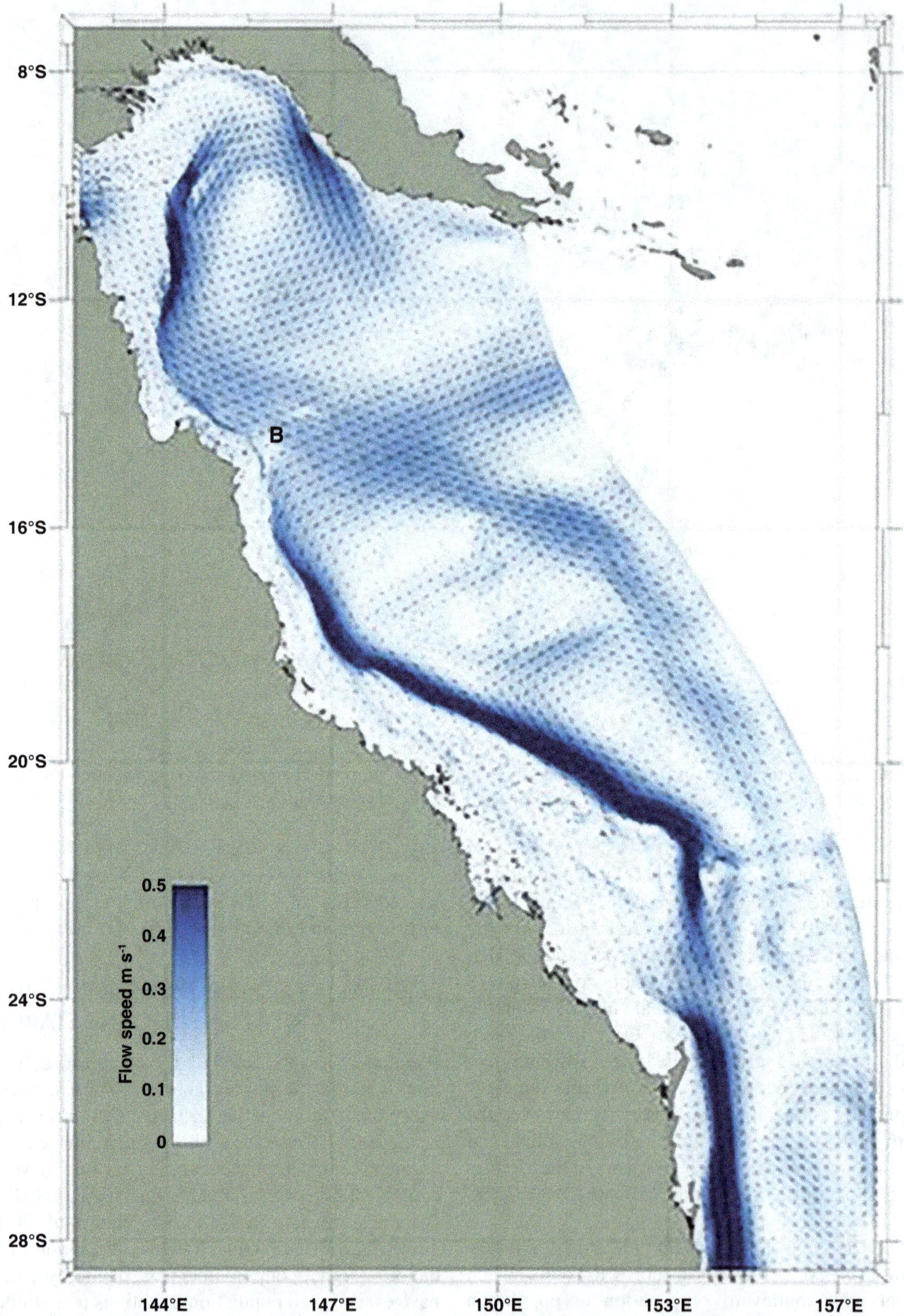

FIGURE 5 Map representing the mean speed in m/s and direction (arrows) at the surface (0–50 m) from the GBR4 eReef model, for the summer period (December to February) averaged from 2010 to 2021. GBR4 eReef hourly files were obtained from https://dapds00.nci.org.au/thredds/catalog/fx3/gbr4_v2/catalog.html. Note the bifurcation (B) of currents at approximately 14°S.

increased frequency and diversity of other disturbances and anthropogenic pressures, in addition to systematic and widespread degradation of reef ecosystems (Ortiz et al. 2018; Bellwood et al. 2019; Mellin et al. 2019). Effective management of population irruptions of CoTS is therefore warranted even if population irruptions are a natural phenomenon. Crucially, minimising coral loss caused by CoTS (e.g., Westcott et al. 2020) is also currently much more tractable than managing or moderating the effects of other large-scale disturbances, such as climate-induced coral bleaching. Emerging threats posed by climate change also provide an increased imperative to try to reverse the ongoing degradation of coral reef ecosystems to maximise the adaptive capacity and resilience of coral reef organisms (Hughes et al. 2003).

20.5.1. Large-Scale Management Interventions

If population irruptions of CoTS are caused or exacerbated by anthropogenic degradation of coastal environments and habitats (e.g., due to over-fishing or eutrophication), then it seems logical that redressing these issues would effectively prevent future population irruptions (Brodie et al. 2005) or at least reduce their frequency of occurrence, intensity of impact, and/or spread through reef systems. There are, however, considerable challenges associated with effectively reversing historical and significant effects of anthropogenic activities in coastal environments (Fidelman et al. 2013; Hughes et al. 2017). It is also not certain that these changes (if achieved) would necessarily prevent or reduce the frequency and intensity of future population irruptions (Pratchett et al. 2014; Westcott et al. 2020). Improvements in water quality (by minimising the run-off of sediments, nutrients, and pollutants) and fisheries management (through both the increasing establishment of no-take marine reserves and establishment and implementation of sustainable harvest limits) are nonetheless critical for improving the resilience and outlook for coral reefs (e.g., Wooldridge 2009) and may have subtle or incremental benefits for moderating the incidence or severity of population irruptions of CoTS (e.g., Babcock et al. 2016a; Westcott et al. 2020).

20.5.2. Local-Scale Management Interventions

The most direct and effective way to control population irruptions of CoTS is manual culling (e.g., Westcott et al. 2020), whereby individual (and mostly adult) starfish are collected or killed. Most contemporary culling programs use lethal injections of toxic or acidic chemicals to kill CoTS *in situ*, which is much less time consuming than physically collecting and removing starfish from the reef environment (Pratchett and Cumming 2019). Further increases in the efficiency and effectiveness of culling have been achieved through the development of the single-shot injection method, whereby modifications to the injecting apparatus, combined with the use of select chemicals (Rivera-Posada et al. 2014), ensured that high levels of mortality were achieved with a single injection. Even so, there are considerable constraints to the extent and effectiveness of culling (Fletcher et al. 2021), especially given that multiple discrete culling events are required in each location to effectively suppress CoTS densities and thereby protect local coral assemblages (Westcott et al. 2020). Recently, the effectiveness of intensive culling has been demonstrated at a broad range of locations across the GBR (Westcott et al. 2020). Significant, larger-scale impacts may also be achieved given suitable investment and capacity and by sustained culling (albeit at different locations) throughout the course of successive population irruptions (Fletcher et al. 2021).

The foremost constraint on the effectiveness of the current CoTS culling program in the GBR is the limited detectability of CoTS, which not only limits the proportion of CoTS that are actually killed during instantaneous culling (thereby explaining why multiple visits are needed to effectively suppress CoTS densities in any given location) but also makes it very difficult to know the current or impending status of population irruptions in different locations. Detectability of CoTS is particularly low for smaller individuals (MacNeil et al. 2017) such that culling has very limited effectiveness until CoTS attain sufficient size and density.

20.6. Key Knowledge Gaps and Research Priorities

Persistent controversy and uncertainty regarding the underlying causes of population irruptions of CoTS on the GBR (Kenchington 1987; Babcock et al. 2016a) reflect persistent gaps in the understanding of their population dynamics and demographic rates (Pratchett et al. 2021c). It is intuitive, for example, that population irruptions arise from increased reproductive success and replenishment, whether caused by increases in larval production or increased rates of larval survival and settlement success (e.g., Endean 1973; Uthicke et al. 2009). The issue is that there is limited temporal data on relevant demographic rates (e.g., settlement rates) necessary to understand the initiation and establishment of population irruptions. There are also conspicuous spatiotemporal patterns in the occurrence of population irruptions of CoTS on the GBR (Pratchett et al. 2014), which highlight the potential importance of extrinsic biophysical processes in causing or exacerbating population irruptions. New research methods and opportunities to detect and document the initiation of renewed population irruptions of CoTS (Uthicke et al. 2018, 2022; Pratchett et al. 2022) should therefore be matched with detailed monitoring of potentially important biophysical processes at relevant spatial and temporal scales, including improved understanding of hydrodynamic conditions within key reef areas in the northern and far northern GBR. It is also important to account for potential changes in the population structure and demography of CoTS due to changing environmental conditions (Uthicke et al. 2013, 2015; Lamare et al. 2014; Caballes et al. 2017; Lang et al. 2022). Given that population irruptions and climate-induced coral bleaching represent the two foremost threats to coral assemblages and, thus, to reef ecosystems (e.g., Mellin et al. 2019), more research is needed to understand the links among marine heatwaves, mass coral bleaching events, and population irruptions of CoTS.

20.6.1. Measuring Settlement Rates

Marked spatiotemporal variation in the abundance of CoTS is almost certainly linked to differential rates of larval supply and settlement (Zann et al. 1987; Uthicke et al. 2009; Wilmes et al. 2018). Notably, large-scale, multi-year studies of settlement have been instrumental in understanding population dynamics for marine organisms, such as fishes (e.g., Sale et al. 1984) and corals (e.g., Hughes et al. 2000). However, comparable research on settlement rates has not yet been conducted for CoTS, mainly due to logistical constraints and challenges associated with sampling, detecting, and identifying CoTS at or after settlement (Doll et al. 2021). Indeed, persistent failures in detecting newly settled individuals in shallow reef habitats on the GBR (e.g., Doherty and Davidson 1988; Johnson et al. 1992) led to the suggestion that CoTS must settle predominantly in deep (30–60 m), inter-reefal habitats (Johnson et al. 1991). This hypothesis was given added weight when it was experimentally shown that CoTS preferentially settle on the calcareous algae *Lithothamnium pseudosorum*, which predominates in deep, non-reefal habitats (Johnson et al. 1991). However, recent work on settlement cues suggests that many calcareous algal species that are prevalent across the depth continuum induce equally high or higher settlement rates, including algal species that primarily occur in shallow reef environments (Doll et al. 2023). These new insights challenge the hypothesis that larval settlement of CoTS predominantly occurs in deep water, and these insights are corroborated by documented settlement (Doll et al. 2021) and newly settled juveniles (Wilmes et al. 2020) at a range of relatively shallow depths (1–18 m). Importantly, new techniques developed to aid in the study of CoTS settlement (Doll et al. 2021) also provide an opportunity to add settlement as a variable in demographic models to resolve key information gaps, such as the stock-recruitment dynamics of this species.

To date, settlement rates of CoTS have been inferred based on the distribution and abundance of 6+-month-old juveniles or adults (e.g., Zann et al. 1987). However, the distribution and abundance of juvenile CoTS may be moderated by variation in post-settlement survival and movement (Wilmes et al. 2018), thereby obscuring patterns established at settlement. Explicitly linking high densities of adult starfish to elevated larval settlement rates is one of the foremost research priorities for advancing the understanding and management of CoTS (Pratchett et al. 2017a, 2021c). In some cases, seemingly sudden population irruptions of CoTS have been attributed to a single mass-settlement event (Zann et al. 1987; Houk and Raubani 2010) or may occur due to accumulation of cryptic juveniles that grow and transition *en masse* when food becomes available (Deaker et al. 2020), but there is also evidence of gradual and sustained increases in the local abundance of CoTS (Pratchett 2005) attributed to progressive accumulation of multiple cohorts over successive years of elevated settlement. Sudden versus protracted increases in population size have different explanations (Pratchett et al. 2014) and highlight the critical importance of assessing demographic processes that lead to initiation and establishment of population irruptions. Quantitative studies of natural settlement will also provide much needed insights into stock-recruitment relationships, which are critical for resolving the factors involved in the initiation of population irruptions (McCallum 1992).

20.6.2. Improved Understanding of Larval Retention and Dispersal

The origination and spread of population irruptions of CoTS are influenced by a combination of both biological (intrinsic) and physical (extrinsic) processes, both of which could benefit from increased research and improved understanding. Most importantly, it is no longer sufficient to consider only the dispersal of larvae from areas with established population irruptions. Rather, hydrodynamic modelling may be critical in identifying reefs or regions with high levels of larval retention (even if only under certain conditions) that may allow for the iterative increases in the local abundance of CoTS over successive years of self-recruitment (Hock et al. 2014; Wooldridge and Brodie 2015). Larval dispersal models for CoTS will also benefit from explicit data on the vertical position of gametes and larvae in the water column at various stages throughout their development, which can have a major bearing on potential retention of or advection away from natal reefs.

Larval dispersal models for CoTS rely very heavily on estimates of hydrodynamic connectivity between reefs (e.g., Dight et al. 1990; Hock et al. 2014), which is relatively well-studied in the central and southern sections of the GBR. However, given the potential importance of reefs in the far northern GBR, the lack of research and understanding of net water circulation in this region (Wolanksi and Lambrechts 2020) greatly constrains an understanding of the potential origination and spread of CoTS outbreaks. Critically, the existing array of oceanographic moorings deployed on the shelf and slope along the GBR to monitor the Western Boundary current of the Coral Sea (https://data.aims.gov.au/moorings/) extends southward from Lizard Island. The mooring array provides invaluable data to calibrate and validate hydrodynamic models of the GBR. There has also been very limited independent research on the physical oceanography in the far northern GBR (Wolanksi and Lambrechts 2020), which is needed to validate hydrodynamic models for the region and greatly increase spatiotemporal resolution in understanding water circulation.

20.6.3. Environmental Change

Changing environmental conditions, especially ocean warming and the increased incidence of severe marine heatwaves, pose a major threat to coral reef ecosystems, leading to observed and projected increases in the occurrence of mass coral bleaching and mortality (e.g., Hughes et al. 2017). Environmental change will also affect key biophysical processes that influence the distribution and abundance of marine species (Bashevkin et al. 2020), including CoTS (Wooldridge and Brodie 2015). Climate change may alter the force and direction of ocean currents, such as the SEC, and the intensity and regularity of ENSO and upwelling events (Collins et al. 2010; Luo and Rothstein 2011; Bashevkin et al. 2020; Figueiredo et al. 2021). These oceanographic changes may have various synergistic or antagonistic physical and biological impacts on the dispersal patterns of larval CoTS and, therefore, the distribution and density of their adult populations and outbreak risk (Bashevkin et al. 2020).

Environmental change will also have direct biological impacts on the performance of CoTS at all life stages (e.g., Uthicke et al. 2013, 2015; Kamya et al. 2016; Lang et al. 2022). Laboratory studies have revealed that temperatures ≥30 °C increase the speed of development, thus reducing the planktonic larval duration of CoTS (Uthicke et al. 2015), which increases the likelihood of larval retention. Other studies have shown reduced larval size and normality but increased juvenile growth at the same temperatures and under near-future ocean acidification (pH 7.6–7.8) conditions (Uthicke et al. 2013; Kamya et al. 2014, 2016; Lamare et al. 2014). At > 30 °C, adult CoTS also experience significant mortality (Lang et al. 2022). These observations indicate that further inevitable environmental change may lead to population declines or possibly redistributions of CoTS to more favourable habitats (Dight et al. 1990; Sill and Dawson 2021). Holistic models that incorporate both biological and biophysical effects of climate change will enable better predictions of the future risk of population irruptions (Wooldridge and Brodie 2015). However, there is also fundamental research needed to understand how CoTS respond to coral bleaching.

REFERENCES

Andutta FP, Kingsford MJ, Wolanski E (2012) 'Sticky water' enables the retention of larvae in a reef mosaic. Est. Coast. Shelf Sci. 101:54–63.

Babcock RC, Dambacher JM, Morello EB, Plagányi ÉE, Hayes KR, Sweatman HPA, Pratchett MS (2016a) Assessing different causes of crown-of-thorns starfish outbreaks and appropriate responses for management on the Great Barrier Reef. PLoS One 11(12):e0169048.

Babcock RC, Milton DA, Pratchett MS (2016b) Relationships between size and reproductive output in the crown-of-thorns starfish. Mar. Biol. 163(11):1–7.

Babcock RC, Mundy CN, Whitehead D (1994) Sperm diffusion models and in situ confirmation of long-distance fertilization in the free-spawning asteroid *Acanthaster planci*. Biol. Bull. 186(1):17–28.

Babcock RC, Plagányi ÉE, Condie SA, Westcott DA, Fletcher CS, Bonin MC, Cameron D (2020) Suppressing the next crown-of-thorns outbreak on the Great Barrier Reef. Coral Reefs 39(5):1233–1244.

Barnes JH (1966) The crown-of-thorns starfish as a destroyer of coral. Aust. Nat. Hist. 15:257–261.

Bashevkin SM, Dibble CD, Dunn RP, Hollarsmith JA, Ng G, Satterthwaite EV, Morgan SG (2020) Larval dispersal in a changing ocean with an emphasis on upwelling regions. Ecosphere 11(1):e03015.

Bellwood DR, Pratchett MS, Morrison TH, Gurney GG, Hughes TP, Álvarez-Romero JG, Day JC, Grantham R, Grech A, Hoey AS, Jones GP (2019) Coral reef conservation in the Anthropocene: Confronting spatial mismatches and prioritizing functions. Biol. Cons. 236:604–615.

Birkeland C (1982) Terrestrial runoff as a cause of outbreaks of *Acanthaster planci* (Echinodermata: Asteroidea). Mar. Biol. 69(2):175–185.

Birkeland C (1989) The Faustian traits of the crown-of-thorns starfish. Am. Sci. 77(2):154–63.

Birkeland C, Lucas J (1990) *Acanthaster planci*: Major management problem of coral reefs. CRC Press, Boca Raton.

Brodie J, Fabricius K, De'ath G, Okaji K (2005) Are increased nutrient inputs responsible for more outbreaks of crown-of-thorns starfish? An appraisal of the evidence. Mar. Poll. Bull. 51(1–4):266–278.

Budden C, Butler I, Wolfe K, Deaker D, Sweatman H, Byrne M (2019) Effect of sublethal predation on reproductive output of the crown-of-thorns starfish *Acanthaster* sp., with an overview of arm damage. Mar. Ecol. Prog. Ser. 629:103–116.

Caballes CF, Messmer V, Raymundo ML, Pratchett MS (2022) Prevalence and severity of sublethal injuries in crown-of-thorns starfish relative to marine reserves in the Great Barrier Reef. Aquat. Cons. 32(6):993–1004.

Caballes CF, Pratchett MS, Kerr AM, Rivera-Posada JA (2016) The role of maternal nutrition on oocyte size and quality, with respect to early larval development in the coral-eating starfish, *Acanthaster planci*. PLoS One 11(6):e0158007.

Caballes CF, Pratchett MS, Raymundo ML, Rivera-Posada JA (2017) Environmental tipping points for sperm motility, fertilization, and embryonic development in the crown-of-thorns starfish. Diversity 9(1):10.

Chesher RH (1969) Destruction of Pacific corals by the sea star *Acanthaster planci*. Science 165(3890):280–283.

Choat J, Russell B, Chin A (2019) The fish assemblages of the Great Barrier Reef: Their diversity and origin. Pages 389–406 In: Hutchings P, Kingsford M, and Hoegh-Guldberg O (Eds) The great barrier reef: Biology, environment and management. CSIRO Publishing, Clayton South, VIC.

Choukroun S, Ridd PV, Brinkman R, McKinna LI (2010) On the surface circulation in the western Coral Sea and residence times in the Great Barrier Reef. J. Geophys. Res. Ocean. 115(C6):C06013.

Clements M, Selvakumaraswamy P, Deaker D, Byrne M (2022) Freshening of Great Barrier Reef waters is deleterious for larval crown-of-thorns starfish, counter to the terrestrial run-off hypothesis. Mar. Ecol. Prog. Ser. 696:1–4.

Collins M, An S-I, Cai W, Ganachaud A, Guilyardi E, Jin F-F, Jochum M, Lengaigne M, Power S, Timmermann A, Vecchi G, Wittenberg A (2010) The impact of global warming on the tropical Pacific Ocean and El Niño. Nat. Geosci. 3:391–397.

Couvelard X, Marchesiello P, Gourdeau L, Lefèvre J (2008) Barotropic zonal jets induced by islands in the southwest Pacific. J. Phys. Oceanogr. 38(10):2185–2204.

Cowan ZL, Pratchett M, Messmer V, Ling S (2017) Known predators of crown-of-thorns starfish (*Acanthaster* spp.) and their role in mitigating, if not preventing, population outbreaks. Diversity 9(1):7.

Dana TF, Newman WA, Fager EW (1972) *Acanthaster* aggregations: Interpreted as primarily responses to natural phenomena. Pac. Sci. 26:355–372.

Deaker DJ, Agüera A, Lin HA, Lawson C, Budden C, Dworjanyn SA, Mos B, Byrne M (2020) The hidden army: Corallivorous crown-of-thorns seastars can spend years as herbivorous juveniles. Biol. Lett. 16(4):20190849.

De'Ath G, Fabricius KE, Sweatman H, Puotinen M (2012) The 27–year decline of coral cover on the Great Barrier Reef and its causes. Proc. Nat. Acad. Sci. USA 109(44):17995–17999.

Dight IJ, James MK, Bode L (1990) Modelling the larval dispersal of *Acanthaster planci*. Coral Reefs 9(3):125–134.

Doherty PJ, Davidson J (1988) Monitoring the distribution and abundance of juvenile *Acanthaster planci* in the central Great Barrier Reef. Pages 131–136 In: Choat JH et al. (eds) Proceedings of the 6th International Coral Reef Symposium, Vol. 2. Townsville.

Doll PC, Caballes CF, Hoey AS, Uthicke S, Ling SD, Pratchett MS (2022) Larval settlement in echinoderms: A review of processes and patterns. Oceanogr. Mar. Biol. Ann. Rev. 60:433–494.

Doll PC, Messmer V, Uthicke S, Doyle JR, Caballes CF, Pratchett MS (2021) DNA-based detection and patterns of larval settlement of the corallivorous crown-of-thorns sea star (*Acanthaster* sp.). Biol. Bull. 241(3):271–285.

Doll PC, Uthicke S, Caballes CF, Diaz-Pulido G, Abdul Wahab MA, Lang BJ, Jeong SY, Pratchett MS (2023) Settlement cue selectivity by larvae of the destructive crown-of-thorns starfish. Biol. Lett. 19:20220399.

Dulvy NK, Freckleton RP, Polunin NVC (2004) Coral reef cascades and the indirect effects of predator removal by exploitation. Ecol. Lett. 7:410–416.

Emslie MJ, Bray P, Cheal AJ, Johns KA, Osborne K, Sinclair-Taylor T, Thompson CA (2020) Decades of monitoring have informed the stewardship and ecological understanding of Australia's Great Barrier Reef. Biol. Cons. 252:108854.

Endean R (1969) Report on investigations made into aspects of the current *Acanthaster planci* (crown-of-thorns) infestations of certain reefs of the Great Barrier Reef. Queensland Department of Primary Industries (Fisheries Branch), Brisbane.

Endean R. (1973) Population explosion of *Acanthaster planci* and associated destruction of hermatypic corals in the Indo-West Pacific region. Pages 389–438. In: Jones Oa, Endean R (eds) Biology and geology of coral reefs, Vol. 2, Biology 1. Academic Press, New York.

Endean R, Stablum W (1975) A study of some aspects of the crown-of-thorns starfish (*Acanthaster planci*) infestations of reefs of Australia's Great Barrier Reef. Atoll Res. Bull. 167:1–76.

Fabricius KE, Okaji K, De'Ath G (2010) Three lines of evidence to link outbreaks of the crown-of-thorns seastar *Acanthaster planci* to the release of larval food limitation. Coral Reefs 29(3):593–605.

Fidelman PI, Leitch AM, Nelson DR (2013) Unpacking multi-level adaptation to climate change in the Great Barrier Reef, Australia. Global Environ. Change 23(4):800–812.

Figueiredo J, Thomas CJ, Deleersnijder E, Lambrechts J, Baird AH, Connolly SR, Hanert E (2021) Global warming decreases connectivity among coral populations. Nat. Clim. Change 12:83–87.

Fletcher CS, Castro-Sanguino C, Condie S, Bozec YM, Hock K, Gladish, DW, Mumby PJ, Westcott DA (2021) Regional-scale modelling capability for assessing crown-of-thorns starfish control strategies on the Great Barrier Reef. Report 5.1 to the Australian Government's National Environmental Science Program (NESP) Tropical Water Quality (TWQ) Hub. Reef and Rainforest Research Centre Limited, Cairns.

Gourdeau L, Kessler WS, Davis RE, Sherman J, Maes C, Kestenare E (2008) Zonal jets entering the Coral Sea. J. Phys. Oceanogr. 38(3):715–725.

Harrison HB, Pratchett MS, Messmer V, Saenz-Agudelo P, Berumen ML (2017) Microsatellites reveal genetic homogeneity among outbreak populations of crown-of-thorns starfish (*Acanthaster* cf. *solaris*) on Australia's Great Barrier Reef. Diversity 9(1):16.

Haszprunar G, Vogler C, Wörheide G (2017) Persistent gaps of knowledge for naming and distinguishing multiple species of crown-of-thorns-seastar in the *Acanthaster planci* species complex. Diversity 9(2):22.

Hock K, Wolff NH, Condie SA, Anthony KR, Mumby PJ (2014) Connectivity networks reveal the risks of crown-of-thorns starfish outbreaks on the Great Barrier Reef. J. Appl. Ecol. 51(5):1188–1196.

Houk P, Bograd S, van Woesik R (2007) The transition zone chlorophyll front can trigger *Acanthaster planci* outbreaks in the Pacific Ocean: Historical confirmation. J. Oceanogr. 63(1):149–154.

Houk P, Raubani J (2010) *Acanthaster planci* outbreaks in Vanuatu coincide with ocean productivity, furthering trends throughout the Pacific Ocean. J. Oceanog. 66(3):435-438.

Hughes TP, Baird AH, Bellwood DR, Card M, Connolly SR, Folke C, Grosberg R, Hoegh-Guldberg O, Jackson JB, Kleypas J, Lough JM, Marshall P, Nyström M, Palumbi SR, Pandolfi JM, Rosen, B, Roughgarden J (2003) Climate change, human impacts, and the resilience of coral reefs. Science 301(5635): 929–933.

Hughes TP, Baird AH, Dinsdale EA, Moltschaniwskyj NA, Pratchett MS, Tanner JE, Willis BL (2000) Supply-side ecology works both ways: The link between benthic adults, fecundity, and larval recruits. Ecology 81(8):2241–9.

Hughes, TP, Kerry JT, Alvarez-Noriega M, Alvarez-Romero JG, Anderson KD, Baird AH, Babcock RC, Beger M, Bellwood DR, Berkelmans Rl, Bridge TC, Butler ER, Dia-Pulido G, Eakin M, Figueira WF, Gilmour JP, Harrison HB, Heron SF, Hoey AS, Hobbs J-PA, Hoogenboom MO, Kennedy EV, Kuo C-Y, Lough JM, Lowe RJ, Liu G, McCulloch MT, Malcolm HA, McWilliam MJ, Pandolfi JM, Pears RJ, Pratchett MS, Schoepf V, Simpson T, Skirving WJ, Sommer B, Torda G, Wachenfeld DR, Willis BL, Wilson SK (2017) Global warming and recurrent mass bleaching of corals. Nature 543:373–377.

Johnson C (1992) Settlement and recruitment of *Acanthaster planci* on the Great Barrier Reef: Questions of process and scale. Aust. J. Mar. Freshw. Res. 43(3):611–627.

Johnson CR, Sutton DC, Olson RR, Giddins R (1991) Settlement of crown-of-thorns starfish: Role of bacteria on surfaces of coralline algae and a hypothesis for deepwater recruitment. Mar. Ecol. Prog. Ser. 11:143–162.

Kamya PZ, Byrne M, Graba-Landry A, Dworjanyn SA (2016) Near-future ocean acidification enhances the feeding rate and development of the herbivorous juveniles of the crown-of-thorns starfish, *Acanthaster planci*. Coral Reefs 35:1241–1251.

Kamya PZ, Dworjanyn SA, Hardy N, Mos B, Uthicke S, Byrne M (2014) Larvae of the coral eating crown-of-thorns starfish, *Acanthaster planci* in a warmer-high CO_2 ocean. Global Change Biol. 20:3365–3376.

Kayal M, Vercelloni J, Lison de Loma T, Bosserelle P, Chancerelle Y, Geoffroy S, Stievenart C, Michonneau F, Penin L, Planes S, Adjeroud M (2012) Predator crown-of-thorns starfish (*Acanthaster planci*) outbreak, mass mortality of corals, and cascading effects on reef fish and benthic communities. PLoS One 7(10):e47363.

Keesing JK, Bradbury RH, DeVantier LM, Riddle MJ, De'Ath G (1992) Geological evidence for recurring outbreaks of the crown-of-thorns starfish: A reassessment from an ecological perspective. Coral Reefs 11(2):79–85.

Kenchington RA (1977) Growth and recruitment of *Acanthaster planci* (L.) on the Great Barrier Reef. Biol. Cons. 11(2):103–18.

Kenchington RA (1987) *Acanthaster planci* and management of the Great Barrier Reef. Bull. Mar. Sci. 41(2):552–560.

Kenyon JC, Aeby GS (2009) Localized outbreak and feeding preferences of the crown-of-thorns seastar *Acanthaster planci* echinodermata, asteroidea) on reefs off Oahu, Hawaii. Bull. Mar. Sci. 84(2):199.

Kroon FJ, Barneche DR, Emslie MJ (2021) Fish predators control outbreaks of Crown-of-Thorns Starfish. Nature Comm. 12(1):6986.

Kroon FJ, Lefèvre CD, Doyle JR, Patel F, Milton G, Severati A, Kenway M, Johansson CL, Schnebert S, Thomas-Hall P, Bonin MC (2020) DNA-based identification of predators of the corallivorous Crown-of-Thorns Starfish (*Acanthaster* cf. *solaris*) from fish faeces and gut contents. Sci. Rep. 10(1):8184.

Lamare M, Pecorino D, Hardy N, Liddy M, Byrne M, Uthicke S (2014) The thermal tolerance of crown-of-thorns (*Acanthaster planci*) embryos and bipinnaria larvae: Implications for spatial and temporal variation in adult populations. Coral Reefs 33:207–219.

Lang BJ, Donelson JM, Caballes CF, Uthicke S, Doll PC, Pratchett MS (2022) Effects of elevated temperature on the performance and survival of pacific crown-of-thorns starfish (*Acanthaster* cf. *solaris*). Mar. Biol. 169(4):1–3.

Lucas JS (1982) Quantitative studies of feeding and nutrition during larval development of the coral reef asteroid *Acanthaster planci* (L.). J. Exp. Mar. Biol. Ecol. 65(2):173–193.

Luo Y, Rothstein LM (2011) Response of the Pacific Ocean circulation to climate change. Atmosphere-Ocean 49(3): 235–244.

MacNeil MA, Chong-Seng KM, Pratchett DJ, Thompson CA, Messmer V, Pratchett MS (2017) Age and growth of an outbreaking *Acanthaster* cf. *solaris* population within the Great Barrier Reef. Diversity 9(1):18.

MacNeil MA, Mellin C, Pratchett MS, Hoey J, Anthony KR, Cheal AJ, Miller I, Sweatman H, Cowan ZL, Taylor S, Moon S (2016) Joint estimation of crown of thorns (*Acanthaster planci*) densities on the Great Barrier Reef. PeerJ 4:e2310.

McCallum HI (1987) Predator regulation of *Acanthaster planci*. J. Theor. Biol. 127(2):207–220.

McCallum HI (1992) Completing the circle: Stock-recruitment relationships and *Acanthaster*. Mar. FW Res. 43(3):653–662.

Mellin C, Matthews S, Anthony KR, Brown SC, Caley MJ, Johns KA, Osborne K, Puotinen M, Thompson A, Wolff NH, Fordham DA (2019) Spatial resilience of the Great Barrier Reef under cumulative disturbance impacts. Global Change Biol. 25(7):2431–2445.

Miller I, Sweatman H, Cheal A, Emslie M, Johns K, Jonker M, Osborne K (2015) Origins and implications of a primary crown-of-thorns starfish outbreak in the southern Great Barrier Reef. J. Mar. Biol. 809624.

Moore RJ (1978) Is *Acanthaster planci* an r-strategist?. Nature 271(5640):56–57.

Moran PJ (1986) The *Acanthaster* phenomenon. Oceanogr. Mar. Biol. Ann. Rev. 24:379–480.

Moran PJ, Bradbury RH, Reichelt RE (1988) Distribution of recent outbreaks of the crown-of-thorns starfish (*Acanthaster planci*) along the Great Barrier Reef: 1985–1986. Coral Reefs 7(3):125–137.

Moran PJ, De'ath G, Baker VJ, Bass DK, Christie CA, Miller IR, Miller-Smith BA, Thompson AA (1992) Pattern of Outbreaks of Crown-of-thorns Starfish (*Acanthaster planci* (L.)) along the Great Barrier Reef since 1966. Aust. J. Mar. Freshwater Res. 43:555–568.

Olson RR (1987) In situ culturing as a test of the larval starvation hypothesis for the crown-of-thorns starfish, *Acanthaster planci*. Limnol. Oceanogr. 32(4):895–904.

Ormond RF, Campbell AC, Head SH, Moore RJ, Rainbow PR, Saunders AP (1973) Formation and breakdown of aggregations of the crown-of-thorns starfish, *Acanthaster planci* (L.). Nature 246(5429):167–169.

Ortiz JC, Wolff NH, Anthony KR, Devlin M, Lewis S, Mumby PJ (2018) Impaired recovery of the Great Barrier Reef under cumulative stress. Sci. Adv. 4(7):eaar6127.

Pandolfi JM (1992) A palaeobiological examination of the geological evidence for recurring outbreaks of the crown-of-thorns starfish, *Acanthaster planci* (L.). Coral Reefs 11(2): 87–93.

Pearson RG, Endean R (1969) A preliminary study of the coral predator *Acanthaster planci* (L.) (Asteroidea) on the Great Barrier Reef. Fish. Notes. Queensland. Dep. Harb. Mar. 3:27–55.

Pratchett MA, Cumming GS (2019) Managing cross-scale dynamics in marine conservation: Pest irruptions and lessons from culling of crown-of-thorns starfish (*Acanthaster* spp.). Biol. Cons. 23:108211.

Pratchett MS (2005) Dynamics of an outbreak population of *Acanthaster planci* at Lizard Island, northern Great Barrier Reef (1995–1999). Coral Reefs 24(3):453–462.

Pratchett MS, Caballes CF, Burn D, Doll PC, Chandler JF, Doyle JR, Uthicke S (2022) Scooter-assisted large area diver-based (SALAD) visual surveys to test for renewed outbreaks of crown-of-thorns starfish (*Acanthaster* cf. *solaris*) in the northern Great Barrier Reef. Technical Report prepared for the Australian Government by the COTS Control Innovation Program. Great Barrier Reef Foundation, Brisbane.

Pratchett MS, Caballes CF, Rivera-Posada JA, Sweatman HPA (2014). Limits to understanding and managing outbreaks of crown-of-thorns starfish (*Acanthaster* spp.). Oceanogr. Mar. Biol. Ann. Rev. 52:133–200.

Pratchett MS, Caballes CF, Wilmes JC, Matthews S, Mellin C, Sweatman HPA, Nadler LE, Brodie J, Thompson CA, Hoey J, Bos AR, Byrne M, Messmer V, Fortunato SAV, Chen CMN, Buck ACE, Babcock RC, Uthicke S (2017a) 30 years of research on crown-of-thorns starfish (1986–2016): Scientific advances and emerging opportunities. Diversity 9(4):41.

Pratchett MS, Dworjanyn S, Mos B, Caballes CF, Thompson CA, Blowes S (2017b) Larval survivorship and settlement of crown-of-thorns starfish (*Acanthaster* cf. *solaris*) at varying algal cell densities. Diversity 9(1):2.

Pratchett MS, Harrison HB, Messmer V, Berumen BL, Saenz-Agudelo P, Chong-Seng KM (2015) Initiation and Spread of Outbreaks of Crown-of-thorns Starfish on the Great Barrier Reef. Caring for Country Project A010388G, Canberra.

Pratchett MS, Heron SF, Mellin C, Cumming GS (2021a) Recurrent mass-bleaching and the potential for ecosystem collapse on Australia's Great Barrier Reef. In: Ecosystem collapse and climate change 2021 (pp. 265–289). Springer, Cham.

Pratchett MS, Nadler LE, Burn D, Lang BJ, Messmer V, Caballes CF (2021b) Reproductive investment and fecundity of Pacific crown-of-thorns starfish (*Acanthaster* cf. *solaris*) on the Great Barrier Reef. Mar. Biol. 168(6):1–8.

Pratchett MS, Caballes CF, Cvitanovic C, Raymundo ML, Babcock RC, Bonin MC, Bozec YM, Burn D, Byrne M, Castro-Sanguino C, Chen CC (2021c) Knowledge gaps in the biology, ecology, and management of the Pacific crown-of-thorns sea star *Acanthaster* sp. on Australia's Great Barrier Reef. Biol. Bull. 241(3):330–346.

Reichelt RE, Bradbury RH, Moran PJ (1990) Distribution of *Acanthaster planci* outbreaks on the Great barrier Reef between 1966 and 1989. Coral Reefs 9:97–103.

Ridgway KR, Benthuysen JA, Steinberg C (2018) Closing the gap between the Coral Sea and the equator: Direct observations of the north Australian western boundary currents. J. Geophys. Res. 123:9212–9231.

Rivera-Posada J, Pratchett MS, Aguilar C, Grand A, Caballes CF (2014) Bile salts and the single-shot lethal injection method for killing crown-of-thorns sea stars (*Acanthaster planci*). Ocean Coast. Manag. 102:383–390.

Rivera-Posada JA, Pratchett M, Cano-Gomez A, Arango-Gomez JD, Owens L (2011) Refined identification of Vibrio bacterial flora from *Acanthasther planci* based on biochemical profiling and analysis of housekeeping genes. Dis. Aquat. Org. 96(2):113–123.

Rogers JG, Pláganyi ÉE, Babcock RC (2017) Aggregation, Allee effects and critical thresholds for the management of the crown-of-thorns starfish *Acanthaster planci*. Mar. Ecol. Prog. Ser. 578:99–114.

Sale PF, Doherty PJ, Eckert GJ, Douglas WA, Ferrell DJ (1984) Large scale spatial and temporal variation in recruitment to fish populations on coral reefs. Oecologia 64(2):191–198.

Sill SR, Dawson TP (2021) Climate change impacts on the ecological dynamics of two coral reef species, the humphead wrasse (*Cheilinus undulatus*) and crown-of-thorns starfish (*Ancanthaster planci*). Ecol. Inform. 65:101399.

Stump R (1996) An investigation to describe the population dynamics of *Acanthaster planci* (L.) around Lizard Island, Cairns section, Great Barrier Reef Marine Park. CRC Reef Research Centre, Townsville.

Sweatman H (2008) No-take reserves protect coral reefs from predatory starfish. Curr. Biol. 18:R598–R599.

Sweatman H, Bass DK, Cheal AJ, Coleman G, Miller IR, Ninio R, Osborne K, Oxley WG, Ryan DAJ, Thompson AA, Tomkins P (1998) Long-term monitoring of the Great Barrier Reef. Status report number 3. Australian Institute of Marine Science. Townsville.

Treml EA, Ford JR, Black KP, Swearer SE (2015) Identifying the key biophysical drivers, connectivity outcomes, and metapopulation consequences of larval dispersal in the sea. Mov. Ecol. 3(1):1–6.

Uthicke S, Lamare M, Doyle JR (2018) eDNA detection of corallivorous seastar (*Acanthaster* cf. *solaris*) outbreaks on the Great Barrier Reef using digital droplet PCR. Coral Reefs. 37:1229–1239.

Uthicke S, Logan M, Liddy M, Francis D, Hardy N, Lamare M (2015) Climate change as an unexpected co-factor promoting coral eating seastar (*Acanthaster planci*) outbreaks. Sci. Rep. 5:8402.

Uthicke S, Pecorino D, Albright R, Negri AP, Cantin N, Liddy M, Dworjanyn S, Kamya P, Byrne M, Lamare M (2013) Impacts of ocean acidification on early life-history stages and settlement of the coral-eating sea star *Acanthaster planci*. PLoS One 8(12):e82938.

Uthicke S, Robson B, Doyle JR, Logan M, Pratchett MS, Lamare M (2022) Developing an effective marine eDNA monitoring: eDNA detection at pre-outbreak densities of corallivorous seastar (*Acanthaster* cf. *solaris*). Sci. Total Environ. 851(1):158143.

Uthicke S, Schaffelke B, Byrne M (2009) A boom–bust phylum? Ecological and evolutionary consequences of density variations in echinoderms. Ecol. Monogr. 79(1):3–24.

Vanhatalo J, Hosack GR, Sweatman H (2017) Spatiotemporal modelling of crown-of-thorns starfish outbreaks on the Great Barrier Reef to inform control strategies. J. Appl. Ecol. 54(1):188–197.

Vine PJ (1973) Crown of thorns (*Acanthaster planci*) plagues: The natural causes theory. Atoll Res. Bull. 166:1–10.

Walbran PD, Henderson RA, Jull AJT, Head MJ (1989) Evidence from sediments of long-term *Acanthaster planci* predation on corals of the Great Barrier Reef. Science 245:847–850.

Webb DJ (2000) Evidence for shallow zonal jets in the South Equatorial Current region of the southwest Pacific. J. Phys. Oceanography 30(4):706–720.

Weber JN, Woodhead PM (1970) Ecological studies of the coral predator *Acanthaster planci* in the South Pacific. Mar. Biol. 6:12–17.

Westcott DA, Fletcher CS, Kroon FJ, Babcock RC, Plagányi EE, Pratchett MS, Bonin MC (2020) Relative efficacy of three approaches to mitigate crown-of-thorns starfish outbreaks on Australia's Great Barrier Reef. Sci. Rep. 10(1):12594.

Wilmes JC, Caballes CF, Cowan ZL, Hoey AS, Lang BJ, Messmer V, Pratchett MS (2018) Contributions of pre-versus post-settlement processes to fluctuating abundance of crown-of-thorns starfishes (*Acanthaster* spp.). Mar. Poll. Bull. 135:332–345.

Wilmes JC, Schultz DJ, Hoey AS, Messmer V, Pratchett MS (2020) Habitat associations of settlement-stage crown-of-thorns starfish on Australia's Great Barrier Reef. Coral Reefs 39(4):1163–1174.

Wolanski E, Lambrechts J (2020) The net water circulation in the far Northern Great Barrier Reef. Est. Coast. Shelf Sci. 235:106569.

Wolanksi E, Kingsford MJ, Lambrechts J, Marmorino G (2024) The physical oceanography of the Great Barrier Reef: A Review. In Wolanski E, Kingsford MJ (eds.), Oceanographic Processes of Coral Reefs. Physical and Biological Links in the Great Barrier Reef. CRC Press, Boca Raton, FL.

Wolfe K, Graba-Landry A, Dworjanyn SA, Byrne M (2015) Larval starvation to satiation: Influence of nutrient regime on the success of *Acanthaster planci*. PLoS One 10(3):e0122010.

Wolfe K, Graba-Landry A, Dworjanyn SA, Byrne M (2017) Superstars: Assessing nutrient thresholds for enhanced larval success of *Acanthaster planci*, a review of the evidence. Mar. Poll. Bull. 116(1–2):307–314.

Wooldridge S, Brodie J, Furnas M (2006) Exposure of inner-shelf reefs to nutrient enriched runoff entering the Great Barrier Reef Lagoon: Post-European changes and the design of water quality targets. Mar. Poll. Bull. 52(11):1467–1479.

Wooldridge SA (2009) Water quality and coral bleaching thresholds: Formalising the linkage for the inshore reefs of the Great Barrier Reef, Australia. Mar. Poll. Bull. 58(5): 745–751.

Wooldridge SA, Brodie JE (2015) Environmental triggers for primary outbreaks of crown-of-thorns starfish on the Great Barrier Reef, Australia. Mar. Poll. Bull. 101:805–815.

Zann L, Brodie J, Berryman C, Naqasima M (1987) Recruitment, ecology, growth and behavior of juvenile *Acanthaster planci* (L.)(Echinodermata: Asteroidea). Bull. Mar. Sci. 41(2):561–575.

21 The Biophysics of Sharks and Rays on the Great Barrier Reef

Andrew Chin, Stacy Bierwagen, Jodie L. Rummer, and Vinay Udyawer

21.1. INTRODUCTION

The Great Barrier Reef (GBR) is home to a wide array of chondrichthyan fishes—sharks, rays, skates, and chimaeras—which are henceforth referred to here as sharks and rays. Sharks and rays are typically mid- to high-level predators, occupying a wide range of GBR habitats and niches where they play a variety of ecological roles including transferring energy between habitats and trophic systems and affecting prey density and behaviour. Sharks and rays can even alter physical habitats and facilitate the survival of other species (Heithaus et al. 2022). Aside from their ecological roles, the sharks and rays of the GBR have important social, cultural, and economic values (Choat et al. 2019). Many of the First Nations Peoples of the GBR have special cultural connections with these species, including ongoing cultural uses and lore and cultural significance dating back more than 40,000 years (Gerhardt 2018). Meanwhile, sharks and rays are among the most sought-after tourist attractions by divers visiting the GBR, and they have also been important components of commercial fisheries.

Although sharks and rays can affect the ecosystems that they inhabit, shark and ray communities themselves are also affected by the physical environment in which they occur. In this chapter, we explore the biophysical relationships between sharks and rays and the habitats and ecological processes of the GBR. The chapter discusses how the physical structure of the GBR, a mosaic of interconnected habitats on a relatively shallow continental shelf, supports the rich diversity of sharks and rays present in the GBR and affects their movements and distributions. The impact of environmental variables in affecting species distribution and habitat is also explored, including that of extreme weather events such as cyclones. We also explore the energetic relationships between the GBR ecosystem and sharks and rays and discuss where sharks derive energy and how this energy flows through the GBR ecosystem. Lastly, we explore how the changing climate and the biophysical environment may affect sharks and rays. This overview of the biophysics of sharks and rays in the GBR not only provides both a synthesis of some of the key knowledge available on the subject to date but also highlights how much remains unknown about this morphologically and functionally diverse group of fishes that are instrumental to the GBR.

21.2. LINKAGES BETWEEN GBR PHYSICAL HABITATS AND SHARKS AND RAYS

21.2.1. Shark and Ray Diversity and Ecological Groupings

There are approximately 138 species of sharks and rays in the GBR including 10 skates, 82 sharks, 42 rays, and 4 chimaeras (Chin et al. 2010; Chin et al. 2023). These species range from small cryptic species, such as the cookie cutter shark (*Isistius brasiliensis* at 40 cm in length), to the whale shark (*Rhincodon typus*) that grows to 12 m in length. This diversity of species, sizes, and life history strategies is reflected in the diversity of their ecology and links to the physical environment.

The GBR is a mosaic of different habitats, each with their own biological communities and a characteristic set of ecological processes (Hutchings et al. 2019). Shark and ray communities are likewise associated with specific habitats and ecological processes. Compagno (1990) introduced the concept of shark and ray 'ecomorphotypes', where shark morphology and life history, and by extension the species present in a community, are shaped by the physical environment and a species' ecological adaptations to that environment. This concept has been applied to the sharks and rays of the GBR, where the recorded 140 species have been placed into six 'ecological groups' (Table 1) (Chin et al. 2010; Chin 2023). Each ecological group encompasses a discrete set of habitats and associated ecological processes, and the sharks and rays occurring in each group are the species that have adapted to these habitats and conditions. For example, the largetooth sawfish (*Pristis pristis*) is typically restricted to turbid freshwater and estuarine environments (Grant et al. 2019). This species has a large, toothed rostrum that, along with associated prey detection sensory modalities, evolved to ambush prey in low visibility water (Wueringer 2012), and juveniles spend their early years in the freshwater environment with fewer elasmobranch competitors and predators and greater prey availability (Whitty et al. 2009). As a result, the extensive freshwater and estuarine habitats of the GBR coast once supported a large sawfish community, and where impacts are low, these habitats continue to support globally significant sawfish populations (Grant et al. 2021). However, species such as the bull shark (*Carcharhinus leucas*) are ecologically flexible and

DOI: 10.1201/9781003320425-24

TABLE 1

Ecological groups of GBR sharks and rays, their respective habitats, and example species.

Ecological group	Habitat types and processes	Species
Freshwater and estuarine	Species occurring in rivers and streams, riparian habitats, mangroves and saltmarsh, and shallow intertidal and subtidal habitats such as seagrass beds and mudflats. Key biophysical processes include coastal runoff, wave- and wind-driven currents, and resuspension, erosion, and deposition of the coastal zone.	4 species including the largetooth sawfish *Pristis pristis*; freshwater whipray *Urogymnus dalyensis*; bull shark *Carcharhinus leucas*; and the spear-tooth shark *Glyphis glyphis*
Coastal and inshore	Species occurring in habitats extending from coastal subtidal habitats to the midshelf platform or ribbon reefs. Habitats include estuaries and bays, subtidal seagrass beds, inshore fringing reefs, rocky shoals, sponge gardens, and other benthic habitats of the GBR lagoon to 30 m depth. Key biophysical processes include coastal runoff, wave- and wind-driven currents and resuspension, GBR lagoonal currents, and circulation patterns that drive inshore productivity.	52 species including the narrow sawfish *Anoxypristis cuspidata*; Australian cownose ray *Rhinoptera neglecta*; pigeye shark *Carcharhinus amboinensis*; tiger shark *Galeocerdo cuvier*; bull shark *Carcharhinus leucas*; blacktip reef shark *Carcharhinus melanopterus*; great hammerhead shark *Sphyrna mokarran*; giant shovelnose ray *Glaucostegus typus*; zebra shark *Stegastoma tigrinum*
Reefal	Species occurring in habitats on and immediately adjacent to midshelf and outer shelf coral reefs, down to a maximum depth of 40 m in the GBR lagoon to 60 m adjacent to the outer shelf reefs. Key biophysical processes include coral reef processes, lagoonal currents, and circulation patterns, (especially in the outer reef zones).	28 species including the bluespotted fantail ray *Taeniura lymma*; ornate eagle ray *Aetomylaeus vespertilio*; blacktip reef shark *Carcharhinus melanopterus*; whitetip reef shark *Triaenodon obesus*; bull shark *Carcharhinus leucas*; grey reef shark *Carcharhinus amblyrhynchos*; epaulette shark *Hemiscyllium ocellatum*; ornate wobbegong *Orectolobus ornatus*; zebra shark *Stegastoma tigrinum*
Shelf	Species occurring in deeper water and seabed habitats between the midshelf and outer reefs, extending to the continental slope edge. Includes waters from the surface to 200 m (approximately the shelf edge) and benthic habitats such as deepwater seagrass beds and Halimeda mounds, rocky shoals, and sponge gardens (40–60 m depth). Key biophysical processes include deep water currents and circulation, upwellings and pelagic influences, deposition.	28 species such as the eastern angel shark *Squatina albipunctata*; short-tail torpedo ray *Torpedo macneilli*; piked spurdog *Squalus megalops*; white shark *Carcharodon carcharias*; pencil shark *Hypogaleus hyugaensis;* tiger shark *Galeocerdo cuvier* spot-tail shark *Carcharhinus sorrah;* great hammerhead shark *Sphyrna mokarran*; argus skate *Dipturus polyommata*
Bathyal	Species occurring in deep water, benthic habitats of the continental slope and beyond, and bathypelagic waters below 1,000 m depth. Key biophysical processes include deep water currents, upwellings, and deposition.	50 species including the argus skate *Dipturus polyommata*; longspine chimaera *Chimaera macrospina*; blackfin ghostshark *Hydrolagus lemures*; bartail spurdog *Squalus notocaudatus*
Pelagic	Species occurring in the epipelagic and mesopelagic zones of open ocean waters extending from the edge of the outer reefs and beyond into the Coral Sea, with vertical depth to 1,000 m. Key biophysical processes include upwellings, water currents, and thermoclines.	13 species including the Oceanic whitetip shark *Carcharhinus longimanus*; blue shark *Prionace glauca*; whale shark *Rhincodon typus*; oceanic manta ray *Manta birostris*

Source: Adapted from Chin et al. (2010) and Chin et al. (2023)

occupy a range of different habitats. As such, these species occur in more than one ecological group.

The six ecological groups and their respective shark and ray species illustrate the connections between the physical environment and GBR sharks and rays. Each ecological group comprises specific habitats with different biophysical characteristics, and the shark and ray community of each ecological group reflects these characteristics. More specific examples of these connections are provided in the following.

21.2.2. Biophysical Relationships Shape Shark and Ray Behaviour and Community Composition

The biophysical relationships between habitats and sharks and rays are illustrated in the ways that the structural and physical characteristics of habitats have shaped shark and ray morphology, physiology, diversity, and population structure. These examples range from the adaptations and behaviours of individual species to the ecological functions provided by habitats that shape entire shark and ray communities.

A species-focused example of this relationship is the epaulette shark (*Hemiscyllium ocellatum*). This small carpet shark resides in shallow reef and seagrass habitats (typically < 5 m deep), including shallow reef lagoons, and forages for invertebrates in very shallow water. The epaulette sharks' morphology and physiology reflect this habitat. The species has a long, slender body form that is very flexible, and the fins are small and rounded, which enable it to squeeze into small cervices between corals and within the reef matrix (Last and Stevens 2009). The fin morphology and musculature also enable the shark to 'walk' out of one tidal pool into another (Porter et al. 2022), giving rise to their name of 'walking sharks' (Dudgeon et al. 2020). However, these morphological adaptations make this small catshark a very poor swimmer that is highly susceptible to predators. To compensate for these limitations, the epaulette shark is mainly nocturnal (Wheeler et al. 2022) and also has adaptations that enable it to exploit coral reef habitats inaccessible to larger predators (Dudgeon et al. 2020). The species has physiological adaptations that enable it to remain in very shallow reef lagoons at low tide during low oxygen conditions when water temperatures rise during the day or at night when photosynthesis declines, dissolved oxygen levels decrease (Wise et al. 1998; Hickey et al. 2012), and PCO_2 levels increase, thereby decreasing water pH (Heinrich et al. 2014, 2015). Here, this species has developed morphological and physiological adaptations and behaviours that are shaped by the biophysical characteristics of its shallow, coral reef habitat.

Biophysical characteristics, such as depth, water currents, and physical habitat complexity, can also shape entire shark and ray population structures and community dynamics. Shallow coastal areas are key nursery habitats for many shark and ray species. These shallow environments include mud and sand flats, mangroves, and seagrass beds, which provide ample prey and shallow water refuge from predators for highly vulnerable neonate and juvenile sharks and rays (Figure 1) (Heupel et al. 2007; Martins et al. 2018; Martins et al. 2020b).

In the GBR, sheltered coastal environments protect seagrass beds and mangroves from strong wave action and currents, which promotes growth and development thereby enhancing the capacity of coastal environments to provide shelter for sharks and rays in vulnerable life stages. Although the role of these habitats in providing energy to growing neonates and juveniles is widely reported, in the GBR, it seems likely that the most important function of these areas is as a refuge from predators. Trophodynamic research in the GBR has shown that two stingray species inhabiting shallow coastal environments derived little energy from mangrove systems (Martins et al. 2022). Moreover, different shark species in shallow foreshore environments exhibited a high level of niche partitioning, which suggests that resources were limited and resulted in different species adopting different foraging strategies (Kinney

FIGURE 1 Shallow habitats provide neonate and juvenile sharks and rays, such as these blacktip reef sharks (*C. melanopterus*), with protection from predators.

Source: Photo credit: Andrew Chin

et al. 2011). Meanwhile, these tidal and subtidal habitats are too shallow for larger predators, so they are inhabited by neonates and juveniles who remain in these areas while growing. For example, the mangroves and shallow flats of Orpheus Island in the Central GBR provide critical habitats for cowtail rays (*Pastinachus ater*) and mangrove whiprays (*Urogymnus granulatus*), which have been recorded as hiding within mangrove roots during high tide when larger predators can access the system (Davy et al. 2015; Martins et al. 2020a; Martins et al. 2022). Similarly, these shallow coastal habitats provide sheltered nursery habitats for the early life stages of blacktip reef sharks (*Carcharhinus melanopterus*) (Chin et al. 2016; George et al. 2019; Bouyoucos et al. 2022) that remain in these areas until reaching a large enough size that enables them to disperse to other habitats, including offshore coral reefs (Chin et al. 2013).

These examples provide an overview as to how biophysical factors shape shark and ray communities. These concepts are explored in more detail in sections *Effects of Biophysical Factors on Shark and Ray Movement and Distribution* and *The Physiological Linkages between Sharks and Rays and Their Environment*. However, sharks and rays themselves can alter the biophysics of the systems that they inhabit, and these interactions are explored here in sections *Sharks and Rays as Habitat Engineers* and *Trophodynamics of GBR Sharks and Rays*.

21.2.3. Sharks and Rays as Habitat Engineers

Biophysical factors can shape shark and ray communities; however, in some instances, sharks and rays themselves can act as 'ecological engineers' that alter the biophysics of their environment. One of the most obvious examples of this effect is bioturbation during stingray feeding. Many stingrays are benthic predators that forage for invertebrates in the substrate. One of their feeding behaviours involves pumping water into the substrate and turning over the sediment, leaving large depressions known as 'feeding pits' behind. This activity can be significant, with one study estimating that 42% of the soft sediment in one study area (equivalent to an estimate of ~42,000 m^2) would be turned over by stingrays every year (O'Shea et al. 2012). Although bioturbation is an important mechanism that can oxygenate sediments, affect pH, and influence chemical and nutrient cycling (Krantzberg 1985; Sarker et al. 2021), the pits themselves create habitat rugosity in areas that may otherwise be uniform (Figure 2), thus creating habitats that are colonised by other organisms such as ostracods and amphipods (O'Shea et al. 2012). Ray-mediated bioturbation can also alter habitats by inhibiting the growth of other organisms. Cownose rays (*Rhinoptera bonasus*) reportedly can inhibit the growth of eelgrass (*Zostera marina*) by disrupting the rhizomes while foraging in the sediment, thus affecting the physical habitat and associated biodiversity (Orth 1975). However, different ray species may have different effects. For example, the Australian whipray (*Himantura australis*) creates larger feeding pits than the cowtail ray (*Pastinachus ater*); however, the cowtail ray forages over a wider area and may thus exert bioturbation influences over a larger area (Crook et al. 2022). Nevertheless, stingrays (and their feeding pits) are commonly found across many different GBR habitats and coastal areas (Crook 2020; Martins et al. 2022) and despite inter-specific differences in foraging behaviours and associated biophysical effects, are likely to have widespread influences on habitats and biodiversity across the GBR.

Sharks can also alter the physical environment through indirect effects. The presence of sharks can alter prey behaviour and feeding strategies (Frid et al. 2008; Rizzari et al. 2014b; Lester et al. 2020), which, in turn, create indirect effects on habitat structure. Some species, such as tiger sharks (*Galeocerdo cuvier*), are top-level predators that exert influence throughout a food web that results in changes to seagrass beds. The presence of tiger sharks has been shown to affect the foraging depth of dugongs (*Dugong dugon*) and whether dugongs forage by cropping or excavating seagrasses (Wirsing et al. 2007). When tiger sharks were present, dugongs foraged on deeper seagrass beds, and because they were warier, dugongs fed by grazing or cropping seagrasses instead of excavating seagrasses and their rhizomes. In turn, these changes altered the nature of disturbance to seagrass meadows and the turnover and succession of seagrass species (Wirsing et al. 2008). Although these studies have not been repeated in the GBR, both tiger sharks and dugongs are present in the GBR, and it seems likely that this dynamic would exist in the GBR where both species co-occur.

FIGURE 2 Stingrays such as the porcupine ray (*Urogymnus asperrimus*) can act as habitat engineers through bioturbation and excavating feeding pits.

Source: Photo credit: Henriette Grimmel | © Save Our Seas Foundation

21.3. TROPHODYNAMICS OF GBR SHARKS AND RAYS

Trophodynamics are 'the dynamics of nutrition and metabolism' (Lindeman 1942) and crucial in understanding how energy flows through ecosystems (Libralato et al. 2014; Bierwagen et al. 2018). Research on shark and ray trophodynamics has been ongoing for some time but only more prominently on the GBR since trophodynamic work on reef fishes was highlighted by Polunin (1996). The scope of diversity in sharks and rays on the GBR makes the description of their trophodynamics a complex task. Hence, this topic is as equally difficult to describe with any definitive assuredness without underpinning elements of inference, particularly when describing ecological interactions of sharks and rays. This is not unexpected in the context of coral reefs, which are widely known for their substantial complexity and diversity. The reason behind inferences of ecological patterns is often due to a lack of specific and time-intensive data needed for these semi-open, complex systems that require research at correct spatial scales (Sale 2002). For species that have a large distribution and/or home range, the geomorphology and scale of the GBR must also be considered, as entire assemblages can change both through latitudes and across the continental shelf. Thus, the range of species, each with different distributions and movement patterns, creates a diverse array of interactions with different levels of trophic connectivity that are difficult to quantify and generalise.

From a physical perspective, energy needs and, thus, energy flows between organisms are dictated by the biomass and availability of basal sources of nutrition derived from coral reefs, which comprise both benthic and pelagic inputs. In general, the GBR is driven by bottom-up processes where nutrients are assimilated between different levels of biomass to form, in the simplest of terms, a resource-limited pyramid structure. Larger-bodied organisms with substantial biomass per individual at the top of the pyramid require the most energy per individual and, therefore, in theory, will be limited in abundance compared to lower trophic levels. On coral reefs, this pyramid is more diffuse and unconsolidated where reef predators, such as sharks and rays, occupy a variety of dietary niche breadths at multiple trophic levels (Wilson et al. 2008) and may also transfer nutrients from the pelagic system to coral reefs.

To make sense of the complex nature of these systems, generalisations are often made to gain some perspective of the ecological groupings of sharks and rays. Research focus also tends to concentrate heavily on the more 'charismatic' and less cryptic species. For instance, reef sharks are often described in the literature as a conglomerate of three main species, namely, grey reef sharks (*C. amblyrhynchos*), white-tip reef sharks (*T. obesus*), and blacktip reef sharks (*C. melanopterus*). Other apex species that occupy higher trophic levels, such as bull sharks (*C. leucas*), hammerheads (*Sphyrna spp.*), and tiger sharks (*G. cuvier*), are often the focus of coral reef research but are less resident and more roving than other reef-associated species. All other sharks and rays are usually classed as benthic or demersal-dwelling species occurring on coastal and shelf ecological groups (Table 1). Many of these species groups, however, occupy vastly different niche breaths and use reef resources differently, thus deriving energy from varying sources. Sharks and rays also occupy different habitats across ontogeny, where nutrients assimilated and rates of consumption as a juvenile may be entirely different than during adult life stages (Dale et al. 2013). Rays are also only just beginning to be studied in terms of ecological research on the GBR, and the description of ecological function in this group of organisms is limited (Crook 2020; Martins et al. 2022).

The discussion of ecological roles of sharks and rays often equates with their importance to the ecosystem and a need to understand what happens in the event of their removal or loss. There is a documented element of top-down control by high-order predators through both consumptive and fear-related behavioural effects (Cortes 1999; Rizzari et al. 2014b; Frisch et al. 2016; Lester et al. 2020). However, evidence of collapse of these controlling effects, effectively creating a cascading breakdown of a balanced system (e.g., trophic cascade), is mostly anecdotal and extremely limited in cases that have actually been able to report this observed effect from a single causal source such as species removal or loss (Ferretti et al. 2010; Ruppert et al. 2013; Grubbs et al. 2016). Irrespective of the lack of *in situ* observations, models attempting to reproduce cascades have shown that top-down control has weak effects on coral reefs (Casey et al. 2017), but these models may also not accurately represent trophic structure appropriately based on our most up-to-date understanding of ecological roles (e.g., use of reef sharks grouping as an apex predator when they are currently considered to be a mesopredator) (Ruppert et al. 2019). Alternatively, the lack of food web linearity in reef systems assumes a great degree of redundancy across trophic levels, which may inherently create more stability when considering the loss of a single species and may explain why ecosystem models are unable to detect significant impacts. These models also do not often account for loss of sharks and rays in lower trophic levels and the wider consequences of the loss of ecological services that they provide, such as bioturbation (Dulvy et al. 2017).

Although there is no question of very high levels of shark diversity on the GBR (MacNeil et al. 2020), there is mixed evidence in terms of abundance of these species (Robbins et al. 2006; Heupel et al. 2009; Rizzari et al. 2014b; Bierwagen 2019) and, therefore, of the influence that they may carry in their ecological roles. For instance, the influence of a species that is highly abundant in isolated remote environments compared to lower abundance in semi-open environments is likely ecologically distinct. Grey reef sharks, which were assumed to be limited in their diet by their mouth gape size, have been known to socially forage in groups for larger prey items (Mourier et al. 2016) and are subject to kleptoparasitism by foraging among other species such as whitetip reef sharks (Labourgade et al. 2020), increasing their ability to successfully capture organisms. These examples illustrate how their overall ecological influence may change between remote and isolated reef systems and the shallow and highly interconnected reefs of the GBR.

Ecological relevance is also difficult to interpret among species occupying similar trophic levels. Some sharks and rays are comparable in carbon and nitrogen enrichment to large-bodied teleosts (such as mesopredators), and other species derive nutrients from entirely different basal sources (Frisch et al. 2016; Roff et al. 2016; Bierwagen 2019; Espinoza et al. 2019). This means that assumed trophic redundancy across reef-dwelling species is not the case in some instances. There has also been little research until recently on scaling up the bioenergetic needs of these animals. Even for the most commonly occurring reef sharks on the GBR such as the grey reef shark (*C. amblyrhynchos*), there is only a paucity of information available regarding the energy needs of the species (Mourier et al. 2016; Dunn et al. 2022). However, recent research has shown that reef sharks are responsible for transferring more nutrients and energy from pelagic sources to reefs than previously considered, which makes their and any mobile marine predator's role in nutrient cycling likely critical for reefs (Williams et al. 2018; Bierwagen 2019).

There are also a great many unknowns in what processes are occurring on reefs and the anomalies that affect our understanding of energy flow in terms of behavioural adaptations that require direct observation. These overall impacts may be significant in terms of effects on ecosystems and are usually unexpected. For example, white sharks (*Carcharodon carcarias*, which are one of the most commonly cited examples of an apex predator) in South Africa have been documented to change distribution as a response to predation events by orcas (*Orcinus orca*), creating a space for the bronze whaler (*Carcharhinus brachyurus*) to increase its presence in coastal areas (Towner et al. 2022). Bonnethead sharks (*Sphyrna tiburo*), known previously as entirely carnivorous, exhibit herbivorous feeding behaviours, with a significant portion of seagrass found in their diet (Leigh et al. 2018).

These examples highlight that although literature has documented some clear examples of shark and ray contributions and performance in terms of trophodynamics, the interpretation of their ecological role and importance to the environment still involves substantial inferences, where the use of such language, particularly in terms of management, should be restrictive. This is especially important considering the risk of overexploitation and significant loss to these groups of species (MacNeil et al. 2020; Sherman et al. 2022). In addition to exploitation, climate change is also expected to significantly change the range distributions of migratory species (Heupel et al. 2019) or pose increased risk to the species whose ranges are extremely limited (e.g., walking sharks) (Gervais et al. 2018), which also has flow-on effects for our current understanding of ecological roles.

21.4. EFFECTS OF BIOPHYSICAL FACTORS ON SHARK AND RAY MOVEMENT AND DISTRIBUTION

Biophysical relationships between the GBR and sharks and rays are bidirectional. Despite having the ability to engineer their physical environment to varying degrees, the distributions and movements of sharks and rays are also largely influenced by a range of biophysical conditions (Espinoza et al. 2014; Rizzari et al. 2014a). The broad ecological and functional diversity of elasmobranchs on the GBR means that these species occupy a range of habitats including freshwater and estuarine systems, shallow coastal habitats, coral reefs, and pelagic environments (Table 1). The sheer diversity of form and function in this group makes the task of identifying key influential factors difficult, but fundamentally, the distribution of sharks and rays on the GBR is influenced by a complex interplay of environmental, ecological, and anthropogenic factors (Ceccarelli et al. 2013; Schlaff et al. 2014).

Broad-scale surveys of sharks along the GBR have revealed that the complexity of hard coral cover, proximity to reef systems, and associated biomass and diversity of prey influence shark and ray diversity and species richness (Espinoza et al. 2014). More specifically within reef systems, the reef slope habitats are areas that have an overall higher abundance and diversity of reef sharks (e.g., grey reef sharks *C. amblyrhynchos*, blacktip reef sharks *C. melanopterus*) across the length of the GBR (Rizzari et al. 2014a). Other biophysical variables, such as salinity and temperature, are strong drivers of seasonal distributions and movements of euryhaline and freshwater species (e.g., freshwater sharks *Glyphis* spp., sawfishes *Pristis* spp.), and they dynamically impact coastal and offshore shark and ray distributions (Yates et al. 2015; Lyon et al. 2017). For example, pregnant sawfishes (*Pristis* spp.) may move upstream into lower salinity environments for parturition, but the pups remain in low salinity habitats for their early years to avoid predators. Meanwhile, prey availability, together with oceanic currents and tidal patterns, drive distributions and migratory behaviours of large pelagic migratory species of sharks and rays on the east coast of Australia (Werry et al. 2014; Armstrong et al. 2016).

21.4.1. Biophysical Drivers of Movement

Although the long-term distribution patterns of marine species reflect broader scale habitat requirements, mobile species such as sharks and rays have the ability to leave an area should the local environmental conditions exceed thresholds for maintaining key physiological processes (Schlaff et al. 2017). A number of biophysical factors influence the movement behaviours of sharks and rays, which act across multiple spatial and temporal scales, and affect different species and different life stages in different ways. These factors include extrinsic parameters (i.e., temperature, salinity, and dissolved oxygen) or ecological dynamics (i.e., predator-prey dynamics, ontogenetic shifts, and reproductive activity) that change or limit how individuals move and select habitats, with extrinsic factors and ecological dynamics often working synergistically (Mull et al. 2022). Here, we discuss a few variables that have been identified as key drivers in the context of shark and ray movements, which may also influence seasonal changes on the movement and distribution of GBR sharks and rays.

21.4.1.1. Water Temperature

Temperature is a key variable that effects the physiology of ectotherms, and, in turn, has a strong influence on the movement patterns of sharks and rays. Individual species may actively seek out and occupy specific temperatures or traverse thermal gradients, which is termed behavioural thermoregulation. In general, seasonal fluctuations in water temperatures are known to be cues for specific movement behaviours in some sharks and rays, which can trigger migratory behaviours (Heupel 2007) or influence residency patterns within specific habitats (Kessel et al. 2014).

21.4.1.2. Salinity

Sharks and rays have different tolerances to water salinity, from euryhaline species that can exploit a wide range of salinities to the more common stenohaline species of sharks and rays that occupy a narrow salinity range (Grant et al. 2021). Fluctuating salinities in coastal and estuarine habitats during the onset of monsoonal runoffs are often a key driver of movements for shark and ray species that inhabit euryhaline environments across northern Australia (e.g., immature bull sharks *Carcharhinus leucas*, speartooth sharks *Glyphis glyphis*; Dwyer et al. 2020). Notwithstanding their osmoregulatory adaptations (see *The physiological Linkages between Sharks and Rays and their Environment*), species-specific behavioural responses to changing salinities help individuals balance the energetic demands of ion regulation with optimal habitat requirements.

21.4.1.3. Dissolved Oxygen

The availability of oxygen in the environment and the species-specific tolerance to varying levels of oxygen depletion in certain habitats drive movements in many shark and ray species. As previously described, species that have a high tolerance to hypoxic conditions, such as the epaulette shark (*H. ocellatum*), have the ability to access and remain in habitats such as shallow reef platforms during large tidal fluctuations (Wise et al. 1998). Meanwhile, the limiting influence of dissolved oxygen also plays a role in restricting access to depth in pelagic systems, with some shark species with lower hypoxic tolerances restricted from accessing deep waters (Abascal et al. 2011).

21.4.1.4. Tidal Phase

Movements and access to shallower habitats are tidally restricted in nearshore environments. In coastal species, tidally driven movements are often associated with foraging tactics (Davy et al. 2015; Martins et al. 2020a); however, for juvenile individuals, access to more complex and shallow habitats (e.g., mangroves) may be more likely driven by predator avoidance tactics (see 21.2.2; George et al. 2019).

21.4.2. Influence of Extreme Weather Events

So far, we have discussed how long-term seasonal changes in biophysical variables may dictate movements and distributions of sharks and rays. However, acute and extreme changes of these variables over much smaller temporal scales also frequently occur across northern Australia in the form of tropical cyclones, floods, and heatwaves. Extreme weather events can have a significant impact on shark movements and distributions and can disrupt the physical and biological conditions of marine ecosystems, which, in turn, can affect the behaviour and distribution of sharks.

As mobile marine predators, most sharks and rays can leave an area should local environmental conditions change sufficiently in the face of extreme weather (Heupel et al. 2003). Individuals that leave, however, will encounter risks associated with finding a new, suitable habitat and may experience increased chances of predation, while those that stay must endure challenging conditions (Schlaff et al. 2014). Whether or not individuals that leave return to their original habitat or relocate permanently often depends on the length and severity of environmental fluctuations (Udyawer et al. 2013). Tropical cyclones on the GBR are common, and in a changing climate cyclones are likely to increase in intensity (Wu et al. 2022) and perhaps frequency. Past tracking work on coastal shark species has identified that elasmobranchs most likely use the abrupt change in barometric pressure as a cue to evacuate shallow habitats during the onset of tropical cyclones (Heupel et al. 2003; Gutowsky et al. 2021). Multi-species studies have shown that sharks have species-specific responses to tropical cyclones, with some species moving into deeper waters, and others staying within shallow habitats (Udyawer et al. 2013; Strickland et al. 2020). In the GBR, tropical cyclone Anthony (Category 2) and Tropical Cyclone Yasi (Category 5) caused three coastal shark species (Australian blacktip *Carcharhinus tilstoni*; spot tail *C. sorrah*; and pigeye sharks *C. ambionensis*) to seek refuge in deeper water before the storms' arrival (Udyawer et al. 2013), thus illustrating how biophysical factors, such as pressure, can trigger shark behaviours during the onset of extreme weather.

Similarly, marine heatwaves can have a significant impact on the distribution of sharks and rays (Espinoza et al. 2011; Matich and Heithaus 2012). Heatwave events can disrupt the physical and biological conditions of marine ecosystems, which, in turn, can affect the behaviour and distribution of species. During heatwave events, prolonged high temperature conditions are likely to cause water temperatures to exceed thermal tolerances (Matich and Heithaus 2012). In addition to exceeding species' thermal tolerance, prolonged heatwave conditions degrade other key biophysical variables within habitats, such as dissolved oxygen and habitat structure, and simultaneously disrupt prey availability, which has a detrimental impact on species that cannot move (Osgood et al. 2021). The effects of temperature and other biophysical variables on shark and ray physiology are explored in more detail in *The Physiological Linkages between Sharks and Their Environment.*

21.5. THE PHYSIOLOGICAL LINKAGES BETWEEN SHARKS AND RAYS AND THEIR ENVIRONMENT

Understanding the physiology of sharks and rays—and their 450-million-year evolutionary history (Miller et al. 2003)—has revealed numerous physiological traits related to life history, metabolism, osmoregulation and ion balance, and stress responses that are unique to this taxon and key to how they function in their environment.

Firstly, unlike most teleosts, sharks and rays are K-strategists, meaning that these species are long-lived and slow-growing, have a late age of sexual maturity and long reproductive cycles, and produce a low number of large, high-quality offspring (Conrath and Musick 2012). This alone has implications toward growth and energy utilisation, considering, for example, maternal size at parturition and maternal nutritional input or matrotrophy (Hussey et al. 2010; Weideli et al. 2019). Indeed, increased reproductive output can be seen with larger mothers (e.g., dusky shark *Carcharhinus obscurus* and spinner shark *C. brevipinna*), and provisioned energy reserves in the form of an enlarged liver can constitute approximately 20% of the total body mass of the neonate (Hussey et al. 2010).

Secondly, sharks and rays exhibit a unique energy metabolism compared to teleost fishes, where they preferentially oxidize ketone bodies and amino acids (i.e., as opposed to fatty acids in teleosts) for ATP production (reviewed in Speers-Roesch and Treberg 2010). Such stark differences in energy metabolism also have implications when considering environmental influences, such as elevated temperatures, hypoxia, and high CO_2. Although an increase in overall metabolic rates (e.g., as estimated via oxygen uptake rates) is the common response amongst ectotherms with elevated temperatures (reviewed in Bouyoucos et al. 2019), sharks and rays may also be metabolically compensating in other ways via unchanged metabolic enzyme activities (Tullis and Baillie 2005), via increasing plasma glucose concentrations (e.g., neonates of *C. melanopterus*; Bouyoucos et al. 2018), or haematologically via changes in haemoglobin concentrations (e.g., increased [Hb], neonates of *Negaprion acutidens*; Bouyoucos et al. 2018; decreased [Hb], neonates of *C. melanopterus* Bouyoucos et al. 2021) or increased temperature sensitivity of haemoglobin-oxygen binding affinity (e.g., neonates of *C. melanopterus*; Bouyoucos et al. 2020a). When epaulette sharks (*H. ocellatum*) are exposed to extreme hypoxia and transition to anaerobic glycolysis, they accumulate lactate but do not increase plasma glucose concentrations, as seen in hypoxia-tolerant teleosts (Routley et al. 2002), and their metabolic enzyme activities (e.g., citrate synthase) are not altered in the red muscle, heart, or brain upon extended exposure to elevated CO_2 (Heinrich et al. 2014). There may also be neuroprotective attributes of ketone bodies during challenging environmental conditions.

Thirdly, this unique energy metabolism in sharks and rays is thought to be related to their capacity to synthesise and accumulate high levels of urea that allows marine sharks and rays to be osmoconforming ionoregulators (i.e., as opposed to osmo- or iono-regulating teleosts). This allows them to maintain their osmolality very close to that of seawater while excreting ions (i.e., such as sodium and chloride, similarly to teleosts) to maintain a different internal ion concentration from that of seawater (Yancey 2001). However, at high levels, urea destabilises proteins, and to thus counteract these normally toxic effects, sharks and rays also co-accumulate methylated amines, such as trimethylamine oxide (TMAO) (Yancey 2001), which protect these protein conformations by binding tightly to water (Liao et al. 2017). Sharks synthesise urea by using their gills to absorb ammonia—the major nitrogenous waste product that other organisms release upon breaking down dietary protein—that has been excreted into the seawater (Wood and Giacomin 2016). This physiological trait that is unique to sharks and rays has implications in terms of coastal areas with high ammonia levels, which are areas that are expected to increase with future development, agricultural run-off, and aquaculture operations, as increased ammonia concentrations could alter habitat use or shift home ranges of species to optimize urea production (Wood and Giacomin 2016).

Finally, a fourth physiological trait that is unique to sharks and rays is related to the primary and secondary stress responses (reviewed by Skomal and Mandelman 2012). Teleosts use cortisol as the key glucocorticoid involved in the stress axis and liberate glucose as the fuel to restore post-stress homeostasis (Wendelaar Bonga 1997), whereas sharks and rays are thought to use 1α-hydroxycorticosterone (1α-OHB) and potentially glucose and ketone bodies (Anderson 2012). Moreover, although teleosts adrenergically swell their red blood cells as part of the secondary stress response to protect intracellular pH and, therefore, oxygen transport, sharks and rays do not (Schwieterman et al. 2021). Indeed, these differences make identifying stress biomarkers and their utility challenging, but advances are underway. Schoen et al. (2021) related changes in 1α-OHB, glucose, and ketone bodies to simulated heatwave conditions in neonates of *C. melanopterus* and determined the ontogenetic differences in 1α-OHB mobilisation between neonates and adults of the same species. However, more work is needed to utilize 1α-OHB as a classic stress biomarker in sharks and rays.

It is evident that studies are needed to identify useful biomarkers that will be meaningful given the new environmental challenges that sharks and rays are facing into the Anthropocene.

The physiology of sharks and rays, like any other taxonomic group, is inextricably linked to their morphology and behaviour, and no species or population can be truly understood without first comprehending its interactions within its environment. Highly mobile species can actively seek out and use environmental situations that best allow their morphological and physiological abilities to function adequately for survival and reproduction, thus enhancing fitness (e.g., *C. leucas*; Bangley et al. 2018; Birkmanis et al. 2020). For example, tiger sharks (*G. cuvier*) swim optimally in 22 °C waters and are therefore most abundant in these temperatures (Payne et al. 2018). In this light, changes in environmental conditions or fishing pressure can alter distribution patterns and range retractions that may impact a species' role ecologically and from a socio-economic perspective (Gallagher and Hammerschlag 2011; Hammerschlag et al. 2019; Kendrick et al. 2019). Even species that are less-mobile can utilise various behaviours (e.g., avoidance, Bouyoucos et al. 2020a) to actively select conditions suited to their physiological capacities, but this may eventually lead to a reduction of performance in warming oceans (Lear et al. 2019). Tolerating a range of environmental temperatures by adjusting metabolic rates to correspond with optimal temperatures, diel patterns, and seasons to, for example, move, forage, or reproduce may be the strategy for other species (e.g., *H. ocellatum*; Gervais et al. 2018; Nay et al. 2021; Wheeler et al. 2022). Moreover, some strategies may be important for certain life history stages but not for others, as reproductive modes (e.g., at least 10 unique strategies, representing one of the most diverse arrays amongst vertebrates; Conrath and Musick 2012) and ontogenetic shifts frequently correspond with changes in habitat requirements and, therefore, often abiotic conditions as well (Wheeler et al. 2020). Consequently, the physiological linkages between these organisms and their environments are inseparable and must be investigated in concert. As such, with 140 species of sharks, skates, rays, and chimeras on the GBR, we could at the very least assume no fewer than

140 unique organism-environment relationships, and these are likely to change across space and time.

The physiological linkages between GBR species and their environments have been investigated with ecological and evolutionary objectives, management prioritisation, conservation outcomes, and since 2010, to understand the effects of climate change (Chin et al. 2010; Pereira Santos et al. 2021; Rummer et al. 2022). Overfishing remains the top threat to sharks and rays worldwide, and many studies have investigated physiological implications of catch-and-release and discard mortality in sharks (e.g., *C. plumbeus*, *C. limbatus*, *G. cuvier*, *C. brevipinna*, and *C. leucas*) using various haematological and stress biomarkers (Whitney et al. 2012), which helps refine management strategies. Although understanding the effects of climate change on sharks and rays remains critical, climate change was only explicitly identified as a threat to sharks and rays in 2010 (Chin et al. 2010); since then, there have still only been fewer than 50 studies on climate change effects on only 10–12 species of sharks and rays, only three of which are found on the GBR (Pereira Santos et al. 2021; Rummer et al. 2022). The available data suggest that of the climate change threats (e.g., ocean warming, acidification, and deoxygenation), ocean warming is likely to have the most pervasive effects on sharks and rays and is also the most well-studied to date (Rummer et al. 2022). Ocean warming and acidification together will affect swimming performance and locomotion (Vilmar and Di Santo 2022), development, and metabolism, but no study to date has addressed all three climate change stressors in combination (Pereira Santos et al. 2021).

The three most well-studied GBR species within a climate change physiology context include *C. melanopterus*, *H. ocellatum*, and the brown banded bamboo shark *Chiloscyllium punctatum*. Neonates of *C. melanopterus* populations have been the most extensively investigated within the context of early life habitat use, where home ranges can be constrained to areas as small as 0.02 km^2 (Bouyoucos et al. 2020a), but it is thought that such small nursery areas are important for food quality and foraging success (Heupel et al. 2007; Chin et al. 2016; Weideli et al. 2019). However, the shallow, warm, periodically hypoxic, and otherwise anthropogenically influenced nursery areas that they inhabit during the first years of life (Mourier and Planes 2013) necessitate physiological tolerance (i.e., metabolism, growth, and escape behaviours) for survival (Bouyoucos et al. 2018; Rummer et al. 2020; Bouyoucos et al. 2020a, 2020b, 2021, 2022; Trujillo et al. 2022). Indeed, although less is known about adult life stages, neonates tolerate a wide range of environmental temperatures and are predicted to be generally unaffected by mid- or end-of-century ocean acidification conditions (Rummer et al. 2020; Bouyoucos and Rummer 2021). Neonates of *C. melanopterus* exhibited elevated haematocrit, which could indicate stress or metabolic compensation, and blood lactate concentrations, which could indicate prolonged recovery, but all other metrics, including foraging behaviour, activity, and lateralisation, were unaffected under simulated ocean acidification conditions (Rummer et al. 2020; Bouyoucos et al. 2020a; Bouyoucos and Rummer 2021).

Meanwhile, the shallow reef flat specialist, *H. ocellatum* (Figure 3), has been investigated across all life stages—developing embryos *in ovo*, neonates, juveniles, and both non- and reproducing adults—across all climate change stressors. With this species still noted as the most hypoxia-tolerant shark that has been investigated to date (Chapman et al. 2011) and robust to ocean acidification conditions across life stages (Heinrich et al. 2014, 2015; Johnson et al. 2016), recent research has focused on the effects of temperature. Early life stages are significantly affected upon acclimation to end-of-century warming scenarios, with shorter incubation periods and faster yolk consumption *in ovo*, higher metabolic costs, smaller sizes and reduced energy reserves at hatching (Wheeler et al. 2021), altered colouration (Gervais et al. 2016), and decreased survival (Gervais et al. 2018). However, juveniles through non-encapsulating females, regardless of sex or body size, exhibit no differences in their upper thermal tolerance limits (Wheeler et al. 2022), which could have implications for distribution patterns of this site-attached mesopredator; however, more work is needed in this realm. Finally, although syntheses suggest that sharks and rays should be robust to future elevated CO_2 conditions, given their evolutionary history (Rosa et al. 2017), as has been found in *H. ocellatum*, studies on *Chiloscyllium punctatum* indicate otherwise. This well-studied species exhibits decreased aerobic metabolism in the brain and increased activity of antioxidant enzymes but still exhibits neuro-oxidative damage as a result (Rosa et al. 2016a). Although the effects on digestion are equivocal (Rosa et al. 2016b), this species does exhibit reductions in body condition and survival, especially when exposed

FIGURE 3 Epaulette shark (*Hemiscyllium ocellatum*) in the shallow reef flats of the southern GBR. This species can tolerate the low oxygen and elevated CO_2 (low pH) conditions that occur during the night-time hours, which is when this species is most active, and can also tolerate the dramatic diel fluctuations in temperature that are experienced in these shallow reef flats.

Source: Photo credit: Kristian Laine for @physioshark

to the combined stressors of elevated CO_2 and temperature (Rosa et al. 2014). These studies on only three GBR species, although thorough, emphasise that there remains a substantial knowledge gap with respect to the 140 species of sharks and rays on the GBR and how they physiologically respond to their environment, especially considering anthropogenic stressors that are increasing in frequency and intensity as we progress through the 21st century. However, these studies are a starting point to identify thresholds, limits, and trigger points for various stressors and potentially reveal adaptations necessary for future climate change conditions and/or determine which species will move to and thrive in more favourable habitats.

21.6. CONCLUSIONS AND FUTURE DIRECTIONS

Accordingly, the sharks and rays of the GBR are diverse and have a wide range of connections to their habitats and the biophysical characteristics of these habitats. Such characteristics shape shark and ray diversity and population structures, and, in turn, these species shape the environment around them. For example, different species select different habitats, and this can vary by life stage, such as juveniles spending their early years in nurseries. Connections to habitat and trophic relationships illustrate how sharks and rays affect and facilitate energy flow between different ecosystems in the GBR. Such characteristics drive species-specific morphological and physiological adaptations and explain how sharks and rays respond to and/or move in response to environmental parameters and major weather events. These characteristics also affect shark and ray physiology, which, in turn, will determine how these species respond to a changing climate. Meanwhile, our understanding of these connections remains limited. Even basic dietary information, which is essential to understanding the trophodynamic relationships between sharks and rays and their foodwebs, is sparse and only known for a handful of species. Similarly, physiological tolerances and performance characteristics are only known for a fraction of shark and ray species and in a rapidly warming world, represent an area that needs urgent attention. There is great scope for further exploration of these research areas, but this comes with a great challenge. Although many shark species in the GBR are relatively secure (Simpfendorfer et al. 2017), some species are at high risk and have already experienced declines (Tobin et al. 2010). As pressures mount on the GBR, it is urgent to better understand these species to help inform their ongoing management and safeguard such important social, cultural, economic, and ecological roles that they play in the World Heritage Area.

ACKNOWLEDGEMENTS

The authors would like to thank Carmen Dobszewicz for their technical assistance.

REFERENCES

Abascal FJ, Quintans M, Ramos-Cartelle A, Mejuto J (2011) Movements and environmental preferences of the shortfin mako, Isurus oxyrinchus, in the southeastern Pacific Ocean. Mar Biol 158:1175–1184

Anderson WG (2012) The endocrinology of 1α-hydroxy-corticosterone in elasmobranch fish: A review. *Comp Biochem Physiol A Mol Integr Physiol* 162:73–80

Armstrong AO, Armstrong AJ, Jaine FRA, Couturier LIE, Fiora K, Uribe-Palomino J, Weeks SJ, Townsend KA, Bennett MB, Richardson AJ (2016) Prey density threshold and tidal influence on reef manta ray foraging at an aggregation site on the Great Barrier Reef. PLoS One 11:e0153393-e0153393

Bangley CW, Paramore L, Shiffman DS, Rulifson RA (2018) Increased abundance and nursery habitat use of the bull shark (Carcharhinus leucas) in response to a changing environment in a warm-temperate estuary. Sci Rep 8:6018

Bierwagen SL (2019) Functional roles of reef sharks on the Great Barrier Reef. Dissertation. James Cook University

Bierwagen SL, Heupel MR, Chin A, Simpfendorfer CA (2018) Trophodynamics as a tool for understanding coral reef ecosystems. Front Mar Sci 5

Birkmanis CA, Freer JJ, Simmons LW, Partridge JC, Sequeira AMM (2020) Future distribution of suitable habitat for pelagic sharks in australia under climate change models. Front Mar Sci 7:570

Bouyoucos IA, Morrison PR, Weideli OC, Jacquesson E, Planes S, Simpfendorfer CA, Brauner CJ, Rummer JL (2020a) Thermal tolerance and hypoxia tolerance are associated in blacktip reef shark (Carcharhinus melanopterus) neonates. J Exp Biol 223

Bouyoucos IA, Rummer JL (2021) Improving "shark park" protections under threat from climate change using the conservation physiology tool box, Chapter 11 in Cooke SJ, Madliger CL, Love OP, and Franklin CE (eds) Conservation physiology: Integrating physiology into animal conservation and management. Oxford University Press. ISBN: 9780198843627. pp. 185–204. doi: 10.1093/oso/9780198843610.003.0011

Bouyoucos IA, Shipley ON, Jones E, Brooks EJ, Mandelman JW (2020b) Wound healing in an elasmobranch fish is not impaired by high-CO2 exposure. J Fish Biol 96:1508–1511

Bouyoucos IA, Simpfendorfer CA, Planes S, Schwieterman GD, Weideli OC, Rummer JL (2022) Thermally insensitive physiological performance allows neonatal sharks to use coastal habitats as nursery areas. Mar Ecol Prog Ser 682:137–152

Bouyoucos IA, Simpfendorfer CA, Rummer JL (2019) Estimating oxygen uptake rates to understand stress in sharks and rays. Rev Fish Biol Fish 29:297–311

Bouyoucos IA, Trujillo JE, Weideli OC, Nakamura N, Mourier J, Planes S, Simpfendorfer CA, Rummer JL (2021) Investigating links between thermal tolerance and oxygen supply capacity in shark neonates from a hyperoxic tropical environment. Sci Tot Environ 782:146854–146854

Bouyoucos IA, Weideli OC, Planes S, Simpfendorfer CA, Rummer JL (2018) Dead tired: Evaluating the physiological status and survival of neonatal reef sharks under stress. Conserv Physiol 6:coy053–coy053

Casey JM, Baird AH, Brandl SJ, Hoogenboom MO, Rizzari JR, Frisch AJ, Mirbach CE, Connolly SR (2017) A test of trophic cascade theory: Fish and benthic assemblages across a predator density gradient on coral reefs. Oecologia 183:161–175

Ceccarelli DM, Frisch AJ, Graham NAJ, Ayling AM, Beger M (2013) Habitat partitioning and vulnerability of sharks in the Great Barrier Reef Marine Park. Rev Fish Biol Fish 24:169–197

Chapman CA, Harahush BK, Renshaw GMC (2011) The physiological tolerance of the grey carpet shark (Chiloscyllium punctatum) and the epaulette shark (Hemiscyllium ocellatum) to anoxic exposure at three seasonal temperatures. Fish Physiol Biochem 37:387–399

Chin A (2023) Checklist of sharks and rays of the Great Barrier Reef World Heritage Area. Shark Search IndoPacific Programme. James Cook University, Townsville

Chin A, Heupel MR, Simpfendorfer CA, Tobin AJ (2013) Ontogenetic movements of juvenile blacktip reef sharks: Evidence of dispersal and connectivity between coastal habitats and coral reefs: Reef shark dispersal and connectivity. Aquat Conserv Mar Freshw Ecosyst 23:468–474

Chin A, Heupel MR, Simpfendorfer CA, Tobin AJ (2016) Population organisation in reef sharks: New variations in coastal habitat use by mobile marine predators. Mar Ecol Prog Ser 544:197–211

Chin A, Kyne PM, Walker TI, McAuley RB (2010) An integrated risk assessment for climate change: Analysing the vulnerability of sharks and rays on Australia's Great Barrier Reef. Glob Change Biol 16:1936–1953

Choat J, Russell B, Chin A (2019) The fish assemblages of the Great Barrier Reef: Their diversity and origin. In Hutchings P and Hoegh-Guldberg O (ed.) The great barrier reef: Biology, environment and management. CSIRO Publishing. pp. 389–406

Compagno LJV (1990) Alternative life-history styles of cartilaginous fishes in time and space. Environ Biol Fishes 28:33–75

Conrath CL, Musick JA (2012) Reproductive biology of elasmobranchs. Biol Sharks Relat 2:291–311

Cortes E (1999) Standardized diet compositions and trophic levels of sharks. ICES J Mar Sci 56:707–717

Crook K (2020) Assessing the functional roles of rays in coastal sandflats. Dissertation, James Cook University, Townsville

Crook KA, Sheaves M, Barnett A (2022) Species-specific foraging behaviors define the functional roles of sympatric stingrays. Limnol Oceanogr 67:219–230

Dale JJ, Drazen JC, Holland KN (2013) Stingray life history trade-offs associated with nursery habitat use inferred from a bioenergetics model. Mar Biol 160:3181–3192

Davy LE, Simpfendorfer CA, Heupel MR (2015) Movement patterns and habitat use of juvenile mangrove whiprays (Himantura granulata). Mar Freshw Res 66:481–492

Dudgeon CL, Corrigan S, Yang L, Allen GR, Erdmann MV, Fahmi, Sugeha HY, White WT, Naylor GJP (2020) Walking, swimming or hitching a ride? Phylogenetics and biogeography of the walking shark genus Hemiscyllium. Mar Freshw Res 71:1107–1117

Dulvy NK, Davidson LN, Kyne PM, Simpfendorfer CA, Harrison LR, Carlson JK, Fordham SV (2017) Ghosts of the coast: Global extinction risk and conservation of sawfishes. Aquat Conserv Mar Freshw Ecosyst 27(1):97–106.

Dunn RE, Bradley D, Heithaus MR, Caselle JE, Papastamatiou YP (2022) Conservation implications of forage base requirements of a marine predator population at carrying capacity. iScience 25:103646

Dwyer RG, Campbell HA, Cramp RL, Burke CL, Micheli-Campbell MA, Pillans RD, Lyon BJ, Franklin CE (2020) Niche partitioning between river shark species is driven by seasonal fluctuations in environmental salinity. Funct Ecol 34:2170–2185

Espinoza M, Cappo M, Heupel MR, Tobin AJ, Simpfendorfer CA (2014) Quantifying shark distribution patterns and species-habitat associations: Implications of marine park zoning. PLoS One 9:e106885.

Espinoza M, Farrugia TJ, Lowe CG (2011) Habitat use, movements and site fidelity of the gray smooth-hound shark (Mustelus californicus, Gill 1863) in a newly restored southern California estuary. J Exp Mar Biol Ecol 401:63–74

Espinoza M, Matley J, Heupel MR, Tobin AJ, Fisk AT, Simpfendorfer CA (2019) Multi-tissue stable isotope analysis reveals resource partitioning and trophic relationships of large reef-associated predators. Mar Ecol Prog Ser 615:159–176

Ferretti F, Worm B, Britten GL, Heithaus MR, Lotze HK (2010) Patterns and ecosystem consequences of shark declines in the ocean. Ecol Lett 13:1055–1071

Frid A, Baker GG, Dill LM (2008) Do shark declines create fear-released systems? Oikos 117:191–201

Frisch AJ, Ireland M, Rizzari JR, Lönnstedt OM, Magnenat KA, Mirbach CE, Hobbs J-PA (2016) Reassessing the trophic role of reef sharks as apex predators on coral reefs. Coral Reefs 35:459–472

Gallagher AJ, Hammerschlag N (2011) Global shark currency: The distribution, frequency, and economic value of shark ecotourism. Curr Issues Tour 14:797–812

George LW, Martins APB, Heupel MR, Simpfendorfer CA (2019) Fine-scale movements of juvenile blacktip reef sharks Carcharhinus melanopterus in a shallow nearshore nursery. Mar Ecol Prog Ser 623:85–97

Gerhardt K (2018) Indigenous knowledge and cultural values of hammerhead sharks in Northern Australia: Report to the National Environmental Science Program, Marine Biodiversity Hub. James Cook University, Townsville, Queensland

Gervais CR, Mourier J, Rummer JL (2016) Developing in warm water: Irregular colouration and patterns of a neonate elasmobranch. Mar Biodivers 46:743–744

Gervais CR, Nay TJ, Renshaw G, Johansen JL, Steffensen JF, Rummer JL (2018) Too hot to handle? Using movement to alleviate effects of elevated temperatures in a benthic elasmobranch, Hemiscyllium ocellatum. Mar Biol 165:162

Grant MI, Kyne PM, Simpfendorfer CA, White WT, Chin A (2019) Categorising use patterns of non-marine environments by elasmobranchs and a review of their extinction risk. Rev Fish Biol Fish 29:689–710

Grant MI, White WT, Amepou Y, Appleyard SA, Baje L, Devloo-Delva F, Feutry P, Ibana D, Jogo DJ, Jogo S, Kyne PM, Mana R, Mapmani N, Nagul A, Roeger D, Simpfendorfer CA, Chin A (2021) Papua New Guinea: A potential refuge for threatened Indo–Pacific river sharks and sawfishes. Front Conserv Sci 2

Grubbs RD, Carlson JK, Romine JG, Curtis TH, McElroy WD, McCandless CT, Cotton CF, Musick JA (2016) Critical assessment and ramifications of a purported marine trophic cascade. Sci Rep 6:20970

Gutowsky LFG, Rider MJ, Roemer RP, Gallagher AJ, Heithaus MR, Cooke SJ, Hammerschlag N (2021) Large sharks exhibit varying behavioral responses to major hurricanes. Estuar Coast Shelf Sci 256:107373

Hammerschlag N, Schmitz OJ, Flecker AS, Lafferty KD, Sih A, Atwood TB, Gallagher AJ, Irschick DJ, Skubel R, Cooke SJ (2019) Ecosystem Function and services of aquatic predators in the Anthropocene. Trends Ecol Evol 34:369–383

Heinrich DDU, Rummer JL, Morash AJ, Watson S-A, Simpfendorfer CA, Heupel MR, Munday PL (2014) A product of its environment: The epaulette shark (Hemiscyllium ocellatum) exhibits physiological tolerance to elevated environmental CO_2. Conserv Physiol 2:cou047

Heinrich DDU, Watson S-A, Rummer JL, Brandl SJ, Simpfendorfer CA, Heupel MR, Munday PL (2015) Foraging behaviour of the epaulette shark Hemiscyllium ocellatum is not affected by ocean acidification. ICES J Mar Sci 73:633–640

Heithaus MR, Dunn RE, Farabaugh NF, Lester E, Madin E, Meekan MG, Papastamatiou YP, Roff G, Vuado JJ, Wirsing AJ (2022) Advances in our understanding of the ecological importance of sharks and their relatives, In: Biology of sharks and their relatives. CRC Press, pp. 487–521

Heupel MR (2007) Exiting Terra Ceia Bay: An examination of cues stimulating migration from a summer nursery area, In: McCandless CT, Kohler NE, Pratt HL (eds) Shark Nursery Grounds of the Gulf of Mexico and the East Coast Waters of the United States, American Fisheries Society, Symposium 50, Bethesda, MD, pp. 265–280

Heupel MR, Carlson JK, Simpfendorfer CA (2007) Shark nursery areas: Concepts, definition, characterization and assumptions. Mar Ecol Prog Ser 337:287–297

Heupel MR, Papastamatiou YP, Espinoza M, Green ME, Simpfendorfer CA (2019) Reef shark science—Key questions and future directions. Front Mar Sci 6:12

Heupel MR, Simpfendorfer CA, Hueter RE (2003) Running before the storm: Blacktip sharks respond to falling barometric pressure associated with Tropical Storm Gabrielle: Response of sharks to a tropical storm. J Fish Biol 63:1357–1363

Heupel MR, Williams AJ, Welch DJ, Ballagh A, Mapstone BD, Carlos G, Davies C, Simpfendorfer CA (2009) Effects of fishing on tropical reef associated shark populations on the Great Barrier Reef. Fish Res 95:350–361

Hickey AJR, Renshaw GMC, Speers-Roesch B, Richards JG, Wang Y, Farrell AP, Brauner CJ (2012) A radical approach to beating hypoxia: Depressed free radical release from heart fibres of the hypoxia-tolerant epaulette shark (Hemiscyllum ocellatum). J Comp Physiol B Biochem Syst Environ Physiol 182:91–100

Hussey NE, MacNeil MA, Fisk AT (2010) The requirement for accurate diet-tissue discrimination factors for interpreting stable isotopes in sharks: Comment on: Stable isotope dynamics in elasmobranch fishes. Hydrobiologia 654:1–5

Hutchings P, Kingsford MJ, Hoegh-Guldberg O (2019) The Great Barrier Reef: Biology, Environment and Management, 2nd edn. CSIRO Publishing

Johnson MS, Kraver DW, Renshaw GMC, Rummer JL (2016) Will ocean acidification affect the early ontogeny of a tropical oviparous elasmobranch (Hemiscyllium ocellatum)? Conserv Physiol 4:1–11

Kendrick GA, Nowicki RJ, Olsen YS, Strydom S, Fraser MW, Sinclair EA, Statton J, Hovey RK, Thomson JA, Burkholder DA, McMahon KM, Kilminster K, Hetzel Y, Fourqurean JW, Heithaus MR, Orth RJ (2019) A systematic review of how multiple stressors from an extreme event drove ecosystem-wide loss of resilience in an iconic seagrass community. Front Mar Sci 6

Kessel ST, Chapman DD, Franks BR, Gedamke T, Gruber SH, Newman JM, White ER, Perkins RG (2014) Predictable temperature-regulated residency, movement and migration in a large, highly mobile marine predator (Negaprion brevirostris). Mar Ecol Prog Ser 514:175–190

Kinney MJ, Hussey NE, Fisk AT, Tobin AJ, Simpfendorfer CA (2011) Communal or competitive?: Stable isotope analysis provides evidence of resource partitioning within a communal shark nursery. Mar Ecol Prog Ser 439:263–276

Krantzberg G (1985) The influence of bioturbation on physical, chemical and biological parameters in aquatic environments: A review. Environ Pollut A 39:99–122

Labourgade P, Ballesta L, Huveneers C, Papastamatiou Y, Mourier J (2020) Heterospecific foraging associations between reef-associated sharks: First evidence of kleptoparasitism in sharks. Ecology 101:1–4

Last PR, Stevens JD (2009) Sharks and rays of Australia. CSIRO Publishing

Lear KO, Whitney NM, Morgan DL, Brewster LR, Whitty JM, Poulakis GR, Scharer RM, Guttridge TL, Gleiss AC (2019) Thermal performance responses in free-ranging elasmobranchs depend on habitat use and body size. Oecologia 191:829–842

Leigh SC, Papastamatiou YP, German DP (2018) Seagrass digestion by a notorious 'carnivore'. Proc R Soc B Biol Sci 285:20181583

Lester EK, Langlois TJ, Simpson SD, McCormick MI, Meekan MG (2020) The hemisphere of fear: The presence of sharks influences the three dimensional behaviour of large mesopredators in a coral reef ecosystem. Oikos 129:731–739

Liao Y-T, Manson AC, DeLyser MR, Noid WG, Cremer PS (2017) Trimethylamine N-oxide stabilizes proteins via a distinct mechanism compared with betaine and glycine. Proc Natl Acad Sci USA 114:2479–2484

Libralato S, Pranovi F, Stergiou KI, Link J (2014) Trophodynamics in marine ecology: 70 years after Lindeman. Mar Ecol Prog Ser 512:1–7

Lindeman RL (1942) The trophic-dynamic aspect of ecology. Ecology 23:399–417

Lyon BJ, Dwyer RG, Pillans RD, Campbell HA, Franklin CE (2017) Distribution, seasonal movements and habitat utilisation of an endangered shark, Glyphis glyphis, from northern Australia. Mar Ecol Prog Ser 573:203–213

MacNeil MA, Chapman DD, Heupel M, Simpfendorfer CA, Heithaus M, Meekan M, Harvey E, Goetze J, Kiszka J, Bond ME (2020) Global status and conservation potential of reef sharks. Nature 583:801–806

Martins APB, Heupel MR, Bierwagen SL, Chin A, Simpfendorfer C (2020a) Diurnal activity patterns and habitat use of juvenile Pastinachus ater in a coral reef flat environment. PLoS One 15:e0228280.

Martins APB, Heupel MR, Bierwagen SL, Chin A, Simpfendorfer CA (2020b) Tidal–diel patterns of movement, activity and habitat use by juvenile mangrove whiprays using towed-float GPS telemetry. Mar Freshw Res 72:534–541

Martins APB, Heupel MR, Chin A, Simpfendorfer CA (2018) Batoid nurseries: Definition, use and importance. Mar Ecol Prog Ser 595:253–267

Martins APB, Matley J, Heupel M, Fisk A, Chin A, Simpfendorfer C (2022) Trophic ecology of sympatric juvenile stingrays within a nursery area. Mar Freshw Res 73:678–688

Matich P, Heithaus MR (2012) Effects of an extreme temperature event on the behavior and age structure of an estuarine top predator, Carcharhinus leucas. Mar Ecol Prog Ser 447:165–178

Miller RF, Cloutier R, Turner S (2003) The oldest articulated chondrichthyan from the Early Devonian period. Nature 425:501–504

Mourier J, Maynard J, Parravicini V, Ballesta L, Clua E, Domeier Michael L, Planes S (2016) Extreme inverted trophic pyramid of reef sharks supported by spawning groupers. Curr Biol 26:2011–2016

Mourier J, Planes S (2013) Direct genetic evidence for reproductive philopatry and associated fine-scale migrations in female blacktip reef sharks (Carcharhinus melanopterus) in French Polynesia. Mol Ecol 22:201–214

Mull CG, Andrzejaczek S, Udyawer V, Dwyer RG (2022) Advances in methods, understanding, and applications of elasmobranch movement ecology biology of sharks and their relatives. CRC Press, pp. 357–399

Nay TJ, Longbottom RJ, Gervais CR, Johansen JL, Steffensen JF, Rummer JL, Hoey AS (2021) Regulate or tolerate: Thermal strategy of a coral reef flat resident, the epaulette shark, Hemiscyllium ocellatum. J Fish Biol 98:723–732

Orth RJ (1975) Destruction of Eelgrass, Zostera marina, by the cownose ray, Rhinoptera bonasus, in the Chesapeake Bay. Chesapeake Sci 16:205–208

Osgood GJ, White ER, Baum JK (2021) Effects of climate-change-driven gradual and acute temperature changes on shark and ray species. J Anim Ecol 90:2547–2559

O'Shea OR, Thums M, Van Keulen M, Meekan M (2012) Bioturbation by stingrays at Ningaloo reef, Western Australia. Mar Freshw Res 63:189–197

Payne NL, Meyer CG, Smith JA, Houghton JDR, Barnett A, Holmes BJ, Nakamura I, Papastamatiou YP, Royer MA, Coffey DM, Anderson JM, Hutchinson MR, Sato K, Halsey LG (2018) Combining abundance and performance data reveals how temperature regulates coastal occurrences and activity of a roaming apex predator. Glob Change Biol 24:1884–1893

Pereira Santos C, Sampaio E, Pereira B, Pegado MR, Borges FO, Wheeler C, Bouyoucos I, Rummer J, Frazão Santos C, Rosa R (2021) Elasmobranch responses to experimental warming, acidification, and oxygen loss—a meta-analysis. Front Mar Sci 8.

Polunin NV (1996) Trophodynamics of reef fisheries productivity. Reef Fisheries 113–135

Porter ME, Hernandez AV, Gervais CR, Rummer JL (2022) Aquatic walking and swimming kinematics of neonate and juvenile epaulette sharks. Integr Comp Biol 62:1710–1724

Rizzari JR, Frisch AJ, Hoey AS, McCormick MI (2014b) Not worth the risk: Apex predators suppress herbivory on coral reefs. Oikos 123:829–836

Rizzari JR, Frisch AJ, Magnenat KA (2014a) Diversity, abundance, and distribution of reef sharks on outer-shelf reefs of the Great Barrier Reef, Australia. Mar Biol 161:2847–2855

Robbins WD, Hisano M, Connolly SR, Choat JH (2006) Ongoing collapse of coral-reef shark populations. Curr Biol 16:2314–2319

Roff G, Doropoulos C, Rogers A, Bozec Y-M, Krueck NC, Aurellado E, Priest M, Birrell C, Mumby PJ (2016) The ecological role of sharks on coral reefs. Trends Ecol Evol 31:395–407

Rosa R, Baptista M, Lopes VM, Pegado MR, Ricardo Paula J, Trübenbach K, Leal MC, Calado R, Repolho T (2014) Early-life exposure to climate change impairs tropical shark survival. Proc R Soc B Biol Sci 281:20141738

Rosa R, Pimentel M, Galan JG, Baptista M, Lopes VM, Couto A, Guerreiro M, Sampaio E, Castro J, Santos C, Calado R, Repolho T (2016b) Deficit in digestive capabilities of bamboo shark early stages under climate change. Mar Biol 163:160

Rosa R, Ricardo Paula J, Sampaio E, Pimentel M, Lopes AR, Baptista M, Guerreiro M, Santos C, Campos D, Almeida-Val VMF, Calado R, Diniz M, Repolho T (2016a) Neuro-oxidative damage and aerobic potential loss of sharks under elevated CO_2 and warming. Mar Biol 163:119

Rosa R, Rummer JL, Munday PL (2017) Biological responses of sharks to ocean acidification. Biol Lett 13:20160796.

Routley MH, Nilsson GE, Renshaw GMC (2002) Exposure to hypoxia primes the respiratory and metabolic responses of the epaulette shark to progressive hypoxia. *Comp Biochem Physiol A Mol Integr Physiol* 131:313–321

Rummer JL, Bouyoucos IA, Mourier J, Nakamura N, Planes S (2020) Responses of a coral reef shark acutely exposed to ocean acidification conditions. Coral Reefs 39:1215–1220

Rummer JL, Bouyoucos IA, Wheeler CR, Santos CP, Rosa R (2022) Climate change and sharks, Ch. 25 in Carrier JC, Simpfendorfer CA, Heithaus MR, Yopak KE (eds), Biology of Sharks and their Relatives, 3rd edn. CRC Press. 840pp. doi: 10.1201/9781003262190

Ruppert JLW, Travers MJ, Smith LL, Fortin M-J, Meekan MG (2013) Caught in the middle: Combined impacts of shark removal and coral loss on the fish communities of coral reefs. PLoS One 8:e74648

Ruppert JLW, Travers MJ, Smith LL, Fortin M-J, Meekan MG, Meeuwig JJ (2019) Demography of reef sharks using ray- and laser-based photogrammetry in the Chagos Archipelago. Coral Reefs 38(5):985–994.

Sale PF (2002) Chapter 16 — The science we need to develop for more effective management. Academic Press, pp. 361–376

Sarker S, Masud-Ul-Alam M, Hossain MS, Rahman Chowdhury S, Sharifuzzaman SM (2021) A review of bioturbation and sediment organic geochemistry in mangroves. Geol J 56:2439–2450

Schlaff AM, Heupel MR, Simpfendorfer CA (2014) Influence of environmental factors on shark and ray movement, behaviour and habitat use: A review. Rev Fish Biol Fish 24:1089–1103

Schlaff AM, Heupel MR, Udyawer V, Simpfendorfer CA (2017) Biological and environmental effects on activity space of a common reef shark on an inshore reef. Mar Ecol Prog Ser 571:169–181

Schoen AN, Bouyoucos IA, Anderson WG, Wheaton CJ, Planes S, Mylniczenko ND, Rummer JL (2021) Simulated heatwave and fishing stressors alter corticosteroid and energy balance in neonate blacktip reef sharks, Carcharhinus melanopterus. Conserv Physiol 9:1–coab067

Schwieterman GD, Rummer JL, Bouyoucos IA, Bushnell PG, Brill RW (2021) A lack of red blood cell swelling in five elasmobranch fishes following air exposure and exhaustive exercise. *Comp Biochem Physiol A Mol Integr Physiol* 258:110978.

Sherman CS, Digel ED, Zubick P, Eged J, Haque AB, Matsushiba JH, Simpfendorfer CA, Sant G, Dulvy NK (2022) High overexploitation risk due to management shortfall in highly traded requiem sharks. Cold Spring Harbor Laboratory Press.

Simpfendorfer C, Chin A, Rigby C, Sherman S, White W (2017) Shark futures: A report card for Australia's sharks and rays. Centre for Sustainable Tropical Fisheries and Aquaculture, James Cook University.

Skomal GB, Mandelman JW (2012) The physiological response to anthropogenic stressors in marine elasmobranch fishes: A review with a focus on the secondary response. *Comp Biochem Physiol A Mol Integr Physiol* 162:146–155

Speers-Roesch B, Treberg JR (2010) The unusual energy metabolism of elasmobranch fishes. *Comp Biochem Physiol A Mol Integr Physiol* 155:417–434

Strickland BA, Massie JA, Viadero N, Santos R, Gastrich KR, Paz V, O'Donnell P, Kroetz AM, Ho DT, Rehage JS, Heithaus MR (2020) Movements of Juvenile bull sharks in response to a major hurricane within a tropical estuarine nursery area. Estaur Coast 43:1144–1157

Tobin A, Simpfendorfer C, Mapleston A, Currey L, Harry A, Welch D, Ballagh A, Chin A, Szczenski N, Schlaff A (2010) A quantitative ecological risk assessment of sharks and finfish of Great Barrier Reef World Heritage Area inshore waters: A tool for fisheries and marine park managers: Identifying species at risk and potential mitigation strategies. Marine and Tropical Sciences Research Facility, Cairns.

Towner AV, Watson RGA, Kock AA, Papastamatiou Y, Sturup M, Gennari E, Baker K, Booth T, Dicken M, Chivell W, Elwen S, Kaschke T, Edwards D, Smale MJ (2022) Fear at the top: Killer whale predation drives white shark absence at South Africa's largest aggregation site. Afr J Mar Sci 44:139–152

Trujillo JE, Bouyoucos I, Rayment WJ, Domenici P, Planes S, Rummer JL, Allan BJM (2022) Escape response kinematics in two species of tropical shark: Short escape latencies and high turning performance. J Exp Biol 225

Tullis A, Baillie M (2005) The metabolic and biochemical responses of tropical whitespotted bamboo shark Chiloscyllium plagiosum to alterations in environmental temperature. J Fish Biol 67:950–968

Udyawer V, Chin A, Knip DM, Simpfendorfer CA, Heupel MR (2013) Variable response of coastal sharks to severe tropical storms: Environmental cues and changes in space use. Mar Ecol Prog Ser 480:171–183

Vilmar M, Di Santo V (2022) Swimming performance of sharks and rays under climate change. Rev Fish Biol Fish 32:765–781

Weideli OC, Bouyoucos IA, Papastamatiou YP, Gauthier M, Rummer JL, Planes S (2019) Same species, different prerequisites: Investigating body condition and foraging success in young reef sharks between an atoll and an island system. Sci Rep 9:1–11

Wendelaar Bonga SE (1997) The stress response in fish. Physiol Rev 77:591–625

Werry JM, Planes S, Berumen ML, Lee KA, Braun CD, Clua E (2014) Reef-fidelity and migration of tiger sharks, Galeocerdo cuvier, across the Coral Sea. PLoS One 9:e83249.

Wheeler CR, Gervais CR, Johnson MS, Vance S, Rosa R, Mandelman JW, Rummer JL (2020) Anthropogenic stressors influence reproduction and development in elasmobranch fishes. Rev Fish Biol Fish 30:373–386

Wheeler CR, Kneebone J, Heinrich D, Strugnell JM, Mandelman JW, Rummer JL (2022) Diel rhythm and thermal independence of metabolic rate in a benthic shark. J Biol Rhythms 37:484–497

Wheeler CR, Rummer JL, Bailey B, Lockwood J, Vance S, Mandelman JW (2021) Future thermal regimes for epaulette sharks (Hemiscyllium ocellatum): Growth and metabolic performance cease to be optimal. Sci Rep 11:454

Whitney NM, Papastamatiou YP, Gleiss AC (2012) Integrative multi-sensor tagging of elasmobranchs: Emerging techniques to quantify behavior, physiology, and ecology. In Carrier J, Musick JA, Heithaus M (eds) Biology of sharks and their relatives. CRC Press.

Whitty JM, Morgan DL, Peverell SC, Thorburn DC, Beatty SJ (2009) Ontogenetic depth partitioning by juvenile freshwater sawfish (Pristis microdon: Pristidae) in a riverine environment. Mar Freshw Res 60:306–316

Williams JJ, Papastamatiou YP, Caselle JE, Bradley D, Jacoby DM (2018) Mobile marine predators: An understudied source of nutrients to coral reefs in an unfished atoll. Proc R Soc B Biol Sci 285:20172456

Wilson SK, Burgess SC, Cheal AJ, Emslie MJ, Fisher R, Miller I, Polunin NVC, Sweatman H (2008) Habitat utilization by coral reef fish: Implications for specialists vs. generalists in a changing environment. J Anim Ecol 77(2):220–228.

Wirsing AJ, Heithaus MR, Dill LM (2007) Fear factor: Do dugongs (Dugong dugon) trade food for safety from tiger sharks (Galeocerdo cuvier). Oecologia 153:1031–1040

Wirsing AJ, Heithaus MR, Frid A, Dill LM (2008) Seascapes of fear: Evaluating sublethal predator effects experienced and generated by marine mammals. Mar Mammal Sci 24:1–15

Wise G, Mulvey JM, Renshaw GMC (1998) Hypoxia tolerance in the epaulette shark (Hemiscyllium ocellatum). The J Exp Zool 281:1–5

Wood CM, Giacomin M (2016) Feeding through your gills and turning a toxicant into a resource: How the dogfish shark scavenges ammonia from its environment. J Exp Biol 219:3218–3226

Wu L, Zhao H, Wang C., Cao J, Liang J (2022) Understanding of the effect of climate change on tropical cyclone intensity: A review. Adv Atmos Sci 39:205–221.

Wueringer BE (2012) Electroreception in Elasmobranchs: Sawfish as a case study. Brain Behav Evol 80:97–107

Yancey PH (2001) Nitrogen compounds as osmolytes. Elsevier Science & Technology, pp. 309–341

Yates PM, Heupel MR, Tobin AJ, Simpfendorfer CA (2015) Ecological drivers of shark distributions along a tropical coastline. PLoS One 10:e0121346.

Section 4

Consequences: Impact on GBR Water and Remediation

22 Impacts of Climate Change Stressors on the Great Barrier Reef

Maria Byrne, Shawna A. Foo, Ana Vila-Concejo, and Kennedy Wolfe

22.1. INTRODUCTION

In a high CO_2 world, coral reefs face a perilous future due to the multitude of co-occurring climate stressors. Climate change is causing sea-level rise, thermal stress, ocean acidification, and an increase in the intensity of storms with associated physical, hydromechanical, and hydrochemical impacts (Albright et al. 2016a; Cheal et al. 2017; Hoegh-Guldberg et al. 2017; Hughes et al. 2017a, 2017b, 2018; Putnam et al. 2017; Tebaldi et al. 2021; NOAA https://tinyurl.com/53z66rdj). Moreover, the rate and magnitude of change of these stressors are well beyond what modern reefs have evolved to cope with over geological timeframes (Pandolfi et al. 2011). Climate-driven warming and marine heatwaves have caused significant coral bleaching and mortality and in some cases, collapse of reef structures and associated communities on the Great Barrier Reef (GBR) and elsewhere (Hughes et al. 2017, 2018). Heatwaves, especially in combination with cyclones, are the climate change stressors of the utmost contemporary concern for the GBR. Accordingly, this iconic reef system has a very uncertain future.

As detailed in several reviews (De'Ath et al. 2012; Ban et al. 2014; Albright et al. 2016b; Mellin et al. 2019; Fabricius et al. 2020; Wolff et al. 2018; Castro-Sanguino et al. 2021), the impacts of climate change on the GBR have been evident for some time. Here, we review the suite of climate change stressors that are impacting coral reefs as single and combined factors and highlight their impacts in GBR-specific case studies.

Understanding the changes that have occurred on coral reefs in response to a changing climate and predicting future prospects for these ecosystems depend on long-term datasets, several of which are available for the GBR. These long-term studies are presented here. Reports of coral cover on well-studied reefs at Lizard Island and Low Isles in the northern GBR provide important insight into disturbance and recovery dynamics (Case Study 1). Long-term changes in the biogeochemistry of the GBR are reviewed (Case Study 2), the ability of reef organisms such as bêche-de-mer sea cucumbers to help buffer these changes through their biology and ecological roles is investigated (Case Study 3), and the resilience of the influential crown-of-thorns sea star is examined (Case Study 4). From reef evolution and climate change perspectives, the dynamics of sand sheets (also known as sand aprons) in the southern GBR and impacts of sea level rise are also considered (Case Study 5). The impacts of climate change on non-coral invertebrates and the dynamics of reef fishes following cyclone disturbance are highlighted (Case Study 6).

22.2. CLIMATE CHANGE STRESSORS

22.2.1. Chemical Processes in a Changing Ocean—Warming and Acidification

The global increase in sea surface temperature (SST) and marine heatwaves are major threats to coral reef ecosystems. The Intergovernmental Panel on Climate Change (IPCC) projections based on the best and worst Representative Concentration Pathway (RCP) scenarios indicate an increase between 0.7 and 1.1 °C by 2050 and 0.8 and 2.7 °C by 2100 (Fox-Kemper et al. 2021). Heatwaves have become more frequent with greater impacts in the northern GBR, while the southern GBR has, for the most part, been less affected (Hughes et al. 2017a, 2017b). Acute punctuated warming events cause temperatures to rise rapidly, often beyond the thermal limits of many reef species. A series of heatwave-driven bleaching events over recent years have been devastating for areas of the north and central GBR (Hughes et al. 2017a, 2017b). Although recent surveys have shown coral recovery, this recovery is dominated by fast-growing *Acropora* (AIMS 2022).

Bleaching mortality causes the collapse of habitat and coral reef structural complexity (Perry and Alvarez-Filip 2019) upon which a great diversity of species depends. Consequently, impacts to reef fishes and invertebrates have been documented, especially for species that depend on coral for food and/or shelter with flow on effects for the ecosystem (Jones et al. 2004; Pratchett et al. 2008; Prezeslawski et al. 2008; Stella et al. 2011; Rice et al. 2019; Froehlich et al. 2021). Corallivores such as *Acanthaster* may have a more concentrated impact on surviving corals by narrowing recovery windows before they go locally extinct (Keesing et al. 2019). The breakdown of dead coral into rubble gives rise to a diversity of small cryptic invertebrates that can rapidly avail themselves of this newly created habitat (Wolfe et al. 2021a, 2023; Stella et al. 2022), which also serves as the nursery habitat for the herbivorous juvenile stage of the crown-of-thorns sea star (Deaker et al. 2020; Wilmes et al. 2020).

In addition to warming, ocean chemistry is being altered by increasing atmospheric CO_2. The ocean surface is the largest sink for anthropogenic CO_2 emissions, which is

DOI: 10.1201/9781003320425-26

causing the ocean to become more acidic, a phenomenon called ocean acidification. Globally, surface ocean pH has decreased from 8.2 in pre-industrial times to the present-day average of 8.07 (Jiang et al. 2019), and IPCC projections based on the best and worst RCP scenarios indicate a drop in pH to 7.6–7.7 by 2050 and 7.4–7.7 by 2100 (Fox-Kemper et al. 2021). Due to upstream processes, the change in ocean chemistry is likely to be steeper on the GBR than currently projected by the IPCC with the largest impacts on inshore reefs influenced by freshwater, land runoff, and coastal nutrient fluxes (Uthicke et al. 2014; Mongin et al. 2016). Data from two monitoring stations, one in the southern GBR at Heron Island and the second at Yongala Reef in the central coastal GBR near the influence of the Burdekin River, show that seawater CO_2 fugacity (*f*CO2) has progressively increased by $>2.0 \pm 0.3$ μatm year^{-1} where carbonate dissolution from the seafloor is unlikely sufficient to buffer the GBR against ocean acidification (Fabricius et al. 2020). Although we have some information on the impacts of ocean acidification on the GBR from these monitoring stations and from coral cores (Wei et al. 2009), ecological modelling (Mongin et al. 2016), and mesocosm experiments (Dove et al. 2013; Kline et al. 2019), our overall understanding of the carbonate chemistry of the waters of the GBR is poor. In particular, the influence of coastal acidification (see Uthicke et al. 2014), offshore dynamics (e.g., upwelling, nutrients), and the inherent variability in reef carbonate systems (Albright et al. 2016a; Wolfe and Roff 2022) are key knowledge gaps in our understanding of the oceanography of the GBR and future responses of the reef to changing climate.

Ocean acidification itself is a multiple stressor, encompassing three main components (reduced pH, increased organism uptake of environmental CO_2, and reduced carbonate mineral saturation), all of which can negatively impact coral reef biota. The relative importance and impact of each component differ depending on the species and differing levels of exposure across environmental gradients (Hoegh-Guldberg 2005; Wolfe et al. 2020). An increased body burden of CO_2 reduces the ability of organisms to regulate their acid-base balance due to hypercapnia, which disrupts many physiological processes (Tresguerres and Hamilton 2017). For example, hypercapnia can disrupt sensory systems in fishes (Munday et al. 2012; McNeil and Sasse 2016). This is suggested to be the main mechanism of effects of ocean acidification on fishes, although this remains an area of discussion (Munday et al. 2012, 2020; Clark et al. 2020).

For the diversity of reef animals that calcify (corals, foraminifera, crustaceans, echinoderms, and molluscs), the reduction in the availability of the calcium carbonate minerals (aragonite and calcite) required to produce shells and skeletons is a major stressor (Byrne and Fitzer 2019). The physiology of calcification is challenged by ocean acidification, and the skeletons and shells that are produced are often smaller and weaker (Byrne and Fitzer 2019). This has been shown for many tropical calcifiers (Fabricius et al. 2011; Byrne et al. 2013; Uthicke et al. 2013b; Doo et al. 2014; Dworjanyn and Byrne 2017; Cornwall et al. 2021). Reef species that calcify in both the pelagic larval and benthic adult stages (e.g., bivalves, sea urchins) are especially vulnerable to ocean acidification (Uthicke et al. 2013b; Dworjanyn and Byrne 2017; Byrne et al. 2018; Byrne and Fitzer 2019; Byrne and Hernandez 2020). For species that have calcified structures not covered by living tissue, such as many gastropods and bivalves, reduced pH can cause etching and dissolution of the skeleton (Harvey et al. 2018; Armstrong et al. 2022). When more energy is needed to make and maintain calcified structures, the energy available for other processes such as growth and reproduction is reduced. As shown at CO_2 vent sites that experience extreme local acidification, the balance between skeleton production and dissolution in projected near future ocean acidification conditions can lead to the complete absence of calcifying invertebrates and shifts in reef structure (Fabricius et al. 2011; Foo et al. 2018). The species that can survive at low pH often remain small (the Lilliput effect) (Garilli et al. 2015). In some cases, moderate warming can counter the negative effects of acidification through stimulation of calcification systems if sufficient energetic resources (food) are available but only up to a limit when thermal stress compounds the negative effect of ocean acidification (Shepard Brennand et al. 2010).

Species that do not calcify as larvae and/or produce a weakly calcified skeleton as adults appear less affected by ocean acidification. This is seen for the most influential sea star on the GBR, *Acanthaster* sp. (crown-of-thorns sea star; COTS). The larvae do not calcify and are less affected by acidification than calcifying larvae, and they are also tolerant of warming (Case Study 3). Acidification and warming enhance growth of the herbivorous and early coralivorous juvenile stages of COTS. Juvenile COTS are resilient to heatwaves that kill corals (Byrne et al. 2023).

Marine invertebrates that do not calcify in either the larval or adult stages, such as non-calcareous sponges and many jellyfishes, are less vulnerable to ocean acidification (Byrne et al. 2018). It is therefore proposed that some coral reefs will transition to sponge-dominated reefs in the future due to climate change (Gonzalez-Rivero et al. 2011; Bell et al. 2013). Small, often cryptic invertebrates that have rapid rates of reproduction, such as worms (e.g., nematodes and polychaetes) and small crustaceans (e.g., copepods and amphipods), may be better able to adapt and persist under future climate scenarios (Sarmento et al. 2015; Wolfe et al. 2021a, 2021b; Timmers et al. 2021).

Warming and acidification can have negative interactive effects on coral reef species (Putnam et al. 2017; Dove et al. 2020). This is evident for crustose coralline algae that are essential in cementing reef structure (Diaz-Pullido et al. 2012) and for foraminifera that are considered as 'living sands' (Doo et al. 2012, 2014). These negative stressor interactions can have direct flow-on impacts for reef infrastructure, including sand production and the formation and maintenance of islands (Hosono et al. 2014; Fellows et al. 2022).

Large benthic foraminifera are among the most important calcifying organisms on coral reefs. Studies at One Tree Island showed foraminifera on the reef flat at densities of thousands per m^2, which produce up to 10,000 g CaCO3 m^{-2} yr^1 (Doo et al. 2012, 2017), and the sand aprons have up to 32% foraminifera content (Fellows et al. 2017). Foraminifera are highly sensitive to warming and acidification with negative consequences for sediment budgets and island stability. For example, at Raine Island in the northern GBR where foraminifera constitute 50% of beach sand (Dawson et al. 2014), climate-driven reduction in the ability of these organisms to calcify has caused a depletion of sand at this most important turtle nesting site in the southern hemisphere. Species interactions are also an important consideration for foraminifera. These organisms often live as epibionts on algae and seagrass, and in this situation, the negative effects of acidification on foraminifera calcification are buffered by the plant, although reproduction remains negatively impacted (Doo et al. 2020; Reymond et al. 2022).

In contrast to the coral-centric view of tropical ecosystems, the reef benthos is dominated by sediment habitats over vast expanses (Snelgrove 1997; Gattuso et al. 1998). Carbonate sediment biomes are less studied than coral-based reef habitats but are the dominant habitat and repository for the vast majority of the calcium carbonate stores of the ecosystem. Reef sediment is largely derived through erosive processes, from biological calcification and structural accretion by corals and other calcifying organisms (e.g., molluscs, foraminifera) to the subsequent breakdown of calcium carbonate materials by physical, chemical, and biological forces (Scoffin 1992; Wolfe et al. 2021b; Fellows et al. 2022). Sediment biomes harbour major ecosystem processes and functions including primary production (Furnas et al. 1990), nutrient cycling and carbon sequestration (Kinsey 1985; Crossland et al. 1991; Hansen et al. 1992), sediment trapping (Scoffin and Tudhope 1985; Snelgrove 1997; Wolanski 2000), and reef biogeochemistry (Andersson and Gledhill 2013; Eyre et al. 2014). However, under low-pH conditions, carbonate sediment can be more sensitive to dissolution than the skeletons of living reef calcifiers due to a range of physical parameters including mineralogy, grain size, and surface area (Walter and Morse 1984; Andersson et al. 2007; Andersson and Mackenzie 2012; Cyronak et al. 2013; Cyronak and Eyre 2016).

Overall, ocean acidification is expected to shift coral reefs from a state of net calcification to net dissolution by the end of the century (Cornwall et al. 2021). Estimates of when coral reefs will cross this tipping point range from as early as 2031 to 2082 (Silverman et al. 2009; Cantin et al. 2010; Eyre et al. 2018). However, variation in local and global data suggests that net-zero calcification may occur anywhere from 2030 to beyond 2150 (Wolfe and Roff 2022) based on a global average decline in net ecosystem calcification of 4.3 ± 1.9% yr^1 (Davis et al. 2021). Reefs in the Pacific, including the GBR, are predicted to fare better than most under future scenarios due to comparably high rates of net carbonate production and low rates of bioerosion (Cornwall et al. 2021).

Our understanding of the biogeochemical changes on the GBR has been gained through long-term datasets from several reefs and a series of manipulative field experiments (Case Study 2). In addition to ocean acidification, biological processes such as calcification, photosynthesis, and respiration influence the metabolism and biogeochemistry of coral reefs. Thus, the species and communities that contribute most to these processes can have disproportionate influences on local reef chemistry, serving to buffer or exacerbate the impacts of ocean acidification in the ecosystem (Case Study 4).

22.2.2. Physical Processes in a Changing Ocean—Storms and Sea Level Rise

Sea-level rise (SLR) is a significant threat to the survival of coral reefs (Storlazzi et al. 2018; Tebaldi et al. 2021; Fellows et al. 2022). This physical change challenges the very existence of the GBR. Coral reefs grow to sea level, and if their growth rate cannot keep up with SLR, then they may drown (Masseline et al. 2020). This is evidenced by the terraces of fossil reefs in deep water along the GBR created by the glacial-interglacial cycle changes in sea level (Woodroffe and Webster 2014; Webster et al. 2018). The dynamics of sand aprons, a ubiquitous feature of lagoon systems, will also be altered by changing climate due to decreasing sources of sediment and SLR (Case Study 5). At present, global mean SLR is 3.7 mm yr^{-1}, and IPPC predictions for the best and worst RCP scenarios indicate that there will be mean increases in sea level between 0.2 and 0.3 m by 2050 and 0.4 and 0.7 m by 2100 (Fox-Kemper et al. 2021). The impacts of tropical storms will be exacerbated by SLR (Case Study 5), which will make even small storms or increases in wave height a major threat as coral cover and complexity decline (Beetham and Kench 2018; Callaghan et al. 2020; Castro-Sanguino et al. 2022) and threaten the existence of coral reef islands (Fellows et al. 2022).

In addition to the direct impacts of heat stress on coral reefs, increased temperature also boosts the energy of cyclones. With a changing climate, fewer tropical cyclones are projected, but their intensity is predicted to increase (Knutson et al. 2010; Cheal et al. 2017). On the GBR there have been more category 4–5 cyclones in the last 15 years than in the previous 30 years (NOAA https://tinyurl.com/53z66rdj). Tropical cyclones are a major driver of extreme wave conditions on the GBR and produce waves almost 10 m in height offshore (Smith et al. 2023). Cyclones are likely to become more destructive, especially to coral reefs already under stress from warming-induced bleaching and from skeletal weakening due to acidification, which exacerbates the negative impacts of SLR on net erosion and reef depth (Perry et al. 2018). Physical abrasion of coral cover, rolling and tossing of coral boulders, and extensive movement of rubble can cause pronounced changes and destruction of the reef architecture, impeding reef recoverability (Woolsey et al. 2012; Kenyon et al. 2020; Kenyon 2021).

In recent years, cyclones have been particularly devastating in the northern GBR where the ability to recover is diminished due to mortality of corals caused by bleaching (Ceccarelli et al. 2016; Hughes et al. 2017a, 2017b, 2018) (Case Study 1,6). Cyclones result in a substantial loss of live habitat forming coral with major impacts on reef communities. On the GBR, the effects of cyclones are patchy and vary among even closely located sites and reefs. The extent of habitat destruction depends on the habitat type, species of hard coral and other habitat formers that make up the reef architecture, reef complexity, water depth, and wave exposure (Woolsey et al. 2012; Callaghan et al. 2020; Castro-Sanguino et al. 2022). Impacts can be highly localised to exposed areas of reefs.

In a 30-year study of sites around Heron Island, significant declines in coral cover were recorded in years when cyclones passed within 200 km of the reef (Connell et al. 1997). These events were followed by a distinct recovery pattern of corals. Cyclones had the greatest impact on branching species by allowing massive coral species to avail themselves of the newly available space (Connell et al. 1997). Over time, the branching species recovered in an eventual return to dominance, which suggests that, outside of the climate change context, cyclones are important in shaping the growth and diversity of coral reef communities (Connell et al. 1997).

Reef fish communities can have variable responses to habitat degradation caused by cyclones, and recent research shows that this is best understood in studies at the species level (Ceccarelli et al. 2016; Gerlach et al. 2021). Reef fishes that have close habitat associations, such as coral-associated damselfishes and gobies, are among the most susceptible groups (Cheal et al. 2002; Pratchett et al. 2008; Brandl et al. 2016; Froehlich et al. 2021). Conversely, invertivorous fishes may benefit from cyclones because their invertebrate food can be revealed from hidden reef spaces during storms, as found on reefs at Lizard Island following a cyclone (Brandl et al. 2016). Populations of rubble-inhabiting invertebrates also increase as reefs break down to rubble (Wolfe et al. 2021b). Shifts in reef fish assemblages have been documented on the GBR following cyclone disturbance, with some species moving to safety in deeper water during major storm events (Ceccarelli et al. 2016). Importantly, lesser-impacted sites can serve as sources of replenishment for degraded sites (Ceccarelli et al. 2016; Gerlach et al. 2021) (Case study 6).

In addition to the physical impacts of cyclones, ocean chemistry is changed by associated heavy rainfall that causes significant freshening of GBR waters. Due to inflow from major rivers, low salinity plumes that are generated by storms can impact near-shore and mid-shelf reefs in the GBR for days or weeks (Schroeder et al. 2012; Wooldridge and Brodie 2015; eReefs: www.ereefs.org.au/) and can have deleterious effects on corals and other reef species. For instance, these plumes cause coral bleaching and mortality (Van Woesik et al. 1995, Kerswell and Jones 2003) and are detrimental to development of marine larvae that coincide with these plumes as shown for the larvae of COTS (Clements et al. 2022). The historic impact of runoff is recorded in coral skeletons (McCulloch et al. 2003) and is seen in the inverse relationship between salinity stress and coral calcification (D'Olivo et al. 2013). Flood runoff when accompanied by terrestrial nutrients and contaminants also reduces water quality (Uthicke et al. 2014; Kroon et al. 2020). In addition to freshening, this has had deleterious impacts on near-shore reefs on the GBR following colonialisation and anthropogenic alterations to the reef river catchments.

22.3. CONCLUSION

Long-term studies clearly show that the GBR has significantly changed from its pre-Anthropocene state. The foundation of the reef is changing with a shift from faster-growing and more fragile branching corals to massive corals and soft corals (Hughes et al. 2018), although local recovery of GBR reefs post 2016–2017 from bleaching mortality has been due to fast-growing branching *Acropora* species (AIMS, 2022).

The reduction in living reef complexity following mortality events (bleaching, cyclones) has cascading effects for a great diversity of reef species, particularly those that depend on coral, with a flow-on impact for trophic cascades. After mass mortality events, there is a shift towards a rubble state and reef communities dominated by small cryptic invertebrates (Wolfe et al. 2021b; Stella et al. 2022).

For the southern GBR, which has been less impacted by bleaching events (Hughes et al. 2017a), reefs in the region have recovered following stress events such as cyclones. Areas that are less impacted by climate change stressors on the GBR, including the southern GBR and deeper-water reefs, have been identified as important areas for reef refugia. Areas such as coral-mangrove environments and seagrasses have also been identified as important refugia for corals and a potential source of stress-resistant corals on the GBR (Camp et al. 2016, 2019). There is currently an emphasis on reef restoration, especially for high-value ecologically and economically important reefs (Howlett et al. 2021).

The future of coral reefs depends on local and global action to reduce stressor impacts (Albright et al. 2016b). Coastal management and the nexus between terrestrial parks and marine parks are important to reduce river run-off and other land impacts that are deleterious for coral reef communities. Although these measures may serve to buy time for coral reefs, ultimately, the only comprehensive solution to reduce the impacts of climate change on the GBR and coral reefs globally is to rapidly decrease anthropogenic emissions of CO_2.

ACKNOWLEDGMENTS

This chapter was supported by grants from the Australian Research Council, DP 220101125 (AVC, MB), DE220100555 (SAF), and FT100100215 (AVC). Figure 5C was taken during the Commonwealth Scientific Industry Research Organisation (CSIRO) Marine National Facility Voyage IN2022_V07, and we acknowledge the help of the facility. We thank Matt Clements for his assistance with the figures.

CASE STUDY 1 – LONG-TERM CORAL REEF DYNAMICS ON THE GREAT BARRIER REEF

Several long-term studies of coral cover, biodiversity and reef island dynamics are available for the GBR. These include a study of a 27-year data set from the long-term monitoring program (LTMP) (De'ath et al. 2012) and studies based at Heron Island, Lizard Island and Low Isles encompassing 30 to 91-year data sets (Connell et al. 1997; Fine et al. 2019; Tanner 2017; Richards et al. 2021). Long-term data sets are rare, and the baseline data established by the 1928 GBR expedition at Low Isles, are particularly important (Figure 1).

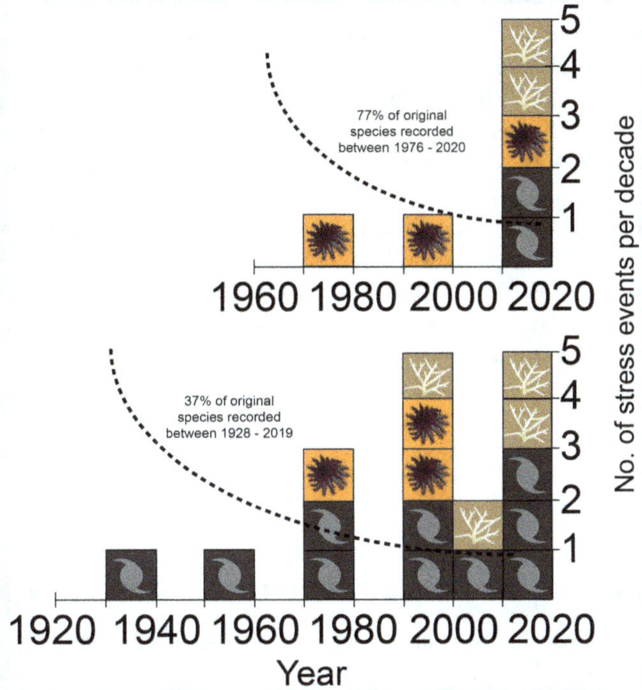

FIGURE 1 Lizard Island (top panel) and Low Isles (bottom panel) have experienced a variety of disturbance events of varying severity with an upward trajectory in recent decades, including COTS outbreaks, cyclones, and coral bleaching. For cyclone categories and tracks on the GBR, see NOAA (https://tinyurl.com/53z66rdj). Because of these compounding stressors, species richness has declined at both reefs.

Source: Figure adapted from Fine et al. (2019) and Richards et al. (2021). Symbols courtesy of the Integration and Application Network, University of Maryland Center for Environmental Science (https://ian.umces.edu/symbols/).

The LTMP study, encompassing the entire GBR found that coral cover declined from 28 to 13.8% over a 27-year period (1985 – 2012). The loss of 50.7% of initial coral cover was largely attributed to cyclones, coral predation by crown of thorns starfish (COTS) and coral bleaching (De'ath et al. 2012). Disease was also implicated in a parallel study of these data (Osborne et al. 2011). These assessments were made prior to marked decline in coral cover caused by subsequent mass bleaching events (2016 – 2022) and it would be of interest to revisit these assessments with respect to recent drivers of coral decline.

Permanent sites monitored at Heron Island showed that the change in coral abundance over 30-years (1962 – 1992) was largely due to storms, with significant damage caused by multiple cyclones impacting recovery in coral cover (Connell et al. 1997). There were marked differences between the impacts of cyclones on coral communities, highlighting the difficulties in drawing conclusions from isolated events. A 50-year study of the same sites (1963 – 2012) showed that some areas were recovering 40 years after major cyclonic disturbance. One site showed genera level recovery after 30 years although species richness did not fully recover. Importantly, this study found that 40 years is insufficient time for corals to recover from major physical disturbance (Tanner 2017). The Heron Island dataset (Tanner and Connell 2022) represents the longest (56 years) photographic record of coral cover in existence.

A study from Lizard Island reef combined newly gathered and historical presence/absence data for coral species diversity across 44-years (1976 – 2020). During this time, the reef experienced three COTS outbreaks, two large cyclones and two bleaching events (Figure 1). While 77% of species were consistently recorded across the four decades, up to 16% of species were at risk of local extinction and reduction (Richards et al. 2021).

The longest temporal data set, spanning 91-years, is from Low Isles in the northern GBR (Fine et al. 2019). The year-long study of Low Isles by the GBR expedition in 1928 provides details of coral reef structure and environmental conditions before the rise of SCUBA. The entire island was mapped with detailed aerial photography allowing the same sites to be identified for study in later years. The environmental conditions around Low Isles have changed since 1928, with average annual sea surface temperature now ~0.7°C higher and seawater pH ~0.1 pH units lower. Flooding occurs more often increasing the sediment and nutrient content of the seawater.

Since 1928, coral cover and species richness has declined, notably in the intertidal. Conversely, subtidal coral communities have not shown as consistent a decline, with similar cover reported in 1928–1929 and 1986–1998. Coral cover has oscillated in accordance with stressors such as a bleaching event in 1998, COTS outbreaks and cyclones (Figure 1), however species richness has declined over time with the 2019 survey indicating that only 37% of species recorded in 1928 were present. In addition, coral colonies were 30% smaller in 2019. These changes have contributed to a reduced diversity of invertebrates. Over the years, the reef has transitioned to a greater proportion of massive corals compared with more sensitive branching species, and with an increase in soft corals relative to hard corals.

These studies from the GBR show a common theme of major phase-shifts resulting from multiple local and global stressors. While these studies differ in survey period and location, the reefs investigated here over the time periods of 27 to 91 years did not return to their initial state after major disturbance, even over long periods of time. With climate change increasing the frequency of stressor events on coral reefs, it is of major concern that disturbance intervals are becoming too short to allow coral communities to recover.

CASE STUDY 2 – DECLINES IN NET REEF CALCIFICATION ON THE GREAT BARRIER REEF

Predicting changes in natural biogeochemical processes on coral reefs relies on long-term datasets that facilitate comparison of modern conditions to historical baselines. Three sites on the GBR (Lizard Island, Davies Reef and One Tree Island) have sufficient data to determine biogeochemical changes over time (Davis et al. 2021). Relative changes in reef biogeochemistry from these datasets vary, with only Lizard Island showing a significant decrease in net calcification of ~75% since the 1970s (Figure 2), which is tightly linked to declines in coral cover over the years (McMahon et al. 2019; Wolfe and Roff 2022).

Shifts from net calcification towards net dissolution on the GBR will be spatially and temporally variable as reefs cycle between impact, degradation and recovery (Wolfe and Roff 2022). Davis et al. (2019) showed that net calcification on a reef flat at One Tree Island decreased by ~66% between 1975 and 2014, but then sharply increased by ~400% in 2017, above the historical baseline (Figure 2). This increase was largely attributed to the recovery of coral cover from 18% in 2014 to 31% in 2017. Impacts that drastically reduce coral cover on the GBR (Case Study 1,6) impair net calcification and reef growth. This highlights the need to consider short-term trends in reef impact and recoverability when predicting long-term trends of change.

Different reef ecosystems will perform and change differently in response to OA. For example, heatwave-induced coral bleaching reduced net calcification by 40–46% at Lizard Island (McMahon et al. 2019) and ~70% in controlled mesocosm experiments (Dove et al. 2013). However, Lantz et al. (2022) showed that net calcification in the sediment-dominated lagoon of Heron Island was maintained during a major bleaching event in February 2020. The somewhat unexpected contributions of reef carbonate sediments (Eyre et al. 2014; Cyronak and Eyre 2016; Lantz et al. 2017a, 2017b; Cyronak et al. 2018), dead reef and epilithic algal matrices (Romanó de Orte et al. 2021), and

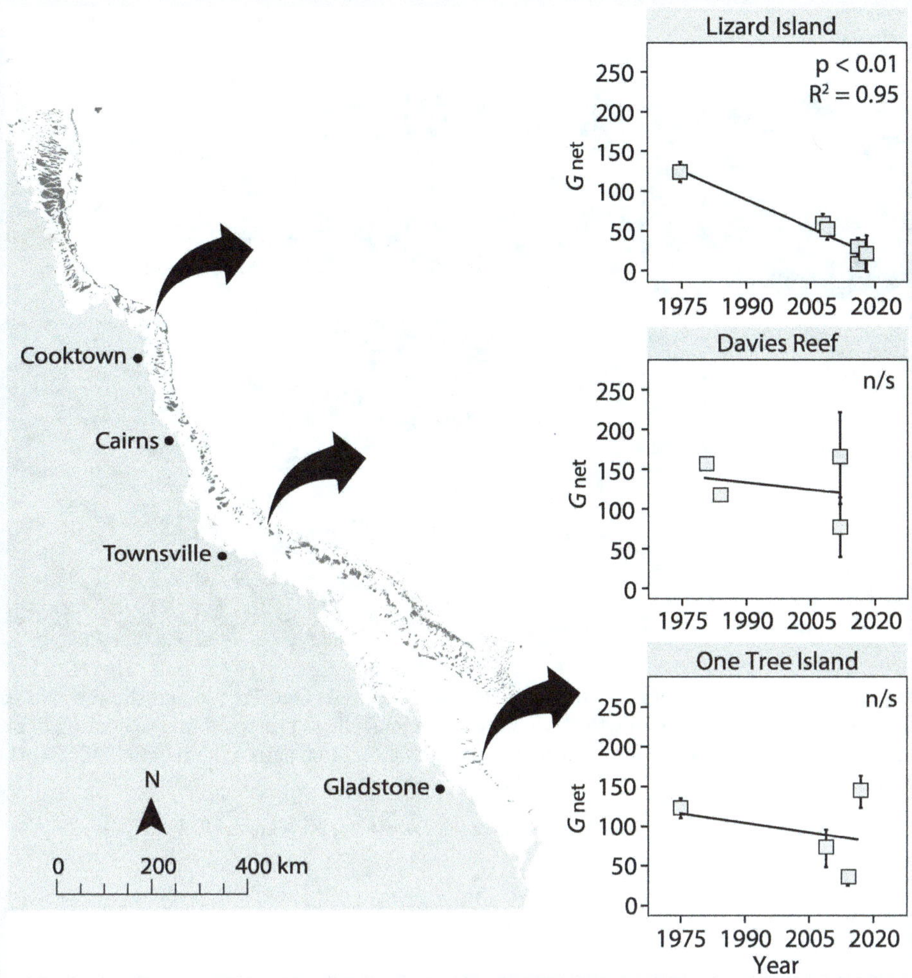

FIGURE 2 Repeated measures data on net reef calcification (G_{net}) available for three locations on the GBR, with only Lizard Island showing significant decline. Reliable projections of shifts to net dissolution are masked by local scale variability in biogeochemistry determined by a suite of factors including season, wave exposure, and coral cover.

Source: Figure adapted from (Wolfe and Roff 2022)

other reef organisms (Wolfe et al. 2018; Case Study 4) to fluxes in biogeochemistry can mask or even buffer declines in ecosystem calcification rates. It is becoming clear that net ecosystem calcification may not be an appropriate proxy of reef growth or health, especially in regions where coral cover is naturally low (e.g., lagoons) or progressively lost following bleaching impacts (Lantz et al. 2022).

A wealth of studies document the negative impacts of ocean warming and acidification on net calcification based on data from controlled experiments, including enclosed-ecosystem flumes deployed on the reef (Kline et al. 2019). Using a water reservoir similar to a large inflatable pool (Figure 3), Albright et al. (2016a) manipulated the seawater chemistry of a section of One Tree Reef to pre-Industrial conditions. Through this reversal, they found that ocean acidification has already reduced net ecosystem calcification by ~7% on this reef. The team repeated this experiment to test ecosystem-scale outcomes under near-future acidification and found that net calcification may decrease by a further ~34% in response to acidification by the end of this century (Albright et al. 2018). This suggests that the impacts and outcomes of ocean acidification, measured through relative changes in sea water carbonate chemistry, on net reef calcification-dissolution processes are non-linear (Wolfe and Roff 2022), and will perhaps amplify as pH continues to decrease on coral reefs (Albright et al. 2018). This is expected as pH is measured on the logarithmic scale, meaning that even small changes in pH have exponential effects in the acidification process.

FIGURE 3 Experimental manipulation of net ecosystem calcification at (A) One Tree Island showed that ocean acidification has already impaired reef calcification and growth, with more extreme declines projected for the coming decades. (B) Diagram of experimental setup, with (B, D, F) pink dye revealing treatment water plume across the reef and (C, E) pH conditions established in a tank *in situ.*

Source: Photo credits: University of Sydney (A), Ken Caldeira (C, D), Aaron Ninokawa (E, F)

CASE STUDY 3 – IMPACTS OF CLIMATE CHANGE STRESSORS ON THE CROWN-OF-THORNS SEA STAR

Acanthaster, a species complex of sea stars called the crown-of-thorns (COTS) (Haszprunar and Spies 2014) are keystone coral predators that cause widespread coral mortality when their populations outbreak on the GBR and elsewhere across the Indo-Pacific. Due to their ecological influence, it is important to understand the impact of climate change stressors on COTS and in context with the sensitivity of their coral prey.

A study that tested temperatures across present-day and ocean warming/heatwave conditions showed that COTS development has a broad thermal optimum with high survival and normal development across 25–32 °C up to three days post fertilisation (Figure 4A), indicating resilience to acute heatwave conditions (Lamare et al. 2013; Uthicke et al. 2013a). By day 10 however, mortality of advanced larvae occurred, likely due to the cumulative effects of stress, showing the importance to consider heatwave duration (Kamya et al. 2014). For juvenile COTS, increased temperature (30 °C) had a positive effect on growth given ad libitum access to food (Figure 4B) (Kamya et al. 2016, 2018). Adult COTS appear to be more sensitive to warming than other life stages (Lang et al. 2022).

In warming-acidification experiments these factors independently enhanced growth of the herbivorous and the early coralivorous juvenile COTS (Kamya et al. 2016, 2017, 2018). The herbivorous juveniles grew more due to the indirect effect of acidification on their food which made crustose coralline algae more nutritious and easier to digest (Figure 4B) and the high CO_2 may also have enhanced algal growth. The coral-eating juveniles grew larger but produced less skeleton (Figure 5A). Under acidification conditions the juveniles invested more in soft tissue growth than calcification. In contrast, corals that had been exposed to a pulse of COTS predation under these conditions (Figure 5B) exhibited high mortality after release from predation pressure. These findings indicate that COTS may benefit from near-future warming and acidification, while their coral prey is more sensitive (Byrne et al. 2023).

While mortality of coral in response to warming indicates that the future of COTS is bleak due to demise of its prey, this sea star has flexibility in its diet as juveniles (Deaker et al. 2020) and adult COTS can avail of non-coral food including bivalves and algae (Byrne, pers obs). They may also survive in cooler deep-water in areas with low coral cover (Figure 5C). It is not clear what COTS are eating at these near mesophotic depths.

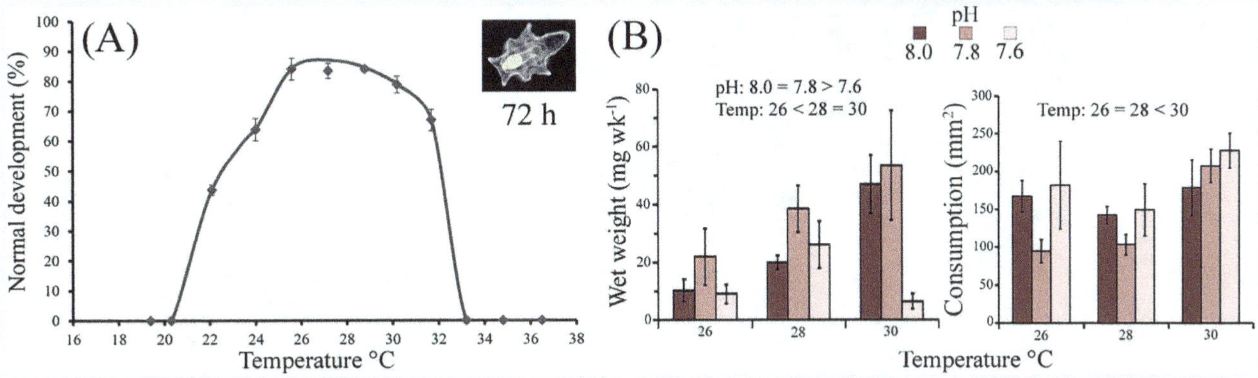

FIGURE 4 (A) Thermal response curve for development to the 72-h larval stage. (B) Impact of warming and acidification as single factors (there was no interaction) on growth (wet weight) and consumption of coralline algae by herbivorous juvenile COTS.

Source: Figure (A) adapted from Lamare et al. (2014), Figure (B) adapted from Kamya et al. (2018)

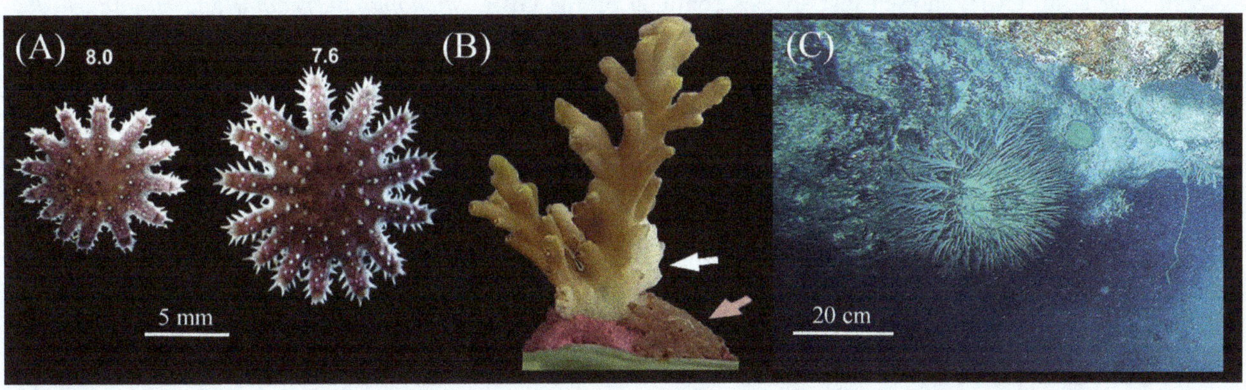

FIGURE 5 (A) Early coral-eating juveniles grew faster under acidification conditions (pH 7.6). (B) Coral and juvenile COTS (pink arrow) showing a feeding scar (white arrow). (C) Drop camera image of an adult COTS at 41 m in the northern GBR (CSIRO Marine National Facility RV Investigator IN2022_V07).

Source: Figures (A-B) adapted from Kamya et al. (2018)

CASE STUDY 4 – BIOGENIC BUFFERING OF REEF WATER CHEMISTRY BY SEA CUCUMBERS

Through their feeding activity, tropical deposit-feeding sea cucumbers influence their sediment biomes by increasing oxygen content and benthic respiration (Hammond 1982; Lee et al. 2018), reducing bacteria and pathogens linked to coral disease (Grayson et al. 2022), and altering the structure of sediment-based communities (Wolfe et al. 2018; Wolfe et al. 2021a). These are all likely to influence reef biogeochemistry. It has been proposed that the carbonate sediment ingested by tropical deposit-feeding sea cucumbers (Figure 6) is partially dissolved in their gut (Hammond 1981). Several studies show that sea cucumber digestive tracts full of sediment are higher in pH than those that are empty (Hammond 1981; Schneider et al. 2013; Wolfe and Byrne 2017). This dissolution process can reduce sediment grain size and buffer the internal chemistry of sea cucumbers through increased total alkalinity (Schneider et al. 2011). Whether this internal change is transferred to the local environment upon defaecation (Figure 6A, B), thereby contributing to the buffering of reef water chemistry, has been explored in several studies on the GBR.

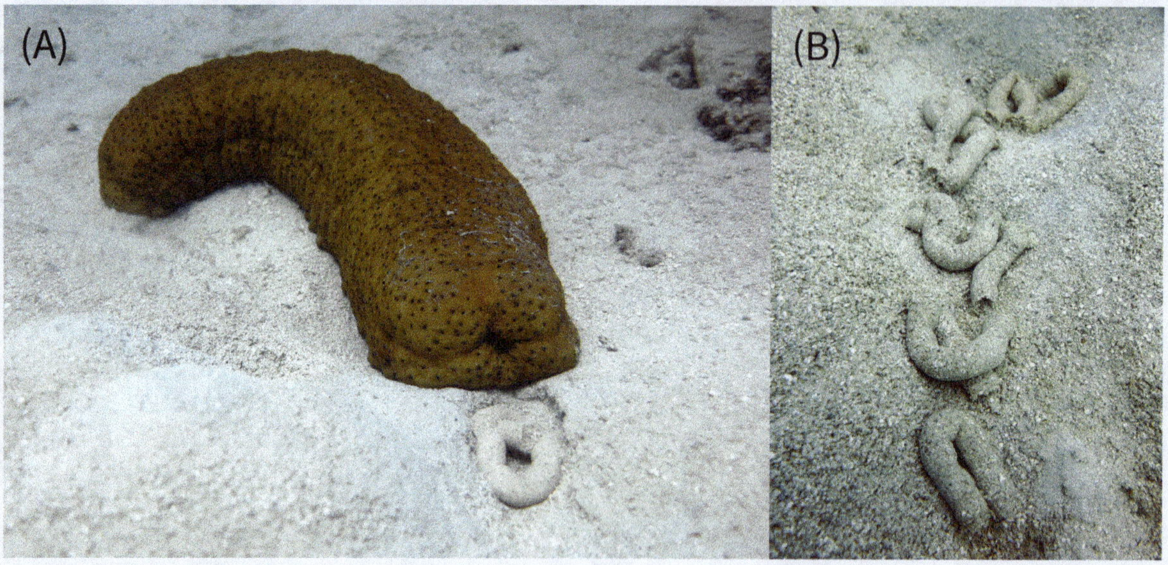

FIGURE 6 Tropical deposit-feeding sea cucumbers can alter reef carbonate chemistry, as shown for (A) *Stichopus herrmanni* under present-day and near-future ocean acidification scenarios, through partial dissolution of sediment and alkalinity released in (A, B) faecal casts.

Schneider et al. (2011) showed that species common on the GBR, *Stichopus herrmanni* and *Holothuria leucospilota*, increased total alkalinity in aquaria at levels that could represent >50% of net nighttime dissolution in their local environment. This highlights the contribution of sea cucumber feeding and defaecation to local scale biogeochemistry on coral reefs today. In context of future scenarios, Wolfe et al. (2018) measured changes in seawater chemistry in aquaria with and without *S. herrmanni* under present-day and near-future ocean acidification (OA) scenarios. In these experiments, *S. herrmanni* buffered changes in seawater chemistry both day and night. Importantly, net dissolution was reduced by ~20% (present-day OA) and ~30% (near-future OA) by *S. herrmanni* at night (Figure 7) when sediment dissolution dominates geochemical processes (Cyronak and Eyre 2016) and calcifying organisms are most susceptible to acidification (Andersson and Gledhill 2013). The buffering role of tropical sea cucumbers in a present and future ocean may be most appreciable at night when rates of photosynthesis and calcification are low (Schneider et al. 2013; Wolfe et al. 2018), but whether this benefits species and communities at the broader reef scale is not known (Vidal-Ramirez and Dove 2016).

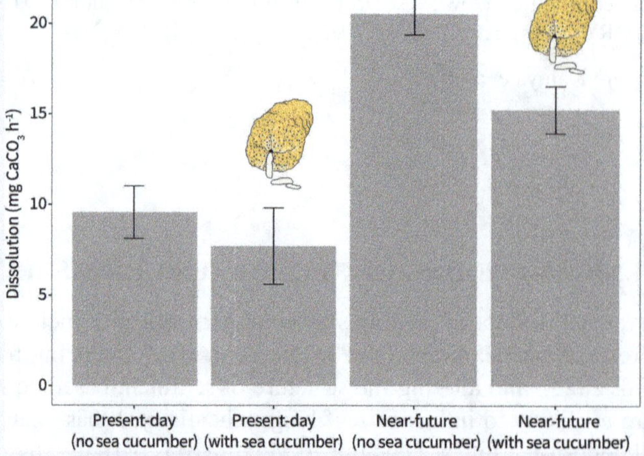

FIGURE 7 The sea cucumber, *Stichopus herrmanni*, reduces (or buffers) the magnitude of nighttime dissolution in mesocosm experiments under present-day and near-future ocean acidification scenarios.

Source: Data from Wolfe et al. (2018)

Over 0.2% of the carbonate sediment processed through the gut into faecal deposits can be dissolved (Hammond 1981; Schneider et al. 2013), which is sufficient to alter the internal chemistry of sea cucumbers and within aquaria (Schneider et al. 2011; Wolfe et al. 2018). While this amount of dissolution seems negligible at the level of the individual, outcomes in nature would be appreciable for diverse high-density populations of sea cucumbers, and on reefs with long seawater residence times, such as One Tree Island, where added alkalinity is retained and benefits to neighbouring corals and other calcifying taxa are enhanced. Alterations to seawater chemistry have been reported for two species of tropical sea cucumber, *S. herrmanni* (Figure 7) and *H. leucospilota* (Schneider et al. 2011; Wolfe et al. 2018), but no significant influence has been determined for *H. atra* (Vidal-Ramirez and Dove 2016). This species has comparatively low sediment processing rates (~3.6 g h^{-1}; Wolfe and Byrne 2017), although their abundance in shallow reefs may result in a high cumulative contribution to reef scale outcomes (Williamson et al. 2021). The largest species with the highest sediment processing rates, such as *Thelenota anax* (~34 g h^{-1}; Hammond et al. 2020), would be expected to have the greatest impact on reef scale chemistry, but this is yet to be examined.

More studies are needed to determine how sea cucumber density, size and processing rates influence biogeochemical outcomes at reef scales, and how likely this buffering process is to mediate the negative impacts of OA on coral reefs. Regardless, this role of tropical sea cucumbers in reef biogeochemistry has important implications for the conservation of sea cucumbers on the GBR, and elsewhere, with respect the harvest of these animals (Wolfe and Byrne 2022). Reef resilience is likely to be enhanced by a high density and diversity of these animals, especially on reefs characterised by local oceanographic processes.

CASE STUDY 5 – DYNAMICS OF THE SAND APRONS AND CLIMATE CHANGE IMPACTS

Coral reefs are formed from the interaction between ecological and physical structures and often grow in rings creating a "bucket" structure as seen for platform reefs such as One Tree Reef in the southern GBR and the Ribbon Reefs in the north. This structure results from the biotic self-organisation of coral reefs, as the reef edge is favoured by living corals because the bottom shear stress is higher, and the diffusive boundary layer between the reef and the water is thinner (Schlager and Purkis 2013), enhancing access to nutrients and decreasing sedimentation. On the lagoon side (the "bucket") shear stress decreases, and sedimentation is enhanced (Figure 8). Sand aprons are ubiquitous features in coral reefs generated by enhanced deposition lagoonward from the reef crest. Sediment produced in the outer areas of the reef is transported by waves and currents and becomes trapped in the lagoon forming sand aprons. This sediment transport is driven by waves and modulated by tides (Hamylton et al. 2016; Vila-Concejo et al. 2014) and controlled with respect to sea level change (Harris et al. 2015) and sediment production.

Platform reef classifications (Hopley 1982) consider the degree of infilling given by sand apron progradation into the lagoon to determine whether a platform reef is juvenile (unfilled), mature (semi infilled), or senile (completely infilled). However, recent decadal studies have shown that the sand aprons at One Tree Reef are remarkably stable and do not seem to be prograding, not even under the influence of tropical cyclones (Vila-Concejo et al. 2013). In addition, numerical modelling has shown that lagoon infilling is a self-limiting process where reef flats attain a constant critical water depth at the threshold for sediment mobilisation (Ortiz and Ashton 2019) and that progradation is limited by lagoonward decrease in shear stress (Rankey 2021). Recent research on the Capricorn Bunker Group reefs shows that sand apron volume can be used as proxy to estimate carbonate production as originally produced by living calcifiers (e.g., corals, foraminifera) as well as and past tide and wave climates during the time of active sand apron progradation (Vila-Concejo et al. 2022). It appears that past carbonate production was twice the present values, indicating the impacts of climate change and the nature of storm frequency and duration.

Coral reefs around the world are changing because of shifts in climatic conditions, and returning reefs to past configurations is no longer an option (Hughes et al. 2017b). The effects of climate change on the main controls for sand apron development – sediment production, hydrodynamic forcing, and accommodation space – are not clear (Figure 8B). Sea-level rise is a major concern. If corals cannot keep up with rising waters this will increase the accommodation space in the lagoon and may drive sand accumulation and sand apron progradation, depending on sediment availability and transport capacity. In the short term, sediment production may increase due to bleaching and coral mortality, likely followed by low sediment production while corals recover or transition to heat resilient ecosystems. Wave dissipation will decrease with the increased depth, in theory enhancing sediment transport capacity, but higher water depths due to rising sea level will mean a decrease in shear stress. There is a need to understand these oceanographic and hydrodynamic processes, and how they will be altered by climate change using ecomorphodynamic models of coral reef evolution driven by biotic/abiotic feedbacks in response to stressors including sea-level rise and waves, as well increasing water temperature and acidification.

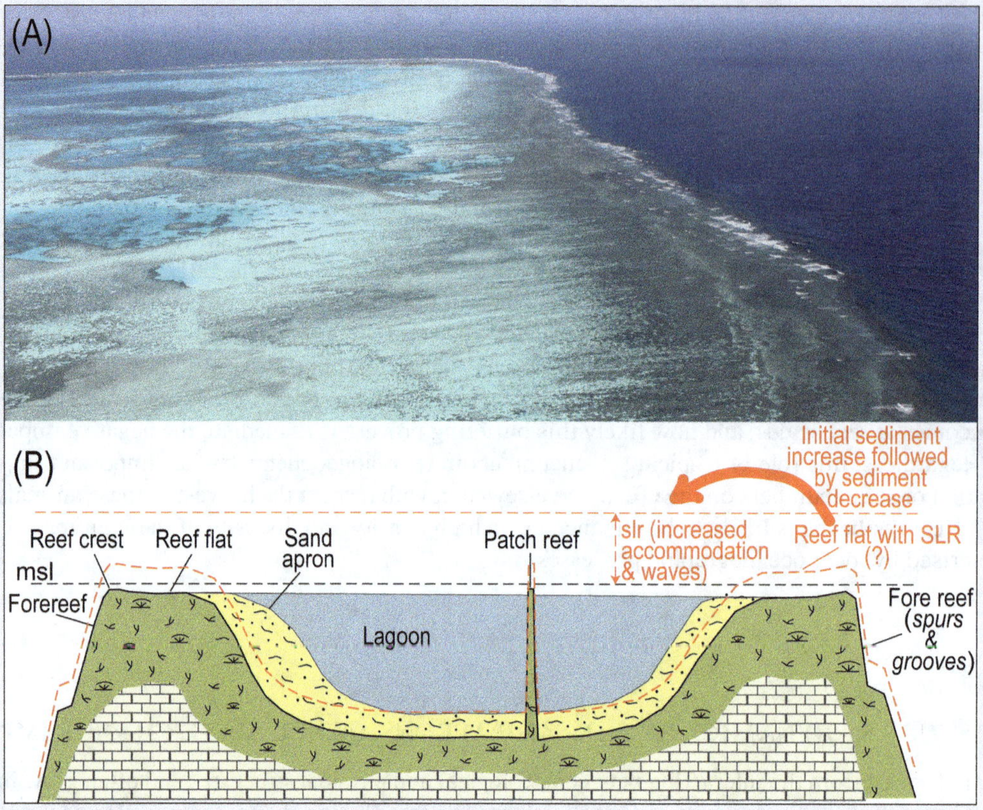

FIGURE 8 (A) The southern sand apron at One Tree Reef showing evidence of sediment transport into the lagoon (right to left in photo). (B) A cross-section of an idealised reef showing potential effects of climate change: SLR will mean increased accommodation space to be infilled with sediments and less wave dissipation; sediment availability will initially increase but will decrease in the long term. msl, mean sea level.

Source: Figure (B) adapted from Fellows et al. (2022)

CASE STUDY 6 – CYCLONE HAMISH IN THE CAPRICORN BUNKER REEFS: IMPACT ON CORALS AND FISHES – LEGACY AND RECOVERY

The southern GBR is occasionally impacted by cyclones, and in March 2009, Category 5 tropical cyclone (TC) Hamish hit the Capricorn Bunker Reef group reducing the coral cover to the lowest levels recorded by the LTMP (AIMS 2022). This was the first major cyclone (≥ category 5) to traverse the southern GBR in 14 years (Woolsey et al. 2012). In this reef group, coral cover has been recorded annually or biannually (2007–2017) at One Tree Island (OTI) (Figure 9). This reef has not experienced the devastation of crown-of-thorns sea star (COTS) outbreaks and escaped the devastating bleaching heat stress that occurred in the northern GBR (Hughes et al. 2017a). This provides an opportunity to focus on cyclone impacts and recovery over 15 years. Other data on the recovery of reefs in the Capricorn Bunker Group (e.g., Broomfield, Lady Musgrave) are available from the LTMP (see Gerlach et al. 2021).

Following TC Hamish, benthic surveys of exposed sites at OTI revealed a large reduction in coral cover with disappearance of laminar corals (0% cover of laminar coral) (Woolsey et al. 2012). Coral cover decreased from 42% to 18–25% and algal cover increased by 30%. Dislodged laminar corals and coral rubble was common at the exposed sites (Woolsey et al. 2012). Channels of coral rubble and sand were common in many places on the exposed side of the reef after the cyclone where the benthos was previously dominated by live coral. This was followed by initial low to no recovery, but over time, near full recovery of coral cover 7–8 year after the cyclone. In the absence of other stressors such as bleaching and COTS outbreaks, the Capricorn Bunker reefs recovered well (Gerlach et al. 2021).

For fishes, populations of the habitat responder, *Pomacentrus coelestis*, a short-lived species of damselfish, increased as they temporarily benefitted from the new habitat. These fishes colonised the newly formed dead coral rubble in great abundance (Figure 9). However, as the corals started to recover the mortality rate of *P. coelestis* increased, along with the availability of shelter for their predators (Gerlach et al. 2021). This pattern of a successful founder group that mature fast and increase in abundance quickly following storm damage may apply to other fishes and is predicted to be more common in the future with increased storminess. In contrast, fish species with slow growth and longer maturation time and that have special habitat requirements for an abundance of live coral may find it difficult to maintain populations. This is especially true for obligate coralivores, such as butterfly fish and obligate coral dwellers such as coral gobies (Pratchett and Berumen 2008; Froehlich et al. 2021). After cyclones the abundance of macro- and meso-predators in the families Serranidae and Lujanidae that seek habitat complexity is also greatly reduced (Jones et al. 2004). The intervals between disturbances, such as cyclones and heat waves-coral bleaching events, may become too short for populations of these diverse fishes to recover.

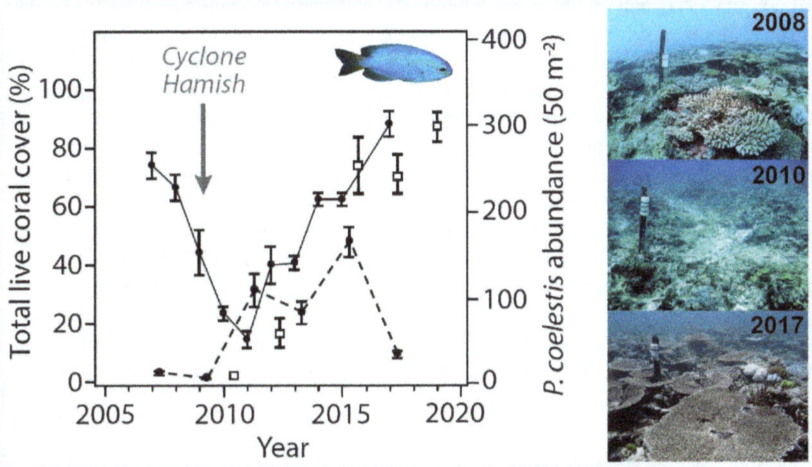

FIGURE 9 Percentage cover of live coral (solid line) and abundance of *Pomacentrus coelestis* (dashed line) on the exposed northeast slope of One Tree Reef monitored annually from 2005–2019. Open squares indicate coral cover at a replicate site 2 km away monitored approximately biennially. Photos show changes in coral at the northeast slope site over 10 years, 2008, prior to and after tropical cyclone Hamish, 2010, and with high coral cover, 2017.

Source: Figure adapted from Gerlach et al. (2021), photos courtesy of M. Kingsford. *P. coelestis* photo courtesy of Reef Life Survey

REFERENCES

AIMS (2022) Annual Summary Report on Coral Reef Condition for 2021/22. Australian Institute of Marine Science, Government of Australia

Albright R, Anthony KRN, Baird M, Beeden R, Byrne M, Collier C, Dove S, Fabricius K, Hoegh-Guldberg O, Kelly RP, Lough J, Mongin M, Munday P, Pears RJ, Russell BD, Tilbrook B, Abal E (2016b). Ocean acidification: Linking science to management solutions using the Great Barrier Reef as a case study. J Environ Manage 182:641–650.

Albright R, Caldeira K, Hosfelt J, Kwiatkowski, L, Maclaren JK, Mason BM, Nebuchina Y, Ninokawa A, Pongratz J, Ricke. KL, Rivlin T, Schneider K, Sesboue M, Shamberger K, Silverman J, Wolfe K, Zhu K, Caldeira K (2016a) Reversal of ocean acidification enhances net coral reef calcification. Nature 531:362–365

Albright R, Takeshita Y, Koweek DA, Ninokawa A, Wolfe K, Rivlin T, Nebuchina Y, Young J, Caldeira K (2018) Carbon dioxide addition to coral reef waters suppresses net community calcification. Nature 555:516–519

Andersson AJ, Bates NR, Mackenzie FT (2007) Dissolution of carbonate sediments under rising pCO(2) and ocean acidification: Observations from Devil's Hole, Bermuda. Aquatic Geochemistry 13:237–264

Andersson AJ, Gledhill D (2013) Ocean acidification and coral reefs: Effects on breakdown, dissolution, and net ecosystem calcification. Ann Rev Mar Sci 5:321–348

Andersson AJ, Mackenzie FT (2012) Revisiting four scientific debates in ocean acidification research. Biogeosciences 9:893–905

Armstrong EJ, Watson SA, Stillman JH, Calosi P (2022) Elevated temperature and carbn dioxide levels alter growth rates and shell composition in the giant clam, *Tridachna squamosa*. Scientific Reports 12:11034

Ban SS, Graham NAJ, Connolly SR (2014) Evidence for multiple stressor interactions and effects on coral reefs. Global Change Biol. 20:681–697

Beetham E, and Kench PS (2018) Predicting wave overtopping thresholds on coral reef-island shorelines with future sea-level rise: Nature Communications 9(1):3997.

Bell JJ, Davy SK, Jones T, Taylor MW, Webster NS (2013) Could some coral reefs become sponge reefs as our climate changes? Global Change Biol 19:2613–2624

Brandl SJ, Emslie MJ, Ceccarelli DM, Richards ZT (2016) Habitat degradation increases functional originality in highly diverse coral reef fish assemblages. Ecosphere 7:e01557

Byrne M, Fitzer S (2019) The impact of environmental acidification on the microstructure and mechanical integrity of marine invertebrate skeletons. Conser Physiol 7:coz062

Byrne M, Hernández JC (2020) Sea urchins in a high CO_2 world: Impacts of climate warming and ocean acidification across life history stages. In: JM Lawrence (ed), Sea Urchins: Biology and Ecology, Fourth Edition, Elsevier B.V., 281–297.

Byrne M, Lamare M, Winter D, Dworjanyn SA, Uthicke S (2013) The stunting effect of a high CO_2 ocean on calcification and development in sea urchin larvae, a synthesis from the tropics to the poles. Phil Trans R Soc B 368:20120439

Byrne M, Ross PM, Dworjanyn SA, Parker L (2018) Larval ecology in the face of changing climate—impacts of ocean warming and ocean acidification. In: TJ Carrier, AM Reitzel, and A Heyland (eds), Evolutionary Ecology of Marine Invertebrate Larvae. Oxford University Press 251–272

Byrne M, Deaker DJ, Gibbs M, Selvakumaraswamy P, Clements M (2023) Juvenile waiting stage crown-of-thorns sea stars are resilient in heatwave conditions that bleach and kill corals. Glob Change Biol 29:6439–6502

Callaghan DP, Mumby PJ, Mason MS (2020) Near-reef and near-shore tropical cyclone wave climate in the Great Barrier Reef with and without reef structure. Coastal Eng 157:103652

Camp EF, Edmondson J, Doheny A, Rumney J, Grima AJ, Huete A, Suggett DJ (2019) Mangrove lagoons of the Great Barrier Reef support coral populations persisting under extreme environmental conditions. Mar Ecol Prog Ser 625:1–14.

Camp EF, Suggett DJ, Gendron G, Jompa J, Manfrino C, Smith DJ (2016) Mangrove and seagrass beds provide different biogeochemical services for corals threatened by climate change. Front Mar Sci 3:52

Cantin NE, Cohen AL, Karnauskas KB, Tarrant AM, McCorkle DG (2010) Ocean warming slows coral growth in the central Red Sea. Science 329:322–325

Castro-Sanguino C, Bozec YM, Callaghan D, Vercelloni J, Rodriguez-Ramirez A, Lopez-Marcano S, Gonzalez-Marrero Y, Puotinen M, Hoegh-Guldberg O, Gonzalez-Rivero, M (2022) Coral composition and bottom-wave metrics improve understanding of the patchiness of cyclone damage on reefs. Science Total Environ 804:150178

Castro-Sanguino C, Ortiz JC, Thompson A, Nicholas H, Magno-Canto MM, Puotinen M, Fabricius KE, Uthicke S (2021) Reef state and performance as indicators of cumulative impacts on coral reefs. Ecol Ind. 123:107335

Ceccarelli DM, Emslie MJ, Richards ZT (2016) Post-disturbance stability of fish assemblages measured at coarse taxonomic resolution masks change at finer scales. PLoS One 11:e0156232

Cheal A, Coleman G, Delean S, Miller I, Osborne K, Sweatman H (2002) Responses of coral and fish assemblages to a severe but short-lived tropical cyclone on the Great Barrier Reef, Australia. Coral Reefs 21:131–142

Cheal A, MacNeil MA, Emslie MJ, Sweatman H (2017) The threat to coral reefs from more intense cyclones under climate change. Global Ch Biol 23:1511–1524

Clark TD, Rqby GD, Roche DG, Binning SA, Speers-Roesch B, Jurfelt F, Sundin J (2020) Ocean acidification does not impair the behaviour of coral reef fishes. Nature 577:370–375

Clements M, Selvakumaraswamy P, Deaker DJ, Byrne M (2022) Freshening of Great Barrier Reef waters is deleterious for larval crown-of-thorns starfish, counter to the terrestrial run-off hypothesis. Mar Ecol Prog Ser 696:1–14

Connell JH, Hughes TP, Wallace CC (1997) A 30-year study of coral abundance, recruitment, and disturbance at several scales in space and time. Ecol Monogr 67:461–488

Cornwall CE, Comeau S, Kornderna, Perry CT, van Hooidonk R, DeCarlo TM, Pratchett MS, Anderson KD, Browne N, Carpenter R, Diaz-Pulido G, D'Olivo DP, Doo SS, Figueiredo J, Fortunato SAV, Kennedy E, Lantz CA, McCulloch MT, González-Rivero M, Schoepf V, Smithers SG, Lowe RJ (2021) Global declines in coral reef calcium carbonate production under ocean acidification and warming. Proc Nat Acad Sci USA 118:e2015265118

Crossland C, Hatcher B, Smith S (1991) Role of coral reefs in global ocean production. Coral Reefs 10:55–64

Cyronak T, Andersson AJ, Langdon C, Albright R, Bates NR, Caldeira K, Carlton R, Corredor JE, Dunbar RB, Enochs I, Erez J, Eyre BD, Gattuso JP, Gledhill D, Kayanne H, Kline DI, Koweek DA, Lantz C, Lazar B, Manzello D, McMahon A, Melendez M, Page HN, Santos IR, Schulz KG, Shaw E, Silverman J, Suzuki A, Teneva L, Watanabe A, Yamamoto S (2018) Taking the metabolic pulse of the world's coral reefs. PLoS One 13:e0190872

Cyronak T, Eyre BD (2016) The synergistic effects of ocean acidification and organic metabolism on calcium carbonate (CaCO3) dissolution in coral reef sediments. Mar Chem 183:1–12

Cyronak T, Santos IR, Eyre BD (2013) Permeable coral reef sediment dissolution driven by elevated pCO(2) and pore water advection. Geophys Res Let 40:4876–4881

Davis KL, Colefax AP, Tucker JP, Kelaher BP, Santos IR (2021). Global coral reef ecosystems exhibit declining calcification and increasing primary productivity. Comm Earth Env 2:1–10

Davis KL, McMahon A, Kelaher B, Shaw E, Santos IR (2019) Fifty years of sporadic coral reef calcification estimates at One Tree Island, Great Barrier Reef: Is it enough to imply long term trends? Front Mar Sci 6.

Dawson JL, Smithers SG, Hua Q (2014) The importance of large benthic foraminifera to reef island sediment budget and dynamics at Raine Island, northern Great Barrier Reef: Geomorphology 222:68–81.

Deaker DJ, Mos B, Lin H-A, Lawson C, Budden C, Dworjanyn SA, Byrne M (2020) Diet flexibility and growth of the early herbivorous juvenile crown-of-thorns sea star, implications for its boom-bust population dynamics. PLoS One, e0236142

De'ath G, Fabricius KE, Sweatman H, Puotinen M (2012) The 27-year decline of coral cover on the Great Barrier Reef and its causes. Proc Nat Acad Sci USA 109:17995–17999.

Diaz-Pulido G, Anthony KRN, Kline DI, Dove S, Hoegh-Guldberg O (2012) Interactions between ocean acidification and warming on the mortality and dissolution of coralline algae. J Phycol 48:32–39

D'Olivo J, McCulloch M, Judd K (2013) Long-term records of coral calcification across the central Great Barrier Reef: Assessing the impacts of river runoff and climate change. Coral Reefs 32:999 – 1012

Doo S, Fujita K, Byrne M, Uthicke S (2014) Fate of calcifying tropical symbiont-bearing large benthic foraminfera: Living sands in a changing ocean. Biol Bull 226:169–186

Doo SS, Hamylton S, Byrne M (2012) Reef-scale assessment of intertidal large benthic Foraminifera populations on One Tree Island and their future carbonate production potential in a warming ocean. Zool Stud 51:1298–1307

Doo SS, Hamylton S, Finfer J, Byrne M (2017) Spatial and temporal variation in reef-scale carbonate storage of large benthic foraminifera: A case study on One Tree Reef. Coral Reefs 36:293–303

Doo SS, Leplastrier A, Graba-Landry A, Harianto JH, Coleman RA, Byrne M (2020) Amelioration of ocean acidification and warming effects through physiological buffering of a macroalgae. Ecol Evol 10:8465–847.

Dove SG, Brown KT, Van Den Heuvel A, Chai A, Hoegh-Guldberg O (2020) Ocean warming and acidification uncouple calcification from calcifier biomass which accelerates coral reef decline. Comm Earth Env 1:55

Dove SG, Kline DI, Pantos O, Angly FE, Tyson GW, Hoegh-Guldberg O (2013) Future reef decalcification under a business-as-usual CO_2 emission scenario. Proc Nat Acad USA 10:15342–15347

Dworjanyn SA, Byrne M (2018) Impacts of ocean acidification on sea urchin growth across the juvenile to mature adult life-stage transition is mitigated by warming. Proc Roy Soc B 285:20172684

Eyre BD, Andersson AJ, Cyronak P (2014) Benthic coral reef calcium carbonate dissolution in an acidifying ocean. Nat Clim Change 4:969–976

Eyre BD, Cyronak P, Drupp P, De Carlo EH, Sachs DP, Andersson AJ (2018) Coral reefs will transition to net dissolving before end of century. Science 359:908–911

Fabricius KE, Langdon C, Uthicke S, Humphrey C, Noonan S, De'ath G, Okazaki R, Muehllehner N, Glas MS. Lough JM (2011) Losers and winners in coral reefs acclimatized to elevated carbon dioxide concentrations. Nat Clim Change 1:165–169

Fabricius KE, Neill C, Ooijen EV, Smith JN, Tilbrook B (2020) Progressive seawater acidification on the Great Barrier Reef continental shelf. Sci Rep 10:18602

Fellowes TE, Anggadi F, Vila-Concejo A, Byrne M, Bruce E, Baker E (2022) Stability of coral reef islands and associated legal maritime zones in a changing ocean. Env Res Let 17:093003

Fellowes TE, Gacutan J, Harris DL, Vila-Concejo A, Webster JM, Byrne M (2017) Patterns of Sediment Transport Using Foraminifera Tracers across Sand Aprons on the Great Barrier Reef. J Coastal Res 33:864–873

Fine M, Hoegh-Guldberg O, Fine M, Dove S (2019) Ecological changes over 90 years at Low Isles on the Great Barrier Reef. Nat Comm10:4409

Foo SA, Byrne M, Ricevuto E, Gambi MC (2018) The carbon dioxide vents of Ischia, Italy, a natural laboratory to assess impacts of ocean acidification on marine ecosystems: An overview of research and comparisons with other vent systems. Oceanog Mar Biol Ann Rev 56:237–310

Fox-Kemper, B, Hewitt, HT, Xiao, C, Aðalgeirsdóttir, G, Drijfhout, SS, Edwards, TL, Golledge, NR, Hemer, M, Kopp, RE, Krinner, G, Mix, A, Notz, D, Nowicki, S, Nurhati, IS, Ruiz, L, Sallée, J-B, Slangen, ABA, Yu, Y, 2021, Ocean, Cryosphere and Sea Level Change, *in* MassonDelmotte, V, Zhai, P, Pirani, A, Connors, SL, Péan, C, Berger, S, Caud, N, Chen, Y, Goldfarb, L, Gomis, MI, Huang, M, Leitzell, K, Lonnoy, E, Matthews, JBR, Maycock, TK, Waterfield, T, Yelekçi, O, Yu, R, Zhou, B, eds., Climate Change 2021: The Physical Science Basis. Contribution of Working Group I to the Sixth Assessment Report of the Intergovernmental Panel on Climate Change, Cambridge University Press.

Froehlich CYM, Klanten OS, Hing ML, Dowton M, Wong MYL (2021) Uneven declines between corals and cryptobenthic fish symbionts from multiple disturbances. Sci Rep 11:16420

Furnas MJ, MitchellAW, Gilmartin M, Revelante M (1990) Phytoplankton biomass and primary production in semi-enclosed reef lagoons of the central Great Barrier Reef, Australia. Coral Reefs 9:1–10

Garilli V, Rodolfo-Metalpa R, Scuderi D, Brusca L, Parrinello D, Rastrick SPS, Foggo A, Twitchett RJ, Hall-Spencer JM, Milazzo M (2015) Physiological advantages of dwarfing in surviving extinctions in high-CO_2 oceans. Nat Clim Chang 5:678–682.

Gattuso JP, Frankignoulle M, Wollast K (1998) Carbon and carbonate metabolism in coastal aquatic ecosystems. Ann Rev Ecol Sys 29:405–434

Gerlach G, Kraemer P, Weist P, Eokelmann L, Kingsford MJ (2021) Impact of cyclones on hard coral and metapopulatiopn structure, connectivity and genetic diversity of coral reef fishes. Coral Reefs 40:999–1011

Gonzalez-Rivero FM, Yakob L, Mumby PJ (2011) The role of sponge competition on coral reef alternative steady states. Ecological Modelling 222:1847–1853

Grayson N, Clements CS, Towner AA, Beatty DS, Hay ME (2022) Did the historic overharvesting of sea cucumbers make coral more susceptible to pathogens? Coral Reefs 41:447–453

Hammond AR, Meyers L, Purcell SW (2020) Not so sluggish: Movement and sediment turnover of the world's heaviest holothuroid, *Thelenota anax*. Mar Biol 167:1–9

Hammond L (1981) An analysis of grain size modification in biogenic carbonate sediments by deposit-feeding holothurians and echinoids (Echinodermata). Limnol Oceanog 26:898–906

Hammond L (1982) Patterns of feeding and activity in deposit-feeding holothurians and echinoids (Echinodermata) from a shallow back-reef lagoon, Discovery Bay, Jamaica. Bull Mar Sci 32:549–571

Hamylton SM, Carvalho RC, Duce S, Roelfsema CM, Vila-Concejo A (2016) Linking pattern to process in reef sediment dynamics at Lady Musgrave Island, southern Great Barrier Reef. Sedimentology 63:1634–1650

Hansen JA, Klumpp DW, Alongi DM, Dayton K, Riddle MJ (1992) Detrital pathways in a coral reef lagoon. 2. Detritus deposition, benthic microbial biomass and production. Mar Biol 113:363–372

Harris DL, Webster JM, Vila-Concejo A, Hua Q, Yokoyama Y, Reimer PJ (2015) Late Holocene sea-level fall and turn-off of reef flat carbonate production: Rethinking bucket fill and coral reef growth models. Geology 43:175–178

Harvey BP, Agostini S, Wada S, Inaba K, Hall-Spencer JM (2018) Dissolution: The Achilles' heel of the triton shell in an acidifying ocean. Front Mar Sci 2018:371

Haszprunar G, Spies M (2014) An integrative approach to the taxonomy of the crown-of-thorns starfish species group (Asteroidea: *Acanthaster*): A review of names and comparison to recent molecular data. Zootaxa 3841:271 – 284

Hoegh-Guldberg O (2005) Low coral cover in a high-CO_2 world. J Geophys Res Oceans 110:C09S06

Hoegh-Guldberg O, Poloczanska ES, Skirving W, Dove S (2017) Coral reef ecosystems under climate change and ocean acidification. Front Mar Sci 4:158

Hopley D (1982) The geomorphology of the great barrier reef. Wiley Interscience.

Hosono T, Lopati P, Makolo F, Kayanne H (2014) Mass culturing of living sands (*Baculogypsina sphaerulata)* to protect island coasts against sea-level rise: J Sea Res 90:121–126.

Howlett L, Camp EF, Edmondson J, Henderson N, Suggett DJ (2021) Coral growth, survivorship and return-on-effort within nurseries at high-value sites on the Great Barrier Reef. PLoS One 16:e0244961

Hughes TP, Barnes ML, Bellwood DR, Cinner JE, Cumming GS, Jackson JBC, Kleypas J, Van De Leemput IA, Lough JM, Morrison TH, Palumbi SR, Van Nes EH, Scheffer M (2017b) Coral reefs in the Anthropocene. Nature 546:82–90

Hughes TP, Kerry JT, Álvarez-Noriega M, Álvarez-Romero JG, Anderson KD, Baird AH, Babcock RC, Beger M, Bellwood DR, Berkelmans R, Bridge TC, Butler IR, Byrne M, Cantin NE, Comeau S, Connolly SR, Cumming GS, Dalton SJ, Diaz-Pulido G, Eakin, CM, Figueira WF, Gilmour JP, Harrison HB, Heron SF, Hoey AS, Hobbs J-P A, Hoogenboom MO, Kennedy EV, Kuo C-Y, Lough, JM, Lowe, RJ, Liu G, McCulloch MT, Malcolm HA, McWilliam MJ, Pandolfi JM, Pears RJ, Pratchett MS, Schoepf V, Simpson T, Skirving WJ, Sommer B, Torda G, Wachenfeld DR, Willis, BL, Wilson SK (2017a) Global warming and recurrent mass bleaching of corals. Nature 543:373–377

Hughes TP, Kerry JT, Baird AH, Connolly SR, Dietzel A, Eakin CM, Liu G (2018) Global warming transforms coral reef assemblages. Nature 556 (7702):492–496

Jiang L-Q, Carter BR, Feely RA, Lauvset SK Olsen A (2019) Surface ocean pH and buffer capacity: Past, present and future. Sci Rep 9:18624

Jones GP, McCormick M, Srinivasan M, Eagle JV (2004) Coral decline threatens fish diversity in marine reserves. Proc Nat Acad Sci USA 101:8251–8253

Kamya PZ, Byrne M, Graba-Landry, A, Dworjanyn SA (2016) Near future ocean acidification enhances the feeding rate and development of the herbivorous juveniles of the Crown-of-Thorns Starfish, *Acanthaster planci*. Coral Reefs 35:1241–1251

Kamya PZ, Byrne M, Hall L, Mos B, Dworjanyn SA (2017) Indirect effects of ocean acidification drive feeding and growth of juvenile crown-of-thorns starfish, *Acanthaster planci*. Proc Roy Soc B. 284:20170778

Kamya PZ, Byrne M, Mos B, Dworjanyn SA (2018) Enhanced performance of juvenile crown-of-thorns starfish, *Acanthaster planci* in a warm-high CO_2 ocean exacerbates poor growth and survival of their coral prey. Coral Reefs 37:751–762

Kamya PZ, Dworjanyn SA, Hardy N, Mos B, Uthicke S, Byrne M (2014). Larvae of the coral eating crown-of-thorns starfish, *Acanthaster planci* in a warmer-high CO_2 ocean. Global Change Biol 20:3365–3376

Keesing JM, Thomson DP, Haywood MDE, Babcock RC (2019) Two time losers: Selective feeding by crown-of-thorns starfish on corals most affected by successive coral-bleaching episodes on western Australian coral reefs. Mar Biol 166:72

Kenyon T (2021) From rubble to reef: The physical and biological dynamics of coral reef rubble beds. PhD thesis, University of Queensland.

Kenyon TM, Doropoulos C, Dove S, Webb GE, Newman SP, Sim CW, Arzan M, Mumby PJ (2020) The effects of rubble mobilisation on coral fragment survival, partial mortality and growth. J Exp Mar Biol Ecol 533:151467

Kerswell AP, Jones RJ (2003) Effects of hypo-osmosis on the coral *Stylophora pistillata*: Nature and cause of low-salinity bleaching. Mar Ecol Prog Ser 253:145 – 154

Kinsey D (1985) Metabolism, calcification and production: I. System level studies. Proceedings of the [5t]h International Coral Reef Congress 4:503–542

Kline DI, Teneva L, Okamoto DK, Schneider K, Caldeira K, Miard T, Chai A, Marker M, Dunbar RB, Mitchell BG, Dove S, Hoegh-Guldberg O (2019) Living coral tissue slows skeletal dissolution related to ocean acidification. Nat Ecol Evol 3:1438–1444

Knutson TR, McBride JL, Chan J, Emanuel K, Holland G, Landsea C, Held I, Kissin JP, Srivastava AK, Sugi M (2010) Tropical cyclones and climate change. Nat Geo Sci 3: 157–163.

Kroon FJ, Berry KLE, Brinkman DL, Kookana R, Leusch F, Melvin SD, Neale PA, Negri AP, Puotinen M, Tsang JJ, van de Merwe JP, Williams M (2020) Sources, presence and potential effects of contaminants of emerging concern in the marine environments of the Great Barrier Reef and Torres Strait, Australia. Sci Total Environ 719:135140

Lamare M, Pecorino D, Hardy N, Liddy M, Byrne M, Uthicke S (2014) The thermal tolerance of crown-of-thorns (*Acanthaster planci*) embryos and bipinnaria larvae: Implications for spatial and temporal variation in adult populations. Coral Reefs 33:207–219

Lang BJ, Donelson JM, Caballes CF, Uthicke S, Doll PC, Pratchett MS (2022) Effects of elevated temperature on the performance and survival of Pacific crown-of-thorns starfish (*Acanthaster* cf. *solaris*). Mar Biol 169:43

Lantz CA, Carpenter RC, Edmunds PJ (2017b) Calcium carbonate (CaCO3) sediment dissolution under elevated concentrations of carbon dioxide (CO2) and nitrate (NO3). J Exp Mar Biol Ecol 495:48–56

Lantz CA, Leggat W, Bergman JL, Fordyce A, Page C, Mesaglio T, Ainsworth TD (2022) Will daytime community calcification reflect reef accretion on future, degraded coral reefs? Biogeosciences 19:891–906

Lantz CA, Schulz KG, Stoltenberg L, Eyre BD (2017a) The short-term combined effects of temperature and organic matter enrichment on permeable coral reef carbonate sediment metabolism and dissolution. Biogeosciences 14:5377–5391

Lee S, Ford AK. Mangubhai S, Wild C, Fers SCA (2018) Effects of sandfish (*Holothuria scabra*) removal on shallow-water sediments in Fiji. PeerJ 6:e4773.

Masseline G, Beetham E, Kench P (2020) Coral reef islands can accrete vertically in response to sea level rise. Sci Adv 6:eaay3656

McCulloch MT, Fallon SJ, Wyndham T, Hendy E, Lough JM, Barnes DJ (2003) Coral record of increased sediment flux to the inner Great Barrier Reef of Australia since European settlement. Nature 6924:727–730

McMahon A, Santos IR, Schulz KG, Scott A, Silverman J, Davis KL, Maher DT (2019) Coral reef calcification and production after the 2016 bleaching event at Lizard Island, Great Barrier Reef. J Geophys Res Oceans 124: 4003–4016

McNeil BI, Sasse TP (2016) Future ocean hypercapnia driven by anthropogenic amplification of the natural CO_2 cycle. Nature, 529(7586):383–386

Mellin, C. Matthews S, Anthony KRN, Brown SC, Caley MJ, Johns KA, Osborne K, Puotinen M, Thompson A, Wolff NH, Fordham DA, MacNeil MA (2019) Spatial resilience of the Great Barrier Reef under cumulative disturbance impacts. Glob Ch Biol 25:2431–2445

Mongin M, Baird ME, Tilbrook B, Matear RJ, Lenton A, Herzfeld M, Wild-Allen K, Skerratt J, Margvelashvili N, Robson BJ, Duarte CM (2016) The exposure of the Great Barrier Reef to ocean acidification. Nat Comm 7:10732

Munday PL, Dixson DL, Welsch MJ, Chivers DP, Domenici P, Grosell M, Heuer RM, Jones GP, McCormick MI, Meekan M, Nilsson, GE, Ravasi T, Watson S-A (2020) Methods matter in repeating ocean acidification studies. Nature 586:E20–E24

Munday PL, McCormick MI, Nilsson GE (2012) Impact of global warming and rising CO_2 levels on coral reef fishes: What hope for the future? J Exp Biol 215:3865–3873

Ortiz AC, Ashton AD (2019) Exploring carbonate reef flat hydrodynamics and potential formation and growth mechanisms for motu. Mar Geol 412:173–186

Osborne K, Dolman AM, Burgess SC, Johns KA (2011) Disturbance and the dynamics of coral cover on the Great Barrier Reef (1995 – 2009). PLoS One 6:e17516

Pandolfi JM, Connolly SR, Marshall DJ, Cohen AL (2011) Projecting coral reef futures under global warming and ocean acidification. Science 333(604):418–422

Perry CT, Alvarez-Filip L (2019) Changing geo-ecological functions of coral reefs in the Anthropocene. Funct Ecol 33:976–988

Perry CT, Alvarez-Filip L, Graham NAJ, Mumby PJ, Wilson SK, Kench PS, Manzello DP, Morgan KM, Slangen ABA, Thomson DP, Januchowski-Hartley F, Smithers SG, Steneck RS, Carlton R, Edinger EN, Enochs LC, Estrada-Saldívar N, Haywood MDE, Kolodziej G, Murphy GN, Pérez-Cervantes E, Suchley A, Valentino L, Boenish R, Wilson M, Macdonald C (2018) Loss of coral reef growth capacity to track future increases in sea level. Nature, 558(7710):396–400

Pratchett MS, Berumen ML (2008) Interspecific variation in distributions and diets of coral reef butterflyfishes (Teleostei:Chaetodontidae). J Fish Biol 7:1730–1747

Pratchett MS, Munday PL, Wilson SK, Graham NAJ, Cinner JE, Bellwood DR, Jones GP, Polunin NVC, McClanahan TR (2008) Effects of climate-induced coral bleaching on coral-reef fishes—Ecological and economic consequences. Ocean Mar Biol Ann Rev 46:251–296

Przeslawski R, Ahyong S, Byrne M, Wörheide G, Hutching P (2008) Beyond corals and fish: The effects of climate change on non-coral benthic invertebrates of tropical reefs. Global Change Biol 14: 2273–2795

Putnam HM, Barott KL, Ainsworth TD, Gates RD (2017) The vulnerability and resilience of reef-building corals. Curr Biol 27:R528–R540

Rankey EC (2021) Platform-top reef sand apron morphodynamics and the half-empty bucket. Sediment Geol 412:105825

Reymond CE, Patel F, Uthicke S (2022) Stable adult growth but reduced asexual fecundity in *Marginopora vertebralis*, under global climate change scenario. J Earth Sciences 33:1400–1410

Rice MM, Ezzat L, Burkepile DE (2019) Corallivory in the Anthropocene: Interactive effects of anthropogenic stressors and corallivory on coral reefs. Front Mar Sci 5:525

Richards ZT, Juszkiewicz DJ, Hoggett A (2021) Spatio-temporal persistence of scleractinian coral species at Lizard Island, Great Barrier Reef. Coral Reefs 40:1369–1378

Romanó de Orte M, Koweek DA, Cyronak T, Takeshita Y, Griffin A, Wolfe K, Szmant A, Whitehead R, Albright R, Caldeira K (2021) Unexpected role of communities colonizing dead coral substrate in the calcification of coral reefs. Limnol Oceanogr 66:1793–1803

Sarmento VC, Souza TP, Esteves AM, Santos PJP (2015) Effects of seawater acidification on a coral reef meiofauna community. Coral Reefs 34:955–966

Schlager W, Purkis S, (2013) Bucket structure in carbonate accumulations of the Maldive, Chagos and Laccadive archipelagos. Int J Earth Sci 102:2225–2238

Schneider K, Silverman J, Kravitz B, Rivlin T, Schneider-Mor A, Barbosa S, Byrne M, Caldeira K (2013) Inorganic carbon turnover caused by digestion of carbonate sands sands and metabolic activity of holothurians. Est Coastal Shelf Sci 133:217–223

Schneider K, Silverman J, Woolsey E, Eriksson H, Byrne M, Caldeira K (2011) Potential influence of sea cucumbers on coral reef CaCO3 budget: A case study at One Tree Reef. J Geophys Res Biogeosci 116:G04032

Schroeder T, Devlin MJ, Brando VE, Dekker AG, Brodie JE, Clementson LA, McKinna L (2012) Inter-annual variability of wet season freshwater plume extent into the Great Barrier Reef lagoon based on satellite coastal ocean colour observations. Mar Poll Bull 65:210–223.

Scoffin TP (1992) Taphonomy of Coral Reefs—a review. Coral Reefs 11:57–77

Scoffin TP, Tudhope SW (1985) Sedimentary environments of the central region of the Great Barrier Reef of Australia. Coral Reefs 4:81–93

Sheppard-Brennand H, Soars N, Dworjanyn SA, Davis AR, Byrne M (2010) Impact of ocean warming and ocean acidification on larval development and calcification in the sea urchin *Tripneustes gratilla*. PlosOne 5:e11372

Silverman JB, Lazar B, Cao L, Caldeira K, Erez J (2009) Coral reefs may start dissolving when atmospheric CO_2 doubles. Geophys Res Let 36:L05606

Smith C, Vila-Concejo A, Salles T (2023) Offshore wave climate of the Great Barrier Reef. Coral Reefs 42:661–676

Snelgrove PV (1997) The importance of marine sediment biodiversity in ecosystem processes. Ambiology 26:578–583

Stella JS, Pratchett MS, Hutchings PA, Jones GP (2011) Coral-associated invertebrates: Diversity, ecological importance and vulnerability to disturbance. Oceanog Mar Biol Ann Rev 49:43–104

Stella JS, Wolfe K, Roff G, Rogers A, Priest M, Golbuu Y, Kenyon, Mumby PJ (2022) Functional and phylogenetic responses of mobile cryptofauna to habitat degradation. J Animal Ecol 2022:00:1–17.

Storlazzi CD, Gingerich SB, van Dongeren AR, Cheriton OM, Swarzenski PW, Quataet E, Voss CI, Field DW, Annamalai H, Piniak GA, McCall R (2018) Most atolls will be uninhabitable by the mid-2[1s]t century because of sea-level rise exacerbating wave-driven flooding. Sci Adv 4:1–10

Tanner JE (2017) Multi-decadal analysis reveals contrasting patterns of resilience and decline in coral assemblages. Coral Reefs 36:1225–1233

Tanner JE, Connell JH (2022) Coral community data Heron Island Great Barrier Reef 1962–2016. Sci Data 9:617

Tebaldi C, Ranasinghe R, Vousdoukas M, Rasmussen DJ, Vega-Westhoff B, Kirezci E, Kopp RE, Sriver R, Mentaschi L (2021) Extreme sea levels at different global warming levels: Nat Clim Change 11:746–751

Timmers MA, Jury CP, Vicente J, Bahr KD, Webb MK, Toonen RJ (2021) Biodiversity of coral reef cryptobiota shuffles but does not decline under the combined stressors of ocean warming and acidification. Proc Nat Acad Sci USA 118:e2103275118

Tresguerres M, Hamilton TJ (2017) Acid–base physiology, neurobiology and behaviour in relation to CO2-induced ocean acidification. J Exp Biol 220:2136–2148

Uthicke S, Furnas M, Lonborg C (2014) Coral reefs on the edge? Carbon chemistry on inshore reefs of the Great Barrier Reef. PLoS One, 9:e109092

Uthicke S, Pecorino D, Albright R, Negri A, Cantin N, Liddy M, Dworjanyn S, Kamya P, Byrne M, Lamare M (2013a) Impacts of ocean acidification on early life-history stages and settlement of the coral-eating sea star *Acanthaster planci* PLoS One 8:e82938

Uthicke S, Soars N, Foo S, Byrne M (2013b). Physiological effects of increased pCO_2 and the effect of parent acclimation on development in the tropical Pacific sea urchin *Echinometra mathaei*. Mar Biol 160:1913–1926

Van Woesik R, De Vantier LM, Glazebrook JS (1995) Effects of Cyclone 'Joy' on nearshore coral communities of the Great Barrier Reef. Mar Ecol Prog Ser 128:261 – 270

Vidal-Ramirez F, Dove S (2016) Diurnal effects of *Holothuria atra* on seawater carbonate chemistry in a sedimentary environment. J Exp Mar Biol Ecol 474:156–163

Vila-Concejo A, Hamylton SM, Webster JM, Duce SJ, Fellowes TE (2022) Lagoon infilling by coral reef sand aprons as a proxy for carbonate sediment productivity. Geology 50:1427–1431

Vila-Concejo A, Harris DL, Power HE, Shannon AM, Webster JM (2014) Sediment transport and mixing depth on a coral reef sand apron. Geomorphology 222:143–154

Vila-Concejo A, Harris DL, Shannon AM, Webster JM, Power HE (2013) Coral reef sediment dynamics: Evidence of sand-apron evolution on a daily and decadal scale. J Coast Res SI:606–611

Walter LM, Morse JW (1984) Reactive surface area of skeletal carbonates during dissolution; effect of grain size. J Sed Res 54:1081–1090

Webster JM, Braga JC, Humblet M, Potts DC, Iryu Y, Yokoyama Y, Fujita K, Bourillot R, Esat TM, Fallon S, Thompson WG, Thomas AL, Kan H, McGregor HV, Hinestrosa G, Obrochta SP, Lougheed BC (2018) Response of the Great Barrier Reef to sea-level and environmental changes over the past 30,000 years. Nat Geosci 11:426–432

Wei G, McCulloch MT, Mortimer G, Deng W, Xie L (2009) Evidence for ocean acidification in the Great Barrier Reef of Australia. Geochem Cosmochem Acta 73:2332–2346

Williamson JES, Duce S, Joyce KE, Raoult V (2021) Putting sea cucumbers on the map: Projected holothurian bioturbation rates on a coral reef scale. Coral Reefs 40:559–569.

Wilmes JC, Schultz DJ, Hoey AS, Messmer V, Pratchett MS (2020) Habitat associations of settlement-stage crown-of-thorns starfish on Australia's Great Barrier Reef. Coral Reefs 39:1163–1174

Wolanski E (2000) Physics–biology links in the Great Barrier Reef. In: Oceanographic Processes of Coral Reefs. pp. 27–38. CRC Press

Wolfe K, Anthony K, Babcock RC, Bay L, Bourne D, Burrows D, Byrne M, Deaker D, Diaz-Pulido G, Frade PR, Gonzalez-Rivero M, Hoey A, Hoogenboom M, McCormick M, Ortiz J-C, Razak T, Richardson AJ, Roff G, Sheppard-Brennand H, Stella J, Thompson A, Watson S-A, Webster N, Audas D, Beeden R, Carver J, Cowlishaw M, Dyer M, Groves P, Horne D, Thiault L, Vains J, Wachenfeld D, Weekers D, Williams G, Mumby PJ (2020) Priority species to support the functional integrity of coral reefs. Oceanog Mar Biol Ann Rev 58:179–318

Wolfe K, Byrne M (2017) Biology and ecology of the vulnerable holothuroid, *Stichopus herrmanni*, on a high-latitude coral reef on the Great Barrier Reef. Coral Reefs 36: 1143–1156

Wolfe K, Byrne M (2022) Overview of the Great Barrier Reef sea cucumber fishery with a focus on vulnerable and endangered species. Biol Cons 266:109451

Wolfe K, Deaker DJ, Graba-Landry A, Champion C, Dove S, Lee R, Byrne M (2021a) Current and future trophic interactions in tropical shallow-reef lagoon habitats. Coral Reefs 40:83–96

Wolfe K, Kenyon TM, Desbiens A, de la Motte K, Mumby PJ (2023) Hierarchical drivers of cryptic biodiversity on coral reefs. Ecol Monogr e1586

Wolfe K, Kenyon TM, Mumby PJ (2021b) The biology and ecology of coral rubble and implications for the future of coral reefs. Coral Reefs. 40:1769–1806

Wolfe K, Roff G (2022) Global predictions of coral reef dissolution in the Anthropocene. Comm Earth Env 3:42

Wolfe K, Vidal-Ramirez F, Dove S, Deaker D, Byrne M (2018) Altered sediment biota and lagoon habitat carbonate dynamics due to sea cucumber bioturbation in a high-pCO(2) environment. Glob Ch Biol 24:465–480

Wolff NH, Mumby PJ, Devlin M, Anthony KR (2018) Vulnerability of the Great Barrier Reef to climate change and local pressures. Glob Cha Biol 24:1978–1991

Woodroffe CD, Webster JM (2014) Coral reefs and sea-level change. Mar Geol 352:248–267.

Wooldridge SA, Brodie JE (2015) Environmental triggers for primary outbreaks of crown-of-thorns starfish on the Great Barrier Reef, Australia. Mar Poll Bull 101:805 – 815

Woolsey E, Bainbridge SJ, Kingsford MJ, Byrne M (2012) Impacts of cyclone Hamish at One Tree Reef: Integrating environmental and benthic habitat data. Mar Biol 159:793–803

23 Selective Breeding and Promotion of Naturally Heat-Tolerant Coral Reef Species

Kate M. Quigley and Jennifer M. Donelson

23.1. INTRODUCTION

Climate change continues to warm and acidify the worlds' oceans—threatening coral reefs with mass bleaching, disease, cyclones, and related mortality. Alongside strong action to reduce carbon emissions, active restoration actions are increasingly investigated and promoted to assist and speed up the adaptation of organisms, in particular, corals, and the reefs that they build, to extreme and ever-increasing heat. One of the actions (Figure 1) being investigated is the enhancement of thermal tolerance using selective breeding (SB), a process by which humans selectively develop phenotypic traits by choosing which males and females sexually reproduce and have offspring together (Kincaid 1993). Producing individuals bred from the same species but from different locations (intra-specific hybrids) through the SB of coral populations has resulted in substantial increases in the heat tolerance of coral reef larvae and juveniles across multiple species (Drury et al. 2022). To continue to assess the feasibility of SB as an active restoration action for rapid adaptation, this chapter explores whether SB can reliably achieve increases in thermal tolerance across a larger number of coral reef genera and species, including coral reef fish. In this chapter, we also highlight the potential trade-offs (growth versus survival versus reproduction) that may exist when selecting for enhanced thermal tolerance in corals and fish (Shulman and Ogden 1987). We additionally emphasize that a shift is needed from the perspective of single species restoration to the restoration of functions or systems.

Despite being one of the best-managed coral reef ecosystems in the world; there is broad scientific consensus that the long-term survival of the Great Barrier Reef is under threat from climate change (Souter et al. 2021), even with booms and busts of recovery. Threats include local impacts in the form of decreased water quality from flood events and run-off of nutrients and sediments, damage due to cyclonic waves, and the global threats of coral bleaching and ocean acidification induced by the warming waters and lower pH associated with climate change (De'ath et al. 2012; Souter et al. 2021). These **cumulative impacts**, both acute and prolonged, mean that in addition to the strong global action needed to reduce carbon emissions and continued management of local pressures (Donovan et al. 2021), bold conservation action may be needed if carbon emissions continue unabated. Important decisions also need to be made about priorities and acceptable risk. Central to this need is that resulting actions must be co-designed and understood with Traditional Owners, reef stakeholders, and the broader community (Quigley et al. 2022). These threats, especially climate change, have the potential to rapidly reduce coral population sizes and cause range shifts or contractions, with negative consequences for genetic diversity through either the rapid or slow erosion from disturbance. Importantly, climate change will also change and/or destabilize oceanographic processes such as hydrodynamics, impacting dispersal and the resulting connectivity of organisms (Wolanski et al. 2024). These will also have profound impacts on the feasibility of management strategies such as SB.

23.2. GREAT BARRIER REEF OCEANOGRAPHY AND DISPERSAL (AND HOW DIFFERENCES IMPACT SPECIES)

Currents influence the dispersal of marine organisms, which go on to influence population structure and their underlying genetics (Kininmonth 2024). However, the **dispersal** of organisms and their connectivity are also influenced by local, fine-scale hydrodynamics within the ocean (e.g., tidal jets, recirculation eddies) that can make particular reefs "source" or "sink" locations for marine organisms (Hock et al. 2017) and are critical for recovery (Gouezo et al. 2021). Thus, biophysical oceanic conditions can influence dispersal and connectivity between populations. The Great Barrier Reef is influenced by the numerous currents that flow through it, enabling the transport of nutrients, sediments, and organisms, including coral and fish larvae. This includes transport across the shelf of the Great Barrier Reef and up and down its length—connecting organisms across the diversity of habitats that make up the Great Barrier Reef (**connectivity**, see Wolanski et al. 2024). The major surface currents that influence the Great Barrier Reef lagoon include the North Queensland Current, East Australian Current and Northern Caledonia Jet (Mongin et al. 2016). On a historical timescale, dispersal and connectivity will have affected the eco-evolutionary processes for species and populations. For example, populations with large amounts of self-recruitment and limited immigration will

DOI: 10.1201/9781003320425-27

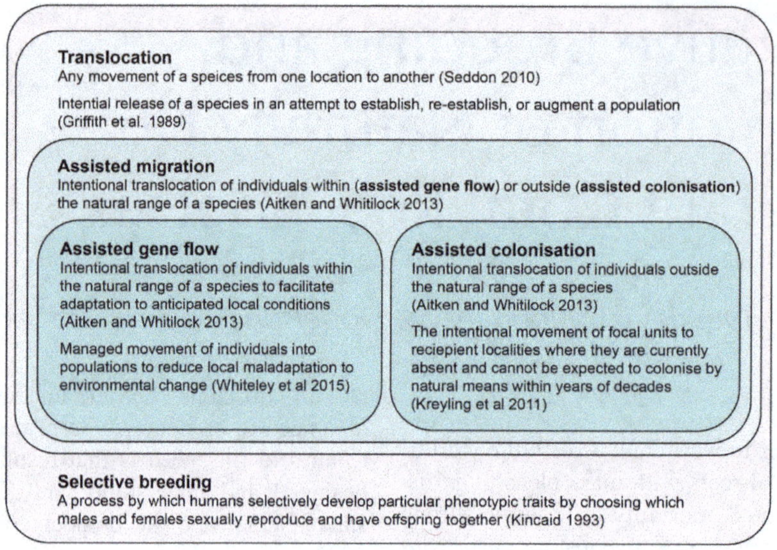

FIGURE 1 Summary of the terms used to describe different forms of the implementation of genetic interventions for conservation. SB can be used to provide organisms for each of these interventions. The classification of terms is broadly based on if individuals are being moved around in (Assisted Gene Flow—AGF)—or outside of (Assisted Colonization—AC)—their current distributions. See also (Bay et al. 2020) for further definitions.

be more likely to have adapted to the local conditions (i.e., **local adaptation**). Accordingly, current and future oceanographic processes will also influence the natural dispersal potential of tolerant and non-heat-tolerant populations and by extension, how successful SB and other conservation methods such as **assisted gene flow** (AGF, Figure 1) will be in the enhancement of additional tolerance on the existing reserve of genetic diversity. Altimetry-derived field data (see Wolanski et al. 2024) and modelled data should shed light on important future conditions, for example, if the East Australian Current has accelerated. Given that connectivity inherently involves different scales of distance, we define them here (for this chapter). Habitat-forming species, such as scleractinean corals, may have a "**localised**" or **small** population structure (**10s of kms**), whereas benthic spawning fishes exhibit less structure across **larger** spatial scales (**100s of kms**) (Richards et al. 2017a; but see Gerlach et al. 2007). Therefore, both connectivity and dispersal are critical to determining the need and effectiveness of SB and AGF.

23.3. SELECTIVE BREEDING AND PROMOTION OF NATURAL HEAT TOLERANCE

The ongoing decline of global ecosystems has prompted large-scale interest in restoration practices, including highly interventionist strategies that incorporate a focus on accelerating or enhancing the evolution of organisms (briefly reviewed in Filbee-Dexter and Smajdor 2019). SB is part of this broader restoration effort, that is, to restore degraded ecosystems to a previous state or a state that provides similar ecosystem services. Marine systems have traditionally lagged behind terrestrial systems in the sense that terrestrial conservation has focused on testing interventions in the lab and field for generations (see commentary in Quigley et al. 2022). However, the United Nations has declared the current decade as the "next decade of restoration", prompting increased interest, funding, and project ideation concerning marine restoration, including restoration aimed at using adaptive principles to restore ecosystems (Baums 2008; Voolstra et al. 2021; Shaver et al. 2022).

Many forms of genetic interventions have been proposed—with varying degrees of ethical debate (see Filbee-Dexter and Smajdor 2019)—and can be classified on a continuum of risk and the level of genetic manipulation (see Hoffmann et al. 2021 for the classification of general strategies). Classically, many populations that have been targeted for conservation efforts have been small, isolated, or inbred (Gomulkiewicz and Holt 1995). Briefly, their classification depends on the level of gene flow, population sizes, evolutionary history of the populations, and future environmental projections. A range of genetic conservation methods have been put forward to "rescue" populations from decline, including **assisted migration** (AM, movement of species within or outside their natural ranges, Figure 1) and AGF as mentioned previously (the movement of organisms within their known ranges) (reviewed in Aitken and Whitlock 2013). A modified version of previous forms of AGF incorporates SB, where individuals are bred, and their resulting offspring are moved (Quigley et al. 2019). The goal of breeding and then movement would be to elicit an impactful (i.e., a fitness benefit), sustainable shift in allele frequencies over time in wild populations (Torda and Quigley 2021). Essential to making an informed decision on whether to use AGF or AM is an understanding of the connectivity of populations within a species, which determines the boundaries of the species range (Figures 1 and 2). In the context of warming induced by climate change, the increased prevalence of heat tolerance is the desired conservation outcome.

Heritable genetic diversity associated with heat tolerance exists in some coral populations (Dixon et al. 2015; Quigley et al. 2020; Weeriyanun et al. 2021; Quigley and van Oppen 2022) and fish populations (Munday et al. 2017) on the Great Barrier Reef. This provides an opportunity to use and amplify natural, enhanced tolerance to facilitate rapid adaptation of populations. AGF involves the intentional movement ("translocation") of warm-adapted adults or their offspring to vulnerable reef locations to prepare these habitats for anticipated warming. Where AGF includes the implementation of SB, offspring of warm-adapted adults are produced from aquarium-based breeding methods. Therefore, these warm-adapted offspring, when deployed to cooler, sensitive reefs at high risk of heat stress, would provide local populations with increased tolerance. Through interbreeding, this should facilitate the spread of heat tolerance genes in and between populations to prepare them for future warming, a process estimated to occur too slowly to keep pace with rates of temperature increase (Quigley et al. 2019). Possessing genetic variation associated with heat tolerance enables populations to withstand higher temperatures that are fast becoming the norm on reefs. AGF therefore holds the potential to facilitate adaptation to heat tolerance for these reef organisms, preparing them for further climate warming.

SB techniques can be used to select a trait of interest and increase the adaptive genetic variation in the resulting offspring. These techniques can be used to reproductively cross coral individuals from different populations within the same species to produce intra-specific hybrid offspring. Recent work by Quigley and colleagues in the field has shown that selectively breeding corals in this way can confer either enhanced or equal fitness in the hybrid offspring, as measured by traits such as survival and growth (Drury et al. 2022). This is assessed by comparing the fitness of selectively bred coral offspring to non-crossed coral individuals from the same populations, referred to as purebred offspring, and using these as a comparative baseline (Quigley et al. 2021). The ability for laboratory SB to be a useful tool in nature is restricted by our knowledge of what limits survival for a particular taxonomic group. Thus, SB will only ever be as good as the traits targeted. The goal is to select for traits that are linked to fitness and/or are the bottleneck for persistence, in which many lessons can be learnt from an aquaculture context. Due to the rich biodiversity on the Great Barrier Reef, we do not have an understanding of the fundamental biology of many species, let alone what traits limit fitness and persistence. Currently, there are over 600 recognized benthic coral and over 1,600 fish species on the Great Barrier Reef (Great Barrier Reef Marine Park Authority 2022). Each have different dispersal capabilities, colonization ability, ecological niches, and tolerances. The selection of any one of them for a breeding program will necessitate a rigorous approach to species selection criteria for any organism of interest, whether it is coral or fish (Madin et al. 2021; Quigley et al. 2022). A potential counterfactual would be to accept range shifts and/or hope that species can acclimate or adapt to the future local conditions and includes the prospect of strong actions to limit warming and curb climate change impacts.

The ability of organisms to withstand the stress of high temperatures (heat tolerance) and maintain performance in a warm environment (thermal performance) are phenotypic traits that vary considerably between even closely related species. This variation arises from the adaptive evolution that species undergo as a result of genetic and environmental forces. Here, we focus on two case studies for SB on the Great Barrier Reef and discuss the research underway, particularly the role that oceanographic processes have on determining their feasibility and likelihood of success.

23.4. INFLUENCE OF OCEANIC PROCESSES FOR SELECTIVE BREEDING IN CORAL REEF-ASSOCIATED SPECIES

Over time, barriers can drive the ecological and evolutionary independence of lineages, leading to patterns of differential performance between populations. Local adaptation, where performance is optimized to the conditions experienced, has occurred through the process of natural selection (Kawecki and Ebert 2004; Savolainen et al. 2013). This is a useful mechanism in the case of SB as it potentially provides populations with enhanced traits maximized from habitats of interest (e.g., more heat-prone or thermally variable locations). SB may also create poor fitness if the bred individuals are moved into environments that create maladapted responses (Tillotson et al. 2019a). Simply observing performance differences between populations that correspond to the environmental conditions experienced does not indicate local adaptation, as performance differences may arise non-genetically (i.e., phenotypic plasticity) or differences may not be driven by heritable genetic variation (Torda et al. 2017). There is also experimental evidence that the heritability of traits can shift with climate change, such that adaptive potential may be constrained under future conditions (Welch and Munday 2017).

The genetic component of differences between populations forms a fundamental basis for AGF and even SB potential. Healthy sources of adaptive alleles must be identified and first protected in their own right and for breeding purposes (Quigley and van Oppen 2022). Once identified, the connectivity between these sources with other sinks must also be determined (Hock et al. 2017). If this connectivity is unknown, then it would be possible to introduce maladapted alleles into receiving populations—potentially leading to decreases in fitness for local coral stock. Corresponding to barriers, it is important to understand if there is considerable exchange of larvae occurring throughout the conservation region of interest. Our understanding of the extent of the exchange of larvae and resulting gene flow is still developing (discussed in Kininmonth 2024). Indeed, marine systems were originally thought to be completely "open" with little genetic structure, an assumption that has been updated in recent years with the incorporation of increasingly powerful genetic tests (reviewed in Jones et al. 2009). Therefore, to understand the potential for the retention and eventual fixation of alleles (i.e., retained within the population), genetic analyses paired with biophysical modelling incorporating oceanic processes and evolutionary modelling are critically needed (Torda and Quigley 2021). Recent assessments of coral marine connectivity have shown an increased prevalence of self-recruitment, structure, and cryptic species diversity in corals and in fish (Gerlach et al. 2007; Jones et al. 2009; Steneck et al. 2009; Quigley et al. 2019). This understanding of connectivity and barriers to dispersal has been greatly enhanced by knowledge on the physical transport of organisms (Wolanski et al. 2024; Kingsford et al. 2024). By extension, this can also aid in our understanding of the transport and retention of **adaptive alleles,** which has consequences for methods such as AGF.

We highlight two examples to show the importance of the interaction of these processes. Figure 2 depicts a reef cluster in the northern Great Barrier Reef: Thetford, Moore, and Elford (top to bottom). These reefs may experience a significant amount of self-seeding of coral or fish larvae, indicated by the inward facing arrows. If the underlying genetic predisposition is not heat-tolerant (blue), then this will serve to re-enforce the local adaptation of these genotypes within the populations. Techniques such as AGF, either paired with SB or with only the movement of naturally heat-tolerant individuals sourced from other locations (AGF *sensu stricto*), can serve to introduce warm-adapted variants or genotypes (in red) into established non-heat-tolerant populations. To underscore the potential usefulness of AGF, the closest reef to Moore (reef centroid to centroid distance) is Elford, at 5.62 km away. If there are at least a few strong connections between these reefs, modelled under a simplified unigene model (with caveats), with a selection (S) coefficient of 0.1, then it would take ~37 generations for any heat tolerance genes to move and then reach fixation in that neighbouring reef under average rates of gene flow for this northern region of the Great Barrier Reef. This could therefore be accelerated with increased, strategic inputs via AGF.

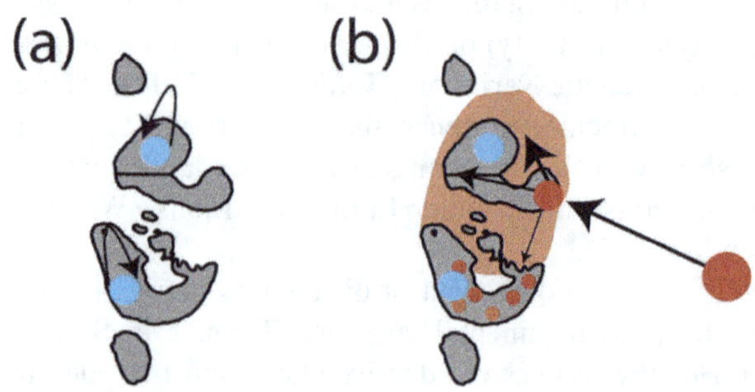

FIGURE 2 The role of connectivity and self-seeding in determining the spread of adaptive variation in a meta-population reef structure. The dots represent theoretical populations with either underlying genetic predisposition that is not heat-tolerant (blue) or that is heat-tolerant (red). Heat-tolerant individuals can be sourced from naturally heat-tolerant stock within a species range (AGF) or selectively bred for heat tolerance and then moved (SB+AGF). The wavefront of the spread of adaptive alleles is symbolised as the lighter red shading.

A second example shows how barriers to connectivity can have unexpected consequences for the retention of adaptive alleles in corals, shown through comparing reef topology in the northern versus the southern Great Barrier Reef. Naturally, the spread of adaptive alleles occurs through accelerating waves that spread from a leading edge (Slatkin 1976). Like ripples in a pond, when multiple waves meet (wave front), this can decelerate the leading edge of the movement of adaptive alleles. Deceleration occurs more rapidly when multiple waves meet and decreases where waves are allowed to travel without barriers. In modelling the movement of adaptive alleles throughout the Great Barrier Reef, acceleration of adaptive alleles through AGF methods occurred more rapidly in the far north, above Lockhart River, where reefs were more distant from one another, allowing the acceleration of "waves" of gene flow, compared to the southern GBR, where reefs are more closely spaced, leading to deceleration of spread (Quigley et al. 2019). Further complexities arise when biophysical properties such as current deceleration and acceleration are factored in. This example underscores how understanding the physical oceanographic processes will influence the rate of success of SB.

The expectation for marine species is that the distributional range infers thermal performance within populations, with the warmest part of the species distribution being "warm adapted". However, this is broadly assumed and for many species, remains untested, and where tested, it has been shown to be more complex, in which heat-tolerant populations have been found across the thermal envelope of some species range of plasticity (Gardiner et al. 2010; Donelson and Munday 2012; Veilleux et al. 2018; Weeriyanun et al. 2021; Quigley and van Oppen 2022). However, populations at the edges of a species range can be smaller or highly isolated due to underlying hydrology and may have resultantly low genetic diversity. This situation represents a risk for the use of SB and its implementation in AGF. For example, if individuals of low genetic diversity with high tolerance are added to populations with high diversity, then this may influence the success of colonization using the selectively bred offspring, potentially leading to failure. This breeding pitfall is caused by the erosion of genetic diversity, sometimes called an extinction vortex, brought on by genetic bottlenecks (Aitken and Whitlock 2013). This may occur in any context of SB, AGF, or AC. Information on connectivity and its subsequent impact on genetic diversity is therefore critical to avoid pitfalls

FIGURE 3 The two study species examined in this chapter, including scleractinean corals (photo: Martin Reef in 2021) (a), and reef-associated and commercially important aquarium fish species (b).

Source: Image credit as follows: author supplied (M. Roman), AIMS 2019 Report (N. Thake)

associated with the application of genetic conservation practices such as SB and AGF.

In the following, we review and summarize our understanding on population differentiation and connectivity, in terms of performance and genetics, for scleractinian corals and teleost reef fish as they relate to oceanographic processes. To create generality rather than species specificity, we establish the understanding through an eco-evolutionary and life history lens. We outline the considerations and challenges for moving forward with AGF and SB in these groups and what aspects require further investigation to move this field forward (Figure 3).

23.4.1. A. Corals

23.4.1.1. Reproduction

Approximately 85% of coral species are hermaphrodites that broadcast spawn (Baird et al. 2009). This is advantageous for methods like SB in that it allows the spawn to be collected from a single colony, which can then be separated into eggs and sperm, manipulated, and fertilized to create known familial "crosses," in effect—selectively breeding. The principles of spawning have been developed over many field seasons of trial-and-error (Harrison et al. 2024), but a series of methodological resources are currently available for spawning, including manuals (Guest et al. 2010), scientific publications (Babcock 1984; Pollock et al. 2017), and databases (Baird et al. 2021). Generally, reproductively ready (gravid) coral colonies are collected from the field and transported back to an experimental or field facility for spawning. Once spawning occurs, bundles are collected at the surface and separated with mesh filters, their concentrations are standardized, and then, they are mixed in specific proportions to create selectively bred crosses. The fertilized embryos are then reared into larvae, and at the appropriate developmental stage, they are cued to metamorphose and settle for further experimentation (for full details on selectively bred methods, see Humanes et al. 2021; Drury et al. 2022). The offspring, either as larva or as juveniles, are tested and analysed for offspring performance for improvements in key traits (Weeriyanun et al. 2021) (e.g., heat tolerance, growth, or bleaching resistance, Quigley et al. 2020a; Quigley and van Oppen 2022; Strader and Quigley 2022) under laboratory or field conditions (Quigley et al. 2021).

23.4.1.2. Dispersal

Like the general theory outlined earlier, dispersal in corals was thought to be largely open, with high connectivity among populations of coral species. However, genetic tools have shown more self-recruitment and retention than initially expected (reviewed in Jones et al. 2009). This change in understanding over time, and the resulting impact that it has on fundamental ecological principles such as population regulation, persistence, and recovery underscores the importance of quantifying the baselines of connectivity both for fundamental knowledge of life-history and for any intervention or management activity. Setting the connectivity baseline (i.e., how connected and how open or closed populations are) is important for identifying how potential adaptive alleles will move through populations naturally and how they will respond to interventions such as AGF after SB has been performed. The search for resilient coral species and populations has been on-going since the first record of differential impacts to bleaching-related stress was recorded between colonies and species (Loya et al. 2001b; Morikawa and Palumbi 2019; Voolstra et al. 2020). Therefore, healthy sources of adaptive alleles must first be identified, protected in their own right, or used for breeding purposes (Quigley and van Oppen 2022). Other methods that use rapid heat tolerance assays (CBASS system, Voolstra et al. 2020) have also been deployed to identify resilient populations that could be sources for heat or generalized stress tolerance. Once identified, the connectivity between these sources with other sinks should be quantified (Hock et al. 2017; Riginos et al. 2019). If this connectivity is unknown, then it would be possible to introduce maladapted alleles into receiving populations—potentially leading to decreases in fitness for local coral stock. Another consideration for corals is how connectivity varies between brooders and spawning coral species (Figure 4). Brooding species, such as pocilloporids, generally recruit close to their natal reefs (Torda et al. 2013; Warner et al. 2016), whereas broadcasting species may travel for many days and potentially hundreds of kilometres (Graham et al. 2008), leading to larger, more contiguous distributions, even in species with high levels of cryptic variation (Cornwell et al. 2021). Therefore, in corals, both life-history categories and how the resulting connectivity is impacted need to be considered before SB programs are developed, but this will likely differ between broadcasting and brooding species.

23.4.1.3. Current Research, Possibilities, and Challenges

The ability of corals to withstand the stress of high temperatures is a phenotypic trait that varies considerably among different species (Loya et al. 2001a; van Woesik et al. 2011) and is largely influenced by the population of symbionts

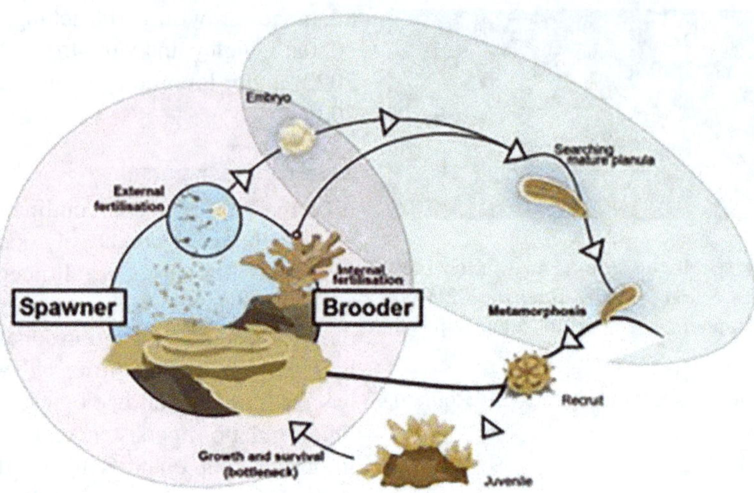

FIGURE 4 Simplified life-history of a scleractinean coral individual, highlighting the reproductive stage (pink) and the dispersal stage (blue), both of which are critical to the implementation and success of SB for AGF. Note that broadcast spawners are thought to disperse more widely compared to brooding coral species.

Source: Modified with permission from (Vanderklift et al. 2020)

hosted (reviewed in Davies et al. 2022). This variation arises from the adaptive evolution that species undergo as a result of genetic and environmental forces. Thus far on the Great Barrier Reef, SB has been undertaken or attempted across 6 species, including *Acropora tenuis* (Quigley et al. 2016), *A. millepora* (van Oppen et al. 2014; Dixon et al. 2015), *A. spathulata* (Quigley et al. 2020a, 2020b), *A. hyacinthus*, *Goniastrea retiformis*, and *Porites lutea* (Macadam et al. *in-prep*). Additional species have been tested in the Red Sea and in the Caribbean (reviewed in Drury et al. 2022).

A recognition of understanding and measuring biological responses outside of the laboratory and in the wild are increasing. Although field trials using selectively bred corals are still in their infancy (van Oppen et al. 2014; Quigley et al. 2016, 2021), their success or failure will depend on several factors tied to oceanic processes. The source or sink dynamics, driven by hydrology, will be critical for success. If many reefs are sinks for example, then the contribution of new alleles will tend to be diluted in each generation. Alternatively, if connectivity is low, then the relative contribution of each larva should be higher in shaping the gene pool at the sink reef (Torda and Quigley 2021). Biophysical models suggest that most (~80%) reefs in the Great Barrier Reef exhibit high local retention, and only ~3% of reefs reach ~50% of reefs (Hock et al. 2017), which is perhaps partly explained by the "sticky water" phenomenon (Andutta et al. 2012). If taken simplistically, then this suggests that connectivity is overall lower than expected, which should increase the probability of success of AGF at localized reefs where offspring are outplanted, but this decreases the probability of success extending outward to other areas. This is in line with estimates on the spread of adaptive variation being quite slow relative to the pace of change needed (Quigley et al. 2019). Overall, this suggests that AGF may have only a limited spatial impact of success beyond direct deployment reefs.

Finally, although the range of reproductive modes are presented simply in two dominant forms (broadcast spawning vs. brooding) compared to the more complex fish (discussed in the following section), there is substantial variability and many unknowns in coral reproduction. Mechanisms such as sperm competition (Willis et al. 1997; Warner et al. 2016; Kitanobo et al. 2022), cryptic diversity in reproductive modes, including the release of sperm but not eggs (spermcasting, Warner et al. 2016), or simultaneous spermcasting with planulation (Smith et al. 2019), and novel spawning behaviours (i.e., daytime spawning, Schmidt-Roach et al. 2014) all suggest that there is more to understand about coral reproduction.

23.4.2. B. Fish

23.4.2.1. Reproduction

There is a diversity of reproductive strategies, reproductive traits, mating systems, and spawning modes for teleost marine fishes (see reviews in Choat 2012; Kuwamura et al. 2020). Coral reef fish are generally iteroparous (multiple spawning events in a lifetime) and undertake sexual reproduction. For the purpose of this chapter, we focus our discussion on four main reproductive modes (combinations of the aforementioned diversity, noting that some species can breed across multiple modes, e.g., Fricke 1980; van den Berghe and Warner 1989) to highlight how they might influence the success of translocation and assistance strategies. The first group that we describe are the fish that form either social groups, which include only a pair of breeding individuals, or monogamous pairs (Figure 5a; pair breeding). This reproductive mode commonly occurs with

FIGURE 5 Current developments in using SB for AGF on the Great Barrier Reef. A selectively bred juvenile coral produced from a northern coral mother and a central coral father and outplanted to Davies Reef (a), first mass deployment of AGF corals on the Great Barrier Reef at Davies Reef (b), second generation mass deployment of AGF corals on enhanced (e.g., high survivorship) substrates, attached on tables above the substratum to reduce sedimentation, at Palm Islands.

Source: All photos author-supplied from M. Marzonie, M. Roman, and M. Panero

benthic reproduction (e.g., demersal spawning) and will often involve some level of parental care (Kuwamura et al. 2020). Species can be either gonochoristic or sequentially hermaphroditic (Kuwamura et al. 2020), which will impact the ease at which SB can be applied. As an alternative, many species form polygamous harems (either polygyny or polyandry), with these species tending to be sequentially hermaphroditic, and often, the direction of sex change will correlate with the dominant sex (Petersen and Fischer 1986). In harems, the dominant individual has a disproportionate genetic contribution to offspring (Figure 5b; Petersen and Fischer 1986), which could facilitate or hinder AGF. Coral reef fish also aggregate to broadcast their pelagic eggs (either transient or resident spawning; Choat 2012), and this can still mean paired reproduction or reproduction involving multiple individuals at one time. One aspect that these three reproductive strategies have in common is the occurrence of a pelagic larval development period (discussed further in the following). To a lesser taxonomic and numerical extent than the previous three reproductive modes, there are pair-forming species that produce offspring that directly develop and do not possess a pelagic larval development period (e.g., the damselfish *Acanthochromis polyacanthus*, some apogonids (most are mouth brooders with pelagic larvae) and syngnathids; Figure 5d). In these cases, embryonic development can occur on benthic substrate, within the oral cavity, or can be incubated in a pouch. Direct development includes a large amount of parental care of the embryos, through larval development until becoming juveniles (Santaquiteria et al. 2021). Reproductive modes that might best lend themselves to SB would include the aggregative pelagic spawning species (synonymous with the previously discussed broadcasting corals) and direct developing species, for which completing the life cycle in captivity is generally easier (DiMaggio et al. 2017).

Reef fish reproduction includes a large range of sexual selection aspects, such as mate choice, male-male or female-female competition, and sperm competition, like in corals (discussed earlier). Processes of sexual selection will differ among species and have the potential to impact any assisted management strategies employed as not all individuals in a population get to reproduce (Fox et al. 2019). This consequently affects the retention and assimilation of adaptive alleles. Mate choice in fish can occur well ahead of reproductive events through the formation of pairs and social groups (Wong et al. 2007) or can shift regularly within an individual's lifetime (Goulet and Goulet 2006; Rueger et al. 2016). Consequently, bottlenecks to contributing genetics to the next generation could be limited prior to mate choice. In these hierarchies, the progression of an individual to be reproductively active is generally limited and relies on the death or eviction of a higher-ranked individual in the group (Wong et al. 2007; Rueger et al. 2018). Even in species that do not form strict social hierarchies, there can be limitations to accessing breeding territories or mates (Fricke 1980; Hobbs et al. 2004). Mate choice also occurs in some aggregative spawning species with courtship occurring between individuals once they reach the spawning site (Samoilys and Squire 1994), and when aggregating species spawn en masse, there is likely to be sperm competition (Erisman et al. 2007; Munoz et al. 2014). Sneaker males, males that pretend to be females and rush into a nesting or spawning site, can also introduce aspects of sperm competition (Fitzpatrick 2020). From genetics research on species hybridization in nature, we know that particular directions of integration can be more common, with the maternal parent of surgeonfish hybrids four times more likely to be *Acanthurus nigricans* than *A. leucosternon* in the eastern Indian Ocean (Marie et al. 2007). This could in part be evidence of the role that sexual selection can play in evolutionary processes.

23.4.2.2. Dispersal

For most coral reef fish, dispersal is predominantly limited to the larval development period (e.g., *Lutjanus sebae*). Some exceptions exist, such as population-mixing, aggregative spawning species that travel long distances to spawning sites (Kaunda-Arara and Rose 2004; Green et al. 2015). For coral reef fish, there is extensive diversity in key aspects that will influence the larval dispersal outcomes, and this is

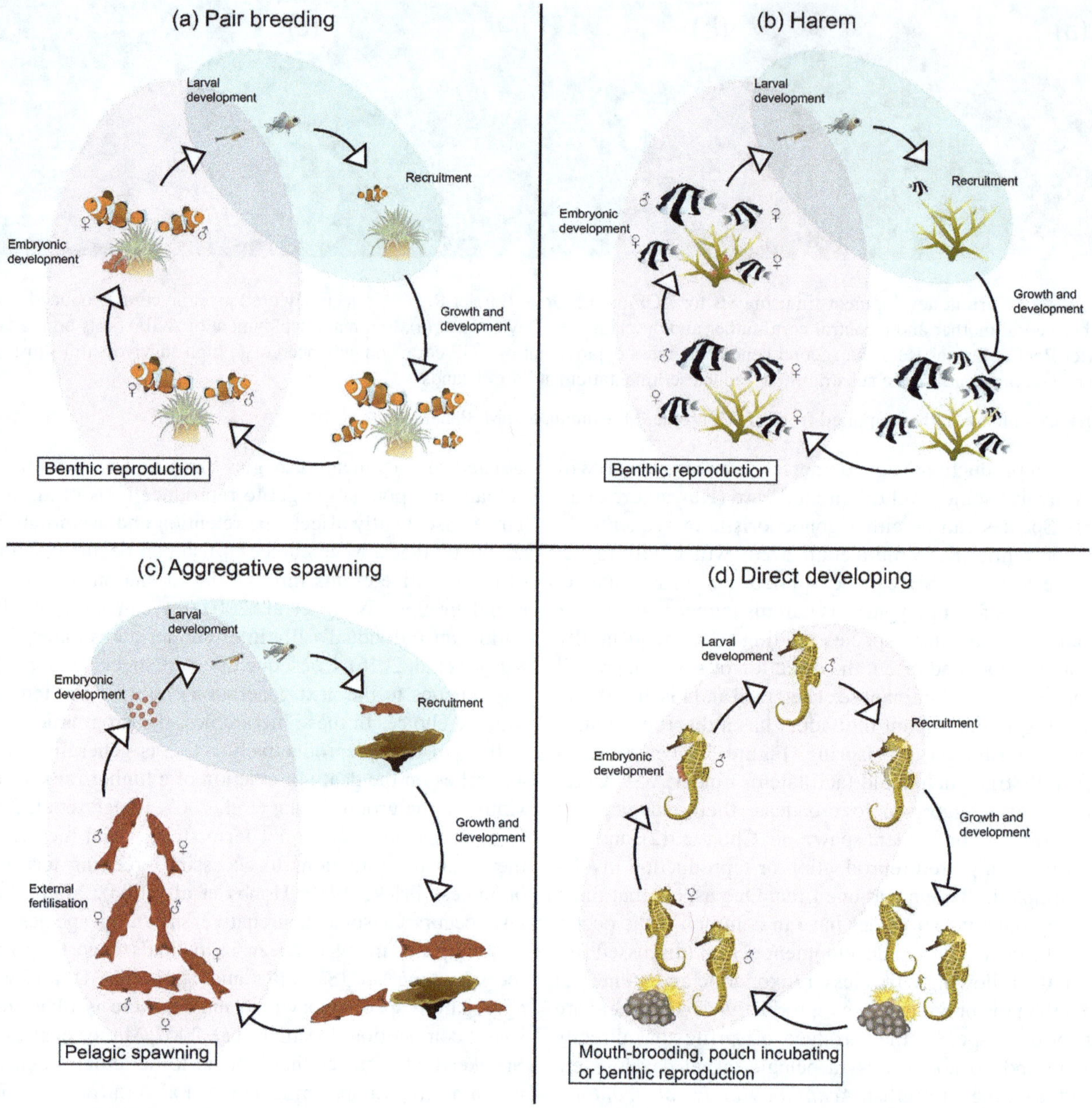

FIGURE 6 Simplified life-history of four diverse modes of reproduction and development found in coral reef fish, with reproductive (pink) and dispersal stages (blue) highlighted. This figure was created with images from the Integration and Application Network (ian.umces.edu/media-library).

Source: Licensed under Attribution-ShareAlike 4.0 International (CC BY-SA 4.0). Eggs and early larvae (alevin): Claire Sbardella; clownfish, whitetail dascyllus, incubating and hatching seahorses, anemone, staghorn coral, plate coral, brain coral and tubastrea: Tracey Saxby; lined seahorse: Jane Thomas; coral trout: Kim Kraeer. Late development larvae created by Jennifer M. Donelson.

not necessarily correlated simply to reproductive mode or pelagic larval duration (Bay et al. 2006; Gerlach et al. 2007; Jones et al. 2009). Instead, a combination of time spent in the plankton, level of development at hatching, the rate at which key physical developments occur during the larval period, sensory capacity and larval behavior, and oceanographic processes will all play a role in determining how many offspring self-recruit and are retained with the parent population and the connectivity between populations (Wolanski et al. 1997; Fisher 2005; Gerlach et al. 2007; Leis et al. 2007; Jones et al. 2009; Andutta et al. 2012; Burgess et al. 2022; Kingsford et al. 2024). As is seen for corals, reef fish species can exhibit high levels of self-recruitment (5–75%, reviewed in Jones et al. 2009; Planes et al. 2009; Harrison et al. 2012), mixed with longer-distance dispersal (e.g., up to 250 km, Williamson et al. 2016, even 570 km, Patterson and Swearer 2007); average dispersal distance can often range from 10 to 190 km (Harrison et al. 2012;

Jones et al. 2015; Williamson et al. 2016; Abesamis et al. 2017; Almany et al. 2017). High levels of self-recruitment would be beneficial for SB and AGF strategies, as this increases the likely retention of beneficial alleles in targeted locations.

The regularity and level of connectivity between populations will be influential to the effectiveness of SB and AGF strategies and whether local adaptation may have occurred in the past for reefs of interest. Connectivity has often been studied from the view of whether high-quality locations can supply and replenish poorer-quality locations (sources and sinks; e.g., Bode et al. 2006; Figueira 2009) or whether marine protected areas can supply nearby fished reefs (e.g., Planes et al. 2009; Harrison et al. 2012; Williamson et al. 2016). An investigation of population genetics shows extensive variation in the spatial scales at which reef fish populations exhibit either differentiation or panmixia (reviewed in Planes 2002; van Herwerden et al. 2006; Evans et al. 2010; Jones et al. 2010). Several studies have detected genetic differentiation at relatively small spatial scales, suggesting that ecological connectivity can be restricted (Gerlach et al. 2007, 2016). However, local scale differentiation may reflect strong selection or age-related mortality—rather than low connectivity—such as in the cardinalfish *Ostorhinchus doederleini* at One Tree Island where strong genetic differences exist on small spatial scales but also shift on short timescales among years (Gerlach et al. 2016). Coral reef fish species that do not have a pelagic larval stage can exhibit local genetic differentiation, which suggests little connectivity among reefs (Doherty et al. 1995; Hoffman et al. 2005; Miller-Sims et al. 2008; but see Bay et al. 2006). Whether high levels of self-recruitment result in locally thermally adapted populations in the warmer parts of the species range is still unknown for most reef fish but represents a critical knowledge gap for future research.

23.4.2.3. Possibilities and Challenges in Fishes

Considering two example fish species that fall within the preferential reproductive modes for SB, proposed here as aggregative spawning species and direct developing species, we can explore the potential for SB. Firstly, the aggregative spawning of the economically and socially important coral grouper, *Plectropomus leopardus*, exhibits high connectivity, limited genetic differentiation across populations on the Great Barrier Reef, and no observable differences in thermal performance from the northern to southern extent (van Herwerden et al. 2006, 2009; Pratchett et al. 2017; Payet et al. 2022). Although adaptive management actions would be valuable from cultural and economic perspectives, the eco-evolutionary history of the species makes it a poor candidate for AGF and SB within the Great Barrier Reef. If gene flow extends beyond the Great Barrier Reef, then alternative options might be AM/AGF from outside this region (e.g., the Coral Sea, Western Australia, van Herwerden et al. 2009; Payet et al. 2022) or hybridization with *Plectropomus maculatus*, which can already occur in nature (van Herwerden et al. 2006; Harrison et al. 2017). However, in both cases, there is no evidence yet that thermal performance would be improved with either approach.

In contrast, the spiny chromis damselfish (*Acanthochromis polyacanthus*) exhibits high levels of population genetic differentiation due to limited dispersal (direct development) and difference in thermal performance along the Great Barrier Reef (Doherty et al. 1995; Miller-Sims et al. 2008; Gardiner et al. 2010; Donelson and Munday 2012; Rodgers et al. 2018). Interestingly, in this species, thermal performance does not entirely align with latitudinal thermal expectations (a pattern also seen in some coral species, Quigley and van Oppen 2022), with fish from the south of the Great Barrier Reef showing enhanced thermal physiology and plasticity (Gardiner et al. 2010; Donelson and Munday 2012; Veilleux et al. 2018). This species has been extensively investigated to explore the potential for thermal plasticity within and across generations, including molecular plasticity. From this, we know that developmental exposure from early life can result in improved thermal performance in future generations (Donelson et al. 2012; Bernal et al. 2022) and expression genes involved in this thermal plasticity (Veilleux et al. 2015; Bernal et al. 2018; Ryu et al. 2018), and this phenotypic change is heritable (Munday et al. 2017). This work has almost entirely focused on fish originating from the central part of the Great Barrier Reef, and for SB, a greater investigation of performance differences across the Great Barrier Reef is needed by focusing on the inclusion of cross-shelf and latitudinal variation (Kingsford et al. 2019).

The fundamental information that we lack for implementation of AGF and SB approaches in reef fish is an understanding of thermal performance and local adaptation on the Great Barrier Reef. This raises an important question prior to measuring thermal performance—what trait(s) limit persistence and regulate fitness? One trait that is historically used as a test of thermal tolerance is the critical thermal maximum, as measured through a rapid rise in water temperature until locomotory activity becomes disorganized (Becker and Genoway 1979). This trait does not seem to be relevant for coral reef fish and would indicate a greater thermal safety margin of substantial thermal tolerance than what exists in reality (Mora and Ospina 2001; Rodgers et al. 2018). Aerobic physiology has also emerged as an influential trait for understanding thermal adaptation in fish with some work identifying a correlation between aerobic scope and other high-level performance (e.g., Pörtner and Knust 2007; Eliason et al. 2011). However, other research shows that aerobic scope does not always provide a holistic indication of thermal performance for growth and reproductive capacity (Gräns et al. 2014; Donelson et al. 2016; Bernal et al. 2018). This point brings us to a significant challenge for SB in reef fish—a set of easily measured and relevant thermal performance traits are not agreed upon in the literature, which is also a challenge for corals (Grottoli et al. 2020).

Considering that reproductive adults and early life development are generally the most thermally sensitive and are expected to be those that will limit the persistence of species in the future due to their importance for regeneration (Pankhurst and Munday 2011; Asch and Erisman 2018), traits focused on these life stages may be the best starting point, which has been the case in corals (Weeriyanun et al. 2021). Combining thermal sensitivity and genetics could also represent a practical way forward, in which detailed knowledge across species that looks for consistencies and differences could allow knowledge expansion more broadly. In fishes, candidate gene approaches have often focused on thermally sensitive physiological functions such as heat shock protein and metabolic genes (Edmunds et al. 2009, 2012; Newton et al. 2012). As technology has become more affordable, full genomic and transcriptomic studies have emerged, which can allow a more holistic understanding of genomic plasticity to environmental warming. For example, recent transcriptomic analysis has been used to explore the response of cardinalfish and damselfish species to a large-scale marine heatwave on the Great Barrier Reef. Although there were species-specific differences, similar molecular pathways were activated and associated with the thermal stress (Bernal et al. 2020). Similar experimental work has also been completed in corals and shows the conservation of heat stress responses across taxa (Strader and Quigley 2022).

It is also important to acknowledge the challenges for AGF, and especially SB, that arise from the large diversity of reproductive modes in reef fish. In particular, there is knowledge of only a few species in terms of breeding and early life development in captivity (Moorhead and Zeng 2010). Direct developers and the species with high levels of self-recruitment present a situation where assisted management actions could provide the greatest value due to limited connectivity and relative ease with captive breeding. However, fish species that humans rely on for food security and livelihoods generally do not fit into this group, and a more pragmatic approach is needed to move forward with AGF and SB discussions. The complexities of sex change, behaviour, social systems, and sexual selection should also be considered. Ultimately, so will the acceptance of range shifts as organisms either move, adapt, or die. The role that sexual selection might play in hindering adaptive thermal plasticity (i.e., when sexually selected traits are not correlated with thermally beneficial traits), whether inheritance differs depending on parental sex, and thus how sequential hermaphroditism could influence adaptation have yet to be contemplated.

A suite of research over the years shows that reef fish productivity and biodiversity may decrease without healthy and diverse habitat-forming species such as corals. It is unclear if the number of reef fish that obligately use corals is higher or lower than expected or if there are proportional declines in abundance and diversity in fish with the loss of coral (Jones et al. 2004; Wilson et al. 2006; Pratchett et al. 2008; Cheal et al. 2017). This raises the question that if reef fish are highly dependent on corals, is there any point in actions to save fish species without simultaneously focusing on the corals? If modelling efforts (Madin et al. 2021) can identify keystone coral species targeted for different conservation outcomes, then there is an opportunity to match efforts in AGF and SB in both corals and fish by allowing for maintenance of symbiotic, mutualistic, and obligate relationships. Alternatively, it could be argued that conservation efforts should be focused only on the true generalists that can live in a range of reef conditions (rocky reef, rubble, complex coral, macro-algae, etc.) or species that are likely to thrive in altered ecosystem states (e.g., herbivores).

23.5. SYNTHESIS AND COMPARISON OF THE TWO CASE STUDIES WITHIN AN ECO/EVO FRAMEWORK

The impact of any conservation effort, especially one aimed at influencing the underlying genetic background of populations and species, will be strongly impacted by the eco-evolutionary history of that organism. Population structure of the species is driven by its connectivity and then often, subsequent high mortality during post-settlement processes. As an example, conservation impacts will depend on the size of source populations (small or large), their connectivity to, and the size of the recipient sink populations (Figure 7) (Hoffmann and Sgró 2011; Hoffmann et al. 2021). In this applied context, AGF is usually recommended for species with large population sizes and limited connectivity (Aitken and Whitlock 2013). Therefore, knowing the population structure, and the sizes of sources and sinks, is critical in determining if action is necessary or likely to lead to conservation success. A broader discussion of AGF in line with other conservation strategies underpinned by a species' adaptive capacity is timely (Thurman et al. 2020, 2022). This work by Thurman and colleagues highlights much of what we have discussed here, namely, that life history, movement, evolutionary potential, and demography are all essential considerations for how we conserve species in a changing climate (Figure 7), but for most species, we are data-deficient.

The two case studies presented here—reef corals and fish—reflect how key differences in ecology and life history may likely impact options for applied conservation action. Some coral species on the Great Barrier Reef likely have substantial evolutionary connectivity and large population sizes—at least relative to other current and projected at-risk locations and species with larger realized distributions—suggesting that what seems feasible with SB and AGF in corals may vary for other species. Modelling and genetic analyses at various spatial scales confirm that in at least the few coral species examined, global connectivity is likely to be larger for some coral species compared to fish (> 42 km in coral compared to 10 s of km in fish, Green et al. 2015; Greiner et al. 2022). The lower connectivity is reflective of the relatively small population sizes with highly localized recruitment. Based on conservation guidelines, the relatively limited connectivity in fish would

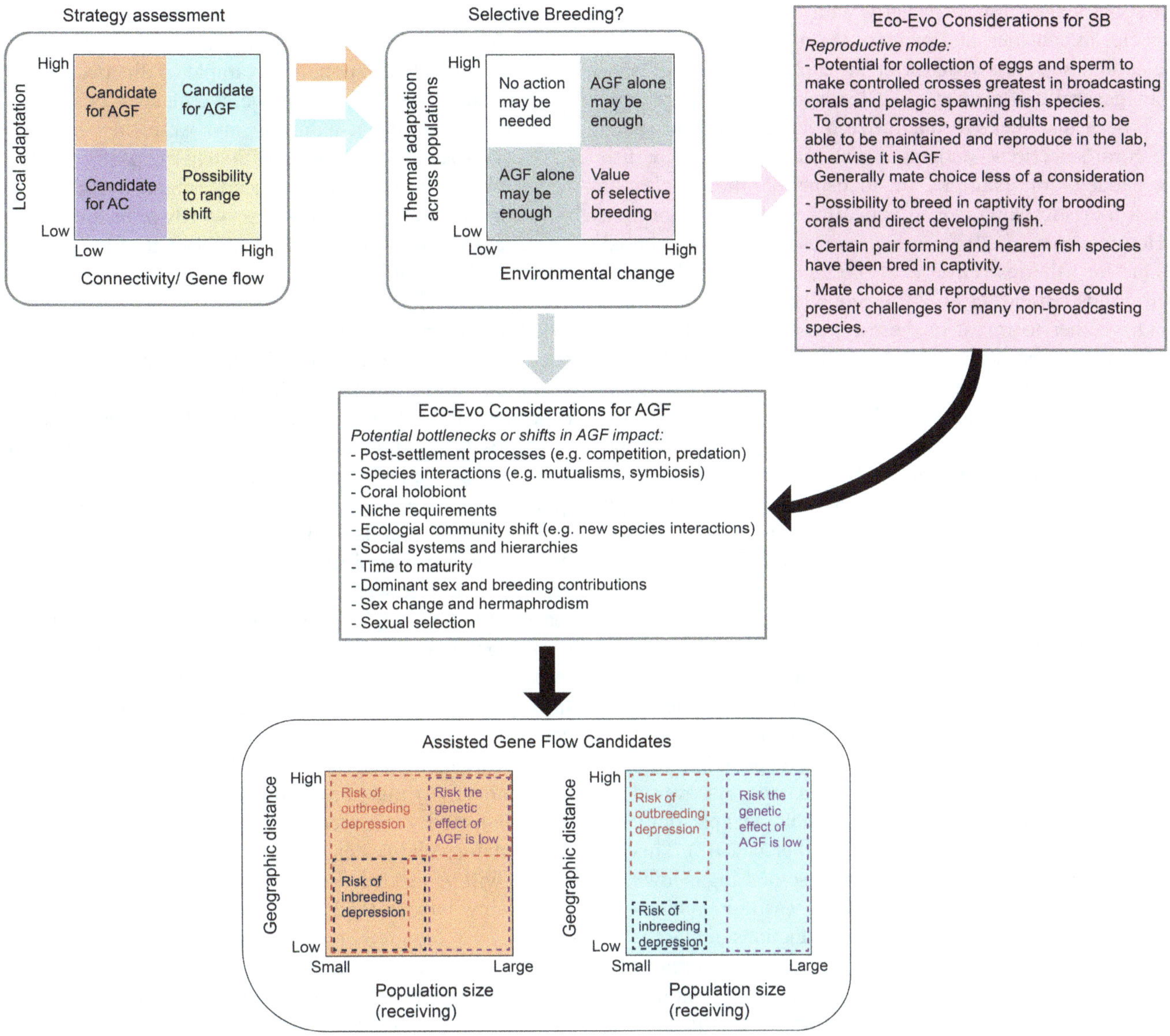

FIGURE 7 Conceptual summary of expectations and considerations when contemplating translocation conservation options for a species. Starting from the top left, knowledge of local adaption (heritable genetic diversity associated with heat tolerance) and connectivity (oceanographic, biological, and ecological) can be used to assess what assisted management strategy might be suitable. Focusing on AGF to maintain populations within the current distribution (orange and blue arrows), the need for SB will relate to whether populations adapted to the expected environmental change are available. If not, SB could potentially create thermally adapted individuals. We suggest that reproductive ecology and evolutionary history should be strong considerations before undertaking SB (pink box). Once thermally adapted individuals can be sourced, there are still further ecological and evolutionary considerations prior to undertaking AGF (white box). Geographic distance between source and receiving populations and receiving population size could result in risks of outbreeding depression, inbreeding depression, and limited impact of AGF (bottom section). This summary builds on previous discussions on this topic including Aitken and Whitlock 2013.

suggest that AGF may be a more viable conservation action relative to corals (Aitken and Whitlock 2013; Hoffmann et al. 2021), but the existence of locally adapted populations is key. Long-distance movement does present increased risk of failure in cases of high local adaptation due to dispersal outside the range where fitness is optimized (Figure 7). We acknowledge that for many reef fish species, AGF would present the same risks and limitations as with corals, as discussed previously in an aquaculture-like setting (Quigley et al. 2021), as they possess panmictic populations and high connectivity.

Given the large diversity of organisms living on coral reefs and the variability in life-history traits, eco-evolutionary processes, and oceanography, which influence resilience to future environmental change—where is the best place to start our efforts if SB is required for coral reef persistence? Researchers, managers, and other stakeholders working with coral reefs face similar difficulties in species selection, especially with the vast diversity of species to consider, and many of them are cryptic. Species selection in corals (Madin et al. 2021) and in reef species generally (Beyer et al. 2018) has taken a "portfolio" approach that seeks to maximize diversity and trait space in coral selection—perhaps a matching approach for fish is needed, for example, by focusing on the species of high touristic value due to appealing or bright coloration (Hemingson et al. 2022) or the species of high commercial value for fisheries (Payet et al. 2022) or functionally critical guilds (Brandl et al. 2019; Siqueira et al. 2021). When and whether to invest in these types of conservation and restoration approaches for fish and corals should take into account multiple eco-evolutionary lenses (e.g., time to reach sexual maturity, longevity, and the potential for behavioral or social specialization) that may result in bottlenecks and limit the benefits of the efforts. Activities that may produce unwanted outcomes need to be considered with formalized risk assessments (weighing risks vs. rewards) and modelled (e.g., Peterson and Bode 2021; Torda and Quigley 2021). These may include the following risks.

- **Intervention was not needed**, thereby wasting precious conservation resources. Taking into account the potential for processes such as phenotypic plasticity when assessing the need for intervention is important. This was completed for the red-cheeked salamander, and it was discovered that salamanders exhibited physiological plasticity to future climate change (Riddell et al. 2018). This suggests that the value in doing something is low.
- **A lack of outcomes**, thereby wasting precious conservation resources. Reintroductions of macropods to areas of the species' native range in mainland Australia that has been impacted by human development have often been unsuccessful. One of the main issues for the survival of reintroduced native species has been the presence of exotic predators such as foxes, feral cats, and dogs (Short et al. 1992), with potentially a similar scenario for some coral reef-associated groups such as fish when differences exist in community composition.
- **Unintended outcomes** due to a lack of consideration of other non-genetic factors. This occurred in the case of supplementary feeding provided to the critically endangered kakapo in an attempt to increase breeding frequency. However, this instead caused females to produce male-dominated broods (as production of male offspring is favoured in "good" quality conditions), resulting in a surplus of males in the population (Clout et al. 2002).
- **Negative outcomes** because it had an unexpected negative outcome on another part of the ecosystem. SB in the lab for one performance phenotype may reduce the overall plastic capacity of individuals via directional selection. This means that actions might risk creating highly thermally adapted species that cannot cope with other aspects of environmental variation. An example of this occurred when Sockeye salmon were collected on their return to spawn naturally and were then hatchery spawned. Offspring of hatchery spawned fish released to the wild produced a maladaptive shift in the timing of spawning, which reduced their reproductive success and the survival of juveniles (Tillotson et al. 2019b).

What is certain is that oceanographic processes will influence the feasibility and effectiveness of any applied conservation action, such as AGF, in the wild. This may be especially true if there is a mismatch between realized connectivity (as shown through genetics or modelling) and assumed connectivity (from life-history characteristics), requiring connectivity assessments at a species-by-species or higher level (Richards et al. 2017b). This will have critical impacts on the feasibility of AGF. Understanding oceanographic processes will also have direct implications for the feasibility of SB itself and will need to be considered from a management perspective. Patterns in connectivity vary by species, and their management plans should also vary. In management strategies that seek to promote population persistence and regeneration or improve resilience through conservation practices such as SB, key ecological processes like connectivity must be considered. Not only high-resolution connectivity resources (e.g., particle tracking models) that match spatial scales of genetics, but also management action, will be needed.

Notably, understanding the boundaries of species ranges—as dictated by connectivity—is challenging. Often the boundaries are not clear-cut and instead present as diffuse boundaries of connectivity and self-seeding that can change over time and under different spatial contexts (e.g., Cowen 1985). The uncertainty in these porous borders is an important ecological theme of research and will be important for applied outcomes that specifically parameterize the levels of diffusion to understand the risks and benefits of any conservation action implemented. One example of this includes the thus far unparameterized influence of near-surface currents of the fringing Coral Sea on the connectivity of the neighbouring Great Barrier Reef (see Wolanski et al. 2024). Variability of these Coral Sea currents should affect the time and magnitude of connectivity/self-seeding across the Great Barrier Reef, especially in the northern region. This is relevant for two reasons. Firstly, modelling suggests that around coral spawning at latitudes greater than approximately 15°S, dispersal of northern coral larvae is generally directed to the Coral Sea (there are high probability densities of larvae in this region) (Quigley et al. 2019). If a high preponderance of heat-tolerant populations can be found in this northern region (for full details see Quigley and van Oppen 2022), then it will be increasingly important to quantify the larval subsidies being "lost" or "gained" from this region depending on the neighbouring influence of near-surface fringing Coral Sea currents. Secondly, the structure of fish assemblages along the Great Barrier Reef is heavily influenced by the angle of currents within the Great Barrier Reef lagoon, but it is unknown how fringing currents likewise influence structure (Benthuysen et al. 2022). Clearly, further investigation is needed to link biophysical oceanographic processes to both fundamental knowledge and applied conservation outcomes.

Accordingly, we highlight the need to understand the speed of adaptive evolution in the wild. This will not only influence the potential success of SB but also inform as to the need for it. There are countless examples of how one or a few actions led to extreme negative consequences for ecosystems (e.g., cane toads and rabbits in Australia; Alves et al. 2022). The *fundamental theorem of natural selection* predicts that the amount of genetic variability related to improved ability to survive and reproduce (fitness) is proportional to the population's rate of adaptive evolution. Recent evidence suggests that some organisms in the wild are evolving two to four times the speed suggested by earlier work (Bitter et al. 2021)—leading to the question on whether breeding, with all its risks, is even needed. Considering these risks, we suggest that planning genetic interventions must be considered within a multi-step approach (Figure 7). This approach and work by others (e.g., Aitken and Whitlock 2013) highlights the critical need to understand connectivity between populations and eco-evolutionary aspects of a species prior to any genetic intervention actions.

REFERENCES

Abesamis RA, Saenz-Agudelo P, Berumen ML, Bode M, Jadloc CRL, Solera LA, Villanoy CL, Bernardo LPC, Alcala AC, Russ GR (2017) Reef-fish larval dispersal patterns validate no-take marine reserve network connectivity that links human communities. Coral Reefs 36:791–801

AIMS (2019) Australian Institute of Marine Science—Report for 2018–19

Aitken SN, Whitlock MC (2013) Assisted gene flow to facilitate local adaptation to climate change. Annu Rev Ecol Evol Syst 44.

Almany GR, Planes S, Thorrold SR, Berumen ML, Bode M, Saenz-Agudelo P, Bonin MC, Frisch AJ, Harrison HB, Messmer V (2017) Larval fish dispersal in a coral-reef seascape. Nat Ecol Evol 1:1–7

Alves JM, Carneiro M, Day JP, Welch JJ, Duckworth JA, Cox TE, Letnic M, Strive T, Ferrand N, Jiggins FM (2022) A single introduction of wild rabbits triggered the biological invasion of Australia. Proc Nat Acad Sci 119:e2122734119

Andutta FP, Kingsford MJ, Wolanski E (2012) 'Sticky water' enables the retention of larvae in a reef mosaic. Estuar Coast Shelf Sci 101:54–63

Asch RG, Erisman B (2018) Spawning aggregations act as a bottleneck influencing climate change impacts on a critically endangered reef fish. Divers Distrib 24:1712–1728

Babcock RC (1984) Reproduction and distribution of two species of *Goniastrea* (Scleractinia) from the Great Barrier Reef Province. Coral Reefs 2:187–195

Baird AH, Guest JR, Edwards AJ, et al. (2021) An Indo-Pacific coral spawning database. Sci Data 8.

Baird AH, Guest JR, Willis BL (2009) Systematic and biogeographical patterns in the reproductive biology of scleractinian corals. Annu Rev Ecol Evol Syst 40:551–571

Baums I (2008) A restoration genetics guide for coral reef conservation. Mol Ecol 17:2796–2811

Bay L, Rocker M, Boström-Einarsson L, Babcock R, Buerger P, Cleves P, Harrison D, Negri A, Quigley K, Randall CJ, van Oppen MJH Webster (2020) Reef Restoration and Adaptation Program, T3: Intervention Technical Summary.

Bay LK, Crozier RH, Caley MJ (2006) The relationship between population genetic structure and pelagic larval duration in coral reef fishes on the Great Barrier Reef. Mar Biol 149:1247–1256

Becker CD, Genoway RG (1979) Evaluation of the critical thermal maximum for determining thermal tolerance of freshwater fish. Environ Biol Fishes 4:245–256

Benthuysen JA, Emslie MJ, Currey-Randall LM, Cheal AJ, Heupel MR (2022) Oceanographic influences on reef fish assemblages along the Great Barrier Reef. Prog Oceanogr 102901

Bernal De Leon M, Donelson JM, Veilleux HD, Ryu T, Munday PL, Ravasi T (2018) Phenotypic and molecular consequences of step-wise temperature increase across generations in a coral reef fish. Mol Ecol 27(22):4516–4528

Bernal MA, Schunter C, Lehmann R, Lightfoot DJ, Allan BJM, Veilleux HD, Rummer JL, Munday PL, Ravasi T (2020) Species-specific molecular responses of wild coral reef fishes during a marine heatwave. Sci Adv 6:eaay3423

Bernal MA, Ravasi T, Rodgers GG, Munday PL, Donelson JM (2022) Plasticity to ocean warming is influenced by transgenerational, reproductive, and developmental exposure in a coral reef fish. Evol Appl 15:249–261

Beyer HL, Kennedy EV, Beger M, Chen CA, Cinner JE, Darling ES, Eakin CM, Gates RD, Heron SF, Knowlton N, Obura DO, Palumbi SR, Possingham HP, Puotinen M, Runting RK, Skirving WJ, Spalding M, Wilson KA, Wood S, Veron JE, Hoegh-Guldberg O (2018) Risk-sensitive planning for conserving coral reefs under rapid climate change. Conserv Lett e12587

Bitter MC, Wong JM, Dam HG, Donelan SC, Kenkel CD, Komoroske LM, Nickols KJ, Rivest EB, Salinas S, Burgess SC (2021) Fluctuating selection and global change: A synthesis and review on disentangling the roles of climate amplitude, predictability and novelty. Proc Roy Soci B 288:20210727

Bode M, Bode L, Armsworth P (2006) Regional scale source-sink structure in a marine metapopulation. Mar Ecol Prog Ser 308:17–25

Brandl SJ, Rasher DB, Côté IM, Casey JM, Darling ES, Lefcheck JS, Duffy JE (2019) Coral reef ecosystem functioning: Eight core processes and the role of biodiversity. Front Ecol Environ 17:445–454

Burgess SC, Bode M, Leis JM, Mason LB (2022) Individual variation in marine larval-fish swimming speed and the emergence of dispersal kernels. Oikos 2022:e08896

Cheal AJ, MacNeil MA, Emslie MJ, Sweatman H (2017) The threat to coral reefs from more intense cyclones under climate change. Glob Chang Biol 23:1511–1524

Choat JH (2012) Spawning aggregations in reef fishes; ecological and evolutionary processes. Reef fish spawning aggregations: Biology, research and management. Springer, pp. 85–116

Clout MN, Elliott GP, Robertson BC (2002) Effects of supplementary feeding on the offspring sex ratio of kakapo: A dilemma for the conservation of a polygynous parrot. Biol Conserv 107:13–18

Cornwell B, Armstrong K, Walker NS, Lippert M, Nestor V, Golbuu Y, Palumbi SR (2021) Widespread variation in heat tolerance and symbiont load are associated with growth tradeoffs in the coral Acropora hyacinthus in Palau. Elife 10:e64790

Cowen RK (1985) Large scale patterns of recruitment by the labrid, *Semicossyphus pulcher*: Causes and implications. J Mar Res 43:719–742.

Davies S, Gamache MH, Howe-Kerr LI, Kriefall NG, Baker AC, Banaszak AT, Bay LK, Bellantuono AJ, Bhattacharya D, Chan CX (2022) Building consensus around the assessment and interpretation of Symbiodiniaceae diversity. PeerJ 11 (2023):e15023

De'ath G, Fabricius KE, Sweatman H, Puotinen M (2012) The 27-year decline of coral cover on the Great Barrier Reef and its causes. Proc Nat Acad Sci 109:17995–17999

DiMaggio MA, Cassiano EJ, Barden KP, Ramee SW, Ohs CL, Watson CA (2017) First record of captive larval culture and metamorphosis of the Pacific blue tang, Paracanthurus hepatus. J World Aquac Soc 48:393–401

Dixon GB, Davies SW, Aglyamova GV, Meyer E, Bay LK, Matz MV (2015a) Genomic determinants of coral heat tolerance across latitudes. Science (1979) 348:1460–1462

Doherty PJ, Planes S, Mather P (1995) Gene flow and larval duration in seven species of fish from the Great Barrier Reef. Ecology 76:2373–2391

Donelson JM, Munday PL (2012) Thermal sensitivity does not determine acclimation capacity for a tropical reef fish. J Anim Ecol 81:1126–1131

Donelson JM, Munday PL, McCormick MI, Pitcher CR (2012) Rapid transgenerational acclimation of a tropical reef fish to climate change. Nat Clim Chang 2:30–32

Donelson JM, Wong M, Booth DJ, Munday PL (2016) Transgenerational plasticity of reproduction depends on rate of warming across generations. Evol Appl 9:1072–1081

Donovan MK, Burkepile DE, Kratochwill C, Shlesinger T, Sully S, Oliver TA, Hodgson G, Freiwald J, van Woesik R (2021) Local conditions magnify coral loss after marine heatwaves. Science (1979) 372:977–980

Drury C, Caruso C, Quigley K (2022) Selective breeding to enhance the adaptive potential of corals. Coral reef conservation and restoration in the omics age. Springer, pp. 71–84

Edmunds RC, Hillersøy G, Momigliano P, van Herwerden L (2009) Classic approach revitalizes genomics: Complete characterization of a candidate gene for thermal adaptation in two coral reef fishes. Mar Genomics 2:215–222

Edmunds RC, Smith-Keune C, van Herwerden L, Fulton CJ, Jerry DR (2012) Exposing local adaptation: Synergistic stressors elicit population-specific lactate dehydrogenase-B (ldh-b) expression profiles in Australian barramundi, *Lates calcarifer*. Aquat Sci 74:171–178

Eliason EJ, Clark TD, Hague MJ, Hanson LM, Gallagher ZS, Jeffries KM, Gale MK, Patterson DA, Hinch SG, Farrell AP (2011) Differences in thermal tolerance among sockeye salmon populations. Science (1979) 332:109–112

Erisman BE, Buckhorn ML, Hastings PA (2007) Spawning patterns in the leopard grouper, Mycteroperca rosacea, in comparison with other aggregating groupers. Mar Biol 151:1849–1861

Evans RD, van Herwerden L, Russ GR, Frisch AJ (2010) Strong genetic but not spatial subdivision of two reef fish species targeted by fishers on the Great Barrier Reef. Fish Res 102:16–25

Figueira WF (2009) Connectivity or demography: Defining sources and sinks in coral reef fish metapopulations. Ecol Model 220:1126–1137

Filbee-Dexter K, Smajdor A (2019) Ethics of assisted evolution in marine conservation. Front Mar Sci 6:20

Fisher R (2005) Swimming speeds of larval coral reef fishes: Impacts on self-recruitment and dispersal. Mar Ecol Prog Ser 285:223–232

Fitzpatrick JL (2020) Sperm competition and fertilization mode in fishes. Philos Trans Roy Soc B 375:20200074

Fox RJ, Fromhage L, Jennions MD (2019) Sexual selection, phenotypic plasticity and female reproductive output. Philos Trans Roy Soc B 374:20180184

Fricke HW (1980) Control of different mating systems in a coral reef fish by one environmental factor. Anim Behav 28:561–569

Gardiner NM, Munday PL, Nilsson GE (2010) Counter-gradient variation in respiratory performance of coral reef fishes at elevated temperatures. PLoS One 5:e13299

Gerlach G, Atema J, Kingsford MJ, Black KP, Miller-Sims V (2007) Smelling home can prevent dispersal of reef fish larvae. Proc Nat Acad Sci 104:858–863

Gerlach G, Atema J, Raupach MJ, Deister F, Müller A, Kingsford MJ (2016) Cryptic species of cardinalfish with evidence for old and new divergence. Coral Reefs 35:437–450

Gomulkiewicz R, Holt RD (1995) When does evolution by natural selection prevent extinction? Evolution (N Y) 201–207

Gouezo M, Wolanski E, Critchell K, Fabricius K, Harrison P, Golbuu Y, Doropoulos C (2021) Modelled larval supply predicts coral population recovery potential following disturbance. Mar Ecol Prog Ser 661:127–145

Goulet D, Goulet TL (2006) Nonindependent mating in a coral reef damselfish: Evidence of mate choice copying in the wild. Behavioral Ecology 17:998–1003

Graham EM, Baird AH, Connolly SR (2008) Survival dynamics of scleractinian coral larvae and implications for dispersal. Coral Reefs 27:529–539

Gräns A, Jutfelt F, Sandblom E, Jönsson E, Wiklander K, Seth H, Olsson C, Dupont S, Ortega-Martinez O, Einarsdottir I (2014) Aerobic scope fails to explain the detrimental effects on growth resulting from warming and elevated CO2 in Atlantic halibut. J Exp Biol 217:711–717

Great Barrier Reef Marine Park Authority (2022)

Green AL, Maypa AP, Almany GR, Rhodes KL, Weeks R, Abesamis RA, Gleason MG, Mumby PJ, White AT (2015) Larval dispersal and movement patterns of coral reef fishes, and implications for marine reserve network design. Biol Rev 90:1215–1247

Greiner A, Andrello M, Darling E, Krkošek M, Fortin M (2022) Limited spatial rescue potential for coral reefs lost to future climate warming. Glob Ecol Biogeography 31(11):2245–2258

Grottoli AG, Toonen RJ, van Woesik R, Vega Thurber R, Warner ME, McLachlan RH, Price JT, Bahr KD, Baums IB, Castillo K, Coffroth MA, Cunning R, Dobson K, Donahue M, Hench JL, Iglesias-Prieto R, Kemp DW, Kenkel CD, Kline DI, Kuffner IB, Matthews JL, Mayfield A, Padilla-Gamino J, Palumbi SR, Voolstra CR, Weis V, Wu HC (2020) Increasing comparability among coral bleaching experiments. Ecol Appl e2262

Guest JR, Heyward AJ, Omori M, Iwao K, Morse A, Boch C (2010) Rearing coral larvae for reef rehabilitation. Reef Rehabil Manual 73–98

Harrison HB, Berumen ML, Saenz-Agudelo P, Salas E, Williamson DH, Jones GP (2017) Widespread hybridization and bidirectional introgression in sympatric species of coral reef fish. Mol Ecol 26(20):5692–5704.

Harrison HB, Williamson DH, Evans RD, Almany GR, Thorrold SR, Russ GR, Feldheim KA, van Herwerden L, Planes S, Srinivasan M (2012) Larval export from marine reserves and the recruitment benefit for fish and fisheries. Curr Biol 22:1023–1028

Harrison PL (2024) Sexual reproduction of reef corals and application to coral restoration. In Wolanski E, Kingsford MJ (eds.), Oceanographic Processes of Coral Reefs. Physical and Biological Links in the Great Barrier Reef. CRC Press.

Hemingson CR, Mihalitsis M, Bellwood DR (2022) Are fish communities on coral reefs becoming less colourful? Glob Chang Biol 28:3321–3332

Hobbs JA, Munday PL, Jones GP (2004) Social induction of maturation and sex determination in a coral reef fish. Proc R Soc Lond B Biol Sci 271:2109–2114

Hock K, Wolff NH, Ortiz JC, Condie SA, Anthony KRN, Blackwell PG, Mumby PJ (2017) Connectivity and systemic resilience of the Great Barrier Reef. PLoS Biol 15:e2003355

Hoffman EA, Kolm N, Berglund A, Arguello JR, Jones AG (2005) Genetic structure in the coral-reef-associated Banggai cardinalfish, Pterapogon kauderni. Mol Ecol 14:1367–1375

Hoffmann AA, Miller AD, Weeks AR (2021) Genetic mixing for population management: From genetic rescue to provenancing. Evol Appl 14:634–652

Hoffmann AA, Sgró CM (2011) Climate change and evolutionary adaptation. Nature 470:479–485

Humanes A, Bythell J, Beauchamp E, Carl M, Craggs J, Edwards A, Golbuu Y, Lachs L, Randle JL, Martinez H (2021) A framework for selectively breeding corals for assisted evolution. Front Mar Sci (8):669995

Jones DB, Jerry DR, McCormick MI, Bay LK (2010) The population genetic structure of a common tropical damselfish on the Great Barrier Reef and eastern Papua New Guinea. Coral Reefs 29:455–467

Jones GP (2015) Mission impossible: Unlocking the secrets of coral reef fish dispersal. Ecol Fishes Coral Reefs:16–27.

Jones GP, Almany GR, Russ GR, Sale PF, Steneck RS, van Oppen MJH, Willis BL (2009) Larval retention and connectivity among populations of corals and reef fishes: History, advances and challenges. Coral Reefs 28:307–325

Jones RJ, Bowyer J, Hoegh-Guldberg O, Blackall LL (2004) Dynamics of a temperature-related coral disease outbreak. Mar Ecol Prog Ser 281:63–77

Kaunda-Arara B, Rose GA (2004) Long-distance movements of coral reef fishes. Coral Reefs 23:410–412

Kawecki TJ, Ebert D (2004) Conceptual issues in local adaptation. Ecol Lett 7:1225–1241

Kincaid HL (1993) Selective breeding and domestication. In: Cloud JG, Thorgaard GH (eds.), Genetic Conservation of Salmonid Fishes. NATO ASI Series, vol 248. Springer.

Kingsford MJ, Spiecker L, Gerlach G (2024) The interactions between larval behaviour and oceanography. In Wolanski E, Kingsford MJ (eds.), Oceanographic Processes of Coral Reefs. Physical and Biological Links in the Great Barrier Reef. CRC Press.

Kingsford MJ, Welch D, O'Callaghan M (2019) Latitudinal and cross-shelf patterns of size, age, growth, and mortality of a tropical damselfish Acanthochromis polyacanthus on the Great Barrier Reef. Diversity (Basel) 11:67

Kininmonth S (2024) Currents modulate the genetic character of marine populations in the Great Barrier Reef. In Wolanski E, Kingsford MJ (eds.), Oceanographic Processes of Coral Reefs. Physical and Biological Links in the Great Barrier Reef. CRC Press.

Kitanobo S, Iwao K, Fukami H, Isomura N, Morita M (2022) First evidence for backcrossing of F1 hybrids in Acropora corals under sperm competition. Sci Rep 12:1–8

Kuwamura T, Sunobe T, Sakai Y, Kadota T, Sawada K (2020) Hermaphroditism in fishes: An annotated list of species, phylogeny, and mating system. Ichthyol Res 67:341–360

Leis JM, Wright KJ, Johnson RN (2007) Behaviour that influences dispersal and connectivity in the small, young larvae of a reef fish. Mar Biol 153:103–117

Loya Y, Sakai K, Nakano Y, Woesik R van (2001a) Coral bleaching: The winners and the losers. Ecol Lett 4:122–131

Loya Y, Sakai K, Yamazato K, Nakano Y, Sambali H, van Woesik R (2001b) Coral bleaching: The winners and the losers. Ecol Lett 4:122–131

Madin J, McWilliam M, Quigley K, Bay L, Bellwood D, Bridge T, Doropoulos C, Fernandes L, Harrison P, Hoey A (2021) A trait-based approach to advance coral reef science. Trends in Ecology & Evolution 31.6 (2016):419–428.

Marie AD, van Herwerden L, Choat JH, Hobbs JP (2007) Hybridization of reef fishes at the Indo-Pacific biogeographic barrier: A case study. Coral Reefs 26:841–850

Miller-Sims VC, Gerlach G, Kingsford MJ, Atema J (2008) Dispersal in the spiny damselfish, Acanthochromis polyacanthus, a coral reef fish species without a larval pelagic stage. Mol Ecol 17:5036–5048

Mongin M, Baird ME, Tilbrook B, Matear RJ, Lenton A, Herzfeld M, Wild-Allen K, Skerratt J, Margvelashvili N, Robson BJ (2016) The exposure of the Great Barrier Reef to ocean acidification. Nat Commun 7:1–8

Moorhead JA, Zeng C (2010) Development of captive breeding techniques for marine ornamental fish: A review. Rev Fish Sci 18:315–343

Mora C, Ospina A (2001) Tolerance to high temperatures and potential impact of sea warming on reef fishes of Gorgona Island (tropical eastern Pacific). Mar Biol 139:765–769

Morikawa MK, Palumbi SR (2019) Using naturally occurring climate resilient corals to construct bleaching-resistant nurseries. Proc Natl Acad Sci U S A 116:10586–10591

Munday PL, Donelson JM, Domingos JA (2017) Potential for adaptation to climate change in a coral reef fish. Glob Chang Biol 23:307–317

Munoz RC, Zgliczynski BJ, Teer BZ, Laughlin JL (2014) Spawning aggregation behavior and reproductive ecology of the giant bumphead parrotfish, Bolbometopon muricatum, in a remote marine reserve. PeerJ 2:e681

Newton JR, de Santis C, Jerry DR (2012) The gene expression response of the catadromous perciform barramundi Lates calcarifer to an acute heat stress. J Fish Biol 81:81–93

Pankhurst NW, Munday PL (2011) Effects of climate change on fish reproduction and early life history stages. Mar Freshw Res 62:1015–1026

Patterson HM, Swearer SE (2007) Long-distance dispersal and local retention of larvae as mechanisms of recruitment in an island population of a coral reef fish. Austral Ecol 32:122–130

Payet SD, Pratchett MS, Saenz-Agudelo P, Berumen ML, DiBattista JD, Harrison HB (2022) Demographic histories shape population genomics of the common coral grouper (*Plectropomus leopardus*). Evol Appl 15:1221–1235

Petersen CW, Fischer EA (1986) Mating system of the hermaphroditic coral-reef fish, Serranus baldwini. Behav Ecol Sociobiol 19:171–178

Peterson K, Bode M (2021) Using ensemble modeling to predict the impacts of assisted migration on recipient ecosystems. Conserv Biol 35:678–687

Planes S (2002) Biogeography and larval dispersal inferred from population genetic analysis. Coral Reef Fishes 201–220

Planes S, Jones GP, Thorrold SR (2009) Larval dispersal connects fish populations in a network of marine protected areas. Proc Nat Acad Sci 106:5693–5697

Pollock FJ, Katz SM, van de Water JAJM, Davies SW, Hein M, Torda G, Matz M v, Beltran VH, Buerger P, Puill-Stephan E, Abrego D, Bourne DG, Willis BL (2017) Coral larvae for restoration and research: A large-scale method for rearing *Acropora millepora* larvae, inducing settlement, and establishing symbiosis. PeerJ 5:e3732

Pörtner HO, Knust R (2007) Climate change affects marine fishes through the oxygen limitation of thermal tolerance. Science (1979) 315:95–97

Pratchett MS, Cameron DS, Donelson J, Evans L, Frisch AJ, Hobday AJ, Hoey AS, Marshall NA, Messmer V, Munday PL (2017) Effects of climate change on coral grouper (Plectropomus spp.) and possible adaptation options. Rev Fish Biol Fish 27:297–316

Pratchett MS, Munday PL, Wilson SK, Graham NAJ, Cinner JE, Bellwood DR, Jones GP, Polunin NVC, McClanahan TR (2008) Effects of climate-induced coral bleaching on coral-reef fishes—ecological and economic consequences. Oceanography Marine Biology, CRC Press, pp. 257–302

Quigley KM, Bay LK, van Oppen MJH (2019) The active spread of adaptive variation for reef resilience. Ecol Evol 9

Quigley KM, Bay LK, van Oppen MJH (2020) Genome-wide SNP analysis reveals an increase in adaptive genetic variation through selective breeding of coral. Mol Ecol. 29(12):2176–2188

Quigley KM, Hein M, Suggett DJ (2022) Translating the ten golden rules of reforestation for coral reef restoration. Conserv Biol 36(4):e13890

Quigley KM, Marzonie M, Ramsby B, Abrego D, Milton G, van Oppen MJH, Bay LK (2021) Variability in fitness trade-offs amongst coral juveniles with mixed genetic backgrounds held in the wild. Front Mar Sci 8:161

Quigley KM, Randall CJ, van Oppen MJH, Bay LK (2020b) Assessing the role of historical temperature regime and algal symbionts on the heat tolerance of coral juveniles. Biol Open.

Quigley KM, van Oppen MJH (2022) Predictive models for the selection of thermally tolerant corals based on offspring survival. Nat Commun 13:1–13

Quigley KM, Willis BL, Bay LK (2016) Maternal effects and symbiodinium community composition drive differential patterns in juvenile survival in the coral acropora tenuis. R Soc Open Sci 3.

Richards Z, Berry O, Underwood J, McMahon K, Travers M, Moore G, Hernawan U, DiBattista J, Evans R, Gilmour J (2017) Ecological connectivity of Kimberley marine communities. Synthesis report of project 1.1.3 prepared for the Kimberley marine research program. WAMSI, Australia

Riddell EA, Odom JP, Damm JD, Sears MW (2018) Plasticity reveals hidden resistance to extinction under climate change in the global hotspot of salamander diversity. Sci Adv 4:eaar5471

Riginos C, Hock K, Matias AM, Mumby PJ, van Oppen MJH, Lukoschek V (2019) Asymmetric dispersal is a critical element of concordance between biophysical dispersal models and spatial genetic structure in Great Barrier Reef corals. Divers Distrib 25:1684–1696

Rodgers GG, Donelson JM, McCormick MI, Munday PL (2018) In hot water: Sustained ocean warming reduces survival of a low-latitude coral reef fish. Mar Biol 165:1–10

Rueger T, Barbasch TA, Wong MYL, Srinivasan M, Jones GP, Buston PM (2018) Reproductive control via the threat of eviction in the clown anemonefish. Proc Roy Soc B 285:20181295

Rueger T, Gardiner NM, Jones GP (2016) Size matters: Male and female mate choice leads to size-assortative pairing in a coral reef cardinalfish. Behav Ecol arw082

Ryu T, Veilleux HD, Donelson JM, Munday PL, Ravasi T (2018) The epigenetic landscape of transgenerational acclimation to ocean warming. Nat Clim Chang 8:504–509

Samoilys MA, Squire LC (1994) Preliminary observations on the spawning behavior of coral trout, Plectropomus leopardus (Pisces: Serranidae), on the Great Barrier Reef. Bull Mar Sci 54:332–342

Santaquiteria A, Siqueira AC, Duarte-Ribeiro E, Carnevale G, White WT, Pogonoski JJ, Baldwin CC, Ortí G, Arcila D, Ricardo B-R (2021) Phylogenomics and historical biogeography of seahorses, dragonets, goatfishes, and allies (Teleostei: Syngnatharia): Assessing factors driving uncertainty in biogeographic inferences. Syst Biol 70:1145–1162

Savolainen O, Lascoux M, Merilä J (2013) Ecological genomics of local adaptation. Nat Rev Genet 14:807

Schmidt-Roach S, Johnston E, Fontana S, Jury CP, Forsman Z (2014) Daytime spawning of Pocillopora species in Kaneohe Bay, Hawai 'i. Galaxea. J Coral Reef Stud 16:11–12

Shaver EC, McLeod E, Hein MY, Palumbi SR, Quigley K, Vardi T, Mumby PJ, Smith D, Montoya-Maya P, Muller EM (2022) A roadmap to integrating resilience into the practice of coral reef restoration. Glob Chang Biol 28(16):4751–4764

Short J, Bradshaw SD, Giles J, Prince RIT, Wilson GR (1992) Reintroduction of macropods (Marsupialia: Macropodoidea) in Australia—a review. Biol Conserv 62:189–204

Shulman MJ, Ogden JC (1987) What controls tropical reef fish populations: Recruitment or benthic mortality? An example in the Caribbean reef fish *Haemulon flavolineatum*. Marine Ecology Progress Series 39:233–242.

Siqueira AC, Morais RA, Bellwood DR, Cowman PF (2021) Planktivores as trophic drivers of global coral reef fish diversity patterns. Proceedings of the National Academy of Sciences 118:e2019404118

Slatkin M (1976) The rate of spread of an advantageous allele in a subdivided population. Population genetics and ecology. Elsevier, pp. 767–780

Smith HA, Moya A, Cantin NE, van Oppen MJH, Torda G (2019) Observations of simultaneous sperm release and larval planulation suggest reproductive assurance in the coral Pocillopora acuta. Front Mar Sci 6:362

Souter, SP, Wicquart J, Logan M, Obura D, Staub F (2021) The sixth status of corals of the world: 2020 report. Global Coral Reef Monitoring Network

Steneck R, Jones GP, Willis BL, Sale PF, Almany GR, van Oppen MJH, Russ GR (2009) Larval retention and connectivity among populations of corals and reef fishes: History, advances and challenges. Coral Reefs: 28:307–325

Strader ME, Quigley KM (2022) The role of gene expression and symbiosis in reef-building coral acquired heat tolerance. Nat Commun 13:1–8

Thurman LL, Gross JE, Mengelt C, Beever EA, Thompson LM, Schuurman GW, Hoving CL, Olden JD (2022) Applying assessments of adaptive capacity to inform natural-resource management in a changing climate. Conservation Biology 36:e13838

Thurman LL, Stein BA, Beever EA, Foden W, Geange SR, Green N, Gross JE, Lawrence DJ, LeDee O, Olden JD (2020) Persist in place or shift in space? Evaluating the adaptive capacity of species to climate change. Front Ecol Environ 18:520–528

Tillotson MD, Barnett HK, Bhuthimethee M, Koehler ME, Quinn TP (2019) Artificial selection on reproductive timing in hatchery salmon drives a phenological shift and potential maladaptation to climate change. Evol Appl 12:1344–1359

Torda G, Donelson JM, Aranda M, Barshis DJ, Bay L, Berumen ML, Bourne DG, Cantin N, Foret S, Matz M (2017) Rapid adaptive responses to climate change in corals. Nat Clim Chang 7:627–636

Torda G, Lundgren P, Willis BL, Oppen MJH (2013) Revisiting the connectivity puzzle of the common coral *Pocillopora damicornis*. Mol Ecol 22:5805–5820

Torda G, Quigley K (2021) Drivers of adaptive capacity in wild populations: Implications for genetic interventions. bioRxiv

van den Berghe EP, Warner RR (1989) The effects of mating system on male mate choice in a coral reef fish. Behav Ecol Sociobiol 24:409–415

van Herwerden L, Choat JH, Dudgeon CL, Carlos G, Newman SJ, Frisch A, van Oppen M (2006) Contrasting patterns of genetic structure in two species of the coral trout Plectropomus (Serranidae) from east and west Australia: Introgressive hybridisation or ancestral polymorphisms. Mol Phylogenet Evol 41:420–435

van Herwerden L, Choat JH, Newman SJ, Leray M, Hillersøy G (2009) Complex patterns of population structure and recruitment of Plectropomus leopardus (Pisces: Epinephelidae) in the Indo-West Pacific: Implications for fisheries management. Mar Biol 156:1595–1607

van Oppen MJH, Puill-Stephan E, Lundgren P, De'ath G, Bay LK (2014) First-generation fitness consequences of interpopulational hybridisation in a Great Barrier Reef coral and its implications for assisted migration management. Coral Reefs 33:607–611.

van Woesik R, Sakai K, Ganase A, Loya Y (2011) Revisiting the winners and the losers a decade after coral bleaching. Mar Ecol Prog Ser 434:67–76

Vanderklift MA, Doropoulos C, Gorman D, Leal I, Minne AJP, Statton J, Steven ADL, Wernberg T (2020) Using propagules to restore coastal marine ecosystems. Front Mar Sci 7:724

Veilleux HD, Ryu T, Donelson JM, Ravasi T, Munday PL (2018) Molecular response to extreme summer temperatures differs between two genetically differentiated populations of a coral reef fish. Front Mar Sci 5:349

Veilleux HD, Ryu T, Donelson JM, van Herwerden L, Seridi L, Ghosheh Y, Berumen ML, Leggat W, Ravasi T, Munday PL (2015) Molecular processes of transgenerational acclimation to a warming ocean. Nat Clim Chang 5:1074–1078

Voolstra CR, Buitrago-López C, Perna G, Cárdenas A, Hume BCC, Rädecker N, Barshis DJ (2020) Standardized short-term acute heat stress assays resolve historical differences in coral thermotolerance across microhabitat reef sites. Global Chang Biol 26:4328–4343

Voolstra CR, Suggett DJ, Peixoto RS, Parkinson JE, Quigley KM, Silveira CB, Sweet M, Muller EM, Barshis DJ, Bourne DG (2021) Extending the natural adaptive capacity of coral holobionts. Nat Rev Earth Environ 2:747–762

Warner PA, Willis BL, van Oppen MJH (2016) Sperm dispersal distances estimated by parentage analysis in a brooding scleractinian coral. Mol Ecol 25:1398–1415

Weeriyanun P, Collins RB, Macadam A, Kiff H, Randle JL, Quigley KM (2021) Predicting selection-response gradients of heat tolerance in a wide-ranging reef-building coral. Journal of Experimental Biology

Welch MJ, Munday PL (2017) Heritability of behavioural tolerance to high CO 2 in a coral reef fish is masked by nonadaptive phenotypic plasticity. Evol Appl 10:682–693

Williamson DH, Harrison HB, Almany GR, Berumen ML, Bode M, Bonin MC, Choukroun S, Doherty PJ, Frisch AJ, Saenz-Agudelo P (2016) Large-scale, multidirectional larval connectivity among coral reef fish populations in the Great Barrier Reef Marine Park. Mol Ecol 25:6039–6054

Willis BL, Babcock RC, Harrison PL, Wallace CC (1997) Experimental hybridization and breeding incompatibilities within the mating systems of mass spawning reef corals. Coral Reefs 16:S53–S65

Wilson SK, Graham NAJ, Pratchett MS, Jones GP, Polunin NVC (2006) Multiple disturbances and the global degradation of coral reefs: Are reef fishes at risk or resilient? Glob Chang Biol 12:2220–2234

Wolanski E, Doherty P, Carleton J (1997) Directional swimming of fish larvae determines connectivity of fish populations on the Great Barrier Reef. Naturwissenschaften 84:262–268

Wolanski E, Kingsford M, Lambrechts J, Marmorino, G (2024) The physical oceanography of the great barrier reef: A review. In Wolanski E, Kingsford MJ (eds.), Oceanographic Processes of Coral Reefs. Physical and Biological Links in the Great Barrier Reef. CRC Press.

Wong MYL, Buston PM, Munday PL, Jones GP (2007) The threat of punishment enforces peaceful cooperation and stabilizes queues in a coral-reef fish. Proc Roy Soc B: Biol Sci 274:1093–1099

24 Coastal Wetland Restoration

Case Studies from Great Barrier Reef Catchments

Nathan J. Waltham, Paula Cartwright, Katie Motson, Marcus Sheaves, and Mike Ronan

24.1. INTRODUCTION

Coastal wetlands are transitional areas between land and sea and include estuaries, rivers, creeks, floodplains, seasonal flowing (dry for parts of the year) channels, and low-lying areas (Elliott and Whitfield 2011; Gillanders and Kingsford 2002; Schuerch et al. 2018). They are dynamic environments, with highly variable physical, hydrological, and biological components, making them truly amazing ecosystems that are vital for the many services they provide (Beck et al. 2001). The coastal seascape (as with landscape theory on land) exists as a mixed set of habitats, including vegetated areas, (e.g., seagrass, mangroves, and saltmarshes), sandy beaches, and mudflats (Barbier et al. 2010; Nagelkerken et al. 2015). However, with expanding coastal urban and industrial development, once pristine coastal and estuarine habitats now include a range of engineered infrastructure features, including roads, seawalls, pontoons, marinas, ports, jetties, and pylons (Browne and Chapman 2014; Bulleri et al. 2005; Chapman 2003; Waltham and Connolly 2011). These anthropogenic encroachments raise concerns about the ability of these habitats to continue providing ecosystem services (Chapman 2003; Ng et al. 2015); however, evidence suggests that these modified habitats can display a decline or an increase or undergo no change in the composition of species colonising them. This raises the fundamental question around the true value of estuaries and whether restoration of these damaged ecosystems back to pristine or near-pristine conditions is possible or even necessary (Beck et al. 2001; Nagelkerken et al. 2015).

Wetlands (defined as all aquatic habitats in the coastal zone to a depth of 6 m at low tide) are among the most vulnerable of coastal ecosystems. Globally, 100% of wetlands are estimated as likely or highly likely to suffer from habitat loss and fragmentation exacerbated by climate change, compared to rainforest ecosystems at 45.3% (Powers and Jetz 2019). By far, coastal industrialisation and urbanisation are the largest contributors to coastal wetland modification (Airoldi et al. 2021; Bugnot et al. 2021), and although recent data have revealed an expansion of wetlands in some places, the net trend shows a decline in coastal wetland extent (Murray et al. 2022). Australia also faces a legacy of degraded freshwater and coastal wetland habitat, despite a small population and a relatively short 250 years of urban, industrial, and agricultural development (Creighton et al. 2016). In the Great Barrier Reef (GBR) catchment, the loss and degradation of wetlands is also reducing the GBR's resilience to other pressures (e.g., slower coral recovery following marine heatwave or cyclone activity) due to ongoing pollutant runoff (Adame et al. 2019; Brodie and Waterhouse 2012; Lewis et al. 2021) and reduced habitat availability for species with freshwater life stages (Arthington et al. 2015; Waltham et al. 2019). This has sparked management targets seeking to maintain and improve the extent and condition of wetlands (State of Queensland 2018). In addition to agriculture and urban development demands, there is increasing pressure to alter wetlands to capitalise on both their carbon sequestration (Alongi et al. 2016; Hagger et al. 2022; Herr et al. 2015; Rogers et al. 2019) and water quality improvement services (McJannet et al. 2012; Wallace and Waltham 2021), with potentially negative biodiversity consequences (Sheaves et al. 2014; Waltham et al. 2019). Successfully addressing these multiple issues will inevitably rely on applying a range of methods, with known efficacy, to restore or create wetlands that can be adapted to a local context.

The coastal and floodplain areas of the GBR region (Figure 1) are spectacular, dynamic, and hold incredible value; however, extensive land-use change has modified these habitats, and agricultural development currently dominates the coastal landscape (Lewis et al. 2021). This land-use change has resulted in considerable negative consequences to catchment hydrology (Lewis et al. 2024; Waterhouse et al. 2024), nearshore coastal dynamics (Wolanski et al. 2024), and connectivity of the GBR floodplains, undermining the functioning of the many diverse coastal and estuarine ecosystems (Sheaves et al. 2024). Urban and industrial development has also expanded along the GBR coast. A review of the spatial extent and distribution of engineered structures (including roads, pontoons, seawalls, marinas, ports, and boat ramps) reveals that more than 10% of the GBR coast's linear extent has been developed (Waltham and Sheaves 2015). Research into the ecosystem services and functionality of these built features is emerging, and although revealing some evidence that these habitats support marine flora and fauna, still more research is needed to understand whether even the most basic ecological processes and traits (such as diet) hold

DOI: 10.1201/9781003320425-28

FIGURE 1 Wetlands in the GBR catchments are dynamic and complex. A) Barratt's lagoon on the Tully floodplain is a permanent freshwater system; B) saltmarsh meadows in Eurimbula National Park Round Hill Reserve; and C) *Rhizophora stylosa* sapling growing on an intertidal muddy bank in Hinchinbrook channel.

Source: Photos credit— Nathan J. Waltham

true when compared to natural habitat settings. There is a growing body of research on the ecology of anthropogenic/urban coastal infrastructure (Evans et al. 2016; Firth et al. 2014), but only a small number of recent studies have been completed in the GBR (Bradley et al. 2023; Waltham and Sheaves 2018).

Official wetland mapping across the GBR catchment has been carried out by the Queensland Government since 2001 (Environmental Protection Agency 2005), with the latest iteration released in 2017 that also includes estimates of pre-clear (pre-European settlement) extent. The mapping program uses a modified version of the Ramsar definition that excludes riparian zones above the saturation level and intermittently covered floodplains that do not meet the hydrophyte and soil criteria (Environmental Protection Agency 2005). According to the Queensland Government's Wetland*Info* website (https://wetla ndinfo.des.qld.gov.au/wetlands/facts -maps/study -area-great-barrier-reef/), in 2017, approximately 90.5% of pre-clear estuarine areas (excluding open water), 96.1% of pre-clear lacustrine, 78.8% of pre-clear palustrine, and 83.5% of pre-clear riverine wetlands remained across the GBR catchment (Canning and Waltham 2021). Between 2001 and 2017, there was a net loss of 7,688 ha across natural wetlands (i.e., excluding artificial/highly modified wetlands), with riverine wetlands accounting for 6,255 ha, estuarine salt flats and saltmarshes accounting for 605 ha, and coastal and subcoastal tree swamps (*Melaleuca* spp. and *Eucalyptus* spp.) accounting for 569 ha and 537 ha on non-floodplains and floodplains, respectively. Much of the decline is attributed to clearing and drainage for urban and agricultural development. Artificial/highly modified wetlands (including dams, ring tanks, and irrigation channels), created primarily for irrigation storage, accounted for the majority of increased wetland, with an additional 21,546 ha (approximately) in 2017 compared with 2001, representing a 15.2% increase. A substantial proportion of the artificial/highly modified wetlands was created through bunding (constructing a wall to exclude saltwater and retain freshwater; Abbott et al. 2020), accounting for 8,299 ha of the increase (Department of Environment & Science 2019b). These statistics do not include wetlands smaller than 1 ha, as they are not mapped. The change in extent and composition of wetlands across the GBR catchment can be explored by sub-catchment with further background information on the Wetland*Info* website.

Despite extensive modification of coastal wetlands and floodplains in the GBR region (Canning and Waltham 2021), there is growing interest in restoring these ecosystems back to some hybrid state (Waltham et al. 2019). Restoration objectives have included improving water quality (Adame et al. 2019; Wallace et al. 2022) and habitat connectivity (Sheaves et al. 2017; Sheaves and Johnston 2010), controlling/removing invasive species and fisheries and increasing the abundance and diversity of other aquatic species (Baker and Sheaves 2005; Sheaves et al. 2012; Sheaves and Molony

2000), and more recently, abating greenhouse gases via restoration of marine vegetation (Cadier et al. 2022; de Paula Costa et al. 2022; Hagger et al. 2022).

In this chapter, a selection of wetland restoration case studies are summarised to outline the challenges and barriers and the knowledge learned from these efforts. These examples are examined in the context of a values-based approach to restoration, where the focus is on understanding the components and processes of interest so that the most appropriate and targeted restoration mitigation response is implemented.

24.2. VALUES-BASED FRAMEWORK

The increased push for the restoration of coastal ecosystems has been met with some optimism by scientists, although to date, many restoration attempts have been unsuccessful, partly due to a lack of scientific guidance, data collection, guidelines, and technical skills to complete these projects (Paschke et al. 2019; Sheaves et al. 2021b; Weinstein et al. 2019). For instance, removing non-native macrophytes is common regardless of whether the goal of the restoration project is to improve water quality or to increase biodiversity (Zedler 2016). If biodiversity outcomes are expected, then the removal of non-native macrophytes could be important, given that excessive organic loads reduce oxygen availability in the water column by altering the net rate difference between consumption and production—an imbalance that can lead to major fish kills (Perna et al. 2012; Waltham and Fixler 2017). However, macrophyte removal efforts might be conducted differently if the goal is to improve water quality, as denitrification (a desired outcome) will decrease if the macrophytes are removed. Therefore, a 'one size fits all' approach can often lead to pervasive outcomes for restoration when intervention is poorly aligned with project objectives and values. Inappropriate application of restoration strategies can be prevented by examining previous outcomes where they have been applied and using this knowledge to assess suitability at future restoration sites.

Values-based management is adaptive (i.e., informed by 'learning-by-doing') based on the intrinsic and existing values and services provided by a system, and it accounts for the different values, management needs, and priorities of stakeholders and beneficiaries (e.g., water quality, fisheries production, and roosting areas). Values-based management ensures that biophysical, social, cultural, and economic values inform management priorities and are the basis for ongoing monitoring and evaluation of management performance. The Whole-of-System Values-Based Framework (herein the Framework) draws explicit connections among the biophysical environment, the beneficiaries of the services provided by the ecosystem, and their values. The Framework uses a holistic management approach to achieve outcomes that consider the biophysical environment alongside social, economic, and cultural outcomes. For example, coastal sand dunes (components) provide the service of protection from storms (process) for people living on the coast (beneficiaries) who may value the coastline for its aesthetic view or habitat for migratory birds (Schlacher et al. 2008). The Framework has been developed by the Queensland Government through the synthesis of other frameworks and tools and with extensive stakeholder consultation. The Framework is a specified integrated approach to catchment management and the protection, maintenance, and restoration of wetland systems (catchment strategy). The Framework proposes that information on natural and human components of a system can be integrated with an understanding of how catchments function. A whole-of-system approach encompasses both terrestrial and aquatic environments within broader catchment environmental processes. Understanding the parts of a catchment, how they work, and the values and services they provide is critical to determining appropriate and integrated management activities. Threats can then be used to prioritise options for management intervention, based on the objectives associated with a management activity. It is intended that the Framework will be used in day-to-day wetland management decisions that will help deliver more effective and sustainable management of the GBR catchment (DEHP 2016). The Framework provides the foundation for establishing a comprehensive approach to prioritising investment.

In the Reef 2050 Plan (GBRMPA 2015), 'coastal habitat' is recognised as supporting the ecological processes of the Reef by providing habitat for biodiversity, the community, economic benefits, and resilience to climate change. The *Wetlands in the Great Barrier Reef Catchments Management Strategy 2016–2021* outlines that on-ground actions, education, and scientific research are necessary to improve the management and repair of wetlands in the GBR catchment. The Reef 2050 Water Quality Improvement Plan 2017–2022 (Reef WQ Plan) identifies the need for targeted catchment repair projects (State of Queensland 2018). Failure to use an integrated, whole-of-system, values-based approach can lead to poorly designed management interventions that generate biophysically or socially inappropriate outcomes or have low buy-in from beneficiaries and other stakeholders because their values were not considered when the management interventions were developed.

24.3. CASE STUDIES

Genuine efforts to restore coastal and marine ecosystems continue to be a major focus for the community, non-government organisations (NGOs), government, industry, indigenous groups, and landholders (Jiang et al. 2015), though success is often limited because coastal wetland ecological processes are complex and driven by multiple interwoven, dynamic causes and effects (Sheaves et al. 2021a). This complexity has limited the acquisition of relevant knowledge of habitat characteristics, species utilisation, and ecosystem dynamics, thereby rendering each restoration project an experiment (Martin et al. 2021;

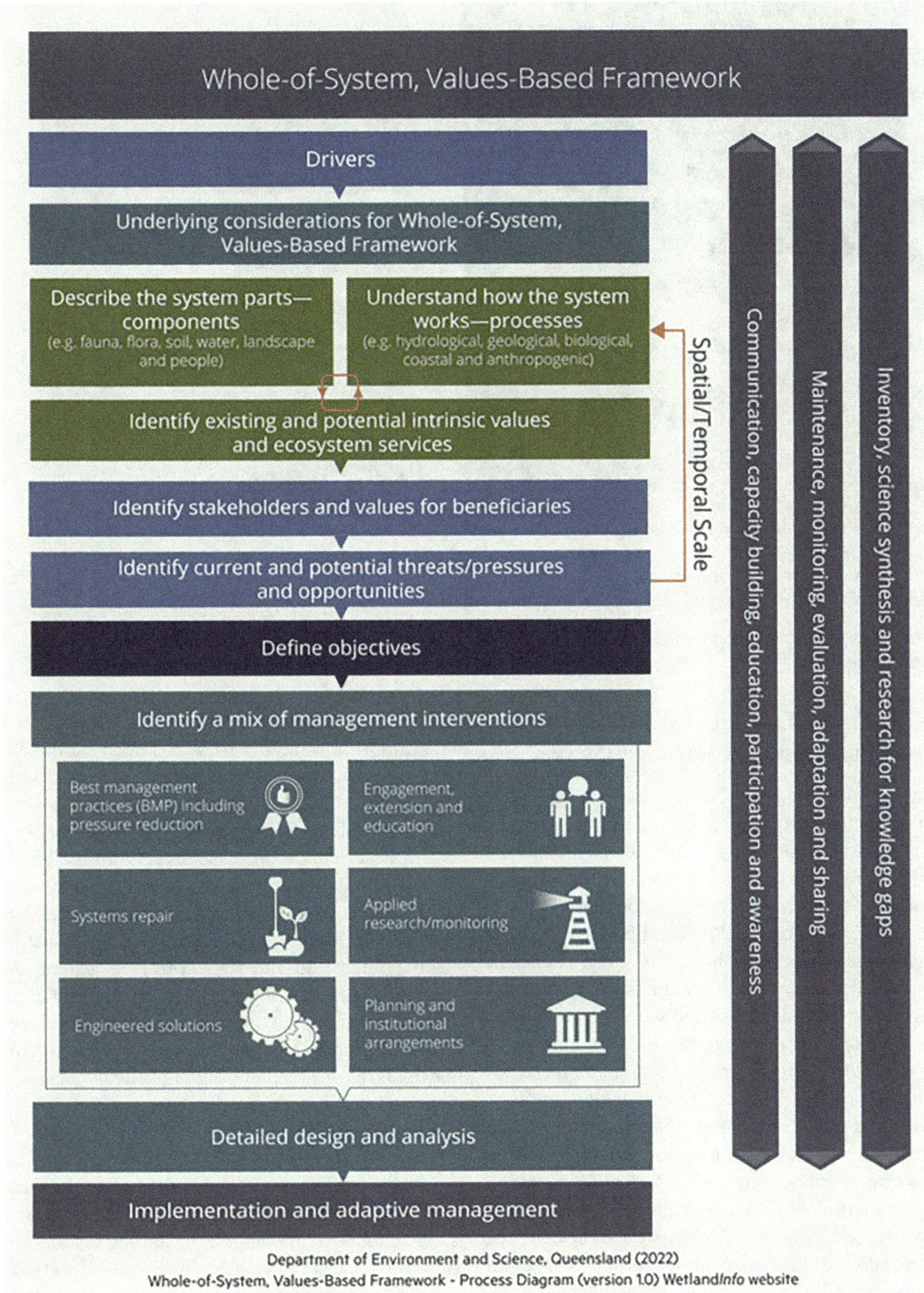

FIGURE 2 The whole-of-system catchment management framework developed for restoration projects in Queensland (DEHP 2016).

Sparks et al. 2017; Waltham et al. 2021a). Three restoration projects are examined to understand the nature of each project, what was learned, and the outcomes in the context of the values-based Framework (Figure 3). This chapter is important and timely given the increasing interest in the restoration of coastal wetland ecosystems driven by national and international recognition of the values and services provided by these habitat resources—for example, the United Nations Decade on Ecosystem Restoration (Waltham et al. 2020b).

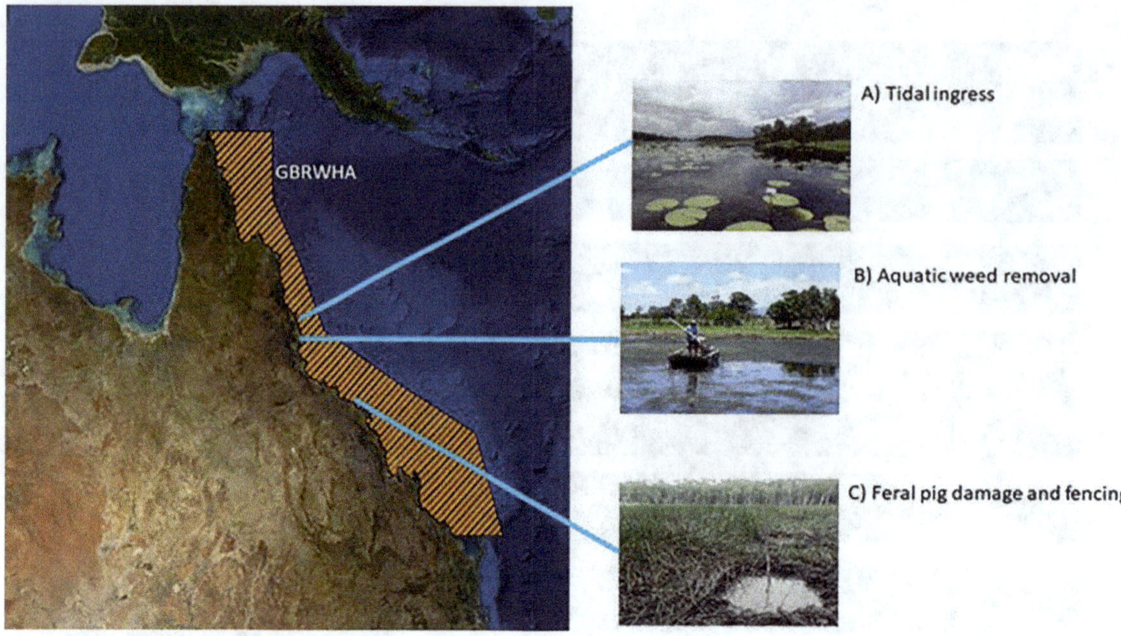

FIGURE 3 Location of three wetland restoration sites: A) Mungalla wetland complex; B) Crooked waterhole, Burdekin floodplain; and C) Eurimbula National Park Round Hill Reserve, with feral pig exclusion to restore wetland values.

Source: Photo credit— Nathan J. Waltham

24.3.1. Case Study: Tidal Bund Wall Removal and Tidal Ingress at Mungalla Station

The removal of tidal bunds to restore coastal habitats has become an attractive proposition for coastal restoration managers in Australia, including in the GBR catchments. However, like many other restoration methods, the efficacy and outcomes of this technique have not been assessed in Australia (Challen and Long 2004). Tidal bund walls were initially constructed in several locations along the GBR coastline to provide stock with ponded pasture as late dry season forage. The exclusion of saline sea water permitted the growth of freshwater macrophytes, namely, Para grass (*Urochloa mutica*), Aleman grass (*Echinochloa polystachya*), and Olive hymenachne (*Hymenachne amplexicaulis*). These macrophytes allow the health and condition of livestock to be maintained throughout the dry season, reducing the cost of supplementary feeding and improving farm resilience to drought (Challen and Long 2004; Hyland 2002). Although this may provide some on-farm benefits, the ponded pastures contribute to a range of environmental processes that depending on context, can be deemed positive or negative. Understanding these processes and the processes of altered systems (e.g., where tidal bunds have been removed to allow seawater ingress) is critical in deciding whether tidal bund removal is a suitable restoration technique for future coastal wetland rehabilitation projects.

Several environmental processes result from ponded pasture construction. The most apparent is that the earth wall becomes a direct barrier to the movement of fish and other aquatic fauna, effectively limiting available habitat areas to complete lifecycles (e.g., mangrove jack). On the central Queensland plains, it has been estimated that more than 16,000 ha of ponded areas occur adjacent to important fish nursery areas (Houston and Duivenvoorden 2002). With excessive aquatic plant growth/high organic load and warm water conditions associated with the bund walls, the stagnant waters can become anoxic (Burrows and Butler 2012). During periods of rainfall, these waters (termed black water) contribute to massive fish kills not just in the pond itself (Butler and Burrows 2007) but extend downstream into estuaries causing more regional problems including fish kills and impact on marine plants (Burrows et al. 2012). Another product of ponded pastures is sediment trapping, particularly of larger sediment particles, in much the same way that the natural wetland vegetation would trap sediment. The provision and loss of these services should be a determinant when deciding whether bund wall removal is a suitable practice for the objective of wetland restoration.

The balance between a degraded and desired wetland state remains a challenge for land managers. Therefore, to maximise the services offered by coastal wetlands restored through bund removal, we examine the introduction of saltwater bores installed at Mungalla, Ingham. Here, we track the impact that pumping (using solar pumping technology) water from a saltwater aquifer has on the immediate receiving freshwater wetland that is persistently in a state of overgrowth by aquatic weeds. Currently, efforts to introduce saltwater from nearby tidal creeks have had limited success in removing freshwater weeds. In this trial, we track the water quality in the bores over time, examine the water quality conditions in the immediate receiving wetland area after salty bore water introduction, and investigate salt accumulation in wetland soils and the ability of salty bore water to destroy freshwater weeds. These small-scale experiments have been designed to understand the potential for saltwater bores to assist with wetland restoration where persistent freshwater weeds maintain this wetland system in a continual overall poor condition. The results here will assist with upscaling operations to maximise restoration return efforts in the complex more broadly.

Mungalla Station (-18.707°, 146.261°) is an 830-ha property located on Palm Creek in the lower Herbert Valley east of Ingham in north Queensland (Figure 2). The Mungalla property is centrally located within the ancestral lands of the Nywaigi people (Grice et al. 2012). The Nywaigi people have been heavily impacted by urban and agricultural development and suffered disconnection from their lands for close to 100 years. In 1999, Mungalla was reclaimed through purchase on behalf of the Nywaigi Aboriginal Land Corporation (NALC), providing an opportunity for the Nywaigi people to re-establish links with their coastal country.

The Mungulla site is mapped as a High Ecological System (HES) with very high conservation values (Wetland*Info*, accessed October 2022). The property includes a large area (~160 ha) of coastal wetland that historically (pre-European settlement) was estuarine towards the coast, merging into palustrine wetlands further inland. These estuarine wetlands would have previously supported *Melaleuca* and mangrove forests on the seaward edge, with saline to hyper-saline vegetated saltmarsh landward. Further inland, saltmarsh would have merged into intertidal grass-sedge wetlands dominated by Bulkuru (*Eleocharis dulcis*) with fringing *Melaleuca* freshwater forest (Blackman et al. 1999). Bunding of the Mungalla grass-sedge wetlands, landward of the saltmarsh, was completed in the mid-1940s as a trackway to provide access across the wetlands. Shortly after, Para grass (*Urochloa mutica*) was introduced into the freshwater ponded area on the landward side of the bund wall. More recently, two other ponded pasture species were introduced—Olive hymenachne (*Hymenachne amplexicaulis*) and Aleman grass (*Echinochloa polystachya*). These two species were released into Queensland in 1988 despite warnings about possible weedy invasion by the Queensland Environmental Protection Agency (Csurhes and Edwards 1998). Subsequently, the two grasses have been locally recognised as weeds, with Olive hymenachne declared a weed of national significance (WONS) only 10 years after its release (Wearne et al. 2010). Two other nationally declared weeds are present in the wetlands, specifically, Salvinia (*Salvinia molesta*) and Water hyacinth (*Eichhornia crassipes*), both of which are spread by spore/seed or fragments that travel downstream.

In 2008, the Commonwealth Scientific Industry Research Organisation (CSIRO), in conjunction with the Nywaigi Traditional Owners, wetland scientists, and managers (James Cook University and Queensland local/state agencies), developed a wetlands management strategy, including the urgent need for restoration of cultural and environmental values within the Mungalla wetlands complex. In 2012 and 2013, CSIRO, assisted by hydrodynamic modelling (Karim et al. 2021), removed a section of the earthen bund wall to allow saltwater intrusion (Figure 4). This dramatically changed the vegetation community from a freshwater-dominated invasive weed community to an open water

FIGURE 4 Mungalla wetland complex: A) Excavator breaching the earth wall in August 2013; B) removing the earth wall to permit tidal ingress; C) the aquatic plant community upstream of the earth wall (approximately July 2013) with extensive invasive freshwater weeds; and D) same photo location after the removal of the earth wall section and re-establishment of native aquatic vegetation (Bulkuru, *Eleocharis dulcis*).

Source: Photo credits M. Nicholas and Nathan J. Waltham

system, with fewer hypoxic events and altered fish assemblage. In the time following bund wall removal, the upstream wetland has oscillated between aquatic weed-free and covered, an apparent cyclical process that is influenced by catchment flow (Karim et al. 2021). Despite the early destruction of aquatic weeds with tidal ingress, the wetland now functions mostly as a freshwater-dominated system, with weeds and limited habitat values, which means that the initial success has been overwhelmed by the wetland returning to a weed-impacted system again (Abbott et al. 2020).

24.3.2. Case Study: Mechanical Weed Removal in Irrigation Channels and Lagoons on the Burdekin Floodplain

The Burdekin catchment consists of Australia's largest sugar cane production region (Davis et al. 2014). Concomitant with the ongoing expansion of sugar cane since the 1980s, the Queensland Government regional ecosystem mapping estimates that 77% of pre-European freshwater wetlands have been lost within the Burdekin, with much of the remaining 23% in a fragmented state (GBRMPA 2015). Remaining wetlands also face altered hydrology, poor water quality (O'Brien et al. 2016), extensive invasive aquatic vegetation (macrophytes; (Waltham and Fixler 2017), and engineered barriers impeding fish migration (Burrows et al. 2012; Perna and Burrows 2005). The Queensland Government and the Commonwealth of Australia both recognise the need to reverse the decline in water quality reaching the GBR (GBRMPA 2015). As part of this response, funding for the 'Delivering biodiversity dividends to the Barratta Creek, Burdekin Catchment' project was created in 2013 to restore ecological function in the creek/floodplain and the Bowling Green Bay Ramsar wetland. Although the broader project covered riparian tree planting, fire management, feral animal management, and extensive training in land management for farmers, the study also examined water quality and the fish community occupying the floodplain's Lochinvar wetland (Waltham and Fixler 2017). In this study, the wetland was treated with herbicides to control invasive aquatic plants (particularly species listed under Australian legislation as WONS)—most notably the water hyacinth (*Eichhornia crassipes*), one of the world's most prevalent invasive aquatic plants (Montiel-Martínez et al. 2015; Toft et al. 2003; Villamagna and Murphy 2010). Water hyacinth occurs in floodplain wetlands across northern Queensland and is particularly invasive in the Burdekin due to permanent water and high nutrients (Waltham et al. 2020a). Clearing this declared WONS has become an obligation for local government agencies and natural resource management groups (Burrows et al. 2012), particularly in the lower Burdekin floodplain. Here, hyacinth growth is fuelled by an extensive water distribution network, with over 1,500 km of channels (in linear extent) delivering irrigation water to sugar cane farms. Tailwaters of the Burdekin (rich in nutrients, sediments, and herbicides) then flow via coastal wetlands to Bowling Green Bay (Davis et al. 2014).

Most of the floodplain has been extensively modified for agricultural expansion and development, including an extensive water irrigation network that channels water from the dam in the upper catchment to farmers across the floodplain. This irrigation network supplements the water supply during the dry months of the year (Davis et al. 2014). Excess water not used on farms is generally allowed to drain back into the irrigation network, either directly or via recycled water pits that do reduce some returned flow to the network. Consequently, the channels and lagoons on the floodplain that were previously ephemeral now permanently hold water that contains high amounts of sediment, nutrients, and herbicides (Davis et al. 2014; O'Brien et al. 2016). With excessive nutrients available, the slow-flowing water has become a haven for extreme aquatic weed growth (Waltham et al. 2020a), particularly floating species that can overwhelm the entire water body. Such overgrowth can reduce water quality, particularly dissolved oxygen concentrations, and in some instances, cause large-scale fish kills (Burrows et al. 2009a; Burrows et al. 2009b; Davis et al. 2014; Waltham et al. 2020a). This excessive aquatic weed growth is a major challenge to the floodplains' ability to sustain past ecosystem services, such as fisheries production. As a result, most of the floodplain's waterways now require continual maintenance, either through aerial spraying of targeted herbicide or direct removal using machinery (Perna and Burrows 2005; Waltham and Fixler 2017).

Crooked Waterhole (Figure 5), a heavily weed-choked creek (common on the Burdekin floodplain), has had a legacy of excessive aquatic weed growth that has contributed to poor water quality and fishery production values (Burrows and Perna 2005). To improve the waterway, a mechanical weed harvester was used in November 2016 (late dry season) to clear around 5 ha of aquatic weeds from an approximate 10-km stretch (linear extent) of the creek. Prior to weed removal, the creek was persistently anoxic, with only one native fish species and two invasive fish species present. Following weed removal, there was a return of diurnal dissolved oxygen fluctuations and 15 native fish species (Waltham et al. 2020c). Despite this initial success, more recent visual spot inspections found substantial weed re-growth. Ongoing weed control to ensure enduring health is a critical maintenance exercise but requires a stable and perpetual funding mechanism (Burrows et al. 2009a). The aquatic weed removal program at the nearby Sheep Station Creek (also on the Burdekin floodplain) has been operating for two decades, funded through an environmental levy paid by farmers, the Burdekin Shire Council, the North Queensland Dry Tropics Natural Resource Management group, and the Lower Burdekin Water Board, totalling approximately AUD $20,000 annually (Waltham et al. 2020a). The funding is made available each year to undertake aquatic weed treatment, ensuring the irrigation network remains open and free-flowing for water movement and fish habitat

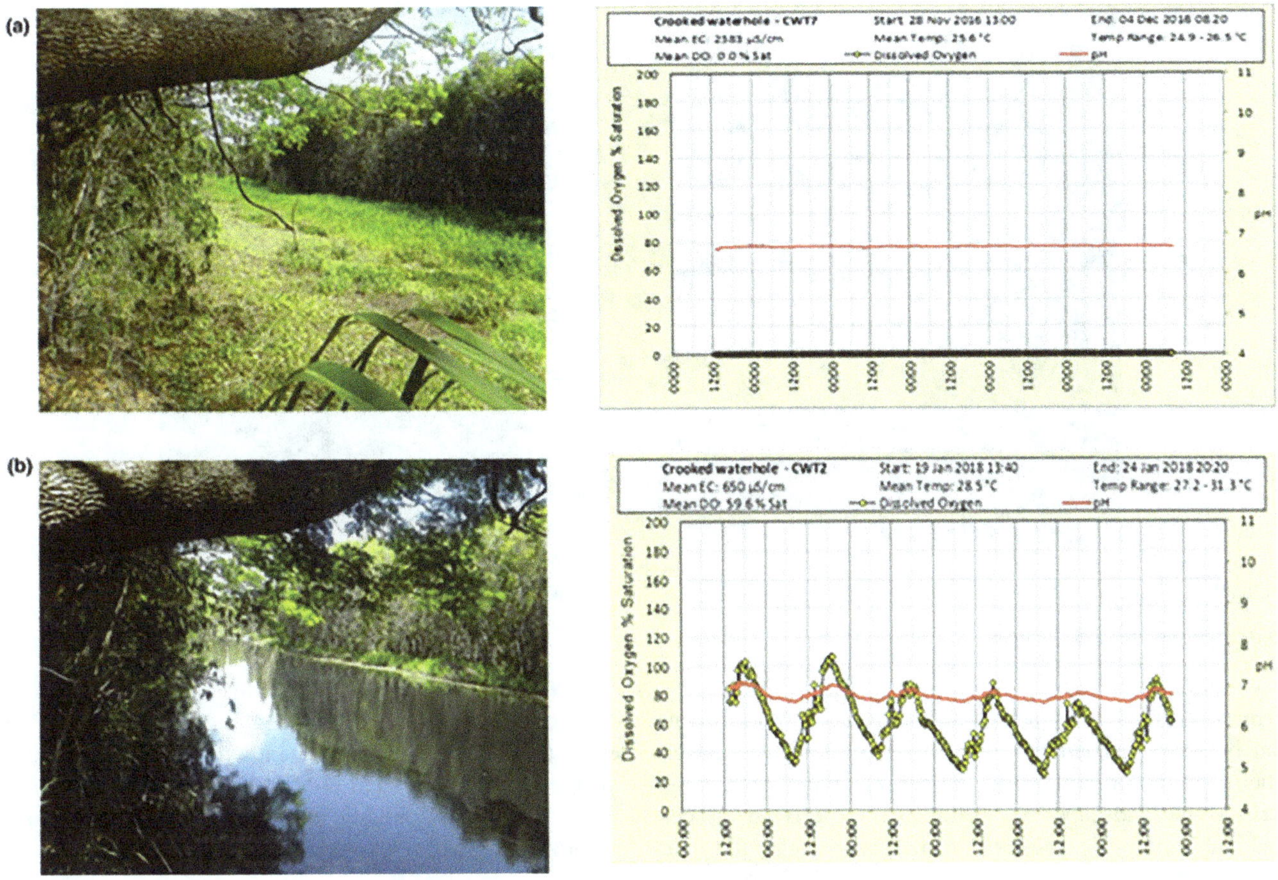

FIGURE 5 Photograph of Crooked Waterhole (A) before major aquatic plant removal and corresponding Hydrolab time-series plot; and (B) Photograph at the same location in January 2018 during the wet season and with ongoing follow-up maintenance works and corresponding Hydrolab time-series plot. Plots show summary statistics during each logging period (about 1 week in duration) for dissolved oxygen and pH.

Source: Photo credit Nathan J. Waltham

provision, though the fish community is not monitored (Veitch et al. 2008).

24.3.3. Case Study: Feral Pig Exclusion Fencing for Wetland Protection and Restoration

Across northern Australia, feral pigs (*Sus scrofa*) have a wide-scale negative impact on wetland vegetation assemblages, water quality, biological communities, and wider ecological processes (Fordham et al. 2008; Krull et al. 2013). Feral pigs have an omnivorous diet supported by foraging or digging plant roots, bulbs, and other belowground vegetation throughout terrestrial and wetland areas (Barrios-Garcia and Ballari 2012). This feeding strategy negatively impacts wetland aquatic vegetation, giving rise to soil erosion, benthic sediment resuspension, reduced water clarity, and eutrophication, which becomes particularly critical in the late dry season. Only a few studies have quantified the negative impacts feral pigs have on coastal wetlands (Doupe et al. 2010; Mitchell and Mayer 1997; Steward et al. 2018; Waltham and Schaffer 2018), limiting the ability of land managers to assess the consequences of feral pig destruction (Fordham et al. 2008) or indeed, other large invasive species (Ens et al. 2016). Strategies focused on reducing or removing feral pigs from the landscape have been employed since their introduction to Australia (Fordham et al. 2006), including poison baiting, aerial shooting, and trapping using specially constructed mesh cages (Ross et al. 2017). Attempts to exclude feral pigs for conservation outcomes have included building exclusion fencing that directly limits feral pig access to essential resources (Figure 6) (Nordberg et al. 2019). The installation of fences around wetlands has only been recently examined in Australia (Doupe et al. 2009; Waltham and Schaffer 2018). The evidence suggests that fences may be less effective in protecting freshwater refugia from pig damage, particularly in ephemeral wetlands that will dry out before the wet season rain reconnects them with primary rivers. The exception to this is where wetlands are more permanent in the landscape (Balcombe et al. 2007), either supported by groundwater springs or that are large enough to survive the dry season. Fencing these more permanent wetlands to protect their ecosystem services and values is therefore probably more critical. In addition, fences may prevent non-target

FIGURE 6 Fencing installed to prevent feral pigs accessing the Eurimbula National Park Round Hill Reserve.

Source: Photo credit S. Jackson

terrestrial fauna from accessing wetlands, a requirement that becomes particularly relevant in the late dry season when wetlands function as regional water points for animals in the landscape. For example, small terrestrial species including birds, snakes, and lizards are generally able to access and subsequently escape fenced wetlands (Ross et al. 2017); however, access or escape from drying wetlands by other species, including macropods or freshwater turtles, can be hindered by fencing (Waltham and Schaffer 2016).

In November 2016, the Burnett Mary Regional Group (BMRG) constructed an exclusion fence surrounding the Round Hill Reserve, a coastal wetland in the upper reaches of Round Hill Creek, located in the Baffle catchment (Figure 6). The primary objective of this restora tion activity was to control feral pig damage directly to the wetland (Figure 6d). Round Hill Creek was identified as a restoration priority because it is an important Fish Habitat Area (FHA) and a Directory of Important Wetlands in Australia (DIWA)-listed wetland (Queensland Government 2018a, b).

Disturbance in Round Hill Reserve is largely attributed to feral pigs and livestock access. Between July 2017 and November 2019 (proximal to the post-wet season and late dry season), surveys of the native vegetation cover and extent were completed using drone aerial analysis and quadrat surveys. Fish communities were examined using fyke nets set to soak overnight, with water sampling also completed during each survey. In addition to the fenced wetland area (Eurimbula National Park (NP), Round Hill Reserve), four additional wetland sites were included, consisting of two sites approximately 60 km to the south and another site located in the Eurimbula NP approximately 20 km to the north, each open to cattle and feral pig access. Although the Eurimbula NP, Round Hill Reserve is the only wetland site here with a conservation fence in place, as part of land tenure arrangements with adjacent landholders, the fence is opened periodically to allow cattle (and presumably feral pigs) to enter the wetland to graze without any apparent monitoring of cattle numbers entering the wetland.

Summary of the results

- Of the five wetlands surveyed, the fenced wetland in Eurimbula NP supported the most diverse floral community, mostly due to the presence of freshwater species that were absent elsewhere. The fenced wetland site is positioned high in the estuary and is therefore the least frequently inundated by tides, and it is more regularly inundated by freshwater flows following rainfall events. The conservation fence at the Eurimbula NP, however, was ineffective at protecting fragile marsh vegetation from trampling. Therefore, the management practice of opening the fence to allow cattle (and presumably feral pigs) to enter the wetland to graze must cease to ensure that the mitigative effects of fencing are effective.
- Water quality measurements for physicochemical parameters (temperature, conductivity, dissolved oxygen, and pH), nutrients, and suspended sediments showed varied results among wetlands and over time. The data are limited because of the nature of the study; however, there was some evidence of seasonal cycling and elevated nutrient concentrations in the post-wet season surveys (but this cannot be confirmed). Dissolved oxygen was at times critically low, below asphyxiation thresholds known for some of the fish species captured, which might also explain the low number of fish

species. Most of the nitrogen present was in the filterable fraction, while TN:TP and TDN:TDP ratios indicated that nitrogen was generally in excess.

- A total of 12 fish species were detected across the wetland sites. The small-bodied fish Gobiidae (true goby) and *Pseudomugil signifier* (Pacific blue-eye) were the most abundant species encountered. Two exotic species, *Oreochromis mossambicus* (Mozambique tilapia) and *Gambusia affinis* (Eastern gambusia) were present but not across all wetland sites. *Macrobrachium rosenbergii* (giant river prawn) was present at only a single wetland site, as were *Metapenaeus bennettae* (greasy-back prawn) and *Metapenaeus macleayi* (school prawn), which are both commercially and recreationally targeted species in Queensland. Species richness was highest at the fenced wetland with 15 species present. Overall, the number and abundance of fish species captured in these wetlands were low when compared to the number and abundance of species known to occur in the region (Sheaves and Johnston 2008). This result is likely due to the limited connectivity between these wetlands and downstream estuaries, with most sites drying out almost completely during the late dry season. These wetland areas may therefore provide limited fish habitat value.
- Twelve herpetofauna species were present across the five wetland sites (five skink, one gecko, five frog, and one snake species). This number is considered low and likely reflects the short survey period and seasonality of herpetofauna, with surveys conducted during cooler winter months. Surveys conducted in warmer months would likely result in higher recorded species richness and abundance. The highest diversity and abundance of herpetofauna were from Round Hill site 3 (RH3), which, paradoxically, is one of the most disturbed sites (with cattle and pig access). The introduced cane toad (*Rhinella marina*) was also detected at three of the five sites.
- Higher waterbird species richness was detected during the post-wet season surveys (March 2018) relative to the dry season surveys (July 2018). Water availability and depth are key factors determining the use of wetlands by waterbirds, and a decrease in water availability during the dry season may explain the reduced abundance recorded in July 2018.

Key conclusions and recommendations

- Restoration projects targeting the threat of feral pig (and cattle) damage to coastal wetlands are important and will continue to be necessary in the GBR catchments in the future. The results presented here highlight that although fencing for wetland vegetation community values has indeed achieved more when compared to wetlands exposed to feral pig damage, in the years following this study, it became apparent that the protection of wetland values was limited by continual access to the fenced wetland by cattle.
- Maintenance of the fence has been also challenging. Routine inspection of fences and the undertaking of general maintenance as required will preserve the investment and ensure that the supporting infrastructure is in working order.
- Finally, conservation fencing for wetland values protection is important and should be implemented in broader GBR catchments where the impact of feral pig damage is persistent and extensive. Aerial shooting and baited traps are still tools used to manage feral pig populations in northern Australia. However, fencing, if designed and constructed to a high standard, could result in a functioning and productive coastal wetland system at a much-reduced cost (Ross et al. 2017). Importantly, a clear understanding and determination of values need to be set to avoid situations where broader expectations are not met (e.g., the movement of terrestrial wildlife across wetland regions). Clearly, allowing cattle to enter wetlands fenced for feral pig control is counterproductive, will continue to impact the broader value and services of these wetlands, and should cease. Under a market-based scheme (e.g., blue carbon or water quality markets) these coastal wetlands could become more valuable as part of climate change mitigation and could potentially earn more generated income relative to using these ecosystems as late dry season cattle feed areas (Waltham et al. 2023).

24.4. LESSONS LEARNED AND FUTURE OPPORTUNITIES

Wetlands and their ecosystem setting collectively contribute to the productivity and function of the coastal zone, providing many services that are valued by humans. It is important to acknowledge that not all wetlands provide the same values or services, for example, one wetland may be primarily valued for its natural features, while others may be considered more important for their cultural value. Restoring wetland services first requires defining lcoal stakeholder values. Once the values are set and agreed to by relevant stakeholders, available management strategies can be developed and implemented. In planning restoration projects, it is important to recognise that stakeholders (beneficiaries) have different and sometimes conflicting views or priorities when valuing coastal wetland ecosystem services (Harris and Heathwaite 2012). For example, placing a high value on services such as freshwater extraction for agriculture can directly undermine cultural ecosystem service values related to aquatic biodiversity (Boulton et al. 2016) and reduce the duration and frequency of water connection across floodplains, which also has biological consequences (Rayner et al. 2009; Sheaves et al. 2024). Ecosystem restoration efforts seem to be most effective when the values of all stakeholders are incorporated, a process best facilitated through discussions that set objectives early (Guerrero et al. 2017; Sheaves et al. 2021b; Zedler 2016). The scale of the project and its desired outcomes are also important, for example, local-scale improvement of fish habitat versus catchment-scale amelioration of agricultural fertiliser loads exported to coastal waters. Focusing on an appropriate scale not only is important for informing technical aspects of the restoration management activities but also ensures that appropriate management bodies are involved (Sheaves et al. 2021b).

The whole-of-system catchment management framework developed by the Queensland Government has already been adopted in GBR catchments. Understanding the restoration project site, including the broader catchment area, how it works, and the values and services it provides, are critical to determining appropriate and integrated management activities. A whole-of-system approach encompasses both terrestrial and aquatic environments within broader catchment processes. When assigning management decisions, decision-makers must take into account the full range of ecological, economic, and social values that a wetland provides. Historically, wetland management decisions have favoured either conversion or management for a single ecosystem service, such as water supply or fish nursery habitats. As wetlands come under increasing pressure and as we further understand the full range of services they provide, the best restoration and management options will address a broader array of services that aligns with the 'wise use principles' of the Ramsar Convention (DEHP 2016).

Funding wetland restoration continues to be a challenge, particularly at the scale needed for meaningful and lasting positive effect (Bayraktarov et al. 2016; Saunders et al. 2022). Government investment into on-ground action will always be important and necessary, but there is growing interest for funding through private corporations. In Australia, the Clean Energy Regulator (Australian Government) prepared a Blue Carbon method to encourage industry investment schemes to fund coastal wetland research into systems such as mangroves and tidal marshes for their greenhouse gas (GHG) mitigation services (Clean Energy Regulator 2021). The only approved method at this time is tidal re-introduction, with Australian Carbon Credit Units awarded for GHG abatement with coastal wetland restoration. Other methods to restore coastal wetland carbon stocks could be developed in the future, including a method for managing ungulate damage to coastal wetlands, if appropriate evidence was available. Total soil organic carbon stocks in the coastal wetlands of the GBR catchment (including mangroves, seagrass, and tidal marshes) represent approximately 9–13% of the Australian total stock (de Paula Costa et al. 2021), indicating potentially large carbon stocks that may be sensitive to disturbance if not managed effectively (e.g., fencing to prevent damage from non-native species). However, there are barriers to the success of blue carbon projects (Stewart-Sinclair et al. 2020). For example, projects need to be cognisant of many things including, catchment hydrology (Abbott et al. 2020), future climate scenarios (which might include substantial immersion of coastal habitats in the next 100 years) (Karim et al. 2021), laws concerning reflooding, how carbon rights are granted and managed, and the liabilities for potential impacts on adjacent land and biota (Bell-James and Lovelock 2019). In addition to carbon, restoration could also deliver improved biodiversity outcomes, or water quality, which unlocks multiple (bundled) investment opportunities for corporate enterprises (Waltham et al. 2021b). Despite these opportunities to fund wetland restoration projects for climate mitigation outcomes, the values-based approach still presents complications, e.g., bund wall removal for carbon outcomes and freshwater weed control could lead to a loss of freshwater habitats (Abbott et al. 2020).

Land-use change has altered floodplain hydrology, limnology, and ground/surface water interface, meaning that floodplains will persist in a modified, hybrid state, providing only a subset of past ecosystem services. For as long as sugar production occurs, floodplains will deliver excessive nutrients to local creeks (Brodie et al. 2016; Waterhouse et al. 2017; Waterhouse et al. 2024), and aquatic invasive plants will continue to require management intervention. To retain the advantages gained by wetland restoration and to avoid returning to a degraded condition, ongoing weed maintenance is required. Without a plan of continual funding to maintain a site after weed removal, restoration sites in the GBR catchment have returned to pre-restoration conditions. A funding model like the Sheep Station Creek maintenance program, that has continued for almost 20 years, may be the best example of the benefits of stable and long-term, funded maintenance (Waltham et al. 2020a). Establishing a similar partnership arrangement is necessary for all floodplains along the GBR coastline.

Fencing wetlands can keep out unwanted threats and prevent damage to coastal wetlands. The evidence presented here illustrates that fencing can provide a strategy forward to control these threats. However, there are still values and services that might be lost in doing so, which require stakeholder consideration during the planning phase. Moreover, adaptations that allow non-target species to move through fences, while keeping out target species, need to be developed. For example, trials to install earthen ramps on either side of exclusion fences have allowed macropods to pass over fences, and removing wire pieces in fences can allow freshwater turtles to pass through (Figure 7). Maintenance of fences is also necessary, particularly after storms where fallen tree branches can damage them or the build-up of debris can pose a fire burn point, compromising their integrity.

FIGURE 7 (A) Rain and flooding cause accumulation of timber and organic material along fence lines. This becomes a major problem for fences during a fire when the timber burns hot causing damage to fences and requiring expensive maintenance work; (B) Fences around wetlands also prevent animal movement, such as kangaroos, from accessing resources in the landscape. In Round Hill Reserve, Queensland Parks and Wildlife Services constructed an earth ramp to trial whether kangaroos would use it to move across fence boundaries by jumping over it, while at the same time preventing access by feral pigs and cattle (photo S Jackson, Queensland Government); and (C) Fences around wetlands will prevent feral pig access; however, for freshwater turtles that move around the landscape, fences pose an obvious barrier. A simple conservation measure that allows turtles to move across floodplains is to remove a small piece of vertical wire, shown here, allowing prospecting turtles to pass through.

Source: Photo credit E. Nordberg; Waltham et al. (2022)

A funding model for coastal wetland restoration in GBR catchments and equally, their long-term maintenance are challenges for practitioners. Government funding for works and maintenance is important and necessary in the future, but achieving the scale needed for positive impact requires investment from private partnerships (i.e., affected stakeholders). These funding pipelines are emerging but will require careful integration and coordination to ensure projects are successful and contribute positively to broader values and services (not simply focused on addressing the threat; Waltham et al. (2019)).

Coordination of wetland restoration efforts, knowledge, and information are shared via the GBR wetlands network, a group with members from Natural Resource Management (NRM), the government, NGOs, universities, and industry. During regular network events, information is presented and shared among members, with a central repository of information, reports, and a database of restoration projects held on the Queensland Government's Wetland*Info* website. Users can also access wetland and coastal spatial databases on the site, which are presented in an interactive setting, to learn more about coastal wetlands in Queensland. New policy updates, technical guidelines, fact sheets, conceptual models, and the latest scientific publications are also available to facilitate information sharing and to build local capacity.

24.5. CONCLUSION

The natural beauty of the GBR's ecosystems is well-known and requires a coordinated and long-term approach to its management and protection. This challenge is exacerbated by the impacts of climate change, cyclones, and frequent, extreme water temperatures that have contributed to several widescale coral bleaching events over the past decade. Efforts to restore and protect the floodplains and coastal wetland ecosystems of the GBR have become even more critical given the multiple ecosystem services they provide to the region. However, although efforts have been made to restore these ecosystems, success is dependent on the many important factors outlined here. These opportunities will continue to be a challenge for managers, NRM, NGOs, and industry, as interest in restoration extends beyond many federal and state policy and strategy initiatives. Funding for these projects and their maintenance long term will likely continue, mostly via market mechanisms and interest from private partner investors. Government still has an important role in extending restoration efforts, but achieving large-scale efforts needs considerably more investment, along with engagement with stakeholders. Such levels of investment require coordination and capacity building, which is underway but requires further focused consideration and resources.

REFERENCES

Abbott BN, Wallace J, Nicholas DM, Karim F, Waltham NJ (2020) Bund removal to re-establish tidal flow, remove aquatic weeds and restore coastal wetland services—North Queensland, Australia. PLoS One 15:e0217531.

Adame MF, Arthington AH, Waltham N, Hasan S, Selles A, Ronan M (2019) Managing threats and restoring wetlands within catchments of the Great Barrier Reef, Australia. Aquatic Conservation: Marine and Freshwater Ecosystems 29:829–839.

Airoldi L, Beck MW, Firth LB, Bugnot AB, Steinberg PD, Dafforn KA (2021) Emerging solutions to return nature to the urban ocean. Annual Review of Marine Science 13:445–477.

Alongi D, Murdiyarso D, Fourqurean J, Kauffman J, Hutahaean A, Crooks S, Lovelock C, Howard J, Herr D, Fortes M (2016) Indonesia's blue carbon: A globally significant and vulnerable sink for seagrass and mangrove carbon. Wetlands Ecology and Management 24:3–13.

Arthington AH, Godfrey PC, Pearson RG, Karim F, Wallace J (2015) Biodiversity values of remnant freshwater floodplain lagoons in agricultural catchments: Evidence for fish of the Wet Tropics bioregion, northern Australia. Aquatic Conservation: Marine and Freshwater Ecosystems 25:336–352.

Baker R, Sheaves M (2005) Redefining the piscivore assemblage of shallow estuarine nursery habitats. Marine Ecology Progress Series 291:197–213.

Balcombe SR, Bunn S, Arthington A, Fawcett J, McKenzie-Smith FJ, Wright A (2007) Fish larvae, growth and biomass relationships in an Australian arid zone river: Links between floodplains and waterholes. Freshwater Biology 52:2385–2398.

Barbier EB, Hacker SD, Kennedy C, Koch EW, Stier AC, Silliman BR (2010) The value of estuarine and coastal ecosystem services. Ecological Monographs 81:169–193.

Barrios-Garcia MN, Ballari SA (2012) Impact of wild boar (Sus scrofa) in its introduced and native range: A review. Biological Invasions 14:2283–2300.

Bayraktarov E, Saunders MI, Abdullah S, Mills M, Beher J, Possingham HP, Mumby PJ, Lovelock CE (2016) The cost and feasibility of marine coastal restoration. Ecological Applications 26:1055–1074.

Beck MW, Heck KL, Able KW, Childers DL, Eggleston DB, Gillanders BM, Halpern B, Hays CG, Hoshino K, Minello TJ, Orth RJ, Sheridan PF, Weinstein MR (2001) The identification, conservation, and management of estuarine and marine nurseries for fish and invertebrates. Bioscience 51:633–641.

Bell-James J, Lovelock CE (2019) Tidal boundaries and climate change mitigation–the curious case of ponded pastures. Australian Property Law Journal 27:114–133.

Blackman JG, Perry TW, Ford GI, Craven SA, Gardiner SJ, De Lai RJ (1999) Characteristics of important wetlands in Queensland. Environment Protection Agency, Queensland Government.

Boulton AJ, Ekebom J, Gislason GM (2016) Integrating ecosystem services into conservation strategies for freshwater and marine habitats: A review. Aquatic Conservation: Marine and Freshwater Ecosystems 26:963–985.

Bradley M, Sheaves M, Waltham NJ (2023) Urban-industrial seascapes can be abundant and dynamic fish habitat. Frontiers in Marine Science 9:1034039.

Brodie J, Waterhouse J (2012) A critical review of environmental management of the 'not so Great' Barrier Reef. Estuarine, Coastal and Shelf Science 104–105:1–22.

Brodie JE, Lewis SE, Collier CJ, Wooldridge S, Bainbridge ZT, Waterhouse J, Rasheed MA, Honchin C, Holmes G, Fabricius K (2016) Setting ecologically relevant targets for river pollutant loads to meet marine water quality requirements for the Great Barrier Reef, Australia: A preliminary methodology and analysis. Ocean & Coastal Management 143:136–147.

Browne MA, Chapman M (2014) Mitigating against the loss of species by adding artificial intertidal pools to existing seawalls. Marine Ecology Progress Series 497:119–129.

Bugnot A, Mayer-Pinto M, Airoldi L, Heery E, Johnston E, Critchley L, Strain E, Morris R, Loke L, Bishop M (2021) Current and projected global extent of marine built structures. Nature Sustainability 4:33–41.

Bulleri F, Chapman M, Underwood A (2005) Intertidal assemblages on seawalls and vertical rocky shores in Sydney Harbour, Australia. Austral Ecology 30:655–667.

Burrows D, Davis A, Knott M (2009a) Survey of the freshwater fishes of the Beylando-Suttor System, Burdekin Catchment, Queensland. Australian Centre for Tropical Freshwater Research, James Cook University, Townsville.

Burrows D, Perna C (2005) Identification of fish habitat and wetland management issues in the Haughton River Catchment, north Queensland. Australian Centre for Tropical Freshwater Research. James Cook University, Townsville.

Burrows D, Perna C, Davis A (2009b) An assessment of fish passage and connectivity to Gladys Lagoon, Burdekin Floodplain, North Queensland. Australian Centre for Tropical Freshwater Research, James Cook University, Townsville.

Burrows DW, Butler B (2012) Preliminary studies of temperature regimes and temperature tolerance of aquatic fauna in freshwater habitats of northern Australia. Australian Centre for Tropical Freshwater Research, James Cook University, Townsville.

Burrows DW, Sheaves M, Johnston R, Dowe JL, Schaffer J (2012) Impact of excess freshwater flow on the lower Barratta Creek and estuary. TropWATER—Centre for tropical water and aquatic ecosystem research, James Cook University, Townsville.

Butler B, Burrows DW (2007) Dissolved oxygen guidelines for freshwater habitats of northern Australia. Australian Centre for Tropical Freshwater Research, James Cook University, Townsville.

Cadier C, Waltham NJ, Canning A, Fry S, Adame MF (2022) Tidal restoration to reduce greenhouse gas emissions from freshwater impounded coastal wetlands. Restoration Ecology:e13829.

Canning AD, Waltham NJ (2021) Ecological impact assessment of climate change and habitat loss on wetland vertebrate assemblages of the Great Barrier Reef catchment and the influence of survey bias. Ecology and Evolution 11:5244–5254.

Challen S, Long P (2004) Fisheries guidelines for managing ponded pastures, Fish Habitat Guidelines. Department of Primary Industries, Brisbane, 27pp.

Chapman MG (2003) Paucity of mobile species on constructed seawalls: Effects of urbanization on biodiversity. Marine Ecology Progress Series 264:21–29.

Clean Energy Regulator (2021) Emissions Reduction Fund: Method development, www.cleanenergyregulator.gov.au/ERF/Pages/Method-development.aspx. Accessed May 5, 2021.

Creighton C, Hobday AJ, Lockwood M, Pecl GT (2016) Adapting management of marine environments to a changing climate: A checklist to guide reform and assess progress. Ecosystems 19:187–219.

Csurhes S and Edwards RH (1998) Potential environmental weeds in Australia. Biodiversity Group, Environment Australia.

Davis AM, Lewis SE, O'Brien DS, Bainbridge ZT, Bentley C, Mueller JF, Brodie JE (2014) Water resource development and high value coastal wetlands on the lower Burdekin floodplain, Australia. In Wolanski E (Ed.), Estuaries of Australia in 2050 and beyond. Springer, New York, pp. 223–245.

de Paula Costa MD, Lovelock CE, Waltham NJ, Moritsch MM, Butler D, Power T, Thomas E, Macreadie PI (2022) Modelling blue carbon farming opportunities at different spatial scales. Journal of Environmental Management 301:113813.

de Paula Costa MD, Lovelock CE, Waltham NJ, Young M, Adame MF, Bryant CV, Butler D, Green D, Rasheed MA, Salinas C, York P, Whitt AA, Macreadie PI (2021) Current and future carbon stocks in coastal wetlands within the Great Barrier Reef catchments. Global Change Biology 27:3257–3271.

DEHP (2016) Wetlands in the Great Barrier Reef Catchment Management Strategy 2016–2021. Department of Environment Heritage and Protection, Brisbane.

Doupe RG, Mitchell J, Knott MJ, Davis AM, Lymbery AJ (2010) Efficacy of exclusion fencing to protect ephemeral floodplain lagoon habitats from feral pigs (Sus scrofa). Wetlands Ecology Management 18:69–78.

Doupe RG, Schaffer J, Knott MJ, Dicky PW (2009) A description of freshwater turtle habitat destruction by feral pigs in tropical north eastern Australia. Herpetological Conservation and Biology 4:331–339.

Elliott M, Whitfield AK (2011) Challenging paradigms in estuarine ecology and management. Estuarine, Coastal and Shelf Science 94:306–314.

Ens E, Daniels C, Nelson E, Roy J, Dixon P (2016) Creating multifunctional landscapes: Using exclusion fences to frame feral ungulate management preferences in remote Aboriginal-owned northern Australia. Biological Conservation 197:235–246.

Environmental Protection Agency (2005) Wetland Mapping and Classification Methodology - A Method to Provide Baseline Mapping and Classification for Wetlands in Queensland, Version 1.2. Brisbane, Australia.

Evans AJ, Firth LB, Hawkins SJ, Morris ES, Goudge H, Moore PJ (2016) Drill-cored rock pools: An effective method of ecological enhancement on artificial structures. Marine and Freshwater Research 67:123–130.

Firth LB, Schofield M, White FJ, Skov MW, Hawkins SJ (2014) Biodiversity in intertidal rock pools: Informing engineering criteria for artificial habitat enhancement in the built environment. Marine Environmental Research 102:122–130.

Fordham DA, Georges A, Brook BW (2008) Indigenous harvest, exotic pig predation and local persistence of a long-lived vertebrate: Managing a tropical freshwater turtle for sustainability and conservation. Journal of Applied Ecology 45:52–62.

Fordham D, Georges A, Corey B, Brook BW (2006) Feral pig predation threatens the indigenous harvest and local persistence of snake-necked turtles in northern Australia. Biological Conservation 133:379–388.

Gillanders BM, Kingsford MJ (2002) Impact of changes in flow of freshwater on estuarine and open coastal habitats and the associated organisms. Oceanography and Marine Biology 40:233–309.

Great Barrier Reef Marine Park Authority (GBRMPA) (2015) Reef 2050 Long term sustainability plan. Great Barrier Reef Marine Park Authority, Commonwealth of Australia 111pp.

Grice AC, Cassady J, Nicholas DM (2012) Indigenous and non-Indigenous knowledge and values combine to support management of Nywaigi lands in the Queensland coastal tropics. Ecological Management & Restoration 13:93–97.

Guerrero AM, Shoo L, Iacona G, Standish RJ, Catterall CP, Rumpff L, de Bie K, White Z, Matzek V, Wilson KA (2017) Using structured decision-making to set restoration objectives when multiple values and preferences exist. Restoration Ecology 25:858–865.

Hagger V, Waltham NJ, Lovelock CE (2022) Opportunities for coastal wetland restoration for blue carbon with co-benefits for biodiversity, coastal fisheries, and water quality. Ecosystem Services 55:101423.

Harris GP, Heathwaite AL (2012) Why is achieving good ecological outcomes in rivers so difficult? Freshwater Biology 57:91–107.

Herr D, Agardy T, Benzaken D, Hicks F, Howard J, Landis E, Soles A, Vegh T (2015) Coastal 'blue' carbon. A revised guide to supporting coastal wetland programs and projects using climate finance and other financial mechanisms. Gland: IUCN, 47pp.

Houston WA, Duivenvoorden LJ (2002) Replacement of littoral native vegetation with the ponded pasture grass Hymenachne amplexicaulis: Effects on plant, macroinvertebrate and fish biodiversity of backwaters in the Fitzroy River, Central Queensland, Australia. Marine and Freshwater Research 53:1235–1244.

Hyland SL (2002) An investigation of the impacts of ponded pastures on Barramundi and other finfish populations in tropical coastal wetlands. In: Industries, D.o.P. (Ed.). Queensland Government, Brisbane.

Jiang TT, Pan JF, Pu XM, Wang B, Pan JI (2015) Current status of coastal wetlands in China: Degradation, restoration, and future management. Estuarine, Coastal and Shelf Science 164:265–275.

Karim F, Wallace J, Abbott BN, Nicholas M, Waltham NJ (2021) Modelling the removal of an earth bund to maximise seawater ingress into a coastal wetland. Estuarine, Coastal and Shelf Science 263:107626.

Krull CR, Choquenot D, Burns BR, Stanley MC (2013) Feral pigs in a temperate rainforest ecosystem: Disturbance and ecological impacts. Biological Invasions 15:2193–2204.

Lewis S, Bainbridge Z, Davis, A, McCloskey G, Bartley R, Turner R (2024). Sediment and nutrient flux from land. In Wolanski E, Kingsford MJ (eds.), Oceanographic Processes of Coral Reefs. Physical and Biological Links in the Great Barrier Reef. CRC Press, Boca Raton, FL.

Lewis SE, Bartley R, Wilkinson SN, Bainbridge ZT, Henderson AE, James CS, Irvine SA, Brodie JE (2021) Land use change in the river basins of the Great Barrier Reef, 1860 to 2019: A foundation for understanding environmental history across the catchment to reef continuum. Marine Pollution Bulletin 166:112193.

Martin S, Sparks EL, Constantin AJ, Cebrian J, Cherry JA (2021) Restoring fringing tidal marshes for ecological function and ecosystem resilience to moderate sea-level rise in the Northern Gulf of Mexico. Environmental Management 67:384–397.

McJannet D, Wallace J, Keen R, Hawdon A, Kemei J (2012) The filtering capacity of a tropical riverine wetland: II. Sediment and nutrient balances. Hydrological Processes 26:53–72.

Mitchell J, Mayer R (1997) Diggings by feral pigs within the Wet Tropics World Heritage Area of north Queensland. Wildlife Research 24:591–601.

Montiel-Martínez A, Ciros-Pérez J, Corkidi G (2015) Littoral zooplankton–water hyacinth interactions: Habitat or refuge? Hydrobiologia 755:173–182.

Murray NJ, Worthington TA, Bunting P, Duce S, Hagger V, Lovelock CE, Lucas R, Saunders MI, Sheaves M, Spalding M, Waltham NJ, Lyons M (2022) High-resolution mapping of losses and gains of Earth's tidal wetlands. Science 376:744–749.

Nagelkerken I, Sheaves M, Baker R, Connolly RM (2015) The seascape nursery: A novel spatial approach to identify and manage nurseries for coastal marine fauna. Fish and Fisheries 16:362–371.

Nordberg EJ, Macdonald S, Zimny G, Hoskins A, Zimny A, Somaweera R, Ferguson J and Perry J (2019) An evaluation of nest predator impacts and the efficacy of plastic meshing on marine turtle nests on the western Cape York Peninsula, Australia. Biological Conservation 238:108201.

Ng CSL, Lim SC, Ong JY, Teo LMS, Chou LM, Chua KE, Tan KS (2015) Enhancing the biodiversity of coastal defence structures: Transplantation of nursery-reared reef biota onto intertidal seawalls. Ecological Engineering 82:480–486.

O'Brien D, Lewis S, Davis A, Gallen C, Smith R, Turner R, Warne M, Turner S, Caswell S, Mueller JF (2016) Spatial and temporal variability in pesticide exposure downstream of a heavily irrigated cropping area: Application of different monitoring techniques. Journal of Agricultural and Food Chemistry 64:3975–3989.

Paschke MW, Perkins LB, Veblen KE (2019) Restoration for multiple use. Restoration Ecology 27:701–704.

Perna C, Burrows DW (2005) Improved dissolved oxygen status following removal of exotic weed mats in important fish habitat lagoons of the tropical Burdekin River floodplain, Australia. Marine Pollution Bulletin 51:138–148.

Perna CN, Cappo M, Pusey BJ, Burrows DW, Pearson RG (2012) Removal of aquatic weeds greatly enhances fish community richness and diversity: An example from the Burdekin River floodplain, Tropical Australia. River Research and Applications 28:1093–1104.

Powers RP, Jetz W (2019) Global habitat loss and extinction risk of terrestrial vertebrates under future land-use-change scenarios. Nature Climate Change 9:323–329.

Queensland Government (2018a) Bustard Bay Wetlands DIWA nationally important wetland—facts and maps. WetlandInfo, Department of Environment and Heritage Protection, Queensland.

Queensland Government (2018b) Seventeen seventy-round hill fish habitat area—facts and maps. WetlandInfo, Department of Environment and Heritage Protection, Queensland, Queensland

Rayner TS, Pusey BJ, Pearson RG (2009) Seasonal flooding, instream habitat structure and fish assemblages in the Mulgrave River, north-east Queensland: Towards a new conceptual framework for understanding fish-habitat dynamics in small tropical rivers. Marine & Freshwater Research 59:97–116.

Rogers K, Saintilan N, Mazumder D, Kelleway JJ (2019) Mangrove dynamics and blue carbon sequestration. Biology Letters 15:20180471.

Ross B, Waltham NJ, Schaffer J, Jaffer T, Whyte S, Perry J, Vanderduys E, Macdonald S, Morgan M, Walsh T, Huerlimann R, LePort A, Burrows D, Dean J (2017) Improving biodiversity outcomes and carbon reduction through feral pig abatement. Balkanu Cape York Development Corporation Ltd Pty, Cairns, p. 172.

Saunders M, Waltham N, Cannard T, Sheppard M, Fischer M, Twomey A, Bishop M, Boody K, Callaghan D, Fulton B (2022) A roadmap for coordinated landscape-scale coastal and marine ecosystem restoration. Rainforest, Reef, Research Center, Ltd Pty, Cairns

Schlacher TA, Schoeman DS, Dugan J, Lastra M, Jones A, Scapini F, McLachlan A (2008) Sandy beach ecosystems: Key features, sampling issues, management challenges and climate change impacts. Marine Ecology 29:70–90.

Schuerch M, Spencer T, Temmerman S, Kirwan ML, Wolff C, Lincke D, McOwen CJ, Pickering MD, Reef R, Vafeidis AT (2018) Future response of global coastal wetlands to sea-level rise. Nature 561:231–234.

Sheaves M, Baker R, Abrantes KG, Connolly RM (2017) Fish biomass in tropical estuaries: Substantial variation in food web structure, sources of nutrition and ecosystem-supporting processes. Estuaries and Coasts 40:580–593.

Sheaves M, Bradley M, Lubitz N, Mattone C, Myers J, Venkataraman A, Waltham N, Winter S (2024) The influence of the spatio-temporal dynamics of fish populations on the outcomes of land-sea connectivity. In Wolanski E, Kingsford MJ (eds.), Oceanographic Processes of Coral Reefs. Physical and Biological Links in the Great Barrier Reef. CRC Press, Boca Raton, FL.

Sheaves M, Brookes J, Coles R, Freckelton M, Groves P, Johnston R, Winberg P (2014) Repair and revitalisation of Australia's tropical estuaries and coastal wetlands: Opportunities and constraints for the reinstatement of lost function and productivity. Marine Policy 47:23–38.

Sheaves M, Johnston R (2008) Influence of marine and freshwater connectivity on the dynamics of subtropical estuarine wetland fish metapopulations. Marine Ecology Progress Series 357:225–243.

Sheaves M, Johnston R (2010) Implications of spatial variability of fish assemblages for monitoring of Australia's tropical estuaries. Aquatic Conservation: Marine and Freshwater Ecosystems 20:348–356.

Sheaves M, Johnston R, Connolly RM, Baker R (2012) Importance of estuarine mangroves to juvenile banana prawns. Estuarine, Coastal and Shelf Science 114:208–219.

Sheaves M, Mattone C, Connolly RM, Hernandez S, Nagelkerken I, Murray N, Ronan M, Waltham NJ, Bradley M (2021a) Ecological constraint mapping: Understanding outcome-limiting bottlenecks for improved environmental decision-making in marine and coastal environments. Frontiers in Marine Science 8. doi: 10.3389/fmars.2021.717448

Sheaves M, Molony B (2000) Short-circuit in the mangrove food chain. Marine Ecology Progress Series 199:97–109.

Sheaves M, Waltham N, Benham C, Bradley M, Mattone C, Diedrich A, Sheaves J, Sheaves A, Hernandez S, Dale P (2021b) Restoration of marine ecosystems: Understanding possible futures for optimal outcomes. Science of the Total Environment 796:48845.

Sparks RE, Blodgett KD, Casper AF, Hagy HM, Lemke MJ, Velho LFM, Rodrigues LC (2017). Why experiment with success? Opportunities and risks in applying assessment and adaptive management to the Emiquon floodplain restoration project. Hydrobiologia 804:177–200.

Steward AL, Negus P, Marshall JC, Clifford SE, Dent C (2018). Assessing the ecological health of rivers when they are dry. Ecological Indicators 85:537–547.

Stewart-Sinclair PJ, Purandare J, Bayraktarov E, Waltham NJ, Reeves SE, Statton J, Sinclair EA, Brown BM, Shribman ZI, Lovelock CE (2020) Blue restoration—building confidence and overcoming barriers. Frontiers in Marine Science 7. doi: 10.3389/fmars.2020.541700

Toft JD, Simenstad CA, Cordell JR, Grimaldo LF (2003) The effects of introduced water hyacinth on habitat structure, invertebrate assemblages, and fish diets. Estuaries 26:746–758.

Veitch V, Tait J, Burrows DW (2008) Fish passage connectivity issues lower Sheep Station Creek: Dry and wet season investigations of water quality and fish assemblages. Australian Centre for Tropical Freshwater Research (08/02), James Cook University, Townsville.

Villamagna AM, Murphy BR (2010) Ecological and socioeconomic impacts of invasive water hyacinth (*Eichhornia crassipes*): A review. Freshwater Biology 55:282–298.

Wallace J, Bueno C, Waltham NJ (2022) Modelling the removal of nitrogen and sediment by a constructed wetland system in north Queensland, Australia. Ecological Engineering 184:106767.

Wallace J, Waltham NJ (2021) On the potential for improving water quality entering the Great Barrier Reef lagoon using constructed wetlands. Marine Pollution Bulletin 170:112627.

Waltham N, Fixler S (2017) Aerial herbicide spray to control invasive water hyacinth (*Eichhornia crassipes*): Water quality concerns fronting fish occupying a tropical floodplain wetland. Tropical Conservation Science 10:1940082917741592.

Waltham NJ, Alcott C, Barbeau MA, Cebrian J, Connolly RM, Deegan LA, Dodds K, Goodridge Gaines LA, Gilby BL, Henderson CJ (2021a) Tidal marsh restoration optimism in a changing climate and urbanizing seascape. Estuaries and Coasts 44:1681–1690.

Waltham NJ, Burrows D, Wegscheidl C, Buelow C, Ronan M, Connolly N, Groves P, Audas D, Creighton C, Sheaves M (2019). Lost floodplain wetland environments and efforts to restore connectivity, habitat and water quality settings on the Great Barrier Reef. Frontiers in Marine Science 6. doi: 10.3389/fmars.2019.00071

Waltham NJ, Coleman L, Buelow C, Fry S, Burrows D (2020a) Restoring fish habitat values on a tropical agricultural floodplain: Learning from two decades of aquatic invasive plant maintenance efforts. Ocean & Coastal Management 198:105355.

Waltham NJ, Connolly RM (2011) Global extent and distribution of artificial, residential waterways in estuaries. Estuarine, Coastal and Shelf Science 94:192–197.

Waltham NJ, Elliott M, Lee SY, Lovelock C, Duarte CM, Buelow C, Simenstad C, Nagelkerken I, Classens L, Wen CCK, Barletta M, Connolly RM, Gillies C, Mitsch WJ, Ogburn MB, Purandare J, Possingham H, Sheaves M (2020b) UN Decade on Ecosystem of Restoration 2021–2030—what chance for success in restoring coastal ecosystems? Frontiers in Marine Science 7. doi: 10.3389/fmars.2020.00071

Waltham NJ, Lovelock C, Buelow C (2023). Blue carbon stocks and cycling in tropical tidal marshes facing grazing pressure. Marine Ecological Progress Series doi:103354/meps14379

Waltham NJ, Pyott M, Buelow C, Wearne L (2020c) Mechanical harvester removes invasive aquatic weeds to restore water quality and fish habitat values on the Burdekin floodplain. Ecological Management & Restoration 21:187–197.

Waltham NJ, Schaffer J (2016) Baseline aquatic assessment of wetlands identified for feral pig fence exclusion, Archer River catchment. James Cook University, Townsville.

Waltham NJ, Schaffer J, Walker S, Perry J, Nordberg E (2022). Simple fence modification increases land movement prospects for freshwater turtles on floodplains. Wildlife Biology:e01012.

Waltham NJ, Schaffer JR (2018) Thermal and asphyxia exposure risk to freshwater fish in feral-pig-damaged tropical wetlands. Journal of Fish Biology 93:723–728.

Waltham NJ, Sheaves M (2015) Expanding coastal urban and industrial seascape in the Great Barrier Reef World Heritage Area: Critical need for coordinated planning and policy. Marine Policy 57:78–84.

Waltham NJ, Sheaves M (2018) Eco-engineering rock pools to a seawall in a tropical estuary: Microhabitat features and fine sediment accumulation. Ecological Engineering 120:631–636.

Waltham NJ, Wegscheidl C, Volders A, Smart JC, Hasan S, Lédée E, Waterhouse J (2021b) Land use conversion to improve water quality in high DIN risk, low-lying sugarcane areas of the Great Barrier Reef catchments. Marine Pollution Bulletin 167:112373.

Waterhouse J, Pearson R, Lewis S, Davis A, Waltham NJ (2024) Great Barrier Reef Ecohydrology. In Wolanski E, Kingsford MJ (eds.), Oceanographic Processes of Coral Reefs. Physical and Biological Links in the Great Barrier Reef. CRC Press, Boca Raton, FL.

Waterhouse J, Schaffelke B, Bartley R, Eberhard R, Brodie J, Star M, Thorburn P, Rolfe J, Ronan M, Taylor B (2017) Scientific consensus statement: A synthesis of the science of land-based water quality impacts on the great barrier reef. State of Queensland, Brisbane.

Wearne LJ, Clarkson J, Grice AC, van Klinken RD, Vitelli JS (2010) The biology of Australian weeds 56. 'Hymenachne amplexicaulis' (Rudge) nees. Plant Protection Quarterly 25:146–161.

Weinstein MP, Hazen R, Litvin SY (2019) Response of Nekton to Tidal Salt Marsh Restoration, a Meta-Analysis of Restoration Trajectories. Wetlands 39:575–585.

Wolanski E, Kingsford M, Lambrechts J, Marmorino G (2024). The physical oceanography of the Great Barrier Reef: A review. In Wolanski E, Kingsford MJ (eds.), Oceanographic Processes of Coral Reefs. Physical and Biological Links in the Great Barrier Reef. CRC Press, Boca Raton, FL.

Zedler JB (2016) What's New in Adaptive Management and Restoration of Coasts and Estuaries? Estuaries and Coasts 40:1–21.

25 Pathways to Improved Water Quality in the GBR Lagoon—Exploring Opportunities for Broadscale Application of Low-Risk Practices in the Lower Burdekin Irrigated Agriculture Areas

Stephen Attard, John G. Connell, and Taha Chaiechi

25.1. INTRODUCTION

Although climate change has an increasingly important influence on the health and bio-diversity of the Great Barrier Reef (GBR), water quality remains a critical factor. The effect that land-based activities, mainly agricultural, have had on water quality of the GBR has been identified as being of major concern for some time (Brodie 1997). Beginning in 2003, great effort has been made to identify 'low-risk practices' for agriculture, along with launching a range of programs that supports their application. Progress has been made towards the water quality (WQ) targets but at a rate that will not achieve them within the required timeframe.

The Lower Burdekin is comprised of two basins, the Burdekin and Haughton, with complimentary irrigation systems. Sugarcane farming dominates the landscape, where between 65,000 and 72,000 ha is harvested annually. The area has been a focus of efforts to reduce nutrient flows to the GBR. The assurance of high radiation and reliable water has led farmers to apply high rates of fertilizer generating high rates of dissolved inorganic nitrogen (DIN) that are transported to the GBR through surface runoff and deep drainage. A series of Water Quality Improvement Plans (WQIPs) beginning in 2003 have attempted to reduce this DIN export to the GBR, and research has been successful to some degree in identifying practices that could achieve this. However, achieving change has been a challenge for a variety of reasons that include difficulties in applying practices developed under research conditions to be applied under practical farm conditions, social constraints on adoption, and an apparent contest between market and environmental protection objectives. Over the years, there have been calls for programs to find new approaches with social research indicating areas that should be considered. Based on these, this chapter attempts to examine how the WQIPs could be reframed to take account of these social factors and the necessary technical factors that still need to mature.

25.2. WATER QUALITY POLLUTANTS AND THEIR FUNCTION

25.2.1. Pollutant Effects on GBR Ecosystems

The main pollutants entering the GBR are sediments of various size fractions, nutrients (nitrogen and phosphorous), and pesticides. Sediments are generated by soil erosion, which grazing and cropping acerbate. The coarse fractions tend to be deposited at the river mouths, with the smaller size fractions (< 16 μm) traveling out into the lagoon, where through turbidity and deposition, they directly affect sea grass, corals, and many other processes on the reef (Bartley et al. 2017; Waterhouse et al. 2017).

The nutrients nitrogen (N) and phosphorous (P) can appear in both particulate and dissolved forms, with the dissolved forms being readily bio-available. Of these, DIN is of greater concern because it is the form in which N is taken up by living organisms as a nutrient. In irrigation systems, this occurs principally in the nitrate (NO_3^-) form (Thorburn et al. 2011). An increase of DIN in the lagoon will promote an increase in phytoplankton biomass, which is food for the larvae of the crown-of-thorns starfish (Furnas et al. 2013). This increases the survival rate of the larvae and is now believed to be responsible for the increased frequency of crown-of-thorns outbreaks, a major cause of coral mortality (De'arth et al. 2012; Fabricius et al. 2010).

High temperature events due to climate change cause coral bleaching, but increased DIN further increases the susceptibility of corals to bleaching from higher temperatures (Fabricius et al. 2013). DIN further affects reef health

DOI: 10.1201/9781003320425-29

in a number of other ways by promoting increased populations of organisms (borers, sponges, bivalves, etc.) that feed on corals (Hutchings et al. 2005), increased coral disease (Havall et al. 2007; Thompson et al. 2014; Pollock et al. 2016), and reduced growth of seagrass and coral because of increased water turbidity (Collier et al. 2016; Petus et al. 2014).

25.2.2. Land-Based Delivery of DIN to the GBR

There is a natural loss of N from the soil to the environment through leaching and erosion, but this reaches a steady state with input of N supplied through rainfall. Anthropogenic N enters the system mainly through fertiliser applied to crops. Accumulation of N in the soil is not significant, meaning that the N applied in excess of what is taken up by crops is closely related to the N lost to the environment (Brodie et al. 2013). The pathways for DIN loss to the GBR are through surface runoff or deep drainage. N is also lost to the environment through N in gaseous forms (denitrification and volatilisation) to the atmosphere, and so this does not enter the GBR. The balance among these three loss paths depends primarily on soil characteristics and crop management.

The river basins that drain into the GBR have been divided into five administrative marine Natural Resource Management (NRM) regions (Figure 1). The most recent estimate of DIN loss from all five catchments into the GBR is 10,967 t/yr. Of this, 5,319 t/yr (48%) is anthropogenic. Sugarcane production is responsible for most of this and contributes 42% of the total DIN to the GBR (McCloskey et al. 2021).

The Burdekin Region is composed of 5 basins: the Black; the Ross; the Haughton; the Burdekin, and the Don. The Burdekin and the Haughton basins, referred to collectively as the Lower Burdekin, are dominated by sugarcane cultivation, and discharge a total of 1,705 t DIN to the GBR, of which 787 t is anthropogenic. The Lower Burdekin thus supplies 14.8% of the total anthropogenic DIN discharged to the GBR, second only to the Herbert basin, which supplies 20.8% (McCloskey et al. 2021).

25.3. LAND BASED ACTIVITY AND MANAGING DIN EXPORT IN THE LOWER BURDEKIN

25.3.1. Irrigation Systems and DIN Pathways

The total area of irrigated land in the Lower Burdekin is approximately 90,000 ha of which about 70,000 is sugarcane that is harvested annually. Other crops, including diverse horticulture (e.g., capsicums, melons, and beans), grains (e.g., soy, mung, and corn), and permanent tree crops (e.g., mango and lychee) are also significant contributors to the local economy. Other sugarcane producing regions largely rely on rainfall with irrigation as a supplement. The Lower Burdekin, is located in the dry tropics, with low summer-dominated rainfall (av. 966 mm/yr) and high evapotranspiration rates (2,057 mm/yr), it would not be able to host cultivated sugarcane without irrigation (State of Queensland 2013). Construction of the Burdekin Falls Dam (1987) granted farmers access to abundant water, which combined with high solar radiation, provides the area with a high yield potential. To realise this potential, farmers have tended to apply fertilizer at rates ~30 to 60 kg N/ha higher than other cane producing regions (Schroeder et al. 2003; Calcino et al. 2010). The average cane yields (~120 t/ha) are at least 30% higher than other cane producing regions, and as a result, the Lower Burdekin is regarded as a premier sugarcane producing region for the country.

The Lower Burdekin is best characterised by its irrigation systems: the Delta and the Burdekin River Irrigation Area (BRIA) (see Figure 2). The Delta lies fully within the Burdekin basin, while the BRIA includes portions of both the Burdekin and Haughton basins. The two areas have quite different soils, farm layouts, and irrigation management, which greatly affect the relative nutrient loss pathways.

The **Delta** was established in the 1880s with a small farm size (median ~56 ha) with short furrow lengths (300–700 m). Irrigation of the ~45,000 ha of sugarcane is achieved predominantly by groundwater from on-farm bores (10–20 m deep). The majority of soils are sandy to sandy loam and, thus, highly permeable, which can result in high deep drainage losses. Extraction of water from the shallow aquifer lowers the water table, and this must be recharged or risk intrusion of salt water from the sea. The Delta irrigation area is managed by Lower Burdekin Water, with a key aspect of its management being to monitor water extraction and balance this with recharge of the aquifer. The **BRIA** area was first established in the 1950s and then expanded to over 49,000 ha following the construction of the Burdekin Falls Dam, with the dominant crop being sugarcane. Farm sizes are larger (median ~94 ha) with a long furrow length, usually between 400 and 1,000 m. Soils tend to be uniform light to heavy clays or duplex soils. These are typically described as high runoff and low drainage. Over the years, gypsum has been used to remove sodium from the profile and improve soakage, but this has also increased the potential for deep drainage losses.

Field trials that measured N balances consistently show that the main loss pathway in the clay-based soils of the BRIA is through surface water runoff, with deep drainage the main loss pathway in the sandy soils of the Delta (Thorburn et al. 2011; Hesp et al. 2011). These studies reported that the amount of N leached from the soils is substantially less in the BRIA (typically < 10% of fertilizer application) than in the Delta (between ~30 and 75% of fertilizer application) and was balanced by high rates of N loss through denitrification in the BRIA. Overall, the

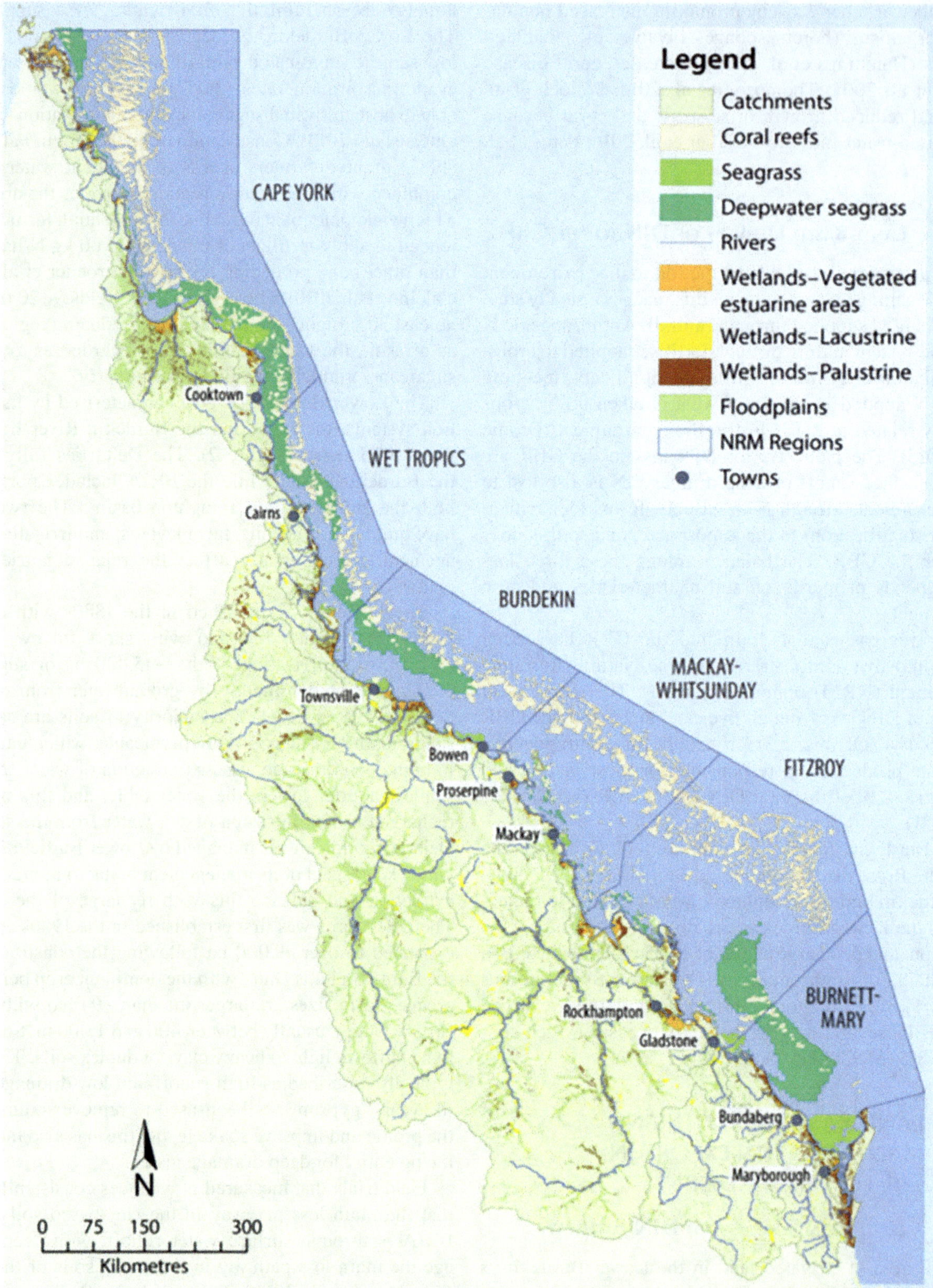

FIGURE 1 The NRM regions that form the basis of initiatives under the Reef Rescue programs (Waterhouse et al. 2017).

FIGURE 2 Map of the Lower Burdekin sugar production area, with overlays indicating the two irrigation management systems: the Delta, managed by Lower Burdekin Water, and the BRIA (or Burdekin Haughton Water Supply Scheme), managed by Sunwater.

export rate of DIN from the Delta (~14 kg/ha/yr) is somewhat higher than that from the BRIA (~10 kg N/ha/yr) (Waterhouse et al. 2018). Cumulatively, the Burdekin River and Haughton River basins export 1,085 t DIN to the GBR each year (Brodie et al. 2017).

25.3.2. Irrigation and Rising Water Table

Although the Delta must manage the water table to prevent salt water intrusion, the BRIA is affected by a rising water table due to the surface water supplied across the area. This puts at risk sugarcane productivity in some areas of the BRIA.

The impact of surface water from the irrigation made available by construction of the Burdekin Falls Dam (1987) was recognised at the time; its rise has closely followed the predictions (Williams et al. 2009). The surface water applied in excess of the land's capacity to drain is estimated to be in the range of 130,000–145,000 ML per year (Williams et al. 2009, State of Queensland 2013; Shaw 2014). As a result, since the dam was constructed, the water table has risen from 10 m to 3 m below the surface 3 m being (the level judged to be sufficient to ensure that the cane root zone will not be affected by salinity or waterlogging) (Figure 3 left). This is variable, and recent studies conducted by Queensland's Department of Regional Development, Manufacturing and Water indicates that there is already 5,000 ha in the BRIA where the water table is at 1.5 m, with a further 2,800 ha at 0.5 m (Figure 3 right). Modelling shows that by 2030, at least 2,300 ha will go out of production. This would potentially reduce cane production by 265,000 t per year (Reef Trust Partnership 2020).

Two approaches can be used to lower the water table, namely, by (a) reducing irrigation system-induced deep drainage and (b) extracting water from the aquifer for drawdown of the water table. The second will have a more rapid effect but will not address the underlying cause of the problem (State of Queensland 2017). There are plans for a 'groundwater mitigation project' that will address leakage from the Sunwater distribution channels but not the on-farm deep drainage from irrigation. Given that approximately 60% of farms are operating at irrigation efficiencies of < 50%, there is much scope for reducing deep drainage by achieving water-efficient irrigation (Waterhouse et al. 2018).

25.4. MANAGING DIN EXPORT FROM THE FIELD

With increasing scrutiny on the role of nutrients on the GBR and recognition of the need to give greater attention to profitability for the sugar industry as a whole to be sustainable, the industry developed a new set of nutrient recommendations to farmers called "SIX EASY STEPS®" (Schroeder et al. 2003). These shifted recommendations from providing regional blanket fertilizer rates to farmers using the 'district potential yield' as a target and then making farm-specific calculations based on this; they also recommended conducting soil analysis, accounting for all forms of nutrient inputs (e.g., growing legume crops and the addition of mill by-products), maintaining records, and reviewing the response of the crop to continuously revise rates. These proposals formed the basis of the sugar industry Best Management Practices (BMP), which were expanded to include soil and irrigation management practices. For

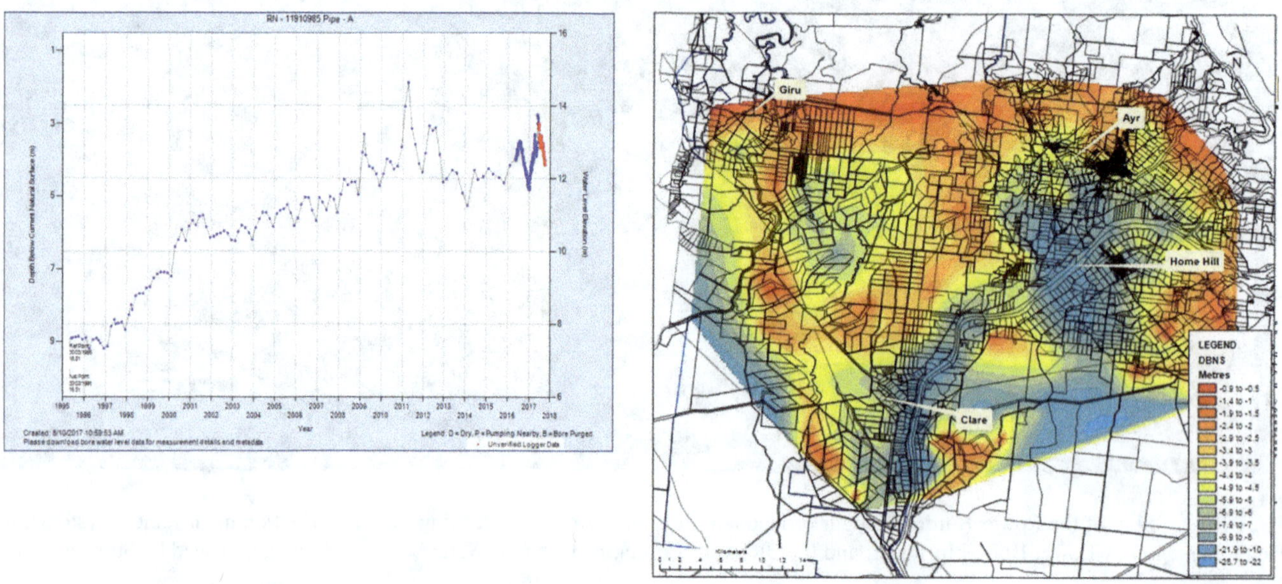

FIGURE 3 Historic rise in the water table (left) since construction of Burdekin Falls Dam and expansion of irrigated agriculture in the BRIA area; the ground water depth (right) is below the soil surface, May to June 2017.

Source: Adapted from the Department of Natural Resources and Mines, Queensland 2017

TABLE 1

Comparison of the 'water quality risk framework' (2017 Reef 2050 WQIP) to the sugar industry BMP (Waterhouse et al. 2018).

Program	Practice levels			
Reef 2050 WQIP	A	B	C	D
Water Quality Risk Framework	*Lowest Risk* Innovative	*Low risk* Best practice	*Moderate risk* Min. Standard	*High risk* Superseded
Industry BMP	Above Industry Standard		Industry Standard	Below Industry Standard

the Burdekin, this produces typical rates of 90–180 kg and 130–220 kg N/ha for plant and ratoon crops, respectively (Calcino et al. 2022).

Early reviews of the BMP indicated that even with universal application across all catchments, they would reduce DIN export by just 29%, (Armour et al. 2009); this was later revised down to be only 12%, (Thorburn and Wilkinson 2011). The BMP persisted as the main strategy for DIN reduction until the 2017 Reef 2050 WQIP. This introduced the 'water quality risk framework', which provided a progressive set of management practices rated in terms of their risk to water quality, from Lowest to High risk, using an A/B/C/D ranking scale (Table 1). The ABCD framework is shown as follows compared to the industry BMP.

Specific practices for each level were defined for four areas of crop management: soil management, pesticide application, fertiliser application, and irrigation. Practices that focus only on fertilizer management, the source of N, are not enough to achieve DIN reduction targets. Irrigation leaches and transports DIN, and so it is essential for this to be managed. This has resulted in a suite of practices for both fertilizer and irrigation management that will provide low-risk cultivation of sugarcane (Waterhouse et al. 2017).

25.4.1. Optimising Fertiliser Management

Each application of fertiliser incurs costs (product, labour, and fuel), and as the crop grows and the canopy closes, sub-surface placement becomes increasingly difficult and impractical. As a result, typical practices involve the application of fertiliser once or twice early in the season, thus leaving it exposed in the soil to runoff and leaching. This can be optimised by the following.

- *Reducing surplus fertiliser* can be accomplished by estimating the potential crop yield for a specific farm and from this, the required N application. The industry recommendation remains based on district yield potential (Calcino et al. 2022; Schroeder et al. 2010).
- *Burying fertiliser* at a minimum depth with compacted soil will reduce loss from irrigation- and rain-induced run-off and loss from denitrification or volatilisation.
- *Splitting fertiliser application* to match crop requirements at different growth stages through multiple, smaller applications will limit the time fertilizer is exposed to leaching and denitrification opportunities.

25.4.2. Optimising Furrow Irrigation Management

Furrow irrigation (Figure 4) is used on more than 95% of the Lower Burdekin sugarcane area. Improving irrigation practice primarily involves better matching application to crop requirements by applying the right amount at the right time. This can be achieved in multiple ways including the following.

- *Matching application to crop requirements.* This includes shutting off irrigation once the soil profile is full.
- *Construction of recycle pits* to catch flows at the end of furrows and making them available to be recycled (Figure 8).
- *Irrigation scheduling* so that sufficient water is provided according to the crop requirement. This can be scheduled but also informed by simple soil moisture sensors or sophisticated crop models so that water is always available in the root zone.
- *Furrow length and flow rate* must be considered to minimise crop waterlogging and losses of excess water to deep drainage and runoff. A paddock design will be required that takes into account infiltration characteristics of the soil, field slope, and furrow length to achieve maximum application efficiency (i.e., optimum inflow rates and irrigation event durations) (Gillies et al. 2017b; Raine and Bakker 1996).

Application of these principles would be equivalent to using low-risk (B) practices, which if applied universally, would reduce DIN loads by an estimated 60–80% (Thorburn and Wilkinson 2011; Waterhouse et al. 2018).

These practices were developed under controlled conditions, and some degree of automation is needed to apply them to real farm conditions. Irrigation management in the Lower Burdekin is still largely a manual activity. At the farm level, due to the number of fields, it can be quite

FIGURE 4 Furrow irrigation is considered low-cost in terms of installation, maintenance, and energy. It requires high labour inputs to assess when water has reached the end of the field and to manage pumps. To increase labour efficiencies, some fields are amalgamated to increase the length of the furrow, but this can also reduce water use efficiencies. To achieve balanced water delivery at the beginning and end of furrows, optimisation of field length, inflow rate, and duration of each irrigation event needs to be conducted.

Source: Image: Tom Major, ABC Rural 2022

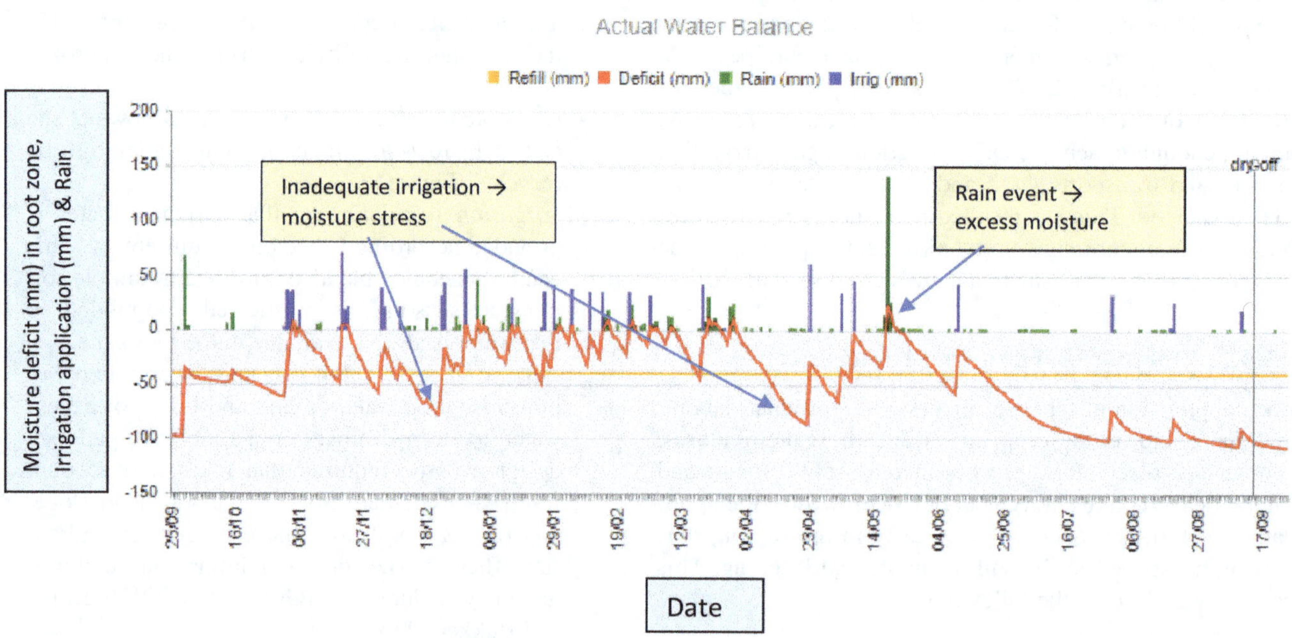

FIGURE 5 Automated furrow irrigation field trial—Delta area of Lower Burdekin. The figure of an IrrigWeb sugarcane water balance, records irrigation, and rain events that illustrates how excess or shortage affects moisture content in the root zone.

Source: Adapted from Wang et al. 2020. Decision support tools such as IrrigWeb are more likely to be integrated into farming practices when they are practical and effortless to access and use. Irrigation records collected by the automation system are simultaneously transferred to IrrigWeb, reducing the need for manual data entry, which is a known barrier to adoption.

FIGURE 6 Sophisticated platforms that display and integrate information from multiple sources promote the use of technology. Being able to effortlessly access and interpret farm records and information, such as weather forecasts, improves the ability of farmers to adopt low-risk practices over the longer term rather than for just the duration of short-term projects. Investing in programs that support continuous improvement through measurement will underpin legacy.

Source: Photograph courtesy of Sugar Research Australia

complex and labour-demanding, and it is a year-round commitment. Furrow irrigated crops require between 15 to 25 irrigations per year. Farmers must take into account the weather (timing and intensity of rainfall events), along with the duration of each crop (i.e., time between crop initiation and harvest varies from 12–16 months for plant crops and 11–14 months for ratoon crops). When multiple fields are irrigated simultaneously, with some at different growth stages, continuous manual checking of each field's progress becomes labour-intensive and time-consuming. As such, most farms resort to operating on a convenient fixed duration for each irrigation event, for example, a 12- or 24-hr irrigation event (Gillies et al. 2019). Although practical from a labour perspective, this results in low irrigation and energy efficiencies.

Early field trials of automation (Figure 6) simply provided fixed times for pump on/shut-off, which is sufficient to shift irrigation efficiency from D to C. More advanced trials to match irrigation to crop need included installation of soil moisture sensors in fields linked to pumps through programs (Wang et al. 2020). These begin to provide B or higher-level practices.

These practices have been piloted in the Lower Burdekin at both Delta and BRIA sites over the last decade or so. Results vary among sites, but they offer a consistent picture that water reduction can often provide an increase in both productivity and profitability; typical results have been 2–5 ML/ha water saved and DIN reductions of around 4.8 kg/ha (Gilles et al. 2017a, Gillies et al. 2017b; Gilles et al. 2019; AgriTech Solutions and NQ Dry Tropics 2020; Wang et al. 2020). If these practices were applied across the Lower Burdekin, then they should achieve at a minimum a DIN reduction of 346 t/yr, which is equivalent to just over half the basin's Reef 2050 WQIP target of 742 t. A reduced volume of water pumped will provide additional savings and reduced CO_2, contributing to additional environmental benefits.

The results from these trials are arguably still conservative. During these trials, real-farm conditions constrained the practices that could be applied or tested. As farmer confidence and understanding develops, so will their willingness to push the decision support systems (and themselves) toward an ever-improving outcome. Such an engagement challenges the boundaries of both the tools and the farmer. As a result, there remains potential for further farm-specific innovation to be made once automation is generally applied. Automation can move beyond simply scheduling pump operation to offering farmers the opportunity to actively control pumping and therefore achieve optimum moisture within the root zone—this would result in minimal stress due to large water deficits (under-irrigation), and no over-irrigation that could lead to deep drainage and runoff (Figure 5). Real-time information will lead to improved crop yields (e.g., irrigating to avoid periods of water-stress) and cost savings (e.g., forgoing irrigation ahead of forecast rain events). As farmers gain real-time feedback on their crop status via multiple lines of evidence, this will encourage further fine-tuning and innovation at an individual irrigation management unit scale.

The role of automation may have first been seen as providing a degree of convenience to the farm operator (Figure 6). However, it is in fact an essential element to apply the suite of low-risk practices to farm conditions. Without automation, it is simply not possible for farmers to apply these practices and manage them manually and arguably, for DIN reduction targets to be reached.

There are alternative irrigation systems to furrow irrigation, specifically, pressurised sub-surface drip irrigation (SSDI) and low-pressure overhead irrigation (OHLP), sometimes termed 'precision irrigation'. These would also allow fertigation (addition of fertilizer through irrigation to match crop needs), which could significantly reduce DIN losses (Attard et al. 2009, Hesp et al. 2011; Attard and O'Donnell 2013). Despite their efficiency, adoption of these systems is held back for multiple reasons including high installation costs (> $10,000/ha for SSDI and > $7,000/ha for OHLP), high energy costs to run the pressurised systems, and a high level of commitment from farmers to manage and maintain the systems.

Transitioning from traditional (manual) to automated irrigation systems entails overcoming the natural hesitancy to change established practices. This has taken place in other contexts with positive results. A South Korean study (Lee 2022) showed that automated irrigation had improved water productivity by over 12.7% and reduced labour requirements by 21.8%. Closer to home, dairy farms in northern Victoria found that the main gains of the automated irrigation were from labour efficiencies (Armstrong and Ho 2011). Labour is already showing to be a source of significant cost savings from automating irrigation in the Lower Burdekin (Gillies et al. 2017a, b, 2019). In addition to the economic assessment of labour reduction is the benefit to the farmers themselves and their family lifestyle. Automation will relieve the farmer of the back-and-forth travel to manage pumps throughout the day and night and enable this when the farmer is off-site and engaged in other activities. The arrival of water-efficient irrigation and the relief of the drudgery of manually managed irrigation should appeal to younger farmers and may thus help to resolve one concern within the local community: succession as the current farmers reach retirement age. Labour shortages, which is a growing problem in shrinking rural communities, will increasingly impact the efficiency of manually operated irrigation systems.

25.5. PROGRAMS FOR DIN REDUCTION FOR THE GBR AND THE LOWER BURDEKIN

25.5.1. Valuing the GBR

The intrinsic value of the GBR is well-accepted and evidenced by its listing as a World Heritage site in 1981. This obliges Australia to ensure its protection enabled through Commonwealth Law, sections 3a and 12 of the Environment Protection and Biodiversity Act 1999. Thus, it is not really necessary to establish a market value for the GBR. This does not mean that non-market goods and services have no value (Chaiechi et al. 2022). In fact, although ecosystem services are not traded in the market, they contribute significantly to human welfare, and accounting for this will help enabling their efficient use and protection.

A large number of studies have been conducted on the value of the GBR using different approaches and focusing in some cases, on specific industries (fishing, tourism, etc.) or specific areas, and these have been reviewed (Stoeckl et al. 2011). It is estimated that the GBR directly contributes $6.4 billion (AUD) to the Australian economy, mainly through the tourism sector (O'Mahoney et al. 2017).

The Total Economic Value (TEV) framework combines both monetary and non-monetary aspects of overall value and sums all the benefits obtained from a resource to assist policy and decision-making processes (Figure 7).

Prominent examples of TEV's application have been by the United Nations in its 'System of Environmental-Economic Accounts' and its application to water resources (United Nations 2012) and by the World Bank for measuring the value of ecosystems (Pagiola et al. 2004). More recently it was used to assess the non-market value of the natural resources for entire Papua New Guinea (PNG; Chaiechi et al. 2022). Application of the TEV framework to the GBR thus includes the financial benefits not only from the reef itself but also the direct and indirect uses of the reef and the surrounding natural environment that generate economic and anthropocentric value.

Oxford Economics, in conducting its study of the potential impact of coral bleaching for the GBR Foundation (Oxford Economics, 2009), used the TEV framework to attempt a comprehensive value of its full 'ecosystem services'. This took into account the various industries such as tourism and fishing, including the GBR's coastal defence function against erosion, particularly in the case of cyclones. The study estimated the total TEV to be $51B per year. It also assessed TEV costs in the event the whole reef was lost to be $38B per year. Although such estimates are not likely to be the final word on the value of the GBR, they nonetheless assist in establishing a clear economic justification for investing in its protection.

25.5.2. Setting Targets for DIN Reduction

Recognition of the impacts of land-based activities on the health of the GBR began in the late 90s (Furnas and Brodie 1996; Furnas et al. 1997; Brodie 1997). Information

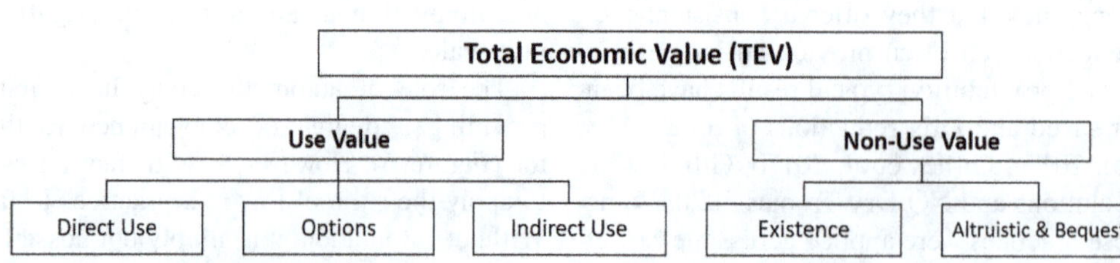

FIGURE 7 TEV approach (WCPA 1998; Munasinghe 1992).

on this was consolidated in a seminal GBR Marine Park Authority (GBRMPA) report (Great Barrier Reef Marine Park Authority 2001), which led to the development of government policies and programs to manage the quality of water for catchment draining into the GBR lagoon. In 2003, the Queensland and Commonwealth Governments launched the ten-year Reef Water Quality Protection Plan (RWQPP), and pursuant to this, each NRM Region developed its WQIP. For the Burdekin Region, the first of these occurred in 2009, which was followed by the now completed 2017–2022 Burdekin WQIP. The regional WQIPs are now nested within the Reef 2050 Long-Term Sustainability Plan (Reef 2050), which continues to direct the overarching strategy for management of the GBR.

The 2003 RWQPP expressed its objectives only in general terms, for example, '*Objective #1: Reduce the load of pollutants from diffuse sources in the water entering the Reef*' (State of Queensland and Commonwealth of Australia 2003). The program was reviewed mid-way in 2009 and from then on, set specific targets for nutrient load reduction, which was then expressed as a 50% reduction of total N entering the reef. The 2009 loads were then taken as the baseline to reduce nutrient levels. The 2013 WQIP defined targets in terms of DIN and extended the time for the target of 50% DIN reduction to 2018, towards an 80% reduction by 2025.

Ongoing research has now provided the 'trigger' points at which pollutant loads would affect reef health. As a result, the pollutant targets could be refined to be 'ecologically relevant' (ERT), that is, to ensure that the quality of water entering the GBR has 'no detrimental effects on the reef's health and resilience'. These are somewhat less than the earlier targets and were applied for the 2017–2022 WQIP. The 2025 target for anthropogenic DIN reduction for the whole GBR is now 3,594 t (equivalent to a 60% reduction). For the Lower Burdekin (i.e., Burdekin and Haughton basins) the target it is a DIN reduction of 742 t (equivalent to a 68% reduction). The ERT targets, however, are those required only to maintain reef health. To promote ecosystem restoration and recovery, stricter targets may need to be considered (Brodie et al. 2017).

25.5.3. Progress towards DIN Reduction Targets

The various programs to improve water quality have now been operating since 2003, that is, almost two decades. Along with the incentives to reduce the capital cost for farmers to purchase equipment and install systems (e.g., automated furrow), the programs have provided other support, such as for extension, etc. Overall, the Australian and Queensland Governments have invested approximately $1.1 B (AUD) to change agricultural practices to improve water quality in the GBR (Australian Government 2020). In the

FIGURE 8 Irrigation-induced run-off captured in recycle pits and then re-used on-farm. Installed sensors can continuously monitor the water height to lower the risk of over-filling and losing water or emptying the pit and causing problems with the pump. As the pit fills, this type of monitoring can allow the automation to turn on the pump and distribute water to a nominated field. Smart sensor technologies and continuous monitoring of the pit's water depth enables lower-risk practices for both the farmer and the environment to be adopted.

Source: Photograph courtesy of Sugar Research Australia

Lower Burdekin alone, up until 2018 there were conservatively 24 programs for improved water quality, with an investment of about $51M (Waterhouse et al. 2018).

The progress that these programs have made towards WQ has been reported through the Paddock to Reef (P2R) program Report Cards, which tracks changes in management practices and from this, estimates the pollutant loads to the GBR and for each NRM region. The Reef Report Card for 2012 and 2013 already noted that the N reduction targets at the time would not be reached based on application of BMP and that new thinking and approaches would be needed (Reef Water Quality Protection Plan Secretariat 2013). This has continued to be reiterated over time (Kroon et al. 2016; Brodie et al. 2019; Eberhard et al. 2021). After almost two decades, the last available Reef Report Card (2020) reports reductions of 27.7% and 34% for the GBR and Burdekin Region, respectively (State of Queensland and Commonwealth of Australia 2022), which is far short of the DIN reduction target of 60% by 2025.

These results reported in the Reef Report Cards are modelled from farmer self-reported applications of BMP, that is, moderate-risk practices. Notably, the confidence level for the data in the 2020 Report Card is rated as two out of five. A comprehensive review of practices in 2016, using multiple sources of data, indicated that a large proportion of the area in the Lower Burdekin was still under high- to moderate-risk practices (D/C). To illustrate this, surface water or runoff was still lost from most blocks (80%), the irrigation volume being applied based on crop requirements rather than 'time' was only 4%, and fertiliser rates are higher than BMP in 40% and 60% of the BRIA and Delta areas, respectively. Fertilizer sales indicated that approximately 2,120 t of N was being applied above crop requirements (Waterhouse et al. 2018). These gross figures suggest that real application of low risk practices might be far lower than the self-reporting indicates. Arguably the application of automation platforms to manage irrigation would provide greatly increased accuracy and real-time information on farm practices and, thus, the estimates of DIN export.

25.5.4. Factors Affecting the Rate of Uptake of Low-Risk Practices

The application of the low-risk practices would appear to provide a triple-win for farmers: reduced rates of inputs and associated costs, improved yield potential, and favourable Internal Rate of Return (IRR). The results would seem to speak for themselves and be enough to provide sufficient encouragement for neighbouring farmers to adopt. However, such a second wave of broadscale adoption has not taken place.

Although the Reef program's aim is to get farmers over the investment hurdle to install low-risk irrigation systems (Figure 8), this does not cover indirect and transition costs. These include the time and effort to attend meetings and obtain information and the application processes themselves. These costs could range from 7–37% over and above any grant funds received (Falconer 2000; Coggan et al. 2014). Although these costs are a hurdle, they are not necessarily prohibitive, as demonstrated by substantial co-investment by farmers working with the WQIPs and then, their later expansion of the practices across their properties (Waterhouse et al. 2018).

The National Environmental Science Program commissioned an assessment of the efficiency and effectiveness of the Reef WQIPs. This identified a range of practical difficulties, including that the multiple programs did not have consistent messages; extension workers focussed on engaged farmers and avoided less-engaged farmers; documents were too complex; the tone of documents was often dictatorial and patronising; and farmers found applying for programs both difficult and disempowering (Farr et al. 2017: Hay and Eagle, 2019).

Other studies have revealed social factors that could inhibit farmers response to the incentives offered in the WQ programs. These found that farmers see 'good farming' as associated with 'high yields' rather than profitability or sustainability (Blackstock et al. 2010) and are influenced by how their peers will judge their actions on this basis (Ferraro et al. 2019). In terms of engagements with agri-environmental programs, farmers had a high distrust of government programs and did not want their practices to be driven by GBR WQ issues. These views are encapsulated by one farmer's statement: *'I'm a results person—I like to see yep, . . . we're growing two tonne per acre better because we did this; Whereas the Reef Rescue—it was all about the reef!'* (Taylor and van Grieken 2015).

The data collected under the National Environmental Science Program were further analysed to provide a more quantitative appreciation of the social factors affecting adoption (Hasan et al. 2020). This showed that for the early adopters of low-risk fertilizer practices, their two most prominent characteristics were that they (a) considered it less essential to maintain good relations with other growers and (b) had attended at least five workshops. At first glance, they appear to be disparate but can in fact refer to the same set of social tensions. At first, they can appear somewhat bland but not when real-life contexts are considered, such as family dinners, local barbecues, and other gatherings where any behaviour that diverges from 'good farming' could attract ridicule and criticism. The workshops can also be understood to have a social function. Attendance at a series of workshops (five) could thus help to form sub-groups of peers where the early adopters can feel mutually supported against negative feedback from the wider peer group. All of this empahsises the importance of understanding and addressing social contexts.

25.5.5. Reframing the WQIPs for Social Inclusion

The practical difficulties of the WQIPs (clarity in documents, extension availability, etc.) can be addressed by making adjustments to the delivery of the programs.

Addressing the social constraints to enable the broadscale adoption of low-risk practices would, however, still remain as a potent hurdle with the current framing of the programs. The objectives of agriculture and conservation have often been perceived as conflicting. A recent study shows that where conservation outcomes can be achieved without loss of market outcomes, the issue should become how to best promote the conservation outcomes so that farmers feel comfortable in pursing both market and non-market (conservation) objectives (Stoeckl et al. 2015).

Behavioural science approaches have been applied to examine adoption of BMP. These consider not so much the hurdles but the contexts that need to be in place for these different objectives to be accommodated. They include the need for the producers to retain a sense of identity and group reinforcement, to have control over their practices, and importantly, for short-term goals to align with their current needs, while still allowing for longer-term goals to reach beyond individual needs (Pickering et al. 2017). In addition, they point to the influence that social networks, norms, and framing can have in achieving change. These have been applied with some success to sugarcane farmers in the Wet Tropics (Pickering et al. 2019).

Thus, it is worthwhile exploring how the WQIPs might be reframed to address these social constraints to adoption of improved practices, which on the face of it, would benefit farm productivity. In their current form, the WQIPs have WQ as the stated objective. This implicitly casts farmers and their practices as the problem. This means that in accepting the WQIP incentives, farmers implicitly must accept this negative characterisation. Reframing the WQIPs would need to focus on factors that avoid this negative framing and align them with farmers' existing attitudes and objectives. It is suggested here that in the case of the Lower Burdekin, this could be achieved through a program to 'improve sugarcane production and productivity', to be achieved through support for automating irrigation and better targeted fertiliser application across the whole area. The use of public funds to support such a 'productivity enhancement program' would thus be justified due to the environmental services that it would provide.

The practice changes in the field for such a 'productivity enhancement program' would be the same as those already incentivised through the past WQIP's, but they would instead emphasise the direct benefits to farmers of improved productivity (higher yields through precise irrigation), lower costs (water, energy, and labour), and the resolution of a rising water table at its root cause. They should also highlight important non-economic benefits to the farming communities as follows: the improved lifestyle, which would make farming a more attractive opportunity for succession to a younger generation of producers within families; solving labour shortages; and addressing public perception that the sugarcane industry is using out-of-date thinking and technology. These could assist in stabilising the shrinking local communities of the Lower Burdekin.

Such a shift would require more than a change in name. Both the structure and processes of a program with a productivity orientation would need to change. The existing WQ-focussed programs rely on decisions by individual producers to adopt low-risk practices against negative peer-pressure. The productivity-orientated program would need to be designed by the production community itself, through a process that would accommodate site specifics, with producers defining their requirements. The design process itself would in effect create a new peer group—one that farmers would see as action-orientated—and which would also have a major role in managing the program. With such a framework, 'adoption' would have already been achieved in the design stage, and peer-pressure would shift to mobilising action to achieve the program objectives in a timely manner. Such approaches to more effective WQIPs are not entirely new. They lay at the core of the recommendations in a NQ Dry Tropics synthesis document and in other unrecorded forums (Waterhouse et al. 2018).

The Burdekin Irrigation Project (Great Barrier Reef Foundation 2023) has successfully implemented aspects of this approach. As a result, adoption of low-risk irrigation management practices underpinned by water-efficient irrigation technology, including automation, has increased significantly during the life of the project, which started in 2021 (Figure 9). By the project completion date, more than 3,700 ha is expected to have been transitioned to automated irrigation, an increase of approximately 450% over a 3.5-year period.

The success of the Burdekin Irrigation Project is evidence that many farmers have the desire to transition to smart, innovative technology that enhances farming practices. The farmer co-investment has not been discussed; however, the cash and in-kind contribution has been significant, which is further evidence that farmers are committed to these improvements. Projects such as this have built a foundation for a sustained legacy after the end of the project funding.

25.6. TRANSFORMATIVE APPROACHES TO ACHIEVING WATER QUALITY

Coming to the end of the 2017–2022 WQIP, the Lower Burdekin struggles to meet the target for DIN reduction to the GBR lagoon. In a contrasting set of initiatives, there has been an increasing number of proposals for a series of greenfield agricultural production areas supported by construction of new dams on the Burdekin River. The major ones include the Hells Gates Dam, Burdekin Falls Dam Raising, Bowen Pipeline, and Urannah Dam. Together, these propose to add a further 140,000 ha cultivated annually, an area that would more than double the area currently cultivated in the Lower Burdekin. DIN loads generally increase linearly with the area of fertilised land (Connelly et al. 2015; Mitchell et al. 2009). Although the proponents state that they would use precision irrigation, it is difficult to imagine that in practice, they would not add significant

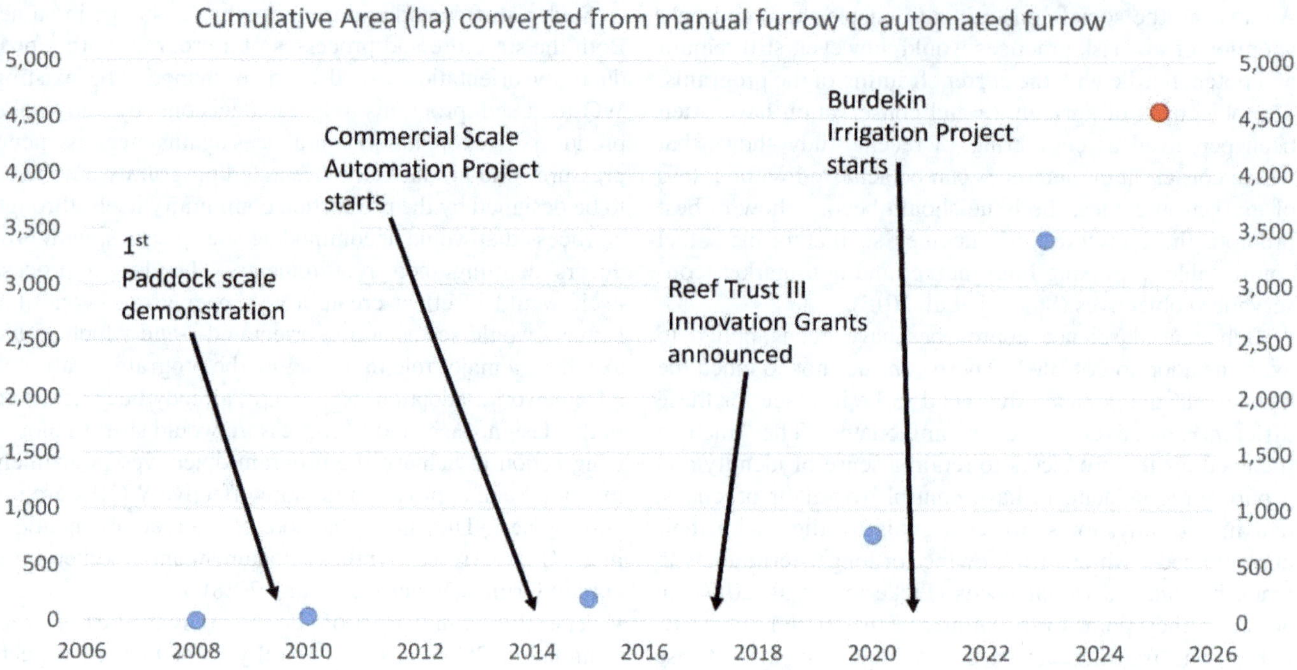

FIGURE 9 Recent increases in the adoption rate of automated irrigation technology.

Basic automation of furrow irrigation, in the Lower Burdekin, has been demonstrated since 2010 with slow uptake (blue dots). Improved platforms and consultation with producers have seen an additional 2,700 ha transitioned to automation, with farmers, involved in the Burdekin Irrigation Project committing to at least an additional 1,000 ha by 2025 (red dot).

Source: Chart courtesy of AgriTech Solutions

pollutant loads to the GBR. These proposals have recently been rejected by the current state and federal governments, but the proponents continue to argue for their construction. This illustrates clearly competing market and environmental protection objectives as noted by Dale et al. (2018), indicating the need for an independent institution to ensure that both objectives are taken into account when major new development project are proposed.

Stricter application of the Environmental Protection Act 1994 (Qld) has been seen by some as one way to achieve changes in farmers practices. However, turning to enforcement can have perverse outcomes. The authors are aware of repeated unwillingness to countenance improved practices and even willingness of some farmers to suppress their innovation for fear that new practices might come to be legislated. Some proponents see stricter enforcement of regulation as a way to rationalise the sugar industry, which is likely to force the exit of smaller producers from the industry (Deane et al. 2018). This might make economic sense, but it would further contribute to the decline in the population of rural communities in North Queensland. Apart from the practical (verification) and political difficulties of robust enforcement, it is generally recognised that although regulations 'can be effective augmentations to a behaviour change strategy, they are rarely fully effective in their own right' (Pickering et al. 2017).

There have been calls that more transformative approaches to reducing pollutant loads on the GBR should be considered (Kroon et al. 2016). These have included a wide range of approaches, for instance, with policy support, shifting land use from sugarcane to less N-dependent broadacre and tree crops (Thorburn and Wilkinson 2013), improving the hydrological performance of riparian and wetland areas so that they provide ecological filtering and buffering of pollutants (Kroon et al. 2013; Lough et al. 2015; Thorburn et al. 2013), and retirement of less productive land from production. These may in fact all have a role to play, but it would seem foolish to turn to them while sugarcane production would continue exporting nutrient loads to the GBR.

Transitioning surface irrigation from the current manual operation to automated systems is now well-proven in practical, economic, and environmental terms. This is an available and sensible strategy that will deliver significant improvement in DIN export to the reef in a relatively short timeframe. This is not without challenge. Transitioning the > 70,000 ha of surface irrigation will still require the design of systems that fit the varied needs of each property by taking into account the soils, farm layout, labour, energy costs, etc., and plans for the property's operation and maintenance should be provided. Transitioning too rapidly without adequate support would lead to poor design, which

FIGURE 10 Shed meetings play an important role by allowing neighbouring farmers to discuss practices and share experiences with new technologies that improve irrigation management. Peer-to-peer interactions are strong influences on the adoption of technology.

Source: Photograph courtesy of Sugar Research Australia

would result in systems with inadequate capacity, high energy costs, and mismatched scheduling. Finally, perhaps the most important factor is a realistic timeframe for the transition. Individual landowners may already have the desire, confidence, and resources to transition quickly, but the majority would still need to become engaged with the new program.

As we move towards the next WQIP, the context for sugarcane production itself is also changing. Energy and fertiliser costs have already risen considerably over the last decade and with the current global disruption, are predicted to continue to rise steeply. The demand for Australian sugar itself may also be negatively affected, firstly, by the expansion and improved productivity of sugar production from competing countries, such as Brazil (de Oliveria Bordonal et al. 2018; Caldarelli and Gillo 2018), and secondly, by the emerging requirement of world markets that agricultural products have green credentials and be traceable. These factors may encourage farmers to be more receptive to practices that both improve profitability and can demonstrate that they are more environmentally sustainable rather than practices that just increase productivity.

Through broadscale adoption of the low-risk 'B' practices, sugarcane would achieve the target reductions in DIN. The barriers to this could be addressed by a similarly transformative approach for future WQIPs that align with producers' concerns and that articulate productivity goals (Figure 10).

25.7. CONCLUSION

Although rising temperatures due to climate change are the main influence on the health of the GBR, water quality remains a significant factor. The health of the GBR, a World Heritage-listed area under the care of Australia, is coming under increasing scrutiny by the UN and other bodies around the world. There are limited opportunities for Australia to counter the trend in water temperature in the GBR, but there are opportunities for Australia to achieve a direct and significant reduction of nutrients flowing from land-based activities onto the GBR.

Irrigators in the Lower Burdekin are in a special position to be able to reduce the DIN loads exported to the GBR. The abundant water for surface irrigation that has driven the high export of nutrient loads also provides levers that producers can now use to manage this. Low-risk practices for fertilizer application and water-efficient irrigation have been identified under commercial field conditions. Application of these practices would achieve a reduction of DIN to below the Environmental Relevant Targets (ERTs) and at the same time, enhance the sustainability of sugarcane production and the communities in the Lower Burdekin.

There is some despair, however, that after several decades of WQIPs to support the application of low-risk practices, a wave of broadscale adoption has not developed. A range of factors have influenced this including 1) focussing on fertilizer rates and less so on irrigation practices, 2) not recognizing that farmers also require convenient and labour-efficient practices, and 3) a degree of social resistance to the low-risk practices due to scepticism and negative branding. These perspectives have persisted along with multiple stakeholders focusing on specific and seemingly competing issues that include crop productivity, water quality, salinization and rising water tables, and in the process attempting to deal with them separately. In fact, all of these issues are driven by the application of irrigation water that activates and mobilises nutrients in the soil.

The past decades of research and piloting of low-risk practices are beginning to bear fruit. And along with these are two other important changes, a) new digital platforms that make it feasible for farmers to apply the low-risk practices and save labor, and b) projects and approaches that emphasise consultation and planning with local communities that see the traditional resistance to water-efficient irrigation decrease. Many external factors affecting the industry in the Lower Burdekin have changed since the early WQIPs. Producers now need to take into account rising costs of power for irrigation and fertilizer inputs. World markets seeking products that have green credentials could become less accessible to the local sugar market.

This chapter argues that with these new factors affecting producers, new accessible practices will be available, and by reframing WQIPs, it should be possible to achieve broadscale committed application of low-risk irrigation practices and thus see nutrient flows to the GBR reduced to limit further damage and eventually enable reef recovery.

ACKNOWLEDGEMENTS

The authors wish to acknowledge two key sources of information used to examine the issues of addressing water quality in the GBR. Firstly, we recognise the Scientific Consensus Statement 2017 Chapters 1, 2, 3, 4, and 5, as compiled by many researchers, and some of whom are also contributing to this book. Secondly, we acknowledge the work and reports of the NQ Dry Tropics, the NRM responsible for the region in which the Lower Burdekin lies.

REFERENCES

ABC Rural 25 Nov 2020, accessed: www.abc.net.au/news/2020-11-25/raising-burdekin-falls-dam-could-be-environmental-disaster/12915820.

AgriTech Solutions and NQ Dry Tropics (2020) Assessment Report of the Reef Trust III: Reef Alliance Project, Burdekin Innovation Program. A report for the Great Barrier Reef Foundation, Brisbane.

Amour JD, Hately IR, Pitt GL (2009) Catchment modelling of sediment, nitrogen and phosphorous nutrient loads with sedNet/ANNEX in the Tully-Murray basin. Marine and Freshwater Research 60:1091–1096.

Armstrong DP, Ho CKM (2011) Economic analysis of automatic flood irrigation for dairy farms in northern Victoria. AFBM Journal 8:1.

Attard S, O'Donnell D (2013) Drip irrigation on sugarcane in the Burdekin region (case study). NQ Dry Tropics, Townsville.

Attard SJ, Inman-Bamber NG, Hesp C (2009) Use of overhead irrigation in the lower Burdekin uses less water and does not penalise yield. Proceedings Australian Society of Sugar Cane Technologists 31:230–239.

Australian Government (2020) Reef 2050 Long Term sustainability Plan. Public Consultation Draft. Australian Government.

Bartley R, Waters D, Turner R, Kroon F, Wilkinson S, Garzon-Garcia A, Kuhnert P, Lewis S, Smith R, Bainbridge Z, Olley J, Brooks A, Burton J, Brodie J, Waterhouse J (2017) Scientific Consensus Statement 2017: A synthesis of the science of land-based water quality impacts on the Great Barrier Reef, Chapter 2: Sources of sediment, nutrients, pesticides and other pollutants to the Great Barrier Reef. State of Queensland, Brisbane.

Blackstock KL, Ingram J, Burton KM, Slee B (2010) Understanding and influencing behaviour change by farmers to improve water quality. Science Total of the Environment. 408:5631–5638.

Brodie J (1997) The water quality status of the Great Barrier Reef World Heritage Area. In: State of the Great Barrier Reef World Heritage Area Workshop, eds. Wachenfeld D, Oliver J, Davis K. Great Barrier Reef Marine Park Authority, Townsville.

Brodie J, Baird M, Waterhouse J, Mongin M, Skerratt J, Robillot C, Smith R, Mann R, Warne M (2017) Development of basin-specific ecologically relevant water quality targets for the Great Barrier Reef. TropWATER Report No. 17/38, James Cook University, Published by the State of Queensland, Brisbane, p. 68.

Brodie J, Grech A, Pressy B, Day J, Dale AP, Morrison T, Wenger A (2019) The future of the Great Barrier Reef: The water quality imperative. In Coasts and estuaries: The future, eds. Wolanski E, Day JW, Elliott M, Ramachandran R. Elsevier, Amsterdam.

Brodie J, Waterhouse J, Schaffelke B, Johnson JE, Kroon F, Thorburn P, Rolfe J, Lewis S, Warne MStJ, Fabricius K, McKenzie L, Devlin M (2013) Reef water quality Scientific Consensus Statement 2013. Department of the Premier and Cabinet, Queensland Government, Brisbane.

BSES Ltd (2010) Burdekin irrigation trial—final report August 2010. Internal report for Rural Water Use Efficiency 4 Project.

Calcino DV, Schroeder BL, Hurney AP (2010) Extension and adoption of "six easy steps' nutrient management program in sugarcane production in North Queensland. Proceedings of the Australian Society of Sugar Cane Technology 27.

Calcino D, Schroeder B, Panitz J, Hurney A, Skocaj D, Wood A, Salter B (2022) Australian sugarcane nutritional manual. Sugar Research Australia Limited. First published in 1994 by BSES Limited.

Caldarelli CE, Gillo L (2018) Expansion of sugarcane industry and its effects on land use in Sao Paulo: Analysis from 200 through 2015. Land Use Policy 76:264–274.

Chaiechi T, Gabriel J, Esparon M, Sivers K (2022) Making nature's value visible. Papua New Guinea Government through the Conservation and Environment Protection Authority. https://www.undp.org/papua-new-guinea/publications/making-natures-value-visibile

Coggan A, van Grieken M, Boulier A, Jardi X (2014) Private transaction costs of participation in water quality improvement programs for Australia's Great Barrier Reef: Extent, causes and policy implications: An abatement perspective. Agriculture Water Management 104:10–20.

Collier CJ, Adams M, Langlois L, Waycott M, O'Brien K, Maxwell P, McKenzie L (2016) Thresholds for morphological response to light reduction for four tropical seagrass species. Ecological Indicators 67:358–366.

Connolly NM, Pearson RG, Loong D, Maughan M, Brodie J (2015) Water quality variation along streams with similar agricultural development but contrasting riparian vegetation. Agriculture, Ecosystems & Environment 213: 11–20.

Dale AP, Vella K, Gooch M, Potts R, Pressy RL, Brodie J, Eberhard R (2018) Avoiding implementation failures in catchment landscapes: A case study in governance of the Great Barrier Reef. Environmental Management 62: 70–81.

de Oliveria Bordonal, Carvalho JLN, Lal R, de Figueiredo EB, de Oliveria BG, La Scala Jr N (2018) Sustainability of sugarcane production in Brazil. A review. Agronomy for Sustainable development 38:13.

Deane F, Wilson C, Rowlings D, Webb J, Mitchell E, Hamman E, Grace P (2018) Sugarcane farming and the Great Barrier Reef: The role of a principled approach to change. Land Use Policy 78:691–698.

De'ath G, Fabricius KE, Sweatman H, Puotinen M (2012) The 27-year decline of coral cover on the Great Barrier Reef and its causes. Proceedings of the National Academy of Sciences 190:17995–17999.

Department of Natural Resources and Mines (2013) Burdekin groundwater management area—rising water tables: Part A—an estimate of impacts of irrigation and water distribution activities on groundwater levels.

Eberhard R, Coggan A, Jarvis D, Hamman E, Taylor B, Baresi U, Vella K, Dean AJ, Deane F, Helmstedt K (2021) Understanding the effectiveness of policy instruments to encourage adoption of farming practices to improve water quality for the Great Barrier Reef. Marine Pollution Bulletin 172:112793.

Fabricius KE, Okaji K, De'ath G (2010) Three lines of evidence to link outbreaks of the crown-of-thorns seastar *Acanthaster planci* to the release of larval food limitation. Coral Reefs 29:593–605.

Fabricius KE, Cséke S, Humphrey C, De'ath G (2013). Does trophic status enhance or reduce the thermal tolerance of scleractinian corals? A review, experiment and conceptual framework. PLoS ONE 8(1):e54399.

Falconer K (2000) Farm level constraints on agri-environmental scheme participation: A transactional perspective. Journal of Rural Studies16:379–394.

Farr M, Eagle L, Hay R, Churchill M (2017) Questionnaire design, sampling strategy and preliminary findings: The Burdekin region. NESP Project 2.1.3 Interim report. Report to the National Environmental Science Program. Reef and Rainforest Research Centre Limited, Cairns, p.124.

Ferraro P, Messer K, Wu S (2019) Applying behavioral insights to improve water security. Choices 32.

Furnas MJ, Brodie J (1996) Current stats of nutrient levels and other water quality parameters in the Great Barrier Reef. In Downstream effects of land-use, eds. Hunter HM, Eyles AG, Rayment GE. Department of Natural Resources, Brisbane, pp. 9–28.

Furnas MJ, Mitchell A, Skuza M (1997) Shelf scale nitrogen and phosphorous budgets for the Central Great Barrier reef (16°–19° S). Proceeding of the 8th International Coral Reef Symposium, Panama, pp. 809–814.

Furnas MJ, O'Brien D, Warne M (2013) Chapter 2: The Redfield Ratio and potential nutrient limitation of phytoplankton in the Great Barrier Reef. In: Assessment of the relative risk of water quality to ecosystems of the Great Barrier Reef: Supporting Studies. A report to the Department of the Environment and Heritage Protection, Queensland Government, Brisbane. TropWATER Report 13/30. Townsville, Australia.

GBRMPA (2001) Great Barrier reef catchment water quality action plan, key issues: water quality & coastal development, a report to ministerial council on targets for pollutant loads. Great Barrier Reef Marine Park Authority, Townsville.

Gillies M, Attard S, Foley J (2017a) Modernisation of furrow irrigation in the sugar industry: Final report project 2014/079. Sugar Research Australia Limited, Brisbane.

Gillies M, Attard S, Foley J, Davis M (2019) Cost-benefit analysis of automated furrow irrigation in the Burdekin. Proceedings of the Australian Society of Sugar Cane Technologists 41:320–327.

Gillies M, Attard S, Jaramillo A, Davis M, Foley J (2017b) Smart automation of furrow irrigation in the sugar industry. Proceedings of the Australian Society of Sugar Cane Technologists 39:320–325.

Great Barrier Reef Foundation (GBRF 2023). https://www.barrierreef.org/what-we-do/reef-trust-partnership/water-quality-improvement/regional-actions/lower-burdekin-water-quality-program

Harvell D, Jordán-Dahlgren E, Merkel S, Rosenberg E, Raymundo L, Smith G, Weil E, Willis B (2007) Coral disease, environmental drivers, and the balance between coral and microbial associates. Oceanography 20:172–195.

Hasan S, Smart JCR, Hay R, Rundle-Thiele S (2020) Changing fertilizer management practices in sugarcane production: Cane grower survey insights. Land 10(2):98.

Hay R, Eagle L, (2019) Findings from a longitudinal study of farmer decision influencers for Best Management Practices, Queensland, Australia. Report to the National Environmental Science Program. Reef and Rainforest Research Centre Limited, Cairns (p. 209).

Hesp C, Attard S, McShane T, Poggio M, Page J (2011) Evaluating alternative irrigation for a greener future. Final Report for the Sugar Research and Development Corporation, Research Project MAF002. CSIRO Sustainable Ecosystems, Brisbane.

Hutchings P, Peyrot-Clausade M, Osnorno A (2005) Influence of land runoff on rates and agents of bioerosion of coral substrates. Marine Pollution Bulletin 51(1):438–447.

Kroon FJ, Thorburn P, Schaffelke B, Whitten S (2016) Towards protecting the Great Barrier Reef from land-based pollution. Global Changing Biology 22:1985–2002.

Kroon FJ, Turner R, Smith R, Warne MStJ, Hunter H, Bartley R, Wilkinson S, Lewis S, Waters D, Carroll C (2013) Chapter 4: Sources of sediment, nutrients, pesticides and other pollutants in the Great Barrier Reef catchment. In 2013 Scientific Consensus Statement. The State of Queensland, Brisbane.

Lee J (2022) Evaluation of automatic irrigation system for rice cultivation and sustainable agriculture water management. Sustainability 14(17):11044.

Lough JM, Lewis SE, Cantin NE (2015) Freshwater impacts in the central Great Barrier Reef: 1648–2011. Coral Reefs 34:739–751.

McCloskey GL, Baheerathan R, Dougall C, Ellis R, Bennett FR, Waters D, Darr S, Fentie, B, Hately LR. Askildsen M (2021) Modelled estimates of dissolved inorganic nitrogen exported to the Great Barrier Reef lagoon. Marine Pollution Bulletin 171(1):112655.

Mitchell A, Reghenzani J, Faithful J, Furnas M, Brodie J (2009) Relationships between land use and nutrient concentrations in streams draining a 'wet-tropics' catchment in northern Australia. Marine and Freshwater Research 60:1097–1108.

Munasinghe M (1992) Biodiversity protection policy. Environmental valuation and distribution issues. Ambiology 21(3):227–236.

O'Mahoney J, Simes R, Redhill D, Heaton K, Atkinson C, Haywood E, Nquyen M (2017) At what price? The economic, social and icon value of the great barrier reef. Deloitte Access Economics, Townsville.

Oxford Economics (2009) Valuing the effects of Great Barrier Reef Bleaching. Report prepared for the Great Barrer Reef Foundation, Brisbane.

Pagiola S, Von Ritter K, Bishop J (2004) Assessing the economic value of ecosystem conservation. In collaboration with the nature conservancy and IUCN—The World Conservation Union. The World Bank, Townsville.

Petus CC, Collier CJ, Devlin M, Rasheed MA, McKenna S (2014) Using MODIS data for understanding changes in seagrass meadow health: A case study in the Great Barrier Reef (Australia). Marine Environmental Research 98:68–85.

Pickering JA, Hong J, Hong D, Kealley M (2017) Applying behavioural science to the Queensland sugar cane industry and its relationship to the Great Barrier Reef. Rural Extension and Innovation Systems Journal 13(2).

Pickering JA, McIntosh T, Moore S, Priwitzer S, Haatera K, Preston G, Hong J, Law I Win, Simmons E, Ruzsicska N, Kealley M (2019) Project cane changer: Using behavioural science to create practice change. Proceedings of the Australian Society of Sugar Cane Technology 41:101–107.

Pollock FJ, Lamb JB, Field SN, Heron SF, Schaffelke B, Shedrawi G, Bourne DG, Willis BL (2016) Correction: Sediment and turbidity associated with offshore dredging increase coral disease prevalence on nearby reefs. PLoS One 11(11), e0165541.

Raine SR, Bakker DM (1996) Increased furrow irrigation efficiency through better design and management of cane fields. Proceedings Australian Society of Sugar Cane Technologists, 18:119–124.

Reef Trust Partnerships, Lower Burdekin Regional Plan (2020) https://www.barrierreef.org/uploads/WQ-RP-Lower-Burdekin-Regional-Plan.pdf.

Reef Water Quality Protection Plan Secretariat (2013) Reef Water Quality Protection Plan 2013.

Rolfe J, Windle J (2016) Benchmarking costs of improving agricultural water management in GBR catchments. Report to the National Environmental Science Program. Reef and Rainforest Research Centre Limited, Cairns, p. 52.

Schroeder BL, Moody PW, Wood AW (2003) Improved nutrient management in the Australian sugar industry. Sugar Research and Development Corporation Project BSS232 Final Report. Bureau of Sugar Experiment Stations, Indooroopilly, Australia.

Schroeder BL, Wood AW, Sefton M, Hurney AP, Skocaj DM, Stainley T, Moody PW (2010) Proceedings Australian Society of Sugar Cane Technologists 32:193–209.

Shaw RJ (1988) Predicted deep drainage loss under dryland and irrigation managements for Burdekin soils. Pages 37–48 in Soils of the Burdekin River Irrigation Area, 1988, Workshop Proceedings, Ayr. Publication QC89003, edited GE Rayment & VE Eldershaw. Queensland Department of Primary Industries, Brisbane.

Shaw RJ (2014) Groundwater in the BRIA and recommendations to achieve sustainable groundwater management under Local management authority approach—Report to the Burdekin River Irrigation Area Irrigator's Committee.

State of Queensland (2013) Burdekin groundwater management area—rising water tables. Part A—an estimate of the impacts of the irrigation and water distribution activities on groundwater levels 2002–2010. State of Queensland, Brisbane.

State of Queensland (2017) Lower Burdekin groundwater strategy project, Discussion Paper. State of Queensland, Brisbane.

State of Queensland and Commonwealth of Australia (2003) Reef Water Quality Protection Plan; for catchments adjacent to the Great Barrier Reef World Heritage Area. Queensland Department of Premier and Cabinet, Brisbane.

State of Queensland and Commonwealth of Australia (2022) In reef water quality report card 2020. State of Queensland and Commonwealth of Australia, Brisbane.

Stoeckl N, Chaiechi T, Farr M, Jarvis D, Álvarez-Romero J, Kennard MJ, Hermoso V, Pressey RL (2015) Co-benefits and trade-offs between agriculture and conservation: A case study in Northern Australia, Biological Conservation 191:478–494.

Stoeckl N, Hicks C C, Mills M, Fabricius K, Esparon M, Kroon F, Costanza R (2011) The economic value of ecosystem services in the GBR: Our state of knowledge. Annals of the New York Academy of Sciences, 1219(1):113–133.

Taylor BM, van Grieken M (2015) Local institutions and farmer participation in agri-environmental schemes. Journal of Rural Studies 37:10–19.

Thompson, A, Schroeder T, Brando VE, Schaffelke B (2014) Coral community responses to declining water quality: Whitsunday Islands, Great Barrier Reef, Australia. Coral Reefs 33(4):923–938.

Thorburn PJ, Biggs JS, Attard SJ, Kemei J (2011) Environmental impacts of irrigated sugarcane production: Nitrogen lost through runoff and leaching. Agriculture, Ecosystems & Environment 144(1):1–12.

Thorburn PJ, Wilkinson SN (2013) Conceptual frameworks for estimating the water quality benefits of improved agricultural management practices in large catchments. Agriculture, Ecosystems & Environment 180:192–209.

United Nations (2012) SEEA-water: System of environmental-economic accounting for water. Statistical Papers, Series F, No. 100. Sales No. E11.XVII.12.

Wang E, Attard S, Linton A, McGlinchey M, Xiang W, Philippa B, Everingham Y (2020) Development of a closed-loop irrigation system for sugarcane farms using the Internet of Things. Computers and Electronics in Agriculture 172:10.

Waterhouse J, Attard S, Rickert A, Buono T, Hunt R (2018) Burdekin Water Quality Improvement Plan Lower Burdekin Implementation Plan: Water quality improvement through improved irrigation management in sugarcane. Phase 1—Knowledge synthesis, evaluation of past programs and identification of preliminary strategies for implementation. NQ Dry Tropics.

Waterhouse J, Brodie J, Tracey D, Smith R, VanderGragt M, Collier C, Petus C, Baird M, Kroon F, Mann R, Sutcliffe T, Waters D, Adame F (2017) 2017 Scientific Consensus Statement: Land use impacts on the Great Barrier Reef water quality and ecosystem condition, Chapter 3: The risk from anthropogenic pollutants to Great Barrier Reef coastal and marine ecosystems. State of Queensland.

Williams J, Stubbs T, Bristow KL (2009) Final Report: The water and salt balances of the Burdekin River Irrigation Area. Importance for strategic planning and institutional arrangements for the entire Lower Burdekin. Prepared for Queensland Canegrowers Pty Ltd, Brisbane.

World Commission on Protected Areas (WCPA) (1998) Economic values of protected areas: Guidelines for protected areas managers (Best Practice Protected Area Guidelines Series No. 2). ed. Adrian Phillips. World Commission on Protected Areas, Cambridge.

26 Raine Island Recovery Project—Intervening at One of the Most Significant Sites on the Great Barrier Reef

Katharine Robertson, John Dawson, and Owen Coffee

26.1. INTRODUCTION

The Great Barrier Reef (GBR) includes more than 900 islands, which are an important part of the GBR ecosystem including nesting habitats for marine turtles and seabirds. Raine Island (11° 35' 25" S, 144° 02' 05" E), a 27-hectare vegetated cay located on a detached reef on the outer edge of the northern GBR, is one of the most important islands on the GBR (Figure 1). The island is highly protected as part of the Raine Island National Park (Scientific), which also incorporates Moulter (11° 24' 33" S, 144° 02' 13" E) and MacLennan (11° 19' 18" S, 143° 48' 41" E) Cays, due to its natural, ecological, cultural, and scientific values. It is

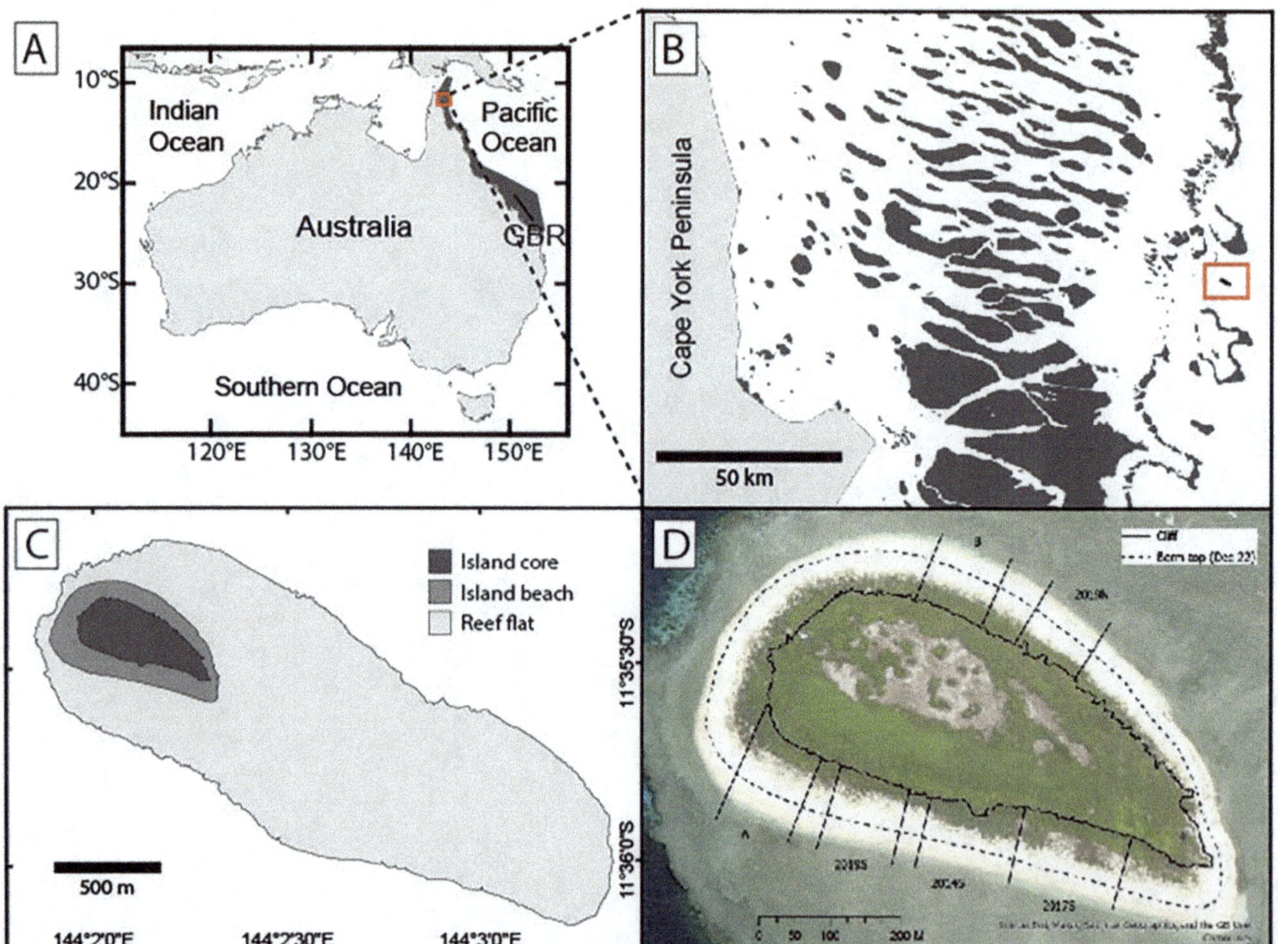

FIGURE 1 Location map of Raine Island. (A) General location within the western Pacific region. (B) Enlarged view of the GBR showing the position of Raine Reef (red box). (C) Enlarged view of Raine Island and the surrounding reef platform showing the stable island core (primary seabird rookery) and the surrounding beach (primary turtle rookery). (D) Map of the study sectors on Raine Island beach including control sector "A" and "B" and reprofiled sectors from 2014, 2017, and 2019.

DOI: 10.1201/9781003320425-30

one of the world's largest remaining rookeries for the internationally endangered green turtle (Seminoff 2004), supporting the nesting for 90% of the northern GBR (nGBR) green turtle population (Limpus et al. 2003). Raine Island is also a significant seabird rookery in the region and is considered one of the most important seabird nesting areas in the GBR World Heritage Area (Batianoff and Cornelius 2005). The Wuthathi People from eastern Cape York Peninsula and the Meriam Nation People from the eastern Torres Strait islands of Ugar, Mer, and Erub have strong cultural connections and assert custodial responsibility for the Raine Island National Park (Scientific). The island and the species that nest on the island are culturally significant to both the Wuthathi People and Meriam Nation People.

On the GBR, there are two genetically distinct nesting populations of green turtles, namely, the southern GBR population and the northern GBR (nGBR) population (Limpus et al. 2003). Despite the size of the nGBR population, Raine Island has been considered a failing turtle rookery since at least the mid-1990s, raising concerns for the future of this population (Limpus et al. 2003; Limpus 2008; Jensen et al. 2016). In the 1970s and 1980s, it was reported that 'vast numbers of hatchlings were crossing the beach throughout the night' and nest emergence success was above 75% (Limpus et al. 2003). However, in the 1996–1997 breeding season, coinciding with observations of tidal inundation throughout the nesting beach, few hatchlings were sighted. It was inferred that the tidal inundation of the nesting beach was causing developing clutches to drown, reducing hatchling numbers (Limpus et al. 2003).

Marine turtles are slow-growing and long-lived with adult green turtles taking several decades to reach maturity. As a result, any impact and/or reduction in population would not be seen at nesting grounds for decades. Recent foraging ground studies have reported a decrease in the proportion of juvenile nGBR green turtles recruiting into feeding areas, supporting observations of limited hatchling production at Raine Island, with very little production of male hatchlings (Fuentes et al. 2010; Jensen et al. 2016, 2018; Booth et al. 2020b). Poor hatchling production, recorded since at least the mid-1990s, in addition to pressures from climate change such as rising sea levels, biased female sex ratios due to rising temperatures, and further reductions in hatching success from extreme nest temperatures are predicted to have devastating consequences to the long-term viability of the nGBR green turtle population (Limpus et al. 2003; Jensen et al. 2018; Booth et al. 2020b).

With these identified issues and other broader threats to the population, it was decided to undertake an adaptive management approach with an objective to improve breeding success in parallel with undertaking further monitoring and research. The Queensland Government initiated the Raine Island Recovery Project which aimed to protect and restore the island's critical habitat to ensure the future of key marine species, including green turtles and seabirds (Coffee and Robertson 2022). The Project was a collaborative partnership among the Queensland Government, the GBR Marine Park Authority, the Wuthathi People and Meriam Nation (Ugar, Mer, Erub) peoples, the company BHP and the GBR Foundation, which aimed to deliver pioneering adaptive management action to restore viable turtle nesting habitat at Raine Island. The Project was not undertaken lightly and represents a paradigm shift for park management to embark on the world's first intervention works at a highly protected, environmentally and culturally significant site (Coffee and Robertson 2022).

Management agencies identified that the beach could be artificially reprofiled using heavy machinery to raise nest levels above the highest tide levels and restore the beach to viable turtle nesting habitat (Coffee and Robertson 2022; Hamann et al. 2022; Smithers and Dawson 2023). By redistributing sand from the top of the beach and beach slope to the lower areas of the swale and back swale (Figure 2), this would increase hatchling production as incubating eggs would no longer be vulnerable to flooding (Smithers and Dawson 2023). In addition, reprofiling low-lying sections of the nesting area was predicted to increase the area utilised by nesting turtles and increase nesting success, as some areas of the swale did not provide enough sand depth for nesting turtles to dig a nest (Coffee and Robertson 2022; Hamann et al. 2022).

26.2. SITE DESCRIPTION

Raine Island (11°35′25″S 144°02′06″E) is a vegetated sand cay that has developed over the past 3–4 kyr. Situated approximately 100 km off the northeast coast of Australia on one of a series of detached reefs (Figure 1), Raine Island is the largest vegetated cay located on the outer GBR and as such, is a prominent habitat for a wide range of fauna. Island geomorphology has been previously described in detail (Dawson and Smithers 2014; Smithers and Dawson 2023). In summary, Raine Island is approximately 820 m long and 440 m wide, although size and shape are variable among seasons and years (Dawson 2021). Maximum elevation of the island is ~8 m above lowest astronomical tide (LAT), attained in the central core, while the contemporary beach (green turtle nesting habitat) is almost exclusively below 4 m LAT (Figure 2). During the previous decade of turtle research undertaken at Raine Island, the contemporary beach was classified into four main sections from the toe of the beach (intersection between the bottom of the beach slope and the reef flat) and the phosphatic cliff including the i) berm; ii) back berm, iii) swale; and iv) back swale (Booth et al. 2020b) (Figure 2).

Raine Island has a unique standing and very high environmental, cultural, and ecological value extending back to before the European settlement of Australia's mainland. However, in recent decades, threats to these unique values have been identified and have provided stimulus for a shift from passive (focusing on maintaining natural integrity to support system resilience) to more adaptive (direct intervention) management strategies. One particular stimulus has been the decline in hatching success of green turtles

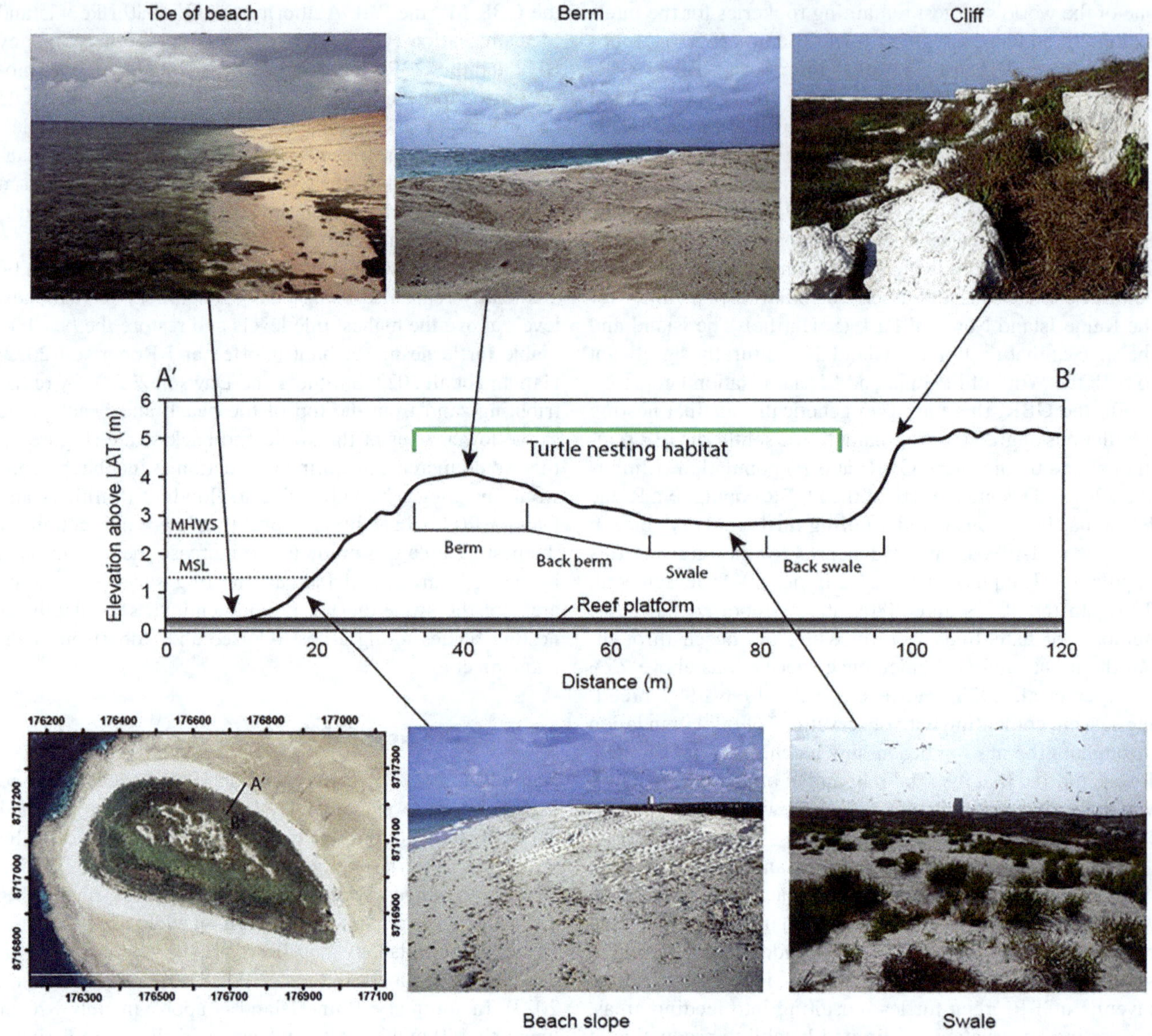

FIGURE 2 Schematics of a typical beach cross-section at Raine Island including photographs of the main geomorphological features. Turtle nesting habitat encompasses the entire berm and swale. The position of the transect is shown in the bottom-left photograph.

observed since the mid-1990s, which has largely been attributed to the tidal inundation of nests in the lower elevated swale regions of the beach (Limpus et al. 2003; Pike et al. 2015) (Figure 3A), although a range of additional factors may contribute to embryo mortality prior to hatching (Booth and Dunstan 2018; Booth et al. 2020a and b; Booth et al. 2021; Laloe et al. 2021).

The most significant adaptive intervention strategy undertaken so far has been the artificial reprofiling of large sections of the beach to raise the elevation of turtle nesting habitat and therefore reduce the number of egg clutches that are drowned during peak high tides or storm surges (Smithers and Dawson 2023) (Figure 3). At Raine Island, estimates of egg chamber depth below the beach surface range from 0.65 m (based on survey data collected during the Raine Island Recovery Project) to 0.82 m (Limpus et al. 2003). This typically means that eggs laid within the lower elevated swales (~2.5–3 m LAT) would be deposited between the mean sea level and the mean high water springs, where they are vulnerable to inundation (Figure 3C).

26.2.1. What Is Beach Reprofiling?

Beach reprofiling, sometimes referred to as "beach scraping", is a "soft engineering" strategy that involves the physical redistribution of sediment from the lower parts of the beach to the upper parts of the beach (Staudt et al. 2021), typically with the aid of heavy machinery (Figure 3B). The engineering strategy mimics the natural process of berm build-up by overtopping waves—a process observed at Raine

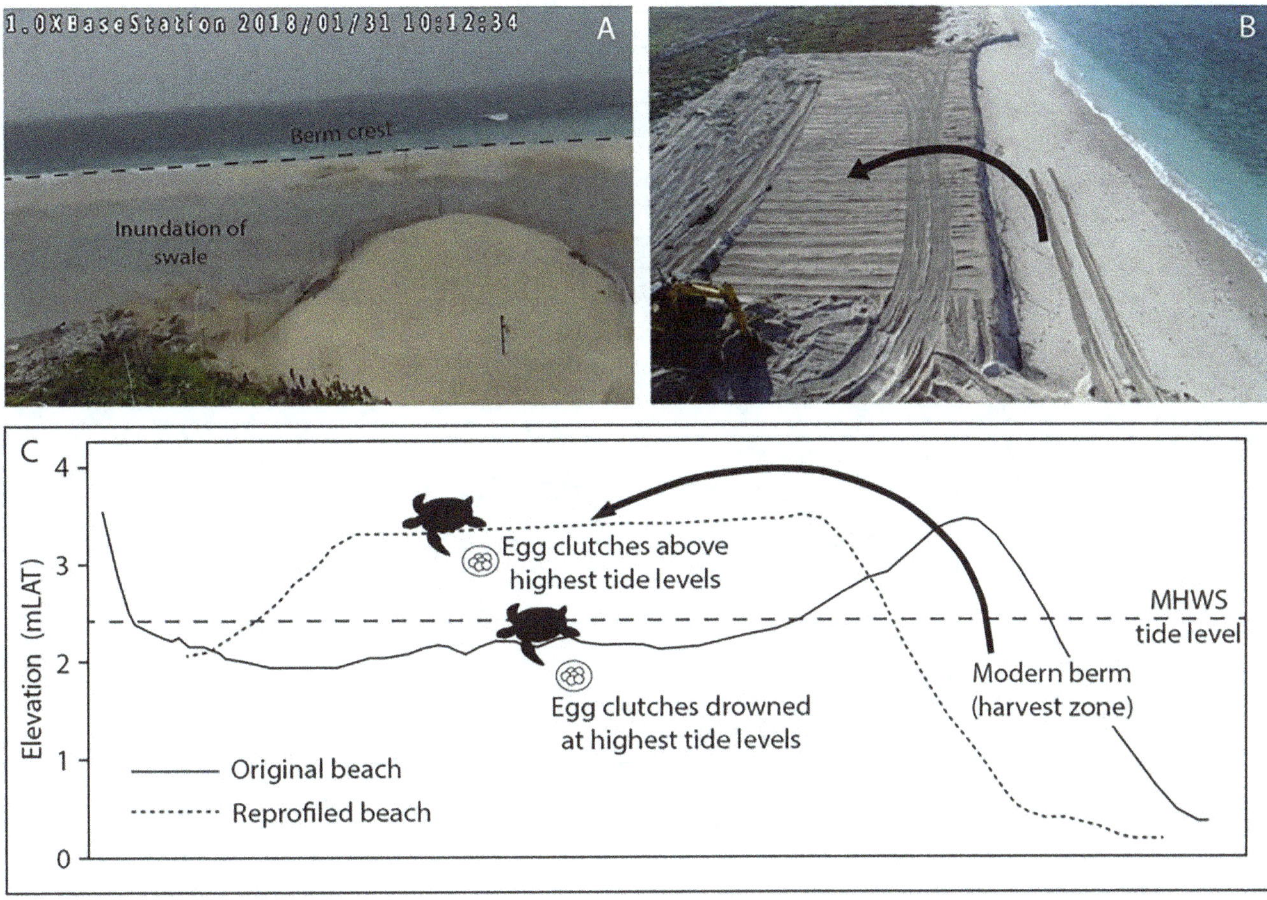

FIGURE 3 Beach reprofiling at Raine Island. (A) Photograph showing peak tidal/storm surge inundation of the swale at the western end of Raine Island in January 2018. (B) Photograph showing an example of mechanical reprofiling of the beach using heavy machinery. (C) Schematic showing nesting beach reprofiling approach. Mean high water spring (MHWS) tide level is 2.4 m above LAT while average turtle egg chamber depth is ~0.65 m below the surface. In this example, much of the swale is raised from about 2 m LAT to approximately 3.5 m LAT (1.1 m above MHWS).

Island (Gourlay and Hacker 1991) and other reef islands (Smithers and Hoeke 2014). It has commonly been used as a strategy for coastal erosion on mainland coastlines (Staudt et al. 2021), but it has been rarely documented in reef island settings (Duvat 2009), likely due to the logistical challenges involved in carrying out such an exercise. Nevertheless, despite the recognition of such challenges, an expert panel agreed that reprofiling the beach was the best management strategy to increase the area of viable nesting habitat on the island and reduce fatal nest inundation. Traditional Owners (the Wuthathi and Meriam Nation People) were involved at all stages (proposal, planning, and implementation) and were strongly supportive of the strategy.

26.3. RESULTS

26.3.1. How Was Beach Reprofiling Implemented?

Beach reprofiling was initially informed by investigations of Raine Island's geomorphological development, coastal morphodynamics, sediment budget, and local hydrodynamic regime (Dawson and Smithers 2020; Smithers and Dawson 2023). During planning, careful consideration was given to minimise negative impacts and to maximise positive outcomes. Particular attention was given to the potential of exposure of underlying beachrock and timing reprofiling to minimise disturbance to nesting birds and turtles and to avoid erosion caused by seasonal redistribution of harvested sand (Dawson 2021; Smithers and Dawson 2023). The work was undertaken at three stages during Phase 1 of the Raine Island Recovery Project (2014–2021). An initial trial commenced in 2014 followed by two subsequent beach reprofiling works in 2017 and 2019 (Figure 1D). Digital elevation models (DEMs) were used to target infill areas to maximise the nesting area benefit from the harvested sand yield. Meanwhile, historical records of island geomorphology (e.g., satellite imagery and aerial photography) were consulted to ensure targeted areas would be stable over the longer term (Dawson and Smithers 2020). Infilled areas were levelled to the height of the berm crest (~4 m LAT), lifting the elevation across most of the swales by 1–1.5 m (Smithers and Dawson 2023, Figures 3 and 4).

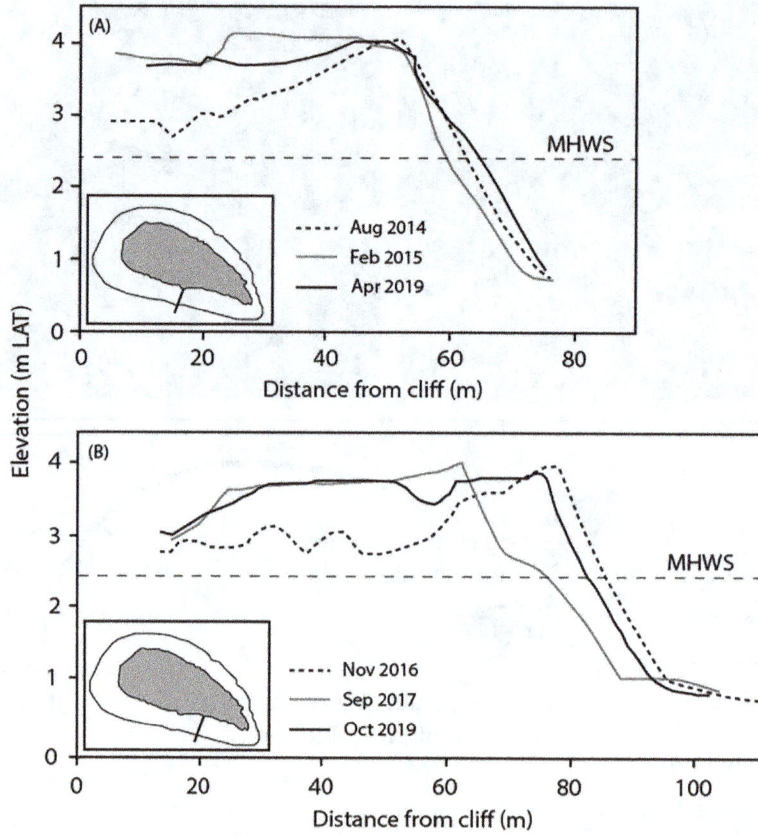

FIGURE 4 Cross-sections of the nesting beach based on topographic surveys prior to reprofiling, soon after, and in 2019.

Stage 1—September 2014

In September 2014, a 150-m section of the southern contemporary beach was reprofiled during which approximately 10,000 m^3 of sand was harvested from the beach slope and transferred into the swale, raising elevations by about 1 m so that the entire section of beach was above 3.5 m LAT (design heights were 4.2 m LAT at the berm gently sloping back toward the cliff) (Figure 4). By the following February, around 75% of the sand harvested from the beach had been replenished by natural redistribution, primarily from the northwest shoreline (Dawson and Smithers 2020; Dawson 2021) (Figure 4). Overall, the reprofiling of the beach in 2014 increased the volume of sand in the sector by approximately 11% and increased the area of available turtle nesting habitat by 400% while also raising nests above peak tidal inundation levels (Dunstan and Robertson 2017; Dawson 2021). The sector was monitored for three nesting seasons, and it was concluded that the results were positive and that the scope of the project could be expanded to other sections of the beach.

Stage 2—September 2017

In September 2017, approximately 14,000 m^3 of sand was harvested from a 170 m section of beach at the eastern end of the southern shoreline and moved up into the adjoining swale (Figures 1D and 4). The method was adjusted slightly based on learning from the initial reprofiling in 2014. Firstly, the top layer of sand was initially stripped and stockpiled to preserve grass runners and seeds, which could be replaced on the surface once the beach had been reprofiled. This was implemented to improve the recovery time of grasses and other plants within the sector, which play an important role in stabilising the surface of the swales. Secondly, infilled areas extended no closer than 10 m from the cliff to protect bird nesting habitat such as that of red-tailed tropicbirds from possible sand drifts. Patterns of natural shoreline recovery were consistent with those observed following initial works in 2014. For example, during the following months, about 6,000 m^3 of sand (~44%) was returned to fill the void produced by harvesting, and by 2019 (just prior to the final reprofiling works), the volume of sand harvested from the beach face in 2017 had been completely replenished by naturally redistributed material from elsewhere on the island (Figure 4).

Stage 3—September 2019

The final stage of beach reprofiling was concluded in September 2019 and involved the harvesting of approximately 16,000 m^3 of sand to infill the swale regions of two

TABLE 1

Anthropogenic modifications of the nesting beach at Raine Island. Values in parentheses indicate percentages of February 2020 island totals.

Date of beach reprofiling	Total length of modified shoreline (m)	Total area of harvested beach (area loss) (m²)	Estimated volume of harvested sand
September 2014	150 (7.1%)	2300 (0.9%)	10,000
September 2017	170 (8.0%)	3400 (1.2%)	14,000
September 2019	240 (11.3%)	3800 (1.4%)	16,000
Total	570 (26.9%)	9500 (3.5%)	40,000

sections of beach, one extending to the west of the 2014 sector and one located on the northern beach (Figure 1D). Each section of reprofiled beach was about 120 m long and extended back to within 10 m of the cliff on the southern beach and approximately 20 m from the cliff on the northern beach. The majority of the material used to infill the swales was harvested from the beach face in the southern sector.

The three stages of reprofiling have transferred a total of ~40,000 m^3 of sand from the beach slopes into the swales and have raised the elevation of the nesting beach by 1–1.5 m. Although this material is not lost from the overall sediment budget (volume), its removal from the beach face has had an impact on the island area. For example, sand harvesting has modified about 570 m of beach (~27% of the total shoreline) and accounts for a total reduction in beach area of approximately 9,500 m^2 (Table 1). However, the effects are small (~1% reduction during each stage) and relatively short-lived as shoreline records confirm that sand harvested from the lower beach is replaced relatively quickly through natural readjustment of the island shoreline (Figure 4). Furthermore, it is difficult to ascertain whether these changes are outside the natural variability (Dawson 2021). Consequently, geomorphological data collected thus far imply that the direct interventions of the project have had negligible impacts on long-term shoreline behaviour and have led to less than a 5% decrease in the proportion of stable shoreline (Dawson 2021).

26.3.2. A Success Story

Rather than witnessing the decline of a vulnerable and world significant green turtle nesting population, the Raine Island Recovery Project delivered pioneering adaptive management strategies. It represents a paradigm shift in parks management by intervening to ensure the preservation of natural and cultural values. To investigate the outcome of the intervention actions, reproductive success and hatchling production were monitored to ascertain seasonal reproductive baselines for the island and compare the non-modified areas of the beach to the reprofiled areas. Nesting females were monitored through nightly nesting success surveys to establish the proportion of the nesting population ashore successfully laying clutches, and some individual turtles were tracked using satellite tags to investigate their reproductive behaviour and use of the reprofiled areas of the beach. Hatchling production was monitored through hatchling counts in sectors and by excavating nests post-development to assess the hatching success of eggs. Complementing the marine turtle research, seabird surveys were conducted to monitor nesting numbers and any impacts to the populations from the management works.

The project has led to a dramatic decrease in the amount of nesting beach that is vulnerable to peak tidal inundation from ~51,000 m^2 (December 2013) to ~16,000 m^2 (December 2019), effectively doubling the area of viable nesting beach at Raine Island from ~35,000 m^2 to ~70,000 m^2 (Dawson and Smithers 2020; Smithers and Dawson 2023). The total area of nesting beach above the inundation threshold of 3.05 m LAT (the sum of MHWS (2.4 m) and nest depths (0.65 m)) improved from 59% in December 2013 to 86% in December 2019 (Figure 5) and from as low as 49% in December 2009 (Dawson and Smithers 2020; Smithers and Dawson 2023). All three stages of beach reprofiling had incremental positive effects on the total area of viable nesting beach at Raine Island (Figure 5).

Nesting success surveys comparing the nesting of turtles ashore within reprofiled sections of the beach with a selection of unmodified control sectors commenced in the 2014–15 breeding season. Following management action conducted in 2017 and 2019, further sectors were added to the comparison in their subsequent nesting seasons (Figure 1D). Except for the 2019–20 and 2020–21 seasons, nesting success in reprofiled areas of the beach was equivalent to or greater than comparable control sectors, demonstrating no negative impact on nesting success from the project's reprofiling actions. In the two seasons, control sector B recorded higher nesting success along the berm possibly due to overtopping waves creating a higher moisture level. Increased nesting success is typically observed following rainfall, and nesting success correlated with sand moisture content.

To understand how individual female turtles utilised the nesting habitat, including the reprofiled sections, and how they responded to the reprofiling actions, 39 green turtles had Argos-linked Fastloc-GPS tags fitted to monitor their

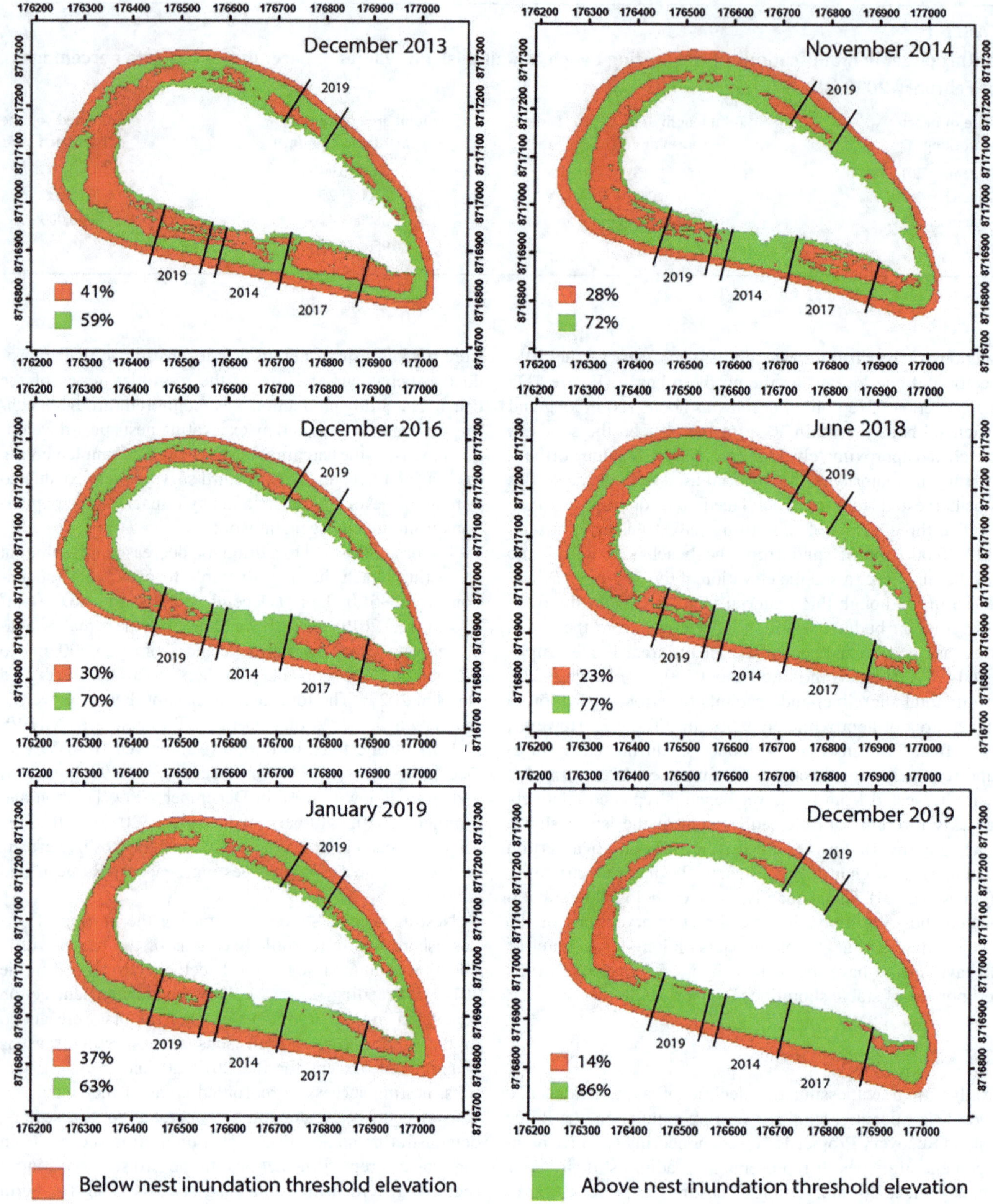

FIGURE 5 DEMs showing areas of Raine Island beach above (green) and below (red) the turtle nest inundation threshold elevation of 3.05 m (peak tidal depth of 2.4 m + typical nest depth of 0.65 m) prior to and following each of the three stages of reprofiling in 2014, 2017, and 2019. Areas on the beach slope have been excluded from calculations as they are not considered viable nesting habitat.

nesting behaviour across the breeding season (Hamann et al. 2022). Tags were fitted to 19 turtles at the start of the 2017–2018 season and 20 turtles at the start of the 2018–2019 season (Hamann et al. 2022). In the 2017–2018 season, almost 60% of clutches were laid on the southern beach, and one-third of all clutches were laid in the reprofiled areas. However, in the 2018–2019 season, the opposite pattern occurred, with 60% of clutches laid on the northern side of the island (Figure 7). In the 2017–2018 season, the 2017 reprofiled area (which was reprofiled only a few months earlier) was a nesting hotspot and contained 28 out of 111 clutches laid by the satellite tagged turtles. In addition, turtles using this reprofiled area had higher nesting success at 70% compared to 36% elsewhere on the island. In 2018–2019, the reprofiled sectors were not nesting hotspots and only 6 of 121 clutches were laid in these sectors. Differential use of the island from season to season indicates that nest site determination by nesting turtles is driven by numerous environmental factors. Nevertheless, the satellite tracking demonstrated that turtles continued to use the reprofiled areas and that intervention works neither negatively impacted nor deterred the green turtles from nesting in these areas. The satellite tracking of individual turtles revealed that immediately after beach reprofiling, nesting success improved in the respective modified sector. This is likely due to the method of reprofiling, as moist sand was reprofiled from the toe of the beach, temporarily increasing moisture content in the nesting habitat and aiding in the construction of viable egg chambers by reducing the likelihood of chamber collapse.

Hatching success of clutches was compared between the unmodified areas and reprofiled areas of nesting habitat to establish whether management interventions improved hatching success for the population. Except for the 2018–19 nesting season, hatching success of clutches laid in reprofiled areas were equivalent to or greater than hatching success recorded in control sectors (Figure 6). Hatching successes have remained variable in both control and reprofiled sectors throughout the monitored seasons of the Project. It is likely that individual variation in hatching success is driven by factors of the nest micro-environment such as nest temperature, sand moisture content, respiratory gas composition, bacterial and fungal pathogens, and proximity to nearby clutches (Booth and Dunstan 2018; Booth et al. 2021 and 2022). However, season wide hatching success is strongly correlated with seasonal nesting population, with highest mean hatching successes typically recorded in low density nesting seasons (Coffee and Robertson 2022).

To determine the effects of reprofiling the nesting beach at Raine Island on hatchling production, we compared the observed numbers of hatchlings per 100 m of beach in reprofiled and non-modified areas. Except for the 2018–19 season, hatchling production in the reprofiled sectors of the nesting beach was equivalent to or greater than hatchling production in the control survey sectors (Coffee and Robertson 2022).

During the project, it has become clear that the issues contributing to elevated hatching failure of eggs and the nGBR green turtle population's low hatchling recruitment are more nuanced than just tidal inundation and the flooding of the nesting habitat. The population's viability still faces significant challenges such as climate change (e.g., impacts to foraging habitat, increases in sea level, and changes in nesting beach sand temperatures). Nevertheless,

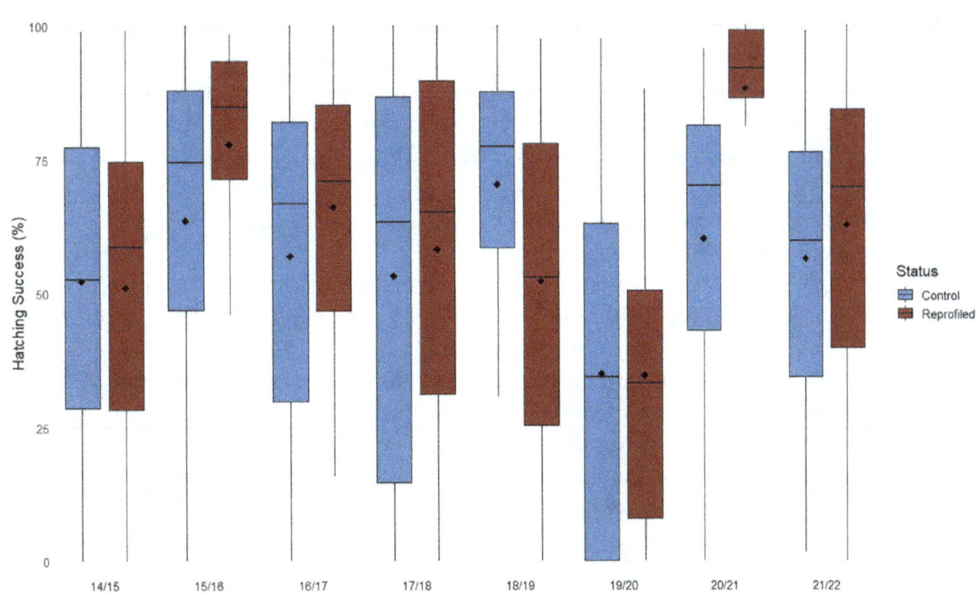

FIGURE 6 A comparison of hatching success between control and reprofiled sectors across the seasons of the project. Diamonds denote sector means, boxes encompass the interquartile range with the median values identified by a horizontal line, and whiskers represent the recorded variation in hatching success for each sector.

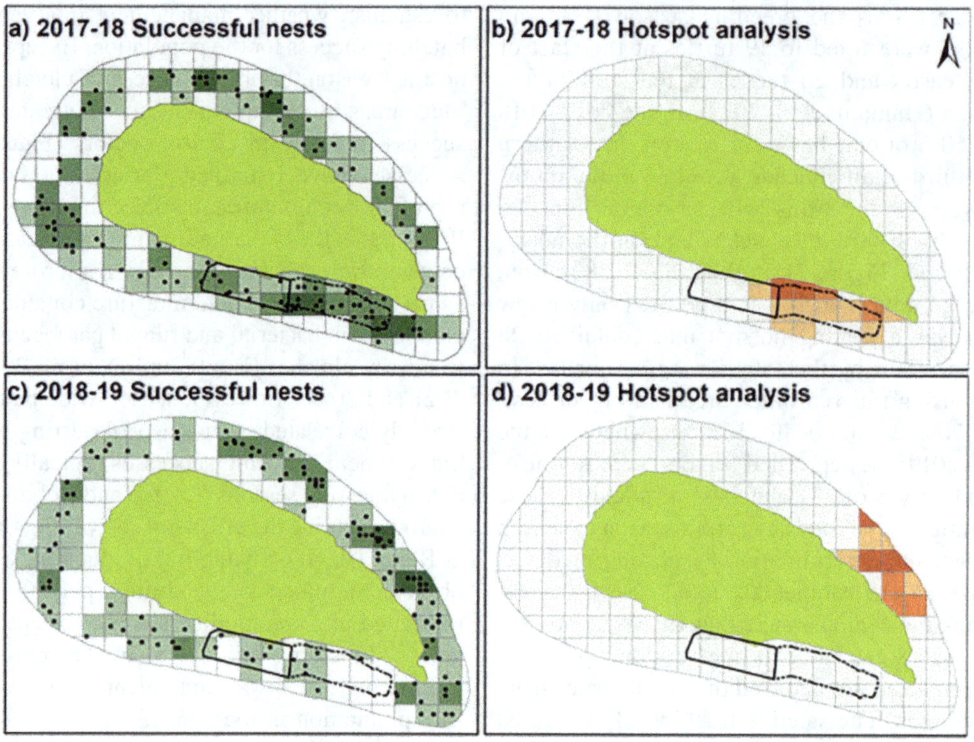

FIGURE 7 Clutch density and hotspot analysis of clutches laid in 2017–18 and 2018–19 by tracked turtles (from Hamann et al. 2022).

the higher hatching success of eggs laid in reprofiled areas is contributing additional hatchlings to the nGBR green turtle population.

Complementing the marine turtle research, seabird surveys were conducted at Raine Island to monitor nesting numbers and any impacts to the populations from the management works. For most bird species that breed on the island, counts have remained stable during the Project, indicating that the sand reprofiling and other management works appear to not have negatively impact the seabirds. Beach reprofiling was conducted outside of the peak seabird nesting season and a majority of the bird nesting occurs above the phosphatic cliff and, therefore, away from the reprofiling works. Lessons learned from the 2014 reprofiling works were applied in stages 2 and 3 to reduce potential impacts on nesting red-tailed tropicbirds and brown noddies. Red-tailed tropicbirds utilise crevices under the cliff for nesting, so the 2017 and 2019 reprofiling works did not extend to the cliff area to protect their nesting habitat from possible sand drifts. The red-tailed tropicbird population has continued fidelity to the island as indicated by banding recoveries. Recaptures of birds that were originally banded in the late 1990s to the late 2000s are regularly recorded each year, indicating that the intervention works have not impacted this species in returning to the island and continuing to nest. Brown noddies, which nest in the vegetated swale area outside of the turtle nesting season, appear to have been impacted from the first reprofiling actions in 2014 as the breeding effort in 2015 and 2016 was lower than in other years. This reduced breeding effort was caused in part by vegetation not recovering on the beach in the 2014 reprofiling area; however, the modified reprofiling method as described above used in stages 2 and 3 sought to allow for a natural recolonisation of grasses, and no impacts to breeding effort were observed after stages 2 and 3 of reprofiling.

26.3.3. Future Projections

Recent studies estimate that it will take about 125 years for natural processes to supply and replenish the volume of sand harvested from the beach slopes. Over this period of time, global sea level rise (SLR) will be a significant threat to Raine Island and to the green turtle and seabird populations that rely on it for nesting habitat. Sea level at Raine Island is rising by ~3.9 mm/yr based on satellite altimetry data (Figure 8), which is slightly above the global mean rate (Horton et al. 2020). Although the strategy used at Raine Island provides a modest buffer (typically about 50–70 cm after settling) to accommodate higher water tables associated with extreme wet weather and/or future sea-level rise, the future viability of Raine Island as a green turtle rookery will depend on the future rates of SLR and its impacts on the supply of sediment from the adjacent reef to offset

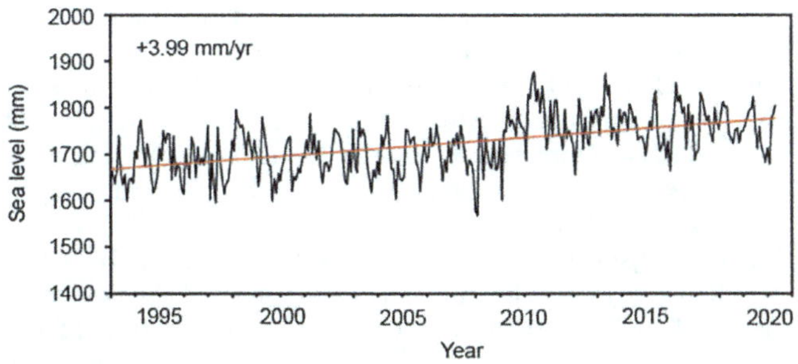

FIGURE 8 Mean monthly regional sea level record at Raine Island based on satellite altimetry data compiled from the IMOS Ocean Current Project.

increased losses through erosion. For example, the most likely projections are currently at between 4.5 mm/yr and 11.5 mm/yr according to the Intergovernmental Panel on Climate Change (IPCC), although some scientists believe that these projections may underestimate the contribution of Antarctic and Greenland ice melt (Garner et al. 2018; Horton et al. 2020). As a result, water depths above the reef flat will substantially increase, allowing larger waves to reach the island beach (Storlazzi et al. 2018). Raine Reef's location on the outer edge of the GBR means that Raine Island is exposed to oceanic waves, and it is thus more vulnerable to these disruptions than islands located on relatively low energy reef flats inside the GBR lagoon. The consequences are increased beach elevation (wave overtopping), accelerating shoreline retreat, and a shrinking of available turtle nesting habitat. More frequent turtle nest inundation would be expected if vertical accretion of the beach lags behind SLR (Tuck et al. 2019).

26.4. CONCLUSION

The Raine Island Recovery Project is an example of how bringing together partners to undertake pro-active adaptive management strategies at one of the most protected sites on the GBR can have positive outcomes. By being prepared to accept the risk of intervening and undertaking such strategies, the Raine Island Recovery Project has implemented actions to increase the resilience of the island to climate change. The intervention action has not negatively impacted the island and the species that nest on it, and the reprofiling interventions have provided a buffer against climate change and SLRs in the future, enabling further actions to take place in attempts to conserve the nGBR green turtle population.

REFERENCES

Batianoff GN, Cornelius NJ (2005) Birds of Raine Island: Population trend, breeding behaviour and nesting habitats. Proceedings of the Royal Society of Queensland 112:1–29.

Booth DT, Archibald-Binge A, Limpus CJ (2020a). The effect of respiratory gases and incubation temperature on early stage embryonic development in sea turtles. PLoS One 15(12).

Booth DT, Dunstan A (2018) A preliminary investigation into the early embryo death syndrome (EEDS) at the world's largest green turtle rookery. PLoS One 13(4):e0195462.

Booth DT, Dunstan A, Bell I, Reina RD, Tedeschi JN (2020b) Low male production at the world's largest green turtle rookery. Marine Ecology Progress Series 653:181–190.

Booth DT, Dunstan A, Robertson K, Tedeschi J, (2021) Egg viability of green turtles nesting on Raine Island, the world's largest nesting aggregation of green turtles. Australian Journal of Zoology 69(1):12–17.

Booth DT, Staines MN, Reina RD (2022) Sand characteristics do not influence hatching success of nests at the world's largest green turtle rookery. Australian Journal of Zoology 69(4):113–124.

Coffee OI, Robertson K (2022) Raine Island Recovery Project: 2021–2022 Season Technical Report. Brisbane: Department of Environment and Science, Queensland Government.

Dawson JL (2021) Multi-decadal shoreline morphodynamics of a shelf-edge reef island, Great Barrier Reef: Implications for future island persistence. Geomorphology 392:107920.

Dawson JL, Smithers SG (2014) Carbonate sediment production, transport, and supply to a coral cay at Raine Reef, Northern Great Barrier Reef, Australia: A facies approach. Journal of Sedimentary Research 84(11):1120–1138.

Dawson JL, Smithers SG (2020) Geomorphological understanding for improved island management at Raine Island: Cay evolution, morphodynamics, and sediment budget. Project Report. College of Science and Engineering, James Cook University, 423p.

Dunstan AJ, Robertson K (2017) Raine Island Recovery Project: 2016–17 Season Technical Report to the Raine Island Scientific Advisory Committee and Raine Island Reference Group. Brisbane: Department of National Parks, Sport and Racing, Queensland Government.

Duvat V (2009) Beach Erosion management in small island developing states: Indian Ocean case studies. WIT Trans. Ecology Environment 126:149–160.

Fuentes MMPB, Hamann M, Limpus CJ (2010) Past, current and future thermal profiles for green turtle nesting grounds: Implications from climate change. Journal of Experimental Marine Biology and Ecology 383:56–64.

Garner AJ, Weiss JL, Parris A, Kopp RE, Horton RM, Overpeck, JT, Horton BP (2018) Evolution of 21st century sea level rise projections. Earth's Future 6(11):1603–1615.

Gourlay MR, Hacker JLF (1991) Raine Island Coastal Processes and Sedimentology. Tech. Rpt. CH40/91. Department of Civil Engineering, The University of Queensland, 150pp.

Hamann M, Shimada T, Duce S, Foster A, Tsz Ying To A, Limpus C (2022) Patterns of nesting behaviour and nesting success for green turtles at Raine Island, Australia. Endangered Species Research 47:217–229.

Horton BP, Khan NS, Cahill N, Lee JSH, Shaw TA, Garner AJ, Kemp AC, Engelhart SE, Rahmstorf S (2020) Estimating global mean sea-level rise and its uncertainties by 2100 and 2300 from an expert survey. NPJ Climate and Atmospheric Science 3(1):18.

Jensen MPJ, Allen CD, Eguchi T, Bell IP, LaCasella E, Hilton WA, Hof CAM, Dutton PH (2018) Environmental warming and feminization of one of the largest sea turtle populations in the world. Current Biology 28:154–159.

Jensen MPJ, Bell I, Limpus CJ, Hamann M, Ambar S, Whap T, David C, FitzSimmons NN (2016) Spatial and temporal genetic variation among size classes of green turtles (*Chelonia mydas*) provides information on oceanic dispersal and population dynamics. Marine Ecology-Progress Series 543:241–256.

Laloe JO, Tedeschi JN, Booth DT, Bell I, Dunstan A, Reina RD, Hays GC (2021) Extreme rainfall events and cooling of sea turtle clutches: Implications in the face of climate warming. Ecology and Evolution 11(1):560–565.

Limpus CJ (2008) A biological review of Australian marine turtles. 2. Green turtle Chelonia mydas (Linnaeus). Queensland Environmental Protection Agency. Availabl e from: https://environment.des.qld.gov.au/wildlife/animals/discovering-wildlife/turtle-watching/publications.

Limpus CJ, Miller JD, Parmenter JC, Limpus DJ (2003) The green turtle, *Chelonia mydas*, population of Raine Island and the northern Great Barrier Reef: 1843–2001. Memoirs-Queensland Museum 49:349–440.

Pike DA, Roznik EA, Bell I (2015) Nest inundation from sea-level rise threatens sea turtle population viability. Royal Society Open Science 2(7):150127.

Seminoff JA (2004) Chelonia mydas. The IUCN Red List of Threatened Species 2004. doi: 10.2305/IUCN.UK.2004.RLTS.T4615A11037468.en.

Smithers SG, Dawson JL (2023) Beach reprofiling to improve reproductive output at the world's largest remaining green turtle rookery: Raine Island, northern Great Barrier Reef. Ocean & Coastal Management 231:106385.

Smither SG, Hoeke RK (2014) Geomorphological impacts of high-latitude storm waves on low-latitude reef islands—Observations of the December 2008 event on Nukutoa, Takuu, Papua New Guinea. Geomorphology 222:106–121.

Staudt F, Gijsman R, Ganal C, Mielck F, Wolbring J, Hass HC, Goseberg N, Schüttrumpf H, Schlurmann T, Schimmels S (2021) The sustainability of beach nourishments: A review of nourishment and environmental monitoring practice. Journal of Coastal Conservation 25(2):34.

Storlazzi CD, Gingerich SB, van Dongeren A, Cheriton OM, Swarzenski PW, Quataert E, Voss CI, Field DW, Annamalai H, Piniak GA, McCall R (2018) Most atolls will be uninhabitable by the mid-21st century because of sea-level rise exacerbating wave-driven flooding. Science Advance 4(4). doi: 10.1126/sciadv.aap9741.

Tuck ME, Kench PS, Ford MR, Masselink G (2019) Physical modelling of the response of reef islands to sea-level rise. Geology 47(9):803–806.

27 An Overview of Environmental Engineering Methods for Reducing Coral Bleaching Stress

Daniel P. Harrison

27.1. INTRODUCTION

Temperature-driven episodic coral bleaching events have increased dramatically in both frequency and severity over the last four decades (Baker et al. 2008; Hughes et al. 2017a and 2017b). Although a multitude of stressors can cause coral bleaching, the sharp contemporary rise in mass coral bleaching events is widely attributed to anthropogenic greenhouse gas emissions and their impact on warming the world's oceans (Hughes et al. 2017b). In the context of the Great Barrier Reef (GBR), the first of the recent contemporary mass bleaching events was recorded during 1998 and 2002, both El Niño years, although historically, there were bleaching events but less frequently (Kamenos and Hennige 2024). The El Niño phase of the El Niño-Southern Oscillation (ENSO) is associated with global positive temperature anomalies and is often attributed responsibility for above-average seasonal sea surface temperatures (SSTs) leading to mass bleaching events (Hoegh-Guldberg 1999; McGowan and Theobald 2017).

The GBR again experienced mass bleaching during the summer of 2015/2016 after a hiatus of a decade and a half, coincident with another strong El Niño (Hughes et al. 2017b). In recent years, the GBR experienced further mass coral bleaching in 2017, 2020, and 2022, with the 2022 event being particularity noteworthy as it was the first to be recorded during a La Niña phase of the ENSO. The increasing threat to the GBR from thermal mass bleaching events and the link to anthropogenic climate change are readily apparent when comparing the global warming trend with the recent mass bleaching events on the GBR (Figure 1). A linear fit to the National Aeronautics and Space Administration (NASA) GISTEMP v4 global surface (land—ocean) temperature index from 1970 onwards (Lenssen et al. 2019) gives an annually increasing temperature trend of ~0.18 °C decade-1. Simply extrapolating the current warming trend another 28 years to 2050 gives a predicted average global temperature increase over the climatological baseline (1951–1980) of +1.43 °C. This contrasts with +0.49 °C in 1998 when one of the first severe contemporary mass bleaching events was recorded. Without intervention, mass bleaching events are expected to continue increasing in both frequency and severity (Hughes et al. 2018b). Future strong El Niño conditions exacerbating an already increased baseline SST have the potential to be especially devastating.

27.2. THE ROLE OF LIGHT AND TEMPERATURE

The primary driver of mass coral bleaching events is positive water column temperature anomalies. A metric commonly used to estimate the thermally induced stress on corals integrates the daily accumulated temperature anomaly above a local climatological maximum monthly mean, plus a threshold (generally 1 °C). This results in a measure with units of Degree Heating Weeks (DHWs; Liu et al. 2013 and 2014; Heron et al. 2016). The DHW metric can be derived from satellite, *in situ* measurements, or ocean model data and exhibits a good correlation with severity of bleaching response in coral communities during recent events on the GBR (Hughes et al. 2018a). If temperature stress is high enough, then this alone may be sufficient to cause a bleaching response, and this has been shown to occur in the dark (Tolleter et al. 2013). However, in the natural environment, the temperature-induced stress is often

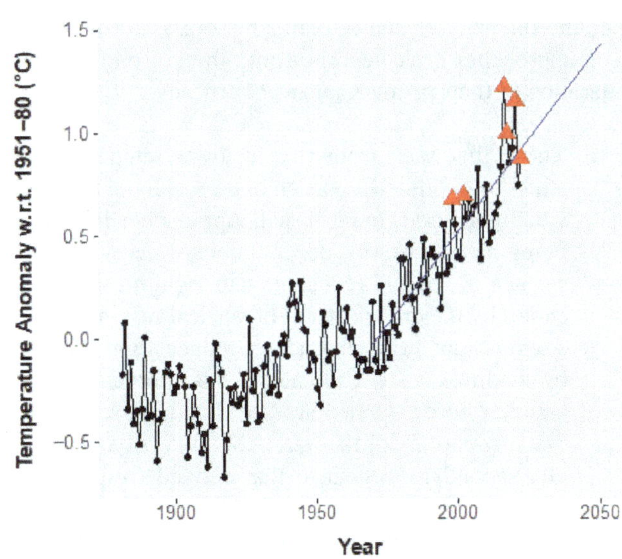

FIGURE 1 Global land-ocean surface temperature anomaly for December-February. Summer temperature anomaly for years with mass bleaching events on the GBR are indicated by red triangles. Blue line represents a linear fit to the data for the period 1970–2022. Temperature data obtained from NASA GISTEMP v4 (Lenssen et al. 2019).

DOI: 10.1201/9781003320425-31

further exacerbated by light (Coles and Jokiel 1978; Lesser 1997; Lesser and Farrell 2004). Stress metrics are now being developed that take into account both temperature and light (Skirving et al. 2018), and mechanistic numerical models include both stressors (Baird et al. 2020b). There are believed to be multiple physiological mechanisms by which excessive light and temperature can lead to cell damage and cause a breakdown in symbiosis between zooxanthellae and coral host (Lesser and Farrell 2004). A large body of literature examines the impacts of shading on the bleaching process using experiments undertaken in aquaria (recently reviewed by Tagliafico et al. 2022). During mass bleaching events, light has been shown to be capable of either exacerbating or alleviating bleaching stress (Leahy et al. 2013; Hoogenboom et al. 2017) depending on whether light levels are high or low. Sea surface exposure to light during a mass bleaching event varies primarily due to atmospheric and meteorological processes (aerosol optical depth, cloud optical depth, and cloud cover). The amount of light penetrating to the depth of coral is impacted by water column processes (ocean surface albedo and turbidity), which are also influenced by meteorological processes (wind and rain). Finally, the amount of light reaching an individual coral is strongly a factor of its depth and microhabitat (e.g., shading by other corals).

27.3. INTERVENTION BY ENVIRONMENTAL ENGINEERING

Given that temperature and light are the primary stressors in contemporary mass bleaching events, attention is being directed towards interventionist approaches to engineer the environmental conditions experienced by corals and mitigate the intensity of one or both stressors. For this review, such approaches are categorised into three groups by consideration of their primary means of efficacy:

a) Those that seek primarily to lower temperature stress by cooling the portion of the water column to which the corals are exposed. Approaches include pumping to or mixing deeper cooler water with the surface ocean and reducing water column stratification by the enhancement of vertical mixing.
b) Actions that primarily act to reduce light stress by shading corals. This category of intervention is assumed to be applied at a local enough scale so that any cooling of the water column is of a generally secondary benefit to that derived from light reduction. An exception might be ponding lagoons whereby shading of a large enough portion of the lagoon at low tide may result in significantly lowered water temperature because of restricted water exchange. Methods that fall into this category include installing physical structures such as shade cloth, increasing the albedo of the upper ocean by 'whitening' it using floating reflective surface films or microbubbles and atmospheric options such as creating a fog or mist using seawater or other obscurants at smaller scales, for example, that of an individual reef or part of a reef.
c) Any 'shading' technology employed on a sufficiently ongoing basis, over a large enough region of ocean, will result in a change to water temperature by altering the shortwave (and other) terms of the surface ocean heat budget. Such regional scale approaches may reduce both light and temperature stress, with potentially synergistic outcomes. Interventions in this category include marine haze brightening and marine cloud brightening, which target the direct (light scattering) and indirect (cloud microphysical interactions) aerosol effects, respectively.

Assuming that they are found to be effective, safe, and socially acceptable, environmental engineering approaches are likely to be desirable as a management tool at both local and regional scales (GBRMPA 2018; Anthony et al. 2019).

27.4. COOLING

27.4.1. Mixing and Pumping: A Description

Oceans are generally characterised by a well-mixed surface layer and then decreasing temperature with depth (e.g., for the GBR, see Walther et al. 2013), with shallow surface waters receiving the bulk of the sun's radiative energy. Vertical mixing on the GBR is driven by a large variety of processes. During doldrum conditions, low wind velocities can lead to reduced physical mixing and increased upper water column heating and stratification. Increased heating under these conditions results from a combination of processes including reduced cloud cover, decreased ocean surface reflectance in the absence of waves, and decreased turbidity in the water column. If calm conditions persist over several days, then water column stratification can develop and contribute to the severity of coral bleaching events (West and Salm 2003; Bainbridge 2017). Paradoxically, upwelling intensity has been shown to be positively correlated with doldrum conditions and positive thermal temperature anomalies on the GBR (Berkelmans et al. 2010). These authors observed that upwelling was enhanced during the doldrum conditions that characterised the 1998 and 2022 mass bleaching events but that the upwelling of cooler water was intermittent and spatially restricted to the central GBR and remained mostly subsurface. Thus, although upwelling of cooler deeper waters may be enhanced during doldrum conditions when coral bleaching risk is highest, the mechanisms of upwelling are such that cooler upwelled water is not distributed widely or consistently enough to shallow coral reefs to avoid the development of mass bleaching events, although it has also been observed that local scale upwelling can reduce the occurrence of bleaching (Randall et al. 2020). Cyclones have occasionally been credited with reducing the overall severity of several mass bleaching events on the GBR by leading to cooler SSTs over large regions of the reef system. This effect could, however, be attributed to either increased cloud cover, enhanced oceanic mixing (Carrigan and Puotinen 2014; Steinberg et al. 2021), or perhaps both acting together.

The aim of mixing or pumping technology is to cool the upper water column where the majority of corals are found by increasing the upward vertical transport of cooler deeper water, ideally from below the thermocline. Lowering the water temperature would reduce the exposure of corals to accumulated thermal stress during a bleaching event. Cooling the upper ocean where corals are at risk of bleaching has the advantage of supporting reef biodiversity and functionality by mitigating the primary environmental stressor during a mass bleaching event. Given the strong correlation between accumulated heat stress and bleaching severity, it is reasonable to assume that the benefit of cooling is well-approximated by any reduction in DHW achieved. Based on bleaching severity observations during the 2016 mass bleaching event, Hughes et al. (2017b) derived a relationship between the fraction of corals bleached and DHW. This regression suggests that reducing the thermal stress from 8 to 4 DHW would on average reduce the fraction of corals bleached from ~80% to ~45%, that is, by approximately half. An important engineering consideration is that the magnitude and duration of cooling required will vary among locations and are very difficult to know in advance. Consider a goal to reduce the accumulated stress by 4 DHW: it may be achieved at low cooling intensity over a longer period (e.g., 0.5 °C for 8 weeks) or higher intensity over a short period (e.g., 2 °C for 2 weeks). However, for any given site, the traditional DHW stress metric will accumulate only when temperatures are above the threshold, which is taken as the maximum monthly mean SST plus 1 °C (Liu et al. 2014). Therefore, a given applicable profile of temperature reduction achieved at one site may not achieve the same reduction in DHW at another.

Strategies for cooling corals by mixing and pumping include bulk pumping of water from depth to the surface, using slow turning impellors to disrupt water column stratification in shallow waters, and creating artificial upwelling by building a subsurface structure such as an 'undersea berm' with the intent of enhancing vertical mixing.

27.4.2. Undersea Berms

A proposal to generate artificial upwelling by the construction of undersea berms, positioned in the path of prevailing deep-water currents, was forwarded to the Reef Restoration and Adaptation Program (B Covens, pers. comm., 9 December 2019). Some inherent challenges are readily identified for such a scheme. Assuming such a structure was permanent and worked as intended, rather than treat intermittent periods of anomalously high SSTs, the installation would permanently alter the local hydrodynamics and circulation. Upwelling along the GBR was first documented by Andrews and Gentien (1982) to be a staged process where the East Australian Current pumps sub-thermocline cooler, nutrient rich water up the slope to the shelf break, from where periodic Eckman pumping moves the upwelled water inshore along the seabed under poleward winds. Shoreward intrusions of cooler bottom water have been shown only occasionally and mostly in summer to penetrate the GBR lagoon or upwards into the surface layer within the channels between reefs (Benthuysen et al. 2016; Wolanski et al. 2024). Ultimately some of the bottom water is transported up onto shallow coral reefs by localised upwelling caused by winds and tides along 'frontal topographic discontinuities', that is, onto the reef structures themselves. The bottom waters of the GBR lagoon are not dominated by consistent strong flowing currents of cold water. If the structure were to be located at a reasonably accessible depth on the continental shelf, then it is only the final stages of upwelling described above that such devices could target for enhancement. Given that the reefs themselves already perform this function, it is not immediately apparent that adding further structures would result in meaningful net additional cooling of surface waters. No scientific literature was identified supporting the efficacy of such a proposal, and when issues around environmental risk and social acceptability are also taken into account, it is not currently considered a viable approach.

27.4.3. Underwater Impellors

It has been proposed that enhanced vertical mixing of the water column could be achieved through submerged, slow-turning, solar powered impellors (Bay et al. 2019; Figure 2). The primary objective of these systems would be to reduce shallow water temperature stress by breaking down water column stratification during summer doldrum conditions. Increased water flow has been shown to be beneficial to corals (Rogers et al. 2016) and may thus provide a potential secondary benefit. To lower water temperatures beyond the immediate flow exit, impellors need to move colder deeper water to mix with surface waters, thus lowering surface temperatures and creating enhanced mixing in the water column. A similar need for destratification is prevalent in artificially created still bodies of water such as lakes and reservoirs, where the necessity is driven by water quality requirements (McAuliffe and Rosich 1989) or the desire to reduce evaporation (Sherman et al. 2021). Historically, destratification of these bodies of water was achieved by releasing plumes of compressed air near the bottom of the reservoir (Kirke 2000). However, technological advances have led to increasing use of slow-turning impellors that can operate 10 times more efficiently in terms of energy (Kirke 2000). Slow speed impellor prototypes have been tested at very low flow velocities of 0.2–0.4 m s^{-1} to produce flow of 1 m^3 s^{-1} using around 200 W of hydraulic power in a weak thermocline (Kirke and El Gezawy 1997). A specific type of Flygt impellor has been demonstrated in real-world applications to generate approximately 1 $m^3 s^{-1} kW^{-1}$ of flow (Kirke 2000).

A trial program (Reef Havens Research Project, www.rrrc.org.au/reef-havens/, 2 April 2023) was funded by the Australian Government to install eight impellors at Moore Reef, a major reef tourism site (Mitchell-Whittington 2017). The project, which is led by the Reef and Rainforest Research Centre, drew criticism from the media after receiving $2.2m AUD of government funding for the project despite it not being endorsed by the Reef 2050 Independent Expert Panel (IEP). The IEP had recommended that the project not be supported or if it were to go ahead, that it be reframed as a research project, which it ultimately was (Lavars 2018). The project intends to target intervention at the scale of 'a snorkelling footprint' around an established moored pontoon (350–500 m^2). The project will monitor flow, temperature, coral light stress, and the depth of stratification that occurs during doldrum conditions and has made background monitoring data publicly available but does not appear to have gone ahead with installation of any impellors (Long and Fisher 2021).

27.4.4. Pumping

An alternative technique to using slow-turning impellors is to pump the cooler water from depth to the surface using pumps and pipes (Baird et al. 2019; Bay et al. 2019). An advantage of this approach is that the location of deeper cooler water and the targeted coral reef can be spatially separated. Baird et al. (2020a) used a hydrodynamic-biogeochemical model (eReefs) to examine cool water injection as a potential intervention on 19 individual reefs in the GBR system. Of these, the Lizard Island reef flats were identified as an optimum target for cool water injection because of longer water residence time and proximity (3 km) to a 40-m deep channel connected to cooler continental slope waters (Baird et al. 2020a).

Another suggested approach is the region-wide application of artificial upwelling to move cooler deeper waters to the surface. A modelling study has been reported that uses a coarse global ocean model to examine a hypothetical scenario of artificial ocean upwelling applied across the entirety of the GBR (Feng et al. 2020). The study assumes an even distribution of upwelling sites at a density of 1 km^{-2} with each pumping water from a depth of 130 or 550 m at a rate of either 1 or 50 $m^3 s^{-1}$. Given that this scenario is unrealistic in the extreme and was conducted with a numerical earth system model that does not capture hydrodynamic processes on the GBR and that all of the GBR lagoon is much shallower than 130 m, this proposal is not considered further.

27.4.5. Feasibility of Mixing and Pumping

The key factors influencing the magnitude and extent of cooling that could be derived from mixing and pumping can readily be identified as the temperature differential between surface and source water, the volume of water pumped, and the volume of receiving waters. The latter in an open hydrodynamic system is not constant and is itself a function of

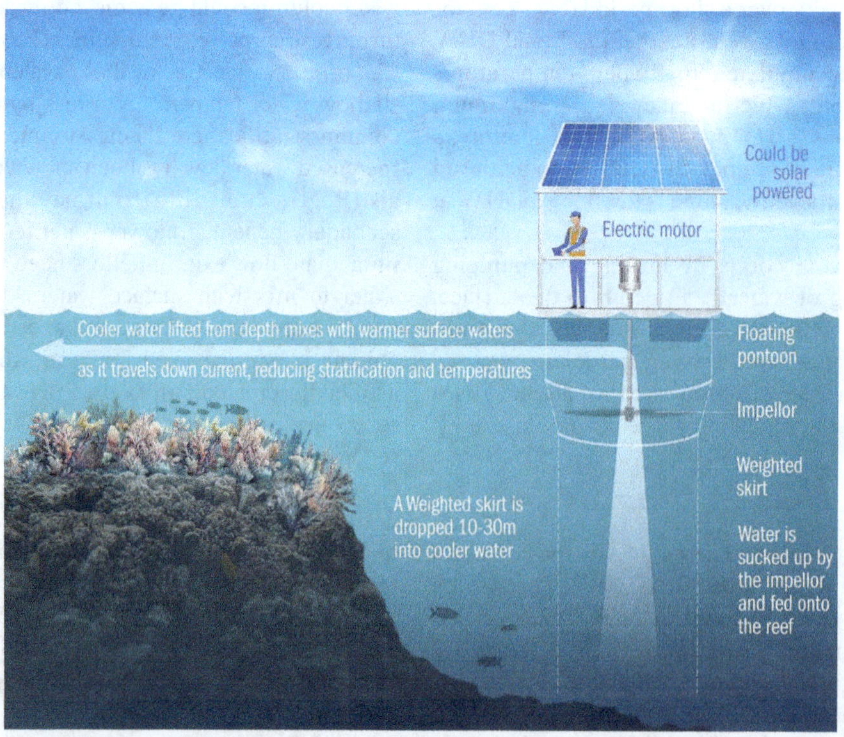

FIGURE 2 A conceptual diagram of a solar-powered underwater impeller system scheduled to be trialled on Moore Reef in the northern sector of the GBR by the Reef Havens Project.

depth, mixing rates, and current velocity. Assuming rapid vertical mixing in the receiving waters, greater depth and faster currents result in lower cooling for a given temperature differential and pumped flow rate. The ideal situation then is transporting water with a high temperature differential to shallow reef habitats with a high water residence time (Baird et al. 2019).

One of the criticisms levelled at the Reef Havens Project was that the modelling by proponents assumed a 3°C temperature difference between the surface and bottom waters, which was described in the IEP review as a 'major departure from reality' (Slezak 2018). At Lizard Island, Baird et al. (2020a) found the average temperature difference during a tidal cycle in February 2017 between surface waters and those at 40 m depth in the channel 3 km away to be approximately 1 °C. These values were derived from a 1-km resolution hydrodynamic model, and although the model does a generally good job at representing temperatures on the GBR, it can underestimate variability with depth (Steinberg et al. 2021). Nevertheless, little of the GBR lagoon is deeper than 40 m, and accessing a temperature differential between surface waters and a 40-m deep channel that is hydrodynamically connected to slope waters could be considered a best-case scenario for a reef within the GBR lagoon. Baird et al. (2020a) did not, however, report an analysis of surface to seabed temperature differentials on the GBR, and in highly stratified conditions such as those targeted by the Reef Havens project, larger differentials may occur.

Baird et al. (2020a) calculated the power required to move 5 m^3 s^{-1} from 40 m depth, 3 km away, through 4 pipes as ~466 kW, which is two orders of magnitude higher than the claimed 1 m^3 s^{-1} kW^{-1} for impellors. Considering the calculation for the Lizard Island scenario, 457 kW of the total energy is for friction losses incurred by pumping the water over this distance. A further 6 kW is expended accelerating the water to its flow velocity through the pipe, and only 2 kW are attributed to overcoming the density difference between waters at depth and those at the surface (Baird et al. 2019). Although slow-turning impellors ducted to the seabed directly below do not incur the majority of acceleration or friction losses, they are limited to vertically transporting water *in situ* and must therefore be co-located over the cool water source. This then poses a potential problem for targeting the dispersion onto recipient corals, which may not be immediately adjacent and may not be downcurrent. Baird et al. (2019) observed that the regions of greatest coral density are typically the exposed reef slopes with fast moving currents that make them less suitable for thermal mitigation by cool-water injections because of very low water residence times.

Ultimately, Baird et al. (2019) found that 5 m^3 s^{-1} of 1 °C cooler water from a 40-m depth pumped continuously for 3 months into an ideal receiving habitat of shallow water with a long residence time was able to cool 97 ha by 0.15 °C or more. This effort generated a reduction in thermal stress of 2 °C-weeks. Transporting sufficient water through pipes to a targeted location is energy-intensive and costly compared with the benefit imparted to only a small footprint of coral. Using impellors located directly above the source water would require orders of magnitude less energy for equivalent vertical water transport. However, there is little peer-reviewed research to ascertain where and under what conditions this approach could be effective. Detailed site-specific high-resolution modelling studies validated with observational data are required to discover whether there are suitable sites with hydrodynamic conditions conducive to benefiting even a small area of reef habitat using this method. The Reef Havens project is collecting and publishing observational datasets to support such an evaluation for the Moore Reef site (Long and Fisher 2021).

27.4.6. Risks of Mixing and Pumping

Risks of mixing and pumping technologies include project risk, risks associated with adding infrastructure to the reef and risk of unintended ecological or environmental impacts. Permanent infrastructure on the GBR is notably at risk of cyclone damage (Do and Kuleshov 2023). Project risks include financial risks if the technology does not perform as intended, including stakeholder and social acceptability risks. The financial risk associated with failure to perform relative to expectations is normally borne by the project operator. On the GBR, mixing and pumping has been suggested as applicable to small footprint sites that generate disproportionally large tourism revenue. It is conceivable that tourism operators may fund mixing and pumping intervention at their own sites. Given that both the Reef Restoration and Adaptation Program and the Reef Havens project have received Australian government funding, it is also possible that further taxpayer funding may be directed towards future research or trials, in which case some financial project risk may be borne by the public. The GBR is a significant source of economic revenue for the Northeast Queensland region that generates income from tourism, fishing, and other industries (Deloitte Access Economics 2017). Any intervention that inadvertently negatively affects the reef's health or limits access to the public could result in significant economic losses for local communities and the broader regional economy. Additionally, there is a potential for conflict to arise over resource allocation among different stakeholders. For instance, conservation groups may advocate for stronger protections for the reef, while industry groups may argue that certain activities are necessary for economic development. Lastly, public opposition to the intervention can lead to social conflict and political pressure that may result in delays or cancellation of the proposed intervention.

Moving cooler water from depth to the surface inherently transports the constituents of the deeper water and mixes them with those of the surface receiving waters (Bauman et al. 2015). Of primary concern is the potential to enhance surface nutrient concentrations. Bay et al. (2019) pointed out a range of nutrification-related risks including eutrophication, destabilisation of internal nutrient dynamics of corals potentially exacerbating thermal sensitivity, blooms of benthic algae or phytoplankton, enhanced survival of crown-of-thorns larvae, and changes in overall water chemistry on the reef (e.g., decreased oxygen). When ambient nutrient levels are low, corals have been observed to effectively assimilate some additional nutrient loading up to a limit after which further loading becomes detrimental (Koop et al. 2001). Studies have also shown that increased nutrients can increase coral growth rates (Koop et al. 2001; Savage 2019; Becker et al. 2021). Speculatively, increased nutrient supply during coral mass bleaching events could improve coral bleaching survivability by increasing plankton concentrations, given that corals under thermal stress in aquaria experiments have been shown to have improved outcomes through heterotrophic feeding (Hoogenboom et al. 2012; Tagliafico et al. 2022). Overall, the effects of additional nutrients are expected to be context-specific (Shantz and Burkepile 2014) and may have positive and negative attributes regarding ecosystem health.

Mixing and pumping is likely to be both costly and energy-intensive, and given that the intervention is aimed at relieving intermittent excessive temperature stress, the overall ecological risk is likely to be manageable. Impacts of pumping and mixing on corals (both negative and positive) must be holistically understood and negative impacts mitigated prior to deployment. Further risks are associated with the required infrastructure, such as contamination to the marine environment, the creation of habitat acting as a stepping-stone for invasive species and pests, and the potential creation of marine debris if infrastructure is damaged or lost, such as during storms. The GBR Marine Park Authority classifies pumping and mixing as a high-risk activity since it requires installation of large infrastructure and facilities (GBRMPA 2018).

27.5. SHADING

There is a large body of experimental evidence that shading corals during periods of excessive temperature can reduce the degree of bleaching stress (Tagliafico et al. 2022). Shading is a direct intervention that could be strategically deployed in summer months, or in response to bleaching forecasts (Coelho et al. 2017). The primary goal of providing shade is to reduce the levels of ultraviolet (UV) light and photosynthetically active radiation (PAR) on the coral holobiont, thereby reducing levels of bleaching and mortality in targeted regions. Shading has been demonstrated to be beneficial to corals under thermal stress in aquaria (Lesser and Farrell 2004; Smith and Birkeland 2007; Hoogenboom et al. 2012; Berg et al. 2020) and can also reduce levels of bleaching in situ (Coelho et al. 2017). In the field, subsurface shading has been shown to reduce the bleaching colour score in a reef pool (Coelho et al. 2017), but the experiment did not extend to the point of inducing mortality so it is not clear how effective it might be under more severe conditions. Shallow coral reefs experiencing reduced light levels due to high turbidity have been proposed as potential coral reef refugia in a warming ocean (Cacciapaglia and van Woesik 2016).

Shade cloths deployed either above or under the water offer an obvious method by which to reduce light and have been trialled on the GBR (Rau et al. 2012) but are not considered further here due to the very limited spatial scales to which they could realistically be applied. Other methods that could temporarily reduce light intensity include 'whitening' the upper ocean with floating reflective surface films or microbubbles and atmospheric techniques such as creating a fog or mist using seawater or other obscurants.

27.5.1. Surface Films

Artificial barriers are commonly employed on the surface of open water storages, such as reservoirs and dams, to prevent evaporation (Waheeb Youssef and Khodzinskaya 2019). One approach is to cover the surface using a monomolecular film (Barnes 2008; Figure 3), such films have been researched for at least 100 years (McJannet et al. 2008). Despite challenges in replicating promising laboratory results in the field (Barnes 2008), development has progressed rapidly, and some outdoor studies have shown the films to be an effective means of evaporation control (Panjabi et al. 2016). An effort is underway to develop surface films which could be deployed to provide shading to reefs under heat and light stress (GBRF 2015). Early laboratory study findings outlined in a press release claim that surface films can be stable, attenuate light by up to 30 percent across the spectrum (including UV), are non-toxic to corals (made from calcium carbonate and biodegradable) and protect some coral species from bleaching (GBRF 2018).

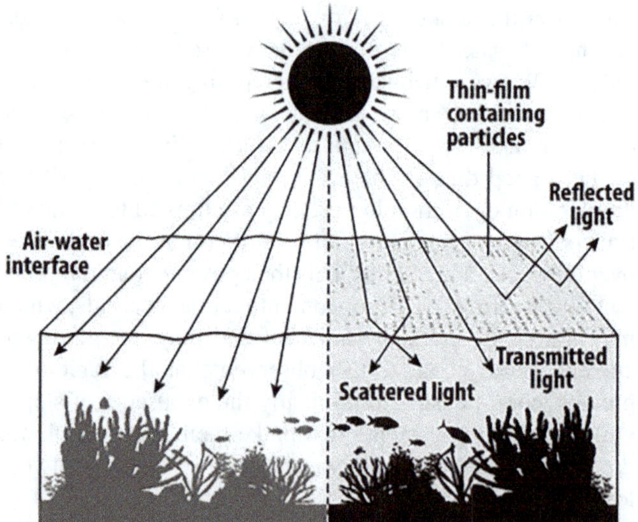

FIGURE 3 Surface films aim to reduce the stress that corals and other photosynthesising organisms experience during heatwave conditions by scattering and reflecting incoming sunlight.

Source: Reproduced with permission (source: Bay et al. 2019)

The maximum effective lifetime of the film in large tank studies is less than 48 hours. Realistic targets for reduction in light levels are estimated to be around 20 percent in the field (Bay et al. 2019). The application of ultra-thin surface films would not require permanent infrastructure and could be intermittent in response to forecast or actual bleaching events. It would likely target high-value reefs with long water retention times to minimise the quantity and rate at which material needed to be deployed. Although research is still underway to develop the final formulation of the surface film product (Bay et al. 2019) the current formulation uses suspended calcium carbonate to act as the light scatterer (Gibbens 2018). Aircraft, vessels and automated buoys have all been suggested as potential deployment methods (GBRF 2018).

There are no reported studies in the literature on the potential effectiveness of surface films for the mitigation of coral bleaching or on their effectiveness in reflecting light. For smaller areas such as the aforementioned 'snorkelling footprint' proposed for underwater impellors, the application might be bunded to retain the film using technology readily available from the offshore oil industry. For more general application, the water residence time on the targeted reef will be a key consideration as for mixing and pumping. There is no information published on the material characteristics, formulation, density, or volume of material required to achieve a suitable coverage per unit area. It is therefore not currently possible to undertake any meaningful evaluation of the feasibility of implementation. Research is required into potential formulations, their stability and lifetime, reflectivity, behaviour under various environmental conditions (e.g., wind and wave climates), effect on other ocean heat budget terms (e.g., evaporation), methods, costs, and strategies of deployment, safety including potential impacts on marine chemical processes, and potential environmental and ecological impacts such as on the feeding habits of fish, plankton, and other marine biota.

27.5.2. Ocean Microbubbles

Microbubbles of air suspended in the ocean surface reflect light in a similar way to droplets of water suspended in air (such as cloud or fog), albeit with less efficiency, due to the absorption of backscattered light in ocean waters (Seitz 2011). The effectiveness of bubbles in changing the reflectivity of the ocean surface is apparent by the high albedo of ship wake (Gatebe et al. 2011), whose enhancement has been suggested as a climate geoengineering strategy (Crook et al. 2016). The deliberate production of large quantities of microbubbles in the upper ocean has been proposed to cool down the planet, a form of global-scale geoengineering (Seitz 2011; Forster et al. 2014). The technique has also been suggested as applicable at local or regional scales (Seitz 2011). Microbubbles in the upper few metres of the ocean are predicted to persist from hours to days. Both the lifetime and reflectance of the microbubbles increase with the concentration of surfactants in the seawater. Microbubbles coated with organic film have reflectance as much as 400 percent higher based on modelling and observations of naturally occurring microbubbles (Zhang et al. 1998). This enhancement may occur naturally in the GBR due to coral-secreted surface-active compounds (Deacon 1979) or potentially through the co-application of biologically benign organic surfactants with the generation of microbubbles (Seitz 2011).

There are multiple methods by which microbubbles could be generated including expansion of air-saturated water through vortex nozzles, ultrasonic excitation, and fluidic oscillation—the last of which may be the most energy-efficient (William et al. 2008). Adding surfactants to the wake of ships could prolong the lifetime of microbubbles in the ships' wake (Crook et al. 2016). There are currently no known projects attempting to develop microbubble technology to shade corals and no studies into the potential efficacy, feasibility, systems engineering, costs, or risks of such a scheme. As with pumping, mixing, and reflective surface films, water residence time over the targeted region would be a key engineering constraint along with the challenge of generating a sufficient quantity of microbubbles.

27.5.3. Misting and Fogging: A Description

Particulate matter suspended in air is termed aerosol; aerosols scatter and absorb radiative energy from the sun and impact the amount of sunlight reaching the Earth's surface, which is known as the aerosol direct effect (Penner et al. 2001). The shading potential of atmospheric aerosols can be observed in naturally occurring phenomena such as smoke, mist, fog, clouds, and haze. A plume of biogenic oil or seawater droplets could be released into the atmosphere over coral reefs to create artificial shading during times of peak bleaching stress. Two approaches are suggested: vaporising biogenic oils to create a 'smoke fog' by borrowing technology from the entertainment industry or military applications; or the spraying of seawater to generate an artificial 'sea fog' or 'mist'. In both cases, the objective would be to improve mortality outcomes over relatively small-scale regions (up to tens km^2) by reducing light stress in a similar manner as shade cloths, surface films, or microbubbles. The terms misting and fogging are both in common use to describe the atomisation of water, while creating obscurants from the vaporisation of oil is mostly referred to as fogging or smoke fog. In the Reef Restoration and Adaptation Program, misting is used to describe the potential use of oils to shade corals, while the use of seawater is described as fogging (Bay et al. 2019).

27.5.4. Smoke Fogging

A variety of commercially available technologies exist for aerosolising oils for applications that range from movie special effects to military purposes. Pulse jet engines are widely used for producing smoke in a variety of uses that require high output rates. One example is in military applications, where smoke generators have been in use since at least World War One to obscure the battlefield or for training purposes (Palacios and Wright 1998). The United States Department of Defense has developed a smoke generator called the M157 that uses a pulse jet engine to produce large amounts of smoke quickly and efficiently. The M157 can be mounted on a trailer or a vehicle and is easily transportable. In 2016, the Russian military was reported to have concealed a naval fleet and perhaps even the city of Severomorsk itself for 3 days using stationary and mobile generators to create an artificial fog (Haines 2016). Another example of the use of pulse jet engines for smoke production is in the entertainment industry. Special effects companies use smoke generators to create realistic smoke and fire effects for movies, air shows, music videos, and live performances. Fogging generators are also widely employed in the pest control industry where the vaporised oil can be used as a carrying medium for insecticides or can form the insecticide itself (Kalita et al. 2013; Rosli et al. 2022).

To block light or reduce visibility, smoke generators heat a fogging oil to vaporisation and then disseminate it into the ambient air. The vapor cools as it enters the air, condensing into small droplets that scatter light. Smoke generators have been used previously in scientific field experiments to examine the cloud microphysical and albedo impacts of additional aerosols (Russell et al. 2013). Due to human health and environmental reasons, it is unlikely that heavily refined mineral oils typically used in military and some other applications would be suitable for use on the GBR. Instead, development of this type of technology for protection of coral reefs would likely aim to develop a fog oil from biogenic oils that already occur naturally in the marine environment.

27.5.5. Seawater Fogging

An alternative to vaporising oil to reflect sunlight is to generate a similar light scattering effect using seawater droplets (Figure 4). Using abundant and naturally occurring seawater to generate the obscurant is likely to carry lower environmental risk and be more acceptable to regulators, stakeholders, and the public than using fog oils. Seawater could be used to produce droplets in the range of 1–20 µm diameter, which would be roughly equivalent to the properties of a heavy sea fog (Sheikh Muhammad et al. 2007; Mazoyer et al. 2022). A relatively large technology base exists for producing droplets consistent with this size range using freshwater for cooling applications ranging from outdoor dining and concerts to industrial, energy production, and dust or fire suppression. The general principle is that water is fed, typically at high pressure, into arrays of nozzles that cause the breakup of the water into a range of droplet sizes dependent on the nozzle technology, orifice size, and pressure used. Dense concentrations of droplets could be dispersed low in the marine boundary layer under low wind conditions. A plume of evaporating seawater droplets would cool the airmass in which the droplets are entrained and suppress further mixing with surrounding air. If fully evaporated, then conservation of mass would lead to salt crystals roughly one-quarter the diameter of the original seawater droplet. However, at typical relative humidity (RH) over the GBR in summer, the droplets would not fully evaporate. At 80% RH, the droplets would reach equilibrium with the surrounding air at around 50% of their sprayed diameter following Köhler theory (Köhler 1936). Evaporation to this level of equilibrium could take some time if the droplet concentrations are high and mixing low and the airmass carrying them becomes locally fully saturated by water vapor. If a fog with these characteristics could be maintained over an

FIGURE 4 The RV Bandicoot testing seawater fogging technology as part of the Reef Restoration and Adaptation Program.
Source: Photographed by Author

area of coral reef, then it could potentially provide cooling by both shading and by lowering sensible heat flux into the ocean surface waters, although it would tend to suppress the transfer of latent heat out of the ocean surface by evaporation (Harrison et al. 2019).

Seawater fogging could conceivably be applied at the scale of an individual reef or a collection of reefs by barges or boats (tens km^2). Creating artificial sea fog is expected to be practical only for short continuous periods (days) when wind speeds are low. The process inputs are seawater and energy. Modelling has not been undertaken to quantify the various components of the ocean-atmosphere heat flux in response to such a perturbation. Although evaporative cooling of the air above the ocean would tend to cool the ocean by adjusting the sensible heat flux, the increased humidity will tend to suppress evaporative cooling of the ocean surface (latent heat flux). The net heat flux change induced by the evaporative cooling is therefore unquantified. Field studies are required to test the level of shading that can be achieved and the feasibility of targeting and maintaining it sufficiently well to mitigate bleaching mortality.

Subject to energy and logistics requirements it may also be possible to disperse sea spray aerosols (SSA) of a similar or slightly smaller size over a much larger area. A regional strategy such as this would target atmospheric SSA concentrations several order of magnitude lower and widely dispersed within the atmospheric boundary layer but maintain them for longer periods and could be termed marine haze brightening. This is described further in the following section.

27.5.6. Risks of Local Shading

As with techniques to cool corals, shading interventions have the potential to inadvertently reduce natural adaptation by protecting corals from stress (Bay et al. 2019). Globally, the natural genetic adaptation of coral reefs does not appear to be keeping pace with climate change. Preventing some mortality while not fully removing the stress may encourage genetic adaptation (Matz et al. 2018). Shading with any of the methods other than shade cloth is likely to be intermittent and require considerable effort and cost to maintain. The effectiveness is subject to prevailing winds and currents, which add randomness and dynamic heterogeneity to the pattern of any stress reduction achieved in both time and space. Understanding the interplay among environmental stress, its manipulation, and mechanisms of adaptation will therefore be critical to managing shading interventions for long-term benefit.

Another potential impact is overshading of light-limited corals or other autotrophic biota such as seagrass. Shading could lead to reduced coral growth rates and reduced rates of translocation of photosynthates from symbionts to corals, potentially also affecting herbivores and higher trophic levels, although if shading was of short duration and intermittent, these impacts are likely to be minimal. Overshading is more likely close to the distribution source for methods that are subject to advection and dispersion. Some undesirable consequences, particularly those related to overshading of light-limited organisms, are less of a concern if intervention is limited to small target footprints as they can be mitigated by prudent site selection and intervention design.

At an individual reef scale, the risk of overshading some areas close to the source, be it with surface films, misting, or fogging, is potentially high, as the shading will be more intense in regions immediately around the release. This impact could be reduced if the source was mobile. Atmospheric fogging carries some additional undesirable impacts such as decreased visual amenity and potential safety considerations to vehicles (aircraft / ships) if visibility is reduced significantly. The use of oils for thermal fogging may have implications for human (Palmer 1990; Harland Bartholomew and Associates 1997) and wildlife health in the vicinity of the generators (Driver et al. 1993), and military grade fogging oils can be harmful to the freshwater aquatic environment in sufficient concentrations (Poston et al. 1986; Cropek et al. 2008). It is likely that some area of exclusion to other marine park users would be required during operations. Any further development of an oil fogging strategy would need to focus on selecting a suitable biogenic oil, perhaps derived from phytoplankton, that would not pose an unacceptable risk and might be socially acceptable. Any oil obscurant would need to be rigorously evaluated for any potential negative impacts to the marine environment.

Some of the risks around using microbubbles for global-scale geoengineering have been outlined by Robock (2011). Of these, most relevant to local-scale application are the potential for adverse impacts if surfactants are added to promote bubble lifetime, the unknown impact of the bubbles themselves on marine biota including corals, and the potential for cooling to affect ocean convection or acidity. All the shading techniques described would require infrastructure to located in the marine park. Shading techniques are likely to be implemented with mobile platforms, although fixed dispensing systems are also a potential implementation method. Mobile infrastructure is subject to the risks outlined above including the inadvertent introduction of pests and invasive species. However, the risk of exposure to storms and cyclones can be more easily mitigated with mobile infrastructure. As with other intervention measures, all risks should be comprehensively evaluated as part of any research and development initiative.

27.6. REGIONAL COOLING AND SHADING

Shading applied continuously or semi continuously over a large enough area will reduce the transfer of radiative energy from the sun into the shallow waters of the GBR lagoon and lead to ocean cooling (Harrison et al. 2019). In theory, this could include regional scale deployment of microbubbles or surface films; however, it does not appear practical to implement either of these techniques at the thousands of square kilometre scales required. Thermal fogging with biogenic oils could possibly be scaled to regional areas, but challenges with regulatory approval and social acceptability are likely to increase with the scale. It is likely impractical to generate continuously a dense seawater fog with large droplets of the type envisioned for reef scale shading over such a large area due to the large quantity of water required to be pumped. Instead, atomisation of seawater at sizes around an order of magnitude smaller (droplet diameter around 0.8 micron) would be the ideal size to target both the direct (light scattering) and indirect (cloud microphysical response) aerosol effects in a process known as marine cloud brightening (MCB). When there are no clouds present and the target is only the aerosol direct effect, the technique could be referred to as marine haze brightening (MHB) or marine sky brightening (MSB), although the term MCB has traditionally been understood to refer to both the aerosol direct and indirect effects.

27.6.1. Marine Cloud Brightening

MCB, like ocean whitening, has been previously considered as a potential geoengineering technique, operated at such large scale as to alter the Earth's radiative budget (Latham 1990; Salter et al. 2008; Latham et al. 2012). These suggestions originate from observations of the impact that additional cloud condensation nuclei (aerosols that activate to become cloud droplets) can have on the brightness (albedo) of low level clouds. Latham et al. (2013) suggested that widespread semi global continuous use of MCB could alleviate coral bleaching by cooling the world's major oceans. Harrison (2018) hypothesised that localized intermittent cloud brightening directly over the GBR could be implemented in response to marine heat wave events. The idea is to shade surface waters over regional-scale areas, downscaling both spatially and in duration the original concept and taking advantage of the restricted water exchange of the relatively shallow GBR lagoon with the ocean. Modelling studies have indicated that reducing shortwave radiative energy by 10–20 W m^{-2} above GBR waters over a summer could lower SST by 0.3–0.7 °C around the mid-shelf after a stabilisation period (Harrison et al. 2019). The combined effects of decreased SSTs and lower light levels are predicted to reduce coral bleaching (Harrison et al. 2019). Ecological modelling suggests that ongoing successful cloud brightening over the GBR could significantly improve the trajectory of coral cover in the future, especially if applied in association with aggressive greenhouse gas reduction strategies (Condie et al. 2021).

The concept is to supply additional sea salt aerosols to regions where droplet density is limited by cloud condensation nuclei (CCN) availability within low-lying marine clouds (Latham 1990). Aerosols can be provided in the form of nano-sized sea salt crystals produced by spraying sub-micron-sized seawater droplets into the atmospheric boundary layer and allowing them to evaporate (Latham et al. 2012). The portion of nano-sized aerosols that reach the cloud condensation level act as additional cloud condensation nuclei. Sea salt crystals have high hygroscopicity and can activate to form cloud droplets immediately on reaching an existing cloud base or when a new boundary layer cloud is formed. Additional CCN affect the droplet size distribution within clouds by increasing the number of droplets while decreasing the mean droplet size (Twomey 1991). The same cloud formed of more numerous droplets with a smaller radius has higher albedo, reflecting a greater portion of incoming solar radiation back into space (Twomey 1991). When seawater is used to generate the aerosol, the cloud albedo effect is enhanced further by the light scattering of the resulting sea salt crystals themselves if they are of sufficient size, regardless of whether a cloud is present or not (Ahlm et al. 2017). A brightened cloud is on average less likely to precipitate and may thus remain in state longer, resulting in further reduction in net radiative forcing (Albrecht 1989). Some other secondary cloud microphysical responses may be perturbed with opposite forcing, potentially offsetting some of the benefit in enhanced reflectivity (Partanen et al. 2012; Wood 2021), particularly if the size of the sea spray aerosol is not optimised (Alterskjær et al. 2012). Globally however, the dominant effect of additional cloud condensation nuclei supply is an increase in cloud albedo (Twomey 1974). A striking example of the increase in cloud albedo resulting from additional CCN in cargo-ship exhaust is evident from satellite imagery, providing in some cases an unambiguous easily observed example of albedo increase (e.g., Figure 5). Visually striking albedo increases of the magnitude illustrated occur typically

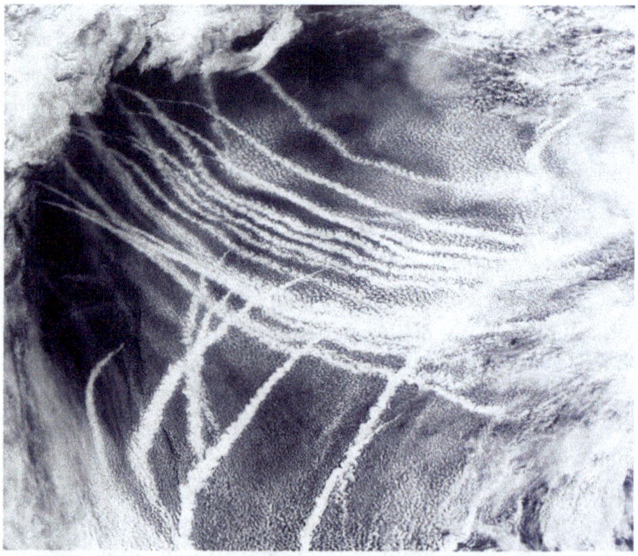

FIGURE 5 Cloud brightening resulting from marine ship exhaust is sometimes visually apparent due to the high albedo change that occurs under certain atmospheric conditions, for example, in this satellite image of the Pacific Ocean captured by the MODIS instrument aboard the Terra satellite operated by NASA. These conditions are not often present on the reef, but atmospheric modelling has indicated that low-level clouds on the reef are amenable to a lower intensity of brightening.

Source: Image credit: NASA

under certain meteorological conditions; with shallow cloudy marine boundary layers with very low background aerosol concentrations, often the albedo increase is more subtle (Christensen et al. 2022).

MCB for the GBR would aim to replicate the natural generation of sea salt aerosol by breaking waves in the ocean; as with waves the process inputs required are only energy and water. The reef itself has been shown to generate sulphur-based aerosols from the biogenic action of corals and algae (through the production of dimethyl sulphide; DMS). These reef-produced aerosols are also hypothesised to be linked to low-level cloud albedo (Jones et al. 2017), although the overall contribution of coral-generated CCN is likely to be small (Fiddes et al. 2022). The benefits of reducing incoming short-wave energy are cumulative over time if the technology is operated during daylight hours each day for continuous periods (Harrison et al. 2019). Scenarios run using the eReefs hydrodynamic model show that over the first several weeks of operation, GBR waters cool until reaching a semi equilibrium state (more correctly, heating is avoided), after which SST stabilises after several weeks at a reduced temperature compared to the control. The amount of time to reach this equilibrium is related to the water residence time within the impacted region and would depend on the size and bathymetry of the area over which MCB is operated (Harrison et al. 2019). MCB has the potential to significantly relieve bleaching stress over large portions of the reef at a regional spatial scale. To be effective and achieve these scales, MCB would need to operate continuously whenever suitable clouds were present over weeks to months, during periods when forecasting algorithms predict a high risk of bleaching. Cloud brightening stations would be required to be spread throughout the target region at appropriate spacing. These could conceivably be hosted on mobile or moored barges, dedicated ships, ships of opportunity, or on islands or platforms. The Reef Restoration and Adaptation Program is undertaking experimental work on the GBR to develop MCB technology and investigate if clouds over the GBR are susceptible to brightening (Tollefson 2021). The spraying apparatus developed to support these experiments is analogous to a snow-making cannon. Each station includes a fan for initial dispersion of the droplets and up to 330 atomising nozzles to produce ~1–3 quadrillion (10^{14}) nano-sized aerosols per second from 1.6 L/s of seawater using a modified version of the effervescent nozzle described by Cooper et al. (2014).

27.6.2. Risks of Regional Shading Techniques

A possible unintended impact of MCB is the potential to change local weather and rainfall patterns, a concern that also exists for cloud seeding operations undertaken to enhance precipitation. At the present time, it is not known whether cloud brightening over the Reef would increase, decrease, or have negligible impact on local or regional rainfall patterns. If SSTs can be successfully reduced by the use of MCB, then ocean-atmosphere fluxes of heat and water will be altered. The impacts of successfully reducing SST may have greater consequences for local weather than the aerosol perturbation to cloud microphysical properties. The potential for impacts on weather beyond the local region and even across jurisdictional boundaries will need to be thoroughly investigated using computational weather models and observational studies.

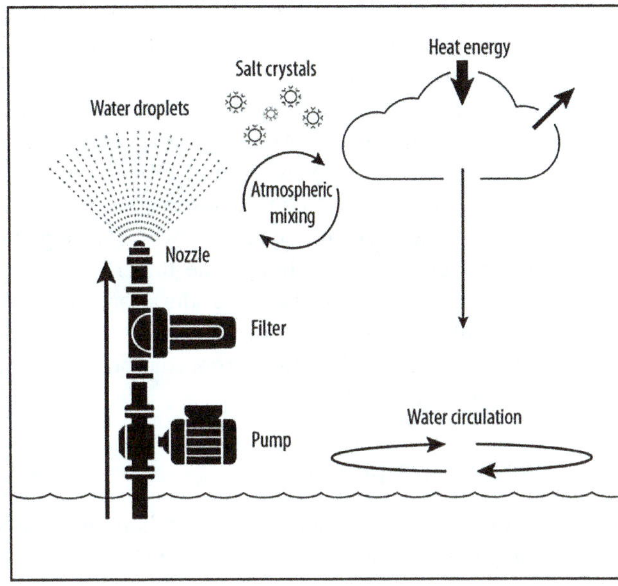

FIGURE 6 MCB aims to increase the portion of incoming solar radiation reflected back into space (i.e., albedo). The basic concept is to atomise seawater to produce droplets that then evaporate to leave behind sea salt crystals, which form effective cloud condensation nuclei.

Source: Figure 27 used with permission from Harrison (2018)

As with other shading techniques, risks apply such as reduced ocean primary productivity or overshading of light limited coral populations. However, shading intensities from a regional approach such as MCB would be at much lower intensities for longer periods than local scale shading techniques. At the regional scale, the magnitude of additional shading may not differ markedly from natural variability due to changes in atmospheric conditions year to year, such as generally higher cloud cover during La Niña years. Successful cooling of ocean waters may increase the capacity of the ocean to absorb CO_2, leading to higher acidity (Robock 2011). Ocean acidification has been identified as a major risk to GBR ecosystems (Eyre et al. 2018).

Regional cloud modification (for the purpose of enhancing rain or snowfall) is extensively and regularly practiced around the globe, often for the enhancement of hydroelectricity generation. In Australia, Snowy Mountains Hydro and Tasmania Hydro have used the technique in an effort to improve energy yields. Legislation in Australia specifically permits cloud seeding activities in the Snowy Mountains alpine National Park (Snowy Mountains Cloud Seeding Act 2014), which could provide a precedent for similar legislation regulating MCB in the future.

Implementation of solar radiation management techniques at the regional scale required to materially impact the future trajectory of coral health in the GBR marine park would represent an unprecedented use of eco-engineering for conservation management. Many of the albedo modification techniques examined here have also been proposed as potential geoengineering technologies to alter the global climate and offset warming caused by anthropogenic greenhouse gas emissions. Although the underlying technology used may be similar, the scale, duration, intermittency, and intent of solar radiation management considered for the GBR fundamentally differs from that proposed for global geoengineering.

27.7. WHERE TO NEXT?

All the proposed interventions reviewed here represent novel management actions for the conservation of coral reefs. Although some of the underlying technologies have been studied in other contexts, there is minimal and, in many cases, no peer reviewed literature relating to their potential efficacy, risks, governance, or social acceptability related to coral reef conservation. As the climate continues to warm, the pressure to find solutions to preserve coral cover and ecosystem functioning to support the many benefits to humankind of the GBR is likely to increase. To support informed decision making, it is incumbent on scientists and engineers to quantitatively investigate these and other novel approaches to mitigating the impacts of anthropogenic climate change on the reef.

Cooling and shading interventions represent transitory perturbations to the dynamic physical environment. Efficacy and risk are dependent on the prevailing conditions at the location and time of implementation. Numerical modelling is required to extrapolate both spatial and temporal performance estimates derived through laboratory and field testing of delivery methods and to aid in planning experiments and interpreting the results. The required modelling effort extends across multiple scales, resolutions, domains, and modelling approaches. Modelling will be required for estimating efficacy, designing and interpreting field experiments, optimising deployment strategies, and quantifying risk. Exacerbating the challenge, atmospheric and meteorological conditions over the reef, which are crucial to many of the proposed interventions, are remarkably understudied. Further research effort should be invested in understanding the meteorological drivers of bleaching and how this forcing may change in the future.

By their nature, techniques for cooling and shading represent a deliberate attempt to temporarily alter the environment of coral reef ecosystems to relieve stress. Technologies have been suggested that are applicable across a vast range of scales from one to hundreds of thousands of km^2 (Figure 7). Eco-engineering approaches are commonly utilised for conservation purposes on land, whereas their application for conservation within the marine environment, especially at the larger spatial scales proposed, have scarce precedent. Accordingly, engagement with stakeholders, the public, and indigenous traditional owners of sea country, combined with robust governance are as critical to achieving successful implementation as is the technical and engineering development of new technologies and approaches.

ACKNOWLEDGEMENT

This work was funded by the Reef Restoration and Adaptation Program a partnership between the Australian Governments Reef Trust and the Great Barrier Reef Foundation.

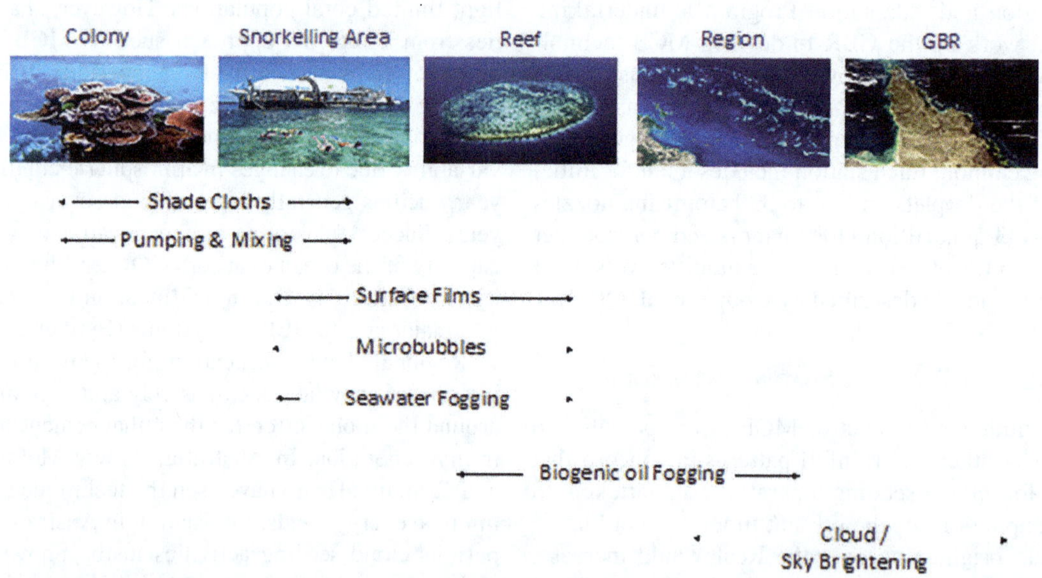

FIGURE 7 The author's estimate of indicative scales over which key cooling and shading technologies could potentially be applicable, given the currently limited available information.

REFERENCES

Ahlm L, Jones A, Stjern CW, Muri H, Kravitz B, Kristjánsson JE (2017) Marine cloud brightening—as effective without clouds. Atmos Chem Phys 17:13071–13087. doi: 10.5194/acp-17-13071-2017

Albrecth B. A. 1989. Aerosols, Cloud Microphysics, and Fractional Cloudiness. *Science*, 245, 1227–1230.

Alterskjær K, Kristjánsson JE, Seland Ø (2012) Sensitivity to deliberate sea salt seeding of marine clouds—observations and model simulations. Atmos Chem Phys 12:2795–2807. doi: 10.5194/acp-12-2795-2012

Andrews JC, Gentien P (1982) Upwelling as a source of nutrients for the Great Barrier Reef ecosystems: A solution to Darwin's Question. Mar Ecol Prog Ser 8:257–269. doi: 10.3354/meps008257

Anthony K, Bowen J, Mead D, Hardisty P (2019) Reef restoration and adaptation program: Intervention analysis and recommendations. A report provided to the Australian Government by the Reef Restoration and Adaptation Program. Australian Institute of Marine Science. https://gbrrestoration.org/reports

Bainbridge SJ (2017) Temperature and light patterns at four reefs along the Great Barrier Reef during the 2015–2016 austral summer: Understanding patterns of observed coral bleaching. J Oper Oceanogr 10:16–29. doi: 10.1080/1755876X.2017.1290863

Baird ME, Green R, Lowe R (2019) Cool water injection. A report provided to the Australian Government by the Reef Restoration and Adaptation Program. Australian Institute of Marine Science. https://gbrrestoration.org/reports

Baird ME, Green R, Lowe R, Mongin M, Bougeot E (2020a) Optimising cool-water injections to reduce thermal stress on coral reefs of the Great Barrier Reef. PLoS One 15:e0239978. doi: 10.1371/journal.pone.0239978

Baird ME, Wild-Allen KA, Parslow J, Mongin M, Robson B, Skerratt J, Rizwi F, Soja-Woźniak M, Jones E, Herzfeld M, Margvelashvili N, Andrewartha J, Langlais C, Adams MP, Cherukuru N, Gustafsson M, Hadley S, Ralph PJ, Rosebrock U, Schroeder T, Laiolo L, Harrison D, Steven ADL (2020b) CSIRO Environmental Modelling Suite (EMS): Scientific description of the optical and biogeochemical models (vB3p0). Geosci Model Dev 13:4503–4553. doi: 10.5194/gmd-13-4503-2020

Baker AC, Glynn PW, Riegl B (2008) Climate change and coral reef bleaching: An ecological assessment of long-term impacts, recovery trends and future outlook. Estuar Coast Shelf Sci 80:435–471. doi: 10.1016/j.ecss.2008.09.003

Barnes GT (2008) The potential for monolayers to reduce the evaporation of water from large water storages. Agric Water Manage 95:339–353. doi: 10.1016/j.agwat.2007.12.003

Bauman SJ, Costa MT, Fong MB, House BM, Perez EM, Tan MH, Thornton AE, Franks PJS (2015) Augmenting the biological pump: The shortcomings of geoengineered upwelling. Oceanography 27:17–23. doi: 10.5670/oceanog.2014.79

Bay L, Rocker M, Boström-Einarsson L, Babcock R, Buerger P, Cleves P, Harrison DP, Negri A, Quigley K, Randall CJ, van Oppen MJH, Webster N (2019) Reef restoration and adaptation program: Intervention technical summary. A report provided to the Australian Government by the Reef Restoration and Adaptation Program. Australian Institute of Marine Science. https://gbrrestoration.org/reports

Becker DM, Putnam HM, Burkepile DE, Adam TC, Vega Thurber R, Silbiger NJ (2021) Chronic low-level nutrient enrichment benefits coral thermal performance in a fore reef habitat. Coral Reefs 40:1637–1655. doi: 10.1007/s00338-021-02138-2

Benthuysen JA, Tonin H, Brinkman R, Herzfeld M, Steinberg C (2016) Intrusive upwelling in the Central Great Barrier Reef. Journal of Geophysical Research: Oceans 121:8395–8416. doi: 10.1002/2016JC012294

Berg JT, David CM, Gabriel MM, Bentlage B (2020) Fluorescence signatures of persistent photosystem damage in the staghorn coral Acropora cf. pulchra (Anthozoa: Scleractinia) during bleaching and recovery. Mar Biol Res 16:643–655. doi: 10.1080/17451000.2021.1875245

Berkelmans R, Weeks SJ, Steinberga CR (2010) Upwelling linked to warm summers and bleaching on the Great Barrier Reef. Limnol Oceanogr 55:2634–2644. doi: 10.4319/lo.2010.55.6.2634

Cacciapaglia C, van Woesik R (2016) Climate-change refugia: Shading reef corals by turbidity. Global Change Biol 22:1145–1154. doi: 10.1111/gcb.13166

Carrigan AD, Puotinen M (2014) Tropical cyclone cooling combats region-wide coral bleaching. Global Change Biol 20:1604–1613. doi: 10.1111/gcb.12541

Christensen MW, Gettelman A, Cermak J, Dagan G, Diamond M, Douglas A, Feingold G, Glassmeier F, Goren T, Grosvenor DP, Gryspeerdt E, Kahn R, Li Z, Ma PL, Malavelle F, McCoy IL, McCoy DT, McFarquhar G, Mülmenstädt J, Pal S, Possner A, Povey A, Quaas J, Rosenfeld D, Schmidt A, Schrödner R, Sorooshian A, Stier P, Toll V, Watson-Parris D, Wood R, Yang M, Yuan T (2022) Opportunistic experiments to constrain aerosol effective radiative forcing. Atmos Chem Phys 22:641–674. doi: 10.5194/acp-22-641-2022

Coelho VR, Fenner D, Caruso C, Bayles BR, Huang Y, Birkeland C (2017) Shading as a mitigation tool for coral bleaching in three common Indo-Pacific species. J Exp Mar Biol Ecol 497:152–163. doi: 10.1016/j.jembe.2017.09.016

Coles SL, Jokiel PL (1978) Synergistic effects of temperature, salinity and light on the hermatypic coral Montipora verrucosa. Mar Biol 49:187–195. doi: 10.1007/BF00391130

Condie SA, Anthony KRN, Babcock RC, Baird ME, Beeden R, Fletcher CS, Gorton R, Harrison D, Hobday AJ, Plagányi ÉE, Westcott DA (2021) Large-scale interventions may delay decline of the Great Barrier Reef. R Soc Open Sci 8:201296. doi: 10.1098/rsos.201296

Cooper G, Foster J, Galbraith L, Jain S, Neukermans A, Ormond B (2014) Preliminary results for salt aerosol production intended for marine cloud brightening, using effervescent spray atomization. Philos Trans A Math Phys Eng Sci 372:20140055. doi: 10.1098/rsta.2014.0055

Crook JA, Jackson LS, Forster PM (2016) Can increasing albedo of existing ship wakes reduce climate change? J Geophys Res Atmos 121:1549–1558. doi: 10.1002/2015JD024201

Cropek DM, Esarey JC, Conner CL, Goran JM, Smith T, Soucek DJ (2008) Toxicological effects of military fog oil obscurant on Daphnia magna and Ceriodaphnia dubia in field and laboratory exposures. Ecotoxicology 17:517–525. doi: 10.1007/s10646-008-0207-z

Deacon EL (1979) The role of coral mucus in reducing the wind drag over coral reefs. Boundary Layer Meteorol 17:517–521. doi: 10.1007/BF00118614

Deloitte Access Economics (2017) At what price? The economic, social and icon value of the Great Barrier Reef. Report. Great Barrier Reef Foundation. https://www2.deloitte.com/content/dam/Deloitte/au/Documents/Economics/deloitte-au-economics-great-barrier-reef-230617.pdf

Do C, Kuleshov Y (2023) Multi-hazard tropical cyclone risk assessment for Australia. Remote Sens 15:795. doi: 10.3390/rs15030795

Driver CJ, Ligotke MW, Downs JL, Tiller BL, Poston TM, Moore JC, Cataldo DA (1993) Environmental and health effects review for obscurant fog oil. ERDEC-CR-071. U. S. Army Chemical and Biological Defence Agency.

Eyre BD, Cyronak T, Drupp P, De Carlo EH, Sachs JP, Andersson AJ (2018) Coral reefs will transition to net dissolving before end of century. Science 359:908–911. doi: 10.1126/science.aao1118

Feng EY, Sawall Y, Wall M, Lebrato M, Fu Y (2020) Modeling coral bleaching mitigation potential of water vertical translocation—an analogue to geoengineered artificial upwelling. Front Mar Sci 7. doi: 10.3389/fmars.2020.556192

Fiddes SL, Woodhouse MT, Utembe S, Schofield R, Alexander SP, Alroe J, Chambers SD, Chen Z, Cravigan L, Dunne E, Humphries RS, Johnson G, Keywood MD, Lane TP, Miljevic B, Omori Y, Protat A, Ristovski Z, Selleck P, Swan HB, Tanimoto H, Ward JP, Williams AG (2022) The contribution of coral-reef-derived dimethyl sulfide to aerosol burden over the Great Barrier Reef: A modelling study. Atmos Chem Phys 22:2419–2445. doi: 10.5194/acp-22-2419-2022

Forster P, Crook JA, Jacksoin LS, Jenkins AKL (2014) Geoengineering Using oceanic microbubbles. American Geophysical Union, Fall Meeting 2014

Gatebe CK, Wilcox E, Poudyal R, Wang J (2011) Effects of ship wakes on ocean brightness and radiative forcing over ocean. Geophys Res Lett 38. doi: 10.1029/2011GL048819

GBRF (2015) Surface films: A feasibility study. Press Release. Great Barrier Reef Foundation. www.barrierreef.org/uploads/Surface%20Films%20Fact%20Sheet.pdf

GBRF (2018) Reef 'Sun Shield' trials show promise to prevent coral bleaching. Press Release. Great Barrier Reef Foundation. www.barrierreef.org/news/news/reef-sun-shield-trials-show-promise-to-prevent-coral-bleaching

GBRMPA (2018) Permit applications for restoration/adaptation projects to improve resilience of habitats in the Great Barrier Reef Marine Park. Great Barrier Reef Marine Park Authority. https://elibrary.gbrmpa.gov.au/jspui/handle/11017/3420

Gibbens S (2018) Floating sunscreen-like film could protect the Great Barrier Reef. National Geographic, March 28, 2018. www.nationalgeographic.com/science/article/sun-shield-film-protection-great-barrier-reef-coral-bleaching-spd

Haines JR (2016) Why is Russia Blowing Smoke (Literally)? The military uses of artificial fog. Foreign Policy Research Institute, August 19, 2016. www.fpri.org/article/2016/08/russia-blowing-smoke-literally-military-uses-artificial-fog/

Harland Bartholomew & Associates (1997) Evaluation of human health risks associated with fog oil training at fort leonard wood, Missouri Preliminary Risk Evaluation Report. U.S. Army Corps of Engineers. https://apps.dtic.mil/sti/citations/ADA326407

Harrison DP (2018) Could localized Marine Cloud Brightening Buy Coral Reefs Time? Abstract [AI44B-1733] presented at 2018 Ocean Sciences Meeting, Portland, Oregon, USA

Harrison DP, Harrison L, Baird M, Utembe S, Schofield R, Escobar Correa R, Mongin M, Rizwi F (2019) Environmental modelling of large scale solar radiation management. A report provided to the Australian Government by the Reef Restoration and Adaptation Program. Australian Institute of Marine Science. https://gbrrestoration.org/reports

Heron SF, Johnston L, Liu G, Geiger EF, Maynard JA, De La Cour JL, Johnson S, Okano R, Benavente D, Burgess TFR, Iguel J, Perez DI, Skirving WJ, Strong AE, Tirak K, Eakin CM (2016) Validation of reef-scale thermal stress satellite products for coral bleaching monitoring. Remote Sens 8:59. doi: 10.3390/rs8010059

Hoegh-Guldberg O (1999) Climate change, coral bleaching and the future of the world's coral reefs. Mar Freshw Res 50:839–866. doi: 10.1071/MF99078

Hoogenboom MO, Campbell DA, Beraud E, DeZeeuw K, Ferrier-Pagès C (2012) Effects of light, food availability and temperature stress on the function of photosystem II and photosystem I of coral symbionts. PLoS One 7:e30167. doi: 10.1371/journal.pone.0030167

Hoogenboom MO, Frank GE, Chase TJ, Jurriaans S, Álvarez-Noriega M, Peterson K, Critchell K, Berry KLE, Nicolet KJ, Ramsby B, Paley AS (2017) Environmental drivers of variation in bleaching severity of acropora species during an extreme thermal anomaly. Front Mar Sci 4:376. doi: 10.3389/fmars.2017.00376

Hughes TP, Anderson KD, Connolly SR, Heron SF, Kerry JT, Lough JM, Baird AH, Baum JK, Berumen ML, Bridge TC, Claar DC, Eakin CM, Gilmour JP, Graham NAJ, Harrison H, Hobbs J-PA, Hoey AS, Hoogenboom M, Lowe RJ, McCulloch MT, Pandolfi JM, Pratchett M, Schoepf V, Torda G, Wilson SK (2018b) Spatial and temporal patterns of mass bleaching of corals in the anthropocene. Science 359:80–83. doi: 10.1126/science.aan8048

Hughes TP, Barnes ML, Bellwood DR, Cinner JE, Cumming GS, Jackson JBC, Kleypas J, van de Leemput IA, Lough JM, Morrison TH, Palumbi SR, van Nes EH, Scheffer M (2017a) Coral reefs in the anthropocene. Nature 546:82. doi: 10.1038/nature22901

Hughes TP, Kerry JT, Álvarez-Noriega M, Álvarez-Romero JG, Anderson KD, Baird AH, Babcock RC, Beger M, Bellwood DR, Berkelmans R, Bridge TC, Butler IR, Byrne M, Cantin NE, Comeau S, Connolly SR, Cumming GS, Dalton SJ, Diaz-Pulido G, Eakin CM, Figueira WF, Gilmour JP, Harrison HB, Heron SF, Hoey AS, Hobbs J-PA, Hoogenboom MO, Kennedy EV, Kuo C-y, Lough JM, Lowe RJ, Liu G, McCulloch MT, Malcolm HA, McWilliam MJ, Pandolfi JM, Pears RJ, Pratchett MS, Schoepf V, Simpson T, Skirving WJ, Sommer B, Torda G, Wachenfeld DR, Willis BL, Wilson SK (2017b) Global warming and recurrent mass bleaching of corals. Nature 543:373–377. doi: 10.1038/nature21707

Hughes TP, Kerry JT, Simpson T (2018a) Large-scale bleaching of corals on the Great Barrier Reef. Ecology 99:501–501. doi: 10.1002/ecy.2092

Jones G, Curran M, Swan H, Deschaseaux E (2017) Dimethylsulfide and coral bleaching: Links to solar radiation, low level cloud and the regulation of seawater temperatures and climate in the Great Barrier Reef. Am J Clim Change 6:328–359. doi: 10.4236/ajcc.2017.62017

Kalita B, Bora S, Sharma AK (2013) Plant essential oils as mosquito repellent-a review. Int J Res Dev Pharm Life Sci 3:741–747. doi: 10.1186/1475-2875-10-S1-S11

Kamenos NA, Hennige SJ (2024) A historical perspective to thermal- and heatwave-induced bleaching on the great barrier reef. in Wolanski E, Kingsford MJ (eds.), Oceanographic Processes of Coral Reefs. Physical and Biological Links in the Great Barrier Reef. CRC Press, Boca Raton, FL.

Kirke BK (2000) Circulation, destratification, mixing and aeration: Why and how? Water:16–21

Kirke BK, El Gezawy A (1997) Design and model tests for an efficient mechanical circulator/aerator for lakes and reservoirs. Water Res 31:1283–1290. doi: 10.1016/S0043-1354(96)00172-8

Köhler H (1936) The nucleus in and the growth of hygroscopic droplets. Trans Faraday Soc 32:1152–1161. doi: 10.1039/TF9363201152

Koop K, Booth D, Broadbent A, Brodie J, Bucher D, Capone D, Coll J, Dennison W, Erdmann M, Harrison P, Hoegh-Guldberg O, Hutchings P, Jones GB, Larkum AWD, O'Neil J, Steven A, Tentori E, Ward S, Williamson J, Yellowlees D (2001) ENCORE: The effect of nutrient enrichment on coral reefs. Synthesis of results and conclusions. Mar Pollut Bull 42:91–120. doi: 10.1016/S0025-326X(00)00181-8

Latham J (1990) Control of global warming? Nature 347:339–340. doi: 10.1038/347339b0

Latham J, Bower K, Choularton T, Coe H, Connolly P, Cooper G, Craft T, Foster J, Gadian A, Galbraith L, Iacovides H, Johnston D, Launder B, Leslie B, Meyer J, Neukermans A, Ormond B, Parkes B, Rasch P, Rush J, Salter S, Stevenson T, Wang H, Wang Q, Wood R (2012) Marine cloud brightening. Philos Trans A Math Phys Eng Sci 370:4217–4262. doi: 10.1098/rsta.2012.0086

Latham J, Kleypas J, Hauser R, Parkes B, Gadian A (2013) Can marine cloud brightening reduce coral bleaching? Atmos Sci Lett 14:214–219. doi: 10.1002/asl2.442

Lavars N (2018) Giant fans, starfish assaults and the half-baked plan to save the Great Barrier Reef. New Atlas, 23 January 2018. https://newatlas.com/giant-fans-starfish-great-barrier-reef-plan/53076/

Leahy SM, Kingsford MJ, Steinberg CR (2013) Do clouds save the Great Barrier Reef? Satellite imagery elucidates the cloud-SST relationship at the local scale. PLoS One 8:e70400. doi: 10.1371/journal.pone.0070400

Lenssen NJL, Schmidt GA, Hansen JE, Menne MJ, Persin A, Ruedy R, Zyss D (2019) Improvements in the GISTEMP uncertainty model. J Geophys Res Atmos 124:6307–6326. doi: 10.1029/2018JD029522

Lesser MP (1997) Oxidative stress causes coral bleaching during exposure to elevated temperatures. Coral Reefs 16:187–192. doi: 10.1007/s003380050073

Lesser MP, Farrell JH (2004) Exposure to solar radiation increases damage to both host tissues and algal symbionts of corals during thermal stress. Coral Reefs 23:367–377. doi: 10.1007/s00338-004-0392-z

Liu G, Heron SF, Eakin CM, Muller-Karger FE, Vega-Rodriguez M, Guild LS, De La Cour JL, Geiger EF, Skirving WJ, Burgess TFR, Strong AE, Harris A, Maturi E, Ignatov A, Sapper J, Li J, Lynds S (2014) Reef-scale thermal stress monitoring of coral ecosystems: New 5-km global products from NOAA Coral Reef watch. Remote Sens 6:11579–11606. doi: 10.3390/rs61111579

Liu G, Rauenzahn JL, Heron SF, Eakin MC, Skirving W, Christensen TRL, Strong AE, Li J (2013) NOAA coral reef watch 50 km satellite sea surface temperature-based decision support system for coral bleaching management. NOAA Technical Report NESDIS 143. https://repository.library.noaa.gov/view/noaa/743

Long S, Fisher E (2021) Reef Havens Research Project (Temperature, Current, CTD casts at Moore Reef) 2018–2021. Dataset. eAtlas. https://eatlas.org.au/pydio/public/gbr-rrrc-reef-havens-metadata-and-images

Matz MV, Treml EA, Aglyamova GV, Bay LK (2018) Potential and limits for rapid genetic adaptation to warming in a Great Barrier Reef coral. PLoS Genet 14:e1007220. doi: 10.1371/journal.pgen.1007220

Mazoyer M, Burnet F, Denjean C (2022) Experimental study on the evolution of droplet size distribution during the fog life cycle. Atmos Chem Phys 22:11305–11321. doi: 10.5194/acp-22-11305-2022

McAuliffe TF, Rosich RS (1989) Review of artificial destratification of water storages in Australia. Research report (Urban Water Research Association of Australia); no. 9. Melbourne and Metropolitan Board of Works, Melbourne. ISBN:1875298096

McGowan H, Theobald A (2017) ENSO weather and coral bleaching on the Great Barrier Reef, Australia. Geophys Res Lett 44:10,601–610,607. doi: 10.1002/2017GL074877

McJannet D, Cook F, Knight J, Burn S (2008) Evaporation reduction by monolayers: Overview, modelling and effectiveness. Urban Water Security Research Alliance Technical Report No. 6. CSIRO: Water for a Healthy Country National Research Flagship. www.urbanwateralliance.org.au/publications/UWSRA-tr6.pdf

Mitchell-Whittington A (2017) Giant $2.2 million fans to cool Great Barrier Reef. Brisbane Times, December 8, 2017. www.brisbanetimes.com.au/national/queensland/giant-2-2m-fans-to-cool-great-barrier-reef-20171207-p4yxi5.html

Palacios CF, Wright BR (1998) Literature survey on materials to improve low temperature fluidity of fog oil. Interim Report TFLRF No. 336. Army US. https://apps.dtic.mil/sti/pdfs/ADA345722.pdf

Palmer WG (1990) Exposure standard for fog oil. Technical report: AD-A-231714/7/XAB; USABRDL-TR-9010. Army Biomedical Research and Development Laboratory. https://apps.dtic.mil/sti/citations/ADA255479

Panjabi K, Rudra R, Goel P (2016) Evaporation retardation by monomolecular layers: An experimental study at the Aji Reservoir (India). Open J Civ Eng 6:346–357. doi: 10.4236/ojce.2016.63029

Partanen A-I, Kokkola H, Romakkaniemi S, Kerminen V-M, Lehtinen KEJ, Bergman T, Arola A, Korhonen H (2012) Direct and indirect effects of sea spray geoengineering and the role of injected particle size. J Geophys Res Atmos 117. doi: 10.1029/2011JD016428

Penner JE, Andreae M, Annegarn H, Barrie L, Feichter J, Hegg D, Jayaraman A, Leaitch R, Murphy D, Nganga J, Pitari G (2001) Aerosols, Their Direct and Indirect Effects Climate Change 2001: The Scientific Basis. Cambridge University Press, Cambridge.

Poston TM, McFadden KM, Bean RM, Clark ML, Thomas BL (1986) Acute toxicity of smoke screen materials to aquatic organisms, white phosphorus-felt, red phosphorus-butyl rubber and SGF (Smoke Generator Fuel) No. 2 Fog Oil. Technical report. https://apps.dtic.mil/sti/citations/ADA167900

Randall CJ, Toth LT, Leichter JJ, Maté JL, Aronson RB (2020) Upwelling buffers climate change impacts on coral reefs of the eastern tropical Pacific. Ecology 101:e02918. doi: 10.1002/ecy.2918

Rau GH, McLeod EL, Hoegh-Guldberg O (2012) The need for new ocean conservation strategies in a high-carbon dioxide world. Nat Clim Change 2:720–724. doi: 10.1038/nclimate1555

Robock A (2011) Bubble, bubble, toil and trouble. Clim Change 105:383–385. doi: 10.1007/s10584-010-0017-1

Rogers JS, Monismith SG, Koweek DA, Torres WI, Dunbar RB (2016) Thermodynamics and hydrodynamics in an atoll reef system and their influence on coral cover. Limnol Oceanogr 61:2191–2206. doi: 10.1002/lno.10365

Rosli SRN, Yong KTL, Mohd Nasir NS, Alikasturi AS, Anuar MR (2022) Application of biodiesel in fogging system with Azadirachta Indica Oil as insecticide towards mosquito and larvae. Mater Sci Forum 1077:153–163. doi: 10.4028/p-6b4rb2

Russell LM, Sorooshian A, Seinfeld JH, Albrecht BA, Nenes A, Ahlm L, Chen Y-C, Coggon M, Craven JS, Flagan RC, Frossard AA, Jonsson H, Jung E, Lin JJ, Metcalf AR, Modini R, Mülmenstädt J, Roberts G, Shingler T, Song S, Wang Z, Wonaschütz A (2013) Eastern pacific emitted aerosol cloud experiment. Bull Am Meteorol Soc 94:709–729. doi: 10.1175/BAMS-D-12-00015.1

Salter S, Sortino G, Latham J (2008) Sea-going hardware for the cloud albedo method of reversing global warming. Philos Trans A Math Phys Eng Sci 366:3989–4006. doi: 10.1098/rsta.2008.0136

Savage C (2019) Seabird nutrients are assimilated by corals and enhance coral growth rates. Sci Rep 9:4284. doi: 10.1038/s41598-019-41030-6

Seitz R (2011) Bright water: Hydrosols, water conservation and climate change. Clim Change 105:365–381. doi: 10.1007/s10584-010-9965-8

Shantz AA, Burkepile DE (2014) Context-dependent effects of nutrient loading on the coral–algal mutualism. Ecology 95:1995–2005. doi: 10.1890/13-1407.1

Sheikh Muhammad S, Flecker B, Leitgeb E, Gebhart M (2007) Characterization of fog attenuation in terrestrial free space optical links. Opt Eng 46:066001. doi: 10.1117/1.2749502

Sherman B, Lemckert C, Zhang H (2021) The impact of artificial destratification on reservoir evaporation. Technical Report No. 35. Urban Water Security Research Alliance CSIRO. www.urbanwateralliance.org.au/publications/UWSRA-tr35.pdf

Skirving W, Enríquez S, Hedley JD, Dove S, Eakin CM, Mason RAB, De La Cour JL, Liu G, Hoegh-Guldberg O, Strong AE, Mumby PJ, Iglesias-Prieto R (2018) Remote sensing of coral bleaching using temperature and light: Progress towards an operational algorithm. Remote Sens 10:18. doi: 10.3390/rs10010018

Slezak M (2018) Millions spent on Great Barrier Reef projects against expert advice. The Guardian, 21 January 2018. www.theguardian.com/environment/2018/jan/21/millions-spent-on-great-barrier-reef-projects-against-expert-advice

Smith LW, Birkeland C (2007) Effects of intermittent flow and irradiance level on back reef Porites corals at elevated seawater temperatures. J Exp Mar Biol Ecol 341:282–294. doi: 10.1016/j.jembe.2006.10.053

Snowy Mountains Cloud Seeding Act (2004) No 19 https://legislation.nsw.gov.au/view/whole/html/inforce/current/act-2004-019

Snowy Mountains Cloud Seeding Act (2004) NSW Government Legislation, Snowy Mountains Cloud Seeding Trial Act 2004 No 19, https://legislation.nsw.gov.au/view/html/inforce/2008-05-20/act-2004-019

Steinberg CR, Benthuysen JA, Klein-SalasEduardo, Cantin NE, Tonin H, Spagnol S, Spillman CM (2021) Oceanographic drivers of bleaching in the GBR: From observations to prediction, Volume 1: Summary of oceanographic conditions during the 2015–17 bleaching years. Report to the National Environmental Science Program. Reef and Rainforest Research Centre Limited.

Tagliafico A, Baker P, Kelaher B, Ellis S, Harrison D (2022) The effects of shade and light on corals in the context of coral bleaching and shading technologies. Front Mar Sci 9. doi: 10.3389/fmars.2022.919382

Tollefson J (2021) Can artificially altered clouds save the Great Barrier Reef? Nature 596. doi: 10.1038/d41586-021-02290-3

Tolleter D, Seneca François O, DeNofrio Jan C, Krediet Cory J, Palumbi Stephen R, Pringle John R, Grossman Arthur R (2013) Coral bleaching independent of photosynthetic activity. Curr Biol 23:1782–1786. doi: 10.1016/j.cub.2013.07.041

Twomey S (1974) Pollution and the planetary albedo. Atmos Environ 8:1251–1256. doi: 10.1016/0004-6981(74)90004-3

Twomey S. 1991. Aerosols, clouds and radiation. *Atmospheric Environment. Part A. General Topics*, 25, 2435–2442

Waheeb Youssef Y, Khodzinskaya A (2019) A review of evaporation reduction methods from water surfaces. In: AP Volkov, T Sultanov, A Adamtsevich (eds) XXII International Scientific Conference "Construction the Formation of Living Environment" (FORM-2019). E3S-Conferences, Tashkent, Uzbekistan

Walther BD, Kingsford MJ, McCulloch MT (2013) Environmental records from Great Barrier Reef Corals: Inshore versus offshore drivers. PLoS One 8:e77091. doi: 10.1371/journal.pone.0077091

West JM, Salm RV (2003) Resistance and resilience to coral bleaching: Implications for coral reef conservation and management. Conserv Biol 17:956–967. www.jstor.org/stable/3588851

William BZ, Vaclav T, Simon B, Himiyage CHB (2008) Microbubble generation. Recent Patents on Engineering 2:1–8. doi: 10.2174/187221208783478598

Wolanski E, Kingsford M, Lambrechts J, Marmorino G (2024) The physical oceanography of the Great Barrier Reef: A review. In Wolanski E, Kingsford MJ (eds.), Oceanographic Processes of Coral Reefs. Physical and Biological Links in the Great Barrier Reef. CRC Press, Boca Raton, FL.

Wood R (2021) Assessing the potential efficacy of marine cloud brightening for cooling earth using a simple heuristic model. Atmos Chem Phys 21:14507–14533. doi: 10.5194/acp-21-14507-2021

Zhang X, Lewis M, Johnson B (1998) Influence of bubbles on scattering of light in the ocean. Appl Opt 37:6525–6536. doi: 10.1364/AO.37.006525

28 Sexual Reproduction of Reef Corals and Application to Coral Restoration

Peter L. Harrison

28.1. INTRODUCTION

Scleractinian reef-building corals are foundation species and ecosystem engineers with critically important roles in the formation, development, and ecological functions of hyperdiverse coral reef ecosystems (Bruno et al. 2007; Harrison and Booth 2007; Knowlton et al. 2021). Reef-building corals are holobionts formed by complex symbioses with photosynthetic Symbiodiniaceae microalgae, Bacteria, Archaea and viruses (Blackall et al. 2015), and these symbioses are important for their ecological success, resilience, and evolution. The mutualism with photosynthetic Symbiodiniaceae combined with heterotrophic feeding by coral polyps provide sufficient productivity and energy for allocations to key life functions of survival, growth and repair of tissues and skeletons, and sexual reproduction (Muscatine 1990; Harrison and Wallace 1990; Anthony and Fabricius 2000; Houlbreque et al. 2004; Matthews et al. 2020). In this chapter, the term 'coral' is used to refer to hard corals from the Order Scleractinia within the Class Anthozoa (Phylum Cnidaria).

Sexual reproduction, larval development, dispersal, settlement, post-settlement survival, and recruitment processes are essential for the maintenance, recovery, and evolution of coral populations, assemblages, and the reef systems they build (reviewed in Fadlallah 1983; Harrison and Wallace 1990; Richmond and Hunter 1990; Baird et al. 2009; Harrison 2011; Gleason and Hoffman 2011; Randall et al. 2020; Banaszak et al. 2023). However, these processes are not sufficiently understood to accurately predict natural recruitment patterns and recovery pathways after major disturbances. This is a consequence of insufficient biological and ecological information on coral reproduction processes, in addition to inadequate understanding of the complex biophysical interactions and dynamic oceanographic processes operating within and between reefs at varying spatial and temporal scales that control larval dispersal and settlement (Pineda et al. 2010; Wolanski et al. 2024). Post-settlement survival and growth also vary widely within and among species with different life histories and in different reef environments (Doropoulos et al. 2016; Ritson-Williams et al. 2016; Gouezo et al. 2020a). This natural complexity is exacerbated by the increasing frequency and severity of anthropogenic disturbances and multiple stressors impacting reef systems, including rapid changes to climate and ocean systems resulting in ocean warming and mass bleaching episodes and altered carbonate saturation states that are overwhelming the natural resilience of reefs (Hoegh-Guldberg et al. 2017; Hughes et al. 2018, 2019; Knowlton et al. 2021). These environmental changes are altering selection pressures on coral populations and their adaptation and evolutionary processes, with uncertain outcomes for future reefs.

This chapter broadly reviews current knowledge and identifies new research required for improved understanding of sexual reproduction among scleractinian hermatypic corals with a focus on the GBR and emerging global patterns. It also examines the ways in which knowledge of coral reproduction is being applied to coral restoration research on the GBR and in other reef regions. Understanding when and how different corals reproduce and, therefore, when gametes and planula larvae are available are essential prerequisites for application to active coral restoration of damaged and highly degraded reefs that are now limited by impaired larval supply. This work is time-critical because halting the accelerating decline in coral communities and enabling reef recovery and future resilience are essential for maintaining the health and viability of the GBR and reefs globally.

28.2. CORAL REPRODUCTION: SEXUAL PATTERNS, SPAWNING AND PLANULATION, ENVIRONMENTAL INFLUENCES, AND REGIONAL PATTERNS OF REPRODUCTION

Scleractinian corals have two primary reproductive pathways resulting in the production of new corals: asexual reproduction including development and budding of genetically identical polyps to enable the formation of larger sessile coral colonies or solitary polyps such as clonal mushroom corals that are sedentary and natural colony fragmentation for some species; and sexual reproduction resulting in genetically diverse offspring (Harrison and Booth 2007; Richmond et al. 2018). Most corals have a biphasic life cycle typically involving a dominant sessile adult polyp and colony phase and a shorter motile planula larval stage that is usually planktonic and dispersive (Fadlallah 1983; Harrison and Wallace 1990; Randall et al. 2020). Planktonic embryos and larvae are typically dispersed away from their natal reef site to some degree by surface wind and waves interacting with complex reef currents and dynamic oceanographic conditions, which interconnect populations on different areas of the reef and between reefs to create metapopulation networks that vary spatially and temporally (Harrison and Wallace 1990; Jones et al. 2009; Gleason and Hoffman 2011; Bode et al.

DOI: 10.1201/9781003320425-32

2018; Wolanski et al. 2024; Kininmonth 2024). The extent to which pelagic planulae disperse is highly variable and strongly influenced by life history traits including broadcast spawning versus brooding modes of development, rates of larval development until settlement competency, larval size and energetics, presence of photosymbionts, and whether larvae are entrained in eddies or large lagoons that restrict dispersal during development or dispersed over longer distances by major currents (e.g., Richmond 1988; Jones et al. 2009; Graham et al. 2008; Figueiredo et al. 2013; Hock et al. 2019; Riginos et al. 2019).

28.2.1. Sexual Patterns

Sexual reproductive patterns have been studied in more than 400 species of Scleractinia in various regions around the world, and the majority are broadcast spawning hermaphrodites (~65%), with fewer species known to be gonochoric (separate sexes) broadcast spawners (~20%) and a smaller number of hermaphroditic or gonochoric species (15%) recorded to brood planula larvae within their polyps before release (Harrison and Wallace 1990; Baird et al. 2009, 2021; Harrison 2011; Randall et al. 2020).

However, not all scleractinians conform to these four separate reproductive categories, as some coral populations and species are known to have mixed sexual patterns (hermaphrodites and gonochores), and some species have both broadcast spawning and brooding modes of development recorded (Harrison 1985; Harrison and Wallace 1990; Harrison 2011; Guest et al. 2012). Furthermore, some brooded larvae are known to be produced asexually in some *Pocillopora* spp. (Stoddart and Black 1985; Schmidt-Roach et al. 2012) and in two species of *Tubastraea* (Ayre and Resing 1986).

Hermaphroditism is the dominant sexual pattern among Scleractinia (~71% of species studied), and most are simultaneous hermaphrodites that develop both ova and spermatozoa within each of the polyps, which mature simultaneously just prior to spawning (Harrison 2011; Figure 1). However, some coral species are sequential hermaphrodites (Glynn and Colley 2009), and some fungiid mushroom corals and other colonial species exhibit protandrous sex change from male to female function as they grow or exhibit bidirectional sex change in response to changing environmental conditions (Rinkevich and Loya 1979; Loya and Sakai 2008; Loya et al. 2009; Eyal-Shaham et al. 2019; Figure 2). Initial sex allocation to male function conforms to sex allocation theory that predicts that patterns of sex change respond to differential male and female fitness as size and energy resources available for allocation to sexual reproduction change over the lifetime of individuals (Charnov 1982; Polikansky 1982). Longer-term studies of two *Ctenactis* spp. fungiids in Japan showed that these species exhibit bidirectional sex change and respond to changing environmental conditions and variable energy costs of sperm versus egg production (Loya and Sakai 2008; Loya et al. 2009). Another longer-term study of *Fungia fungites* corals in Japan recorded male, female, and hermaphroditic individuals in the population, and isolated females continuously released planulae over multiple months even when males were absent, possibly resulting from self-fertilization, asexual planulae production, or extended sperm storage (Eyal-Shaham et al. 2020). Further research on other apparently gonochoric coral species is likely to reveal additional species with labile sex allocation and sex change (Harrison 2011), which highlights the need for detailed long-term monitoring of coral gametogenic and breeding cycles within individuals from a broader range of taxa, reef regions, and different reef habitats subject to variable environmental conditions.

FIGURE 1 Most coral species studied so far are hermaphroditic broadcast spawners, releasing mature eggs and sperm in buoyant bundles from each gravid polyp in the colony, for example, *Acropora valida* (left) and *Platygyra* cf. *daedalea* colony (right).

Source: Photos: Peter Harrison

FIGURE 2 Fungiid mushroom corals typically release either eggs or sperm within a breeding season.

Source: Photo: Peter Harrison

An unusual sexual pattern (termed pseudo-gynodioecy) was recorded in *Galaxea* spp. populations on the GBR, which have female colonies that spawn red pigmented eggs and hermaphroditic colonies that develop functional sperm and white pseudo-eggs filled with large storage granules that were not able to be fertilized (Harrison 1988). The same sexual pattern has been reported for *Galaxea fascicularis* from Japan (Okubo et al. 2013) and from Taiwan, where the white eggs were able to be fertilized, and, therefore, these populations have a gynodioecious sexual pattern (Keshavmurphy et al. 2012).

Fewer strictly gonochoric coral species (~26% of species studied) have been reported, and these species have separate female and male colonies or solitary polyps throughout their life (Harrison 2011). Some primarily gonochoric coral species such as *Porites* spp. also have a low incidence of hermaphroditic polyps or colonies within some populations studied; where this is relatively rare, they are considered to exhibit stable gonochorism (Giese and Pearse 1974; Harrison and Wallace 1990; Figure 3). However, mixed sexual patterns have been reported in more than 12 coral species where different populations are either gonochoric or have mixtures of males, females, and hermaphrodites (Harrison and Wallace 1990; Harrison 2011; Guest et al. 2012; Santiago-Valentin et al. 2015; Abecia et al. 2016).

Strong systematic trends in sexual patterns and some other reproductive characteristics are evident in many species, genera, and some families of Scleractinia (Harrison 1985, 2011; Baird et al. 2009). However, exceptions to these systematic trends occur among some other families, which indicates that contrasting sexual patterns have evolved within a range of families (Harrison 2011). The dominance of hermaphroditism among corals led Szmant (1986) to suggest that this might be the ancestral pattern for Scleractinia. Conversely, Harrison (1990) considered that gonochorism was likely to be the ancestral sexual pattern in Scleractinia, based on the more primitive conical sperm morphology and ultrastructural features in gonochoric species compared with hermaphroditic corals (Harrison 1985, 1990; Harrison and Jamieson 1999). Subsequent phylogenetic analyses have indicated that gonochorism is likely to be the ancestral condition among anthozoan hexacorals including scleractinians (Daly et al. 2003; Kerr et al. 2011).

28.2.2. Mode of Larval Development

Historically, scleractinian corals were typically considered to brood planulae within their polyps as most early studies focused on brooding species leading to the 'dogma of viviparity' (Hyman 1940; Wells 1956). Early reports of corals spawning gametes were either overlooked or dismissed as aberrant stress responses and therefore thought to not be representative of typical reproductive patterns (reviewed in Harrison and Wallace 1990; Harrison 2011). However, the discovery of mass coral spawning on the GBR in the early 1980s (Harrison et al. 1984; Willis et al. 1985; Babcock et al. 1986) combined with increasing global research (Shlesinger and Loya 1985; Simpson 1985) quickly overturned this misconception and showed that broadcast spawning and external fertilization was the predominant mode of reproduction in corals (Harrison and Wallace 1990; Richmond and Hunter 1990).

Global research over the past few decades has contributed substantial new information on sexual reproduction in corals from many reef regions and has further increased

FIGURE 3 *Porites* spp. colonies are primarily gonochoric and release either sperm or eggs.

Source: Photo: Peter Harrison

the dominance of known broadcast spawning species. Of the more than 400 scleractinian species studied globally by 2010, ~83% were recorded to spawn gametes, and only 14% of coral species were known to be brooding corals, with a small number of species (3%) having mixed modes of development (Harrison 2011). Some species have consistent modes of development in some reef regions but mixed modes of development in other regions, for example, colonies of the merulinid brain coral identified as *Goniastrea aspera* are broadcast spawners on the GBR (Babcock 1984; Babcock et al. 1986) but have both spawning and brooding modes of development in Japan (Sakai 1997; Nozawa and Harrison 2005). Similarly, the vast majority of coral species studied so far on the GBR are broadcast spawners (~93%) with comparatively few species known to brood larvae (~7%) on the GBR.

Many scleractinian families have relatively consistent modes of development, with broadcast spawning dominating in most Acroporidae, Merulinidae, Fungiidae, and some other groups studied to date, although some brooding species are also recorded in these families (Harrison 1985, 2011; Baird et al. 2009). Mixed modes of development have been recorded in 13 families, including Dendrophylliidae, Agariciidae, and Poritidae in which spawning is more common than brooding and Pocilloporidae where brooding is more common (Baird et al. 2009; Harrison 2011; Madin et al. 2016).

Both spawning and brooding development modes have also been reported in different *Pocillopora* species. *Pocillopora damicornis* on the GBR reproduces sexually via broadcast spawning of clouds of eggs and sperm over a few hours after sunrise and asexually via production and release of brooded larvae around full moon periods (Tanner 1996; Schmidt-Roach et al. 2012, 2014). *Pocillopora acuta* on the GBR releases brooded larvae (Marshall and Stephenson 1933) that are asexually produced (Schmidt-Roach et al. 2014), with sperm released in the early morning during the peak planulation periods (Smith et al. 2019). Both modes of development have been reported in *Pocillopora verrucosa* populations from different regions; Schmidt-Roach et al. (2014) concluded that true *P. verrucosa* is a broadcast spawning species. Only gamete release has been observed in colonies confirmed to be *P. meandrina*, *P. eydouxi*, and *P. elegans* (Schmidt-Roach et al. 2012, 2014).

Mixed modes of development have now been recorded in a wide range of other species, genera, and many families (Harrison and Wallace 1990; Baird et al. 2009; Harrison 2011; Kerr et al. 2011), confirming the earlier conclusion that the mode of development is a relatively plastic and variable life history trait in Scleractinia compared with sexual patterns (Harrison 1985). Shlesinger et al. (1998) concluded that viviparous brooding is a derived reproductive trait and that the ancestral mode of development in scleractinian and other Hexacorallia is broadcast spawning. Phylogenetic analyses indicate that the mode of reproduction has evolved at up to four times the rate of sexual patterns (Kerr et al. 2011). Approximately 48% of Atlantic species studied to date are brooding species, whereas brooders account for only about 10% of the Indo-Pacific species that have been studied (Kerr et al. 2011).

Most broadcast spawning scleractinian species produce eggs without symbionts and therefore need to acquire these during early post-settlement stages. However, eggs of the *Porites*, *Montipora*, *Anacropora*, and *Pocillopora* species studied so far all have Symbiodiniaceae in their eggs. Baird et al. (2009) noted ~90% of brooding species have vertical transmission of Symbiodiniaceae dinoflagellate endosymbionts from parent tissues to eggs and planulae; in contrast, *Isopora* brooding species produce planulae without photosymbionts (Kojis 1986a). Larvae with photosymbionts are potentially autotrophic and can therefore presumably supplement energy acquired from their eggs via photosynthetic production, which may enable longer planktonic dispersal periods (Richmond 1988; Serrano et al. 2016). Survival of photosymbiotic larvae is also likely to be enhanced after metamorphosis into the primary polyp and the onset of calcification as these processes have high energy demands.

Given the range of sexual patterns and modes of development evident among corals studied to date, it is highly likely that a greater diversity of reproductive characteristics will be found when a broader range of coral taxa from the GBR and other reef regions are studied in detail.

28.2.3. Reproductive Cycles

Scleractinian corals exhibit a range of gametogenic cycles that culminate in maturation of eggs and sperm within polyps prior to spawning and external or internal fertilization. Most broadcast spawning hermaphroditic corals studied to date have a single cycle of egg and sperm production and a single spawning season each year, whereas brooding corals may have multiple cycles of gamete development and release larvae frequently throughout the year or seasonally, with peak planulation at different lunar phases (Jokiel et al. 1985; Harrison and Wallace 1990; Richmond and Hunter 1990; Tanner 1996; Baird et al. 2009; Harrison 2011). Oogenesis requires larger investments of energy and resources and usually occurs over longer time periods from about five to more than nine or ten months, whereas spermatogenesis usually occurs over fewer months.

Some broadcast spawning hermaphroditic colonies are known to have two gametogenic cycles and spawning periods annually, including some *Montipora* spp. colonies on the Central GBR (Stobart et al. 1992) and in Palau (Gouezo et al. 2020b). Two gametogenic cycles and likely biannual spawning have been reported in some species from equatorial Madang (Oliver et al. 1988), in two *Platygyra* spp. from Kenya (Mangubhai and Harrison 2008a), and in *Diploria labrinthiformis* populations from Curacao (Chamberland et al. 2016). Therefore, as more coral populations and species are studied, it is likely that additional records of multiple gametogenic cycles and biannual spawning events within hermaphroditic broadcast spawning colonies will be reported.

Some gonochoric broadcast spawners also have multiple cycles of egg or sperm development and release each year, whereas other gonochoric species have extended periods of oogenesis over 14 to 15 months and overlapping cycles of egg production (Harrison and Wallace 1990; Randall et al. 2020). Variable numbers of gametogenic cycles have also been documented among some brooding coral species. *Isopora palifera* colonies from Heron Island on the southern GBR have a single gametogenic cycle each year, whereas colonies at Lizard Island on the northern GBR breed year-round, and colonies at Lae in northern Papua New Guinea have six gametogenic cycles per year (Kojis 1986a, 1986b). *Isopora cuneata* colonies at Heron Island have two consecutive gametogenic cycles each year (Kojis 1986a).

28.2.4. Fecundity and Stress

Sexual reproduction is energetically costly, particularly for egg production where investment of energy-rich compounds and other resources may occur over many months each year. As with all living organisms, corals have limited energy resources and interstitial cells that must be shared among different life functions including essential tissue maintenance and repair, growth of tissues and calcium carbonate exoskeletons, and allocation to sexual and asexual reproduction (Harrison and Wallace 1990). Coral ecophysiology and energy balance are influenced by the coral holobiont responses to varying environmental conditions on reefs and their 'fundamental niche' limits in which they can survive, grow, and reproduce (Hoogenboom and Connolly 2009). These responses are also strongly influenced by stressors including climate and ocean change (Sweet and Brown 2016) and other anthropogenic disturbances (Richmond 1993; Richmond et al. 2018).

Allocation of resources to sexual reproduction and fecundity varies among different coral groups and morphologies (Alvarez-Noriega et al. 2016), over different life stages, and in response to varying environmental conditions and impacts of stressors (Harrison and Wallace 1990; Leuzinger et al. 2012; Randall et al. 2020). Therefore, sexual reproduction and fecundity are important and highly sensitive indicators of the overall health of the coral holobiont (Harrison and Wallace 1990; Richmond et al. 2018), as they integrate the outcomes from heterotrophic and autotrophic energy supply to corals and the resources available for investing in sexual production of gametes and larvae during each cycle.

Resource allocation to survival and growth during early life stages enables corals to grow large enough to then divert some resources to sexual reproduction and determines important life history parameters such as the onset of sexual reproduction and numbers and quality of progeny produced over the lifetime of corals (Harrison and Wallace 1990; Hall and Hughes 1996; Randall et al. 2020). Some brooding corals become reproductive within the first few years of life and at small colony sizes (Rinkevich and Loya 1979; Szmant-Froelich et al. 1985; Szmant 1986; Kojis 1986a, Harrison and Wallace 1990). For broadcast spawning corals, colony size and age at first reproduction has been clearly established for cohorts of *Acropora tenuis* grown from larvae directly settled onto reef areas during larval restoration trials in the Philippines, with sexual reproduction occurring in some colonies at age 2–3 years (dela Cruz and Harrison 2017; Harrison et al. 2021). Estimated ages of first reproduction for some other spawning coral species range from 4–8 years, with some merulinid colonies as small as 1.6–3 cm in diameter recorded to be sexually reproductive (Kojis and Quinn 1981; Babcock 1984; Wallace 1985; Harrison and Wallace 1990; Sakai 1998; Bonilla et al. 2021).

Scleractinian corals studied to date are iteroparous and have multiple reproductive cycles during their lifetime, with annual reproductive allocations divided among different numbers and sizes of eggs and their energy-provisioning, numbers and quality of sperm produced, and numbers and sizes of brooded planula larvae (Harrison and Wallace 1990). Fecundity is often used to estimate reproductive effort and is usually expressed as eggs or planulae produced per polyp or colony each cycle or annually, but this does not account for allocation and costs of sperm production or the quality of eggs, sperm, and planulae.

Fecundity can vary among polyps within a colony, including in rapidly growing branch tips of acroporid and some pocilloporid species where the newly developing polyps can be sterile, and in polyps adjacent to this zone that produce only sperm or fewer eggs (Rinkevich and Loya 1979; Wallace 1985). Fecundity can also vary over the lifetime of a coral, with colony fecundity increasing as the size and age of colonies increase (Babcock 1991; Hall and Hughes 1996), and fecundity can vary with depth, reef location, and microhabitat (Tan et al. 2016; Pratchett et al. 2019). Detailed studies of populations of three *Acropora* spp. along the length of the GBR showed that fecundity and the proportion of colonies that are reproductive in populations can vary spatially and temporally and that fecundity was highly correlated with subsequent recruitment patterns (Hughes et al. 2000). Gametogenesis and fecundity can also vary in response to asexual fragmentation (Szmant-Froelich et al. 1985; Kojis and Quinn 1985; Okubo et al. 2007; Randall et al. 2020).

28.2.5. Environmental Regulation of Reproduction and Synchrony

Seasonal reproductive cycles and synchronisation of coral spawning are thought to be strongly influenced and entrained by a range of environmental factors that interact with endogenous clock processes to create important biological rhythms to maximise reproductive success. Environmental cues that have been correlated with reproductive patterns include seasonal changes in sea temperature, photoperiod, lunar cycles (moonlight), diel light and dark cycles that act as the final triggers for spawning or planulae release (Babcock et al. 1986; Harrison and Wallace 1990; Randall et al. 2020; Sakai et al. 2020). Periodic changes in these environmental factors can act as proximate cues that progressively synchronise gamete production and maturation in separate colonies or solitary corals within populations through physiological and biochemical changes inducing molecular changes in cells and tissues to entrain biological rhythms (Vize 2009; Kaniewska et al. 2015; Hoadley et al. 2016; Rosenberg et al. 2019). Some of these environmental changes can also act as ultimate factors that exert strong evolutionary selective pressures that align spawning and larval production periods with changing environmental conditions to optimise survival and reproductive success (Babcock et al. 1986; Oliver et al. 1988; Harrison 2011). For example, most corals studied so far spawn after dark (Baird et al. 2009; Harrison 2011), and in *A. millepora* colonies, nocturnal spawning is influenced by biochemical responses to altered light and dark cycles and nocturnal illumination that influence gene expression create signalling cascades that affect cells and gamete release (Vize 2009; Kaniewska et al. 2015). However, nocturnal spawning may also act as an ultimate selective pressure by reducing predation rates by avoiding daytime visual predators (Chamberland et al. 2016), and synchronised nocturnal mass spawning events potentially further enhance survival through predator satiation (Harrison et al. 1984; Babcock et al. 1986; McCormick 2003).

Increased spawning synchrony promotes higher reproductive success through increased concentrations of gametes, which leads to higher rates of fertilization and cross-fertilization (Oliver and Babcock 1992; Willis et al. 1997; Levitan et al. 2004; Nozawa et al. 2015) and increased larval production for recruitment. Alignment of spawning seasons with periods of warming or high sea temperatures (Willis et al. 1985; Harrison and Wallace 1990; Babcock et al. 1994; Nozawa 2012; Keith et al. 2016; Sakai et al. 2020), changes in solar radiation in equatorial regions (van Woesik et al. 2006; Brady et al. 2009), calmer weather (van Woesik 2009; Sakai et al. 2020), low rainfall (Harrison et al. 1984; Mendes and Woodley 2002; Scott et al. 2013), and during neap tidal phases that reduce water movement and therefore sperm dilution (Babcock et al. 1986) have all been proposed as factors potentially enhancing reproductive synchrony and success.

Sea temperature changes strongly influence coral physiological processes, and rising sea surface temperatures (SST) in late spring and summer are thought to stimulate the maturation of gametes in many corals on the GBR and in some other regions (Willis et al. 1985; Harrison and Wallace 1990; Keith et al. 2016; Lin and Nozawa 2023). Mass coral spawning periods on the GBR can occur one lunar cycle earlier in populations on inshore reefs compared with reefs further offshore, corresponding to rising sea temperatures in late spring that occur more quickly on inshore reefs (Harrison et al. 1984; Willis et al. 1985; Babcock et al. 1986). Similarly in Taiwan, spawning patterns among two *Acropora* and two merulinid brain coral species over ten years showed consistent patterns of spawning occurring one month earlier at reef areas that warmed more quickly compared to reef areas that had delayed sea temperature rises (Lin and Nozawa 2023). Keith et al. (2016) analysed spawning periods among *Acropora* spp. assemblages from 34 Indo-Pacific reefs and concluded that rapid increases in SST rather than high mean SST were the best predictor of major spawning events in *Acropora* assemblages, supporting the original hypotheses of environmental factors influencing mass coral spawning on the GBR. Sakai et al. (2020)

found that high SST and low wind speeds were correlated with earlier spawning events among *Acropora* corals on seven reefs in Japan and on the GBR.

Given the wide range of spawning periods and extended breeding seasons recorded for some *Acropora* species, rising or high SST is not a universal predictor of spawning for all *Acropora* species. Future increases in sea temperatures resulting from rapid climate change and ocean warming may alter coral reproductive phenology including the seasonal timing of gametogenesis and synchrony of gamete maturation within some populations, potentially reducing fertilization rates and reproductive success (Baird et al. 2009; Shlesinger and Loya 2019).

In contrast to the main annual late spring and summer spawning periods of many GBR corals, many species on Western Australian (WA) reefs experience similar annual sea temperature cycles to those on the GBR but spawn during large synchronous mass spawning events in autumn, with additional secondary spawning events of some colonies within some populations in spring and summer (Simpson 1985; Babcock et al. 1994; Rosser and Gilmour 2008; Foster et al. 2018). In equatorial regions where annual temperature ranges are small, changes in solar radiation have been correlated with reproductive cycles (van Woesik et al. 2006; Brady et al. 2009). However, more recent work has confirmed that coral assemblages in equatorial Palau have extended spawning periods over 9 months of the year with different environmental cues correlating with spawning periods of different species (Gouezo et al. 2020b). Given the diversity of environmental cues that correlate with seasonal reproductive cycles in corals and that correlation does not prove causality (Harrison and Wallace 1990), it is clear that further research and manipulative experiments are required to elucidate the key relationships between different environmental cues and their influence on gametogenic cycles, gamete maturation, and spawning seasons in corals from different reef regions.

Lunar cycles of gamete spawning or planulation have been well-documented for a range of corals from many families (Harrison and Wallace 1990; Baird et al. 2009; Foster and Gilmour 2018), and experimental manipulation of night irradiance has been shown to synchronise these cycles in some species (Jokiel et al. 1985; Hunter 1988; Kaniewska et al. 2015). In contrast, planulation cycles in the brooding coral *Stylophora pistillata* from the Red Sea do not conform to strict circalunar 29.5 day cycles, and they instead have an ~30 day periodicity, which is a circatrigintran rhythm not tightly aligned with lunar phase and irradiance intensities (Linden et al. 2018). Some corals are sensitive to moonlight via photosensitive molecules within their tissues (Gorbunov and Falkowski 2002; Levy et al. 2007; Kaniewska et al. 2015). These photoreceptors are likely to be cryptochromes or opsins that are thought to entrain circadian rhythms to promote spawning synchrony (Vize 2009; Hoadley et al. 2016).

The lunar period of spawning is remarkably precise and predictable in some corals, whereas in other species the lunar period of spawning can vary between years. For example, longer-term observations of spawning periods on the central GBR have shown that the lunar night of mass spawning among many *Acropora* spp. often occurs between 4 and 6 nights after a full moon (nAFM), but in some years and in different GBR regions, they can vary from these lunar periods, whereas merulinid brain corals and some other taxa generally have more consistent lunar nights of spawning (Harrison et al. 1984; Willis et al. 1985; Babcock et al. 1986; Harrison and Wallace 1990; Baird et al. 2009, 2021). Similar trends were recorded over a seven-year period for spawning corals in Taiwan and Japan (Lin and Nozawa 2017). Spawning of the Caribbean *Orbicella annularis* species complex colonies consistently occurred between 4–7 days after a full moon (AFM) over eight years (Levitan et al. 2011), and relatively stable lunar periodicity of spawning has been recorded in *Diploria labrinthiformis* colonies from 10 to 13 nAFM (Chamberland et al. 2017). Sakai et al. (2020) noted that high wind speed and rough seas in the period immediately prior to predicted spawning periods might delay the night of spawning, which aligns with anecdotal reports and long-term observations of delayed spawning on some GBR reefs during periods of strong winds (P. Harrison, pers. obs.).

Lin and Nozawa (2017) suggested that *Acropora* spp. may be more sensitive to changes in environmental cues and respond through an hourglass regulatory biological clock mechanism that triggers spawning after a period of exposure to stimuli, whereas merulinid corals may have an oscillator regulatory mechanism in which the biological clock is entrained by environmental cues and is therefore more stable and less responsive to unpredictable changes. As these coral groups belong to different scleractinian clades with divergent evolutionary pathways from about 380 MYA (Campoy et al. 2020), it will be interesting to determine if these different regulatory mechanisms for spawning synchrony are evident in related coral taxa, and are a conserved trait. Wolstenholme et al. (2018) proposed that a close association between the third quarter period within a lunar synodic rhythm and the zero declination of the moon over the equator (and periods of low atmospheric pressure) may influence the timing of synchronous coral spawning events.

Spawning periods also vary interannually in response to incomplete alignment of annual seasonal changes and 29.5 day lunar cycles such that full moon periods become progressively offset by about 10–12 calendar days between years. Willis et al. (1985) noted that split-spawning over two consecutive lunar cycles occurred during some years on the GBR resulting from a shift in interannual spawning periods from 12 to 13 lunar cycles. Longer-term records of interannual spawning periods on the GBR (Baird et al. 2009, 2021) and in WA reefs (Foster et al. 2018) have also shown that coral populations split-spawn every few years when the full moon coincides with the first week of the usual spawning calendar month or when the full moon occurs in the last week of the previous month. Split-spawning therefore maintains coral broadcast spawning periods within appropriate environmental periods but varies in relation to the numbers of lunar cycles elapsed between annual spawning events and the calendar months of spawning in some years.

Most corals studied to date spawn after dark, and many show highly conserved spawning times that are more predictable than the lunar night of spawning (Harrison and Wallace 1990; Baird et al. 2009, 2021). Nocturnal spawning has been experimentally shown to be controlled by the onset of darkness and time elapsed after sunset (Babcock 1984; Hayashibara et al. 2004; Brady et al. 2009). Lin and Nozawa (2017) showed that moonlight suppresses spawning in *Dipsastraea speciosa* on reefs in Taiwan and that spawning is triggered by the darkness interval occurring after sunset and before moonrise and requires at least a few nights of dark periods after a full moon, which may be a spawning trigger in coral species that consistently spawn in the fourth quarter lunar phase.

Relatively few coral species have been observed to spawn during daylight (Harrison and Wallace 1990), but daytime sperm release has been observed in a range of gonochoric corals (Marshall and Stevenson 1933; Babcock et al. 1986; Mangubhai et al. 2007; Eyal-Shaham et al. 2019). Daytime spawning has also been recorded in the hermaphroditic corals *Diploria labrinthiformis* (Muller and Vermeij 2011; Chamberland et al. 2016) and *Goniastrea favulus* (Kojis and Quinn 1981; Babcock et al. 1986). It is likely that some other coral species will be recorded spawning during daylight hours when a greater range of corals are studied, and for gonochoric species in particular, periods of low tides and reduced currents may be important for reducing sperm dilution and enhancing fertilization rates.

28.2.6. Spawning Synchrony and Regional Patterns

Spawning is often synchronised among polyps within each colony and among different colonies of each species and most often occurs during darkness some nights after full moon periods in spring and summer, although in some species and reef regions, the peak spawning period occurs during other periods and lunar phases. Synchronous spawning results in high concentrations of eggs and sperm of the same species in the water column or floating on the sea surface, and the high sperm concentrations can result in high rates of cross-fertilization. Outcrossing of gametes from different genetically distinct corals promotes increased genetic diversity in the embryos and larvae and allows for increased adaptive potential and evolution of the next generations of corals. Cross-fertilization is essential for rapid evolution of populations and is increasingly important for enhancing resilience in response to rapidly changing reef environments and increasing climate change impacts.

In many regions, coral spawning events also include synchronous spawning of different species on the same night or over a few nights; however, the extent of multi-species spawning synchrony varies greatly in different reef regions and among different groups of corals (Harrison and Wallace 1990; Richmond and Hunter 1990; Harrison and Booth 2007; Baird et al. 2009; Harrison 2011). Global information shows a continuum of reproductive patterns and levels of spawning synchrony among coral assemblages in different regions (Harrison 2011), including some unusual patterns that are not entirely predictable based on current understanding of phylogenetic, latitudinal, ecophysiological, and environmental influences on coral reproduction.

Accordingly, reproductive patterns range from highly synchronised mass coral spawning events to relatively asynchronous spawning patterns. Mass coral spawning events involving many (but not all) species from a wide range of genera and families occur on a few nights each year after full moon periods on the GBR and some WA reefs (Harrison et al. 1984; Willis et al. 1985; Simpson 1985; Babcock et al. 1986, 1994; Rosser and Gilmour 2008; Baird et al. 2011; Gilmour et al. 2016; Foster et al. 2018; Baird et al. 2021), with emerging evidence of similar events in some other Indo-Pacific regions. Intermediate levels of multi-species spawning synchrony occur on many reef systems where some populations spawn synchronously, but other species spawn over different months or lunar phases (Harrison and Wallace 1990; Richmond and Hunter 1990; Baird et al. 2009; Harrison 2011). In some reef regions, spawning may be synchronised within populations of each species but is asynchronous between many or most species studied to date, resulting in different species spawning at different lunar cycles or in different months or seasons over much more extended spawning seasons (Shlesinger and Loya 1985; Szmant 1986; Oliver et al. 1988; Shlesinger et al. 1998; Mangubhai and Harrison 2008b; Gouezo et al. 2020b). Asynchronous spawning has been recorded within and between populations of a range of *Acropora*, merulinid, and *Porites* species in equatorial Maldives, with small 'dribble' spawning releases recorded over multiple nights and lunar phases and periodic larger-scale multi-species spawning resulting in occasional coral spawn slicks at the sea surface (Harrison and Hakeem 2012). Therefore, determining the extent of spawning synchrony within populations and among coral species and assemblages requires detailed research and repeated monitoring of gravid reproductive corals and ideally involves direct spawning observations over multiple nights (Harrison 2011).

28.2.7. Coral Reproduction on the GBR

Coral reproductive patterns have been most extensively studied on the GBR, particularly in the Central GBR region. Traditional Owners and custodians of sea country along the GBR will have observed coral spawn slicks and spawning periods of other reef organisms for many tens of thousands of years. The first published records of coral reproduction on the GBR resulted from the Great Barrier Reef Expedition in 1928–1929 based at Low Isles, with *Pavona cactus* colonies observed spawning at dawn and *Pocillopora acuta* releasing planulae periodically (Marshall and Stephenson 1933). Increased interest in coral reproduction in the late 1970s (Kojis and Quinn 1980, 1981) and particularly in the early 1980s culminated in the discovery of mass coral spawning on the GBR in 1981 with many additional species recorded spawning during 1982–1986 (Harrison et al. 1984; Willis et al. 1985; Babcock et al. 1986; Oliver et al. 1988).

More than 140 coral species from a wide range of coral families and genera were found to participate in mass coral spawning events on the GBR during the initial phase of research, with most colonies recorded to spawn over a few nights following full moon periods in the austral spring (October) and summer (November, December) each year, and coincident spawning of more than 20–30 species was recorded on peak spawning nights (often 4–7 nAFM) at many reefs including reefs more than 1,200 km apart (Harrison et al. 1984; Willis et al. 1985; Babcock et al. 1986; Oliver et al. 1988; Harrison and Wallace 1990). Although many colonies within populations had mature eggs and sperm and spawned synchronously during these periods, varying proportions of each population were recorded to have split-spawning across lunar cycles between years, and gametes were not present before or after mass spawning periods in some colonies of some populations (op. cit.). Hughes et al. (2000) found that the proportion of reproductively mature colonies within populations of three *Acropora* spp. sampled in six regions spanning multiple regions of the GBR varied substantially between years.

Subsequent detailed records and observations by researchers and members of the public in many parts of the GBR have continued to add substantial information about the degree of synchrony and split-spawning within and among populations on different reefs (Hughes et al. 2000; Baird et al. 2002) and timing of reproduction, and much of this has been integrated into the Indo-Pacific coral spawning database (Baird et al. 2021).

Synthesising the available information on GBR coral reproduction is difficult due to the large volume of information, ongoing research, and new observations each year. However, based on current knowledge, more than 180 coral species that have been studied on the GBR are confirmed to be broadcast spawning corals, and most but certainly not all of these species spawn once per year over one or a few nights during spectacular multi-species mass spawning events. In contrast, only about 15 species have been recorded to brood and release planulae on the GBR, including two *Pocillopora* spp. and two *Tubastraea* spp. that have asexual production of brooded planulae (Ayre and Resing 1986; Schmidt-Roach et al. 2014). The majority of broadcast spawning species recorded spawning after full moon periods during mass spawning events are from families Acroporidae and Merulinidae, with some Lobophylliidae, Fungiidae, Poritidae, Agariciidae, Euphylliidae, Dendrophylliidae, and a few species in other families (records primarily from Harrison et al. 1984; Willis et al. 1985; Babcock et al. 1986; Oliver et al. 1988; Harrison and Wallace 1990; Baird et al. 2009; Harrison 2011; Baird et al. 2021 and references cited therein).

Mass coral spawning events from healthy coral assemblages can result in trillions of eggs and sperm being released (Doropoulos et al. 2019a), and when mass spawning coincides with periods of calm weather and oceanographic conditions, large coral spawn slicks accumulate at the sea surface, which are then gradually dispersed by currents, wind, and wave action (Harrison et al. 1984; Babcock et al. 1986; Oliver and Willis 1987; Oliver et al. 1992; Baird et al. 2009; Randall et al. 2020; Figure 4). These slicks provide access to vast numbers of eggs and embryos from which small samples can be collected for mass larval production on reefs and coral restoration of damaged or degraded reefs in the GBR and other regions (Heyward et al. 2002; Doropoulos et al. 2019b; Harrison et al. 2021; Harrison and dela Cruz 2022; Banaszak et al. 2023).

Although substantial research is continuing on the GBR, the mode of development and sexual patterns of less than half of the > 400 coral species recorded on the GBR are known; therefore, much additional research is needed in the future to enable these ecologically complex and essential reef-building assemblages to be more fully understood. Additionally, the dominance of mass spawning records tends to obscure the fact that some other coral species on the GBR are known to spawn or release planulae at other lunar phases or over some months after the peak spawning periods, and a few species have multiple or biannual reproductive cycles each year (Marshall and Stephenson 1933; Harriott 1983; Harrison et al. 1984; Willis et al. 1985; Babcock et al. 1986; Kojis 1986a, 1986b; Oliver et al. 1988; Harrison and Wallace 1990; Stobart et al. 1992; Tanner 1996; Ayre and Hughes 2000; Hughes et al. 2000; Baird et al. 2002, 2021; Wolstenholme 2004), reflecting the diversity of taxa and life histories in these species-rich assemblages.

Synchronous spawning within populations is obviously advantageous for maximising fertilization and outcrossing, but the coincidence of spawning of so many coral species across a diverse range of taxa on the GBR and in some other reef regions indicates that these spawning periods, lunar phases, and tidal windows are likely to enhance reproductive success, with multi-species spawning resulting from evolutionary selective pressures operating separately or in combination to constrain successful reproduction in many taxa (Babcock et al. 1986; Harrison and Wallace 1990). Additionally, a range of other reef invertebrates are known to spawn at similar lunar phases and times coinciding with peak coral spawning periods on the GBR and around new moon periods (Babcock et al. 1986; Hutchings and Howitt 1988; Babcock and Mundy 1992) and in some other reef regions (Van Veghel 1993; Bouwmeester et al. 2016), indicating that these periods are important for reproductive success in a wide range of reef organisms.

The oceanographic conditions during and following mass coral spawning events and smaller spawning events at other times have important consequences for initial sperm dilution, probability of fertilization, and likelihood of slick development (Figure 4), and they drive larval dispersal patterns and the extent of connectivity and gene flow between populations on different reefs (Kininmonth 2024). Currents and wind direction vary substantially in space and time resulting in complex connectivity and metapopulation patterns among corals and other reef organisms that are difficult to model

FIGURE 4 Large coral spawn slicks accumulate at the sea surface after mass spawning events during periods of calm weather. Surface slick at Heron Island, Southern GBR (left) and near Lizard Island, Northern GBR (right).

Source: Photos: Peter Harrison

effectively (Bode et al. 2018; Kininmonth 2023). Although some biophysical models indicate that some reefs are more likely to act as source reefs while others tend to be sinks, source and sink reefs are likely to vary seasonally and annually in response to variable wind and oceanographic conditions and in response to climate change (Bode et al. 2018; Hock et al. 2019; Riginos et al. 2019; Kininmonth 2024).

28.2.8. Other Regional Patterns of Coral Reproduction and Varying Synchrony

The wide range of reproductive patterns and degree of spawning synchrony or temporal partitioning within coral assemblages in other reef regions has been reviewed in detail previously (Szmant 1986; Harrison and Wallace 1990; Richmond and Hunter 1990; Baird et al. 2009; Harrison 2011). Extensive research on coral reproduction has continued and expanded in many reef regions during the last decade, with reproductive traits summarised in the coral trait database (Madin et al. 2016) and Indo-Pacific coral spawning records summarised in the coral spawning database (Baird et al. 2021). Given the large volume and complexity of these global data, a detailed review of global patterns in each of these regions is beyond the scope of this chapter. However, this rapidly expanding global knowledge of coral reproductive patterns and timing in many reef regions provides opportunities for improved reef management and larval-based restoration to be applied on many damaged reef systems in future.

Currently expanding research on coral reproduction in some reef regions including the GBR e.g., Reef Restoration and Adaptation Program (RRAP), Philippines, Maldives, French Polynesia, and Caribbean is shifting from pure discovery science to research that can be directly applied to scaling up sexual-based breeding and larval propagation for coral restoration (Randall et al. 2020; Bostrom-Einarsson et al. 2020; Harrison et al. 2021; McLeod et al. 2022; Harrison and dela Cruz 2022; Banaszak et al. 2023). This includes substantial innovative research on determining enhanced heat tolerance in adult broodstock and their progeny (Howells et al. 2021; Quigley and Donelson 2024), maximising fertilization rates and genetic diversity (Baums et al. 2022), hybridization (Chan et al. 2018, 2019), cryopreservation of gametes and embryos (Hagendorn et al. 2021), larval development and nutrition (Rodd et al. 2022), probiotic research (Rosado et al. 2018), post-settlement survival and growth, and the onset of sexual reproduction to re-establish resilient breeding populations on damaged and highly degraded reefs (dela Cruz and Harrison 2017; Harrison et al. 2021). These advances are briefly reviewed in the following section of this chapter.

28.3. ADVANCES IN CORAL RESTORATION AND BREEDING FOR LARVAL-BASED PROPAGATION

The accelerating loss of live corals on the GBR and most other reef systems globally in recent decades has resulted in increased interest in active coral restoration to stimulate recovery of damaged and chronically degraded reefs. Coral restoration has a relatively short history of development and application on reefs over recent decades compared with terrestrial ecological restoration research and practice, which has been well established for much longer periods. Therefore, much of the initial research on coral restoration has had to focus on improved understanding of asexual and sexual reproductive processes and timing of gamete and larval release (reviewed above), and smaller-scale testing of the wide range of rapidly evolving methods for application to restoring damaged reefs at larger scales in the future (Edwards 2010; Omori 2019; Vaughan 2021).

28.3.1. Asexual Restoration Methods

Most coral restoration projects globally have focused on asexual methods of propagation, primarily using coral fragmentation, nursery phase rearing, and outplanting on reefs, which is also referred to as coral gardening (reviewed in Harriott and Harrison 1984; Rinkevich 1995, 2005; Edwards 2010; Young et al. 2012; Lirman and Schopmeyer 2016; Omori 2019; Bostrom-Einarsson et al. 2020). Many of the early coral fragmentation and outplanting projects were relatively small-scale, short-term, and with limited monitoring to quantify and report on outcomes and impacts from bleaching events and disease outbreaks (Bostrom-Einarsson et al. 2020). More recent research has focused on increasing the scale and effectiveness of outplanting by using a wider range of genotypes (Baums et al. 2019), multi-hectare scales of outplanting (Omori 2019; Vaughan 2021), development of new methods including novel Coralclip rapid attachment devices (Suggett et al. 2020), and larger-scale outplanting on engineered structures deployed above rubble reef areas (Williams et al. 2019). Additional research is examining the use of micro-fragmentation and 'reskinning' of polyps and tissues of slow growing massive corals to increase the rates of growth for restoration (Page et al. 2018; Vaughan 2021). These and other recent advances are showing promising results for increasing the success rate and cost-effectiveness of some asexual propagation methods on some reefs.

28.3.2. Sexual Reproduction and Larval Production for Restoration

The alternative approach to coral restoration is based on sexual reproduction and larger-scale larval production. This approach aims to restore genetically diverse, resilient breeding populations on damaged and badly degraded reef areas (Harrison et al. 2021; Banaszak et al. 2023). Sexual-based propagation has some inherent advantages over asexual propagation methods including increased genetic diversity and adaptive capacity in restored populations. Furthermore, selecting coral broodstock with traits such as increased heat tolerance for mass larval culture and application of interspecific hybridization may increase coral fitness and enable larger-scale and sustainable reef restoration outcomes (van Oppen et al. 2017; Randall et al. 2020; Banaszak et al. 2023). However, most corals have life histories with Type III survivorship, characterised by high levels of mortality in early life stages (Wilson and Harrison 2005; Ritson-Williams et al. 2009; Doropoulos et al. 2016; dela Cruz and Harrison 2017, 2020; Harrison et al. 2021). Therefore, substantial ongoing research is examining new methods to increase larval survival, routine production of millions of larvae in reef-based larval culture pools or on vessels (Harrison et al. 2019; Doropoulos et al. 2019b; Suzuki et al. 2020; Harrison et al. 2021; Harrison and dela Cruz 2022), and reducing post-settlement mortality to offset this low early survival 'bottleneck' and the ongoing loss of adult breeding corals on many reef systems (Randall et al. 2020; Banaszak et al. 2023). All of these approaches require relatively predictable access to coral spawn and larvae, which is reliant on increased global knowledge of sexual reproductive processes, larval development, and seasonal patterns of reproduction, as noted in the previous section.

The two main sexual reproductive approaches to coral restoration are larval-based restoration to increase larval supply and settlement directly on damaged reefs with limited natural larval supply (Edwards et al. 2015; dela Cruz and Harrison 2017; Harrison et al. 2021), and coral reseeding in which larvae are initially settled onto settlement surfaces or engineered devices prior to deployment on reefs (Guest et al. 2014; Chamberland et al. 2017; Randall et al. 2021; McLeod et al. 2022; Banaszak et al. 2023). The larval restoration approach was conceived following the discovery of mass coral spawning on the GBR in 1981 (Harrison et al. 1984) and a subsequent mass bleaching event in early 1982, which highlighted the need for novel approaches to managing and restoring damaged reefs (Heyward et al. 2002; Harrison and dela Cruz 2022). Larval restoration involves five phases: 1. determining the presence of mature gametes in healthy breeding coral populations and assemblages, 2. collecting coral spawn and mixing gametes to enhance cross-fertilization and genetic diversity, 3. mass larval culture, 4. larval release or coral reseeding onto coral-depleted and degraded reef areas lacking sufficient natural larval supply for natural recovery, and 5. long-term monitoring of restoration outcomes compared with control sites without cultured larval supply (methods detailed in Harrison et al. 2016, 2021; dela Cruz and Harrison 2017, 2020; Doropoulos et al. 2019b; Harrison and dela Cruz 2022; McLeod et al. 2022 for GBR and Philippines projects; Banaszak et al. 2023 for global projects).

Initial small-scale trials of larval restoration showed that capturing small volumes of multi-species coral spawn enabled successful larval cultures and significantly increased larval settlement on larval-enhanced reef plots compared with control plots on reefs in WA (Heyward et al. 2002), Japan (Suzuki et al. 2012, 2020; Omori 2019), and Palau (Edwards et al. 2015). Subsequent larger-scale larval production and settlement on coral-depleted and algal phase-shifted reef areas in the Philippines and GBR have clearly demonstrated that larval restoration can be successfully applied at small scales to rapidly re-establish reef coral dominance even on highly degraded reef areas and produce new breeding populations of *Acropora* spp. and some merulinid brain corals within two to five years, which have become a source of gametes for successive larval restoration trials (dela Cruz and Harrison 2017, 2020; Harrison et al. 2021; Harrison and dela Cruz 2022; Harrison and colleagues, unpubl. Data; Figure 5). Increased abundance of small pomacentrids and corallivore chaetodontids has

FIGURE 5 High densities of three-year-old breeding colonies of *A. tenuis* in a restored population grown from larvae that were settled directly onto badly degraded reef areas in Northern Luzon, Philippines. Colonies of other species recruited naturally.

Source: Harrison et al. 2021, Photo: Peter Harrison

been recorded on larval restoration plots coinciding with the growth and increasing dominance of restored coral populations (Harrison et al. 2021). Re-establishing breeding populations also enables renewed larval supply and dispersal to other reef areas via local reef and ocean currents, thereby significantly increasing the scale and cost-effectiveness of restoration outcomes over time (dela Cruz and Harrison 2017; Harrison et al. 2021). Multi-hectare deployment of larvae has been achieved in the Philippines (Harrison et al. 2019; Dunbabin et al. 2020), and current research on the GBR (RRAP) and in some other reef regions is focusing on upscaling routine production of 10s-100s millions of larvae for larval restoration at multi-hectare scales.

Ongoing research is examining innovative approaches to larger-scale reef-based mass larval culture of tens of millions of larvae (Harrison et al. 2019, 2020, 2021; Doropoulos et al. 2019b; Suzuki et al. 2020; Harrison and dela Cruz 2022). New methods for deploying competent larvae are being trialled using a range of approaches, including larval cloud releases of millions of larvae that are dispersed by complex local currents or concentrating larvae from reef culture pools and delivering them via pipes and autonomous vehicles, and under mesh enclosures to reduce dispersal away from target restoration areas (Harrison et al. 2019, 2020, 2021; Dunbabin et al. 2020; Mou et al. 2022; Harrison and dela Cruz 2022).

The larval reseeding approach also relies on larger-scale larval production and has mostly focused on ex situ laboratory-based collection of coral spawn from gravid corals collected from reefs just prior to predicted spawning periods. Larval culture is usually managed in aquaculture tanks (Pollock et al. 2017), followed by settling competent larvae onto engineered devices and holding the settled spat and juvenile corals in seawater tanks during nursery phases that extend over various periods ranging from weeks to months before the juvenile corals are deployed onto reef sites (Petersen et al. 2005; Guest et al. 2014; Chamberland et al. 2017; Baria et al. 2019; Randall et al. 2020, 2021, 2023; Banaszak et al. 2023). These extended nursery periods increase initial production costs but enable laboratory control of environmental conditions and testing of innovative settlement substrata designed to enhance post-settlement survival of spat (op. cit.; McLeod et al. 2022). Pre-deployment settlement of competent larvae onto settlement surfaces and devices within floating larval culture pools on reefs prior to outplanting is also showing promise for potential larger-scale application in future (Suzuki et al. 2020; Miller et al. 2022; Harrison and dela Cruz 2022).

28.3.3. Flexibility and Limitations of Coral Restoration

The variety of coral restoration techniques now available create a suite of tools that can be selectively applied to successfully restore coral populations and assemblages in a wide range of reef environments, depths, and scales under different hydrodynamic conditions and reef status. However, all restoration approaches have limitations; therefore, it is important to clearly define the aims of the restoration intervention, the potential for success, the funding and other resources available, and the scale and outcomes required before commencing any work. Furthermore, coral restoration should not be prioritized on reefs that are naturally recovering from disturbance through sufficient natural larval supply and recruitment. However, if the reef system is unlikely to recover naturally, then intervention may be warranted.

A key issue in the initial restoration evaluation decision framework that should be carefully considered before investing resources and effort to initiate coral restoration on any reef area is, are the current and predicted near-future environmental conditions on the target reefs suitable for coral survival, growth, and reproduction, that is, are these damaged or degraded reef sites potentially 'recoverable' (*sensu* Harrison et al. 2021)? If the reef system is now too badly polluted for propagule or fragment survival or frequently disturbed by blast fishing, mass bleaching, high sediment loads, and other disturbances, then further reef management is required before restoration is feasible. Conversely, if the reef area is recoverable and suitable for intervention, then the inherent limitations and cost-effectiveness of each technique need to be evaluated.

For larval-based restoration, adverse reef states or oceanographic conditions at the reef site may preclude certain approaches, and variations in environmental conditions over small spatial scales on reefs can greatly affect performance and survival of recruits deployed on coral seeding devices (Randall et al. 2023). Oceanographic conditions can also strongly influence restoration outcomes when larvae are released directly onto degraded reefs. For example, fewer larvae are likely to settle onto target restoration areas after release if the reef habitat is subjected to strong currents or turbulent wave action that rapidly disperse larvae. In contrast, larvae are more likely to be retained on sheltered and topographically complex reef sites with lower current flow during their settlement period resulting in a simple and efficient way to rapidly increase localised larval supply and recruitment for larger scale coral restoration outcomes.

28.4. CONCLUSIONS

The large and growing body of research on scleractinian coral reproduction and recruitment has provided a wealth of data on many critically important aspects of coral life cycles and continues to reveal surprising complexity and variation in reproductive biology, spawning patterns, larval dispersal, and recruitment processes that shape the evolution and resilience of coral assemblages that create the foundations of coral reefs. Variation in reproductive patterns between coral taxa and between regions and the large number of species (particularly some gonochoric and brooding taxa) and coral reef regions that have not been sufficiently studied highlight the urgent need for additional research on coral reproduction to better understand the adaptive potential and resilience of corals to increasing stressors that threaten reef function and future existence in a rapidly changing environment.

In addition to continued expansion of this discovery science, it is essential to harness this knowledge to increase the scale and effectiveness of coral restoration processes to reduce the impacts of mass coral bleaching and chronic human impacts that interact with natural disturbances to accelerate the loss of corals on many reefs. Active coral restoration can buy time for maintaining and recovering coral assemblages on impacted reefs with impaired natural larval supply and enhance adaptive potential of surviving remnant populations while there is still sufficient genetic diversity present. Large-scale multidisciplinary teams working with local communities, traditional owners, businesses, government departments, and other key partners need to rapidly apply current knowledge to increase the scales and effectiveness of coral restoration methods and outcomes while momentum builds for meaningful and effective global action on managing greenhouse gas emissions and direct human impacts that threaten to destroy these extraordinary hyperdiverse and immensely valuable ecosystems.

ACKNOWLEDGEMENTS

Sincere thanks to Alicia Forbes for collating and summarising recent coral reproduction references and GBR coral reproduction data, Lauren Hasson for collating cited references and GBR coral reproductive data, Dr. Dexter dela Cruz and Dr. Marine Gouezo for their detailed reviews, and Barbara Harrison for editing assistance. Thanks also to Professors Eric Wolanski and Mike Kingsford for their support and encouragement to complete this chapter and the many researchers who continue to contribute important new information on coral reproduction and restoration practices around the world.

REFERENCES

Abecia JED, Guest JR, Villanueva RD (2016) Geographical variation in reproductive biology is obscured by the species problem: A new record of brooding in *Porites cylindrica*, or misidentification? Invertebrate Biology 135:58–67.

Alvarez-Noriega M, Baird AH, Dornelas M, Madin JS, Cumbo VR, Connolly SR (2016) Fecundity and the demographic strategies of coral morphologies. Ecology 97:3485–3493.

Anthony KRN, Fabricius KE (2000) Shifting roles of heterotrophy and autotrophy in coral energetics under varying turbidity. Journal of Experimental Marine Biology and Ecology 252:221–253.

Ayre DJ, Hughes TP (2000) Genotypic diversity and gene flow in brooding and spawning corals along the Great Barrier Reef, Australia. Evolution 54:1590–1605.

Ayre DJ, Resing JM (1986) Sexual and asexual production of planulae in reef corals. Marine Biology 90:187–190.

Babcock RC (1984) Reproduction and distribution of two species of *Goniastrea* (Scleractinia) from the Great Barrier Reef province. Coral Reefs 2:187–195.

Babcock RC (1991) Comparative demography of three species of scleractinian corals using age- and size-dependent classifications. Ecological Monographs 61:225–244.

Babcock RC, Bull GD, Harrison PL, Heyward AJ, Oliver JK, Wallace CC, Willis BL (1986) Synchronous spawning of 105 scleractinian coral species on the Great Barrier Reef. Marine Biology 90:379–394.

Babcock RC, Mundy CN (1992) Reproductive biology, spawning and field fertilization rates of *Acanthaster planci*. Australian Journal of Marine and Freshwater Research 43(3):525–533.

Babcock RC, Willis BL, Simpson CJ (1994) Mass spawning of corals on a high latitude coral reef. Coral Reefs 13:161–169.

Baird AH, Blakeway DR, Hurley TJ, Stoddart JA (2011) Seasonality of coral reproduction in the Dampier Archipelago, northern Western Australia. Marine Biology 158(2):275–285.

Baird AH, Guest JR, Edwards AJ, Bauman AG, Bouwmeester J, Mera H, et al. (2021) An Indo-Pacific coral spawning database. Scientific Data 8:35–9.

Baird AH, Guest JR, Willis BL (2009) Systematic and biogeographical patterns in the reproductive biology of Scleractinian corals. Annual Review of Ecology, Evolution, and Systematics 40:551–571.

Baird AH, Marshall PA, Wolstenholme J (2002) Latitudinal variation in the reproduction of *Acropora* in the Coral Sea. Proc. 9th International Coral Reef Symposium 1:385–389.

Banaszak AT, Marhaver KL, Miller MW, Hartmann AC, Albright R, Hagedorn M, Harrison PL, Latijnhouwers RW, Quiroz SM, Pizarro V, Chamberland VF (2023) Applying coral breeding to reef restoration: Best practiced, knowledge gaps, and priority actions in a rapidly evolving field. Restoration Ecology:1–17.

Baria-Rodriguez MVB, dela Cruz DW, Dizon RM, Yap YT, Villanueva RD (2019) Performance and cost-effectiveness of sexually produced *Acropora granulosa* juveniles compared with asexually generated coral fragments in restoring degraded reef areas. Aquatic Conservation: Marine and Freshwater Ecosystems 29:891–900.

Baums IB, Baker AC, Davies SW, Grottoli AG, Kenkel CD, Kitchen SA, Kuffner IB, LaJeunesse TC, Matz MV, Miller MW (2019) Considerations for maximizing the adaptive potential of restored coral populations in the western Atlantic. Ecological Applications 29:e01978.

Baums IB, Chamberland VF, Locatelli NS, Conn T (2022) Maximizing genetic diversity in coral restoration projects. In: Van Oppen MJH and Aranda M (Editors) Coral reef conservation and restoration in the 'omics' age. Springer International Publishing

Blackall LL, Wilson B, van Oppen MJH (2015) Coral—The world's most diverse symbiotic ecosystem. Marine Ecology 24:5330–5347.

Bode M, Bode L, Choukroun S, James, MK, Mason LB (2018) Resilient reefs may exist, but can larval dispersal models find them? PLoS Biology 16(8):e2005964–e2005964.

Bonilla KG, Guest JR, dela Cruz DW, Baria-Rodriguez MV (2021) Onset of sexual maturity of sexually propagated and wild colonies of the massive coral *Favites abdita* in northwestern Philippines. Invertebrate Reproduction and Development 65(3):201–209.

Boström-Einarsson L, Babcock RC, Bayraktarov E, Ceccarelli D, Cook N, Ferse SCA, et al. (2020) Coral restoration—a systematic review of current methods, successes, failures and future directions. PLoS One 15:e0226631.

Bouwmeester J, Gatins R, Giles EC, Sinclair-Taylor TH, Berumen ML (2016) Spawning of coral reef invertebrates and a second spawning season for scleractinian corals in the central Red Sea. Invertebrate Biology 135:273–284.

Brady AK, Hilton JD, Vize PD (2009) Coral spawn timing is a direct response to solar light cycles and is not an entrained circadian response. Coral Reefs 28:677–680.

Bruno JF, Selig ER, Casey ES, Page CA, Willis BL, Harvell CD, Sweatman H, Melendy AM (2007) Thermal stress and coral cover as drivers of coral disease outbreaks. PLoS Biology 5(6):e124.

Campoy AN, Addamo AM, Machordom A, Meade A, Rivadeneira MM, Hernández CE, Venditti C (2020) The origin and correlated evolution of symbiosis and coloniality in scleractinian corals. Frontiers in Marine Science 7:461.

Chamberland VF, Petersen D, Guest JR, Petersen U, Brittsan M, Vermeij MJA (2017) New seeding approach reduces cost and time to outplant sexually propagated corals for reef restoration. Scientific Report 7:1–12.

Chamberland VF, Petersen D, Latijnhouwers KRW, Snowden S, Mueller B, Vermeij MJA (2016) Four-year-old Caribbean Acropora colonies reared from field-collected gametes are sexually mature. Bulletin of Marine Science 92:1–3.

Chan WY, Hoffmann AA, van Oppen MJ (2019) Hybridization as a conservation management tool. Conservation Letters 12:e12652.

Chan WY, Peplow LM, Menéndez P, Hoffmann AA, van Oppen MJ (2018) Interspecific hybridization may provide novel opportunities for coral reef restoration. Frontiers in Marine Science 5:160.

Charnov EL (1982) The Theory of Sex Allocation. Princeton Monograph in Population Biology. Princeton University Press 18:355.

Daly M, Fautin, DG, Cappola VA (2003) Systematics of the Hexacorallia (Cnidaria: Anthozoa). Zoological Journal of the Linnean Society 139(3):419–437.

dela Cruz DW, Harrison PL (2017) Enhanced larval supply and recruitment can replenish reef corals on degraded reefs. Scientific Reports 7:1385–13.

dela Cruz DW, Harrison PL (2020) Enhancing coral recruitment through assisted mass settlement of cultured coral larvae. PLoS One 15:1–21

Doropoulos C, Elzinga J, ter Hofstede R, van Koningsveld M, Babcock RC (2019a) Optimizing industrial-scale coral reef restoration: Comparing harvesting wild coral spawn slicks and transplanting gravid adult colonies. Restoration Ecology 27:758–767.

Doropoulos C, Roff G, Bozec YM, Zupan M, Werminghausen J, Mumby PJ (2016) Characterizing the ecological trade-offs throughout the early ontogeny of coral recruitment. Ecology Monogram 86:20–44.

Doropoulos C, Vons F, Elzinga J, ter Hofstede R, Salee K, et al. (2019b) Testing industrial-scale coral restoration techniques: Harvesting and culturing wild coral-spawn slicks. Frontiers Marine Science 6:658.

Dunbabin M, Manley J, Harrison PL (2020) Uncrewed maritime systems for coral reef conservation. Global Oceans 2020. Singapore: U.S. Gulf Coast, 1–6pp.

Edwards AJ, ed (2010) Reef Rehabilitation Manual. St Lucia: Coral Reef Targeted Research & Capacity Building or Management Program, 166pp.

Edwards AJ, Guest JR, Heyward AJ, Villanueva RD, Baria MV, Bollozos ISF, et al. (2015) Direct seeding of mass-cultured coral larvae is not an effective option for reef rehabilitation. Marine Ecology Progress Series 525:105–116.

Eyal-Shaham L, Eyal G, Sakai Z, Nozawa Y, Harii S, Sinniger F, Bronstein O, Ben-Zvi O, Shlesinger T, Loya Y (2019) Repetitive sex change in the stony coral *Herpolitha limax* across a wide geographic range. Scientific Reports 9:1–12.

Eyal-Shaham L, Eyal G, Ben-Zvi O, Sakai K, Harii S, Sinniger F, et al. (2020) A unique reproductive strategy in the mushroom coral *Fungia fungites*. Coral Reefs 39:1793–1804.

Fadlallah YH (1983) Sexual reproduction, development and larval biology in scleractinian corals. Coral Reefs 2:129–150.

Figueiredo J, Baird AH, Connolly SR (2013) Synthesizing larval competence dynamics and reef-scale retention reveals a high potential for self-recruitment in corals. Ecology 94(3):650–659.

Foster T, Gilmour J (2018) Reproduction of brooding corals at Scott Reef, Western Australia. Matters 2297–8240.

Foster T, Heyward AJ, Gilmour JP (2018) Split spawning realigns coral reproduction with optimal environmental windows. Nature Communication 9:718–8.

Giese AC, Pearse JS (1974) Introduction: General principles. In: Giese AC and Pearse JS (Editors). Reproduction of Marine Invertebrates, I. Academic Press, New York, pp. 1–49.

Gilmour JP, Underwood JN, Howells EJ, Gates E, Heyward AJ, Johnson N (2016) Biannual spawning and temporal reproductive isolation in Acropora corals. PLoS One 11(3):e0150916-e015916.

Gleason DF, Hoffman DK (2011) Coral larvae: From gametes to recruits. Journal of Experimental Marine Biology and Ecology 408:42–57.

Glynn PW, Colley SB (2009) Survival of brooding and broadcasting reef corals following large scale disturbances: Is there any hope for broadcasting species during global warming? In: Proceedings of the 11th International Coral Reef Symposium 1:361–365.

Gorbunov MY, Falkowski PG (2002) Photoreceptors in the Cnidarian hosts allow symbiotic corals to sense blue moonlight. Limnology and Oceanography 47(1):309–315.

Gouezo M, Doropoulos C, Fabricius K, Olsudong D, Nestor V, Kurihara H, Golbuu Y and Harrison P (2020b) Multispecific coral spawning events and extended breeding periods on an equatorial reef. Coral Reefs 39:1107–1123.

Gouezo M, Olsudong D, Fabricius K, Harrison P, Golbuu Y, Doropoulos C (2020a) Relative roles of biological and physical processes influencing coral recruitment during the lag phase of reef community recovery. Scientific Reports 10:2471

Graham EM, Baird AH, Connolly SR (2008) Survival dynamics of scleractinian coral larvae and implications for dispersal. Coral Reefs 27:529–539.

Guest JR, Baird AH, Goh BPL, Chou LM (2012) Sexual systems in scleractinian corals: An unusual pattern in the reef-building species *Diploastrea heliopora*. Coral Reefs 31:705–713.

Guest JR, Baria MV, Gomez ED, et al. (2014) Closing the circle: Is it feasible to rehabilitate reefs with sexually propagated corals? Coral Reefs 33:45–55

Hagedorn M, Page CA, O'Neil KL, O'Neil KL, Flores DM, Tichy L, et al. (2021) Assisted gene flow using cryopreserved sperm in critically endangered coral. Proceedings of the National Academy of Science 118:e2110559118.

Hall VR, Hughes TP (1996) Reproductive strategies of modular organisms: Comparative studies of reef-building corals. Ecology 77(3):950–963.

Harriott VJ (1983) Reproductive ecology of four scleractinian species at Lizard Island, Great Barrier Reef. Coral Reefs 2:9–18.

Harriott VJ, Harrison PL (1984) Methods to accelerate the recolonization of corals in damaged reef systems. Final report to the Great Barrier Reef Marine Park Authority, Townsville, p. 58.

Harrison P, Hakeem AAA (2012) Asynchronous and multispecific spawning patterns among Maldivian equatorial reef corals.

Proceedings of the Twelfth International Coral Reef Society Conference.

Harrison P, Villanueva R, dela Cruz D (2016) Coral reef restoration using mass coral larval reseeding. Australian Centre for International Agricultural Research, Bruce, ACT, 60pp. www.aciar.gov.au/project/FIS-2014-063.

Harrison, PL (1985) Sexual characteristics of scleractinian corals: Systematic and evolutionary implications. Proceedings of the Fifth International Coral Reef Symposium 4:337–342.

Harrison PL (1988) Pseudo-gynodioecy: An unusual breeding system in the scleractinian coral *Galaxea fascicularia.* Proceedings of the Sixth International Coral Reef Symposium 699–705.

Harrison PL (1990) Sperm morphology and fertilization strategies in scleractinian corals. Advances in Invertebrate Reproduction 5:299–304.

Harrison PL (2011) Sexual reproduction of Scleractinian corals. Coral Reefs: An Ecosystem in Transition 59–85.

Harrison PL, Babcock RC, Bull GD, Oliver JK, Wallace CC and Willis BL (1984) Mass spawning in tropical corals. Science 223:1186–1189.

Harrison PL, Booth DJ (2007) Coral reefs: Naturally dynamic and increasingly disturbed ecosystems. Chapter 13. In: Connell SD and Gillanders BM (Editors), Marine Ecology, Oxford University Press, Melbourne, pp. 316–377.

Harrison PL, dela Cruz D (2022) Methods for Restoring Damaged Reefs Using Coral Larval Restoration. Technical Report, Southern Cross University, Lismore, p. 35.

Harrison PL, dela Cruz D, Cabaitan P, Cameron K, Ligson C, Horan K, et al. (2019) Restoring damaged coral reefs using mass coral larval reseeding. Report to Australian Centre for International Agricultural Research, Bruce, ACT, p. 55.

Harrison PL, Chartrand K, Suggett D. (2020) Scaling up coral restoration using innovative *Symbiodinium* coculture and mass larval supply on reefs. Final Report for Coral Abundance Challenge Project to Queensland Department of Environment and Science, p. 30.

Harrison PL, dela Cruz DW, Cameron KA, Cabaitan PC (2021) Increased coral larval supply enhances recruitment for coral and fish habitat restoration. Frontiers in Marine Science 8:1–22.

Harrison PL, Jamieson BGM (1999) Cnidaria and Ctenophora. In: Jamieson BGM (Editor), Reproductive Biology of the Invertebrates, Volume IX Part A, Progress in Male Gamete Ultrastructure and Phylogeny, Oxford-IBH 21–95.

Harrison PL, Wallace CC (1990) Reproduction, dispersal and recruitment of scleractinian corals. Chapter 7. In: Dubinsky Z (Editor), Coral Reef Ecosystems, Ecosystems of the World 25:133–207. Elsevier Science Publishers, Amsterdam.

Hayashibara T, Iwao K, Omori M (2004) Induction and control of spawning in Okinawan staghorn corals. Coral Reefs 23(3):406–409.

Heyward AJ, Smith LD, Rees M, Field SN (2002) Enhancement of coral recruitment by *in situ* mass culture of coral larvae. Marine Ecology Progress Series 230:113–118.

Hoadley KD, Vize PD, Pyott SJ (2016) Current understanding of the circadian clock within Cnidaria. In: Goffredo S and Dubinsky Z (Editors) The Cnidaria, Past, Present and Future. Springer, 511–520.

Hock K, Doropoulos C, Gorton R, Condie SA, Mumby PJ (2019) Split spawning increases robustness of coral larval supply and inter-reef connectivity. Nature Communications 10:3463.

Hoegh-Guldberg O, Mumby PJ, Hooten AJ, Steneck RS, Greenfield P, Gomez E, et al. (2017) Coral reefs under rapid climate change and ocean acidification. Science 318:1737–1742.

Hoogenboom MO, Connolly SR (2009) Defining fundamental niche dimensions of corals: Synergistic effects of colony size, light, and flow. Ecology 90:767–780.

Houlbrèque FE, Tambutté C, Richards C, Ferrier-Pagès C (2004) Importance of a micro diet for scleractinian corals. Marine Ecology Progress Series 282:151–160.

Howells EJ, Abrego D, Liew YJ, Burt JA, Meyer E, Aranda M (2021) Enhancing the heat tolerance of reef-building corals to future warming. Science Advances 7(34).

Hughes TP, Baird AH, Dinsdale EA, Moltschaniwskyj NA, Pratchett MS, Tanner JE, Willis BL (2000) Supply side ecology works both ways: The link between benthic adults, fecundity, and larval recruits. Ecology 81:2241–2249.

Hughes, TP, Kerry JT, Baird AH, Connolly SR, Chase TJ, Dietzel A, Hill T, Hoey AS, Hoogenboom MO, Jacobson M, Kerswell A, Madin JS, Mieog A, Paley AS, Pratchett MS, Torda G, Woods RM (2019) Global warming impairs stock-recruitment dynamics of corals. Nature 568:387–390.

Hughes TP, Kerry JT, Baird AH, et al. (2018) Global warming transforms coral reef assemblages. Nature 556:492–496.

Hunter CL (1988) Environmental cues controlling spawning in two Hawaiian corals, *Montipora verrucosa* and *M. dilitata.* Proceedings of the 6th International Coral Reef Symposium 2:727–732.

Hutchings P, Howitt L (1988) Swarming of polychaetes on the Great Barrier Reef. Proceedings of the Sixth International Coral Reef Symposium, Townsville 2:739–744.

Hyman LH (1940) The invertebrates: Protozoa through Ctenophora, McGraw-Hill, New York 1:726.

Jokiel PL, Ito RY, Liu PM (1985) Night irradiance and synchronization of lunar release of planula larvae in the reef coral *Pocillopora damicornis.* Marine Biology 88:167 – 174.

Jones GP, Almay GR, Russ GR, Sale PF, Steneck RS, van Oppen MJH, Willis BL (2009) Larval retention and connectivity among populations of corals and reef fishes: History, advances and challenges. Coral Reefs 28:307–325.

Kaniewska P, Alon S, Karako-Lampert S, Hoegh-Guldberg O, Levy O (2015) Signalling cascades and the importance of moonlight in coral broadcast mass spawning. eLife 4:e09991.

Keith SA, Maynard JA, Edwards AJ, Guest JR, et al. (2016) Coral mass spawning predicted by rapid seasonal rise in ocean temperature. Proceedings of the Royal Society B: Biological Science 283:20160011.

Kerr AM, Baird AH, Hughes TP (2011) Correlated evolution of sex and reproductive mode in corals (Anthozoa: Scleractinia). Proceedings of the Royal Society. B, Biological Sciences 278(1702):75–81.

Keshavmurthy S, Hsu CM, Kuo CY, Denis V, Leung JKL, Fontana S, et al. (2012) Larval development of fertilized "Pseudo-Gynodioecious" eggs suggests a sexual pattern of gynodioecy in *Galaxea fascicularis* (Scleractinia: Euphyllidae). Zoological studies 51:143–149.

Kininmonth S (2024) Currents modulate the genetic character of marine populations in the Great Barrier Reef. In Wolanski E and Kingsford MJ (Editors), Oceanographic Processes of Coral Reefs. Physical and Biological Links in the Great Barrier Reef. CRC Press, Boca Raton, FL.

Knowlton N, Grottoli AG, Kleypas J, Obura D, Corcoran E, et al. (2021) Rebuilding coral reefs: A decadal grand challenge. International Coral Reef Society and Future Earth Coasts, 56 pp. https://doi.org/10.53642/NRKY9386

Kojis BL (1986a) Sexual reproduction in *Acropora* (*Isopora*) species (Coelenterata: Scleractinia) I. *A. cuneata* and *A. palifera* on Heron Island reef, Great Barrier Reef. Marine Biology 91:291–309.

Kojis BL (1986b) Sexual reproduction in *Acropora* (*Isopora*) (Coelenterata: Scleractinia) II. Latitudinal variation in *A. palifera* from the Great Barrier Reef and Papua New Guinea. Marine Biology 91:311–318.

Kojis BL, Quinn NJ (1980) Mode and timing of sexual reproduction in some members of the hermatypic coral family Faviidae. American Zoologist 20:819.

Kojis BL, Quinn NJ (1981) Aspects of sexual reproduction and larval development in the shallow water hermatypic coral, *Goniastrea australensis* (Edwards and Haime, 1857). Bulletin of Marine Science 31:558–573.

Kojis BL, Quinn NJ (1985) Puberty in *Goniastrea favulus*: Age or size limited? Proceedings 5th International Coral Reef Congress, 4:289 – 293.

Leuzinger S, Willis BL, Anthony KRN (2012) Energy allocation in a reef coral under varying resource availability. Marine biology 159(1):177–186.

Levitan DR, Fogarty ND, Jara J, Lotterhos KE, Knowlton N (2011) Genetic, spatial, and temporal components of precise spawning synchrony in reef building corals of the *Montastraea annularis* species complex. Evolution 65:1254–1270.

Levitan DR, Fukami H, Jara J, Kline D, McGovern TM, McGhee KE, Swanson CA, Knowlton N (2004) Mechanisms of reproductive isolation among sympatric broadcast-spawning corals of the *Montastraea annularis* species complex. Evolution 58:308–323.

Levy O, Appelbaum L, Leggat W, Gothlif Y, Hayward DC, Miller DJ, Hoegh-Guldberg O (2007) Light-responsive cryptochromes from a simple multicellular animal, the coral *Acropora millepora*. Science 318:467–470.

Lin CH, Nozawa, Y (2017) Variability of spawning time (lunar day) in *Acropora* versus merulinid corals: A 7-yr record of *in situ* coral spawning in Taiwan. Coral Reefs 36:1269–1278.

Lin CH, Nozawa Y (2023) The influence of seawater temperature on the timing of coral spawning. Coral Reefs 42(2):417–426.

Linden B, Huisman J, Rinkevich B (2018) Circatrigintan instead of lunar periodicity of larval release in a brooding coral species. Scientific Reports 8:5668.

Lirman D, Schopmeyer S (2016) Ecological solutions to reef degradation: Optimizing coral reef restoration in the Caribbean and Western Atlantic. PeerJ 4:e2597.

Loya Y, Sakai K (2008) Bidirectional sex change in mushroom stony corals. Proceedings of the Royal Society B: Biological Sciences 275:2335–2343.

Loya Y, Sakai K, Heyward A (2009) Reproductive patterns of fungiid corals in Okinawa, Japan. Galaxea. Journal of Coral Reef Studies 11:119 – 129.

Madin JS, Anderson KD, Andreasen MH, Bridge TCL, et al. (2016) The Coral Trait Database, a curated database of trait information for coral species from the global oceans. Sci Data 3:160017

Mangubhai S, Harris A, Graham NAJ (2007) Synchronous daytime spawning of the solitary coral *Fungia danai* (Fungiidae) in the Chagos Archipelago, central Indian Ocean. Coral reefs 26(1):15–15.

Mangubhai S, Harrison PL (2008a) Gametogenesis, spawning and fecundity of *Platygyra daedalea* (Scleractinia) on equatorial reefs in Kenya. Coral Reefs 27:117 – 122.

Mangubhai S, Harrison PL (2008b) Asynchronous coral spawning patterns on equatorial reefs in Kenya. Marine Ecology Progress Series 360:85–96.

Marshall SM, Stephenson TA (1933) The breeding of reef animals. Part 1. The corals. Scientific Reports Great Barrier Reef Expedition 29(3):219–245.

Matthews JL, Raina JB, Kahlke T, Seymour JR, van Oppen MJ, Suggett D J (2020) Symbiodiniaceae-bacteria interactions: Rethinking metabolite exchange in reef-building corals as multi-partner metabolic networks. Environmental Microbiology 22:1675–1687.

McCormick MI (2003) Consumption of Coral Propagules after Mass Spawning Enhances Larval Quality of Damselfish through Maternal Effects. Oecologia 136(1):37–45.

McLeod IM, Hein MY, Babcock R, Bay L, Bourne DG, Cook N, et al. (2022) Coral restoration and adaptation in Australia: The first five years. PLoS One 17(11):e0273325.

Mendes JM, Woodley JD (2002) Timing of reproduction in *Montastrea annularis*: Relationship to environmental variables. Marine Ecology Progress Series 227:241–251.

Miller MW, Latijnhouwers KR, Bickel A, Mendoza-Quiroz S, Schick M, Burton K, Banaszak AT (2022) Settlement yields in large-scale in situ culture of Caribbean coral larvae for restoration. Restoration Ecology 30:e13512.

Mou S, Tsai D, Dunbabin M (2022) Reconfigurable robots for scaling reef restoration. arXiv.org.

Muller E, Vermeij MJ (2011) Day time spawning of a Caribbean coral. Coral reefs 30(4):1147–1147.

Muscatine L (1990) The role of symbiotic algae in carbon and energy flux in coral reefs. In: Dubinsky Z (Editor). Ecosystems of the World: Coral Reefs, Elsevier, Amsterdam, pp. 75–87.

Nozawa Y (2012) Annual variation in the timing of coral spawning in a high-latitude environment: Influence of temperature. Biology Bulletin 222:192–202.

Nozawa Y, Harrison PL (2005) Temporal settlement patterns of larvae of the broadcast spawning reef coral *Favites chinensis* and the broadcast spawning and brooding reef coral *Goniastrea aspera* from Okinawa, Japan. Coral Reefs 24:274–282.

Nozawa Y, Isomura N, Fukami H (2015) Influence of sperm dilution and gamete contact time on the fertilization rate of scleractinian corals. Coral reefs 34(4):1199–1206.

Okubo N, Isomura N, Motokawa T, Hidaka M (2007) Possible self fertilization in the brooding coral *Acropora* (Isopora) *brueggemanni*. Zoology Science 24:277–280.

Okubo N, Mezaki T, Nozawa Y, Nakano Y and others (2013) Comparative embryology of eleven species of stony corals (Scleractinia). PLoS One 8:e84115.

Oliver J, Babcock RC (1992). Aspects of the fertilization ecology of broadcast spawning corals: Sperm dilution effects and in situ measurements of fertilization. Biology Bulletin 183:409–417.

Oliver JK, Babcock RC, Harrison PL, Willis BL (1988) Geographic extent of mass coral spawning: Clues to ultimate causal factors. Proceedings of the 6th International Coral Reef Symposium, Australia 2:803–810.

Oliver JK, King BA, Willis BL, Babcock RC, Wolanski E (1992) Dispersal of coral larvae from a lagoonal reef. II. Comparisons between model predictions and observed concentrations. Continental Shelf Research 12:873–889.

Oliver JK, Willis BL (1987) Coral-spawn slicks in the Great Barrier Reef: Preliminary observations. Marine Biology 94:521–529.

Omori, M (2019) Coral restoration research and technical developments: What we have learned so far. Marine Biology Research 15:377–409.

Page CA, Muller EM, Vaughan DE (2018) Micro fragmenting for the successful restoration of slow growing massive corals. Ecological Engineering 123:86–94.

Petersen D, Laterveer M, Schuhmacher H (2005) Spatial and temporal variation in larval settlement of reef building corals in mariculture. Aquaculture 249:317–327.

Pineda J, Porri F, Starczak V, Blythe J (2010) Causes of decoupling between larval supply and settlement and consequences for understanding recruitment and population connectivity. Journal of Experimental Marine Biology and Ecology 392:9–21

Policansky D (1982) Sex change in plants and animals. Annual Review of Ecology, Evolution and Systematics 13:471–495.

Pollock FJ, Katz SM, van de Water JA, Davies SW, Hein M, Torda G, et al. (2017) Coral larvae for restoration and research: A large-scale method for rearing *Acropora millepora* larvae, inducing settlement, and establishing symbiosis. PeerJ 5:e3732.

Pratchett MS, Hoey AS, Tan CH, Kuo CY, Bauman AG, et al. (2019) Spatial and Temporal Variation in Fecundity of *Acropora* spp. in the Northern Great Barrier Reef. Diversity 11:60

Quigley KM, Donelson JM (2024) Selective breeding and promotion of naturally heat-tolerant coral reef species. In Wolanski E and Kingsford MJ (Editors), Oceanographic Processes of Coral Reefs. Physical and Biological Links in the Great Barrier Reef. CRC Press, Boca Raton, FL.

Randall CJ, Giuliano C, Heyward AJ, Negri AP (2021) Enhancing coral survival on deployment devices with microrefugia. Frontiers in Marine Science 8:662263.

Randall CJ, Giuliano C, Allen C, Bickel A, Miller M, Negri AP (2023) Site mediates performance in a coral-seeding trial. Restoration Ecology 31:e13745.

Randall CJ, Negri AP, Quigley KM, Foster T, Ricardo GF, Webster NS, Bay LK, Harrison PL, Babcock RC, Heyward AJ (2020) Sexual production of corals for reef restoration in the Anthropocene. Marine Ecology Progress Series 635:203–232.

Richmond R, Hunter C (1990) Reproduction and recruitment of corals: Comparisons among the Caribbean, the Tropical Pacific, and the Red Sea. Marine Ecology Progress Series 60:185–203.

Richmond RH (1988) Competency and dispersal potential of planula larvae of a spawning versus a brooding coral. In: Proceedings of the 6th International Coral Reef Symposium, Townsville 2:827–831.

Richmond RH (1993) Coral reefs: Present problems and future concerns resulting from anthropogenic disturbance. American Zoologist 33:524–536.

Richmond RH, Tisthammer KH, Spies NP (2018) The effects of anthropogenic stressors on reproduction and recruitment of corals and reef organisms. Frontiers in Marine Science 5:226.

Riginos C, Hock K, Matias AM, Mumby PJ, van Oppen MJH, Lukoschek, V (2019) Asymmetric dispersal is a critical element of concordance between biophysical dispersal models and spatial genetic structure in Great Barrier Reef corals. Diversity & Distributions 25:1684–1696.

Rinkevich B (1995) Restoration strategies for coral reefs damaged by recreational activities: The use of sexual and asexual recruits. Restoration Ecology 3:241–251.

Rinkevich B (2005) Conservation of coral reefs through active restoration measures: Recent approaches and last decade progress. Environmental Science and Technology 39:4333–4342.

Rinkevich B, Loya Y (1979) The reproduction of the Red Sea coral *Stylophora pistillata*. II. Synchronization in breeding and seasonality of planulae shedding. Marine Ecology Progress Series 1:145–152.

Ritson-Williams R, Arnold S, Fogarty ND, Steneck RS, Vermeij M, Paul VJ (2009) New perspectives on ecological mechanisms affecting coral recruitment on reefs. Smiths. Contributions in Marine Science 38:437–457.

Ritson-Williams R, Arnold S N, Paul V J (2016) Patterns of larval settlement preferences and post-settlement survival for seven Caribbean corals. Marine Ecology Progress Series 548:127–138.

Rodd C, Whalan S, Humphrey C, Harrison PL (2022) Enhancing coral settlement through a novel larval feeding protocol. Frontiers in Marine Science 9:918232.

Rosado PM, Leite DC, Duarte GA, Chaloub RM, Jospin G, Nones da Rocha U, et al. (2018) Marine probiotics: Increasing coral resistance to bleaching through microbiome manipulation. The ISME Journal 13:921–936.

Rosenberg Y, Doniger T, Harii S, Sinniger F, Levy O (2019) Demstifying circalunar and diel rhythmicity in *Acropora digitifera* under constant dim light. iScience 22:477–488.

Rosser NL, Gilmour JP (2008) New insights into patterns of coral spawning on western Australian reefs. Coral Reefs 27:345–349.

Sakai K (1997) Gametogenesis, spawning, and planula brooding by the reef coral *Goniastrea aspera* (Scleractinia) in Okinawa, Japan. Marine Ecology Progress Series 151:67–72.

Sakai K (1998) Delayed maturation in the colonial coral *Goniastrea aspera* (Scleractinia): Whole-colony mortality, colony growth and polyp egg production. Researches on Population Ecology 40:287–292.

Sakai Y, Hatta M, Furukawa S, Kawata M, et al. (2020) Environmental factors explain spawning day deviation from full moon in the scleractinian coral Acropora. Biology Letters 16:20190760.

Santiago-Valentín JD, Rodríguez-Troncoso AP, Carpizo-Ituarte E, Benítez-Villalobos F, Torres-Hernández P, López-Pérez A (2015) Reproductive pattern of the reef-building coral *Pavona gigantea* (Scleractinia: Agariciidae) off southwestern Mexico. Ciencias Marinas 41(3):233–246.

Schmidt-Roach S, Miller KJ, Lundgren P, Andreakis N (2014) With eyes wide open: A revision of species within and closely related to the *Pocillopora damicornis* species complex (Scleractinia; Pocilloporidae) using morphology and genetics. Zoological journal of the Linnean Society 170(1):1–33.

Schmidt-Roach S, Miller KJ, Woolsey E, Gerlach G, Baird AH (2012) Broadcast spawning by *Pocillopora* species on the Great Barrier Reef. PLoS One 7:e50847.

Scott A, Harrison PL, Brooks LO (2013) Reduced salinity decreases the fertilization success and larval survival of two scleractinian coral species. Marine Environmental Research 1–5.

Serrano XM, Baums IB, Smith TB, Jones RJ, Shearer TL, Baker AC (2016) Long distance dispersal and vertical gene flow in the Caribbean brooding coral *Porites astreoides*. Scientific reports 6(1):21619–21619.

Shlesinger Y, Goulet TL, Loya Y (1998) Reproductive patterns of scleractinian corals in the northern Red Sea. Marine Biology 132:691–701.

Shlesinger Y, Loya Y (1985) Coral community reproductive patterns: Red Sea versus the Great Barrier Reef. Science 228:1333–1335.

Shlesinger T, Loya Y (2019) Breakdown in spawning synchrony: A silent threat to coral persistence. Science 365: 1002–1007.

Simpson CJ (1985) Mass spawning of scleractinian corals in the Dampier archipelago and the implications for management of coral reefs in Western Australia. Department of Conservation and Environment Bulletin 244:1–35.

Smith HA, Moya A, Cantin NE, Van Oppen MJ, Torda G (2019) Observations of Simultaneous Sperm Release and Larval

Planulation Suggest Reproductive Assurance in the Coral *Pocillopora acuta*. Frontiers in Marine Science 6:362.

Stobart B, Babcock RC, Willis BL (1992) Biannual spawning of three species of scleractinian coral from the Great Barrier Reef. In: Richmond RH (Editor) Proc 7th International Coral Reef Symposium 1:494–499.

Stoddart JA, Black R (1985) Cycles of gametogenesis and planulation in the corals *Pocillopora damicornis*. Marine Ecology Progress Series 23:153–164.

Suggett DJ, Edmondson J, Howlett L, Camp EF (2020) Coralclip®: A low-cost solution for rapid and targeted out-planting of coral at scale. Restoration Ecology 28:289–296.

Suzuki G, Arakaki S, Suzuki K, Lehisa Y, Hayashibara T (2012) What is the optimal density of larval seeding in Acropora corals? Fisheries Science 78:801–808.

Suzuki G, Okada W, Yasutake Y, Yamamoto H, Tanita I, et al. (2020) Enhancing coral larval supply and seedling production using a special bundle collection system "coral larval cradle" for large-scale coral restoration. Restoration Ecology 28:1172–1182.

Sweet MJ, Brown BE (2016) Coral responses to anthropogenic stress in the 21st century—an eco physiological perspective. Oceanography Marine Biology Annual Review 54:271–314.

Szmant AM (1986) Reproductive ecology of Caribbean reef corals. Coral Reefs 5:43–53.

Szmant-Froelich A (1985) The effect of colony size on the reproductive ability of the Caribbean coral *Montastrea annularis* (Ellis and Solander) [In: C. Gabrie and B. Salvat (Editors)]. Proceeding 5th International Coral Reef Symposium 4:295–300.

Tan CH, Pratchett MS, Bay LK, Graham EM, Baird AH (2016) Spatial and temporal variation in fecundity among populations of *Acropora millepora* on the Great Barrier Reef. Marine ecology Progress Series 561:147–153.

Tanner JE (1996) Seasonality and lunar periodicity in the reproduction of pocilloporid corals. Coral Reefs 15:59–66.

van Oppen MJ, Gates RD, Blackall LL, Cantin N, Chakravarti LJ, et al. (2017) Shifting paradigms in restoration of the world's coral reefs. Global Change Biology 23:3437–3448.

Van Veghel MLJ (1993) Multiple species spawning on Curacao reefs. Bulletin of Marine Science 52:10171021.

van Woesik R (2009) Calm before the spawn: Global coral spawning patterns are explained by regional wind fields. Proceedings of the Royal Society B: Biological Sciences 277:715–722.

van Woesik R, Lacharmoise F, Köksal S (2006) Annual cycles of solar insolation predict spawning times of Caribbean corals. Ecology Letters 9:390–398.

Vaughan DE (2021) Active Coral Restoration. Techniques for a Changing Planet. JR Ross, Plantation, p. 609.

Vize PD (2009) Transcriptome analysis of the circadian regulatory network in the coral, *Acropora millepora*. Biology Bulletin 216:131–137.

Wallace CC (1985) Reproduction, recruitment and fragmentation in nine sympatric species of the coral genus Acropora. Marine Biology 88:217–233.

Wells JW (1956) Scleractinia. In: Moore RC (Editor) Treatise on Invertebrate Palaeontology. Pt. F. Coelenterata. Geological Society of America, University of Kansas Press, Lawrence, 328–444.

Williams SL, Sur C, Janetski N, Hollarsmith JA, Rapi S, Barron L, Heatwole SJ, et al. (2019) Large-scale coral reef rehabilitation after blast fishing in Indonesia. Restoration Ecology 27(2):447–456.

Willis BL, Babcock RC, Harrison PL, Oliver JK (1985) Patterns in the mass spawning of corals on the Great Barrier Reef from 1981 to 1984. Proceedings in the Fifth International Coral Reef Congress, Tahiti 4:343–348.

Willis BL, Babcock RC, Harrison PL, Wallace CC (1997) Experimental hybridization and breeding incompatibilities within the mating systems of mass spawning reef corals. Coral Reefs 16:S53–S65.

Wilson J, Harrison P (2005) Post-settlement mortality and growth of newly settled reef corals in a subtropical environment. Coral Reefs 24:418–421.

Wolanski E, Kingsford M, Lambrechts J, Marmorino G (2024) The physical oceanography of the Great Barrier Reef: A review. In Wolanski E and Kingsford MJ (Editor), Oceanographic Processes of Coral Reefs. Physical and Biological Links in the Great Barrier Reef. CRC Press, Boca Raton, FL.

Wolstenholme J, Nozawa Y, Byrne M, Burke W (2018) Timing of mass spawning in corals: Potential influence of the coincidence of lunar factors and associated changes in atmospheric pressure from northern and southern hemisphere case studies. Invertebrate Reproduction and Development 62:98–108.

Wolstenholme JK (2004) Temporal reproductive isolation and gametic compatibility are evolutionary mechanisms in the *Acropora humilis* species group (Cnidaria; Scleractinia). Marine Biology 144:567–582.

Young CN, Schopmeyer SA, Lirman D (2012) A review of reef restoration and coral propagation using the threatened genus *Acropora* in the Caribbean and Western Atlantic. Bulletin of Marine Science 88:1075–1098.

29 Great Barrier Reef Biophysics
A Synthesis of Challenges and Opportunities

Eric Wolanski and Michael J. Kingsford

29.1. INTRODUCTION

The Great Barrier Reef (GBR; Figure 1), together with its 424,000 km^2 catchment comprising 35 rivers, is enormous. It is nearly 2,000 km long, it has about 2,500 reefs of various sizes and shapes, and these reefs are scattered in diverse ways in different regions of the continental shelf. The shelf width and depth vary with latitude between 30 and 200 km, and its mean depth also varies with latitude between 30 and 100 m. It borders the Coral Sea with depths of 2,000–4,000 m.

The oceanography of the GBR is influenced by mesoscale turbulent oceanic jets in the Coral Sea (with water speeds generally < 1 m s^{-1}) that reach the shelf break. The GBR waters are also affected by large semi-diurnal tides with a maximum range at the shelf break of typically near 3 m but approximately 3.5 m in the southern GBR, a large tidal range at the coast, and commonly moderate to strong wind. During the Austral wet season, there are occasional cyclones and seasonal rains that create significant pulses of freshwater in the nearshore environment where inner reefs are located. The wet season usually occurs between January and March. Physical processes and source of nutrient inputs have a profound influence on biological processes (Figure 2). Although the resulting nutrient input is weak by global standards, there is a significant and biologically important upwelling along the shelf break. Furthermore, additional nutrient input comes from rivers. Currents are complex in the maze of reefs that make up the GBR. Individual reefs and islands generate significant sub-mesoscale turbulence and secondary flows. The reef matrix reshapes the tidal and the mean currents through it and around it, and all of this determines the connectivity of larvae and other particles between reefs (Kingsford and Wolanski 2024).

This book focuses on the links between the biology/ecology and the hydrodynamics/oceanography. These links control or at least modulate the population dynamics of most fauna and flora of the GBR. To understand these links, experts were invited to review the state of knowledge in their field of expertise concerning several themes, specifically, (1) what is the GBR oceanography and how does it influence the ecology? Here, we cover large- and small-scale aspects of GBR oceanography, microbiology, and geomorphology and the resulting connectivity. (2) What is the land-sea connectivity, and what are the ecological consequences? (3) How does biophysical oceanography control the population dynamics of a multitude of organisms including seagrasses, mangroves, charismatic nektonic taxa such as dugongs and turtles, and the larvae of crown-of-thorns starfish, reef and coastal fish, and planktonic jellyfish? (4) What are the impacts of human activities and climate change on GBR organisms, and what potential is there for remediation measures that have a focused impact? The multiple natural threats are real and well documented, but the remediation measures are still in their infancy in terms of research and implementation. Initiatives have included the reseeding of corals and fish, the genetic manipulation of corals, the restoration of wetlands, the upgrade of irrigation infrastructure, the engineering of beach morphology for nesting turtles, and engineering proposals to combat climate change.

In synthesizing the content of the book, we have taken the following approach: (1) the present GBR as a point in time, that is, the millennial origins of the GBR to its present state; (2) the present GBR in space; (3) the biophysical aspects of the GBR including links with the mainland and connectivity of organisms over different spatial scales; (4) in the face of increasing perturbations that affect the GBR and its flora and flora, what remediation measures can be taken?

29.2. THE PRESENT GBR AS A POINT IN TIME

The GBR has evolved in time from natural processes. At 12,000 years BP, the reef as we know it did not exist. About 5,000 years ago, the shelf that formed the foundation of the GBR was fully immersed. Sea levels have varied to present. The GBR is still evolving from natural processes and the impact of climate change and anthropogenic effects. What we see at present is the GBR as a point in time of an evolving body. The authors report important findings on this evolution as follows.

29.2.1. Past Physical Patterns Generated the Present Genetic Character of Populations

Over thousands of years the reef-building scleractinian corals have shaped the water flows in the GBR. At the same time, the water flow has also shaped the distribution of

DOI: 10.1201/9781003320425-33

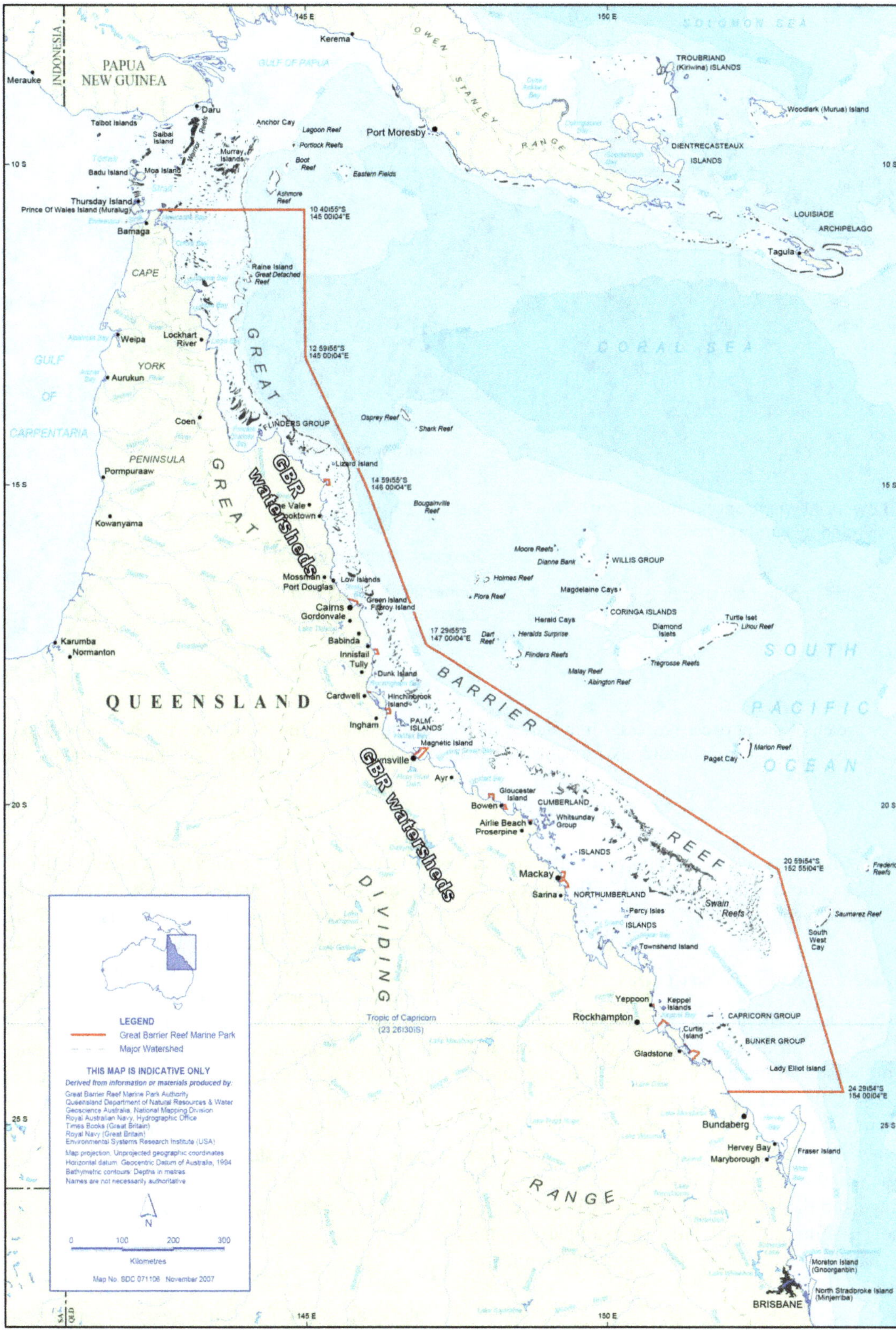

FIGURE 1 The GBR and its catchments. The red line shows the seaward boundary of the GBR Marine Park. The land east of the Great Dividing Range is the GBR watersheds. The base map is courtesy of Geospatial Services, Great Barrier Reef Marine Park Authority © Commonwealth of Australia (GBRMPA) 2022.

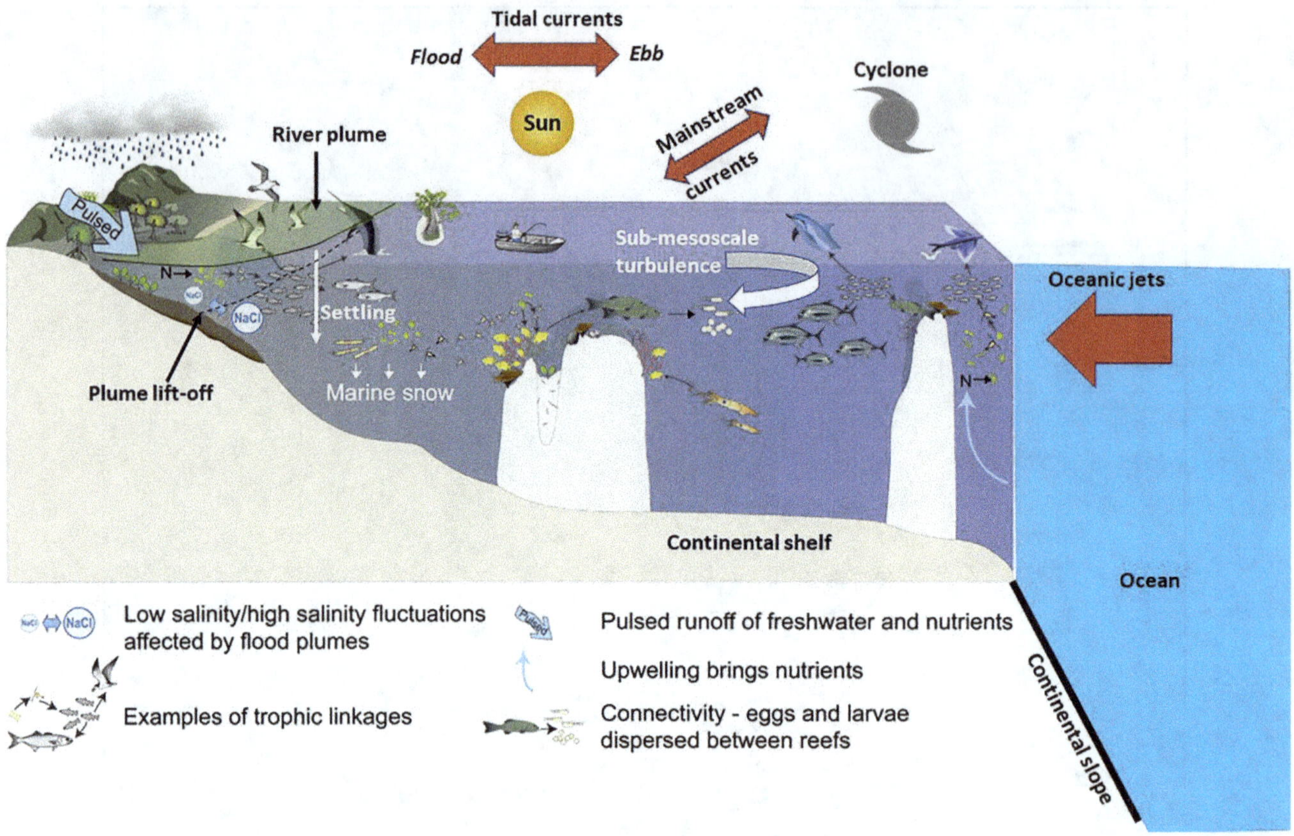

FIGURE 2 A sketch of the physical processes influencing the continental shelf of the GBR from inshore to offshore. Coral reefs, sources of nutrients, and the dominant currents are indicated. Some key organisms of the GBR food chain are shown. Modified from Kingsford and Welch (2007).

these organisms. Kininmonth (2024) shows that the genetic character of the populations in the central GBR reflects the past biophysical patterns.

29.2.2. Reef Island Geomorphology

Coral reef islands are continuously changing and are shaped over time by interrelating biological, meteorological, and geological processes (Hamylton et al. 2024). In low-lying carbonate islands, the reef flats shape the mangrove forests that form on top of them, and beach geomorphology influences turtle nesting. Questions remain unanswered on how these biological and geological links will behave because of climate change and the implications for reef islands. Torres Strait Islander communities provide an example on how people may have to increasingly deal with environmental changes.

29.2.3. Many Points in Time and Space Are Revealed by Coral Cores

There are surprisingly few coral-based paleoclimate records, particularly for pre-instrumental times and regions outside the inshore central GBR (Webb 2024). Some datasets show multidecadal oscillations that suggest greater amplitude El Niño-Southern Oscillation (ENSO) variability since the 19th century. The best multi-century record of palaeo-temperature in the GBR is out of step with the regional average, suggesting key missing knowledge in past GBR climate controls. A better understanding is needed on how local factors affect contemporary coral archives. The GBR is too large and complex to be interpreted as a single environment. Anthropogenic components of change observed over the last centuries cannot yet be identified, isolated, and quantified for better predictive modelling for the future. Furthermore, at geologic time scales, terrigenous sediment is stored on the shelf during sea level lowstand, and it is exported seaward to the Coral Sea during transgression (Dunbar et al. 2000).

29.2.4. The History of Severe Tropical Cyclones

Duke et al. (2024) found that severe tropical cyclones (STC) have occurred widely and caused considerable damage to shoreline mangroves. There has been a notable increase in both the number and intensity of STCs over the recent decade (2010–2019) – with 7 during this decade compared to the 0–2 recorded earlier. During the most recent decade, 38,913 ha of mangroves were severely damaged, which is up 13-fold from the 2,926 ha of previous decades. This damage was greater than concurrent impacts on terrestrial forests nearby. Both forest types notably recovered naturally, but damaged mangrove canopies took longer to recover. More severe storm damage further impedes natural regenerative processes. Several sites in the GBR area were characterised by repeated impacts from 2–3 STCs during the 35-year study period. These impacts were often accumulative, resulting in proportionately delayed periods of recovery. The unabated continuation of such threats is trending inevitably towards the collapse of mangrove forest ecosystems already threatened by rising sea levels and other factors. Such monitoring studies need to be expanded to mangroves on islands. Similar perturbations have affected the architecture of coral reefs and the organisms associated with them (e.g., Woolsey et al. 2012; Byrne et al. 2024), and there is a concern that the frequency and intensity of cyclones will increase with a changing climate.

29.2.5. The History of Coral Bleaching

The history of coral bleaching in the GBR is reviewed by Kamenos and Hennige (2024). Mass coral bleaching events during the last 20 years have caused major concern over the future of the GBR coral reefs. Regular systematic scientific observations on the GBR began only in the late 1970s and reveal that marine heatwaves caused mass bleaching in 1982, 1998, 2002, 2016, 2017, and 2020. In the geological record, there is also evidence of possible global mass bleaching events ~20 kyr BP and prior to the Industrial Revolution, with more recent individual bleaching and heat stress events identified on the GBR in at least 1877 and 1929. In the observational record, there is evidence of increased GBR bleaching frequency since at least the 1990s, and the proxy record indicates that this may have begun in the 1800s, a period when the Industrial Revolution was reaching its peak. These patterns highlight that marine heatwaves are pushing coral populations into repeated bleaching cycles. The concern is that with an increase in the frequency of bleaching events, the benthic assemblages may not have time to recover.

29.2.6. The Irruptions of Crown-of-Thorns Starfish

Much scientific knowledge has been gained from studying the irruptions of crown-of-thorns starfish in the GBR (Pratchett et al. 2024). There have been four well-documented population irruptions of the western Pacific crown-of-thorns starfish (CoTS; *Acanthaster* cf. *solaris*) since the 1960s. These irruptions have had a major deleterious effect on the biological structure of coral reefs on the GBR. Understanding and management of CoTS has much improved since, together with increased data on the spatiotemporal incidence of population irruptions and knowledge of the tolerance and dispersal of the larval stages. Most critically, the location for the initiation of population irruptions is becoming increasingly resolved. However, the key biophysical processes that influence the initiation and spread of population irruptions are still poorly understood.

29.2.7. The Cumulative Impact of Climate Change Stressors

Climate change stressors impact coral reefs both as single and combined factors as is documented in GBR-specific case studies (Byrne et al. 2024). The GBR faces a perilous future in a high CO_2 world due to a multitude of co-occurring climate stressors, including sea-level rise, thermal stress, ocean acidification, and an increase in the intensity of storms. Long-term datasets on well-studied reefs exist at Lizard Island (44 years) and Low Isles (91 years) and show the responses to disturbance and recovery dynamics with respect to coral and non-coral reef invertebrates. The resilience to climate change stressors of sea cucumbers and crown-of-thorns starfish has positive and negative implications for reefs, respectively. A key result is that the impact of these various stressors may be cumulative. From reef evolution and climate change perspectives, impacts on sand production and coral islands from storms and sea-level rise highlight concerns for the persistence of reef islands.

29.3. THE PRESENT GBR IN SPACE

The GBR is not a closed ecosystem. It is impacted by throughflows of oceanic waters and coastal drainages, together with associated sediments, nutrients, fauna and flora, and pollutants.

29.3.1. Pan-Oceanic Pollutants

Pollutants reach the GBR from the Coral Sea, and control measures are inadequate to minimise such pollution (Hoogenboom et al. 2024). Plastic pollution is ubiquitous, but we still cannot predict how much is present at a given location at a given time and who are the polluters. Solving plastic pollution will involve legislation based on a legally binding global plastic treaty. At present, there is no such legally binding treaty. At the local scale, the management of plastic pollution is challenging because of the small size of the municipal area, while the available data describing plastic transport and sources are at a much larger spatial scale. Another pollutant of concern is washwater from Exhaust Gas Cleaning Systems (EGCS) on ships (i.e., basically 'scrubbers' used to comply with the sulphur limit of 0.50 mass per cent concentration (m/m) in fuel oil as required by the International Convention for the Prevention of Pollution from Ships (MARPOL) Annex VI and Australia's domestic law). EGCS remove sulphur from the exhaust gas using seawater, and they discharge the washwater overboard. Five hundred ships that use scrubbers operate within the GBR. This washwater is more acidic than seawater and contains nitrates, particulate matter, polycyclic aromatic hydrocarbons (PAHs), and heavy metals that are toxic to marine organisms. For predictive biophysical models to be operational, research is needed to quantify the breakdown of these compounds in GBR waters and how compounds bind with microbes and suspended particles.

29.3.2. Animal Migration

The GBR is a leaky entity to many highly mobile taxa and highlights the need to consider Indo-Pacific pathways for many species. Animals migrate annually in considerable numbers in and out of the GBR. These include (Figure 3) shorebirds that migrate from Papua New Guinea and Russia (Bird Life International 2022) and seabirds that nest on Raine Island and nearby cays and migrate up to 4,000 km in the tropical western Pacific Ocean (Dobbs 2005). Humpback and GBR dwarf minke whales migrate every year from Antarctica to the northern GBR in winter against the prevailing net currents and with the currents on the way back (Andrews-Goff et al. 2018; Ramirez-Flores 2019). Green and loggerhead turtles that have rookeries in the GBR migrate in the Pacific Ocean before returning many years later (Hamann et al. 2024). Black marlin billfish disperse in the Coral Sea after spawning in the outer GBR and at Cape Bowling Green, their only coastal spawning ground (Domeier and Speare 2012). The ornate spiny lobster takes three generations to migrate from Torres Strait to Vietnam and return to Torres Strait (Dao et al. 2015). There is a net export of GBR coral reef fish to Lord Howe Island located in the Pacific Ocean 1,153 km southeast of the southern GBR (Hobbs 2010). Dugongs have occasional small migrations from Torres Strait to the Gulf of Papua (Marsh and Cleguer 2024; see also www.seaturtle.org/tracking/index.shtml?project_id=549).

29.4. BIOPHYSICAL OCEANOGRAPHY OF THE PRESENT GBR

29.4.1. The GBR Physical Oceanography

Like all continental shelves worldwide, the water circulation over the GBR continental shelf is strongly influenced by the circulation in the adjoining Coral Sea. Commonly, shelf waters have a gentle slope, with few shoals, islands, and reefs, and as a result, the flow field over the continental shelf is smooth. Field data reveal that, in contrast, the reefs of the GBR generate a bioengineered, turbulent flow field, and this sub-mesoscale turbulence is enhanced by the mesoscale turbulence in the adjoining Coral Sea (Wolanski et al. 2024). The currents among reefs are dominated by strong tidal currents, eddies, jets, boundary layers, topographically driven upwelling and downwelling, shear zones, stagnation zones, sticky water, boundary mixing, reef-generated turbulence, and topographically steered flows at all scales. The

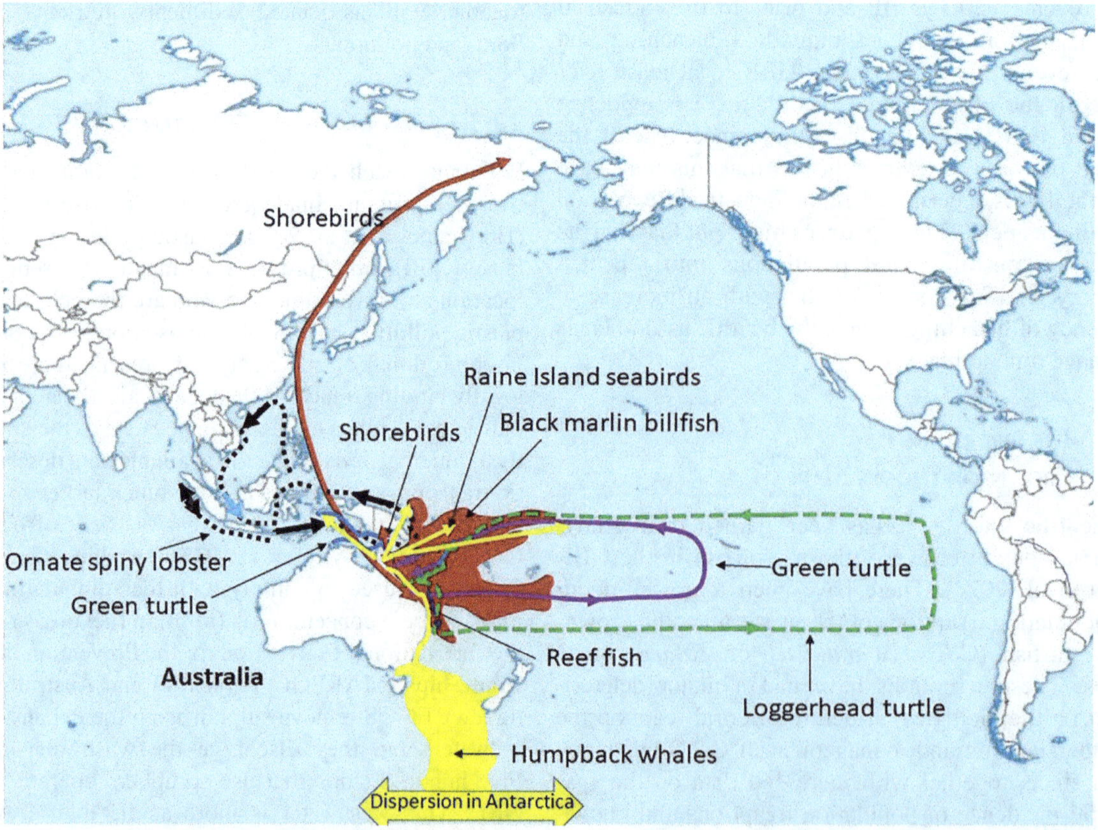

FIGURE 3 A sketch of the ecological watershed of the GBR showing either the approximate migration routes or the dispersal areas according to the availability of field data.

net circulation varies strongly from north to south following the topography and the mesoscale turbulent jets in the Coral Sea, and furthermore, it varies seasonally and interannually with the wind. River floods during the wet season are sporadic events that generate transient river plumes that transport fine sediment and nutrients over long distances. A key finding is that small-scale processes control, through feedback processes, the large-scale water circulation in the GBR.

29.4.2. Micro-Scale Processes Also Modulate the Large Scales

Micro-scale processes play a significant role in the GBR biophysics (Raina and Seymour 2024). Physical processes affect the activity and ecology of microorganisms on the GBR. In turn, these microorganisms play a key role in the cycling of carbon, nitrogen, and sulfuric compounds and their emission of volatile compounds, which can affect atmospheric processes. A change is needed in the way that reef microorganisms are studied. This will enable more accurate predictions of their impact on ecosystem function and health.

29.4.3. The GBR Ecohydrology

The biophysical connections between the GBR and the adjacent river catchments are critical to the function of the GBR ecosystem and in the survival of the plants and animals that depend on various parts of the system as part of their life cycles (Waterhouse et al. 2024). Human development has led to declining water quality in catchment waterways and increases in the loads of pollutants. In turn, these degrade the GBR. Future changes to connectivity in the GBR are predicted. Key criteria for success in managing complex catchment to reef landscapes are (i) the lead role of the community, (ii) the need for a systems approach, (iii) the need for targeted, cost-effective, and sustainable long-term investment, and (iv) building knowledge and maintaining consensus and commitment in the face of politically motivated interference. All of these criteria are challenging to implement.

29.4.4. The Human Impact Through Land Use

The changing delivery of sediment and nutrients from the land because of land use is quantified by Lewis et al. (2024). This delivery is a threat to the health and productivity of key ecosystems of the GBR lagoon, from the burying of live corals to increased water turbidity and eutrophication. Reliable estimates are now available of sediment and nutrient loads and how they have changed over time under catchment development for each of the 35 basins of the GBR catchment area for sediment, particulate nitrogen and phosphorus, and dissolved inorganic nitrogen. Various policy instruments and management options are available to reduce the loads of sediment and nutrients delivered to the GBR lagoon. They have different efficacies, and their implementation depends on political will.

29.4.5. Pesticides: The Proposed Jellyfish Watch

Templeman and Williams (2024) comment on the state of knowledge concerning the impact of pesticides in the GBR. Pesticides are heavily used in the GBR catchment area, but the monitoring of their effects on the GBR has been problematic. Thus, the ecological impact of pesticides in the GBR is still unknown. Ecological studies suggest that the spatial distribution of pesticides and pesticide metabolites across the GBR and their risks to key biota can be mapped by the large benthic medusa *Cassiopea* sp. at both small and large scales. Thus, for the GBR, the 'mussel watch' used worldwide to monitor pollution of coastal waters should instead be 'jellyfish watch'.

29.4.6. Connectivity for Fish

Fish use and connect the reefs themselves, inter-reefal habitats, and coastal and terrestrial systems (Sheaves et al. 2024). This functional interlinking enables the disparate components from estuaries to offshore reefs to work together, allowing the operation of a range of vital ecological processes across space and time. Some land-sea linkages are threatened by land use. Measures are needed to maintain robust fish-mediated land-sea connectivity in the future. This in-depth review further confirms the 'Blue Highway' hypothesis (Cappo and Kelley 2001) that reefs are not self-contained biological islands. Instead, the GBR is a profoundly interconnected system from the reef to the 'inter-reef' and to the mangroves. This, in turn, has significant implications for reef zoning plans.

29.4.7. Limited Connectivity for Seagrass

Seagrasses are widely distributed in the GBR shelf (Tol et al. 2024). The currents disperse the seagrass propagules, and this is critical to the distribution, structure, and resilience of seagrass populations. The dispersion and connectivity of seagrass depends strongly on the wind drift coefficient. This coefficient was found to depend on the seagrass species, size of the seagrass fragments, and wind speed. The GBR seagrass meadows have a higher level of bounding, that is, significantly less connectivity between seagrass meadows than previously inferred. Research is needed on the loss of seagrass from above average temperatures and how much the compromising of such an important foundation species will affect the connectivity and survivorship of dependent species such as dugongs.

29.4.8. Dugong Biophysics

The dugong, a medium-sized coastal marine mammal and a seagrass community specialist, has globally significant populations in the GBR (Marsh and Cleguer 2024). Dugongs are strong swimmers that perceive their environment through touch, hydrodynamic reception, and hearing; vision and taste are complementary sensory modalities. They have strong spatial cognition. Dugongs also apparently rely on socially transmitted knowledge learned from their mothers during their years of calf dependency. They occasionally migrate, although not for long distances, and the trigger for this is unknown. The most profound effects of climate change on dugongs in the GBR are likely to be changes in the distribution, species composition, and biomass of seagrass habitats, which will affect dugong fecundity and mortality. Dugongs are now largely confined to two hot spots, one in Harvey Bay south of the GBR and the other in Torres Strait in the far north, with small remnant populations in between (Marsh et al. 2019). There is a genetic discontinuity for dugongs in the Whitsundays that may be due to the local oceanography and to variation in the distribution of seagrass, their primary food source. Multidisciplinary research is much needed to progress understanding of the interactions between dugongs and their biophysical environment.

29.4.9. Jellyfish Biophysics

The GBR is home to a diverse array of jellyfish (Subphylum Medusozoa) that interact with their environments in a variety of ways (Schlaefer and Kingsford 2024). Jellyfish perceive their environments with multiple senses, which can include eyes with lenses. Furthermore, they are highly mobile, and they can propel themselves by rowing, jetting, or sailing. In open waters, jellyfish can orient themselves to swim counter to currents and maintain blooms. In structured environments such as mangroves and reefs, jellyfish exhibit sophisticated behavioural responses that can include navigating and selectively avoiding strong currents. Accordingly, most jellyfish are far from passive drifters, and local populations at small spatial scales are likely to be common. This is particularly true of the deadly stinger that affects communities in the GBR.

29.4.10. Turtle Biophysics

Of the six turtle species found the in the GBR, four have nesting rookeries in the region, and all but one of these have a pelagic-ocean dispersal phase (Hamann et al. 2024). Turtles are a spectacular example of how organisms interact with currents at large spatial scales (thousands of kilometres). After emerging from the nest, hatchling turtles move quickly down the beach and then navigate coastal currents and tides to disperse away from nesting beaches using combinations of natural cues and behaviour. Flatback turtle hatchlings remain within the GBR coastal zone, whereas hatchlings of the other three species aim to make their way into offshore currents to begin an oceanic journey. The route and duration of the oceanic journey are unknown but are likely to comprise a variety of active swimming and passive drifting and can take a decade to complete. On completion of the oceanic journey, the post-dispersal turtles arrive and settle at coastal habitats to begin the neritic-benthic stage of their lives. However, important knowledge gaps remain.

29.4.11. The Biophysics of Reef Fish Larvae

GBR fish larvae can swim well and often at speeds that are greater than local net currents (Kingsford et al. 2024). In addition, they can orientate to sound, detect reefs using olfaction, and even use a sun compass and magnetic senses to orientate to swim towards reefs. The halo of sensory information does not extend further than approximately 5–10 km from reefs. The distances that larvae can detect sensory cues may vary, however, by species and the size of larvae. Some larvae disperse widely, but a high percentage stay close to home and often close to the natal reef from which they were spawned. Tightly packed reefs may minimize dispersal and favour retention of fish larvae through the 'sticky water' effect. The field of sensory biology is elucidating imprinting processes of natal environments and chemicals from conspecifics. The challenge is to incorporate these behaviours into biophysical models.

29.4.12. Crown-of-Thorns Starfish Biophysics Modelling

Larval dispersal drives the population dynamics of a considerable proportion of species on coral reefs, including most reef fishes and scleractinian corals (Pratchett et al. 2024). Biophysical models of larval dispersal offer insight into larval dispersal patterns at large spatial and temporal scales, that is, the GBR connectivity matrix. Such models need improvement. A biophysical model is developed to enable potentially more successful control decisions for the crown-of-thorns starfish (CoTS) in the GBR through a network theory approach. This approach identifies the best reefs to cull CoTS by considering the limited budget, the field data on CoTS abundance and population dynamics, the connectivity with adjacent reefs, the tourism potential, and the fishing potential of the targeted and connected reefs.

29.4.13. The Biophysics of Sharks and Rays

Sharks and rays play a key role in the ecology of the GBR (Chin et al. 2024). These animals have a wide range of connections between their habitats and their biophysical characteristics, and this shapes their diversity, population structures, and use of habitats. Sharks and rays affect and facilitate energy flows between different ecosystems in the GBR. They are bioengineers that help to ventilate soft sediment. They move in response to environmental parameters and major weather events, and, in turn, this will determine how these species respond to a changing climate. Some species are at an elevated risk from exploitation. Research is needed to help inform the management of these species that have such important social, cultural, economic, and ecological importance.

29.5. REMEDIATION MEASURES

29.5.1. Selective Breeding of Coral and Fish

Quigley and Donelson (2024) discuss the pros and cons of using selective breeding and the assisted movement of species as conservation tools on the GBR. They present two case studies of Scleractinean corals and coral reef fishes. These methods could be used in an eco-evolutionary framework to promote resilience of the ecosystem. They question whether it is realistically possible to restore one species without the other. They show how an understanding of oceanographic processes helps overcome issues of the scalability of these methods and the choice of source reefs.

29.5.2. Restoring Wetlands

Waltham et al. (2024) demonstrate the need and discuss the practicalities of restoring and protecting the floodplains and coastal wetland ecosystems of the GBR. There have been efforts to restore these ecosystems with various levels of success. Common problems are degradation by feral pigs and swift re-invasion by weeds after clearing the sites. Opportunities for restoration efforts will increase as interest grows in federal and state policy and strategy initiatives and in private partner investors. Further coordination and capacity building, through appropriate levels of funding, are needed.

29.5.3. Remediation for Irrigated Agriculture and Fertiliser Practices

Attard et al. (2024) report on the status of the 90,000 ha of irrigated agriculture in the Lower Burdekin, which produce high nutrient loads that enter the GBR. They focus on dissolved inorganic nitrogen (DIN) and its impact on the health of the GBR, the land-based practices that create the loads, and the mechanisms by which they are delivered to the GBR. Work in the Lower Burdekin over the past two decades has identified improved irrigation and fertilizer practices to reduce the export of DIN, while at the same time improving sugarcane productivity. A series of Water Quality Improvement Plans (WQIPs) have supported uptake of these practices by farmers by providing information, training, and incentives. The WQIPs have confirmed the viability of the improved practices, but they fall very far short of broadscale adoption. To promote better adoption by the farming community, these plans could be reframed as adopting water-efficient irrigation in the Lower Burdekin.

29.5.4. Remediation of Beaches for Sea Turtles

Raine Island supports 90% of the northern GBR green turtle nesting stock. However, its viability as a turtle rookery has become problematic since the mid-1990's due to geomorphic changes leading to tidal inundation of the nesting beach, which, in turn, led to low hatchling numbers. Robertson et al. (2024) described the remediation measures to restore viable nesting habitat on the island to increase hatchling numbers. The key intervention action was to mechanically raise the nesting beach level through reprofiling so that the beach and natural nest depth was above the peak tidal water table. Associated with the reprofiling actions, significant increases have been recorded in turtle hatchling production and emergence compared to the non-modified areas of the island. Intervention and restoration have delivered positive outcomes for Raine Island and the green turtle population, and this may need to be continued in the future due to rising seas.

29.5.5. Environmental Engineering to Reduce Coral Bleaching

A range of environmental engineering approaches is available to mitigate the stress on coral during marine heatwaves to reduce the severity of coral bleaching (Harrison 2024a), but each approach has pros and cons. Methods to directly reduce heat stress, to delay bleaching onset, and to thus reduce both the severity and mortality from bleaching include shading, piping, pumping cooler water from depth, enhancing water column mixing to remove stratification, 'whitening' the upper ocean with floating reflective surface films or microbubbles, and atmospheric options such as creating a fog or mist using seawater or other obscurants. Regional scale interventions include marine haze brightening and marine cloud brightening, which target the direct and indirect aerosol effects of atmospheric aerosols, respectively.

29.5.6. Sexual Reproduction of Reef Corals and Application to Coral Restoration

There is increasing knowledge of when and how different corals reproduce and, therefore, of when gametes and planula larvae are available (Harrison 2024b). These are essential prerequisites for active coral restoration of damaged and highly degraded reefs that are now limited by impaired larval supply. This can help maintain and recover coral assemblages and enhance the adaptive potential of the surviving remnant populations while there is still sufficient genetic diversity present.

29.6. CONCLUSIONS

29.6.1. The Importance of small-Scale Physical Processes in Generating Connectivity

The currents in the GBR reef matrix are dominated by strong tidal currents, eddies and jets, boundary layers, topographically driven upwelling and downwelling, shear zones, stagnation zones, sticky water, boundary mixing, turbulence, and topographically steered flows at all scales. These small-scale processes control, through feedback processes, the large-scale net currents in the GBR reef matrix (Wolanski et al. 2024). The scale of these processes starts at scales equal to or even smaller than the reef passages, which is commonly 200 m. However, most GBR-wide oceanographic models have a mesh size of at least 1 km in the reef matrix. At a 1-km resolution, these models yield similar tides but are quite different from the same models with a mesh size of 250 m, that is, these 1-km resolution models are unreliable (Saint-Amand et al. 2023). Thus, the resolution of the present GBR-wide models needs to be significantly increased to make them realistic for biophysical applications.

A rule for accepting such models should be that they must reproduce the observed flows within the reef matrix; this necessitates a much smaller mesh size than used at present. The flows within the reef matrix appear chaotic at first glance because there are many vortices and jets (Figure 4). However, because these vortices are generated by the topography and because the topography does not vary, they show deterministic patterns, which is demonstrated by the two Landsat images of Figure 4. These two images have a time difference of two years, but one can track most vortices between the two images, which are imbedded in a net westward current, as if one sees the flow developing with a time lag of maybe a few tens of minutes instead of two years. Thus, the flow within the reef matrix is not chaotic, with chaos implying that the behaviour is unpredictable. Instead, the flow is predictable and thus amendable to biophysical modelling. Therefore, there is no justification to use coarse resolution models with a mesh size larger than 150 m near reefs; coarser meshes erase the very processes generating the connectivity (Saint-Amand et al. 2023). In addition, the models must be modified to also include the water flux over the reef driven by the waves breaking on the outer reef crest.

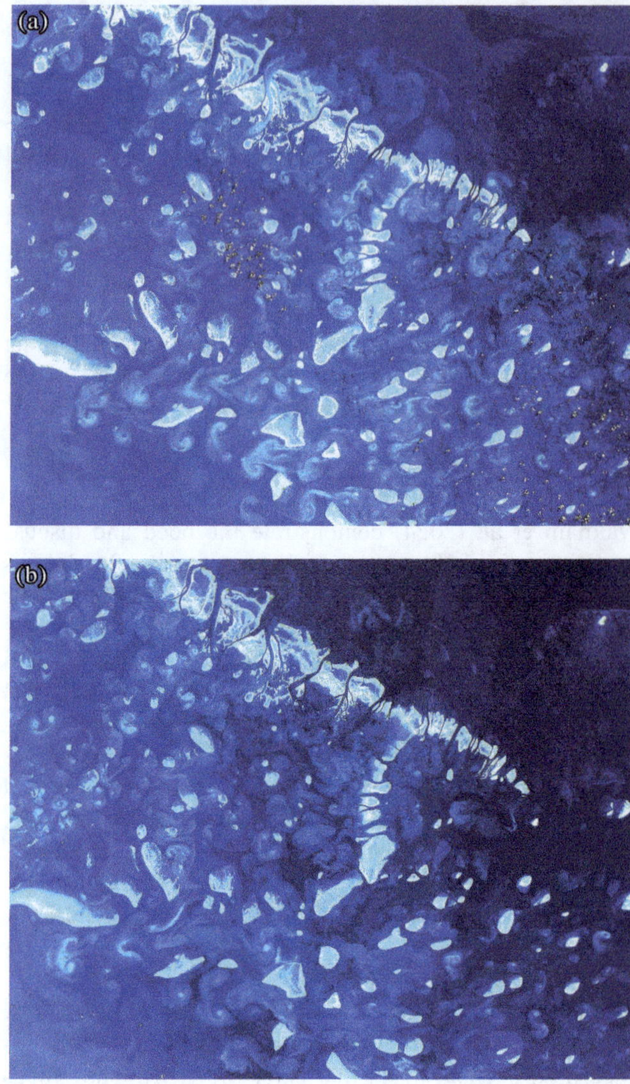

FIGURE 4 True colour Landsat-8 images of the southern region of the southern GBR, with the Swain Reefs on the right-hand side, on (a) September 1, 2019 and (b) August 21, 2021. Images kindly provided by George Marmorino.

29.6.2. The Importance of the Mesoscale Turbulence in the Coral Sea

There is a need to include a more realistic forcing by the Coral Sea in the present GBR models. The currents in the Coral Sea are dominated by mesoscale turbulent oceanic jets (see Animation 2 in Wolanski et al. 2024). When these jets meet the shelf break, they generate events of intense *cross-shelf* flows on the continental shelf, such as during the 1995 Coral Spawning Experiment at Lizard Island (Animation 1). This *cross-shelf* flow carried water from the shelf break all the way to the coast. It was generated by an oceanic jet that impinged on this area during the field study (Figure 5). Present GBR biophysical models overlook these jets by smoothening out the oceanic inflow. As a result, they predict a connectivity that is *longshelf* in the area (Figure 5). Such events of swift cross-shelf transport completely alter the nature of source and sink reefs, specifically, the whole connectivity matrix. They also enable the transport of fish larvae from the reef to coastal mangroves, which Sheaves et al. (2024) reveal. Thus, the GBR oceanography models need to be forced at their ocean open boundary by reliable Coral Sea current data. The present Coral Sea models are still unreliable for providing these data because even with some gentle nudging towards the altimetry data, these models do not reproduce well the mesoscale turbulent jets in the Coral Sea (Wolanski et al. 2021). This is because these jets are not generated by a fixed topography (in this case, they could be predicted), but by the unresolved, wicked 'turbulence problem' in the open ocean (Eckert 2019). The GBR models will yield more reliable data if they simply use for their ocean open boundary for the well-mixed surface layer the altimetry-measured currents in the adjoining Coral Sea.

29.6.3. The Importance of Animal Behaviour

Another challenge is a more accurate incorporation of animal behaviour into biophysical models. Currently the knowledge of marine animal orientation and migrations

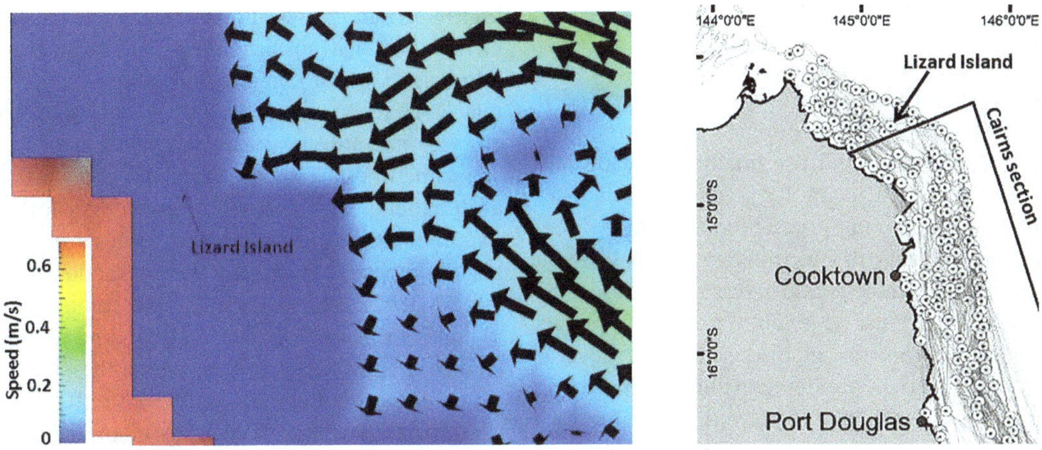

FIGURE 5 (Left) The altimetry-derived near-surface currents in the Coral Sea adjoining the northern GBR during the 1995 Coral Spawning Experiment whose data are shown in Animation 1. The colour bar shows the speed in m/s. (Right) The predicted mean connectivity network around Lizard Island is *longshelf* (adapted from Kininmonth 2024).

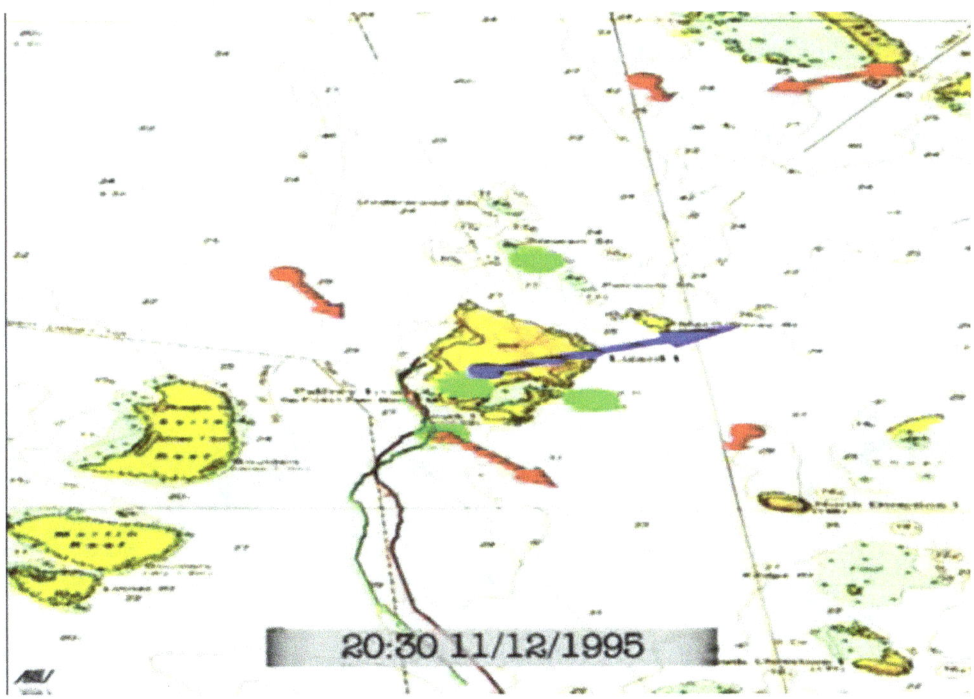

ANIMATION 1 A visualisation of the oceanographic data from the 1995 Coral Spawning Experiment. The blue arrow shows the wind at Lizard Island, the other arrows show the currents, and the solid lines show the trajectories of two parachute drogues deployed near the surface over the coral during mass coral spawning. A *cross-shelf* shoreward transport (hence, connectivity) is evidenced.

Available online at https://www.routledge.com/9781032340692

with changing conditions is crude. It is clear from the examples that we have given that understanding the GBR ecosystem can reach a higher level of robust predictions through a better integration of the biological and oceanographic knowledge. The biophysics are a major determinant of the population dynamics of many species on the GBR, from coral, fish, and jellyfish and charismatic animals such as dugongs, whales, and sea turtles to important habitat-changing species such as the CoTS. The biophysical science behind other impacts, such as that of severe tropical cyclones and coral bleaching, also needs support in the hope that it will lead to finding solutions such as prevention measures or restoration methods adequate for implementation at the large scales of the GBR.

29.6.4. Remediation

Some remediation measures, such as improved irrigation and fertiliser practices, are proven technology and if adopted, would go a long way to mitigate the degradation of the GBR by detrimental land use practices. Beach restoration has been a proven technology at Raine Island to increase green turtle hatchling production and emergence. With rising mean sea levels, this technology may have to be used more frequently at Raine Island and at other sites in the GBR, but consideration must be given to potential impacts on other sympatric organisms such as sea birds. Several other remediation initiatives are being evaluated, including bioengineering measures on land (e.g., wetland restoration) and at sea (e.g., environmental engineering measures to reduce coral bleaching stress, selective breeding of coral and fish, and active coral restoration). All of this research is much needed to explore options for the adaptation of biota and landforms to climate change and other anthropogenic impacts such as land use.

29.7. KEY MESSAGES FROM AUTHORS

There were recuring messages from our contributors as follows. (1) Biophysical modelling where physical oceanography is combined with a knowledge of behavioural ecology and the distribution and evolution of foundation species that form habitats (e.g., corals and seagrass) is providing new insights to complex ecosystems. (2) From an algorithmic point of view, oceanographic models will be improved with the incorporation of 3D and new approaches to deal with the highly turbulent flows at all scales of Coral Sea waters and GBR waters in a mosaic of reefs. (3) The incorporation of animal behaviour into biophysical models is still at a rudimentary level, where the 'rules' are often simple, age/size related changes are not always incorporated, and the assumptions are sometimes based on other species. Furthermore, there are challenges in getting complex algorithms to work. (4) Satellite telemetry and acoustic arrays are providing data to test hypotheses concerning the interactions of currents with large marine organisms, such as fishes and megafauna that includes sharks, sea turtles, dugongs and whales. (5) Climate change is a wicked problem, where the biota, currents, and even geomorphology are likely to be affected by increased perturbations such as bleaching, changing water chemistry, increased frequency of tropical cyclones, and inundation of shallow habitats. (6) There are interventionist options for addressing problems such as coral bleaching, acid sea, and changing coastlines, but consideration has to be given to the distinction between gaining knowledge and our ability to facilitate changes at the relevant spatial scales. (7) There is no quick fix to some of the grand challenges, such as managing coastal resources in the face of multiple perturbations, as management requires consistency and often substantial financial investment to succeed.

For managers, a developing problem is that the physical environment is changing due to human activities, thereby changing the physics-biology links that determine the ecosystem functioning and health. Two examples are described. The first example is the coastal waters of Lucinda, 96 km northwest of Townsville. In 1956, there was a large intertidal sandy area cut by an inshore shore-parallel channel that remained unchanged until 1977 (Figure 6). In 1979 a

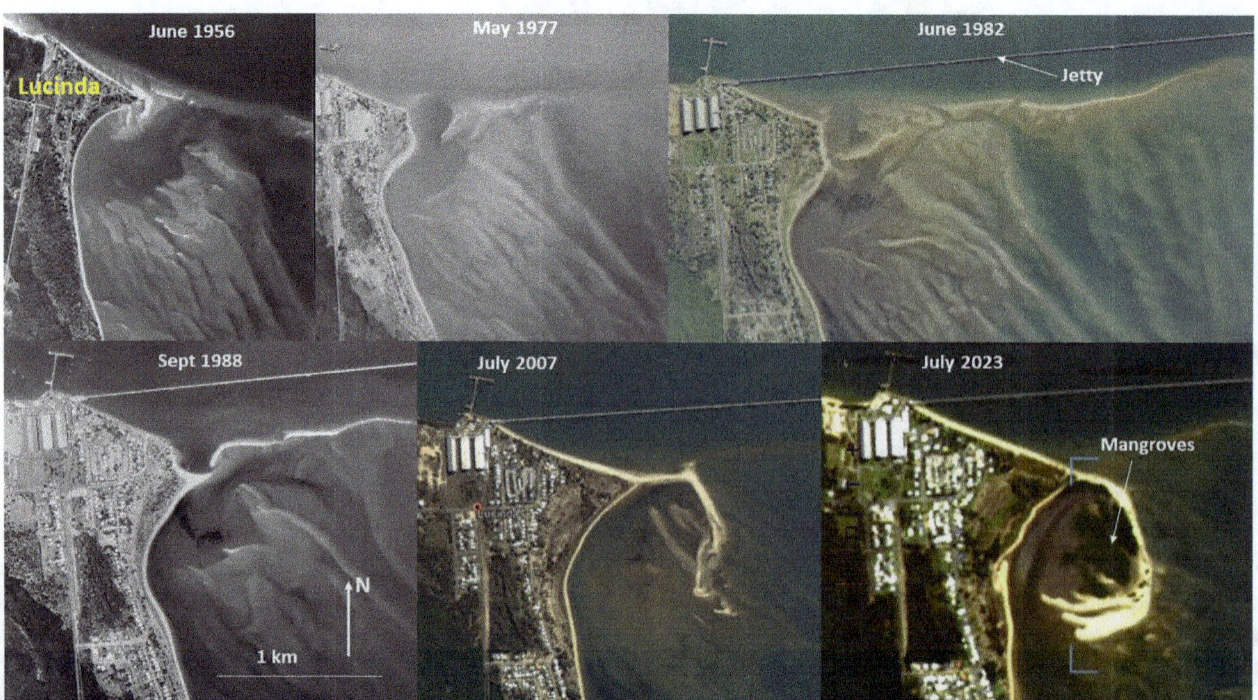

FIGURE 6 Aerial and satellite images of the coastal waters of Lucinda from 1956 to 2023.

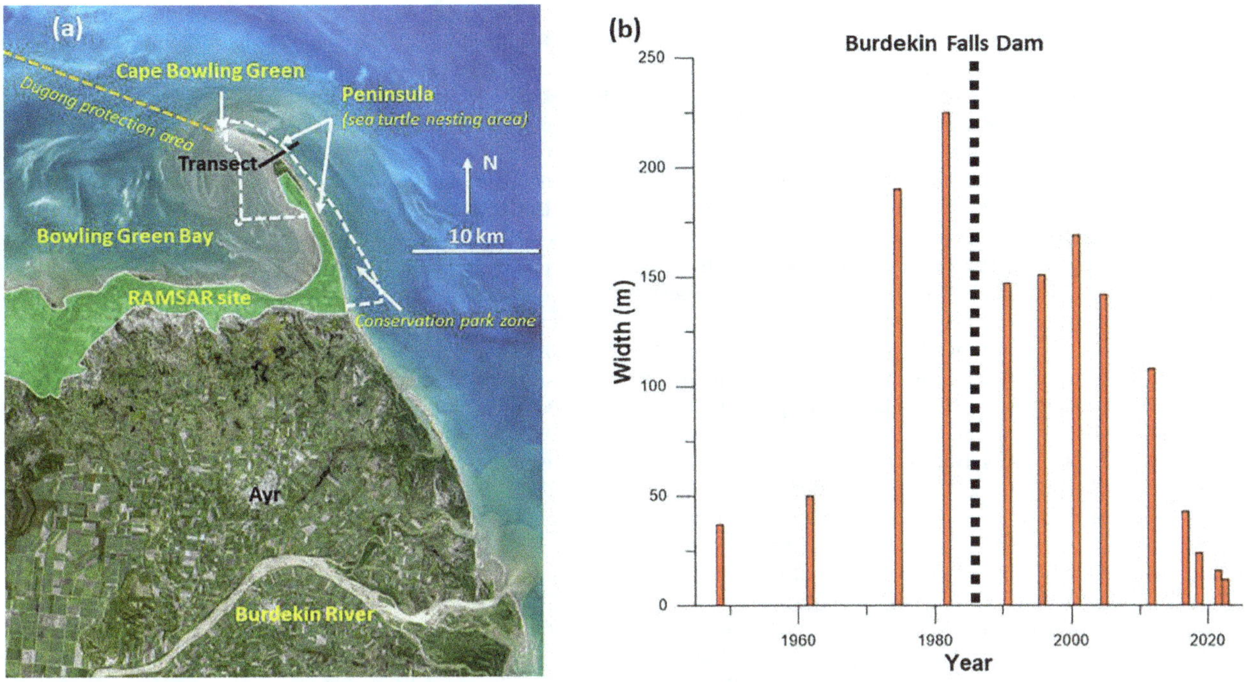

FIGURE 7 (a) A Landsat-8 image of the Cape Bowling Green peninsula, Bowling Green Bay, and the Burdekin River Delta. (b) A time-series plot of the width of the peninsula at high tide at the transect shown in (a) from 1948 to 2022. The data were obtained from aerial and satellite images. The dotted line shows the year when the construction of the Burdekin Falls dam was completed.

5.76-km long sugar loading pier was constructed, and its concrete and steel pylons cut across the inshore channel. Three years later, in 1982, the pylons had interfered sufficiently with the currents that a cross-shore sand bar had developed, which was still discontinuous as it was cut by small passages. By 1988, the sand bar was continuous with no passages. It then started to curve southward, against the dominant wind, as in the 2007 image. By 2023, it had curved shoreward to form a lagoon populated by mangroves. This environment now supports mosquitoes upwind and in closed proximity to Lucinda, potentially posing a public health risk.

The second example is Cape Bowling Green (the Cape) located 63 km east of Townsville (Figure 7). The Cape has a long sandy peninsula that is a sea turtle nesting area, and much of the Cape's coastal waters are a conservation park zone (Wolanski and Hopper 2022). The Cape shelters Bowling Green Bay that is a dugong protection area. An annual spawning aggregation of the black marlin billfish occurs near the Cape, the only such aggregation in the GBR in coastal waters. The bay is fringed by extensive wetlands that are listed by Ramsar as a wetland of international importance. The peninsula is maintained by the sand discharge of the Burdekin River. The Burdekin Falls dam was built in 1987, and it captures water and the riverine sand from approximately 88% of the catchment (Wolanski and Hopper 2022; Lewis et al. 2024). This decrease of the riverine sand supply to the coast results in the coast being starved of sediment. The upper peninsula appears to respond to these human-induced changes as it was accreting (widening) pre-dam, and it is eroding post-dam (thinning; Figure 7). Already in late 2022, there had been wave overtopping over the peninsula at the transect site. This suggests that the peninsula could breach in the foreseeable future, which would decrease the shelter for Bowling Green Bay. This could endanger the billfish aggregation and degrade the Ramsar-listed wetlands. Further, there are proposals for more dams in the Burdekin catchment that would decrease even more the riverine sand flux to the coast and thus exacerbate coastal erosion.

Managers must deal with such complex issues of a changing biophysical environment. Our hope is that this book will expand biophysical approaches to address challenges in a changing world and facilitate innovative ideas to contribute to our knowledge of the GBR and other environments worldwide.

REFERENCES

Andrews-Goff V, Bestley S, Gales NJ, Laverick SM, Paton D, Polanowski AM, Schmitt NT, Double MC (2018) Humpback whale migrations to Antarctic summer foraging grounds through the southwest Pacific Ocean. Scientific Reports 8:12333.

Attard S, Connell JG, Chaiechi T (2024) Pathways to improved water quality in the GBR lagoon – exploring opportunities for broadscale application of low-risk practices in the Lower Burdekin irrigated agriculture areas. In Wolanski

E, Kingsford MJ (eds.) Oceanographic Processes of Coral Reefs. Physical and Biological Links in the Great Barrier Reef. CRC Press, Boca Raton, FL.

BirdLife International (2022) http://datazone.birdlife.org/userfiles/file/sowb/flyways/8_East_Asia_Australasia_Factsheet.pdf

Bode M, Stewart O, Choukroun SM (2024) Incorporating biophysical larval dispersal simulations into coral reef conservation decision-making. In Wolanski E, Kingsford MJ (eds.) Oceanographic Processes of Coral Reefs. Physical and Biological Links in the Great Barrier Reef. CRC Press, Boca Raton, FL.

Byrne M, Foo SA, Vila-Concejo A, Wolfe K (2024) Impacts of climate change stressors on the Great Barrier Reef. In Wolanski E, Kingsford MJ (eds.) Oceanographic Processes of Coral Reefs. Physical and Biological Links in the Great Barrier Reef. CRC Press, Boca Raton, FL.

Cappo M, Kelley R (2001) Connectivity in the Great Barrier Reef World Heritage Area – An overview of pathways and processes, pp. 161–187. In Wolanski E (ed.) Oceanographic Processes of Coral Reefs. Physical and Biological Links in the Great Barrier Reef. CRC Press, Boca Raton, FL.

Chin A, Bierwagen S, Rummer JL, Udyawer V (2024) The biophysics of sharks and rays on the Great Barrier Reef. In Wolanski E, Kingsford MJ (eds.) Oceanographic Processes of Coral Reefs. Physical and Biological Links in the Great Barrier Reef. CRC Press, Boca Raton, FL.

Dao HT, Smith-KEune C, Wolanski E, Jones CM, Jerry DR (2015) Oceanographic currents and local ecological knowledge indicate, and genetics does not refute, a contemporary pattern of larval dispersal for the ornate spiny lobster, Panulirus ornatus in the South-East Asian archipelago. PLoS One 10(5):e0124568.

Dobbs KA (2005) Recoveries of seabirds banded between 1978 and 1987 at Raine Island, MacLennan and Moulter cays and sandbanks No. 7 and 8, Northern Great Barrier Reef, Australia. Corella 29(3):65–72.

Domeier ML, Speare P (2012) Dispersal of adult black marlin (Istiompax indica) from a Great Barrier Reef spawning aggregation. PLoS One 7(2):e31629.

Duke N, Canning A, Mackenzie J (2024) More intense severe tropical cyclones in recent decades cause greater impacts on mangroves bordering Australia's Great Barrier Reef. In Wolanski E, Kingsford MJ (eds.) Oceanographic Processes of Coral Reefs. Physical and Biological Links in the Great Barrier Reef. CRC Press, Boca Raton, FL.

Dunbar G, Dickens G, Carter RM (2000) Sediment flux across the Great Barrier Reef Shelf to the Queensland trough over the last 300ky. Sedimentary Geology 133:49–92.

Eckert M (2019) The turbulence problem. A persistent riddle in historical perspective. Springer, Dordrecht, 107pp.

Hamann, M, Limpus CJ, Kophamel S (2024) Dispersal and connectivity of marine turtles in the Great Barrier Reef and links to the South Pacific Ocean. In Wolanski E, Kingsford MJ (eds.) Oceanographic Processes of Coral Reefs. Physical and Biological Links in the Great Barrier Reef. CRC Press, Boca Raton, FL.

Hamylton SM, McLean R, Bell IP, Thomas C (2024) Biological and geological links on coral reef islands. In Wolanski E, Kingsford MJ (eds.) Oceanographic Processes of Coral Reefs. Physical and Biological Links in the Great Barrier Reef. CRC Press, Boca Raton, FL.

Harrison D (2024a) An overview of environmental engineering methods for reducing coral bleaching stress. In Wolanski E, Kingsford MJ (eds.) Oceanographic Processes of Coral Reefs. Physical and Biological Links in the Great Barrier Reef. CRC Press, Boca Raton, FL.

Harrison P (2024b) Sexual reproduction of reef corals and application to coral restoration. In Wolanski E, Kingsford MJ (eds.) Oceanographic Processes of Coral Reefs. Physical and Biological Links in the Great Barrier Reef. CRC Press, Boca Raton, FL.

Hobbs J-PA (2010) Poleward range expansion of a tropical coral reef fish (Centropyge flavissima) to Lord Howe Island, Australia. Marine Biodiversity Records 3:e118.

Hoogenboom MO, Osipova L, Nordborg M, Schlaefer J, Critchell K (2024) Dispersal and environmental impacts of pan-oceanic contaminants. In Wolanski E, Kingsford MJ (eds.) Oceanographic Processes of Coral Reefs. Physical and Biological Links in the Great Barrier Reef. CRC Press, Boca Raton, FL.

Kamenos NA, Hennige SJ (2024) A historical perspective to thermal- and heatwave-induced bleaching on the Great Barrier Reef. In Wolanski E, Kingsford MJ (eds.) Oceanographic Processes of Coral Reefs. Physical and Biological Links in the Great Barrier Reef. CRC Press, Boca Raton, FL.

Kingsford MJ, Spiecker L, Gerlach G (2024) The interactions between larval behaviour and oceanography. In Wolanski E, Kingsford MJ (eds.) Oceanographic Processes of Coral Reefs. Physical and Biological Links in the Great Barrier Reef. CRC Press, Boca Raton, FL.

Kingsford MJ, Welch DJ (2007) Vulnerability of pelagic systems of the Great Barrier Reef to climate change. In Johnson JE, Marshall PA (eds.) 'Climate change and the Great Barrier Reef: A vulnerability Assessment'. Great Barrier Reef Marine Park Authority and the Australia Green House Office in the Department of the Environment and Water Resources, pp. 555–592.

Kingsford MJ, Wolanski E (2024) The emergence of biophysical sciences for the Great Barrier Reef. In Wolanski E, Kingsford MJ (eds.) Oceanographic Processes of Coral Reefs. Physical and Biological Links in the Great Barrier Reef. CRC Press, Boca Raton, FL.

Kininmonth S (2024) Currents modulate the genetic character of marine populations in the Great Barrier Reef. In Wolanski E, Kingsford MJ (eds.) Oceanographic Processes of Coral Reefs. Physical and Biological Links in the Great Barrier Reef. CRC Press, Boca Raton, FL.

Lewis S, McCloskey G, Bainbridge Z, DAvis A, BaRtley R, Turner R (2024) Sediment and nutrient flux from land. In Wolanski E, Kingsford MJ (eds.) Oceanographic Processes of Coral Reefs. Physical and Biological Links in the Great Barrier Reef. CRC Press, Boca Raton, FL.

Marsh H, Cleguer C (2024) Interactions between dugong biology and the biophysical determinants of their environment: A review. In Wolanski E, Kingsford MJ (eds.) Oceanographic Processes of Coral Reefs. Physical and Biological Links in the Great Barrier Reef. CRC Press, Boca Raton, FL.

Marsh H, Hagihara R, Hodgson A, Rankin R, Sobtzick S (2019) Monitoring dugongs within the Reef 2050 Integrated Monitoring and Reporting Program: Final report of the Dugong Team in the Megafauna Expert Group, Great Barrier Reef Marine Park Authority, Townsville. https://elibrary.gbrmpa.gov.au/jspui/retrieve/7c09af33-c541-4f26-9c98-6a791f0706f3/RIMReP%20Dugong%20report.pdf

Pratchett MS, Chandler J, Choukroun SM, Doll PC, Lang BJ, Kwong S, Chen CCM, EMslie M, Caballes CF, Uthicke S, Matthews S (2024) Biophysical processes involved in the initiation and spread of population irruptions of crown-of-thorns starfish on the Great Barrier Reef. In Wolanski E, Kingsford MJ (eds.) Oceanographic Processes of Coral Reefs. Physical and Biological Links in the Great Barrier Reef. CRC Press, Boca Raton, FL.

Quigley KM, Donelson JM (2024) Selective breeding and promotion of naturally heat-tolerant coral reef species. In Wolanski E, Kingsford MJ (eds.) Oceanographic Processes of Coral Reefs. Physical and Biological Links in the Great Barrier Reef. CRC Press, Boca Raton, FL.

Raina J-B, Seymour JR (2024) From the microscale to the reef: The role of microorganisms in the chemical ecology and gaseous emissions of the Great Barrier Reef. In Wolanski E, Kingsford MJ (eds.) Oceanographic Processes of Coral Reefs. Physical and Biological Links in the Great Barrier Reef. CRC Press, Boca Raton, FL.

Ramirez-Flores O, Birtles A, Pazmino D, Zenger KR, Van-Herwerden L (2019) Dwarf minke whales from the South Pacific share a matrilineal lineage distinct from *Balaenoptera acutorostrata* and *B. a. scammony*. Australian Mammalogy https://doi.org/10.1071/AM18015

Robertson K, Dawson J, Coffee O (2024) Raine Island Recovery Project – Intervening at one of the most significant sites on the Great Barrier Reef. In Wolanski E, Kingsford MJ (eds.) Oceanographic Processes of Coral Reefs. Physical and Biological Links in the Great Barrier Reef. CRC Press, Boca Raton, FL.

Saint-Amand A, Lambrechts J, Hanert E (2023) Biophysical models resolution affects coral connectivity estimates. Scientific Reports 13(1):9414.

Schlaefer JA, Kingsford MJ (2024) Bio-physical interactions of jellyfish on the Great Barrier Reef. In Wolanski E, Kingsford MJ (eds.) Oceanographic Processes of Coral Reefs. Physical and Biological Links in the Great Barrier Reef. CRC Press, Boca Raton, FL.

Sheaves M, Bradley M, Lubitz N, Mattone C, Myers J, VenkAtaramaN A, Waltham N, Winter S (2024) The influence of the spatio-temporal dynamics of fish populations on the outcomes of land-sea connectivity. In Wolanski E, Kingsford MJ (eds.) Oceanographic Processes of Coral Reefs. Physical and Biological Links in the Great Barrier Reef. CRC Press, Boca Raton, FL.

Templeman MA, Williams CD (2024) Jellyfish: A Window into pesticide distribution and risks on the Great Barrier Reef. In Wolanski E, Kingsford MJ (eds.) Oceanographic Processes of Coral Reefs. Physical and Biological Links in the Great Barrier Reef. CRC Press, Boca Raton, FL.

Tol SJ, Carter A, York PH, Grech A, Situ R, Coles RG (2024) Estimates of wind drift coefficient to inform biophysical models of seagrass dispersal in the Great Barrier Reef. In Wolanski E, Kingsford MJ (eds.) Oceanographic Processes of Coral Reefs. Physical and Biological Links in the Great Barrier Reef. CRC Press, Boca Raton, FL.

Waltham NJ, Cartwright P, Motson K, Sheaves M, Ronan M (2024) Coastal wetland restoration: Case studies from Great Barrier Reef catchments. In Wolanski E, Kingsford MJ (eds.) Oceanographic Processes of Coral Reefs. Physical and Biological Links in the Great Barrier Reef. CRC Press, Boca Raton, FL.

Waterhouse J, PearSon R, Lewis S, DAvis A, Waltham N (2024) Great Barrier Reef ecohydrology. In Wolanski E, Kingsford MJ (eds.) Oceanographic Processes of Coral Reefs. Physical and Biological Links in the Great Barrier Reef. CRC Press, Boca Raton, FL.

Webb GE (2024) Advances in understanding climate change on the Great Barrier Reef using coral-based proxies. In Wolanski E, Kingsford MJ (eds.) Oceanographic Processes of Coral Reefs. Physical and Biological Links in the Great Barrier Reef. CRC Press, Boca Raton, FL.

Wolanski E, Hopper C (2022) Dams and climate change accelerate channel avulsion and coastal erosion and threaten Ramsar-listed wetlands in the largest Great Barrier Reef watershed. Ecohydrology and Hydrobiology 22:197–212.

Wolanski E, Kingsford MJ, Lambrechts J, Marmorino G (2024) The physical oceanography of the Great Barrier Reef: A review. In Wolanski E, Kingsford MJ (eds.) Oceanographic Processes of Coral Reefs. Physical and Biological Links in the Great Barrier Reef. CRC Press, Boca Raton, FL.

Wolanski E, Richmond RH, Golbuu Y (2021) Oceanographic chaos and its role in larval self-recruitment and connectivity among fish populations in Micronesia. Estuarine, Coastal and Shelf Science 259:107461.

Woolsey E, Bainbridge SJ, Kingsford MJ, Byrne M (2012) Impacts of cyclone Hamish at One Tree Reef: Integrating environmental and benthic habitat data. Marine Biology 159:793–803.

Index

Note: Page numbers in *italic* indicate a figure and page numbers in **bold** indicate a table on the corresponding page.

D

E

F

For Product Safety Concerns and Information please contact our EU
representative GPSR@taylorandfrancis.com
Taylor & Francis Verlag GmbH, Kaufingerstraße 24, 80331 München, Germany

www.ingramcontent.com/pod-product-compliance
Lightning Source LLC
LaVergne TN
LVHW080118011026
849160LV00013B/1127